FORMULAS FROM GEOMETRY

area A perimeter P circumference C volume V curved surface area S altitude h radius r

RIGHT TRIANGLE

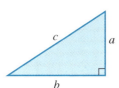

Pythagorean Theorem: $c^2 = a^2 + b^2$

TRIANGLE

$A = \frac{1}{2}bh$ $P = a + b + c$

$h = \dfrac{\sqrt{3}}{2}s$ $A = \dfrac{\sqrt{3}}{4}s^2$

RECTANGLE

$A = lw$ $P = 2l + 2w$

PARALLELOGRAM

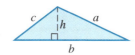

$A = bh$

TRAPEZOID

$A = \frac{1}{2}(a + b)h$

CIRCLE

$A = \pi r^2$ $C = 2\pi r$

CIRCULAR SECTOR

$A = \frac{1}{2}r^2\theta$ $s = r\theta$

CIRCULAR RING

$A = \pi(R^2 - r^2)$

RECTANGULAR BOX

$V = lwh$ $S = 2(hl + lw + hw)$

SPHERE

$V = \frac{4}{3}\pi r^3$ $S = 4\pi r^2$

RIGHT CIRCULAR CYLINDER

$V = \pi r^2 h$ $S = 2\pi rh$

RIGHT CIRCULAR CONE

$V = \frac{1}{3}\pi r^2 h$ $S = \pi r\sqrt{r^2 + h^2}$

FRUSTUM OF A CONE

$V = \frac{1}{3}\pi h(r^2 + rR + R^2)$

PRISM

$V = Bh$ with B the area of the base

ANALYTIC GEOMETRY

DISTANCE FORMULA

$$d(P_1, P_2) = \sqrt{(x_2 - x_1)^2 + (y_2 - y_1)^2}$$

EQUATION OF A CIRCLE

$$(x - h)^2 + (y - k)^2 = r^2$$

SLOPE m OF A LINE

$$m = \frac{y_2 - y_1}{x_2 - x_1}$$

GRAPH OF A QUADRATIC FUNCTION

$$y = ax^2, \, a > 0 \qquad y = ax^2 + bx + c, \, a > 0$$

POINT-SLOPE FORM OF A LINE

$$y - y_1 = m(x - x_1)$$

CONSTANTS

$\pi \approx 3.14159$

$e \approx 2.71828$

CONVERSIONS

1 centimeter $\approx$ 0.3937 inch

1 meter $\approx$ 3.2808 feet

1 kilometer $\approx$ 0.6214 mile

1 gram $\approx$ 0.0353 ounce

1 kilogram $\approx$ 2.2046 pounds

1 liter $\approx$ 0.2642 gallon

1 milliliter $\approx$ 0.0381 fluid ounce

1 joule $\approx$ 0.7376 foot-pound

1 newton $\approx$ 0.2248 pound

1 lumen $\approx$ 0.0015 watt

1 acre = 43,560 square feet

SLOPE-INTERCEPT FORM OF A LINE

$$y = mx + b$$

INTERCEPT FORM OF A LINE

$$\frac{x}{a} + \frac{y}{b} = 1 \quad (a \neq 0, b \neq 0)$$

CLASSIC ELEVENTH EDITION

ALGEBRA AND TRIGONOMETRY

WITH ANALYTIC GEOMETRY

EARL W. SWOKOWSKI

JEFFERY A. COLE

Anoka Ramsey Community College

THOMSON

BROOKS/COLE

Australia • Canada • Mexico • Singapore • Spain
United Kingdom • United States

Algebra and Trigonometry with Analytic Geometry
Classic Eleventh Edition

Earl W. Swokowski and Jeffery A. Cole

Acquisitions Editor: *John-Paul Ramin*
Assistant Editor: *Katherine Brayton*
Editorial Assistant: *Leata Holloway*
Technology Project Manager: *Earl Perry*
Senior Marketing Manager: *Karin Sandberg*
Marketing Assistant: *Erin Mitchell*
Managing Communications Project Manager: *Bryan Vann*
Senior Project Manager, Editorial Production: *Janet Hill*
Senior Art Director: *Vernon Boes*

Senior Buyer: *Karen Hunt*
Production Service: *Lifland et al., Bookmakers*
Text Designers: *Andrew Ogus/Rokusek*
Production/Art Editor: *Sally Lifland*
Illustrator: *Scientific Illustrators*
Cover Designer: *Tani Hasegawa*
Cover Printer: *Coral Graphics Services*
Compositor: *Better Graphics, Inc.*
Printer: *Transcontinental Interglobe*

Printed in Canada

2 3 4 5 6 7 09 08 07 06 05

For more information about our products,
contact us at:
Thomson Learning Academic Resource Center
1-800-423-0563

For permission to use material from this text or product,
submit a request online at **http://www.thomsonrights.com**.
Any additional questions about permissions can be
submitted by email to **thomsonrights@thomson.com**.

Library of Congress Control Number: 2005921681

ISBN 0-534-40469-3

Thomson Brooks/Cole
10 Davis Drive
Belmont, CA 94002-3098
USA

Asia (Includes India)
Thomson Learning
5 Shenton Way
#01-01 UIC Building
Singapore 068808

Australia/New Zealand
Thomson Learning
102 Dodds Street
Southbank, Victoria 3006
Australia

Canada
Nelson
1120 Birchmount Road
Toronto, Ontario M1K 5G4
Canada

Europe/Middle East/Africa
Thomson Learning
High Holborn House
50/51 Bedford Row
London WC1R 4LR
United Kingdom

Latin America
Thomson Learning
Seneca, 53
Colonia Polanco
11560 Mexico
D.F. Mexico

Spain (Includes Portugal)
Thomson Paraninfo
Calle Magallanes, 25
28015 Madrid, Spain

To the memory of Earl W. Swokowski

CONTENTS

Preface   viii

CHAPTER 1 *Fundamental Concepts of Algebra*   1

   1.1 Real Numbers   2
   1.2 Exponents and Radicals   16
   1.3 Algebraic Expressions   27
   1.4 Fractional Expressions   40
      Chapter 1 Review Exercises   49
      Chapter 1 Discussion Exercises   51

CHAPTER 2 *Equations and Inequalities*   53

   2.1 Equations   54
   2.2 Applied Problems   61
   2.3 Quadratic Equations   73
   2.4 Complex Numbers   87
   2.5 Other Types of Equations   94
   2.6 Inequalities   102
   2.7 More on Inequalities   111
      Chapter 2 Review Exercises   119
      Chapter 2 Discussion Exercises   122

CHAPTER 3 *Functions and Graphs*   123

   3.1 Rectangular Coordinate Systems   124
   3.2 Graphs of Equations   130
   3.3 Lines   140
   3.4 Definition of Function   155
   3.5 Graphs of Functions   171
   3.6 Quadratic Functions   185
   3.7 Operations on Functions   197
      Chapter 3 Review Exercises   205
      Chapter 3 Discussion Exercises   210

CHAPTER **4** *Polynomial and Rational Functions*   213

4.1 Polynomial Functions of Degree Greater Than 2   214
4.2 Properties of Division   222
4.3 Zeros of Polynomials   229
4.4 Complex and Rational Zeros of Polynomials   241
4.5 Rational Functions   248
4.6 Variation   265
Chapter 4 Review Exercises   272
Chapter 4 Discussion Exercises   275

CHAPTER **5** *Inverse, Exponential, and Logarithmic Functions*   277

5.1 Inverse Functions   278
5.2 Exponential Functions   287
5.3 The Natural Exponential Function   299
5.4 Logarithmic Functions   308
5.5 Properties of Logarithms   323
5.6 Exponential and Logarithmic Equations   330
Chapter 5 Review Exercises   342
Chapter 5 Discussion Exercises   345

CHAPTER **6** *The Trigonometric Functions*   347

6.1 Angles   348
6.2 Trigonometric Functions of Angles   358
6.3 Trigonometric Functions of Real Numbers   375
6.4 Values of the Trigonometric Functions   393
6.5 Trigonometric Graphs   400
6.6 Additional Trigonometric Graphs   412
6.7 Applied Problems   420
Chapter 6 Review Exercises   433
Chapter 6 Discussion Exercises   439

CHAPTER **7**   *Analytic Trigonometry*   441

7.1	Verifying Trigonometric Identities	442
7.2	Trigonometric Equations	447
7.3	The Addition and Subtraction Formulas	457
7.4	Multiple-Angle Formulas	467
7.5	Product-to-Sum and Sum-to-Product Formulas	477
7.6	The Inverse Trigonometric Functions	482
	Chapter 7 Review Exercises	496
	Chapter 7 Discussion Exercises	499

CHAPTER **8**   *Applications of Trigonometry*   501

8.1	The Law of Sines	502
8.2	The Law of Cosines	512
8.3	Vectors	522
8.4	The Dot Product	536
8.5	Trigonometric Form for Complex Numbers	546
8.6	De Moivre's Theorem and nth Roots of Complex Numbers	552
	Chapter 8 Review Exercises	557
	Chapter 8 Discussion Exercises	560

CHAPTER **9**   *Systems of Equations and Inequalities*   563

9.1	Systems of Equations	564
9.2	Systems of Linear Equations in Two Variables	573
9.3	Systems of Inequalities	582
9.4	Linear Programming	590
9.5	Systems of Linear Equations in More Than Two Variables	598
9.6	The Algebra of Matrices	614
9.7	The Inverse of a Matrix	623
9.8	Determinants	628
9.9	Properties of Determinants	634
9.10	Partial Fractions	642
	Chapter 9 Review Exercises	648
	Chapter 9 Discussion Exercises	651

CHAPTER **10** *Sequences, Series, and Probability*   653

10.1 Infinite Sequences and Summation Notation   654
10.2 Arithmetic Sequences   664
10.3 Geometric Sequences   671
10.4 Mathematical Induction   680
10.5 The Binomial Theorem   686
10.6 Permutations   695
10.7 Distinguishable Permutations and
         Combinations   702
10.8 Probability   709
         Chapter 10 Review Exercises   723
         Chapter 10 Discussion Exercises   725

CHAPTER **11** *Topics from Analytic Geometry*   727

11.1 Parabolas   728
11.2 Ellipses   737
11.3 Hyperbolas   750
11.4 Plane Curves and Parametric Equations   762
11.5 Polar Coordinates   772
11.6 Polar Equations of Conics   786
         Chapter 11 Review Exercises   792
         Chapter 11 Discussion Exercises   794

Appendixes   797

I   Common Graphs and Their Equations   798
II   A Summary of Graph Transformations   800
III Graphs of Trigonometric Functions and
         Their Inverses   802
IV   Values of the Trigonometric Functions of
         Special Angles on a Unit Circle   804

Answers to Selected Exercises   A1

Index of Applications   A79

Index   A84

PREFACE

The classic edition of *Algebra and Trigonometry with Analytic Geometry* is a special version of the eleventh edition of the same title. It has been written for professors seeking to teach a traditional course that requires only a scientific calculator. Both editions improve upon the tenth edition in several ways. This edition includes over 100 new examples and exercises, many of these resulting from suggestions of users and reviewers of the tenth edition. All have been incorporated without sacrificing the mathematical soundness that has been paramount to the success of this text.

Below is a brief overview of the chapters, followed by a short description of the College Algebra course that I teach at Anoka Ramsey Community College, and then a list of the general features of the text.

Overview

Chapter 1   This chapter contains a summary of some basic algebra topics. Students should be familiar with much of this material, but also challenged by some of the exercises that prepare them for calculus.

Chapter 2   Equations and inequalities are solved algebraically in this chapter. Students will extend their knowledge of these topics; for example, they have worked with the quadratic formula, but will be asked to relate it to factoring and work with coefficients that are not real numbers (see Examples 10 and 11 in Section 2.3).

Chapter 3   Two-dimensional graphs and functions are introduced in this chapter. See the new Example 8 in Section 3.5 for a topical application (taxes) that relates tables, formulas, and graphs.

Chapter 4   This chapter begins with a discussion of polynomial functions and some polynomial theory. A thorough treatment of rational functions is given in Section 4.5. This is followed by a section on variation (formerly Section 3.9), which now includes graphs of rational functions.

Chapter 5   Inverse functions (formerly Section 3.8) is the first topic of discussion, followed by several sections that deal with exponential and logarithmic functions. Modeling an exponential function is given additional attention in this chapter (see Example 8 in Section 5.2), as well as in Chapter 9.

Chapter 6   Angles are the first topic in this chapter. Next, the trigonometric functions are introduced using a right triangle approach and then defined in terms of a unit circle. Basic trigonometric identities appear throughout the chapter. The chapter concludes with sections on trigonometric graphs and applied problems.

Chapter 7   This chapter consists mostly of trigonometric identities, formulas, and equations. The last section contains definitions, properties, and applications of the inverse trigonometric functions.

Chapter 8   The law of sines and the law of cosines are used to solve oblique triangles. Vectors are then introduced and used in applications. The last two sections relate the trigonometric functions and complex numbers.

Chapter 9   Systems of inequalities and linear programming immediately follow solving systems by substitution and elimination. Next, matrices are introduced and used to solve systems. This chapter concludes with a discussion of determinants and partial fractions.

Chapter 10   This chapter begins with a discussion of sequences. Mathematical induction and the binomial theorem are next, followed by counting topics (see Example 3 in Section 10.7 for an example involving both combinations and permutations). The last section is about probability and includes topics such as odds and expected value.

Chapter 11   Sections on the parabola, ellipse, and hyperbola begin this chapter. Two different ways of representing functions are given in the next sections on parametric equations and polar coordinates.

My Course

At Anoka Ramsey Community College in Coon Rapids, Minnesota, College Algebra I is a one-semester 3-credit course. For students intending to take Calculus, this course is followed by a one-semester 4-credit course, College Algebra II and Trigonometry. This course also serves as a terminal math course for many students.

   The sections covered in College Algebra I are

   3.1–3.7, 4.1, 4.5 (part), 4.6, 5.1–5.6, 9.1–9.4, 10.1–10.3, and 10.5–10.8.

Chapters 1 and 2 are used as review material in some classes, and the remaining sections are taught in the following course.

Features

Illustrations   Brief demonstrations of the use of definitions, laws, and theorems are provided in the form of illustrations.

Charts   Charts give students easy access to summaries of properties, laws, graphs, relationships, and definitions. These charts often contain simple illustrations of the concepts that are being introduced.

Examples   Titled for easy reference, all examples provide detailed solutions of problems similar to those that appear in exercise sets. Many examples include graphs, charts, or tables to help the student understand procedures and solutions.

Step-by-Step Explanations   In order to help students follow them more easily, many of the solutions in examples contain step-by-step explanations.

Discussion Exercises   Each chapter ends with several exercises that are suitable for small-group discussions. These exercises range from easy to difficult and from theoretical to application-oriented.

Checks   The solutions to some examples are explicitly checked, to remind students to verify that their solutions satisfy the conditions of the problems.

Applications   To arouse student interest and to help students relate the exercises to current real-life situations, applied exercises have been titled. One look at the Index of Applications in the back of the book reveals the wide array of topics. Many professors have indicated that the applications constitute one of the strongest features of the text.

Exercises   Exercise sets begin with routine drill problems and gradually progress to more difficult problems. An ample number of exercises contain graphs and tabular data; others require the student to find a mathematical model for the given data. Many of the new exercises require the student to understand the conceptual relationship of an equation and its graph.

Applied problems generally appear near the end of an exercise set, to allow students to gain confidence in working with the new ideas that have been presented before they attempt problems that require greater analysis and synthesis of these ideas. Review exercises at the end of each chapter may be used to prepare for examinations.

Guidelines   Boxed guidelines enumerate the steps in a procedure or technique to help students solve problems in a systematic fashion.

Warnings   Interspersed throughout the text are warnings to alert students to common mistakes.

Text Art   Forming a total art package that is second to none, figures and graphs have been computer-generated for accuracy, using the latest technology. Colors are employed to distinguish between different parts of figures. For example, the graph of one function may be shown in blue and that of a second function in red. Labels are the same color as the parts of the figure they identify.

Text Design   The text has been designed to ensure that discussions are easy to follow and important concepts are highlighted. Color is used pedagogically to clarify complex graphs and to help students visualize applied problems. Previous adopters of the text have confirmed that the text strikes a very appealing balance in terms of color use.

Endpapers   The endpapers in the front and back of the text provide useful summaries from algebra, geometry, and trigonometry.

Appendixes   Appendix I, "Common Graphs and Their Equations," is a pictorial summary of graphs and equations that students commonly encounter in precalculus mathematics. Appendix II, "A Summary of Graph Transformations," is an illustrative synopsis of the basic graph transformations discussed in the text: shifting, stretching, compressing, and reflecting. Appendix III, "Graphs of Trigonometric Functions and Their Inverses," contains graphs, domains, and ranges of the six trigonometric functions and their inverses. Appendix IV, "Values of the Trigonometric Functions of Special Angles on a Unit Circle," is a full-page reference for the most common angles on a unit circle—valuable for students who are trying to learn the basic trigonometric functions values.

Answer Section   The answer section at the end of the text provides answers for most of the odd-numbered exercises, as well as answers for all chapter review exercises. Considerable thought and effort were devoted to making this section a learning device for the student instead of merely a place to check answers. For instance, proofs are given for mathematical induction problems. Numerical answers for many exercises are stated in both an exact and an approximate form. Graphs, proofs, and hints are included whenever appropriate. Author-prepared solutions and answers ensure a high degree of consistency among the text, the solutions manuals, and the answers.

Teaching Tools for the Instructor

Instructor's Solutions Manual by Jeffery A. Cole (ISBN 0-534-40470-7)   This author-prepared manual includes answers to all exercises and detailed solutions to most exercises. The manual has been thoroughly reviewed for accuracy.

Test Bank (ISBN 0-534-49453-6)   The *Test Bank* includes multiple tests per chapter as well as final exams. The tests are made up of a combination of multiple-choice, free-response, true/false, and fill-in-the-blank questions.

Text-Specific Videotapes (ISBN 0-534-49471-4)   Text-specific videotape sets, available at no charge to qualified adopters of the text, feature 10- to 20-minute problem-solving lessons that cover each section of every chapter. The particular text examples covered in the videos are designated by a videocassette/CD icon (shown to the left).

Instructor's Resource CD-ROM for Swokowski/Cole's Precalculus Series (ISBN 0-534-49513-3) This CD-ROM provides the instructor with dynamic media tools for teaching Precalculus or College Algebra. Animations and PowerPoint® lecture slides, combined with supplements in electronic format, are all included on this CD-ROM.

iLrn Instructor Version (ISBN 0-534-28052-8)   Efficient and versatile, iLrn gives you the power to transform the teaching and learning experience. iLrn Instructor Version is made up of two components, iLrn Testing and iLrn Tutorial. iLrn Testing is an internet-ready, text-specific testing suite that allows instructors to customize exams and track student progress in an accessible, browser-based format. iLrn offers full algorithmic generation of problems and free-response mathematics. iLrn Tutorial is a text-specific, interactive tutorial software program that is delivered via the web (at http://ilrn.com) and is offered in both student and instructor versions. Like iLrn Testing, it is browser-based, making it an intuitive mathematical guide even for students with little technological proficiency. iLrn Tutorial allows students to work with real math notation in real time, providing instant analysis and feedback. The tracking program built into the instructor version of the software enables instructors to carefully monitor student progress. The complete integration of the testing, tutorial, and course management components simplifies your routine tasks. Results flow automatically to your gradebook, and you can easily communicate with individuals, sections, or entire courses.

   Plus, a personalized study plan can be generated by the course-specific diagnostic built into iLrn. With a personalized study plan, students can focus their time where they need it the most, creating a positive learning environment and paving a pathway to success in their mathematics course. Additional features you'll find when using iLrn:

- You can manipulate and format your tests with the "rich text format" (RTF) conversion tool. RTF conversion allows you to open tests in most word processors, like Microsoft Word, for further formatting and customization.

- iLrn offers more problem types to provide greater variety and more diverse challenges in your tests.

The iLrn Tutorial interface effectively engages students and helps them learn math concepts faster.

WebTutor ToolBox on WebCT or Blackboard   Preloaded with content and available free via access code when packaged with this text, WebTutor ToolBox for WebCT or Blackboard pairs all the content of this text's rich Book Companion Website with all the sophisticated course management functionality of a WebCT or Blackboard product. You can assign materials (including online quizzes) and have the results flow *automatically* to your gradebook. ToolBox is ready to use as soon as you log on—or you can customize its preloaded content by uploading images and other resources, adding web links, or creating your own practice materials. Students have access only to student resources on the website. Instructors can enter an access code to reach password-protected instructor resources.

Learning Tools for the Student

Student Solutions Manual by Jeffery A. Cole (ISBN 0-534-40471-5)   This author-prepared manual provides worked-out solutions for all the odd-numbered exercises, as well as strategies for solving additional exercises.

Thomson Brooks/Cole Mathematics Website (http://mathematics.brookscole.com)   Visit us on the web for access to a wealth of free learning resources, including graphing calculator tutorials, tools to address math anxiety, Historical Notes, an extensive glossary of math terms, career information, and more.

iLrn Tutorial Student Version   A complete content mastery resource offered online, iLrn Tutorial allows students to work with real math notation in real time, with unlimited practice problems, instant analysis and feedback, and streaming video to illustrate key concepts. And live, online, text-specific tutorial help is just a click away with vMentor™, accessed seamlessly through iLrn Tutorial. The iLrn Tutorial access card located inside the cover of the text provides an access code for students to utilize when logging into iLrn.

 Interactive Video SkillBuilder CD-ROM   Think of it as portable office hours! Packaged with the text, the Interactive Video SkillBuilder CD-ROM contains video instruction covering each chapter of the text. The particular text examples covered in the videos are designated by a videocassette/CD icon (shown to the left). The problems worked during each video lesson are shown first so that students can try working them before watching the solution. To help students evaluate their progress, each section contains a 10-question Web quiz (the results of which can be e-mailed to the instructor), and each chapter contains a chapter test, with the answers to each problem on each test. This CD-ROM also includes MathCue tutorial and quizzing software, featuring a SkillBuilder that presents problems to solve and evaluates answers with step-by-step explanations; a Quiz function that enables students to generate quiz problems keyed to problem types from each section of the book; a Chapter Test that provides many problems keyed to problem types from each chapter; and a Solution Finder that allows students to enter their own basic problems and receive step-by-step help as if they were working with a tutor. English/Spanish closed caption translations can be selected for display along with the video instruction.

Acknowledgments

Many changes for this edition are due to the following individuals, who reviewed the manuscript and/or made suggestions to increase the usefulness of the text for the student:

Jean H. Bevis, Georgia State University
David Boliver, University of Central Oklahoma
Randall Dorman, Cochise College
Karen Hinz, Anoka Ramsey Community College
Sudhir Goel, Valdosta State University
John W. Horton, Sr., St. Petersburg College
Robert Jajcay, Indiana State University
Conrad D. Krueger, San Antonio College

Susan McLoughlin, Union County College
Lakshmi Nigam, Quinnipiac University
Wesley J. Orser, Clark College
Don E. Soash, Hillsborough Community College
Thomas A. Tredon, Lord Fairfax Community College
Fred Worth, Henderson State University

In addition, I thank Mike Rosenborg, of Canyonville (Oregon) Christian Academy, and Anna Fox, accuracy checkers for the *Instructor's Solutions Manual*.

I am thankful for the excellent cooperation of the staff of Brooks/Cole, especially Acquisitions Editor John-Paul Ramin, for his helpful advice and support throughout the project. Rachael Sturgeon and Katherine Brayton managed the excellent ancillary package that accompanies the text. Sally Lifland, Gail Magin, Denise Throckmorton, and Jeannie Yost, all of Lifland et al., Bookmakers, saw the book through all the stages of production, took exceptional care in seeing that no inconsistencies occurred, and offered many helpful suggestions. The late George Morris, of Scientific Illustrators, created the mathematically precise art package and updated all the art through several editions. This tradition of excellence is carried on by his son Brian.

In addition to all the persons named here, I would like to express my sincere gratitude to the many students and teachers who have helped shape my views on mathematics education. Please feel free to write to me about any aspect of this text—I value your opinion.

Jeffery A. Cole

Fundamental Concepts of Algebra

1.1   Real Numbers

1.2   Exponents and Radicals

1.3   Algebraic Expressions

1.4   Fractional Expressions

The word *algebra* comes from *ilm al-jabr w'al muqabala,* the title of a book written in the ninth century by the Arabian mathematician al-Khworizimi. The title has been translated as the science of restoration and reduction, which means transposing and combining similar terms (of an equation). The Latin transliteration of al-jabr led to the name of the branch of mathematics we now call algebra.

In algebra we use symbols or letters—such as a, b, c, d, x, y—to denote arbitrary numbers. This general nature of algebra is illustrated by the many formulas used in science and industry. As you proceed through this text and go on either to more advanced courses in mathematics or to fields that employ mathematics, you will become more and more aware of the importance and the power of algebraic techniques.

1.1

Real Numbers

Real numbers are used throughout mathematics, and you should be acquainted with symbols that represent them, such as

$$1, \quad 73, \quad -5, \quad \tfrac{49}{12}, \quad \sqrt{2}, \quad 0, \quad \sqrt[3]{-85}, \quad 0.33333\ldots, \quad 596.25,$$

and so on. The **positive integers,** or **natural numbers,** are

$$1, \quad 2, \quad 3, \quad 4, \quad \ldots.$$

The **whole numbers** (or *nonnegative integers*) are the natural numbers combined with the number 0. The **integers** are often listed as follows:

$$\ldots, \quad -4, \quad -3, \quad -2, \quad -1, \quad 0, \quad 1, \quad 2, \quad 3, \quad 4, \quad \ldots$$

Throughout this text lowercase letters $a, b, c, x, y, \ldots$ represent arbitrary real numbers (also called *variables*). If a and b denote the same real number, we write $a = b$, which is read "a **is equal to** b" and is called an **equality.** The notation $a \neq b$ is read "a **is *not*** equal to b."

If a, b, and c are integers and $c = ab$, then a and b are **factors,** or **divisors,** of c. For example, since

$$6 = 2 \cdot 3 = (-2)(-3) = 1 \cdot 6 = (-1)(-6),$$

we know that $1, -1, 2, -2, 3, -3, 6$, and -6 are factors of 6.

A positive integer p different from 1 is **prime** if its only positive factors are 1 and p. The first few primes are 2, 3, 5, 7, 11, 13, 17, and 19. The **Fundamental Theorem of Arithmetic** states that every positive integer different from 1 can be expressed as a product of primes in one and only one way (except for order of factors). Some examples are

$$12 = 2 \cdot 2 \cdot 3, \quad 126 = 2 \cdot 3 \cdot 3 \cdot 7, \quad 540 = 2 \cdot 2 \cdot 3 \cdot 3 \cdot 3 \cdot 5.$$

A **rational number** is a real number that can be expressed in the form a/b, where a and b are integers and $b \neq 0$. Note that every integer a is a rational number, since it can be expressed in the form $a/1$. Every real number can be expressed as a decimal, and the decimal representations for rational numbers are either *terminating* or *nonterminating and repeating*. For example, we can show by using the arithmetic process of division that

$$\tfrac{5}{4} = 1.25 \quad \text{and} \quad \tfrac{177}{55} = 3.2181818\ldots,$$

where the digits 1 and 8 in the representation of $\frac{177}{55}$ repeat indefinitely (sometimes written $3.2\overline{18}$).

Real numbers that are not rational are **irrational numbers.** Decimal representations for irrational numbers are always *nonterminating and nonrepeating.* One common irrational number, denoted by π, is the ratio of the circumference of a circle to its diameter. We sometimes use the notation $\pi \approx 3.1416$ to indicate that π **is approximately equal to** 3.1416.

There is no *rational* number b such that $b^2 = 2$, where b^2 denotes $b \cdot b$. However, there is an *irrational* number, denoted by $\sqrt{2}$ (the **square root** of 2), such that $\left(\sqrt{2}\right)^2 = 2$.

The system of **real numbers** consists of all rational and irrational numbers. Relationships among the types of numbers used in algebra are illustrated in the diagram in Figure 1, where a line connecting two rectangles means that the numbers named in the higher rectangle include those in the lower rectangle. The complex numbers, discussed in Section 2.4, contain all real numbers.

Figure 1   Types of numbers used in algebra

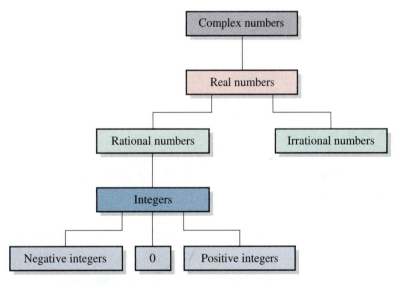

The real numbers are **closed relative to the operation of addition** (denoted by $+$); that is, to every pair a, b of real numbers there corresponds exactly one real number $a + b$ called the **sum** of a and b. The real numbers are also **closed relative to multiplication** (denoted by $\cdot$); that is, to every pair a, b of real numbers there corresponds exactly one real number $a \cdot b$ (also denoted by ab) called the **product** of a and b.

Important properties of addition and multiplication of real numbers are listed in the following chart.

Properties of Real Numbers

Terminology	General case	Meaning
(1) Addition is **commutative.**	$a + b = b + a$	Order is immaterial when adding two numbers.
(2) Addition is **associative.**	$a + (b + c) = (a + b) + c$	Grouping is immaterial when adding three numbers.
(3) 0 is the **additive identity.**	$a + 0 = a$	Adding 0 to any number yields the same number.
(4) $-a$ is the **additive inverse,** or **negative,** of a.	$a + (-a) = 0$	Adding a number and its negative yields 0.
(5) Multiplication is **commutative.**	$ab = ba$	Order is immaterial when multiplying two numbers.
(6) Multiplication is **associative.**	$a(bc) = (ab)c$	Grouping is immaterial when multiplying three numbers.
(7) 1 is the **multiplicative identity.**	$a \cdot 1 = a$	Multiplying any number by 1 yields the same number.
(8) If $a \neq 0$, $\dfrac{1}{a}$ is the **multiplicative inverse,** or **reciprocal,** of a.	$a\left(\dfrac{1}{a}\right) = 1$	Multiplying a nonzero number by its reciprocal yields 1.
(9) Multiplication is **distributive** over addition.	$a(b + c) = ab + ac$ and $(a + b)c = ac + bc$	Multiplying a number and a sum of two numbers is equivalent to multiplying each of the two numbers by the number and then adding the products.

Since $a + (b + c)$ and $(a + b) + c$ are always equal, we may use $a + b + c$ to denote this real number. We use abc for either $a(bc)$ or $(ab)c$. Similarly, if four or more real numbers a, b, c, d are added or multiplied, we may write $a + b + c + d$ for their sum and $abcd$ for their product, regardless of how the numbers are grouped or interchanged.

The distributive properties are useful for finding products of many types of expressions involving sums. The next example provides one illustration.

EXAMPLE 1   Using distributive properties

If p, q, r, and s denote real numbers, show that

$$(p + q)(r + s) = pr + ps + qr + qs.$$

SOLUTION   We use both of the distributive properties listed in (9) of the preceding chart:

$$
\begin{aligned}
(p + q)(r + s) &\\
&= p(r + s) + q(r + s) &&\text{second distributive property, with } c = r + s\\
&= (pr + ps) + (qr + qs) &&\text{first distributive property}\\
&= pr + ps + qr + qs &&\text{remove parentheses}
\end{aligned}
$$

The following are basic properties of equality.

Properties of Equality	If $a = b$ and c is any real number, then
	$\quad$ **(1)** $a + c = b + c$
	$\quad$ **(2)** $ac = bc$

Properties 1 and 2 state that the same number may be added to both sides of an equality, and both sides of an equality may be multiplied by the same number. We will use these properties extensively throughout the text to help find solutions of equations.

The next result can be proved.

Products Involving Zero	**(1)** $a \cdot 0 = 0$ for every real number a.
	(2) If $ab = 0$, then either $a = 0$ or $b = 0$.

When we use the word *or* as we do in (2), we mean that *at least* one of the factors a and b is 0. We will refer to (2) as the *zero factor theorem* in future work.

Some properties of negatives are listed in the following chart.

Properties of Negatives

Property	Illustration
(1) $-(-a) = a$	$-(-3) = 3$
(2) $(-a)b = -(ab) = a(-b)$	$(-2)3 = -(2 \cdot 3) = 2(-3)$
(3) $(-a)(-b) = ab$	$(-2)(-3) = 2 \cdot 3$
(4) $(-1)a = -a$	$(-1)3 = -3$

The reciprocal $\dfrac{1}{a}$ of a nonzero real number a is often denoted by a^{-1}, as in the next chart.

Notation for Reciprocals

Definition	Illustrations
If $a \neq 0$, then $a^{-1} = \dfrac{1}{a}$.	$2^{-1} = \dfrac{1}{2}$ $\left(\dfrac{3}{4}\right)^{-1} = \dfrac{1}{3/4} = \dfrac{4}{3}$

Note that if $a \neq 0$, then

$$a \cdot a^{-1} = a\left(\frac{1}{a}\right) = 1.$$

The operations of **subtraction** $(-)$ and **division** $(\div)$ are defined as follows.

Subtraction and Division

Definition	Meaning	Illustration
$a - b = a + (-b)$	To subtract one number from another, add the negative.	$3 - 7 = 3 + (-7)$
$a \div b = a \cdot \left(\dfrac{1}{b}\right)$ $= a \cdot b^{-1}; b \neq 0$	To divide one number by a nonzero number, multiply by the reciprocal.	$3 \div 7 = 3 \cdot \left(\dfrac{1}{7}\right)$ $= 3 \cdot 7^{-1}$

We use either a/b or $\dfrac{a}{b}$ for $a \div b$ and refer to a/b as the **quotient of a and b** or the **fraction a over b.** The numbers a and b are the **numerator** and **denominator,** respectively, of a/b. Since 0 has no multiplicative inverse, a/b is not defined if $b = 0$; that is, *division by zero is not defined.* It is for this reason that the real numbers are not closed relative to division. Note that

$$1 \div b = \frac{1}{b} = b^{-1} \quad \text{if} \quad b \neq 0.$$

The following properties of quotients are true, provided all denominators are nonzero real numbers.

Properties of Quotients

Property	Illustration
(1) $\dfrac{a}{b} = \dfrac{c}{d}$ if $ad = bc$	$\dfrac{2}{5} = \dfrac{6}{15}$ because $2 \cdot 15 = 5 \cdot 6$
(2) $\dfrac{ad}{bd} = \dfrac{a}{b}$	$\dfrac{2 \cdot 3}{5 \cdot 3} = \dfrac{2}{5}$
(3) $\dfrac{a}{-b} = \dfrac{-a}{b} = -\dfrac{a}{b}$	$\dfrac{2}{-5} = \dfrac{-2}{5} = -\dfrac{2}{5}$
(4) $\dfrac{a}{b} + \dfrac{c}{b} = \dfrac{a + c}{b}$	$\dfrac{2}{5} + \dfrac{9}{5} = \dfrac{2 + 9}{5} = \dfrac{11}{5}$
(5) $\dfrac{a}{b} + \dfrac{c}{d} = \dfrac{ad + bc}{bd}$	$\dfrac{2}{5} + \dfrac{4}{3} = \dfrac{2 \cdot 3 + 5 \cdot 4}{5 \cdot 3} = \dfrac{26}{15}$
(6) $\dfrac{a}{b} \cdot \dfrac{c}{d} = \dfrac{ac}{bd}$	$\dfrac{2}{5} \cdot \dfrac{7}{3} = \dfrac{2 \cdot 7}{5 \cdot 3} = \dfrac{14}{15}$
(7) $\dfrac{a}{b} \div \dfrac{c}{d} = \dfrac{a}{b} \cdot \dfrac{d}{c} = \dfrac{ad}{bc}$	$\dfrac{2}{5} \div \dfrac{7}{3} = \dfrac{2}{5} \cdot \dfrac{3}{7} = \dfrac{6}{35}$

Real numbers may be represented by points on a line l such that to each real number a there corresponds exactly one point on l and to each point P on l there corresponds one real number. This is called a **one-to-one correspondence.** We first choose an arbitrary point O, called the **origin,** and associate with it the real number 0. Points associated with the integers are then determined by laying off successive line segments of equal length on either side of O, as illustrated in Figure 2. The point corresponding to a rational number, such as $\frac{23}{5}$, is obtained by subdividing these line segments. Points associated with certain irrational numbers, such as $\sqrt{2}$, can be found by construction (see Exercise 45).

Figure 2

The number a that is associated with a point A on l is the **coordinate** of A. We refer to these coordinates as a **coordinate system** and call l a **coordinate line** or a **real line.** A direction can be assigned to l by taking the **positive direction** to the right and the **negative direction** to the left. The positive direction is noted by placing an arrowhead on l, as shown in Figure 2.

The numbers that correspond to points to the right of O in Figure 2 are **positive real numbers.** Numbers that correspond to points to the left of O are **negative real numbers.** *The real number* 0 *is neither positive nor negative.*

Note the difference between a negative real number and the *negative of* a real number. In particular, the negative of a real number a can be positive. For example, if a is negative, say $a = -3$, then the negative of a is $-a = -(-3) = 3$, which is positive. In general, we have the following relationships.

Relationships Between a and $-a$	**(1)** If a is positive, then $-a$ is negative. **(2)** If a is negative, then $-a$ is positive.

In the following chart we define the notions of **greater than** and **less than** for real numbers a and b. The symbols $>$ and $<$ are **inequality signs,** and the expressions $a > b$ and $a < b$ are called **(strict) inequalities.**

Greater Than or Less Than

Notation	Definition	Terminology
$a > b$	$a - b$ is positive	a is greater than b
$a < b$	$a - b$ is negative	a is less than b

If points A and B on a coordinate line have coordinates a and b, respectively, then $a > b$ is equivalent to the statement "A is to the *right* of B," whereas $a < b$ is equivalent to "A is to the *left* of B."

ILLUSTRATION **Greater Than (>) and Less Than (<)**

- $5 > 3$, since $5 - 3 = 2$ is positive.
- $-6 < -2$, since $-6 - (-2) = -6 + 2 = -4$ is negative.
- $\frac{1}{3} > 0.33$, since $\frac{1}{3} - 0.33 = \frac{1}{3} - \frac{33}{100} = \frac{1}{300}$ is positive.
- $7 > 0$, since $7 - 0 = 7$ is positive.
- $-4 < 0$, since $-4 - 0 = -4$ is negative.

The next law enables us to compare, or *order*, any two real numbers.

Trichotomy Law	If a and b are real numbers, then exactly one of the following is true: $$a = b, \quad a > b, \quad \text{or} \quad a < b$$

We refer to the **sign** of a real number as positive if the number is positive, or negative if the number is negative. Two real numbers have *the same sign* if both are positive or both are negative. The numbers have *opposite signs* if one is positive and the other is negative. The following results about the signs of products and quotients of two real numbers a and b can be proved using properties of negatives and quotients.

Laws of Signs

(1) If a and b have the same sign, then ab and $\dfrac{a}{b}$ are positive.

(2) If a and b have opposite signs, then ab and $\dfrac{a}{b}$ are negative.

The **converses*** of the laws of signs are also true. For example, if a quotient is negative, then the numerator and denominator have opposite signs.

The notation $a \geq b$, read "a **is greater than or equal to** b," means that either $a > b$ or $a = b$ (but not both). For example, $a^2 \geq 0$ for every real number a. The symbol $a \leq b$, which is read "a **is less than or equal to** b," means that either $a < b$ or $a = b$. Expressions of the form $a \geq b$ and $a \leq b$ are called **nonstrict inequalities,** since a may be equal to b. As with the equality symbol, we may negate any inequality symbol by putting a slash through it— that is, $\not>$ means not greater than.

An expression of the form $a < b < c$ is called a **continued inequality** and means that both $a < b$ *and* $b < c$; we say "b **is between** a and c." Similarly, the expression $c > b > a$ means that both $c > b$ and $b > a$.

ILLUSTRATION **Ordering Three Real Numbers**

- $1 < 5 < \frac{11}{2}$ - $-4 < \frac{2}{3} < \sqrt{2}$ - $3 > -6 > -10$

There are other types of inequalities. For example, $a < b \leq c$ means both $a < b$ and $b \leq c$. Similarly, $a \leq b < c$ means both $a \leq b$ and $b < c$. Finally, $a \leq b \leq c$ means both $a \leq b$ and $b \leq c$.

EXAMPLE 2   **Determining the sign of a real number**

If $x > 0$ and $y < 0$, determine the sign of $\dfrac{x}{y} + \dfrac{y}{x}$.

SOLUTION   Since x is a positive number and y is a negative number, x and y have opposite signs. Thus, both x/y and y/x are negative. The sum of two negative numbers is a negative number, so

$$\text{the sign of} \quad \frac{x}{y} + \frac{y}{x} \quad \text{is negative.}$$

---

*If a theorem is written in the form "if P, then Q," where P and Q are mathematical statements called the *hypothesis* and *conclusion,* respectively, then the *converse* of the theorem has the form "if Q, then P." If both the theorem and its converse are true, we often write "P if and only if Q" (denoted P iff Q).

Figure 3

$|-4| = 4$ $|4| = 4$

If a is an integer, then it is the coordinate of some point A on a coordinate line, and the symbol $|a|$ denotes the number of units between A and the origin, without regard to direction. The nonnegative number $|a|$ is called the *absolute value of a*. Referring to Figure 3, we see that for the point with coordinate -4 we have $|-4| = 4$. Similarly, $|4| = 4$. In general, *if a is negative, we change its sign to find $|a|$; if a is nonnegative, then $|a| = a$.* The next definition extends this concept to every real number.

Definition of Absolute Value	The **absolute value** of a real number a, denoted by $|a|$, is defined as follows.
	(1) If $a \geq 0$, then $|a| = a$.
	(2) If $a < 0$, then $|a| = -a$.

Since a is negative in part (2) of the definition, $-a$ represents a *positive* real number. Some special cases of this definition are given in the following illustration.

ILLUSTRATION   **The Absolute Value Notation $|a|$**

■ $|3| = 3$, since $3 > 0$.
■ $|-3| = -(-3)$, since $-3 < 0$. Thus, $|-3| = 3$.
■ $|2 - \sqrt{2}| = 2 - \sqrt{2}$, since $2 - \sqrt{2} > 0$.
■ $|\sqrt{2} - 2| = -(\sqrt{2} - 2)$, since $\sqrt{2} - 2 < 0$.
  Thus, $|\sqrt{2} - 2| = 2 - \sqrt{2}$.

In the preceding illustration, $|3| = |-3|$ and $|2 - \sqrt{2}| = |\sqrt{2} - 2|$. In general, we have the following:

$$|a| = |-a|, \text{ for every real number } a$$

EXAMPLE 3   **Removing an absolute value symbol**

If $x < 1$, rewrite $|x - 1|$ without using the absolute value symbol.

SOLUTION   If $x < 1$, then $x - 1 < 0$; that is, $x - 1$ is negative. Hence, by part (2) of the definition of absolute value,

$$|x - 1| = -(x - 1) = -x + 1 = 1 - x.$$

Figure 4

$5 = |7 - 2| = |2 - 7|$

We shall use the concept of absolute value to define the distance between any two points on a coordinate line. First note that the distance between the points with coordinates 2 and 7, shown in Figure 4, equals 5 units. This distance is the difference obtained by subtracting the smaller (leftmost) coordinate from the larger (rightmost) coordinate ($7 - 2 = 5$). If we use absolute values, then, since $|7 - 2| = |2 - 7|$, it is unnecessary to be concerned about the order of subtraction. This fact motivates the next definition.

Definition of the Distance Between Points on a Coordinate Line	Let a and b be the coordinates of two points A and B, respectively, on a co-ordinate line. The **distance between A and B,** denoted by $d(A, B)$, is defined by $$d(A, B) = |b - a|.$$

The number $d(A, B)$ is the length of the line segment AB.

Since $d(B, A) = |a - b|$ and $|b - a| = |a - b|$, we see that

$$d(A, B) = d(B, A).$$

Note that the distance between the origin O and the point A is

$$d(O, A) = |a - 0| = |a|,$$

which agrees with the geometric interpretation of absolute value illustrated in Figure 4. The formula $d(A, B) = |b - a|$ is true regardless of the signs of a and b, as illustrated in the next example.

EXAMPLE 4 **Finding distances between points**

Figure 5

Let A, B, C, and D have coordinates -5, -3, 1, and 6, respectively, on a co-ordinate line, as shown in Figure 5. Find $d(A, B)$, $d(C, B)$, $d(O, A)$, and $d(C, D)$.

SOLUTION Using the definition of the distance between points on a coordinate line, we obtain the distances:

$$d(A, B) = |-3 - (-5)| = |-3 + 5| = |2| = 2$$
$$d(C, B) = |-3 - 1| = |-4| = 4$$
$$d(O, A) = |-5 - 0| = |-5| = 5$$
$$d(C, D) = |6 - 1| = |5| = 5$$

The concept of absolute value has uses other than finding distances between points; it is employed whenever we are interested in the magnitude or numerical value of a real number without regard to its sign.

In the next section we shall discuss the *exponential notation a^n*, where a is a real number (called the *base*) and n is an integer (called an *exponent*). In particular, for base 10 we have

$$10^0 = 1, \quad 10^1 = 10, \quad 10^2 = 10 \cdot 10 = 100, \quad 10^3 = 10 \cdot 10 \cdot 10 = 1000,$$

and so on. For negative exponents we use the reciprocal of the corresponding positive exponent, as follows:

$$10^{-1} = \frac{1}{10^1} = \frac{1}{10}, \quad 10^{-2} = \frac{1}{10^2} = \frac{1}{100}, \quad 10^{-3} = \frac{1}{10^3} = \frac{1}{1000}$$

We can use this notation to write any finite decimal representation of a real number as a sum of the following type:

$$437.56 = 4(100) + 3(10) + 7(1) + 5\left(\tfrac{1}{10}\right) + 6\left(\tfrac{1}{100}\right)$$
$$= 4(10^2) + 3(10^1) + 7(10^0) + 5(10^{-1}) + 6(10^{-2})$$

In the sciences it is often necessary to work with very large or very small numbers and to compare the relative magnitudes of very large or very small quantities. We usually represent a large or small positive number a in *scientific form*, using the symbol $\times$ to denote multiplication.

Scientific Form	$a = c \times 10^n$, where $1 \le c < 10$ and n is an integer

The distance a ray of light travels in one year is approximately 5,900,000,000,000 miles. This number may be written in scientific form as 5.9×10^{12}. The positive exponent 12 indicates that the decimal point should be moved 12 places to the *right*. The notation works equally well for small numbers. The weight of an oxygen molecule is estimated to be

0.000 000 000 000 000 000 000 053 gram,

or, in scientific form, 5.3×10^{-23} gram. The negative exponent indicates that the decimal point should be moved 23 places to the *left*.

ILLUSTRATION **Scientific Form**

- $513 = 5.13 \times 10^2$
- $93{,}000{,}000 = 9.3 \times 10^7$
- $0.000\,000\,000\,43 = 4.3 \times 10^{-10}$

- $7.3 = 7.3 \times 10^0$
- $20{,}700 = 2.07 \times 10^4$
- $0.000\,648 = 6.48 \times 10^{-4}$

Figure 6

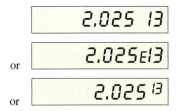

Many calculators use scientific form in their display panels. For the number $c \times 10^n$, the 10 is suppressed and the exponent is often shown preceded by the letter E. For example, to find $(4{,}500{,}000)^2$ on a scientific calculator, we could enter the integer 4,500,000 and press the $\boxed{x^2}$ (or squaring) key, obtaining a display similar to one of those in Figure 6. We would translate this as 2.025×10^{13}. Thus,

$$(4{,}500{,}000)^2 = 20{,}250{,}000{,}000{,}000.$$

Calculators may also use scientific form in the entry of numbers. The user's manual for your calculator should give specific details.

Before we conclude this section, we should briefly consider the issue of rounding off results. Applied problems often include numbers that are obtained by various types of measurements and, hence, are *approximations* to exact values. Such answers should be rounded off, since the final result of a calculation cannot be more accurate than the data that have been used. For example, if the length and width of a rectangle are measured to two-decimal-place accuracy, we cannot expect more than two-decimal-place accuracy in the calculated value of the area of the rectangle. For purely *mathematical* work, if values of the length and width of a rectangle are given, we assume that the dimensions are *exact,* and no rounding off is required.

If a number a is written in scientific form as $a = c \times 10^n$ for $1 \le c < 10$ and if c is rounded off to k decimal places, then we say that a is accurate (or has been rounded off) to $k + 1$ **significant figures, or digits.** For example, 37.2638 rounded to 5 significant figures is 3.7264×10^1, or 37.264; to 3 significant figures, 3.73×10^1, or 37.3; and to 1 significant figure, 4×10^1, or 40.

1.1   *Exercises*

Exer. 1–2: If $x < 0$ and $y > 0$, determine the sign of the real number.

1 (a) xy (b) x^2y (c) $\dfrac{x}{y} + x$ (d) $y - x$

2 (a) $\dfrac{x}{y}$ (b) xy^2 (c) $\dfrac{x - y}{xy}$ (d) $y(y - x)$

Exer. 3–6: Replace the symbol □ with either <, >, or = to make the resulting statement true.

3 (a) $-7 \;\square\; -4$ (b) $\dfrac{\pi}{2} \;\square\; 1.57$ (c) $\sqrt{225} \;\square\; 15$

4 (a) $-3 \;\square\; -5$ (b) $\dfrac{\pi}{4} \;\square\; 0.8$ (c) $\sqrt{289} \;\square\; 17$

5 (a) $\frac{1}{11} \;\square\; 0.09$ (b) $\frac{2}{3} \;\square\; 0.6666$ (c) $\frac{22}{7} \;\square\; \pi$

6 (a) $\frac{1}{7} \;\square\; 0.143$ (b) $\frac{5}{6} \;\square\; 0.833$ (c) $\sqrt{2} \;\square\; 1.4$

Exer. 7–8: Express the statement as an inequality.

7 (a) x is negative.

(b) y is nonnegative.

(c) q is less than or equal to π.

(d) d is between 4 and 2.

(e) t is not less than 5.

(f) The negative of z is not greater than 3.

(g) The quotient of p and q is at most 7.

(h) The reciprocal of w is at least 9.

(i) The absolute value of x is greater than 7.

8 (a) b is positive.

(b) s is nonpositive.

(c) w is greater than or equal to -4.

(d) c is between $\frac{1}{5}$ and $\frac{1}{3}$.

(e) p is not greater than -2.

(f) The negative of m is not less than -2.

(g) The quotient of r and s is at least $\frac{1}{5}$.

(h) The reciprocal of f is at most 14.

(i) The absolute value of x is less than 4.

Exer. 9–14: Rewrite the number without using the absolute value symbol, and simplify the result.

9 (a) $|-3 - 2|$ (b) $|-5| - |2|$ (c) $|7| + |-4|$

10 (a) $|-11 + 1|$ (b) $|6| - |-3|$ (c) $|8| + |-9|$

11 (a) $(-5)|3 - 6|$ (b) $|-6|/(-2)$ (c) $|-7| + |4|$

12 (a) $(4)|6 - 7|$ (b) $5/|-2|$ (c) $|-1| + |-9|$

13 (a) $|4 - \pi|$ (b) $|\pi - 4|$ (c) $|\sqrt{2} - 1.5|$

14 (a) $|\sqrt{3} - 1.7|$ (b) $|1.7 - \sqrt{3}|$ (c) $|\frac{1}{5} - \frac{1}{3}|$

Exer. 15–18: The given numbers are coordinates of points A, B, and C, respectively, on a coordinate line. Find the distance.

(a) $d(A, B)$ (b) $d(B, C)$

(c) $d(C, B)$ (d) $d(A, C)$

15 $3, 7, -5$ **16** $-6, -2, 4$

17 $-9, 1, 10$ **18** $8, -4, -1$

Exer. 19–24: The two given numbers are coordinates of points A and B, respectively, on a coordinate line. Express the indicated statement as an inequality involving the absolute value symbol.

19 x,   7; $d(A, B)$ is less than 5

20 x, $-\sqrt{2}$; $d(A, B)$ is greater than 1

21 x, -3; $d(A, B)$ is at least 8

22 x,   4; $d(A, B)$ is at most 2

23 4, x; $d(A, B)$ is not greater than 3

24 -2, x; $d(A, B)$ is not less than 2

Exer. 25–32: Rewrite the expression without using the absolute value symbol, and simplify the result.

25 $|3 + x|$ if $x < -3$ **26** $|5 - x|$ if $x > 5$

27 $|2 - x|$ if $x < 2$ **28** $|7 + x|$ if $x \geq -7$

29 $|a - b|$ if $a < b$ **30** $|a - b|$ if $a > b$

31 $|x^2 + 4|$ **32** $|-x^2 - 1|$

Exer. 33–40: Replace the symbol $\square$ with either $=$ or $\neq$ to make the resulting statement true for all real numbers a, b, c, and d, whenever the expressions are defined.

33 $\dfrac{ab + ac}{a} \square b + ac$ **34** $\dfrac{ab + ac}{a} \square b + c$

35 $\dfrac{b + c}{a} \square \dfrac{b}{a} + \dfrac{c}{a}$ **36** $\dfrac{a + c}{b + d} \square \dfrac{a}{b} + \dfrac{c}{d}$

37 $(a \div b) \div c \square a \div (b \div c)$

38 $(a - b) - c \square a - (b - c)$

39 $\dfrac{a - b}{b - a} \square -1$ **40** $-(a + b) \square -a + b$

Exer. 41–42: Approximate the real-number expression to four decimal places.

41 (a) $|3.2^2 - \sqrt{3.15}|$

(b) $\sqrt{(15.6 - 1.5)^2 + (4.3 - 5.4)^2}$

42 (a) $\dfrac{3.42 - 1.29}{5.83 + 2.64}$

(b) π^3

Exer. 43–44: Approximate the real-number expression. Express the answer in scientific notation accurate to four significant figures.

43 (a) $\dfrac{1.2 \times 10^3}{3.1 \times 10^2 + 1.52 \times 10^3}$

(b) $(1.23 \times 10^{-4}) + \sqrt{4.5 \times 10^3}$

44 (a) $\sqrt{|3.45 - 1.2 \times 10^4| + 10^5}$

(b) $(1.791 \times 10^2) \times (9.84 \times 10^3)$

45 The point on a coordinate line corresponding to $\sqrt{2}$ may be determined by constructing a right triangle with sides of length 1, as shown in the figure. Determine the points that correspond to $\sqrt{3}$ and $\sqrt{5}$, respectively. (*Hint:* Use the Pythagorean theorem.)

Exercise 45

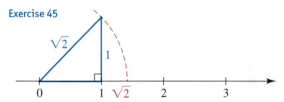

46 A circle of radius 1 rolls along a coordinate line in the positive direction, as shown in the figure. If point P is initially at the origin, find the coordinate of P after one, two, and ten complete revolutions.

Exercise 46

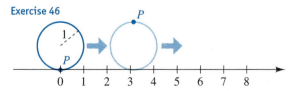

47 Geometric proofs of properties of real numbers were first given by the ancient Greeks. In order to establish the distributive property $a(b + c) = ab + ac$ for positive real numbers a, b, and c, find the area of the rectangle shown in the figure on the next page in two ways.

Exercise 47

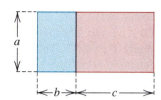

48 Rational approximations to square roots can be found using a formula discovered by the ancient Babylonians. Let x_1 be the first rational approximation for $\sqrt{n}$. If we let

$$x_2 = \frac{1}{2}\left(x_1 + \frac{n}{x_1}\right),$$

then x_2 will be a better approximation for $\sqrt{n}$, and we can repeat the computation with x_2 replacing x_1. Starting with $x_1 = \frac{3}{2}$, find the next two rational approximations for $\sqrt{2}$.

Exer. 49–50: Express the number in scientific form.

49 (a) 427,000   **(b)** 0.000 000 098   **(c)** 810,000,000

50 (a) 85,200   **(b)** 0.000 005 5   **(c)** 24,900,000

Exer. 51–52: Express the number in decimal form.

51 (a) 8.3×10^5 **(b)** 2.9×10^{-12} **(c)** 5.63×10^8

52 (a) 2.3×10^7 **(b)** 7.01×10^{-9} **(c)** 1.23×10^{10}

53 Mass of a hydrogen atom The mass of a hydrogen atom is approximately

0.000 000 000 000 000 000 000 001 7 gram.

Express this number in scientific form.

54 Mass of an electron The mass of an electron is approximately 9.1×10^{-31} kilogram. Express this number in decimal form.

55 Light year In astronomy, distances to stars are measured in light years. One light year is the distance a ray of light travels in one year. If the speed of light is approximately 186,000 miles per second, estimate the number of miles in one light year.

56 Milky Way galaxy

(a) Astronomers have estimated that the Milky Way galaxy contains 100 billion stars. Express this number in scientific form.

(b) The diameter d of the Milky Way galaxy is estimated as 100,000 light years. Express d in miles. (Refer to Exercise 55.)

57 Avogadro's number The number of hydrogen atoms in a mole is Avogadro's number, 6.02×10^{23}. If one mole of the gas has a mass of 1.01 grams, estimate the mass of a hydrogen atom.

58 Fish population The population dynamics of many fish are characterized by extremely high fertility rates among adults and very low survival rates among the young. A mature halibut may lay as many as 2.5 million eggs, but only 0.00035% of the offspring survive to the age of 3 years. Use scientific form to approximate the number of offspring that live to age 3.

59 Frames in a movie film The longest movie ever made is a 1970 British film that runs for 48 hours. Assuming that the film speed is 24 frames per second, approximate the total number of frames in this film. Express your answer in scientific form.

60 Large prime numbers The number $2^{44,497} - 1$ is prime. At the time that this number was determined to be prime, it took one of the world's fastest computers about 60 days to verify that it was prime. This computer was capable of performing 2×10^{11} calculations per second. Use scientific form to estimate the number of calculations needed to perform this computation. (More recently, in 1992, $2^{756,839} - 1$, a number containing 227,832 digits, was shown to be prime.)

61 Tornado pressure When a tornado passes near a building, there is a rapid drop in the outdoor pressure and the indoor pressure does not have time to change. The resulting difference is capable of causing an outward pressure of 1.4 lb/in^2 on the walls and ceiling of the building.

(a) Calculate the force in pounds exerted on 1 square foot of a wall.

(b) Estimate the tons of force exerted on a wall that is 8 feet high and 40 feet wide.

62 Cattle population A rancher has 750 head of cattle consisting of 400 adults (aged 2 or more years), 150 yearlings, and 200 calves. The following information is known about this particular species. Each spring an adult female gives birth to a single calf, and 75% of these calves will survive the first year. The yearly survival percentages for yearlings and adults are 80% and 90%, respectively. The male-female ratio is one in all age classes. Estimate the population of each age class

(a) next spring   **(b)** last spring

1.2

Exponents and Radicals

If n is a positive integer, the exponential notation a^n, defined in the following chart, represents the product of the real number a with itself n times. We refer to a^n as ***a* to the *n*th power** or, simply, *a to the n*. The positive integer n is called the **exponent,** and the real number a is called the **base.**

Exponential Notation

General case (n is any positive integer)	Special cases
$a^n = \underbrace{a \cdot a \cdot a \cdot \,\cdots\, \cdot a}_{n \text{ factors of } a}$	$a^1 = a$ $a^2 = a \cdot a$ $a^3 = a \cdot a \cdot a$ $a^6 = a \cdot a \cdot a \cdot a \cdot a \cdot a$

The next illustration contains several numerical examples of exponential notation.

ILLUSTRATION   **The Exponential Notation a^n**

- $5^4 = 5 \cdot 5 \cdot 5 \cdot 5 = 625$
- $\left(\frac{1}{2}\right)^5 = \frac{1}{2} \cdot \frac{1}{2} \cdot \frac{1}{2} \cdot \frac{1}{2} \cdot \frac{1}{2} = \frac{1}{32}$
- $(-3)^3 = (-3)(-3)(-3) = -27$
- $\left(-\frac{1}{3}\right)^4 = \left(-\frac{1}{3}\right)\left(-\frac{1}{3}\right)\left(-\frac{1}{3}\right)\left(-\frac{1}{3}\right) = \left(\frac{1}{9}\right)\left(\frac{1}{9}\right) = \frac{1}{81}$

It is important to note that if n is a positive integer, then an expression such as $3a^n$ means $3(a^n)$, *not* $(3a)^n$. The real number 3 is the **coefficient** of a^n in the expression $3a^n$. Similarly, $-3a^n$ means $(-3)a^n$, *not* $(-3a)^n$.

ILLUSTRATION   **The Notation ca^n**

- $5 \cdot 2^3 = 5 \cdot 8 = 40$
- $-5 \cdot 2^3 = -5 \cdot 8 = -40$
- $-2^4 = -(2^4) = -16$
- $3(-2)^3 = 3(-2)(-2)(-2) = 3(-8) = -24$

We next extend the definition of a^n to nonpositive exponents.

Zero and Negative (Nonpositive) Exponents

Definition ($a \neq 0$)	Illustrations
$a^0 = 1$	$3^0 = 1, \qquad \left(-\sqrt{2}\right)^0 = 1$
$a^{-n} = \dfrac{1}{a^n}$	$5^{-3} = \dfrac{1}{5^3}, \qquad (-3)^{-5} = \dfrac{1}{(-3)^5}$

If m and n are positive integers, then

$$a^m a^n = \underbrace{a \cdot a \cdot a \cdot \,\cdots\, \cdot a}_{m \text{ factors of } a} \cdot \underbrace{a \cdot a \cdot a \cdot \,\cdots\, \cdot a}_{n \text{ factors of } a}.$$

Since the total number of factors of a on the right is $m + n$, this expression is equal to a^{m+n}; that is,

$$a^m a^n = a^{m+n}.$$

We can extend this formula to $m \le 0$ or $n \le 0$ by using the definitions of the zero exponent and negative exponents. This gives us law 1, stated in the next chart.

To prove law 2, we may write, for m and n positive,

$$(a^m)^n = \underbrace{a^m \cdot a^m \cdot a^m \cdot \,\cdots\, \cdot a^m}_{n \text{ factors of } a^m}$$

and count the number of times a appears as a factor on the right-hand side. Since $a^m = a \cdot a \cdot a \cdot \,\cdots\, \cdot a$, with a occurring as a factor m times, and since the number of such groups of m factors is n, the total number of factors of a is $m \cdot n$. Thus,

$$(a^m)^n = a^{mn}.$$

The cases $m \le 0$ and $n \le 0$ can be proved using the definition of nonpositive exponents. The remaining three laws can be established in similar fashion by counting factors. In laws 4 and 5 we assume that denominators are not 0.

Laws of Exponents for Real Numbers a and b and Integers m and n

Law	Illustration
(1) $a^m a^n = a^{m+n}$	$2^3 \cdot 2^4 = 2^{3+4} = 2^7 = 128$
(2) $(a^m)^n = a^{mn}$	$(2^3)^4 = 2^{3 \cdot 4} = 2^{12} = 4096$
(3) $(ab)^n = a^n b^n$	$(20)^3 = (2 \cdot 10)^3 = 2^3 \cdot 10^3 = 8 \cdot 1000 = 8000$
(4) $\left(\dfrac{a}{b}\right)^n = \dfrac{a^n}{b^n}$	$\left(\dfrac{2}{5}\right)^3 = \dfrac{2^3}{5^3} = \dfrac{8}{125}$
(5) (a) $\dfrac{a^m}{a^n} = a^{m-n}$	$\dfrac{2^5}{2^3} = 2^{5-3} = 2^2 = 4$
(b) $\dfrac{a^m}{a^n} = \dfrac{1}{a^{n-m}}$	$\dfrac{2^3}{2^5} = \dfrac{1}{2^{5-3}} = \dfrac{1}{2^2} = \dfrac{1}{4}$

We usually use 5(a) if $m > n$ and 5(b) if $m < n$.

We can extend laws of exponents to obtain rules such as $(abc)^n = a^n b^n c^n$ and $a^m a^n a^p = a^{m+n+p}$. Some other examples of the laws of exponents are given in the next illustration.

ILLUSTRATION   **Laws of Exponents**

- $x^5 x^6 x^2 = x^{5+6+2} = x^{13}$
- $(y^5)^7 = y^{5 \cdot 7} = y^{35}$

- $(3st)^4 = 3^4 s^4 t^4 = 81 s^4 t^4$
- $\left(\dfrac{p}{2}\right)^5 = \dfrac{p^5}{2^5} = \dfrac{p^5}{32}$

- $\dfrac{c^8}{c^3} = c^{8-3} = c^5$
- $\dfrac{u^3}{u^8} = \dfrac{1}{u^{8-3}} = \dfrac{1}{u^5}$

To **simplify** an expression involving powers of real numbers means to change it to an expression in which each real number appears only once and all exponents are positive. *We shall assume that denominators always represent nonzero real numbers.*

EXAMPLE 1   **Simplifying expressions containing exponents**

Use laws of exponents to simplify each expression:

(a) $(3x^3 y^4)(4xy^5)$ **(b)** $(2a^2 b^3 c)^4$ **(c)** $\left(\dfrac{2r^3}{s}\right)^2 \left(\dfrac{s}{r^3}\right)^3$ **(d)** $(u^{-2} v^3)^{-3}$

SOLUTION

(a) $(3x^3 y^4)(4xy^5) = (3)(4)x^3 xy^4 y^5$ rearrange factors

$\qquad = 12x^4 y^9$ law 1

(b) $\quad (2a^2 b^3 c)^4 = 2^4 (a^2)^4 (b^3)^4 c^4$ law 3

$\qquad = 16a^8 b^{12} c^4$ law 2

(c) $\left(\dfrac{2r^3}{s}\right)^2 \left(\dfrac{s}{r^3}\right)^3 = \dfrac{(2r^3)^2}{s^2} \cdot \dfrac{s^3}{(r^3)^3}$ law 4

$\qquad = \dfrac{2^2 (r^3)^2}{s^2} \cdot \dfrac{s^3}{(r^3)^3}$ law 3

$\qquad = \left(\dfrac{4r^6}{s^2}\right)\left(\dfrac{s^3}{r^9}\right)$ law 2

$\qquad = 4\left(\dfrac{r^6}{r^9}\right)\left(\dfrac{s^3}{s^2}\right)$ rearrange factors

$\qquad = 4\left(\dfrac{1}{r^3}\right)(s)$ laws 5(b) and 5(a)

$\qquad = \dfrac{4s}{r^3}$ rearrange factors

(d) $\quad (u^{-2} v^3)^{-3} = (u^{-2})^{-3} (v^3)^{-3}$ law 3

$\qquad = u^6 v^{-9}$ law 2

$\qquad = \dfrac{u^6}{v^9}$ definition of a^{-n}

The following theorem is useful for problems that involve negative exponents.

Theorem on Negative Exponents	(1) $\dfrac{a^{-m}}{b^{-n}} = \dfrac{b^{n}}{a^{m}}$	(2) $\left(\dfrac{a}{b}\right)^{-n} = \left(\dfrac{b}{a}\right)^{n}$

PROOFS   Using properties of negative exponents and quotients, we obtain

(1) $\dfrac{a^{-m}}{b^{-n}} = \dfrac{1/a^{m}}{1/b^{n}} = \dfrac{1}{a^{m}} \cdot \dfrac{b^{n}}{1} = \dfrac{b^{n}}{a^{m}}$

(2) $\left(\dfrac{a}{b}\right)^{-n} = \dfrac{a^{-n}}{b^{-n}} = \dfrac{b^{n}}{a^{n}} = \left(\dfrac{b}{a}\right)^{n}$

EXAMPLE 2   **Simplifying expressions containing negative exponents**

Simplify:

(a) $\dfrac{8x^{3}y^{-5}}{4x^{-1}y^{2}}$ (b) $\left(\dfrac{u^{2}}{2v}\right)^{-3}$

SOLUTION   We apply the theorem on negative exponents and the laws of exponents.

(a) $\dfrac{8x^{3}y^{-5}}{4x^{-1}y^{2}} = \dfrac{8x^{3}}{4y^{2}} \cdot \dfrac{y^{-5}}{x^{-1}}$ rearrange quotients so that negative exponents are in one fraction

$ = \dfrac{8x^{3}}{4y^{2}} \cdot \dfrac{x^{1}}{y^{5}}$ theorem on negative exponents (1)

$ = \dfrac{2x^{4}}{y^{7}}$ law 1 of exponents

(b) $\left(\dfrac{u^{2}}{2v}\right)^{-3} = \left(\dfrac{2v}{u^{2}}\right)^{3}$ theorem on negative exponents (2)

$ = \dfrac{2^{3}v^{3}}{(u^{2})^{3}}$ laws 4 and 3 of exponents

$ = \dfrac{8v^{3}}{u^{6}}$ law 2 of exponents

We next define the **principal nth root** $\sqrt[n]{a}$ of a real number a.

Definition of $\sqrt[n]{a}$	Let n be a positive integer greater than 1, and let a be a real number.
	(1) If $a = 0$, then $\sqrt[n]{a} = 0$.
	(2) If $a > 0$, then $\sqrt[n]{a}$ is the *positive* real number b such that $b^n = a$.
	(3) **(a)** If $a < 0$ and n is odd, then $\sqrt[n]{a}$ is the *negative* real number b such that $b^n = a$.
	(b) If $a < 0$ and n is even, then $\sqrt[n]{a}$ is not a real number.

Complex numbers, discussed in Section 2.4, are needed to define $\sqrt[n]{a}$ if $a < 0$ and n is an *even* positive integer, because for all real numbers b, $b^n \geq 0$ whenever n is even.

If $n = 2$, we write $\sqrt{a}$ instead of $\sqrt[2]{a}$ and call $\sqrt{a}$ the **principal square root** of a or, simply, the **square root** of a. The number $\sqrt[3]{a}$ is the (principal) **cube root** of a.

ILLUSTRATION   **The Principal nth Root $\sqrt[n]{a}$**

- ■ $\sqrt{16} = 4$, since $4^2 = 16$.
- ■ $\sqrt[5]{\frac{1}{32}} = \frac{1}{2}$, since $\left(\frac{1}{2}\right)^5 = \frac{1}{32}$.
- ■ $\sqrt[3]{-8} = -2$, since $(-2)^3 = -8$.
- ■ $\sqrt[4]{-16}$ is not a real number.

Note that $\sqrt{16} \neq \pm 4$, since, by definition, roots of positive real numbers are positive. The symbol $\pm$ is read "plus or minus."

To complete our terminology, the expression $\sqrt[n]{a}$ is a **radical,** the number a is the **radicand,** and n is the **index** of the radical. The symbol $\sqrt{}$ is called a **radical sign.**

If $\sqrt{a} = b$, then $b^2 = a$; that is, $\left(\sqrt{a}\right)^2 = a$. If $\sqrt[3]{a} = b$, then $b^3 = a$, or $\left(\sqrt[3]{a}\right)^3 = a$. Generalizing this pattern gives us property 1 in the next chart.

Properties of $\sqrt[n]{a}$ (n is a positive integer)

Property	Illustrations	
(1) $\left(\sqrt[n]{a}\right)^n = a$ if $\sqrt[n]{a}$ is a real number	$\left(\sqrt{5}\right)^2 = 5$,	$\left(\sqrt[3]{-8}\right)^3 = -8$
(2) $\sqrt[n]{a^n} = a$ if $a \geq 0$	$\sqrt{5^2} = 5$,	$\sqrt[3]{2^3} = 2$
(3) $\sqrt[n]{a^n} = a$ if $a < 0$ and n is odd	$\sqrt[3]{(-2)^3} = -2$,	$\sqrt[5]{(-2)^5} = -2$
(4) $\sqrt[n]{a^n} = |a|$ if $a < 0$ and n is even	$\sqrt{(-3)^2} = |-3| = 3$,	$\sqrt[4]{(-2)^4} = |-2| = 2$

If $a \geq 0$, then property 4 reduces to property 2. We also see from property 4 that

$$\sqrt{x^2} = |x|$$

for every real number x. In particular, if $x \geq 0$, then $\sqrt{x^2} = x$; however, if $x < 0$, then $\sqrt{x^2} = -x$, which is positive.

The three laws listed in the next chart are true for positive integers m and n, *provided the indicated roots **exist**—*that is, provided the roots are real numbers.

Laws of Radicals

Law	Illustrations
(1) $\sqrt[n]{ab} = \sqrt[n]{a}\,\sqrt[n]{b}$	$\sqrt{50} = \sqrt{25 \cdot 2} = \sqrt{25}\,\sqrt{2} = 5\sqrt{2}$
	$\sqrt[3]{-108} = \sqrt[3]{(-27)(4)} = \sqrt[3]{-27}\,\sqrt[3]{4} = -3\sqrt[3]{4}$
(2) $\sqrt[n]{\dfrac{a}{b}} = \dfrac{\sqrt[n]{a}}{\sqrt[n]{b}}$	$\sqrt[3]{\dfrac{5}{8}} = \dfrac{\sqrt[3]{5}}{\sqrt[3]{8}} = \dfrac{\sqrt[3]{5}}{2}$
(3) $\sqrt[m]{\sqrt[n]{a}} = \sqrt[mn]{a}$	$\sqrt{\sqrt[3]{64}} = \sqrt[2(3)]{64} = \sqrt[6]{2^6} = 2$

The radicands in laws 1 and 2 involve products and quotients. Care must be taken if sums or differences occur in the radicand. The following chart contains two particular warnings concerning commonly made mistakes.

 Warning!

If $a \neq 0$ and $b \neq 0$	Illustration
(1) $\sqrt{a^2 + b^2} \neq a + b$	$\sqrt{3^2 + 4^2} = \sqrt{25} = 5 \neq 3 + 4 = 7$
(2) $\sqrt{a + b} \neq \sqrt{a} + \sqrt{b}$	$\sqrt{4 + 9} = \sqrt{13} \neq \sqrt{4} + \sqrt{9} = 5$

If c is a real number and c^n occurs as a factor in a radical of index n, then we can remove c from the radicand if the sign of c is taken into account. For example, if $c > 0$ or if $c < 0$ and n is *odd,* then

$$\sqrt[n]{c^n d} = \sqrt[n]{c^n}\,\sqrt[n]{d} = c\sqrt[n]{d},$$

provided $\sqrt[n]{d}$ exists. If $c < 0$ and n is *even,* then

$$\sqrt[n]{c^n d} = \sqrt[n]{c^n}\,\sqrt[n]{d} = |c|\sqrt[n]{d},$$

provided $\sqrt[n]{d}$ exists.

ILLUSTRATION **Removing nth Powers from $\sqrt[n]{}$**

■ $\sqrt[5]{x^7} = \sqrt[5]{x^5 \cdot x^2} = \sqrt[5]{x^5}\,\sqrt[5]{x^2} = x\sqrt[5]{x^2}$

■ $\sqrt[3]{x^7} = \sqrt[3]{x^6 \cdot x} = \sqrt[3]{(x^2)^3 x} = \sqrt[3]{(x^2)^3}\,\sqrt[3]{x} = x^2\sqrt[3]{x}$

■ $\sqrt{x^2 y} = \sqrt{x^2}\,\sqrt{y} = |x|\sqrt{y}$

■ $\sqrt{x^6} = \sqrt{(x^3)^2} = |x^3|$

■ $\sqrt[4]{x^6 y^3} = \sqrt[4]{x^4 \cdot x^2 y^3} = \sqrt[4]{x^4}\,\sqrt[4]{x^2 y^3} = |x|\sqrt[4]{x^2 y^3}$

Note: To avoid considering absolute values, *in examples and exercises involving radicals in this chapter, we shall assume that all letters—a, b, c, d, x, y,* *and so on—that appear in radicands represent positive real numbers, unless otherwise specified.*

As shown in the preceding illustration and in the following examples, if the index of a radical is n, then we rearrange the radicand, isolating a factor of the form p^n, where p may consist of several letters. We then remove $\sqrt[n]{p^n} = p$ from the radical, as previously indicated. Thus, in Example 3(b) the index of the radical is 3 and we rearrange the radicand into *cubes,* obtaining a factor p^3, with $p = 2xy^2z$. In part (c) the index of the radical is 2 and we rearrange the radicand into *squares,* obtaining a factor p^2, with $p = 3a^3b^2$.

To *simplify a radical* means to remove factors from the radical until no factor in the radicand has an exponent greater than or equal to the index of the radical and the index is as low as possible.

EXAMPLE 3 Removing factors from radicals

Simplify each radical (all letters denote positive real numbers):

(a) $\sqrt[3]{320}$ (b) $\sqrt[3]{16x^3y^8z^4}$ (c) $\sqrt{3a^2b^3}\,\sqrt{6a^5b}$

SOLUTION

(a) $\sqrt[3]{320} = \sqrt[3]{64 \cdot 5}$ factor out the largest cube in 320

$\qquad\qquad = \sqrt[3]{4^3}\,\sqrt[3]{5}$ law 1 of radicals

$\qquad\qquad = 4\sqrt[3]{5}$ property 2 of $\sqrt[n]{\ }$

(b) $\sqrt[3]{16x^3y^8z^4} = \sqrt[3]{(2^3x^3y^6z^3)(2y^2z)}$ rearrange radicand into cubes

$\qquad\qquad\qquad = \sqrt[3]{(2xy^2z)^3(2y^2z)}$ laws 2 and 3 of exponents

$\qquad\qquad\qquad = \sqrt[3]{(2xy^2z)^3}\,\sqrt[3]{2y^2z}$ law 1 of radicals

$\qquad\qquad\qquad = 2xy^2z\,\sqrt[3]{2y^2z}$ property 2 of $\sqrt[n]{\ }$

(c) $\sqrt{3a^2b^3}\,\sqrt{6a^5b} = \sqrt{3a^2b^3 \cdot 2 \cdot 3a^5b}$ law 1 of radicals

$\qquad\qquad\qquad = \sqrt{(3^2a^6b^4)(2a)}$ rearrange radicand into squares

$\qquad\qquad\qquad = \sqrt{(3a^3b^2)^2(2a)}$ laws 2 and 3 of exponents

$\qquad\qquad\qquad = \sqrt{(3a^3b^2)^2}\,\sqrt{2a}$ law 1 of radicals

$\qquad\qquad\qquad = 3a^3b^2\sqrt{2a}$ property 2 of $\sqrt[n]{\ }$ ◢

If the denominator of a quotient contains a factor of the form $\sqrt[n]{a^k}$, with $k < n$ and $a > 0$, then multiplying the numerator and denominator by $\sqrt[n]{a^{n-k}}$ will eliminate the radical from the denominator, since

$$\sqrt[n]{a^k}\,\sqrt[n]{a^{n-k}} = \sqrt[n]{a^{k+n-k}} = \sqrt[n]{a^n} = a.$$

This process is called **rationalizing a denominator.** Some special cases are listed in the following chart.

Rationalizing Denominators of Quotients ($a > 0$)

Factor in denominator	Multiply numerator and denominator by	Resulting factor
$\sqrt{a}$	$\sqrt{a}$	$\sqrt{a}\,\sqrt{a} = \sqrt{a^2} = a$
$\sqrt[3]{a}$	$\sqrt[3]{a^2}$	$\sqrt[3]{a}\,\sqrt[3]{a^2} = \sqrt[3]{a^3} = a$
$\sqrt[3]{a^3}$	$\sqrt[3]{a^4}$	$\sqrt[3]{a^3}\,\sqrt[3]{a^4} = \sqrt[3]{a^7} = a$

The next example illustrates this technique.

EXAMPLE 4   Rationalizing denominators

Rationalize each denominator:

(a) $\dfrac{1}{\sqrt{5}}$ (b) $\dfrac{1}{\sqrt[3]{x}}$ (c) $\sqrt{\dfrac{2}{3}}$ (d) $\sqrt[5]{\dfrac{x}{y^2}}$

SOLUTION

(a) $\dfrac{1}{\sqrt{5}} = \dfrac{1}{\sqrt{5}}\dfrac{\sqrt{5}}{\sqrt{5}} = \dfrac{\sqrt{5}}{\sqrt{5^2}} = \dfrac{\sqrt{5}}{5}$

(b) $\dfrac{1}{\sqrt[3]{x}} = \dfrac{1}{\sqrt[3]{x}}\dfrac{\sqrt[3]{x^2}}{\sqrt[3]{x^2}} = \dfrac{\sqrt[3]{x^2}}{\sqrt[3]{x^3}} = \dfrac{\sqrt[3]{x^2}}{x}$

(c) $\sqrt{\dfrac{2}{3}} = \dfrac{\sqrt{2}}{\sqrt{3}} = \dfrac{\sqrt{2}}{\sqrt{3}}\dfrac{\sqrt{3}}{\sqrt{3}} = \dfrac{\sqrt{2\cdot3}}{\sqrt{3^2}} = \dfrac{\sqrt{6}}{3}$

(d) $\sqrt[5]{\dfrac{x}{y^2}} = \dfrac{\sqrt[5]{x}}{\sqrt[5]{y^2}} = \dfrac{\sqrt[5]{x}}{\sqrt[5]{y^2}}\dfrac{\sqrt[5]{y^3}}{\sqrt[5]{y^3}} = \dfrac{\sqrt[5]{xy^3}}{\sqrt[5]{y^5}} = \dfrac{\sqrt[5]{xy^3}}{y}$

If we use a calculator to find decimal approximations of radicals, there is no advantage in rationalizing denominators, such as $1/\sqrt{5} = \sqrt{5}/5$ or $\sqrt{2/3} = \sqrt{6}/3$, as we did in Example 4(a) and (c). However, for *algebraic* simplifications, changing expressions to such forms is sometimes desirable. Similarly, in advanced mathematics courses such as calculus, changing $1/\sqrt[3]{x}$ to $\sqrt[3]{x^2}/x$, as in Example 4(b), could make a problem *more* complicated. In such courses it is simpler to work with the expression $1/\sqrt[3]{x}$ than with its rationalized form.

We next use radicals to define *rational exponents*.

Definition of Rational Exponents	Let m/n be a rational number, where n is a positive integer greater than 1. If a is a real number such that $\sqrt[n]{a}$ exists, then

$$\textbf{(1)} \quad a^{1/n} = \sqrt[n]{a}$$

$$\textbf{(2)} \quad a^{m/n} = \left(\sqrt[n]{a}\right)^m = \sqrt[n]{a^m}$$

$$\textbf{(3)} \quad a^{m/n} = (a^{1/n})^m = (a^m)^{1/n}$$

When evaluating $a^{m/n}$ in (2), we usually use $\left(\sqrt[n]{a}\right)^m$; that is, we take the nth root of a first and then raise that result to the mth power, as shown in the following illustration.

ILLUSTRATION　**The Exponential Notation $a^{m/n}$**

- $x^{1/3} = \sqrt[3]{x}$　　■ $x^{3/5} = \left(\sqrt[5]{x}\right)^3 = \sqrt[5]{x^3}$
- $125^{2/3} = \left(\sqrt[3]{125}\right)^2 = \left(\sqrt[3]{5^3}\right)^2 = 5^2 = 25$
- $\left(\frac{32}{243}\right)^{3/5} = \left(\sqrt[5]{\frac{32}{243}}\right)^3 = \left(\sqrt[5]{\left(\frac{2}{3}\right)^5}\right)^3 = \left(\frac{2}{3}\right)^3 = \frac{8}{27}$

The laws of exponents are true for rational exponents and also for *irrational* exponents, such as $3^{\sqrt{2}}$ or 5^π, considered in Chapter 5.

To simplify an expression involving rational powers of letters that represent real numbers, we change it to an expression in which each letter appears only once and all exponents are positive. As we did with radicals, we shall assume that all letters represent positive real numbers unless otherwise specified.

EXAMPLE 5　**Simplifying rational powers**

Simplify:

(a) $(-27)^{2/3}(4)^{-5/2}$　　**(b)** $(r^2 s^6)^{1/3}$　　**(c)** $\left(\dfrac{2x^{2/3}}{y^{1/2}}\right)^2 \left(\dfrac{3x^{-5/6}}{y^{1/3}}\right)$

SOLUTION

(a)　　$(-27)^{2/3}(4)^{-5/2} = \left(\sqrt[3]{-27}\right)^2 \left(\sqrt{4}\right)^{-5}$　definition of rational exponents

$= (-3)^2 (2)^{-5}$　take roots

$= \dfrac{(-3)^2}{2^5}$　definition of negative exponents

$= \dfrac{9}{32}$　take powers

(b)　　$(r^2 s^6)^{1/3} = (r^2)^{1/3}(s^6)^{1/3}$　law 3 of exponents

$= r^{2/3} s^2$　law 2 of exponents

(c)　$\left(\dfrac{2x^{2/3}}{y^{1/2}}\right)^2 \left(\dfrac{3x^{-5/6}}{y^{1/3}}\right) = \left(\dfrac{4x^{4/3}}{y}\right)\left(\dfrac{3x^{-5/6}}{y^{1/3}}\right)$　laws of exponents

$= \dfrac{(4 \cdot 3)x^{4/3 - 5/6}}{y^{1 + (1/3)}}$　law 1 of exponents

$= \dfrac{12x^{8/6 - 5/6}}{y^{4/3}}$　common denominator

$= \dfrac{12x^{1/2}}{y^{4/3}}$　simplify

Rational exponents are useful for problems involving radicals that do not have the same index, as illustrated in the next example.

EXAMPLE 6   **Combining radicals**

Change to an expression containing one radical of the form $\sqrt[n]{a^m}$:

(a) $\sqrt[3]{a}\,\sqrt{a}$ (b) $\dfrac{\sqrt[4]{a}}{\sqrt[3]{a^2}}$

SOLUTION   Introducing rational exponents, we obtain

(a) $\sqrt[3]{a}\,\sqrt{a} = a^{1/3}a^{1/2} = a^{(1/3)+(1/2)} = a^{5/6} = \sqrt[6]{a^5}$

(b) $\dfrac{\sqrt[4]{a}}{\sqrt[3]{a^2}} = \dfrac{a^{1/4}}{a^{2/3}} = a^{(1/4)-(2/3)} = a^{-5/12} = \dfrac{1}{a^{5/12}} = \dfrac{1}{\sqrt[12]{a^5}}$

In Exercises 1.2, whenever an index of a radical is even (or a rational exponent m/n with n even is employed), assume that the letters that appear in the radicand denote positive real numbers unless otherwise specified.

1.2   *Exercises*

Exer. 1–10: Express the number in the form a/b, where a and b are integers.

1 $\left(-\tfrac{2}{3}\right)^4$

2 $(-3)^3$

3 $\dfrac{2^{-3}}{3^{-2}}$

4 $\dfrac{2^0 + 0^2}{2 + 0}$

5 $-2^4 + 3^{-1}$

6 $\left(-\tfrac{3}{2}\right)^4 - 2^{-4}$

7 $16^{-3/4}$

8 $9^{5/2}$

9 $(-0.008)^{2/3}$

10 $(0.008)^{-2/3}$

Exer. 11–46: Simplify.

11 $\left(\tfrac{1}{2}x^4\right)(16x^5)$

12 $(-3x^{-2})(4x^4)$

13 $\dfrac{(2x^3)(3x^2)}{(x^2)^3}$

14 $\dfrac{(2x^2)^3}{4x^4}$

15 $\left(\tfrac{1}{6}a^5\right)(-3a^2)(4a^7)$

16 $(-4b^3)\left(\tfrac{1}{6}b^2\right)(-9b^4)$

17 $\dfrac{(6x^3)^2}{(2x^2)^3} \cdot (3x^2)^0$

18 $\dfrac{(3y^3)(2y^2)^2}{(y^4)^3} \cdot (y^3)^0$

19 $(3u^7v^3)(4u^4v^{-5})$

20 $(x^2yz^3)(-2xz^2)(x^3y^{-2})$

21 $(8x^4y^{-3})\left(\tfrac{1}{2}x^{-5}y^2\right)$

22 $\left(\dfrac{4a^2b}{a^3b^2}\right)\left(\dfrac{5a^2b}{2b^4}\right)$

23 $\left(\tfrac{1}{3}x^4y^{-3}\right)^{-2}$

24 $(-2xy^2)^5\left(\dfrac{x^7}{8y^3}\right)$

25 $(3y^3)^4(4y^2)^{-3}$

26 $(-3a^2b^{-5})^3$

27 $(-2r^4s^{-3})^{-2}$

28 $(2x^2y^{-5})(6x^{-3}y)\left(\tfrac{1}{3}x^{-1}y^3\right)$

29 $(5x^2y^{-3})(4x^{-5}y^4)$

30 $(-2r^2s)^5(3r^{-1}s^3)^2$

31 $\left(\dfrac{3x^5y^4}{x^0y^{-3}}\right)^2$

32 $(4a^2b)^4\left(\dfrac{-a^3}{2b}\right)^2$

33 $(4a^{3/2})(2a^{1/2})$

34 $(-6x^{7/5})(2x^{8/5})$

35 $(3x^{5/6})(8x^{2/3})$

36 $(8r)^{1/3}(2r^{1/2})$

37 $(27a^6)^{-2/3}$

38 $(25z^4)^{-3/2}$

39 $(8x^{-2/3})x^{1/6}$

40 $(3x^{1/2})(-2x^{5/2})$

41 $\left(\dfrac{-8x^3}{y^{-6}}\right)^{2/3}$

42 $\left(\dfrac{-y^{3/2}}{y^{-1/3}}\right)^3$

43 $\left(\dfrac{x^6}{9y^{-4}}\right)^{-1/2}$

44 $\left(\dfrac{c^{-4}}{16d^8}\right)^{3/4}$

45 $\dfrac{(x^6y^3)^{-1/3}}{(x^4y^2)^{-1/2}}$

46 $a^{4/3}a^{-3/2}a^{1/6}$

Exer. 47–52: Rewrite the expression using rational exponents.

47 $\sqrt[4]{x^3}$

48 $\sqrt[3]{x^5}$

49 $\sqrt[3]{(a+b)^2}$

50 $\sqrt{a} + \sqrt{b}$

51 $\sqrt{x^2 + y^2}$

52 $\sqrt[3]{r^3 - s^3}$

Exer. 53–56: Rewrite the expression using a radical.

53 (a) $4x^{3/2}$ (b) $(4x)^{3/2}$

54 (a) $4 + x^{3/2}$ (b) $(4 + x)^{3/2}$

55 (a) $8 - y^{1/3}$ (b) $(8 - y)^{1/3}$

56 (a) $8y^{1/3}$ (b) $(8y)^{1/3}$

Exer. 57–80: Simplify the expression, and rationalize the denominator when appropriate.

57 $\sqrt{81}$ 58 $\sqrt[3]{-125}$

59 $\sqrt[5]{-64}$ 60 $\sqrt[4]{256}$

61 $\dfrac{1}{\sqrt[3]{2}}$ 62 $\sqrt{\dfrac{1}{7}}$

63 $\sqrt{9x^{-4}y^6}$ 64 $\sqrt{16a^8b^{-2}}$

65 $\sqrt[3]{8a^6b^{-3}}$ 66 $\sqrt[4]{81r^5s^8}$

67 $\sqrt{\dfrac{3x}{2y^3}}$ 68 $\sqrt{\dfrac{1}{3x^3y}}$

69 $\sqrt[3]{\dfrac{2x^4y^4}{9x}}$ 70 $\sqrt[3]{\dfrac{3x^2y^5}{4x}}$

71 $\sqrt[4]{\dfrac{5x^8y^3}{27x^2}}$ 72 $\sqrt[4]{\dfrac{x^7y^{12}}{125x}}$

73 $\sqrt[5]{\dfrac{5x^7y^2}{8x^3}}$ 74 $\sqrt[5]{\dfrac{3x^{11}y^3}{9x^2}}$

75 $\sqrt[6]{(3x^5y^{-2})^4}$ 76 $\sqrt[6]{(2u^{-3}v^4)^6}$

77 $\sqrt[5]{\dfrac{8x^3}{y^4}}\sqrt[5]{\dfrac{4x^4}{y^2}}$ 78 $\sqrt{5xy^7}\sqrt{10x^3y^3}$

79 $\sqrt[3]{3t^4v^2}\sqrt[3]{-9t^{-1}v^4}$ 80 $\sqrt[3]{(2r - s)^3}$

Exer. 81–84: Simplify the expression, assuming x and y may be negative.

81 $\sqrt{x^6y^4}$ 82 $\sqrt{x^4y^{10}}$

83 $\sqrt[4]{x^8(y - 1)^{12}}$ 84 $\sqrt[4]{(x + 2)^{12}y^4}$

Exer. 85–90: Replace the symbol $\square$ with either $=$ or $\neq$ to make the resulting statement true, whenever the expression has meaning. Give a reason for your answer.

85 $(a^r)^2 \ \square \ a^{(r^2)}$ 86 $(a^2 + 1)^{1/2} \ \square \ a + 1$

87 $a^xb^y \ \square \ (ab)^{xy}$ 88 $\sqrt{a^r} \ \square \ (\sqrt{a})^r$

89 $\sqrt[n]{\dfrac{1}{c}} \ \square \ \dfrac{1}{\sqrt[n]{c}}$ 90 $a^{1/k} \ \square \ \dfrac{1}{a^k}$

Exer. 91–92: In evaluating negative numbers raised to fractional powers, it may be necessary to evaluate the root and integer power separately. For example, $(-3)^{2/5}$ can be evaluated successfully as $[(-3)^{1/5}]^2$ or $[(-3)^2]^{1/5}$, whereas an error message might otherwise appear. Approximate the real-number expression to four decimal places.

91 (a) $(-3)^{2/5}$ (b) $(-5)^{4/3}$

92 (a) $(-1.2)^{3/7}$ (b) $(-5.08)^{7/3}$

Exer. 93–94: Approximate the real-number expression to four decimal places.

93 (a) $\sqrt{\pi + 1}$ (b) $\sqrt[3]{15.1} + 5^{1/4}$

94 (a) $(2.6 - 1.9)^{-2}$ (b) $5^{\sqrt{7}}$

95 **Savings account** One of the oldest banks in the United States is the Bank of America, founded in 1812. If \$200 had been deposited at that time into an account that paid 4% annual interest, then 180 years later the amount would have grown to $200(1.04)^{180}$ dollars. Approximate this amount to the nearest cent.

96 **Viewing distance** On a clear day, the distance d (in miles) that can be seen from the top of a tall building of height h (in feet) can be approximated by $d \doteq 1.2\sqrt{h}$. Approximate the distance that can be seen from the top of the Chicago Sears Tower, which is 1454 feet tall.

97 **Length of a halibut** The length-weight relationship for Pacific halibut can be approximated by the formula $L = 0.46\sqrt[3]{W}$, where W is in kilograms and L is in meters. The largest documented halibut weighed 230 kilograms. Estimate its length.

98 **Weight of a whale** The length-weight relationship for the sei whale can be approximated by $W = 0.0016L^{2.43}$, where W is in tons and L is in feet. Estimate the weight of a whale that is 25 feet long.

99 **Weight lifters' handicaps** O'Carroll's formula is used to handicap weight lifters. If a lifter who weighs b kilograms lifts w kilograms of weight, then the handicapped weight W is given by

$$W = \frac{w}{\sqrt[3]{b - 35}}.$$

Suppose two lifters weighing 75 kilograms and 120 kilograms lift weights of 180 kilograms and 250 kilograms, respectively. Use O'Carroll's formula to determine the superior weight lifter.

100 **Body surface area** A person's body surface area S (in square feet) can be approximated by

$$S = (0.1091)w^{0.425}h^{0.725},$$

where height h is in inches and weight w is in pounds.

(a) Estimate S for a person 6 feet tall weighing 175 pounds.

(b) If a person is 5 feet 6 inches tall, what effect does a 10% increase in weight have on S?

101 Men's weight The average weight W (in pounds) for men with height h between 64 and 79 inches can be approximated using the formula $W = 0.1166h^{1.7}$. Construct a table for W by letting $h = 64, 65, \ldots, 79$. Round all weights to the nearest pound.

Height	Weight	Height	Weight
64		72	
65		73	
66		74	
67		75	
68		76	
69		77	
70		78	
71		79	

102 Women's weight The average weight W (in pounds) for women with height h between 60 and 75 inches can be approximated using the formula $W = 0.1049h^{1.7}$. Construct a table for W by letting $h = 60, 61, \ldots, 75$. Round all weights to the nearest pound.

Height	Weight	Height	Weight
60		68	
61		69	
62		70	
63		71	
64		72	
65		73	
66		74	
67		75	

1.3
Algebraic Expressions

We sometimes use the notation and terminology of sets to describe mathematical relationships. A **set** is a collection of objects of some type, and the objects are called **elements** of the set. Capital letters $R, S, T, \ldots$ are often used to denote sets, and lowercase letters $a, b, x, y, \ldots$ usually represent elements of sets. Throughout this book, $\mathbb{R}$ denotes the set of real numbers and $\mathbb{Z}$ denotes the set of integers.

Two sets S and T are **equal,** denoted by $S = T$, if S and T contain exactly the same elements. We write $S \neq T$ if S and T are not equal. Additional notation and terminology are listed in the following chart.

Notation or terminology	Meaning	Illustrations
$a \in S$ $a \notin S$	a is an element of S a is not an element of S	$3 \in \mathbb{Z}$ $\frac{3}{5} \notin \mathbb{Z}$
S is a **subset** of T	Every element of S is an element of T	$\mathbb{Z}$ is a subset of $\mathbb{R}$
Constant	A letter or symbol that represents a *specific* element of a set	$5, -\sqrt{2}, \pi$
Variable	A letter or symbol that represents *any* element of a set	Let x denote any real number

We usually use letters near the end of the alphabet, such as x, y, and z, for variables and letters near the beginning of the alphabet, such as a, b, and c, for constants. Throughout this text, unless otherwise specified, variables represent real numbers.

If the elements of a set S have a certain property, we sometimes write $S = \{x: \}$ and state the property describing the variable x in the space after the colon. The expression involving the braces and colon is read "the set of all x such that . . . ," where we complete the phrase by stating the desired property.

$\{x \mid x > 3\}$ is an equivalent notation.

For example, $\{x: x > 3\}$ is read "the set of all x such that x is greater than 3."

For finite sets, we sometimes list all the elements of the set within braces. Thus, if the set T consists of the first five positive integers, we may write $T = \{1, 2, 3, 4, 5\}$. When we describe sets in this way, the order used in listing the elements is irrelevant, so we could also write $T = \{1, 3, 2, 4, 5\}$, $T = \{4, 3, 2, 5, 1\}$, and so on.

If we begin with any collection of variables and real numbers, then an **algebraic expression** is the result obtained by applying additions, subtractions, multiplications, divisions, powers, or the taking of roots to this collection. If specific numbers are substituted for the variables in an algebraic expression, the resulting number is called the **value** of the expression for these numbers. The **domain** of an algebraic expression consists of all real numbers that may represent the variables. Thus, unless otherwise specified, *we assume that the domain consists of the real numbers that, when substituted for the variables, do not make the expression meaningless, in the sense that denominators cannot equal zero and roots always exist.* Two illustrations are given in the following chart.

Algebraic Expressions

Illustration	Domain	Typical value
$x^3 - 5x + \dfrac{6}{\sqrt{x}}$	all $x > 0$	At $x = 4$: $$4^3 - 5(4) + \frac{6}{\sqrt{4}} = 64 - 20 + 3 = 47$$
$\dfrac{2xy + (3/x^2)}{\sqrt[3]{y} - 1}$	all $x \neq 0$ and all $y \neq 1$	At $x = 1$ and $y = 9$: $$\frac{2(1)(9) + (3/1^2)}{\sqrt[3]{9} - 1} = \frac{18 + 3}{\sqrt[3]{8}} = \frac{21}{2}$$

If x is a variable, then a **monomial** in x is an expression of the form ax^n, where a is a real number and n is a nonnegative integer. A **binomial** is a sum of two monomials, and a **trinomial** is a sum of three monomials. A *polynomial in x* is a sum of any number of monomials in x. Another way of stating this is as follows.

Definition of Polynomial	A **polynomial in x** is a sum of the form

$$a_n x^n + a_{n-1} x^{n-1} + \cdots + a_1 x + a_0,$$

where n is a nonnegative integer and each coefficient a_k is a real number. If $a_n \neq 0$, then the polynomial is said to have **degree n.**

Each expression $a_k x^k$ in the sum is a **term** of the polynomial. If a coefficient a_k is zero, we usually delete the term $a_k x^k$. The coefficient a_k of the highest power of x is called the **leading coefficient** of the polynomial.

The following chart contains specific illustrations of polynomials.

Polynomials

Example	Leading coefficient	Degree
$3x^4 + 5x^3 + (-7)x + 4$	3	4
$x^8 + 9x^2 + (-2)x$	1	8
$-5x^2 + 1$	-5	2
$7x + 2$	7	1
8	8	0

By definition, two polynomials are **equal** if and only if they have the same degree and the coefficients of like powers of x are equal. If all the coefficients of a polynomial are zero, it is called the **zero polynomial** and is denoted by 0. However, by convention, the degree of the zero polynomial is *not* zero but, instead, is undefined. If c is a *nonzero real number,* then c is a polynomial of degree 0. Such polynomials (together with the zero polynomial) are **constant polynomials.**

If a coefficient of a polynomial is negative, we usually use a minus sign between appropriate terms. To illustrate,

$$3x^2 + (-5)x + (-7) = 3x^2 - 5x - 7.$$

We may also consider polynomials in variables other than x. For example, $\frac{2}{5}z^2 - 3z^7 + 8 - \sqrt{5}z^4$ is a polynomial in z of degree 7. We often arrange the terms of a polynomial in order of decreasing powers of the variable; thus, we write

$$\frac{2}{5}z^2 - 3z^7 + 8 - \sqrt{5}z^4 = -3z^7 - \sqrt{5}z^4 + \frac{2}{5}z^2 + 8.$$

We may regard a polynomial in x as an algebraic expression obtained by employing a finite number of additions, subtractions, and multiplications involving x. If an algebraic expression contains divisions or roots involving a variable x, then it is not a polynomial in x.

ILLUSTRATION   **Nonpolynomials**

- $\dfrac{1}{x} + 3x$ ■ $\dfrac{x - 5}{x^2 + 2}$ ■ $3x^2 + \sqrt{x} - 2$

Since polynomials represent real numbers, we may use the properties described in Section 1.1. In particular, if additions, subtractions, and multiplications are carried out with polynomials, we may simplify the results by using properties of real numbers, as demonstrated in the following examples.

EXAMPLE 1   **Adding and subtracting polynomials**

(a) Find the sum: $(x^3 + 2x^2 - 5x + 7) + (4x^3 - 5x^2 + 3)$

(b) Find the difference: $(x^3 + 2x^2 - 5x + 7) - (4x^3 - 5x^2 + 3)$

SOLUTION

(a) To obtain the sum of any two polynomials in x, we may add coefficients of like powers of x.

$$
\begin{aligned}
(x^3 &+ 2x^2 - 5x + 7) + (4x^3 - 5x^2 + 3) \\
&= x^3 + 2x^2 - 5x + 7 + 4x^3 - 5x^2 + 3 && \text{remove parentheses} \\
&= (1 + 4)x^3 + (2 - 5)x^2 - 5x + (7 + 3) && \text{add coefficients of like} \\
& && \text{powers of } x \\
&= 5x^3 - 3x^2 - 5x + 10 && \text{simplify}
\end{aligned}
$$

The grouping in the first step was shown for completeness. You may omit this step after you become proficient with such manipulations.

(b) When subtracting polynomials, we first remove parentheses, noting that the minus sign preceding the second pair of parentheses changes the sign of *each* term of that polynomial.

$$
\begin{aligned}
(x^3 &+ 2x^2 - 5x + 7) - (4x^3 - 5x^2 + 3) \\
&= x^3 + 2x^2 - 5x + 7 - 4x^3 + 5x^2 - 3 && \text{remove parentheses} \\
&= (1 - 4)x^3 + (2 + 5)x^2 - 5x + (7 - 3) && \text{add coefficients of like} \\
& && \text{powers of } x \\
&= -3x^3 + 7x^2 - 5x + 4 && \text{simplify} \quad ◢
\end{aligned}
$$

EXAMPLE 2   **Multiplying binomials**

Find the product: $(4x + 5)(3x - 2)$

SOLUTION   Since $3x - 2 = 3x + (-2)$, we may proceed as in Example 1 of Section 1.1:

$$
\begin{aligned}
(4x &+ 5)(3x - 2) \\
&= (4x)(3x) + (4x)(-2) + (5)(3x) + (5)(-2) && \text{distributive properties} \\
&= 12x^2 - 8x + 15x - 10 && \text{multiply} \\
&= 12x^2 + 7x - 10 && \text{simplify} \quad ◢
\end{aligned}
$$

Calculator check for Example 2: Store 17 in a memory location and show that the original expression and the final expression both equal 3577.

After becoming proficient working problems of the type in Example 2, you may wish to perform the first two steps mentally and proceed directly to the final form.

In the next example we illustrate different methods for finding the product of two polynomials.

EXAMPLE 3   Multiplying polynomials

Find the product: $(x^2 + 5x - 4)(2x^3 + 3x - 1)$

SOLUTION

Method 1   We begin by using a distributive property, treating the polynomial $2x^3 + 3x - 1$ as a single real number:

$$(x^2 + 5x - 4)(2x^3 + 3x - 1)$$
$$= x^2(2x^3 + 3x - 1) + 5x(2x^3 + 3x - 1) - 4(2x^3 + 3x - 1)$$

We next use another distributive property three times and simplify the result, obtaining

$$(x^2 + 5x - 4)(2x^3 + 3x - 1)$$
$$= 2x^5 + 3x^3 - x^2 + 10x^4 + 15x^2 - 5x - 8x^3 - 12x + 4$$
$$= 2x^5 + 10x^4 - 5x^3 + 14x^2 - 17x + 4.$$

Note that the three monomials in the first polynomial were multiplied by each of the three monomials in the second polynomial, giving us a total of nine terms.

Method 2   We list the polynomials vertically and multiply, leaving spaces for powers of x that have zero coefficients, as follows:

$$
\begin{array}{l}
2x^3 \quad + \quad 3x \quad - 1 \\
\underline{\quad x^2 \quad + \quad 5x \quad - 4} \\
2x^5 \qquad\qquad + 3x^3 \quad - \quad x^2 \qquad\qquad\quad = \quad x^2(2x^3 + 3x - 1) \\
\qquad\quad 10x^4 \qquad\qquad + 15x^2 \quad - \quad 5x \qquad = \quad 5x(2x^3 + 3x - 1) \\
\underline{\qquad\qquad\qquad - 8x^3 \qquad\qquad\quad - 12x \quad + 4} = -4(2x^3 + 3x - 1) \\
2x^5 \quad + 10x^4 \quad - 5x^3 \quad + 14x^2 \quad - 17x \quad + 4 = \text{sum of the above}
\end{array}
$$

In practice, we would omit the reasons (equalities) listed on the right in the last four lines.   ◢

We may consider polynomials in more than one variable. For example, a polynomial in *two* variables, x and y, is a finite sum of terms, each of the form $ax^m y^k$ for some real number a and nonnegative integers m and k. An example is

$$3x^4 y + 2x^3 y^5 + 7x^2 - 4xy + 8y - 5.$$

Other polynomials may involve three variables—such as x, y, z—or, for that matter, *any* number of variables. Addition, subtraction, and multiplication are performed using properties of real numbers, just as for polynomials in one variable.

The next example illustrates division of a polynomial by a monomial.

EXAMPLE 4 Dividing a polynomial by a monomial

Express as a polynomial in x and y:

$$\frac{6x^2y^3 + 4x^3y^2 - 10xy}{2xy}$$

SOLUTION

$$\frac{6x^2y^3 + 4x^3y^2 - 10xy}{2xy} = \frac{6x^2y^3}{2xy} + \frac{4x^3y^2}{2xy} - \frac{10xy}{2xy} \qquad \text{divide each term by } 2xy$$

$$= 3xy^2 + 2x^2y - 5 \qquad \text{simplify}$$

The products listed in the next chart occur so frequently that they deserve special attention. You can check the validity of each formula by multiplication. In (2) and (3), we use either the top sign on both sides or the bottom sign on both sides. Thus, (2) is actually *two* formulas:

$$(x + y)^2 = x^2 + 2xy + y^2 \qquad \text{and} \qquad (x - y)^2 = x^2 - 2xy + y^2$$

Similarly, (3) represents two formulas.

Product Formulas

Formula	Illustration
(1) $(x + y)(x - y) = x^2 - y^2$	$(2a + 3)(2a - 3) = (2a)^2 - 3^2 = 4a^2 - 9$
(2) $(x \pm y)^2 = x^2 \pm 2xy + y^2$	$(2a - 3)^2 = (2a)^2 - 2(2a)(3) + (3)^2$
	$\qquad\qquad = 4a^2 \quad 12a \mid 9$
(3) $(x \pm y)^3 = x^3 \pm 3x^2y + 3xy^2 \pm y^3$	$(2a + 3)^3 = (2a)^3 + 3(2a)^2(3) + 3(2a)(3)^2 + (3)^3$
	$\qquad\qquad = 8a^3 + 36a^2 + 54a + 27$

Several other illustrations of the product formulas are given in the next example.

EXAMPLE 5 Using product formulas

Find the product:

(a) $\left(2r^2 - \sqrt{s}\right)\left(2r^2 + \sqrt{s}\right)$ **(b)** $\left(\sqrt{c} + \dfrac{1}{\sqrt{c}}\right)^2$ **(c)** $(2a - 5b)^3$

SOLUTION

(a) We use product formula 1, with $x = 2r^2$ and $y = \sqrt{s}$:

$$\left(2r^2 - \sqrt{s}\right)\left(2r^2 + \sqrt{s}\right) = (2r^2)^2 - \left(\sqrt{s}\right)^2$$
$$= 4r^4 - s$$

(b) We use product formula 2, with $x = \sqrt{c}$ and $y = \dfrac{1}{\sqrt{c}}$:

$$\left(\sqrt{c} + \frac{1}{\sqrt{c}}\right)^2 = \left(\sqrt{c}\right)^2 + 2 \cdot \sqrt{c} \cdot \frac{1}{\sqrt{c}} + \left(\frac{1}{\sqrt{c}}\right)^2$$

$$= c + 2 + \frac{1}{c}$$

Note that the last expression is *not* a polynomial.

(c) We use product formula 3, with $x = 2a$ and $y = 5b$:

$$(2a - 5b)^3 = (2a)^3 - 3(2a)^2(5b) + 3(2a)(5b)^2 - (5b)^3$$
$$= 8a^3 - 60a^2b + 150ab^2 - 125b^3$$

If a polynomial is a product of other polynomials, then each polynomial in the product is a **factor** of the original polynomial. **Factoring** is the process of expressing a sum of terms as a product. For example, since $x^2 - 9 = (x + 3)(x - 3)$, the polynomials $x + 3$ and $x - 3$ are factors of $x^2 - 9$.

Factoring is an important process in mathematics, since it may be used to reduce the study of a complicated expression to the study of several simpler expressions. For example, properties of the polynomial $x^2 - 9$ can be determined by examining the factors $x + 3$ and $x - 3$. As we shall see in Chapter 2, another important use for factoring is in finding solutions of equations.

We shall be interested primarily in **nontrivial factors** of polynomials— that is, factors that contain polynomials of positive degree. However, if the coefficients are restricted to *integers,* then we usually remove a common integral factor from each term of the polynomial. For example,

$$4x^2y + 8z^3 = 4(x^2y + 2z^3).$$

A polynomial with coefficients in some set S of numbers is **prime,** or **irreducible** over S, if it cannot be written as a product of two polynomials of positive degree with coefficients in S. A polynomial may be irreducible over one set S but not over another. For example, $x^2 - 2$ is irreducible over the rational numbers, since it cannot be expressed as a product of two polynomials of positive degree that have *rational* coefficients. However, $x^2 - 2$ is *not* irreducible over the real numbers, since we can write

$$x^2 - 2 = \left(x + \sqrt{2}\right)\left(x - \sqrt{2}\right).$$

Similarly, $x^2 + 1$ is irreducible over the real numbers, but, as we shall see in Section 2.4, not over the complex numbers.

Every polynomial $ax + b$ of degree 1 is irreducible.

Before we factor a polynomial, we must specify the number system (or set) from which the coefficients of the factors are to be chosen. In this chapter we shall use the rule that *if a polynomial has integral coefficients, then the factors should be polynomials with integral coefficients*. To **factor a polynomial** means to express it as a product of irreducible polynomials.

The **greatest common factor (gcf)** of an expression is the product of the factors that appear in each term, with each of these factors raised to the smallest nonzero exponent appearing in any term. In factoring polynomials, it is advisable to first factor out the gcf, as shown in the following illustration.

ILLUSTRATION **Factored Polynomials**

- $8x^2 + 4xy = 4x(2x + y)$
- $25x^2 + 25x - 150 = 25(x^2 + x - 6) = 25(x + 3)(x - 2)$
- $4x^5y - 9x^3y^3 = x^3y(4x^2 - 9y^2) = x^3y(2x + 3y)(2x - 3y)$

It is usually difficult to factor polynomials of degree greater than 2. In simple cases, the following factoring formulas may be useful. Each formula can be verified by multiplying the factors on the right-hand side of the equals sign. It can be shown that the factors $x^2 + xy + y^2$ and $x^2 - xy + y^2$ in the difference and sum of two cubes, respectively, are irreducible over the real numbers.

Factoring Formulas

Formula	Illustration
(1) Difference of two squares: $x^2 - y^2 = (x + y)(x - y)$	$9a^2 - 16 = (3a)^2 - (4)^2 = (3a + 4)(3a - 4)$
(2) Difference of two cubes: $x^3 - y^3 = (x - y)(x^2 + xy + y^2)$	$\begin{aligned} 8a^3 - 27 &= (2a)^3 - (3)^3 \\ &= (2a - 3)[(2a)^2 + (2a)(3) + (3)^2] \\ &= (2a - 3)(4a^2 + 6a + 9) \end{aligned}$
(3) Sum of two cubes: $x^3 + y^3 = (x + y)(x^2 - xy + y^2)$	$\begin{aligned} 125a^3 + 1 &= (5a)^3 + (1)^3 \\ &= (5a + 1)[(5a)^2 - (5a)(1) + (1)^2] \\ &= (5a + 1)(25a^2 - 5a + 1) \end{aligned}$

Several other illustrations of the use of factoring formulas are given in the next two examples.

EXAMPLE 6   **Difference of two squares**

Factor each polynomial:

(a) $25r^2 - 49s^2$ (b) $81x^4 - y^4$ (c) $16x^4 - (y - 2z)^2$

SOLUTION

(a) We apply the difference of two squares formula, with $x = 5r$ and $y = 7s$:

$$25r^2 - 49s^2 = (5r)^2 - (7s)^2 = (5r + 7s)(5r - 7s)$$

(b) We write $81x^4 = (9x^2)^2$ and $y^4 = (y^2)^2$ and apply the difference of two squares formula twice:

$$81x^4 - y^4 = (9x^2)^2 - (y^2)^2$$
$$= (9x^2 + y^2)(9x^2 - y^2)$$
$$= (9x^2 + y^2)[(3x)^2 - (y)^2]$$
$$= (9x^2 + y^2)(3x + y)(3x - y)$$

(c) We write $16x^4 = (4x^2)^2$ and apply the difference of two squares formula:

$$16x^4 - (y - 2z)^2 = (4x^2)^2 - (y - 2z)^2$$
$$= [(4x^2) + (y - 2z)][(4x^2) - (y - 2z)]$$
$$= (4x^2 + y - 2z)(4x^2 - y + 2z)$$

EXAMPLE 7   **Sum and difference of two cubes**

Factor each polynomial:

(a) $a^3 + 64b^3$ (b) $8c^6 - 27d^9$

SOLUTION

(a) We apply the sum of two cubes formula, with $x = a$ and $y = 4b$:

$$a^3 + 64b^3 = a^3 + (4b)^3$$
$$= (a + 4b)[a^2 - a(4b) + (4b)^2]$$
$$= (a + 4b)(a^2 - 4ab + 16b^2)$$

(b) We apply the difference of two cubes formula, with $x = 2c^2$ and $y = 3d^3$:

$$8c^6 - 27d^9 = (2c^2)^3 - (3d^3)^3$$
$$= (2c^2 - 3d^3)[(2c^2)^2 + (2c^2)(3d^3) + (3d^3)^2]$$
$$= (2c^2 - 3d^3)(4c^4 + 6c^2d^3 + 9d^6)$$

A factorization of a trinomial $px^2 + qx + r$, where p, q, and r are integers, must be of the form

$$px^2 + qx + r = (ax + b)(cx + d),$$

where a, b, c, and d are integers. It follows that

$$ac = p, \quad bd = r, \quad \text{and} \quad ad + bc = q.$$

Only a limited number of choices for a, b, c, and d satisfy these conditions. If none of the choices work, then $px^2 + qx + r$ is irreducible. Trying the various possibilities, as depicted in the next example, is called the **method of trial and error.** This method is also applicable to trinomials of the form $px^2 + qxy + ry^2$, in which case the factorization must be of the form $(ax + by)(cx + dy)$.

EXAMPLE 8   **Factoring a trinomial by trial and error**

Factor $6x^2 - 7x - 3$.

SOLUTION If we write

$$6x^2 - 7x - 3 = (ax + b)(cx + d),$$

then the following relationships must be true:

$$ac = 6, \quad bd = -3, \quad \text{and} \quad ad + bc = -7$$

If we assume that a and c are both positive, then all possible values are given in the following table:

a	1	6	2	3
c	6	1	3	2

Thus, if $6x^2 - 7x - 3$ is factorable, then one of the following is true:

$$6x^2 - 7x - 3 = (x + b)(6x + d)$$
$$6x^2 - 7x - 3 = (6x + b)(x + d)$$
$$6x^2 - 7x - 3 = (2x + b)(3x + d)$$
$$6x^2 - 7x - 3 = (3x + b)(2x + d)$$

We next consider all possible values for b and d. Since $bd = -3$, these are as follows:

b	1	−1	3	−3
d	−3	3	−1	1

Trying various (possibly all) values, we arrive at $b = -3$ and $d = 1$; that is,

$$6x^2 - 7x - 3 = (2x - 3)(3x + 1).$$

As a check, you should multiply the final factorization to see whether the original polynomial is obtained.

The method of trial and error illustrated in Example 8 can be long and tedious if the coefficients of the polynomial are large and have many prime factors. We will show a factoring method in Section 2.3 that can be used to factor any trinomial of the form of the one in Example 8—regardless of the size of the coefficients. For simple cases, it is often possible to arrive at the correct choice rapidly.

EXAMPLE 9   **Factoring polynomials**

Factor:

(a) $12x^2 - 36xy + 27y^2$ **(b)** $4x^4y - 11x^3y^2 + 6x^2y^3$

SOLUTION

(a) Since each term has 3 as a factor, we begin by writing

$$12x^2 - 36xy + 27y^2 = 3(4x^2 - 12xy + 9y^2).$$

A factorization of $4x^2 - 12xy + 9y^2$ as a product of two first-degree polynomials must be of the form

$$4x^2 - 12xy + 9y^2 = (ax + by)(cx + dy),$$

with $ac = 4,$ $bd = 9,$ and $ad + bc = -12.$

Using the method of trial and error, as in Example 8, we obtain

$$4x^2 - 12xy + 9y^2 = (2x - 3y)(2x - 3y) = (2x - 3y)^2.$$

Thus, $12x^2 - 36xy + 27y^2 = 3(4x^2 - 12xy + 9y^2) = 3(2x - 3y)^2.$

(b) Since each term has x^2y as a factor, we begin by writing

$$4x^4y - 11x^3y^2 + 6x^2y^3 = x^2y(4x^2 - 11xy + 6y^2).$$

By trial and error, we obtain the factorization

$$4x^4y - 11x^3y^2 + 6x^2y^3 = x^2y(4x - 3y)(x - 2y).$$

If a sum contains four or more terms, it may be possible to group the terms in a suitable manner and then find a factorization by using distributive properties. This technique, called **factoring by grouping,** is illustrated in the next example.

EXAMPLE 10 Factoring by grouping

Factor:

(a) $4ac + 2bc - 2ad - bd$ **(b)** $3x^3 + 2x^2 - 12x - 8$

(c) $x^2 - 16y^2 + 10x + 25$

SOLUTION

(a) We group the first two terms and the last two terms and then proceed as follows:

$$4ac + 2bc - 2ad - bd = (4ac + 2bc) - (2ad + bd)$$
$$= 2c(2a + b) - d(2a + b)$$

At this stage we have not factored the given expression because the right-hand side has the form

$$2ck - dk \quad \text{with } k = 2a + b.$$

However, if we factor out k, then

$$2ck - dk = (2c - d)k = (2c - d)(2a + b).$$

Hence,

$$4ac + 2bc - 2ad - bd = 2c(2a + b) - d(2a + b)$$
$$= (2c - d)(2a + b).$$

Note that if we factor $2ck - dk$ as $k(2c - d)$, then the last expression is $(2a + b)(2c - d)$.

(b) We group the first two terms and the last two terms and then proceed as follows:

$$3x^3 + 2x^2 - 12x - 8 = (3x^3 + 2x^2) - (12x + 8)$$
$$= x^2(3x + 2) - 4(3x + 2)$$
$$= (x^2 - 4)(3x + 2)$$

Finally, using the difference of two squares formula for $x^2 - 4$, we obtain the factorization:

$$3x^3 + 2x^2 - 12x - 8 = (x + 2)(x - 2)(3x + 2)$$

(c) First we rearrange and group terms, and then we apply the difference of two squares formula, as follows:

$$x^2 - 16y^2 + 10x + 25 = (x^2 + 10x + 25) - 16y^2$$
$$= (x + 5)^2 - (4y)^2$$
$$= [(x + 5) + 4y][(x + 5) - 4y]$$
$$= (x + 4y + 5)(x - 4y + 5)$$

1.3   *Exercises*

Exer. 1–44: Express as a polynomial.

1 $(3x^3 + 4x^2 - 7x + 1) + (9x^3 - 4x^2 - 6x)$

2 $(7x^3 + 2x^2 - 11x) + (-3x^3 - 2x^2 + 5x - 3)$

3 $(4x^3 + 5x - 3) - (3x^3 + 2x^2 + 5x - 7)$

4 $(6x^3 - 2x^2 + x - 2) - (8x^2 - x - 2)$

5 $(2x + 5)(3x - 7)$ 6 $(3x - 4)(2x + 9)$

7 $(5x + 7y)(3x + 2y)$ 8 $(4x - 3y)(x - 5y)$

9 $(2u + 3)(u - 4) + 4u(u - 2)$

10 $(3u - 1)(u + 2) + 7u(u + 1)$

11 $(3x + 5)(2x^2 + 9x - 5)$ 12 $(7x - 4)(x^3 - x^2 + 6)$

13 $(t^2 + 2t - 5)(3t^2 - t + 2)$

14 $(r^2 - 8r - 2)(-r^2 + 3r - 1)$

15 $(x + 1)(2x^2 - 2)(x^3 + 5)$ 16 $(2x - 1)(x^2 - 5)(x^3 - 1)$

17 $\dfrac{8x^2y^3 - 10x^3y}{2x^2y}$ 18 $\dfrac{6a^3b^3 - 9a^2b^2 + 3ab^4}{3ab^2}$

19 $\dfrac{3u^3v^4 - 2u^5v^2 + (u^2v^2)^2}{u^3v^2}$ 20 $\dfrac{6x^2yz^3 - xy^2z}{xyz}$

21 $(2x + 3y)(2x - 3y)$ 22 $(5x + 4y)(5x - 4y)$

23 $(x^2 + 2y)(x^2 - 2y)$ 24 $(3x + y^3)(3x - y^3)$

25 $(x^2 + 9)(x^2 - 4)$ 26 $(x^2 + 1)(x^2 - 16)$

27 $(3x + 2y)^2$ 28 $(5x - 4y)^2$

29 $(x^2 - 3y^2)^2$ 30 $(2x^2 + 5y^2)^2$

31 $(x + 2)^2(x - 2)^2$ 32 $(x + y)^2(x - y)^2$

33 $\left(\sqrt{x} + \sqrt{y}\right)\left(\sqrt{x} - \sqrt{y}\right)$

34 $\left(\sqrt{x} + \sqrt{y}\right)^2\left(\sqrt{x} - \sqrt{y}\right)^2$

35 $(x^{1/3} - y^{1/3})(x^{2/3} + x^{1/3}y^{1/3} + y^{2/3})$

36 $(x^{1/3} + y^{1/3})(x^{2/3} - x^{1/3}y^{1/3} + y^{2/3})$

37 $(x - 2y)^3$ 38 $(x + 3y)^3$

39 $(2x + 3y)^3$ 40 $(3x - 4y)^3$

41 $(a + b - c)^2$ 42 $(x^2 + x + 1)^2$

43 $(2x + y - 3z)^2$ 44 $(x - 2y + 3z)^2$

Exer. 45–102: Factor the polynomial.

45 $rs + 4st$ 46 $4u^2 - 2uv$

47 $3a^2b^2 - 6a^2b$ 48 $10xy + 15xy^2$

49 $3x^2y^3 - 9x^3y^2$ 50 $16x^5y^2 + 8x^3y^3$

51 $15x^3y^5 - 25x^4y^2 + 10x^6y^4$ 52 $121r^3s^4 + 77r^2s^4 - 55r^4s^3$

53 $8x^2 - 53x - 21$ 54 $7x^2 + 10x - 8$

55 $x^2 + 3x + 4$ 56 $3x^2 - 4x + 2$

57 $6x^2 + 7x - 20$ 58 $12x^2 - x - 6$

59 $12x^2 - 29x + 15$ 60 $21x^2 + 41x + 10$

61 $4x^2 - 20x + 25$ 62 $9x^2 + 24x + 16$

63 $25z^2 + 30z + 9$ 64 $16z^2 - 56z + 49$

65 $45x^2 + 38xy + 8y^2$ 66 $50x^2 + 45xy - 18y^2$

67 $36r^2 - 25t^2$ 68 $81r^2 - 16t^2$

69 $z^4 - 64w^2$ 70 $9y^4 - 121x^2$

71 $x^4 - 4x^2$ 72 $x^3 - 25x$

73 $x^2 + 25$ 74 $4x^2 + 9$

75 $75x^2 - 48y^2$ 76 $64x^2 - 36y^2$

77 $64x^3 + 27$ 78 $125x^3 - 8$

79 $64x^3 - y^6$ 80 $216x^9 + 125y^3$

81 $343x^3 + y^9$ 82 $x^6 - 27y^3$

83 $125 - 27x^3$ 84 $x^3 + 64$

85 $2ax - 6bx + ay - 3by$ 86 $2ay^2 - axy + 6xy - 3x^2$

87 $3x^3 + 3x^2 - 27x - 27$ 88 $5x^3 + 10x^2 - 20x - 40$

89 $x^4 + 2x^3 - x - 2$ 90 $x^4 - 3x^3 + 8x - 24$

91 $a^3 - a^2b + ab^2 - b^3$ 92 $6w^8 + 17w^4 + 12$

93 $a^6 - b^6$ 94 $x^8 - 16$

95 $x^2 + 4x + 4 - 9y^2$ 96 $x^2 - 4y^2 - 6x + 9$

97 $y^2 - x^2 + 8y + 16$ 98 $y^2 + 9 - 6y - 4x^2$

99 $y^6 + 7y^3 - 8$

100 $8c^6 + 19c^3 - 27$

101 $x^{16} - 1$

102 $4x^3 + 4x^2 + x$

Exer. 103–104: The ancient Greeks gave geometric proofs of the factoring formulas for the difference of two squares and the difference of two cubes. Establish the formula for the special case described.

103 Find the areas of regions I and II in the figure to establish the difference of two squares formula for the special case $x > y$.

Exercise 103

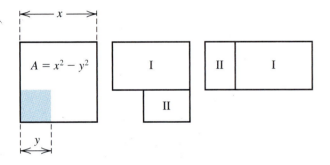

104 Find the volumes of boxes I, II, and III in the figure to establish the difference of two cubes formula for the special case $x > y$.

Exercise 104

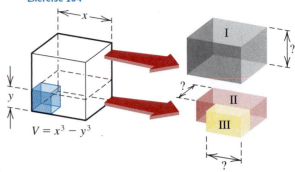

$V = x^3 - y^3$

105 Calorie requirements The basal energy requirement for an individual indicates the minimum number of calories necessary to maintain essential life-sustaining processes such as circulation, regulation of body temperature, and respiration. Given a person's sex, weight w (in kilograms), height h (in centimeters), and age y (in years), we can estimate the basal energy requirement in calories using the following formulas, where C_f and C_m are the calories necessary for females and males, respectively:

$$C_f = 66.5 + 13.8w + 5h - 6.8y$$
$$C_m = 655 + 9.6w + 1.9h - 4.7y$$

(a) Determine the basal energy requirements first for a 25-year-old female weighing 59 kilograms who is 163 centimeters tall and then for a 55-year-old male weighing 75 kilograms who is 178 centimeters tall.

(b) Discuss why, in both formulas, the coefficient for y is negative but the other coefficients are positive.

1.4

Fractional Expressions

A **fractional expression** is a quotient of two algebraic expressions. As a special case, a **rational expression** is a quotient p/q of two *polynomials p and q*. Since division by zero is not allowed, the domain of p/q consists of all real numbers except those that make the denominator zero. Two illustrations are given in the chart.

Rational Expressions

Quotient	Denominator is zero if	Domain
$\dfrac{6x^2 - 5x + 4}{x^2 - 9}$	$x = \pm 3$	All $x \neq \pm 3$
$\dfrac{x^3 - 3x^2y + 4y^2}{y - x^3}$	$y = x^3$	All x and y such that $y \neq x^3$

In most of our work we will be concerned with rational expressions in which both numerator and denominator are polynomials in only one variable.

Since the variables in a rational expression represent real numbers, we may use the properties of quotients in Section 1.1, replacing the letters a, b, c, and d with polynomials. The following property is of particular importance, where $bd \neq 0$:

$$\frac{ad}{bd} = \frac{a}{b} \cdot \frac{d}{d} = \frac{a}{b} \cdot 1 = \frac{a}{b}$$

We sometimes describe this simplification process by saying that *a common nonzero factor in the numerator and denominator of a quotient may be canceled.* In practice, we usually show this cancellation by means of a slash through the common factor, as in the following illustration, where all denominators are assumed to be nonzero.

ILLUSTRATION   **Canceled Common Factors**

- $\dfrac{a\cancel{d}}{b\cancel{d}} = \dfrac{a}{b}$
- $\dfrac{m\cancel{n}}{\cancel{n}pq} = \dfrac{m}{pq}$
- $\dfrac{pq\cancel{r}}{\cancel{r}pv} = \dfrac{q}{v}$

A rational expression is *simplified,* or *reduced to lowest terms,* if the numerator and denominator have no common polynomial factors of positive degree and no common integral factors greater than 1. To simplify a rational expression, we factor both the numerator and the denominator into prime factors and then, assuming the factors in the denominator are not zero, cancel common factors, as in the following illustration.

ILLUSTRATION   **Simplified Rational Expressions**

- $$\frac{3x^2 - 5x - 2}{x^2 - 4} = \frac{(3x + 1)(x - 2)}{(x + 2)(x - 2)} \overset{\text{if } x \neq 2}{=} \frac{3x + 1}{x + 2} \qquad \text{if } x \neq 2/3$$

- $$\frac{2 - x - 3x^2}{6x^2 - x - 2} = \frac{-(3x^2 + x - 2)}{6x^2 - x - 2} = -\frac{(3x - 2)(x + 1)}{(3x - 2)(2x + 1)} \overset{\downarrow}{=} -\frac{x + 1}{2x + 1}$$

- $$\frac{(x^2 + 8x + 16)(x - 5)}{(x^2 - 5x)(x^2 - 16)} = \frac{(x + 4)^{\overset{1}{2}}(x - 5)}{x(x - 5)(x + 4)(x - 4)} \overset{\text{if } x \neq 5,\, x \neq -4}{=} \frac{x + 4}{x(x - 4)}$$

As shown in the next example, when simplifying a product or quotient of rational expressions, we often use properties of quotients to obtain one rational expression. Then we factor the numerator and denominator and cancel common factors, as we did in the preceding illustration.

EXAMPLE 1 **Products and quotients of rational expressions**

Perform the indicated operation and simplify:

(a) $\dfrac{x^2 - 6x + 9}{x^2 - 1} \cdot \dfrac{2x - 2}{x - 3}$ (b) $\dfrac{x + 2}{2x - 3} \div \dfrac{x^2 - 4}{2x^2 - 3x}$

SOLUTION

(a) $\dfrac{x^2 - 6x + 9}{x^2 - 1} \cdot \dfrac{2x - 2}{x - 3} = \dfrac{(x^2 - 6x + 9)(2x - 2)}{(x^2 - 1)(x - 3)}$ property of quotients

$= \dfrac{(x - 3)^{\overset{1}{2}} \cdot 2(x - 1)}{(x + 1)(x - 1)(x - 3)}$ factor all polynomials

if $x \neq 3, x \neq 1$

$\overset{\downarrow}{=} \dfrac{2(x - 3)}{x + 1}$ cancel common factors

(b) $\dfrac{x + 2}{2x - 3} \div \dfrac{x^2 - 4}{2x^2 - 3x} = \dfrac{x + 2}{2x - 3} \cdot \dfrac{2x^2 - 3x}{x^2 - 4}$ property of quotients

$= \dfrac{(x + 2)x(2x - 3)}{(2x - 3)(x + 2)(x - 2)}$ property of quotients; factor all polynomials

if $x \neq -2, x \neq 3/2$

$\overset{\downarrow}{=} \dfrac{x}{x - 2}$ cancel common factors

To add or subtract two rational expressions, we usually find a *common denominator* and use the following properties of quotients:

$$\frac{a}{d} + \frac{c}{d} = \frac{a + c}{d} \qquad \text{and} \qquad \frac{a}{d} - \frac{c}{d} = \frac{a - c}{d}$$

If the denominators of the expressions are not the same, we may obtain a common denominator by multiplying the numerator and denominator of each fraction by a suitable expression. We usually use the **least common denominator (lcd)** of the two quotients. To find the lcd, we factor each denominator into primes and then form the product of the different prime factors, using the *largest* exponent that appears with each prime factor. Let us begin with a numerical example of this technique.

EXAMPLE 2 **Adding fractions using the lcd**

Express as a simplified rational number:

$$\frac{7}{24} + \frac{5}{18}$$

SOLUTION   The prime factorizations of the denominators 24 and 18 are $24 = 2^3 \cdot 3$ and $18 = 2 \cdot 3^2$. To find the lcd, we form the product of the different prime factors, using the largest exponent associated with each factor. This gives us $2^3 \cdot 3^2$. We now change each fraction to an equivalent fraction with denominator $2^3 \cdot 3^2$ and add:

$$\frac{7}{24} + \frac{5}{18} = \frac{7}{2^3 \cdot 3} + \frac{5}{2 \cdot 3^2}$$

$$= \frac{7}{2^3 \cdot 3} \cdot \frac{3}{3} + \frac{5}{2 \cdot 3^2} \cdot \frac{2^2}{2^2}$$

$$= \frac{21}{2^3 \cdot 3^2} + \frac{20}{2^3 \cdot 3^2}$$

$$= \frac{41}{2^3 \cdot 3^2}$$

$$= \frac{41}{72}$$

The method for finding the lcd for rational expressions is analogous to the process illustrated in Example 2. The only difference is that we use factorizations of polynomials instead of integers.

EXAMPLE 3   Sums and differences of rational expressions

Perform the operations and simplify:

$$\frac{6}{x(3x - 2)} + \frac{5}{3x - 2} - \frac{2}{x^2}$$

SOLUTION   The denominators are already in factored form. The lcd is $x^2(3x - 2)$. To obtain three quotients having the denominator $x^2(3x - 2)$, we multiply the numerator and denominator of the first quotient by x, those of the second by x^2, and those of the third by $3x - 2$, which gives us

$$\frac{6}{x(3x - 2)} + \frac{5}{3x - 2} - \frac{2}{x^2} = \frac{6}{x(3x - 2)} \cdot \frac{x}{x} + \frac{5}{3x - 2} \cdot \frac{x^2}{x^2} - \frac{2}{x^2} \cdot \frac{3x - 2}{3x - 2}$$

$$= \frac{6x}{x^2(3x - 2)} + \frac{5x^2}{x^2(3x - 2)} - \frac{2(3x - 2)}{x^2(3x - 2)}$$

$$= \frac{6x + 5x^2 - 2(3x - 2)}{x^2(3x - 2)}$$

$$= \frac{5x^2 + 4}{x^2(3x - 2)}.$$

EXAMPLE 4 Simplifying sums of rational expressions

Perform the operations and simplify:

$$\frac{2x + 5}{x^2 + 6x + 9} + \frac{x}{x^2 - 9} + \frac{1}{x - 3}$$

SOLUTION We begin by factoring denominators:

$$\frac{2x + 5}{x^2 + 6x + 9} + \frac{x}{x^2 - 9} + \frac{1}{x - 3} = \frac{2x + 5}{(x + 3)^2} + \frac{x}{(x + 3)(x - 3)} + \frac{1}{x - 3}$$

Since the lcd is $(x + 3)^2(x - 3)$, we multiply the numerator and denominator of the first quotient by $x - 3$, those of the second by $x + 3$, and those of the third by $(x + 3)^2$ and then add:

$$\frac{(2x + 5)(x - 3)}{(x + 3)^2(x - 3)} + \frac{x(x + 3)}{(x + 3)^2(x - 3)} + \frac{(x + 3)^2}{(x + 3)^2(x - 3)}$$

$$= \frac{(2x^2 - x - 15) + (x^2 + 3x) + (x^2 + 6x + 9)}{(x + 3)^2(x - 3)}$$

$$= \frac{4x^2 + 8x - 6}{(x + 3)^2(x - 3)} = \frac{2(2x^2 + 4x - 3)}{(x + 3)^2(x - 3)}$$

A **complex fraction** is a quotient in which the numerator and/or the denominator is a fractional expression. Certain problems in calculus require simplifying complex fractions of the type given in the next example.

EXAMPLE 5 Simplifying a complex fraction

Simplify the complex fraction:

$$\frac{\dfrac{2}{x + 3} - \dfrac{2}{a + 3}}{x - a}$$

SOLUTION We change the numerator of the given expression into a single quotient and then use a property for simplifying quotients:

$$\frac{\dfrac{2}{x + 3} - \dfrac{2}{a + 3}}{x - a} = \frac{\dfrac{2(a + 3) - 2(x + 3)}{(x + 3)(a + 3)}}{x - a} \quad \text{\textcolor{red}{combine fractions in the numerator}}$$

$$= \frac{2a - 2x}{(x + 3)(a + 3)} \cdot \frac{1}{x - a} \quad \text{\textcolor{red}{simplify; property of quotients}}$$

$$= \frac{2(a - x)}{(x + 3)(a + 3)(x - a)} \quad \text{\textcolor{red}{factor } 2a - 2x \text{; property of quotients}}$$

$$\underset{\text{if } x \neq a}{\overset{\downarrow}{=}} -\frac{2}{(x + 3)(a + 3)} \quad \text{\textcolor{red}{replace } \dfrac{a - x}{x - a} \text{ with } -1}$$

An alternative method is to multiply the numerator and denominator of the given expression by $(x + 3)(a + 3)$, the lcd of the numerator and denominator, and then simplify the result.   ◢

Some quotients that are not rational expressions contain denominators of the form $a + \sqrt{b}$ or $\sqrt{a} + \sqrt{b}$; as in the next example, these quotients can be simplified by multiplying the numerator and denominator by the **conjugate** $a - \sqrt{b}$ or $\sqrt{a} - \sqrt{b}$, respectively. Of course, if $a - \sqrt{b}$ appears, multiply by $a + \sqrt{b}$ instead.

EXAMPLE 6   Rationalizing a denominator

Rationalize the denominator:

$$\frac{1}{\sqrt{x} + \sqrt{y}}$$

SOLUTION

$$\frac{1}{\sqrt{x} + \sqrt{y}} = \frac{1}{\sqrt{x} + \sqrt{y}} \cdot \frac{\sqrt{x} - \sqrt{y}}{\sqrt{x} - \sqrt{y}} \quad \text{multiply numerator and denominator by the conjugate of } \sqrt{x} + \sqrt{y}$$

$$= \frac{\sqrt{x} - \sqrt{y}}{\left(\sqrt{x}\right)^2 - \left(\sqrt{y}\right)^2} \quad \text{property of quotients and difference of squares}$$

$$= \frac{\sqrt{x} - \sqrt{y}}{x - y} \quad \text{law of radicals}$$

◢

In calculus it is sometimes necessary to rationalize the *numerator* of a quotient, as shown in the following example.

EXAMPLE 7   Rationalizing a numerator

If $h \neq 0$, rationalize the numerator of

$$\frac{\sqrt{x + h} - \sqrt{x}}{h}.$$

SOLUTION

$$\frac{\sqrt{x + h} - \sqrt{x}}{h} = \frac{\sqrt{x + h} - \sqrt{x}}{h} \cdot \frac{\sqrt{x + h} + \sqrt{x}}{\sqrt{x + h} + \sqrt{x}} \quad \text{multiply numerator and denominator by the conjugate of } \sqrt{x + h} - \sqrt{x}$$

$$= \frac{\left(\sqrt{x + h}\right)^2 - \left(\sqrt{x}\right)^2}{h\left(\sqrt{x + h} + \sqrt{x}\right)} \quad \text{property of quotients and difference of squares}$$

$$= \frac{(x + h) - x}{h\left(\sqrt{x + h} + \sqrt{x}\right)} \quad \text{law of radicals}$$

$$= \frac{\not{h}}{\not{h}\left(\sqrt{x + h} + \sqrt{x}\right)} \quad \text{simplify}$$

$$= \frac{1}{\sqrt{x + h} + \sqrt{x}} \quad \text{cancel } h \neq 0$$

(continued)

It may seem as though we have accomplished very little, since radicals occur in the denominator. In calculus, however, it is of interest to determine what is true if h is very close to zero. Note that if we use the *given* expression we obtain the following:

$$\text{If} \quad h \approx 0, \quad \text{then} \quad \frac{\sqrt{x+h} - \sqrt{x}}{h} \approx \frac{\sqrt{x+0} - \sqrt{x}}{0} = \frac{0}{0},$$

a meaningless expression. If we use the *rationalized* form, however, we obtain the following information:

$$\text{If} \quad h \approx 0, \quad \text{then} \quad \frac{\sqrt{x+h} - \sqrt{x}}{h} = \frac{1}{\sqrt{x+h} + \sqrt{x}}$$

$$\approx \frac{1}{\sqrt{x} + \sqrt{x}} = \frac{1}{2\sqrt{x}}.$$

Certain problems in calculus require simplifying expressions of the type given in the next example.

EXAMPLE 8 **Simplifying a fractional expression**

Simplify, if $h \neq 0$:

$$\frac{\dfrac{1}{(x+h)^2} - \dfrac{1}{x^2}}{h}$$

SOLUTION

$$\frac{\dfrac{1}{(x+h)^2} - \dfrac{1}{x^2}}{h} = \frac{\dfrac{x^2 - (x+h)^2}{(x+h)^2 x^2}}{h} \qquad \text{combine quotients in numerator}$$

$$= \frac{x^2 - (x^2 + 2xh + h^2)}{(x+h)^2 x^2} \cdot \frac{1}{h} \qquad \text{square } x+h; \text{ property of quotients}$$

$$= \frac{x^2 - x^2 - 2xh - h^2}{(x+h)^2 x^2 h} \qquad \text{remove parentheses}$$

$$= \frac{-\cancel{h}(2x+h)}{(x+h)^2 x^2 \cancel{h}} \qquad \text{simplify; factor out } -h$$

$$= -\frac{2x+h}{(x+h)^2 x^2} \qquad \text{cancel } h \neq 0$$

Problems of the type given in the next example also occur in calculus.

EXAMPLE 9 **Simplifying a fractional expression**

Simplify:

$$\frac{3x^2(2x+5)^{1/2} - x^3\left(\frac{1}{2}\right)(2x+5)^{-1/2}(2)}{[(2x+5)^{1/2}]^2}$$

SOLUTION   One way to simplify the expression is as follows:

$$\frac{3x^2(2x+5)^{1/2} - x^3\left(\frac{1}{2}\right)(2x+5)^{-1/2}(2)}{[(2x+5)^{1/2}]^2}$$

$$= \frac{3x^2(2x+5)^{1/2} - \dfrac{x^3}{(2x+5)^{1/2}}}{2x+5} \qquad \text{definition of negative exponents}$$

$$= \frac{\dfrac{3x^2(2x+5) - x^3}{(2x+5)^{1/2}}}{2x+5} \qquad \text{combine terms in numerator}$$

$$= \frac{6x^3 + 15x^2 - x^3}{(2x+5)^{1/2}} \cdot \frac{1}{2x+5} \qquad \text{property of quotients}$$

$$= \frac{5x^3 + 15x^2}{(2x+5)^{3/2}} \qquad \text{simplify}$$

$$= \frac{5x^2(x+3)}{(2x+5)^{3/2}} \qquad \text{factor numerator}$$

An alternative simplification is to eliminate the negative power, $-\frac{1}{2}$, in the given expression, as follows:

$$\frac{3x^2(2x+5)^{1/2} - x^3\left(\frac{1}{2}\right)(2x+5)^{-1/2}(2)}{[(2x+5)^{1/2}]^2} \cdot \frac{(2x+5)^{1/2}}{(2x+5)^{1/2}} \qquad \begin{array}{l}\text{multiply numerator and} \\ \text{denominator by } (2x+5)^{1/2}\end{array}$$

$$= \frac{3x^2(2x+5) - x^3}{(2x+5)(2x+5)^{1/2}} \qquad \begin{array}{l}\text{property of quotients and} \\ \text{law of exponents}\end{array}$$

The remainder of the simplification is similar.

A third method of simplification is to first factor out the gcf. In this case, the common factors are x and $(2x+5)$, and the smallest exponents are 2 and $-\frac{1}{2}$, respectively. Thus, the gcf is $x^2(2x+5)^{-1/2}$, and we factor the numerator and simplify as follows:

$$\frac{x^2(2x+5)^{-1/2}[3(2x+5)^1 - x]}{(2x+5)^1} = \frac{x^2(5x+15)}{(2x+5)^{3/2}} = \frac{5x^2(x+3)}{(2x+5)^{3/2}}$$

One of the problems in calculus is determining the values of x that make the numerator equal to zero. The simplified form helps us answer this question with relative ease—the values are 0 and -3. ◢

1.4   *Exercises*

Exer. 1–4: Write the expression as a simplified rational number.

1 $\dfrac{3}{50} + \dfrac{7}{30}$

2 $\dfrac{4}{63} + \dfrac{5}{42}$

3 $\dfrac{5}{24} - \dfrac{3}{20}$

4 $\dfrac{11}{54} - \dfrac{7}{72}$

Exer. 5–48: Simplify the expression.

5 $\dfrac{2x^2 + 7x + 3}{2x^2 - 7x - 4}$
$(2x+1)(x-4)$

6 $\dfrac{2x^2 + 9x - 5}{3x^2 + 17x + 10}$

7 $\dfrac{y^2 - 25}{y^3 - 125}$

8 $\dfrac{y^2 - 9}{y^3 + 27}$

$$x^3 - y^3 = (x-y)(x^2 + xy + y^2)$$
$$x^3 + y^3 = (x+y)(x^2 - xy + y^2)$$

9 $\dfrac{12 + r - r^2}{r^3 + 3r^2}$

10 $\dfrac{10 + 3r - r^2}{r^4 + 2r^3}$

11 $\dfrac{9x^2 - 4}{3x^2 - 5x + 2} \cdot \dfrac{9x^4 - 6x^3 + 4x^2}{27x^4 + 8x}$

12 $\dfrac{4x^2 - 9}{2x^2 + 7x + 6} \cdot \dfrac{4x^4 + 6x^3 + 9x^2}{8x^7 - 27x^4}$

13 $\dfrac{5a^2 + 12a + 4}{a^4 - 16} \div \dfrac{25a^2 + 20a + 4}{a^2 - 2a}$

14 $\dfrac{a^3 - 8}{a^2 - 4} \div \dfrac{a}{a^3 + 8}$

15 $\dfrac{6}{x^2 - 4} - \dfrac{3x}{x^2 - 4}$

16 $\dfrac{15}{x^2 - 9} - \dfrac{5x}{x^2 - 9}$

17 $\dfrac{2}{3s + 1} - \dfrac{9}{(3s + 1)^2}$

18 $\dfrac{4}{(5s - 2)^2} + \dfrac{s}{5s - 2}$

19 $\dfrac{2}{x} + \dfrac{3x + 1}{x^2} - \dfrac{x - 2}{x^3}$

20 $\dfrac{5}{x} - \dfrac{2x - 1}{x^2} + \dfrac{x + 5}{x^3}$

21 $\dfrac{3t}{t + 2} + \dfrac{5t}{t - 2} - \dfrac{40}{t^2 - 4}$

22 $\dfrac{t}{t + 3} + \dfrac{4t}{t - 3} - \dfrac{18}{t^2 - 9}$

23 $\dfrac{4x}{3x - 4} + \dfrac{8}{3x^2 - 4x} + \dfrac{2}{x}$

24 $\dfrac{12x}{2x + 1} - \dfrac{3}{2x^2 + x} + \dfrac{5}{x}$

25 $\dfrac{2x}{x + 2} - \dfrac{8}{x^2 + 2x} + \dfrac{3}{x}$

26 $\dfrac{5x}{2x + 3} - \dfrac{6}{2x^2 + 3x} + \dfrac{2}{x}$

27 $\dfrac{p^4 + 3p^3 - 8p - 24}{p^3 - 2p^2 - 9p + 18}$

28 $\dfrac{2ac + bc - 6ad - 3bd}{6ac + 2ad + 3bc + bd}$

29 $3 + \dfrac{5}{u} + \dfrac{2u}{3u + 1}$

30 $4 + \dfrac{2}{u} - \dfrac{3u}{u + 5}$

31 $\dfrac{2x + 1}{x^2 + 4x + 4} - \dfrac{6x}{x^2 - 4} + \dfrac{3}{x - 2}$

32 $\dfrac{2x + 6}{x^2 + 6x + 9} + \dfrac{5x}{x^2 - 9} + \dfrac{7}{x - 3}$

33 $\dfrac{\dfrac{b}{a} - \dfrac{a}{b}}{\dfrac{1}{a} - \dfrac{1}{b}}$

34 $\dfrac{\dfrac{1}{x + 2} - 3}{\dfrac{4}{x} - x}$

35 $\dfrac{\dfrac{x}{y^2} - \dfrac{y}{x^2}}{\dfrac{1}{y^2} - \dfrac{1}{x^2}}$

36 $\dfrac{\dfrac{r}{s} + \dfrac{s}{r}}{\dfrac{r^2}{s^2} - \dfrac{s^2}{r^2}}$

37 $\dfrac{y^{-1} + x^{-1}}{(xy)^{-1}}$

38 $\dfrac{y^{-2} - x^{-2}}{y^{-2} + x^{-2}}$

39 $\dfrac{\dfrac{5}{x + 1} + \dfrac{2x}{x + 3}}{\dfrac{x}{x + 1} + \dfrac{7}{x + 3}}$

40 $\dfrac{\dfrac{3}{w} - \dfrac{6}{2w + 1}}{\dfrac{5}{w} + \dfrac{8}{2w + 1}}$

41 $\dfrac{\dfrac{3}{x - 1} - \dfrac{3}{a - 1}}{x - a}$

42 $\dfrac{\dfrac{x + 2}{x} - \dfrac{a + 2}{a}}{x - a}$

43 $\dfrac{(x + h)^2 - 3(x + h) - (x^2 - 3x)}{h}$

44 $\dfrac{(x + h)^3 + 5(x + h) - (x^3 + 5x)}{h}$

45 $\dfrac{\dfrac{1}{(x + h)^3} - \dfrac{1}{x^3}}{h}$

46 $\dfrac{\dfrac{1}{x + h} - \dfrac{1}{x}}{h}$

47 $\dfrac{\dfrac{4}{3x + 3h - 1} - \dfrac{4}{3x - 1}}{h}$

48 $\dfrac{\dfrac{5}{2x + 2h + 3} - \dfrac{5}{2x + 3}}{h}$

Exer. 49–54: Rationalize the denominator.

49 $\dfrac{\sqrt{t} + 5}{\sqrt{t} - 5}$

50 $\dfrac{\sqrt{t} - 4}{\sqrt{t} + 4}$

51 $\dfrac{81x^2 - 16y^2}{3\sqrt{x} - 2\sqrt{y}}$

52 $\dfrac{16x^2 - y^2}{2\sqrt{x} - \sqrt{y}}$

53 $\dfrac{1}{\sqrt[3]{a} - \sqrt[3]{b}}$ (*Hint:* Multiply numerator and denominator by $\sqrt[3]{a^2} + \sqrt[3]{ab} + \sqrt[3]{b^2}$.)

54 $\dfrac{1}{\sqrt[3]{x} + \sqrt[3]{y}}$

Exer. 55–60: Rationalize the numerator.

55 $\dfrac{\sqrt{a} - \sqrt{b}}{a^2 - b^2}$

56 $\dfrac{\sqrt{b} + \sqrt{c}}{b^2 - c^2}$

57 $\dfrac{\sqrt{2(x + h) + 1} - \sqrt{2x + 1}}{h}$

58 $\dfrac{\sqrt{x} - \sqrt{x + h}}{h\sqrt{x}\sqrt{x + h}}$

59 $\dfrac{\sqrt{1 - x - h} - \sqrt{1 - x}}{h}$

60 $\dfrac{\sqrt[3]{x + h} - \sqrt[3]{x}}{h}$ (*Hint:* Compare with Exercise 53.)

Exer. 61–64: Express as a sum of terms of the form ax^r, where r is a rational number.

61 $\dfrac{4x^2 - x + 5}{x^{2/3}}$

62 $\dfrac{x^2 + 4x - 6}{\sqrt{x}}$

63 $\dfrac{(x^2 + 2)^2}{x^5}$

64 $\dfrac{\left(\sqrt{x} - 3\right)^2}{x^3}$

Exer. 65–68: Express as a quotient.

65 $x^{-3} + x^2$

66 $x^{-4} - x$

67 $x^{-1/2} - x^{3/2}$

68 $x^{-2/3} + x^{7/3}$

Exer. 69–82: Simplify the expression.

69 $(2x^2 - 3x + 1)(4)(3x + 2)^3(3) + (3x + 2)^4(4x - 3)$

70 $(6x - 5)^3(2)(x^2 + 4)(2x) + (x^2 + 4)^2(3)(6x - 5)^2(6)$

71 $(x^2 - 4)^{1/2}(3)(2x + 1)^2(2) + (2x + 1)^3\left(\frac{1}{2}\right)(x^2 - 4)^{-1/2}(2x)$

72 $(3x + 2)^{1/3}(2)(4x - 5)(4) + (4x - 5)^2\left(\frac{1}{3}\right)(3x + 2)^{-2/3}(3)$

73 $(3x + 1)^6\left(\frac{1}{2}\right)(2x - 5)^{-1/2}(2) + (2x - 5)^{1/2}(6)(3x + 1)^5(3)$

74 $(x^2 + 9)^4\left(-\frac{1}{3}\right)(x + 6)^{-4/3} + (x + 6)^{-1/3}(4)(x^2 + 9)^3(2x)$

75 $\dfrac{(6x + 1)^3(27x^2 + 2) - (9x^3 + 2x)(3)(6x + 1)^2(6)}{(6x + 1)^6}$

76 $\dfrac{(x^2 - 1)^4(2x) - x^2(4)(x^2 - 1)^3(2x)}{(x^2 - 1)^8}$

77 $\dfrac{(x^2 + 2)^3(2x) - x^2(3)(x^2 + 2)^2(2x)}{[(x^2 + 2)^3]^2}$

78 $\dfrac{(x^2 - 5)^4(3x^2) - x^3(4)(x^2 - 5)^3(2x)}{[(x^2 - 5)^4]^2}$

79 $\dfrac{(x^2 + 4)^{1/3}(3) - (3x)\left(\frac{1}{3}\right)(x^2 + 4)^{-2/3}(2x)}{[(x^2 + 4)^{1/3}]^2}$

80 $\dfrac{(1 - x^2)^{1/2}(2x) - x^2\left(\frac{1}{2}\right)(1 - x^2)^{-1/2}(-2x)}{[(1 - x^2)^{1/2}]^2}$

81 $\dfrac{(4x^2 + 9)^{1/2}(2) - (2x + 3)\left(\frac{1}{2}\right)(4x^2 + 9)^{-1/2}(8x)}{[(4x^2 + 9)^{1/2}]^2}$

82
$\dfrac{(3x + 2)^{1/2}\left(\frac{1}{3}\right)(2x + 3)^{-2/3}(2) - (2x + 3)^{1/3}\left(\frac{1}{2}\right)(3x + 2)^{-1/2}(3)}{[(3x + 2)^{1/2}]^2}$

CHAPTER 1 REVIEW EXERCISES

1 Express as a simplified rational number:

(a) $\left(\frac{2}{3}\right)\left(-\frac{5}{8}\right)$ (b) $\frac{3}{4} + \frac{6}{5}$ (c) $\frac{5}{8} - \frac{6}{7}$ (d) $\frac{3}{4} \div \frac{6}{5}$

2 Replace the symbol $\square$ with either $<$, $>$, or $=$ to make the resulting statement true.

(a) $-0.1 \,\square\, -0.001$ (b) $\sqrt{9} \,\square\, -3$

(c) $\frac{1}{6} \,\square\, 0.166$

3 Express the statement as an inequality.

(a) x is negative.

(b) a is between $\frac{1}{2}$ and $\frac{1}{3}$.

(c) The absolute value of x is not greater than 4.

4 Rewrite without using the absolute value symbol, and simplify:

(a) $|-7|$ (b) $\dfrac{|-5|}{-5}$ (c) $|3^{-1} - 2^{-1}|$

5 If points A, B, and C on a coordinate line have coordinates -8, 4, and -3, respectively, find the distance:

(a) $d(A, C)$ (b) $d(C, A)$ (c) $d(B, C)$

6 Determine whether the expression is true for all values of the variables, whenever the expression is defined.

(a) $(x + y)^2 = x^2 + y^2$ (b) $\dfrac{1}{\sqrt{x + y}} = \dfrac{1}{\sqrt{x}} + \dfrac{1}{\sqrt{y}}$

(c) $\dfrac{1}{\sqrt{c} - \sqrt{d}} = \dfrac{\sqrt{c} + \sqrt{d}}{c - d}$

7 Express the number in scientific form.

(a) 93,700,000,000   (b) 0.000 004 02

8 Express the number in decimal form.

(a) 6.8×10^7 (b) 7.3×10^{-4}

Exer. 9–10: Rewrite the expression without using the absolute value symbol, and simplify the result.

9 $|x + 3|$ if $x \le -3$

10 $|(x - 2)(x - 3)|$ if $2 < x < 3$

Exer. 11–12: Express the number in the form a/b, where a and b are integers.

11 $-3^2 + 2^0 + 27^{-2/3}$

12 $\left(\frac{1}{2}\right)^0 - 1^2 + 16^{-3/4}$

Exer. 13–38: Simplify the expression, and rationalize the denominator when appropriate.

13 $(3a^2b)^2(2ab^3)$

14 $\dfrac{6r^3y^2}{2r^5y}$

15 $\dfrac{(3x^2y^{-3})^{-2}}{x^{-5}y}$

16 $\left(\dfrac{a^{2/3}b^{3/2}}{a^2b}\right)^6$

17 $(-2p^2q)^3\left(\dfrac{p}{4q^2}\right)^2$

18 $c^{-4/3}c^{3/2}c^{1/6}$

19 $\left(\dfrac{xy^{-1}}{\sqrt{z}}\right)^4 \div \left(\dfrac{x^{1/3}y^2}{z}\right)^3$

20 $\left(\dfrac{-64x^3}{z^6y^9}\right)^{2/3}$

21 $[(a^{2/3}b^{-2})^3]^{-1}$

22 $\dfrac{(3u^2v^5w^{-4})^3}{(2uv^{-3}w^2)^4}$

23 $\dfrac{r^{-1} + s^{-1}}{(rs)^{-1}}$

24 $(u + v)^3(u + v)^{-2}$

25 $s^{5/2}s^{-4/3}s^{-1/6}$

26 $x^{-2} - y^{-1}$

27 $\sqrt[3]{(x^4y^{-1})^6}$

28 $\sqrt[3]{8x^5y^3z^4}$

29 $\dfrac{1}{\sqrt[3]{4}}$

30 $\sqrt{\dfrac{a^2b^3}{c}}$

31 $\sqrt[3]{4x^2y}\,\sqrt[3]{2x^5y^2}$

32 $\sqrt[4]{(-4a^3b^2c)^2}$

33 $\dfrac{1}{\sqrt{t}}\left(\dfrac{1}{\sqrt{t}} - 1\right)$

34 $\sqrt{\sqrt[3]{(c^3d^6)^4}}$

35 $\dfrac{\sqrt{12x^4y}}{\sqrt{3x^2y^5}}$

36 $\sqrt[3]{(a + 2b)^3}$

37 $\sqrt[3]{\dfrac{1}{2\pi^2}}$

38 $\sqrt[3]{\dfrac{x^2}{9y}}$

Exer. 39–42: Rationalize the denominator.

39 $\dfrac{1 - \sqrt{x}}{1 + \sqrt{x}}$

40 $\dfrac{1}{\sqrt{a} + \sqrt{a - 2}}$

41 $\dfrac{81x^2 - y^2}{3\sqrt{x} + \sqrt{y}}$

42 $\dfrac{3 + \sqrt{x}}{3 - \sqrt{x}}$

Exer. 43–58: Express as a polynomial.

43 $(3x^3 - 4x^2 + x - 7) + (x^4 - 2x^3 + 3x^2 + 5)$

44 $(4z^4 - 3z^2 + 1) - z(z^3 + 4z^2 - 4)$

45 $(x + 4)(x + 3) - (2x - 1)(x - 5)$

46 $(4x - 5)(2x^2 + 3x - 7)$

47 $(3y^3 - 2y^2 + y + 4)(y^2 - 3)$

48 $(3x + 2)(x - 5)(5x + 4)$

49 $(a - b)(a^3 + a^2b + ab^2 + b^3)$

50 $\dfrac{9p^4q^3 - 6p^2q^4 + 5p^3q^2}{3p^2q^2}$

51 $(3a - 5b)(2a + 7b)$

52 $(4r^2 - 3s)^2$

53 $(13a^2 + 4b)(13a^2 - 4b)$

54 $(a^3 - a^2)^2$

55 $(2a + b)^3$

56 $(c^2 - d^2)^3$

57 $(3x + 2y)^2(3x - 2y)^2$

58 $(a + b + c + d)^2$

Exer. 59–74: Factor the polynomial.

59 $60xw + 70w$

60 $2r^4s^3 - 8r^2s^5$

61 $28x^2 + 4x - 9$

62 $16a^4 + 24a^2b^2 + 9b^4$

63 $2wy + 3yx - 8wz - 12zx$

64 $2c^3 - 12c^2 + 3c - 18$

65 $8x^3 + 64y^3$

66 $u^3v^4 - u^6v$

67 $p^8 - q^8$

68 $x^4 - 8x^3 + 16x^2$

69 $w^6 + 1$

70 $3x + 6$

71 $x^2 + 36$

72 $x^2 - 49y^2 - 14x + 49$

73 $x^5 - 4x^3 + 8x^2 - 32$

74 $4x^4 + 12x^3 + 20x^2$

Exer. 75–86: Simplify the expression.

75 $\dfrac{6x^2 - 7x - 5}{4x^2 + 4x + 1}$

76 $\dfrac{r^3 - t^3}{r^2 - t^2}$

77 $\dfrac{6x^2 - 5x - 6}{x^2 - 4} \div \dfrac{2x^2 - 3x}{x + 2}$

78 $\dfrac{2}{4x - 5} - \dfrac{5}{10x + 1}$

79 $\dfrac{7}{x + 2} + \dfrac{3x}{(x + 2)^2} - \dfrac{5}{x}$

80 $\dfrac{x + x^{-2}}{1 + x^{-2}}$

81 $\dfrac{1}{x} - \dfrac{2}{x^2 + x} - \dfrac{3}{x + 3}$

82 $(a^{-1} + b^{-1})^{-1}$

83 $\dfrac{x + 2 - \dfrac{3}{x + 4}}{\dfrac{x}{x + 4} + \dfrac{1}{x + 4}}$

84 $\dfrac{\dfrac{x}{x + 2} - \dfrac{4}{x + 2}}{x - 3 - \dfrac{6}{x + 2}}$

85 $(x^2 + 1)^{3/2}(4)(x + 5)^3 + (x + 5)^4(\tfrac{3}{2})(x^2 + 1)^{1/2}(2x)$

86 $\dfrac{(4 - x^2)(\tfrac{1}{3})(6x + 1)^{-2/3}(6) - (6x + 1)^{1/3}(-2x)}{(4 - x^2)^2}$

87 **Red blood cells in a body** The body of an average person contains 5.5 liters of blood and about 5 million red blood cells per cubic millimeter of blood. Given that $1\,\text{L} = 10^6\,\text{mm}^3$, estimate the number of red blood cells in an average person's body.

88 **Heartbeats in a lifetime** A healthy heart beats 70 to 90 times per minute. Estimate the number of heartbeats in the lifetime of an individual who lives to age 80.

89 **Body surface area** At age 2 years, a typical boy is 86 centimeters tall and weighs 13 kilograms. Use the DuBois and DuBois formula, $S = (0.007184)w^{0.425}h^{0.725}$, where w is weight and h is height, to find the body surface area S (in square meters).

90 **Adiabatic expansion** A gas is said to expand *adiabatically* if there is no loss or gain of heat. The formula for the adiabatic expansion of air is $pv^{-1.4} = c$, where p is the pressure, v is the volume, and c is a constant. If, at a certain instant, the pressure is 40 dyne/cm^2 and the volume is 60 cm^3, find the value of c (a *dyne* is the unit of force in the cgs system).

CHAPTER 1 DISCUSSION EXERCISES

1 **Surface area of a tank** You know that a spherical tank holds 10,000 gallons of water. What do you need to know to determine the surface area of the tank? Estimate the surface area of the tank.

2 Determine the conditions under which $\sqrt{a^2 + b^2} = a + b$.

3 Show that the sum of squares $x^2 + 25$ can be factored by adding and subtracting a particular term and following the method demonstrated in Example 10(c) of Section 1.3.

4 What is the difference between the expressions $\dfrac{1}{x + 1}$ and $\dfrac{x - 1}{x^2 - 1}$?

5 Write the quotient of two arbitrary second-degree polynomials in x, and evaluate the quotient with several large values of x. What general conclusion can you reach about such quotients?

6 Simplify the expression $\dfrac{3x^2 - 5x - 2}{x^2 - 4}$. Now evaluate both expressions with a value of x ($x \neq \pm 2$). Discuss what this evaluation proves (or doesn't) and what your simplification proves (or doesn't).

7 **Party trick** To guess your partner's age and height, have him/her do the following:

1 Write down his/her age.

2 Multiply it by 2.

3 Add 5.

4 Multiply this sum by 50.

5 Subtract 365.

6 Add his/her height (in inches).

7 Add 115.

The first two digits of the result equal his/her age, and the last two digits equal his/her height. Explain why this is true.

8 **Circuits problem** In a particular circuits problem, the output voltage is defined by

$$V_{\text{out}} = I_{\text{in}}\left(-\frac{RXi}{R - Xi}\right),$$

where $I_{\text{in}} = \dfrac{V_{\text{in}}}{Z_{\text{in}}}$ and $Z_{\text{in}} = \dfrac{R^2 - X^2 - 3RXi}{R - Xi}$. Find a formula for V_{out} in terms of V_{in} when R is equal to X.

2

Equations and Inequalities

2.1 Equations

2.2 Applied Problems

2.3 Quadratic Equations

2.4 Complex Numbers

2.5 Other Types of Equations

2.6 Inequalities

2.7 More on Inequalities

Methods for solving equations date back to the Babylonians (2000 B.C.), who described equations in words instead of the variables—x, y, and so on—that we use today. Major advances in finding solutions of equations then took place in Italy in the sixteenth century and continued throughout the world well into the nineteenth century. In modern times, computers are used to approximate solutions of very complicated equations.

Inequalities that involve variables have now attained the same level of importance as equations, and they are used extensively in applications of mathematics. In this chapter we shall discuss several methods for solving basic equations and inequalities.

2.1

Equations

An **equation** (or **equality**) is a statement that two quantities or expressions are equal. Equations are employed in every field that uses real numbers. As an illustration, the equation

$$d = rt, \quad \text{or} \quad \text{distance} = (\text{rate})(\text{time}),$$

is used in solving problems involving an object moving at a constant rate of speed. If the rate r is 45 mi/hr (miles per hour), then the distance d (in miles) traveled after time t (in hours) is given by

$$d = 45t.$$

For example, if $t = 2$ hr, then $d = 45 \cdot 2 = 90$ mi. If we wish to find how long it takes the object to travel 75 miles, we let $d = 75$ and *solve* the equation

$$75 = 45t \quad \text{or, equivalently,} \quad 45t = 75.$$

Dividing both sides of the last equation by 45, we obtain

$$t = \tfrac{75}{45} = \tfrac{5}{3}.$$

Thus, if $r = 45$ mi/hr, then the time required to travel 75 miles is $1\frac{2}{3}$ hours, or 1 hour and 40 minutes.

Note that the equation $d = rt$ contains three variables: d, r, and t. In much of our work in this chapter we shall consider equations that contain only one variable. The following chart applies to a variable x, but any other variable may be considered. The abbreviations LS and RS in the second illustration stand for the equation's left side and right side, respectively.

Terminology	Definition	Illustration
Equation in x	A statement of equality involving one variable, x	$x^2 - 5 = 4x$
Solution, or **root**, of an equation in x	A number b that yields a true statement when substituted for x	5 is a solution of $x^2 - 5 = 4x$, since substitution gives us LS: $5^2 - 5 = 25 - 5 = 20$ and RS: $4 \cdot 5 = 20$, and $20 = 20$ is a true statement.
A number b **satisfies** an equation in x	b is a solution of the equation	5 satisfies $x^2 - 5 = 4x$.
Equivalent equations	Equations that have exactly the same solutions	$2x + 1 = 7$ $2x = 7 - 1$ $2x = 6$ $x = 3$
Solve an equation in x	Find all solutions of the equation	To solve $(x + 3)(x - 5) = 0$, set each factor equal to 0: $x + 3 = 0, x - 5 = 0$, obtaining the solutions -3 and 5.

An **algebraic equation** in x contains only algebraic expressions such as polynomials, rational expressions, radicals, and so on. An equation of this type is called a **conditional equation** if there are numbers in the domains of the expressions that are not solutions. For example, the equation $x^2 = 9$ is conditional, since the number $x = 4$ (and others) is not a solution. If *every* number in the domains of the expressions in an algebraic equation is a solution, the equation is called an **identity.**

Sometimes it is difficult to determine whether an equation is conditional or an identity. An identity will often be indicated when, after properties of real numbers are applied, an equation of the form $p = p$ is obtained, where p is some expression. To illustrate, if we multiply both sides of the equation

$$\frac{x}{x^2 - 4} = \frac{x}{(x + 2)(x - 2)}$$

by $x^2 - 4$, we obtain $x = x$. This alerts us to the fact that we may have an identity on our hands; it does not, however, prove anything. A standard method for verifying that an equation is an identity is to show, using properties of real numbers, that the expression which appears on one side of the given equation can be transformed into the expression which appears on the other side of the given equation. That is easy to do in the preceding illustration, since we know that $x^2 - 4 = (x + 2)(x - 2)$. Of course, to show that an equation is not an identity, we need only find one real number in the domain of the variable that fails to satisfy the original equation.

The most basic equation in algebra is the *linear equation,* defined in the next chart, where a and b denote real numbers.

Terminology	Definition	Illustration
Linear equation in x	An equation that can be written in the form $ax + b = 0$, where $a \neq 0$	$4x + 5 = 0$ $4x = -5$ $x = -\frac{5}{4}$

The illustration in the preceding chart indicates a typical method of solving a linear equation. Following the same procedure, we see that

$$\text{if} \quad ax + b = 0, \quad \text{then} \quad x = -\frac{b}{a},$$

provided $a \neq 0$. Thus, a linear equation has exactly one solution.

We sometimes solve an equation by making a list of equivalent equations, each in some sense simpler than the preceding one, ending the list with an equation from which the solutions can be easily obtained. We often simplify an equation by adding the same expression to both sides or subtracting the same expression from both sides. We can also multiply or divide both sides of an equation by an expression that represents a *nonzero* real number. In the following examples, the phrases in color indicate how an equivalent equation was obtained from the preceding equation. To shorten these phrases we have, as in Example 1, used "add 7" instead of the more accurate but lengthy *add 7 to both sides.* Similarly, "subtract $2x$" is used for *subtract $2x$ from both sides,* and "divide by 4" means *divide both sides by 4.*

EXAMPLE 1 **Solving a linear equation**

Solve the equation $6x - 7 = 2x + 5$.

SOLUTION The equations in the following list are equivalent:

$$6x - 7 = 2x + 5 \qquad \text{given}$$
$$(6x - 7) + 7 = (2x + 5) + 7 \qquad \text{add 7}$$
$$6x = 2x + 12 \qquad \text{simplify}$$
$$6x - 2x = (2x + 12) - 2x \qquad \text{subtract } 2x$$
$$4x = 12 \qquad \text{simplify}$$
$$\frac{4x}{4} = \frac{12}{4} \qquad \text{divide by 4}$$
$$x = 3 \qquad \text{simplify}$$

Check $x = 3$ LS: $6(3) - 7 = 18 - 7 = 11$
                   RS: $2(3) + 5 = 6 + 5 = 11$

Since $11 = 11$ is a true statement, $x = 3$ checks as a solution.

As indicated in the preceding example, we often check a solution by substituting it into the given equation. Such checks may detect errors introduced through incorrect manipulations or mistakes in arithmetic.

We say that the equation given in Example 1 *has the solution $x = 3$.* Similarly, we would say that the equation $x^2 = 4$ *has solutions $x = 2$ and $x = -2$.*

The next example illustrates that a seemingly complicated equation may simplify to a linear equation.

EXAMPLE 2 **Solving an equation**

Solve the equation $(8x - 2)(3x + 4) = (4x + 3)(6x - 1)$.

SOLUTION The equations in the following list are equivalent:

$$(8x - 2)(3x + 4) = (4x + 3)(6x - 1) \qquad \text{given}$$
$$24x^2 + 26x - 8 = 24x^2 + 14x - 3 \qquad \text{multiply factors}$$
$$26x - 8 = 14x - 3 \qquad \text{subtract } 24x^2$$
$$12x - 8 = -3 \qquad \text{subtract } 14x$$
$$12x = 5 \qquad \text{add 8}$$
$$x = \tfrac{5}{12} \qquad \text{divide by 12}$$

Hence, the solution of the given equation is $\frac{5}{12}$.

We did not check the preceding solution because each step yields an equivalent equation; however, when you are working exercises or taking a test, it is always a good idea to check answers to guard against errors.

If an equation contains rational expressions, we often eliminate denominators by multiplying both sides by the lcd of these expressions. If we multiply both sides by an expression that equals zero for some value of x, then the resulting equation may *not* be equivalent to the original equation, as illustrated in the following example.

EXAMPLE 3   An equation with no solutions

Solve the equation $\dfrac{3x}{x-2} = 1 + \dfrac{6}{x-2}$.

SOLUTION

$$\frac{3x}{x-2} = 1 + \frac{6}{x-2} \qquad \text{given}$$

$$\left(\frac{3x}{x-2}\right)(x-2) = (1)(x-2) + \left(\frac{6}{x-2}\right)(x-2) \qquad \text{multiply by } x-2$$

$$3x = (x-2) + 6 \qquad \text{simplify}$$

$$3x = x + 4 \qquad \text{simplify}$$

$$2x = 4 \qquad \text{subtract } x$$

$$x = 2 \qquad \text{divide by 2}$$

Check $x = 2$ LS: $\dfrac{3(2)}{(2)-2} = \dfrac{6}{0}$

Since division by 0 is not permissible, $x = 2$ is not a solution. Hence, *the given equation has no solutions.*

   In the process of solving an equation, we may obtain, as a *possible* solution, a number that is *not* a solution of the given equation. Such a number is called an **extraneous solution** or **extraneous root** of the given equation. In Example 3, $x = 2$ is an extraneous solution (root) of the given equation.
   The following guidelines may also be used to solve the equation in Example 3. In this case, observing guideline 2 would make it unnecessary to check the extraneous solution $x = 2$.

Guidelines for Solving an Equation Containing Rational Expressions	*1* Determine the lcd of the rational expressions.
	2 Find the values of the variable that make the lcd zero. These are *not* solutions, because they yield at least one zero denominator when substituted into the given equation.
	3 Multiply each term of the equation by the lcd and simplify, thereby eliminating all of the denominators.
	4 Solve the equation obtained in guideline 3.
	5 The solutions of the given equation are the solutions found in guideline 4, with the exclusion of the values found in guideline 2.

We shall follow these guidelines in the next example.

EXAMPLE 4   An equation containing rational expressions

Solve the equation $\dfrac{3}{2x - 4} - \dfrac{5}{x + 3} = \dfrac{2}{x - 2}$.

SOLUTION

Guideline 1   Rewriting the denominator $2x - 4$ as $2(x - 2)$, we see that the lcd of the three rational expressions is $2(x - 2)(x + 3)$.

Guideline 2   The values of x that make the lcd $2(x - 2)(x + 3)$ zero are 2 and -3, so these numbers cannot be solutions of the equation.

Guideline 3   Multiplying each term of the equation by the lcd and simplifying gives us the following:

$$\frac{3}{2(x - 2)}2(x - 2)(x + 3) - \frac{5}{x + 3}2(x - 2)(x + 3) = \frac{2}{x - 2}2(x - 2)(x + 3)$$

$$3(x + 3) - 10(x - 2) = 4(x + 3) \qquad \textcolor{red}{\text{cancel like factors}}$$

$$3x + 9 - 10x + 20 = 4x + 12 \qquad \textcolor{red}{\text{multiply factors}}$$

Guideline 4   We solve the last equation obtained in guideline 3.

$$3x - 10x - 4x = 12 - 9 - 20 \qquad \textcolor{red}{\text{subtract } 4x, 9, \text{ and } 20}$$

$$-11x = -17 \qquad \textcolor{red}{\text{combine like terms}}$$

$$x = \tfrac{17}{11} \qquad \textcolor{red}{\text{divide by } -11}$$

Guideline 5   Since $\frac{17}{11}$ is not included among the values (2 and -3) that make the lcd zero (guideline 2), we see that $x = \frac{17}{11}$ is a solution of the given equation.

We shall not check the solution $x = \frac{17}{11}$ by substitution, because the arithmetic involved is complicated. It is simpler to carefully check the algebraic manipulations used in each step. However, a calculator check is recommended.

Formulas involving several variables occur in many applications of mathematics. Sometimes it is necessary to solve for a specific variable in terms of the remaining variables that appear in the formula, as the next two examples illustrate.

Figure 1

Celsius   Fahrenheit
scale   scale

100   212

C   F

0   32

−100   −148

EXAMPLE 5   Relationship between temperature scales

The Celsius and Fahrenheit temperature scales are shown on the thermometer in Figure 1. The relationship between the temperature readings C and F is given by $C = \frac{5}{9}(F - 32)$. Solve for F.

SOLUTION   To *solve for* F we must obtain a formula that has F by itself on one side of the equals sign and does not have F on the other side. We may do this as follows:

$$C = \tfrac{5}{9}(F - 32) \quad \text{given}$$

$$\tfrac{9}{5}C = F - 32 \quad \text{multiply by } \tfrac{9}{5}$$

$$\tfrac{9}{5}C + 32 = F \quad \text{add 32}$$

$$F = \tfrac{9}{5}C + 32 \quad \text{equivalent equation}$$

We can make a simple check of our result in Example 5 as follows. Start with $C = \tfrac{5}{9}(F - 32)$ and substitute 212 (an arbitrary choice) for F to obtain 100 for C. Now let $C = 100$ in $F = \tfrac{9}{5}C + 32$ to get $F = 212$. Again, this check does not *prove* we are correct, but certainly lends credibility to our result.

EXAMPLE 6   Resistors connected in parallel

In electrical theory, the formula

$$\frac{1}{R} = \frac{1}{R_1} + \frac{1}{R_2}$$

is used to find the total resistance R when two resistors R_1 and R_2 are connected in parallel, as illustrated in Figure 2. Solve for R_1.

SOLUTION   We first multiply both sides of the given equation by the lcd of the three fractions and then solve for R_1, as follows:

$$\frac{1}{R} = \frac{1}{R_1} + \frac{1}{R_2} \qquad \text{given}$$

$$\frac{1}{R} \cdot RR_1R_2 = \frac{1}{R_1} \cdot RR_1R_2 + \frac{1}{R_2} \cdot RR_1R_2 \quad \text{multiply by the lcd, } RR_1R_2$$

$$R_1R_2 = RR_2 + RR_1 \qquad \text{cancel common factors}$$

$$R_1R_2 - RR_1 = RR_2 \qquad \text{collect terms with } R_1 \text{ on one side}$$

$$R_1(R_2 - R) = RR_2 \qquad \text{factor out } R_1$$

$$R_1 = \frac{RR_2}{R_2 - R} \qquad \text{divide by } R_2 - R$$

An alternative method of solution is to first solve for $\dfrac{1}{R_1}$:

$$\frac{1}{R} = \frac{1}{R_1} + \frac{1}{R_2} \quad \text{given}$$

$$\frac{1}{R_1} + \frac{1}{R_2} = \frac{1}{R} \quad \text{equivalent equation}$$

$$\frac{1}{R_1} = \frac{1}{R} - \frac{1}{R_2} \quad \text{subtract } \frac{1}{R_2}$$

$$\frac{1}{R_1} = \frac{R_2 - R}{RR_2} \quad \text{combine fractions}$$

If two nonzero numbers are equal, then so are their reciprocals. Hence,

$$R_1 = \frac{RR_2}{R_2 - R}.$$

Figure 2

2.1 *Exercises*

Exer. 1–44: Solve the equation.

1 $-3x + 4 = -1$

2 $2x - 2 = -9$

3 $4x - 3 = -5x + 6$

4 $5x - 4 = 2(x - 2)$

5 $4(2y + 5) = 3(5y - 2)$

6 $6(2y + 3) - 3(y - 5) = 0$

7 $\frac{1}{5}x + 2 = 3 - \frac{2}{7}x$

8 $\frac{5}{3}x - 1 = 4 + \frac{2}{3}x$

9 $0.3(3 + 2x) + 1.2x = 3.2$

10 $1.5x - 0.7 = 0.4(3 - 5x)$

11 $\frac{3 + 5x}{5} = \frac{4 - x}{7}$

12 $\frac{2x - 9}{4} = 2 + \frac{x}{12}$

13 $\frac{13 + 2x}{4x + 1} = \frac{3}{4}$

14 $\frac{3}{7x - 2} = \frac{9}{3x + 1}$

15 $8 - \frac{5}{x} = 2 + \frac{3}{x}$

16 $\frac{3}{y} + \frac{6}{y} - \frac{1}{y} = 11$

17 $(3x - 2)^2 = (x - 5)(9x + 4)$

18 $(x + 5)^2 + 3 = (x - 2)^2$

19 $(5x - 7)(2x + 1) - 10x(x - 4) = 0$

20 $(2x + 9)(4x - 3) = 8x^2 - 12$

21 $\frac{3x + 1}{6x - 2} = \frac{2x + 5}{4x - 13}$

22 $\frac{5x + 2}{10x - 3} = \frac{x - 8}{2x + 3}$

23 $\frac{2}{5} + \frac{4}{10x + 5} = \frac{7}{2x + 1}$

24 $\frac{-5}{3x - 9} + \frac{4}{x - 3} = \frac{5}{6}$

25 $\frac{3}{2x - 4} - \frac{5}{3x - 6} = \frac{3}{5}$

26 $\frac{9}{2x + 6} - \frac{7}{5x + 15} = \frac{2}{3}$

27 $2 - \frac{5}{3x - 7} = 2$

28 $\frac{6}{2x + 11} + 5 = 5$

29 $\frac{1}{2x - 1} = \frac{4}{8x - 4}$

30 $\frac{4}{5x + 2} - \frac{12}{15x + 6} = 0$

31 $\frac{7}{y^2 - 4} - \frac{4}{y + 2} = \frac{5}{y - 2}$

32 $\frac{4}{2u - 3} + \frac{10}{4u^2 - 9} = \frac{1}{2u + 3}$

33 $(x + 3)^3 - (3x - 1)^2 = x^3 + 4$

34 $(x - 1)^3 = (x + 1)^3 - 6x^2$

35 $\frac{9x}{3x - 1} = 2 + \frac{3}{3x - 1}$

36 $\frac{2x}{2x + 3} + \frac{6}{4x + 6} = 5$

37 $\frac{1}{x + 4} + \frac{3}{x - 4} = \frac{3x + 8}{x^2 - 16}$

38 $\frac{2}{2x + 3} + \frac{4}{2x - 3} = \frac{5x + 6}{4x^2 - 9}$

39 $\frac{4}{x + 2} + \frac{1}{x - 2} = \frac{5x - 6}{x^2 - 4}$

40 $\frac{2}{2x + 5} + \frac{3}{2x - 5} = \frac{10x + 5}{4x^2 - 25}$

41 $\frac{2}{2x + 1} - \frac{3}{2x - 1} = \frac{-2x + 7}{4x^2 - 1}$

42 $\frac{3}{2x + 5} + \frac{4}{2x - 5} = \frac{14x + 3}{4x^2 - 25}$

43 $\frac{5}{2x + 3} + \frac{4}{2x - 3} = \frac{14x + 3}{4x^2 - 9}$

44 $\frac{-3}{x + 4} + \frac{7}{x - 4} = \frac{-5x + 4}{x^2 - 16}$

Exer. 45–50: Show that the equation is an identity.

45 $(4x - 3)^2 - 16x^2 = 9 - 24x$

46 $(3x - 4)(2x + 1) + 5x = 6x^2 - 4$

47 $\frac{x^2 - 9}{x + 3} = x - 3$

48 $\frac{x^3 + 8}{x + 2} = x^2 - 2x + 4$

49 $\frac{3x^2 + 8}{x} = \frac{8}{x} + 3x$

50 $\frac{49x^2 - 25}{7x - 5} = 7x + 5$

Exer. 51–52: For what value of c is the number a a solution of the equation?

51 $4x + 1 + 2c = 5c - 3x + 6;$ $a = -2$

52 $3x - 2 + 6c = 2c - 5x + 1;$ $a = 4$

Exer. 53–54: Determine whether the two equations are equivalent.

53 (a) $\dfrac{7x}{x-5} = \dfrac{42}{x-5}$, $x = 6$

(b) $\dfrac{7x}{x-5} = \dfrac{35}{x-5}$, $x = 5$

54 (a) $\dfrac{8x}{x-7} = \dfrac{72}{x-7}$, $x = 9$

(b) $\dfrac{8x}{x-7} = \dfrac{56}{x-7}$, $x = 7$

Exer. 55–56: Determine values for a and b such that $\frac{5}{3}$ is a solution of the equation.

55 $ax + b = 0$

56 $ax^2 + bx = 0$

Exer. 57–58: Determine which equation is not equivalent to the equation preceding it.

57 $\quad x^2 - x - 2 = x^2 - 4$

$\quad (x+1)(x-2) = (x+2)(x-2)$

$\quad\quad x + 1 = x + 2$

$\quad\quad\quad 1 = 2$

58 $\quad\quad 5x + 6 = 4x + 3$

$\quad x^2 + 5x + 6 = x^2 + 4x + 3$

$\quad (x+2)(x+3) = (x+1)(x+3)$

$\quad\quad x + 2 = x + 1$

$\quad\quad\quad 2 = 1$

Exer. 59–62: Solve the formula for the specified variable.

59 $EK + L = D - TK$ for K

60 $CD + C = PC + N$ for C

61 $M = \dfrac{Q+1}{Q}$ for Q

62 $\beta = \dfrac{\alpha}{1-\alpha}$ for α

Exer. 63–76: The formula occurs in the indicated application. Solve for the specified variable.

63 $I = Prt$ for P (simple interest)

64 $C = 2\pi r$ for r (circumference of a circle)

65 $A = \frac{1}{2}bh$ for h (area of a triangle)

66 $V = \frac{1}{3}\pi r^2 h$ for h (volume of a cone)

67 $F = g\dfrac{mM}{d^2}$ for m (Newton's law of gravitation)

68 $R = \dfrac{V}{I}$ for I (Ohm's law in electrical theory)

69 $P = 2l + 2w$ for w (perimeter of a rectangle)

70 $A = P + Prt$ for r (principal plus interest)

71 $A = \frac{1}{2}(b_1 + b_2)h$ for b_1 (area of a trapezoid)

72 $s = \frac{1}{2}gt^2 + v_0 t$ for v_0 (distance an object falls)

73 $S = \dfrac{p}{q + p(1-q)}$ for q (Amdahl's law for supercomputers)

74 $S = 2(lw + hw + hl)$ for h (surface area of a rectangular box)

75 $\dfrac{1}{f} = \dfrac{1}{p} + \dfrac{1}{q}$ for q (lens equation)

76 $\dfrac{1}{R} = \dfrac{1}{R_1} + \dfrac{1}{R_2} + \dfrac{1}{R_3}$ for R_2 (three resistors connected in parallel)

2.2
Applied Problems

Equations are often used to solve *applied problems*—that is, problems that involve applications of mathematics to other fields. Because of the unlimited variety of applied problems, it is difficult to state specific rules for finding solutions. The following guidelines may be helpful, provided the problem can be formulated in terms of an equation in one variable.

<table>
<tr><td>**Guidelines for Solving Applied Problems**</td><td>

1 If the problem is stated in writing, read it carefully several times and think about the given facts, together with the unknown quantity that is to be found.

2 Introduce a letter to denote the unknown quantity. This is one of the most crucial steps in the solution. Phrases containing words such as *what, find, how much, how far,* or *when* should alert you to the unknown quantity.

3 If appropriate, draw a picture and label it.

4 List the known facts, together with any relationships that involve the unknown quantity. A relationship may be described by an equation in which written statements, instead of letters or numbers, appear on one or both sides of the equals sign.

5 After analyzing the list in guideline 4, formulate an equation that describes precisely what is stated in words.

6 Solve the equation formulated in guideline 5.

7 Check the solutions obtained in guideline 6 by referring to the original statement of the problem. Verify that the solution agrees with the stated conditions.

</td></tr>
</table>

The use of these guidelines is illustrated in the next example.

EXAMPLE 1   Test average

A student in an algebra course has test scores of 64 and 78. What score on a third test will give the student an average of 80?

SOLUTION

Guideline 1   Read the problem at least one more time.

Guideline 2   The unknown quantity is the score on the third test, so we let

$$x = \text{score on the third test.}$$

Guideline 3   A picture or diagram is unnecessary for this problem.

Guideline 4   Known facts are scores of 64 and 78 on the first two tests. A relationship that involves x is the average score of 64, 78, and x. Thus,

$$\text{average score} = \frac{64 + 78 + x}{3}.$$

Guideline 5   Since the average score in guideline 4 is to be 80, we consider the equation

$$\frac{64 + 78 + x}{3} = 80.$$

Guideline 6 We solve the equation formulated in guideline 5:

$$64 + 78 + x = 80 \cdot 3 \qquad \text{multiply by 3}$$
$$142 + x = 240 \qquad \text{simplify}$$
$$x = 98 \qquad \text{subtract 142}$$

Guideline 7 Check If the three test scores are 64, 78, and 98, then the average is

$$\frac{64 + 78 + 98}{3} = \frac{240}{3} = 80,$$

as desired.

In the remaining examples, try to identify the explicit guidelines that are used in the solutions.

EXAMPLE 2   **Calculating a presale price**

A clothing store holding a clearance sale advertises that all prices have been discounted 20%. If a shirt is on sale for $28, what was its presale price?

SOLUTION   Since the unknown quantity is the presale price, we let

$$x = \text{presale price.}$$

We next note the following facts:

$$0.20x = \text{discount of 20\% on presale price}$$
$$28 = \text{sale price}$$

The sale price is determined as follows:

$$(\text{presale price}) - (\text{discount}) = \text{sale price}$$

Translating the last equation into symbols and then solving gives us

$$x - 0.20x = 28 \qquad \text{formulate an equation}$$
$$0.80x = 28 \qquad \text{subtract } 0.20x \text{ from } 1x$$
$$x = \frac{28}{0.80} = 35. \qquad \text{divide by 0.80}$$

The presale price was $35.

Check   If a $35 shirt is discounted 20%, then the discount (in dollars) is $(0.20)(35) = 7$ and the sale price is $35 - 7$, or $28.

Banks and other financial institutions pay interest on investments. Usually this interest is *compounded* (as described in Section 5.2); however, if money is invested or loaned for a short period of time, *simple interest* may be paid, using the following formula.

Simple Interest Formula	If a sum of money P (the **principal**) is invested at a simple interest rate r (expressed as a decimal), then the **simple interest** I at the end of t years is $$I = Prt.$$

The following table illustrates simple interest for three cases.

Principal P	Interest rate r	Number of years t	Interest $I = Prt$
$1000	$8\% = 0.08$	1	$1000(0.08)(1) = \$80$
$2000	$6\% = 0.06$	$1\frac{1}{2}$	$2000(0.06)(1.5) = \$180$
$3200	$5\frac{1}{2}\% = 0.055$	2	$3200(0.055)(2) = \$352$

EXAMPLE 3 **Investing money in two stocks**

An investment firm has $100,000 to invest for a client and decides to invest it in two stocks, A and B. The expected annual rate of return, or simple interest, for stock A is 15%, but there is some risk involved, and the client does not wish to invest more than $50,000 in this stock. The annual rate of return on the more stable stock B is anticipated to be 10%. Determine whether there is a way of investing the money so that the annual interest is

(a) $12,000   **(b)** $13,000

SOLUTION   The annual interest is given by $I = Pr$, which comes from the simple interest formula $I = Prt$ with $t = 1$. If we let x denote the amount invested in stock A, then $100,000 - x$ will be invested in stock B. This leads to the following equalities:

$$x = \text{amount invested in stock A at 15\%}$$
$$100,000 - x = \text{amount invested in stock B at 10\%}$$
$$0.15x = \text{annual interest from stock A}$$
$$0.10(100,000 - x) = \text{annual interest from stock B}$$

Adding the interest from both stocks, we obtain

$$\text{total annual interest} = 0.15x + 0.10(100,000 - x).$$

Simplifying the right-hand side gives us

$$\text{total annual interest} = 10,000 + 0.05x. \qquad (*)$$

(a) The total annual interest is $12,000 if

$$10,000 + 0.05x = 12,000 \qquad \text{from (*)}$$
$$0.05x = 2000 \qquad \text{subtract 10,000}$$
$$x = \frac{2000}{0.05} = 40,000. \qquad \text{divide by 0.05}$$

Thus, $40,000 should be invested in stock A, and the remaining $60,000 should be invested in stock B. Since the amount invested in stock A is not more than $50,000, this manner of investing the money meets the requirement of the client.

Check If $40,000 is invested in stock A and $60,000 in stock B, then the total annual interest is

$$40,000(0.15) + 60,000(0.10) = 6000 + 6000 = 12,000.$$

(b) The total annual interest is $13,000 if

$$10,000 + 0.05x = 13,000 \qquad \text{from (*)}$$
$$0.05x = 3000 \qquad \text{subtract 10,000}$$
$$x = \frac{3000}{0.05} = 60,000. \qquad \text{divide by 0.05}$$

Thus, $60,000 should be invested in stock A and the remaining $40,000 in stock B. This plan does *not* meet the client's requirement that no more than $50,000 be invested in stock A. Hence, the firm cannot invest the client's money in stocks A and B such that the total annual interest is $13,000. ◼

In certain applications, it is necessary to combine two substances to obtain a prescribed mixture, as illustrated in the next two examples.

EXAMPLE 4   Mixing chemicals

A chemist has 10 milliliters of a solution that contains a 30% concentration of acid. How many milliliters of pure acid must be added in order to increase the concentration to 50%?

SOLUTION Since the unknown quantity is the amount of pure acid to add, we let

$$x = \text{number of mL of pure acid to be added.}$$

To help visualize the problem, let us draw a picture, as in Figure 1, and attach appropriate labels.

(continued)

Figure 1

Original 30% mixture   Pure acid   New 50% mixture

	Original 30% mixture	Pure acid	New 50% mixture
Total amount of solution:	10 mL	x mL	$10 + x$ mL
Amount of pure acid:	$0.30(10) = 3$ mL	$1.00(x) = x$ mL	$0.50(10 + x)$ mL

Since we can express the amount of pure acid in the final solution as either $3 + x$ (from the first two beakers) or $0.50(10 + x)$, we obtain the equation

$$3 + x = 0.50(10 + x).$$

We now solve for x:

$$3 + x = 5 + 0.5x \qquad \text{multiply factors}$$
$$0.5x = 2 \qquad \text{subtract } 0.5x \text{ and } 3$$
$$x = \frac{2}{0.5} = 4 \qquad \text{divide by } 0.5$$

Hence, 4 milliliters of pure acid should be added to the original solution.

✔ **Check** If 4 milliliters of acid is added to the original solution, then the new solution contains 14 milliliters, 7 milliliters of which is pure acid. This is the desired 50% concentration. ◢

EXAMPLE 5 **Replacing antifreeze**

A radiator contains 8 quarts of a mixture of water and antifreeze. If 40% of the mixture is antifreeze, how much of the mixture should be drained and replaced by pure antifreeze so that the resultant mixture will contain 60% antifreeze?

SOLUTION Let

$$x = \text{number of qt of mixture to be drained.}$$

Since there were 8 quarts in the original 40% mixture, we may depict the problem as in Figure 2.

Figure 2

Original 40% mixture,
less amount drained   Pure antifreeze   New 60% mixture

Total amount:	$(8 - x)$ qt	x qt	8 qt
Amount of pure antifreeze:	$0.40(8 - x)$ qt	$1.00(x) = x$ qt	$0.60(8) = 4.8$ qt

Since the number of quarts of pure antifreeze in the final mixture can be expressed as either $0.40(8 - x) + x$ or 4.8, we obtain the equation

$$0.40(8 - x) + x = 4.8.$$

We now solve for x:

$$3.2 - 0.4x + x = 4.8 \qquad \text{multiply factors}$$
$$0.6x = 1.6 \qquad \text{combine } x \text{ terms and subtract 3.2}$$
$$x = \frac{1.6}{0.6} = \frac{16}{6} = \frac{8}{3} \qquad \text{divide by 0.6}$$

Thus, $\frac{8}{3}$ quarts should be drained from the original mixture.

✔ **Check** Let us first note that the amount of antifreeze in the original 8-quart mixture was 0.4(8), or 3.2 quarts. In draining $\frac{8}{3}$ quarts of the original 40% mixture, we lose $0.4\left(\frac{8}{3}\right)$ quarts of antifreeze, and so $3.2 - 0.4\left(\frac{8}{3}\right)$ quarts of antifreeze remain after draining. If we then add $\frac{8}{3}$ quarts of pure antifreeze, the amount of antifreeze in the final mixture is

$$3.2 - 0.4\left(\tfrac{8}{3}\right) + \tfrac{8}{3} = 4.8 \text{ qt.}$$

This number, 4.8, is 60% of 8.   ◢

EXAMPLE 6 Comparing times traveled by cars

Two cities are connected by means of a highway. A car leaves city B at 1:00 P.M. and travels at a constant rate of 40 mi/hr toward city C. Thirty minutes later, another car leaves B and travels toward C at a constant rate of 55 mi/hr. If the lengths of the cars are disregarded, at what time will the second car reach the first car?

SOLUTION Let t denote the number of hours after 1:00 P.M. traveled by the first car. Since the second car leaves B at 1:30 P.M., it has traveled $\frac{1}{2}$ hour less than the first. This leads to the following table.

(continued)

Car	Rate (mi/hr)	Hours traveled	Miles traveled
First car	40	t	$40t$
Second car	55	$t - \frac{1}{2}$	$55\left(t - \frac{1}{2}\right)$

The schematic drawing in Figure 3 illustrates possible positions of the cars t hours after 1:00 P.M. The second car reaches the first car when the number of miles traveled by the two cars is equal—that is, when

$$55\left(t - \tfrac{1}{2}\right) = 40t.$$

Figure 3

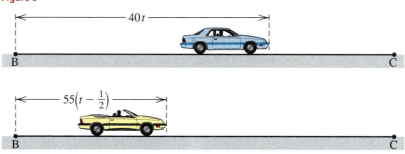

We now solve for t:

$$55t - \tfrac{55}{2} = 40t \qquad \text{multiply factors}$$
$$15t = \tfrac{55}{2} \qquad \text{subtract } 40t \text{ and add } \tfrac{55}{2}$$
$$t = \tfrac{55}{30} = \tfrac{11}{6} \qquad \text{divide by 15}$$

Thus, t is $1\frac{5}{6}$ hours or, equivalently, 1 hour 50 minutes after 1:00 P.M. Consequently, the second car reaches the first at 2:50 P.M.

✔ **Check** At 2:50 P.M. the first car has traveled for $1\frac{5}{6}$ hours, and its distance from B is $40\left(\frac{11}{6}\right) = \frac{220}{3}$ mi. At 2:50 P.M. the second car has traveled for $1\frac{1}{3}$ hours and is $55\left(\frac{4}{3}\right) = \frac{220}{3}$ mi from B. Hence, they are together at 2:50 P.M. ◢

EXAMPLE 7 **Constructing a grain-elevator hopper**

Figure 4

A grain-elevator hopper is to be constructed as shown in Figure 4, with a right circular cylinder of radius 2 feet and altitude h feet on top of a right circular cone whose altitude is one-half that of the cylinder. What value of h will make the total volume V of the hopper 500 ft^3?

SOLUTION   If V_{cylinder} and V_{cone} denote the volumes (in ft³) and h_{cylinder} and h_{cone} denote the heights (in feet) of the cylinder and cone, respectively, then, using the formulas for volume stated on the endpapers at the front of the text, we obtain the following:

$$V_{\text{cylinder}} = \pi r^2 h_{\text{cylinder}} = \pi(2)^2 h = 4\pi h$$

$$V_{\text{cone}} = \tfrac{1}{3}\pi r^2 h_{\text{cone}} = \tfrac{1}{3}\pi(2)^2\left(\tfrac{1}{2}h\right) = \tfrac{2}{3}\pi h$$

Since the total volume V of the hopper is to be 500 ft³, we must have

$$4\pi h + \tfrac{2}{3}\pi h = 500 \qquad\qquad V_{\text{cylinder}} + V_{\text{cone}} = V_{\text{total}}$$

$$12\pi h + 2\pi h = 1500 \qquad\qquad \text{multiply by 3}$$

$$14\pi h = 1500 \qquad\qquad \text{combine terms}$$

$$h = \frac{1500}{14\pi} \approx 34.1 \text{ ft.} \qquad \text{divide by } 14\pi$$

■

EXAMPLE 8   Time required to do a job

Two pumps are available for filling a gasoline storage tank. Pump A, used alone, can fill the tank in 3 hours, and pump B, used alone, can fill it in 4 hours. If both pumps are used simultaneously, how long will it take to fill the tank?

SOLUTION   Let t denote the number of hours needed for A and B to fill the tank if used simultaneously. It is convenient to introduce the *part* of the tank filled in 1 hour as follows:

$$\tfrac{1}{3} = \text{part of the tank filled by A in 1 hr}$$

$$\tfrac{1}{4} = \text{part of the tank filled by B in 1 hr}$$

$$\frac{1}{t} = \text{part of the tank filled by A } and \text{ B in 1 hr}$$

Using the fact that

$$\left(\begin{array}{c}\text{part filled by}\\ \text{A in 1 hr}\end{array}\right) + \left(\begin{array}{c}\text{part filled by}\\ \text{B in 1 hr}\end{array}\right) = \left(\begin{array}{c}\text{part filled by}\\ \text{A } and \text{ B in 1 hr}\end{array}\right),$$

we obtain

$$\frac{1}{3} + \frac{1}{4} = \frac{1}{t}, \qquad \text{or} \qquad \frac{7}{12} = \frac{1}{t}.$$

Taking the reciprocal of each side of the last equation gives us $t = \frac{12}{7}$. Thus, if pumps A and B are used simultaneously, the tank will be filled in $1\frac{5}{7}$ hours, or approximately 1 hour 43 minutes.

■

2.2 Exercises

1 Test scores A student in an algebra course has test scores of 75, 82, 71, and 84. What score on the next test will raise the student's average to 80?

2 Final class average Before the final exam, a student has test scores of 72, 80, 65, 78, and 60. If the final exam counts as one-third of the final grade, what score must the student receive in order to have a final average of 76?

3 Gross pay A worker's take-home pay is $492, after deductions totaling 40% of the gross pay have been subtracted. What is the gross pay?

4 Cost of dining out A couple does not wish to spend more than $70 for dinner at a restaurant. If a sales tax of 6% is added to the bill and they plan to tip 15% after the tax has been added, what is the most they can spend for the meal?

5 Intelligence quotient A person's intelligence quotient (IQ) is determined by multiplying the quotient of his or her mental age and chronological age by 100.

  (a) Find the IQ of a 12-year-old child whose mental age is 15.

  (b) Find the mental age of a person 15 years old whose IQ is 140.

6 Earth's surface area Water covers 70.8%, or about 361×10^6 km^2, of the earth's surface. Approximate the total surface area of the earth.

7 Cost of insulation The cost of installing insulation in a particular two-bedroom home is $1080. Present monthly heating costs average $60, but the insulation is expected to reduce heating costs by 10%. How many months will it take to recover the cost of the insulation?

8 Overtime pay A workman's basic hourly wage is $10, but he receives one and a half times his hourly rate for any hours worked in excess of 40 per week. If his paycheck for the week is $595, how many hours of overtime did he work?

9 Savings accounts An algebra student has won $100,000 in a lottery and wishes to deposit it in savings accounts in two financial institutions. One account pays 8% simple interest, but deposits are insured only to $50,000. The second account pays 6.4% simple interest, and deposits are insured up to $100,000. Determine whether the money can be deposited so that it is fully insured and earns annual interest of $7500.

10 Municipal funding A city government has approved the construction of a $50 million sports arena. Up to $30 million will be raised by selling bonds that pay simple interest at a rate of 12% annually. The remaining amount (up to $40 million) will be obtained by borrowing money from an insurance company at a simple interest rate of 10%. Determine whether the arena can be financed so that the annual interest is $5.2 million.

11 Movie attendance Six hundred people attended the premiere of a motion picture. Adult tickets cost $5, and children were admitted for $2. If box office receipts totaled $2400, how many children attended the premiere?

12 Hourly pay A consulting engineer's time is billed at $60 per hour, and her assistant's is billed at $20 per hour. A customer received a bill for $580 for a certain job. If the assistant worked 5 hours less than the engineer, how much time did each bill on the job?

13 Preparing a glucose solution In a certain medical test designed to measure carbohydrate tolerance, an adult drinks 7 ounces of a 30% glucose solution. When the test is administered to a child, the glucose concentration must be decreased to 20%. How much 30% glucose solution and how much water should be used to prepare 7 ounces of 20% glucose solution?

14 Preparing eye drops A pharmacist is to prepare 15 milliliters of special eye drops for a glaucoma patient. The eye-drop solution must have a 2% active ingredient, but the pharmacist only has 10% solution and 1% solution in stock. How much of each type of solution should be used to fill the prescription?

15 Preparing an alloy British sterling silver is a copper-silver alloy that is 7.5% copper by weight. How many grams of pure copper and how many grams of British sterling silver should be used to prepare 200 grams of a copper-silver alloy that is 10% copper by weight?

16 Drug concentration Theophylline, an asthma medicine, is to be prepared from an elixir with a drug concentration of 5 mg/mL and a cherry-flavored syrup that is to be added to hide the taste of the drug. How much of each must be used to prepare 100 milliliters of solution with a drug concentration of 2 mg/mL?

17 Walking rates Two children, who are 224 meters apart, start walking toward each other at the same instant at rates of 1.5 m/sec and 2 m/sec, respectively (see the figure).

(a) When will they meet?

(b) How far will each have walked?

Exercise 17

1.5 m/s   2 m/s

224 m

18 Running rates A runner starts at the beginning of a runners' path and runs at a constant rate of 6 mi/hr. Five minutes later a second runner begins at the same point, running at a rate of 8 mi/hr and following the same course. How long will it take the second runner to reach the first?

19 Snowplow speed At 6 A.M. a snowplow, traveling at a constant speed, begins to clear a highway leading out of town. At 8 A.M. an automobile begins traveling the highway at a speed of 30 mi/hr and reaches the plow 30 minutes later. Find the speed of the snowplow.

20 Two-way radio range Two children own two-way radios that have a maximum range of 2 miles. One leaves a certain point at 1:00 P.M., walking due north at a rate of 4 mi/hr. The other leaves the same point at 1:15 P.M., traveling due south at 6 mi/hr. When will they be unable to communicate with one another?

21 Rowing rate A boy can row a boat at a constant rate of 5 mi/hr in still water, as indicated in the figure. He rows upstream for 15 minutes and then rows downstream, returning to his starting point in another 12 minutes.

Exercise 21

5 mi/hr   Upstream net speed = $5 - x$ mi/hr

x mi/hr

5 mi/hr   Downstream net speed = $5 + x$ mi/hr

x mi/hr

(a) Find the rate of the current.

(b) Find the total distance traveled.

22 Gas mileage A salesperson purchased an automobile that was advertised as averaging 25 mi/gal in the city and 40 mi/gal on the highway. A recent sales trip that covered 1800 miles required 51 gallons of gasoline. Assuming that the advertised mileage estimates were correct, how many miles were driven in the city?

23 Distance to a target A bullet is fired horizontally at a target, and the sound of its impact is heard 1.5 seconds later. If the speed of the bullet is 3300 ft/sec and the speed of sound is 1100 ft/sec, how far away is the target?

24 Jogging rates A woman begins jogging at 3:00 P.M., running due north at a 6-minute-mile pace. Later, she reverses direction and runs due south at a 7-minute-mile pace. If she returns to her starting point at 3:45 P.M., find the total number of miles run.

25 Fencing a region A farmer plans to use 180 feet of fencing to enclose a rectangular region, using part of a straight river bank instead of fencing as one side of the rectangle, as shown in the figure on the next page. Find the area of the region if the length of the side parallel to the river bank is

(a) twice the length of an adjacent side.

(b) one-half the length of an adjacent side.

(c) the same as the length of an adjacent side.

Exercise 25

26 House dimensions Shown in the figure is a cross section of a design for a two-story home. The center height h of the second story has not yet been determined. Find h such that the second story will have the same cross-sectional area as the first story.

Exercise 26

27 Window dimensions A stained-glass window is being designed in the shape of a rectangle surmounted by a semicircle, as shown in the figure. The width of the window is to be 3 feet, but the height h is yet to be determined. If 24 ft^2 of glass is to be used, find the height h.

Exercise 27

28 Drainage ditch dimensions Every cross section of a drainage ditch is an isosceles trapezoid with a small base of 3 feet and a height of 1 foot, as shown in the figure. Determine the width of the larger base that would give the ditch a cross-sectional area of 5 ft^2.

Exercise 28

29 Constructing a silo A large grain silo is to be constructed in the shape of a circular cylinder with a hemisphere attached to the top (see the figure). The diameter of the silo is to be 30 feet, but the height is yet to be determined. Find the height h of the silo that will result in a capacity of $11{,}250\pi$ ft^3.

Exercise 29

30 Dimensions of a cone The wafer cone shown in the figure is to hold 8 in^3 of ice cream when filled to the bottom. The diameter of the cone is 2 inches, and the top of the ice cream has the shape of a hemisphere. Find the height h of the cone.

Exercise 30

31 Lawn mowing rates It takes a boy 90 minutes to mow the lawn, but his sister can mow it in 60 minutes. How long would it take them to mow the lawn if they worked together, using two lawn mowers?

32 Filling a swimming pool With water from one hose, a swimming pool can be filled in 8 hours. A second, larger hose used alone can fill the pool in 5 hours. How long would it take to fill the pool if both hoses were used simultaneously?

33 Delivering newspapers It takes a girl 45 minutes to deliver the newspapers on her route; however, if her brother helps, it takes them only 20 minutes. How long would it take her brother to deliver the newspapers by himself?

34 Emptying a tank A water tank can be emptied by using one pump for 5 hours. A second, smaller pump can empty the tank in 8 hours. If the larger pump is started at 1:00 P.M., at what time should the smaller pump be started so that the tank will be emptied at 5:00 P.M.?

35 Grade point average (GPA) A college student has finished 48 credit hours with a GPA of 2.75. To get into the program she wishes to enter, she must have a GPA of 3.2. How many additional credit hours of 4.0 work will raise her GPA to 3.2?

36 Ohm's law In electrical theory, Ohm's law states that $I = V/R$, where I is the current in amperes, V is the electromotive force in volts, and R is the resistance in ohms. In a certain circuit $V = 110$ and $R = 50$. If V and R are to be changed by the same numerical amount, what change in them will cause I to double?

37 Air temperature Below the cloud base, the air temperature T (in °F) at height h (in feet) can be approximated by the equation $T = T_0 - \left(\frac{5.5}{1000}\right)h$, where T_0 is the temperature at ground level.

(a) Determine the air temperature at a height of 1 mile if the ground temperature is 70°F.

(b) At what altitude is the temperature freezing?

38 Height of a cloud The height h (in feet) of the cloud base can be estimated using $h = 227(T - D)$, where T is the ground temperature and D is the dew point.

(a) If the temperature is 70°F and the dew point is 55°F, find the height of the cloud base.

(b) If the dew point is 65°F and the cloud base is 3500 feet, estimate the ground temperature.

39 A cloud's temperature The temperature T within a cloud at height h (in feet) above the cloud base can be approximated using the equation $T = B - \left(\frac{3}{1000}\right)h$, where B is the temperature of the cloud at its base. Determine the temperature at 10,000 feet in a cloud with a base temperature of 55°F and a base height of 4000 feet. **Note:** For an interesting application involving the three preceding exercises, see Exercise 6 in the Discussion Exercises at the end of the chapter.

40 Bone-height relationship Archeologists can determine the height of a human without having a complete skeleton. If an archeologist finds only a humerus, then the height of the individual can be determined by using a simple linear relationship. (The humerus is the bone between the shoulder and the elbow.) For a female, if x is the length of the humerus (in centimeters), then her height h (in centimeters) can be determined using the formula $h = 65 + 3.14x$. For a male, $h = 73.6 + 3.0x$ should be used.

(a) A female skeleton having a 30-centimeter humerus is found. Find the woman's height at death.

(b) A person's height will typically decrease by 0.06 centimeter each year after age 30. A complete male skeleton is found. The humerus is 34 centimeters, and the man's height was 174 centimeters. Determine his approximate age at death.

2.3

Quadratic Equations

A toy rocket is launched vertically upward from level ground, as illustrated in Figure 1. If its initial speed is 120 ft/sec and the only force acting on it is gravity, then the rocket's height h (in feet) above the ground after t seconds is given by

$$h = -16t^2 + 120t.$$

Some values of h for the first 7 seconds of flight are listed in the following table.

t (sec)	0	1	2	3	4	5	6	7
h (ft)	0	104	176	216	224	200	144	56

Figure 1

h

We see from the table that, as it ascended, the rocket was 180 feet above the ground at some time between $t = 2$ and $t = 3$. As it descended, the rocket was 180 feet above the ground at some time between $t = 5$ and $t = 6$. To find the exact values of t for which $h = 180$ ft, we must solve the equation

$$180 = -16t^2 + 120t,$$

or

$$16t^2 - 120t + 180 = 0.$$

As indicated in the next chart, an equation of this type is called a *quadratic equation* in t. After developing a formula for solving such equations, we will return to this problem in Example 13 and find the exact times at which the rocket was 180 feet above the ground.

Terminology	Definition	Illustrations
Quadratic equation in x	An equation that can be written in the form $ax^2 + bx + c = 0$, where $a \neq 0$	$4x^2 = 8 - 11x$ $x(3 + x) = 5$ $4x = x^2$

To enable us to solve many types of equations, we will make use of the next theorem.

Zero Factor Theorem	If p and q are algebraic expressions, then $$pq = 0 \quad \text{if and only if} \quad p = 0 \quad \text{or} \quad q = 0.$$

The zero factor theorem can be extended to any number of algebraic expressions—that is,

$$pqr = 0 \quad \text{if and only if} \quad p = 0 \quad \text{or} \quad q = 0 \quad \text{or} \quad r = 0,$$

and so on. It follows that if $ax^2 + bx + c$ can be written as a product of two first-degree polynomials, then solutions can be found by setting each factor equal to 0, as illustrated in the next two examples. This technique is called the **method of factoring.**

EXAMPLE 1 Solving an equation by factoring

Solve the equation $3x^2 = 10 - x$.

SOLUTION   To use the method of factoring, *it is essential that only the number* 0 *appear on one side of the equation.* Thus, we proceed as follows:

$$3x^2 = 10 - x \quad \text{given}$$
$$3x^2 + x - 10 = 0 \quad \text{add } x - 10$$
$$(3x - 5)(x + 2) = 0 \quad \text{factor}$$
$$3x - 5 = 0, \quad x + 2 = 0 \quad \text{zero factor theorem}$$
$$x = \tfrac{5}{3}, \qquad x = -2 \quad \text{solve for } x$$

Hence, the solutions of the given equation are $\frac{5}{3}$ and -2. ◢

EXAMPLE 2   **Solving an equation by factoring**

Solve the equation $x^2 + 16 = 8x$.

SOLUTION   We proceed as in Example 1:

$$x^2 + 16 = 8x \quad \text{given}$$
$$x^2 - 8x + 16 = 0 \quad \text{subtract } 8x$$
$$(x - 4)(x - 4) = 0 \quad \text{factor}$$
$$x - 4 = 0, \quad x - 4 = 0 \quad \text{zero factor theorem}$$
$$x = 4, \qquad x = 4 \quad \text{solve for } x$$

Thus, the given quadratic equation has one solution, 4. ◢

Since $x - 4$ appears as a factor twice in the previous solution, we call 4 a **double root** or **root of multiplicity 2** of the equation $x^2 + 16 = 8x$.

If a quadratic equation has the form $x^2 = d$ for some number $d > 0$, then $x^2 - d = 0$ or, equivalently,

$$\left(x + \sqrt{d}\right)\left(x - \sqrt{d}\right) = 0.$$

Setting each factor equal to zero gives us the solutions $-\sqrt{d}$ and $\sqrt{d}$. We frequently use the symbol $\pm\sqrt{d}$ (*plus or minus* $\sqrt{d}$) to represent both $\sqrt{d}$ and $-\sqrt{d}$. Thus, for $d > 0$, we have proved the following result. (The case $d < 0$ requires the system of complex numbers discussed in Section 2.4.)

A Special Quadratic Equation	If $x^2 = d$, then $x = \pm\sqrt{d}$.

Note on Notation: It is common practice to allow one variable to represent more than one value, as in $x = \pm 3$. A more descriptive notation is $x_{1,2} = \pm 3$, implying that $x_1 = 3$ and $x_2 = -3$.

The process of solving $x^2 = d$ as indicated in the preceding box is referred to as *taking the square root of both sides of the equation.* Note that if

$d > 0$ we obtain both a positive square root and a negative square root, not just the principal square root defined in Section 1.2.

EXAMPLE 3 **Solving equations of the form $x^2 = d$**

Solve the equations:

(a) $x^2 = 5$ **(b)** $(x + 3)^2 = 5$

SOLUTION

(a)

$$x^2 = 5 \qquad \text{given}$$
$$x = \pm\sqrt{5} \qquad \text{take the square root}$$

Thus, the solutions are $\sqrt{5}$ and $-\sqrt{5}$.

(b)

$$(x + 3)^2 = 5 \qquad \text{given}$$
$$x + 3 = \pm\sqrt{5} \qquad \text{take the square root}$$
$$x = -3 \pm \sqrt{5} \qquad \text{subtract 3}$$

Thus, the solutions are $-3 + \sqrt{5}$ and $-3 - \sqrt{5}$.

In the work that follows we will replace an expression of the form $x^2 + kx$ by $(x + d)^2$, where k and d are real numbers. This procedure, called **completing the square** for $x^2 + kx$, calls for adding $(k/2)^2$, as described in the next box. (The same procedure is used for $x^2 - kx$.)

Completing the Square	To complete the square for $x^2 + kx$ or $x^2 - kx$, add $\left(\dfrac{k}{2}\right)^2$; that is, *add the square of half the coefficient of x.* $$\text{(1)} \quad x^2 + kx + \left(\frac{k}{2}\right)^2 = \left(x + \frac{k}{2}\right)^2$$ $$\text{(2)} \quad x^2 - kx + \left(\frac{k}{2}\right)^2 = \left(x - \frac{k}{2}\right)^2$$

EXAMPLE 4 **Completing the square**

Determine the value or values of d that complete the square for each expression. Write the trinomial and the square of the binomial it represents.

(a) $x^2 - 3x + d$ **(b)** $x^2 + dx + 64$

SOLUTION

(a) The square of half the coefficient of x is $\left(-\frac{3}{2}\right)^2 = \frac{9}{4}$. Thus, $d = \frac{9}{4}$ and

$$x^2 - 3x + \tfrac{9}{4} = \left(x - \tfrac{3}{2}\right)^2.$$

(b) If $(x + c)^2 = x^2 + dx + 64$, then $x^2 + 2cx + c^2 = x^2 + dx + 64$, so c^2 must equal 64 and $2c$ must equal d. Hence, c must equal 8 or -8, and since $d = 2c$, d could equal 16 or -16. So we could have

$$x^2 + 16x + 64 = (x + 8)^2 \quad \text{or} \quad x^2 - 16x + 64 = (x - 8)^2. \qquad \blacksquare$$

In the next example we solve a quadratic equation by completing a square.

EXAMPLE 5   Solving a quadratic equation by completing the square

Solve the equation $x^2 - 5x + 3 = 0$.

SOLUTION   It is convenient to first rewrite the equation so that only terms involving x are on the left-hand side, as follows:

$$x^2 - 5x + 3 = 0 \qquad\qquad\qquad \text{given}$$

$$x^2 - 5x = -3 \qquad\qquad\qquad \text{subtract 3}$$

$$x^2 - 5x + \left(\tfrac{5}{2}\right)^2 = -3 + \left(\tfrac{5}{2}\right)^2 \qquad \text{complete the square,}$$
$$\text{adding } \left(\tfrac{5}{2}\right)^2 \text{ to } both \text{ sides}$$

$$\left(x - \tfrac{5}{2}\right)^2 = \tfrac{13}{4} \qquad\qquad\qquad \text{equivalent equation}$$

$$x - \tfrac{5}{2} = \pm\sqrt{\tfrac{13}{4}} \qquad\qquad\qquad \text{take the square root}$$

$$x = \frac{5}{2} \pm \frac{\sqrt{13}}{2} = \frac{5 \pm \sqrt{13}}{2} \qquad \text{add } \tfrac{5}{2}$$

Thus, the solutions of the equation are $\left(5 + \sqrt{13}\right)/2 \approx 4.3$ and $\left(5 - \sqrt{13}\right)/2 \approx 0.7$. $\qquad \blacksquare$

In Example 5, we solved a quadratic equation of the form $ax^2 + bx + c = 0$ with $a = 1$. If $a \neq 1$, we can solve the quadratic equation by adding a step to the procedure used in the preceding example. After rewriting the equation so that only terms involving x are on the left-hand side,

$$ax^2 + bx = -c,$$

we divide both sides by a, obtaining

$$x^2 + \frac{b}{a}x = -\frac{c}{a}.$$

We then complete the square by adding $\left(\dfrac{b}{2a}\right)^2$ to both sides. This technique is used in the proof of the following important formula.

Quadratic Formula

If $a \neq 0$, the roots of $ax^2 + bx + c = 0$ are given by

$$x = \frac{-b \pm \sqrt{b^2 - 4ac}}{2a}.$$

The quadratic formula gives us two solutions of the equation

$$ax^2 + bx + c = 0.$$

They are $x = x_1, x_2$, where

$$x_1 = \frac{-b + \sqrt{b^2 - 4ac}}{2a}$$

and

$$x_2 = \frac{-b - \sqrt{b^2 - 4ac}}{2a}.$$

PROOF   We shall assume that $b^2 - 4ac \geq 0$ so that $\sqrt{b^2 - 4ac}$ is a real number. (The case in which $b^2 - 4ac < 0$ will be discussed in the next section.) Let us proceed as follows:

$$ax^2 + bx + c = 0 \qquad \text{given}$$

$$ax^2 + bx = -c \qquad \text{subtract } c$$

$$x^2 + \frac{b}{a}x = -\frac{c}{a} \qquad \text{divide by } a$$

$$x^2 + \frac{b}{a}x + \left(\frac{b}{2a}\right)^2 = \left(\frac{b}{2a}\right)^2 - \frac{c}{a} \qquad \text{complete the square}$$

$$\left(x + \frac{b}{2a}\right)^2 = \frac{b^2 - 4ac}{4a^2} \qquad \text{equivalent equation}$$

$$x + \frac{b}{2a} = \pm\sqrt{\frac{b^2 - 4ac}{4a^2}} \qquad \text{take the square root}$$

$$x = -\frac{b}{2a} \pm \sqrt{\frac{b^2 - 4ac}{4a^2}} \qquad \text{subtract } \frac{b}{2a}$$

We may write the radical in the last equation as

$$\pm\sqrt{\frac{b^2 - 4ac}{4a^2}} = \pm\frac{\sqrt{b^2 - 4ac}}{\sqrt{(2a)^2}} = \pm\frac{\sqrt{b^2 - 4ac}}{|2a|}.$$

Since $|2a| = 2a$ if $a > 0$ or $|2a| = -2a$ if $a < 0$, we see that in all cases

$$x = -\frac{b}{2a} \pm \frac{\sqrt{b^2 - 4ac}}{2a} = \frac{-b \pm \sqrt{b^2 - 4ac}}{2a}. \qquad \blacksquare$$

Note that if the quadratic formula is executed properly, it is unnecessary to check the solutions.

The number $b^2 - 4ac$ under the radical sign in the quadratic formula is called the **discriminant** of the quadratic equation. The discriminant can be used to determine the nature of the roots of the equation, as in the following chart.

Value of the discriminant $b^2 - 4ac$	Nature of the roots of $ax^2 + bx + c = 0$
Positive value	Two real and unequal roots
0	One root of multiplicity 2
Negative value	No real root

The discriminant in the next two examples is positive. In Example 8 the discriminant is 0.

EXAMPLE 6   Using the quadratic formula

Solve the equation $4x^2 + x - 3 = 0$.

SOLUTION   Let $a = 4$, $b = 1$, and $c = -3$ in the quadratic formula:

$$x = \frac{-1 \pm \sqrt{(1)^2 - 4(4)(-3)}}{2(4)} \qquad x = \frac{-b \pm \sqrt{b^2 - 4ac}}{2a}$$

$$= \frac{-1 \pm \sqrt{49}}{8} \qquad \text{simplify the discriminant}$$

$$= \frac{-1 \pm 7}{8} \qquad \sqrt{49} = 7$$

Hence, the solutions are

$$x = \frac{-1 + 7}{8} = \frac{3}{4} \quad \text{and} \quad x = \frac{-1 - 7}{8} = -1.$$

Example 6 can also be solved by factoring. Writing $(4x - 3)(x + 1) = 0$ and setting each factor equal to zero gives us $x = \frac{3}{4}$ and $x = -1$.

EXAMPLE 7   Using the quadratic formula

Solve the equation $2x(3 - x) = 3$.

SOLUTION   To use the quadratic formula, we must write the equation in the form $ax^2 + bx + c = 0$. The following equations are equivalent:

$$2x(3 - x) = 3 \qquad \text{given}$$
$$6x - 2x^2 = 3 \qquad \text{multiply factors}$$
$$-2x^2 + 6x - 3 = 0 \qquad \text{subtract 3}$$
$$2x^2 - 6x + 3 = 0 \qquad \text{multiply by } -1$$

We now let $a = 2$, $b = -6$, and $c = 3$ in the quadratic formula, obtaining

$$x = \frac{-(-6) \pm \sqrt{(-6)^2 - 4(2)(3)}}{2(2)} = \frac{6 \pm \sqrt{12}}{4} = \frac{6 \pm 2\sqrt{3}}{4}.$$

Note that

$$\frac{3 \pm \sqrt{3}}{2} \neq \frac{3}{2} \pm \sqrt{3}.$$

*The 2 in the denominator must be divided into **both** terms of the numerator, so*

$$\frac{3 \pm \sqrt{3}}{2} = \frac{3}{2} \pm \frac{1}{2}\sqrt{3}.$$

Since 2 is a factor of the numerator and denominator, we can simplify the last fraction as follows:

$$\frac{2(3 \pm \sqrt{3})}{2 \cdot 2} = \frac{3 \pm \sqrt{3}}{2}$$

Hence, the solutions are

$$\frac{3 + \sqrt{3}}{2} \approx 2.37 \quad \text{and} \quad \frac{3 - \sqrt{3}}{2} \approx 0.63.$$

The following example illustrates the case of a double root.

EXAMPLE 8   Using the quadratic formula

Solve the equation $9x^2 - 30x + 25 = 0$.

SOLUTION   Let $a = 9$, $b = -30$, and $c = 25$ in the quadratic formula:

$$x = \frac{-(-30) \pm \sqrt{(-30)^2 - 4(9)(25)}}{2(9)} \qquad x = \frac{-b \pm \sqrt{b^2 - 4ac}}{2a}$$

$$= \frac{30 \pm \sqrt{900 - 900}}{18} \qquad \text{simplify}$$

$$= \frac{30 \pm 0}{18} = \frac{5}{3}$$

Consequently, the equation has one (double) root, $\frac{5}{3}$. ◢

EXAMPLE 9   Clearing an equation of fractions

Solve the equation $\dfrac{2x}{x - 3} + \dfrac{5}{x + 3} = \dfrac{36}{x^2 - 9}$.

SOLUTION   Using the guidelines stated in Section 2.1 for solving an equation containing rational expressions, we multiply by the lcd, $(x + 3)(x - 3)$, remembering that, by guideline 2, the numbers (-3 and 3) that make the lcd zero cannot be solutions. Thus, we proceed as follows:

$$\frac{2x}{x - 3} + \frac{5}{x + 3} = \frac{36}{x^2 - 9} \qquad \text{given}$$

$$2x(x + 3) + 5(x - 3) = 36 \qquad \text{multiply by the lcd, } (x + 3)(x - 3)$$

$$2x^2 + 6x + 5x - 15 - 36 = 0 \qquad \text{multiply factors and subtract 36}$$

$$2x^2 + 11x - 51 = 0 \qquad \text{simplify}$$

$$(2x + 17)(x - 3) = 0 \qquad \text{factor}$$

$$2x + 17 = 0, \quad x - 3 = 0 \qquad \text{zero factor theorem}$$

$$x = -\tfrac{17}{2}, \quad x = 3 \qquad \text{solve for } x$$

Since $x = 3$ cannot be a solution, we see that $x = -\frac{17}{2}$ is the only solution of the given equation. ◢

The next example shows how the quadratic formula can be used to help factor trinomials.

EXAMPLE 10   Factoring with the quadratic formula

Factor the polynomial $21x^2 - 13x - 20$.

SOLUTION   We solve the associated quadratic equation,

$$21x^2 - 13x - 20 = 0,$$

by using the quadratic formula:

$$x = \frac{-(-13) \pm \sqrt{(-13)^2 - 4(21)(-20)}}{2(21)}$$

$$= \frac{13 \pm \sqrt{169 + 1680}}{42} = \frac{13 \pm \sqrt{1849}}{42}$$

$$= \frac{13 \pm 43}{42} = \frac{56}{42}, -\frac{30}{42} = \frac{4}{3}, -\frac{5}{7}$$

We now write the equation as a product of linear factors, both of the form $(x - \text{solution})$:

$$\left(x - \tfrac{4}{3}\right)\left(x - \left(-\tfrac{5}{7}\right)\right) = 0.$$

Eliminate the denominators by multiplying both sides by $3 \cdot 7$:

$$3 \cdot 7\left(x - \tfrac{4}{3}\right)\left(x + \tfrac{5}{7}\right) = 0 \cdot 3 \cdot 7$$

$$3\left(x - \tfrac{4}{3}\right) \cdot 7\left(x + \tfrac{5}{7}\right) = 0$$

$$(3x - 4)(7x + 5) = 0$$

The left side is the desired factoring—that is,

$$21x^2 - 13x - 20 = (3x - 4)(7x + 5).$$

In the next example, we use the quadratic formula to solve an equation that contains more than one variable.

EXAMPLE 11   Using the quadratic formula

Solve $y = x^2 - 6x + 5$ for x, where $x \leq 3$.

SOLUTION   The equation can be written in the form

$$x^2 - 6x + 5 - y = 0,$$

so it is a quadratic equation in x with coefficients $a = 1$, $b = -6$, and

(continued)

$c = 5 - y$. Notice that y is considered to be a constant since we are solving for the variable x. Now we use the quadratic formula:

$$x = \frac{-(-6) \pm \sqrt{(-6)^2 - 4(1)(5 - y)}}{2(1)} \qquad x = \frac{-b \pm \sqrt{b^2 - 4ac}}{2a}$$

$$= \frac{6 \pm \sqrt{16 + 4y}}{2} \qquad \text{simplify } b^2 - 4ac$$

$$= \frac{6 \pm \sqrt{4}\sqrt{4 + y}}{2} \qquad \text{factor out } \sqrt{4}$$

$$= \frac{6 \pm 2\sqrt{4 + y}}{2} \qquad \sqrt{4} = 2$$

$$= 3 \pm \sqrt{4 + y} \qquad \text{divide 2 into } both \text{ terms}$$

Since $\sqrt{4 + y}$ is nonnegative, $3 + \sqrt{4 + y}$ is greater than or equal to 3 and $3 - \sqrt{4 + y}$ is less than or equal to 3. Because the given restriction is $x \leq 3$, we have

$$x = 3 - \sqrt{4 + y}.$$

Many applied problems lead to quadratic equations. One is illustrated in the following example.

Figure 2

EXAMPLE 12   Constructing a rectangular box

A box with a square base and no top is to be made from a square piece of tin by cutting out a 3-inch square from each corner and folding up the sides. If the box is to hold 48 in³, what size piece of tin should be used?

SOLUTION   We begin by drawing the picture in Figure 2, letting x denote the unknown length of the side of the piece of tin. Subsequently, each side of the base of the box will have length $x - 3 - 3 = x - 6$.

Since the area of the base of the box is $(x - 6)^2$ and the height is 3, we obtain

$$\text{volume of box} = 3(x - 6)^2.$$

Since the box is to hold 48 in³,

$$3(x - 6)^2 = 48.$$

We now solve for x:

$$(x - 6)^2 = 16 \qquad \text{divide by 3}$$

$$x - 6 = \pm 4 \qquad \text{take the square root}$$

$$x = 6 \pm 4 \qquad \text{add 6}$$

Consequently,

$$x = 10 \quad \text{or} \quad x = 2.$$

✔ **Check**   Referring to Figure 2, we see that $x = 2$ is unacceptable, since no box is possible in this case. However, if we begin with a 10-inch square of tin, cut out 3-inch corners, and fold, we obtain a box having dimensions 4 inches, 4 inches, and 3 inches. The box has the desired volume of 48 in³. Thus, a 10-inch square is the answer to the problem.   ◢

As illustrated in Example 12, even though an equation is formulated correctly, it is possible to arrive at meaningless solutions because of the physical nature of a given problem. Such solutions should be discarded. For example, we would not accept the answer -7 years for the age of an individual or $\sqrt{50}$ for the number of automobiles in a parking lot.

In the next example we solve the applied problem discussed at the beginning of this section.

EXAMPLE 13   **Finding the height of a toy rocket**

The height above ground h (in feet) of a toy rocket, t seconds after it is launched, is given by $h = -16t^2 + 120t$. When will the rocket be 180 feet above the ground?

SOLUTION Using $h = -16t^2 + 120t$, we obtain the following:

$$180 = -16t^2 + 120t \qquad \text{let } h = 180$$

$$16t^2 - 120t + 180 = 0 \qquad \text{add } 16t^2 - 120t$$

$$4t^2 - 30t + 45 = 0 \qquad \text{divide by 4}$$

Note that the equation is quadratic in t, *so the quadratic formula is solved for* t.

Applying the quadratic formula with $a = 4$, $b = -30$, and $c = 45$ gives us

$$t = \frac{-(-30) \pm \sqrt{(-30)^2 - 4(4)(45)}}{2(4)}$$

$$= \frac{30 \pm \sqrt{180}}{8} = \frac{30 \pm 6\sqrt{5}}{8} = \frac{15 \pm 3\sqrt{5}}{4}.$$

Hence, the rocket is 180 feet above the ground at the following times:

$$t = \frac{15 - 3\sqrt{5}}{4} \approx 2.07 \text{ sec}$$

$$t = \frac{15 + 3\sqrt{5}}{4} \approx 5.43 \text{ sec}$$ ◢

2.3 *Exercises*

Exer. 1–14: Solve the equation by factoring.

1 $6x^2 + x - 12 = 0$

2 $4x^2 + x - 14 = 0$

3 $15x^2 - 12 = -8x$

4 $15x^2 - 14 = 29x$

5 $2x(4x + 15) = 27$

6 $x(3x + 10) = 77$

7 $75x^2 + 35x - 10 = 0$

8 $48x^2 + 12x - 90 = 0$

9 $12x^2 + 60x + 75 = 0$

10 $4x^2 - 72x + 324 = 0$

11 $\dfrac{2x}{x + 3} + \dfrac{5}{x} - 4 = \dfrac{18}{x^2 + 3x}$

12 $\dfrac{5x}{x - 2} + \dfrac{3}{x} + 2 = \dfrac{-6}{x^2 - 2x}$

13 $\dfrac{5x}{x - 3} + \dfrac{4}{x + 3} = \dfrac{90}{x^2 - 9}$

14 $\dfrac{3x}{x - 2} + \dfrac{1}{x + 2} = \dfrac{-4}{x^2 - 4}$

Exer. 15–16: Determine whether the two equations are equivalent.

15 (a) $x^2 = 16,\ x = 4$ (b) $x = \sqrt{9},\ x = 3$

16 (a) $x^2 = 25,\ x = 5$ (b) $x = \sqrt{64},\ x = 8$

Exer. 17–24: Solve the equation by using the special quadratic equation on page 75.

17 $x^2 = 169$

18 $x^2 = 361$

19 $25x^2 = 9$

20 $16x^2 = 49$

21 $(x - 3)^2 = 17$

22 $(x + 4)^2 = 31$

23 $4(x + 2)^2 = 11$

24 $9(x - 1)^2 = 7$

Exer. 25–26: Determine the value or values of d that complete the square for the expression.

25 (a) $x^2 + 9x + d$ (b) $x^2 - 8x + d$

(c) $x^2 + dx + 36$ (d) $x^2 + dx + \frac{49}{4}$

26 (a) $x^2 + 13x + d$ (b) $x^2 - 6x + d$

(c) $x^2 + dx + 25$ (d) $x^2 + dx + \frac{81}{4}$

Exer. 27–30: Solve by completing the square. (*Note:* See the discussion after Example 5 for help in solving Exercises 29 and 30.)

27 $x^2 + 6x + 7 = 0$

28 $x^2 - 8x + 11 = 0$

29 $4x^2 - 12x - 11 = 0$

30 $4x^2 + 20x + 13 = 0$

Exer. 31–44: Solve by using the quadratic formula.

31 $6x^2 - x = 2$

32 $5x^2 + 13x = 6$

33 $x^2 + 4x + 2 = 0$

34 $x^2 - 6x - 3 = 0$

35 $2x^2 - 3x - 4 = 0$

36 $3x^2 + 5x + 1 = 0$

37 $\frac{3}{2}z^2 - 4z - 1 = 0$

38 $\frac{5}{3}s^2 + 3s + 1 = 0$

39 $\dfrac{5}{w^2} - \dfrac{10}{w} + 2 = 0$

40 $\dfrac{x + 1}{3x + 2} = \dfrac{x - 2}{2x - 3}$

41 $4x^2 + 81 = 36x$

42 $24x + 9 = -16x^2$

43 $\dfrac{5x}{x^2 + 9} = -1$

44 $\frac{1}{7}x^2 + 1 = \frac{4}{7}x$

Exer. 45–48: Use the quadratic formula to factor the expressions.

45 $x^2 + x - 30$

46 $x^2 + 7x$

47 $12x^2 - 16x - 3$

48 $15x^2 + 34x - 16$

Exer. 49–50: Use the quadratic formula to solve the equation for (a) x in terms of y and (b) y in terms of x.

49 $4x^2 - 4xy + 1 - y^2 = 0$ 50 $2x^2 - xy = 3y^2 + 1$

Exer. 51–54: Solve for the specified variable.

51 $K = \frac{1}{2}mv^2$ for v (kinetic energy)

52 $F = g\dfrac{mM}{d^2}$ for d (Newton's law of gravitation)

53 $A = 2\pi r(r + h)$ for r (surface area of a closed cylinder)

54 $s = \frac{1}{2}gt^2 + v_0t$ for t (distance an object falls)

55 Velocity of a gas When a hot gas exits a cylindrical smoke-stack, its velocity varies throughout a circular cross section of the smokestack, with the gas near the center of the cross section having a greater velocity than the gas near the perimeter. This phenomenon can be described by the formula

$$V = V_{max}\left[1 - \left(\frac{r}{r_0}\right)^2\right],$$

where V_{max} is the maximum velocity of the gas, r_0 is the radius of the smokestack, and V is the velocity of the gas at a distance r from the center of the circular cross section. Solve this formula for r.

56 Density of the atmosphere For altitudes h up to 10,000 meters, the density D of the earth's atmosphere (in kg/m^3) can be approximated by the formula

$$D = 1.225 - (1.12 \times 10^{-4})h + (3.24 \times 10^{-9})h^2.$$

Approximate the altitude if the density of the atmosphere is 0.74 kg/m^3.

57 Dimensions of a tin can A manufacturer of tin cans wishes to construct a right circular cylindrical can of height 20 centimeters and capacity 3000 cm^3 (see the figure). Find the inner radius r of the can.

Exercise 57

20 cm

CORN

r

58 Constructing a rectangular box Refer to Example 12. A box with an open top is to be constructed by cutting 3-inch squares from the corners of a rectangular sheet of tin whose length is twice its width. What size sheet will produce a box having a volume of 60 in^3?

59 Baseball toss A baseball is thrown straight upward with an initial speed of 64 ft/sec. The number of feet s above the ground after t seconds is given by the equation $s = -16t^2 + 64t$.

(a) When will the baseball be 48 feet above the ground?

(b) When will it hit the ground?

60 Braking distance The distance that a car travels between the time the driver makes the decision to hit the brakes and the time the car actually stops is called the braking distance. For a certain car traveling v mi/hr, the braking distance d (in feet) is given by $d = v + (v^2/20)$.

(a) Find the braking distance when v is 55 mi/hr.

(b) If a driver decides to brake 120 feet from a stop sign, how fast can the car be going and still stop by the time it reaches the sign?

61 Temperature of boiling water The temperature T (in °C) at which water boils is related to the elevation h (in meters above sea level) by the formula

$$h = 1000(100 - T) + 580(100 - T)^2$$

for $95 \le T \le 100$.

(a) At what elevation does water boil at a temperature of 98°C?

(b) The elevation of Mt. Everest is approximately 8840 meters. Estimate the temperature at which water boils at the top of this mountain. (*Hint:* Use the quadratic formula with $x = 100 - T$.)

62 Coulomb's law A particle of charge -1 is located on a coordinate line at $x = -2$, and a particle of charge -2 is located at $x = 2$, as shown in the figure. If a particle of charge $+1$ is located at a position x between -2 and 2, Coulomb's law in electrical theory asserts that the net force F acting on this particle is given by

$$F = \frac{-k}{(x + 2)^2} + \frac{2k}{(2 - x)^2}$$

for some constant $k > 0$. Determine the position at which the net force is zero.

Exercise 62

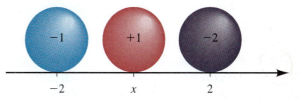

-1 $+1$ -2

-2 x 2

63 Dimensions of a sidewalk A rectangular plot of ground having dimensions 26 feet by 30 feet is surrounded by a walk of uniform width. If the area of the walk is 240 ft^2, what is its width?

64 Designing a poster A 24-by-36-inch sheet of paper is to be used for a poster, with the shorter side at the bottom. The margins at the sides and top are to have the same width, and the bottom margin is to be twice as wide as the other margins. Find the width of the margins if the printed area is to be 661.5 in^2.

65 Fencing a garden A square vegetable garden is to be tilled and then enclosed with a fence. If the fence costs $1 per foot and the cost of preparing the soil is $0.50 per ft^2, determine the size of the garden that can be enclosed for $120.

66 Fencing a region A farmer plans to enclose a rectangular region, using part of his barn for one side and fencing for the other three sides. If the side parallel to the barn is to be twice the length of an adjacent side, and the area of the region is to be 128 ft^2, how many feet of fencing should be purchased?

67 Planning a freeway The boundary of a city is a circle of diameter 5 miles. As shown in the figure, a straight highway runs through the center of the city from A to B. The highway department is planning to build a 6-mile-long freeway from A to a point P on the outskirts and then to B. Find the distance from A to P. (*Hint: APB* is a right triangle.)

Exercise 67

68 City expansion The boundary of a city is a circle of diameter 10 miles. Within the last decade, the city has grown in area by approximately 16π mi^2 (about 50 mi^2). Assuming the city was always circular in shape, find the corresponding change in distance from the center of the city to the boundary.

69 Distance between airplanes An airplane flying north at 200 mi/hr passed over a point on the ground at 2:00 P.M. Another airplane at the same altitude passed over the point at 2:30 P.M., flying east at 400 mi/hr (see the figure).

   (a) If t denotes the time in hours after 2:30 P.M., express the distance d between the airplanes in terms of t.

   (b) At what time after 2:30 P.M. were the airplanes 500 miles apart?

Exercise 69

70 Two-way radio range Two surveyors with two-way radios leave the same point at 9:00 A.M., one walking due south at 4 mi/hr and the other due west at 3 mi/hr. How long can they communicate with one another if each radio has a maximum range of 2 miles?

71 Constructing a pizza box A pizza box with a square base is to be made from a rectangular sheet of cardboard by cutting six 1-inch squares from the corners and the middle sections and folding up the sides (see the figure). If the area of the base is to be 144 in^2, what size piece of cardboard should be used?

Exercise 71

72 Constructing wire frames Two square wire frames are to be constructed from a piece of wire 100 inches long. If the area enclosed by one frame is to be one-half the area enclosed by the other, find the dimensions of each frame. (Disregard the thickness of the wire.)

73 Canoeing rate The speed of the current in a stream is 5 mi/hr. It takes a canoeist 30 minutes longer to paddle 1.2 miles upstream than to paddle the same distance downstream. What is the canoeist's rate in still water?

74 Height of a cliff When a rock is dropped from a cliff into an ocean, it travels approximately $16t^2$ feet in t seconds. If the splash is heard 4 seconds later and the speed of sound is 1100 ft/sec, approximate the height of the cliff.

75 Quantity discount A company sells running shoes to dealers for $40 per pair if less than 50 pairs are ordered. If 50 or more pairs are ordered (up to 600), the price per pair is reduced at a rate of $0.04 times the number ordered. How many pairs can a dealer purchase for $8400?

76 Price of a CD player When a popular brand of CD player is priced at $300 per unit, a store sells 15 units per week. Each time the price is reduced by $10, however, the sales increase by 2 per week. What selling price will result in weekly revenues of $7000?

77 Dimensions of an oil drum A closed right circular cylindrical oil drum of height 4 feet is to be constructed so that the total surface area is 10π ft². Find the diameter of the drum.

78 Dimensions of a vitamin tablet The rate at which a tablet of vitamin C begins to dissolve depends on the surface area of the tablet. One brand of tablet is 2 centimeters long and is in the shape of a cylinder with hemispheres of diameter 0.5 centimeter attached to both ends, as shown in the figure. A second brand of tablet is to be manufactured in the shape of a right circular cylinder of altitude 0.5 centimeter.

(a) Find the diameter of the second tablet so that its surface area is equal to that of the first tablet.

(b) Find the volume of each tablet.

Exercise 78

Exer. 79–80: During a nuclear explosion, a fireball will be produced having a maximum volume V_0. For temperatures below 2000 K and a given explosive force, the volume V of the fireball t seconds after the explosion can be estimated using the given formula. (Note that the kelvin is abbreviated as K, not °K.) Approximate t when V is 95% of V_0.

79 $V/V_0 = 0.8197 + 0.007752t + 0.0000281t^2$
(20-kiloton explosion)

80 $V/V_0 = 0.831 + 0.00598t + 0.0000919t^2$
(10-megaton explosion)

Exer. 81–82: When computations are carried out on a calculator, the quadratic formula will not always give accurate results if b^2 is large in comparison to ac, because one of the roots will be close to zero and difficult to approximate.

(a) Use the quadratic formula to approximate the roots of the given equation.

(b) To obtain a better approximation for the root near zero, rationalize the numerator to change

$$x = \frac{-b \pm \sqrt{b^2 - 4ac}}{2a} \quad \text{to} \quad x = \frac{2c}{-b \mp \sqrt{b^2 - 4ac}},$$

and use the second formula.

81 $x^2 + 4,500,000x - 0.96 = 0$

82 $x^2 - 73,000,000x + 2.01 = 0$

2.4
Complex Numbers

Complex numbers are needed to find solutions of equations that cannot be solved using only the set $\mathbb{R}$ of real numbers. The following chart illustrates several simple quadratic equations and the types of numbers required for solutions.

Equation	Solutions	Type of numbers required
$x^2 = 9$	$3, -3$	Integers
$x^2 = \frac{9}{4}$	$\frac{3}{2}, -\frac{3}{2}$	Rational numbers
$x^2 = 5$	$\sqrt{5}, -\sqrt{5}$	Irrational numbers
$x^2 = -9$	?	Complex numbers

The solutions of the first three equations in the chart are in $\mathbb{R}$; however, since squares of real numbers are never negative, $\mathbb{R}$ does not contain the solutions of $x^2 = -9$. To solve this equation, we need the **complex number system** $\mathbb{C}$, which contains both $\mathbb{R}$ and numbers whose squares are negative.

We begin by introducing the **imaginary unit,** denoted by i, which has the following properties.

Properties of i	$i = \sqrt{-1}, \qquad i^2 = -1$

Because its square is negative, the letter i does not represent a real number. It is a new mathematical entity that will enable us to obtain $\mathbb{C}$. Since i, together with $\mathbb{R}$, is to be contained in $\mathbb{C}$, we must consider products of the form bi for a real number b and also expressions of the form $a + bi$ for real numbers a and b. The next chart provides definitions we shall use.

Terminology	Definition	Example(s)
Complex number	$a + bi$, where a and b are real numbers and $i^2 = -1$	$3, 2 + i, 2i$
Imaginary number	$a + bi$ with $b \neq 0$	$3 + 2i, -4i$
Pure imaginary number	bi with $b \neq 0$	$-4i, \sqrt{3}\,i, i$
Equality	$a + bi = c + di$ if and only if $a = c$ and $b = d$	$x + yi = 3 + 4i$ iff $x = 3$ and $y = 4$
Sum	$(a + bi) + (c + di) = (a + c) + (b + d)i$	see Example 1(a)
Product	$(a + bi)(c + di) = (ac - bd) + (ad + bc)i$	see Example 1(b)

Note that the pure imaginary numbers are a subset of the imaginary numbers and the imaginary numbers are a subset of the complex numbers. We use the phrase *nonreal complex number* interchangeably with *imaginary number*.

It is not necessary to memorize the definitions of addition and multiplication of complex numbers given in the preceding chart. Instead, *we may treat all symbols as having properties of real numbers, with exactly one exception: We replace i^2 by* -1. Thus, for the product $(a + bi)(c + di)$ we simply use the distributive laws and the fact that

$$(bi)(di) = bdi^2 = bd(-1) = -bd.$$

EXAMPLE 1   Addition and multiplication of complex numbers

Express in the form $a + bi$, where a and b are real numbers:

(a) $(3 + 4i) + (2 + 5i)$ **(b)** $(3 + 4i)(2 + 5i)$

SOLUTION

(a) $(3 + 4i) + (2 + 5i) = (3 + 2) + (4 + 5)i = 5 + 9i$

(b) $(3 + 4i)(2 + 5i) = (3 + 4i)(2) + (3 + 4i)(5i)$
$$= 6 + 8i + 15i + 20i^2$$
$$= 6 + 23i + 20(-1)$$
$$= -14 + 23i$$

The set $\mathbb{R}$ of real numbers may be identified with the set of complex numbers of the form $a + 0i$. It is also convenient to denote the complex number $0 + bi$ by bi. Thus,

$$(a + 0i) + (0 + bi) = (a + 0) + (0 + b)i = a + bi.$$

Hence, we may regard $a + bi$ as the sum of two complex numbers a and bi (that is, $a + 0i$ and $0 + bi$). For the complex number $a + bi$, we call a the **real part** and b the **imaginary part.**

EXAMPLE 2   Equality of complex numbers

Find the values of x and y, where x and y are real numbers:

$$(2x - 4) + 9i = 8 + 3yi$$

SOLUTION   We begin by equating the real parts and the imaginary parts of each side of the equation:

$$2x - 4 = 8 \quad \text{and} \quad 9 = 3y$$

Since $2x - 4 = 8$, $2x = 12$ and $x = 6$. Since $9 = 3y$, $y = 3$. The values of x and y that make the complex numbers equal are

$$x = 6 \quad \text{and} \quad y = 3.$$

With complex numbers, we are now able to solve an equation such as $x^2 = -9$. Specifically, since

$$(3i)(3i) = 3^2i^2 = 9(-1) = -9,$$

we see that one solution is $3i$ and another is $-3i$.

In the next chart we define the difference of complex numbers and multiplication of a complex number by a real number.

Terminology	Definition
Difference	$(a + bi) - (c + di) = (a - c) + (b - d)i$
Multiplication by a real number k	$k(a + bi) = ka + (kb)i$

If we are asked to write an expression in the form $a + bi$, the form $a - di$ is acceptable, since $a - di = a + (-d)i$.

EXAMPLE 3 Operations with complex numbers

Express in the form $a + bi$, where a and b are real numbers:

(a) $4(2 + 5i) - (3 - 4i)$ (b) $(4 - 3i)(2 + i)$

(c) $i(3 - 2i)^2$ (d) i^{51}

SOLUTION

(a) $4(2 + 5i) - (3 - 4i) = 8 + 20i - 3 + 4i = 5 + 24i$

(b) $(4 - 3i)(2 + i) = 8 - 6i + 4i - 3i^2 = 11 - 2i$

(c) $i(3 - 2i)^2 = i(9 - 12i + 4i^2) = i(5 - 12i) = 5i - 12i^2 = 12 + 5i$

(d) Taking successive powers of i, we obtain

$$i^1 = i, \quad i^2 = -1, \quad i^3 = -i, \quad i^4 = 1,$$

and then the cycle starts over:

$$i^5 = i, \quad i^6 = i^2 = -1, \quad \text{and so on.}$$

In particular,

$$i^{51} = i^{48}i^3 = (i^4)^{12}i^3 = (1)^{12}i^3 = (1)(-i) = -i.$$

The following concept has important uses in working with complex numbers.

Definition of the Conjugate of a Complex Number	If $z = a + bi$ is a complex number, then its **conjugate,** denoted by $\bar{z}$, is $a - bi$.

Since $a - bi = a + (-bi)$, it follows that the conjugate of $a - bi$ is

$$a - (-bi) = a + bi.$$

Therefore, $a + bi$ and $a - bi$ are conjugates of each other. Some properties of conjugates are given in Exercises 55–60.

ILLUSTRATION Conjugates

Complex number	Conjugate
5 + 7i	5 − 7i
5 − 7i	5 + 7i
4i	−4i
3	3

The following two properties are consequences of the definitions of the sum and the product of complex numbers.

Properties of conjugates	Illustration
$(a + bi) + (a - bi) = 2a$	$(4 + 3i) + (4 - 3i) = 4 + 4 = 2 \cdot 4$
$(a + bi)(a - bi) = a^2 + b^2$	$(4 + 3i)(4 - 3i) = 4^2 - (3i)^2 = 4^2 - 3^2 i^2 = 4^2 + 3^2$

Note that *the sum and the product of a complex number and its conjugate are real numbers.* Conjugates are useful for finding the **multiplicative inverse** of $a + bi$, $1/(a + bi)$, or for simplifying the quotient of two complex numbers. As illustrated in the next example, we may think of these types of simplifications as merely *rationalizing the denominator,* since we are multiplying the quotient by the conjugate of the denominator divided by itself.

EXAMPLE 4   **Quotients of complex numbers**

Express in the form $a + bi$, where a and b are real numbers:

(a) $\dfrac{1}{9 + 2i}$ (b) $\dfrac{7 - i}{3 - 5i}$

SOLUTION

(a) $\dfrac{1}{9 + 2i} = \dfrac{1}{9 + 2i} \cdot \dfrac{9 - 2i}{9 - 2i} = \dfrac{9 - 2i}{81 + 4} = \dfrac{9}{85} - \dfrac{2}{85} i$

(b) $\dfrac{7 - i}{3 - 5i} = \dfrac{7 - i}{3 - 5i} \cdot \dfrac{3 + 5i}{3 + 5i} = \dfrac{21 + 35i - 3i - 5i^2}{9 + 25}$

$= \dfrac{26 + 32i}{34} = \dfrac{13}{17} + \dfrac{16}{17} i$

If p is a positive real number, then the equation $x^2 = -p$ has solutions in $\mathbb{C}$. One solution is $\sqrt{p}\, i$, since

$$\left(\sqrt{p}\, i\right)^2 = \left(\sqrt{p}\right)^2 i^2 = p(-1) = -p.$$

Similarly, $-\sqrt{p}\, i$ is also a solution.

The definition of $\sqrt{-r}$ in the next chart is motivated by $\left(\sqrt{r}\, i\right)^2 = -r$ for $r > 0$. When using this definition, take care *not* to write $\sqrt{ri}$ when $\sqrt{r}\, i$ is intended.

Terminology	Definition	Illustrations
Principal square root $\sqrt{-r}$ for $r > 0$	$\sqrt{-r} = \sqrt{r}\, i$	$\sqrt{-9} = \sqrt{9}\, i = 3i$ $\sqrt{-5} = \sqrt{5}\, i$ $\sqrt{-1} = \sqrt{1}\, i = i$

The radical sign must be used with caution when the radicand is negative. For example, the formula $\sqrt{a}\,\sqrt{b} = \sqrt{ab}$, which holds for positive real numbers, is *not* true when a and b are both negative, as shown below:

$$\sqrt{-3}\,\sqrt{-3} = \left(\sqrt{3}\,i\right)\left(\sqrt{3}\,i\right) = \left(\sqrt{3}\right)^2 i^2 = 3(-1) = -3$$

But
$$\sqrt{(-3)(-3)} = \sqrt{9} = 3.$$

Hence,
$$\sqrt{-3}\,\sqrt{-3} \neq \sqrt{(-3)(-3)}.$$

If only *one* of a or b is negative, then $\sqrt{a}\,\sqrt{b} = \sqrt{ab}$. In general, we shall not apply laws of radicals if radicands are negative. Instead, we shall change the form of radicals before performing any operations, as illustrated in the next example.

EXAMPLE 5 **Working with square roots of negative numbers**

Express in the form $a + bi$, where a and b are real numbers:

$$\left(5 - \sqrt{-9}\right)\left(-1 + \sqrt{-4}\right)$$

SOLUTION First we use the definition $\sqrt{-r} = \sqrt{r}\,i$, and then we simplify:

$$
\begin{aligned}
\left(5 - \sqrt{-9}\right)\left(-1 + \sqrt{-4}\right) &= \left(5 - \sqrt{9}\,i\right)\left(-1 + \sqrt{4}\,i\right) \\
&= (5 - 3i)(-1 + 2i) \\
&= -5 + 10i + 3i - 6i^2 \\
&= -5 + 13i + 6 = 1 + 13i
\end{aligned}
$$

In Section 2.3 we stated that if the discriminant $b^2 - 4ac$ of the quadratic equation $ax^2 + bx + c = 0$ is negative, then there are no real roots of the equation. In fact, the solutions of the equation are two *imaginary* numbers. Moreover, the solutions are conjugates of each other, as shown in the next example.

EXAMPLE 6 **A quadratic equation with complex solutions**

Solve the equation $5x^2 + 2x + 1 = 0$.

SOLUTION Applying the quadratic formula with $a = 5, b = 2$, and $c = 1$, we see that

$$
\begin{aligned}
x &= \frac{-2 \pm \sqrt{2^2 - 4(5)(1)}}{2(5)} \\
&= \frac{-2 \pm \sqrt{-16}}{10} = \frac{-2 \pm 4i}{10} = \frac{-1 \pm 2i}{5} = -\frac{1}{5} \pm \frac{2}{5}i.
\end{aligned}
$$

Thus, the solutions of the equation are $-\frac{1}{5} + \frac{2}{5}i$ and $-\frac{1}{5} - \frac{2}{5}i$.

EXAMPLE 7 **An equation with complex solutions**

Solve the equation $x^3 - 1 = 0$.

Difference of two cubes:

$$a^3 - b^3 = (a - b)(a^2 + ab + b^2)$$

SOLUTION   Using the difference of two cubes factoring formula with $a = x$ and $b = 1$, we write $x^3 - 1 = 0$ as

$$(x - 1)(x^2 + x + 1) = 0.$$

Setting each factor equal to zero and solving the resulting equations, we obtain the solutions

$$1, \quad \frac{-1 \pm \sqrt{1 - 4}}{2} = \frac{-1 \pm \sqrt{3}\,i}{2}$$

or, equivalently,

$$1, \quad -\frac{1}{2} + \frac{\sqrt{3}}{2}i, \quad -\frac{1}{2} - \frac{\sqrt{3}}{2}i.$$

Since the number 1 is called the **unit real number** and the given equation may be written as $x^3 = 1$, we call these three solutions the **cube roots of unity.**

In Section 1.3 we mentioned that $x^2 + 1$ is irreducible over the *real* numbers. However, if we factor over the *complex* numbers, then $x^2 + 1$ may be factored as follows:

$$x^2 + 1 = (x + i)(x - i)$$

2.4   Exercises

Exer. 1–34: Write the expression in the form $a + bi$, where a and b are real numbers.

1 $(5 - 2i) + (-3 + 6i)$

2 $(-5 + 7i) + (4 + 9i)$

3 $(7 - 6i) - (-11 - 3i)$

4 $(-3 + 8i) - (2 + 3i)$

5 $(3 + 5i)(2 - 7i)$

6 $(-2 + 6i)(8 - i)$

7 $(1 - 3i)(2 + 5i)$

8 $(8 + 2i)(7 - 3i)$

9 $(5 - 2i)^2$

10 $(6 + 7i)^2$

11 $i(3 + 4i)^2$

12 $i(2 - 7i)^2$

13 $(3 + 4i)(3 - 4i)$

14 $(4 + 9i)(4 - 9i)$

15 i^{43}

16 i^{92}

17 i^{73}

18 i^{66}

19 $\dfrac{3}{2 + 4i}$

20 $\dfrac{5}{2 - 7i}$

21 $\dfrac{1 - 7i}{6 - 2i}$

22 $\dfrac{2 + 9i}{-3 - i}$

23 $\dfrac{-4 + 6i}{2 + 7i}$

24 $\dfrac{-3 - 2i}{5 + 2i}$

25 $\dfrac{4 - 2i}{-5i}$

26 $\dfrac{-2 + 6i}{3i}$

27 $(2 + 5i)^3$

28 $(3 - 2i)^3$

29 $\left(2 - \sqrt{-4}\right)\left(3 - \sqrt{-16}\right)$

30 $\left(-3 + \sqrt{-25}\right)\left(8 - \sqrt{-36}\right)$

31 $\dfrac{4 + \sqrt{-81}}{7 - \sqrt{-64}}$

32 $\dfrac{5 - \sqrt{-121}}{1 + \sqrt{-25}}$

33 $\dfrac{\sqrt{-36}\,\sqrt{-49}}{\sqrt{-16}}$

34 $\dfrac{\sqrt{-25}}{\sqrt{-16}\,\sqrt{-81}}$

Exer. 35–38: Find the values of x and y, where x and y are real numbers.

35 $4 + (x + 2y)i = x + 2i$ **36** $(x - y) + 3i = 7 + yi$

37 $(2x - y) - 16i = 10 + 4yi$

38 $8 + (3x + y)i = 2x - 4i$

Exer. 39–54: Find the solutions of the equation.

39 $x^2 - 6x + 13 = 0$ **40** $x^2 - 2x + 26 = 0$

41 $x^2 + 4x + 13 = 0$ **42** $x^2 + 8x + 17 = 0$

43 $x^2 - 5x + 20 = 0$ **44** $x^2 + 3x + 6 = 0$

45 $4x^2 + x + 3 = 0$ **46** $-3x^2 + x - 5 = 0$

47 $x^3 + 125 = 0$ **48** $x^3 - 27 = 0$

49 $x^4 = 256$ **50** $x^4 = 81$

51 $4x^4 + 25x^2 + 36 = 0$ **52** $27x^4 + 21x^2 + 4 = 0$

53 $x^3 + 3x^2 + 4x = 0$

54 $8x^3 - 12x^2 + 2x - 3 = 0$

Exer. 55–60: Verify the property.

55 $\overline{z + w} = \bar{z} + \bar{w}$ **56** $\overline{z - w} = \bar{z} - \bar{w}$

57 $\overline{z \cdot w} = \bar{z} \cdot \bar{w}$ **58** $\overline{z/w} = \bar{z}/\bar{w}$

59 $\bar{z} = z$ if and only if z is real.

60 $\overline{z^2} = (\bar{z})^2$

2.5

Other Types of Equations

The equations considered in previous sections are inadequate for many problems. For example, in applications it is often necessary to consider powers x^k with $k > 2$. Some equations involve absolute values or radicals. In this section we give examples of equations of these types that can be solved using elementary methods.

EXAMPLE 1 **Solving an equation containing an absolute value**

Solve the equation $|x - 5| = 3$.

SOLUTION If a and b are real numbers with $b > 0$, then $|a| = b$ if and only if $a = b$ or $a = -b$. Hence, if $|x - 5| = 3$, then either

$$x - 5 = 3 \quad \text{or} \quad x - 5 = -3.$$

Solving for x gives us

$$x = 5 + 3 = 8 \quad \text{or} \quad x = 5 - 3 = 2.$$

Thus, the given equation has two solutions, 8 and 2.

For an equation such as

$$2|x - 5| + 3 = 11,$$

we first isolate the absolute value expression by subtracting 3 and dividing by 2 to obtain

$$|x - 5| = \frac{11 - 3}{2} = 4,$$

and then we proceed as in Example 1.

If an equation is in factored form *with zero on one side,* then we may obtain solutions by setting each factor equal to zero. For example, if p, q, and r are expressions in x and if $pqr = 0$, then either $p = 0$, $q = 0$, or $r = 0$. In the next example we factor by grouping terms.

EXAMPLE 2 Solving an equation using grouping

Solve the equation $x^3 + 2x^2 - x - 2 = 0$.

SOLUTION

$$x^3 + 2x^2 - x - 2 = 0 \qquad \text{given}$$
$$x^2(x + 2) - 1(x + 2) = 0 \qquad \text{group terms}$$
$$(x^2 - 1)(x + 2) = 0 \qquad \text{factor out } x + 2$$
$$(x + 1)(x - 1)(x + 2) = 0 \qquad \text{factor } x^2 - 1$$
$$x + 1 = 0, \quad x - 1 = 0, \quad x + 2 = 0 \qquad \text{zero factor theorem}$$
$$x = -1, \qquad x = 1, \qquad x = -2 \qquad \text{solve for } x$$

EXAMPLE 3 Solving an equation containing rational exponents

Solve the equation $x^{3/2} = x^{1/2}$.

SOLUTION

$$x^{3/2} = x^{1/2} \qquad \text{given}$$
$$x^{3/2} - x^{1/2} = 0 \qquad \text{subtract } x^{1/2}$$
$$x^{1/2}(x - 1) = 0 \qquad \text{factor out } x^{1/2}$$
$$x^{1/2} = 0, \quad x - 1 = 0 \qquad \text{zero factor theorem}$$
$$x = 0, \qquad x = 1 \qquad \text{solve for } x$$

In Example 3 it would have been incorrect to divide both sides of the equation $x^{3/2} = x^{1/2}$ by $x^{1/2}$, obtaining $x = 1$, since the solution $x = 0$ would be lost. In general, *avoid dividing both sides of an equation by an expression that contains variables*—always *factor* instead.

If an equation involves radicals or fractional exponents, we often raise both sides to a positive power. The solutions of the new equation always contain the solutions of the given equation. For example, the solutions of

$$2x - 3 = \sqrt{x + 6}$$

are also solutions of

$$(2x - 3)^2 = \left(\sqrt{x + 6}\right)^2.$$

*Raising both sides of an equation to an **odd** power can introduce imaginary solutions. For example, cubing both sides of $x = 1$ gives us $x^3 = 1$, which is equivalent to $x^3 - 1 = 0$. This equation has three solutions, of which two are imaginary (see Example 7 in Section 2.4).*

In some cases the new equation has *more* solutions than the given equation. To illustrate, if we are given the equation $x = 3$ and we square both sides, we obtain $x^2 = 9$. Note that the given equation $x = 3$ has only one solution, 3, but the new equation $x^2 = 9$ has two solutions, 3 and -3. Any solution of the new equation that is not a solution of the given equation is an extraneous solution. Since extraneous solutions may occur, *it is absolutely essential to check all solutions obtained after raising both sides of an equation to an even power.* Such checks are unnecessary if both sides are raised to an *odd* power, because in this case extraneous (real number) solutions are not introduced.

EXAMPLE 4 Solving an equation containing a radical

Solve the equation $\sqrt[3]{x^2 - 1} = 2$.

SOLUTION

$$\sqrt[3]{x^2 - 1} = 2 \qquad \text{given}$$
$$\left(\sqrt[3]{x^2 - 1}\right)^3 = 2^3 \qquad \text{cube both sides}$$
$$x^2 - 1 = 8 \qquad \text{property of } \sqrt[n]{}$$
$$x^2 = 9 \qquad \text{add 1}$$
$$x = \pm 3 \qquad \text{take the square root}$$

Thus, the given equation has two solutions, 3 and -3. Except to detect algebraic errors, a check is unnecessary, since we raised both sides to an odd power.

In the last solution we used the phrase *cube both sides* of $\sqrt[3]{x^2 - 1} = 2$. In general, for the equation $x^{m/n} = a$, where x is a real number, we raise both sides to the power n/m (the reciprocal of m/n) to solve for x. If m is odd, we obtain $x = a^{n/m}$, but if m is even, we have $x = \pm a^{n/m}$. If n is even, extraneous solutions may occur—for example, if $x^{3/2} = -8$, then $x = (-8)^{2/3} = \left(\sqrt[3]{-8}\right)^2 = (-2)^2 = 4$. However, 4 is not a solution of $x^{3/2} = -8$ since $4^{3/2} = 8$, not -8.

ILLUSTRATION **Solving $x^{m/n} = a$, m odd, x real**

Equation	**Solution**
■ $x^{3/1} = 64$	$x = 64^{1/3} = \sqrt[3]{64} = 4$
■ $x^{3/2} = 64$	$x = 64^{2/3} = \left(\sqrt[3]{64}\right)^2 = 4^2 = 16$

ILLUSTRATION **Solving $x^{m/n} = a$, m even, x real**

Equation	**Solution**
■ $x^{4/1} = 16$	$x = \pm 16^{1/4} = \pm \sqrt[4]{16} = \pm 2$
■ $x^{2/3} = 16$	$x = \pm 16^{3/2} = \pm \left(\sqrt{16}\right)^3 = \pm 4^3 = \pm 64$

In the next two examples, before we raise both sides of the equation to a power, we *isolate a radical*—that is, we consider an equivalent equation in which only the radical appears on one side.

EXAMPLE 5 Solving an equation containing a radical

Solve the equation $3 + \sqrt{3x + 1} = x$.

SOLUTION

$$3 + \sqrt{3x + 1} = x \qquad \text{given}$$
$$\sqrt{3x + 1} = x - 3 \qquad \text{isolate the radical}$$
$$\left(\sqrt{3x + 1}\right)^2 = (x - 3)^2 \qquad \text{square both sides}$$
$$3x + 1 = x^2 - 6x + 9 \qquad \text{simplify}$$
$$x^2 - 9x + 8 = 0 \qquad \text{subtract } 3x + 1$$
$$(x - 1)(x - 8) = 0 \qquad \text{factor}$$
$$x - 1 = 0, \quad x - 8 = 0 \qquad \text{zero factor theorem}$$
$$x = 1, \qquad x = 8 \qquad \text{solve for } x$$

We raised both sides to an even power, so checks are required.

✔ **Check** $x = 1$ LS: $3 + \sqrt{3(1) + 1} = 3 + \sqrt{4} = 3 + 2 = 5$
RS: 1

Since $5 \neq 1$, $x = 1$ is not a solution.

✔ **Check** $x = 8$ LS: $3 + \sqrt{3(8) + 1} = 3 + \sqrt{25} = 3 + 5 = 8$
RS: 8

Since $8 = 8$ is a true statement, $x = 8$ is a solution.
Hence, the given equation has one solution, $x = 8$.   ◢

In order to solve an equation involving several radicals, it may be necessary to raise both sides to powers two or more times, as in the next example.

EXAMPLE 6   **Solving an equation containing radicals**

Solve the equation $\sqrt{2x - 3} - \sqrt{x + 7} + 2 = 0$.

SOLUTION

$$\sqrt{2x - 3} - \sqrt{x + 7} + 2 = 0 \qquad \text{given}$$
$$\sqrt{2x - 3} = \sqrt{x + 7} - 2 \qquad \text{isolate } \sqrt{2x - 3}$$
$$2x - 3 = (x + 7) - 4\sqrt{x + 7} + 4 \qquad \text{square both sides}$$
$$x - 14 = -4\sqrt{x + 7} \qquad \text{isolate the radical term}$$
$$x^2 - 28x + 196 = 16(x + 7) \qquad \text{square both sides}$$
$$x^2 - 28x + 196 = 16x + 112 \qquad \text{multiply factors}$$
$$x^2 - 44x + 84 = 0 \qquad \text{subtract } 16x + 112$$
$$(x - 42)(x - 2) = 0 \qquad \text{factor}$$
$$x - 42 = 0, \quad x - 2 = 0 \qquad \text{zero factor theorem}$$
$$x = 42, \qquad x = 2 \qquad \text{solve for } x$$

A check is required, since both sides were raised to an even power.

✔ **Check** $x = 42$ LS: $\sqrt{84 - 3} - \sqrt{42 + 7} + 2 = 9 - 7 + 2 = 4$
RS: 0

Since $4 \neq 0$, $x = 42$ is not a solution.

✔ **Check** $x = 2$ LS: $\sqrt{4 - 3} - \sqrt{2 + 7} + 2 = 1 - 3 + 2 = 0$
RS: 0

Since $0 = 0$ is a true statement, $x = 2$ is a solution.
Hence, the given equation has one solution, $x = 2$.   ◢

An equation is of **quadratic type** if it can be written in the form

$$au^2 + bu + c = 0,$$

where $a \neq 0$ and u is an expression in some variable. If we find the solutions in terms of u, then the solutions of the given equation can be obtained by referring to the specific form of u.

EXAMPLE 7   **Solving an equation of quadratic type**

Solve the equation $x^{2/3} + x^{1/3} - 6 = 0$.

SOLUTION   Since $x^{2/3} = (x^{1/3})^2$, the form of the equation suggests that we let $u = x^{1/3}$, as in the second line below:

$$
\begin{aligned}
x^{2/3} + x^{1/3} - 6 &= 0 &&\text{given} \\
u^2 + u - 6 &= 0 &&\text{let } u = x^{1/3} \\
(u + 3)(u - 2) &= 0 &&\text{factor} \\
u + 3 = 0, \quad u - 2 &= 0 &&\text{zero factor theorem} \\
u = -3, \quad u &= 2 &&\text{solve for } u \\
x^{1/3} = -3, \quad x^{1/3} &= 2 &&u = x^{1/3} \\
x = -27, \quad x &= 8 &&\text{cube both sides}
\end{aligned}
$$

A check is unnecessary, since we did not raise both sides to an even power. Hence, the given equation has two solutions, -27 and 8.

An alternative method is to factor the left side of the given equation as follows:

$$x^{2/3} + x^{1/3} - 6 = (x^{1/3} + 3)(x^{1/3} - 2)$$

By setting each factor equal to 0, we obtain the solutions.

EXAMPLE 8   **Solving an equation of quadratic type**

Solve the equation $x^4 - 3x^2 + 1 = 0$.

SOLUTION   Since $x^4 = (x^2)^2$, the form of the equation suggests that we let $u = x^2$, as in the second line below:

$$
\begin{aligned}
x^4 - 3x^2 + 1 &= 0 &&\text{given} \\
u^2 - 3u + 1 &= 0 &&\text{let } u = x^2 \\
u &= \frac{3 \pm \sqrt{9 - 4}}{2} = \frac{3 \pm \sqrt{5}}{2} &&\text{quadratic formula} \\
x^2 &= \frac{3 \pm \sqrt{5}}{2} &&u = x^2 \\
x &= \pm\sqrt{\frac{3 \pm \sqrt{5}}{2}} &&\text{take the square root}
\end{aligned}
$$

Thus, there are four solutions:

$$\sqrt{\frac{3 + \sqrt{5}}{2}}, \quad -\sqrt{\frac{3 + \sqrt{5}}{2}}, \quad \sqrt{\frac{3 - \sqrt{5}}{2}}, \quad -\sqrt{\frac{3 - \sqrt{5}}{2}}$$

Using a calculator, we obtain the approximations ± 1.62 and ± 0.62. A check is unnecessary because we did not raise both sides of an equation to an even power.

EXAMPLE 9   Determining the route of a ferry

Figure 1

A passenger ferry makes trips from a town to an island community that is 7 miles downshore from the town and 3 miles off a straight shoreline. As shown in Figure 1, the ferry travels along the shoreline to some point and then proceeds directly to the island. If the ferry travels 12 mi/hr along the shoreline and 10 mi/hr as it moves out to sea, determine the routes that have a travel time of 45 minutes.

SOLUTION   Let x denote the distance traveled along the shoreline. This leads to the sketch in Figure 2, where d is the distance from a point on the shoreline to the island. Refer to the indicated right triangle:

$$
\begin{aligned}
d^2 &= (7 - x)^2 + 3^2 && \text{Pythagorean theorem} \\
&= 49 - 14x + x^2 + 9 && \text{square terms} \\
&= x^2 - 14x + 58 && \text{simplify}
\end{aligned}
$$

Figure 2

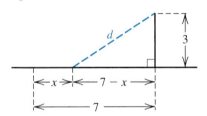

Taking the square root of both sides and noting that $d > 0$, we obtain

$$d = \sqrt{x^2 - 14x + 58}.$$

Using distance = (rate)(time) or, equivalently, time = (distance)/(rate) gives us the following table.

	Along the shoreline	Away from shore
Distance (mi)	x	$\sqrt{x^2 - 14x + 58}$
Rate (mi/hr)	12	10
Time (hr)	$\dfrac{x}{12}$	$\dfrac{\sqrt{x^2 - 14x + 58}}{10}$

The time for the complete trip is the sum of the two expressions in the last row of the table. Since the rate is in mi/hr, we must, for consistency, express this time (45 minutes) as $\frac{3}{4}$ hour. Thus, we have the following:

$$
\begin{aligned}
\frac{x}{12} + \frac{\sqrt{x^2 - 14x + 58}}{10} &= \frac{3}{4} && \text{total time for trip} \\
\frac{\sqrt{x^2 - 14x + 58}}{10} &= \frac{3}{4} - \frac{x}{12} && \text{subtract } \frac{x}{12} \\
6\sqrt{x^2 - 14x + 58} &= 45 - 5x && \text{multiply by the lcd, 60} \\
6\sqrt{x^2 - 14x + 58} &= 5(9 - x) && \text{factor} \\
36(x^2 - 14x + 58) &= 25(9 - x)^2 && \text{square both sides} \\
36x^2 - 504x + 2088 &= 2025 - 450x + 25x^2 && \text{multiply terms} \\
11x^2 - 54x + 63 &= 0 && \text{simplify}
\end{aligned}
$$

(continued)

$$(x - 3)(11x - 21) = 0 \qquad \text{factor}$$

$$x - 3 = 0, \quad 11x - 21 = 0 \qquad \text{zero factor theorem}$$

$$x = 3, \qquad x = \frac{21}{11} \qquad \text{solve for } x$$

A check verifies that these numbers are also solutions of the original equation. Hence, there are two possible routes with a travel time of 45 minutes: the ferry may travel along the shoreline either 3 miles or $\frac{21}{11} \approx 1.9$ miles before proceeding to the island. ◢

2.5   *Exercises*

Exer. 1–50: Solve the equation.

1 $|x + 4| = 11$

2 $|x - 5| = 2$

3 $|3x - 2| + 3 = 7$

4 $2|5x + 2| - 1 = 5$

5 $3|x + 1| - 2 = -11$

6 $|x - 2| + 5 = 5$

7 $9x^3 - 18x^2 - 4x + 8 = 0$

8 $3x^3 - 4x^2 - 27x + 36 = 0$

9 $4x^4 + 10x^3 = 6x^2 + 15x$

10 $15x^5 - 20x^4 = 6x^3 - 8x^2$

11 $y^{3/2} = 5y$

12 $y^{4/3} = -3y$

13 $\sqrt{7 - 5x} = 8$

14 $\sqrt{2x - 9} = \frac{1}{3}$

15 $2 + \sqrt[3]{1 - 5t} = 0$

16 $\sqrt[3]{6 - s^2} + 5 = 0$

17 $\sqrt[5]{2x^2 + 1} - 2 = 0$

18 $\sqrt[4]{2x^2 - 1} = x$

19 $\sqrt{7 - x} = x - 5$

20 $\sqrt{3 - x} - x = 3$

21 $3\sqrt{2x - 3} + 2\sqrt{7 - x} = 11$

22 $\sqrt{2x + 15} - 2 = \sqrt{6x + 1}$

23 $x = 4 + \sqrt{4x - 19}$

24 $x = 3 + \sqrt{5x - 9}$

25 $x + \sqrt{5x + 19} = -1$

26 $x - \sqrt{-7x - 24} = -2$

27 $\sqrt{7 - 2x} - \sqrt{5 + x} = \sqrt{4 + 3x}$

28 $4\sqrt{1 + 3x} + \sqrt{6x + 3} = \sqrt{-6x - 1}$

29 $\sqrt{11 + 8x} + 1 = \sqrt{9 + 4x}$

30 $2\sqrt{x} - \sqrt{x - 3} = \sqrt{5 + x}$

31 $\sqrt{2\sqrt{x + 1}} = \sqrt{3x - 5}$

32 $\sqrt{5\sqrt{x}} = \sqrt{2x - 3}$

33 $\sqrt{1 + 4\sqrt{x}} = \sqrt{x} + 1$

34 $\sqrt{x + 1} = \sqrt{x - 1}$

35 $x^4 - 25x^2 + 144 = 0$

36 $2x^4 - 10x^2 + 8 = 0$

37 $5y^4 - 7y^2 + 1 = 0$

38 $3y^4 - 5y^2 + 1 = 0$

39 $36x^{-4} - 13x^{-2} + 1 = 0$

40 $x^{-2} - 2x^{-1} - 35 = 0$

41 $3x^{2/3} + 4x^{1/3} - 4 = 0$

42 $2y^{1/3} - 3y^{1/6} + 1 = 0$

43 $6w - 23w^{1/2} + 20 = 0$

44 $2x^{-2/3} - 7x^{-1/3} - 15 = 0$

45 $\left(\dfrac{t}{t + 1}\right)^2 - \dfrac{2t}{t + 1} - 8 = 0$

46 $6u^{-1/2} - 13u^{-1/4} + 6 = 0$

47 $27x^3 = (x + 5)^3$

48 $16x^4 = (x - 4)^4$

49 $\sqrt[3]{x} = 2\sqrt[4]{x}$ (*Hint:* Raise both sides to the least common multiple of 3 and 4.)

50 $\sqrt{x + 3} = \sqrt[4]{2x + 6}$

Exer. 51–52: Find the real solutions of the equation.

51 (a) $x^{5/3} = 32$

(b) $x^{4/3} = 16$

(c) $x^{2/3} = -36$

(d) $x^{3/4} = 125$

(e) $x^{3/2} = -27$

52 (a) $x^{3/5} = -27$

(b) $x^{2/3} = 25$

(c) $x^{4/3} = -49$

(d) $x^{3/2} = 27$

(e) $x^{3/4} = -8$

Exer. 53–56: Solve for the specified variable.

53 $T = 2\pi\sqrt{\dfrac{l}{g}}$ for l (period of a pendulum)

54 $d = \frac{1}{2}\sqrt{4R^2 - C^2}$ for C (segments of circles)

55 $S = \pi r\sqrt{r^2 + h^2}$ for h (surface area of a cone)

56 $\omega = \dfrac{1}{\sqrt{LC}}$ for C (alternating-current circuits)

57 **Ladder height** The recommended distance d that a ladder should be placed away from a vertical wall is 25% of its length L. Approximate the height h that can be reached by relating h as a percentage of L.

Exercise 57

58 **Nuclear experiments** Nuclear experiments performed in the ocean vaporize large quantities of salt water. Salt boils and turns into vapor at 1738 K. After being vaporized by a 10-megaton force, the salt takes at least 8–10 seconds to cool enough to crystallize. The amount of salt A that has crystallized t seconds after an experiment is sometimes calculated using $A = k\sqrt{t/T}$, where k and T are constants. Solve this equation for t.

59 **Windmill power** The power P (in watts) generated by a windmill that has efficiency E is given by the formula $P = 0.31ED^2V^3$, where D is the diameter (in feet) of the windmill blades and V is the wind velocity (in ft/sec). Approximate the wind velocity necessary to generate 10,000 watts if $E = 42\%$ and $D = 10$.

60 **Withdrawal resistance of nails** The *withdrawal resistance* of a nail indicates its holding strength in wood. A formula that is used for bright common nails is $P = 15{,}700S^{5/2}RD$, where P is the maximum withdrawal resistance (in pounds), S is the specific gravity of the wood at 12% moisture content, R is the radius of the nail (in inches), and D is the depth (in inches) that the nail has penetrated the wood. A 6d (six-penny) bright, common nail of length 2 inches and diameter 0.113 inch is driven completely into a piece of Douglas fir. If it requires a maximum force of 380 pounds to remove the nail, approximate the specific gravity of Douglas fir.

61 **The effect of price on demand** The demand for a commodity usually depends on its price. If other factors do not affect the demand, then the quantity Q purchased at price P (in cents) is given by $Q = kP^{-c}$, where k and c are positive constants. If $k = 10^5$ and $c = \frac{1}{2}$, find the price that will result in the purchase of 5000 items.

62 **The urban heat island** Urban areas have higher average air temperatures than rural areas, as a result of the presence of buildings, asphalt, and concrete. This phenomenon has become known as the *urban heat island*. The temperature difference T (in °C) between urban and rural areas near Montreal, with a population P between 1000 and 1,000,000, can be described by the formula $T = 0.25P^{1/4}/\sqrt{v}$, where v is the average wind speed (in mi/hr) and $v \geq 1$. If $T = 3$ and $v = 5$, find P.

63 **Dimensions of a sand pile** As sand leaks out of a certain container, it forms a pile that has the shape of a right circular cone whose altitude is always one-half the diameter d of the base. What is d at the instant at which 144 cm³ of sand has leaked out?

Exercise 63

64 Inflating a weather balloon The volume of a spherical weather balloon is $10\frac{2}{3}$ ft³. In order to lift a transmitter and meteorological equipment, the balloon is inflated with an additional $25\frac{1}{3}$ ft³ of helium. How much does its diameter increase?

65 The cube rule in political science The cube rule in political science is an empirical formula that is said to predict the percentage y of seats in the U.S. House of Representatives that will be won by a political party from the popular vote for the party's presidential candidate. If x denotes the percentage of the popular vote for a party's presidential candidate, then the cube rule states that

$$y = \frac{x^3}{x^3 + (1 - x)^3}.$$

What percentage of the popular vote will the presidential candidate need in order for the candidate's party to win 60% of the House seats?

66 Dimensions of a conical cup A conical paper cup is to have a height of 3 inches. Find the radius of the cone that will result in a surface area of 6π in².

67 Installing a power line A power line is to be installed across a river that is 1 mile wide to a town that is 5 miles downstream (see the figure). It costs $7500 per mile to lay the cable underwater and $6000 per mile to lay it overland. Determine how the cable should be installed if $35,000 has been allocated for this project.

Exercise 67

68 Calculating human growth Adolphe Quetelet (1796–1874), the director of the Brussels Observatory from 1832 to 1874, was the first person to attempt to fit a mathematical expression to human growth data. If h denotes height in meters and t denotes age in years, Quetelet's formula for males in Brussels can be expressed as

$$h + \frac{h}{h_M - h} = at + \frac{h_0 + t}{1 + \frac{4}{3}t},$$

with $h_0 = 0.5$, the height at birth; $h_M = 1.684$, the final adult male height; and $a = 0.545$.

(a) Find the expected height of a 12-year-old male.

(b) At what age should 50% of the adult height be reached?

2.6

Inequalities

An **inequality** is a statement that two quantities or expressions are not equal. It may be the case that one quantity is less than ($<$), less than or equal to ($\leq$), greater than ($>$), or greater than or equal to ($\geq$) another quantity. Consider the inequality

$$2x + 3 > 11,$$

where x is a variable. As illustrated in the following table, certain numbers yield true statements when substituted for x, and others yield false statements.

x	$2x + 3 > 11$	**Conclusion**
3	$9 > 11$	False statement
4	$11 > 11$	False statement
5	$13 > 11$	True statement
6	$15 > 11$	True statement

If a true statement is obtained when a number b is substituted for x, then b is a **solution** of the inequality. Thus, $x = 5$ is a solution of $2x + 3 > 11$

since $13 > 11$ is true, but $x = 3$ is not a solution since $9 > 11$ is false. To **solve** an inequality means to find *all* solutions. Two inequalities are **equivalent** if they have exactly the same solutions.

Most inequalities have an infinite number of solutions. To illustrate, the solutions of the inequality

$$2 < x < 5$$

consist of *every* real number x between 2 and 5. We call this set of numbers an **open interval** and denote it by $(2, 5)$. The **graph** of the open interval $(2, 5)$ is the set of all points on a coordinate line that lie between—but do not include—the points corresponding to $x = 2$ and $x = 5$. The graph is represented by shading an appropriate part of the axis, as shown in Figure 1. We refer to this process as **sketching the graph** of the interval. The numbers 2 and 5 are called the **endpoints** of the interval $(2, 5)$. The parentheses in the notation $(2, 5)$ and in Figure 1 are used to indicate that the endpoints of the interval are not included.

If we wish to include an endpoint, we use a bracket instead of a parenthesis. For example, the solutions of the inequality $2 \le x \le 5$ are denoted by $[2, 5]$ and are referred to as a **closed interval.** The graph of $[2, 5]$ is sketched in Figure 2, where brackets indicate that endpoints are included. We shall also consider **half-open intervals** $[a, b)$ and $(a, b]$ and **infinite intervals,** as described in the following chart. The symbol ∞ (read "infinity") used for infinite intervals is merely a notational device and does *not* represent a real number.

Figure 1

Figure 2

Intervals

Notation	Inequality	Graph
(1) (a, b)	$a < x < b$	
(2) $[a, b]$	$a \le x \le b$	
(3) $[a, b)$	$a \le x < b$	
(4) $(a, b]$	$a < x \le b$	
(5) (a, ∞)	$x > a$	
(6) $[a, \infty)$	$x \ge a$	
(7) $(-\infty, b)$	$x < b$	
(8) $(-\infty, b]$	$x \le b$	
(9) $(-\infty, \infty)$	$-\infty < x < \infty$	

Methods for solving inequalities in x are similar to those used for solving equations. In particular, we often use properties of inequalities to replace a given inequality with a list of equivalent inequalities, ending with an inequality from which solutions are easily obtained. The properties in the following chart can be proved for real numbers a, b, c, and d.

Properties of Inequalities

Property	Illustration
(1) If $a < b$ and $b < c$, then $a < c$.	$2 < 5$ and $5 < 9$, so $2 < 9$.
(2) If $a < b$, then $a + c < b + c$ and $a - c < b - c$.	$2 < 7$, so $2 + 3 < 7 + 3$ and $2 - 3 < 7 - 3$.
(3) If $a < b$ and $c > 0$, then $ac < bc$ and $\dfrac{a}{c} < \dfrac{b}{c}$.	$2 < 5$ and $3 > 0$, so $2 \cdot 3 < 5 \cdot 3$ and $\dfrac{2}{3} < \dfrac{5}{3}$.
(4) If $a < b$ and $c < 0$, then $ac > bc$ and $\dfrac{a}{c} > \dfrac{b}{c}$.	$2 < 5$ and $-3 < 0$, so $2(-3) > 5(-3)$ and $\dfrac{2}{-3} > \dfrac{5}{-3}$.

Reverse the inequality when multiplying or dividing by a negative number.

It is important to remember that multiplying or dividing both sides of an inequality by a negative real number *reverses* the inequality sign (see property 4). Properties similar to those above are true for other inequalities and for $\leq$ and $\geq$. Thus, if $a > b$, then $a + c > b + c$; if $a \geq b$ and $c < 0$, then $ac \leq bc$; and so on.

If x represents a real number, then, by property 2, adding or subtracting the same expression containing x on both sides of an inequality yields an equivalent inequality. By property 3, we may multiply or divide both sides of an inequality by an expression containing x if we are certain that the expression is positive for all values of x under consideration. To illustrate, multiplication or division by $x^4 + 3x^2 + 5$ would be permissible, since this expression is always positive. If we multiply or divide both sides of an inequality by an expression that is always negative, such as $-7 - x^2$, then, by property 4, the inequality is reversed.

In examples we shall describe solutions of inequalities by means of intervals and also represent them graphically.

EXAMPLE 1   **Solving an inequality**

Solve the inequality $-3x + 4 < 11$.

SOLUTION

$$-3x + 4 < 11 \qquad \text{given}$$
$$(-3x + 4) - 4 < 11 - 4 \qquad \text{subtract 4}$$
$$-3x < 7 \qquad \text{simplify}$$
$$\frac{-3x}{-3} > \frac{7}{-3} \qquad \begin{array}{l}\text{divide by } -3; \\ \text{reverse the inequality sign}\end{array}$$
$$x > -\tfrac{7}{3} \qquad \text{simplify}$$

Figure 3

Thus, the solutions of $-3x + 4 < 11$ consist of all real numbers x such that $x > -\frac{7}{3}$. This is the interval $\left(-\frac{7}{3}, \infty\right)$ sketched in Figure 3.

EXAMPLE 2 **Solving an inequality**

Solve the inequality $4x - 3 < 2x + 5$.

SOLUTION

$$
\begin{array}{ll}
4x - 3 < 2x + 5 & \text{given} \\
(4x - 3) + 3 < (2x + 5) + 3 & \text{add 3} \\
4x < 2x + 8 & \text{simplify} \\
4x - 2x < (2x + 8) - 2x & \text{subtract } 2x \\
2x < 8 & \text{simplify} \\
\dfrac{2x}{2} < \dfrac{8}{2} & \text{divide by 2} \\
x < 4 & \text{simplify}
\end{array}
$$

Figure 4

Hence, the solutions of the given inequality consist of all real numbers x such that $x < 4$. This is the interval $(-\infty, 4)$ sketched in Figure 4.

EXAMPLE 3 **Solving an inequality**

Solve the inequality $-6 < 2x - 4 < 2$.

SOLUTION A real number x is a solution of the given inequality if and only if it is a solution of *both* of the inequalities

$$-6 < 2x - 4 \qquad \text{and} \qquad 2x - 4 < 2.$$

This first inequality is solved as follows:

$$
\begin{array}{ll}
-6 < 2x - 4 & \text{given} \\
-6 + 4 < (2x - 4) + 4 & \text{add 4} \\
-2 < 2x & \text{simplify} \\
\dfrac{-2}{2} < \dfrac{2x}{2} & \text{divide by 2} \\
-1 < x & \text{simplify} \\
x > -1 & \text{equivalent inequality}
\end{array}
$$

The second inequality is then solved:

$$
\begin{array}{ll}
2x - 4 < 2 & \text{given} \\
2x < 6 & \text{add 4} \\
x < 3 & \text{divide by 2}
\end{array}
$$

Thus, x is a solution of the given inequality if and only if *both*

$$x > -1 \qquad \text{and} \qquad x < 3;$$

that is,

$$-1 < x < 3.$$

(continued)

Figure 5

Hence, the solutions are all numbers in the open interval $(-1, 3)$ sketched in Figure 5.

An alternative (and shorter) method is to solve both inequalities simultaneously—that is, solve the continued inequality:

$$
\begin{array}{lll}
-6 < 2x - 4 < 2 & & \text{given} \\
-6 + 4 < 2x \quad < 2 + 4 & & \text{add 4} \\
-2 < 2x \quad < 6 & & \text{simplify} \\
-1 < \ x \quad < 3 & & \text{divide by 2}
\end{array}
$$

EXAMPLE 4 Solving a continued inequality

Solve the continued inequality $-5 \le \dfrac{4 - 3x}{2} < 1$.

SOLUTION   A number x is a solution of the given inequality if and only if

$$
-5 \le \frac{4 - 3x}{2} \quad \text{and} \quad \frac{4 - 3x}{2} < 1.
$$

We can either work with each inequality separately or solve both inequalities simultaneously, as follows (keep in mind that our goal is to isolate x):

$$
\begin{array}{lll}
-5 \le \dfrac{4 - 3x}{2} < 1 & & \text{given} \\[2mm]
-10 \le 4 - 3x < 2 & & \text{multiply by 2} \\[2mm]
-10 - 4 \le \quad -3x < 2 - 4 & & \text{subtract 4} \\[2mm]
-14 \le \quad -3x < -2 & & \text{simplify} \\[2mm]
\dfrac{-14}{-3} \ge \dfrac{-3x}{-3} > \dfrac{-2}{-3} & & \text{divide by } -3; \text{ reverse} \\
& & \text{the inequality signs} \\[2mm]
\tfrac{14}{3} \ge \quad x \quad > \tfrac{2}{3} & & \text{simplify} \\[2mm]
\tfrac{2}{3} < \quad x \quad \le \tfrac{14}{3} & & \text{equivalent inequality}
\end{array}
$$

Figure 6

Thus, the solutions of the inequality are all numbers in the half-open interval $\left(\tfrac{2}{3}, \tfrac{14}{3}\right]$ sketched in Figure 6.

EXAMPLE 5 Solving a rational inequality

Solve the inequality $\dfrac{1}{x - 2} > 0$.

Figure 7

SOLUTION   Since the numerator is positive, the fraction is positive if and only if the denominator, $x - 2$, is also positive. Thus, $x - 2 > 0$ or, equivalently, $x > 2$, and the solutions are all numbers in the infinite interval $(2, \infty)$ sketched in Figure 7.

Figure 8

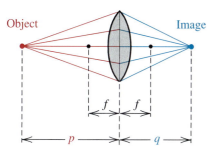

Object   Image

f f

p q

EXAMPLE 6   Using a lens formula

As illustrated in Figure 8, if a convex lens has focal length f centimeters and if an object is placed a distance p centimeters from the lens with $p > f$, then the distance q from the lens to the image is related to p and f by the formula

$$\frac{1}{p} + \frac{1}{q} = \frac{1}{f}.$$

If $f = 5$ cm, how close must the object be to the lens for the image to be more than 12 centimeters from the lens?

SOLUTION   Since $f = 5$, the given formula may be written as

$$\frac{1}{p} + \frac{1}{q} = \frac{1}{5}.$$

We wish to determine the values of q such that $q > 12$. Let us first solve the equation for q:

$$5q + 5p = pq \qquad \text{multiply by the lcd, } 5pq$$
$$q(5 - p) = -5p \qquad \text{collect } q \text{ terms on one side and factor}$$
$$q = -\frac{5p}{5 - p} = \frac{5p}{p - 5} \qquad \text{divide by } 5 - p$$

To solve the inequality $q > 12$, we proceed as follows:

$$\frac{5p}{p - 5} > 12 \qquad\qquad q = \frac{5p}{p - 5}$$
$$5p > 12(p - 5) \qquad \text{allowable, since } p > f \text{ implies } p - 5 > 0$$
$$-7p > -60 \qquad \text{multiply factors and collect } p \text{ terms on one side}$$
$$p < \tfrac{60}{7} \qquad \text{divide by } -7; \text{ reverse the inequality}$$

Figure 9

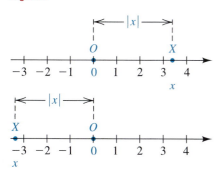

Combining the last inequality with the fact that p is greater than 5, we obtain the solution

$$5 < p < \tfrac{60}{7}. \qquad\qquad ◢$$

If a point X on a coordinate line has coordinate x, as shown in Figure 9, then X is to the right of the origin O if $x > 0$ and to the left of O if $x < 0$. From Section 1.1, the distance $d(O, X)$ between O and X is the *nonnegative* real number given by

$$d(O, X) = |x - 0| = |x|.$$

It follows that the solutions of an inequality such as $|x| < 3$ consist of the coordinates of all points whose distance from O is less than 3. This is the open interval $(-3, 3)$ sketched in Figure 10. Thus,

Figure 10

$$|x| < 3 \quad \text{is equivalent to} \quad -3 < x < 3.$$

Similarly, for $|x| > 3$, the distance between O and a point with coordinate x is greater than 3; that is,

$$|x| > 3 \quad \text{is equivalent to} \quad x < -3 \text{ or } x > 3.$$

Figure 11

The graph of the solutions to $|x| > 3$ is sketched in Figure 11. We often use the **union symbol** $\cup$ and write

$$(-\infty, -3) \cup (3, \infty)$$

to denote all real numbers that are in either $(-\infty, -3)$ or $(3, \infty)$.

The notation

$$(-\infty, 2) \cup (2, \infty)$$

represents the set of all real numbers except 2.

The **intersection symbol** $\cap$ is used to denote the elements that are *common* to two sets. For example,

$$(-\infty, 3) \cap (-3, \infty) = (-3, 3),$$

since the intersection of $(-\infty, 3)$ and $(-3, \infty)$ consists of all real numbers x such that both $x < 3$ *and* $x > -3$.

The preceding discussion may be generalized to obtain the following properties of absolute values.

Properties of Absolute Values ($b > 0$)	
	(1) $|a| < b$ is equivalent to $-b < a < b$.
	(2) $|a| > b$ is equivalent to $a < -b$ or $a > b$.

In the next example we use property 1 with $a = x - 3$ and $b = 0.5$.

EXAMPLE 7 **Solving an inequality containing an absolute value**

Solve the inequality $|x - 3| < 0.5$.

SOLUTION

$|x - 3| < 0.5$	given
$-0.5 < x - 3 < 0.5$	property 1
$-0.5 + 3 < (x - 3) + 3 < 0.5 + 3$	isolate x by adding 3
$2.5 < x < 3.5$	simplify

Figure 12

Thus, the solutions are the real numbers in the open interval $(2.5, 3.5)$. The graph is sketched in Figure 12. ◢

In the next example we use property 2 with $a = 2x + 3$ and $b = 9$.

EXAMPLE 8 Solving an inequality containing an absolute value

Solve the inequality $|2x + 3| > 9$.

SOLUTION

$$
\begin{array}{lll}
|2x + 3| > 9 & & \text{given} \\
2x + 3 < -9 \quad \text{or} \quad 2x + 3 > 9 & & \text{property 2} \\
2x < -12 \quad \text{or} \quad 2x > 6 & & \text{subtract 3} \\
x < -6 \quad \text{or} \quad x > 3 & & \text{divide by 2}
\end{array}
$$

Figure 13

Consequently, the solutions of the inequality $|2x + 3| > 9$ consist of the numbers in $(-\infty, -6) \cup (3, \infty)$. The graph is sketched in Figure 13.

The trichotomy law in Section 1.1 states that for any real numbers a and b exactly one of the following is true:

$$a > b, \qquad a < b, \qquad \text{or} \qquad a = b$$

Thus, after solving $|2x + 3| > 9$ in Example 8, we readily obtain the solutions for $|2x + 3| < 9$ and $|2x + 3| = 9$—namely, $(-6, 3)$ and $\{-6, 3\}$, respectively. Note that the union of these three sets of solutions is necessarily the set $\mathbb{R}$ of real numbers.

When using the notation $a < x < b$, we must have $a < b$. Thus, *it is incorrect to write the solution $x < -6$ or $x > 3$ (in Example 8) as $3 < x < -6$.* Another misuse of inequality notation is to write $a < x > b$, since when several inequality symbols are used in one expression, *they must point in the same direction.*

2.6 *Exercises*

1 Given $-7 < -3$, determine the inequality obtained if

(a) 5 is added to both sides

(b) 4 is subtracted from both sides

(c) both sides are multiplied by $\frac{1}{3}$

(d) both sides are multiplied by $-\frac{1}{3}$

2 Given $4 > -5$, determine the inequality obtained if

(a) 7 is added to both sides

(b) -5 is subtracted from both sides

(c) both sides are divided by 6

(d) both sides are divided by -6

Exer. 3–12: Express the inequality as an interval, and sketch its graph.

3 $x < -2$

4 $x \leq 5$

5 $x \geq 4$

6 $x > -3$

7 $-2 < x \leq 4$

8 $-3 \leq x < 5$

9 $3 \leq x \leq 7$

10 $-3 < x < -1$

11 $5 > x \geq -2$

12 $-3 \geq x > -5$

Exer. 13–20: Express the interval as an inequality in the variable x.

13 $(-5, 8]$

14 $[0, 4)$

15 $[-4, -1]$

16 $(3, 7)$

17 $[4, \infty)$

18 $(-3, \infty)$

19 $(-\infty, -5)$

20 $(-\infty, 2]$

Exer. 21–70: Solve the inequality, and express the solutions in terms of intervals whenever possible.

21 $3x - 2 > 14$

22 $2x + 5 \leq 7$

23 $-2 - 3x \geq 2$

24 $3 - 5x < 11$

25 $2x + 5 < 3x - 7$

26 $x - 8 > 5x + 3$

27 $9 + \frac{1}{3}x \geq 4 - \frac{1}{2}x$

28 $\frac{1}{4}x + 7 \leq \frac{1}{3}x - 2$

29 $-3 < 2x - 5 < 7$

30 $4 \geq 3x + 5 > -1$

31 $3 \leq \dfrac{2x - 3}{5} < 7$

32 $-2 < \dfrac{4x + 1}{3} \leq 0$

33 $4 > \dfrac{2 - 3x}{7} \geq -2$

34 $5 \geq \dfrac{6 - 5x}{3} > 2$

35 $0 \leq 4 - \frac{1}{3}x < 2$

36 $-2 < 3 + \frac{1}{4}x \leq 5$

37 $(2x - 3)(4x + 5) \leq (8x + 1)(x - 7)$

38 $(x - 3)(x + 3) \geq (x + 5)^2$

39 $(x - 4)^2 > x(x + 12)$

40 $2x(6x + 5) < (3x - 2)(4x + 1)$

41 $\dfrac{4}{3x + 2} \geq 0$

42 $\dfrac{3}{2x + 5} \leq 0$

43 $\dfrac{-2}{4 - 3x} > 0$

44 $\dfrac{-3}{2 - x} < 0$

45 $\dfrac{2}{(1 - x)^2} > 0$

46 $\dfrac{4}{x^2 + 4} < 0$

47 $|x| < 3$

48 $|x| \leq 7$

49 $|x| \geq 5$

50 $|-x| > 2$

51 $|x + 3| < 0.01$

52 $|x - 4| \leq 0.03$

53 $|x + 2| + 0.1 \geq 0.2$

54 $|x - 3| - 0.3 > 0.1$

55 $|2x + 5| < 4$

56 $|3x - 7| \geq 5$

57 $-\frac{1}{3}|6 - 5x| + 2 \geq 1$

58 $2|-11 - 7x| - 2 > 10$

59 $|7x + 2| > -2$

60 $|6x - 5| \leq -2$

61 $|3x - 9| > 0$

62 $|5x + 2| \leq 0$

63 $\left|\dfrac{2 - 3x}{5}\right| \geq 2$

64 $\left|\dfrac{2x + 5}{3}\right| < 1$

65 $\dfrac{3}{|5 - 2x|} < 2$

66 $\dfrac{2}{|2x + 3|} \geq 5$

67 $-2 < |x| < 4$

68 $1 < |x| < 5$

69 $1 < |x - 2| < 4$

70 $2 < |2x - 1| < 3$

Exer. 71–72: Solve part (a) and use that answer to determine the answers to parts (b) and (c).

71 (a) $|x + 5| = 3$ (b) $|x + 5| < 3$

(c) $|x + 5| > 3$

72 (a) $|x - 3| < 2$ (b) $|x - 3| = 2$

(c) $|x - 3| > 2$

Exer. 73–76: Express the statement in terms of an inequality involving an absolute value.

73 The weight w of a wrestler must be within 2 pounds of 148 pounds.

74 The radius r of a ball bearing must be within 0.01 centimeter of 1 centimeter.

75 The difference of two temperatures T_1 and T_2 within a chemical mixture must be between 5°C and 10°C.

76 The arrival time t of train B must be at least 5 minutes different from the 4:00 P.M. arrival time of train A.

77 **Temperature scales** Temperature readings on the Fahrenheit and Celsius scales are related by the formula $C = \frac{5}{9}(F - 32)$. What values of F correspond to the values of C such that $30 \leq C \leq 40$?

78 Hooke's law According to Hooke's law, the force F (in pounds) required to stretch a certain spring x inches beyond its natural length is given by $F = (4.5)x$ (see the figure). If $10 \le F \le 18$, what are the corresponding values for x?

Exercise 78

Natural length

Stretched x inches

x

79 Ohm's law Ohm's law in electrical theory states that if R denotes the resistance of an object (in ohms), V the potential difference across the object (in volts), and I the current that flows through it (in amperes), then $R = V/I$. If the voltage is 110, what values of the resistance will result in a current that does not exceed 10 amperes?

80 Electrical resistance If two resistors R_1 and R_2 are connected in parallel in an electrical circuit, the net resistance R is given by

$$\frac{1}{R} = \frac{1}{R_1} + \frac{1}{R_2}.$$

If $R_1 = 10$ ohms, what values of R_2 will result in a net resistance of less than 5 ohms?

81 Linear magnification Shown in the figure is a simple magnifier consisting of a convex lens. The object to be magnified is positioned so that the distance p from the lens is less than the focal length f. The linear magnification M is the ratio of the image size to the object size. It is shown in physics that $M = f/(f - p)$. If $f = 6$ cm, how far should the object be placed from the lens so that its image appears at least three times as large? (Compare with Example 6.)

Exercise 81

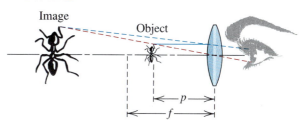

Image

Object

p

f

82 Drug concentration To treat arrhythmia (irregular heartbeat), a drug is fed intravenously into the bloodstream. Suppose that the concentration c of the drug after t hours is given by $c = 3.5t/(t + 1)$ mg/L. If the minimum therapeutic level is 1.5 mg/L, determine when this level is exceeded.

83 Business expenditure A construction firm is trying to decide which of two models of a crane to purchase. Model A costs \$50,000 and requires \$4000 per year to maintain. Model B has an initial cost of \$40,000 and a maintenance cost of \$5500 per year. For how many years must model A be used before it becomes more economical than B?

84 Buying a car A consumer is trying to decide whether to purchase car A or car B. Car A costs \$10,000 and has an mpg rating of 30, and insurance is \$550 per year. Car B costs \$12,000 and has an mpg rating of 50, and insurance is \$600 per year. Assume that the consumer drives 15,000 miles per year and that the price of gas remains constant at \$1.25 per gallon. Based only on these facts, determine how long it will take for the total cost of car B to become less than that of car A.

85 Decreasing height A person's height will typically decrease by 0.024 inch each year after age 30.

(a) If a woman was 5 feet 9 inches tall at age 30, predict her height at age 70.

(b) A 50-year-old man is 5 feet 6 inches tall. Determine an inequality for the range of heights (in inches) that this man will experience between the ages of 30 and 70.

2.7

More on Inequalities

To solve an inequality involving polynomials of degree greater than 1, we shall express each polynomial as a product of linear factors $ax + b$ and/or irreducible quadratic factors $ax^2 + bx + c$. If any such factor is not zero in an interval, then it is either positive throughout the interval or negative throughout the interval. Hence, if we choose any k in the interval and if the factor is positive (or negative) for $x = k$, then it is positive (or negative) throughout the interval. The value of the factor at $x = k$ is called a **test value** of the factor at the test number k. This concept is exhibited in the following example.

EXAMPLE 1   **Solving a quadratic inequality**

Solve the inequality $2x^2 - x < 3$.

SOLUTION   To use test values, *it is essential to have 0 on one side of the inequality sign.* Thus, we proceed as follows:

$$2x^2 - x < 3 \qquad \text{given}$$
$$2x^2 - x - 3 < 0 \qquad \text{make one side 0}$$
$$(x + 1)(2x - 3) < 0 \qquad \text{factor}$$

The factors $x + 1$ and $2x - 3$ are zero at -1 and $\frac{3}{2}$, respectively. The corresponding points on a coordinate line (see Figure 1) determine the nonintersecting intervals

$$(-\infty, -1), \quad \left(-1, \tfrac{3}{2}\right), \quad \text{and} \quad \left(\tfrac{3}{2}, \infty\right).$$

We may find the signs of $x + 1$ and $2x - 3$ in each interval by using a test value taken from each interval. To illustrate, if we choose $k = -10$ in $(-\infty, -1)$, the values of both $x + 1$ and $2x - 3$ are negative, and hence they are negative throughout $(-\infty, -1)$. A similar procedure for the remaining two intervals gives us the following *sign chart,* where the term *resulting sign* in the last row refers to the sign obtained by applying laws of signs to the product of the factors. Note that the resulting sign is positive or negative according to whether the number of negative signs of factors is even or odd, respectively.

Figure 1

Interval	$(-\infty, -1)$	$\left(-1, \frac{3}{2}\right)$	$\left(\frac{3}{2}, \infty\right)$
Sign of $x + 1$	$-$	$+$	$+$
Sign of $2x - 3$	$-$	$-$	$+$
Resulting sign	$+$	$-$	$+$

Sometimes it is convenient to represent the signs of $x + 1$ and $2x - 3$ by using a coordinate line and a *sign diagram,* of the type illustrated in Figure 2. The vertical lines indicate where the factors are zero, and signs of factors are shown above the coordinate line. The resulting signs are shown in red.

Figure 2

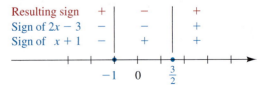

The solutions of $(x + 1)(2x - 3) < 0$ are the values of x for which the product of the factors is *negative*—that is, where the resulting sign is negative. This corresponds to the open interval $\left(-1, \frac{3}{2}\right)$.

Back on page 74, we discussed the zero factor theorem, which dealt with *equalities*. It is a common mistake to extend this theorem to *inequalities*. The following warning shows this incorrect extension applied to the inequality in Example 1.

 Warning!

$(x + 1)(2x - 3) < 0$ is ***not*** equivalent to $x + 1 < 0$ or $2x - 3 < 0$

In future examples we will use either a sign chart or a sign diagram, but not both. When working exercises, you should choose the method of solution with which you feel most comfortable.

EXAMPLE 2 Solving a quadratic inequality

Solve the inequality $-3x^2 < -21x + 30$.

SOLUTION

$$-3x^2 < -21x + 30 \qquad \text{given}$$
$$-3x^2 + 21x - 30 < 0 \qquad \text{make one side 0}$$
$$x^2 - 7x + 10 > 0 \qquad \text{divide by the common factor } -3; \text{ reverse the inequality}$$
$$(x - 2)(x - 5) > 0 \qquad \text{factor}$$

The factors are zero at 2 and 5. The corresponding points on a coordinate line (see Figure 3) determine the nonintersecting intervals

$$(-\infty, 2), \quad (2, 5), \quad \text{and} \quad (5, \infty).$$

As in Example 1, we may use test values from each interval to obtain the following sign chart.

Figure 3

Interval	$(-\infty, 2)$	$(2, 5)$	$(5, \infty)$
Sign of $x - 2$	$-$	$+$	$+$
Sign of $x - 5$	$-$	$-$	$+$
Resulting sign	$+$	$-$	$+$

The solutions of $(x - 2)(x - 5) > 0$ are the values of x for which the resulting sign is *positive*. Thus, the solution of the given inequality is the union $(-\infty, 2) \cup (5, \infty)$.

EXAMPLE 3 Using a sign diagram to solve an inequality

Solve the inequality $\dfrac{(x + 2)(3 - x)}{(x + 1)(x^2 + 1)} \leq 0$.

SOLUTION   Since 0 is already on the right side of the inequality and the left side is factored, we may proceed directly to the sign diagram in Figure 4, where the vertical lines indicate the zeros (-2, -1, and 3) of the factors.

Figure 4

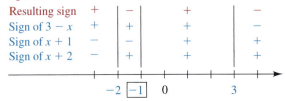

The frame around the -1 indicates that -1 makes a factor in the denominator of the original inequality equal to 0. Since the quadratic factor $x^2 + 1$ is always positive, it has no effect on the sign of the quotient and hence may be omitted from the diagram.

The various signs of the factors can be found using test values. Alternatively, we need only remember that as x increases, the sign of a linear factor $ax + b$ changes from negative to positive if the coefficient a of x is positive, and the sign changes from positive to negative if a is negative.

To determine where the quotient is less than or equal to 0, we first note from the sign diagram that it is *negative* for numbers in $(-2, -1) \cup (3, \infty)$. Since the quotient is 0 at $x = -2$ and $x = 3$, the numbers -2 and 3 are also solutions and must be *included* in our solution. Lastly, the quotient is *undefined* at $x = -1$, so -1 must be *excluded* from our solution. Thus, the solutions of the given inequality are given by

$$[-2, -1) \cup [3, \infty).$$

 EXAMPLE 4   Using a sign diagram to solve an inequality

Solve the inequality $\dfrac{(2x + 1)^2(x - 1)}{x(x^2 - 1)} \geq 0$.

SOLUTION   Rewriting the inequality as

$$\frac{(2x + 1)^2(x - 1)}{x(x + 1)(x - 1)} \geq 0,$$

we see that $x - 1$ is a factor of both the numerator and the denominator. Thus, *assuming that $x - 1 \neq 0$* (that is, $x \neq 1$), we may cancel this factor and reduce our search for solutions to the case

$$\frac{(2x + 1)^2}{x(x + 1)} \geq 0 \qquad \text{and} \qquad x \neq 1.$$

We next observe that this quotient is 0 if $2x + 1 = 0$ $\left(\text{that is, if } x = -\frac{1}{2}\right)$. Hence, $-\frac{1}{2}$ is a solution. To find the remaining solutions, we construct the sign

Figure 5

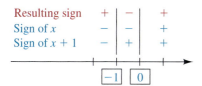

diagram in Figure 5. We do not include $(2x + 1)^2$ in the sign diagram, since this expression is always positive if $x \neq -\frac{1}{2}$ and so has no effect on the sign of the quotient. Referring to the resulting sign and remembering that $-\frac{1}{2}$ is a solution but 1 is *not* a solution, we see that the solutions of the given inequality are given by

$$(-\infty, -1) \cup \left\{-\tfrac{1}{2}\right\} \cup (0, 1) \cup (1, \infty).$$

EXAMPLE 5   Using a sign diagram to solve an inequality

Solve the inequality $\dfrac{x + 1}{x + 3} \leq 2$.

SOLUTION   A common mistake in solving such an inequality is to first multiply both sides by $x + 3$. If we did so, we would have to consider two cases, since $x + 3$ may be positive or negative (assuming $x + 3 \neq 0$), and we might have to reverse the inequality. A simpler method is to first obtain an equivalent inequality that has 0 on the right side and proceed from there:

$$\frac{x + 1}{x + 3} \leq 2 \quad \text{given}$$

$$\frac{x + 1}{x + 3} - 2 \leq 0 \quad \text{make one side 0}$$

$$\frac{x + 1 - 2(x + 3)}{x + 3} \leq 0 \quad \text{combine into one fraction}$$

$$\frac{-x - 5}{x + 3} \leq 0 \quad \text{simplify}$$

$$\frac{x + 5}{x + 3} \geq 0 \quad \text{multiply by } -1$$

Note that the direction of the inequality is changed in the last step, since we multiplied by a negative number. This multiplication was performed for convenience, so that all factors would have positive coefficients of x.

The factors $x + 5$ and $x + 3$ are 0 at $x = -5$ and $x = -3$, respectively. This leads to the sign diagram in Figure 6, where the signs are determined as in previous examples. We see from the diagram that the resulting sign, and hence the sign of the quotient, is positive in $(-\infty, -5) \cup (-3, \infty)$. The quotient is 0 at $x = -5$ (include -5) and undefined at $x = -3$ (exclude -3). Hence, the solution of $(x + 5)/(x + 3) \geq 0$ is $(-\infty, -5] \cup (-3, \infty)$.

Figure 6

(continued)

An alternative method of solution is to begin by multiplying both sides of the given inequality by $(x + 3)^2$, *assuming that* $x \neq -3$. In this case, $(x + 3)^2 > 0$ and the multiplication is permissible; however, after the resulting inequality is solved, the value $x = -3$ must be excluded. ◢

EXAMPLE 6 Determining minimum therapeutic levels

For a drug to have a beneficial effect, its concentration in the bloodstream must exceed a certain value, which is called the *minimum therapeutic level*. Suppose that the concentration c (in mg/L) of a particular drug t hours after it is taken orally is given by

$$c = \frac{20t}{t^2 + 4}.$$

If the minimum therapeutic level is 4 mg/L, determine when this level is exceeded.

SOLUTION   The minimum therapeutic level, 4 mg/L, is exceeded if $c > 4$. Thus, we must solve the inequality

$$\frac{20t}{t^2 + 4} > 4.$$

Since $t^2 + 4 > 0$ for every t, we may multiply both sides by $t^2 + 4$ and proceed as follows:

$$20t > 4t^2 + 16 \qquad \text{allowable, since } t^2 + 4 > 0$$

$$-4t^2 + 20t - 16 > 0 \qquad \text{make one side } 0$$

$$t^2 - 5t + 4 < 0 \qquad \text{divide by the common factor } -4$$

$$(t - 1)(t - 4) < 0 \qquad \text{factor}$$

The factors in the last inequality are 0 when $t = 1$ and $t = 4$. These are the times at which c is *equal to* 4. As in previous examples, we may use a sign chart or sign diagram (with $t \geq 0$) to show that $(t - 1)(t - 4) < 0$ for every t in the interval $(1, 4)$. Hence, the minimum therapeutic level is exceeded if $1 < t < 4$. ◢

Some basic properties of inequalities were stated at the beginning of the last section. The following additional properties are helpful for solving certain inequalities. Proofs of the properties are given after the chart.

Additional Properties of Inequalities

Property	Illustration
(1) If $0 < a < b$, then $\dfrac{1}{a} > \dfrac{1}{b}$.	If $0 < \dfrac{1}{x} < 4$, then $\dfrac{1}{1/x} > \dfrac{1}{4}$, or $x > \dfrac{1}{4}$.
(2) If $0 < a < b$, then $0 < a^2 < b^2$.	If $0 < \sqrt{x} < 4$, then $0 < (\sqrt{x})^2 < 4^2$, or $0 < x < 16$.
(3) If $0 < a < b$, then $0 < \sqrt{a} < \sqrt{b}$.	If $0 < x^2 < 4$, then $0 < \sqrt{x^2} < \sqrt{4}$, or $0 < |x| < 2$.

PROOFS

(1) If $0 < a < b$, then multiplying by $1/(ab)$ yields

$$a \cdot \frac{1}{ab} < b \cdot \frac{1}{ab}, \quad \text{or} \quad \frac{1}{b} < \frac{1}{a}; \quad \text{that is,} \quad \frac{1}{a} > \frac{1}{b}.$$

(2) If $0 < a < b$, then multiplying by a yields $a \cdot a < a \cdot b$ and multiplying by b yields $b \cdot a < b \cdot b$, so $a^2 < ab < b^2$ and hence $a^2 < b^2$.

(3) If $0 < a < b$, then $b - a > 0$ or, equivalently,

$$\left(\sqrt{b} + \sqrt{a}\right)\left(\sqrt{b} - \sqrt{a}\right) > 0.$$

Dividing both sides of the last inequality by $\sqrt{b} + \sqrt{a}$, we obtain $\sqrt{b} - \sqrt{a} > 0$; that is, $\sqrt{b} > \sqrt{a}$.

2.7 *Exercises*

Exer. 1–40: Solve the inequality, and express the solutions in terms of intervals whenever possible.

1 $(3x + 1)(5 - 10x) > 0$

2 $(2 - 3x)(4x - 7) \geq 0$

3 $(x + 2)(x - 1)(4 - x) \leq 0$

4 $(x - 5)(x + 3)(-2 - x) < 0$

5 $x^2 - x - 6 < 0$

6 $x^2 + 4x + 3 \geq 0$

7 $x^2 - 2x - 5 > 3$

8 $x^2 - 4x - 17 \leq 4$

9 $x(2x + 3) \geq 5$

10 $x(3x - 1) \leq 4$

11 $6x - 8 > x^2$

12 $x + 12 \leq x^2$

13 $x^2 < 16$

14 $x^2 > 9$

15 $25x^2 - 9 < 0$

16 $25x^2 - 9x < 0$

17 $16x^2 \geq 9x$

18 $16x^2 > 9$

19 $x^4 + 5x^2 \geq 36$

20 $x^4 + 15x^2 < 16$

21 $x^3 + 2x^2 - 4x - 8 \geq 0$

22 $2x^3 - 3x^2 - 2x + 3 \leq 0$

23 $\dfrac{x^2(x + 2)}{(x + 2)(x + 1)} \leq 0$

24 $\dfrac{(x^2 + 1)(x - 3)}{x^2 - 9} \geq 0$

25 $\dfrac{x^2 - x}{x^2 + 2x} \leq 0$

26 $\dfrac{(x + 3)^2(2 - x)}{(x + 4)(x^2 - 4)} \leq 0$

27 $\dfrac{x - 2}{x^2 - 3x - 10} \geq 0$

28 $\dfrac{x + 5}{x^2 - 7x + 12} \leq 0$

29 $\dfrac{-3x}{x^2 - 9} > 0$

30 $\dfrac{2x}{16 - x^2} < 0$

31 $\dfrac{x + 1}{2x - 3} > 2$

32 $\dfrac{x - 2}{3x + 5} \leq 4$

33 $\dfrac{1}{x - 2} \geq \dfrac{3}{x + 1}$

34 $\dfrac{2}{2x + 3} \leq \dfrac{2}{x - 5}$

35 $\dfrac{4}{3x - 2} \leq \dfrac{2}{x + 1}$

36 $\dfrac{3}{5x + 1} \geq \dfrac{1}{x - 3}$

37 $\dfrac{x}{3x - 5} \leq \dfrac{2}{x - 1}$

38 $\dfrac{x}{2x - 1} \geq \dfrac{3}{x + 2}$

39 $x^3 > x$

40 $x^4 \geq x^2$

Exer. 41–42: As a particle moves along a straight path, its speed v (in cm/sec) at time t (in seconds) is given by the equation. For what subintervals of the given time interval $[a, b]$ will its speed be at least k cm/sec?

41 $v = t^3 - 3t^2 - 4t + 20; \quad [0, 5]; \quad k = 8$

42 $v = t^4 - 4t^2 + 10; \qquad [1, 6]; \quad k = 10$

43 Vertical leap record *Guinness Book of World Records* reports that German shepherds can make vertical leaps of over 10 feet when scaling walls. If the distance s (in feet) off the ground after t seconds is given by the equation $s = -16t^2 + 24t + 1$, for how many seconds is the dog more than 9 feet off the ground?

44 Height of a projected object If an object is projected vertically upward from ground level with an initial velocity of 320 ft/sec, then its distance s above the ground after t seconds is given by $s = -16t^2 + 320t$. For what values of t will the object be more than 1536 feet above the ground?

45 Braking distance The braking distance d (in feet) of a certain car traveling v mi/hr is given by the equation $d = v + (v^2/20)$. Determine the velocities that result in braking distances of less than 75 feet.

46 Gas mileage The number of miles M that a certain compact car can travel on 1 gallon of gasoline is related to its speed v (in mi/hr) by

$$M = -\tfrac{1}{30}v^2 + \tfrac{5}{2}v \quad \text{for} \quad 0 < v < 70.$$

For what speeds will M be at least 45?

47 Salmon propagation For a particular salmon population, the relationship between the number S of spawners and the number R of offspring that survive to maturity is given by the formula $R = 4500S/(S + 500)$. Under what conditions is $R > S$?

48 Population density The population density D (in people/mi²) in a large city is related to the distance x from the center of the city by $D = 5000x/(x^2 + 36)$. In what areas of the city does the population density exceed 400 people/mi²?

49 Weight in space After an astronaut is launched into space, the astronaut's weight decreases until a state of weightlessness is achieved. The weight of a 125-pound astronaut at an altitude of x kilometers above sea level is given by

$$W = 125\left(\dfrac{6400}{6400 + x}\right)^2.$$

At what altitudes is the astronaut's weight less than 5 pounds?

50 Lorentz contraction formula The Lorentz contraction formula in relativity theory relates the length L of an object moving at a velocity of v mi/sec with respect to an observer to its length L_0 at rest. If c is the speed of light, then

$$L^2 = L_0^2\left(1 - \dfrac{v^2}{c^2}\right).$$

For what velocities will L be less than $\tfrac{1}{2}L_0$? State the answer in terms of c.

51 Aircraft's landing speed In the design of certain small turbo-prop aircraft, the landing speed V (in ft/sec) is determined by the formula $W = 0.00334V^2S$, where W is the gross weight (in pounds) of the aircraft and S is the surface area (in ft²) of the wings. If the gross weight of the aircraft is between 7500 pounds and 10,000 pounds and $S = 210$ ft², determine the range of the landing speeds in miles per hour.

CHAPTER 2 REVIEW EXERCISES

Exer. 1–24: Solve the equation.

1 $\dfrac{3x+1}{5x+7} = \dfrac{6x+11}{10x-3}$

2 $2 - \dfrac{1}{x} = 1 + \dfrac{4}{x}$

3 $\dfrac{2}{x+5} - \dfrac{3}{2x+1} = \dfrac{5}{6x+3}$

4 $\dfrac{7}{x-2} - \dfrac{6}{x^2-4} = \dfrac{3}{2x+4}$

5 $\dfrac{1}{\sqrt{x}} - 2 = \dfrac{1-2\sqrt{x}}{\sqrt{x}}$

6 $2x^2 + 5x - 12 = 0$

7 $x(3x+4) = 5$

8 $\dfrac{x}{3x+1} = \dfrac{x-1}{2x+3}$

9 $(x-2)(x+1) = 3$

10 $4x^4 - 33x^2 + 50 = 0$

11 $x^{2/3} - 2x^{1/3} - 15 = 0$

12 $20x^3 + 8x^2 - 35x - 14 = 0$

13 $5x^2 = 2x - 3$

14 $x^2 + \tfrac{1}{3}x + 2 = 0$

15 $6x^4 + 29x^2 + 28 = 0$

16 $x^4 - 3x^2 + 1 = 0$

17 $|4x - 1| = 7$

18 $2|2x + 1| + 1 = 19$

19 $\dfrac{1}{x} + 6 = \dfrac{5}{\sqrt{x}}$

20 $\sqrt[3]{4x - 5} - 2 = 0$

21 $\sqrt{7x + 2} + x = 6$

22 $\sqrt{x + 4} = \sqrt[4]{6x + 19}$

23 $\sqrt{3x + 1} - \sqrt{x + 4} = 1$

24 $x^{4/3} = 16$

Exer. 25–26: Solve the equation by completing the square.

25 $3x^2 - 12x + 3 = 0$

26 $x^2 + 10x + 38 = 0$

Exer. 27–44: Solve the inequality, and express the solutions in terms of intervals whenever possible.

27 $(x - 3)^2 \le 0$

28 $10 - 7x < 4 + 2x$

29 $-\dfrac{1}{2} < \dfrac{2x + 3}{5} < \dfrac{3}{2}$

30 $(3x - 1)(10x + 4) \ge (6x - 5)(5x - 7)$

31 $\dfrac{6}{10x + 3} < 0$

32 $|4x + 7| < 21$

33 $2|3 - x| + 1 > 5$

34 $-2|x - 3| + 1 \ge -5$

35 $|16 - 3x| \ge 5$

36 $2 < |x - 6| < 4$

37 $10x^2 + 11x > 6$

38 $x(x - 3) \le 10$

39 $\dfrac{x^2(3 - x)}{x + 2} \le 0$

40 $\dfrac{x^2 - x - 2}{x^2 + 4x + 3} \le 0$

41 $\dfrac{3}{2x + 3} < \dfrac{1}{x - 2}$

42 $\dfrac{x + 1}{x^2 - 25} \le 0$

43 $x^3 > x^2$

44 $(x^2 - x)(x^2 - 5x + 6) < 0$

Exer. 45–50: Solve for the specified variable.

45 $P + N = \dfrac{C + 2}{C}$ for C

46 $A = B\sqrt[3]{\dfrac{C}{D} - E}$ for D

47 $V = \tfrac{4}{3}\pi r^3$ for r (volume of a sphere)

48 $F = \dfrac{\pi P R^4}{8VL}$ for R (Poiseuille's law for fluids)

49 $c = \sqrt{4h(2R - h)}$ for h (base of a circular segment)

50 $V = \tfrac{1}{3}\pi h(r^2 + R^2 + rR)$ for r (volume of a frustum of a cone)

Exer. 51–56: Express in the form $a + bi$, where a and b are real numbers.

51 $(7 + 5i) - (-8 + 3i)$

52 $(4 + 2i)(-5 + 4i)$

53 $(3 + 8i)^2$

54 $\dfrac{1}{9 - \sqrt{-4}}$

55 $\dfrac{6 - 3i}{2 + 7i}$

56 $\dfrac{20 - 8i}{4i}$

57 Rule of 90 In a particular teachers' union, a teacher may retire when the teacher's age plus the teacher's years of service is at least 90. If a 37-year-old teacher has 15 years of service, at what age will this teacher be eligible to retire? Make reasonable assumptions.

58 Electrical resistance When two resistors R_1 and R_2 are connected in parallel, the net resistance R is given by $1/R = (1/R_1) + (1/R_2)$. If $R_1 = 5$ ohms, what value of R_2 will make the net resistance 2 ohms?

59 Investment income An investor has a choice of two investments: a bond fund and a stock fund. The bond fund yields 7.186% interest annually, which is nontaxable at both the federal and state levels. Suppose the investor pays federal income tax at a rate of 28% and state income tax at a rate of 7%. Determine what the annual yield must be on the taxable stock fund so that the two funds pay the same amount of net interest income to the investor.

60 Gold and silver mixture A ring that weighs 80 grams is made of gold and silver. By measuring the displacement of the ring in water, it has been determined that the ring has a volume of 5 cm³. Gold weighs 19.3 g/cm³, and silver weighs 10.5 g/cm³. How many grams of gold does the ring contain?

61 Preparing hospital food A hospital dietitian wishes to prepare a 10-ounce meat-vegetable dish that will provide 7 grams of protein. If an ounce of the vegetable portion supplies $\frac{1}{2}$ gram of protein and an ounce of meat supplies 1 gram of protein, how much of each should be used?

62 Preparing a bactericide A solution of ethyl alcohol that is 75% alcohol by weight is to be used as a bactericide. The solution is to be made by adding water to a 95% ethyl alcohol solution. How many grams of each should be used to prepare 400 grams of the bactericide?

63 Solar heating A large solar heating panel requires 120 gallons of a fluid that is 30% antifreeze. The fluid comes in either a 50% solution or a 20% solution. How many gallons of each should be used to prepare the 120-gallon solution?

64 Fuel consumption A boat has a 10-gallon gasoline tank and travels at 20 mi/hr with a fuel consumption of 16 mi/gal when operated at full throttle in still water. The boat is moving upstream into a 5-mi/hr current. How far upstream can the boat travel and return on 10 gallons of gasoline if it is operated at full throttle during the entire trip?

65 Train travel A high-speed train makes a 400-mile nonstop run between two major cities in $5\frac{1}{2}$ hours. The train travels 100 mi/hr in the country, but safety regulations require that it travel only 25 mi/hr when passing through smaller, intermediate cities. How many hours are spent traveling through the smaller cities?

66 Windspeed An airplane flew with the wind for 30 minutes and returned the same distance in 45 minutes. If the cruising speed of the airplane was 320 mi/hr, what was the speed of the wind?

67 Passing speed An automobile 20 feet long overtakes a truck 40 feet long that is traveling at 50 mi/hr (see the figure). At what constant speed must the automobile travel in order to pass the truck in 5 seconds?

Exercise 67

68 Filling a bin An extruder can fill an empty bin in 2 hours, and a packaging crew can empty a full bin in 5 hours. If a bin is half full when an extruder begins to fill it and a crew begins to empty it, how long will it take to fill the bin?

69 Gasoline mileage A sales representative for a company estimates that her automobile gasoline consumption averages 28 mpg on the highway and 22 mpg in the city. A recent trip covered 627 miles, and 24 gallons of gasoline was used. How much of the trip was spent driving in the city?

70 City expansion The longest drive to the center of a square city from the outskirts is 10 miles. Within the last decade the city has expanded in area by 50 mi². Assuming the city has always been square in shape, find the corresponding change in the longest drive to the center of the city.

71 Dimensions of a cell membrane The membrane of a cell is a sphere of radius 6 microns. What change in the radius will increase the surface area of the membrane by 25%?

72 Highway travel A north-south highway intersects an east-west highway at a point P. An automobile crosses P at 10 A.M., traveling east at a constant rate of 20 mi/hr. At the same instant another automobile is 2 miles north of P, traveling south at 50 mi/hr.

(a) Find a formula for the distance d between the automobiles t hours after 10:00 A.M.

(b) At approximately what time will the automobiles be 104 miles apart?

73 Fencing a kennel A kennel owner has 270 feet of fencing material to be used to divide a rectangular area into 10 equal pens, as shown in the figure. Find dimensions that would allow 100 ft² for each pen.

Exercise 73

74 Dimensions of an aquarium An open-topped aquarium is to be constructed with 6-foot-long sides and square ends, as shown in the figure.

(a) Find the height of the aquarium if the volume is to be 48 ft³.

(b) Find the height if 44 ft² of glass is to be used.

Exercise 74

6′

75 Dimensions of a pool The length of a rectangular pool is to be four times its width, and a sidewalk of width 6 feet will surround the pool. If a total area of 1440 ft² has been set aside for construction, what are the dimensions of the pool?

76 Dimensions of a bath A contractor wishes to design a rectangular sunken bath with 40 ft² of bathing area. A 1-foot-wide tile strip is to surround the bathing area. The total length of the tiled area is to be twice the width. Find the dimensions of the bathing area.

77 Population growth The population P (in thousands) of a small town is expected to increase according to the formula

$$P = 15 + \sqrt{3t + 2},$$

where t is time in years. When will the population be 20,000?

78 Boyle's law Boyle's law for a certain gas states that if the temperature is constant, then $pv = 200$, where p is the pressure (in lb/in²) and v is the volume (in in³). If $25 \le v \le 50$, what is the corresponding range for p?

79 Sales commission A recent college graduate has job offers for a sales position in two computer firms. Job A pays $25,000 per year plus 5% commission. Job B pays only $20,000 per year, but the commission rate is 10%. How much yearly business must the salesman do for the second job to be more lucrative?

80 Speed of sound The speed of sound in air at 0°C (or 273 K) is 1087 ft/sec, but this speed increases as the temperature rises. The speed v of sound at temperature T in K is given by $v = 1087\sqrt{T/273}$. At what temperatures does the speed of sound exceed 1100 ft/sec?

81 Period of a pendulum If the length of the pendulum in a grandfather clock is l centimeters, then its period T (in seconds) is given by $T = 2\pi\sqrt{l/g}$, where g is a gravitational constant. If, under certain conditions, $g = 980$ and $98 \le l \le 100$, what is the corresponding range for T?

82 Orbit of a satellite For a satellite to maintain an orbit of altitude h kilometers, its velocity (in km/sec) must equal $626.4/\sqrt{h + R}$, where $R = 6372$ km is the radius of the earth. What velocities will result in orbits with an altitude of more than 100 kilometers from the earth's surface?

83 Fencing a region There is 100 feet of fencing available to enclose a rectangular region. For what widths will the fenced region contain at least 600 ft²?

84 **Planting an apple orchard** The owner of an apple orchard estimates that if 24 trees are planted per acre, then each mature tree will yield 600 apples per year. For each additional tree planted per acre, the number of apples produced by each tree decreases by 12 per year. How many trees should be planted per acre to obtain at least 16,416 apples per year?

85 **Apartment rentals** A real estate company owns 180 efficiency apartments, which are fully occupied when the rent is $300 per month. The company estimates that for each $10 increase in rent, 5 apartments will become unoccupied. What rent should be charged in order to pay the monthly bills, which total $54,400?

CHAPTER 2 DISCUSSION EXERCISES

1 When we factor the sum or difference of cubes, $x^3 \pm y^3$, is the factor $(x^2 \mp xy + y^2)$ ever factorable over the real numbers?

2 What is the average of the two solutions of the arbitrary quadratic equation $ax^2 + bx + c = 0$? Discuss how this knowledge can help you easily check the solutions to a quadratic equation.

3 (a) Find an expression of the form $p + qi$ for the multiplicative inverse of $\dfrac{a + bi}{c + di}$, where a, b, c, and d are real numbers.

   (b) Does the expression you found apply to real numbers of the form a/c?

   (c) Are there any restrictions on your answer for part (a)?

4 In solving the inequality $\dfrac{x - 1}{x - 2} \geq 3$, what is wrong with employing $x - 1 \geq 3(x - 2)$ as a first step?

5 Consider the inequality $ax^2 + bx + c \geq 0$, where a, b, and c are real numbers with $a \neq 0$. Suppose the associated equality $ax^2 + bx + c = 0$ has discriminant D. Categorize the solutions of the inequality according to the signs of a and D.

6 **Freezing level in a cloud** Refer to Exercises 37–39 in Section 2.2.

   (a) Approximate the height of the freezing level in a cloud if the ground temperature is 80°F and the dew point is 68°F.

   (b) Find a formula for the height h of the freezing level in a cloud for ground temperature G and dew point D.

7 Explain why you should not try to solve one of these equations.

$$\sqrt{2x - 3} + \sqrt{x + 5} = 0$$
$$\sqrt[3]{2x - 3} + \sqrt[3]{x + 5} = 0$$

8 Solve the equation

$$\sqrt{x} = cx - 2/c$$

for x, where $c = 2 \times 10^{500}$. Discuss why one of your positive solutions is extraneous.

3

Functions and Graphs

3.1 Rectangular Coordinate Systems

3.2 Graphs of Equations

3.3 Lines

3.4 Definition of Function

3.5 Graphs of Functions

3.6 Quadratic Functions

3.7 Operations on Functions

The mathematical term *function* (or its Latin equivalent) dates back to the late seventeenth century, when calculus was in the early stages of development. This important concept is now the backbone of advanced courses in mathematics and is indispensable in every field of science.

In this chapter we study properties of functions using algebraic and graphical methods that include plotting points, determining symmetries, and making horizontal and vertical shifts. These techniques are adequate for obtaining rough sketches of graphs that help us understand properties of functions; modern-day methods, however, employ sophisticated computer software and advanced mathematics to generate extremely accurate graphical representations of functions.

3.1

Rectangular Coordinate Systems

In Section 1.1 we discussed how to assign a real number (coordinate) to each point on a line. We shall now show how to assign an **ordered pair** (a, b) of real numbers to each point in a plane. Although we have also used the notation (a, b) to denote an open interval, there is little chance for confusion, since it should always be clear from our discussion whether (a, b) represents a point or an interval.

We introduce a **rectangular,** or **Cartesian,* coordinate system** in a plane by means of two perpendicular coordinate lines, called **coordinate axes,** that intersect at the **origin** O, as shown in Figure 1. We often refer to the horizontal line as the **x-axis** and the vertical line as the **y-axis** and label them x and y, respectively. The plane is then a **coordinate plane,** or an **xy-plane.** The coordinate axes divide the plane into four parts called the **first, second, third,** and **fourth quadrants,** labeled I, II, III, and IV, respectively (see Figure 1). Points on the axes do not belong to any quadrant.

Each point P in an xy-plane may be assigned an ordered pair (a, b), as shown in Figure 1. We call a the **x-coordinate** (or **abscissa**) of P, and b the **y-coordinate** (or **ordinate**). We say that P *has coordinates* (a, b) and refer to the *point* (a, b) or the *point* $P(a, b)$. Conversely, every ordered pair (a, b) determines a point P with coordinates a and b. We **plot a point** by using a dot, as illustrated in Figure 2.

Figure 1

Figure 2

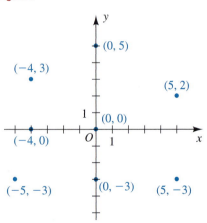

We may use the following formula to find the distance between two points in a coordinate plane.

Distance Formula	The distance $d(P_1, P_2)$ between any two points $P_1(x_1, y_1)$ and $P_2(x_2, y_2)$ in a coordinate plane is $$d(P_1, P_2) = \sqrt{(x_2 - x_1)^2 + (y_2 - y_1)^2}.$$

*The term *Cartesian* is used in honor of the French mathematician and philosopher René Descartes (1596–1650), who was one of the first to employ such coordinate systems.

Figure 3

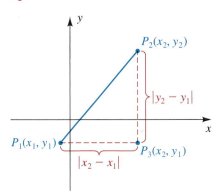

PROOF If $x_1 \neq x_2$ and $y_1 \neq y_2$, then, as illustrated in Figure 3, the points P_1, P_2, and $P_3(x_2, y_1)$ are vertices of a right triangle. By the Pythagorean theorem,

$$[d(P_1, P_2)]^2 = [d(P_1, P_3)]^2 + [d(P_3, P_2)]^2.$$

From the figure we see that

$$d(P_1, P_3) = |x_2 - x_1| \quad \text{and} \quad d(P_3, P_2) = |y_2 - y_1|.$$

Since $|a|^2 = a^2$ for every real number a, we may write

$$[d(P_1, P_2)]^2 = (x_2 - x_1)^2 + (y_2 - y_1)^2.$$

Taking the square root of each side of the last equation and using the fact that $d(P_1, P_2) \geq 0$ gives us the distance formula.

If $y_1 = y_2$, the points P_1 and P_2 lie on the same horizontal line, and

$$d(P_1, P_2) = |x_2 - x_1| = \sqrt{(x_2 - x_1)^2}.$$

Similarly, if $x_1 = x_2$, the points are on the same vertical line, and

$$d(P_1, P_2) = |y_2 - y_1| = \sqrt{(y_2 - y_1)^2}.$$

These are special cases of the distance formula.

Although we referred to the points shown in Figure 3, our proof is independent of the positions of P_1 and P_2. ◢

When applying the distance formula, note that $d(P_1, P_2) = d(P_2, P_1)$ and, hence, the order in which we subtract the x-coordinates and the y-coordinates of the points is immaterial. We may think of the distance between two points as the length of the hypotenuse of a right triangle.

Figure 4

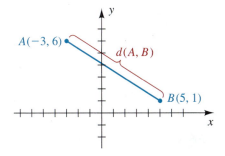

EXAMPLE 1 **Finding the distance between points**

Plot the points $A(-3, 6)$ and $B(5, 1)$, and find the distance $d(A, B)$.

SOLUTION The points are plotted in Figure 4. By the distance formula,

$$d(A, B) = \sqrt{[5 - (-3)]^2 + (1 - 6)^2}$$
$$= \sqrt{8^2 + (-5)^2}$$
$$= \sqrt{64 + 25} = \sqrt{89} \approx 9.43. \qquad ◢$$

EXAMPLE 2 **Showing that a triangle is a right triangle**

(a) Plot $A(-1, -3)$, $B(6, 1)$, and $C(2, -5)$, and show that triangle ABC is a right triangle.

(b) Find the area of triangle ABC.

Figure 5

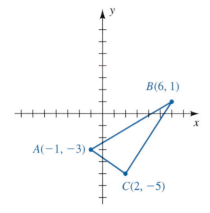

SOLUTION

(a) The points are plotted in Figure 5. From geometry, triangle ABC is a right triangle if the sum of the squares of two of its sides is equal to the square of the remaining side. By the distance formula,

$$d(A, B) = \sqrt{(6 + 1)^2 + (1 + 3)^2} = \sqrt{49 + 16} = \sqrt{65}$$
$$d(B, C) = \sqrt{(2 - 6)^2 + (-5 - 1)^2} = \sqrt{16 + 36} = \sqrt{52}$$
$$d(A, C) = \sqrt{(2 + 1)^2 + (-5 + 3)^2} = \sqrt{9 + 4} = \sqrt{13}.$$

Since $d(A, B) = \sqrt{65}$ is the largest of the three values, the condition to be satisfied is

$$[d(A, B)]^2 = [d(B, C)]^2 + [d(A, C)]^2.$$

Substituting the values found using the distance formula, we obtain

$$[d(A, B)]^2 = \left(\sqrt{65}\right)^2 = 65$$

and $[d(B, C)]^2 + [d(A, C)]^2 = \left(\sqrt{52}\right)^2 + \left(\sqrt{13}\right)^2 = 52 + 13 = 65.$
Thus, the triangle is a right triangle with hypotenuse AB.

(b) The area of a triangle with base b and altitude h is $\frac{1}{2}bh$. Referring to Figure 5, we let

$$b = d(B, C) = \sqrt{52} \quad \text{and} \quad h = d(A, C) = \sqrt{13}.$$

Hence, the area of triangle ABC is

$$\tfrac{1}{2}bh = \tfrac{1}{2}\sqrt{52}\sqrt{13} = \tfrac{1}{2} \cdot 2\sqrt{13}\sqrt{13} = 13.$$

Area of a triangle:

$$A = \tfrac{1}{2}bh$$

EXAMPLE 3 **Applying the distance formula**

Given $A(1, 7)$, $B(-3, 2)$, and $C\left(4, \tfrac{1}{2}\right)$, prove that C is on the perpendicular bisector of segment AB.

SOLUTION The points A, B, C and the *perpendicular bisector l* are illustrated in Figure 6. From plane geometry, l can be characterized by either of the following conditions:

(1) l is the line perpendicular to segment AB at its midpoint.

(2) l is the set of all points equidistant from the endpoints of segment AB.

We shall use condition 2 to show that C is on l by verifying that

$$d(A, C) = d(B, C).$$

Figure 6

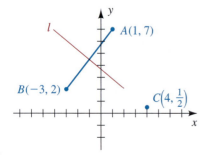

We apply the distance formula:

$$d(A, C) = \sqrt{(4 - 1)^2 + \left(\tfrac{1}{2} - 7\right)^2} = \sqrt{3^2 + \left(-\tfrac{13}{2}\right)^2} = \sqrt{9 + \tfrac{169}{4}} = \sqrt{\tfrac{205}{4}}$$

$$d(B, C) = \sqrt{[4 - (-3)]^2 + \left(\tfrac{1}{2} - 2\right)^2} = \sqrt{7^2 + \left(-\tfrac{3}{2}\right)^2} = \sqrt{49 + \tfrac{9}{4}} = \sqrt{\tfrac{205}{4}}$$

Thus, C is equidistant from A and B, and the verification is complete.

EXAMPLE 4 Finding a formula that describes a perpendicular bisector

Given $A(1, 7)$ and $B(-3, 2)$, find a formula that expresses the fact that an arbitrary point $P(x, y)$ is on the perpendicular bisector l of segment AB.

SOLUTION By condition 2 of Example 3, $P(x, y)$ is on l if and only if $d(A, P) = d(B, P)$; that is,

$$\sqrt{(x - 1)^2 + (y - 7)^2} = \sqrt{[x - (-3)]^2 + (y - 2)^2}.$$

To obtain a simpler formula, let us square both sides and simplify terms of the resulting equation, as follows:

$$(x - 1)^2 + (y - 7)^2 = [x - (-3)]^2 + (y - 2)^2$$
$$x^2 - 2x + 1 + y^2 - 14y + 49 = x^2 + 6x + 9 + y^2 - 4y + 4$$
$$-2x + 1 - 14y + 49 = 6x + 9 - 4y + 4$$
$$-8x - 10y = -37$$
$$8x + 10y = 37$$

Note that, in particular, the last formula is true for the coordinates of the point $C\left(4, \frac{1}{2}\right)$ in Example 3, since if $x = 4$ and $y = \frac{1}{2}$, substitution in $8x + 10y$ gives us

$$8 \cdot 4 + 10 \cdot \tfrac{1}{2} = 37.$$

In Example 9 of Section 3.3, we will find a formula for the perpendicular bisector of a segment using condition 1 of Example 3. ◢

We can find the midpoint of a line segment by using the following formula.

Midpoint Formula	The midpoint M of the line segment from $P_1(x_1, y_1)$ to $P_2(x_2, y_2)$ is $$\left(\frac{x_1 + x_2}{2}, \frac{y_1 + y_2}{2}\right).$$

PROOF The lines through P_1 and P_2 parallel to the y-axis intersect the x-axis at $A_1(x_1, 0)$ and $A_2(x_2, 0)$. From plane geometry, the line through the midpoint M parallel to the y-axis bisects the segment A_1A_2 at point M_1 (see Figure 7). If $x_1 < x_2$, then $x_2 - x_1 > 0$, and hence $d(A_1, A_2) = x_2 - x_1$. Since M_1 is halfway from A_1 to A_2, the x-coordinate of M_1 is equal to the x-coordinate of A_1 plus one-half the distance from A_1 to A_2; that is,

$$x\text{-coordinate of } M_1 = x_1 + \tfrac{1}{2}(x_2 - x_1).$$

(continued)

Figure 7

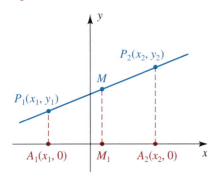

The expression on the right side of the last equation simplifies to

$$\frac{x_1 + x_2}{2}.$$

This quotient is the *average* of the numbers x_1 and x_2. It follows that the x-coordinate of M is also $(x_1 + x_2)/2$. Similarly, the y-coordinate of M is $(y_1 + y_2)/2$. These formulas hold for all positions of P_1 and P_2. ◢

To apply the midpoint formula, it may suffice to remember that

the x-coordinate of the midpoint = the *average* of the x-coordinates,

and that

the y-coordinate of the midpoint = the *average* of the y-coordinates.

EXAMPLE 5   Finding a midpoint

Find the midpoint M of the line segment from $P_1(-2, 3)$ to $P_2(4, -2)$, and verify that $d(P_1, M) = d(P_2, M)$.

Figure 8

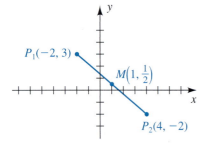

SOLUTION   By the midpoint formula, the coordinates of M are

$$\left(\frac{-2 + 4}{2}, \frac{3 + (-2)}{2}\right), \quad \text{or} \quad \left(1, \frac{1}{2}\right).$$

The three points P_1, P_2, and M are plotted in Figure 8. By the distance formula,

$$d(P_1, M) = \sqrt{(1 + 2)^2 + \left(\tfrac{1}{2} - 3\right)^2} = \sqrt{9 + \tfrac{25}{4}}$$

$$d(P_2, M) = \sqrt{(1 - 4)^2 + \left(\tfrac{1}{2} + 2\right)^2} = \sqrt{9 + \tfrac{25}{4}}.$$

Hence, $d(P_1, M) = d(P_2, M)$. ◢

3.1   *Exercises*

1 Plot the points $A(5, -2)$, $B(-5, -2)$, $C(5, 2)$, $D(-5, 2)$, $E(3, 0)$, and $F(0, 3)$ on a coordinate plane.

2 Plot the points $A(-3, 1)$, $B(3, 1)$, $C(-2, -3)$, $D(0, 3)$, and $E(2, -3)$ on a coordinate plane. Draw the line segments AB, BC, CD, DE, and EA.

3 Plot the points $A(0, 0)$, $B(1, 1)$, $C(3, 3)$, $D(-1, -1)$, and $E(-2, -2)$. Describe the set of all points of the form (a, a), where a is a real number.

4 Plot the points $A(0, 0)$, $B(1, -1)$, $C(3, -3)$, $D(-1, 1)$, and $E(-3, 3)$. Describe the set of all points of the form $(a, -a)$, where a is a real number.

Exer. 5–6: Find the coordinates of the points A–F.

5

6

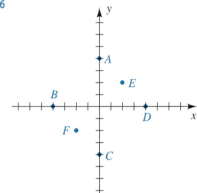

Exer. 7–8: Describe the set of all points $P(x, y)$ in a coordinate plane that satisfy the given condition.

7 (a) $x = -2$ (b) $y = 3$ (c) $x \geq 0$

(d) $xy > 0$ (e) $y < 0$ (f) $x = 0$

8 (a) $y = -2$ (b) $x = -4$ (c) $x/y < 0$

(d) $xy = 0$ (e) $y > 1$ (f) $y = 0$

**Exer. 9–14: (a) Find the distance $d(A, B)$ between A and B.
(b) Find the midpoint of the segment AB.**

9 $A(4, -3)$, $B(6, 2)$ 10 $A(-2, -5)$, $B(4, 6)$

11 $A(-5, 0)$, $B(-2, -2)$ 12 $A(6, 2)$, $B(6, -2)$

13 $A(7, -3)$, $B(3, -3)$ 14 $A(-4, 7)$, $B(0, -8)$

Exer. 15–16: Show that the triangle with vertices A, B, and C is a right triangle, and find its area.

15

16

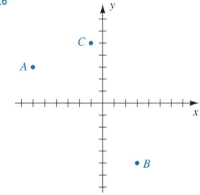

17 Show that $A(-4, 2)$, $B(1, 4)$, $C(3, -1)$, and $D(-2, -3)$ are vertices of a square.

18 Show that $A(-4, -1)$, $B(0, -2)$, $C(6, 1)$, and $D(2, 2)$ are vertices of a parallelogram.

19 Given $A(-3, 8)$, find the coordinates of the point B such that $C(5, -10)$ is the midpoint of segment AB.

20 Given $A(5, -8)$ and $B(-6, 2)$, find the point on segment AB that is three-fourths of the way from A to B.

Exer. 21–22: Prove that C is on the perpendicular bisector of segment AB.

21 $A(-4, -3)$, $B(6, 1)$, $C(5, -11)$

22 $A(-3, 2)$, $B(5, -4)$, $C(7, 7)$

Exer. 23–24: Find a formula that expresses the fact that an arbitrary point $P(x, y)$ is on the perpendicular bisector l of segment AB.

23 $A(-4, -3), B(6, 1)$ 24 $A(-3, 2), B(5, -4)$

25 Find a formula that expresses the fact that $P(x, y)$ is a distance 5 from the origin. Describe the set of all such points.

26 Find a formula that states that $P(x, y)$ is a distance $r > 0$ from a fixed point $C(h, k)$. Describe the set of all such points.

27 Find all points on the y-axis that are a distance 6 from $P(5, 3)$.

28 Find all points on the x-axis that are a distance 5 from $P(-2, 4)$.

29 Find the point with coordinates of the form $(2a, a)$ that is in the third quadrant and is a distance 5 from $P(1, 3)$.

30 Find all points with coordinates of the form (a, a) that are a distance 3 from $P(-2, 1)$.

31 For what values of a is the distance between $P(a, 3)$ and $Q(5, 2a)$ greater than $\sqrt{26}$?

32 Given $A(-2, 0)$ and $B(2, 0)$, find a formula not containing radicals that expresses the fact that the sum of the distances from $P(x, y)$ to A and to B, respectively, is 5.

33 Prove that the midpoint of the hypotenuse of any right triangle is equidistant from the vertices. (*Hint:* Label the vertices of the triangle $O(0, 0)$, $A(a, 0)$, and $B(0, b)$.)

34 Prove that the diagonals of any parallelogram bisect each other. (*Hint:* Label three of the vertices of the parallelogram $O(0, 0)$, $A(a, b)$, and $C(0, c)$.)

3.2

Graphs of Equations

Graphs are often used to illustrate changes in quantities. A graph in the business section of a newspaper may show the fluctuation of the Dow-Jones average during a given month; a meteorologist might use a graph to indicate how the air temperature varied throughout a day; a cardiologist employs graphs (electrocardiograms) to analyze heart irregularities; an engineer or physicist may turn to a graph to illustrate the manner in which the pressure of a confined gas increases as the gas is heated. Such visual aids usually reveal the behavior of quantities more readily than a long table of numerical values.

Two quantities are sometimes related by means of an equation or formula that involves two variables. In this section we discuss how to represent such an equation geometrically, by a graph in a coordinate plane. The graph may then be used to discover properties of the quantities that are not evident from the equation alone. The following chart introduces the basic concept of the graph of an equation in two variables x and y. Of course, other letters can also be used for the variables.

Terminology	Definition	Illustration
Solution of an equation in x and y	An ordered pair (a, b) that yields a true statement if $x = a$ and $y = b$	$(2, 3)$ is a solution of $y^2 = 5x - 1$, since substituting $x = 2$ and $y = 3$ gives us LS: $3^2 = 9$ RS: $5(2) - 1 = 10 - 1 = 9$.

For each solution (a, b) of an equation in x and y there is a point $P(a, b)$ in a coordinate plane. The set of all such points is called the **graph** of the equation. To *sketch the graph of an equation,* we illustrate the significant features of the graph in a coordinate plane. In simple cases, a graph can be sketched by plotting few, if any, points. For a complicated equation, plotting points may give very little information about the graph. In such cases, methods of calculus or computer graphics are often employed. Let us begin with a simple example.

EXAMPLE 1 Sketching a simple graph by plotting points

Sketch the graph of the equation $y = 2x - 1$.

SOLUTION We wish to find the points (x, y) in a coordinate plane that correspond to the solutions of the equation. It is convenient to list coordinates of several such points in a table, where for each x we obtain the value for y from $y = 2x - 1$:

Figure 1

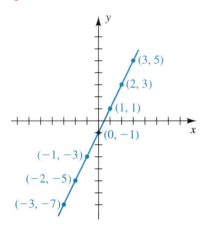

x	-3	-2	-1	0	1	2	3
y	-7	-5	-3	-1	1	3	5

The points with these coordinates appear to lie on a line, and we can sketch the graph in Figure 1. Ordinarily, the few points we have plotted would not be enough to illustrate the graph of an equation; however, in this elementary case we can be reasonably sure that the graph is a line. In the next section we will establish this fact.

It is impossible to sketch the entire graph in Example 1, because we can assign values to x that are numerically as large as desired. Nevertheless, we call the drawing in Figure 1 *the graph of the equation* or *a sketch of the graph.* In general, the sketch of a graph should illustrate its essential features so that the remaining (unsketched) parts are self-evident. For instance, in Figure 1, the **end behavior**—the pattern of the graph as x assumes large positive and negative values (that is, the shape of the right and left ends)—is apparent to the reader.

If a graph terminates at some point (as would be the case for a half-line or line segment), we place a dot at the appropriate *endpoint* of the graph. As a final general remark, *if ticks on the coordinate axes are not labeled* (as in Figure 1), *then each tick represents one unit.* We shall label ticks only when different units are used on the axes. For *arbitrary* graphs, where units of measurement are irrelevant, we omit ticks completely (see, for example, Figures 5 and 6).

EXAMPLE 2 Sketching the graph of an equation

Sketch the graph of the equation $y = x^2 - 3$.

Figure 2

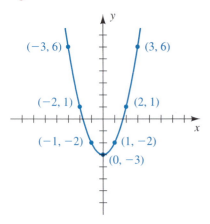

SOLUTION   Substituting values for x and finding the corresponding values of y using $y = x^2 - 3$, we obtain a table of coordinates for several points on the graph:

x	-3	-2	-1	0	1	2	3
y	6	1	-2	-3	-2	1	6

Larger values of $|x|$ produce larger values of y. For example, the points $(4, 13)$, $(5, 22)$, and $(6, 33)$ are on the graph, as are $(-4, 13)$, $(-5, 22)$, and $(-6, 33)$. Plotting the points given by the table and drawing a smooth curve through these points (in the order of increasing values of x) gives us the sketch in Figure 2.

The graph in Figure 2 is a **parabola,** and the y-axis is the **axis of the parabola.** The lowest point $(0, -3)$ is the **vertex** of the parabola, and we say that the parabola *opens upward.* If we invert the graph, then the parabola *opens downward* and the vertex is the highest point on the graph. In general, the graph of *any* equation of the form $y = ax^2 + c$ with $a \neq 0$ is a parabola with vertex $(0, c)$, opening upward if $a > 0$ or downward if $a < 0$. If $c = 0$, the equation reduces to $y = ax^2$ and the vertex is at the origin $(0, 0)$. Parabolas may also open to the right or to the left (see Example 4) or in other directions.

We shall use the following terminology to describe where the graph of an equation in x and y intersects the x-axis or the y-axis.

Intercepts of the Graph of an Equation in x and y

Terminology	Definition	Graphical interpretation	How to find
x-**intercepts**	The x-coordinates of points where the graph intersects the x-axis		Let $y = 0$ and solve for x. Here, a and c are x-intercepts.
y-**intercepts**	The y-coordinates of points where the graph intersects the y-axis		Let $x = 0$ and solve for y. Here, b is the y-intercept.

An x-intercept is sometimes referred to as a *zero* of the graph of an equation or as a *root* of an equation.

EXAMPLE 3   Finding x-intercepts and y-intercepts

Find the x- and y-intercepts of the graph of $y = x^2 - 3$.

SOLUTION   The graph is sketched in Figure 2 (Example 2). We find the intercepts as stated in the preceding chart.

(1) x-intercepts:

$$y = x^2 - 3 \qquad \text{given}$$
$$0 = x^2 - 3 \qquad \text{let } y = 0$$
$$x^2 = 3 \qquad \text{equivalent equation}$$
$$x = \pm\sqrt{3} \approx \pm 1.73 \qquad \text{take the square root}$$

Thus, the x-intercepts are $-\sqrt{3}$ and $\sqrt{3}$. The points at which the graph crosses the x-axis are $\left(-\sqrt{3}, 0\right)$ and $\left(\sqrt{3}, 0\right)$.

(2) y-intercepts:

$$y = x^2 - 3 \qquad \text{given}$$
$$y = 0 - 3 = -3 \qquad \text{let } x = 0$$

Thus, the y-intercept is -3, and the point at which the graph crosses the y-axis is $(0, -3)$.

If the coordinate plane in Figure 2 is folded along the y-axis, the graph that lies in the left half of the plane coincides with that in the right half, and we say that *the graph is symmetric with respect to the y-axis*. A graph is symmetric with respect to the y-axis provided that the point $(-x, y)$ is on the graph whenever (x, y) is on the graph. The graph of $y = x^2 - 3$ in Example 2 has this property, since substitution of $-x$ for x yields the same equation:

$$y = (-x)^2 - 3 = x^2 - 3$$

This substitution is an application of symmetry test 1 in the following chart. Two other types of symmetry and the appropriate tests are also listed. The graphs of $x = y^2$ and $4y = x^3$ in the illustration column are discussed in Examples 4 and 5, respectively.

Symmetries of Graphs of Equations in x and y

Terminology	Graphical interpretation	Test for symmetry	Illustration
The graph is symmetric with respect to the y-axis.	*(graph showing points $(-x, y)$ and (x, y))*	**(1)** Substitution of $-x$ for x leads to the same equation.	*(parabola $y = x^2 - 3$)*
The graph is symmetric with respect to the x-axis.	*(graph showing points (x, y) and $(x, -y)$)*	**(2)** Substitution of $-y$ for y leads to the same equation.	*(sideways parabola $x = y^2$)*
The graph is symmetric with respect to the origin.	*(graph showing points (x, y) and $(-x, -y)$)*	**(3)** Simultaneous substitution of $-x$ for x and $-y$ for y leads to the same equation.	*(cubic $4y = x^3$)*

If a graph is symmetric with respect to an axis, it is sufficient to determine the graph in half of the coordinate plane, since we can sketch the remainder of the graph by taking a *mirror image,* or *reflection,* through the appropriate axis.

EXAMPLE 4 **A graph that is symmetric with respect to the x-axis**

Sketch the graph of the equation $y^2 = x$.

SOLUTION Since substitution of $-y$ for y does not change the equation, the graph is symmetric with respect to the x-axis (see symmetry test 2). Hence, if the point (x, y) is on the graph, then the point $(x, -y)$ is on the graph. Thus, it

Figure 3

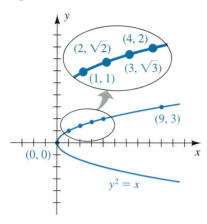

$y^2 = x$

is sufficient to find points with nonnegative y-coordinates and then reflect through the x-axis. The equation $y^2 = x$ is equivalent to $y = \pm\sqrt{x}$. The y-coordinates of points *above* the x-axis (y is *positive*) are given by $y = \sqrt{x}$, whereas the y-coordinates of points *below* the x-axis (y is *negative*) are given by $y = -\sqrt{x}$. Coordinates of some points on the graph are listed below. The graph is sketched in Figure 3.

x	0	1	2	3	4	9
y	0	1	$\sqrt{2} \approx 1.4$	$\sqrt{3} \approx 1.7$	2	3

The graph is a parabola that opens to the right, with its vertex at the origin. In this case, the x-axis is the axis of the parabola.

EXAMPLE 5   **A graph that is symmetric with respect to the origin**

Sketch the graph of the equation $4y = x^3$.

SOLUTION   If we simultaneously substitute $-x$ for x and $-y$ for y, then

$$4(-y) = (-x)^3 \qquad \text{or, equivalently,} \qquad -4y = -x^3.$$

Multiplying both sides by -1, we see that the last equation has the same solutions as the equation $4y = x^3$. Hence, from symmetry test 3, the graph is symmetric with respect to the origin—and if the point (x, y) is on the graph, then the point $(-x, -y)$ is on the graph. The following table lists coordinates of some points on the graph.

Figure 4

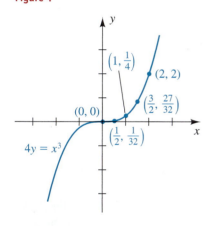

$4y = x^3$

x	0	$\frac{1}{2}$	1	$\frac{3}{2}$	2	$\frac{5}{2}$
y	0	$\frac{1}{32}$	$\frac{1}{4}$	$\frac{27}{32}$	2	$\frac{125}{32}$

Because of the symmetry, we can see that the points $\left(-1, -\frac{1}{4}\right)$, $(-2, -2)$, and so on, are also on the graph. The graph is sketched in Figure 4.

Figure 5

$(x - h)^2 + (y - k)^2 = r^2$

If $C(h, k)$ is a point in a coordinate plane, then a circle with center C and radius $r > 0$ consists of all points in the plane that are r units from C. As shown in Figure 5, a point $P(x, y)$ is on the circle provided $d(C, P) = r$ or, by the distance formula,

$$\sqrt{(x - h)^2 + (y - k)^2} = r.$$

The above equation is equivalent to the following equation, which we will refer to as the **standard equation of a circle.**

Standard Equation of a Circle with Center (h, k) and Radius r	$(x - h)^2 + (y - k)^2 = r^2$

Figure 6

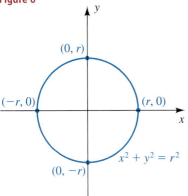

If $h = 0$ and $k = 0$, this equation reduces to $x^2 + y^2 = r^2$, which is an equation of a circle of radius r with center at the origin (see Figure 6). If $r = 1$, we call the graph a **unit circle.**

EXAMPLE 6   Finding an equation of a circle

Find an equation of the circle that has center $C(-2, 3)$ and contains the point $D(4, 5)$.

SOLUTION   The circle is shown in Figure 7. Since D is on the circle, the radius r is $d(C, D)$. By the distance formula,

$$r = \sqrt{(4 + 2)^2 + (5 - 3)^2} = \sqrt{36 + 4} = \sqrt{40}.$$

Using the standard equation of a circle with $h = -2$, $k = 3$, and $r = \sqrt{40}$, we obtain

$$(x + 2)^2 + (y - 3)^2 = 40.$$

By squaring terms and simplifying the last equation, we may write it as

$$x^2 + y^2 + 4x - 6y - 27 = 0.$$

Figure 7

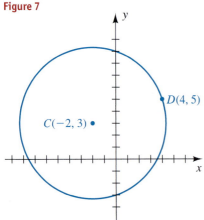

As in the solution to Example 6, squaring terms of an equation of the form $(x - h)^2 + (y - k)^2 = r^2$ and simplifying leads to an equation of the form

$$x^2 + y^2 + ax + by + c = 0,$$

where a, b, and c are real numbers. Conversely, if we begin with this equation, it is always possible, by *completing squares,* to obtain an equation of the form

$$(x - h)^2 + (y - k)^2 = d.$$

This method will be illustrated in Example 7. If $d > 0$, the graph is a circle with center (h, k) and radius $r = \sqrt{d}$. If $d = 0$, the graph consists of only the point (h, k). Finally, if $d < 0$, the equation has no real solutions, and hence there is no graph.

EXAMPLE 7   Finding the center and radius of a circle

Find the center and radius of the circle with equation

$$3x^2 + 3y^2 - 12x + 18y = 9.$$

SOLUTION   Since it is easier to complete the square if the coefficients of x^2 and y^2 are 1, we begin by dividing the given equation by 3, obtaining

$$x^2 + y^2 - 4x + 6y = 3.$$

Next, we rewrite the equation as follows, where the underscored spaces represent numbers to be determined:

$$(x^2 - 4x + \underline{\ \ }) + (y^2 + 6y + \underline{\ \ }) = 3 + \underline{\ \ } + \underline{\ \ }$$

We then complete the squares for the expressions within parentheses, taking care to add the appropriate numbers to *both* sides of the equation. To complete the square for an expression of the form $x^2 + ax$, we add the square of half the coefficient of x (that is, $(a/2)^2$) to both sides of the equation. Similarly, for $y^2 + by$, we add $(b/2)^2$ to both sides. In this example, $a = -4$, $b = 6$, $(a/2)^2 = (-2)^2 = 4$, and $(b/2)^2 = 3^2 = 9$. These additions lead to

$$(x^2 - 4x + \underline{4}\,) + (y^2 + 6y + \underline{9}\,) = 3 + \underline{4} + \underline{9} \qquad \text{\textcolor{red}{completing the squares}}$$

$$(x - 2)^2 + (y + 3)^2 = 16. \qquad \text{\textcolor{red}{equivalent equation}}$$

Comparing the last equation with the standard equation of a circle, we see that $h = 2$ and $k = -3$ and conclude that the circle has center $(2, -3)$ and radius $\sqrt{16} = 4$.   ◢

   In some applications it is necessary to work with only one-half of a circle—that is, a **semicircle.** The next example indicates how to find equations of semicircles for circles with centers at the origin.

EXAMPLE 8   Finding equations of semicircles

Find equations for the upper half, lower half, right half, and left half of the circle $x^2 + y^2 = 81$.

SOLUTION   The graph of $x^2 + y^2 = 81$ is a circle of radius 9 with center at the origin (compare with Figure 6). To find equations for the upper and lower halves, we solve for y in terms of x:

$$x^2 + y^2 = 81 \qquad \text{given}$$

$$y^2 = 81 - x^2 \qquad \text{subtract } x^2$$

$$y = \pm\sqrt{81 - x^2} \qquad \text{take the square root}$$

Since $\sqrt{81 - x^2} \geq 0$, it follows that the upper half of the circle has the equation $y = \sqrt{81 - x^2}$ (y is positive) and the lower half is given by $y = -\sqrt{81 - x^2}$ (y is negative), as illustrated in Figure 8(a) and (b).

(continued)

Figure 8

(a) $y = \sqrt{81 - x^2}$

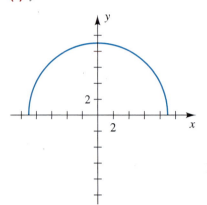

(b) $y = -\sqrt{81 - x^2}$

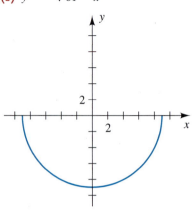

(c) $x = \sqrt{81 - y^2}$

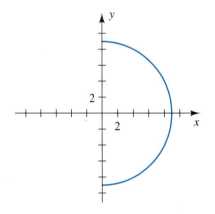

(d) $x = -\sqrt{81 - y^2}$

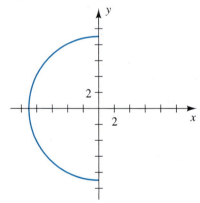

Similarly, to find equations for the right and left halves, we solve $x^2 + y^2 = 81$ for x in terms of y, obtaining

$$x = \pm\sqrt{81 - y^2}.$$

Since $\sqrt{81 - y^2} \geq 0$, it follows that the right half of the circle has the equation $x = \sqrt{81 - y^2}$ (x is positive) and the left half is given by $x = -\sqrt{81 - y^2}$ (x is negative), as illustrated in Figure 8(c) and (d).

3.2 *Exercises*

Exer. 1–20: Sketch the graph of the equation, and label the x- and y-intercepts.

1 $y = 2x - 3$

2 $y = 3x + 2$

3 $y = -x + 1$

4 $y = -2x - 3$

5 $y = -4x^2$

6 $y = \frac{1}{3}x^2$

7 $y = 2x^2 - 1$

8 $y = -x^2 + 2$

9 $x = \frac{1}{4}y^2$

10 $x = -2y^2$

11 $x = -y^2 + 3$

12 $x = 2y^2 - 4$

13 $y = -\frac{1}{2}x^3$

14 $y = \frac{1}{2}x^3$

15 $y = x^3 - 8$

16 $y = -x^3 + 1$

17 $y = \sqrt{x}$

18 $y = \sqrt{-x}$

19 $y = \sqrt{x} - 4$

20 $y = \sqrt{x - 4}$

Exer. 21–22: Use tests for symmetry to determine which graphs in the indicated exercises are symmetric with respect to (a) the y-axis, (b) the x-axis, and (c) the origin.

21 The odd-numbered exercises in 1–20

22 The even-numbered exercises in 1–20

Exer. 23–34: Sketch the graph of the circle or semicircle.

23 $x^2 + y^2 = 11$

24 $x^2 + y^2 = 7$

25 $(x + 3)^2 + (y - 2)^2 = 9$

26 $(x - 4)^2 + (y + 2)^2 = 4$

27 $(x + 3)^2 + y^2 = 16$

28 $x^2 + (y - 2)^2 = 25$

29 $4x^2 + 4y^2 = 25$

30 $9x^2 + 9y^2 = 1$

31 $y = -\sqrt{16 - x^2}$

32 $y = \sqrt{4 - x^2}$

33 $x = \sqrt{9 - y^2}$

34 $x = -\sqrt{25 - y^2}$

Exer. 35–46: Find an equation of the circle that satisfies the stated conditions.

35 Center $C(2, -3)$, radius 5

36 Center $C(-4, 1)$, radius 3

37 Center $C\left(\frac{1}{4}, 0\right)$, radius $\sqrt{5}$

38 Center $C\left(\frac{3}{4}, -\frac{2}{3}\right)$, radius $3\sqrt{2}$

39 Center $C(-4, 6)$, passing through $P(1, 2)$

40 Center at the origin, passing through $P(4, -7)$

41 Center $C(-3, 6)$, tangent to the y-axis

42 Center $C(4, -1)$, tangent to the x-axis

43 Tangent to both axes, center in the second quadrant, radius 4

44 Tangent to both axes, center in the fourth quadrant, radius 3

45 Endpoints of a diameter $A(4, -3)$ and $B(-2, 7)$

46 Endpoints of a diameter $A(-5, 2)$ and $B(3, 6)$

Exer. 47–56: Find the center and radius of the circle with the given equation.

47 $x^2 + y^2 - 4x + 6y - 36 = 0$

48 $x^2 + y^2 + 8x - 10y + 37 = 0$

49 $x^2 + y^2 + 4y - 117 = 0$

50 $x^2 + y^2 - 10x + 18 = 0$

51 $2x^2 + 2y^2 - 12x + 4y - 15 = 0$

52 $9x^2 + 9y^2 + 12x - 6y + 4 = 0$

53 $x^2 + y^2 + 4x - 2y + 5 = 0$

54 $x^2 + y^2 - 6x + 4y + 13 = 0$

55 $x^2 + y^2 - 2x - 8y + 19 = 0$

56 $x^2 + y^2 + 4x + 6y + 16 = 0$

Exer. 57–60: Find equations for the upper half, lower half, right half, and left half of the circle.

57 $x^2 + y^2 = 36$

58 $(x + 3)^2 + y^2 = 64$

59 $(x - 2)^2 + (y + 1)^2 = 49$

60 $(x - 3)^2 + (y - 5)^2 = 4$

Exer. 61–64: Find an equation for the circle or semicircle.

61

62

63

64

Exer. 65–66: Determine whether the point P is inside, outside, or on the circle with center C and radius r.

65 (a) $P(2, 3)$, $C(4, 6)$, $r = 4$

(b) $P(4, 2)$, $C(1, -2)$, $r = 5$

(c) $P(-3, 5)$, $C(2, 1)$, $r = 6$

66 (a) $P(3, 8)$, $C(-2, -4)$, $r = 13$

   (b) $P(-2, 5)$, $C(3, 7)$, $r = 6$

   (c) $P(1, -2)$, $C(6, -7)$, $r = 7$

Exer. 67–68: For the given circle, find (a) the x-intercepts and (b) the y-intercepts.

67 $x^2 + y^2 - 4x - 6y + 4 = 0$

68 $x^2 + y^2 - 10x + 4y + 13 = 0$

69 Find an equation of the circle that is concentric (has the same center) with $x^2 + y^2 + 4x - 6y + 4 = 0$ and passes through $P(2, 6)$.

70 **Radio broadcasting ranges** The signal from a radio station has a circular range of 50 miles. A second radio station, located 100 miles east and 80 miles north of the first station, has a range of 80 miles. Are there locations where signals can be received from both radio stations? Explain your answer.

71 A circle C_1 of radius 5 has its center at the origin. Inside this circle there is a first-quadrant circle C_2 of radius 2 that is tangent to C_1. The y-coordinate of the center of C_2 is 2. Find the x-coordinate of the center of C_2.

72 A circle C_1 of radius 5 has its center at the origin. Outside this circle is a first-quadrant circle C_2 of radius 2 that is tangent to C_1. The y-coordinate of the center of C_2 is 3. Find the x-coordinate of the center of C_2.

Exer. 73–76: Express, in interval form, the x-values such that $y_1 < y_2$. Assume all points of intersection are shown on the interval $(-\infty, \infty)$.

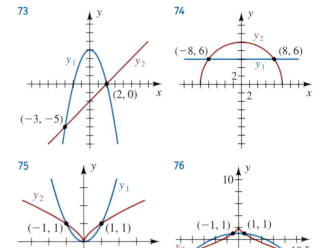

73

74

75

76

3.3

Lines

One of the basic concepts in geometry is that of a *line*. In this section we will restrict our discussion to lines that lie in a coordinate plane. This will allow us to use algebraic methods to study their properties. Two of our principal objectives may be stated as follows:

(1) Given a line l in a coordinate plane, find an equation whose graph corresponds to l.

(2) Given an equation of a line l in a coordinate plane, sketch the graph of the equation.

The following concept is fundamental to the study of lines.

Definition of Slope of a Line	Let l be a line that is not parallel to the y-axis, and let $P_1(x_1, y_1)$ and $P_2(x_2, y_2)$ be distinct points on l. The **slope m** of l is

$$m = \frac{y_2 - y_1}{x_2 - x_1}.$$

If l is parallel to the y-axis, then the slope of l is not defined.

The Greek letter Δ (delta) is used in mathematics to denote "change in." Thus, we can think of the slope m as

$$m = \frac{\Delta y}{\Delta x} = \frac{\text{change in } y}{\text{change in } x}.$$

Figure 1

(a) Positive slope (line rises)

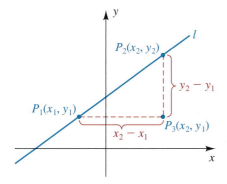

(b) Negative slope (line falls)

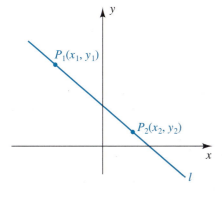

Typical points P_1 and P_2 on a line l are shown in Figure 1. The numerator $y_2 - y_1$ in the formula for m is the vertical change in direction from P_1 to P_2 and may be positive, negative, or zero. The denominator $x_2 - x_1$ is the horizontal change from P_1 to P_2, and it may be positive or negative, but never zero, because l is not parallel to the y-axis if a slope exists. In Figure 1(a) the slope is positive, and we say that the line *rises*. In Figure 1(b) the slope is negative, and the line *falls*.

In finding the slope of a line it is immaterial which point we label as P_1 and which as P_2, since

$$\frac{y_2 - y_1}{x_2 - x_1} = \frac{y_2 - y_1}{x_2 - x_1} \cdot \frac{(-1)}{(-1)} = \frac{y_1 - y_2}{x_1 - x_2}.$$

If the points are labeled so that $x_1 < x_2$, as in Figure 1, then $x_2 - x_1 > 0$, and hence the slope is positive, negative, or zero, depending on whether $y_2 > y_1$, $y_2 < y_1$, or $y_2 = y_1$, respectively.

The definition of slope is independent of the two points that are chosen on l. If other points $P_1'(x_1', y_1')$ and $P_2'(x_2', y_2')$ are used, then, as in Figure 2, the triangle with vertices P_1', P_2', and $P_3'(x_2', y_1')$ is similar to the triangle with vertices P_1, P_2, and $P_3(x_2, y_1)$. Since the ratios of corresponding sides of similar triangles are equal,

$$\frac{y_2 - y_1}{x_2 - x_1} = \frac{y_2' - y_1'}{x_2' - x_1'}.$$

Figure 2

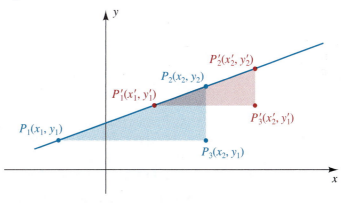

EXAMPLE 1 Finding slopes

Sketch the line through each pair of points, and find its slope m:

(a) $A(-1, 4)$ and $B(3, 2)$ **(b)** $A(2, 5)$ and $B(-2, -1)$

(c) $A(4, 3)$ and $B(-2, 3)$ **(d)** $A(4, -1)$ and $B(4, 4)$

SOLUTION The lines are sketched in Figure 3. We use the definition of slope to find the slope of each line.

(continued)

Figure 3

(a) $m = -\frac{1}{2}$

(b) $m = \frac{3}{2}$

(c) $m = 0$

(d) m undefined

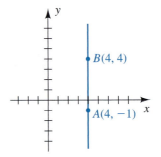

(a) $m = \dfrac{2 - 4}{3 - (-1)} = \dfrac{-2}{4} = -\dfrac{1}{2}$

(b) $m = \dfrac{5 - (-1)}{2 - (-2)} = \dfrac{6}{4} = \dfrac{3}{2}$

(c) $m = \dfrac{3 - 3}{-2 - 4} = \dfrac{0}{-6} = 0$

(d) The slope is undefined because the line is parallel to the y-axis. Note that if the formula for m is used, the denominator is zero. ◢

EXAMPLE 2 Sketching a line with a given slope

Sketch a line through $P(2, 1)$ that has

(a) slope $\frac{5}{3}$ **(b)** slope $-\frac{5}{3}$

SOLUTION   If the slope of a line is a/b and b is positive, then for every change of b units in the horizontal direction, the line rises or falls $|a|$ units, depending on whether a is positive or negative, respectively.

(a) If $P(2, 1)$ is on the line and $m = \frac{5}{3}$, we can obtain another point on the line by starting at P and moving 3 units to the right and 5 units *upward*. This gives us the point $Q(5, 6)$, and the line is determined as in Figure 4(a).

(b) If $P(2, 1)$ is on the line and $m = -\frac{5}{3}$, we move 3 units to the right and 5 units *downward,* obtaining the line through $Q(5, -4)$, as in Figure 4(b).

Figure 4
(a) $m = \frac{5}{3}$ **(b)** $m = -\frac{5}{3}$

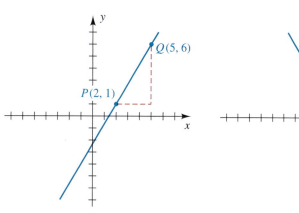

The diagram in Figure 5 indicates the slopes of several lines through the origin. The line that lies on the *x*-axis has slope $m = 0$. If this line is rotated about O in the *counterclockwise* direction (as indicated by the blue arrow), the slope is positive and increases, reaching the value 1 when the line bisects the first quadrant and continuing to increase as the line gets closer to the *y*-axis. If we rotate the line of slope $m = 0$ in the *clockwise* direction (as indicated by the red arrow), the slope is negative, reaching the value -1 when the line bisects the second quadrant and becoming large and negative as the line gets closer to the *y*-axis.

Figure 5

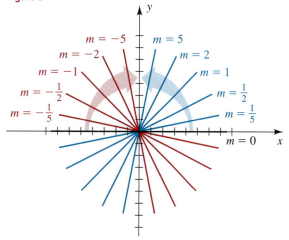

Lines that are horizontal or vertical have simple equations, as indicated in the following chart.

Terminology	Definition	Graph	Equation	Slope
Horizontal line	A line parallel to the x-axis	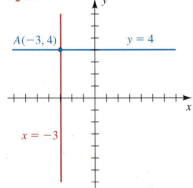	$y = b$ y-intercept is b	Slope is 0
Vertical line	A line parallel to the y-axis		$x = a$ x-intercept is a	Slope is undefined

Figure 6

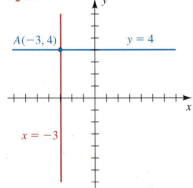

A common error is to regard the graph of $y = b$ as consisting of only the one point $(0, b)$. If we express the equation in the form $0 \cdot x + y = b$, we see that the value of x is immaterial; thus, the graph of $y = b$ consists of the points (x, b) for *every* x and hence is a horizontal line. Similarly, the graph of $x = a$ is the vertical line consisting of all points (a, y), where y is a real number.

EXAMPLE 3   **Finding equations of horizontal and vertical lines**

Find an equation of the line through $A(-3, 4)$ that is parallel to

(a) the x-axis   **(b)** the y-axis

SOLUTION   The two lines are sketched in Figure 6. As indicated in the preceding chart, the equations are $y = 4$ for part (a) and $x = -3$ for part (b).

Figure 7

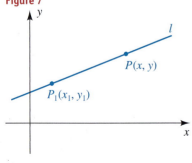

Let us next find an equation of a line l through a point $P_1(x_1, y_1)$ with slope m. If $P(x, y)$ is any point with $x \neq x_1$ (see Figure 7), then P is on l if and only if the slope of the line through P_1 and P is m—that is, if

$$\frac{y - y_1}{x - x_1} = m.$$

This equation may be written in the form

$$y - y_1 = m(x - x_1).$$

Note that (x_1, y_1) is a solution of the last equation, and hence the points on l are precisely the points that correspond to the solutions. This equation for l is referred to as the **point-slope form.**

Point-Slope Form for the Equation of a Line	An equation for the line through the point (x_1, y_1) with slope m is $$y - y_1 = m(x - x_1).$$

The point-slope form is only one possibility for an equation of a line. There are many equivalent equations. We sometimes simplify the equation obtained using the point-slope form to either

$$ax + by = c \qquad \text{or} \qquad ax + by + d = 0,$$

where a, b, and c are integers with no common factor, $a > 0$, and $d = -c$.

EXAMPLE 4 **Finding an equation of a line through two points**

Find an equation of the line through $A(1, 7)$ and $B(-3, 2)$.

Figure 8

Figure 9

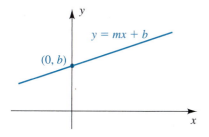

SOLUTION The line is sketched in Figure 8. The formula for the slope m gives us

$$m = \frac{7 - 2}{1 - (-3)} = \frac{5}{4}.$$

We may use the coordinates of either A or B for (x_1, y_1) in the point-slope form. Using $A(1, 7)$ gives us

$$y - 7 = \tfrac{5}{4}(x - 1) \qquad \text{point-slope form}$$
$$4(y - 7) = 5(x - 1) \qquad \text{multiply by 4}$$
$$4y - 28 = 5x - 5 \qquad \text{multiply factors}$$
$$-5x + 4y = 23 \qquad \text{subtract } 5x \text{ and add } 28$$
$$5x - 4y = -23 \qquad \text{multiply by } -1$$

The last equation is one of the desired forms for an equation of a line. Another is $5x - 4y + 23 = 0$.

The point-slope form for the equation of a line may be rewritten as $y = mx - mx_1 + y_1$, which is of the form

$$y = mx + b$$

with $b = -mx_1 + y_1$. The real number b is the y-intercept of the graph, as indicated in Figure 9. Since the equation $y = mx + b$ displays the slope m and

y-intercept b of l, it is called the **slope-intercept form** for the equation of a line. Conversely, if we start with $y = mx + b$, we may write

$$y - b = m(x - 0).$$

Comparing this equation with the point-slope form, we see that the graph is a line with slope m and passing through the point $(0, b)$. We have proved the following result.

Slope-Intercept Form for the Equation of a Line	The graph of $y = mx + b$ is a line having slope m and y-intercept b.

EXAMPLE 5   Expressing an equation in slope-intercept form

Express the equation $2x - 5y = 8$ in slope-intercept form.

SOLUTION   Our goal is to solve the given equation for y to obtain the form $y = mx + b$. We may proceed as follows:

$$2x - 5y = 8 \qquad \text{given}$$
$$-5y = -2x + 8 \qquad \text{subtract } 2x$$
$$y = \left(\frac{-2}{-5}\right)x + \left(\frac{8}{-5}\right) \qquad \text{divide by } -5$$
$$y = \tfrac{2}{5}x + \left(-\tfrac{8}{5}\right) \qquad \text{equivalent equation}$$

The last equation is the slope-intercept form $y = mx + b$ with slope $m = \tfrac{2}{5}$ and y-intercept $b = -\tfrac{8}{5}$.   ◢

It follows from the point-slope form that every line is a graph of an equation

$$ax + by = c,$$

where a, b, and c are real numbers and a and b are not both zero. We call such an equation a **linear equation** in x and y. Let us show, conversely, that the graph of $ax + by = c$, with a and b not both zero, is always a line. If $b \neq 0$, we may solve for y, obtaining

$$y = \left(-\frac{a}{b}\right)x + \frac{c}{b},$$

which, by the slope-intercept form, is an equation of a line with slope $-a/b$ and y-intercept c/b. If $b = 0$ but $a \neq 0$, we may solve for x, obtaining $x = c/a$, which is the equation of a vertical line with x-intercept c/a. This discussion establishes the following result.

General Form for the Equation of a Line	The graph of a linear equation $ax + by = c$ is a line, and conversely, every line is the graph of a linear equation.

For simplicity, we use the terminology *the line $ax + by = c$* rather than *the line with equation $ax + by = c$*.

EXAMPLE 6 Sketching the graph of a linear equation

Sketch the graph of $2x - 5y = 8$.

SOLUTION We know from the preceding discussion that the graph is a line, so it is sufficient to find two points on the graph. Let us find the x- and y-intercepts by substituting $y = 0$ and $x = 0$, respectively, in the given equation, $2x - 5y = 8$.

x-intercept: If $y = 0$, then $2x = 8$, or $x = 4$.

y-intercept: If $x = 0$, then $-5y = 8$, or $y = -\frac{8}{5}$.

Plotting the points $(4, 0)$ and $\left(0, -\frac{8}{5}\right)$ and drawing a line through them gives us the graph in Figure 10.

Figure 10

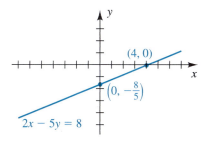

The following theorem specifies the relationship between **parallel lines** (lines in a plane that do not intersect) and slope.

Theorem on Slopes of Parallel Lines	Two nonvertical lines are parallel if and only if they have the same slope.

PROOF Let l_1 and l_2 be distinct lines of slopes m_1 and m_2, respectively. If the y-intercepts are b_1 and b_2 (see Figure 11), then, by the slope-intercept form, the lines have equations

$$y = m_1x + b_1 \quad \text{and} \quad y = m_2x + b_2.$$

The lines intersect at some point (x, y) if and only if the values of y are equal for some x—that is, if

$$m_1x + b_1 = m_2x + b_2,$$

or

$$(m_1 - m_2)x = b_2 - b_1.$$

The last equation can be solved for x if and only if $m_1 - m_2 \neq 0$. We have shown that the lines l_1 and l_2 intersect if and only if $m_1 \neq m_2$. Hence, they do *not* intersect (are parallel) if and only if $m_1 = m_2$.

Figure 11

EXAMPLE 7 Finding an equation of a line parallel to a given line

Find an equation of the line through $P(5, -7)$ that is parallel to the line $6x + 3y = 4$.

SOLUTION We first express the given equation in slope-intercept form:

$$
\begin{array}{ll}
6x + 3y = 4 & \text{given} \\
3y = -6x + 4 & \text{subtract } 6x \\
y = -2x + \tfrac{4}{3} & \text{divide by 3}
\end{array}
$$

The last equation is in slope-intercept form, $y = mx + b$, with slope $m = -2$ and y-intercept $\tfrac{4}{3}$. Since parallel lines have the same slope, the required line also has slope -2. Using the point $P(5, -7)$ gives us the following:

$$
\begin{array}{ll}
y - (-7) = -2(x - 5) & \text{point-slope form} \\
y + 7 = -2x + 10 & \text{simplify} \\
y = -2x + 3 & \text{subtract 7}
\end{array}
$$

The last equation is in slope-intercept form and shows that the parallel line we have found has y-intercept 3. This line and the given line are sketched in Figure 12.

As an alternative solution, we might use the fact that lines of the form $6x + 3y = k$ have the same slope as the given line and hence are parallel to it. Substituting $x = 5$ and $y = -7$ into the equation $6x + 3y = k$ gives us $6(5) + 3(-7) = k$ or, equivalently, $k = 9$. The equation $6x + 3y = 9$ is equivalent to $y = -2x + 3$. ◢

Figure 12

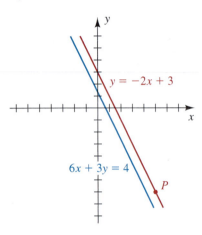

$y = -2x + 3$

$6x + 3y = 4$

P

If the slopes of two nonvertical lines are not the same, then the lines are not parallel and intersect at exactly one point.

The next theorem gives us information about **perpendicular lines** (lines that intersect at a right angle).

Theorem on Slopes of Perpendicular Lines	Two lines with slope m_1 and m_2 are perpendicular if and only if $$m_1 m_2 = -1.$$

Figure 13

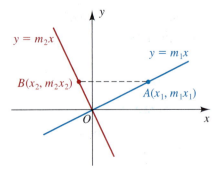

$y = m_2 x$

$y = m_1 x$

$B(x_2, m_2 x_2)$

$A(x_1, m_1 x_1)$

O

PROOF For simplicity, let us consider the special case of two lines that intersect at the origin O, as illustrated in Figure 13. Equations of these lines are $y = m_1 x$ and $y = m_2 x$. If, as in the figure, we choose points $A(x_1, m_1 x_1)$ and $B(x_2, m_2 x_2)$ different from O on the lines, then the lines are perpendicular if and only if angle AOB is a right angle. Applying the Pythagorean theorem, we know that angle AOB is a right angle if and only if

$$[d(A, B)]^2 = [d(O, B)]^2 + [d(O, A)]^2$$

or, by the distance formula,

$$(x_2 - x_1)^2 + (m_2 x_2 - m_1 x_1)^2 = x_2^2 + (m_2 x_2)^2 + x_1^2 + (m_1 x_1)^2.$$

Squaring terms, simplifying, and factoring gives us

$$-2m_1 m_2 x_1 x_2 - 2x_1 x_2 = 0$$

$$-2x_1 x_2 (m_1 m_2 + 1) = 0.$$

Since both x_1 and x_2 are not zero, we may divide both sides by $-2x_1 x_2$, obtaining $m_1 m_2 + 1 = 0$. Thus, the lines are perpendicular if and only if $m_1 m_2 = -1$.

The same type of proof may be given if the lines intersect at *any* point (a, b).

A convenient way to remember the conditions on slopes of perpendicular lines is to note that m_1 and m_2 must be *negative reciprocals* of each other— that is, $m_1 = -1/m_2$ and $m_2 = -1/m_1$.

We can visualize the result of the last theorem as follows. Draw a triangle as in Figure 14; the line containing its hypotenuse has slope $m_1 = b/a$. Now rotate the triangle 90° as in Figure 15. The line now has slope $m_2 = a/(-b)$, the negative reciprocal of m_1.

Figure 14

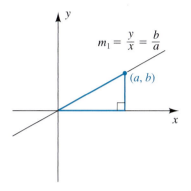

$$m_1 = \frac{y}{x} = \frac{b}{a}$$

(a, b)

Figure 15

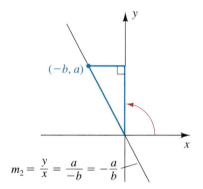

$(-b, a)$

$$m_2 = \frac{y}{x} = \frac{a}{-b} = -\frac{a}{b}$$

EXAMPLE 8 Finding an equation of a line perpendicular to a given line

Find the slope-intercept form for the line through $P(5, -7)$ that is perpendicular to the line $6x + 3y = 4$.

SOLUTION We considered the line $6x + 3y = 4$ in Example 7 and found that its slope is -2. Hence, the slope of the required line is the negative reciprocal $-[1/(-2)]$, or $\frac{1}{2}$. Using $P(5, -7)$ gives us the following:

$$y - (-7) = \tfrac{1}{2}(x - 5) \qquad \text{point-slope form}$$

$$y + 7 = \tfrac{1}{2}x - \tfrac{5}{2} \qquad \text{simplify}$$

$$y = \tfrac{1}{2}x - \tfrac{19}{2} \qquad \text{put in slope-intercept form}$$

The last equation is in slope-intercept form and shows that the perpendicular line has y-intercept $-\frac{19}{2}$. This line and the given line are sketched in Figure 16.

Figure 16

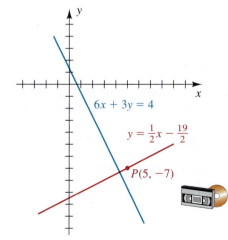

$6x + 3y = 4$

$y = \frac{1}{2}x - \frac{19}{2}$

$P(5, -7)$

EXAMPLE 9 Finding an equation of a perpendicular bisector

Given $A(-3, 1)$ and $B(5, 4)$, find the general form of the perpendicular bisector l of the line segment AB.

Figure 17

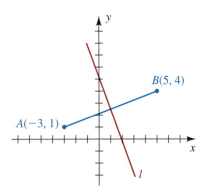

SOLUTION The line segment AB and its perpendicular bisector l are shown in Figure 17. We calculate the following, where M is the midpoint of AB:

Coordinates of M: $\left(\dfrac{-3+5}{2}, \dfrac{1+4}{2}\right) = \left(1, \dfrac{5}{2}\right)$ midpoint formula

Slope of AB: $\dfrac{4-1}{5-(-3)} = \dfrac{3}{8}$ slope formula

Slope of l: $-\dfrac{1}{\frac{3}{8}} = -\dfrac{8}{3}$ negative reciprocal of $\frac{3}{8}$

Using the point $M\left(1, \frac{5}{2}\right)$ and slope $-\frac{8}{3}$ gives us the following equivalent equations for l:

$$y - \tfrac{5}{2} = -\tfrac{8}{3}(x-1) \quad \text{point-slope form}$$
$$6y - 15 = -16(x-1) \quad \text{multiply by the lcd, 6}$$
$$6y - 15 = -16x + 16 \quad \text{multiply}$$
$$16x + 6y = 31 \quad \text{put in general form}$$

Two variables x and y are **linearly related** if $y = ax + b$, where a and b are real numbers and $a \neq 0$. Linear relationships between variables occur frequently in applied problems. The following example gives one illustration.

EXAMPLE 10   **Relating air temperature to altitude**

The relationship between the air temperature T (in °F) and the altitude h (in feet above sea level) is approximately linear for $0 \le h \le 20{,}000$. If the temperature at sea level is 60°, an increase of 5000 feet in altitude lowers the air temperature about 18°.

(a) Express T in terms of h, and sketch the graph on an hT-coordinate system.

(b) Approximate the air temperature at an altitude of 15,000 feet.

(c) Approximate the altitude at which the temperature is 0°.

SOLUTION

(a) If T is linearly related to h, then

$$T = ah + b$$

for some constants a and b (a represents the slope and b the T-intercept). Since $T = 60°$ when $h = 0$ ft (sea level), the T-intercept is 60, and the temperature T for $0 \le h \le 20{,}000$ is given by

$$T = ah + 60.$$

From the given data, we note that when the altitude $h = 5000$ ft, the temperature $T = 60° - 18° = 42°$. Hence, we may find a as follows:

$$42 = a(5000) + 60 \qquad \text{let } T = 42 \text{ and } h = 5000$$
$$a = \frac{42 - 60}{5000} = -\frac{9}{2500} \qquad \text{solve for } a$$

Figure 18

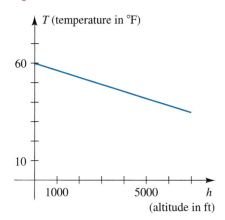

Substituting for a in $T = ah + 60$ gives us the following formula for T:

$$T = -\frac{9}{2500}h + 60$$

The graph is sketched in Figure 18, with different scales on the axes.

(b) Using the last formula for T obtained in part (a), we find that the temperature (in °F) when $h = 15{,}000$ is

$$T = -\frac{9}{2500}(15{,}000) + 60 = -54 + 60 = 6.$$

(c) To find the altitude h that corresponds to $T = 0°$, we proceed as follows:

$$T = -\frac{9}{2500}h + 60 \qquad \text{from part (a)}$$

$$0 = -\frac{9}{2500}h + 60 \qquad \text{let } T = 0$$

$$\frac{9}{2500}h = 60 \qquad \text{add } \frac{9}{2500}h$$

$$h = 60 \cdot \frac{2500}{9} \qquad \text{multiply by } \frac{2500}{9}$$

$$h = \frac{50{,}000}{3} \approx 16{,}667 \text{ ft} \qquad \text{simplify and approximate}$$

A **mathematical model** is a mathematical description of a problem. For our purposes, these descriptions will be graphs and equations. In the last example, the equation $T = -\frac{9}{2500}h + 60$ *models* the relationship between air temperature and altitude.

3.3 Exercises

Exer. 1–6: Sketch the line through A and B, and find its slope m.

1 $A(-3, 2)$, $B(5, -4)$

2 $A(4, -1)$, $B(-6, -3)$

3 $A(2, 5)$, $B(-7, 5)$

4 $A(5, -1)$, $B(5, 6)$

5 $A(-3, 2)$, $B(-3, 5)$

6 $A(4, -2)$, $B(-3, -2)$

Exer. 7–10: Use slopes to show that the points are vertices of the specified polygon.

7 $A(-3, 1), B(5, 3), C(3, 0), D(-5, -2)$;   parallelogram

8 $A(2, 3), B(5, -1), C(0, -6), D(-6, 2)$;   trapezoid

9 $A(6, 15), B(11, 12), C(-1, -8), D(-6, -5)$;   rectangle

10 $A(1, 4), B(6, -4), C(-15, -6)$;   right triangle

11 If three consecutive vertices of a parallelogram are $A(-1, -3)$, $B(4, 2)$, and $C(-7, 5)$, find the fourth vertex.

12 Let $A(x_1, y_1)$, $B(x_2, y_2)$, $C(x_3, y_3)$, and $D(x_4, y_4)$ denote the vertices of an arbitrary quadrilateral. Show that the line segments joining midpoints of adjacent sides form a parallelogram.

Exer. 13–14: Sketch the graph of $y = mx$ for the given values of m.

13 $m = 3, -2, \frac{2}{3}, -\frac{1}{4}$

14 $m = 5, -3, \frac{1}{2}, -\frac{1}{3}$

Exer. 15–16: Sketch the graph of the line through P for each value of m.

15 $P(3, 1)$; $m = \frac{1}{2}, -1, -\frac{1}{5}$

16 $P(-2, 4)$; $m = 1, -2, -\frac{1}{2}$

Exer. 17–18: Write equations of the lines.

17

(2, −3)

18

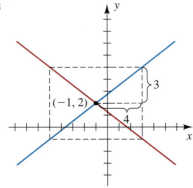

(−1, 2)

Exer. 19–20: Sketch the graphs of the lines on the same coordinate plane.

19 $y = x + 3$, $y = x + 1$, $y = -x + 1$

20 $y = -2x - 1$, $y = -2x + 3$, $y = \frac{1}{2}x + 3$

Exer. 21–32: Find a general form of an equation of the line through the point A that satisfies the given condition.

21 $A(5, -2)$

   (a) parallel to the y-axis

   (b) perpendicular to the y-axis

22 $A(-4, 2)$

   (a) parallel to the x-axis

   (b) perpendicular to the x-axis

23 $A(5, -3)$; slope -4 24 $A(-1, 4)$; slope $\frac{2}{3}$

25 $A(4, 0)$;   slope -3 26 $A(0, -2)$; slope 5

27 $A(4, -5)$; through $B(-3, 6)$

28 $A(-1, 6)$; x-intercept 5

29 $A(2, -4)$; parallel to the line $5x - 2y = 4$

30 $A(-3, 5)$; parallel to the line $x + 3y = 1$

31 $A(7, -3)$; perpendicular to the line $2x - 5y = 8$

32 $A(4, 5)$;   perpendicular to the line $3x + 2y = 7$

Exer. 33–36: Find the slope-intercept form of the line that satisfies the given conditions.

33 x-intercept 4, y-intercept -3

34 x-intercept -5, y-intercept -1

35 Through $A(5, 2)$ and $B(-1, 4)$

36 Through $A(-2, 1)$ and $B(3, 7)$

Exer. 37–38: Find a general form of an equation for the perpendicular bisector of the segment AB.

37 $A(3, -1), B(-2, 6)$ 38 $A(4, 2), B(-2, 10)$

Exer. 39–40: Find an equation for the line that bisects the given quadrants.

39 II and IV   40 I and III

Exer. 41–44: Use the slope-intercept form to find the slope and y-intercept of the given line, and sketch its graph.

41 $2x = 15 - 3y$ 42 $7x = -4y - 8$

43 $4x - 3y = 9$ 44 $x - 5y = -15$

Exer. 45–46: Find an equation of the line shown in the figure.

45 (a)   (b)

$m = -\frac{1}{2}$

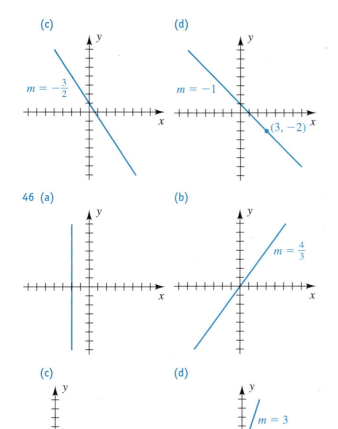

(c)

$m = -\dfrac{3}{2}$

(d)

$m = -1$

$(3, -2)$

46 (a)

(b)

$m = \dfrac{4}{3}$

(c)

$m = \dfrac{1}{3}$

(d)

$m = 3$

$(-2, -5)$

Exer. 47–48: If a line l has nonzero x- and y-intercepts a and b, respectively, then its *intercept form* is

$$\frac{x}{a} + \frac{y}{b} = 1.$$

Find the intercept form for the given line.

47 $4x - 2y = 6$ 48 $x - 3y = -2$

49 Find an equation of the circle that has center $C(3, -2)$ and is tangent to the line $y = 5$.

50 Find an equation of the line that is tangent to the circle $x^2 + y^2 = 25$ at the point $P(3, 4)$.

51 **Fetal growth** The growth of a fetus more than 12 weeks old can be approximated by the formula $L = 1.53t - 6.7$, where L is the length (in centimeters) and t is the age (in weeks). Prenatal length can be determined by ultrasound. Approximate the age of a fetus whose length is 28 centimeters.

52 **Estimating salinity** Salinity of the ocean refers to the amount of dissolved material found in a sample of seawater. Salinity S can be estimated from the amount C of chlorine in seawater using $S = 0.03 + 1.805C$, where S and C are measured by weight in parts per thousand. Approximate C if S is 0.35.

53 **Weight of a humpback whale** The expected weight W (in tons) of a humpback whale can be approximated from its length L (in feet) by using $W = 1.70L - 42.8$ for $30 \le L \le 50$.

(a) Estimate the weight of a 40-foot humpback whale.

(b) If the error in estimating the length could be as large as 2 feet, what is the corresponding error for the weight estimate?

54 **Growth of a blue whale** Newborn blue whales are approximately 24 feet long and weigh 3 tons. Young whales are nursed for 7 months, and by the time of weaning they often are 53 feet long and weigh 23 tons. Let L and W denote the length (in feet) and the weight (in tons), respectively, of a whale that is t months of age.

(a) If L and t are linearly related, express L in terms of t.

(b) What is the daily increase in the length of a young whale? (Use 1 month = 30 days.)

(c) If W and t are linearly related, express W in terms of t.

(d) What is the daily increase in the weight of a young whale?

55 **Baseball stats** Suppose a major league baseball player has hit 5 home runs in the first 14 games, and he keeps up this pace throughout the 162-game season.

(a) Express the number y of home runs in terms of the number x of games played.

(b) How many home runs will the player hit for the season?

56 Cheese production A cheese manufacturer produces 18,000 pounds of cheese from January 1 through March 24. Suppose that this rate of production continues for the remainder of the year.

(a) Express the number y of pounds of cheese produced in terms of the number x of the day in a 365-day year.

(b) Predict, to the nearest pound, the number of pounds produced for the year.

57 Childhood weight A baby weighs 10 pounds at birth, and three years later the child's weight is 30 pounds. Assume that childhood weight W (in pounds) is linearly related to age t (in years).

(a) Express W in terms of t.

(b) What is W on the child's sixth birthday?

(c) At what age will the child weigh 70 pounds?

(d) Sketch, on a tW-plane, a graph that shows the relationship between W and t for $0 \le t \le 12$.

58 Loan repayment A college student receives an interest-free loan of $8250 from a relative. The student will repay $125 per month until the loan is paid off.

(a) Express the amount P (in dollars) remaining to be paid in terms of time t (in months).

(b) After how many months will the student owe $5000?

(c) Sketch, on a tP-plane, a graph that shows the relationship between P and t for the duration of the loan.

59 Vaporizing water The amount of heat H (in joules) required to convert one gram of water into vapor is linearly related to the temperature T (in °C) of the atmosphere. At 10°C this conversion requires 2480 joules, and each increase in temperature of 15°C lowers the amount of heat needed by 40 joules. Express H in terms of T.

60 Aerobic power In exercise physiology, aerobic power P is defined in terms of maximum oxygen intake. For altitudes up to 1800 meters, aerobic power is optimal—that is, 100%. Beyond 1800 meters, P decreases linearly from the maximum of 100% to a value near 40% at 5000 meters.

(a) Express aerobic power P in terms of altitude h (in meters) for $1800 \le h \le 5000$.

(b) Estimate aerobic power in Mexico City (altitude: 2400 meters), the site of the 1968 Summer Olympic Games.

61 Urban heat island The urban heat island phenomenon has been observed in Tokyo. The average temperature was 13.5°C in 1915, and since then has risen 0.032°C per year.

(a) Assuming that temperature T (in °C) is linearly related to time t (in years) and that $t = 0$ corresponds to 1915, express T in terms of t.

(b) Predict the average temperature in the year 2000.

62 Rising ground temperature In 1870 the average ground temperature in Paris was 11.8°C. Since then it has risen at a nearly constant rate, reaching 13.5°C in 1969.

(a) Express the temperature T (in °C) in terms of time t (in years), where $t = 0$ corresponds to the year 1870 and $0 \le t \le 99$.

(b) During what year was the average ground temperature 12.5°C?

63 Business expenses The owner of an ice cream franchise must pay the parent company $1000 per month plus 5% of the monthly revenue R. Operating cost of the franchise includes a fixed cost of $2600 per month for items such as utilities and labor. The cost of ice cream and supplies is 50% of the revenue.

(a) Express the owner's monthly expense E in terms of R.

(b) Express the monthly profit P in terms of R.

(c) Determine the monthly revenue needed to break even.

64 Drug dosage Pharmacological products must specify recommended dosages for adults and children. Two formulas for modification of adult dosage levels for young children are

$$\text{Cowling's rule:} \quad y = \tfrac{1}{24}(t + 1)a$$

and $\quad$ Friend's rule: $\quad y = \tfrac{2}{25}ta,$

where a denotes adult dose (in milligrams) and t denotes the age of the child (in years).

(a) If $a = 100$, graph the two linear equations on the same coordinate plane for $0 \le t \le 12$.

(b) For what age do the two formulas specify the same dosage?

65 Video game In the video game shown in the figure, an airplane flies from left to right along the path given by $y = 1 + (1/x)$ and shoots bullets in the tangent direction at creatures placed along the x-axis at $x = 1, 2, 3, 4$.

Exercise 65

From calculus, the slope of the tangent line to the path at $P(1, 2)$ is $m = -1$ and at $Q\left(\frac{3}{2}, \frac{5}{3}\right)$ is $m = -\frac{4}{9}$. Determine whether a creature will be hit if bullets are shot when the airplane is at

(a) P (b) Q

66 **Temperature scales** The relationship between the temperature reading F on the Fahrenheit scale and the temperature reading C on the Celsius scale is given by $C = \frac{5}{9}(F - 32)$.

(a) Find the temperature at which the reading is the same on both scales.

(b) When is the Fahrenheit reading twice the Celsius reading?

67 **Vertical wind shear** Vertical wind shear occurs when wind speed varies at different heights above the ground. Wind shear is of great importance to pilots during takeoffs and landings. If the wind speed is v_1 at height h_1 and v_2 at height h_2, then the average wind shear s is given by the slope formula

$$s = \frac{v_2 - v_1}{h_2 - h_1}.$$

If the wind speed at ground level is 22 mi/hr and s has been determined to be 0.07, find the wind speed 185 feet above the ground.

68 **Vertical wind shear** In the study of vertical wind shear, the formula

$$\frac{v_1}{v_2} = \left(\frac{h_1}{h_2}\right)^P$$

is sometimes used, where P is a variable that depends on the terrain and structures near ground level. In Montreal, the average daytime value for P with north winds over 29 mi/hr was determined to be 0.13. If a 32 mi/hr north wind is measured 20 feet above the ground, approximate the average wind shear (see Exercise 67) between 20 feet and 200 feet.

Exer. 69–70: The given points were found using empirical methods. Determine whether they lie on the same line $y = ax + b$, and if so, find the values of a and b.

69 $A(-1.3, -1.3598)$, $B(-0.55, -1.11905)$,

$C(1.2, -0.5573)$, $D(3.25, 0.10075)$

70 $A(-0.22, 1.6968)$, $B(-0.12, 1.6528)$,

$C(1.3, 1.028)$ $D(1.45, 0.862)$

3.4

Definition of Function

The notion of **correspondence** occurs frequently in everyday life. Some examples are given in the following illustration.

ILLUSTRATION   **Correspondence**

■ To each book in a library there corresponds the number of pages in the book.

■ To each human being there corresponds a birth date.

■ If the temperature of the air is recorded throughout the day, then to each instant of time there corresponds a temperature.

Each correspondence in the previous illustration involves two sets, D and E. In the first illustration, D denotes the set of books in a library and E the set of positive integers. To each book x in D there corresponds a positive integer y in E—namely, the number of pages in the book.

Figure 1

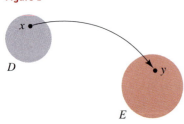

We sometimes depict correspondences by diagrams of the type shown in Figure 1, where the sets D and E are represented by points within regions in a plane. The curved arrow indicates that the element y of E corresponds to the element x of D. The two sets may have elements in common. As a matter of fact, we often have $D = E$. It is important to note that *to each x in D there corresponds exactly one y in E.* However, the same element of E may correspond to different elements of D. For example, two books may have the same number of pages, two people may have the same birthday, and the temperature may be the same at different times.

In most of our work, D and E will be sets of numbers. To illustrate, let both D and E denote the set $\mathbb{R}$ of real numbers, and to each real number x let us assign its square x^2. This gives us a correspondence from $\mathbb{R}$ to $\mathbb{R}$.

Each of our illustrations of a correspondence is a *function,* which we define as follows.

Definition of Function	A **function** f from a set D to a set E is a correspondence that assigns to each element x of D exactly one element y of E.

*For many cases, we can simply remember that the **domain** is the set of x-values and the **range** is the set of y-values.*

The element x of D is the **argument** of f. The set D is the **domain** of the function. The element y of E is the **value** of f at x (or the **image** of x under f) and is denoted by $f(x)$, read "f of x." The **range** of f is the subset R of E consisting of all possible values $f(x)$ for x in D. Note that there may be elements in the set E that are not in the range R of f.

Consider the diagram in Figure 2. The curved arrows indicate that the elements $f(w)$, $f(z)$, $f(x)$, and $f(a)$ of E correspond to the elements w, z, x, and a of D. *To each element in D there is assigned exactly one function value in E;* however, different elements of D, such as w and z in Figure 2, may have the same value in E.

The symbols

Figure 2

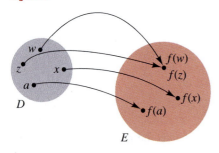

$$D \xrightarrow{f} E, \qquad f: D \to E, \qquad \text{and}$$

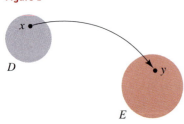

signify that f is a function from D to E, and we say that f **maps** D *into* E. Initially, the notations f and $f(x)$ may be confusing. Remember that f is used to represent the function. It is neither in D nor in E. However, $f(x)$ is an element of the range R—the element that the function f assigns to the element x, which is in the domain D.

Two functions f and g from D to E are **equal,** and we write

$$f = g \quad \text{provided} \quad f(x) = g(x) \quad \text{for every } x \text{ in } D.$$

For example, if $g(x) = \frac{1}{2}(2x^2 - 6) + 3$ and $f(x) = x^2$ for every x in $\mathbb{R}$, then $g = f$.

EXAMPLE 1   Finding function values

Let f be the function with domain $\mathbb{R}$ such that $f(x) = x^2$ for every x in $\mathbb{R}$.

(a) Find $f(-6)$, $f(\sqrt{3})$, $f(a + b)$, and $f(a) + f(b)$, where a and b are real numbers.

(b) What is the range of f?

SOLUTION

(a) We find values of f by substituting for x in the equation $f(x) = x^2$:

$$f(-6) = (-6)^2 = 36$$
$$f(\sqrt{3}) = (\sqrt{3})^2 = 3$$
$$f(a + b) = (a + b)^2 = a^2 + 2ab + b^2$$
$$f(a) + f(b) = a^2 + b^2$$

Note that, in general,

$$f(a + b) \neq f(a) + f(b).$$

(b) By definition, the range of f consists of all numbers of the form $f(x) = x^2$ for x in $\mathbb{R}$. Since the square of every real number is nonnegative, the range is contained in the set of all nonnegative real numbers. Moreover, every nonnegative real number c is a value of f, since $f(\sqrt{c}) = (\sqrt{c})^2 = c$. Hence, the range of f is the set of all nonnegative real numbers. ◢

If a function is defined as in Example 1, the symbols used for the function and variable are immaterial; that is, expressions such as $f(x) = x^2$, $f(s) = s^2$, $g(t) = t^2$, and $k(r) = r^2$ all define the same function. This is true because if a is any number in the domain, then the same value a^2 is obtained regardless of which expression is employed.

In the remainder of our work, the phrase *f is a function* will mean that the domain and range are sets of real numbers. If a function is defined by means of an expression, as in Example 1, and the domain D is not stated, then we will consider D to be the totality of real numbers x such that $f(x)$ is real. This is sometimes called the **implied domain** of f. To illustrate, if $f(x) = \sqrt{x - 2}$, then the implied domain is the set of real numbers x such that $\sqrt{x - 2}$ is real—that is, $x - 2 \geq 0$, or $x \geq 2$. Thus, the domain is the infinite interval $[2, \infty)$. If x is in the domain, we say that f is *defined at x* or that $f(x)$ *exists.* If a set S is contained in the domain, *f is defined on S.* The terminology *f is un-defined at x* means that x is not in the domain of f.

EXAMPLE 2 **Finding function values**

Let $g(x) = \dfrac{\sqrt{4 + x}}{1 - x}$.

(a) Find the domain of g.

(b) Find $g(5)$, $g(-2)$, $g(-a)$, and $-g(a)$.

SOLUTION

(a) The expression $\sqrt{4 + x}/(1 - x)$ is a real number if and only if the radi-cand $4 + x$ is nonnegative and the denominator $1 - x$ is not equal to 0. Thus, $g(x)$ exists if and only if

$$4 + x \geq 0 \quad \text{and} \quad 1 - x \neq 0$$

or, equivalently,

$$x \geq -4 \quad \text{and} \quad x \neq 1.$$

We may express the domain in terms of intervals as $[-4, 1) \cup (1, \infty)$.

(b) To find values of g, we substitute for x:

$$g(5) = \frac{\sqrt{4 + 5}}{1 - 5} = \frac{\sqrt{9}}{-4} = -\frac{3}{4}$$

$$g(-2) = \frac{\sqrt{4 + (-2)}}{1 - (-2)} = \frac{\sqrt{2}}{3}$$

$$g(-a) = \frac{\sqrt{4 + (-a)}}{1 - (-a)} = \frac{\sqrt{4 - a}}{1 + a}$$

$$-g(a) = -\frac{\sqrt{4 + a}}{1 - a} = \frac{\sqrt{4 + a}}{a - 1}$$

Figure 3

MENU	
Hamburger	$1.69
French fries	$0.99
Soda	$0.79

Functions are commonplace in everyday life and show up in a variety of forms. For instance, the menu in a restaurant (Figure 3) can be considered to be a function f from a set of items to a set of prices. Note that f is given in a table format. Here $f(\text{Hamburger}) = 1.69$, $f(\text{French fries}) = 0.99$, and $f(\text{Soda}) = 0.79$.

An example of a function given by a rule can be found in the federal tax tables (Figure 4). Specifically, in 2003, for a single person with a taxable in-come of $120,000, the tax due was given by the rule

$14,010.00 plus 28% of the amount over $68,800.

Figure 4

2003 Tax Rate Schedules

Schedule X –Use if your Filing status is **single**

If the amount on Form 1040, line 40, is: Over–	But not over–	Enter on Form 1040, line 41	of the amount over–
$0	$7,000	- - - - - - - 10%	$0
7,000	28,400	$700.00 + 15%	7,000
28,400	68,800	$3,910.00 + 25%	28,400
68,800	143,500	14,010.00 + 28%	68,800
143,500	311,950	34,926.00 + 33%	143,500
311,950	- - - - - - -	90,514.50 + 35%	311,950

In this case, the tax would be

$$\$14,010.00 + 0.28(\$120,000 - \$68,800) = \$28,346.$$

Graphs are often used to describe the variation of physical quantities. For example, a scientist may use the graph in Figure 5 to indicate the temperature T of a certain solution at various times t during an experiment. The sketch shows that the temperature increased gradually for time $t = 0$ to time $t = 5$, did not change between $t = 5$ and $t = 8$, and then decreased rapidly from $t = 8$ to $t = 9$.

Similarly, if f is a function, we may use a graph to indicate the change in $f(x)$ as x varies through the domain of f. Specifically, we have the following definition.

Figure 5

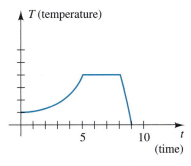

Definition of Graph of a Function	The **graph of a function** f is the graph of the equation $y = f(x)$ for x in the domain of f.

We often attach the label $y = f(x)$ to a sketch of the graph. If $P(a, b)$ is a point on the graph, then the y-coordinate b is the function value $f(a)$, as illustrated in Figure 6 on the next page. The figure displays the domain of f (the set of possible values of x) and the range of f (the corresponding values of y). Although we have pictured the domain and range as closed intervals, they may be infinite intervals or other sets of real numbers.

Since there is exactly one value $f(a)$ for each a in the domain of f, only *one* point on the graph of f has x-coordinate a. In general, we may use the following graphical test to determine whether a graph is the graph of a function.

Vertical Line Test	The graph of a set of points in a coordinate plane is the graph of a function if every vertical line intersects the graph in at most one point.

Figure 6

Thus, *every vertical line intersects the graph of a function in at most one point.* Consequently, the graph of a function cannot be a figure such as a circle, in which a vertical line may intersect the graph in more than one point.

The *x*-intercepts of the graph of a function *f* are the solutions of the equation $f(x) = 0$. These numbers are called the **zeros** of the function. The *y*-intercept of the graph is $f(0)$, if it exists.

EXAMPLE 3   Sketching the graph of a function

Let $f(x) = \sqrt{x - 1}$.

(a) Sketch the graph of *f*.

(b) Find the domain and range of *f*.

SOLUTION

(a) By definition, the graph of *f* is the graph of the equation $y = \sqrt{x - 1}$. The following table lists coordinates of several points on the graph.

x	1	2	3	4	5	6
$y = f(x)$	0	1	$\sqrt{2} \approx 1.4$	$\sqrt{3} \approx 1.7$	2	$\sqrt{5} \approx 2.2$

Figure 7

Range: [0, ∞)

$y = \sqrt{x - 1}$

Domain: [1, ∞)

Plotting points, we obtain the sketch shown in Figure 7. Note that the *x*-intercept is 1 and there is no *y*-intercept.

(b) Referring to Figure 7, note that the domain of *f* consists of all real numbers *x* such that $x \geq 1$ or, equivalently, the interval $[1, \infty)$. The range of *f* is the set of all real numbers *y* such that $y \geq 0$ or, equivalently, $[0, \infty)$.

The **square root function,** defined by $f(x) = \sqrt{x}$, has a graph similar to the one in Figure 7, but the endpoint is at (0, 0). The *y*-value of a point on this graph is the number displayed on a calculator when a square root is requested. This graphical relationship may help you remember that $\sqrt{9}$ is 3 and that $\sqrt{9}$ is *not* ±3. Similarly, $f(x) = x^2$, $f(x) = x^3$, and $f(x) = \sqrt[3]{x}$ are often referred to as the **squaring function,** the **cubing function,** and the **cube root function,** respectively.

In Example 3, as *x* increases, the function value $f(x)$ also increases, and we say that the graph of *f rises* (see Figure 7). A function of this type is said to be *increasing*. For certain functions, $f(x)$ decreases as *x* increases. In this case the graph *falls,* and *f* is a *decreasing* function. In general, we shall consider functions that increase or decrease on an interval *I,* as described in the following chart, where x_1 and x_2 denote numbers in *I.*

Increasing, Decreasing, and Constant Functions

Terminology	Definition	Graphical interpretation
f is **increasing** on an interval *I*	$f(x_1) < f(x_2)$ whenever $x_1 < x_2$	
f is **decreasing** on an interval *I*	$f(x_1) > f(x_2)$ whenever $x_1 < x_2$	
f is **constant** on an interval *I*	$f(x_1) = f(x_2)$ for every x_1 and x_2	

An example of an *increasing function* is the **identity function,** whose equation is $f(x) = x$ and whose graph is the line through the origin with slope 1. An example of a *decreasing function* is $f(x) = -x$, an equation of the line through the origin with slope -1. If $f(x) = c$ for every real number x, then *f* is called a *constant function.*

We shall use the phrases *f is increasing* and *f(x) is increasing* interchangeably. We shall do the same with the terms *decreasing* and *constant*.

EXAMPLE 4　**Using a graph to find domain, range, and where a function increases or decreases**

Let $f(x) = \sqrt{9 - x^2}$.

(a) Sketch the graph of *f*.

(b) Find the domain and range of *f*.

(c) Find the intervals on which *f* is increasing or is decreasing.

SOLUTION

(a) By definition, the graph of *f* is the graph of the equation $y = \sqrt{9 - x^2}$. We know from our work with circles in Section 3.2 that the graph of $x^2 + y^2 = 9$ is a circle of radius 3 with center at the origin. Solving the equation $x^2 + y^2 = 9$ for *y* gives us $y = \pm\sqrt{9 - x^2}$. It follows that the graph of *f* is the *upper half* of the circle, as illustrated in Figure 8.

(b) Referring to Figure 8, we see that the domain of *f* is the closed interval $[-3, 3]$, and the range of *f* is the interval $[0, 3]$.

(c) The graph rises as *x* increases from -3 to 0, so *f* is increasing on the closed interval $[-3, 0]$. Thus, as shown in the preceding chart, if $x_1 < x_2$ in $[-3, 0]$, then $f(x_1) < f(x_2)$ (note that *possibly* $x_1 = -3$ or $x_2 = 0$).

The graph falls as *x* increases from 0 to 3, so *f* is decreasing on the closed interval $[0, 3]$. In this case, the chart indicates that if $x_1 < x_2$ in $[0, 3]$, then $f(x_1) > f(x_2)$ (note that *possibly* $x_1 = 0$ or $x_2 = 3$).

Figure 8

Of special interest in calculus is a problem of the following type.

Problem: Find the slope of the secant line through the points *P* and *Q* shown in Figure 9.

Figure 9

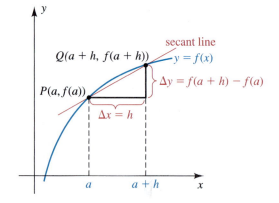

The slope m_{PQ} is given by

$$m_{PQ} = \frac{\Delta y}{\Delta x} = \frac{f(a + h) - f(a)}{h}.$$

The last expression (with $h \neq 0$) is commonly called a **difference quotient.** Let's take a look at the algebra involved in simplifying a difference quotient. (See Discussion Exercise 5 at the end of the chapter for a related problem.)

EXAMPLE 5   Simplifying a difference quotient

Simplify the difference quotient

$$\frac{f(x + h) - f(x)}{h}$$

using the function $f(x) = x^2 + 6x - 4$.

SOLUTION

$$\frac{f(x + h) - f(x)}{h} = \frac{[(x + h)^2 + 6(x + h) - 4] - [x^2 + 6x - 4]}{h}$$

definition of f

$$= \frac{(x^2 + 2xh + h^2 + 6x + 6h - 4) - (x^2 + 6x - 4)}{h}$$

expand numerator

$$= \frac{(x^2 + 2xh + h^2 + 6x + 6h - 4) - (x^2 + 6x - 4)}{h}$$

subtract terms

$$= \frac{2xh + h^2 + 6h}{h}$$

simplify

$$= \frac{h(2x + h + 6)}{h}$$

factor out h

$$= 2x + h + 6$$

cancel $h \neq 0$

The following type of function is one of the most basic in algebra.

Definition of Linear Function	A function f is a **linear function** if $$f(x) = ax + b,$$ where x is any real number and a and b are constants.

The graph of f in the preceding definition is the graph of $y = ax + b$, which, by the slope-intercept form, is a line with slope a and y-intercept b.

Thus, *the graph of a linear function is a line.* Since $f(x)$ exists for every x, the domain of f is $\mathbb{R}$. As illustrated in the next example, if $a \neq 0$, then the range of f is also $\mathbb{R}$.

EXAMPLE 6 **Sketching the graph of a linear function**

Let $f(x) = 2x + 3$.

(a) Sketch the graph of f.

(b) Find the domain and range of f.

(c) Determine where f is increasing or is decreasing.

SOLUTION

(a) Since $f(x)$ has the form $ax + b$, with $a = 2$ and $b = 3$, f is a linear function. The graph of $y = 2x + 3$ is the line with slope 2 and y-intercept 3, illustrated in Figure 10.

(b) We see from the graph that x and y may be any real numbers, so both the domain and the range of f are $\mathbb{R}$.

(c) Since the slope a is positive, the graph of f rises as x increases; that is, $f(x_1) < f(x_2)$ whenever $x_1 < x_2$. Thus, f is increasing throughout its domain. ◢

In applications it is sometimes necessary to determine a specific linear function from given data, as in the next example.

EXAMPLE 7 **Finding a linear function**

If f is a linear function such that $f(-2) = 5$ and $f(6) = 3$, find $f(x)$, where x is any real number.

SOLUTION By the definition of linear function, $f(x) = ax + b$, where a and b are constants. Moreover, the given function values tell us that the points $(-2, 5)$ and $(6, 3)$ are on the graph of f—that is, on the line $y = ax + b$ illustrated in Figure 11. The slope a of this line is

$$a = \frac{5 - 3}{-2 - 6} = \frac{2}{-8} = -\frac{1}{4},$$

and hence $f(x)$ has the form

$$f(x) = -\tfrac{1}{4}x + b.$$

To find the value of b, we may use the fact that $f(6) = 3$, as follows:

$$f(6) = -\tfrac{1}{4}(6) + b \quad \text{let } x = 6 \text{ in } f(x) = -\tfrac{1}{4}x + b$$
$$3 = -\tfrac{3}{2} + b \quad f(6) = 3$$
$$b = 3 + \tfrac{3}{2} = \tfrac{9}{2} \quad \text{solve for } b$$

Figure 10

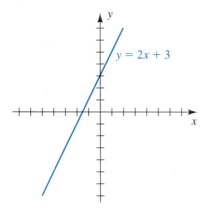

$y = 2x + 3$

Figure 11

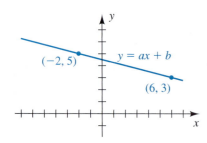

$(-2, 5)$

$y = ax + b$

$(6, 3)$

Thus, the linear function satisfying $f(-2) = 5$ and $f(6) = 3$ is

$$f(x) = -\tfrac{1}{4}x + \tfrac{9}{2}.$$

Many formulas that occur in mathematics and the sciences determine functions. For instance, the formula $A = \pi r^2$ for the area A of a circle of radius r assigns to each positive real number r exactly one value of A. This determines a function f such that $f(r) = \pi r^2$, and we may write $A = f(r)$. The letter r, which represents an arbitrary number from the domain of f, is called an **independent variable.** The letter A, which represents a number from the range of f, is a **dependent variable,** since its value depends on the number assigned to r. If two variables r and A are related in this manner, we say that A *is a function of r.* In applications, the independent variable and dependent variable are sometimes referred to as the **input variable** and **output variable,** respectively. As another example, if an automobile travels at a uniform rate of 50 mi/hr, then the distance d (miles) traveled in time t (hours) is given by $d = 50t$, and hence *the distance d is a function of time t.*

EXAMPLE 8 Expressing the volume of a tank as a function of its radius

A steel storage tank for propane gas is to be constructed in the shape of a right circular cylinder of altitude 10 feet with a hemisphere attached to each end. The radius r is yet to be determined. Express the volume V (in ft^3) of the tank as a function of r (in feet).

Figure 12

SOLUTION   The tank is illustrated in Figure 12. We may find the volume of the cylindrical part of the tank by multiplying the altitude 10 by the area πr^2 of the base of the cylinder. This gives us

$$\text{volume of cylinder} = 10(\pi r^2) = 10\pi r^2.$$

The two hemispherical ends, taken together, form a sphere of radius r. Using the formula for the volume of a sphere, we obtain

$$\text{volume of the two ends} = \tfrac{4}{3}\pi r^3.$$

Thus, the volume V of the tank is

$$V = \tfrac{4}{3}\pi r^3 + 10\pi r^2.$$

This formula expresses V as a function of r. In factored form,

$$V(r) = \tfrac{1}{3}\pi r^2(4r + 30) = \tfrac{2}{3}\pi r^2(2r + 15).$$

EXAMPLE 9 Expressing a distance as a function of time

Two ships leave port at the same time, one sailing west at a rate of 17 mi/hr and the other sailing south at 12 mi/hr. If t is the time (in hours) after their departure, express the distance d between the ships as a function of t.

Figure 13

SOLUTION   To help visualize the problem, we begin by drawing a picture and labeling it, as in Figure 13. By the Pythagorean theorem,

$$d^2 = a^2 + b^2, \quad \text{or} \quad d = \sqrt{a^2 + b^2}.$$

Since distance = (rate)(time) and the rates are 17 and 12, respectively,

$$a = 17t \quad \text{and} \quad b = 12t.$$

Substitution in $d = \sqrt{a^2 + b^2}$ gives us

$$d = \sqrt{(17t)^2 + (12t)^2} = \sqrt{289t^2 + 144t^2} = \sqrt{433t^2} \approx (20.8)t. \quad \text{◢}$$

Ordered pairs can be used to obtain an alternative approach to functions. We first observe that a function f from D to E determines the following set W of ordered pairs:

$$W = \{(x, f(x)): x \text{ is in } D\}$$

Thus, W consists of all ordered pairs such that the first number x is in D and the second number is the function value $f(x)$. In Example 1, where $f(x) = x^2$, W is the set of all ordered pairs of the form (x, x^2). It is important to note that, *for each x, there is exactly one ordered pair (x, y) in W having x in the first position.*

Conversely, if we begin with a set W of ordered pairs such that each x in D appears exactly once in the first position of an ordered pair, then W determines a function. Specifically, for each x in D there is exactly one pair (x, y) in W, and by letting y correspond to x, we obtain a function with domain D. The range consists of all real numbers y that appear in the second position of the ordered pairs.

It follows from the preceding discussion that the next statement could also be used as a definition of function.

Alternative Definition of Function	A **function** with domain D is a set W of ordered pairs such that, for each x in D, there is exactly one ordered pair (x, y) in W having x in the first position.

In terms of the preceding definition, the ordered pairs $\left(x, \sqrt{x-1}\right)$ determine the function of Example 3 given by $f(x) = \sqrt{x-1}$. Note, however, that if

$$W = \{(x, y): x^2 = y^2\},$$

then W is *not* a function, since for a given x there may be more than one pair in W with x in the first position. For example, if $x = 2$, then both $(2, 2)$ and $(2, -2)$ are in W.

As a reference aid, some common graphs and their equations are listed in Appendix I. Many of these graphs are graphs of functions.

3.4 *Exercises*

1 If $f(x) = -x^2 - x - 4$, find $f(-2)$, $f(0)$, and $f(4)$.

2 If $f(x) = -x^3 - x^2 + 3$, find $f(-3)$, $f(0)$, and $f(2)$.

3 If $f(x) = \sqrt{x - 4} - 3x$, find $f(4)$, $f(8)$, and $f(13)$.

4 If $f(x) = \dfrac{x}{x - 3}$, find $f(-2)$, $f(0)$, and $f(3)$.

Exer. 5–10: If a and h are real numbers, find

(a) $f(a)$ (b) $f(-a)$ (c) $-f(a)$ (d) $f(a + h)$

(e) $f(a) + f(h)$ (f) $\dfrac{f(a + h) - f(a)}{h}$, if $h \neq 0$

5 $f(x) = 5x - 2$ **6** $f(x) = 3 - 4x$

7 $f(x) = -x^2 + 4$ **8** $f(x) = 3 - x^2$

9 $f(x) = x^2 - x + 3$ **10** $f(x) = 2x^2 + 3x - 7$

Exer. 11–14: If a is a positive real number, find

(a) $g\left(\dfrac{1}{a}\right)$ (b) $\dfrac{1}{g(a)}$ (c) $g(\sqrt{a})$ (d) $\sqrt{g(a)}$

11 $g(x) = 4x^2$ **12** $g(x) = 2x - 5$

13 $g(x) = \dfrac{2x}{x^2 + 1}$ **14** $g(x) = \dfrac{x^2}{x + 1}$

Exer. 15–16: Explain why the graph is or is not the graph of a function.

15

16

Exer. 17–18: Determine the domain D and range R of the function shown in the figure.

17

18

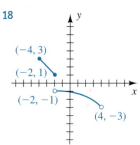

Exer. 19–20: For the graph of the function f sketched in the figure, determine

(a) the domain   (b) the range   (c) $f(1)$

(d) all x such that $f(x) = 1$

(e) all x such that $f(x) > 1$

19

20

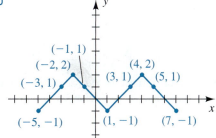

Exer. 21–32: Find the domain of f.

21 $f(x) = \sqrt{2x + 7}$ **22** $f(x) = \sqrt{8 - 3x}$

23 $f(x) = \sqrt{9 - x^2}$ **24** $f(x) = \sqrt{x^2 - 25}$

25 $f(x) = \dfrac{x + 1}{x^3 - 4x}$ **26** $f(x) = \dfrac{4x}{6x^2 + 13x - 5}$

27 $f(x) = \dfrac{\sqrt{2x - 3}}{x^2 - 5x + 4}$

28 $f(x) = \dfrac{\sqrt{4x - 3}}{x^2 - 4}$

29 $f(x) = \dfrac{x - 4}{\sqrt{x - 2}}$

30 $f(x) = \dfrac{1}{(x - 3)\sqrt{x + 3}}$

31 $f(x) = \sqrt{x + 2} + \sqrt{2 - x}$

32 $f(x) = \sqrt{(x - 2)(x - 6)}$

Exer. 33–34: (a) Find the domain D and range R of f. (b) Find the intervals on which f is increasing, is decreasing, or is constant.

33

34

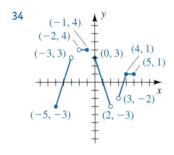

35 Sketch the graph of a function that is increasing on $(-\infty, -3]$ and $[2, \infty)$ and is decreasing on $[-3, 2]$.

36 Sketch the graph of a function that is decreasing on $(-\infty, -2]$ and $[1, 4]$ and is increasing on $[-2, 1]$ and $[4, \infty)$.

Exer. 37–46: (a) Sketch the graph of f. (b) Find the domain D and range R of f. (c) Find the intervals on which f is increasing, is decreasing, or is constant.

37 $f(x) = 3x - 2$

38 $f(x) = -2x + 3$

39 $f(x) = 4 - x^2$

40 $f(x) = x^2 - 1$

41 $f(x) = \sqrt{x + 4}$

42 $f(x) = \sqrt{4 - x}$

43 $f(x) = -2$

44 $f(x) = 3$

45 $f(x) = -\sqrt{36 - x^2}$

46 $f(x) = \sqrt{16 - x^2}$

Exer. 47–48: Simplify the difference quotient
$$\dfrac{f(2 + h) - f(2)}{h} \text{ if } h \neq 0.$$

47 $f(x) = x^2 - 3x$

48 $f(x) = -2x^2 + 3$

Exer. 49–50: Simplify the difference quotient
$$\dfrac{f(x + h) - f(x)}{h} \text{ if } h \neq 0.$$

49 $f(x) = x^2 + 5$

50 $f(x) = 1/x^2$

Exer. 51–52: Simplify the difference quotient $\dfrac{f(x) - f(a)}{x - a}$ if $x \neq a$.

51 $f(x) = \sqrt{x - 3}$ (*Hint:* Rationalize the numerator.)

52 $f(x) = x^3 - 2$

Exer. 53–54: If a linear function f satisfies the given conditions, find $f(x)$.

53 $f(-3) = 1$ and $f(3) = 2$

54 $f(-2) = 7$ and $f(4) = -2$

Exer. 55–64: Determine whether the set W of ordered pairs is a function in the sense of the alternative definition of function on page 166.

55 $W = \{(x, y): 2y = x^2 + 5\}$

56 $W = \{(x, y): x = 3y + 2\}$

57 $W = \{(x, y): x^2 + y^2 = 4\}$

58 $W = \{(x, y): y^2 - x^2 = 1\}$

59 $W = \{(x, y): y = 3\}$

60 $W = \{(x, y): x = 3\}$

61 $W = \{(x, y): xy = 0\}$

62 $W = \{(x, y): x + y = 0\}$

63 $W = \{(x, y): |y| = |x|\}$

64 $W = \{(x, y): y < x\}$

65 **Constructing a box** From a rectangular piece of cardboard having dimensions 20 inches × 30 inches, an open box is to be made by cutting out an identical square of area x^2 from each corner and turning up the sides (see the figure). Express the volume V of the box as a function of x.

Exercise 65

66 Constructing a storage tank Refer to Example 8. A steel storage tank for propane gas is to be constructed in the shape of a right circular cylinder of altitude 10 feet with a hemisphere attached to each end. The radius r is yet to be determined. Express the surface area S of the tank as a function of r.

67 Dimensions of a building A small office unit is to contain 500 ft^2 of floor space. A simplified model is shown in the figure.

(a) Express the length y of the building as a function of the width x.

(b) If the walls cost $100 per running foot, express the cost C of the walls as a function of the width x. (Disregard the wall space above the doors and the thickness of the walls.)

Exercise 67

68 Dimensions of an aquarium An aquarium of height 1.5 feet is to have a volume of 6 ft^3. Let x denote the length of the base and y the width (see the figure).

(a) Express y as a function of x.

(b) Express the total number S of square feet of glass needed as a function of x.

Exercise 68

69 Skyline ordinance A city council is proposing a new skyline ordinance. It would require the setback S for any building from a residence to be a minimum of 100 feet, plus an additional 6 feet for each foot of height above 25 feet. Find a linear function for S in terms of h.

Exercise 69

70 Energy tax A proposed energy tax T on gasoline, which would affect the cost of driving a vehicle, is to be computed by multiplying the number x of gallons of gasoline that you buy by 125,000 (the number of BTUs per gallon of gasoline) and then multiplying the total BTUs by the tax—34.2 cents per million BTUs. Find a linear function for T in terms of x.

71 Childhood growth For children between ages 6 and 10, height y (in inches) is frequently a linear function of age t (in years). The height of a certain child is 48 inches at age 6 and 50.5 inches at age 7.

(a) Express y as a function of t.

(b) Sketch the line in part (a), and interpret the slope.

(c) Predict the height of the child at age 10.

72 Radioactive contamination It has been estimated that 1000 curies of a radioactive substance introduced at a point on the surface of the open sea would spread over an area of 40,000 km^2 in 40 days. Assuming that the area covered by the radioactive substance is a linear function of time t and is always circular in shape, express the radius r of the contamination as a function of t.

73 Distance to a hot-air balloon A hot-air balloon is released at 1:00 P.M. and rises vertically at a rate of 2 m/sec. An observation point is situated 100 meters from a point on the ground directly below the balloon (see the figure). If t denotes the time (in seconds) after 1:00 P.M., express the distance d between the balloon and the observation point as a function of t.

Exercise 73

74 Triangle ABC is inscribed in a semicircle of diameter 15 (see the figure).

(a) If x denotes the length of side AC, express the length y of side BC as a function of x. (*Hint:* Angle ACB is a right angle.)

(b) Express the area $\mathcal{A}$ of triangle ABC as a function of x, and state the domain of this function.

Exercise 74

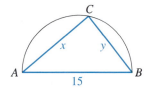

75 Distance to the earth From an exterior point P that is h units from a circle of radius r, a tangent line is drawn to the circle (see the figure). Let y denote the distance from the point P to the point of tangency T.

(a) Express y as a function of h. (*Hint:* If C is the center of the circle, then PT is perpendicular to CT.)

(b) If r is the radius of the earth and h is the altitude of a space shuttle, then y is the maximum distance to the earth that an astronaut can see from the shuttle. In particular, if $h = 200$ mi and $r \approx 4000$ mi, approximate y.

Exercise 75

76 Length of a tightrope The figure illustrates the apparatus for a tightrope walker. Two poles are set 50 feet apart, but the point of attachment P for the rope is yet to be determined.

(a) Express the length L of the rope as a function of the distance x from P to the ground.

(b) If the total walk is to be 75 feet, determine the distance from P to the ground.

Exercise 76

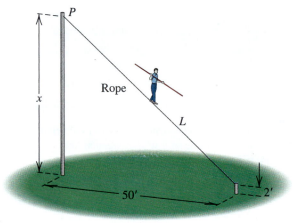

77 Airport runway The relative positions of an aircraft runway and a 20-foot-tall control tower are shown in the figure. The beginning of the runway is at a perpendicular distance of 300 feet from the base of the tower. If x denotes the distance an airplane has moved down the runway, express the distance d between the airplane and the top of the control tower as a function of x.

Exercise 77

78 Destination time A man in a rowboat that is 2 miles from the nearest point A on a straight shoreline wishes to reach a house located at a point B that is 6 miles farther down the shoreline (see the figure). He plans to row to a point P that is between A and B and is x miles from the house, and then he will walk the remainder of the distance. Suppose he can row at a rate of 3 mi/hr and can walk at a rate of 5 mi/hr. If T is the total time required to reach the house, express T as a function of x.

Exercise 78

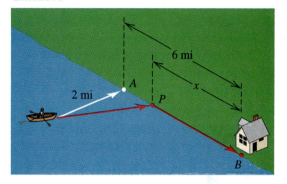

3.5

Graphs of Functions

In this section we discuss aids for sketching graphs of certain types of functions. In particular, a function f is called **even** if $f(-x) = f(x)$ for every x in its domain. In this case, the equation $y = f(x)$ is not changed if $-x$ is substituted for x, and hence, from symmetry test 1 of Section 3.2, the graph of an even function is symmetric with respect to the y-axis.

A function f is called **odd** if $f(-x) = -f(x)$ for every x in its domain. If we apply symmetry test 3 of Section 3.2 to the equation $y = f(x)$, we see that the graph of an odd function is symmetric with respect to the origin.

These facts are summarized in the first two columns of the next chart.

Even and Odd Functions

Terminology	Definition	Illustration	Type of symmetry of graph
f is an *even* **function.**	$f(-x) = f(x)$ for every x in the domain.	$y = f(x) = x^2$	with respect to the y-axis
f is an *odd* **function.**	$f(-x) = -f(x)$ for every x in the domain.	$y = f(x) = x^3$	with respect to the origin

EXAMPLE 1 Determining whether a function is even or odd

Determine whether f is even, odd, or neither even nor odd.

(a) $f(x) = 3x^4 - 2x^2 + 5$ (b) $f(x) = 2x^5 - 7x^3 + 4x$

(c) $f(x) = x^3 + x^2$

SOLUTION In each case the domain of f is $\mathbb{R}$. To determine whether f is even or odd, we begin by examining $f(-x)$, where x is any real number.

(a) $f(-x) = 3(-x)^4 - 2(-x)^2 + 5$ substitute $-x$ for x in $f(x)$

$\qquad\quad = 3x^4 - 2x^2 + 5$ simplify

$\qquad\quad = f(x)$ definition of f

Since $f(-x) = f(x)$, f is an even function.

(b) $f(-x) = 2(-x)^5 - 7(-x)^3 + 4(-x)$ substitute $-x$ for x in $f(x)$

$\qquad\quad = -2x^5 + 7x^3 - 4x$ simplify

$\qquad\quad = -(2x^5 - 7x^3 + 4x)$ factor out -1

$\qquad\quad = -f(x)$ definition of f

Since $f(-x) = -f(x)$, f is an odd function.

(c) $f(-x) = (-x)^3 + (-x)^2$ substitute $-x$ for x in $f(x)$

$\qquad\quad = -x^3 + x^2$ simplify

Since $f(-x) \neq f(x)$, and $f(-x) \neq -f(x)$ (note that $-f(x) = -x^3 - x^2$), the function f is neither even nor odd.

In the next example we consider the **absolute value function** f, defined by $f(x) = |x|$.

EXAMPLE 2 Sketching the graph of the absolute value function

Let $f(x) = |x|$.

(a) Determine whether f is even or odd.

(b) Sketch the graph of f.

(c) Find the intervals on which f is increasing or is decreasing.

Figure 1

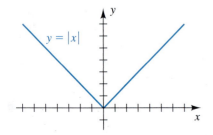

$y = |x|$

SOLUTION

(a) The domain of f is $\mathbb{R}$, because the absolute value of x exists for every real number x. If x is in $\mathbb{R}$, then

$$f(-x) = |-x| = |x| = f(x).$$

Thus, f is an even function, since $f(-x) = f(x)$.

(b) Since f is even, its graph is symmetric with respect to the y-axis. If $x \geq 0$, then $|x| = x$, and therefore the first quadrant part of the graph coincides with the line $y = x$. Sketching this half-line and using symmetry gives us Figure 1.

(c) Referring to the graph, we see that f is decreasing on $(-\infty, 0]$ and is increasing on $[0, \infty)$.

If we know the graph of $y = f(x)$, it is easy to sketch the graphs of

$$y = f(x) + c \quad \text{and} \quad y = f(x) - c$$

for any positive real number c. As in the next chart, for $y = f(x) + c$, we add c to the y-coordinate of each point on the graph of $y = f(x)$. This *shifts* the graph of f *upward* a distance c. For $y = f(x) - c$ with $c > 0$, we subtract c from each y-coordinate, thereby shifting the graph of f a distance c *downward*. These are called **vertical shifts** of graphs.

Vertically Shifting the Graph of $y = f(x)$

Equation	$y = f(x) + c$ with $c > 0$	$y = f(x) - c$ with $c > 0$
Effect on graph	The graph of f is shifted vertically upward a distance c.	The graph of f is shifted vertically downward a distance c.
Graphical interpretation		

EXAMPLE 3 Vertically shifting a graph

Sketch the graph of f:

(a) $f(x) = x^2$ **(b)** $f(x) = x^2 + 4$ **(c)** $f(x) = x^2 - 4$

SOLUTION We shall sketch all graphs on the same coordinate plane.

(a) Since

$$f(-x) = (-x)^2 = x^2 = f(x),$$

the function f is even, and hence its graph is symmetric with respect to the y-axis. Several points on the graph of $y = x^2$ are $(0, 0)$, $(1, 1)$, $(2, 4)$, and $(3, 9)$. Drawing a smooth curve through these points and reflecting through the y-axis gives us the sketch in Figure 2. The graph is a parabola with vertex at the origin and opening upward.

(continued)

Figure 2

Figure 2 (repeated)

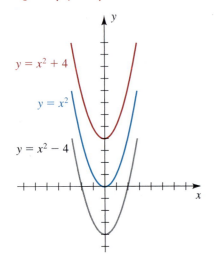

(b) To sketch the graph of $y = x^2 + 4$, we add 4 to the y-coordinate of each point on the graph of $y = x^2$; that is, we shift the graph in part (a) upward 4 units, as shown in the figure.

(c) To sketch the graph of $y = x^2 - 4$, we decrease the y-coordinates of $y = x^2$ by 4; that is, we shift the graph in part (a) downward 4 units.

We can also consider **horizontal shifts** of graphs. Specifically, if $c > 0$, consider the graphs of $y = f(x)$ and $y = g(x) = f(x - c)$ sketched on the same coordinate plane, as illustrated in the next chart. Since

$$g(a + c) = f([a + c] - c) = f(a),$$

we see that the point with x-coordinate a on the graph of $y = f(x)$ has the same y-coordinate as the point with x-coordinate $a + c$ on the graph of $y = g(x) = f(x - c)$. This implies that the graph of $y = g(x) = f(x - c)$ can be obtained by shifting the graph of $y = f(x)$ *to the right* a distance c. Similarly, the graph of $y = h(x) = f(x + c)$ can be obtained by shifting the graph of f *to the left* a distance c, as shown in the chart.

Horizontally Shifting the Graph of $y = f(x)$

Equation	Effect on graph	Graphical interpretation
$y = g(x)$ $= f(x - c)$ with $c > 0$	The graph of f is shifted horizontally to the *right* a distance c.	*graph showing $y = f(x)$ with point (a, b) and $f(a)$, and $y = g(x) = f(x - c)$ with point $(a + c, b)$ and $g(a + c)$; interval $c > 0$ from a to $a + c$*
$y = h(x)$ $= f(x + c)$ with $c > 0$	The graph of f is shifted horizontally to the *left* a distance c.	*graph showing $y = h(x) = f(x + c)$ with point $(a - c, b)$ and $h(a - c)$, and $y = f(x)$ with point (a, b) and $f(a)$; interval $c > 0$ from $a - c$ to a*

Horizontal and vertical shifts are also referred to as *translations*.

Figure 3

EXAMPLE 4   Horizontally shifting a graph

Sketch the graph of f:

(a) $f(x) = (x - 4)^2$ **(b)** $f(x) = (x + 2)^2$

SOLUTION   The graph of $y = x^2$ is sketched in Figure 3.

(a) Shifting the graph of $y = x^2$ to the right 4 units gives us the graph of $y = (x - 4)^2$, shown in the figure.

(b) Shifting the graph of $y = x^2$ to the left 2 units leads to the graph of $y = (x + 2)^2$, shown in the figure.

To obtain the graph of $y = cf(x)$ for some real number c, we may *multiply* the y-coordinates of points on the graph of $y = f(x)$ by c. For example, if $y = 2f(x)$, we double the y-coordinates; or if $y = \frac{1}{2}f(x)$, we multiply each y-coordinate by $\frac{1}{2}$. This procedure is referred to as **vertically stretching** the graph of f (if $c > 1$) or **vertically compressing** the graph (if $0 < c < 1$) and is summarized in the following chart.

Vertically Stretching or Compressing the Graph of $y = f(x)$

Equation	$y = cf(x)$ with $c > 1$	$y = cf(x)$ with $0 < c < 1$
Effect on graph	The graph of f is stretched vertically by a factor c.	The graph of f is compressed vertically by a factor $1/c$.
Graphical interpretation		

EXAMPLE 5 **Vertically stretching or compressing a graph**

Sketch the graph of the equation:

(a) $y = 4x^2$ **(b)** $y = \frac{1}{4}x^2$

Figure 4

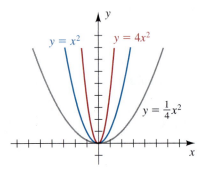

SOLUTION

(a) To sketch the graph of $y = 4x^2$, we may refer to the graph of $y = x^2$ in Figure 4 and multiply the y-coordinate of each point by 4. This stretches the graph of $y = x^2$ vertically by a factor 4 and gives us a narrower parabola that is sharper at the vertex, as illustrated in the figure.

(b) The graph of $y = \frac{1}{4}x^2$ may be sketched by multiplying the y-coordinates of points on the graph of $y = x^2$ by $\frac{1}{4}$. This compresses the graph of $y = x^2$ vertically by a factor $1/\frac{1}{4} = 4$ and gives us a wider parabola that is flatter at the vertex, as shown in Figure 4. ◢

Replacing y with $-y$ reflects the graph of $y = f(x)$ through the x-axis.

We may obtain the graph of $y = -f(x)$ by multiplying the y-coordinate of each point on the graph of $y = f(x)$ by -1. Thus, every point (a, b) on the graph of $y = f(x)$ that lies above the x-axis determines a point $(a, -b)$ on the graph of $y = -f(x)$ that lies below the x-axis. Similarly, if (c, d) lies below the x-axis (that is, $d < 0$), then $(c, -d)$ lies above the x-axis. The graph of $y = -f(x)$ is a **reflection** of the graph of $y = f(x)$ through the x-axis.

Figure 5

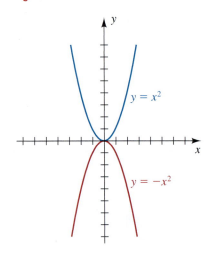

EXAMPLE 6 **Reflecting a graph through the x-axis**

Sketch the graph of $y = -x^2$.

SOLUTION The graph may be found by plotting points; however, since the graph of $y = x^2$ is familiar to us, we sketch it as in Figure 5 and then multiply the y-coordinates of points by -1. This procedure gives us the reflection through the x-axis indicated in the figure. ◢

Sometimes it is useful to compare the graphs of $y = f(x)$ and $y = f(cx)$ if $c \neq 0$. In this case the function values $f(x)$ for

$$a \leq x \leq b$$

are the same as the function values $f(cx)$ for

$$a \leq cx \leq b \qquad \text{or, equivalently,} \qquad \frac{a}{c} \leq x \leq \frac{b}{c}.$$

This implies that the graph of f is **horizontally compressed** (if $c > 1$) or **horizontally stretched** (if $0 < c < 1$), as summarized in the following chart.

Horizontally Compressing or Stretching the Graph of $y = f(x)$

Equation	Effect on graph	Graphical interpretation
$y = f(cx)$ with $c > 1$	The graph of f is compressed horizontally by a factor c.	$y = f(x)$ $y = f(cx)$ with $c > 1$ $\left(\frac{a}{c}, b\right)$ (a, b)
$y = f(cx)$ with $0 < c < 1$	The graph of f is stretched horizontally by a factor $1/c$.	$y = f(x)$ $y = f(cx)$ with $0 < c < 1$ (a, b) $\left(\frac{a}{c}, b\right)$

Replacing x with −x reflects the graph of y = f(x) through the y-axis.

If $c < 0$, then the graph of $y = f(cx)$ may be obtained by reflecting the graph of $y = f(|c|x)$ through the y-axis. For example, to sketch the graph of $y = f(-2x)$, we reflect the graph of $y = f(2x)$ through the y-axis. As a special case, the graph of $y = f(-x)$ is a **reflection** of the graph of $y = f(x)$ through the y-axis.

Functions are sometimes described by more than one expression, as in the next examples. We call such functions **piecewise-defined functions.**

Figure 6

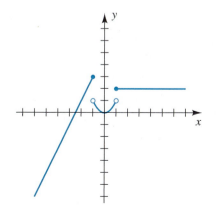

EXAMPLE 7   Sketching the graph of a piecewise-defined function

Sketch the graph of the function f if

$$f(x) = \begin{cases} 2x + 5 & \text{if } x \leq -1 \\ x^2 & \text{if } |x| < 1 \\ 2 & \text{if } x \geq 1 \end{cases}$$

SOLUTION   If $x \leq -1$, then $f(x) = 2x + 5$ and the graph of f coincides with the line $y = 2x + 5$ and is represented by the portion of the graph to the left of the line $x = -1$ in Figure 6. The small dot indicates that the point $(-1, 3)$ is on the graph.

(continued)

Figure 6 (repeated)

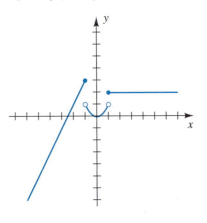

If $|x| < 1$ (or, equivalently, $-1 < x < 1$), we use x^2 to find values of f, and therefore this part of the graph of f coincides with the parabola $y = x^2$, as indicated in the figure. Note that the points $(-1, 1)$ and $(1, 1)$ are *not* on the graph.

Finally, if $x \geq 1$, the values of f are always 2. Thus, the graph of f for $x \geq 1$ is the horizontal half-line in Figure 6.

Note: When you finish sketching the graph of a piecewise-defined function, check that it passes the vertical line test.

It is a common misconception to think that if you move up to a higher tax bracket, *all* your income is taxed at the higher rate. The following example of a graph of a piecewise-defined function helps dispell that notion.

EXAMPLE 8 **Application using a piecewise-defined function**

Sketch a graph of the 2003 Tax Rate Schedule X, shown in Figure 7. Let x represent the taxable income and T represent the amount of tax. (Assume the domain is the set of nonnegative real numbers.)

Figure 7

2003 Tax Rate Schedules

Schedule X –Use if your Filing status is **single**

If the amount on Form 1040, line 40, is: Over–	But not over–	Enter on Form 1040, line 41	of the amount over–
$0	$7,000	-------- 10%	$0
7,000	28,400	**$700.00 + 15%**	7,000
28,400	68,800	**$3,910.00 + 25%**	28,400
68,800	143,500	14,010.00 + 28%	68,800
143,500	311,950	34,926.00 + 33%	143,500
311,950	-------	90,514.50 + 35%	311,950

SOLUTION The tax table can be represented by a piecewise-defined function as follows:

$$T(x) = \begin{cases} 0 & \text{if} & x \leq 0 \\ 0.10x & \text{if} & 0 < x \leq 7000 \\ 700.00 + 0.15(x - 7000) & \text{if} & 7000 < x \leq 28,400 \\ 3910.00 + 0.25(x - 28,400) & \text{if} & 28,400 < x \leq 68,800 \\ 14,010.00 + 0.28(x - 68,800) & \text{if} & 68,800 < x \leq 143,500 \\ 34,926.00 + 0.33(x - 143,500) & \text{if} & 143,500 < x \leq 311,950 \\ 90,514.50 + 0.35(x - 311,950) & \text{if} & x > 311,950 \end{cases}$$

Note that the assignment for the 15% tax bracket is *not* $0.15x$, but 10% of the first $7000 in taxable income plus 15% of the amount *over* $7000; that is,

$$0.10(7000) + 0.15(x - 7000) = 700.00 + 0.15(x - 7000).$$

The other pieces can be established in a similar fashion. The graph of T is shown in Figure 8; note that the slope of each piece represents the tax rate.

Figure 8

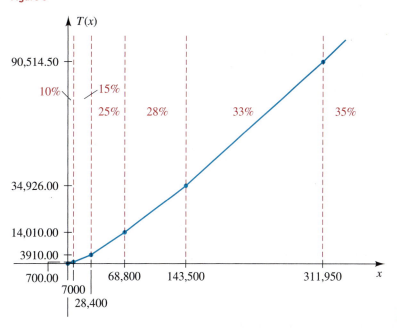

If x is a real number, we define the symbol $[\![x]\!]$ as follows:

$$[\![x]\!] = n, \quad \text{where } n \text{ is the greatest integer such that } n \leq x$$

If we identify $\mathbb{R}$ with points on a coordinate line, then n is the first integer to the *left* of (or *equal* to) x.

ILLUSTRATION   **The Symbol $[\![x]\!]$**

- $[\![0.5]\!] = 0$ ▪ $[\![1.8]\!] = 1$ ▪ $[\![\sqrt{5}]\!] = 2$
- $[\![3]\!] = 3$ ▪ $[\![-3]\!] = -3$ ▪ $[\![-2.7]\!] = -3$
- $[\![-\sqrt{3}]\!] = -2$ ▪ $[\![-0.5]\!] = -1$

The **greatest integer function** f is defined by $f(x) = [\![x]\!]$.

EXAMPLE 9   **Sketching the graph of the greatest integer function**

Sketch the graph of the greatest integer function.

Figure 9

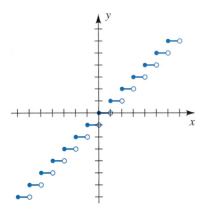

SOLUTION The x- and y-coordinates of some points on the graph may be listed as follows:

Values of x	$f(x) = [\![x]\!]$
.	.
.	.
.	.
$-2 \leq x < -1$	-2
$-1 \leq x < 0$	-1
$0 \leq x < 1$	0
$1 \leq x < 2$	1
$2 \leq x < 3$	2
.	.
.	.
.	.

Whenever x is between successive integers, the corresponding part of the graph is a segment of a horizontal line. Part of the graph is sketched in Figure 9. The graph continues indefinitely to the right and to the left.

The next example involves absolute values.

Figure 10

(a)

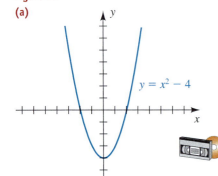

$y = x^2 - 4$

EXAMPLE 10 **Sketching the graph of an equation containing an absolute value**

Sketch the graph of $y = |x^2 - 4|$.

SOLUTION The graph of $y = x^2 - 4$ was sketched in Figure 2 and is resketched in Figure 10(a). We note the following facts:

(1) If $x \leq -2$ or $x \geq 2$, then $x^2 - 4 \geq 0$, and hence $|x^2 - 4| = x^2 - 4$.

(2) If $-2 < x < 2$, then $x^2 - 4 < 0$, and hence $|x^2 - 4| = -(x^2 - 4)$.

It follows from (1) that the graphs of $y = |x^2 - 4|$ and $y = x^2 - 4$ coincide for $|x| \geq 2$. We see from (2) that if $|x| < 2$, then the graph of $y = |x^2 - 4|$ is the reflection of the graph of $y = x^2 - 4$ through the x-axis. This gives us the sketch in Figure 10(b).

(b)

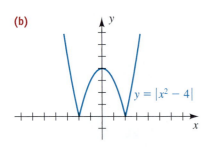

$y = |x^2 - 4|$

Graphing $y = f(|x|)$

In general, if the graph of $y = f(x)$ contains a point $P(c, -d)$ with d positive, then the graph of $y = |f(x)|$ contains the point $Q(c, d)$—that is, Q is the reflection of P through the x-axis. Points with nonnegative y-values are the same for the graphs of $y = f(x)$ and $y = |f(x)|$.

Later in this text and in calculus, you will encounter functions such as

$$g(x) = \ln |x| \qquad \text{and} \qquad h(x) = \sin |x|.$$

Both functions are of the form $y = f(|x|)$. The effect of substituting $|x|$ for x can be described as follows: If the graph of $y = f(x)$ contains a point $P(c, d)$

with c positive, then the graph of $y = f(|x|)$ contains the point $Q(-c, d)$—that is, Q is the reflection of P through the y-axis. Points on the y-axis ($x = 0$) are the same for the graphs of $y = f(x)$ and $y = f(|x|)$. Points with negative x-values on the graph of $y = f(x)$ are not on the graph of $y = f(|x|)$, since the result of the absolute value is always nonnegative.

The processes of shifting, stretching, compressing, and reflecting a graph may be collectively termed *transforming* a graph, and the resulting graph is called a **transformation** of the original graph. A graphical summary of the types of transformations encountered in this section appears in Appendix II.

3.5   Exercises

Exer. 1–2: Suppose f is an even function and g is an odd function. Complete the table, if possible.

1

x	-2	2
$f(x)$		7
$g(x)$		-6

2

x	-3	3
$f(x)$		-5
$g(x)$		15

Exer. 3–12: Determine whether f is even, odd, or neither even nor odd.

3 $f(x) = 5x^3 + 2x$

4 $f(x) = |x| - 3$

5 $f(x) = 3x^4 + 2x^2 - 5$

6 $f(x) = 7x^5 - 4x^3$

7 $f(x) = 8x^3 - 3x^2$

8 $f(x) = 12$

9 $f(x) = \sqrt{x^2 + 4}$

10 $f(x) = 3x^2 - 5x + 1$

11 $f(x) = \sqrt[3]{x^3 - x}$

12 $f(x) = x^3 - \dfrac{1}{x}$

Exer. 13–26: Sketch, on the same coordinate plane, the graphs of f for the given values of c. (Make use of symmetry, shifting, stretching, compressing, or reflecting.)

13 $f(x) = |x| + c$; $c = -3, 1, 3$

14 $f(x) = |x - c|$; $c = -3, 1, 3$

15 $f(x) = -x^2 + c$; $c = -4, 2, 4$

16 $f(x) = 2x^2 - c$; $c = -4, 2, 4$

17 $f(x) = 2\sqrt{x} + c$; $c = -3, 0, 2$

18 $f(x) = \sqrt{9 - x^2} + c$; $c = -3, 0, 2$

19 $f(x) = \frac{1}{2}\sqrt{x - c}$; $c = -2, 0, 3$

20 $f(x) = -\frac{1}{2}(x - c)^2$; $c = -2, 0, 3$

21 $f(x) = c\sqrt{4 - x^2}$; $c = -2, 1, 3$

22 $f(x) = (x + c)^3$; $c = -2, 1, 2$

23 $f(x) = cx^3$; $c = -\frac{1}{3}, 1, 2$

24 $f(x) = (cx)^3 + 1$; $c = -1, 1, 4$

25 $f(x) = \sqrt{cx} - 1$; $c = -1, \frac{1}{9}, 4$

26 $f(x) = -\sqrt{16 - (cx)^2}$; $c = 1, \frac{1}{2}, 4$

Exer. 27–32: If the point P is on the graph of a function f, find the corresponding point on the graph of the given function.

27 $P(0, 5)$; $y = f(x + 2) - 1$

28 $P(3, -1)$; $y = 2f(x) + 4$

29 $P(3, -2)$; $y = 2f(x - 4) + 1$

30 $P(-2, 4)$; $y = \frac{1}{2}f(x - 3) + 3$

31 $P(3, 9)$; $y = \frac{1}{3}f(\frac{1}{2}x) - 1$

32 $P(-2, 1)$; $y = -3f(2x) - 5$

Exer. 33–40: Explain how the graph of the function compares to the graph of $y = f(x)$. For example, for the equation $y = 2f(x + 3)$, the graph of f is shifted 3 units to the left and stretched vertically by a factor of 2.

33 $y = f(x - 2) + 3$

34 $y = 3f(x - 1)$

35 $y = f(-x) - 2$

36 $y = -f(x + 4)$

37 $y = -\frac{1}{2} f(x)$

38 $y = f\left(\frac{1}{2}x\right) - 3$

39 $y = -2f\left(\frac{1}{3}x\right)$

40 $y = \frac{1}{3}| f(x) |$

Exer. 41–42: The graph of a function f with domain $[0, 4]$ is shown in the figure. Sketch the graph of the given equation.

41

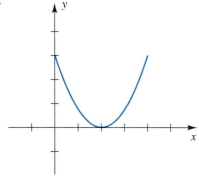

(a) $y = f(x + 3)$ (b) $y = f(x - 3)$

(c) $y = f(x) + 3$ (d) $y = f(x) - 3$

(e) $y = -3f(x)$ (f) $y = -\frac{1}{3} f(x)$

(g) $y = f\left(-\frac{1}{2}x\right)$ (h) $y = f(2x)$

(i) $y = -f(x + 2) - 3$ (j) $y = f(x - 2) + 3$

(k) $y = | f(x) |$ (l) $y = f(| x |)$

42

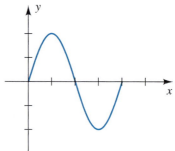

(a) $y = f(x - 2)$ (b) $y = f(x + 2)$

(c) $y = f(x) - 2$ (d) $y = f(x) + 2$

(e) $y = -2f(x)$ (f) $y = -\frac{1}{2} f(x)$

(g) $y = f(-2x)$ (h) $y = f\left(\frac{1}{2}x\right)$

(i) $y = -f(x + 4) - 2$ (j) $y = f(x - 4) + 2$

(k) $y = | f(x) |$ (l) $y = f(| x |)$

Exer. 43–46: The graph of a function f is shown, together with graphs of three other functions (a), (b), and (c). Use properties of symmetry, shifts, and reflecting to find equations for graphs (a), (b), and (c) in terms of f.

43

44

45

46

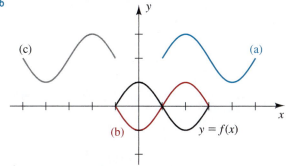

Exer. 47–52: Sketch the graph of f.

47 $f(x) = \begin{cases} 3 & \text{if } x \le -1 \\ -2 & \text{if } x > -1 \end{cases}$

48 $f(x) = \begin{cases} -1 & \text{if } x \text{ is an integer} \\ -2 & \text{if } x \text{ is not an integer} \end{cases}$

49 $f(x) = \begin{cases} 3 & \text{if } x < -2 \\ -x + 1 & \text{if } |x| \le 2 \\ -3 & \text{if } x > 2 \end{cases}$

50 $f(x) = \begin{cases} -2x & \text{if } x < -1 \\ x^2 & \text{if } -1 \le x < 1 \\ -2 & \text{if } x \ge 1 \end{cases}$

51 $f(x) = \begin{cases} x + 2 & \text{if } x \le -1 \\ x^3 & \text{if } |x| < 1 \\ -x + 3 & \text{if } x \ge 1 \end{cases}$

52 $f(x) = \begin{cases} x - 3 & \text{if } x \le -2 \\ -x^2 & \text{if } -2 < x < 1 \\ -x + 4 & \text{if } x \ge 1 \end{cases}$

Exer. 53–54: The symbol $[\![x]\!]$ denotes values of the greatest integer function. Sketch the graph of f.

53 (a) $f(x) = [\![x - 3]\!]$ (b) $f(x) = [\![x]\!] - 3$

    (c) $f(x) = 2[\![x]\!]$ (d) $f(x) = [\![2x]\!]$

    (e) $f(x) = [\![-x]\!]$

54 (a) $f(x) = [\![x + 2]\!]$ (b) $f(x) = [\![x]\!] + 2$

    (c) $f(x) = \frac{1}{2}[\![x]\!]$ (d) $f(x) = [\![\frac{1}{2}x]\!]$

    (e) $f(x) = -[\![-x]\!]$

Exer. 55–56: Explain why the graph of the equation is not the graph of a function.

55 $x = y^2$ 56 $x = -|y|$

Exer. 57–58: For the graph of $y = f(x)$ shown in the figure, sketch the graph of $y = |f(x)|$.

57

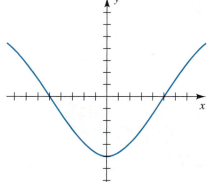

58

[Graph showing a sinusoidal wave on xy-axes]

Exer. 59–62: Sketch the graph of the equation.

59 $y = |9 - x^2|$

60 $y = |x^3 - 1|$

61 $y = |\sqrt{x} - 1|$

62 $y = ||x| - 1|$

63 Let $y = f(x)$ be a function with domain $D = [-2, 6]$ and range $R = [-4, 8]$. Find the domain D and range R for each function. Assume $f(2) = 8$ and $f(6) = -4$.

(a) $y = -2f(x)$

(b) $y = f(\tfrac{1}{2}x)$

(c) $y = f(x - 3) + 1$

(d) $y = f(x + 2) - 3$

(e) $y = f(-x)$

(f) $y = -f(x)$

(g) $y = f(|x|)$

(h) $y = |f(x)|$

64 Let $y = f(x)$ be a function with domain $D = [-6, -2]$ and range $R = [-10, -4]$. Find the domain D and range R for each function.

(a) $y = \tfrac{1}{2}f(x)$

(b) $y = f(2x)$

(c) $y = f(x - 2) + 5$

(d) $y = f(x + 4) - 1$

(e) $y = f(-x)$

(f) $y = -f(x)$

(g) $y = f(|x|)$

(h) $y = |f(x)|$

65 Income tax rates A certain country taxes the first $20,000 of an individual's income at a rate of 15%, and all income over $20,000 is taxed at 20%. Find a piecewise-defined function T that specifies the total tax on an income of x dollars.

66 Property tax rates A certain state taxes the first $500,000 in property value at a rate of 1%; all value over $500,000 is taxed at 1.25%. Find a piecewise-defined function T that specifies the total tax on a property valued at x dollars.

67 Royalty rates A certain paperback sells for $12. The author is paid royalties of 10% on the first 10,000 copies sold, 12.5% on the next 5000 copies, and 15% on any additional copies. Find a piecewise-defined function R that specifies the total royalties if x copies are sold.

68 Electricity rates An electric company charges its customers $0.0577 per kilowatt-hour (kWh) for the first 1000 kWh used, $0.0532 for the next 4000 kWh, and $0.0511 for any kWh over 5000. Find a piecewise-defined function C for a customer's bill of x kWh.

69 Car rental charges There are two car rental options available for a four-day trip. Option I is $29.95 per day, with 200 free miles and $0.25 per mile for each additional mile. Option II is $39.95 per day, with a charge of $0.15 per mile.

(a) Determine the cost of a 500-mile trip for both options.

(b) Model the data with a cost function for each four-day option.

(c) Determine the mileages at which each option is preferable.

70 Traffic flow Cars are crossing a bridge that is 1 mile long. Each car is 12 feet long and is required to stay a distance of at least d feet from the car in front of it (see figure).

(a) Show that the largest number of cars that can be on the bridge at one time is $[\![5280/(12 + d)]\!]$, where $[\![\ ]\!]$ denotes the greatest integer function.

(b) If the velocity of each car is v mi/hr, show that the maximum traffic flow rate F (in cars/hr) is given by $F = [\![5280v/(12 + d)]\!]$.

Exercise 70

3.6

Quadratic Functions

Figure 1

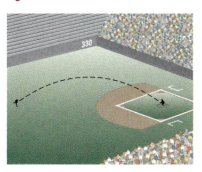

If $a \neq 0$, then the graph of $y = ax^2$ is a parabola with vertex at the origin $(0, 0)$, a vertical axis, opening upward if $a > 0$ or downward if $a < 0$ (see, for example, Figures 4 and 5 in Section 3.5). In this section we show that the graph of an equation of the form

$$y = ax^2 + bx + c$$

can be obtained by vertical and/or horizontal shifts of the graph of $y = ax^2$ and hence is also a parabola. An important application of such equations is to describe the trajectory, or path, of an object near the surface of the earth when the only force acting on the object is gravitational attraction. To illustrate, if an outfielder on a baseball team throws a ball into the infield, as illustrated in Figure 1, and if air resistance and other outside forces are negligible, then the path of the ball is a parabola. If suitable coordinate axes are introduced, then the path coincides with the graph of the equation $y = ax^2 + bx + c$ for some a, b, and c. We call the function determined by this equation a *quadratic function*.

Definition of Quadratic Function	A function f is a **quadratic function** if $$f(x) = ax^2 + bx + c,$$ where a, b, and c are real numbers with $a \neq 0$.

Figure 2

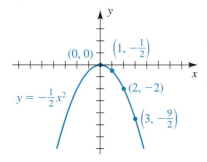

If $b = c = 0$ in the preceding definition, then $f(x) = ax^2$, and the graph is a parabola with vertex at the origin. If $b = 0$ and $c \neq 0$, then

$$f(x) = ax^2 + c,$$

and, from our discussion of vertical shifts in Section 3.5, the graph is a parabola with vertex at the point $(0, c)$ on the y-axis. The following example contains specific illustrations.

Figure 3

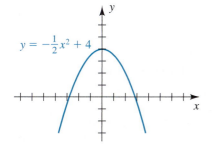

EXAMPLE 1   Sketching the graph of a quadratic function

Sketch the graph of f if

(a) $f(x) = -\frac{1}{2}x^2$ **(b)** $f(x) = -\frac{1}{2}x^2 + 4$

SOLUTION

(a) Since f is even, the graph of f (that is, of $y = -\frac{1}{2}x^2$) is symmetric with respect to the y-axis. It is similar in shape to but wider than the parabola $y = -x^2$, sketched in Figure 5 of Section 3.5. Several points on the graph are $(0, 0)$, $\left(1, -\frac{1}{2}\right)$, $(2, -2)$, and $\left(3, -\frac{9}{2}\right)$. Plotting and using symmetry, we obtain the sketch in Figure 2.

(b) To find the graph of $y = -\frac{1}{2}x^2 + 4$, we shift the graph of $y = -\frac{1}{2}x^2$ upward a distance 4, obtaining the sketch in Figure 3.

If $f(x) = ax^2 + bx + c$ and $b \neq 0$, then, by completing the square, we can change the form to

$$f(x) = a(x - h)^2 + k$$

for some real numbers h and k. This technique is illustrated in the next example.

EXAMPLE 2 **Expressing a quadratic function as $f(x) = a(x - h)^2 + k$**

If $f(x) = 3x^2 + 24x + 50$, express $f(x)$ in the form $a(x - h)^2 + k$.

SOLUTION 1 Before completing the square, *it is essential that we factor out the coefficient of x^2 from the first two terms of $f(x)$*, as follows:

$$\begin{aligned} f(x) &= 3x^2 + 24x + 50 && \text{given} \\ &= 3(x^2 + 8x + \quad) + 50 && \text{factor out 3 from } 3x^2 + 24x \end{aligned}$$

We now complete the square for the expression $x^2 + 8x$ within the parentheses by adding the square of half the coefficient of x—that is, $\left(\frac{8}{2}\right)^2$, or 16. However, if we add 16 to the expression within parentheses, then, because of the factor 3, we are actually adding 48 to $f(x)$. Hence, we must compensate by subtracting 48:

$$\begin{aligned} f(x) &= 3(x^2 + 8x + \quad) + 50 && \text{given} \\ &= 3(x^2 + 8x + 16) + (50 - 48) && \text{complete the square for } x^2 + 8x \\ &= 3(x + 4)^2 + 2 && \text{equivalent equation} \end{aligned}$$

The last expression has the form $a(x - h)^2 + k$ with $a = 3$, $h = -4$, and $k = 2$.

SOLUTION 2 We begin by dividing both sides by the coefficient of x^2.

$$\begin{aligned} f(x) &= 3x^2 + 24x + 50 && \text{given} \\ \frac{f(x)}{3} &= x^2 + 8x + \frac{50}{3} && \text{divide by 3} \\ &= x^2 + 8x \underline{+ 16} + \frac{50}{3} \underline{-16} && \begin{array}{l}\text{add and subtract 16, the number that} \\ \text{completes the square for } x^2 + 8x\end{array} \\ &= (x + 4)^2 + \frac{2}{3} && \text{equivalent equation} \\ f(x) &= 3(x + 4)^2 + 2 && \text{multiply by 3} \end{aligned}$$

If $f(x) = ax^2 + bx + c$, then, by completing the square as in Example 2, we see that the graph of f is the same as the graph of an equation of the form

$$y = a(x - h)^2 + k.$$

The graph of this equation can be obtained from the graph of $y = ax^2$ shown in Figure 4(a) by means of a horizontal and a vertical shift, as follows. First, as in Figure 4(b), we obtain the graph of $y = a(x - h)^2$ by shifting the graph of $y = ax^2$ either to the left or to the right, depending on the sign of h (the figure illustrates the case with $h > 0$). Next, as in Figure 4(c), we shift the graph in (b) vertically a distance $|k|$ (the figure illustrates the case with $k > 0$). It follows that *the graph of a quadratic function is a parabola with a vertical axis.*

Figure 4

(a)

(b)

(c)

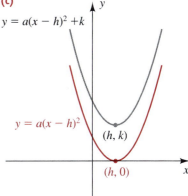

The sketch in Figure 4(c) illustrates one possible graph of the equation $y = ax^2 + bx + c$. If $a > 0$, the point (h, k) is the lowest point on the parabola, and the function f has a **minimum value** $f(h) = k$. If $a < 0$, the parabola opens downward, and the point (h, k) is the highest point on the parabola. In this case, the function f has a **maximum value** $f(h) = k$.

We have obtained the following result.

Standard Equation of a Parabola with Vertical Axis	The graph of the equation $$y = a(x - h)^2 + k$$ for $a \neq 0$ is a parabola that has vertex $V(h, k)$ and a vertical axis. The parabola opens upward if $a > 0$ or downward if $a < 0$.

For convenience, we often refer to the *parabola* $y = ax^2 + bx + c$ when considering the graph of this equation.

EXAMPLE 3   Finding a standard equation of a parabola

Express $y = 2x^2 - 6x + 4$ as a standard equation of a parabola with a vertical axis. Find the vertex and sketch the graph.

Figure 5

Figure 6

Figure 7

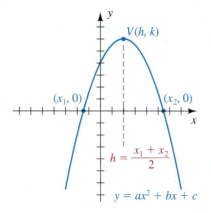

SOLUTION

$$y = 2x^2 - 6x + 4 \qquad \text{given}$$
$$= 2(x^2 - 3x + \quad) + 4 \qquad \text{factor out 2 from } 2x^2 - 6x$$
$$= 2\left(x^2 - 3x + \tfrac{9}{4}\right) + \left(4 - \tfrac{9}{2}\right) \qquad \text{complete the square for } x^2 - 3x$$
$$= 2\left(x - \tfrac{3}{2}\right)^2 - \tfrac{1}{2} \qquad \text{equivalent equation}$$

The last equation has the form of the standard equation of a parabola with $a = 2$, $h = \tfrac{3}{2}$, and $k = -\tfrac{1}{2}$. Hence, the vertex $V(h, k)$ of the parabola is $V\!\left(\tfrac{3}{2}, -\tfrac{1}{2}\right)$. Since $a = 2 > 0$, the parabola opens upward.

To find the y-intercept of the graph of $y = 2x^2 - 6x + 4$, we let $x = 0$, obtaining $y = 4$. To find the x-intercepts, we let $y = 0$ and solve the equation $2x^2 - 6x + 4 = 0$ or the equivalent equation $2(x - 1)(x - 2) = 0$, obtaining $x = 1$ and $x = 2$. Plotting the vertex and using the x- and y-intercepts provides enough points for a reasonably accurate sketch (see Figure 5). ◢

EXAMPLE 4 Finding a standard equation of a parabola

Express $y = -x^2 - 2x + 8$ as a standard equation of a parabola with a vertical axis. Find the vertex and sketch the graph.

SOLUTION

$$y = -x^2 - 2x + 8 \qquad \text{given}$$
$$= -(x^2 + 2x + \quad) + 8 \qquad \text{factor out } -1 \text{ from } -x^2 - 2x$$
$$= -(x^2 + 2x + 1) + (8 + 1) \qquad \text{complete the square for } x^2 + 2x$$
$$= -(x + 1)^2 + 9 \qquad \text{equivalent equation}$$

This is the standard equation of a parabola with $h = -1$, $k = 9$, and hence the vertex is $(-1, 9)$. Since $a = -1 < 0$, the parabola opens downward.

The y-intercept of the graph of $y = -x^2 - 2x + 8$ is the constant term, 8. To find the x-intercepts, we solve $-x^2 - 2x + 8 = 0$ or, equivalently, $x^2 + 2x - 8 = 0$. Factoring gives us $(x + 4)(x - 2) = 0$, and hence the intercepts are $x = -4$ and $x = 2$. Using this information gives us the sketch in Figure 6. ◢

If a parabola $y = ax^2 + bx + c$ has x-intercepts x_1 and x_2, as illustrated in Figure 7 for the case $a < 0$, then the axis of the parabola is the vertical line $x = (x_1 + x_2)/2$ through the midpoint of $(x_1, 0)$ and $(x_2, 0)$. Therefore, the x-coordinate h of the vertex (h, k) is $h = (x_1 + x_2)/2$. Some special cases are illustrated in Figures 5 and 6.

In the following example we find an equation of a parabola from given data.

EXAMPLE 5 Finding an equation of a parabola with a given vertex

Find an equation of a parabola that has vertex $V(2, 3)$ and a vertical axis and passes through the point $(5, 1)$.

Figure 8

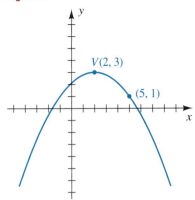

SOLUTION   Figure 8 shows the vertex V, the point $(5, 1)$, and a possible position of the parabola. Using the standard equation

$$y = a(x - h)^2 + k$$

with $h = 2$ and $k = 3$ gives us

$$y = a(x - 2)^2 + 3.$$

To find a, we use the fact that $(5, 1)$ is on the parabola and so is a solution of the last equation. Thus,

$$1 = a(5 - 2)^2 + 3, \quad \text{or} \quad a = -\tfrac{2}{9}.$$

Hence, an equation for the parabola is

$$y = -\tfrac{2}{9}(x - 2)^2 + 3.$$

The next theorem gives us a simple formula for locating the vertex of a parabola.

Theorem for Locating the Vertex of a Parabola	The vertex of the parabola $y = ax^2 + bx + c$ has x-coordinate $$-\frac{b}{2a}.$$

PROOF   Let us begin by writing $y = ax^2 + bx + c$ as

$$y = a\left(x^2 + \frac{b}{a}x + \quad\right) + c.$$

Next we complete the square by adding $\left(\dfrac{1}{2}\dfrac{b}{a}\right)^2$ to the expression within parentheses:

$$y = a\left(x^2 + \frac{b}{a}x + \frac{b^2}{4a^2}\right) + \left(c - \frac{b^2}{4a}\right)$$

Note that if $b^2/(4a^2)$ is added *inside* the parentheses, then, because of the factor a on the *outside,* we have actually added $b^2/(4a)$ to y. Therefore, we must compensate by subtracting $b^2/(4a)$. The last equation may be written

$$y = a\left(x + \frac{b}{2a}\right)^2 + \left(c - \frac{b^2}{4a}\right).$$

This is the equation of a parabola that has vertex (h, k) with $h = -b/(2a)$ and $k = c - b^2/(4a)$.

It is unnecessary to remember the formula for the y-coordinate of the vertex of the parabola in the preceding result. Once the x-coordinate has been found, we can calculate the y-coordinate by substituting $-b/(2a)$ for x in the equation of the parabola.

EXAMPLE 6   Finding the vertex of a parabola

Find the vertex of the parabola $y = 2x^2 - 6x + 4$.

SOLUTION   We considered this parabola in Example 3 and found the vertex by completing the square. We shall use the vertex formula with $a = 2$ and $b = -6$, obtaining the x-coordinate

$$\frac{-b}{2a} = \frac{-(-6)}{2(2)} = \frac{6}{4} = \frac{3}{2}.$$

We next find the y-coordinate by substituting $\frac{3}{2}$ for x in the given equation:

$$y = 2\left(\tfrac{3}{2}\right)^2 - 6\left(\tfrac{3}{2}\right) + 4 = -\tfrac{1}{2}$$

Thus, the vertex is $\left(\frac{3}{2}, -\frac{1}{2}\right)$ (see Figure 5).

Since the graph of $f(x) = ax^2 + bx + c$ for $a \neq 0$ is a parabola, we can use the vertex formula to help find the maximum or minimum value of a quadratic function. Specifically, since the x-coordinate of the vertex V is $-b/(2a)$, the y-coordinate of V is the function value $f(-b/(2a))$. Moreover, since the parabola opens downward if $a < 0$ and upward if $a > 0$, this function value is the maximum or minimum value, respectively, of f. We may summarize these facts as follows.

Theorem on the Maximum or Minimum Value of a Quadratic Function	If $f(x) = ax^2 + bx + c$, where $a \neq 0$, then $f\left(-\dfrac{b}{2a}\right)$ is (1) the maximum value of f if $a < 0$ (2) the minimum value of f if $a > 0$

We shall use this theorem in the next example.

EXAMPLE 7   Finding the maximum value of a quadratic function

A long rectangular sheet of metal, 12 inches wide, is to be made into a rain gutter by turning up two sides so that they are perpendicular to the sheet. How many inches should be turned up to give the gutter its greatest capacity?

Figure 9

SOLUTION The gutter is illustrated in Figure 9. If x denotes the number of inches turned up on each side, the width of the base of the gutter is $12 - 2x$ inches. The capacity will be greatest when the cross-sectional area of the rectangle with sides of lengths x and $12 - 2x$ has its greatest value. Letting $f(x)$ denote this area, we have

$$f(x) = x(12 - 2x)$$
$$= 12x - 2x^2$$
$$= -2x^2 + 12x,$$

which has the form $f(x) = ax^2 + bx + c$ with $a = -2$, $b = 12$, and $c = 0$. Since f is a quadratic function and $a = -2 < 0$, it follows from the preceding theorem that the maximum value of f occurs at

$$x = -\frac{b}{2a} = -\frac{12}{2(-2)} = 3.$$

Thus, 3 inches should be turned up on each side to achieve maximum capacity.

As an alternative solution, we may note that the graph of the function $f(x) = x(12 - 2x)$ has x-intercepts at $x = 0$ and $x = 6$. Hence, the average of the intercepts,

$$x = \frac{0 + 6}{2} = 3,$$

is the x-coordinate of the vertex of the parabola and the value that yields the maximum capacity.

When working with quadratic functions, we are often most interested in finding the vertex and the x-intercepts. Typically, a given quadratic function closely resembles one of the three forms listed in the following chart.

Relationship Between Quadratic Function Forms and Their Vertex and x-intercepts

Form	Vertex (h, k)	x-intercepts (if there are any)
(1) $y = f(x) = a(x - h)^2 + k$	h and k as in the form	$x = h \pm \sqrt{-k/a}$ (see below)
(2) $y = f(x) = a(x - x_1)(x - x_2)$	$h = \dfrac{x_1 + x_2}{2}, \quad k = f(h)$	$x = x_1, x_2$
(3) $y = f(x) = ax^2 + bx + c$	$h = -\dfrac{b}{2a}, \quad k = f(h)$	$x = -\dfrac{b}{2a} \pm \dfrac{\sqrt{b^2 - 4ac}}{2a}$ (see below)

If the radicands in (1) or (3) are negative, then there are no x-intercepts. To find the x-intercepts with form (1), use the special quadratic equation on page 75. If you have a quadratic function in form (3) and want to find the vertex and the x-intercepts, it may be best to first find the x-intercepts by using the quadratic formula. Then you can easily obtain the x-coordinate of the vertex, h, since

$$-\frac{b}{2a} \pm \frac{\sqrt{b^2 - 4ac}}{2a} = h \pm \frac{\sqrt{b^2 - 4ac}}{2a}.$$

Of course, if the function in form (3) is easily factorable, it is not necessary to use the quadratic formula.

We will discuss parabolas further in a later chapter.

3.6 Exercises

Exer. 1–4: Find the standard equation of any parabola that has vertex V.

1 $V(-3, 1)$ 2 $V(4, -2)$

3 $V(0, -3)$ 4 $V(-2, 0)$

Exer. 5–12: Express $f(x)$ in the form $a(x - h)^2 + k$.

5 $f(x) = -x^2 - 4x - 8$ 6 $f(x) = x^2 - 6x + 11$

7 $f(x) = 2x^2 - 12x + 22$ 8 $f(x) = 5x^2 + 20x + 17$

9 $f(x) = -3x^2 - 6x - 5$

10 $f(x) = -4x^2 + 16x - 13$

11 $f(x) = -\frac{3}{4}x^2 + 9x - 34$ 12 $f(x) = \frac{2}{5}x^2 - \frac{12}{5}x + \frac{23}{5}$

Exer. 13–22: (a) Use the quadratic formula to find the zeros of f. (b) Find the maximum or minimum value of $f(x)$. (c) Sketch the graph of f.

13 $f(x) = x^2 - 4x$ 14 $f(x) = -x^2 - 6x$

15 $f(x) = -12x^2 + 11x + 15$

16 $f(x) = 6x^2 + 7x - 24$

17 $f(x) = 9x^2 + 24x + 16$ 18 $f(x) = -4x^2 + 4x - 1$

19 $f(x) = x^2 + 4x + 9$ 20 $f(x) = -3x^2 - 6x - 6$

21 $f(x) = -2x^2 + 20x - 43$

22 $f(x) = 2x^2 - 4x - 11$

Exer. 23–26: Find the standard equation of the parabola shown in the figure.

23

24

25

$V(-2, 4)$

26

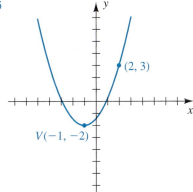

$(2, 3)$

$V(-1, -2)$

Exer. 27–28: Find an equation of the form

$$y = a(x - x_1)(x - x_2)$$

of the parabola shown in the figure. See the chart on page 191.

27

$(2, 4)$

28

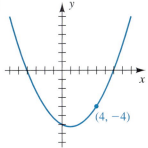

$(4, -4)$

Exer. 29–34: Find the standard equation of a parabola that has a vertical axis and satisfies the given conditions.

29 Vertex $(0, -2)$, passing through $(3, 25)$

30 Vertex $(0, 5)$, passing through $(2, -3)$

31 Vertex $(3, 5)$, x-intercept 0

32 Vertex $(4, -7)$, x-intercept -4

33 x-intercepts -3 and 5, highest point has y-coordinate 4

34 x-intercepts 8 and 0, lowest point has y-coordinate -48

Exer. 35–36: Find the maximum vertical distance d between the parabola and the line for the green region.

35

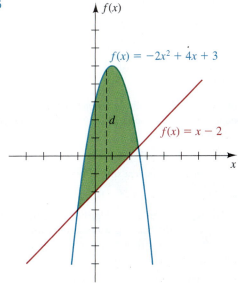

$f(x) = -2x^2 + 4x + 3$

$f(x) = x - 2$

d

36

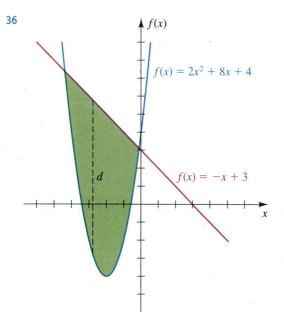

$f(x) = 2x^2 + 8x + 4$

$f(x) = -x + 3$

d

Exer. 37–38: Ozone occurs at all levels of the earth's atmosphere. The density of ozone varies both seasonally and latitudinally. At Edmonton, Canada, the density $D(h)$ of ozone (in 10^{-3} cm/km) for altitudes h between 20 kilometers and 35 kilometers was determined experimentally. For each $D(h)$ and season, approximate the altitude at which the density of ozone is greatest.

37 $D(h) = -0.058h^2 + 2.867h - 24.239$ (autumn)

38 $D(h) = -0.078h^2 + 3.811h - 32.433$ (spring)

39 Infant growth rate The growth rate y (in pounds per month) of an infant is related to present weight x (in pounds) by the formula $y = cx(21 - x)$, where c is a positive constant and $0 < x < 21$. At what weight does the maximum growth rate occur?

40 Gasoline mileage The number of miles M that a certain automobile can travel on one gallon of gasoline at a speed of v mi/hr is given by

$$M = -\tfrac{1}{30}v^2 + \tfrac{5}{2}v \qquad \text{for } 0 < v < 70.$$

(a) Find the most economical speed for a trip.

(b) Find the largest value of M.

41 Height of a projectile An object is projected vertically upward from the top of a building with an initial velocity of 144 ft/sec. Its distance $s(t)$ in feet above the ground after t seconds is given by the equation

$$s(t) = -16t^2 + 144t + 100.$$

(a) Find its maximum distance above the ground.

(b) Find the height of the building.

42 Flight of a projectile An object is projected vertically upward with an initial velocity of v_0 ft/sec, and its distance $s(t)$ in feet above the ground after t seconds is given by the formula $s(t) = -16t^2 + v_0t$.

(a) If the object hits the ground after 12 seconds, find its initial velocity v_0.

(b) Find its maximum distance above the ground.

43 Find two positive real numbers whose sum is 40 and whose product is a maximum.

44 Find two real numbers whose difference is 40 and whose product is a minimum.

45 Constructing cages One thousand feet of chain-link fence is to be used to construct six animal cages, as shown in the figure.

(a) Express the width y as a function of the length x.

(b) Express the total enclosed area A of the cages as a function of x.

(c) Find the dimensions that maximize the enclosed area.

Exercise 45

46 Fencing a field A farmer wishes to put a fence around a rectangular field and then divide the field into three rectangular plots by placing two fences parallel to one of the sides. If the farmer can afford only 1000 yards of fencing, what dimensions will give the maximum rectangular area?

47 Leaping animals Flights of leaping animals typically have parabolic paths. The figure on the next page illustrates a frog jump superimposed on a coordinate plane. The length of the leap is 9 feet, and the maximum height off the ground is 3 feet. Find a standard equation for the path of the frog.

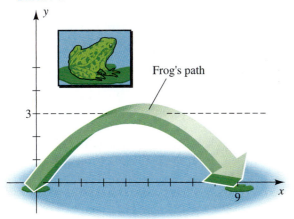

Frog's path

48 The human cannonball In the 1940s, the human cannonball stunt was performed regularly by Emmanuel Zacchini for The Ringling Brothers and Barnum & Bailey Circus. The tip of the cannon rose 15 feet off the ground, and the total horizontal distance traveled was 175 feet. When the cannon is aimed at an angle of 45°, an equation of the parabolic flight (see the figure) has the form $y = ax^2 + x + c$.

(a) Use the given information to find an equation of the flight.

(b) Find the maximum height attained by the human cannonball.

Exercise 48

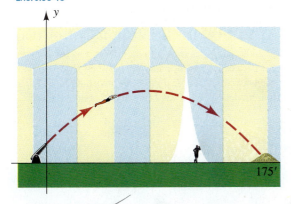

175'

49 Shape of a suspension bridge One section of a suspension bridge has its weight uniformly distributed between twin towers that are 400 feet apart and rise 90 feet above the horizontal roadway (see the figure). A cable strung between the tops of the towers has the shape of a parabola, and its center point is 10 feet above the roadway. Suppose coordinate axes are introduced, as shown in the figure.

Exercise 49

(a) Find an equation for the parabola.

(b) Nine equally spaced vertical cables are used to support the bridge (see the figure). Find the total length of these supports.

50 Designing a highway Traffic engineers are designing a stretch of highway that will connect a horizontal highway with one having a 20% grade (that is, slope $\frac{1}{5}$), as illustrated in the figure. The smooth transition is to take place over a horizontal distance of 800 feet, with a parabolic piece of highway used to connect points A and B. If the equation of the parabolic segment is of the form $y = ax^2 + bx + c$, it can be shown that the slope of the tangent line at the point $P(x, y)$ on the parabola is given by $m = 2ax + b$.

(a) Find an equation of the parabola that has a tangent line of slope 0 at A and $\frac{1}{5}$ at B.

(b) Find the coordinates of B.

Exercise 50

51 Parabolic doorway A doorway has the shape of a parabolic arch and is 9 feet high at the center and 6 feet wide at the base. If a rectangular box 8 feet high must fit through the doorway, what is the maximum width the box can have?

52 Wire rectangle A piece of wire 24 inches long is bent into the shape of a rectangle having width x and length y.

(a) Express y as a function of x.

(b) Express the area A of the rectangle as a function of x.

(c) Show that the area A is greatest if the rectangle is a square.

53 Quantity discount A company sells running shoes to dealers at a rate of $40 per pair if fewer than 50 pairs are ordered. If a dealer orders 50 or more pairs (up to 600), the price per pair is reduced at a rate of 4 cents times the number ordered. What size order will produce the maximum amount of money for the company?

54 Group discount A travel agency offers group tours at a rate of $60 per person for the first 30 participants. For larger groups—up to 90—each person receives a $0.50 discount for every participant in excess of 30. For example, if 31 people participate, then the cost per person is $59.50. Determine the size of the group that will produce the maximum amount of money for the agency.

55 Cable TV fee A cable television firm presently serves 5000 households and charges $20 per month. A marketing survey indicates that each decrease of $1 in the monthly charge will result in 500 new customers. Let $R(x)$ denote the total monthly revenue when the monthly charge is x dollars.

(a) Determine the revenue function R.

(b) Sketch the graph of R and find the value of x that results in maximum monthly revenue.

56 Apartment rentals A real estate company owns 180 efficiency apartments, which are fully occupied when the rent is $300 per month. The company estimates that for each $10 increase in rent, 5 apartments will become unoccupied. What rent should be charged so that the company will receive the maximum monthly income?

57 Crest vertical curves When engineers plan highways, they must design hills so as to ensure proper vision for drivers. Hills are referred to as *crest vertical curves*. Crest vertical curves change the slope of a highway. Engineers use a parabolic shape for a highway hill, with the vertex located at the top of the crest. Two roadways with different slopes are to be connected with a parabolic crest curve. The highway passes through the points $A(-800, -48)$, $B(-500, 0)$, $C(0, 40)$, $D(500, 0)$, and $E(800, -48)$, as shown in the figure. The roadway is linear between A and B, parabolic between B and D, and then linear between D and E. Find a piecewise-defined function f that models the roadway between the points A and E.

Exercise 57

58 Sag vertical curves Refer to Exercise 57. Valleys or dips in highways are referred to as *sag vertical curves*. Sag vertical curves are also modeled using parabolas. Two roadways with different grades meeting at a sag curve need to be connected. The highway passes through the points $A\left(-500, 243\frac{1}{3}\right)$, $B(0, 110)$, $C(750, 10)$, $D(1500, 110)$, and $E\left(2000, 243\frac{1}{3}\right)$, as shown in the figure. The roadway is linear between A and B, parabolic between B and D, and linear between D and E. Find a piecewise-defined function f that models the roadway between the points A and E.

Exercise 58

3.7

Operations on Functions

Functions are often defined using sums, differences, products, and quotients of various expressions. For example, if

$$h(x) = x^2 + \sqrt{5x + 1},$$

we may regard $h(x)$ as a sum of values of the functions f and g given by

$$f(x) = x^2 \quad \text{and} \quad g(x) = \sqrt{5x + 1}.$$

We call h the *sum* of f and g and denote it by $f + g$. Thus,

$$h(x) = (f + g)(x) = x^2 + \sqrt{5x + 1}.$$

In general, if f and g are *any* functions, we use the terminology and notation given in the following chart.

While it is true that

$$(f + g)(x) = f(x) + g(x),$$

remember that, in general,

$$f(a + b) \quad \neq \quad f(a) + f(b).$$

Sum, Difference, Product, and Quotient of Functions

Terminology	Function value
sum $f + g$	$(f + g)(x) = f(x) + g(x)$
difference $f - g$	$(f - g)(x) = f(x) - g(x)$
product fg	$(fg)(x) = f(x)g(x)$
quotient $\dfrac{f}{g}$	$\left(\dfrac{f}{g}\right)(x) = \dfrac{f(x)}{g(x)}, \ g(x) \neq 0$

    The domains of $f + g$, $f - g$, and fg are the intersection I of the domains of f and g—that is, the numbers that are *common* to both domains. The domain of f/g is the subset of I consisting of all x in I such that $g(x) \neq 0$.

EXAMPLE 1　Finding function values of $f + g$, $f - g$, fg, and f/g

If $f(x) = 3x - 2$ and $g(x) = x^3$, find $(f + g)(2)$, $(f - g)(2)$, $(fg)(2)$, and $(f/g)(2)$.

SOLUTION　Since $f(2) = 3(2) - 2 = 4$ and $g(2) = 2^3 = 8$, we have

$$(f + g)(2) = f(2) + g(2) = 4 + 8 = 12$$

$$(f - g)(2) = f(2) - g(2) = 4 - 8 = -4$$

$$(fg)(2) = f(2)g(2) = (4)(8) = 32$$

$$\left(\frac{f}{g}\right)(2) = \frac{f(2)}{g(2)} = \frac{4}{8} = \frac{1}{2}.$$

EXAMPLE 2　Finding $(f + g)(x)$, $(f - g)(x)$, $(fg)(x)$, and $(f/g)(x)$

If $f(x) = \sqrt{4 - x^2}$ and $g(x) = 3x + 1$, find $(f + g)(x)$, $(f - g)(x)$, $(fg)(x)$, and $(f/g)(x)$, and state the domains of the respective functions.

SOLUTION The domain of f is the closed interval $[-2, 2]$, and the domain of g is $\mathbb{R}$. The intersection of these domains is $[-2, 2]$, which is the domain of $f + g, f - g$, and fg. For the domain of f/g, we exclude each number x in $[-2, 2]$ such that $g(x) = 3x + 1 = 0$ $\left(\text{namely, } x = -\frac{1}{3}\right)$. Thus, we have the following:

$$(f + g)(x) = \sqrt{4 - x^2} + (3x + 1), \qquad -2 \le x \le 2$$

$$(f - g)(x) = \sqrt{4 - x^2} - (3x + 1), \qquad -2 \le x \le 2$$

$$(fg)(x) = \sqrt{4 - x^2}\,(3x + 1), \qquad -2 \le x \le 2$$

$$\left(\frac{f}{g}\right)(x) = \frac{\sqrt{4 - x^2}}{3x + 1}, \qquad -2 \le x \le 2 \text{ and } x \ne -\frac{1}{3}$$

A function f is a **polynomial function** if $f(x)$ is a polynomial—that is, if

$$f(x) = a_n x^n + a_{n-1} x^{n-1} + \cdots + a_1 x + a_0,$$

where the coefficients $a_0, a_1, \ldots, a_n$ are real numbers and the exponents are nonnegative integers. A polynomial function may be regarded as a sum of functions whose values are of the form cx^k, where c is a real number and k is a nonnegative integer. Note that the quadratic functions considered in the previous section are polynomial functions.

An **algebraic function** is a function that can be expressed in terms of finite sums, differences, products, quotients, or roots of polynomial functions.

ILLUSTRATION **Algebraic Function**

■ $\quad f(x) = 5x^4 - 2\sqrt[3]{x} + \dfrac{x(x^2 + 5)}{\sqrt{x^3 + \sqrt{x}}}$

Functions that are not algebraic are **transcendental.** The exponential and logarithmic functions considered in Chapter 5 are examples of transcendental functions.

In the remainder of this section we shall discuss how two functions f and g may be used to obtain the *composite functions* $f \circ g$ and $g \circ f$ (read "f circle g" and "g circle f," respectively). Functions of this type are very important in calculus. The function $f \circ g$ is defined as follows.

Definition of Composite Function	The **composite function** $f \circ g$ of two functions f and g is defined by $$(f \circ g)(x) = f(g(x)).$$ The domain of $f \circ g$ is the set of all x in the domain of g such that $g(x)$ is in the domain of f.

*A number x is in the domain of ($f \circ g$)(x) if and only if **both** g(x) **and** f(g(x)) are defined.*

Figure 1

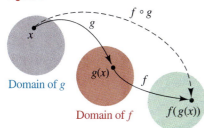

Figure 1 is a schematic diagram that illustrates relationships among f, g, and $f \circ g$. Note that for x in the domain of g, *first we find $g(x)$* (which must be in the domain of f) and then, *second, we find $f(g(x))$*.

For the composite function $g \circ f$, we reverse this order, first finding $f(x)$ and second finding $g(f(x))$. The domain of $g \circ f$ is the set of all x in the domain of f such that $f(x)$ is in the domain of g.

Since the notation $g(x)$ is read "g of x," we sometimes say that *g is a function of x*. For the composite function $f \circ g$, the notation $f(g(x))$ is read "f of g of x," and we could regard f as a function of $g(x)$. In this sense, *a composite function is a function of a function* or, more precisely, a function of another function's values.

EXAMPLE 3 Finding composite functions

Let $f(x) = x^2 - 1$ and $g(x) = 3x + 5$.

(a) Find $(f \circ g)(x)$ and the domain of $f \circ g$.

(b) Find $(g \circ f)(x)$ and the domain of $g \circ f$.

(c) Find $f(g(2))$ in two different ways: first using the functions f and g separately and second using the composite function $f \circ g$.

SOLUTION

(a)

$$(f \circ g)(x) = f(g(x)) \qquad \text{definition of } f \circ g$$
$$= f(3x + 5) \qquad \text{definition of } g$$
$$= (3x + 5)^2 - 1 \qquad \text{definition of } f$$
$$= 9x^2 + 30x + 24 \qquad \text{simplify}$$

The domain of both f and g is $\mathbb{R}$. Since for each x in $\mathbb{R}$ (the domain of g), the function value $g(x)$ is in $\mathbb{R}$ (the domain of f), the domain of $f \circ g$ is also $\mathbb{R}$. Note that *both $g(x)$ and $f(g(x))$* are defined for all real numbers.

(b)

$$(g \circ f)(x) = g(f(x)) \qquad \text{definition of } g \circ f$$
$$= g(x^2 - 1) \qquad \text{definition of } f$$
$$= 3(x^2 - 1) + 5 \qquad \text{definition of } g$$
$$= 3x^2 + 2 \qquad \text{simplify}$$

Since for each x in $\mathbb{R}$ (the domain of f), the function value $f(x)$ is in $\mathbb{R}$ (the domain of g), the domain of $g \circ f$ is $\mathbb{R}$. Note that *both $f(x)$ and $g(f(x))$* are defined for all real numbers.

(c) To find $f(g(2))$ using $f(x) = x^2 - 1$ and $g(x) = 3x + 5$ separately, we may proceed as follows:

$$g(2) = 3(2) + 5 = 11$$
$$f(g(2)) = f(11) = 11^2 - 1 = 120$$

To find $f(g(2))$ using $f \circ g$, we refer to part (a), where we found

$$(f \circ g)(x) = f(g(x)) = 9x^2 + 30x + 24.$$

(continued)

Hence,

$$f(g(2)) = 9(2)^2 + 30(2) + 24$$
$$= 36 + 60 + 24 = 120.$$

Note that in Example 3, $f(g(x))$ and $g(f(x))$ are not always the same; that is, $f \circ g \neq g \circ f$.

If two functions f and g both have domain $\mathbb{R}$, then the domain of $f \circ g$ and $g \circ f$ is also $\mathbb{R}$. This was illustrated in Example 3. The next example shows that the domain of a composite function may differ from those of the two given functions.

EXAMPLE 4   Finding composite functions

Let $f(x) = x^2 - 16$ and $g(x) = \sqrt{x}$.

(a) Find $(f \circ g)(x)$ and the domain of $f \circ g$.

(b) Find $(g \circ f)(x)$ and the domain of $g \circ f$.

SOLUTION   We first note that the domain of f is $\mathbb{R}$ and the domain of g is the set of all nonnegative real numbers—that is, the interval $[0, \infty)$. We may proceed as follows.

(a)
$$(f \circ g)(x) = f(g(x)) \qquad \text{definition of } f \circ g$$
$$= f(\sqrt{x}) \qquad \text{definition of } g$$
$$= (\sqrt{x})^2 - 16 \qquad \text{definition of } f$$
$$= x - 16 \qquad \text{simplify}$$

If we consider only the final expression, $x - 16$, we might be led to believe that the domain of $f \circ g$ is $\mathbb{R}$, since $x - 16$ is defined for every real number x. However, this is not the case. By definition, the domain of $f \circ g$ is the set of all x in $[0, \infty)$ (the domain of g) such that $g(x)$ is in $\mathbb{R}$ (the domain of f). Since $g(x) = \sqrt{x}$ is in $\mathbb{R}$ for every x in $[0, \infty)$, it follows that the domain of $f \circ g$ is $[0, \infty)$. Note that *both* $g(x)$ *and* $f(g(x))$ are defined for x in $[0, \infty)$.

(b)
$$(g \circ f)(x) = g(f(x)) \qquad \text{definition of } g \circ f$$
$$= g(x^2 - 16) \qquad \text{definition of } f$$
$$= \sqrt{x^2 - 16} \qquad \text{definition of } g$$

By definition, the domain of $g \circ f$ is the set of all x in $\mathbb{R}$ (the domain of f) such that $f(x) = x^2 - 16$ is in $[0, \infty)$ (the domain of g). The statement "$x^2 - 16$ is in $[0, \infty)$" is equivalent to each of the inequalities

$$x^2 - 16 \geq 0, \qquad x^2 \geq 16, \qquad |x| \geq 4.$$

Thus, the domain of $g \circ f$ is the union $(-\infty, -4] \cup [4, \infty)$. Note that *both* $f(x)$ *and* $g(f(x))$ are defined for x in $(-\infty, -4] \cup [4, \infty)$. Also note that this domain is different from the domains of both f and g.

The next example illustrates how special values of composite functions may sometimes be obtained from tables.

EXAMPLE 5   **Finding composite function values from tables**

Several values of two functions f and g are listed in the following tables.

x	1	2	3	4
$f(x)$	3	4	2	1

x	1	2	3	4
$g(x)$	4	1	3	2

Find $(f \circ g)(2)$, $(g \circ f)(2)$, $(f \circ f)(2)$, and $(g \circ g)(2)$.

SOLUTION   Using the definition of composite function and referring to the tables above, we obtain

$$(f \circ g)(2) = f(g(2)) = f(1) = 3$$
$$(g \circ f)(2) = g(f(2)) = g(4) = 2$$
$$(f \circ f)(2) = f(f(2)) = f(4) = 1$$
$$(g \circ g)(2) = g(g(2)) = g(1) = 4.$$

In some applied problems it is necessary to express a quantity y as a function of time t. The following example illustrates that it is often easier to introduce a third variable x, express x as a function of t (that is, $x = g(t)$), express y as a function of x (that is, $y = f(x)$), and finally form the composite function given by $y = f(x) = f(g(t))$.

EXAMPLE 6   **Using a composite function
to find the volume of a balloon**

A meteorologist is inflating a spherical balloon with helium gas. If the radius of the balloon is changing at a rate of 1.5 cm/sec, express the volume V of the balloon as a function of time t (in seconds).

SOLUTION   Let x denote the radius of the balloon. If we assume that the radius is 0 initially, then after t seconds

$$x = 1.5t. \quad \text{radius of balloon after } t \text{ seconds}$$

To illustrate, after 1 second, the radius is 1.5 centimeters; after 2 seconds, it is 3.0 centimeters; after 3 seconds, it is 4.5 centimeters; and so on.

Next we write

$$V = \tfrac{4}{3}\pi x^3. \quad \text{volume of a sphere of radius } x$$

This gives us a composite function relationship in which V is a function of x, and x is a function of t. By substitution, we obtain

$$V = \tfrac{4}{3}\pi x^3 = \tfrac{4}{3}\pi(1.5t)^3 = \tfrac{4}{3}\pi\left(\tfrac{3}{2}t\right)^3 = \tfrac{4}{3}\pi\left(\tfrac{27}{8}t^3\right).$$

Simplifying, we obtain the following formula for V as a function of t:

$$V(t) = \tfrac{9}{2}\pi t^3$$

If f and g are functions such that

$$y = f(u) \quad \text{and} \quad u = g(x),$$

then substituting for u in $y = f(u)$ yields

$$y = f(g(x)).$$

For certain problems in calculus we *reverse* this procedure; that is, given $y = h(x)$ for some function h, we find a *composite function form* $y = f(u)$ and $u = g(x)$ such that $h(x) = f(g(x))$.

EXAMPLE 7 Finding a composite function form

Express $y = (2x + 5)^8$ as a composite function form.

SOLUTION Suppose, for a real number x, we wanted to evaluate the expression $(2x + 5)^8$ by using a calculator. We would first calculate the value of $2x + 5$ and then raise the result to the eighth power. This suggests that we let

$$u = 2x + 5 \quad \text{and} \quad y = u^8,$$

which is a composite function form for $y = (2x + 5)^8$. ◢

The method used in the preceding example can be extended to other functions. In general, suppose we are given $y = h(x)$. To choose the *inside* expression $u = g(x)$ in a composite function form, ask the following question: If a calculator were being used, which part of the expression $h(x)$ would be evaluated first? This often leads to a suitable choice for $u = g(x)$. After choosing u, refer to $h(x)$ to determine $y = f(u)$. The following illustration contains typical problems.

ILLUSTRATION   **Composite Function Forms**

Function value	Choice for $u = g(x)$	Choice for $y = f(u)$
■ $y = (x^3 - 5x + 1)^4$	$u = x^3 - 5x + 1$	$y = u^4$
■ $y = \sqrt{x^2 - 4}$	$u = x^2 - 4$	$y = \sqrt{u}$
■ $y = \dfrac{2}{3x + 7}$	$u = 3x + 7$	$y = \dfrac{2}{u}$

The composite function form is never unique. For example, consider the first expression in the preceding illustration:

$$y = (x^3 - 5x + 1)^4$$

If n is any nonzero integer, we could choose

$$u = (x^3 - 5x + 1)^n \quad \text{and} \quad y = u^{4/n}.$$

Thus, there are an *unlimited* number of composite function forms. Generally, our goal is to choose a form such that the expression for y is simple, as we did in the illustration.

3.7 *Exercises*

Exer. 1–2: Find
(a) $(f + g)(3)$ (b) $(f - g)(3)$
(c) $(fg)(3)$ (d) $(f/g)(3)$

1   $f(x) = x + 3$, $g(x) = x^2$

2   $f(x) = -x^2$, $g(x) = 2x - 1$

Exer. 3–8: Find
(a) $(f + g)(x)$, $(f - g)(x)$, $(fg)(x)$, and $(f/g)(x)$
(b) the domain of $f + g$, $f - g$, and fg
(c) the domain of f/g

3   $f(x) = x^2 + 2$, $g(x) = 2x^2 - 1$

4   $f(x) = x^2 + x$, $g(x) = x^2 - 3$

5   $f(x) = \sqrt{x + 5}$, $g(x) = \sqrt{x + 5}$

6   $f(x) = \sqrt{3 - 2x}$, $g(x) = \sqrt{x + 4}$

7   $f(x) = \dfrac{2x}{x - 4}$, $g(x) = \dfrac{x}{x + 5}$

8   $f(x) = \dfrac{x}{x - 2}$, $g(x) = \dfrac{3x}{x + 4}$

Exer. 9–10: Find
(a) $(f \circ g)(x)$ (b) $(g \circ f)(x)$
(c) $(f \circ f)(x)$ (d) $(g \circ g)(x)$

9   $f(x) = 2x - 1$, $g(x) = -x^2$

10   $f(x) = 3x^2$, $g(x) = x - 1$

Exer. 11–20: Find
(a) $(f \circ g)(x)$ (b) $(g \circ f)(x)$
(c) $f(g(-2))$ (d) $g(f(3))$

11   $f(x) = 2x - 5$, $g(x) = 3x + 7$

12   $f(x) = 5x + 2$, $g(x) = 6x - 1$

13   $f(x) = 3x^2 + 4$, $g(x) = 5x$

14   $f(x) = 3x - 1$, $g(x) = 4x^2$

15   $f(x) = 2x^2 + 3x - 4$, $g(x) = 2x - 1$

16   $f(x) = 5x - 7$, $g(x) = 3x^2 - x + 2$

17   $f(x) = 4x$, $g(x) = 2x^3 - 5x$

18   $f(x) = x^3 + 2x^2$, $g(x) = 3x$

19   $f(x) = |x|$, $g(x) = -7$

20   $f(x) = 5$, $g(x) = x^2$

Exer. 21–34: Find (a) $(f \circ g)(x)$ and the domain of $f \circ g$ and (b) $(g \circ f)(x)$ and the domain of $g \circ f$.

21   $f(x) = x^2 - 3x$, $g(x) = \sqrt{x + 2}$

22   $f(x) = \sqrt{x - 15}$, $g(x) = x^2 + 2x$

23   $f(x) = x^2 - 4$, $g(x) = \sqrt{3x}$

24   $f(x) = -x^2 + 1$, $g(x) = \sqrt{x}$

25   $f(x) = \sqrt{x - 2}$, $g(x) = \sqrt{x + 5}$

26   $f(x) = \sqrt{3 - x}$, $g(x) = \sqrt{x + 2}$

27   $f(x) = \sqrt{3 - x}$, $g(x) = \sqrt{x^2 - 16}$

28   $f(x) = x^3 + 5$, $g(x) = \sqrt[3]{x - 5}$

29   $f(x) = \dfrac{3x + 5}{2}$, $g(x) = \dfrac{2x - 5}{3}$

30   $f(x) = \dfrac{1}{x - 1}$, $g(x) = x - 1$

31   $f(x) = x^2$, $g(x) = \dfrac{1}{x^3}$

32   $f(x) = \dfrac{x}{x - 2}$, $g(x) = \dfrac{3}{x}$

33   $f(x) = \dfrac{x - 1}{x - 2}$, $g(x) = \dfrac{x - 3}{x - 4}$

34   $f(x) = \dfrac{x + 2}{x - 1}$, $g(x) = \dfrac{x - 5}{x + 4}$

Exer. 35–36: Solve the equation $(f \circ g)(x) = 0$.

35   $f(x) = x^2 - 2$, $g(x) = x + 3$

36   $f(x) = x^2 - x - 2$, $g(x) = 2x - 1$

37 Several values of two functions f and g are listed in the following tables:

x	5	6	7	8	9
$f(x)$	8	7	6	5	4

x	5	6	7	8	9
$g(x)$	7	8	6	5	4

If possible, find

(a) $(f \circ g)(6)$ (b) $(g \circ f)(6)$ (c) $(f \circ f)(6)$

(d) $(g \circ g)(6)$ (e) $(f \circ g)(9)$

38 Several values of two functions T and S are listed in the following tables:

t	0	1	2	3	4
$T(t)$	2	3	1	0	5

x	0	1	2	3	4
$S(x)$	1	0	3	2	5

If possible, find

(a) $(T \circ S)(1)$ (b) $(S \circ T)(1)$ (c) $(T \circ T)(1)$

(d) $(S \circ S)(1)$ (e) $(T \circ S)(4)$

39 If $D(t) = \sqrt{400 + t^2}$ and $R(x) = 20x$, find $(D \circ R)(x)$.

40 If $S(r) = 4\pi r^2$ and $D(t) = 2t + 5$, find $(S \circ D)(t)$.

41 If f is an odd function and g is an even function, is fg even, odd, or neither even nor odd?

42 There is one function with domain $\mathbb{R}$ that is both even and odd. Find that function.

43 **Payroll functions** Let the social security tax function SSTAX be defined as $SSTAX(x) = 0.0715x$, where $x \geq 0$ is the weekly income. Let ROUND2 be the function that rounds a number to two decimal places. Find the value of $(ROUND2 \circ SSTAX)(437.21)$.

44 **Computer science functions** Let the function CHR be defined by CHR(65) = "A", CHR(66) = "B", ..., CHR(90) = "Z". Then let the function ORD be defined by ORD("A") = 65, ORD("B") = 66, ..., ORD("Z") = 90. Find

(a) $(CHR \circ ORD)("C")$ (b) $CHR(ORD("A") + 3)$

45 **Spreading fire** A fire has started in a dry open field and is spreading in the form of a circle. If the radius of this circle increases at the rate of 6 ft/min, express the total fire area A as a function of time t (in minutes).

46 **Dimensions of a balloon** A spherical balloon is being inflated at a rate of $\frac{9}{2}\pi$ ft^3/min. Express its radius r as a function of time t (in minutes), assuming that $r = 0$ when $t = 0$.

47 **Dimensions of a sand pile** The volume of a conical pile of sand is increasing at a rate of 243π ft^3/min, and the height of the pile always equals the radius r of the base. Express r as a function of time t (in minutes), assuming that $r = 0$ when $t = 0$.

48 **Diagonal of a cube** The diagonal d of a cube is the distance between two opposite vertices. Express d as a function of the edge x of the cube. (*Hint:* First express the diagonal y of a face as a function of x.)

49 **Altitude of a balloon** A hot-air balloon rises vertically from ground level as a rope attached to the base of the balloon is released at the rate of 5 ft/sec (see the figure). The pulley that releases the rope is 20 feet from a platform where passengers board the balloon. Express the altitude h of the balloon as a function of time t.

Exercise 49

50 **Tightrope walker** Refer to Exercise 76 of Section 3.4. Starting at the lowest point, the tightrope walker moves up the rope at a steady rate of 2 ft/sec. If the rope is attached 30 feet up the pole, express the height h of the walker above the ground as a function of time t. (*Hint:* Let d denote the total distance traveled along the wire. First express d as a function of t, and then h as a function of d.)

51 Airplane take-off Refer to Exercise 77 of Section 3.4. When the airplane is 500 feet down the runway, it has reached a speed of 150 ft/sec (or about 102 mi/hr), which it will maintain until take-off. Express the distance d of the plane from the control tower as a function of time t (in seconds). (*Hint:* In the figure, first write x as a function of t.)

52 Cable corrosion A 100-foot-long cable of diameter 4 inches is submerged in seawater. Because of corrosion, the surface area of the cable decreases at the rate of 750 in² per year. Express the diameter d of the cable as a function of time t (in years). (Disregard corrosion at the ends of the cable.)

Exer. 53–60: Find a composite function form for y.

53 $y = (x^2 + 3x)^{1/3}$

54 $y = \sqrt[4]{x^4 - 16}$

55 $y = \dfrac{1}{(x - 3)^4}$

56 $y = 4 + \sqrt{x^2 + 1}$

57 $y = (x^4 - 2x^2 + 5)^5$

58 $y = \dfrac{1}{(x^2 + 3x - 5)^3}$

59 $y = \dfrac{\sqrt{x + 4} - 2}{\sqrt{x + 4} + 2}$

60 $y = \dfrac{\sqrt[3]{x}}{1 + \sqrt[3]{x}}$

61 If $f(x) = \sqrt{x} - 1$ and $g(x) = x^3 + 1$, approximate $(f \circ g)(0.0001)$. In order to avoid calculating a zero value for $(f \circ g)(0.0001)$, rewrite the formula for $f \circ g$ as

$$\frac{x^3}{\sqrt{x^3 + 1} + 1}.$$

62 If $f(x) = \dfrac{x^3}{x^2 + x + 2}$ and $g(x) = \left(\sqrt{3x} - x^3\right)^{3/2}$, approximate

$$\frac{(f + g)(1.12) - (f/g)(1.12)}{[(f \circ f)(5.2)]^2}.$$

CHAPTER 3 REVIEW EXERCISES

1 Describe the set of all points (x, y) in a coordinate plane such that $y/x < 0$.

2 Show that the triangle with vertices $A(3, 1)$, $B(-5, -3)$, and $C(4, -1)$ is a right triangle, and find its area.

3 Given $P(-5, 9)$ and $Q(-8, -7)$, find

(a) the distance $d(P, Q)$

(b) the midpoint of the segment PQ

(c) a point R such that Q is the midpoint of PR

4 Find all points on the y-axis that are a distance 13 from $P(12, 6)$.

5 For what values of a is the distance between $P(a, 1)$ and $Q(-2, a)$ less than 3?

6 Find an equation of the circle that has center $C(7, -4)$ and passes through $P(-3, 3)$.

7 Find an equation of the circle that has endpoints of a diameter $A(8, 10)$ and $B(-2, -14)$.

8 Find an equation for the left half of the circle given by $(x + 2)^2 + y^2 = 9$.

9 Find the slope of the line through $C(11, -5)$ and $D(-8, 6)$.

10 Show that $A(-3, 1)$, $B(1, -1)$, $C(4, 1)$, and $D(3, 5)$ are vertices of a trapezoid.

11 Find an equation of the line through $A\left(\frac{1}{2}, -\frac{1}{3}\right)$ that is

(a) parallel to the line $6x + 2y + 5 = 0$

(b) perpendicular to the line $6x + 2y + 5 = 0$

12 Express $8x + 3y - 24 = 0$ in slope-intercept form.

13 Find an equation of the circle that has center $C(-5, -1)$ and is tangent to the line $x = 4$.

14 Find an equation of the line that has x-intercept -3 and passes through the center of the circle that has equation $x^2 + y^2 - 4x + 10y + 26 = 0$.

15 Find a general form of an equation of the line through $P(4, -3)$ with slope 5.

16 Given $A(-1, 2)$ and $B(3, -4)$, find a general form of an equation for the perpendicular bisector of segment AB.

Exer. 17–18: Find the center and radius of the circle with the given equation.

17 $x^2 + y^2 - 12y + 31 = 0$

18 $4x^2 + 4y^2 + 24x - 16y + 39 = 0$

19 If $f(x) = \dfrac{x}{\sqrt{x+3}}$, find

(a) $f(1)$ (b) $f(-1)$ (c) $f(0)$ (d) $f(-x)$

(e) $-f(x)$ (f) $f(x^2)$ (g) $[f(x)]^2$

Exer. 20–21: Find *the sign* of $f(4)$ without actually finding $f(4)$.

20 $f(x) = \dfrac{-32(x^2 - 4)}{(9 - x^2)^{5/3}}$

21 $f(x) = \dfrac{-2(x^2 - 20)(5 - x)}{(6 - x^2)^{4/3}}$

22 Find the domain and range of f if

(a) $f(x) = \sqrt{3x - 4}$ (b) $f(x) = \dfrac{1}{(x+3)^2}$

Exer. 23–24: Find $\dfrac{f(a+h) - f(a)}{h}$ if $h \neq 0$.

23 $f(x) = -x^2 + x + 5$

24 $f(x) = \dfrac{1}{x+2}$

25 Find a linear function f such that $f(1) = 2$ and $f(3) = 7$.

26 Determine whether f is even, odd, or neither even nor odd.

(a) $f(x) = \sqrt[3]{x^3 + 4x}$ (b) $f(x) = \sqrt[3]{3x^2 - x^3}$

(c) $f(x) = \sqrt[3]{x^4 + 3x^2 + 5}$

Exer. 27–40: Sketch the graph of the equation, and label the x- and y-intercepts.

27 $x + 5 = 0$ **28** $2y - 7 = 0$

29 $2y + 5x - 8 = 0$ **30** $x = 3y + 4$

31 $9y + 2x^2 = 0$ **32** $3x - 7y^2 = 0$

33 $y = \sqrt{1 - x}$ **34** $y = (x - 1)^3$

35 $y^2 = 16 - x^2$

36 $x^2 + y^2 + 4x - 16y + 64 = 0$

37 $x^2 + y^2 - 8x = 0$ **38** $x = -\sqrt{9 - y^2}$

39 $y = (x - 3)^2 - 2$ **40** $y = -x^2 - 2x + 3$

41 Find the center of the small circle.

Exercise 41

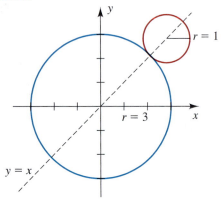

42 Explain how the graph of $y = -f(x - 2)$ compares to the graph of $y = f(x)$.

Exer. 43–52: (a) Sketch the graph of f. (b) Find the domain D and range R of f. (c) Find the intervals on which f is increasing, is decreasing, or is constant.

43 $f(x) = \dfrac{1 - 3x}{2}$ **44** $f(x) = 1000$

45 $f(x) = |x + 3|$ **46** $f(x) = -\sqrt{10 - x^2}$

47 $f(x) = 1 - \sqrt{x + 1}$ **48** $f(x) = \sqrt{2 - x}$

49 $f(x) = 9 - x^2$ **50** $f(x) = x^2 + 6x + 16$

51 $f(x) = \begin{cases} x^2 & \text{if } x < 0 \\ 3x & \text{if } 0 \leq x < 2 \\ 6 & \text{if } x \geq 2 \end{cases}$ **52** $f(x) = 1 + 2[\![x]\!]$

53 Sketch the graphs of the following equations, making use of shifting, stretching, or reflecting:

(a) $y = \sqrt{x}$ (b) $y = \sqrt{x + 4}$

(c) $y = \sqrt{x} + 4$ (d) $y = 4\sqrt{x}$

(e) $y = \tfrac{1}{4}\sqrt{x}$ (f) $y = -\sqrt{x}$

54 The graph of a function f with domain $[-3, 3]$ is shown in the figure. Sketch the graph of the given equation.

(a) $y = f(x - 2)$

(b) $y = f(x) - 2$

(c) $y = f(-x)$

(d) $y = f(2x)$

(e) $y = f(\frac{1}{2}x)$

(f) $y = |f(x)|$

(g) $y = f(|x|)$

Exercise 54

56

57

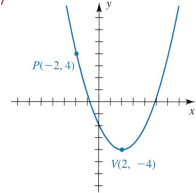

Exer. 55–58: Find an equation for the graph shown in the figure.

55

58

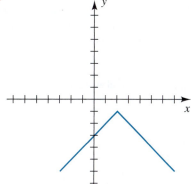

Exer. 59–62: Find the maximum or minimum value of $f(x)$.

59 $f(x) = 5x^2 + 30x + 49$

60 $f(x) = -3x^2 + 30x - 82$

61 $f(x) = -12(x + 1)^2 - 37$

62 $f(x) = 3(x + 2)(x - 10)$

63 Express the function $f(x) = -2x^2 + 12x - 14$ in the form $a(x - h)^2 + k$.

64 Find the standard equation of a parabola with a vertical axis that has vertex $V(3, -2)$ and passes through $(5, 4)$.

65 If $f(x) = \sqrt{4 - x^2}$ and $g(x) = \sqrt{x}$, find the domain of

(a) fg (b) f/g

66 If $f(x) = 8x - 1$ and $g(x) = \sqrt{x - 2}$, find

(a) $(f \circ g)(2)$ (b) $(g \circ f)(2)$

Exer. 67–68: Find (a) $(f \circ g)(x)$ and (b) $(g \circ f)(x)$.

67 $f(x) = 2x^2 - 5x + 1$, $g(x) = 3x + 2$

68 $f(x) = \sqrt{3x + 2}$, $g(x) = 1/x^2$

Exer. 69–70: Find (a) $(f \circ g)(x)$ and the domain of $f \circ g$ and (b) $(g \circ f)(x)$ and the domain of $g \circ f$.

69 $f(x) = \sqrt{25 - x^2}$, $g(x) = \sqrt{x - 3}$

70 $f(x) = \dfrac{x}{3x + 2}$, $g(x) = \dfrac{2}{x}$

71 Find a composite function form for $y = \sqrt[3]{x^2 - 5x}$.

72 **Wheelchair ramp** The Americans with Disabilities Act of 1990 guarantees all persons the right of accessibility of public accommodations. Providing access to a building often involves building a wheelchair ramp. Ramps should have approximately 1 inch of vertical rise for every 12–20 inches of horizontal run. If the base of an exterior door is located 3 feet above a sidewalk, determine the range of appropriate lengths for a wheelchair ramp.

73 **Discus throw** Based on Olympic records, the winning distance for the discus throw can be approximated by the equation $d = 181 + 1.065t$, where d is in feet and $t = 0$ corresponds to the year 1948.

(a) Predict the winning distance for the Summer Olympics in the year 2008.

(b) Estimate the Olympic year in which the winning distance will be 265 feet.

74 **House appreciation** Six years ago a house was purchased for $89,000. This year it was appraised at $125,000. Assume that the value V of the house after its purchase is a linear function of time t (in years).

(a) Express V in terms of t.

(b) How many years after the purchase date was the house worth $103,000?

75 **Temperature scales** The freezing point of water is 0°C, or 32°F, and the boiling point is 100°C, or 212°F.

(a) Express the Fahrenheit temperature F as a linear function of the Celsius temperature C.

(b) What temperature increase in °F corresponds to an increase in temperature of 1°C?

76 **Gasoline mileage** Suppose the cost of driving an automobile is a linear function of the number x of miles driven and that gasoline costs $1.25 per gallon. A certain automobile presently gets 20 mi/gal, and a tune-up that will improve gasoline mileage by 10% costs $50.

(a) Express the cost C_1 of driving without a tune-up in terms of x.

(b) Express the cost C_2 of driving with a tune-up in terms of x.

(c) How many miles must the automobile be driven after a tune-up to make the cost of the tune-up worthwhile?

77 **Constructing a storage shelter** An open rectangular storage shelter, consisting of two 4-foot-wide vertical sides and a flat roof, is to be attached to an existing structure, as illustrated in the figure. The flat roof is made of tin and costs $5 per square foot, and the two sides are made of plywood costing $2 per square foot.

(a) If $400 is available for construction, express the length y as a function of the height x.

(b) Express the volume V inside the shelter as a function of x.

Exercise 77

78 Constructing a cylindrical container A company plans to manufacture a container having the shape of a right circular cylinder, open at the top, and having a capacity of 24π in³. If the cost of the material for the bottom is \$0.30/in² and that for the curved sides is \$0.10/in², express the total cost C of the material as a function of the radius r of the base of the container.

79 Filling a pool A cross section of a rectangular pool of dimensions 80 feet by 40 feet is shown in the figure. The pool is being filled with water at a rate of 10 ft³/min.

Exercise 79

(a) Express the volume V of the water in the pool as a function of time t.

(b) Express V as a function of the depth h at the deep end for $0 \le h \le 6$ and then for $6 < h \le 9$.

(c) Express h as a function of t for $0 \le h \le 6$ and then for $6 < h \le 9$.

80 Filtering water Suppose 5 in³ of water is poured into a conical filter and subsequently drips into a cup, as shown in the figure. Let x denote the height of the water in the filter, and let y denote the height of the water in the cup.

(a) Express the radius r shown in the figure as a function of x. (*Hint:* Use similar triangles.)

(b) Express the height y of the water in the cup as a function of x. (*Hint:* What is the sum of the two volumes shown in the figure?)

Exercise 80

81 Frustum of a cone The shape of the first spacecraft in the Apollo program was a frustum of a right circular cone—a solid formed by truncating a cone by a plane parallel to its base. For the frustum shown in the figure, the radii a and b have already been determined.

Exercise 81

(a) Use similar triangles to express y as a function of h.

(b) Derive a formula for the volume of the frustum as a function of h.

(c) If $a = 6$ ft and $b = 3$ ft, for what value of h is the volume of the frustum 600 ft³?

82 Distance between ships At 1:00 P.M. ship A is 30 miles due south of ship B and is sailing north at a rate of 15 mi/hr. If ship B is sailing west at a rate of 10 mi/hr, find the time at which the distance d between the ships is minimal (see the figure).

Exercise 82

83 Dimensions of a race track The interior of a half-mile race track consists of a rectangle with semicircles at two opposite ends. Find the dimensions that will maximize the area of the rectangle.

84 Vertical leaps When a particular basketball player leaps straight up for a dunk, the player's distance $f(t)$ (in feet) off the floor after t seconds is given by the formula $f(t) = -\frac{1}{2}gt^2 + 16t$, where g is a gravitational constant.

(a) If $g = 32$, find the player's hang time—that is, the total number of seconds that the player is in the air.

(b) Find the player's vertical leap—that is, the maximum distance of the player's feet from the floor.

(c) On the moon, $g = \frac{32}{6}$. Rework parts (a) and (b) for the player on the moon.

85 Trajectory of a rocket A rocket is fired up a hillside, following a path given by $y = -0.016x^2 + 1.6x$. The hillside has slope $\frac{1}{5}$, as illustrated in the figure.

(a) Where does the rocket land?

(b) Find the maximum height of the rocket *above the ground*.

Exercise 85

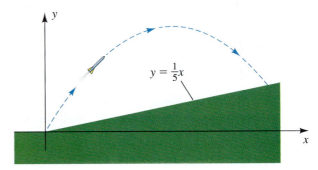

CHAPTER 3 DISCUSSION EXERCISES

1 Compare the graphs of $y = \sqrt[3]{x}$, $y = \sqrt{x}$, $y = x$, $y = x^2$, and $y = x^3$ on the interval $0 \le x \le 2$. Write a generalization based on what you find out about graphs of equations of the form $y = x^{p/q}$, where $x \ge 0$ and p and q are positive integers.

2 Write an expression for $g(x)$ if the graph of g is obtained from the graph of $f(x) = \frac{1}{2}x - 3$ by reflecting f about the

(a) x-axis   **(b)** y-axis

(c) line $y = 2$ **(d)** line $x = 3$

3 Consider the graph of $g(x) = \sqrt{f(x)}$, where f is given by $f(x) = ax^2 + bx + c$. Discuss the general shape of g, including its domain and range. Discuss the advantages and disadvantages of graphing g as a composition of the functions $h(x) = \sqrt{x}$ and $f(x)$. (*Hint:* You may want to use the following expressions for f: $x^2 - 2x - 8$, $-x^2 + 2x + 8$, $x^2 - 2x + 2$, $-x^2 + 2x - 2$.)

4 Simplify the difference quotient in Exercises 49 and 50 of Section 3.4 for an arbitrary quadratic function of the form $f(x) = ax^2 + bx + c$.

5 Refer to Example 5 in Section 3.4. Geometrically, what does the expression $2x + h + 6$ represent on the graph of f? What do you think it represents if $h = 0$?

6 The midpoint formula could be considered to be the "halfway" formula since it gives us the point that is $\frac{1}{2}$ of the distance from the point $P(x_1, y_1)$ to the point $Q(x_2, y_2)$. Develop an "m-nth way" formula that gives the point $R(x_3, y_3)$ that is m/n of the distance from P to Q (assume m and n are positive integers with $m < n$).

7 Consider the graphs of equations of the quadratic form $y = ax^2 + bx + c$ that have two x-intercepts. Let d denote the distance from the axis of the parabola to either of the x-intercepts, and let h denote the value of the y-coordinate of the vertex. Explore the relationship between d and h for several specific equations, and then develop a formula for this relationship.

8 **Billing for service** A common method of billing for service calls is to charge a flat fee plus an additional fee for each quarter-hour spent on the call. Create a function for a washer repair company that charges $40 plus $20 for each quarter-hour or portion thereof—for example, a 30-minute repair call would cost $80, while a 31-minute repair call would cost $100. The input to your function is any positive integer. (*Hint:* See Exercise 54(e) of Section 3.5.)

9 **Density of the ozone layer** The density D (in 10^{-3} cm/km) of the ozone layer at altitudes x between 3 and 15 kilometers during winter at Edmonton, Canada, was determined experimentally to be

$$D = 0.0833x^2 - 0.4996x + 3.5491.$$

Express x as a function of D.

10 **Precipitation in Minneapolis** The average monthly precipitation in inches in Minneapolis is listed in the table. Model these data with a piecewise function f that is first quadratic and then linear.

Month	Precipitation
Jan.	0.7
Feb.	0.8
Mar.	1.5
Apr.	1.9
May	3.2
June	4.0
July	3.3
Aug.	3.2
Sept.	2.4
Oct.	1.6
Nov.	1.4
Dec.	0.9

4

Polynomial and Rational Functions

4.1 Polynomial Functions of Degree Greater Than 2

4.2 Properties of Division

4.3 Zeros of Polynomials

4.4 Complex and Rational Zeros of Polynomials

4.5 Rational Functions

4.6 Variation

Polynomial functions are the most basic functions in mathematics, because they are defined only in terms of addition, subtraction, and multiplication. In applications it is often necessary to sketch their graphs and to find (or approximate) their zeros. In the first part of this chapter we discuss results that are useful in obtaining this information. We then turn our attention to quotients of polynomial functions—that is, rational functions.

4.1

Polynomial Functions of Degree Greater Than 2

If f is a polynomial function with real coefficients of degree n, then

$$f(x) = a_n x^n + a_{n-1} x^{n-1} + \cdots + a_1 x + a_0,$$

with $a_n \neq 0$. The special cases listed in the following chart were previously discussed.

Degree of f	Form of $f(x)$	Graph of f (with y-intercept a_0)
0	$f(x) = a_0$	A horizontal line
1	$f(x) = a_1 x + a_0$	A line with slope a_1
2	$f(x) = a_2 x^2 + a_1 x + a_0$	A parabola with a vertical axis

Figure 1

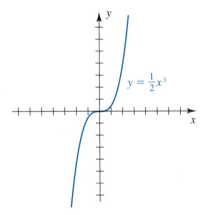

$y = \frac{1}{2}x^3$

In this section we shall discuss graphs of polynomial functions of degree greater than 2. All polynomial functions are **continuous functions**—that is, their graphs can be drawn without any breaks.

If f has degree n *and all the coefficients except a_n are zero,* then

$$f(x) = ax^n \quad \text{for some} \quad a = a_n \neq 0.$$

In this case, if $n = 1$, the graph of f is a line through the origin. If $n = 2$, the graph is a parabola with vertex at the origin. Two illustrations with $n = 3$ (**cubic polynomials**) are given in the next example.

EXAMPLE 1 **Sketching graphs of $y = ax^3$**

Sketch the graph of f if

(a) $f(x) = \frac{1}{2}x^3$ (b) $f(x) = -\frac{1}{2}x^3$

SOLUTION

(a) The following table lists several points on the graph of $y = \frac{1}{2}x^3$.

x	0	$\frac{1}{2}$	1	$\frac{3}{2}$	2	$\frac{5}{2}$
y	0	$\frac{1}{16} \approx 0.06$	$\frac{1}{2}$	$\frac{27}{16} \approx 1.7$	4	$\frac{125}{16} \approx 7.8$

Figure 2

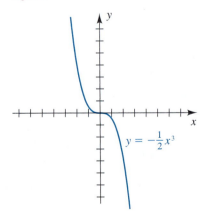

$y = -\frac{1}{2}x^3$

Since f is an odd function, the graph of f is symmetric with respect to the origin, and hence points such as $\left(-\frac{1}{2}, -\frac{1}{16}\right)$ and $\left(-1, -\frac{1}{2}\right)$ are also on the graph. The graph is sketched in Figure 1.

(b) If $y = -\frac{1}{2}x^3$, the graph can be obtained from that in part (a) by multiplying all y-coordinates by -1 (that is, by reflecting the graph in part (a) through the x-axis). This gives us the sketch in Figure 2.

If $f(x) = ax^n$ and n is an *odd* positive integer, then f is an odd function and the graph of f is symmetric with respect to the origin, as illustrated in Figures 1 and 2. For $a > 0$, the graph is similar in shape to that in Figure 1; however, as either n or a increases, the graph rises more rapidly for $x > 1$. If $a < 0$, we reflect the graph through the x-axis, as in Figure 2.

If $f(x) = ax^n$ and n is an *even* positive integer, then f is an even function and the graph of f is symmetric with respect to the y-axis, as illustrated in Figure 3 for the case $a = 1$ and $n = 4$. Note that as the exponent increases, the graph becomes flatter at the origin. It also rises more rapidly for $x > 1$. If $a < 0$, we reflect the graph through the x-axis. Also note that the graph *intersects* the x-axis at the origin, but it does not *cross* the x-axis (change sign).

Figure 3

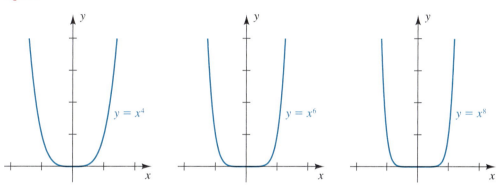

A complete analysis of graphs of polynomial functions of degree greater than 2 requires methods that are used in calculus. As the degree increases, the graphs usually become more complicated. They always have a smooth appearance, however, with a number of high points and low points, such as P, Q, R, and S in Figure 4. Such points are sometimes called **turning points** for the graph. It should be noted that an n-degree polynomial has at most $n - 1$ turning points. Each function value (y-coordinate) corresponding to a high or low point is called an **extremum** of the function f. At an extremum, f changes from an increasing function to a decreasing function, or vice versa.

The intermediate value theorem specifies another important property of polynomial functions.

Figure 4

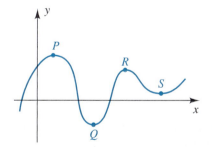

Intermediate Value Theorem for Polynomial Functions	If f is a polynomial function and $f(a) \neq f(b)$ for $a < b$, then f takes on every value between $f(a)$ and $f(b)$ in the interval $[a, b]$.

The intermediate value theorem for polynomial functions states that if w is any number between $f(a)$ and $f(b)$, there is at least one number c between a and b such that $f(c) = w$. If we regard the graph of f as extending continuously

Figure 5

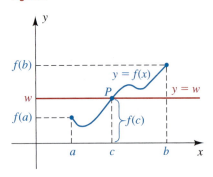

from the point $(a, f(a))$ to the point $(b, f(b))$, as illustrated in Figure 5, then for any number w between $f(a)$ and $f(b)$, the horizontal line $y = w$ intersects the graph in at least one point P. The x-coordinate c of P is a number such that $f(c) = w$.

A consequence of the intermediate value theorem is that if $f(a)$ and $f(b)$ have opposite signs (one positive and one negative), there is at least one number c between a and b such that $f(c) = 0$; that is, f has a **zero** at c. Thus, if the point $(a, f(a))$ lies below the x-axis and the point $(b, f(b))$ lies above the x-axis, or vice versa, the graph crosses the x-axis at least once between $x = a$ and $x = b$, as illustrated in Figure 6.

Figure 6

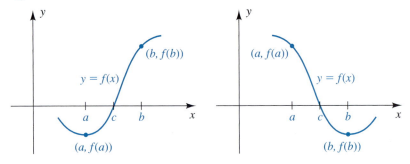

EXAMPLE 2   Using the intermediate value theorem

Show that $f(x) = x^5 + 2x^4 - 6x^3 + 2x - 3$ has a zero between 1 and 2.

SOLUTION Substituting 1 and 2 for x gives us the following function values:

$$f(1) = 1 + 2 - 6 + 2 - 3 = -4$$
$$f(2) = 32 + 32 - 48 + 4 - 3 = 17$$

Since $f(1)$ and $f(2)$ have opposite signs ($f(1) = -4 < 0$ and $f(2) = 17 > 0$), we see that $f(c) = 0$ for at least one real number c between 1 and 2. ◢

Example 2 illustrates a method for locating real zeros of polynomials. By using *successive approximations,* we can approximate each zero at any degree of accuracy by locating it in smaller and smaller intervals.

If c and d are *successive* at real zeros of $f(x)$—that is, there are no other zeros between c and d—then $f(x)$ *does not change sign on the interval* (c, d). Thus, if we choose any number k such that $c < k < d$ and if $f(k)$ is positive, then $f(x)$ is positive throughout (c, d). Similarly, if $f(k)$ is negative, then $f(x)$ is negative throughout (c, d). We shall call $f(k)$ a **test value** for $f(x)$ on the interval (c, d). Test values may also be used on infinite intervals of the form $(-\infty, a)$ or (a, ∞), provided that $f(x)$ has no zeros on these intervals. The use of test values in graphing is similar to the technique used for inequalities in Section 2.7.

EXAMPLE 3   Sketching the graph of a polynomial function of degree 3

Let $f(x) = x^3 + x^2 - 4x - 4$. Find all values of x such that $f(x) > 0$ and all x such that $f(x) < 0$, and then sketch the graph of f.

SOLUTION   We may factor $f(x)$ as follows:

$$
\begin{aligned}
f(x) &= x^3 + x^2 - 4x - 4 && \text{given} \\
&= (x^3 + x^2) + (-4x - 4) && \text{group terms} \\
&= x^2(x + 1) - 4(x + 1) && \text{factor out } x^2 \text{ and } -4 \\
&= (x^2 - 4)(x + 1) && \text{factor out } (x + 1) \\
&= (x + 2)(x - 2)(x + 1) && \text{difference of squares}
\end{aligned}
$$

Figure 7

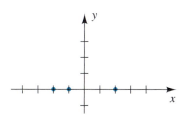

We see from the last equation that the zeros of $f(x)$ (the x-intercepts of the graph) are -2, -1, and 2. The corresponding points on the graph (see Figure 7) divide the x-axis into four parts, and we consider the open intervals

$$(-\infty, -2), \quad (-2, -1), \quad (-1, 2), \quad (2, \infty).$$

As in our work with inequalities in Section 2.7, the sign of $f(x)$ in each of these intervals can be determined by using a sign chart. The graph of f lies above the x-axis for values of x such that $f(x) > 0$, and it lies below the x-axis for all x such that $f(x) < 0$.

Interval	$(-\infty, -2)$	$(-2, -1)$	$(-1, 2)$	$(2, \infty)$
Sign of $x + 2$	$-$	$+$	$+$	$+$
Sign of $x + 1$	$-$	$-$	$+$	$+$
Sign of $x - 2$	$-$	$-$	$-$	$+$
Sign of $f(x)$	$-$	$+$	$-$	$+$
Position of graph	Below x-axis	Above x-axis	Below x-axis	Above x-axis

Figure 8

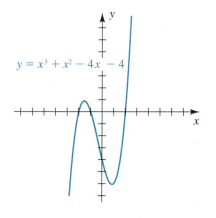

$y = x^3 + x^2 - 4x - 4$

Referring to the sign of $f(x)$ in the chart, we conclude that

$$f(x) > 0 \quad \text{if } x \text{ is in } (-2, -1) \cup (2, \infty)$$

and $$f(x) < 0 \quad \text{if } x \text{ is in } (-\infty, -2) \cup (-1, 2).$$

Using this information leads to the sketch in Figure 8. To find the turning points on the graph, it would be necessary to use a computational device or methods developed in calculus.

The graph of every polynomial function of degree 3 has an appearance similar to that of Figure 8, or it has an inverted version of that graph if the coefficient of x^3 is negative. Sometimes, however, the graph may have only one x-intercept or the shape may be elongated, as in Figures 1 and 2.

EXAMPLE 4 **Sketching the graph of a polynomial function of degree 4**

Let $f(x) = x^4 - 4x^3 + 3x^2$. Find all values of x such that $f(x) > 0$ and all x such that $f(x) < 0$, and then sketch the graph of f.

SOLUTION We begin by factoring $f(x)$:

$$\begin{aligned} f(x) &= x^4 - 4x^3 + 3x^2 && \text{given} \\ &= x^2(x^2 - 4x + 3) && \text{factor out } x^2 \\ &= x^2(x - 1)(x - 3) && \text{factor } x^2 - 4x + 3 \end{aligned}$$

Next, we construct the sign diagram in Figure 9, where the vertical lines indicate the zeros 0, 1, and 3 of the factors. Since the factor x^2 is always positive if $x \neq 0$, it has no effect on the sign of the product and hence may be omitted from the diagram.

Figure 9

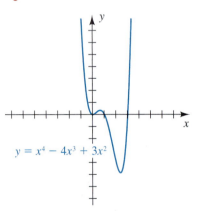

Figure 10

Referring to the sign of $f(x)$ in the diagram, we see that

$$f(x) > 0 \quad \text{if } x \text{ is in } (-\infty, 0) \cup (0, 1) \cup (3, \infty)$$

and $\quad f(x) < 0 \quad$ if x is in $(1, 3)$.

Note that the sign of $f(x)$ does not change at $x = 0$. Making use of these facts leads to the sketch in Figure 10.

In the next example we construct a graph of a polynomial knowing only its sign.

EXAMPLE 5 **Sketch the graph of a polynomial knowing its sign**

Given the sign diagram in Figure 11, sketch a possible graph of the polynomial f.

Figure 11

Figure 12

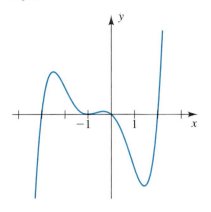

SOLUTION Since the sign of $f(x)$ is *negative* in the interval $(-\infty, -3)$, the graph of f must be *below* the x-axis, as shown in Figure 12. In the interval $(-3, -1)$, the sign of $f(x)$ is *positive,* so the graph of f is *above* the x-axis.

The sign of $f(x)$ is also *positive* in the next interval, $(-1, 0)$. Thus, the graph of f must touch the x-axis at the x-intercept -1 and then remain *above* the x-axis. (The graph of f is *tangent* to the x-axis at $x = -1$.)

In the interval $(0, 2)$, the sign of $f(x)$ is *negative,* so the graph of f is *below* the x-axis. Lastly, the sign of $f(x)$ is *positive* in the interval $(2, \infty)$, and the graph of f is *above* the x-axis.

In the last example we used the function

$$f(x) = (x + 3)(x + 1)^2(x)(x - 2).$$

Note how the graph of f relates to the solutions of the following inequalities.

Inequality	Solution	Position of graph in relation to the x-axis
(1) $f(x) > 0$	$(-3, -1) \cup (-1, 0) \cup (2, \infty)$	Above
(2) $f(x) \geq 0$	$[-3, 0] \cup [2, \infty)$	Above or on
(3) $f(x) < 0$	$(-\infty, -3) \cup (0, 2)$	Below
(4) $f(x) \leq 0$	$(-\infty, -3] \cup \{-1\} \cup [0, 2]$	Below or on

Notice that every real number must be in the solution to either inequality (1) or inequality (4)—the same can be said for inequalities (2) and (3).

4.1 *Exercises*

Exer. 1–4: Sketch the graph of f for the indicated value of c or a.

1 $f(x) = 2x^3 + c$

   (a) $c = 3$ (b) $c = -3$

2 $f(x) = -2x^3 + c$

   (a) $c = -2$ (b) $c = 2$

3 $f(x) = ax^3 + 2$

   (a) $a = 2$ (b) $a = -\frac{1}{3}$

4 $f(x) = ax^3 - 3$

   (a) $a = -2$ (b) $a = \frac{1}{4}$

Exer. 5–10: Use the intermediate value theorem to show that f has a zero between a and b.

5 $f(x) = x^3 - 4x^2 + 3x - 2$; $a = 3,$ $b = 4$

6 $f(x) = 2x^3 + 5x^2 - 3$; $a = -3,$ $b = -2$

7 $f(x) = -x^4 + 3x^3 - 2x + 1$; $a = 2,$ $b = 3$

8 $f(x) = 2x^4 + 3x - 2$; $a = \frac{1}{2},$ $b = \frac{3}{4}$

9 $f(x) = x^5 + x^3 + x^2 + x + 1$; $a = -\frac{1}{2},$ $b = -1$

10 $f(x) = x^5 - 3x^4 - 2x^3 + 3x^2 - 9x - 6$; $a = 3,$ $b = 4$

Exer. 11–12: Match each graph with an equation.

11

(a)

(b)

(c)

(d)

(A) $f(x) = x(x - 2)^2$

(B) $f(x) = -x^2(x - 2)$

(C) $f(x) = (x + 1)(x - 1)(x - 2)$

(D) $f(x) = (x + 1)(x - 1)^2(x - 2)$

12

(a)

(b)

(c)

(d)

(A) $f(x) = x^2(x - 1)$

(B) $f(x) = -x(x + 2)^2$

(C) $f(x) = (x + 2)(x + 1)(x - 3)$

(D) $f(x) = (x + 2)^2(x + 1)(x - 1)$

Exer. 13–28: Find all values of x such that $f(x) > 0$ and all x such that $f(x) < 0$, and sketch the graph of f.

13 $f(x) = \frac{1}{4}x^3 - 2$

14 $f(x) = -\frac{1}{9}x^3 - 3$

15 $f(x) = -\frac{1}{16}x^4 + 1$

16 $f(x) = x^5 + 1$

17 $f(x) = x^4 - 4x^2$

18 $f(x) = 9x - x^3$

19 $f(x) = -x^3 + 3x^2 + 10x$

20 $f(x) = x^4 + 3x^3 - 4x^2$

21 $f(x) = \frac{1}{6}(x + 2)(x - 3)(x - 4)$

22 $f(x) = -\frac{1}{8}(x + 4)(x - 2)(x - 6)$

23 $f(x) = x^3 + 2x^2 - 4x - 8$

24 $f(x) = x^3 - 3x^2 - 9x + 27$

25 $f(x) = x^4 - 6x^2 + 8$

26 $f(x) = -x^4 + 12x^2 - 27$

27 $f(x) = x^2(x + 2)(x - 1)^2(x - 2)$

28 $f(x) = x^3(x + 1)^2(x - 2)(x - 4)$

Exer. 29–30: Sketch the graph of a polynomial given the sign diagram.

29

Sign of $f(x)$ $+$ $|$ $-$ $|$ $-$ $|$ $+$ $|$ $-$

-4 0 1 3

30

Sign of $f(x)$ $+$ $|$ $+$ $|$ $-$ $|$ $+$ $|$ $-$

-3 -2 0 2

31 (a) Sketch a graph of

$$f(x) = (x - a)(x - b)(x - c),$$

where $a < 0 < b < c$.

(b) What is the y-intercept? *(continued)*

(c) What is the solution to $f(x) < 0$?

(d) What is the solution to $f(x) \geq 0$?

32 (a) Sketch a graph of
$$f(x) = (x - a)^2(x - b)(x - c),$$
where $a < b < 0 < c$.

(b) What is the y-intercept?

(c) What is the solution to $f(x) > 0$?

(d) What is the solution to $f(x) \leq 0$?

33 Let $f(x)$ be a polynomial such that the coefficient of every odd power of x is 0. Show that f is an even function.

34 Let $f(x)$ be a polynomial such that the coefficient of every even power of x is 0. Show that f is an odd function.

35 If $f(x) = 3x^3 - kx^2 + x - 5k$, find a number k such that the graph of f contains the point $(-1, 4)$.

36 If $f(x) = kx^3 + x^2 - kx + 2$, find a number k such that the graph of f contains the point $(2, 12)$.

37 If one zero of $f(x) = x^3 - 2x^2 - 16x + 16k$ is 2, find two other zeros.

38 If one zero of $f(x) = x^3 - 3x^2 - kx + 12$ is -2, find two other zeros.

39 **A Legendre polynomial** The third-degree Legendre polynomial $P(x) = \frac{1}{2}(5x^3 - 3x)$ occurs in the solution of heat transfer problems in physics and engineering. Find all values of x such that $P(x) > 0$ and all x such that $P(x) < 0$, and sketch the graph of P.

40 **A Chebyshev polynomial** The fourth-degree Chebyshev polynomial $f(x) = 8x^4 - 8x^2 + 1$ occurs in statistical studies. Find all values of x such that $f(x) > 0$. (*Hint:* Let $z = x^2$, and use the quadratic formula.)

41 **Constructing a box** From a rectangular piece of cardboard having dimensions 20 inches × 30 inches, an open box is to be made by cutting out identical squares of area x^2 from each corner and turning up the sides (see Exercise 65 of Section 3.4).

(a) Show that the volume of the box is given by the function $V(x) = x(20 - 2x)(30 - 2x)$.

(b) Find all positive values of x such that $V(x) > 0$, and sketch the graph of V for $x > 0$.

42 **Constructing a crate** The frame for a shipping crate is to be constructed from 24 feet of 2 × 2 lumber (see the figure).

(a) If the crate is to have square ends of side x feet, express the outer volume V of the crate as a function of x (disregard the thickness of the lumber).

(b) Sketch the graph of V for $x > 0$.

Exercise 42

43 **Determining temperatures** A meteorologist determines that the temperature T (in °F) for a certain 24-hour period in winter was given by the formula $T = \frac{1}{20}t(t - 12)(t - 24)$ for $0 \leq t \leq 24$, where t is time in hours and $t = 0$ corresponds to 6 A.M.

(a) When was $T > 0$, and when was $T < 0$?

(b) Sketch the graph of T.

(c) Show that the temperature was 32°F sometime between 12 noon and 1 P.M. (*Hint:* Use the intermediate value theorem.)

44 **Deflections of diving boards** A diver stands at the very end of a diving board before beginning a dive (see the figure).

Exercise 44

The deflection d of the board at a position s feet from the stationary end is given by $d = cs^2(3L - s)$ for $0 \le s \le L$, where L is the length of the board and c is a positive constant that depends on the weight of the diver and on the physical properties of the board. Suppose the board is 10 feet long.

(a) If the deflection at the end of the board is 1 foot, find c.

(b) Show that the deflection is $\frac{1}{2}$ foot somewhere between $s = 6.5$ and $s = 6.6$.

45 **Deer population** A herd of 100 deer is introduced onto a small island. At first the herd increases rapidly, but eventually food resources dwindle and the population declines. Suppose that the number $N(t)$ of deer after t years is given by $N(t) = -t^4 + 21t^2 + 100$, where $t > 0$.

(a) Determine the values of t for which $N(t) > 0$, and sketch the graph of N.

(b) Does the population become extinct? If so, when?

46 **Deer population** Refer to Exercise 45. It can be shown by means of calculus that the rate R (in deer per year) at which the deer population changes at time t is given by $R = -4t^3 + 42t$.

(a) When does the population cease to grow?

(b) Determine the positive values of t for which $R > 0$.

47 (a) Construct a table containing the values of the fourth-degree polynomials

$$f(x) = 2x^4,$$
$$g(x) = 2x^4 - 5x^2 + 1,$$
$$h(x) = 2x^4 + 5x^2 - 1,$$

and

$$k(x) = 2x^4 - x^3 + 2x,$$

when $x = \pm 20, \pm 40,$ and ± 60.

(b) As $|x|$ becomes large, how do the values for each function compare?

(c) Which term has the greatest influence on each function's value when $|x|$ is large?

4.2

Properties of Division

In this section we use $f(x)$, $g(x)$, and so on, to denote polynomials in x. If $g(x)$ is a factor of $f(x)$, then $f(x)$ is **divisible** by $g(x)$. For example, $x^4 - 16$ is divisible by $x^2 - 4$, by $x^2 + 4$, by $x + 2$, and by $x - 2$.

The polynomial $x^4 - 16$ is not divisible by $x^2 + 3x + 1$; however, we can use the process called **long division** to find a *quotient* and a *remainder,* as in the following illustration, where we have inserted terms with zero coefficients.

ILLUSTRATION   **Long Division of Polynomials**

$$
\begin{array}{r}
\overbrace{x^2 - 3x + 8}^{\text{quotient}} \\
x^2 + 3x + 1 \overline{)\, x^4 + 0x^3 + 0x^2 + 0x - 16} \\
\underline{x^4 + 3x^3 + x^2} \qquad\qquad x^2(x^2 + 3x + 1) \\
-3x^3 - x^2 \qquad\qquad \text{subtract} \\
\underline{-3x^3 - 9x^2 - 3x} \qquad -3x(x^2 + 3x + 1) \\
8x^2 + 3x - 16 \quad \text{subtract} \\
\underline{8x^2 + 24x + 8} \qquad 8(x^2 + 3x + 1) \\
-21x - 24 \quad \text{subtract} \\
\underbrace{}_{\text{remainder}}
\end{array}
$$

The long division process ends when we arrive at a polynomial (the remainder) that either is 0 or has smaller degree than the divisor. The result of the long division in the preceding illustration can be written

$$\frac{x^4 - 16}{x^2 + 3x + 1} = (x^2 - 3x + 8) + \left(\frac{-21x - 24}{x^2 + 3x + 1}\right).$$

Multiplying both sides of this equation by $x^2 + 3x + 1$, we obtain

$$x^4 - 16 = (x^2 + 3x + 1)(x^2 - 3x + 8) + (-21x - 24).$$

This example illustrates the following theorem.

Division Algorithm for Polynomials	If $f(x)$ and $p(x)$ are polynomials and if $p(x) \neq 0$, then there exist unique polynomials $q(x)$ and $r(x)$ such that $$f(x) = p(x) \cdot q(x) + r(x),$$ where either $r(x) = 0$ or the degree of $r(x)$ is less than the degree of $p(x)$. The polynomial $q(x)$ is the **quotient,** and $r(x)$ is the **remainder** in the division of $f(x)$ by $p(x)$.

A useful special case of the division algorithm for polynomials occurs if $f(x)$ is divided by $x - c$, where c is a real number. If $x - c$ is a factor of $f(x)$, then

$$f(x) = (x - c)q(x)$$

for some quotient $q(x)$, and the remainder $r(x)$ is 0. If $x - c$ is not a factor of $f(x)$, then the degree of the remainder $r(x)$ is less than the degree of $x - c$, and hence $r(x)$ must have degree 0. This means that the remainder is a nonzero number. Consequently, for every $x - c$ we have

$$f(x) = (x - c)q(x) + d,$$

where the remainder d is a real number (possibly $d = 0$). If we substitute c for x, we obtain

$$\begin{aligned} f(c) &= (c - c)q(c) + d \\ &= 0 \cdot q(c) + d \\ &= 0 + d = d. \end{aligned}$$

This proves the following theorem.

Remainder Theorem	If a polynomial $f(x)$ is divided by $x - c$, then the remainder is $f(c)$.

EXAMPLE 1 Using the remainder theorem

If $f(x) = x^3 - 3x^2 + x + 5$, use the remainder theorem to find $f(2)$.

SOLUTION According to the remainder theorem, $f(2)$ is the remainder when $f(x)$ is divided by $x - 2$. By long division,

$$
\begin{array}{r}
x^2 - x - 1 \\
x - 2 \overline{\smash{\big)}\ x^3 - 3x^2 + x + 5} \\
\end{array}
$$

$x^3 - 2x^2$	$x^2(x - 2)$
$-x^2 + x$	subtract
$-x^2 + 2x$	$-x(x - 2)$
$-x + 5$	subtract
$-x + 2$	$(-1)(x - 2)$
3	subtract

Hence, $f(2) = 3$. We may check this fact by direct substitution:

$$f(2) = 2^3 - 3(2)^2 + 2 + 5 = 3$$

We shall use the remainder theorem to prove the following important result.

Factor Theorem	A polynomial $f(x)$ has a factor $x - c$ if and only if $f(c) = 0$.

PROOF By the remainder theorem,

$$f(x) = (x - c)q(x) + f(c)$$

for some quotient $q(x)$.

If $f(c) = 0$, then $f(x) = (x - c)q(x)$; that is, $x - c$ is a factor of $f(x)$. Conversely, if $x - c$ is a factor of $f(x)$, then the remainder upon division of $f(x)$ by $x - c$ must be 0, and hence, by the remainder theorem, $f(c) = 0$.

The factor theorem is useful for finding factors of polynomials, as illustrated in the next example.

EXAMPLE 2 Using the factor theorem

Show that $x - 2$ is a factor of $f(x) = x^3 - 4x^2 + 3x + 2$.

SOLUTION Since $f(2) = 8 - 16 + 6 + 2 = 0$, we see from the factor theorem that $x - 2$ is a factor of $f(x)$. Another method of solution would be to divide $f(x)$ by $x - 2$ and show that the remainder is 0. The quotient in the division would be another factor of $f(x)$.

EXAMPLE 3 Finding a polynomial with prescribed zeros

Find a polynomial $f(x)$ of degree 3 that has zeros 2, -1, and 3.

SOLUTION By the factor theorem, $f(x)$ has factors $x - 2$, $x + 1$, and $x - 3$. Thus,

$$f(x) = a(x - 2)(x + 1)(x - 3),$$

where any nonzero value may be assigned to a. If we let $a = 1$ and multiply, we obtain

$$f(x) = x^3 - 4x^2 + x + 6.$$

To apply the remainder theorem it is necessary to divide a polynomial $f(x)$ by $x - c$. The method of **synthetic division** may be used to simplify this work. The following guidelines state how to proceed. The method can be justified by a careful (and lengthy) comparison with the method of long division.

Guidelines for Synthetic Division of
$$a_n x^n + a_{n-1} x^{n-1} + \cdots + a_1 x + a_0$$
by $x - c$

1 Begin with the following display, supplying zeros for any missing coefficients in the given polynomial.

$$\underline{c\,|}\ a_n \quad a_{n-1} \quad a_{n-2} \quad \cdots \quad a_1 \quad a_0$$

$$\rule{6cm}{0.4pt}$$

$$a_n$$

2 Multiply a_n by c, and place the product ca_n underneath a_{n-1}, as indicated by the arrow in the following display. (This arrow, and others, is used only to clarify these guidelines and will not appear in *specific* synthetic divisions.) Next find the sum $b_1 = a_{n-1} + ca_n$, and place it below the line as shown.

$$\underline{c\,|}\ a_n \quad a_{n-1} \quad a_{n-2} \quad \cdots \quad\quad\quad a_1 \quad a_0$$
$$ca_n \quad cb_1 \quad cb_2 \quad \cdots \quad\quad cb_{n-2} \quad cb_{n-1}$$
$$a_n \quad b_1 \quad b_2 \quad \cdots \quad\quad b_{n-2} \quad b_{n-1} \quad r$$

3 Multiply b_1 by c, and place the product cb_1 underneath a_{n-2}, as indicated by the second arrow. Proceeding, we next find the sum $b_2 = a_{n-2} + cb_1$ and place it below the line as shown.

4 Continue this process, as indicated by the arrows, until the final sum $r = a_0 + cb_{n-1}$ is obtained. The numbers

$$a_n, \quad b_1, \quad b_2, \quad \ldots, \quad b_{n-2}, \quad b_{n-1}$$

are the coefficients of the quotient $q(x)$; that is,

$$q(x) = a_n x^{n-1} + b_1 x^{n-2} + \cdots + b_{n-2} x + b_{n-1},$$

and r is the remainder.

Synthetic division does not replace long division; it is merely a faster method and is applicable only when the divisor is of the form $x - c$.

The following examples illustrate synthetic division for some special cases.

EXAMPLE 4 **Using synthetic division to find a quotient and remainder**

Use synthetic division to find the quotient $q(x)$ and remainder r if the polynomial $2x^4 + 5x^3 - 2x - 8$ is divided by $x + 3$.

SOLUTION Since the divisor is $x + 3 = x - (-3)$, the value of c in the expression $x - c$ is -3. Hence, the synthetic division takes this form:

$$
\begin{array}{r|rrrrr}
-3 & 2 & 5 & 0 & -2 & -8 \\
   &   & -6 & 3 & -9 & 33 \\
\hline
   & 2 & -1 & 3 & -11 & 25
\end{array}
$$

coefficients   remainder
of quotient

As we have indicated, the first four numbers in the third row are the coefficients of the quotient $q(x)$, and the last number is the remainder r. Thus,

$$q(x) = 2x^3 - x^2 + 3x - 11 \qquad \text{and} \qquad r = 25.$$

Synthetic division can be used to find values of polynomial functions, as illustrated in the next example.

EXAMPLE 5 **Using synthetic division to find values of a polynomial**

If $f(x) = 3x^5 - 38x^3 + 5x^2 - 1$, use synthetic division to find $f(4)$.

SOLUTION By the remainder theorem, $f(4)$ is the remainder when $f(x)$ is divided by $x - 4$. Dividing synthetically, we obtain

$$
\begin{array}{r|rrrrrr}
4 & 3 & 0 & -38 & 5 & 0 & -1 \\
  &   & 12 & 48 & 40 & 180 & 720 \\
\hline
  & 3 & 12 & 10 & 45 & 180 & 719
\end{array}
$$

coefficients   remainder
of quotient

Consequently, $f(4) = 719$.

Synthetic division may be used to help find zeros of polynomials. By the method illustrated in the preceding example, $f(c) = 0$ if and only if the remainder in the synthetic division by $x - c$ is 0.

EXAMPLE 6 **Using synthetic division to find zeros of a polynomial**

Show that -11 is a zero of the polynomial

$$f(x) = x^3 + 8x^2 - 29x + 44.$$

SOLUTION Dividing synthetically by $x - (-11) = x + 11$ gives us

*The quotient gives us the **depressed equation**,*

$$x^2 - 3x + 4 = 0, \rightarrow$$

which can be used to find the remaining zeros of f.

$$\begin{array}{r|rrrr} -11 & 1 & 8 & -29 & 44 \\ & & -11 & 33 & -44 \\ \hline & 1 & -3 & 4 & 0 \end{array}$$

$\underbrace{}_{\substack{\text{coefficients} \\ \text{of quotient}}}$ $\underbrace{}_{\text{remainder}}$

Thus, $f(-11) = 0$, and -11 is a zero of f.

Example 6 shows that the number -11 is a solution of the equation $x^3 + 8x^2 - 29x + 44 = 0$. In Section 4.4 we shall use synthetic division to find rational solutions of equations.

At this stage you should recognize that the following three statements are equivalent for a polynomial function f whose graph is the graph of the equation $y = f(x)$.

equivalent statements for $f(a) = b$

- **(1)** The point (a, b) is on the graph of f.
- **(2)** The value of f at $x = a$ equals b; that is, $f(a) = b$.
- **(3)** If $f(x)$ is divided by $x - a$, then the remainder is b.

Furthermore, if b is equal to 0, then the next four statements are also equivalent.

additional equivalent statements for $f(a) = 0$

- **(1)** The number a is a zero of the function f.
- **(2)** The point $(a, 0)$ is on the graph of f; that is, a is an x-intercept.
- **(3)** The number a is a solution of the equation $f(x) = 0$.
- **(4)** The binomial $x - a$ is a factor of the polynomial $f(x)$.

You should become familiar with these statements—so familiar that if you know one of them is true, you can easily recall and apply any appropriate equivalent statement.

4.2 Exercises

Exer. 1–8: Find the quotient and remainder if $f(x)$ is divided by $p(x)$.

1 $f(x) = 2x^4 - x^3 - 3x^2 + 7x - 12;$ $p(x) = x^2 - 3$

2 $f(x) = 3x^4 + 2x^3 - x^2 - x - 6;$ $p(x) = x^2 + 1$

3 $f(x) = 3x^3 + 2x - 4;$ $p(x) = 2x^2 + 1$

4 $f(x) = 3x^3 - 5x^2 - 4x - 8;$ $p(x) = 2x^2 + x$

5 $f(x) = 7x + 2;$ $p(x) = 2x^2 - x - 4$

6 $f(x) = -5x^2 + 3;$ $p(x) = x^3 - 3x + 9$

7 $f(x) = 9x + 4;$ $p(x) = 2x - 5$

8 $f(x) = 7x^2 + 3x - 10;$ $p(x) = x^2 - x + 10$

Exer. 9–12: Use the remainder theorem to find $f(c)$.

9 $f(x) = 3x^3 - x^2 + 5x - 4;$ $c = 2$

10 $f(x) = 2x^3 + 4x^2 - 3x - 1;$ $c = 3$

11 $f(x) = x^4 - 6x^2 + 4x - 8;$ $c = -3$

12 $f(x) = x^4 + 3x^2 - 12;$ $c = -2$

Exer. 13–16: Use the factor theorem to show that $x - c$ is a factor of $f(x)$.

13 $f(x) = x^3 + x^2 - 2x + 12$; $c = -3$

14 $f(x) = x^3 + x^2 - 11x + 10$; $c = 2$

15 $f(x) = x^{12} - 4096$; $c = -2$

16 $f(x) = x^4 - 3x^3 - 2x^2 + 5x + 6$; $c = 2$

Exer. 17–20: Find a polynomial $f(x)$ with leading coefficient 1 and having the given degree and zeros.

17 degree 3; zeros $-2, 0, 5$

18 degree 3; zeros $\pm 2, 3$

19 degree 4; zeros $-2, \pm 1, 4$

20 degree 4; zeros $-3, 0, 1, 5$

Exer. 21–28: Use synthetic division to find the quotient and remainder if the first polynomial is divided by the second.

21 $2x^3 - 3x^2 + 4x - 5$; $x - 2$

22 $3x^3 - 4x^2 - x + 8$; $x + 4$

23 $x^3 - 8x - 5$; $x + 3$

24 $5x^3 - 6x^2 + 15$; $x - 4$

25 $3x^5 + 6x^2 + 7$; $x + 2$

26 $-2x^4 + 10x - 3$; $x - 3$

27 $4x^4 - 5x^2 + 1$; $x - \frac{1}{2}$

28 $9x^3 - 6x^2 + 3x - 4$; $x - \frac{1}{3}$

Exer. 29–34: Use synthetic division to find $f(c)$.

29 $f(x) = 2x^3 + 3x^2 - 4x + 4$; $c = 3$

30 $f(x) = -x^3 + 4x^2 + x$; $c = -2$

31 $f(x) = 0.3x^3 + 0.04x - 0.034$; $c = -0.2$

32 $f(x) = 8x^5 - 3x^2 + 7$; $c = \frac{1}{2}$

33 $f(x) = x^2 + 3x - 5$; $c = 2 + \sqrt{3}$

34 $f(x) = x^3 - 3x^2 - 8$; $c = 1 + \sqrt{2}$

Exer. 35–38: Use synthetic division to show that c is a zero of $f(x)$.

35 $f(x) = 3x^4 + 8x^3 - 2x^2 - 10x + 4$; $c = -2$

36 $f(x) = 4x^3 - 9x^2 - 8x - 3$; $c = 3$

37 $f(x) = 4x^3 - 6x^2 + 8x - 3$; $c = \frac{1}{2}$

38 $f(x) = 27x^4 - 9x^3 + 3x^2 + 6x + 1$; $c = -\frac{1}{3}$

Exer. 39–40: Find all values of k such that $f(x)$ is divisible by the given linear polynomial.

39 $f(x) = kx^3 + x^2 + k^2x + 3k^2 + 11$; $x + 2$

40 $f(x) = k^2x^3 - 4kx + 3$; $x - 1$

Exer. 41–42: Show that $x - c$ is not a factor of $f(x)$ for any real number c.

41 $f(x) = 3x^4 + x^2 + 5$ 42 $f(x) = -x^4 - 3x^2 - 2$

43 Find the remainder if the polynomial

$$3x^{100} + 5x^{85} - 4x^{38} + 2x^{17} - 6$$

is divided by $x + 1$.

Exer. 44–46: Use the factor theorem to verify the statement.

44 $x - y$ is a factor of $x^n - y^n$ for every positive integer n.

45 $x + y$ is a factor of $x^n - y^n$ for every positive even integer n.

46 $x + y$ is a factor of $x^n + y^n$ for every positive odd integer n.

47 Let $P(x, y)$ be a first-quadrant point on $y = 6 - x$, and consider the vertical line segment PQ shown in the figure.

(a) If PQ is rotated about the y-axis, determine the volume V of the resulting cylinder.

(b) For what point $P(x, y)$ with $x \neq 1$ is the volume V in part (a) the same as the volume of the cylinder of radius 1 and altitude 5 shown in the figure?

Exercise 47

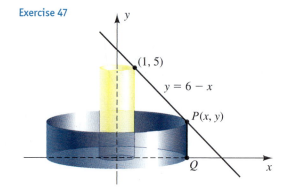

48 Strength of a beam The strength of a rectangular beam is directly proportional to the product of its width and the square of the depth of a cross section (see the figure). A beam of width 1.5 feet has been cut from a cylindrical log of radius 1 foot. Find the width of a second rectangular beam of equal strength that could have been cut from the log.

Exercise 48

49 Parabolic arch An arch has the shape of the parabola $y = 4 - x^2$. A rectangle is fit under the arch by selecting a point (x, y) on the parabola (see the figure).

Exercise 49

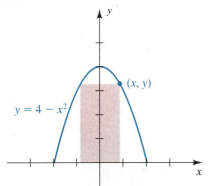

(a) Express the area A of the rectangle in terms of x.

(b) If $x = 1$, the rectangle has base 2 and height 3. Find the base of a second rectangle that has the same area.

50 Dimensions of a capsule An aspirin tablet in the shape of a right circular cylinder has height $\frac{1}{3}$ centimeter and radius $\frac{1}{2}$ centimeter. The manufacturer also wishes to market the aspirin in capsule form. The capsule is to be $\frac{3}{2}$ centimeters long, in the shape of a right circular cylinder with hemispheres attached at both ends (see the figure).

(a) If r denotes the radius of a hemisphere, find a formula for the volume of the capsule.

(b) Find the radius of the capsule so that its volume is equal to that of the tablet.

Exercise 50

4.3
Zeros of Polynomials

The **zeros of a polynomial** $f(x)$ are the solutions of the equation $f(x) = 0$. Each real zero is an x-intercept of the graph of f. In applied fields, calculators and computers are usually used to find or approximate zeros. Before using a calculator, however, it is worth knowing what type of zeros to expect. Some questions we could ask are

(1) How many zeros of $f(x)$ are real? imaginary?

(2) How many real zeros of $f(x)$ are positive? negative?

(3) How many real zeros of $f(x)$ are rational? irrational?

(4) Are the real zeros of $f(x)$ large or small in value?

In this and the following section we shall discuss results that help answer some of these questions. These results form the basis of the *theory of equations*.

The factor and remainder theorems can be extended to the system of complex numbers. Thus, a complex number $c = a + bi$ is a zero of a polynomial $f(x)$ if and only if $x - c$ is a factor of $f(x)$. Except in special cases, zeros of polynomials are very difficult to find. For example, there are no obvious zeros of $f(x) = x^5 - 3x^4 + 4x^3 - 4x - 10$. Although we have no formula that can be used to find the zeros, the next theorem states that there is at *least* one zero c, and hence, by the factor theorem, $f(x)$ has a factor of the form $x - c$.

Fundamental Theorem of Algebra	If a polynomial $f(x)$ has positive degree and complex coefficients, then $f(x)$ has at least one complex zero.

The standard proof of this theorem requires results from an advanced field of mathematics called *functions of a complex variable*. A prerequisite for studying this field is a strong background in calculus. The first proof of the fundamental theorem of algebra was given by the German mathematician Carl Friedrich Gauss (1777–1855), who is considered by many to be the greatest mathematician of all time.

As a special case of the fundamental theorem of algebra, if all the coefficients of $f(x)$ are real, then $f(x)$ has at least one complex zero. If $a + bi$ is a complex zero, it may happen that $b = 0$, in which case the number a is a real zero.

The fundamental theorem of algebra enables us, at least in theory, to express every polynomial $f(x)$ of positive degree as a product of polynomials of degree 1, as in the next theorem.

Complete Factorization Theorem for Polynomials	If $f(x)$ is a polynomial of degree $n > 0$, then there exist n complex numbers $c_1, c_2, \ldots, c_n$ such that $$f(x) = a(x - c_1)(x - c_2) \cdots (x - c_n),$$ where a is the leading coefficient of $f(x)$. Each number c_k is a zero of $f(x)$.

PROOF If $f(x)$ has degree $n > 0$, then, by the fundamental theorem of algebra, $f(x)$ has a complex zero c_1. Hence, by the factor theorem, $f(x)$ has a factor $x - c_1$; that is,

$$f(x) = (x - c_1)f_1(x),$$

where $f_1(x)$ is a polynomial of degree $n - 1$. If $n - 1 > 0$, then, by the same argument, $f_1(x)$ has a complex zero c_2 and therefore a factor $x - c_2$. Thus,

$$f_1(x) = (x - c_2)f_2(x),$$

where $f_2(x)$ is a polynomial of degree $n - 2$. Hence,

$$f(x) = (x - c_1)(x - c_2)f_2(x).$$

Continuing this process, after n steps we arrive at a polynomial $f_n(x)$ of degree 0. Thus, $f_n(x) = a$ for some nonzero number a, and we may write

$$f(x) = a(x - c_1)(x - c_2)\cdots(x - c_n),$$

where each complex number c_k is a zero of $f(x)$. The leading coefficient of the polynomial on the right-hand side in the last equation is a, and therefore a is the leading coefficient of $f(x)$. ◪

ILLUSTRATION   **Complete Factorization Theorem for Polynomials**

A Polynomial $f(x)$	A Factored Form of $f(x)$	Zeros of $f(x)$
■ $3x^2 - (12 + 6i)x + 24i$	$3(x - 4)(x - 2i)$	$4, 2i$
■ $-6x^3 - 2x^2 - 6x - 2$	$-6\left(x + \frac{1}{3}\right)(x + i)(x - i)$	$-\frac{1}{3}, \pm i$
■ $5x^3 - 30x^2 + 65x$	$5(x - 0)[x - (3 + 2i)][x - (3 - 2i)]$	$0, 3 \pm 2i$
■ $\frac{2}{3}x^3 + 8x^2 - \frac{2}{3}x - 8$	$\frac{2}{3}(x + 12)(x + 1)(x - 1)$	$-12, \pm 1$

We may now prove the following.

Theorem on the Maximum Number of Zeros of a Polynomial	A polynomial of degree $n > 0$ has at most n different complex zeros.

PROOF   We will give an indirect proof; that is, we will suppose $f(x)$ has *more* than n different complex zeros and show that this supposition leads to a contradiction. Let us choose $n + 1$ of the zeros and label them $c_1, c_2, \ldots, c_n$, and c. We may use the c_k to obtain the factorization indicated in the statement of the complete factorization theorem for polynomials. Substituting c for x and using the fact that $f(c) = 0$, we obtain

$$0 = a(c - c_1)(c - c_2)\cdots(c - c_n).$$

However, each factor on the right-hand side is different from zero because $c \neq c_k$ for every k. Since the product of nonzero numbers cannot equal zero, we have a contradiction. ◪

EXAMPLE 1 Finding a polynomial with prescribed zeros

Find a polynomial $f(x)$ in factored form that has degree 3; has zeros 2, -1, and 3; and satisfies $f(1) = 5$.

SOLUTION By the factor theorem, $f(x)$ has factors $x - 2$, $x + 1$, and $x - 3$. No other factors of degree 1 exist, since, by the factor theorem, another linear factor $x - c$ would produce a fourth zero of $f(x)$, contrary to the preceding theorem. Hence, $f(x)$ has the form

$$f(x) = a(x - 2)(x + 1)(x - 3)$$

for some number a. Since $f(1) = 5$, we see that

$$5 = a(1 - 2)(1 + 1)(1 - 3) \quad \text{let } x = 1 \text{ in } f(x)$$

$$5 = a(-1)(2)(-2) \quad \text{simplify}$$

$$a = \tfrac{5}{4} \quad \text{solve for } a$$

Consequently,

$$f(x) = \tfrac{5}{4}(x - 2)(x + 1)(x - 3).$$

If we multiply the factors, we obtain the polynomial

$$f(x) = \tfrac{5}{4}x^3 - 5x^2 + \tfrac{5}{4}x + \tfrac{15}{2}.$$

The numbers $c_1, c_2, \ldots, c_n$ in the complete factorization theorem are not necessarily all different. To illustrate, $f(x) = x^3 + x^2 - 5x + 3$ has the factorization

$$f(x) = (x + 3)(x - 1)(x - 1).$$

If a factor $x - c$ occurs m times in the factorization, then c is a **zero of multiplicity** m of the polynomial $f(x)$, or a **root of multiplicity** m of the equation $f(x) = 0$. In the preceding display, 1 is a zero of multiplicity 2, and -3 is a zero of multiplicity 1.

If c is a real zero of $f(x)$ of multiplicity m, then $f(x)$ has the factor $(x - c)^m$ and the graph of f has an x-intercept c. The general shape of the graph at $(c, 0)$ depends on whether m is an odd integer or an even integer. If m is odd, then $(x - c)^m$ changes sign as x increases through c, and hence the graph of f crosses the x-axis at $(c, 0)$, as indicated in the first row of the following chart. The figures in the chart do not show the complete graph of f, but only its general shape near $(c, 0)$. If m is even, then $(x - c)^m$ does not change sign at c and the graph of f near $(c, 0)$ has the appearance of one of the two figures in the second row.

Factor of $f(x)$	General shape of the graph of f near $(c, 0)$	
$(x - c)^m$, with m odd and $m \neq 1$	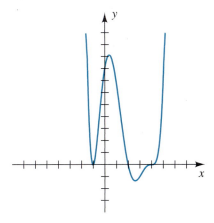	
$(x - c)^m$, with m even		

EXAMPLE 2   Finding multiplicities of zeros

Find the zeros of the polynomial $f(x) = \frac{1}{16}(x - 2)(x - 4)^3(x + 1)^2$, state the multiplicity of each, and then sketch the graph of f.

SOLUTION   We see from the factored form that $f(x)$ has three distinct zeros, 2, 4, and -1. The zero 2 has multiplicity 1, the zero 4 has multiplicity 3, and the zero -1 has multiplicity 2. Note that $f(x)$ has degree 6.

The x-intercepts of the graph of f are the real zeros -1, 2, and 4. Since the multiplicity of -1 is an even integer, the graph intersects, but does not cross, the x-axis at $(-1, 0)$. Since the multiplicities of 2 and 4 are odd, the graph crosses the x-axis at $(2, 0)$ and $(4, 0)$. (Note that the graph is "flatter" at 4 than at 2.) The y-intercept is $f(0) = \frac{1}{16}(-2)(-4)^3(1)^2 = 8$. The graph is shown in Figure 1.

Figure 1

If $f(x) = a(x - c_1)(x - c_2) \cdots (x - c_n)$ is a polynomial of degree n, then the n complex numbers $c_1, c_2, \ldots, c_n$ are zeros of $f(x)$. Counting a zero of multiplicity m as m zeros tells us that $f(x)$ has at least n zeros (not necessarily all different). Combining this fact with the fact that $f(x)$ has at most n zeros gives us the next result.

Theorem on the Exact Number of Zeros of a Polynomial	If $f(x)$ is a polynomial of degree $n > 0$ and if a zero of multiplicity m is counted m times, then $f(x)$ has precisely n zeros.

Notice how the polynomial of degree 6 in Example 2 relates to the last theorem. The multiplicities are 1, 3, and 2, so f has precisely $1 + 3 + 2 = 6$ zeros.

EXAMPLE 3   Finding the zeros of a polynomial

Express $f(x) = x^5 - 4x^4 + 13x^3$ as a product of linear factors, and find the five zeros of $f(x)$.

SOLUTION   We begin by factoring out x^3:

$$f(x) = x^3(x^2 - 4x + 13)$$

By the quadratic formula, the zeros of the polynomial $x^2 - 4x + 13$ are

$$\frac{-(-4) \pm \sqrt{(-4)^2 - 4(1)(13)}}{2(1)} = \frac{4 \pm \sqrt{-36}}{2} = \frac{4 \pm 6i}{2} = 2 \pm 3i.$$

Hence, by the factor theorem, $x^2 - 4x + 13$ has factors $x - (2 + 3i)$ and $x - (2 - 3i)$, and we obtain the factorization

$$f(x) = x \cdot x \cdot x \cdot (x - 2 - 3i)(x - 2 + 3i).$$

Since $x - 0$ occurs as a factor three times, the number 0 is a zero of multiplicity 3, and the five zeros of $f(x)$ are 0, 0, 0, $2 + 3i$, and $2 - 3i$. ◢

We next show how to use *Descartes' rule of signs* to obtain information about the zeros of a polynomial $f(x)$ with real coefficients. In the statement of the rule we assume that the terms of $f(x)$ are arranged in order of decreasing powers of x and that terms with zero coefficients are deleted. We also assume that the **constant term**—that is, the term that does not contain x—is different from 0. We say there is a **variation of sign** in $f(x)$ if two consecutive coefficients have opposite signs. To illustrate, the polynomial $f(x)$ in the following illustration has *three* variations of sign, as indicated by the braces—one variation from $2x^5$ to $-7x^4$, a second from $-7x^4$ to $3x^2$, and a third from $6x$ to -5.

ILLUSTRATION   **Variations of Sign in $f(x) = 2x^5 - 7x^4 + 3x^2 + 6x - 5$**

$\qquad\qquad$ + to −$\qquad$ − to +$\qquad$ no variation$\qquad$ + to −

■ $\quad f(x) = 2x^5 \quad - 7x^4 \quad + 3x^2 \quad + 6x \quad - 5$

Descartes' rule also refers to the variations of sign in $f(-x)$. Using the previous illustration, note that

$$f(-x) = 2(-x)^5 - 7(-x)^4 + 3(-x)^2 + 6(-x) - 5$$
$$= -2x^5 - 7x^4 + 3x^2 - 6x - 5.$$

Hence, as indicated in the next illustration, there are *two* variations of sign in $f(-x)$—one from $-7x^4$ to $3x^2$ and a second from $3x^2$ to $-6x$.

ILLUSTRATION   **Variations of Sign in $f(-x)$ if $f(x) = 2x^5 - 7x^4 + 3x^2 + 6x - 5$**

	no variation		$-$ to $+$	$+$ to $-$	no variation	
■ $f(-x) = -2x^5$		$- 7x^4$	$+ 3x^2$	$- 6x$		$- 5$

We may state Descartes' rule as follows.

Descartes' Rule of Signs	Let $f(x)$ be a polynomial with real coefficients and a nonzero constant term.
	(1) The number of *positive* real zeros of $f(x)$ either is equal to the number of variations of sign in $f(x)$ or is less than that number by an even integer.
	(2) The number of *negative* real zeros of $f(x)$ either is equal to the number of variations of sign in $f(-x)$ or is less than that number by an even integer.

A proof of Descartes' rule will not be given.

EXAMPLE 4   Using Descartes' rule of signs

Discuss the number of possible positive and negative real solutions and imaginary solutions of the equation $f(x) = 0$, where

$$f(x) = 2x^5 - 7x^4 + 3x^2 + 6x - 5.$$

SOLUTION   The polynomial $f(x)$ is the one given in the two previous illustrations. Since there are three variations of sign in $f(x)$, the equation has either three positive real solutions or one positive real solution.

Since $f(-x)$ has two variations of sign, the equation has either two negative solutions or no negative solution. Because $f(x)$ has degree 5, there are a total of 5 solutions. The solutions that are not positive or negative real numbers are imaginary numbers. The following table summarizes the various possibilities that can occur for solutions of the equation.

(continued)

Number of positive real solutions	3	3	1	1
Number of negative real solutions	2	0	2	0
Number of imaginary solutions	0	2	2	4
Total number of solutions	5	5	5	5

Descartes' rule stipulates that the constant term of the polynomial $f(x)$ is different from 0. If the constant term is 0, as in the equation

$$x^4 - 3x^3 + 2x^2 - 5x = 0,$$

we factor out the lowest power of x, obtaining

$$x(x^3 - 3x^2 + 2x - 5) = 0.$$

Thus, one solution is $x = 0$, and we apply Descartes' rule to the polynomial $x^3 - 3x^2 + 2x - 5$ to determine the nature of the remaining three solutions.

When applying Descartes' rule, we count roots of multiplicity k as k roots. For example, given $x^2 - 2x + 1 = 0$, the polynomial $x^2 - 2x + 1$ has two variations of sign, and hence the equation has either two positive real roots or none. The factored form of the equation is $(x - 1)^2 = 0$, and hence 1 is a root of multiplicity 2.

We next discuss the *bounds* for the real zeros of a polynomial $f(x)$ that has real coefficients. By definition, a real number b is an **upper bound** for the zeros if no zero is greater than b. A real number a is a **lower bound** for the zeros if no zero is less than a. Thus, if r is any real zero of $f(x)$, then $a \leq r \leq b$; that is, r is in the closed interval $[a, b]$, as illustrated in Figure 2. Note that upper and lower bounds are not unique, since any number greater than b is also an upper bound and any number less than a is also a lower bound.

Figure 2

We may use synthetic division to find upper and lower bounds for the zeros of $f(x)$. Recall that if we divide $f(x)$ synthetically by $x - c$, the third row in the division process contains the coefficients of the quotient $q(x)$ together with the remainder $f(c)$. The following theorem indicates how this third row may be used to find upper and lower bounds for the real solutions.

Theorem on Bounds for Real Zeros of Polynomials	Suppose that $f(x)$ is a polynomial with real coefficients and a positive leading coefficient and that $f(x)$ is divided synthetically by $x - c$.

(1) If $c > 0$ and if all numbers in the third row of the division process are either positive or zero, then c is an upper bound for the real zeros of $f(x)$.

(2) If $c < 0$ and if the numbers in the third row of the division process are alternately positive and negative (and a 0 in the third row is considered to be either positive or negative), then c is a lower bound for the real zeros of $f(x)$.

EXAMPLE 5   Finding bounds for the solutions of an equation

Find upper and lower bounds for the real solutions of the equation $f(x) = 0$, where $f(x) = 2x^3 + 5x^2 - 8x - 7$.

SOLUTION   We divide $f(x)$ synthetically by $x - 1$ and $x - 2$.

$$\begin{array}{r|rrrr} 1 & 2 & 5 & -8 & -7 \\ & & 2 & 7 & -1 \\ \hline & 2 & 7 & -1 & -8 \end{array} \qquad \begin{array}{r|rrrr} 2 & 2 & 5 & -8 & -7 \\ & & 4 & 18 & 20 \\ \hline & 2 & 9 & 10 & 13 \end{array}$$

The third row of the synthetic division by $x - 1$ contains negative numbers, and hence part (1) of the theorem on bounds for real zeros of polynomials does not apply. However, since all numbers in the third row of the synthetic division by $x - 2$ are positive, it follows from part (1) that 2 is an upper bound for the real solutions of the equation. This fact is also evident if we express the division by $x - 2$ in the division algorithm form

$$2x^3 + 5x^2 - 8x - 7 = (x - 2)(2x^2 + 9x + 10) + 13,$$

for if $x > 2$, then the right-hand side of the equation is positive (why?), and hence $f(x)$ is not zero.

We now find a lower bound. After some trial-and-error attempts using $x - (-1)$, $x - (-2)$, and $x - (-3)$, we see that synthetic division of f by $x - (-4)$ gives us

$$\begin{array}{r|rrrr} -4 & 2 & 5 & -8 & -7 \\ & & -8 & 12 & -16 \\ \hline & 2 & -3 & 4 & -23 \end{array}$$

Since the numbers in the third row are alternately positive and negative, it follows from part (2) of the preceding theorem that -4 is a lower bound for the

(continued)

Figure 3

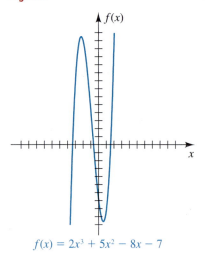

$$f(x) = 2x^3 + 5x^2 - 8x - 7$$

real solutions. This can also be proved by expressing the division by $x + 4$ in the form

$$2x^3 + 5x^2 - 8x - 7 = (x + 4)(2x^2 - 3x + 4) - 23,$$

for if $x < -4$, then the right-hand side of this equation is negative (why?), and hence $f(x)$ is not zero.

Since lower and upper bounds for the real solutions are -4 and 2, respectively, it follows that all real solutions are in the closed interval $[-4, 2]$.

The graph of f in Figure 3 shows that the three zeros of f are in the intervals $[-4, -3]$, $[-1, 0]$, and $[1, 2]$, respectively. ◢

EXAMPLE 6 **Finding a polynomial from a graph**

Shown in Figure 4 are all the zeros of a polynomial function f.

(a) Find a factored form for f that has minimal degree.

(b) Assuming the leading coefficient of f is 1, find the y-intercept.

SOLUTION

(a) The zero at $x = -2$ must have a multiplicity that is an even number, since f does not change sign at $x = -2$. The zero at $x = 1$ must have an odd multiplicity of 3 or greater, since f changes sign at $x = 1$ and levels off. The zero at $x = 3$ is of multiplicity 1, since f changes sign and does not level off. Thus, a factored form of f is

$$f(x) = a(x + 2)^m(x - 1)^n(x - 3)^1.$$

Because we desire the function having minimal degree, we let $m = 2$ and $n = 3$, obtaining

$$f(x) = a(x + 2)^2(x - 1)^3(x - 3),$$

which is a sixth-degree polynomial.

(b) If the leading coefficient of f is to be 1, then, from the complete factorization theorem for polynomials, we know that the value of a is 1. To find the y-intercept, we let $x = 0$ and compute $f(0)$:

$$f(0) = 1(0 + 2)^2(0 - 1)^3(0 - 3) = 1(4)(-1)(-3) = 12$$

Hence, the y-intercept is 12. ◢

Figure 4

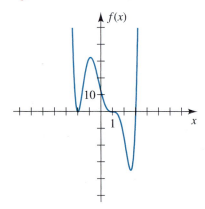

4.3 Exercises

Exer. 1–6: Find a polynomial $f(x)$ of degree 3 that has the indicated zeros and satisfies the given condition.

1 $-1, 2, 3;$ $f(-2) = 80$

2 $-5, 2, 4;$ $f(3) = -24$

3 $-4, 3, 0;$ $f(2) = -36$

4 $-3, -2, 0;$ $f(-4) = 16$

5 $-2i, 2i, 3;$ $f(1) = 20$

6 $-3i, 3i, 4;$ $f(-1) = 50$

7 Find a polynomial $f(x)$ of degree 4 with leading coefficient 1 such that both -4 and 3 are zeros of multiplicity 2, and sketch the graph of f.

8 Find a polynomial $f(x)$ of degree 4 with leading coefficient 1 such that both -5 and 2 are zeros of multiplicity 2, and sketch the graph of f.

9 Find a polynomial $f(x)$ of degree 6 such that 0 and 3 are both zeros of multiplicity 3 and $f(2) = -24$. Sketch the graph of f.

10 Find a polynomial $f(x)$ of degree 7 such that -2 and 2 are both zeros of multiplicity 2, 0 is a zero of multiplicity 3, and $f(-1) = 27$. Sketch the graph of f.

11 Find the third-degree polynomial function whose graph is shown in the figure.

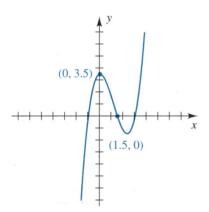

12 Find the fourth-degree polynomial function whose graph is shown in the figure.

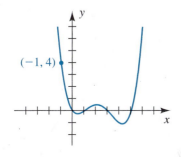

Exer. 13–14: Find the polynomial function of degree 3 whose graph is shown in the figure.

13

14

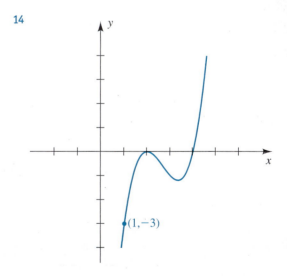

Exer. 15–22: Find the zeros of $f(x)$, and state the multiplicity of each zero.

15 $f(x) = x^2(3x + 2)(2x - 5)^3$

16 $f(x) = x(x + 1)^4(3x - 7)^2$

17 $f(x) = 4x^5 + 12x^4 + 9x^3$

18 $f(x) = (4x^2 - 5)^2$

19 $f(x) = (x^2 + x - 12)^3(x^2 - 9)^2$

20 $f(x) = (6x^2 + 7x - 5)^4(4x^2 - 1)^2$

21 $f(x) = x^4 + 7x^2 - 144$

22 $f(x) = x^4 + 21x^2 - 100$

Exer. 23–26: Show that the number is a zero of $f(x)$ of the given multiplicity, and express $f(x)$ as a product of linear factors.

23 $f(x) = x^4 + 7x^3 + 13x^2 - 3x - 18;$ -3 (multiplicity 2)

24 $f(x) = x^4 - 9x^3 + 22x^2 - 32;$ 4 (multiplicity 2)

25 $f(x) = x^6 - 4x^5 + 5x^4 - 5x^2 + 4x - 1;$
1 (multiplicity 5)

26 $f(x) = x^5 + x^4 - 6x^3 - 14x^2 - 11x - 3;$
-1 (multiplicity 4)

Exer. 27–34: Use Descartes' rule of signs to determine the number of possible positive, negative, and nonreal complex solutions of the equation.

27 $4x^3 - 6x^2 + x - 3 = 0$

28 $5x^3 - 6x - 4 = 0$

29 $4x^3 + 2x^2 + 1 = 0$

30 $3x^3 - 4x^2 + 3x + 7 = 0$

31 $3x^4 + 2x^3 - 4x + 2 = 0$

32 $2x^4 - x^3 + x^2 - 3x + 4 = 0$

33 $x^5 + 4x^4 + 3x^3 - 4x + 2 = 0$

34 $2x^6 + 5x^5 + 2x^2 - 3x + 4 = 0$

Exer. 35–40: Applying the theorem on bounds for real zeros of polynomials, determine the smallest and largest integers that are upper and lower bounds, respectively, for the real solutions of the equation.

35 $x^3 - 4x^2 - 5x + 7 = 0$

36 $2x^3 - 5x^2 + 4x - 8 = 0$

37 $x^4 - x^3 - 2x^2 + 3x + 6 = 0$

38 $2x^4 - 9x^3 - 8x - 10 = 0$

39 $2x^5 - 13x^3 + 2x - 5 = 0$

40 $3x^5 + 2x^4 - x^3 - 8x^2 - 7 = 0$

Exer. 41–42: Find a factored form for a polynomial function f that has a minimal degree. Assume that the intercept values are integers.

41

42

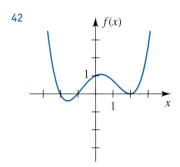

Exer. 43–44: (a) Find a factored form for a polynomial function f that has minimal degree. Assume that the intercept values are integers. (b) If the leading coefficient of f is a, find the y-intercept.

43 $a = 1$

44 $a = -1$

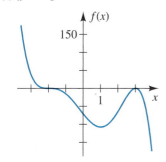

Exer. 45–48: Is there a polynomial of the given degree n whose graph contains the indicated points?

45 $n = 4$;
$(-2, 0), (0, -24), (1, 0), (3, 0), (2, 0), (-1, -52)$

46 $n = 5$;
$(0, 0), (-3, 0), (-1, 0), (2, 0), (3, 0), (-2, 5), (1, 2)$

47 $n = 3$;
$(1.1, -49.815), (2, 0), (3.5, 25.245), (5.2, 0),$
$(6.4, -29.304), (10.1, 0)$

48 $n = 4$;
$(1.25, 0), (2, 0), (2.5, 56.25), (3, 128.625), (6.5, 0),$
$(9, -307.75), (10, 0)$

49 Using limited data A scientist has limited data on the temperature T (in °C) during a 24-hour period. If t denotes time in hours and $t = 0$ corresponds to midnight, find the fourth-degree polynomial that fits the information in the following table.

t (hours)	0	5	12	19	24
T (°C)	0	0	10	0	0

50 Lagrange interpolation polynomial A polynomial $f(x)$ of degree 3 with zeros at c_1, c_2, and c_3 and with $f(c) = 1$ for $c_2 < c < c_3$ is a third-degree *Lagrange interpolation polynomial*. Find an explicit formula for $f(x)$ in terms of c_1, c_2, c_3, and c.

4.4

Complex and Rational Zeros of Polynomials

Example 3 of the preceding section illustrates an important fact about polynomials with real coefficients: The two complex zeros $2 + 3i$ and $2 - 3i$ of $x^5 - 4x^4 + 13x^3$ are conjugates of each other. The relationship is not accidental, since the following general result is true.

Theorem on Conjugate Pair Zeros of a Polynomial	If a polynomial $f(x)$ of degree $n > 1$ has real coefficients and if $z = a + bi$ with $b \neq 0$ is a complex zero of $f(x)$, then the conjugate $\bar{z} = a - bi$ is also a zero of $f(x)$.

A proof is left as a discussion exercise at the end of the chapter.

EXAMPLE 1   **Finding a polynomial with prescribed zeros**

Find a polynomial $f(x)$ of degree 4 that has real coefficients and zeros $2 + i$ and $-3i$.

SOLUTION   By the theorem on conjugate pair zeros of a polynomial, $f(x)$ must also have zeros $2 - i$ and $3i$. Applying the factor theorem, we find that $f(x)$ has the following factors:

$$x - (2 + i), \quad x - (2 - i), \quad x - (-3i), \quad x - (3i)$$

(continued)

Multiplying these four factors gives us

$$f(x) = [x - (2 + i)][x - (2 - i)](x + 3i)(x - 3i)$$
$$= (x^2 - 4x + 5)(x^2 + 9) \qquad (*)$$
$$= x^4 - 4x^3 + 14x^2 - 36x + 45.$$

Note that in $(*)$ the symbol i does not appear. This is not a coincidence, since if $a + bi$ is a zero of a polynomial with real coefficients, then $a - bi$ is also a zero and we can multiply the associated factors as follows:

$$[x - (a + bi)][x - (a - bi)] = x^2 - 2ax + a^2 + b^2$$

In Example 1 we have $a = 2$ and $b = 1$, so $-2a = -4$ and $a^2 + b^2 = 5$ and the associated quadratic factor is $x^2 - 4x + 5$. This resulting quadratic factor will always have real coefficients, as stated in the next theorem.

Theorem on Expressing a Polynomial as a Product of Linear and Quadratic Factors	Every polynomial with real coefficients and positive degree n can be expressed as a product of linear and quadratic polynomials with real coefficients such that the quadratic factors are irreducible over $\mathbb{R}$.

PROOF   Since $f(x)$ has precisely n complex zeros $c_1, c_2, \ldots, c_n$, we may write

$$f(x) = a(x - c_1)(x - c_2) \cdots (x - c_n),$$

where a is the leading coefficient of $f(x)$. Of course, some of the zeros may be real. In such cases we obtain the linear factors referred to in the statement of the theorem.

If a zero c_k is not real, then, by the theorem on conjugate pair zeros of a polynomial, the conjugate $\overline{c}_k$ is also a zero of $f(x)$ and hence must be one of the numbers $c_1, c_2, \ldots, c_n$. This implies that both $x - c_k$ and $x - \overline{c}_k$ appear in the factorization of $f(x)$. If those factors are multiplied, we obtain

$$(x - c_k)(x - \overline{c}_k) = x^2 - (c_k + \overline{c}_k)x + c_k\overline{c}_k,$$

which has *real* coefficients, since $c_k + \overline{c}_k$ and $c_k\overline{c}_k$ are real numbers. Thus, if c_k is a complex zero, then the product $(x - c_k)(x - \overline{c}_k)$ is a quadratic polynomial that is irreducible over $\mathbb{R}$. This completes the proof.

EXAMPLE 2   **Expressing a polynomial as a product of linear and quadratic factors**

Express $x^5 - 4x^3 + x^2 - 4$ as a product of

(a) linear and quadratic polynomials with real coefficients that are irreducible over $\mathbb{R}$

(b) linear polynomials

SOLUTION

(a) $x^5 - 4x^3 + x^2 - 4$

$$= (x^5 - 4x^3) + (x^2 - 4) \qquad \text{group terms}$$

$$= x^3(x^2 - 4) + 1(x^2 - 4) \qquad \text{factor out } x^3$$

$$= (x^3 + 1)(x^2 - 4) \qquad \text{factor out } (x^2 - 4)$$

$$= (x + 1)(x^2 - x + 1)(x + 2)(x - 2) \qquad \text{factor as the sum of cubes}$$
$$\text{and the difference of squares}$$

Using the quadratic formula, we see that the polynomial $x^2 - x + 1$ has the complex zeros

$$\frac{-(-1) \pm \sqrt{(-1)^2 - 4(1)(1)}}{2(1)} = \frac{1 \pm \sqrt{3}\,i}{2} = \frac{1}{2} \pm \frac{\sqrt{3}}{2}i,$$

and hence is irreducible over $\mathbb{R}$. Thus, the desired factorization is

$$(x + 1)(x^2 - x + 1)(x + 2)(x - 2).$$

(b) Since the polynomial $x^2 - x + 1$ in part (a) has zeros $\frac{1}{2} \pm \left(\sqrt{3}/2\right)i$, it follows from the factor theorem that the polynomial has factors

$$x - \left(\frac{1}{2} + \frac{\sqrt{3}}{2}i\right) \quad \text{and} \quad x - \left(\frac{1}{2} - \frac{\sqrt{3}}{2}i\right).$$

Substituting in the factorization found in part (a), we obtain the following complete factorization into linear polynomials:

$$(x + 1)\left(x - \frac{1}{2} - \frac{\sqrt{3}}{2}i\right)\left(x - \frac{1}{2} + \frac{\sqrt{3}}{2}i\right)(x + 2)(x - 2)$$

We previously pointed out that it is generally very difficult to find the zeros of a polynomial of high degree. If all the coefficients are integers, however, there is a method for finding the *rational* zeros, if they exist. The method is a consequence of the following result.

Theorem on Rational Zeros of a Polynomial	If the polynomial $$f(x) = a_n x^n + a_{n-1} x^{n-1} + \cdots + a_1 x + a_0$$ has *integer* coefficients and if c/d is a rational zero of $f(x)$ such that c and d have no common prime factor, then **(1)** the numerator c of the zero is a factor of the constant term a_0 **(2)** the denominator d of the zero is a factor of the leading coefficient a_n

PROOF   Assume that $c > 0$. (The proof for $c < 0$ is similar.) Let us show that c is a factor of a_0. The case $c = 1$ is trivial, since 1 is a factor of *any*

(continued)

number. Thus, suppose $c \neq 1$. In this case $c/d \neq 1$, for if $c/d = 1$, we obtain $c = d$, and since c and d have no prime factor in common, this implies that $c = d = 1$, a contradiction. Hence, in the following discussion we have $c \neq 1$ and $c \neq d$.

Since $f(c/d) = 0$,

$$a_n\frac{c^n}{d^n} + a_{n-1}\frac{c^{n-1}}{d^{n-1}} + \cdots + a_1\frac{c}{d} + a_0 = 0.$$

We multiply by d^n and then add $-a_0d^n$ to both sides:

$$a_nc^n + a_{n-1}c^{n-1}d + \cdots + a_1cd^{n-1} = -a_0d^n$$
$$c(a_nc^{n-1} + a_{n-1}c^{n-2}d + \cdots + a_1d^{n-1}) = -a_0d^n$$

The last equation shows that c is a factor of the integer a_0d^n. Since c and d have no common factor, c is a factor of a_0. A similar argument may be used to prove that d is a factor of a_n.

As an aid in listing the possible rational zeros, remember the following quotient:

$$\text{Possible rational zeros} = \frac{\text{factors of the constant term } a_0}{\text{factors of the leading coefficient } a_n}$$

The theorem on rational zeros of a polynomial may be applied to equations with rational coefficients by merely multiplying both sides of the equation by the lcd of all the coefficients to obtain an equation with integral coefficients.

EXAMPLE 3   **Showing a polynomial has no rational zeros**

Show that $f(x) = x^3 - 4x - 2$ has no rational zeros.

SOLUTION   If $f(x)$ has a rational zero c/d such that c and d have no common prime factor, then, by the theorem on rational zeros of a polynomial, c is a factor of the constant term -2 and hence is either 2 or -2 (which we write as ± 2) or ± 1. The denominator d is a factor of the leading coefficient 1 and hence is ± 1. Thus, the only possibilities for c/d are

$$\frac{\pm 1}{\pm 1} \quad \text{and} \quad \frac{\pm 2}{\pm 1} \qquad \text{or, equivalently,} \qquad \pm 1 \quad \text{and} \quad \pm 2.$$

Substituting each of these numbers for x, we obtain

$$f(1) = -5, \quad f(-1) = 1, \quad f(2) = -2, \quad \text{and} \quad f(-2) = -2.$$

Since $f(\pm 1) \neq 0$ and $f(\pm 2) \neq 0$, it follows that $f(x)$ has no rational zeros.

E X A M P L E 4   **Finding the rational solutions of an equation**

Find all rational solutions of the equation

$$3x^4 + 14x^3 + 14x^2 - 8x - 8 = 0.$$

SOLUTION   The problem is equivalent to finding the rational zeros of the polynomial on the left-hand side of the equation. If c/d is a rational zero and c and d have no common factor, then c is a factor of the constant term -8 and d is a factor of the leading coefficient 3. All possible choices are listed in the following table.

Choices for the numerator c	$\pm 1, \pm 2, \pm 4, \pm 8$
Choices for the denominator d	$\pm 1, \pm 3$
Choices for c/d	$\pm 1, \pm 2, \pm 4, \pm 8, \pm \frac{1}{3}, \pm \frac{2}{3}, \pm \frac{4}{3}, \pm \frac{8}{3}$

We can reduce the number of choices by finding upper and lower bounds for the real solutions; however, we shall not do so here. It is necessary to determine which of the choices for c/d, if any, are zeros. We see by substitution that neither 1 nor -1 is a solution. If we divide synthetically by $x + 2$, we obtain

$$
\begin{array}{r|rrrrr}
-2 & 3 & 14 & 14 & -8 & -8 \\
  &   & -6 & -16 & 4 & 8 \\
\hline
  & 3 & 8 & -2 & -4 & 0
\end{array}
$$

This result shows that -2 is a zero. Moreover, the synthetic division provides the coefficients of the quotient in the division of the polynomial by $x + 2$. Hence, we have the following factorization of the given polynomial:

$$(x + 2)(3x^3 + 8x^2 - 2x - 4)$$

The remaining solutions of the equation must be zeros of the second factor, so we use that polynomial to check for solutions. *Do not* use the polynomial in the original equation. (Note that $\pm \frac{8}{3}$ are no longer candidates, since the numerator must be a factor of 4.) Again proceeding by trial and error, we ultimately find that synthetic division by $x + \frac{2}{3}$ gives us the following result:

$$
\begin{array}{r|rrrr}
-\frac{2}{3} & 3 & 8 & -2 & -4 \\
  &   & -2 & -4 & 4 \\
\hline
  & 3 & 6 & -6 & 0
\end{array}
$$

Therefore, $-\frac{2}{3}$ is also a zero.

Using the coefficients of the quotient, we know that the remaining zeros are solutions of the equation $3x^2 + 6x - 6 = 0$. Dividing both sides by 3

(continued)

gives us the equivalent equation $x^2 + 2x - 2 = 0$. By the quadratic formula, this equation has solutions

$$\frac{-2 \pm \sqrt{2^2 - 4(1)(-2)}}{2(1)} = \frac{-2 \pm \sqrt{12}}{2} = \frac{-2 \pm 2\sqrt{3}}{2} = -1 \pm \sqrt{3}.$$

Hence, the given polynomial has two rational roots, -2 and $-\frac{2}{3}$, and two irrational roots, $-1 + \sqrt{3} \approx 0.732$ and $-1 - \sqrt{3} \approx -2.732$. ▧

EXAMPLE 5 **Finding the radius of a grain silo**

A grain silo has the shape of a right circular cylinder with a hemisphere attached to the top. If the total height of the structure is 30 feet, find the radius of the cylinder that results in a total volume of 1008π ft³.

SOLUTION Let x denote the radius of the cylinder as shown in Figure 1. The volume of the cylinder is $\pi r^2 h = \pi x^2(30 - x)$, and the volume of the hemisphere is $\frac{2}{3}\pi r^3 = \frac{2}{3}\pi x^3$, so we solve for x as follows:

$$\pi x^2(30 - x) + \tfrac{2}{3}\pi x^3 = 1008\pi \qquad \text{total volume is } 1008\pi$$

$$3x^2(30 - x) + 2x^3 = 3024 \qquad \text{multiply by } \frac{3}{\pi}$$

$$90x^2 - x^3 = 3024 \qquad \text{simplify}$$

$$x^3 - 90x^2 + 3024 = 0 \qquad \text{equivalent equation}$$

Since the leading coefficient of the polynomial on the left-hand side of the last equation is 1, any rational root has the form $c/1 = c$, where c is a factor of 3024. If we factor 3024 into primes, we find that $3024 = 2^4 \cdot 3^3 \cdot 7$. It follows that some of the positive factors of 3024 are

$$1, \quad 2, \quad 3, \quad 4, \quad 6, \quad 7, \quad 8, \quad 9, \quad 12, \quad \ldots.$$

To help us decide which of these numbers to test first, let us make a rough estimate of the radius by assuming that the silo has the shape of a right circular cylinder of height 30 feet. In that case, the volume would be $\pi r^2 h = 30\pi r^2$. Since this volume should be close to 1008π, we see that

$$30r^2 = 1008, \qquad \text{or} \qquad r^2 = 1008/30 = 33.6.$$

This suggests that we use 6 in our first synthetic division, as follows:

$$
\begin{array}{r|rrrr}
6 & 1 & -90 & 0 & 3024 \\
  &   & 6 & -504 & -3024 \\
\hline
  & 1 & -84 & -504 & 0
\end{array}
$$

Thus, 6 is a solution of the equation $x^3 - 90x^2 + 3024 = 0$.

The remaining two solutions of the equation can be found by solving the depressed equation $x^2 - 84x - 504 = 0$. These zeros are approximately -5.62 and 89.62—neither of which satisfies the conditions of the problem. Hence, the desired radius is 6 feet. ▧

Figure 1

4.4 *Exercises*

Exer. 1–10: A polynomial $f(x)$ with real coefficients and leading coefficient 1 has the given zero(s) and degree. Express $f(x)$ as a product of linear and quadratic polynomials with real coefficients that are irreducible over $\mathbb{R}$.

1 $3 + 2i$;   degree 2

2 $-4 + 3i$;   degree 2

3 $2, -2 - 5i$;   degree 3

4 $-3, 1 - 7i$;   degree 3

5 $-1, 0, 3 + i$;   degree 4

6 $0, 2, -2 - i$;   degree 4

7 $4 + 3i, -2 + i$;   degree 4

8 $3 + 5i, -1 - i$;   degree 4

9 $0, -2i, 1 - i$;   degree 5

10 $0, 3i, 4 + i$;   degree 5

Exer. 11–14: Show that the equation has no rational root.

11 $x^3 + 3x^2 - 4x + 6 = 0$

12 $3x^3 - 4x^2 + 7x + 5 = 0$

13 $x^5 - 3x^3 + 4x^2 + x - 2 = 0$

14 $2x^5 + 3x^3 + 7 = 0$

Exer. 15–24: Find all solutions of the equation.

15 $x^3 - x^2 - 10x - 8 = 0$

16 $x^3 + x^2 - 14x - 24 = 0$

17 $2x^3 - 3x^2 - 17x + 30 = 0$

18 $12x^3 + 8x^2 - 3x - 2 = 0$

19 $x^4 + 3x^3 - 30x^2 - 6x + 56 = 0$

20 $3x^5 - 10x^4 - 6x^3 + 24x^2 + 11x - 6 = 0$

21 $6x^5 + 19x^4 + x^3 - 6x^2 = 0$

22 $6x^4 + 5x^3 - 17x^2 - 6x = 0$

23 $8x^3 + 18x^2 + 45x + 27 = 0$

24 $3x^3 - x^2 + 11x - 20 = 0$

Exer. 25–26: Find a factored form with integer coefficients of the polynomial f shown in the figure.

25 $f(x) = 6x^5 - 23x^4 + 24x^3 + x^2 - 12x + 4$

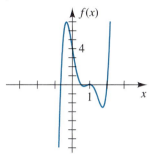

26 $f(x) = -6x^5 + 5x^4 + 14x^3 - 8x^2 - 8x + 3$

27 Does there exist a polynomial of degree 3 with real coefficients that has zeros 1, -1, and i? Justify your answer.

28 The polynomial $f(x) = x^3 - ix^2 + 2ix + 2$ has the complex number i as a zero; however, the conjugate $-i$ of i is not a zero. Why doesn't this result contradict the theorem on conjugate pair zeros of a polynomial?

29 If n is an odd positive integer, prove that a polynomial of degree n with real coefficients has at least one real zero.

30 If a polynomial of the form

$$x^n + a_{n-1}x^{n-1} + \cdots + a_1x + a_0,$$

where each a_k is an integer, has a rational root r, show that r is an integer and is a factor of a_0.

31 **Constructing a box** From a rectangular piece of cardboard having dimensions 20 inches $\times$ 30 inches, an open box is to be made by removing squares of area x^2 from each corner and turning up the sides. (See Exercise 41 of Section 4.1.)

(a) Show that there are two boxes that have a volume of 1000 in³.

(b) Which box has the smaller surface area?

32 Constructing a crate The frame for a shipping crate is to be constructed from 24 feet of 2 × 2 lumber. Assuming the crate is to have square ends of length x feet, determine the value(s) of x that result(s) in a volume of 4 ft³. (See Exercise 42 of Section 4.1.)

33 A right triangle has area 30 ft² and a hypotenuse that is 1 foot longer than one of its sides.

(a) If x denotes the length of this side, then show that $2x^3 + x^2 - 3600 = 0$.

(b) Show that there is a positive root of the equation in part (a) and that this root is less than 13.

(c) Find the lengths of the sides of the triangle.

34 Constructing a storage tank A storage tank for propane gas is to be constructed in the shape of a right circular cylinder of altitude 10 feet with a hemisphere attached to each end. Determine the radius x so that the resulting volume is 27π ft³. (See Example 8 of Section 3.4.)

35 Constructing a storage shelter A storage shelter is to be constructed in the shape of a cube with a triangular prism forming the roof (see the figure). The length x of a side of the cube is yet to be determined.

(a) If the total height of the structure is 6 feet, show that its volume V is given by $V = x^3 + \frac{1}{2}x^2(6 - x)$.

(b) Determine x so that the volume is 80 ft³.

Exercise 35

36 Designing a tent A canvas camping tent is to be constructed in the shape of a pyramid with a square base. An 8-foot pole will form the center support, as illustrated in the figure. Find the length x of a side of the base so that the total amount of canvas needed for the sides and bottom is 384 ft².

Exercise 36

4.5

Rational Functions

A function f is a **rational function** if

$$f(x) = \frac{g(x)}{h(x)},$$

where $g(x)$ and $h(x)$ are polynomials. The domain of f consists of all real numbers *except* the zeros of the denominator $h(x)$.

ILLUSTRATION **Rational Functions and Their Domains**

■ $f(x) = \dfrac{1}{x - 2}$;   *domain:* all *x except x* = 2

■ $f(x) = \dfrac{5x}{x^2 - 9}$;   *domain:* all *x except x* = ± 3

■ $f(x) = \dfrac{x^3 - 8}{x^2 + 4}$;   *domain:* all real numbers *x*

Previously we simplified rational expressions as follows:

$$\frac{x^2 - 4}{x - 2} = \frac{(x + 2)(x - 2)}{x - 2} \overset{\text{if } x \neq 2}{=} \frac{x + 2}{1} = x + 2$$

Figure 1

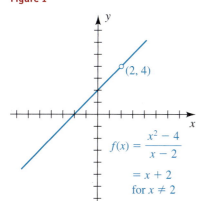

$$f(x) = \frac{x^2 - 4}{x - 2}$$
$$= x + 2$$
$$\text{for } x \neq 2$$

If we let $f(x) = \dfrac{x^2 - 4}{x - 2}$ and $g(x) = x + 2$, then the domain of *f* is all *x* except *x* = 2 and the domain of *g* is all real numbers. These domains and the above simplification suggest that the graphs of *f* and *g* are the same except for *x* = 2. What happens to the graph of *f* at *x* = 2? There is a *hole* in the graph—that is, a single point is missing. To find the *y*-value of the hole, we can substitute 2 for *x* in the reduced function, which is simply *g*(2) = 4. A graph of *f* is shown in Figure 1.

We now turn our attention to rational functions that do not have a common factor in the numerator and the denominator.

When sketching the graph of a rational function *f*, it is important to answer the following two questions.

Question 1 What can be said of the function values *f*(*x*) when *x* is close to (but not equal to) a zero of the denominator?

Question 2 What can be said of the function values *f*(*x*) when *x* is large positive or when *x* is large negative?

As we shall see, if *a* is a zero of the denominator, one of several situations often occurs. These are shown in Figure 2, where we have used notations from the following chart.

Notation	Terminology
$x \rightarrow a^-$	*x* approaches *a* from the left (through values *less* than *a*).
$x \rightarrow a^+$	*x* approaches *a* from the right (through values *greater* than *a*).
$f(x) \rightarrow \infty$	*f*(*x*) increases without bound (can be made as large positive as desired).
$f(x) \rightarrow -\infty$	*f*(*x*) decreases without bound (can be made as large negative as desired).

Figure 2

$f(x) \to \infty$ as $x \to a^-$

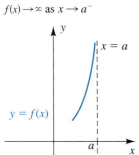

$f(x) \to \infty$ as $x \to a^+$

$f(x) \to -\infty$ as $x \to a^-$

$f(x) \to -\infty$ as $x \to a^+$

The symbols ∞ (read "infinity") and $-\infty$ (read "minus infinity") do not represent real numbers; they simply specify certain types of behavior of functions and variables.

The dashed line $x = a$ in Figure 2 is called a *vertical asymptote,* as in the following definition.

Definition of Vertical Asymptote	The line $x = a$ is a **vertical asymptote** for the graph of a function f if $$f(x) \to \infty \quad \text{or} \quad f(x) \to -\infty$$ as x approaches a from either the left or the right.

Thus, the answer to Question 1 is that if a is a zero of the denominator of a rational function f, then the graph of f *may* have a vertical asymptote $x = a$. There are rational functions where this is *not* the case (as in Figure 1 of this section). If the numerator and denominator have no common factor, then f *must* have a vertical asymptote $x = a$.

Let us next consider Question 2. For x *large positive* or *large negative,* the graph of a rational function may look like one of those in Figure 3, where the notation

$$f(x) \to c \quad \text{as} \quad x \to \infty$$

is read "$f(x)$ approaches c as x increases without bound" or "$f(x)$ approaches c as x approaches infinity," and the notation

$$f(x) \to c \quad \text{as} \quad x \to -\infty$$

is read "$f(x)$ approaches c as x decreases without bound."

Figure 3

$f(x) \to c$ as $x \to \infty$

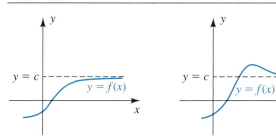

$f(x) \to c$ as $x \to -\infty$

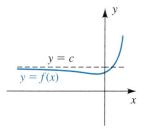

We call the dashed line in Figure 3 a *horizontal asymptote,* as in the next definition.

Definition of Horizontal Asymptote	The line $y = c$ is a **horizontal asymptote** for the graph of a function f if $$f(x) \to c \quad \text{as} \quad x \to \infty \quad \text{or as} \quad x \to -\infty.$$

Thus, the answer to Question 2 is that $f(x)$ *may* be very close to some number c when x is large positive or large negative; that is, the graph of f may have a horizontal asymptote $y = c$. There are rational functions where this is *not* the case (as in Examples 2(c) and 9).

Note that, as in the second and fourth sketches in Figure 3, the graph of f may cross a horizontal asymptote.

In the next example we find the asymptotes for the graph of a simple rational function.

EXAMPLE 1 Sketching the graph of a rational function

Sketch the graph of f if

$$f(x) = \frac{1}{x - 2}.$$

SOLUTION Let us begin by considering Question 1, stated at the beginning of this section. The denominator $x - 2$ is zero at $x = 2$. If x is close to 2 and $x > 2$, then $f(x)$ is large positive, as indicated in the following table.

x	2.1	2.01	2.001	2.0001	2.00001
$\dfrac{1}{x - 2}$	10	100	1000	10,000	100,000

(continued)

Since we can make $1/(x - 2)$ as large as desired by taking x close to 2 (and $x > 2$), we see that

$$f(x) \to \infty \quad \text{as} \quad x \to 2^+.$$

If $f(x)$ is close to 2 and $x < 2$, then $f(x)$ is large negative; for example, $f(1.9999) = -10,000$ and $f(1.99999) = -100,000$. Thus,

$$f(x) \to -\infty \quad \text{as} \quad x \to 2^-.$$

The line $x = 2$ is a vertical asymptote for the graph of f, as illustrated in Figure 4.

We next consider Question 2. The following table lists some approximate values for $f(x)$ when x is large and positive.

Figure 4

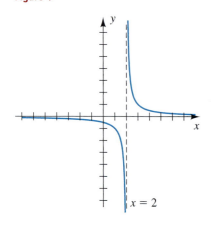

$x = 2$

x	100	1000	10,000	100,000	1,000,000
$\dfrac{1}{x - 2}$ **(approx.)**	0.01	0.001	0.0001	0.00001	0.000001

We may describe this behavior of $f(x)$ by writing

$$f(x) \to 0 \quad \text{as} \quad x \to \infty.$$

Similarly, $f(x)$ is close to 0 when x is large negative; for example, $f(-100,000) \approx -0.00001$. Thus,

$$f(x) \to 0 \quad \text{as} \quad x \to -\infty.$$

The line $y = 0$ (the x-axis) is a horizontal asymptote, as shown in Figure 4.

Plotting the points $(1, -1)$ and $(3, 1)$ helps give us a rough sketch of the graph. ◢

The function considered in Example 1, $f(x) = 1/(x - 2)$, closely resembles one of the simplest rational functions, the **reciprocal function.** The reciprocal function has equation $f(x) = 1/x$, vertical asymptote $x = 0$ (the y-axis), and horizontal asymptote $y = 0$ (the x-axis). The graph of the reciprocal function (shown in Appendix I) is the graph of a *hyperbola* (discussed later in the text). Note that we can obtain the graph of $y = 1/(x - 2)$ by shifting the graph of $y = 1/x$ to the right 2 units.

The following theorem is useful for finding the horizontal asymptote for the graph of a rational function.

Theorem on Horizontal Asymptotes	Let $f(x) = \dfrac{a_n x^n + a_{n-1} x^{n-1} + \cdots + a_1 x + a_0}{b_k x^k + b_{k-1} x^{k-1} + \cdots + b_1 x + b_0}$, where $a_n \neq 0$ and $b_k \neq 0$. **(1)** If $n < k$, then the x-axis (the line $y = 0$) is the horizontal asymptote for the graph of f. **(2)** If $n = k$, then the line $y = a_n/b_k$ (the ratio of leading coefficients) is the horizontal asymptote for the graph of f. **(3)** If $n > k$, the graph of f has no horizontal asymptote. Instead, either $f(x) \to \infty$ or $f(x) \to -\infty$ as $x \to \infty$ or as $x \to -\infty$.

Proofs for each part of this theorem may be patterned after the solutions in the next example. Concerning part (3), if $q(x)$ is the quotient obtained by dividing the numerator by the denominator, then $f(x) \to \infty$ if $q(x) \to \infty$ or $f(x) \to -\infty$ if $q(x) \to -\infty$.

EXAMPLE 2   **Finding horizontal asymptotes**

Find the horizontal asymptote for the graph of f, if it exists.

(a) $f(x) = \dfrac{3x - 1}{x^2 - x - 6}$ 　　　　 **(b)** $f(x) = \dfrac{5x^2 + 1}{3x^2 - 4}$

(c) $f(x) = \dfrac{2x^4 - 3x^2 + 5}{x^2 + 1}$

SOLUTION

(a) The degree of the numerator, 1, is less than the degree of the denominator, 2, so, by part (1) of the theorem on horizontal asymptotes, the x-axis is a horizontal asymptote. To verify this directly, we divide the numerator and denominator of the quotient by x^2 (since 2 is the highest power on x in the denominator), obtaining

$$f(x) = \frac{\dfrac{3x - 1}{x^2}}{\dfrac{x^2 - x - 6}{x^2}} = \frac{\dfrac{3}{x} - \dfrac{1}{x^2}}{1 - \dfrac{1}{x} - \dfrac{6}{x^2}} \quad \text{for} \quad x \neq 0.$$

If x is large positive or large negative, then $3/x$, $1/x^2$, $1/x$, and $6/x^2$ are close to 0, and hence

$$f(x) \approx \frac{0 - 0}{1 - 0 - 0} = \frac{0}{1} = 0.$$

Thus,

$$f(x) \to 0 \quad \text{as} \quad x \to \infty \quad \text{or as} \quad x \to -\infty.$$

Since $f(x)$ is the y-coordinate of a point on the graph, the last statement means that the line $y = 0$ (that is, the x-axis) is a horizontal asymptote.

(continued)

(b) If $f(x) = (5x^2 + 1)/(3x^2 - 4)$, then the numerator and denominator have the same degree, 2, and the leading coefficients are 5 and 3, respectively. Hence, by part (2) of the theorem on horizontal asymptotes, the line $y = \frac{5}{3}$ is the horizontal asymptote. We could also show that $y = \frac{5}{3}$ is the horizontal asymptote by dividing the numerator and denominator of $f(x)$ by x^2, as in part (a).

(c) The degree of the numerator, 4, is greater than the degree of the denominator, 2, so, by part (3) of the theorem on horizontal asymptotes, the graph has no horizontal asymptote. If we use long division, we obtain

$$f(x) = 2x^2 - 5 + \frac{10}{x^2 + 1}.$$

As either $x \to \infty$ or $x \to -\infty$, the quotient $2x^2 - 5$ increases without bound and $10/(x^2 + 1) \to 0$. Hence, $f(x) \to \infty$ as $x \to \infty$ or as $x \to -\infty$. ✔

We next list some guidelines for sketching the graph of a rational function. Their use will be illustrated in Examples 3, 6, and 7.

Guidelines for Sketching the Graph of a Rational Function	Assume that $f(x) = \dfrac{g(x)}{h(x)}$, where $g(x)$ and $h(x)$ are polynomials that have no common factor.

1 Find the x-intercepts—that is, the real zeros of the numerator $g(x)$—and plot the corresponding points on the x-axis.

2 Find the real zeros of the denominator $h(x)$. For each real zero a, sketch the vertical asymptote $x = a$ with dashes.

3 Find the y-intercept $f(0)$, if it exists, and plot the point $(0, f(0))$ on the y-axis.

4 Apply the theorem on horizontal asymptotes. If there is a horizontal asymptote $y = c$, sketch it with dashes.

5 If there is a horizontal asymptote $y = c$, determine whether it intersects the graph. The x-coordinates of the points of intersection are the solutions of the equation $f(x) = c$. Plot these points, if they exist.

6 Sketch the graph of f in each of the regions in the xy-plane determined by the vertical asymptotes in guideline 2. If necessary, use the sign of specific function values to tell whether the graph is above or below the x-axis or the horizontal asymptote. Use guideline 5 to decide whether the graph approaches the horizontal asymptote from above or below.

In the following examples our main objective is to determine the general shape of the graph, paying particular attention to how the graph approaches the asymptotes. We will plot only a few points, such as those corresponding to the *x*-intercepts and *y*-intercept or the intersection of the graph with a horizontal asymptote.

EXAMPLE 3   Sketching the graph of a rational function

Sketch the graph of *f* if

$$f(x) = \frac{3x + 4}{2x - 5}.$$

SOLUTION   We follow the guidelines.

Figure 5

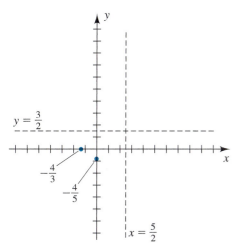

Guideline 1   To find the *x*-intercepts we find the zeros of the numerator. Solving $3x + 4 = 0$ gives us $x = -\frac{4}{3}$, and we plot the point $\left(-\frac{4}{3}, 0\right)$ on the *x*-axis, as shown in Figure 5.

Guideline 2   The denominator has zero $\frac{5}{2}$, so the line $x = \frac{5}{2}$ is a vertical asymptote. We sketch this line with dashes, as in Figure 5.

Guideline 3   The *y*-intercept is $f(0) = -\frac{4}{5}$, and we plot the point $\left(0, -\frac{4}{5}\right)$ in Figure 5.

Guideline 4   The numerator and denominator of $f(x)$ have the same degree, 1. The leading coefficients are 3 and 2, so by part (2) of the theorem on horizontal asymptotes, the line $y = \frac{3}{2}$ is a horizontal asymptote. We sketch the line with dashes in Figure 5.

Guideline 5   The *x*-coordinates of the points where the graph intersects the horizontal asymptote $y = \frac{3}{2}$ are solutions of the equation $f(x) = \frac{3}{2}$. We solve this equation as follows:

$$\frac{3x + 4}{2x - 5} = \frac{3}{2} \qquad \text{let } f(x) = \frac{3}{2}$$

$$2(3x + 4) = 3(2x - 5) \qquad \text{multiply by } 2(2x - 5)$$

$$6x + 8 = 6x - 15 \qquad \text{multiply}$$

$$8 = -15 \qquad \text{subtract } 6x$$

Figure 6

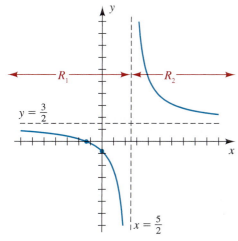

Since $8 \neq -15$ for any value of *x*, this result indicates that the graph of *f* does *not* intersect the horizontal asymptote. As an aid in sketching, we can now think of the horizontal asymptote as a boundary that cannot be crossed.

Guideline 6   The vertical asymptote in Figure 5 divides the *xy*-plane into two regions:

$$R_1: \quad \text{the region to the left of } x = \frac{5}{2}$$

$$R_2: \quad \text{the region to the right of } x = \frac{5}{2}$$

For R_1, we have the two points $\left(-\frac{4}{3}, 0\right)$ and $\left(0, -\frac{4}{5}\right)$ that the graph of *f* must pass through, as well as the two asymptotes that the graph must approach. This portion of *f* is shown in Figure 6.

(continued)

Figure 7

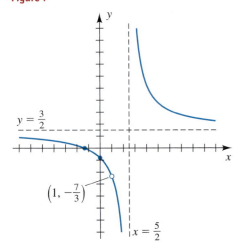

For R_2, the graph must again approach the two asymptotes. Since the graph cannot cross the x-axis (there is no x-intercept in R_2), it must be above the horizontal asymptote, as shown in Figure 6.

EXAMPLE 4   Sketching a graph that has a hole

Sketch the graph of g if

$$g(x) = \frac{(3x + 4)(x - 1)}{(2x - 5)(x - 1)}.$$

SOLUTION   The domain of g is all real numbers except $\frac{5}{2}$ and 1. If g is reduced, we obtain the function f in the previous example. The only difference between the graphs of f and g is that g has a hole at $x = 1$. Since $f(1) = -\frac{7}{3}$, we need only make a hole on the graph in Figure 6 to obtain the graph of g in Figure 7.

EXAMPLE 5   Finding an equation of a rational function satisfying prescribed conditions

Find an equation of a rational function f that satisfies the following conditions:

$$x\text{-intercept: 4, vertical asymptote: } x = -2,$$

$$\text{horizontal asymptote: } y = -\tfrac{3}{5}, \text{ and a hole at } x = 1$$

SOLUTION   An x-intercept of 4 implies that $x - 4$ must be a factor in the numerator, and a vertical asymptote of $x = -2$ implies that $x + 2$ is a factor in the denominator. So we can start with the form

$$\frac{x - 4}{x + 2}.$$

The horizontal asymptote is $y = -\frac{3}{5}$. We can multiply the numerator by -3 and the denominator by 5 to get the form

$$\frac{-3(x - 4)}{5(x + 2)}.$$

(Do *not* write $(-3x - 4)/(5x + 2)$, since that would change the x-intercept and the vertical asymptote.) Lastly, since there is a hole at $x = 1$, we must have a factor of $x - 1$ in both the numerator and the denominator. Thus, an equation for f is

$$f(x) = \frac{-3(x - 4)(x - 1)}{5(x + 2)(x - 1)} \quad \text{or, equivalently,} \quad f(x) = \frac{-3x^2 + 15x - 12}{5x^2 + 5x - 10}.$$

EXAMPLE 6 Sketching the graph of a rational function

Sketch the graph of f if

$$f(x) = \frac{x-1}{x^2 - x - 6}.$$

SOLUTION It is useful to express both numerator and denominator in factored form. Thus, we begin by writing

$$f(x) = \frac{x-1}{x^2 - x - 6} = \frac{x-1}{(x+2)(x-3)}.$$

Figure 8

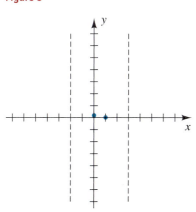

Guideline 1 To find the x-intercepts we find the zeros of the numerator. Solving $x - 1 = 0$ gives us $x = 1$, and we plot the point $(1, 0)$ on the x-axis, as shown in Figure 8.

Guideline 2 The denominator has zeros -2 and 3. Hence, the lines $x = -2$ and $x = 3$ are vertical asymptotes; we sketch them with dashes, as in Figure 8.

Guideline 3 The y-intercept is $f(0) = \frac{1}{6}$, and we plot the point $\left(0, \frac{1}{6}\right)$, shown in Figure 8.

Guideline 4 The degree of the numerator of $f(x)$ is less than the degree of the denominator, so, by part (1) of the theorem on horizontal asymptotes, the x-axis is the horizontal asymptote.

Guideline 5 The points where the graph intersects the horizontal asymptote (the x-axis) found in guideline 4 correspond to the x-intercepts. We already plotted the point $(1, 0)$ in guideline 1.

Guideline 6 The vertical asymptotes in Figure 8 divide the xy-plane into three regions:

R_1: the region to the left of $x = -2$

R_2: the region between $x = -2$ and $x = 3$

R_3: the region to the right of $x = 3$

For R_1, we have $x < -2$. There are only two choices for the shape of the graph of f in R_1: as $x \to -\infty$, the graph approaches the x-axis either from above or from below. To determine which choice is correct, we will examine the *sign* of a typical function value in R_1. Choosing -10 for x, we use the factored form of $f(x)$ to find the sign of $f(-10)$ (this process is similar to the one used in Section 2.7):

$$f(-10) = \frac{(-)}{(-)(-)} = -$$

The negative value of $f(-10)$ indicates that the graph approaches the horizontal asymptote from *below* as $x \to -\infty$. Moreover, as $x \to -2^-$, the graph

(continued)

extends *downward;* that is, $f(x) \to -\infty$. A sketch of f on R_1 is shown in Figure 9(a).

Figure 9

(a)

(b)

(c)

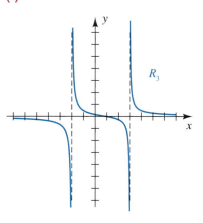

In R_2, we have $-2 < x < 3$, and the graph crosses the x-axis at $x = 1$. Since, for example, $f(0)$ is positive, it follows that the graph lies *above* the x-axis if $-2 < x < 1$. Thus, as $x \to -2^+$, the graph extends *upward;* that is, $f(x) \to \infty$. Since $f(2)$ can be shown to be negative, the graph lies *below* the x-axis if $1 < x < 3$. Hence, as $x \to 3^-$, the graph extends *downward;* that is, $f(x) \to -\infty$. A sketch of f on R_2 is shown in Figure 9(b).

Finally, in R_3, $x > 3$, and the graph does not cross the x-axis. Since, for example, $f(10)$ can be shown to be positive, the graph lies *above* the x-axis. It follows that $f(x) \to \infty$ as $x \to 3^+$ and that the graph approaches the horizontal asymptote from *above* as $x \to \infty$. The graph of f is sketched in Figure 9(c).

 EXAMPLE 7   **Sketching the graph of a rational function**

Sketch the graph of f if

$$f(x) = \frac{x^2}{x^2 - x - 2}.$$

SOLUTION   Factoring the denominator gives us

$$f(x) = \frac{x^2}{x^2 - x - 2} = \frac{x^2}{(x + 1)(x - 2)}.$$

We again follow the guidelines.

Figure 10

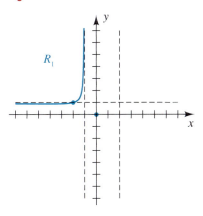

Guideline 1   To find the *x*-intercepts we find the zeros of the numerator. Solving $x^2 = 0$ gives us $x = 0$, and we plot the point $(0, 0)$ on the *x*-axis, as shown in Figure 10.

Guideline 2   The denominator has zeros -1 and 2. Hence, the lines $x = -1$ and $x = 2$ are vertical asymptotes, and we sketch them with dashes, as in Figure 10.

Guideline 3   The *y*-intercept is $f(0) = 0$. This gives us the same point $(0, 0)$ found in guideline 1.

Guideline 4   The numerator and denominator of $f(x)$ have the same degree, and the leading coefficients are both 1. Hence, by part (2) of the theorem on horizontal asymptotes, the line $y = \frac{1}{1} = 1$ is a horizontal asymptote. We sketch the line with dashes, as in Figure 10.

Guideline 5   The *x*-coordinates of the points where the graph intersects the horizontal asymptote $y = 1$ are solutions of the equation $f(x) = 1$. We solve this equation as follows:

$$\frac{x^2}{x^2 - x - 2} = 1 \qquad \text{\color{red}let } f(x) = 1$$

$$x^2 = x^2 - x - 2 \quad \text{\color{red}multiply by } x^2 - x - 2$$

$$x = -2 \qquad \text{\color{red}subtract } x^2 \text{ and add } x$$

This result indicates that the graph intersects the horizontal asymptote $y = 1$ *only* at $x = -2$; hence, we plot the point $(-2, 1)$ shown in Figure 10.

Guideline 6   The vertical asymptotes in Figure 10 divide the *xy*-plane into three regions:

R_1:   the region to the left of $x = -1$

R_2:   the region between $x = -1$ and $x = 2$

R_3:   the region to the right of $x = 2$

For R_1, let us first consider the portion of the graph that corresponds to $-2 < x < -1$. From the point $(-2, 1)$ on the horizontal asymptote, the graph must extend *upward* as $x \to -1^-$ (it cannot extend downward, since there is no *x*-intercept between $x = -2$ and $x = -1$). As $x \to -\infty$, there will be a low point on the graph between $y = 0$ and $y = 1$, and then the graph will approach the horizontal asymptote $y = 1$ from *below*. It is difficult to see where the low point occurs in Figure 10 because the function values are very close to one another. Using calculus, it can be shown that the low point is $\left(-4, \frac{8}{9}\right)$.

In R_2, we have $-1 < x < 2$, and the graph intersects the *x*-axis at $x = 0$. Since the function does not cross the horizontal asymptote in this region, we know that the graph extends *downward* as $x \to -1^+$ and as $x \to 2^-$, as shown in Figure 11(a).

(continued)

Figure 11

(a)

(b)

(c)

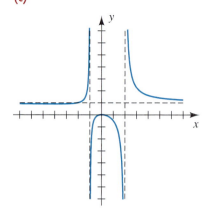

In R_3, the graph approaches the horizontal asymptote $y = 1$ (from either above or below) as $x \to \infty$. Furthermore, the graph must extend *upward* as $x \to 2^+$ because there are no x-intercepts in R_3. This implies that as $x \to \infty$, the graph approaches the horizontal asymptote from *above,* as in Figure 11(b).

The graph of f is sketched in Figure 11(c). ◢

In the remaining solutions we will not formally write down each guideline.

EXAMPLE 8   Sketching the graph of a rational function

Sketch the graph of f if

$$f(x) = \frac{2x^4}{x^4 + 1}.$$

Figure 12

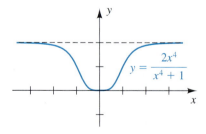

$$y = \frac{2x^4}{x^4 + 1}$$

SOLUTION   Note that since $f(-x) = f(x)$, the function is even, and hence the graph is symmetric with respect to the y-axis.

The graph intersects the x-axis at $(0, 0)$. Since the denominator of $f(x)$ has no real zero, the graph has no vertical asymptote.

The numerator and denominator of $f(x)$ have the same degree. Since the leading coefficients are 2 and 1, respectively, the line $y = \frac{2}{1} = 2$ is the horizontal asymptote. The graph does not cross the horizontal asymptote $y = 2$, since the equation $f(x) = 2$ has no real solution.

Plotting the points $(1, 1)$ and $\left(2, \frac{32}{17}\right)$ and making use of symmetry leads to the sketch in Figure 12. ◢

An **oblique asymptote** for a graph is a line $y = ax + b$, with $a \ne 0$, such that the graph approaches this line as $x \to \infty$ or as $x \to -\infty$. (If the graph is a line, we consider it to be its own asymptote.) If the rational function

$f(x) = g(x)/h(x)$ for polynomials $g(x)$ and $h(x)$ and *if the degree of $g(x)$ is one greater than the degree of $h(x)$*, then the graph of f has an oblique asymptote. To find this oblique asymptote we may use long division to express $f(x)$ in the form

$$f(x) = \frac{g(x)}{h(x)} = (ax + b) + \frac{r(x)}{h(x)},$$

where either $r(x) = 0$ or the degree of $r(x)$ is less than the degree of $h(x)$. From part (1) of the theorem on horizontal asymptotes,

$$\frac{r(x)}{h(x)} \to 0 \quad \text{as} \quad x \to \infty \quad \text{or as} \quad x \to -\infty.$$

Consequently, $f(x)$ approaches the line $y = ax + b$ as x increases or decreases without bound; that is, $y = ax + b$ is an oblique asymptote.

EXAMPLE 9   Finding an oblique asymptote

Find all the asymptotes and sketch the graph of f if

$$f(x) = \frac{x^2 - 9}{2x - 4}.$$

SOLUTION   A vertical asymptote occurs if $2x - 4 = 0$ (that is, if $x = 2$).
The degree of the numerator of $f(x)$ is greater than the degree of the denominator. Hence, by part (3) of the theorem on horizontal asymptotes, there is no *horizontal* asymptote; but since the degree of the numerator, 2, is *one* greater than the degree of the denominator, 1, the graph has an *oblique* asymptote. By long division we obtain

$$
\begin{array}{r}
\tfrac{1}{2}x + 1 \\
2x - 4 \overline{\smash{)}\, x^2 - 9 } \\
\underline{x^2 - 2x} \qquad \left(\tfrac{1}{2}x\right)(2x - 4) \\
2x - 9 \qquad \text{subtract} \\
\underline{2x - 4} \qquad (1)(2x - 4) \\
-5 \qquad \text{subtract}
\end{array}
$$

Therefore,

$$\frac{x^2 - 9}{2x - 4} = \left(\frac{1}{2}x + 1\right) - \frac{5}{2x - 4}.$$

As we indicated in the discussion preceding this example, the line $y = \frac{1}{2}x + 1$ is an oblique asymptote. This line and the vertical asymptote $x = 2$ are sketched with dashes in Figure 13.
The x-intercepts of the graph are the solutions of $x^2 - 9 = 0$ and hence are 3 and -3. The y-intercept is $f(0) = \frac{9}{4}$. The corresponding points are plotted in Figure 13. We may now show that the graph has the shape indicated in Figure 14.

Figure 13

Figure 14

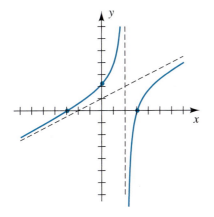

In Example 9, the graph of f approaches the line $y = \frac{1}{2}x + 1$ *asymptotically* as $x \to \infty$ or as $x \to -\infty$. Graphs of rational functions may approach different types of curves asymptotically. For example, if

$$f(x) = \frac{x^4 - x}{x^2} = x^2 - \frac{1}{x},$$

then for large values of $|x|$, $1/x \approx 0$ and hence $f(x) \approx x^2$. Thus, the graph of f approaches the parabola $y = x^2$ asymptotically as $x \to \infty$ or as $x \to -\infty$. In general, if $f(x) = g(x)/h(x)$ and if $q(x)$ is the quotient obtained by dividing $g(x)$ by $h(x)$, then the graph of f approaches the graph of $y = q(x)$ asymptotically as $x \to \infty$ or as $x \to -\infty$.

Graphs of rational functions may become increasingly complicated as the degrees of the polynomials in the numerator and denominator increase. Techniques developed in calculus are very helpful in achieving a more thorough treatment of such graphs.

Formulas that represent physical quantities may determine rational functions. For example, consider Ohm's law in electrical theory, which states that $I = V/R$, where R is the resistance (in ohms) of a conductor, V is the potential difference (in volts) across the conductor, and I is the current (in amperes) that flows through the conductor. The resistance of certain alloys approaches zero as the temperature approaches absolute zero (approximately $-273°C$), and the alloy becomes a *superconductor* of electricity. If the voltage V is fixed, then, for such a superconductor,

$$I = \frac{V}{R} \to \infty \quad \text{as} \quad R \to 0^+;$$

that is, as R approaches 0, the current increases without bound. Superconductors allow very large currents to be used in generating plants and motors. They also have applications in experimental high-speed ground transportation, where the strong magnetic fields produced by superconducting magnets enable trains to levitate so that there is essentially no friction between the wheels and the track. Perhaps the most important use for superconductors is in circuits for computers, because such circuits produce very little heat.

4.5 *Exercises*

Exer. 1–2: (a) Sketch the graph of f. (b) Find the domain D and range R of f. (c) Find the intervals on which f is increasing or is decreasing.

1 $f(x) = \dfrac{4}{x}$

2 $f(x) = \dfrac{1}{x^2}$

Exer. 3–4: Identify any vertical asymptotes, horizontal asymptotes, and holes.

3 $f(x) = \dfrac{-2(x + 5)(x - 6)}{(x - 3)(x - 6)}$

4 $f(x) = \dfrac{2(x + 4)(x + 2)}{5(x + 2)(x - 1)}$

Exer. 5–6: All asymptotes, intercepts, and holes of a rational function f are labeled in the figure. Sketch a graph of f and find a formula for f.

5

6

Exer. 7–32: Sketch the graph of f.

7 $f(x) = \dfrac{3}{x-4}$

8 $f(x) = \dfrac{-3}{x+3}$

9 $f(x) = \dfrac{-3x}{x+2}$

10 $f(x) = \dfrac{4x}{2x-5}$

11 $f(x) = \dfrac{4x-1}{2x+3}$

12 $f(x) = \dfrac{5x+3}{3x-7}$

13 $f(x) = \dfrac{(4x-1)(x-2)}{(2x+3)(x-2)}$

14 $f(x) = \dfrac{(5x+3)(x+1)}{(3x-7)(x+1)}$

15 $f(x) = \dfrac{x-2}{x^2-x-6}$

16 $f(x) = \dfrac{x+1}{x^2+2x-3}$

17 $f(x) = \dfrac{-4}{(x-2)^2}$

18 $f(x) = \dfrac{2}{(x+1)^2}$

19 $f(x) = \dfrac{x-3}{x^2-1}$

20 $f(x) = \dfrac{x+4}{x^2-4}$

21 $f(x) = \dfrac{2x^2-2x-4}{x^2+x-12}$

22 $f(x) = \dfrac{-3x^2-3x+6}{x^2-9}$

23 $f(x) = \dfrac{-x^2-x+6}{x^2+3x-4}$

24 $f(x) = \dfrac{x^2-3x-4}{x^2+x-6}$

25 $f(x) = \dfrac{3x^2-3x-36}{x^2+x-2}$

26 $f(x) = \dfrac{2x^2+4x-48}{x^2+3x-10}$

27 $f(x) = \dfrac{-2x^2+10x-12}{x^2+x}$

28 $f(x) = \dfrac{2x^2+8x+6}{x^2-2x}$

29 $f(x) = \dfrac{x-1}{x^3-4x}$

30 $f(x) = \dfrac{x^2-2x+1}{x^3-9x}$

31 $f(x) = \dfrac{-3x^2}{x^2+1}$

32 $f(x) = \dfrac{x^2-4}{x^2+1}$

Exer. 33–36: Find the oblique asymptote, and sketch the graph of f.

33 $f(x) = \dfrac{x^2-x-6}{x+1}$

34 $f(x) = \dfrac{2x^2-x-3}{x-2}$

35 $f(x) = \dfrac{8-x^3}{2x^2}$

36 $f(x) = \dfrac{x^3+1}{x^2-9}$

Exer. 37–44: Simplify $f(x)$, and sketch the graph of f.

37 $f(x) = \dfrac{2x^2+x-6}{x^2+3x+2}$

38 $f(x) = \dfrac{x^2-x-6}{x^2-2x-3}$

39 $f(x) = \dfrac{x-1}{1-x^2}$

40 $f(x) = \dfrac{x+2}{x^2-4}$

41 $f(x) = \dfrac{x^2+x-2}{x+2}$

42 $f(x) = \dfrac{x^3-2x^2-4x+8}{x-2}$

43 $f(x) = \dfrac{x^2+4x+4}{x^2+3x+2}$

44 $f(x) = \dfrac{(x^2+x)(2x-1)}{(x^2-3x+2)(2x-1)}$

Exer. 45–48: Find an equation of a rational function f that satisfies the given conditions.

45 vertical asymptote: $x = 4$
horizontal asymptote: $y = -1$
x-intercept: 3

46 vertical asymptotes: $x = -2, x = 0$
horizontal asymptote: $y = 0$
x-intercept: 2; $f(3) = 1$

47 vertical asymptotes: $x = -3, x = 1$
horizontal asymptote: $y = 0$
x-intercept: -1; $f(0) = -2$
hole at $x = 2$

48 vertical asymptotes: $x = -1$, $x = 3$
horizontal asymptote: $y = 2$
x-intercepts: -2, 1; hole at $x = 0$

49 **A container for radioactive waste** A cylindrical container for storing radioactive waste is to be constructed from lead. This container must be 6 inches thick. The volume of the outside cylinder shown in the figure is to be 16π ft³.

(a) Express the height h of the inside cylinder as a function of the inside radius r.

(b) Show that the inside volume $V(r)$ is given by

$$V(r) = \pi r^2 \left[\frac{16}{(r + 0.5)^2} - 1 \right].$$

(c) What values of r must be excluded in part (b)?

Exercise 49

6"

6"

h

6"

50 **Drug dosage** Young's rule is a formula that is used to modify adult drug dosage levels for young children. If a denotes the adult dosage (in milligrams) and if t is the age of the child (in years), then the child's dose y is given by the equation $y = ta/(t + 12)$. Sketch the graph of this equation for $t > 0$ and $a = 100$.

51 **Salt concentration** Salt water of concentration 0.1 pound of salt per gallon flows into a large tank that initially contains 50 gallons of pure water.

(a) If the flow rate of salt water into the tank is 5 gal/min, find the volume $V(t)$ of water and the amount $A(t)$ of salt in the tank after t minutes.

(b) Find a formula for the salt concentration $c(t)$ (in lb/gal) after t minutes.

(c) Discuss the variation of $c(t)$ as $t \to \infty$.

52 **Amount of rainfall** The total number of inches $R(t)$ of rain during a storm of length t hours can be approximated by

$$R(t) = \frac{at}{t + b},$$

where a and b are positive constants that depend on the geographical locale.

(a) Discuss the variation of $R(t)$ as $t \to \infty$.

(b) The intensity I of the rainfall (in in./hr) is defined by $I = R(t)/t$. If $a = 2$ and $b = 8$, sketch the graph of R and I on the same coordinate plane for $t > 0$.

53 **Salmon propagation** For a particular salmon population, the relationship between the number S of spawners and the number R of offspring that survive to maturity is given by the formula

$$R = \frac{4500S}{S + 500}.$$

(a) Under what conditions is $R > S$?

(b) Find the number of spawners that would yield 90% of the greatest possible number of offspring that survive to maturity.

(c) Work part (b) with 80% replacing 90%.

(d) Compare the results for S and R (in terms of percentage increases) from parts (b) and (c).

54 **Population density** The population density D (in people/mi²) in a large city is related to the distance x (in miles) from the center of the city by

$$D = \frac{5000x}{x^2 + 36}.$$

(a) What happens to the density as the distance from the center of the city changes from 20 miles to 25 miles?

(b) What eventually happens to the density?

(c) In what areas of the city does the population density exceed 400 people/mi²?

55 Let $f(x)$ be the polynomial

$$(x + 3)(x + 2)(x + 1)(x)(x - 1)(x - 2)(x - 3).$$

(a) Describe the graph of $g(x) = f(x)/f(x)$.

(b) Describe the graph of $h(x) = g(x)p(x)$, where $p(x)$ is a polynomial function.

56 Refer to Exercise 55.

(a) Describe the graph of $y = f(x)$.

(b) Describe the graph of $k(x) = 1/f(x)$.

57 Grade point average (GPA)

(a) A student has finished 48 credit hours with a GPA of 2.75. How many additional credit hours y at 4.0 will raise the student's GPA to some desired value x? (Determine y as a function of x.)

(b) Create a table of values for x and y, starting with $x = 2.8$ and using increments of 0.2.

(c) Graph the function in part (a).

(d) What is the vertical asymptote of the graph in part (c)?

(e) Explain the practical significance of the value $x = 4$.

4.6

Variation

In some scientific investigations, the terminology of *variation* or *proportion* is used to describe relationships between variable quantities. In the following chart, k is a nonzero real number called a **constant of variation** or a **constant of proportionality.**

Terminology	General formula	Illustration
y **varies directly** as x, or y is **directly proportional** to x	$y = kx$	$C = 2\pi r$, where C is the circumference of a circle, r is the radius, and $k = 2\pi$
y **varies inversely** as x, or y is **inversely proportional** to x	$y = \dfrac{k}{x}$	$I = \dfrac{110}{R}$, where I is the current in an electrical circuit, R is the resistance, and $k = 110$ is the voltage

The variable x in the chart can also represent a power. For example, the formula $A = \pi r^2$ states that the area A of a circle varies directly as the *square* of the radius r, where π is the constant of variation. Similarly, the formula $V = \frac{4}{3}\pi r^3$ states that the volume V of a sphere is directly proportional to the *cube* of the radius. In this case the constant of proportionality is $\frac{4}{3}\pi$.

In general, graphs of variables related by *direct variation* resemble graphs of **power functions** of the form $y = x^n$ with $n > 0$ (such as $y = \sqrt{x}$ or $y = x^2$ for nonnegative x-values, as shown in Figure 1). With direct variation, as one variable increases, so does the other variable. An example of two quantities that are directly related is the number of miles run and the number of calories burned.

Figure 1

As x increases, y increases,
or as x decreases, y decreases

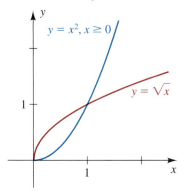

Figure 2

As x increases, y decreases,
or as x decreases, y increases

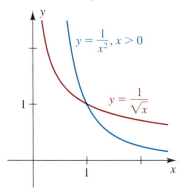

Graphs of variables related by *inverse variation* resemble graphs of power functions of the form $y = x^n$ with $n < 0$ (such as $y = 1/\sqrt{x}$ or $y = 1/x^2$ for positive x-values, as shown in Figure 2). In this case, as one variable increases, the other variable decreases. An example of two quantities that are inversely related is the number of inches of rainfall and the number of grass fires.

EXAMPLE 1 Directly proportional variables

Suppose a variable q is directly proportional to a variable z.

(a) If $q = 12$ when $z = 5$, determine the constant of proportionality.

(b) Find the value of q when $z = 7$ and sketch a graph of this relationship.

SOLUTION   Since q is directly proportional to z,

$$q = kz,$$

where k is a constant of proportionality.

(a) Substituting $q = 12$ and $z = 5$ gives us

$$12 = k \cdot 5, \qquad \text{or} \qquad k = \tfrac{12}{5}.$$

(b) Since $k = \tfrac{12}{5}$, the formula $q = kz$ has the specific form

$$q = \tfrac{12}{5}z.$$

Thus, when $z = 7$,

$$q = \tfrac{12}{5} \cdot 7 = \tfrac{84}{5} = 16.8.$$

Figure 3 illustrates the relationship of the variables q and z—a simple linear relationship.

Figure 3

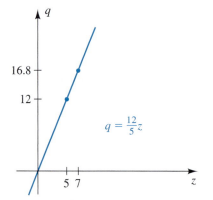

The following guidelines may be used to solve applied problems that involve variation or proportion.

Guidelines for Solving Variation Problems	*1* Write a *general* formula that involves the variables and a constant of variation (or proportion) k.
	2 Find the value of k in guideline 1 by using the initial data given in the statement of the problem.
	3 Substitute the value of k found in guideline 2 into the formula of guideline 1, obtaining a *specific* formula that involves the variables.
	4 Use the new data to solve the problem.

We shall follow these guidelines in the solution of the next example.

EXAMPLE 2 **Pressure and volume as inversely proportional quantities**

If the temperature remains constant, the pressure of an enclosed gas is inversely proportional to the volume. The pressure of a certain gas within a spherical balloon of radius 9 inches is 20 lb/in². If the radius of the balloon increases to 12 inches, approximate the new pressure of the gas. Sketch a graph of the relationship between the pressure and the volume.

SOLUTION

Guideline 1   If we denote the pressure by P (in lb/in²) and the volume by V (in in³), then since P is inversely proportional to V,

$$P = \frac{k}{V}$$

for some constant of proportionality k.

Guideline 2   We find the constant of proportionality k in guideline 1. Since the volume V of a sphere of radius r is $V = \frac{4}{3}\pi r^3$, the initial volume of the balloon is $V = \frac{4}{3}\pi(9)^3 = 972\pi$ in³. This leads to the following:

$$20 = \frac{k}{972\pi} \qquad\qquad \textcolor{red}{P = 20 \text{ when } V = 972\pi}$$

$$k = 20(972\pi) = 19{,}440\pi \quad \textcolor{red}{\text{solve for } k}$$

Guideline 3   Substituting $k = 19{,}440\pi$ into $P = k/V$, we find that the pressure corresponding to any volume V is given by

$$P = \frac{19{,}440\pi}{V}.$$

(continued)

Guideline 4　If the new radius of the balloon is 12 inches, then

$$V = \tfrac{4}{3}\pi(12)^3 = 2304\pi \text{ in}^3.$$

Substituting this number for V in the formula obtained in guideline 3 gives us

$$P = \frac{19{,}440\pi}{2304\pi} = \frac{135}{16} = 8.4375.$$

Thus, the pressure decreases to approximately 8.4 lb/in² when the radius increases to 12 inches.

Figure 4 illustrates the relationship of the variables P and V for $V > 0$. Since $P = 19{,}440\pi/V$ and $V = \tfrac{4}{3}\pi r^3$, we can show that $(P \circ V)(r) = 14{,}580/r^3$, so we could also say that P is inversely proportional to r^3. Note that this is a graph of a simple rational function.

Figure 4

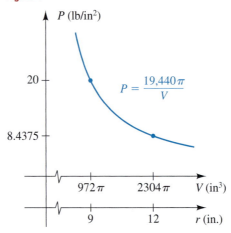

There are other types of variation. If x, y, and z are variables and $y = kxz$ for some real number k, we say that y *varies directly as the product of x and z* or y **varies jointly as x and z**. If $y = k(x/z)$, then y *varies directly as x and inversely as z*. As a final illustration, if a variable w varies directly as the product of x and the cube of y and inversely as the square of z, then

$$w = k\frac{xy^3}{z^2},$$

where k is a constant of proportionality. Graphs of equations for these types of variation will not be considered in this text.

EXAMPLE 3 **Combining several types of variation**

A variable w varies directly as the product of u and v and inversely as the square of s.

(a) If $w = 20$ when $u = 3$, $v = 5$, and $s = 2$, find the constant of variation.

(b) Find the value of w when $u = 7$, $v = 4$, and $s = 3$.

SOLUTION   A general formula for w is

$$w = k\frac{uv}{s^2},$$

where k is a constant of variation.

(a) Substituting $w = 20$, $u = 3$, $v = 5$, and $s = 2$ gives us

$$20 = k\frac{3 \cdot 5}{2^2}, \quad \text{or} \quad k = \frac{80}{15} = \frac{16}{3}.$$

(b) Since $k = \frac{16}{3}$, the specific formula for w is

$$w = \frac{16}{3}\frac{uv}{s^2}.$$

Thus, when $u = 7$, $v = 4$, and $s = 3$,

$$w = \frac{16}{3}\frac{7 \cdot 4}{3^2} = \frac{448}{27} \approx 16.6.$$

In the next example we again follow the guidelines stated in this section.

EXAMPLE 4   **Finding the support load of a rectangular beam**

The weight that can be safely supported by a beam with a rectangular cross section varies directly as the product of the width and square of the depth of the cross section and inversely as the length of the beam. If a 2-inch by 4-inch beam that is 8 feet long safely supports a load of 500 pounds, what weight can be safely supported by a 2-inch by 8-inch beam that is 10 feet long? (Assume that the width is the *shorter* dimension of the cross section.)

SOLUTION

Guideline 1   If the width, depth, length, and weight are denoted by w, d, l, and W, respectively, then a general formula for W is

$$W = k\frac{wd^2}{l},$$

where k is a constant of variation.

Guideline 2   To find the value of k in guideline 1, we see from the given data that

$$500 = k\frac{2(4^2)}{8}, \quad \text{or} \quad k = 125.$$

Guideline 3   Substituting $k = 125$ into the formula of guideline 1 gives us the specific formula

$$W = 125\frac{wd^2}{l}.$$

Guideline 4   To answer the question, we substitute $w = 2$, $d = 8$, and $l = 10$ into the formula found in guideline 3, obtaining

$$W = 125 \cdot \frac{2 \cdot 8^2}{10} = 1600 \text{ lb.}$$

4.6 Exercises

Exer. 1–12: Express the statement as a formula that involves the given variables and a constant of proportionality k, and then determine the value of k from the given conditions.

1 u is directly proportional to v. If $v = 30$, then $u = 12$.

2 s varies directly as t. If $t = 10$, then $s = 18$.

3 r varies directly as s and inversely as t. If $s = -2$ and $t = 4$, then $r = 7$.

4 w varies directly as z and inversely as the square root of u. If $z = 2$ and $u = 9$, then $w = 6$.

5 y is directly proportional to the square of x and inversely proportional to the cube of z. If $x = 5$ and $z = 3$, then $y = 25$.

6 q is inversely proportional to the sum of x and y. If $x = 0.5$ and $y = 0.7$, then $q = 1.4$.

7 z is directly proportional to the product of the square of x and the cube of y. If $x = 7$ and $y = -2$, then $z = 16$.

8 r is directly proportional to the product of s and v and inversely proportional to the cube of p. If $s = 2$, $v = 3$, and $p = 5$, then $r = 40$.

9 y is directly proportional to x and inversely proportional to the square of z. If $x = 4$ and $z = 3$, then $y = 16$.

10 y is directly proportional to x and inversely proportional to the sum of r and s. If $x = 3$, $r = 5$, and $s = 7$, then $y = 2$.

11 y is directly proportional to the square root of x and inversely proportional to the cube of z. If $x = 9$ and $z = 2$, then $y = 5$.

12 y is directly proportional to the square of x and inversely proportional to the square root of z. If $x = 5$ and $z = 16$, then $y = 10$.

13 **Liquid pressure** The pressure P acting at a point in a liquid is directly proportional to the distance d from the surface of the liquid to the point.

(a) Express P as a function of d by means of a formula that involves a constant of proportionality k.

(b) In a certain oil tank, the pressure at a depth of 2 feet is 118 lb/ft². Find the value of k in part (a).

(continued)

(c) Find the pressure at a depth of 5 feet for the oil tank in part (b).

(d) Sketch a graph of the relationship between P and d for $d \geq 0$.

14 **Hooke's law** Hooke's law states that the force F required to stretch a spring x units beyond its natural length is directly proportional to x.

(a) Express F as a function of x by means of a formula that involves a constant of proportionality k.

(b) A weight of 4 pounds stretches a certain spring from its natural length of 10 inches to a length of 10.3 inches. Find the value of k in part (a).

(c) What weight will stretch the spring in part (b) to a length of 11.5 inches?

(d) Sketch a graph of the relationship between F and x for $x \geq 0$.

15 **Electrical resistance** The electrical resistance R of a wire varies directly as its length l and inversely as the square of its diameter d.

(a) Express R in terms of l, d, and a constant of variation k.

(b) A wire 100 feet long of diameter 0.01 inch has a resistance of 25 ohms. Find the value of k in part (a).

(c) Sketch a graph of the relationship between R and d for $l = 100$ and $d > 0$.

(d) Find the resistance of a wire made of the same material that has a diameter of 0.015 inch and is 50 feet long.

16 **Intensity of illumination** The intensity of illumination I from a source of light varies inversely as the square of the distance d from the source.

(a) Express I in terms of d and a constant of variation k.

(b) A searchlight has an intensity of 1,000,000 candlepower at a distance of 50 feet. Find the value of k in part (a).

(c) Sketch a graph of the relationship between I and d for $d > 0$.

(d) Approximate the intensity of the searchlight in part (b) at a distance of 1 mile.

17 Period of a pendulum The period P of a simple pendulum—that is, the time required for one complete oscillation—is directly proportional to the square root of its length l.

(a) Express P in terms of l and a constant of proportionality k.

(b) If a pendulum 2 feet long has a period of 1.5 seconds, find the value of k in part (a).

(c) Find the period of a pendulum 6 feet long.

18 Dimensions of a human limb A circular cylinder is sometimes used in physiology as a simple representation of a human limb.

(a) Express the volume V of a cylinder in terms of its length L and the square of its circumference C.

(b) The formula obtained in part (a) can be used to approximate the volume of a limb from length and circumference measurements. Suppose the (average) circumference of a human forearm is 22 centimeters and the average length is 27 centimeters. Approximate the volume of the forearm to the nearest cm^3.

19 Period of a planet Kepler's third law states that the period T of a planet (the time needed to make one complete revolution about the sun) is directly proportional to the $\frac{3}{2}$ power of its average distance d from the sun.

(a) Express T as a function of d by means of a formula that involves a constant of proportionality k.

(b) For the planet Earth, $T = 365$ days and $d = 93$ million miles. Find the value of k in part (a).

(c) Estimate the period of Venus if its average distance from the sun is 67 million miles.

20 Range of a projectile It is known from physics that the range R of a projectile is directly proportional to the square of its velocity v.

(a) Express R as a function of v by means of a formula that involves a constant of proportionality k.

(b) A motorcycle daredevil has made a jump of 150 feet. If the speed coming off the ramp was 70 mi/hr, find the value of k in part (a).

(c) If the daredevil can reach a speed of 80 mi/hr coming off the ramp and maintain proper balance, estimate the possible length of the jump.

21 Automobile skid marks The speed V at which an automobile was traveling before the brakes were applied can sometimes be estimated from the length L of the skid marks. Assume that V is directly proportional to the square root of L.

(a) Express V as a function of L by means of a formula that involves a constant of proportionality k.

(b) For a certain automobile on a dry surface, $L = 50$ ft when $V = 35$ mi/hr. Find the value of k in part (a).

(c) Estimate the initial speed of the automobile in part (b) if the skid marks are 150 feet long.

22 Coulomb's law Coulomb's law in electrical theory states that the force F of attraction between two oppositely charged particles varies directly as the product of the magnitudes Q_1 and Q_2 of the charges and inversely as the square of the distance d between the particles.

(a) Find a formula for F in terms of Q_1, Q_2, d, and a constant of variation k.

(b) What is the effect of reducing the distance between the particles by a factor of one-fourth?

23 Threshold weight Threshold weight W is defined to be that weight beyond which risk of death increases significantly. For middle-aged males, W is directly proportional to the third power of the height h.

(a) Express W as a function of h by means of a formula that involves a constant of proportionality k.

(b) For a 6-foot male, W is about 200 pounds. Find the value of k in part (a).

(c) Estimate, to the nearest pound, the threshold weight for an individual who is 5 feet 6 inches tall.

24 The ideal gas law The ideal gas law states that the volume V that a gas occupies is directly proportional to the product of the number n of moles of gas and the temperature T (in K) and is inversely proportional to the pressure P (in atmospheres).

(a) Express V in terms of n, T, P, and a constant of proportionality k.

(b) What is the effect on the volume if the number of moles is doubled and both the temperature and the pressure are reduced by a factor of one-half?

25 Poiseuille's law Poiseuille's law states that the blood flow rate F (in L/min) through a major artery is directly proportional to the product of the fourth power of the radius r of the artery and the blood pressure P.

(a) Express F in terms of P, r, and a constant of proportionality k.

(b) During heavy exercise, normal blood flow rates sometimes triple. If the radius of a major artery increases by 10%, approximately how much harder must the heart pump?

26 Trout population Suppose 200 trout are caught, tagged, and released in a lake's general population. Let T denote the number of tagged fish that are recaptured when a sample of n trout are caught at a later date. The validity of the mark-recapture method for estimating the lake's total trout population is based on the assumption that T is directly proportional to n. If 10 tagged trout are recovered from a sample of 300, estimate the total trout population of the lake.

27 Radioactive decay of radon gas When uranium disintegrates into lead, one step in the process is the radioactive decay of radium into radon gas. Radon enters through the soil into home basements, where it presents a health hazard if inhaled. In the simplest case of radon detection, a sample of air with volume V is taken. After equilibrium has been established, the radioactive decay D of the radon gas is counted with efficiency E over time t. The radon concentration C present in the sample of air varies directly as the product of D and E and inversely as the product of V and t.

For a fixed radon concentration C and time t, find the change in the radioactive decay count D if V is doubled and E is reduced by 20%.

28 Radon concentration Refer to Exercise 27. Find the change in the radon concentration C if D increases by 30%, t increases by 60%, V decreases by 10%, and E remains constant.

29 Density at a point A thin flat plate is situated in an xy-plane such that the density d (in lb/ft^2) at the point $P(x, y)$ is inversely proportional to the square of the distance from the origin. What is the effect on the density at P if the x- and y-coordinates are each multiplied by $\frac{1}{3}$?

30 Temperature at a point A flat metal plate is positioned in an xy-plane such that the temperature T (in °C) at the point (x, y) is inversely proportional to the distance from the origin. If the temperature at the point $P(3, 4)$ is 20°C, find the temperature at the point $Q(24, 7)$.

Exer. 31–34: Examine the expression for the given set of data points of the form (x, y). Find the constant of variation and a formula that describes how y varies with respect to x.

31 y/x; $\{(0.6, 0.72), (1.2, 1.44), (4.2, 5.04), (7.1, 8.52), (9.3, 11.16)\}$

32 xy; $\{(0.2, -26.5), (0.4, -13.25), (0.8, -6.625), (1.6, -3.3125), (3.2, -1.65625)\}$

33 x^2y; $\{(0.16, -394.53125), (0.8, -15.78125), (1.6, -3.9453125), (3.2, -0.986328125)\}$

34 y/x^3; $\{(0.11, 0.00355377), (0.56, 0.46889472), (1.2, 4.61376), (2.4, 36.91008)\}$

CHAPTER 4 REVIEW EXERCISES

Exer. 1–6: Find all values of x such that $f(x) > 0$ and all x such that $f(x) < 0$, and sketch the graph of f.

1 $f(x) = (x + 2)^3$

2 $f(x) = x^6 - 32$

3 $f(x) = -\frac{1}{4}(x + 2)(x - 1)^2(x - 3)$

4 $f(x) = 2x^2 + x^3 - x^4$

5 $f(x) = x^3 + 2x^2 - 8x$

6 $f(x) = \frac{1}{15}(x^5 - 20x^3 + 64x)$

7 If $f(x) = x^3 - 5x^2 + 7x - 9$, use the intermediate value theorem for polynomial functions to prove that there is a real number a such that $f(a) = 100$.

8 Prove that the equation $x^5 - 3x^4 - 2x^3 - x + 1 = 0$ has a solution between 0 and 1.

Exer. 9–10: Find the quotient and remainder if $f(x)$ is divided by $p(x)$.

9 $f(x) = 3x^5 - 4x^3 + x + 5$; $p(x) = x^3 - 2x + 7$

10 $f(x) = 4x^3 - x^2 + 2x - 1$; $p(x) = x^2$

11 If $f(x) = -4x^4 + 3x^3 - 5x^2 + 7x - 10$, use the remainder theorem to find $f(-2)$.

12 Use the factor theorem to show that $x - 3$ is a factor of $f(x) = 2x^4 - 5x^3 - 4x^2 + 9$.

Exer. 13–14: Use synthetic division to find the quotient and remainder if $f(x)$ is divided by $p(x)$.

13 $f(x) = 6x^5 - 4x^2 + 8$; $p(x) = x + 2$

14 $f(x) = 2x^3 + 5x^2 - 2x + 1$; $p(x) = x - \sqrt{2}$

Exer. 15–16: A polynomial $f(x)$ with real coefficients has the indicated zero(s) and degree and satisfies the given condition. Express $f(x)$ as a product of linear and quadratic polynomials with real coefficients that are irreducible over $\mathbb{R}$.

15 $-3 + 5i, -1$; degree 3; $f(1) = 4$

16 $1 - i, 3, 0$;   degree 4; $f(2) = -1$

17 Find a polynomial $f(x)$ of degree 7 with leading coefficient 1 such that -3 is a zero of multiplicity 2 and 0 is a zero of multiplicity 5, and sketch the graph of f.

18 Show that 2 is a zero of multiplicity 3 of the polynomial $f(x) = x^5 - 4x^4 - 3x^3 + 34x^2 - 52x + 24$, and express $f(x)$ as a product of linear factors.

Exer. 19–20: Find the zeros of $f(x)$, and state the multiplicity of each zero.

19 $f(x) = (x^2 - 2x + 1)^2(x^2 + 2x - 3)$

20 $f(x) = x^6 + 2x^4 + x^2$

Exer. 21–22: (a) Use Descartes' rule of signs to determine the number of possible positive, negative, and nonreal complex solutions of the equation. (b) Find the smallest and largest integers that are upper and lower bounds, respectively, for the real solutions of the equation.

21 $2x^4 - 4x^3 + 2x^2 - 5x - 7 = 0$

22 $x^5 - 4x^3 + 6x^2 + x + 4 = 0$

23 Show that $7x^6 + 2x^4 + 3x^2 + 10$ has no real zero.

Exer. 24–26: Find all solutions of the equation.

24 $x^4 + 9x^3 + 31x^2 + 49x + 30 = 0$

25 $16x^3 - 20x^2 - 8x + 3 = 0$

26 $x^4 - 7x^2 + 6 = 0$

Exer. 27–28: Find an equation for the sixth-degree polynomial f shown in the figure.

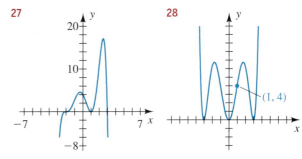

29 Identify any vertical asymptotes, horizontal asymptotes, intercepts, and holes for $f(x) = \dfrac{4(x + 2)(x - 1)}{3(x + 2)(x - 5)}$.

Exer. 30–39: Sketch the graph of f.

30 $f(x) = \dfrac{-2}{(x + 1)^2}$

31 $f(x) = \dfrac{1}{(x - 1)^3}$

32 $f(x) = \dfrac{3x^2}{16 - x^2}$

33 $f(x) = \dfrac{x}{(x + 5)(x^2 - 5x + 4)}$

34 $f(x) = \dfrac{x^3 - 2x^2 - 8x}{-x^2 + 2x}$

35 $f(x) = \dfrac{x^2 - 2x + 1}{x^3 - x^2 + x - 1}$

36 $f(x) = \dfrac{3x^2 + x - 10}{x^2 + 2x}$

37 $f(x) = \dfrac{-2x^2 - 8x - 6}{x^2 - 6x + 8}$

38 $f(x) = \dfrac{x^2 + 2x - 8}{x + 3}$

39 $f(x) = \dfrac{x^4 - 16}{x^3}$

40 Find an equation of a rational function f that satisfies the given conditions.

> vertical asymptote: $x = -3$
> horizontal asymptote: $y = \frac{3}{2}$
> x-intercept: 5
> hole at $x = 2$

41 Suppose y is directly proportional to the cube root of x and inversely proportional to the square of z. Find the constant of proportionality if $y = 6$ when $x = 8$ and $z = 3$.

42 Suppose y is inversely proportional to the square of x. Sketch a graph of this relationship for $x > 0$, given that $y = 18$ when $x = 4$. Include a point for $x = 12$.

43 Deflection of a beam A horizontal beam l feet long is supported at one end and unsupported at the other end (see the figure). If the beam is subjected to a uniform load and if y denotes the deflection of the beam at a position x feet from the supported end, then it can be shown that

$$y = cx^2(x^2 - 4lx + 6l^2),$$

where c is a positive constant that depends on the weight of the load and the physical properties of the beam.

(a) If the beam is 10 feet long and the deflection at the unsupported end of the beam is 2 feet, find c.

(b) Show that the deflection is 1 foot somewhere between $x = 6.1$ and $x = 6.2$.

Exercise 43

44 Elastic cylinder A rectangle made of elastic material is to be made into a cylinder by joining edge AD to edge BC, as shown in the figure. A wire of fixed length l is placed along the diagonal of the rectangle to support the structure. Let x denote the height of the cylinder.

(a) Express the volume V of the cylinder in terms of x.

(b) For what positive values of x is $V > 0$?

Exercise 44

45 Determining temperatures A meteorologist determines that the temperature T (in °F) for a certain 24-hour period in winter was given by the formula $T = \frac{1}{20}t(t - 12)(t - 24)$ for $0 \le t \le 24$, where t is time in hours and $t = 0$ corresponds to 6 A.M. At what time(s) was the temperature 32°F?

46 Deer propagation A herd of 100 deer is introduced onto a small island. Assuming the number $N(t)$ of deer after t years is given by $N(t) = -t^4 + 21t^2 + 100$ (for $t > 0$), determine when the herd size exceeds 180.

47 Threshold response curve In biochemistry, the general threshold response curve is the graph of an equation

$$R = \frac{kS^n}{S^n + a^n},$$

where R is the chemical response when the level of the substance being acted on is S and a, k, and n are positive constants. An example is the removal rate R of alcohol from the bloodstream by the liver when the blood alcohol concentration is S.

(a) Find an equation of the horizontal asymptote for the graph.

(b) In the case of alcohol removal, $n = 1$ and a typical value of k is 0.22 gram per liter per minute. What is the interpretation of k in this setting?

48 Oil spill clean-up The cost $C(x)$ of cleaning up x percent of an oil spill that has washed ashore increases greatly as x approaches 100. Suppose that

$$C(x) = \frac{20x}{101 - x} \text{ (thousand dollars)}.$$

(a) Compare $C(100)$ to $C(90)$.

(b) Sketch the graph of C for $0 < x < 100$.

49 Telephone calls In a certain county, the average number of telephone calls per day between any two cities is directly proportional to the product of their populations and inversely proportional to the square of the distance between them. Cities A and B are 25 miles apart and have populations of 10,000 and 5000, respectively. Telephone records indicate an average of 2000 calls per day between the two cities. Estimate the average number of calls per day between city A and another city of 15,000 people that is 100 miles from A.

50 Power of a wind rotor The power P generated by a wind rotor is directly proportional to the product of the square of the area A swept out by the blades and the third power of the wind velocity v. Suppose the diameter of the circular area swept out by the blades is 10 feet, and $P = 3000$ watts when $v = 20$ mi/hr. Find the power generated when the wind velocity is 30 mi/hr.

CHAPTER 4 DISCUSSION EXERCISES

1 Compare the domain, range, number of x-intercepts, and general shape of even-degreed polynomials and odd-degreed polynomials.

2 When using synthetic division, could you use a complex number c rather than a real number in $x - c$?

3 Discuss how synthetic division can be used to help find the quotient and remainder when $4x^3 - 8x^2 - 11x + 9$ is divided by $2x + 3$. Discuss how synthetic division can be used with any linear factor of the form $ax + b$.

4 Draw (by hand) a graph of a polynomial function of degree 3 that has x-intercepts 1, 2, and 3, has a y-intercept of 6, and passes through the point $(-1, 25)$. Can you actually have the graph you just drew?

5 How many different points do you need to specify a polynomial of degree n?

6 Prove the theorem on conjugate pair zeros of a polynomial. (*Hint:* For an arbitrary polynomial f, examine the conjugates of both sides of the equation $f(z) = 0$.)

7 Give an example of a rational function that has a common factor in the numerator and denominator, but does *not* have a hole in its graph. Discuss, in general, how this can happen.

8 (a) Can the graph of $f(x) = \dfrac{ax + b}{cx + d}$ (where $ax + b \neq cx + d$) cross its horizontal asymptote? If yes, then where?

(b) Can the graph of $f(x) = \dfrac{ax^2 + bx + c}{dx^2 + ex + f}$ (assume there are no like factors) cross its horizontal asymptote? If yes, then where?

9 **Gambling survival formula** An empirical formula for the bankroll B (in dollars) that is needed to survive a gambling session with confidence C (a percent expressed as a decimal) is given by the formula

$$B = \frac{GW}{29.3 + 53.1E - 22.7C},$$

where G is the number of games played in the session, W is the wager per game, and E is the player's edge on the game (expressed as a decimal).

(a) Approximate the bankroll needed for a player who plays 500 games per hour for 3 hours at \$5 per game with a -5% edge, provided the player wants a 95% chance of surviving the 3-hour session.

(b) Discuss the validity of the formula; a table and graph may help.

10 Multiply three consecutive integers together and then add the second integer to that product. Use synthetic division to help prove that the sum is the cube of an integer, and determine which integer.

5

Inverse, Exponential, and Logarithmic Functions

5.1 Inverse Functions

5.2 Exponential Functions

5.3 The Natural Exponential Function

5.4 Logarithmic Functions

5.5 Properties of Logarithms

5.6 Exponential and Logarithmic Equations

Exponential and logarithmic functions are transcendental functions, since they cannot be defined in terms of only addition, subtraction, multiplication, division, and rational powers of a variable x, as is the case for the algebraic functions considered in previous chapters. Such functions are of major importance in mathematics and have applications in almost every field of human endeavor. They are especially useful in the fields of chemistry, biology, physics, and engineering, where they help describe the manner in which quantities in nature grow or decay. As we shall see in this chapter, there is a close relationship between specific exponential and logarithmic functions—they are inverse functions of each other.

5.1

Inverse Functions

A function f may have the same value for different numbers in its domain. For example, if $f(x) = x^2$, then $f(2) = 4$ and $f(-2) = 4$, but $2 \neq -2$. For *the inverse of a function* to be defined, it is essential that different numbers in the domain *always* give different values of f. Such functions are called *one-to-one functions*.

Definition of One-to-One Function	A function f with domain D and range R is a **one-to-one function** if either of the following equivalent conditions is satisfied: **(1)** Whenever $a \neq b$ in D, then $f(a) \neq f(b)$ in R. **(2)** Whenever $f(a) = f(b)$ in R, then $a = b$ in D.

Figure 1

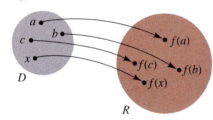

The arrow diagram in Figure 1 illustrates a one-to-one function. Note that each function value in the range R corresponds to *exactly one* element in the domain D. The function illustrated in Figure 2 of Section 3.4 is not one-to-one, since $f(w) = f(z)$, but $w \neq z$.

EXAMPLE 1   Determining whether a function is one-to-one

(a) If $f(x) = 3x + 2$, prove that f is one-to-one.

(b) If $g(x) = x^2 - 3$, prove that g is not one-to-one.

SOLUTION

(a) We shall use condition 2 of the preceding definition. Thus, suppose that $f(a) = f(b)$ for some numbers a and b in the domain of f. This gives us

$$
\begin{array}{ll}
3a + 2 = 3b + 2 & \text{definition of } f(x) \\
3a = 3b & \text{subtract 2} \\
a = b & \text{divide by 3}
\end{array}
$$

Since we have concluded that a must equal b, f is one-to-one.

(b) Showing that a function *is* one-to-one requires a *general* proof, as in part (a). To show that g is *not* one-to-one we need only find two distinct real numbers in the domain that produce the same function value. For example, $-1 \neq 1$, but $g(-1) = g(1)$. In fact, since g is an even function, $g(-a) = g(a)$ for every real number a.

Figure 2

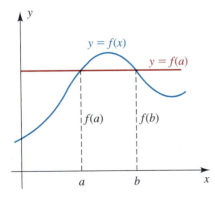

If we know the graph of a function f, it is easy to determine whether f is one-to-one. For example, the function whose graph is sketched in Figure 2 is not one-to-one, since $a \neq b$, but $f(a) = f(b)$. Note that the horizontal line $y = f(a)$ (or $y = f(b)$) intersects the graph in more than one point. In general, we may use the following graphical test to determine whether a function is one-to-one.

Horizontal Line Test	A function f is one-to-one if and only if every horizontal line intersects the graph of f in at most one point.

Let's apply the horizontal line test to the functions in Example 1.

EXAMPLE 2   Using the horizontal line test

Use the horizontal line test to determine if the function is one-to-one.

(a) $f(x) = 3x + 2$

(b) $g(x) = x^2 - 3$

SOLUTION

(a) The graph of $f(x) = 3x + 2$ is a line with y-intercept 2 and slope 3, as shown in Figure 3. We see that any horizontal line intersects the graph of f in at most one point. Thus, f is one-to-one.

Figure 3   **Figure 4**

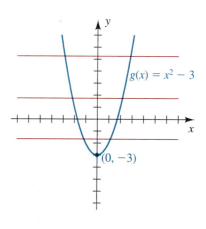

(b) The graph of $g(x) = x^2 - 3$ is a parabola opening upward with vertex $(0, -3)$, as shown in Figure 4. In this case, any horizontal line with equation $y = k$, where $k > -3$, will intersect the graph of g in two points. Thus, g is *not* one-to-one.

We may surmise from Example 2 that every increasing function or decreasing function passes the horizontal line test. Hence, we obtain the following result.

Theorem: Increasing or Decreasing Functions Are One-to-One	**(1)** A function that is increasing throughout its domain is one-to-one. **(2)** A function that is decreasing throughout its domain is one-to-one.

Let f be a one-to-one function with domain D and range R. Thus, for each number y in R, there is *exactly one* number x in D such that $y = f(x)$, as illustrated by the arrow in Figure 5(a). We may, therefore, define a function g from R to D by means of the following rule:

$$x = g(y)$$

As in Figure 5(b), g *reverses the correspondence given by f.* We call g the *inverse function* of f, as in the next definition.

Figure 5

(a) $y = f(x)$

(b) $x = g(y)$

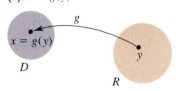

Definition of Inverse Function	Let f be a one-to-one function with domain D and range R. A function g with domain R and range D is the **inverse function** of f, provided the following condition is true for every x in D and every y in R: $$y = f(x) \quad \text{if and only if} \quad x = g(y)$$

Remember that for the inverse of a function f to be defined, *it is absolutely essential that f be one-to-one.* The following theorem, stated without proof, is useful to verify that a function g is the inverse of f.

Theorem on Inverse Functions	Let f be a one-to-one function with domain D and range R. If g is a function with domain R and range D, then g is the inverse function of f if and only if both of the following conditions are true:
	(1) $g(f(x)) = x$ for every x in D
	(2) $f(g(y)) = y$ for every y in R

Conditions 1 and 2 of the preceding theorem are illustrated in Figure 6(a) and (b), respectively, where the blue arrow indicates that f is a function from D to R and the red arrow indicates that g is a function from R to D.

Figure 6

(a) First f, then g **(b)** First g, then f

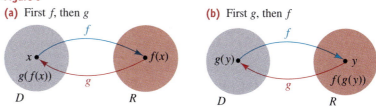

Note that in Figure 6(a) we first apply f to the number x in D, obtaining the function value $f(x)$ in R, and then apply g to $f(x)$, obtaining the number $g(f(x))$ in D. Condition 1 of the theorem states that $g(f(x)) = x$ for every x; that is, *g reverses* the correspondence given by f.

In Figure 6(b) we use the opposite order for the functions. We first apply g to the number y in R, obtaining the function value $g(y)$ in D, and then apply f to $g(y)$, obtaining the number $f(g(y))$ in R. Condition 2 of the theorem states that $f(g(y)) = y$ for every y; that is, *f reverses* the correspondence given by g.

If a function f has an inverse function g, we often denote g by f^{-1}. The -1 used in this notation should not be mistaken for an exponent; that is,

$$f^{-1}(y) \text{ does not mean } 1/[f(y)].$$

The reciprocal $1/[f(y)]$ may be denoted by $[f(y)]^{-1}$. It is important to remember the following facts about the domain and range of f and f^{-1}.

Domain and Range of f and f^{-1}	domain of f^{-1} = range of f
	range of f^{-1} = domain of f

When we discuss functions, we often let x denote an arbitrary number in the domain. Thus, for the inverse function f^{-1}, we may wish to consider $f^{-1}(x)$, *where x is in the domain R of f^{-1}*. In this event, the two conditions in the theorem on inverse functions are written as follows:

(1) $f^{-1}(f(x)) = x$ for every x in the domain of f

(2) $f(f^{-1}(x)) = x$ for every x in the domain of f^{-1}

Figure 6 contains a hint for finding the inverse of a one-to-one function in certain cases: If possible, *we solve the equation $y = f(x)$ for x in terms of y,* obtaining an equation of the form $x = g(y)$. If the two conditions $g(f(x)) = x$ and $f(g(x)) = x$ are true for every x in the domains of f and g, respectively, then g is the required inverse function f^{-1}. The following guidelines summarize this procedure; in guideline 2, in anticipation of finding f^{-1}, we write $x = f^{-1}(y)$ instead of $x = g(y)$.

Guidelines for Finding f^{-1} in Simple Cases

1 Verify that f is a one-to-one function throughout its domain.

2 Solve the equation $y = f(x)$ for x in terms of y, obtaining an equation of the form $x = f^{-1}(y)$.

3 Verify the following two conditions:

  (a) $f^{-1}(f(x)) = x$ for every x in the domain of f

  (b) $f(f^{-1}(x)) = x$ for every x in the domain of f^{-1}

The success of this method depends on the nature of the equation $y = f(x)$, since we must be able to solve for x in terms of y. For this reason, we include the phrase *in simple cases* in the title of the guidelines. We shall follow these guidelines in the next three examples.

EXAMPLE 3   **Finding the inverse of a function**

Let $f(x) = 3x - 5$. Find the inverse function of f.

SOLUTION

Guideline 1   The graph of the linear function f is a line of slope 3, and hence f is increasing throughout $\mathbb{R}$. Thus, f is one-to-one and the inverse function f^{-1} exists. Moreover, since the domain and range of f are $\mathbb{R}$, the same is true for f^{-1}.

Guideline 2   Solve the equation $y = f(x)$ for x:

$$y = 3x - 5 \quad \text{let } y = f(x)$$

$$x = \frac{y + 5}{3} \quad \text{solve for } x \text{ in terms of } y$$

We now formally let $x = f^{-1}(y)$; that is,

$$f^{-1}(y) = \frac{y + 5}{3}.$$

Since the symbol used for the variable is immaterial, we may also write

$$f^{-1}(x) = \frac{x + 5}{3},$$

where x is in the domain of f^{-1}.

Guideline 3 Since the domain and range of both f and f^{-1} are $\mathbb{R}$, we must verify conditions (a) and (b) for every real number x. We proceed as follows:

(a)
$$f^{-1}(f(x)) = f^{-1}(3x - 5) \qquad \text{definition of } f$$
$$= \frac{(3x - 5) + 5}{3} \qquad \text{definition of } f^{-1}$$
$$= x \qquad \text{simplify}$$

(b)
$$f(f^{-1}(x)) = f\left(\frac{x + 5}{3}\right) \qquad \text{definition of } f^{-1}$$
$$= 3\left(\frac{x + 5}{3}\right) - 5 \qquad \text{definition of } f$$
$$= x \qquad \text{simplify}$$

These verifications prove that the inverse function of f is given by

$$f^{-1}(x) = \frac{x + 5}{3}.$$

EXAMPLE 4   Finding the inverse of a function

Let $f(x) = x^2 - 3$ for $x \geq 0$. Find the inverse function of f.

SOLUTION

Figure 7

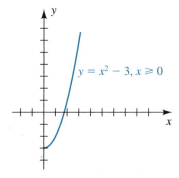

$y = x^2 - 3, x \geq 0$

Guideline 1 The graph of f is sketched in Figure 7. The domain of f is $[0, \infty)$, and the range is $[-3, \infty)$. Since f is increasing, it is one-to-one and hence has an inverse function f^{-1} with domain $[-3, \infty)$ and range $[0, \infty)$.

Guideline 2 We consider the equation

$$y = x^2 - 3$$

and solve for x, obtaining

$$x = \pm\sqrt{y + 3}.$$

Since x is nonnegative, we reject $x = -\sqrt{y + 3}$ and let

$$f^{-1}(y) = \sqrt{y + 3} \qquad \text{or, equivalently,} \qquad f^{-1}(x) = \sqrt{x + 3}.$$

(Note that if the function f had domain $x \leq 0$, we would choose the function $f^{-1}(x) = -\sqrt{x + 3}$.)

Guideline 3 We verify conditions (a) and (b) for x in the domains of f and f^{-1}, respectively.

(a) $f^{-1}(f(x)) = f^{-1}(x^2 - 3)$
$$= \sqrt{(x^2 - 3) + 3} = \sqrt{x^2} = x \text{ for } x \geq 0$$

(b) $f(f^{-1}(x)) = f(\sqrt{x + 3})$
$$= (\sqrt{x + 3})^2 - 3 = (x + 3) - 3 = x \text{ for } x \geq -3$$

Thus, the inverse function is given by

$$f^{-1}(x) = \sqrt{x + 3} \quad \text{for } x \geq -3.$$

Figure 8

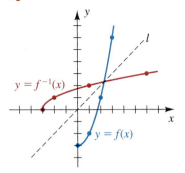

Note that the graphs of f and f^{-1} intersect on the line $y = x$.

Figure 9

Figure 10

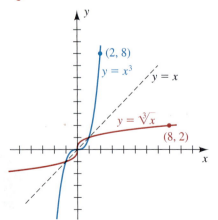

There is an interesting relationship between the graph of a function f and the graph of its inverse function f^{-1}. We first note that $b = f(a)$ is equivalent to $a = f^{-1}(b)$. These equations imply that *the point (a, b) is on the graph of f if and only if the point (b, a) is on the graph of f^{-1}.*

As an illustration, in Example 4 we found that the functions f and f^{-1} given by

$$f(x) = x^2 - 3 \qquad \text{and} \qquad f^{-1}(x) = \sqrt{x + 3}$$

are inverse functions of each other, provided that x is suitably restricted. Some points on the graph of f are $(0, -3)$, $(1, -2)$, $(2, 1)$, and $(3, 6)$. Corresponding points on the graph of f^{-1} are $(-3, 0)$, $(-2, 1)$, $(1, 2)$, and $(6, 3)$. The graphs of f and f^{-1} are sketched on the same coordinate plane in Figure 8. If the page is folded along the line $y = x$ that bisects quadrants I and III (as indicated by the dashes in the figure), then the graphs of f and f^{-1} coincide. The two graphs are *reflections* of each other through the line $y = x$, or are *symmetric* with respect to this line. This is typical of the graph of every function f that has an inverse function f^{-1} (see Exercise 50).

EXAMPLE 5   The relationship between the graphs of f and f^{-1}

Let $f(x) = x^3$. Find the inverse function f^{-1} of f, and sketch the graphs of f and f^{-1} on the same coordinate plane.

SOLUTION   The graph of f is sketched in Figure 9. Note that f is an odd function, and hence the graph is symmetric with respect to the origin.

Guideline 1   Since f is increasing throughout its domain $\mathbb{R}$, it is one-to-one and hence has an inverse function f^{-1}.

Guideline 2   We consider the equation

$$y = x^3$$

and solve for x by taking the cube root of each side, obtaining

$$x = y^{1/3} = \sqrt[3]{y}.$$

We now let

$$f^{-1}(y) = \sqrt[3]{y} \qquad \text{or, equivalently,} \qquad f^{-1}(x) = \sqrt[3]{x}.$$

Guideline 3   We verify conditions (a) and (b):

(a) $f^{-1}(f(x)) = f^{-1}(x^3) = \sqrt[3]{x^3} = x$ for every x in $\mathbb{R}$

(b) $f(f^{-1}(x)) = f(\sqrt[3]{x}) = (\sqrt[3]{x})^3 = x$ for every x in $\mathbb{R}$

The graph of f^{-1} $\left(\text{that is, the graph of the equation } y = \sqrt[3]{x}\right)$ may be obtained by reflecting the graph in Figure 9 through the line $y = x$, as shown in Figure 10. Three points on the graph of f^{-1} are $(0, 0)$, $(1, 1)$, and $(8, 2)$.

5.1 Exercises

Exer. 1–2: If possible, find (a) $f^{-1}(5)$ and (b) $g^{-1}(6)$.

1

x	2	4	6
$f(x)$	3	5	9

x	1	3	5
$g(x)$	6	2	6

2

t	0	3	5
$f(t)$	2	5	6

t	1	2	4
$g(t)$	3	6	6

Exer. 3–4: Determine if the graph is a graph of a one-to-one function.

3 (a)   (b)   (c)

4 (a)   (b)   (c)

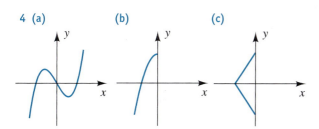

Exer. 5–16: Determine whether the function f is one-to-one.

5 $f(x) = 3x - 7$

6 $f(x) = \dfrac{1}{x - 2}$

7 $f(x) = x^2 - 9$

8 $f(x) = x^2 + 4$

9 $f(x) = \sqrt{x}$

10 $f(x) = \sqrt[3]{x}$

11 $f(x) = |x|$

12 $f(x) = 3$

13 $f(x) = \sqrt{4 - x^2}$

14 $f(x) = 2x^3 - 4$

15 $f(x) = \dfrac{1}{x}$

16 $f(x) = \dfrac{1}{x^2}$

Exer. 17–20: Use the theorem on inverse functions to prove that f and g are inverse functions of each other, and sketch the graphs of f and g on the same coordinate plane.

17 $f(x) = 3x - 2$; $g(x) = \dfrac{x + 2}{3}$

18 $f(x) = x^2 + 5, x \le 0$; $g(x) = -\sqrt{x - 5}, x \ge 5$

19 $f(x) = -x^2 + 3, x \ge 0$; $g(x) = \sqrt{3 - x}, x \le 3$

20 $f(x) = x^3 - 4$; $g(x) = \sqrt[3]{x + 4}$

Exer. 21–24: Determine the domain and range of f^{-1} for the given function without actually finding f^{-1}. Hint: First find the domain and range of f.

21 $f(x) = -\dfrac{2}{x - 1}$

22 $f(x) = \dfrac{5}{x + 3}$

23 $f(x) = \dfrac{4x + 5}{3x - 8}$

24 $f(x) = \dfrac{2x - 7}{9x + 1}$

Exer. 25–42: Find the inverse function of f.

25 $f(x) = 3x + 5$

26 $f(x) = 7 - 2x$

27 $f(x) = \dfrac{1}{3x - 2}$

28 $f(x) = \dfrac{1}{x + 3}$

29 $f(x) = \dfrac{3x + 2}{2x - 5}$

30 $f(x) = \dfrac{4x}{x - 2}$

31 $f(x) = 2 - 3x^2, x \le 0$

32 $f(x) = 5x^2 + 2, x \ge 0$

33 $f(x) = 2x^3 - 5$

34 $f(x) = -x^3 + 2$

35 $f(x) = \sqrt{3 - x}$

36 $f(x) = \sqrt{4 - x^2}, 0 \le x \le 2$

37 $f(x) = \sqrt[3]{x} + 1$

38 $f(x) = (x^3 + 1)^5$

39 $f(x) = x$

40 $f(x) = -x$

41 $f(x) = x^2 - 6x, x \ge 3$

42 $f(x) = x^2 - 4x + 3, x \le 2$

Exer. 43–44: Let $h(x) = 4 - x$. Use h, the table, and the graph to evaluate the expression.

x	2	3	4	5	6
$f(x)$	-1	0	1	2	3

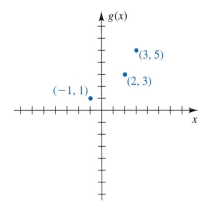

43 (a) $(g^{-1} \circ f^{-1})(2)$ **(b)** $(g^{-1} \circ h)(3)$

   (c) $(h^{-1} \circ f \circ g^{-1})(3)$

44 (a) $(g \circ f^{-1})(-1)$ **(b)** $(f^{-1} \circ g^{-1})(3)$

   (c) $(h^{-1} \circ g^{-1} \circ f)(6)$

Exer. 45–48: The graph of a one-to-one function f is shown. **(a)** Use the reflection property to sketch the graph of f^{-1}. **(b)** Find the domain D and range R of the function f. **(c)** Find the domain D_1 and range R_1 of the inverse function f^{-1}.

45

46

47

48

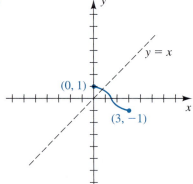

49 (a) Prove that the function defined by $f(x) = ax + b$ (a linear function) for $a \neq 0$ has an inverse function, and find $f^{-1}(x)$.

   (b) Does a constant function have an inverse? Explain.

50 Show that the graph of f^{-1} is the reflection of the graph of f through the line $y = x$ by verifying the following conditions:

(1) If $P(a, b)$ is on the graph of f, then $Q(b, a)$ is on the graph of f^{-1}.

(2) The midpoint of line segment PQ is on the line $y = x$.

(3) The line PQ is perpendicular to the line $y = x$.

51 Verify that $f(x) = f^{-1}(x)$ if

(a) $f(x) = -x + b$ (b) $f(x) = \dfrac{ax + b}{cx - a}$ for $c \neq 0$

(c) $f(x)$ has the following graph:

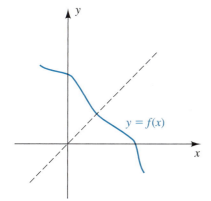

52 Let n be any positive integer. Find the inverse function of f if

(a) $f(x) = x^n$ for $x \geq 0$

(b) $f(x) = x^{m/n}$ for $x \geq 0$ and m any positive integer

53 **Ventilation requirements** Ventilation is an effective way to improve indoor air quality. In nonsmoking restaurants, air circulation requirements (in ft^3/min) are given by the function $V(x) = 35x$, where x is the number of people in the dining area.

(a) Determine the ventilation requirements for 23 people.

(b) Find $V^{-1}(x)$. Explain the significance of V^{-1}.

(c) Use V^{-1} to determine the maximum number of people that should be in a restaurant having a ventilation capability of 2350 ft^3/min.

54 **Radio stations** The table lists the total numbers of radio stations in the United States for certain years.

Year	Number
1950	2773
1960	4133
1970	6760
1980	8566
1990	10,819

(a) Determine a linear function $f(x) = ax + b$ that models these data, where x is the year. Plot f and the data on the same coordinate axes.

(b) Find $f^{-1}(x)$. Explain the significance of f^{-1}.

(c) Use f^{-1} to predict the year in which there were 7744 radio stations. Compare it with the true value, which is 1975.

5.2

Exponential Functions

Previously, we considered functions having terms of the form

$$\text{variable base}^{\text{constant power}},$$

such as x^2, $0.2x^{1.3}$, and $8x^{2/3}$. We now turn our attention to functions having terms of the form

$$\text{constant base}^{\text{variable power}},$$

such as 2^x, $(1.04)^{4x}$, and 3^{-x}. Let us begin by considering the function f defined by

$$f(x) = 2^x,$$

Figure 1

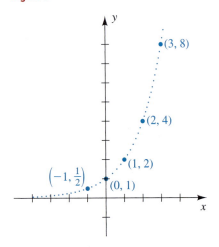

where x is restricted to *rational* numbers. (Recall that if $x = m/n$ for integers m and n with $n > 0$, then $2^x = 2^{m/n} = \left(\sqrt[n]{2} \right)^m$.) Coordinates of several points on the graph of $y = 2^x$ are listed in the following table.

x	-10	-3	-2	-1	0	1	2	3	10
$y = 2^x$	$\frac{1}{1024}$	$\frac{1}{8}$	$\frac{1}{4}$	$\frac{1}{2}$	1	2	4	8	1024

Other values of y for x rational, such as $2^{1/3}$, $2^{-9/7}$, and $2^{5.143}$, can be approximated with a calculator. We can show algebraically that if x_1 and x_2 are rational numbers such that $x_1 < x_2$, then $2^{x_1} < 2^{x_2}$. Thus, f is an increasing function, and its graph rises. Plotting points leads to the sketch in Figure 1, where the small dots indicate that only the points with *rational* x-coordinates are on the graph. There is a *hole* in the graph whenever the x-coordinate of a point is irrational.

To extend the domain of f to all real numbers, it is necessary to define 2^x for every *irrational* exponent x. To illustrate, if we wish to define 2^π, we could use the nonterminating decimal representing $3.1415926\ldots$ for π and consider the following *rational* powers of 2:

$$2^3, \quad 2^{3.1}, \quad 2^{3.14}, \quad 2^{3.141}, \quad 2^{3.1415}, \quad 2^{3.14159}, \quad \ldots$$

It can be shown, using calculus, that each successive power gets closer to a unique real number, denoted by 2^π. Thus,

$$2^x \rightarrow 2^\pi \quad \text{as} \quad x \rightarrow \pi, \quad \text{with } x \text{ rational.}$$

The same technique can be used for any other irrational power of 2. To sketch the graph of $y = 2^x$ with x *real,* we replace the holes in the graph in Figure 1 with points, and we obtain the graph in Figure 2. The function f defined by $f(x) = 2^x$ for every real number x is called the **exponential function** *with base 2.*

Let us next consider *any* base a, where a is a positive real number different from 1. As in the preceding discussion, to each real number x there corresponds exactly one positive number a^x such that the laws of exponents are true. Thus, as in the following chart, we may define a function f whose domain is $\mathbb{R}$ and range is the set of positive real numbers.

Figure 2

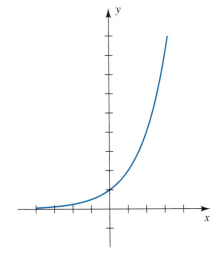

Terminology	Definition	Graph of f for $a > 1$	Graph of f for $0 < a < 1$
Exponential function f with base a	$f(x) = a^x$ for every x in $\mathbb{R}$, where $a > 0$ and $a \neq 1$		

The graphs in the chart show that if $a > 1$, then f is increasing on $\mathbb{R}$, and if $0 < a < 1$, then f is decreasing on $\mathbb{R}$. (These facts can be proved using calculus.) The graphs merely indicate the *general* appearance—the *exact* shape depends on the value of a. Note, however, that since $a^0 = 1$, the y-intercept is 1 for every a.

Note that if $a > 1$, then $a = 1 + d$ ($d > 0$) and the base a in $y = a^x$ can be thought of as representing multiplication by more than 100% as x increases by 1, so the function is increasing. For example, if $a = 1.15$, then $y = (1.15)^x$ can be considered to be a 15% per year growth function. More details on this concept appear later.

If $a > 1$, then as x *decreases* through negative values, the graph of f approaches the x-axis (see the third column in the chart). Thus, the x-axis is a *horizontal asymptote*. As x increases through positive values, the graph rises rapidly. This type of variation is characteristic of the **exponential law of growth,** and f is sometimes called a **growth function.**

If $0 < a < 1$, then as x *increases,* the graph of f approaches the x-axis asymptotically (see the last column in the chart). This type of variation is known as **exponential decay.**

When considering a^x we exclude the cases $a \leq 0$ and $a = 1$. Note that if $a < 0$, then a^x is not a real number for many values of x such as $\frac{1}{2}, \frac{3}{4}$, and $\frac{11}{6}$. If $a = 0$, then $a^0 = 0^0$ is undefined. Finally, if $a = 1$, then $a^x = 1$ for every x, and the graph of $y = a^x$ is a horizontal line.

The graph of an exponential function f is either increasing throughout its domain or decreasing throughout its domain. Thus, f is one-to-one by the theorem on page 280. Combining this result with the definition of a one-to-one function (see page 278) gives us parts (1) and (2) of the following theorem.

Theorem: Exponential Functions Are One-to-One	The exponential function f given by $$f(x) = a^x \quad \text{for} \quad 0 < a < 1 \quad \text{or} \quad a > 1$$ is one-to-one. Thus, the following equivalent conditions are satisfied for real numbers x_1 and x_2. $\quad$ **(1)** If $x_1 \neq x_2$, then $a^{x_1} \neq a^{x_2}$. $\quad$ **(2)** If $a^{x_1} = a^{x_2}$, then $x_1 = x_2$.

When using this theorem as a reason for a step in the solution to an example, we will state that *exponential functions are one-to-one.*

ILLUSTRATION $\quad$ **Exponential Functions Are One-to-One**

■ $\quad$ If $7^{3x} = 7^{2x+5}$, then $3x = 2x + 5$, or $x = 5$.

In the following example we solve a simple *exponential equation*—that is, an equation in which the variable appears in an exponent.

EXAMPLE 1 **Solving an exponential equation**

Solve the equation $3^{5x-8} = 9^{x+2}$.

SOLUTION

$3^{5x-8} = 9^{x+2}$	given
$3^{5x-8} = (3^2)^{x+2}$	express both sides with the same base
$3^{5x-8} = 3^{2x+4}$	law of exponents
$5x - 8 = 2x + 4$	exponential functions are one-to-one
$3x = 12$	subtract $2x$ and add 8
$x = 4$	divide by 3

Note that the solution in Example 1 depended on the fact that the base 9 could be written as 3 to some power. We will consider only exponential equations of this type for now, but we will solve more general exponential equations later in the chapter.

In the next two examples we sketch the graphs of several different exponential functions.

EXAMPLE 2 **Sketching graphs of exponential functions**

If $f(x) = \left(\frac{3}{2}\right)^x$ and $g(x) = 3^x$, sketch the graphs of f and g on the same coordinate plane.

Figure 3

SOLUTION   Since $\frac{3}{2} > 1$ and $3 > 1$, each graph *rises* as x increases. The following table displays coordinates for several points on the graphs.

x	-2	-1	0	1	2	3	4
$y = \left(\frac{3}{2}\right)^x$	$\frac{4}{9} \approx 0.4$	$\frac{2}{3} \approx 0.7$	1	$\frac{3}{2}$	$\frac{9}{4} \approx 2.3$	$\frac{27}{8} \approx 3.4$	$\frac{81}{16} \approx 5.1$
$y = 3^x$	$\frac{1}{9} \approx 0.1$	$\frac{1}{3} \approx 0.3$	1	3	9	27	81

Plotting points and being familiar with the general graph of $y = a^x$ leads to the graphs in Figure 3.

Example 2 illustrates the fact that if $1 < a < b$, then $a^x < b^x$ for positive values of x and $b^x < a^x$ for negative values of x. In particular, since $\frac{3}{2} < 2 < 3$, the graph of $y = 2^x$ in Figure 2 lies between the graphs of f and g in Figure 3.

Figure 4

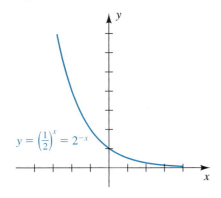

$y = \left(\frac{1}{2}\right)^x = 2^{-x}$

Figure 5

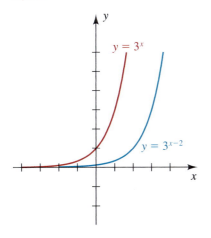

$y = 3^x$

$y = 3^{x-2}$

Figure 6

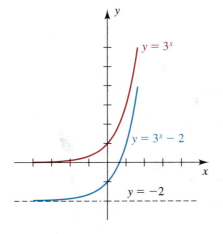

$y = 3^x$

$y = 3^x - 2$

$y = -2$

EXAMPLE 3 Sketching the graph of an exponential function

Sketch the graph of the equation $y = \left(\frac{1}{2}\right)^x$.

SOLUTION Since $0 < \frac{1}{2} < 1$, the graph *falls* as x increases. Coordinates of some points on the graph are listed in the following table.

x	-3	-2	-1	0	1	2	3
$y = \left(\frac{1}{2}\right)^x$	8	4	2	1	$\frac{1}{2}$	$\frac{1}{4}$	$\frac{1}{8}$

The graph is sketched in Figure 4. Since $\left(\frac{1}{2}\right)^x = (2^{-1})^x = 2^{-x}$, the graph is the same as the graph of the equation $y = 2^{-x}$. Note that the graph is a reflection through the y-axis of the graph of $y = 2^x$ in Figure 2. ◢

Equations of the form $y = a^u$, where u is some expression in x, occur in applications. The next two examples illustrate equations of this form.

EXAMPLE 4 Shifting graphs of exponential functions

Sketch the graph of the equation:

(a) $y = 3^{x-2}$ **(b)** $y = 3^x - 2$

SOLUTION

(a) The graph of $y = 3^x$, sketched in Figure 3, is resketched in Figure 5. From the discussion of horizontal shifts in Section 3.5, we can obtain the graph of $y = 3^{x-2}$ by shifting the graph of $y = 3^x$ two units to the right, as shown in Figure 5.

The graph of $y = 3^{x-2}$ can also be obtained by plotting several points and using them as a guide to sketch an exponential-type curve.

(b) From the discussion of vertical shifts in Section 3.5, we can obtain the graph of $y = 3^x - 2$ by shifting the graph of $y = 3^x$ two units downward, as shown in Figure 6. Note that the y-intercept is -1 and the line $y = -2$ is a horizontal asymptote for the graph. ◢

EXAMPLE 5 Finding an equation of an exponential function satisfying prescribed conditions

Find an exponential function of the form $f(x) = ba^{-x} + c$ that has horizontal asymptote $y = -2$, y-intercept 16, and x-intercept 2.

SOLUTION The horizontal asymptote of the graph of an exponential function of the form $f(x) = ba^{-x}$ is the x-axis—that is, $y = 0$. Since the desired horizontal asymptote is $y = -2$, we must have $c = -2$, so $f(x) = ba^{-x} - 2$.

Because the y-intercept is 16, $f(0)$ must equal 16. But $f(0) = ba^{-0} - 2 = b - 2$, so $b - 2 = 16$ and $b = 18$. Thus, $f(x) = 18a^{-x} - 2$.

(continued)

Figure 7

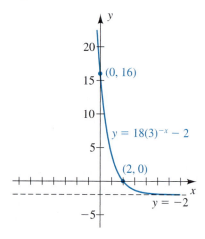

Lastly, we find the value of a:

$$f(x) = 18a^{-x} - 2 \qquad \text{given form of } f$$
$$0 = 18(a)^{-2} - 2 \qquad f(2) = 0 \text{ since 2 is the } x\text{-intercept}$$
$$2 = 18 \cdot \frac{1}{a^2} \qquad \text{add 2; definition of negative exponent}$$
$$a^2 = 9 \qquad \text{multiply by } a^2/2$$
$$a = \pm 3 \qquad \text{take square root}$$

Since a must be positive, we have

$$f(x) = 18(3)^{-x} - 2.$$

Figure 7 shows a graph of f that satisfies all of the conditions in the problem statement. Note that $f(x)$ could be written in the equivalent form

$$f(x) = 18\left(\tfrac{1}{3}\right)^x - 2.$$

The bell-shaped graph of the function in the next example is similar to a *normal probability curve* used in statistical studies.

EXAMPLE 6 Sketching a bell-shaped graph

Figure 8

If $f(x) = 2^{-x^2}$, sketch the graph of f.

SOLUTION If we rewrite $f(x)$ as

$$f(x) = \frac{1}{2^{(x^2)}},$$

we see that as x increases through positive values, $f(x)$ decreases rapidly; hence the graph approaches the x-axis asymptotically. Since x^2 is smallest when $x = 0$, the maximum value of f is $f(0) = 1$. Since f is an even function, the graph is symmetric with respect to the y-axis. Some points on the graph are $(0, 1)$, $\left(1, \tfrac{1}{2}\right)$, and $\left(2, \tfrac{1}{16}\right)$. Plotting and using symmetry gives us the sketch in Figure 8.

APPLICATION Bacterial Growth

Exponential functions may be used to describe the growth of certain populations. As an illustration, suppose it is observed experimentally that the number of bacteria in a culture doubles every day. If 1000 bacteria are present at the start, then we obtain the following table, where t is the time in days and $f(t)$ is the bacteria count at time t.

t (time in days)	0	1	2	3	4
$f(t)$ (bacteria count)	1000	2000	4000	8000	16,000

Figure 9

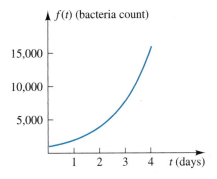

It appears that $f(t) = (1000)2^t$. With this formula we can predict the number of bacteria present at any time t. For example, at $t = 1.5 = \frac{3}{2}$,

$$f(t) = (1000)2^{3/2} \approx 2828.$$

The graph of f is sketched in Figure 9.

APPLICATION   Radioactive Decay

Certain physical quantities *decrease* exponentially. In such cases, if a is the base of the exponential function, then $0 < a < 1$. One of the most common examples of exponential decrease is the decay of a radioactive substance, or isotope. The **half-life** of an isotope is the time it takes for one-half the original amount in a given sample to decay. The half-life is the principal characteristic used to distinguish one radioactive substance from another. The polonium isotope ^{210}Po has a half-life of approximately 140 days; that is, given any amount, one-half of it will disintegrate in 140 days. If 20 milligrams of ^{210}Po is present initially, then the following table indicates the amount remaining after various intervals of time.

Figure 10

t (time in days)	0	140	280	420	560
$f(t)$ (mg remaining)	20	10	5	2.5	1.25

The sketch in Figure 10 illustrates the exponential nature of the disintegration.

Other radioactive substances have much longer half-lives. In particular, a by-product of nuclear reactors is the radioactive plutonium isotope ^{239}Pu, which has a half-life of approximately 24,000 years. It is for this reason that the disposal of radioactive waste is a major problem in modern society.

APPLICATION   Compound Interest

Compound interest provides a good illustration of exponential growth. If a sum of money P, the *principal,* is invested at a *simple* interest rate r, then the interest at the end of one interest period is the product Pr when r is expressed as a decimal. For example, if $P = \$1000$ and the interest rate is 9% per year, then $r = 0.09$, and the interest at the end of one year is $\$1000(0.09)$, or $\$90$.

If the interest is reinvested with the principal at the end of the interest period, then the new principal is

$$P + Pr \quad \text{or, equivalently,} \quad P(1 + r).$$

Note that to find the new principal we may multiply the original principal by $(1 + r)$. In the preceding example, the new principal is $\$1000(1.09)$, or $\$1090$.

After another interest period has elapsed, the new principal may be found by multiplying $P(1 + r)$ by $(1 + r)$. Thus, the principal after two interest periods is $P(1 + r)^2$. If we continue to reinvest, the principal after three periods is $P(1 + r)^3$; after four it is $P(1 + r)^4$; and, in general, the amount A accumulated after k interest periods is

$$A = P(1 + r)^k.$$

Interest accumulated by means of this formula is **compound interest.** Note that A is expressed in terms of an exponential function with base $1 + r$. The interest period may be measured in years, months, weeks, days, or any other suitable unit of time. When applying the formula for A, remember that *r is the interest rate per interest period expressed as a decimal.* For example, if the rate is stated as 6% *per year compounded monthly,* then the rate per month is $\frac{6}{12}\%$ or, equivalently, 0.5%. Thus, $r = 0.005$ and k is the number of months. If $100 is invested at this rate, then the formula for A is

$$A = 100(1 + 0.005)^k = 100(1.005)^k.$$

In general, we have the following formula.

Compound Interest Formula	$$A = P\left(1 + \frac{r}{n}\right)^{nt},$$ where P = principal r = annual interest rate expressed as a decimal n = number of interest periods per year t = number of years P is invested A = amount after t years.

The next example illustrates a special case of the compound interest formula.

EXAMPLE 7   Using the compound interest formula

Suppose that $1000 is invested at an interest rate of 9% compounded monthly. Find the new amount of principal after 5 years, after 10 years, and after 15 years. Illustrate graphically the growth of the investment.

SOLUTION   Applying the compound interest formula with $r = 9\% = 0.09$, $n = 12$, and $P = \$1000$, we find that the amount after t years is

$$A = 1000\left(1 + \frac{0.09}{12}\right)^{12t} = 1000(1.0075)^{12t}.$$

Substituting $t = 5$, 10, and 15 and using a calculator, we obtain the following table.

Note that when working with monetary values, we use = instead of ≈ and round to two decimal places.

Number of years	Amount
5	$A = \$1000(1.0075)^{60} = \$1565.68$
10	$A = \$1000(1.0075)^{120} = \$2451.36$
15	$A = \$1000(1.0075)^{180} = \$3838.04$

Figure 11

Compound interest: $A = 1000(1.0075)^{12t}$

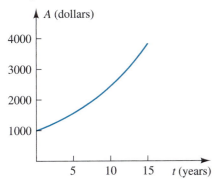

The exponential nature of the increase is indicated by the fact that during the first five years, the growth in the investment is $565.68; during the second five-year period, the growth is $885.68; and during the last five-year period, it is $1386.68.

The sketch in Figure 11 illustrates the growth of $1000 invested over a period of 15 years.

EXAMPLE 8   Finding an exponential model

In 1938, a federal law establishing a minimum wage was enacted, and the wage was set at $0.25 per hour; the wage had risen to $5.15 per hour by 1997. Find a simple exponential function of the form $y = ab^t$ that models the federal minimum wage for 1938–1997.

SOLUTION

$$y = ab^t \qquad \text{given}$$
$$0.25 = ab^0 \qquad \text{let } t = 0 \text{ for 1938}$$
$$0.25 = a \qquad b^0 = 1$$
$$y = 0.25b^t \qquad \text{replace } a \text{ with } 0.25$$
$$5.15 = 0.25b^{59} \qquad t = 1997 - 1938 = 59$$
$$b^{59} = \frac{5.15}{0.25} = 20.6 \qquad \text{divide by } 0.25$$
$$b = \sqrt[59]{20.6} \qquad \text{take 59th root}$$
$$b \approx 1.0526 \qquad \text{approximate}$$

We obtain the model $y = 0.25(1.0526)^t$, which indicates that the federal minimum wage rose about 5.26% per year from 1938 to 1997. A graph of the model is shown in Figure 12. Do you think this model will hold true through the year 2010?

Figure 12

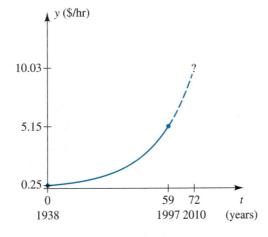

5.2   *Exercises*

Exer. 1–10: Solve the equation.

1   $7^{x+6} = 7^{3x-4}$

2   $6^{7-x} = 6^{2x+1}$

3   $3^{2x+3} = 3^{(x^2)}$

4   $9^{(x^2)} = 3^{3x+2}$

5   $2^{-100x} = (0.5)^{x-4}$

6   $\left(\frac{1}{2}\right)^{6-x} = 2$

7   $4^{x-3} = 8^{4-x}$

8   $27^{x-1} = 9^{2x-3}$

9   $4^x \cdot \left(\frac{1}{2}\right)^{3-2x} = 8 \cdot (2^x)^2$

10   $9^{2x} \cdot \left(\frac{1}{3}\right)^{x+2} = 27 \cdot (3^x)^{-2}$

11   Sketch the graph of f if $a = 2$.

   (a) $f(x) = a^x$

   (b) $f(x) = -a^x$

   (c) $f(x) = 3a^x$

   (d) $f(x) = a^{x+3}$

   (e) $f(x) = a^x + 3$

   (f) $f(x) = a^{x-3}$

   (g) $f(x) = a^x - 3$

   (h) $f(x) = a^{-x}$

   (i) $f(x) = \left(\dfrac{1}{a}\right)^x$

   (j) $f(x) = a^{3-x}$

12   Work Exercise 11 if $a = \frac{1}{2}$.

Exer. 13–24: Sketch the graph of f.

13   $f(x) = \left(\frac{2}{5}\right)^{-x}$

14   $f(x) = \left(\frac{2}{5}\right)^{x}$

15   $f(x) = 5\left(\frac{1}{2}\right)^x + 3$

16   $f(x) = 8(4)^{-x} - 2$

17   $f(x) = -\left(\frac{1}{2}\right)^x + 4$

18   $f(x) = -3^x + 9$

19   $f(x) = 2^{|x|}$

20   $f(x) = 2^{-|x|}$

21   $f(x) = 3^{1-x^2}$

22   $f(x) = 2^{-(x+1)^2}$

23   $f(x) = 3^x + 3^{-x}$

24   $f(x) = 3^x - 3^{-x}$

Exer. 25–28: Find an exponential function of the form $f(x) = ba^x$ or $f(x) = ba^x + c$ that has the given graph.

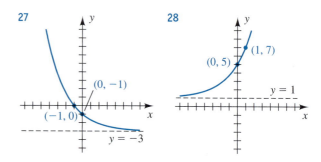

Exer. 29–30: Find an exponential function of the form $f(x) = ba^x$ that has the given y-intercept and passes through the point P.

29   y-intercept 8; $P(3, 1)$

30   y-intercept 6; $P\left(2, \frac{3}{32}\right)$

Exer. 31–32: Find an exponential function of the form $f(x) = ba^{-x} + c$ that has the given horizontal asymptote and y-intercept and passes through point P.

31   $y = 32$; y-intercept 212; $P(2, 112)$

32   $y = 72$; y-intercept 425; $P(1, 248.5)$

33   **Elk population** One hundred elk, each 1 year old, are introduced into a game preserve. The number $N(t)$ alive after t years is predicted to be $N(t) = 100(0.9)^t$. Estimate the number alive after

   (a) 1 year   (b) 5 years   (c) 10 years

34 Drug dosage A drug is eliminated from the body through urine. Suppose that for an initial dose of 10 milligrams, the amount $A(t)$ in the body t hours later is given by $A(t) = 10(0.8)^t$.

(a) Estimate the amount of the drug in the body 8 hours after the initial dose.

(b) What percentage of the drug still in the body is eliminated each hour?

35 Bacterial growth The number of bacteria in a certain culture increased from 600 to 1800 between 7:00 A.M. and 9:00 A.M. Assuming growth is exponential, the number $f(t)$ of bacteria t hours after 7:00 A.M. is given by $f(t) = 600(3)^{t/2}$.

(a) Estimate the number of bacteria in the culture at 8:00 A.M., 10:00 A.M., and 11:00 A.M.

(b) Sketch the graph of f for $0 \le t \le 4$.

36 Newton's law of cooling According to Newton's law of cooling, the rate at which an object cools is directly proportional to the difference in temperature between the object and the surrounding medium. The face of a household iron cools from 125° to 100° in 30 minutes in a room that remains at a constant temperature of 75°. From calculus, the temperature $f(t)$ of the face after t hours of cooling is given by $f(t) = 50(2)^{-2t} + 75$.

(a) Assuming $t = 0$ corresponds to 1:00 P.M., approximate to the nearest tenth of a degree the temperature of the face at 2:00 P.M., 3:30 P.M., and 4:00 P.M.

(b) Sketch the graph of f for $0 \le t \le 4$.

37 Radioactive decay The radioactive bismuth isotope ^{210}Bi has a half-life of 5 days. If there is 100 milligrams of ^{210}Bi present at $t = 0$, then the amount $f(t)$ remaining after t days is given by $f(t) = 100(2)^{-t/5}$.

(a) How much ^{210}Bi remains after 5 days? 10 days? 12.5 days?

(b) Sketch the graph of f for $0 \le t \le 30$.

38 Light penetration in an ocean An important problem in oceanography is to determine the amount of light that can penetrate to various ocean depths. The Beer-Lambert law asserts that the exponential function given by $I(x) = I_0 c^x$ is a model for this phenomenon (see the figure). For a certain location, $I(x) = 10(0.4)^x$ is the amount of light (in calories/cm²/sec) reaching a depth of x meters.

(a) Find the amount of light at a depth of 2 meters.

(b) Sketch the graph of I for $0 \le x \le 5$.

Exercise 38

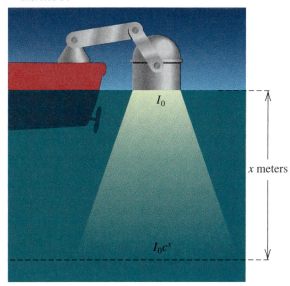

39 Decay of radium The half-life of radium is 1600 years. If the initial amount is q_0 milligrams, then the quantity $q(t)$ remaining after t years is given by $q(t) = q_0 2^{kt}$. Find k.

40 Dissolving salt in water If 10 grams of salt is added to a quantity of water, then the amount $q(t)$ that is undissolved after t minutes is given by $q(t) = 10\left(\frac{4}{5}\right)^t$. Sketch a graph that shows the value $q(t)$ at any time from $t = 0$ to $t = 10$.

41 Compound interest If \$1000 is invested at a rate of 12% per year compounded monthly, find the principal after

(a) 1 month (b) 6 months

(c) 1 year (d) 20 years

42 Compound interest If a savings fund pays interest at a rate of 10% per year compounded semiannually, how much money invested now will amount to \$5000 after 1 year?

43 Automobile trade-in value If a certain make of automobile is purchased for C dollars, its trade-in value $V(t)$ at the end of t years is given by $V(t) = 0.78C(0.85)^{t-1}$. If the original cost is \$10,000, calculate, to the nearest dollar, the value after

(a) 1 year (b) 4 years (c) 7 years

44 Real estate appreciation If the value of real estate increases at a rate of 5% per year, after t years the value V of a house purchased for P dollars is $V = P(1.05)^t$. A graph for the value of a house purchased for $80,000 in 1986 is shown in the figure. Approximate the value of the house, to the nearest $1000, in the year 2010.

Exercise 44

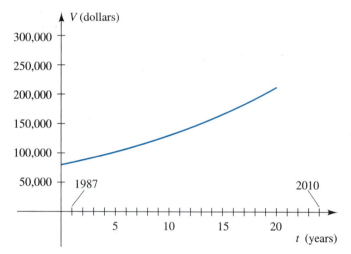

45 Manhattan Island The Island of Manhattan was sold for $24 in 1626. How much would this amount have grown to by 2006 if it had been invested at 6% per year compounded quarterly?

46 Credit-card interest A certain department store requires its credit-card customers to pay interest on unpaid bills at the rate of 18% per year compounded monthly. If a customer buys a television set for $500 on credit and makes no payments for one year, how much is owed at the end of the year?

47 Depreciation The declining balance method is an accounting method in which the amount of depreciation taken each year is a fixed percentage of the present value of the item. If y is the value of the item in a given year, the depreciation taken is ay for some depreciation rate a with $0 < a < 1$, and the new value is $(1 - a)y$.

(a) If the initial value of the item is y_0, show that the value after n years of depreciation is $(1 - a)^n y_0$.

(b) At the end of T years, the item has a salvage value of s dollars. The taxpayer wishes to choose a depreciation rate such that the value of the item after T years will equal the salvage value (see the figure). Show that $a = 1 - \sqrt[T]{s/y_0}$.

Exercise 47

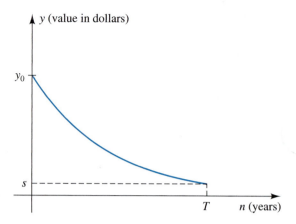

48 Language dating Glottochronology is a method of dating a language at a particular stage, based on the theory that over a long period of time linguistic changes take place at a fairly constant rate. Suppose that a language originally had N_0 basic words and that at time t, measured in millennia (1 millennium = 1000 years), the number $N(t)$ of basic words that remain in common use is given by $N(t) = N_0(0.805)^t$.

(a) Approximate the percentage of basic words lost every 100 years.

(b) If $N_0 = 200$, sketch the graph of N for $0 \le t \le 5$.

Exer. 49–52: Some lending institutions calculate the monthly payment M on a loan of L dollars at an interest rate r (expressed as a decimal) by using the formula

$$M = \frac{Lrk}{12(k - 1)},$$

where $k = [1 + (r/12)]^{12t}$ and t is the number of years that the loan is in effect.

49 Home mortgage

(a) Find the monthly payment on a 30-year $90,000 home mortgage if the interest rate is 12%.

(b) Find the total interest paid on the loan in part (a).

50 Home mortgage Find the largest 25-year home mortgage that can be obtained at an interest rate of 10% if the monthly payment is to be $800.

51 Car loan An automobile dealer offers customers no-down-payment 3-year loans at an interest rate of 15%. If a customer can afford to pay $220 per month, find the price of the most expensive car that can be purchased.

52 Business loan The owner of a small business decides to finance a new computer by borrowing $3000 for 2 years at an interest rate of 12.5%.

(a) Find the monthly payment.

(b) Find the total interest paid on the loan.

Exer. 53–54: Approximate the function at the value of x to four decimal places.

53 (a) $f(x) = 13^{\sqrt{x+1.1}}$, $x = 3$

(b) $g(x) = \left(\frac{5}{42}\right)^{-x}$, $x = 1.43$

(c) $h(x) = (2^x + 2^{-x})^{2x}$, $x = 1.06$

54 (a) $f(x) = 2^{\sqrt[3]{1-x}}$, $x = 2.5$

(b) $g(x) = \left(\frac{2}{25} + x\right)^{-3x}$, $x = 2.1$

(c) $h(x) = \dfrac{3^{-x} + 5}{3^x - 16}$, $x = \sqrt{2}$

55 Cost of a stamp The price of a first-class stamp was 3¢ in 1958 and 37¢ in 2002 (it was 2¢ in 1885). Find a simple ex-

ponential function of the form $y = ab^t$ that models the cost of a first-class stamp for 1958–2002, and predict its value for 2010.

56 Consumer Price Index The CPI is the most widely used measure of inflation. In 1970, the CPI was 37.8, and in 2000, the CPI was 168.8. This means that an urban consumer who paid $37.80 for a market basket of consumer goods and services in 1970 would have needed $168.80 for similar goods and services in 2000. Find a simple exponential function of the form $y = ab^t$ that models the CPI for 1970–2000, and predict its value for 2010.

57 Inflation comparisons In 1974, Johnny Miller won 8 tournaments on the PGA tour and accumulated $353,022 in official season earnings. In 1999, Tiger Woods accumulated $6,616,585 with a similar record.

(a) Suppose the monthly inflation rate from 1974 to 1999 was 0.0025 (3%/yr). Use the compound interest formula to estimate the equivalent value of Miller's winnings in the year 1999. Compare your answer with that from an inflation calculation on the web (e.g., bls.gov/cpi/home.htm).

(b) Find the annual interest rate needed for Miller's winnings to be equivalent in value to Woods's winnings.

(c) What type of function did you use in part (a)? part (b)?

5.3

The Natural Exponential Function

The *compound interest formula* discussed in the preceding section is

$$A = P\left(1 + \frac{r}{n}\right)^{nt},$$

where P is the principal invested, r is the annual interest rate (expressed as a decimal), n is the number of interest periods per year, and t is the number of years that the principal is invested. The next example illustrates what happens if the rate and total time invested are fixed, but the *interest period* is varied.

EXAMPLE 1   Using the compound interest formula

Suppose $1000 is invested at a compound interest rate of 9%. Find the new amount of principal after one year if the interest is compounded quarterly, monthly, weekly, daily, hourly, and each minute.

SOLUTION   If we let $P = \$1000$, $t = 1$, and $r = 0.09$ in the compound interest formula, then

$$A = \$1000\left(1 + \frac{0.09}{n}\right)^{n}$$

(continued)

for n interest periods per year. The values of n we wish to consider are listed in the following table, where we have assumed that there are 365 days in a year and hence $(365)(24) = 8760$ hours and $(8760)(60) = 525{,}600$ minutes. (In many business transactions an investment year is considered to be only 360 days.)

Interest period	Quarter	Month	Week	Day	Hour	Minute
n	4	12	52	365	8760	525,600

Using the compound interest formula (and a calculator), we obtain the amounts given in the following table.

Interest period	Amount after one year
Quarter	$\$1000\left(1 + \dfrac{0.09}{4}\right)^4 = \$1093.08$
Month	$\$1000\left(1 + \dfrac{0.09}{12}\right)^{12} = \$1093.81$
Week	$\$1000\left(1 + \dfrac{0.09}{52}\right)^{52} = \$1094.09$
Day	$\$1000\left(1 + \dfrac{0.09}{365}\right)^{365} = \$1094.16$
Hour	$\$1000\left(1 + \dfrac{0.09}{8760}\right)^{8760} = \$1094.17$
Minute	$\$1000\left(1 + \dfrac{0.09}{525{,}600}\right)^{525{,}600} = \$1094.17$

Note that, in the preceding example, after we reach an interest period of one hour, the number of interest periods per year has no effect on the final amount. If interest had been compounded each *second,* the result would still be $\$1094.17$. (Some decimal places *beyond* the first two *do* change.) Thus, the amount approaches a fixed value as n increases. Interest is said to be **compounded continuously** if the number n of time periods per year increases without bound.

If we let $P = 1$, $r = 1$, and $t = 1$ in the compound interest formula, we obtain

$$A = \left(1 + \frac{1}{n}\right)^n.$$

The expression on the right-hand side of the equation is important in calculus. In Example 1 we considered a similar situation: as n increased, A approached a limiting value. The same phenomenon occurs for this formula, as illustrated by the following table.

n	Approximation to $\left(1 + \dfrac{1}{n}\right)^{n}$
1	2.00000000
10	2.59374246
100	2.70481383
1000	2.71692393
10,000	2.71814593
100,000	2.71826824
1,000,000	2.71828047
10,000,000	2.71828169
100,000,000	2.71828181
1,000,000,000	2.71828183

In calculus it is shown that as n increases without bound, the value of the expression $[1 + (1/n)]^{n}$ approaches a certain irrational number, denoted by e. The number e arises in the investigation of many physical phenomena. An approximation is $e \approx 2.71828$. Using the notation we developed for rational functions in Section 4.5, we denote this fact as follows.

The Number e	If n is a positive integer, then $$\left(1 + \frac{1}{n}\right)^{n} \to e \approx 2.71828 \quad \text{as} \quad n \to \infty.$$

In the following definition we use e as a base for an important exponential function.

Definition of the Natural Exponential Function	The **natural exponential function** f is defined by $$f(x) = e^{x}$$ for every real number x.

The natural exponential function is one of the most useful functions in advanced mathematics and applications. Since $2 < e < 3$, the graph of $y = e^{x}$

Figure 1

lies between the graphs of $y = 2^x$ and $y = 3^x$, as shown in Figure 1. Scientific calculators have an $\boxed{e^x}$ key for approximating values of the natural exponential function.

APPLICATION   Continuously Compounded Interest

The compound interest formula is

$$A = P\left(1 + \frac{r}{n}\right)^{nt}.$$

If we let $1/k = r/n$, then $k = n/r$, $n = kr$, and $nt = krt$, and we may rewrite the formula as

$$A = P\left(1 + \frac{1}{k}\right)^{krt} = P\left[\left(1 + \frac{1}{k}\right)^{k}\right]^{rt}.$$

For continuously compounded interest we let n (the number of interest periods per year) increase without bound, denoted by $n \rightarrow \infty$ or, equivalently, by $k \rightarrow \infty$. Using the fact that $[1 + (1/k)]^k \rightarrow e$ as $k \rightarrow \infty$, we see that

$$P\left[\left(1 + \frac{1}{k}\right)^{k}\right]^{rt} \rightarrow P[e]^{rt} = Pe^{rt} \quad \text{as} \quad k \rightarrow \infty.$$

This result gives us the following formula.

Continuously Compounded Interest Formula	$A = Pe^{rt},$
	where $P = $ principal
	$r = $ annual interest rate expressed as a decimal
	$t = $ number of years P is invested
	$A = $ amount after t years.

The next example illustrates the use of this formula.

EXAMPLE 2 Using the continuously compounded interest formula

Suppose $20,000 is deposited in a money market account that pays interest at a rate of 8% per year compounded continuously. Determine the balance in the account after 5 years.

SOLUTION Applying the formula for continuously compounded interest with $P = 20,000$, $r = 0.08$, and $t = 5$, we have

$$A = Pe^{rt} = 20,000e^{0.08(5)} = 20,000e^{0.4}.$$

Using a calculator, we find that $A = \$29,836.49$.

The continuously compounded interest formula is just one specific case of the following law.

Law of Growth (or Decay) Formula	Let q_0 be the value of a quantity q at time $t = 0$ (that is, q_0 is the initial amount of q). If q changes instantaneously at a rate proportional to its current value, then $$q = q(t) = q_0 e^{rt},$$ where $r > 0$ is the rate of growth (or $r < 0$ is the rate of decay) of q.

EXAMPLE 3 Predicting the population of a city

The population of a city in 1970 was 153,800. Assuming that the population increases continuously at a rate of 5% per year, predict the population of the city in the year 2010.

SOLUTION We apply the growth formula $q = q_0 e^{rt}$ with initial population $q_0 = 153,800$, rate of growth $r = 0.05$, and time $t = 2010 - 1970 = 40$ years. Thus, a prediction for the population of the city in the year 2010 is

$$153,800e^{(0.05)(40)} = 153,800e^2 \approx 1,136,437.$$

The function f in the next example is important in advanced applications of mathematics.

EXAMPLE 4   **Sketching a graph involving two exponential functions**

Sketch the graph of f if

$$f(x) = \frac{e^x + e^{-x}}{2}.$$

SOLUTION   Note that f is an even function, because

$$f(-x) = \frac{e^{-x} + e^{-(-x)}}{2} = \frac{e^{-x} + e^x}{2} = f(x).$$

Thus, the graph is symmetric with respect to the y-axis. Using a calculator, we obtain the following approximations of $f(x)$.

x	0	0.5	1.0	1.5	2.0
$f(x)$ (approx.)	1	1.13	1.54	2.35	3.76

Plotting points and using symmetry with respect to the y-axis gives us the sketch in Figure 2. The graph *appears* to be a parabola; however, this is not actually the case.   ◢

Figure 2

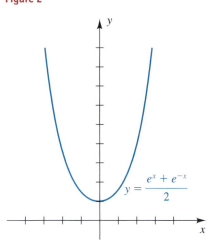

$$y = \frac{e^x + e^{-x}}{2}$$

APPLICATION   **Flexible Cables**

The function f of Example 4 occurs in applied mathematics and engineering, where it is called the **hyperbolic cosine function.** This function can be used to describe the shape of a uniform flexible cable or chain whose ends are supported from the same height, such as a telephone or power line cable (see Figure 3). If we introduce a coordinate system, as indicated in the figure, then it can be shown that an equation that corresponds to the shape of the cable is

$$y = \frac{a}{2}(e^{x/a} + e^{-x/a}),$$

where a is a real number. The graph is called a **catenary,** after the Latin word for *chain.* The function in Example 4 is the special case in which $a = 1$.

Figure 3

APPLICATION   **Radiotherapy**

Exponential functions play an important role in the field of *radiotherapy,* the treatment of tumors by radiation. The fraction of cells in a tumor that survive a treatment, called the *surviving fraction,* depends not only on the energy and nature of the radiation, but also on the depth, size, and characteristics of the tumor itself. The exposure to radiation may be thought of as a number of potentially damaging events, where at least one *hit* is required to kill a tumor cell. For instance, suppose that each cell has exactly one *target* that must be hit. If k denotes the average target size of a tumor cell and if x is the number of damaging events (the *dose*), then the surviving fraction $f(x)$ is given by

$$f(x) = e^{-kx}.$$

This is called the *one target–one hit surviving fraction.*

Suppose next that each cell has n targets and that each target must be hit once for the cell to die. In this case, the *n target–one hit surviving fraction* is given by

$$f(x) = 1 - (1 - e^{-kx})^n.$$

The graph of f may be analyzed to determine what effect increasing the dosage x will have on decreasing the surviving fraction of tumor cells. Note that $f(0) = 1$; that is, if there is no dose, then all cells survive. As an example, if $k = 1$ and $n = 2$, then

$$f(x) = 1 - (1 - e^{-x})^2$$
$$= 1 - (1 - 2e^{-x} + e^{-2x})$$
$$= 2e^{-x} - e^{-2x}.$$

A complete analysis of the graph of f requires calculus. The graph is sketched in Figure 4. The *shoulder* on the curve near the point $(0, 1)$ represents the threshold nature of the treatment—that is, a small dose results in very little tumor cell elimination. Note that for a large x, an increase in dosage has little effect on the surviving fraction. To determine the ideal dose to administer to a patient, specialists in radiation therapy must also take into account the number of healthy cells that are killed during a treatment.

Problems of the type illustrated in the next example occur in the study of calculus.

Figure 4

Surviving fraction of tumor cells after a radiation treatment

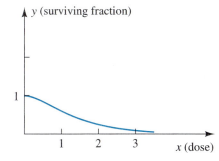

EXAMPLE 5   Finding zeros of a function involving exponentials

If $f(x) = x^2(-2e^{-2x}) + 2xe^{-2x}$, find the zeros of f.

SOLUTION   We may factor $f(x)$ as follows:

$$f(x) = 2xe^{-2x} - 2x^2e^{-2x} \quad \text{given}$$
$$= 2xe^{-2x}(1 - x) \quad \text{factor out } 2xe^{-2x}$$

To find the zeros of f, we solve the equation $f(x) = 0$. Since $e^{-2x} > 0$ for every x, we see that $f(x) = 0$ if and only if $x = 0$ or $1 - x = 0$. Thus, the zeros of f are 0 and 1.

5.3 *Exercises*

Exer. 1–4: Use the graph of $y = e^x$ to help sketch the graph of f.

1 (a) $f(x) = e^{-x}$ (b) $f(x) = -e^x$

2 (a) $f(x) = e^{2x}$ (b) $f(x) = 2e^x$

3 (a) $f(x) = e^{x+4}$ (b) $f(x) = e^x + 4$

4 (a) $f(x) = e^{-2x}$ (b) $f(x) = -2e^x$

Exer. 5–6: If P dollars is deposited in a savings account that pays interest at a rate of $r\%$ per year compounded continuously, find the balance after t years.

5 $P = 1000$, $r = 8\frac{1}{4}$, $t = 5$

6 $P = 100$, $r = 12\frac{1}{2}$, $t = 10$

Exer. 7–8: How much money, invested at an interest rate of $r\%$ per year compounded continuously, will amount to A dollars after t years?

7 $A = 100{,}000$, $r = 11$, $t = 18$

8 $A = 15{,}000$, $r = 9.5$, $t = 4$

Exer. 9–10: An investment of P dollars increased to A dollars in t years. If interest was compounded continuously, find the interest rate. (*Hint:* Use trial and error.)

9 $A = 13{,}464$, $P = 1000$, $t = 20$

10 $A = 890.20$, $P = 400$, $t = 16$

Exer. 11–12: Solve the equation.

11 $e^{(x^2)} = e^{7x-12}$ 12 $e^{3x} = e^{2x-1}$

Exer. 13–16: Find the zeros of f.

13 $f(x) = xe^x + e^x$

14 $f(x) = -x^2 e^{-x} + 2xe^{-x}$

15 $f(x) = x^3(4e^{4x}) + 3x^2 e^{4x}$

16 $f(x) = x^2(2e^{2x}) + 2xe^{2x} + e^{2x} + 2xe^{2x}$

Exer. 17–18: Simplify the expression.

17 $\dfrac{(e^x + e^{-x})(e^x + e^{-x}) - (e^x - e^{-x})(e^x - e^{-x})}{(e^x + e^{-x})^2}$

18 $\dfrac{(e^x - e^{-x})^2 - (e^x + e^{-x})^2}{(e^x + e^{-x})^2}$

19 **Crop growth** An exponential function W such that $W(t) = W_0 e^{kt}$ for $k > 0$ describes the first month of growth for crops such as maize, cotton, and soybeans. The function value $W(t)$ is the total weight in milligrams, W_0 is the weight on the day of emergence, and t is the time in days. If, for a species of soybean, $k = 0.2$ and $W_0 = 68$ mg, predict the weight at the end of 30 days.

20 **Crop growth** Refer to Exercise 19. It is often difficult to measure the weight W_0 of a plant from when it first emerges from the soil. If, for a species of cotton, $k = 0.21$ and the weight after 10 days is 575 milligrams, estimate W_0.

21 **U.S. population growth** The 1980 population of the United States was approximately 227 million, and the population has been growing continuously at a rate of 0.7% per year. Predict the population $N(t)$ in the year 2010 if this growth trend continues.

22 **Population growth in India** The 1985 population estimate for India was 762 million, and the population has been growing continuously at a rate of about 2.2% per year. Assuming that this rapid growth rate continues, estimate the population $N(t)$ of India in the year 2010.

23 **Longevity of halibut** In fishery science, a cohort is the collection of fish that results from one annual reproduction. It is usually assumed that the number of fish $N(t)$ still alive after t years is given by an exponential function. For Pacific halibut, $N(t) = N_0 e^{-0.2t}$, where N_0 is the initial size of the cohort. Approximate the percentage of the original number still alive after 10 years.

24 **Radioactive tracer** The radioactive tracer ^{51}Cr can be used to locate the position of the placenta in a pregnant woman. Often the tracer must be ordered from a medical laboratory. If A_0 units (microcuries) are shipped, then because of the radioactive decay, the number of units $A(t)$ present after t days is given by $A(t) = A_0 e^{-0.0249t}$.

(a) If 35 units are shipped and it takes 2 days for the tracer to arrive, approximately how many units will be available for the test?

(b) If 35 units are needed for the test, approximately how many units should be shipped?

25 **Blue whale population growth** In 1978, the population of blue whales in the southern hemisphere was thought to number 5000. Since whaling has been outlawed and an abundant food supply is available, the population $N(t)$ is expected to grow exponentially according to the formula $N(t) = 5000e^{0.0036t}$, where t is in years and $t = 0$ corresponds to 1978. Predict the population in the year 2010.

26 **Halibut growth** The length (in centimeters) of many common commercial fish t years old can be approximated by a von Bertalanffy growth function having an equation of the form $f(t) = a(1 - be^{-kt})$, where a, b, and k are constants.

(a) For Pacific halibut, $a = 200$, $b = 0.956$, and $k = 0.18$. Estimate the length of a 10-year-old halibut.

(b) Use the graph of f to estimate the maximum attainable length of the Pacific halibut.

27 **Atmospheric pressure** Under certain conditions the atmospheric pressure p (in inches) at altitude h feet is given by $p = 29e^{-0.000034h}$. What is the pressure at an altitude of 40,000 feet?

28 **Polonium isotope decay** If we start with c milligrams of the polonium isotope ^{210}Po, the amount remaining after t days may be approximated by $A = ce^{-0.00495t}$. If the initial amount is 50 milligrams, approximate, to the nearest hundredth, the amount remaining after

(a) 30 days   (b) 180 days   (c) 365 days

29 **Growth of children** The Jenss model is generally regarded as the most accurate formula for predicting the height of preschool children. If y is height (in centimeters) and x is age (in years), then

$$y = 79.041 + 6.39x - e^{3.261 - 0.993x}$$

for $\frac{1}{4} \le x \le 6$. From calculus, the rate of growth R (in cm/year) is given by $R = 6.39 + 0.993e^{3.261 - 0.993x}$. Find the height and rate of growth of a typical 1-year-old child.

30 **Particle velocity** A very small spherical particle (on the order of 5 microns in diameter) is projected into still air with an initial velocity of v_0 m/sec, but its velocity decreases because of drag forces. Its velocity t seconds later is given by $v(t) = v_0 e^{-at}$ for some $a > 0$, and the distance $s(t)$ the particle travels is given by

$$s(t) = \frac{v_0}{a}(1 - e^{-at}).$$

The stopping distance is the total distance traveled by the particle.

(a) Find a formula that approximates the stopping distance in terms of v_0 and a.

(b) Use the formula in part (a) to estimate the stopping distance if $v_0 = 10$ m/sec and $a = 8 \times 10^5$.

31 **Minimum wage** In 1971 the minimum wage in the United States was $1.60 per hour. Assuming that the rate of inflation is 5% per year, find the equivalent minimum wage in the year 2010.

32 **Land value** In 1867 the United States purchased Alaska from Russia for $7,200,000. There is 586,400 square miles of land in Alaska. Assuming that the value of the land increases continuously at 3% per year and that land can be purchased at an equivalent price, determine the price of 1 acre in the year 2010. (One square mile is equivalent to 640 acres.)

Exer. 33–34: The *effective yield* (or effective annual interest rate) for an investment is the simple interest rate that would yield at the end of one year the same amount as is yielded by the compounded rate that is actually applied. Approximate, to the nearest 0.01%, the effective yield corresponding to an interest rate of $r\%$ per year compounded (a) quarterly and (b) continuously.

33 $r = 7$ 34 $r = 12$

35 **Probability density function** In statistics, the probability density function for the normal distribution is defined by

$$f(x) = \frac{1}{\sigma\sqrt{2\pi}} e^{-z^2/2} \quad \text{with} \quad z = \frac{x - \mu}{\sigma},$$

where μ and σ are real numbers (μ is the *mean* and σ^2 is the *variance* of the distribution). Sketch the graph of f for the case $\sigma = 1$ and $\mu = 0$.

5.4

Logarithmic Functions

In Section 5.2 we observed that the exponential function given by $f(x) = a^x$ for $0 < a < 1$ or $a > 1$ is one-to-one. Hence, f has an inverse function f^{-1} (see Section 5.1). This inverse of the exponential function with base a is called the **logarithmic function with base a** and is denoted by **$\log_a$**. Its values are written $\log_a (x)$ or $\log_a x$, read "the logarithm of x with base a." Since, by the definition of an inverse function f^{-1},

$$y = f^{-1}(x) \qquad \text{if and only if} \qquad x = f(y),$$

the definition of $\log_a$ may be expressed as follows.

Definition of $\log_a$	Let a be a positive real number different from 1. The **logarithm of x with base a** is defined by $$y = \log_a x \qquad \text{if and only if} \qquad x = a^y$$ for every $x > 0$ and every real number y.

Note that the two equations in the definition are equivalent. We call the first equation the **logarithmic form** and the second the **exponential form.** You should strive to become an expert in changing each form into the other. The following diagram may help you achieve this goal.

Observe that when forms are changed, *the bases of the logarithmic and exponential forms are the same.* The number y (that is, $\log_a x$) corresponds to the exponent in the exponential form. In words, $\log_a x$ is *the exponent to which the base a must be raised to obtain x.* This is what people are referring to when they say "Logarithms are exponents."

The following illustration contains examples of equivalent forms.

ILLUSTRATION **Equivalent Forms**

Logarithmic form	Exponential form
■ $\log_5 u = 2$	$5^2 = u$
■ $\log_b 8 = 3$	$b^3 = 8$
■ $r = \log_p q$	$p^r = q$
■ $w = \log_4 (2t + 3)$	$4^w = 2t + 3$
■ $\log_3 x = 5 + 2z$	$3^{5+2z} = x$

The next example contains an application that involves changing from an exponential form to a logarithmic form.

EXAMPLE 1 **Changing exponential form to logarithmic form**

The number N of bacteria in a certain culture after t hours is given by $N = (1000)2^t$. Express t as a logarithmic function of N with base 2.

SOLUTION $N = (1000)2^t$ given

$$\frac{N}{1000} = 2^t \qquad \text{isolate the exponential expression}$$

$$t = \log_2 \frac{N}{1000} \qquad \text{change to logarithmic form} \qquad \blacksquare$$

Some special cases of logarithms are given in the next example.

EXAMPLE 2 **Finding logarithms**

Find the number, if possible.

(a) $\log_{10} 100$ **(b)** $\log_2 \frac{1}{32}$ **(c)** $\log_9 3$ **(d)** $\log_7 1$ **(e)** $\log_3 (-2)$

SOLUTION   In each case we are given $\log_a x$ and must find the exponent y such that $a^y = x$. We obtain the following.

(a) $\log_{10} 100 = 2$ because $10^2 = 100$.

(b) $\log_2 \frac{1}{32} = -5$ because $2^{-5} = \frac{1}{32}$.

(c) $\log_9 3 = \frac{1}{2}$ because $9^{1/2} = 3$.

(d) $\log_7 1 = 0$ because $7^0 = 1$.

(e) $\log_3 (-2)$ is not possible because $3^y \neq -2$ for any real number y. $\qquad \blacksquare$

The following general properties follow from the interpretation of $\log_a x$ as an exponent.

Property of $\log_a x$	Reason	Illustration
(1) $\log_a 1 = 0$	$a^0 = 1$	$\log_3 1 = 0$
(2) $\log_a a = 1$	$a^1 = a$	$\log_{10} 10 = 1$
(3) $\log_a a^x = x$	$a^x = a^x$	$\log_2 8 = \log_2 2^3 = 3$
(4) $a^{\log_a x} = x$	as follows	$5^{\log_5 7} = 7$

The reason for property 4 follows directly from the definition of $\log_a$, since

$$\text{if} \quad y = \log_a x, \quad \text{then} \quad x = a^y, \quad \text{or} \quad x = a^{\log_a x}.$$

The logarithmic function with base a is the inverse of the exponential function with base a, so the graph of $y = \log_a x$ can be obtained by reflecting the graph of $y = a^x$ through the line $y = x$ (see Section 5.1). This procedure is illustrated in Figure 1 for the case $a > 1$. Note that the x-intercept of the graph is 1, the domain is the set of positive real numbers, the range is $\mathbb{R}$, and the

Figure 1

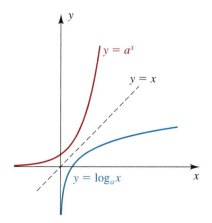

y-axis is a vertical asymptote. Logarithms with base $0 < a < 1$ are seldom used, so we will not emphasize their graphs.

We see from Figure 1 that if $a > 1$, then $\log_a x$ is increasing on $(0, \infty)$ and hence is one-to-one by the theorem on page 280. Combining this result with parts (1) and (2) of the definition of one-to-one function on page 278 gives us the following theorem, which can also be proved if $0 < a < 1$.

Theorem: Logarithmic Functions Are One-to-One	The logarithmic function with base a is one-to-one. Thus, the following equivalent conditions are satisfied for positive real numbers x_1 and x_2. **(1)** If $x_1 \neq x_2$, then $\log_a x_1 \neq \log_a x_2$. **(2)** If $\log_a x_1 = \log_a x_2$, then $x_1 = x_2$.

When using this theorem as a reason for a step in the solution to an example, we will state that *logarithmic functions are one-to-one*.

In the following example we solve a simple *logarithmic equation*—that is, an equation involving a logarithm of an expression that contains a variable. Extraneous solutions may be introduced when logarithmic equations are solved. Hence, we must check solutions of logarithmic equations to make sure that we are taking logarithms of *only positive real numbers;* otherwise, a logarithmic function is not defined.

EXAMPLE 3   Solving a logarithmic equation

Solve the equation $\log_6 (4x - 5) = \log_6 (2x + 1)$.

SOLUTION

$$\log_6 (4x - 5) = \log_6 (2x + 1) \qquad \text{given}$$
$$4x - 5 = 2x + 1 \qquad \text{logarithmic functions are one-to-one}$$
$$2x = 6 \qquad \text{subtract } 2x; \text{ add } 5$$
$$x = 3 \qquad \text{divide by 2}$$

✔ **Check $x = 3$**   LS: $\log_6 (4 \cdot 3 - 5) = \log_6 7$
                    RS: $\log_6 (2 \cdot 3 + 1) = \log_6 7$

Since $\log_6 7 = \log_6 7$ is a true statement, $x = 3$ is a solution.   ◢

When we check the solution $x = 3$ in Example 3, it is not required that the solution be positive. But it is required that the two expressions, $4x - 5$ and $2x + 1$, be positive after we substitute 3 for x. If we extend our idea of *argument* from variables to expressions, then when checking solutions, we can simply remember that *arguments must be positive*.

In the next example we use the definition of logarithm to solve a logarithmic equation.

EXAMPLE 4   Solving a logarithmic equation

Solve the equation $\log_4 (5 + x) = 3$.

SOLUTION

$$\log_4 (5 + x) = 3 \qquad \text{given}$$
$$5 + x = 4^3 \qquad \text{change to exponential form}$$
$$x = 59 \qquad \text{solve for } x$$

 Check $x = 59$ LS: $\log_4 (5 + 59) = \log_4 64 = \log_4 4^3 = 3$

RS: 3

Since $3 = 3$ is a true statement, $x = 59$ is a solution.

We next sketch the graph of a specific logarithmic function.

EXAMPLE 5   Sketching the graph of a logarithmic function

Sketch the graph of f if $f(x) = \log_3 x$.

SOLUTION   We will describe three methods for sketching the graph.

Method 1   Since the functions given by $\log_3 x$ and 3^x are inverses of each other, we proceed as we did for $y = \log_a x$ in Figure 1; that is, we first sketch the graph of $y = 3^x$ and then reflect it through the line $y = x$. This gives us the sketch in Figure 2. Note that the points $(-1, 3^{-1})$, $(0, 1)$, $(1, 3)$, and $(2, 9)$ on the graph of $y = 3^x$ reflect into the points $(3^{-1}, -1)$, $(1, 0)$, $(3, 1)$, and $(9, 2)$ on the graph of $y = \log_3 x$.

Figure 2

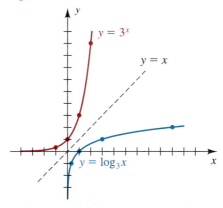

Method 2   We can find points on the graph of $y = \log_3 x$ by letting $x = 3^k$, where k is a real number, and then applying property 3 of logarithms on page 309, as follows:

$$y = \log_3 x = \log_3 3^k = k$$

(continued)

Using this formula, we obtain the points on the graph listed in the following table.

$x = 3^k$		3^{-3}	3^{-2}	3^{-1}	3^0	3^1	3^2	3^3
$y = \log_3 x = k$		-3	-2	-1	0	1	2	3

Figure 3

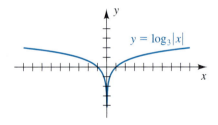

This gives us the same points obtained using the first method.

Method 3   We can sketch the graph of $y = \log_3 x$ by sketching the graph of the equivalent exponential form $x = 3^y$.

Before proceeding, let's plot one more point on $y = \log_3 x$ in Figure 2. If we let $x = 5$, then $y = \log_3 5$ (see Figure 3). (We see that $\log_3 5$ is a number between 1 and 2; we'll be able to better approximate $\log_3 5$ in Section 5.6.) Now on the graph of $y = 3^x$ we have the point $(x, y) = (\log_3 5, 5)$, so $5 = 3^{\log_3 5}$, which illustrates property 4 of logarithms on page 309 and reinforces the claim that *logarithms are exponents*.

As in the following examples, we often wish to sketch the graph of $f(x) = \log_a u$, where u is some expression involving x.

Figure 4

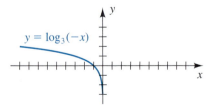

EXAMPLE 6   Sketching the graph of a logarithmic function

Sketch the graph of f if $f(x) = \log_3 |x|$ for $x \neq 0$.

SOLUTION   The graph is symmetric with respect to the y-axis, since

$$f(-x) = \log_3 |-x| = \log_3 |x| = f(x).$$

If $x > 0$, then $|x| = x$ and the graph coincides with the graph of $y = \log_3 x$ sketched in Figure 2. Using symmetry, we reflect that part of the graph through the y-axis, obtaining the sketch in Figure 4.

Alternatively, we may think of this function as $g(x) = \log_3 x$ with $|x|$ substituted for x (refer to the discussion on page 180). Since all points on the graph of g have positive x-coordinates, we can obtain the graph of f by combining g with the reflection of g through the y-axis.

Figure 5

EXAMPLE 7   Reflecting the graph of a logarithmic function

Sketch the graph of f if $f(x) = \log_3 (-x)$.

SOLUTION   The domain of f is the set of negative real numbers, since $\log_3 (-x)$ exists only if $-x > 0$ or, equivalently, $x < 0$. We can obtain the graph of f from the graph of $y = \log_3 x$ by replacing each point (x, y) in Figure 2 by $(-x, y)$. This is equivalent to reflecting the graph of $y = \log_3 x$ through the y-axis. The graph is sketched in Figure 5.

Another method is to change $y = \log_3 (-x)$ to the exponential form $3^y = -x$ and then sketch the graph of $x = -3^y$.

EXAMPLE 8 **Shifting graphs of logarithmic equations**

Sketch the graph of the equation:

(a) $y = \log_3 (x - 2)$ **(b)** $y = \log_3 x - 2$

SOLUTION

(a) The graph of $y = \log_3 x$ was sketched in Figure 2 and is resketched in Figure 6. From the discussion of horizontal shifts in Section 3.5, we can obtain the graph of $y = \log_3 (x - 2)$ by shifting the graph of $y = \log_3 x$ two units to the right, as shown in Figure 6.

(b) From the discussion of vertical shifts in Section 3.5, the graph of the equation $y = \log_3 x - 2$ can be obtained by shifting the graph of $y = \log_3 x$ two units downward, as shown in Figure 7. Note that the x-intercept is given by $\log_3 x = 2$, or $x = 3^2 = 9$.

Figure 6

Figure 7

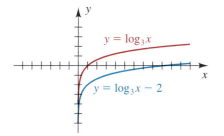

EXAMPLE 9 **Reflecting the graph of a logarithmic function**

Sketch the graph of f if $f(x) = \log_3 (2 - x)$.

SOLUTION If we write

$$f(x) = \log_3 (2 - x) = \log_3 [-(x - 2)],$$

then, by applying the same technique used to obtain the graph of the equation $y = \log_3 (-x)$ in Example 7 (with x replaced by $x - 2$), we see that the graph of f is the reflection of the graph of $y = \log_3 (x - 2)$ through the vertical line $x = 2$. This gives us the sketch in Figure 8.

Another method is to change $y = \log_3 (2 - x)$ to the exponential form $3^y = 2 - x$ and then sketch the graph of $x = 2 - 3^y$.

Figure 8

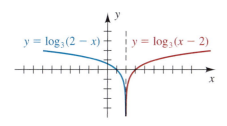

Before electronic calculators were invented, logarithms with base 10 were used for complicated numerical computations involving products, quotients, and powers of real numbers. Base 10 was used because it is well suited for numbers that are expressed in scientific form. Logarithms with base 10 are called **common logarithms.** The symbol $\log x$ is used as an abbreviation for $\log_{10} x$, just as $\sqrt{}$ is used as an abbreviation for $\sqrt[2]{}$.

Definition of Common Logarithm	$\log x = \log_{10} x$ for every $x > 0$

Since inexpensive calculators are now available, there is no need for common logarithms as a tool for computational work. Base 10 does occur in applications, however, and hence many calculators have a $\boxed{\text{LOG}}$ key, which can be used to approximate common logarithms.

The natural exponential function is given by $f(x) = e^x$. The logarithmic function with base e is called the **natural logarithmic function.** The symbol **ln x** (read "ell-en of x") is an abbreviation for $\log_e x$, and we refer to it as the **natural logarithm of x.** Thus, *the natural logarithmic function and the natural exponential function are inverse functions of each other.*

Definition of Natural Logarithm	$\ln x = \log_e x$ for every $x > 0$

Most calculators have a key labeled $\boxed{\text{LN}}$, which can be used to approximate natural logarithms. The next illustration gives several examples of equivalent forms involving common and natural logarithms.

ILLUSTRATION   **Equivalent Forms**

Logarithmic form	**Exponential form**
■ $\log x = 2$	$10^2 = x$
■ $\log z = y + 3$	$10^{y+3} = z$
■ $\ln x = 2$	$e^2 = x$
■ $\ln z = y + 3$	$e^{y+3} = z$

To find x when given $\log x$ or $\ln x$, we may use the $\boxed{10^x}$ key or the $\boxed{e^x}$ key, respectively, on a calculator, as in the next example. If your calculator has an $\boxed{\text{INV}}$ key (for inverse), you may enter x and successively press $\boxed{\text{INV}}$ $\boxed{\text{LOG}}$ or $\boxed{\text{INV}}$ $\boxed{\text{LN}}$.

EXAMPLE 10   Solving a simple logarithmic equation

Find x if

(a) $\log x = 1.7959$ **(b)** $\ln x = 4.7$

SOLUTION

(a) Changing $\log x = 1.7959$ to its equivalent exponential form gives us

$$x = 10^{1.7959}.$$

Evaluating the last expression to three-decimal-place accuracy yields

$$x \approx 62.503.$$

(b) Changing $\ln x = 4.7$ to its equivalent exponential form gives us

$$x = e^{4.7} \approx 109.95.$$

The following chart lists common and natural logarithmic forms for the properties on page 309.

Logarithms with base a	Common logarithms	Natural logarithms
(1) $\log_a 1 = 0$	$\log 1 = 0$	$\ln 1 = 0$
(2) $\log_a a = 1$	$\log 10 = 1$	$\ln e = 1$
(3) $\log_a a^x = x$	$\log 10^x = x$	$\ln e^x = x$
(4) $a^{\log_a x} = x$	$10^{\log x} = x$	$e^{\ln x} = x$

The last property for natural logarithms allows us to write the number a as $e^{\ln a}$, so the exponential function $f(x) = a^x$ can be written as $f(x) = (e^{\ln a})^x$ or as $f(x) = e^{x \ln a}$. Many calculators compute an exponential regression model of the form $y = ab^x$. If an exponential model with base e is desired, we can write the model

$$y = ab^x \quad \text{as} \quad y = ae^{x \ln b}.$$

ILLUSTRATION   **Converting to Base e Expressions**

- 3^x is equivalent to $e^{x \ln 3}$
- x^3 is equivalent to $e^{3 \ln x}$
- $4 \cdot 2^x$ is equivalent to $4 \cdot e^{x \ln 2}$

Figure 9

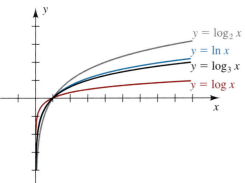

$y = \log_2 x$
$y = \ln x$
$y = \log_3 x$
$y = \log x$

Figure 9 shows four logarithm graphs with base $a > 1$. Note that for $x > 1$, as the base of the logarithm increases, the graphs increase more slowly (they are more horizontal). This makes sense when we consider the graphs of the inverses of these functions: $y = 2^x$, $y = e^x$, $y = 3^x$, and $y = 10^x$. Here, for $x > 0$, as the base of the exponential expression increases, the graphs increase faster (they are more vertical).

The next four examples illustrate applications of common and natural logarithms.

EXAMPLE 11   **The Richter scale**

On the Richter scale, the magnitude R of an earthquake of intensity I is given by

$$R = \log \frac{I}{I_0},$$

where I_0 is a certain minimum intensity.

(a) If the intensity of an earthquake is $1000 I_0$, find R.

(b) Express I in terms of R and I_0.

SOLUTION

(a) $R = \log \dfrac{I}{I_0}$ given

$ = \log \dfrac{1000I_0}{I_0}$ let $I = 1000I_0$

$ = \log 1000$ cancel I_0

$ = \log 10^3$ $1000 = 10^3$

$ = 3$ $\log 10^x = x$ for every x

From this result we see that a tenfold increase in intensity results in an increase of 1 in magnitude (if 1000 were changed to 10,000, then 3 would change to 4).

(b) $R = \log \dfrac{I}{I_0}$ given

$ \dfrac{I}{I_0} = 10^R$ change to exponential form

$ I = I_0 \cdot 10^R$ multiply by I_0

EXAMPLE 12   **Newton's law of cooling**

Newton's law of cooling states that the rate at which an object cools is directly proportional to the difference in temperature between the object and its surrounding medium. Newton's law can be used to show that under certain conditions the temperature T (in °C) of an object at time t (in hours) is given by $T = 75e^{-2t}$. Express t as a function of T.

SOLUTION $T = 75e^{-2t}$ given

$ e^{-2t} = \dfrac{T}{75}$ isolate the exponential expression

$ -2t = \ln \dfrac{T}{75}$ change to logarithmic form

$ t = -\dfrac{1}{2} \ln \dfrac{T}{75}$ divide by -2

EXAMPLE 13   **Approximating a doubling time**

Assume that a population is growing continuously at a rate of 4% per year. Approximate the amount of time it takes for the population to double its size—that is, its **doubling time.**

SOLUTION   Note that an initial population size is not given. Not knowing the initial size does not present a problem, however, since we wish only to determine the time needed to obtain a population size *relative* to the initial population size. Using the growth formula $q = q_0 e^{rt}$ with $r = 0.04$ gives us

$2q_0 = q_0 e^{0.04t}$ let $q = 2q_0$

$2 = e^{0.04t}$ divide by q_0 ($q_0 \neq 0$)

$$0.04t = \ln 2 \qquad \text{change to logarithmic form}$$

$$t = 25 \ln 2 \approx 17.3 \text{ yr.} \quad \text{multiply by } \frac{1}{0.04} = 25$$

The fact that q_0 did not have any effect on the answer indicates that the doubling time for a population of 1000 is the same as the doubling time for a population of 1,000,000 or any other reasonable initial population.

From the last example we may obtain a general formula for the doubling time of a population—namely,

$$rt = \ln 2 \qquad \text{or, equivalently,} \qquad t = \frac{\ln 2}{r}.$$

Since $\ln 2 \approx 0.69$, we see that the doubling time t for a growth of this type is approximately $0.69/r$. Because the numbers 70 and 72 are close to 69 but have more divisors, some resources refer to this doubling relationship as the **rule of 70** or the **rule of 72.** As an illustration of the rule of 72, if the growth rate of a population is 8%, then it takes about $72/8 = 9$ years for the population to double. More precisely, this value is

$$\frac{\ln 2}{8} \cdot 100 \approx 8.7 \text{ yr.}$$

EXAMPLE 14 **Determining the half-life of a radioactive substance**

A physicist finds that an unknown radioactive substance registers 2000 counts per minute on a Geiger counter. Ten days later the substance registers 1500 counts per minute. Using calculus, it can be shown that after t days the amount of radioactive material, and hence the number of counts per minute $N(t)$, is directly proportional to e^{ct} for some constant c. Determine the half-life of the substance.

SOLUTION Since $N(t)$ is directly proportional to e^{ct},

$$N(t) = ke^{ct},$$

where k is a constant. Letting $t = 0$ and using $N(0) = 2000$, we obtain

$$2000 = ke^{c \cdot 0} = k \cdot 1 = k.$$

Hence, the formula for $N(t)$ may be written

$$N(t) = 2000e^{ct}.$$

Since $N(10) = 1500$, we may determine c as follows:

$$1500 = 2000e^{c \cdot 10} \qquad \text{let } t = 10 \text{ in } N(t)$$

$$\tfrac{3}{4} = e^{10c} \qquad \text{isolate the exponential expression}$$

$$10c = \ln \tfrac{3}{4} \qquad \text{change to logarithmic form}$$

$$c = \tfrac{1}{10} \ln \tfrac{3}{4} \qquad \text{divide by 10}$$

(continued)

Finally, since the half-life corresponds to the time t at which $N(t)$ is equal to 1000, we have the following:

$$1000 = 2000e^{ct} \qquad \text{let } N(t) = 1000$$

$$\tfrac{1}{2} = e^{ct} \qquad \text{isolate the exponential expression}$$

$$ct = \ln \tfrac{1}{2} \qquad \text{change to logarithmic form}$$

$$t = \frac{1}{c}\ln\frac{1}{2} \qquad \text{divide by } c$$

$$= \frac{1}{\tfrac{1}{10}\ln\tfrac{3}{4}}\ln\frac{1}{2} \qquad c = \tfrac{1}{10}\ln\tfrac{3}{4}$$

$$\approx 24 \text{ days} \qquad \text{approximate}$$

5.4 Exercises

Exer. 1–2: Change to logarithmic form.

1 (a) $4^3 = 64$　　(b) $4^{-3} = \frac{1}{64}$　　(c) $t^r = s$

　　(d) $3^x = 4 - t$　　(e) $5^{7t} = \dfrac{a+b}{a}$　　(f) $(0.7)^t = 5.3$

2 (a) $3^5 = 243$　　(b) $3^{-4} = \frac{1}{81}$　　(c) $c^p = d$

　　(d) $7^x = 100p$　　(e) $3^{-2x} = \dfrac{P}{F}$　　(f) $(0.9)^t = \frac{1}{2}$

Exer. 3–4: Change to exponential form.

3 (a) $\log_2 32 = 5$　　　　(b) $\log_3 \frac{1}{243} = -5$

　　(c) $\log_t r = p$　　　　(d) $\log_3 (x + 2) = 5$

　　(e) $\log_2 m = 3x + 4$　　(f) $\log_b 512 = \frac{3}{2}$

4 (a) $\log_3 81 = 4$　　　　(b) $\log_4 \frac{1}{256} = -4$

　　(c) $\log_v w = q$　　　　(d) $\log_6 (2x - 1) = 3$

　　(e) $\log_4 p = 5 - x$　　(f) $\log_a 343 = \frac{3}{4}$

Exer. 5–10: Solve for t using logarithms with base a.

5 $2a^{t/3} = 5$　　　　　**6** $3a^{4t} = 10$

7 $K = H - Ca^t$　　　　**8** $F = D + Ba^t$

9 $A = Ba^{Ct} + D$　　　**10** $L = Ma^{t/N} - P$

Exer. 11–12: Change to logarithmic form.

11 (a) $10^5 = 100,000$　　(b) $10^{-3} = 0.001$

　　(c) $10^x = y + 1$　　　(d) $e^7 = p$

　　(e) $e^{2t} = 3 - x$

12 (a) $10^4 = 10,000$　　　(b) $10^{-2} = 0.01$

　　(c) $10^x = 38z$　　　　(d) $e^4 = D$

　　(e) $e^{0.1t} = x + 2$

Exer. 13–14: Change to exponential form.

13 (a) $\log x = 50$　　　　(b) $\log x = 20t$

　　(c) $\ln x = 0.1$　　　　(d) $\ln w = 4 + 3x$

　　(e) $\ln (z - 2) = \frac{1}{6}$

14 (a) $\log x = -8$　　　　(b) $\log x = y - 2$

　　(c) $\ln x = \frac{1}{2}$　　　　(d) $\ln z = 7 + x$

　　(e) $\ln (t - 5) = 1.2$

Exer. 15–16: Find the number, if possible.

15 (a) $\log_5 1$ (b) $\log_3 3$ (c) $\log_4 (-2)$

   (d) $\log_7 7^2$ (e) $3^{\log_3 8}$ (f) $\log_5 125$

   (g) $\log_4 \frac{1}{16}$

16 (a) $\log_8 1$ (b) $\log_9 9$ (c) $\log_5 0$

   (d) $\log_6 6^7$ (e) $5^{\log_5 4}$ (f) $\log_3 243$

   (g) $\log_2 128$

Exer. 17–18: Find the number.

17 (a) $10^{\log 3}$ (b) $\log 10^5$ (c) $\log 100$

   (d) $\log 0.0001$ (e) $e^{\ln 2}$ (f) $\ln e^{-3}$

   (g) $e^{2 + \ln 3}$

18 (a) $10^{\log 7}$ (b) $\log 10^{-6}$ (c) $\log 100{,}000$

   (d) $\log 0.001$ (e) $e^{\ln 8}$ (f) $\ln e^{2/3}$

   (g) $e^{1 + \ln 5}$

Exer. 19–34: Solve the equation.

19 $\log_4 x = \log_4 (8 - x)$

20 $\log_3 (x + 4) = \log_3 (1 - x)$

21 $\log_5 (x - 2) = \log_5 (3x + 7)$

22 $\log_7 (x - 5) = \log_7 (6x)$

23 $\log x^2 = \log (-3x - 2)$

24 $\ln x^2 = \ln (12 - x)$

25 $\log_3 (x - 4) = 2$

26 $\log_2 (x - 5) = 4$

27 $\log_9 x = \frac{3}{2}$ 28 $\log_4 x = -\frac{3}{2}$

29 $\ln x^2 = -2$ 30 $\log x^2 = -4$

31 $e^{2 \ln x} = 9$ 32 $e^{-\ln x} = 0.2$

33 $e^{x \ln 3} = 27$ 34 $e^{x \ln 2} = 0.25$

35 Sketch the graph of f if $a = 4$:

   (a) $f(x) = \log_a x$ (b) $f(x) = -\log_a x$

   (c) $f(x) = 2 \log_a x$ (d) $f(x) = \log_a (x + 2)$

   (e) $f(x) = (\log_a x) + 2$ (f) $f(x) = \log_a (x - 2)$

   (g) $f(x) = (\log_a x) - 2$ (h) $f(x) = \log_a |x|$

   (i) $f(x) = \log_a (-x)$ (j) $f(x) = \log_a (3 - x)$

   (k) $f(x) = |\log_a x|$ (l) $f(x) = \log_{1/a} x$

36 Work Exercise 35 if $a = 5$.

Exer. 37–42: Sketch the graph of f.

37 $f(x) = \log (x + 10)$ 38 $f(x) = \log (x + 100)$

39 $f(x) = \ln |x|$ 40 $f(x) = \ln |x - 1|$

41 $f(x) = \ln e + x$ 42 $f(x) = \ln (e + x)$

Exer. 43–44: Find a logarithmic function of the form $f(x) = \log_a x$ for the given graph.

43

44

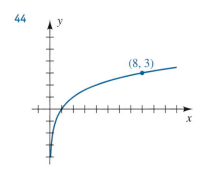

Exer. 45–50: Shown in the figure is the graph of a function f. Express $f(x)$ in terms of F.

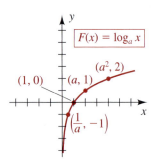

$F(x) = \log_a x$

$(1, 0)$ $(a, 1)$ $(a^2, 2)$ $\left(\frac{1}{a}, -1\right)$

45

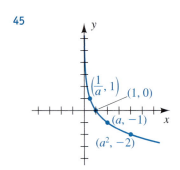

$\left(\frac{1}{a}, 1\right)$ $(1, 0)$ $(a, -1)$ $(a^2, -2)$

46

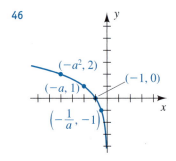

$(-a^2, 2)$ $(-1, 0)$ $(-a, 1)$ $\left(-\frac{1}{a}, -1\right)$

47

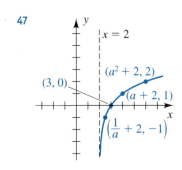

$x = 2$ $(a^2 + 2, 2)$ $(3, 0)$ $(a + 2, 1)$ $\left(\frac{1}{a} + 2, -1\right)$

48

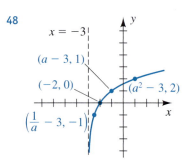

$x = -3$ $(a - 3, 1)$ $(-2, 0)$ $(a^2 - 3, 2)$ $\left(\frac{1}{a} - 3, -1\right)$

49

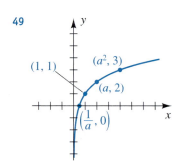

$(1, 1)$ $(a^2, 3)$ $(a, 2)$ $\left(\frac{1}{a}, 0\right)$

50

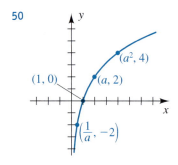

$(a^2, 4)$ $(1, 0)$ $(a, 2)$ $\left(\frac{1}{a}, -2\right)$

Exer. 51–52: Approximate x to three significant figures.

51 (a) $\log x = 3.6274$ (b) $\log x = 0.9469$

    (c) $\log x = -1.6253$ (d) $\ln x = 2.3$

    (e) $\ln x = 0.05$ (f) $\ln x = -1.6$

52 (a) $\log x = 1.8965$ (b) $\log x = 4.9680$

    (c) $\log x = -2.2118$ (d) $\ln x = 3.7$

    (e) $\ln x = 0.95$ (f) $\ln x = -5$

53 Finding a growth rate Change $f(x) = 1000(1.05)^x$ to an exponential function with base e and approximate the growth rate of f.

54 Finding a decay rate Change $f(x) = 100(\frac{1}{2})^x$ to an exponential function with base e and approximate the decay rate of f.

55 Radium decay If we start with q_0 milligrams of radium, the amount q remaining after t years is given by the formula $q = q_0(2)^{-t/1600}$. Express t in terms of q and q_0.

56 Bismuth isotope decay The radioactive bismuth isotope ^{210}Bi disintegrates according to $Q = k(2)^{-t/5}$, where k is a constant and t is the time in days. Express t in terms of Q and k.

57 Electrical circuit A schematic of a simple electrical circuit consisting of a resistor and an inductor is shown in the figure. The current I at time t is given by the formula $I = 20e^{-Rt/L}$, where R is the resistance and L is the inductance. Solve this equation for t.

Exercise 57

58 Electrical condenser An electrical condenser with initial charge Q_0 is allowed to discharge. After t seconds the charge Q is $Q = Q_0e^{kt}$, where k is a constant. Solve this equation for t.

59 Richter scale Use the Richter scale formula $R = \log (I/I_0)$ to find the magnitude of an earthquake that has an intensity

(a) 100 times that of I_0

(b) 10,000 times that of I_0

(c) 100,000 times that of I_0

60 Richter scale Refer to Exercise 59. The largest recorded magnitudes of earthquakes have been between 8 and 9 on the Richter scale. Find the corresponding intensities in terms of I_0.

61 Sound intensity The loudness of a sound, as experienced by the human ear, is based on its intensity level. A formula used for finding the intensity level α (in decibels) that corresponds to a sound intensity I is $\alpha = 10 \log (I/I_0)$, where I_0 is a special value of I agreed to be the weakest sound that can be detected by the ear under certain conditions. Find α if

(a) I is 10 times as great as I_0

(b) I is 1000 times as great as I_0

(c) I is 10,000 times as great as I_0 (This is the intensity level of the average voice.)

62 Sound intensity Refer to Exercise 61. A sound intensity level of 140 decibels produces pain in the average human ear. Approximately how many times greater than I_0 must I be in order for α to reach this level?

63 U.S. population growth The population $N(t)$ (in millions) of the United States t years after 1980 may be approximated by the formula $N(t) = 227e^{0.007t}$. When will the population be twice what it was in 1980?

64 Population growth in India The population $N(t)$ (in millions) of India t years after 1985 may be approximated by the formula $N(t) = 762e^{0.022t}$. When will the population reach 1.5 billion?

65 Children's weight The Ehrenberg relation

$$\ln W = \ln 2.4 + (1.84)h$$

is an empirically based formula relating the height h (in meters) to the average weight W (in kilograms) for children 5 through 13 years old.

(a) Express W as a function of h that does not contain ln.

(b) Estimate the average weight of an 8-year-old child who is 1.5 meters tall.

66 Continuously compounded interest If interest is compounded continuously at the rate of 10% per year, approximate the number of years it will take an initial deposit of $6000 to grow to $25,000.

67 Air pressure The air pressure $p(h)$ (in lb/in^2) at an altitude of h feet above sea level may be approximated by the formula $p(h) = 14.7e^{-0.0000385h}$. At approximately what altitude h is the air pressure

(a) 10 lb/in^2?

(b) one-half its value at sea level?

68 Vapor pressure A liquid's vapor pressure P (in lb/in^2), a measure of its volatility, is related to its temperature T (in °F) by the Antoine equation

$$\log P = a + \frac{b}{c + T},$$

where a, b, and c are constants. Vapor pressure increases rapidly with an increase in temperature. Express P as a function of T.

69 Elephant growth The weight W (in kilograms) of a female African elephant at age t (in years) may be approximated by

$$W = 2600(1 - 0.51e^{-0.075t})^3.$$

(a) Approximate the weight at birth.

(b) Estimate the age of a female African elephant weighing 1800 kilograms by using (1) the accompanying graph and (2) the formula for W.

Exercise 69

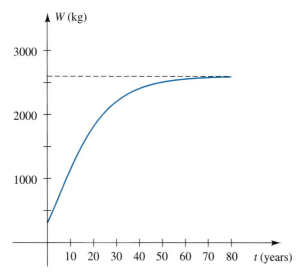

70 Coal consumption A country presently has coal reserves of 50 million tons. Last year 6.5 million tons of coal was consumed. Past years' data and population projections suggest that the rate of consumption R (in million tons/year) will increase according to the formula $R = 6.5e^{0.02t}$, and the total amount T (in million tons) of coal that will be used in t years is given by the formula $T = 325(e^{0.02t} - 1)$. If the country uses only its own resources, when will the coal reserves be depleted?

71 Urban population density An urban density model is a formula that relates the population density D (in thousands/mi^2) to the distance x (in miles) from the center of the city. The formula $D = ae^{-bx}$ for the central density a and coefficient of decay b has been found to be appropriate for many large U.S. cities. For the city of Atlanta in 1970, $a = 5.5$ and $b = 0.10$. At approximately what distance was the population density 2000 per square mile?

72 Brightness of stars Stars are classified into categories of brightness called magnitudes. The faintest stars, with light flux L_0, are assigned a magnitude of 6. Brighter stars of light flux L are assigned a magnitude m by means of the formula

$$m = 6 - 2.5 \log \frac{L}{L_0}.$$

(a) Find m if $L = 10^{0.4}L_0$.

(b) Solve the formula for L in terms of m and L_0.

73 Radioactive iodine decay Radioactive iodine ^{131}I is frequently used in tracer studies involving the thyroid gland. The substance decays according to the formula $A(t) = A_0a^{-t}$, where A_0 is the initial dose and t is the time in days. Find a, assuming the half-life of ^{131}I is 8 days.

74 Radioactive contamination Radioactive strontium ^{90}Sr has been deposited in a large field by acid rain. If sufficient amounts make their way through the food chain to humans, bone cancer can result. It has been determined that the radioactivity level in the field is 2.5 times the safe level S. ^{90}Sr decays according to the formula

$$A(t) = A_0e^{-0.0239t},$$

where A_0 is the amount currently in the field and t is the time in years. For how many years will the field be contaminated?

75 Walking speed In a survey of 15 cities ranging in population P from 300 to 3,000,000, it was found that the average walking speed S (in ft/sec) of a pedestrian could be approximated by $S = 0.05 + 0.86 \log P$.

(a) How does the population affect the average walking speed?

(b) For what population is the average walking speed 5 ft/sec?

76 Computer chips For manufacturers of computer chips, it is important to consider the fraction F of chips that will fail after t years of service. This fraction can sometimes be approximated by the formula $F = 1 - e^{-ct}$, where c is a positive constant.

(a) How does the value of c affect the reliability of a chip?

(b) If $c = 0.125$, after how many years will 35% of the chips have failed?

Exer. 77–78: Approximate the function at the value of x to four decimal places.

77 (a) $f(x) = \ln(x + 1) + e^x, \quad x = 2$

(b) $g(x) = \dfrac{(\log x)^2 - \log x}{4}, \quad x = 3.97$

78 (a) $f(x) = \log(2x^2 + 1) - 10^{-x}, \quad x = 1.95$

(b) $g(x) = \dfrac{x - 3.4}{\ln x + 4}, \quad x = 0.55$

79 Cholesterol level in women Studies relating serum cholesterol level to coronary heart disease suggest that a risk factor is the ratio x of the total amount C of cholesterol in the blood to the amount H of high-density lipoprotein cholesterol in the blood. For a female, the lifetime risk R of having a heart attack can be approximated by the formula

$$R = 2.07 \ln x - 2.04 \quad \text{provided} \quad 0 \le R \le 1.$$

For example, if $R = 0.65$, then there is a 65% chance that a woman will have a heart attack over an average lifetime. Calculate R for a female with $C = 242$ and $H = 78$.

80 Cholesterol level in men Refer to Exercise 79. For a male, the risk can be approximated by the formula $R = 1.36 \ln x - 1.19$. Calculate R for a male with $C = 287$ and $H = 65$.

5.5

Properties of Logarithms

In the preceding section we observed that $\log_a x$ can be interpreted as an exponent. Thus, it seems reasonable to expect that the laws of exponents can be used to obtain corresponding laws of logarithms. This is demonstrated in the proofs of the following laws, which are fundamental for all work with logarithms.

Laws of Logarithms	If u and w denote positive real numbers, then
	(1) $\log_a(uw) = \log_a u + \log_a w$
	(2) $\log_a\left(\dfrac{u}{w}\right) = \log_a u - \log_a w$
	(3) $\log_a(u^c) = c \log_a u \quad$ for every real number c

PROOFS For all three proofs, let

$$r = \log_a u \qquad \text{and} \qquad s = \log_a w.$$

The equivalent exponential forms are

$$u = a^r \qquad \text{and} \qquad w = a^s.$$

We now proceed as follows:

(1)
$$uw = a^r a^s \qquad\qquad \text{definition of } u \text{ and } w$$
$$uw = a^{r+s} \qquad\qquad \text{law 1 of exponents}$$
$$\log_a (uw) = r + s \qquad\qquad \text{change to logarithmic form}$$
$$\log_a (uw) = \log_a u + \log_a w \qquad\qquad \text{definition of } r \text{ and } s$$

(2)
$$\frac{u}{w} = \frac{a^r}{a^s} \qquad\qquad \text{definition of } u \text{ and } w$$
$$\frac{u}{w} = a^{r-s} \qquad\qquad \text{law 5(a) of exponents}$$
$$\log_a \left(\frac{u}{w}\right) = r - s \qquad\qquad \text{change to logarithmic form}$$
$$\log_a \left(\frac{u}{w}\right) = \log_a u - \log_a w \qquad\qquad \text{definition of } r \text{ and } s$$

(3)
$$u^c = (a^r)^c \qquad\qquad \text{definition of } u$$
$$u^c = a^{cr} \qquad\qquad \text{law 2 of exponents}$$
$$\log_a (u^c) = cr \qquad\qquad \text{change to logarithmic form}$$
$$\log_a (u^c) = c \log_a u \qquad\qquad \text{definition of } r$$

The laws of logarithms for the special cases $a = 10$ (common logs) and $a = e$ (natural logs) are written as shown in the following chart.

Common logarithms	**Natural logarithms**
(1) $\log (uw) = \log u + \log w$	**(1)** $\ln (uw) = \ln u + \ln w$
(2) $\log \left(\dfrac{u}{w}\right) = \log u - \log w$	**(2)** $\ln \left(\dfrac{u}{w}\right) = \ln u - \ln w$
(3) $\log (u^c) = c \log u$	**(3)** $\ln (u^c) = c \ln u$

As indicated by the following warning, there are no laws for expressing $\log_a (u + w)$ or $\log_a (u - w)$ in terms of simpler logarithms.

◤ *Warning!* ◤

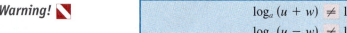

$$\log_a (u + w) \neq \log_a u + \log_a w$$
$$\log_a (u - w) \neq \log_a u - \log_a w$$

The following examples illustrate uses of the laws of logarithms.

EXAMPLE 1 **Using laws of logarithms**

Express $\log_a \dfrac{x^3 \sqrt{y}}{z^2}$ in terms of logarithms of x, y, and z.

SOLUTION We write $\sqrt{y}$ as $y^{1/2}$ and use laws of logarithms:

$$
\begin{aligned}
\log_a \dfrac{x^3 \sqrt{y}}{z^2} &= \log_a (x^3 y^{1/2}) - \log_a z^2 && \text{law 2} \\
&= \log_a x^3 + \log_a y^{1/2} - \log_a z^2 && \text{law 1} \\
&= 3 \log_a x + \tfrac{1}{2} \log_a y - 2 \log_a z && \text{law 3}
\end{aligned}
$$

Note that if a term with a positive exponent (such as x^3) is in the numerator of the original expression, it will have a positive coefficient in the expanded form, and if it is in the denominator (such as z^2), it will have a negative coefficient in the expanded form. ◢

EXAMPLE 2 **Using laws of logarithms**

Express as one logarithm:

$$
\tfrac{1}{3} \log_a (x^2 - 1) - \log_a y - 4 \log_a z
$$

SOLUTION We apply the laws of logarithms as follows:

$$
\begin{aligned}
\tfrac{1}{3} \log_a (x^2 - 1) &- \log_a y - 4 \log_a z \\
&= \log_a (x^2 - 1)^{1/3} - \log_a y - \log_a z^4 && \text{law 3} \\
&= \log_a \sqrt[3]{x^2 - 1} - (\log_a y + \log_a z^4) && \text{algebra} \\
&= \log_a \sqrt[3]{x^2 - 1} - \log_a (yz^4) && \text{law 1} \\
&= \log_a \dfrac{\sqrt[3]{x^2 - 1}}{yz^4} && \text{law 2}
\end{aligned}
$$

◢

EXAMPLE 3 **Solving a logarithmic equation**

Solve the equation $\log_5 (2x + 3) = \log_5 11 + \log_5 3$.

SOLUTION

$$
\begin{aligned}
\log_5 (2x + 3) &= \log_5 11 + \log_5 3 && \text{given} \\
\log_5 (2x + 3) &= \log_5 (11 \cdot 3) && \text{law 1 of logarithms} \\
2x + 3 &= 33 && \text{logarithmic functions are one-to-one} \\
x &= 15 && \text{solve for } x
\end{aligned}
$$

✔ **Check** $x = 15$ LS: $\log_5 (2 \cdot 15 + 3) = \log_5 33$
$\phantom{\text{Check } x = 15 \quad}$ RS: $\log_5 11 + \log_5 3 = \log_5 (11 \cdot 3) = \log_5 33$

Since $\log_5 33 = \log_5 33$ is a true statement, $x = 15$ is a solution. ◢

The laws of logarithms were proved for logarithms of *positive* real numbers u and w. If we apply these laws to equations in which u and w are expressions involving a variable, then extraneous solutions may occur. Answers should therefore be substituted for the variable in u and w to determine whether these expressions are defined.

EXAMPLE 4 Solving a logarithmic equation

Solve the equation $\log_2 x + \log_2 (x + 2) = 3$.

SOLUTION

$$\log_2 x + \log_2 (x + 2) = 3 \qquad \text{given}$$
$$\log_2 [x(x + 2)] = 3 \qquad \text{law 1 of logarithms}$$
$$x(x + 2) = 2^3 \qquad \text{change to exponential form}$$
$$x^2 + 2x - 8 = 0 \qquad \text{multiply and set equal to 0}$$
$$(x - 2)(x + 4) = 0 \qquad \text{factor}$$
$$x - 2 = 0, \quad x + 4 = 0 \qquad \text{zero factor theorem}$$
$$x = 2, \qquad x = -4 \qquad \text{solve for } x$$

✔ **Check** $x = 2$ LS: $\log_2 2 + \log_2 (2 + 2) = 1 + \log_2 4$
$$= 1 + \log_2 2^2 = 1 + 2 = 3$$

RS: 3

Since $3 = 3$ is a true statement, $x = 2$ is a solution.

✔ **Check** $x = -4$ LS: $\log_2 (-4) + \log_2 (-4 + 2)$

Since logarithms of negative numbers are undefined, $x = -4$ is not a solution.

◢

EXAMPLE 5 Solving a logarithmic equation

Solve the equation $\ln (x + 6) - \ln 10 = \ln (x - 1) - \ln 2$.

SOLUTION

$$\ln (x + 6) - \ln (x - 1) = \ln 10 - \ln 2 \qquad \text{rearrange terms}$$
$$\ln \left(\frac{x + 6}{x - 1} \right) = \ln \frac{10}{2} \qquad \text{law 2 of logarithms}$$
$$\frac{x + 6}{x - 1} = 5 \qquad \text{ln is one-to-one}$$
$$x + 6 = 5x - 5 \qquad \text{multiply by } x - 1$$
$$x = \tfrac{11}{4} \qquad \text{solve for } x$$

✔ **Check**   Since both $\ln (x + 6)$ and $\ln (x - 1)$ are defined at $x = \frac{11}{4}$ (they are logarithms of positive real numbers) and since our algebraic steps are correct, it follows that $\frac{11}{4}$ is a solution of the given equation. ◢

EXAMPLE 6 Shifting the graph of a logarithmic equation

Sketch the graph of $y = \log_3 (81x)$.

SOLUTION We may rewrite the equation as follows:

$$
\begin{aligned}
y &= \log_3 (81x) && \text{given} \\
&= \log_3 81 + \log_3 x && \text{law 1 of logarithms} \\
&= \log_3 3^4 + \log_3 x && 81 = 3^4 \\
&= 4 + \log_3 x && \log_a a^x = x
\end{aligned}
$$

Thus, we can obtain the graph of $y = \log_3 (81x)$ by vertically shifting the graph of $y = \log_3 x$ in Figure 2 in Section 5.4 upward four units. This gives us the sketch in Figure 1.

Figure 1

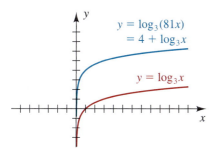

$y = \log_3(81x)$
$= 4 + \log_3 x$

$y = \log_3 x$

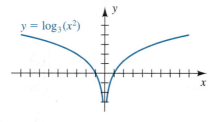

EXAMPLE 7 Sketching graphs of logarithmic equations

Sketch the graph of the equation:

(a) $y = \log_3 (x^2)$ **(b)** $y = 2 \log_3 x$

SOLUTION

(a) Since $x^2 = |x|^2$, we may rewrite the given equation as

$$y = \log_3 |x|^2.$$

Using law 3 of logarithms, we have

$$y = 2 \log_3 |x|.$$

We can obtain the graph of $y = 2 \log_3 |x|$ by multiplying the y-coordinates of points on the graph of $y = \log_3 |x|$ in Figure 4 of Section 5.4 by 2. This gives us the graph in Figure 2(a).

Figure 2

(a)

$y = \log_3 (x^2)$

(b)

$y = 2 \log_3 x$

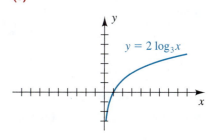

(b) If $y = 2 \log_3 x$, then x must be positive. Hence, the graph is identical to that part of the graph of $y = 2 \log_3 |x|$ in Figure 2(a) that lies to the right of the y-axis. This gives us Figure 2(b).

EXAMPLE 8 A relationship between selling price and demand

In the study of economics, the demand D for a product is often related to its selling price p by an equation of the form

$$\log_a D = \log_a c - k \log_a p,$$

where a, c, and k are positive constants.

(a) Solve the equation for D.

(b) How does increasing or decreasing the selling price affect the demand?

SOLUTION

(a) $\log_a D = \log_a c - k \log_a p$ given

$\log_a D = \log_a c - \log_a p^k$ law 3 of logarithms

$\log_a D = \log_a \dfrac{c}{p^k}$ law 2 of logarithms

$D = \dfrac{c}{p^k}$ $\log_a$ is one-to-one

(b) If the price p is increased, the denominator p^k in $D = c/p^k$ will also increase and hence the demand D for the product will decrease. If the price is decreased, then p^k will decrease and the demand D will increase. ◢

5.5 Exercises

Exer. 1–8: Express in terms of logarithms of x, y, z, or w.

1 (a) $\log_4 (xz)$ (b) $\log_4 (y/x)$ (c) $\log_4 \sqrt[3]{z}$

2 (a) $\log_3 (xyz)$ (b) $\log_3 (xz/y)$ (c) $\log_3 \sqrt[5]{y}$

3 $\log_a \dfrac{x^3 w}{y^2 z^4}$

4 $\log_a \dfrac{y^5 w^2}{x^4 z^3}$

5 $\log \dfrac{\sqrt[3]{z}}{x\sqrt{y}}$

6 $\log \dfrac{\sqrt{y}}{x^4 \sqrt[3]{z}}$

7 $\ln \sqrt[4]{\dfrac{x^7}{y^5 z}}$

8 $\ln x \sqrt[3]{\dfrac{y^4}{z^5}}$

Exer. 9–16: Write the expression as one logarithm.

9 (a) $\log_3 x + \log_3 (5y)$ (b) $\log_3 (2z) - \log_3 x$

   (c) $5 \log_3 y$

10 (a) $\log_4 (3z) + \log_4 x$ (b) $\log_4 x - \log_4 (7y)$

   (c) $\frac{1}{3} \log_4 w$

11 $2 \log_a x + \frac{1}{3} \log_a (x - 2) - 5 \log_a (2x + 3)$

12 $5 \log_a x - \frac{1}{2} \log_a (3x - 4) - 3 \log_a (5x + 1)$

13 $\log (x^3 y^2) - 2 \log x \sqrt[3]{y} - 3 \log \left(\dfrac{x}{y}\right)$

14 $2 \log \dfrac{y^3}{x} - 3 \log y + \dfrac{1}{2} \log x^4 y^2$

15 $\ln y^3 + \frac{1}{3} \ln (x^3 y^6) - 5 \ln y$

16 $2 \ln x - 4 \ln (1/y) - 3 \ln (xy)$

Exer. 17–34: Solve the equation.

17 $\log_6 (2x - 3) = \log_6 12 - \log_6 3$

18 $\log_4 (3x + 2) = \log_4 5 + \log_4 3$

19 $2 \log_3 x = 3 \log_3 5$

20 $3 \log_2 x = 2 \log_2 3$

21 $\log x - \log (x + 1) = 3 \log 4$

22 $\log (x + 2) - \log x = 2 \log 4$

23 $\ln (-4 - x) + \ln 3 = \ln (2 - x)$

24 $\ln x + \ln (x + 6) = \frac{1}{2} \ln 9$

25 $\log_2 (x + 7) + \log_2 x = 3$

26 $\log_6 (x + 5) + \log_6 x = 2$

27 $\log_3 (x + 3) + \log_3 (x + 5) = 1$

28 $\log_3 (x - 2) + \log_3 (x - 4) = 2$

29 $\log (x + 3) = 1 - \log (x - 2)$

30 $\log (57x) = 2 + \log (x - 2)$

31 $\ln x = 1 - \ln (x + 2)$

32 $\ln x = 1 + \ln (x + 1)$

33 $\log_3 (x - 2) = \log_3 27 - \log_3 (x - 4) - 5^{\log_5 1}$

34 $\log_2 (x + 3) = \log_2 (x - 3) + \log_3 9 + 4^{\log_4 3}$

Exer. 35–46: Sketch the graph of f.

35 $f(x) = \log_3 (3x)$

36 $f(x) = \log_4 (16x)$

37 $f(x) = 3 \log_3 x$

38 $f(x) = \frac{1}{3} \log_3 x$

39 $f(x) = \log_3 (x^2)$

40 $f(x) = \log_2 (x^2)$

41 $f(x) = \log_2 (x^3)$

42 $f(x) = \log_3 (x^3)$

43 $f(x) = \log_2 \sqrt{x}$

44 $f(x) = \log_2 \sqrt[3]{x}$

45 $f(x) = \log_3 \left(\frac{1}{x} \right)$

46 $f(x) = \log_2 \left(\frac{1}{x} \right)$

Exer. 47–50: Shown in the figure is the graph of a function f. Express f(x) as one logarithm with base 2.

47

48

49

50

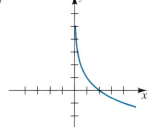

51 **Volume and decibels** When the volume control on a stereo system is increased, the voltage across a loudspeaker changes from V_1 to V_2, and the decibel increase in gain is given by

$$\text{db} = 20 \log \frac{V_2}{V_1}.$$

Find the decibel increase if the voltage changes from 2 volts to 4.5 volts.

52 **Volume and decibels** Refer to Exercise 51. What voltage ratio k is needed for a +20 decibel gain? for a +40 decibel gain?

53 Pareto's law Pareto's law for capitalist countries states that the relationship between annual income x and the number y of individuals whose income exceeds x is

$$\log y = \log b - k \log x,$$

where b and k are positive constants. Solve this equation for y.

54 Price and demand If p denotes the selling price (in dollars) of a commodity and x is the corresponding demand (in number sold per day), then the relationship between p and x is sometimes given by $p = p_0 e^{-ax}$, where p_0 and a are positive constants. Express x as a function of p.

55 Wind velocity If v denotes the wind velocity (in m/sec) at a height of z meters above the ground, then under certain conditions $v = c \ln (z/z_0)$, where c is a positive constant and z_0 is the height at which the velocity is zero. Sketch the graph of this equation on a zv-plane for $c = 0.5$ and $z_0 = 0.1$ m.

56 Eliminating pollution If the pollution of Lake Erie were stopped suddenly, it has been estimated that the level y of pollutants would decrease according to the formula $y = y_0 e^{-0.3821t}$, where t is the time in years and y_0 is the pollutant level at which further pollution ceased. How many years would it take to clear 50% of the pollutants?

57 Reaction to a stimulus Let R denote the reaction of a subject to a stimulus of strength x. There are many possibilities for R and x. If the stimulus x is saltiness (in grams of salt per liter), R may be the subject's estimate of how salty the solution tasted, based on a scale from 0 to 10. One relationship between R and x is given by the Weber-Fechner formula, $R(x) = a \log (x/x_0)$, where a is a positive constant and x_0 is called the threshold stimulus.

(a) Find $R(x_0)$.

(b) Find a relationship between $R(x)$ and $R(2x)$.

58 Electron energy The energy $E(x)$ of an electron after passing through material of thickness x is given by the equation $E(x) = E_0 e^{-x/x_0}$, where E_0 is the initial energy and x_0 is the radiation length.

(a) Express, in terms of E_0, the energy of an electron after it passes through material of thickness x_0.

(b) Express, in terms of x_0, the thickness at which the electron loses 99% of its initial energy.

59 Ozone layer One method of estimating the thickness of the ozone layer is to use the formula

$$\ln I_0 - \ln I = kx,$$

where I_0 is the intensity of a particular wavelength of light from the sun before it reaches the atmosphere, I is the intensity of the same wavelength after passing through a layer of ozone x centimeters thick, and k is the absorption constant of ozone for that wavelength. Suppose for a wavelength of 3176×10^{-8} cm with $k \approx 0.39$, I_0/I is measured as 1.12. Approximate the thickness of the ozone layer to the nearest 0.01 centimeter.

60 Ozone layer Refer to Exercise 59. Approximate the percentage decrease in the intensity of light with a wavelength of 3176×10^{-8} centimeter if the ozone layer is 0.24 centimeter thick.

5.6

Exponential and Logarithmic Equations

In this section we shall consider various types of exponential and logarithmic equations and their applications. When solving an equation involving exponential expressions with constant bases and variables appearing in the exponent(s), we often *equate the logarithms of both sides* of the equation. When we do so, the variables in the exponent become multipliers, and the resulting equation is usually easier to solve. We will refer to this step as simply "take log of both sides."

EXAMPLE 1   Solving an exponential equation

Solve the equation $3^x = 21$.

SOLUTION

$$3^x = 21 \qquad \text{given}$$

$$\log(3^x) = \log 21 \qquad \text{take log of both sides}$$

$$x \log 3 = \log 21 \qquad \text{law 3 of logarithms}$$

$$x = \frac{\log 21}{\log 3} \qquad \text{divide by log 3}$$

We could also have used natural logarithms to obtain

$$x = \frac{\ln 21}{\ln 3}.$$

Using a calculator gives us the approximate solution $x \approx 2.77$. A partial check is to note that since $3^2 = 9$ and $3^3 = 27$, the number x such that $3^x = 21$ must be between 2 and 3, somewhat closer to 3 than to 2.

We could also have solved the equation in Example 1 by changing the exponential form $3^x = 21$ to logarithmic form, as we did in Section 5.4, obtaining

$$x = \log_3 21.$$

This is, in fact, the solution of the equation; however, since calculators typically have keys only for log and ln, we cannot approximate $\log_3 21$ directly. The next theorem gives us a simple *change of base formula* for finding $\log_b u$ if $u > 0$ and b is *any* logarithmic base.

Theorem: Change of Base Formula	If $u > 0$ and if a and b are positive real numbers different from 1, then $$\log_b u = \frac{\log_a u}{\log_a b}.$$

PROOF We begin with the equivalent equations

$$w = \log_b u \qquad \text{and} \qquad b^w = u$$

and proceed as follows:

$$b^w = u \qquad \text{given}$$

$$\log_a b^w = \log_a u \qquad \text{take } \log_a \text{ of both sides}$$

$$w \log_a b = \log_a u \qquad \text{law 3 of logarithms}$$

$$w = \frac{\log_a u}{\log_a b} \qquad \text{divide by } \log_a b$$

Since $w = \log_b u$, we obtain the formula.

The following special case of the change of base formula is obtained by letting $u = a$ and using the fact that $\log_a a = 1$:

$$\log_b a = \frac{1}{\log_a b}$$

The change of base formula is sometimes confused with law 2 of logarithms. The first of the following warnings could be remembered with the phrase "a quotient of logs is *not* the log of the quotient."

 Warning!

$$\frac{\log_a u}{\log_a b} \neq \log_a \frac{u}{b}; \qquad \frac{\log_a u}{\log_a b} \neq \log_a (u - b)$$

The most frequently used special cases of the change of base formula are those for $a = 10$ (common logarithms) and $a = e$ (natural logarithms), as stated in the next box.

Special Change of Base Formulas	(1) $\log_b u = \dfrac{\log_{10} u}{\log_{10} b} = \dfrac{\log u}{\log b}$ (2) $\log_b u = \dfrac{\log_e u}{\log_e b} = \dfrac{\ln u}{\ln b}$

Next, we will rework Example 1 using a change of base formula.

EXAMPLE 2 **Using a change of base formula**

Solve the equation $3^x = 21$.

SOLUTION We proceed as follows:

$$3^x = 21 \qquad \text{given}$$
$$x = \log_3 21 \qquad \text{change to logarithmic form}$$
$$= \frac{\log 21}{\log 3} \qquad \text{special change of base formula 1}$$

Another method is to use special change of base formula 2, obtaining

$$x = \frac{\ln 21}{\ln 3}.$$

Logarithms with base 2 are used in computer science. The next example indicates how to approximate logarithms with base 2 using change of base formulas.

EXAMPLE 3 **Approximating a logarithm with base 2**

Approximate $\log_2 5$ using

(a) common logarithms   (b) natural logarithms

SOLUTION Using special change of base formulas 1 and 2, we obtain the following:

(a) $\log_2 5 = \dfrac{\log 5}{\log 2} \approx 2.322$

(b) $\log_2 5 = \dfrac{\ln 5}{\ln 2} \approx 2.322$

EXAMPLE 4 **Solving an exponential equation**

Solve the equation $5^{2x+1} = 6^{x-2}$.

SOLUTION We can use either common or natural logarithms. Using common logarithms gives us the following:

$5^{2x+1} = 6^{x-2}$	given
$\log\,(5^{2x+1}) = \log\,(6^{x-2})$	take log of both sides
$(2x + 1)\log 5 = (x - 2)\log 6$	law 3 of logarithms
$2x \log 5 + \log 5 = x \log 6 - 2 \log 6$	multiply
$2x \log 5 - x \log 6 = -\log 5 - 2 \log 6$	get all terms with x on one side
$x(\log 5^2 - \log 6) = -(\log 5 + \log 6^2)$	factor, and use law 3 of logarithms
$x = -\dfrac{\log\,(5 \cdot 36)}{\log \frac{25}{6}}$	solve for x, and use laws of logarithms

Substituting $-\log 180/\log \frac{25}{6} \approx -3.64$ for x in both 5^{2x+1} and 6^{x-2} gives us the approximate value 0.00004. We deduce from this that the graphs of $y = 5^{2x+1}$ and $y = 6^{x-2}$ intersect at approximately $(-3.64, 0.00004)$.

EXAMPLE 5 **Solving an exponential equation**

Solve the equation $\dfrac{5^x - 5^{-x}}{2} = 3$.

SOLUTION	$\dfrac{5^x - 5^{-x}}{2} = 3$	given
	$5^x - 5^{-x} = 6$	multiply by 2
	$5^x - \dfrac{1}{5^x} = 6$	definition of negative exponent
	$5^x(5^x) - \dfrac{1}{5^x}(5^x) = 6(5^x)$	multiply by the lcd, 5^x
	$(5^x)^2 - 6(5^x) - 1 = 0$	simplify and subtract $6(5^x)$

(continued)

Note that $(5^x)^2$ can be written as 5^{2x}.

We recognize this form of the equation as a quadratic in 5^x and proceed as follows:

$$(5^x)^2 - 6(5^x) - 1 = 0 \qquad \text{law of exponents}$$

$$5^x = \frac{6 \pm \sqrt{36 + 4}}{2} \qquad \text{quadratic formula}$$

$$5^x = 3 \pm \sqrt{10} \qquad \text{simplify}$$

$$5^x = 3 + \sqrt{10} \qquad 5^x > 0, \text{ but } 3 - \sqrt{10} < 0$$

$$\log 5^x = \log\left(3 + \sqrt{10}\right) \qquad \text{take log of both sides}$$

$$x \log 5 = \log\left(3 + \sqrt{10}\right) \qquad \text{law 3 of logarithms}$$

$$x = \frac{\log\left(3 + \sqrt{10}\right)}{\log 5} \qquad \text{divide by log 5}$$

An approximation is $x \approx 1.13$.

EXAMPLE 6 Solving an equation involving logarithms

Solve the equation $\log \sqrt[3]{x} = \sqrt{\log x}$ for x.

SOLUTION

$$\log x^{1/3} = \sqrt{\log x} \qquad \sqrt[n]{x} = x^{1/n}$$

$$\tfrac{1}{3}\log x = \sqrt{\log x} \qquad \log x^r = r \log x$$

$$\tfrac{1}{9}(\log x)^2 = \log x \qquad \text{square both sides}$$

$$(\log x)^2 = 9 \log x \qquad \text{multiply by 9}$$

$$(\log x)^2 - 9 \log x = 0 \qquad \text{make one side 0}$$

$$(\log x)(\log x - 9) = 0 \qquad \text{factor out log } x$$

$$\log x = 0, \quad \log x - 9 = 0 \qquad \text{set each factor equal to 0}$$

$$\log x = 9 \qquad \text{add 9}$$

$$x = 10^0 = 1 \quad \text{or} \quad x = 10^9 \qquad \log_{10} x = a \Longleftrightarrow x = 10^a$$

✔ **Check** $x = 1$ LS: $\log \sqrt[3]{1} = \log 1 = 0$
  RS: $\sqrt{\log 1} = \sqrt{0} = 0$

✔ **Check** $x = 10^9$ LS: $\log \sqrt[3]{10^9} = \log 10^3 = 3$
  RS: $\sqrt{\log 10^9} = \sqrt{9} = 3$

The equation has two solutions, 1 and 1 billion.

The function $y = 2/(e^x + e^{-x})$ is called the **hyperbolic secant function.** In the next example we solve this equation for x in terms of y. Under suitable restrictions, this gives us the inverse function.

EXAMPLE 7 **Finding an inverse hyperbolic function**

Solve $y = 2/(e^x + e^{-x})$ for x in terms of y.

SOLUTION

$$y = \frac{2}{e^x + e^{-x}} \qquad \text{given}$$

$$ye^x + ye^{-x} = 2 \qquad \text{multiply by } e^x + e^{-x}$$

$$ye^x + \frac{y}{e^x} = 2 \qquad \text{definition of negative exponent}$$

$$ye^x(e^x) + \frac{y}{e^x}(e^x) = 2(e^x) \qquad \text{multiply by the lcd, } e^x$$

$$y(e^x)^2 - 2e^x + y = 0 \qquad \text{simplify and subtract } 2e^x$$

We recognize this form of the equation as a quadratic in e^x with coefficients $a = y$, $b = -2$, and $c = y$. Note that we are solving for e^x, not x.

$$e^x = \frac{-(-2) \pm \sqrt{(-2)^2 - 4(y)(y)}}{2(y)} \qquad \text{quadratic formula}$$

$$= \frac{2 \pm \sqrt{4 - 4y^2}}{2y} \qquad \text{simplify}$$

$$= \frac{2 \pm \sqrt{4}\sqrt{1 - y^2}}{2y} \qquad \text{factor out } \sqrt{4}$$

$$e^x = \frac{1 \pm \sqrt{1 - y^2}}{y} \qquad \text{cancel a factor of 2}$$

$$x = \ln \frac{1 \pm \sqrt{1 - y^2}}{y} \qquad \text{take ln of both sides}$$

For the blue curve $y = f(x)$ in Figure 1, the inverse function is

$$y = f^{-1}(x) = \ln \frac{1 + \sqrt{1 - x^2}}{x},$$

shown in blue in Figure 2. Notice the domain and range relationships. For the red curve $y = g(x)$ in Figure 1, the inverse function is

$$y = g^{-1}(x) = \ln \frac{1 - \sqrt{1 - x^2}}{x},$$

shown in red in Figure 2. Since the hyperbolic secant is not one-to-one, it cannot have one simple equation for its inverse.

Figure 1

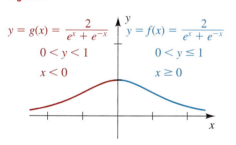

$y = g(x) = \dfrac{2}{e^x + e^{-x}}$
$0 < y < 1$
$x < 0$

$y = f(x) = \dfrac{2}{e^x + e^{-x}}$
$0 < y \le 1$
$x \ge 0$

Figure 2

$y = f^{-1}(x) = \ln \dfrac{1 + \sqrt{1 - x^2}}{x}$
$0 < x \le 1$
$y \ge 0$

$y = g^{-1}(x) = \ln \dfrac{1 - \sqrt{1 - x^2}}{x}$
$0 < x < 1$
$y < 0$

The inverse hyperbolic secant is part of the equation of the curve called a **tractrix.** The curve is associated with Gottfried Wilhelm von Leibniz's (1646–1716) solution to the question "What is the path of an object dragged along a horizontal plane by a string of constant length when the end of the string not joined to the object moves along a straight line in the plane?"

EXAMPLE 8 Approximating light penetration in an ocean

The Beer-Lambert law states that the amount of light I that penetrates to a depth of x meters in an ocean is given by $I = I_0 c^x$, where $0 < c < 1$ and I_0 is the amount of light at the surface.

(a) Solve for x in terms of common logarithms.

(b) If $c = \frac{1}{4}$, approximate the depth at which $I = 0.01 I_0$. (This determines the photic zone where photosynthesis can take place.)

SOLUTION

(a) $I = I_0 c^x$ given

$\dfrac{I}{I_0} = c^x$ isolate the exponential expression

$x = \log_c \dfrac{I}{I_0}$ change to logarithmic form

$= \dfrac{\log (I/I_0)}{\log c}$ special change of base formula 1

(b) Letting $I = 0.01 I_0$ and $c = \frac{1}{4}$ in the formula for x obtained in part (a), we have

$$x = \frac{\log (0.01 I_0/I_0)}{\log \frac{1}{4}}$$ substitute for I and c

$$= \frac{\log (0.01)}{\log 1 - \log 4}$$ cancel I_0; law 2 of logarithms

$$= \frac{\log 10^{-2}}{0 - \log 4}$$ property of logarithms

$$= \frac{-2}{-\log 4}$$ $\log 10^x = x$

$$= \frac{2}{\log 4}.$$ simplify

An approximation is $x \approx 3.32$ m.

EXAMPLE 9 **Comparing light intensities**

If a beam of light that has intensity I_0 is projected vertically downward into water, then its intensity $I(x)$ at a depth of x meters is $I(x) = I_0 e^{-1.4x}$ (see Figure 3). At what depth is the intensity one-half its value at the surface?

SOLUTION   At the surface, $x = 0$, and the intensity is

$$I(0) = I_0 e^0$$

$$= I_0.$$

Figure 3

We wish to find the value of x such that $I(x) = \frac{1}{2}I_0$. This leads to the following:

$$
\begin{aligned}
I(x) &= \tfrac{1}{2}I_0 && \text{desired intensity} \\
I_0 e^{-1.4x} &= \tfrac{1}{2}I_0 && \text{formula for } I(x) \\
e^{-1.4x} &= \tfrac{1}{2} && \text{divide by } I_0 \, (I_0 \neq 0) \\
-1.4x &= \ln \tfrac{1}{2} && \text{change to logarithmic form} \\
x &= \frac{\ln \tfrac{1}{2}}{-1.4} && \text{divide by } -1.4
\end{aligned}
$$

An approximation is $x \approx 0.495$ m.

EXAMPLE 10 A logistic curve

A **logistic curve** is the graph of an equation of the form

$$y = \frac{k}{1 + be^{-cx}},$$

where k, b, and c are positive constants. Such curves are useful for describing a population y that grows rapidly initially, but whose growth rate decreases after x reaches a certain value. In a famous study of the growth of protozoa by Gause, a population of *Paramecium caudata* was found to be described by a logistic equation with $c = 1.1244$, $k = 105$, and x the time in days.

(a) Find b if the initial population was 3 protozoa.

(b) In the study, the maximum growth rate took place at $y = 52$. At what time x did this occur?

(c) Show that after a long period of time, the population described by any logistic curve approaches the constant k.

SOLUTION

(a) Letting $c = 1.1244$ and $k = 105$ in the logistic equation, we obtain

$$y = \frac{105}{1 + be^{-1.1244x}}.$$

We now proceed as follows:

$$3 = \frac{105}{1 + be^{0}} = \frac{105}{1 + b} \qquad y = 3 \text{ when } x = 0$$

$$1 + b = 35 \qquad \text{multiply by } \frac{1 + b}{3}$$

$$b = 34 \qquad \text{solve for } b$$

(b) Using the fact that $b = 34$ leads to the following:

$$52 = \frac{105}{1 + 34e^{-1.1244x}} \qquad \text{let } y = 52 \text{ in part (a)}$$

$$1 + 34e^{-1.1244x} = \frac{105}{52} \qquad \text{multiply by } \frac{1 + 34e^{-1.1244x}}{52}$$

$$e^{-1.1244x} = \left(\frac{105}{52} - 1\right) \cdot \frac{1}{34} = \frac{53}{1768} \qquad \text{isolate } e^{-1.1244x}$$

$$-1.1244x = \ln \frac{53}{1768} \qquad \text{change to logarithmic form}$$

$$x = \frac{\ln \frac{53}{1768}}{-1.1244} \approx 3.12 \text{ days} \qquad \text{divide by } -1.1244$$

(c) As $x \to \infty$, $e^{-cx} \to 0$. Hence,

$$y = \frac{k}{1 + be^{-cx}} \longrightarrow \frac{k}{1 + b \cdot 0} = k.$$

A sketch of the logistic curve that has equation $y = 105/(1 + 34e^{-1.1244x})$ is shown in Figure 4.

Figure 4

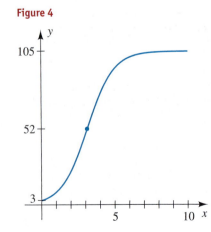

5.6   *Exercises*

Exer. 1–4: Find the exact solution and a two-decimal-place approximation for it by using (a) the method of Example 1 and (b) the method of Example 2.

1 $5^x = 8$

2 $4^x = 3$

3 $3^{4-x} = 5$

4 $\left(\frac{1}{3}\right)^x = 100$

Exer. 5–8: Estimate using the change of base formula.

5 $\log_5 6$

6 $\log_2 20$

7 $\log_9 0.2$

8 $\log_6 \frac{1}{2}$

Exer. 9–10: Evaluate using the change of base formula (without a calculator).

9 $\dfrac{\log_5 16}{\log_5 4}$

10 $\dfrac{\log_7 243}{\log_7 3}$

Exer. 11–24: Find the exact solution, using common logarithms, and a two-decimal-place approximation of each solution, when appropriate.

11 $3^{x+4} = 2^{1-3x}$

12 $4^{2x+3} = 5^{x-2}$

13 $2^{2x-3} = 5^{x-2}$

14 $3^{2-3x} = 4^{2x+1}$

15 $2^{-x} = 8$

16 $2^{-x^2} = 5$

17 $\log x = 1 - \log(x - 3)$

18 $\log(5x + 1) = 2 + \log(2x - 3)$

19 $\log(x^2 + 4) - \log(x + 2) = 2 + \log(x - 2)$

20 $\log(x - 4) - \log(3x - 10) = \log(1/x)$

21 $5^x + 125(5^{-x}) = 30$

22 $3(3^x) + 9(3^{-x}) = 28$

23 $4^x - 3(4^{-x}) = 8$

24 $2^x - 6(2^{-x}) = 6$

Exer. 25–32: Solve the equation without using a calculator.

25 $\log(x^2) = (\log x)^2$

26 $\log \sqrt{x} = \sqrt{\log x}$

27 $\log(\log x) = 2$

28 $\log \sqrt{x^3 - 9} = 2$

29 $x^{\sqrt{\log x}} = 10^8$

30 $\log(x^3) = (\log x)^3$

31 $e^{2x} + 2e^x - 15 = 0$

32 $e^x + 4e^{-x} = 5$

Exer. 33–34: Solve the equation.

33 $\log_3 x - \log_9(x + 42) = 0$

34 $\log_4 x + \log_8 x = 1$

Exer. 35–38: Use common logarithms to solve for x in terms of y.

35 $y = \dfrac{10^x + 10^{-x}}{2}$

36 $y = \dfrac{10^x - 10^{-x}}{2}$

37 $y = \dfrac{10^x - 10^{-x}}{10^x + 10^{-x}}$

38 $y = \dfrac{10^x + 10^{-x}}{10^x - 10^{-x}}$

Exer. 39–42: Use natural logarithms to solve for x in terms of y.

39 $y = \dfrac{e^x - e^{-x}}{2}$

40 $y = \dfrac{e^x + e^{-x}}{2}$

41 $y = \dfrac{e^x + e^{-x}}{e^x - e^{-x}}$

42 $y = \dfrac{e^x - e^{-x}}{e^x + e^{-x}}$

Exer. 43–44: Sketch the graph of f, and use the change of base formula to approximate the y-intercept.

43 $f(x) = \log_2(x + 3)$

44 $f(x) = \log_3(x + 5)$

Exer. 45–46: Sketch the graph of f, and use the change of base formula to approximate the x-intercept.

45 $f(x) = 4^x - 3$

46 $f(x) = 3^x - 6$

Exer. 47–50: Chemists use a number denoted by pH to describe quantitatively the acidity or basicity of solutions. By definition, pH $= -\log[\text{H}^+]$, where $[\text{H}^+]$ is the hydrogen ion concentration in moles per liter.

47 Approximate the pH of each substance.

(a) vinegar: $[\text{H}^+] \approx 6.3 \times 10^{-3}$

(b) carrots: $[\text{H}^+] \approx 1.0 \times 10^{-5}$

(c) sea water: $[\text{H}^+] \approx 5.0 \times 10^{-9}$

48 Approximate the hydrogen ion concentration $[\text{H}^+]$ of each substance.

(a) apples: pH ≈ 3.0

(b) beer: pH ≈ 4.2

(c) milk: pH ≈ 6.6

49 A solution is considered basic if $[\text{H}^+] < 10^{-7}$ or acidic if $[\text{H}^+] > 10^{-7}$. Find the corresponding inequalities involving pH.

50 Many solutions have a pH between 1 and 14. Find the corresponding range of $[\text{H}^+]$.

51 Compound interest Use the compound interest formula to determine how long it will take for a sum of money to double if it is invested at a rate of 6% per year compounded monthly.

52 Compound interest Solve the compound interest formula

$$A = P\left(1 + \frac{r}{n}\right)^{nt}$$

for t by using natural logarithms.

53 Photic zone Refer to Example 8. The most important zone in the sea from the viewpoint of marine biology is the photic zone, in which photosynthesis takes place. The photic zone ends at the depth where about 1% of the surface light penetrates. In very clear waters in the Caribbean, 50% of the light at the surface reaches a depth of about 13 meters. Estimate the depth of the photic zone.

54 Photic zone In contrast to the situation described in the previous exercise, in parts of New York harbor, 50% of the surface light does not reach a depth of 10 centimeters. Estimate the depth of the photic zone.

55 Drug absorption If a 100-milligram tablet of an asthma drug is taken orally and if none of the drug is present in the body when the tablet is first taken, the total amount A in the bloodstream after t minutes is predicted to be

$$A = 100[1 - (0.9)^t] \qquad \text{for} \qquad 0 \le t \le 10.$$

(a) Sketch the graph of the equation.

(b) Determine the number of minutes needed for 50 milligrams of the drug to have entered the bloodstream.

56 Drug dosage A drug is eliminated from the body through urine. Suppose that for a dose of 10 milligrams, the amount $A(t)$ remaining in the body t hours later is given by $A(t) = 10(0.8)^t$ and that in order for the drug to be effective, at least 2 milligrams must be in the body.

(a) Determine when 2 milligrams is left in the body.

(b) What is the half-life of the drug?

57 Genetic mutation The basic source of genetic diversity is mutation, or changes in the chemical structure of genes. If a gene mutates at a constant rate m and if other evolutionary forces are negligible, then the frequency F of the original gene after t generations is given by $F = F_0(1 - m)^t$, where F_0 is the frequency at $t = 0$.

(a) Solve the equation for t using common logarithms.

(b) If $m = 5 \times 10^{-5}$, after how many generations does $F = \frac{1}{2}F_0$?

58 Employee productivity Certain learning processes may be illustrated by the graph of an equation of the form $f(x) = a + b(1 - e^{-cx})$, where a, b, and c are positive constants. Suppose a manufacturer estimates that a new employee can produce five items the first day on the job. As the employee becomes more proficient, the daily production increases until a certain maximum production is reached. Suppose that on the nth day on the job, the number $f(n)$ of items produced is approximated by

$$f(n) = 3 + 20(1 - e^{-0.1n}).$$

(a) Estimate the number of items produced on the fifth day, the ninth day, the twenty-fourth day, and the thirtieth day.

(b) Sketch the graph of f from $n = 0$ to $n = 30$. (Graphs of this type are called *learning curves* and are used frequently in education and psychology.)

(c) What happens as n increases without bound?

59 Height of trees The growth in height of trees is frequently described by a logistic equation. Suppose the height h (in feet) of a tree at age t (in years) is

$$h = \frac{120}{1 + 200e^{-0.2t}},$$

as illustrated by the graph in the figure.

(a) What is the height of the tree at age 10?

(b) At what age is the height 50 feet?

Exercise 59

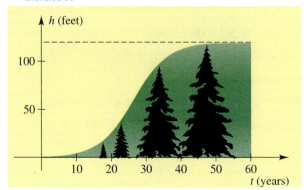

60 Employee productivity Manufacturers sometimes use empirically based formulas to predict the time required to produce the nth item on an assembly line for an integer n. If $T(n)$ denotes the time required to assemble the nth item and T_1 denotes the time required for the first, or prototype, item, then typically $T(n) = T_1 n^{-k}$ for some positive constant k.

(a) For many airplanes, the time required to assemble the second airplane, $T(2)$, is equal to $(0.80)T_1$. Find the value of k.

(b) Express, in terms of T_1, the time required to assemble the fourth airplane.

(c) Express, in terms of $T(n)$, the time $T(2n)$ required to assemble the $(2n)$th airplane.

61 Vertical wind shear Refer to Exercises 67–68 in Section 3.3. If v_0 is the wind speed at height h_0 and if v_1 is the wind speed at height h_1, then the vertical wind shear can be described by the equation

$$\frac{v_0}{v_1} = \left(\frac{h_0}{h_1}\right)^P,$$

where P is a constant. During a one-year period in Montreal, the maximum vertical wind shear occurred when the winds at the 200-foot level were 25 mi/hr while the winds at the 35-foot level were 6 mi/hr. Find P for these conditions.

62 Vertical wind shear Refer to Exercise 61. The average vertical wind shear is given by the equation

$$s = \frac{v_1 - v_0}{h_1 - h_0}.$$

Suppose that the velocity of the wind increases with increasing altitude and that all values for wind speeds taken at the 35-foot and 200-foot altitudes are greater than 1 mi/hr. Does increasing the value of P produce larger or smaller values of s?

Exer. 63–64: An economist suspects that the following data points lie on the graph of $y = c2^{kx}$, where c and k are constants. If the data points have three-decimal-place accuracy, is this suspicion correct?

63 (0, 4), (1, 3.249), (2, 2.639), (3, 2.144)

64 (0, −0.3), (0.5, −0.345), (1, −0.397), (1.5, −0.551), (2, −0.727)

Exer. 65–66: It is suspected that the following data points lie on the graph of $y = c \log (kx + 10)$, where c and k are constants. If the data points have three-decimal-place accuracy, is this suspicion correct?

65 (0, 1.5), (1, 1.619), (2, 1.720), (3, 1.997)

66 (0, 0.7), (1, 0.782), (2, 0.847), (3, 0.900), (4, 0.945)

Exer. 67–68: Approximate the function at the value of x to four decimal places.

67 $h(x) = \log_4 x - 2 \log_8 1.2x;$ $x = 5.3$

68 $h(x) = 3 \log_3 (2x - 1) + 7 \log_2 (x + 0.2);$ $x = 52.6$

69 Human memory A group of elementary students were taught long division over a one-week period. Afterward, they were given a test. The average score was 85. Each week thereafter, they were given an equivalent test, without any review. Let $n(t)$ represent the average score after $t \geq 0$ weeks. Determine which function best models the situation.

(1) $n(t) = 85e^{t/3}$

(2) $n(t) = 70 + 10 \ln (t + 1)$

(3) $n(t) = 86 - e^t$

(4) $n(t) = 85 - 15 \ln (t + 1)$

70 Cooling A jar of boiling water at 212°F is set on a table in a room with a temperature of 72°F. If $T(t)$ represents the temperature of the water after t hours, determine which function best models the situation.

(1) $T(t) = 212 - 50t$

(2) $T(t) = 140e^{-t} + 72$

(3) $T(t) = 212e^{-t}$

(4) $T(t) = 72 + 10 \ln (140t + 1)$

71 Ozone layer Measurements made by the total ozone mapping spectrometer aboard NASA's Nimbus-7 weather satellite showed that ozone levels in the stratosphere are decreasing at a higher rate than anticipated. In April 1993, ozone levels in most of the northern hemisphere measured 11% below levels for April 1992. According to some experts, the cause of this rapid depletion may be the volcanic eruption of Mount Pinatubo in the Philippines in June 1991, rather than the thinning of the ozone layer. Assuming that the ozone level in April 1992 is the normal level and that the 1993 rate of decrease continues, predict the year in which the ozone layer will be 50% of its normal level.

72 Skin cancer Ultraviolet-B radiation from the sun is the primary cause of nonmelanotic skin cancer. Much of this radiation is filtered out by ozone in the atmosphere. In Toronto, the amount of ultraviolet-B radiation increased by 7% from the summer of 1988 to the summer of 1989. If this increase in ultraviolet light is permanent, it may cause the likelihood of contracting skin cancer over a lifetime to increase by the same percentage. If this rate of increase continues, approximate how many years it will take for the likelihood of contracting skin cancer in Toronto to double.

CHAPTER 5 REVIEW EXERCISES

1 Is $f(x) = 2x^3 - 5$ a one-to-one function?

2 The graph of a function f with domain $[-3, 3]$ is shown in the figure. Sketch the graph of $y = f^{-1}(x)$.

Exercise 2

Exercise 5

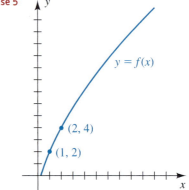

6 Suppose f and g are one-to-one functions such that $f(2) = 7$, $f(4) = 2$, and $g(2) = 5$. Find the value, if possible.

(a) $(g \circ f^{-1})(7)$ (b) $(f \circ g^{-1})(5)$

(c) $(f^{-1} \circ g^{-1})(5)$ (d) $(g^{-1} \circ f^{-1})(2)$

Exer. 3–4: (a) Find $f^{-1}(x)$. (b) Sketch the graphs of f and f^{-1} on the same coordinate plane.

3 $f(x) = 10 - 15x$ **4** $f(x) = 9 - 2x^2, \ x \le 0$

5 Refer to the figure to determine each of the following:

(a) $f(1)$ (b) $(f \circ f)(1)$ (c) $f^{-1}(4)$

(d) all x such that $f(x) = 4$

(e) all x such that $f(x) > 4$

Exer. 7–22: Sketch the graph of f.

7 $f(x) = 3^{x+2}$ **8** $f(x) = \left(\frac{3}{5}\right)^x$

9 $f(x) = \left(\frac{3}{2}\right)^{-x}$ **10** $f(x) = 3^{-2x}$

11 $f(x) = 3^{-x^2}$ **12** $f(x) = 1 - 3^{-x}$

13 $f(x) = e^{x/2}$ **14** $f(x) = \frac{1}{2}e^x$

15 $f(x) = e^{x-2}$ **16** $f(x) = e^{2-x}$

17 $f(x) = \log_6 x$ **18** $f(x) = \log_6 (36x)$

19 $f(x) = \log_4 (x^2)$ **20** $f(x) = \log_4 \sqrt[3]{x}$

21 $f(x) = \log_2 (x + 4)$ **22** $f(x) = \log_2 (4 - x)$

Exer. 23–24: Evaluate without using a calculator.

23 (a) $\log_2 \frac{1}{16}$ (b) $\log_\pi 1$ (c) $\ln e$

(d) $6^{\log_6 4}$ (e) $\log 1{,}000{,}000$ (f) $10^{3 \log 2}$

(g) $\log_4 2$

24 (a) $\log_5 \sqrt[3]{5}$ (b) $\log_5 1$ (c) $\log 10$

(d) $e^{\ln 5}$ (e) $\log \log 10^{10}$ (f) $e^{2 \ln 5}$

(g) $\log_{27} 3$

Exer. 25–44: Solve the equation without using a calculator.

25 $2^{3x-1} = \frac{1}{2}$

26 $8^{2x} \cdot \left(\frac{1}{4}\right)^{x-2} = 4^{-x} \cdot \left(\frac{1}{2}\right)^{2-x}$

27 $\log \sqrt{x} = \log (x - 6)$

28 $\log_8 (x - 5) = \frac{2}{3}$

29 $\log_4 (x + 1) = 2 + \log_4 (3x - 2)$

30 $2 \ln (x + 3) - \ln (x + 1) = 3 \ln 2$

31 $\ln (x + 2) = \ln e^{\ln 2} - \ln x$ 32 $\log \sqrt[4]{x + 1} = \frac{1}{2}$

33 $2^{5-x} = 6$ 34 $3^{(x^2)} = 7$

35 $2^{5x+3} = 3^{2x+1}$

36 $\log_3 (3x) = \log_3 x + \log_3 (4 - x)$

37 $\log_4 x = \sqrt[3]{\log_4 x}$ 38 $e^{x + \ln 4} = 3e^x$

39 $10^{2 \log x} = 5$ 40 $e^{\ln (x+1)} = 3$

41 $x^2(-2xe^{-x^2}) + 2xe^{-x^2} = 0$ 42 $e^x + 2 = 8e^{-x}$

43 (a) $\log x^2 = \log (6 - x)$ (b) $2 \log x = \log (6 - x)$

44 (a) $\ln (e^x)^2 = 16$ (b) $\ln e^{(x^2)} = 16$

45 Express $\log x^4 \sqrt[3]{y^2/z}$ in terms of logarithms of x, y, and z.

46 Express $\log (x^2/y^3) + 4 \log y - 6 \log \sqrt{xy}$ as   one logarithm.

47 Find an exponential function that has y-intercept 6 and passes through the point $(1, 8)$.

48 Sketch the graph of $f(x) = \log_3(x + 2)$.

Exer. 49–50: Use common logarithms to solve the equation for x in terms of y.

49 $y = \dfrac{1}{10^x + 10^{-x}}$

50 $y = \dfrac{1}{10^x - 10^{-x}}$

Exer. 51–52: Approximate x to three significant figures.

51 (a) $x = \ln 6.6$ (b) $\log x = 1.8938$

(c) $\ln x = -0.75$

52 (a) $x = \log 8.4$ (b) $\log x = -2.4260$

(c) $\ln x = 1.8$

Exer. 53–54: (a) Find the domain and range of the function. (b) Find the inverse of the function and its domain and range.

53 $y = \log_2 (x + 1)$ 54 $y = 2^{3-x} - 2$

55 **Bacteria growth** The number of bacteria in a certain culture at time t (in hours) is given by $Q(t) = 2(3^t)$, where $Q(t)$ is measured in thousands.

(a) What is the number of bacteria at $t = 0$?

(b) Find the number of bacteria after 10 minutes, 30 minutes, and 1 hour.

56 **Compound interest** If $1000 is invested at a rate of 12% per year compounded quarterly, what is the principal after one year?

57 **Radioactive iodine decay** Radioactive iodine ^{131}I, which is frequently used in tracer studies involving the thyroid gland, decays according to $N = N_0(0.5)^{t/8}$, where N_0 is the initial dose and t is the time in days.

(a) Sketch the graph of the equation if $N_0 = 64$.

(b) Find the half-life of ^{131}I.

58 **Trout population** A pond is stocked with 1000 trout. Three months later, it is estimated that 600 remain. Find a formula of the form $N = N_0 a^{ct}$ that can be used to estimate the number of trout remaining after t months.

59 **Continuously compounded interest** Ten thousand dollars is invested in a savings fund in which interest is compounded continuously at the rate of 11% per year.

(a) When will the account contain $35,000?

(b) How long does it take for money to double in the account?

60 **Ben Franklin's will** In 1790, Ben Franklin left $4000 with instructions that it go to the city of Philadelphia in 200 years. It was worth about $2 million at that time. Approximate the annual interest rate for the growth.

61 Electrical current The current $I(t)$ in a certain electrical circuit at time t is given by $I(t) = I_0 e^{-Rt/L}$, where R is the resistance, L is the inductance, and I_0 is the initial current at $t = 0$. Find the value of t, in terms of L and R, for which $I(t)$ is 1% of I_0.

62 Sound intensity The sound intensity level formula is $\alpha = 10 \log (I/I_0)$.

(a) Solve for I in terms of α and I_0.

(b) Show that a one-decibel rise in the intensity level α corresponds to a 26% increase in the intensity I.

63 Fish growth The length L of a fish is related to its age by means of the von Bertalanffy growth formula

$$L = a(1 - be^{-kt}),$$

where a, b, and k are positive constants that depend on the type of fish. Solve this equation for t to obtain a formula that can be used to estimate the age of a fish from a length measurement.

64 Earthquake area in the West In the western United States, the area A (in mi^2) affected by an earthquake is related to the magnitude R of the quake by the formula

$$R = 2.3 \log (A + 3000) - 5.1.$$

Solve for A in terms of R.

65 Earthquake area in the East Refer to Exercise 64. For the eastern United States, the area-magnitude formula has the form

$$R = 2.3 \log (A + 34,000) - 7.5.$$

If A_1 is the area affected by an earthquake of magnitude R in the West and A_2 is the area affected by a similar quake in the East, find a formula for A_1/A_2 in terms of R.

66 Earthquake area in the Central states Refer to Exercise 64. For the Rocky Mountain and Central states, the area-magnitude formula has the form

$$R = 2.3 \log (A + 14,000) - 6.6.$$

If an earthquake has magnitude 4 on the Richter scale, estimate the area A of the region that will feel the quake.

67 Atmospheric pressure Under certain conditions, the atmospheric pressure p at altitude h is given by the formula $p = 29e^{-0.000034h}$. Express h as a function of p.

68 Rocket velocity A rocket of mass m_1 is filled with fuel of initial mass m_2. If frictional forces are disregarded, the total mass m of the rocket at time t after ignition is related to its upward velocity v by $v = -a \ln m + b$, where a and b are constants. At ignition time $t = 0$, $v = 0$ and $m = m_1 + m_2$. At burnout, $m = m_1$. Use this information to find a formula, in terms of one logarithm, for the velocity of the rocket at burnout.

69 Earthquake frequency Let n be the average number of earthquakes per year that have magnitudes between R and $R + 1$ on the Richter scale. A formula that approximates the relationship between n and R is

$$\log n = 7.7 - (0.9)R.$$

(a) Solve the equation for n in terms of R.

(b) Find n if $R = 4$, 5, and 6.

70 Earthquake energy The energy E (in ergs) released during an earthquake of magnitude R may be approximated by using the formula

$$\log E = 11.4 + (1.5)R.$$

(a) Solve for E in terms of R.

(b) Find the energy released during the famous Alaskan quake of 1964, which measured 8.4 on the Richter scale.

71 Radioactive decay A certain radioactive substance decays according to the formula $q(t) = q_0 e^{-0.0063t}$, where q_0 is the initial amount of the substance and t is the time in days. Approximate the half-life of the substance.

72 Children's growth The Count Model is a formula that can be used to predict the height of preschool children. If h is height (in centimeters) and t is age (in years), then

$$h = 70.228 + 5.104t + 9.222 \ln t$$

for $\frac{1}{4} \le t \le 6$. From calculus, the rate of growth R (in cm/year) is given by $R = 5.104 + (9.222/t)$. Predict the height and rate of growth of a typical 2-year-old.

73 Electrical circuit The current I in a certain electrical circuit at time t is given by

$$I = \frac{V}{R}(1 - e^{-Rt/L}),$$

where V is the electromotive force, R is the resistance, and L is the inductance. Solve the equation for t.

74 Carbon 14 dating The technique of carbon 14 (^{14}C) dating is used to determine the age of archaeological and geological specimens. The formula $T = -8310 \ln x$ is sometimes used to predict the age T (in years) of a bone fossil, where x is the percentage (expressed as a decimal) of ^{14}C still present in the fossil.

(a) Estimate the age of a bone fossil that contains 4% of the ^{14}C found in an equal amount of carbon in present-day bone.

(b) Approximate the percentage of ^{14}C present in a fossil that is 10,000 years old.

75 Population of Kenya Based on present birth and death rates, the population of Kenya is expected to increase according to the formula $N = 21.4e^{0.033t}$, with N in millions and $t = 0$ corresponding to 1989. How many years will it take for the population to double?

76 Language history Refer to Exercise 48 of Section 5.2. If a language originally had N_0 basic words of which $N(t)$ are still in use, then $N(t) = N_0(0.805)^t$, where time t is measured in millennia. After how many years are one-half the basic words still in use?

CHAPTER 5 DISCUSSION EXERCISES

1 (a) Sketch the graph of $f(x) = -(x - 1)^3 + 1$ along with the graph of $y = f^{-1}(x)$.

(b) Discuss what happens to the graph of $y = f^{-1}(x)$ (in general) as the graph of $y = f(x)$ is increasing or is decreasing.

(c) What can you conclude about the intersection points of the graphs of a function and its inverse?

2 Find the inverse function of $f(x) = \dfrac{9x}{\sqrt{x^2 + 1}}$ and identify

any asymptotes of the graph of f^{-1}. How do they relate to the asymptotes of the graph of f?

3 Shown in the figure is a graph of $f(x) = (\ln x)/x$ for $x > 0$. The maximum value of $f(x)$ occurs at $x = e$.

(a) The integers 2 and 4 have the unusual property that $2^4 = 4^2$. Show that if $x^y = y^x$ for positive real numbers x and y, then $(\ln x)/x = (\ln y)/y$.

(b) Use the graph of f to explain why many pairs of real numbers satisfy the equation $x^y = y^x$.

Exercise 3

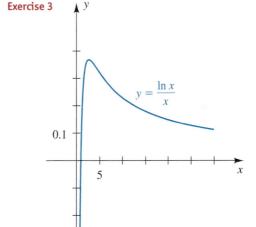

$y = \dfrac{\ln x}{x}$

4 Refer to Exercise 70 of Section 5.4. Discuss how to solve this exercise *without* the use of the formula for the total amount T. Proceed with your solution, and compare your answer to the answer arrived at using the formula for T.

5 Since $y = \log_3 (x^2)$ is equivalent to $y = 2 \log_3 x$ by law 3 of logarithms, why aren't the graphs in Figure 2(a) and (b) of Section 5.5 the same?

6 (a) Compare the growth of the functions $f(x) = (1.085)^x$ and $g(x) = e^{0.085x}$, discuss what they could represent, and explain the difference between the two functions.

(b) Now suppose you are investing money at 8.5% per year compounded monthly. How would a graph of this growth compare with the two graphs in part (a)?

7 Salary increases Suppose you started a job at $40,000 per year. In 5 years, you are scheduled to be making $60,000 per year. Determine the annual exponential rate of increase that describes this situation. Assume that the same exponential rate of increase will continue for 40 years. Using the rule of 70 (page 317), mentally estimate your annual salary in 40 years, and compare the estimate to an actual computation.

8 Energy release Consider these three events:

(1) On May 18, 1980, the volcanic eruption of Mount St. Helens in Washington released approximately 1.7×10^{18} joules of energy.

(2) When a 1-megaton nuclear bomb detonates, it releases about 4×10^{15} joules of energy.

(3) The 1989 San Francisco earthquake registered 7.1 on the Richter scale.

(a) Make some comparisons (i.e., how many of one event is equivalent to another) in terms of energy released. (*Hint:* Refer to Exercise 70 in Chapter 5 Review Exercises.) *Note:* The atomic bombs dropped in World War II were 1-kiloton bombs (1000 1-kiloton bombs = 1 1-megaton bomb).

(b) What reading on the Richter scale would be equivalent to the Mount St. Helens eruption? Has there ever been a reading that high?

9 Discuss how many solutions the equation

$$\log_5 x + \log_7 x = 11$$

has. Solve the equation using the change of base formula.

6

The Trigonometric Functions

6.1 Angles

6.2 Trigonometric Functions of Angles

6.3 Trigonometric Functions of Real Numbers

6.4 Values of the Trigonometric Functions

6.5 Trigonometric Graphs

6.6 Additional Trigonometric Graphs

6.7 Applied Problems

Trigonometry was invented over 2000 years ago by the Greeks, who needed precise methods for measuring angles and sides of triangles. In fact, the word *trigonometry* was derived from the two Greek words *trigonon* (triangle) and *metria* (measurement). This chapter begins with a discussion of angles and how they are measured. We next introduce the trigonometric functions by using ratios of sides of a right triangle. After extending the domains of the trigonometric functions to arbitrary angles and real numbers, we consider their graphs and graphing techniques that make use of amplitudes, periods, and phase shifts. The chapter concludes with a section on applied problems.

6.1

Angles

Figure 1

Figure 2

Coterminal angles

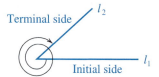

In geometry an **angle** is defined as the set of points determined by two rays, or half-lines, l_1 and l_2, having the same endpoint O. If A and B are points on l_1 and l_2, as in Figure 1, we refer to **angle *AOB*** (denoted $\angle AOB$). An angle may also be considered as two finite line segments with a common endpoint.

In trigonometry we often interpret angles as rotations of rays. Start with a fixed ray l_1, having endpoint O, and rotate it about O, in a plane, to a position specified by ray l_2. We call l_1 the **initial side**, l_2 the **terminal side**, and O the **vertex** of $\angle AOB$. The amount or direction of rotation is not restricted in any way. We might let l_1 make several revolutions in either direction about O before coming to position l_2, as illustrated by the curved arrows in Figure 2. Thus, many different angles have the same initial and terminal sides. Any two such angles are called **coterminal angles.** A **straight angle** is an angle whose sides lie on the same straight line but extend in opposite directions from its vertex.

If we introduce a rectangular coordinate system, then the **standard position** of an angle is obtained by taking the vertex at the origin and letting the initial side l_1 coincide with the positive x-axis. If l_1 is rotated in a *counterclockwise* direction to the terminal position l_2, then the angle is considered **positive.** If l_1 is rotated in a *clockwise* direction, the angle is **negative.** We often denote angles by lowercase Greek letters such as α (*alpha*), β (*beta*), γ (*gamma*), θ (*theta*), ϕ (*phi*), and so on. Figure 3 contains sketches of two positive angles, α and β, and a negative angle, γ. If the terminal side of an angle in standard position is in a certain quadrant, we say that the *angle* is in that quadrant. In Figure 3, α is in quadrant III, β is in quadrant I, and γ is in quadrant II. An angle is called a **quadrantal angle** if its terminal side lies on a coordinate axis.

Figure 3 Standard position of an angle

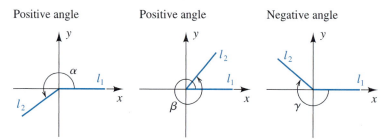

One unit of measurement for angles is the **degree.** The angle in standard position obtained by one complete revolution in the counterclockwise direction has measure 360 degrees, written 360°. Thus, an angle of measure 1 degree (1°) is obtained by $\frac{1}{360}$ of one complete counterclockwise revolution. In Figure 4, several angles measured in degrees are shown in standard position on rectangular coordinate systems. Note that the first three are quadrantal angles.

Figure 4

Throughout our work, a notation such as $\theta = 60°$ specifies an angle θ whose measure is 60°. We also refer to *an angle of* 60° or *a* 60° *angle,* instead of using the more precise (but cumbersome) phrase *an angle having measure* 60°.

EXAMPLE 1 **Finding coterminal angles**

If $\theta = 60°$ is in standard position, find two positive angles and two negative angles that are coterminal with θ.

SOLUTION The angle θ is shown in standard position in the first sketch in Figure 5. To find positive coterminal angles, we may add 360° or 720° (or any other positive integer multiple of 360°) to θ, obtaining

$$60° + 360° = 420° \quad \text{and} \quad 60° + 720° = 780°.$$

These coterminal angles are also shown in Figure 5.

To find negative coterminal angles, we may add −360° or −720° (or any other negative integer multiple of 360°), obtaining

$$60° + (-360°) = -300° \quad \text{and} \quad 60° + (-720°) = -660°,$$

as shown in the last two sketches in Figure 5.

Figure 5

A **right angle** is half of a straight angle and has measure 90°. The following chart contains definitions of other special types of angles.

Terminology	Definition	Illustration
acute angle θ	$0° < \theta < 90°$	12°; 37°
obtuse angle θ	$90° < \theta < 180°$	95°; 157°
complementary angles α, β	$\alpha + \beta = 90°$	20°, 70°; 7°, 83°
supplementary angles α, β	$\alpha + \beta = 180°$	115°, 65°; 18°, 162°

If smaller measurements than the degree are required, we can use tenths, hundredths, or thousandths of degrees. Alternatively, we can divide the degree into 60 equal parts, called **minutes** (denoted by '), and each minute into 60 equal parts, called **seconds** (denoted by "). Thus, $1° = 60'$, and $1' = 60''$. The notation $\theta = 73°56'18''$ refers to an angle θ that has measure 73 degrees, 56 minutes, 18 seconds.

EXAMPLE 2   **Finding complementary angles**

Find the angle that is complementary to θ:
(a) $\theta = 25°43'37''$ (b) $\theta = 73.26°$

SOLUTION   We wish to find $90° - \theta$. It is convenient to write 90° as an equivalent measure, $89°59'60''$.

(a)
$$90° \quad = 89°59'60''$$
$$\underline{\quad \theta \quad = 25°43'37''}$$
$$90° - \theta = 64°16'23''$$

(b)
$$90° \quad = 90.00°$$
$$\underline{\quad \theta \quad = 73.26°}$$
$$90° - \theta = 16.74°$$

Figure 6
Central angle θ

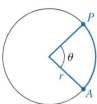

Degree measure for angles is used in applied areas such as surveying, navigation, and the design of mechanical equipment. In scientific applications that require calculus, it is customary to employ *radian measure*. To define an angle of radian measure 1, we consider a circle of any radius r. A **central angle** of a circle is an angle whose vertex is at the center of the circle. If θ is the central angle shown in Figure 6, we say that the **arc AP** (denoted $\overset{\frown}{AP}$) of the circle **subtends** θ or that θ **is subtended by** $\overset{\frown}{AP}$. If the length of $\overset{\frown}{AP}$ is equal to the radius r of the circle, then θ has a measure of one radian, as in the next definition.

Definition of Radian Measure	**One radian** is the measure of the central angle of a circle subtended by an arc equal in length to the radius of the circle.

If we consider a circle of radius r, then an angle α whose measure is 1 radian intercepts an arc AP of length r, as illustrated in Figure 7(a). The angle β in Figure 7(b) has radian measure 2, since it is subtended by an arc of length $2r$. Similarly, γ in (c) of the figure has radian measure 3, since it is subtended by an arc of length $3r$.

(a) $\alpha = 1$ radian

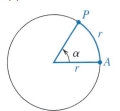

(b) $\beta = 2$ radians

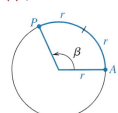

(c) $\gamma = 3$ radians

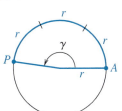

(d) $360° = 2\pi \approx 6.28$ radians

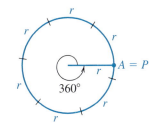

To find the radian measure corresponding to 360°, we must find the number of times that a circular arc of length r can be laid off along the circumference (see Figure 7(d)). This number is not an integer or even a rational number. Since the circumference of the circle is $2\pi r$, the number of times r units can be laid off is 2π. Thus, an angle of measure 2π radians corresponds to the degree measure 360°, and we write 360° = 2π radians. This result gives us the following relationships.

Relationships Between Degrees and Radians	
	(1) $180° = \pi$ radians
	(2) $1° = \dfrac{\pi}{180}$ radian ≈ 0.0175 radian
	(3) 1 radian $= \left(\dfrac{180°}{\pi}\right) \approx 57.2958°$

When radian measure of an angle is used, no units will be indicated. Thus, if an angle has radian measure 5, we write $\theta = 5$ instead of $\theta = 5$ *radians.* There should be no confusion as to whether radian or degree measure is being used, since if θ has *degree* measure 5°, we write $\theta = 5°$, and *not* $\theta = 5$.

The next chart illustrates how to change from one angular measure to another.

Changing Angular Measures

To change	Multiply by	Illustration
degrees to radians	$\dfrac{\pi}{180°}$	$150° = 150°\left(\dfrac{\pi}{180°}\right) = \dfrac{5\pi}{6}$ $225° = 225°\left(\dfrac{\pi}{180°}\right) = \dfrac{5\pi}{4}$
radians to degrees	$\dfrac{180°}{\pi}$	$\dfrac{7\pi}{4} = \dfrac{7\pi}{4}\left(\dfrac{180°}{\pi}\right) = 315°$ $\dfrac{\pi}{3} = \dfrac{\pi}{3}\left(\dfrac{180°}{\pi}\right) = 60°$

We may use the techniques illustrated in the preceding chart to obtain the following table, which displays the corresponding radian and degree measures of special angles.

Radians	0	$\dfrac{\pi}{6}$	$\dfrac{\pi}{4}$	$\dfrac{\pi}{3}$	$\dfrac{\pi}{2}$	$\dfrac{2\pi}{3}$	$\dfrac{3\pi}{4}$	$\dfrac{5\pi}{6}$	π	$\dfrac{7\pi}{6}$	$\dfrac{5\pi}{4}$	$\dfrac{4\pi}{3}$	$\dfrac{3\pi}{2}$	$\dfrac{5\pi}{3}$	$\dfrac{7\pi}{4}$	$\dfrac{11\pi}{6}$	2π
Degrees	0°	30°	45°	60°	90°	120°	135°	150°	180°	210°	225°	240°	270°	300°	315°	330°	360°

Several of these special angles, in radian measure, are shown in standard position in Figure 8.

Figure 8

EXAMPLE 3   **Changing radians to degrees, minutes, and seconds**

If $\theta = 3$, approximate θ in terms of degrees, minutes, and seconds.

SOLUTION

$$3 \text{ radians} = 3\left(\frac{180°}{\pi}\right) \qquad \text{multiply by } \frac{180°}{\pi}$$

$$\approx 171.8873° \qquad \text{approximate}$$

$$= 171° + (0.8873)(60') \qquad 1° = 60'$$

$$= 171° + 53.238' \qquad \text{multiply}$$

$$= 171° + 53' + (0.238)(60'') \qquad 1' = 60''$$

$$= 171°53' + 14.28'' \qquad \text{multiply}$$

$$\approx 171°53'14'' \qquad \text{approximate}$$

EXAMPLE 4   **Expressing minutes and seconds as decimal degrees**

Express $19°47'23''$ as a decimal, to the nearest ten-thousandth of a degree.

SOLUTION   Since $1' = \left(\frac{1}{60}\right)°$ and $1'' = \left(\frac{1}{60}\right)' = \left(\frac{1}{3600}\right)°$,

$$19°47'23'' = 19° + \left(\frac{47}{60}\right)° + \left(\frac{23}{3600}\right)°$$

$$\approx 19° + 0.7833° + 0.0064°$$

$$= 19.7897°.$$

The next result specifies the relationship between the length of a circular arc and the central angle that it subtends.

Formula for the Length of a Circular Arc	If an arc of length s on a circle of radius r subtends a central angle of radian measure θ, then $$s = r\theta.$$

A mnemonic device for remembering $s = r\theta$ is SRO (<u>S</u>tanding <u>R</u>oom <u>O</u>nly).

Figure 9

(a)

(b)

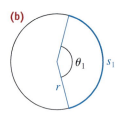

PROOF A typical arc of length s and the corresponding central angle θ are shown in Figure 9(a). Figure 9(b) shows an arc of length s_1 and central angle θ_1. If radian measure is used, then, from plane geometry, the ratio of the lengths of the arcs is the same as the ratio of the angular measures; that is,

$$\frac{s}{s_1} = \frac{\theta}{\theta_1}.$$

If we consider the special case in which θ_1 has radian measure 1, then, from the definition of radian, $s_1 = r$ and the above equation becomes

$$\frac{s}{r} = \frac{\theta}{1}, \quad \text{or} \quad s = r\theta. \quad \blacksquare$$

Notice that if $\theta = 2\pi$, then the formula for the length of a circular arc becomes $s = r(2\pi)$, which is simply the formula for the circumference of a circle, $C = 2\pi r$.

The next formula is proved in a similar manner.

Formula for the Area of a Circular Sector	If θ is the radian measure of a central angle of a circle of radius r and if A is the area of the circular sector determined by θ, then $$A = \tfrac{1}{2}r^2\theta.$$

PROOF If A and A_1 are the areas of the sectors in Figures 9(a) and 9(b), respectively, then, from plane geometry,

$$\frac{A}{A_1} = \frac{\theta}{\theta_1}, \quad \text{or} \quad A = \frac{A_1}{\theta_1}\theta.$$

If we consider the special case $\theta_1 = 2\pi$, then $A_1 = \pi r^2$ and

$$A = \frac{\pi r^2}{2\pi}\theta = \frac{1}{2}r^2\theta. \quad \blacksquare$$

When using the preceding formulas, it is important to remember to use the radian measure of θ rather than the degree measure, as illustrated in the next example.

EXAMPLE 5 **Using the circular arc and sector formulas**

In Figure 10, a central angle θ is subtended by an arc 10 centimeters long on a circle of radius 4 centimeters.

Figure 10

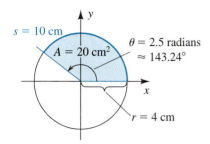

$s = 10$ cm

$A = 20$ cm^2

$\theta = 2.5$ radians
$\approx 143.24°$

$r = 4$ cm

(a) Approximate the measure of θ in degrees.

(b) Find the area of the circular sector determined by θ.

SOLUTION We proceed as follows:

(a)
$$s = r\theta \qquad \text{length of a circular arc formula}$$
$$\theta = \frac{s}{r} \qquad \text{solve for } \theta$$
$$= \frac{10}{4} = 2.5 \qquad \text{let } s = 10, \, r = 4$$

This is the *radian* measure of θ. Changing to degrees, we have

$$\theta = 2.5\left(\frac{180°}{\pi}\right) = \frac{450°}{\pi} \approx 143.24°.$$

(b)
$$A = \tfrac{1}{2}r^2\theta \qquad \text{area of a circular sector formula}$$
$$= \tfrac{1}{2}(4)^2(2.5) \qquad \text{let } r = 4, \, \theta = 2.5 \text{ radians}$$
$$= 20 \text{ cm}^2 \qquad \text{multiply}$$

Figure 11

P

The **angular speed** of a wheel that is rotating at a constant rate is the angle generated in one unit of time by a line segment from the center of the wheel to a point P on the circumference (see Figure 11). The **linear speed** of a point P on the circumference is the distance that P travels per unit of time.

EXAMPLE 6 **Finding angular and linear speeds**

Suppose that a machine contains a wheel of diameter 3 feet, rotating at a rate of 1600 rpm (revolutions per minute).

(a) Find the angular speed of the wheel.

(b) Find the linear speed of a point P on the circumference of the wheel.

SOLUTION

(a) Let O denote the center of the wheel, and let P be a point on the circumference. Because the number of revolutions per minute is 1600 and because each revolution generates an angle of 2π radians, the angle generated by the line segment OP in one minute has radian measure $(1600)(2\pi)$; that is,

$$\text{angular speed} = (1600)(2\pi) = 3200\pi \text{ radians per minute.}$$

Note that the diameter of the wheel is irrelevant in finding the angular speed.

(b) The linear speed of P is the distance it travels per minute. We may find this distance by using the formula $s = r\theta$, with $r = \tfrac{3}{2}$ ft and $\theta = 3200\pi$. Thus,

$$s = \tfrac{3}{2}(3200\pi) = 4800\pi \text{ ft,}$$

and consequently the linear speed of P is 4800π ft/min. To the nearest integer, this is approximately 15,080 ft/min, or 171.36 mi/hr. Unlike the angular speed, the linear speed *is* dependent on the diameter of the wheel.

6.1 *Exercises*

Exer. 1–4: If the given angle is in standard position, find two positive coterminal angles and two negative coterminal angles.

1 (a) $120°$ (b) $135°$ (c) $-30°$

2 (a) $240°$ (b) $315°$ (c) $-150°$

3 (a) $620°$ (b) $\dfrac{5\pi}{6}$ (c) $-\dfrac{\pi}{4}$

4 (a) $570°$ (b) $\dfrac{2\pi}{3}$ (c) $-\dfrac{5\pi}{4}$

Exer. 5–6: Find the angle that is complementary to θ.

5 (a) $\theta = 5°17'34''$ (b) $\theta = 32.5°$

6 (a) $\theta = 63°4'15''$ (b) $\theta = 82.73°$

Exer. 7–8: Find the angle that is supplementary to θ.

7 (a) $\theta = 48°51'37''$ (b) $\theta = 136.42°$

8 (a) $\theta = 152°12'4''$ (b) $\theta = 15.9°$

Exer. 9–12: Find the exact radian measure of the angle.

9 (a) $150°$ (b) $-60°$ (c) $225°$

10 (a) $120°$ (b) $-135°$ (c) $210°$

11 (a) $450°$ (b) $72°$ (c) $100°$

12 (a) $630°$ (b) $54°$ (c) $95°$

Exer. 13–16: Find the exact degree measure of the angle.

13 (a) $\dfrac{2\pi}{3}$ (b) $\dfrac{11\pi}{6}$ (c) $\dfrac{3\pi}{4}$

14 (a) $\dfrac{5\pi}{6}$ (b) $\dfrac{4\pi}{3}$ (c) $\dfrac{11\pi}{4}$

15 (a) $-\dfrac{7\pi}{2}$ (b) 7π (c) $\dfrac{\pi}{9}$

16 (a) $-\dfrac{5\pi}{2}$ (b) 9π (c) $\dfrac{\pi}{16}$

Exer. 17–20: Express θ in terms of degrees, minutes, and seconds, to the nearest second.

17 $\theta = 2$ **18** $\theta = 1.5$

19 $\theta = 5$ **20** $\theta = 4$

Exer. 21–24: Express the angle as a decimal, to the nearest ten-thousandth of a degree.

21 $37°41'$ **22** $83°17'$

23 $115°26'27''$ **24** $258°39'52''$

Exer. 25–28: Express the angle in terms of degrees, minutes, and seconds, to the nearest second.

25 $63.169°$ **26** $12.864°$

27 $310.6215°$ **28** $81.7238°$

Exer. 29–30: If a circular arc of the given length s subtends the central angle θ on a circle, find the radius of the circle.

29 $s = 10$ cm, $\theta = 4$ **30** $s = 3$ km, $\theta = 20°$

Exer. 31–32: (a) Find the length of the arc of the colored sector in the figure. (b) Find the area of the sector.

31   **32**

Exer. 33–34: (a) Find the radian and degree measures of the central angle θ subtended by the given arc of length s on a circle of radius r. (b) Find the area of the sector determined by θ.

33 $s = 7$ cm, $r = 4$ cm   **34** $s = 3$ ft, $r = 20$ in.

Exer. 35–36: (a) Find the length of the arc that subtends the given central angle θ on a circle of diameter d. (b) Find the area of the sector determined by θ.

35 $\theta = 50°$, $d = 16$ m   **36** $\theta = 2.2$, $d = 120$ cm

37 Measuring distances on Earth The distance between two points A and B on Earth is measured along a circle having center C at the center of Earth and radius equal to the distance from C to the surface (see the figure). If the diameter of Earth is approximately 8000 miles, approximate the distance between A and B if angle ACB has the indicated measure:

(a) 60°   (b) 45°   (c) 30°   (d) 10°   (e) 1°

Exercise 37

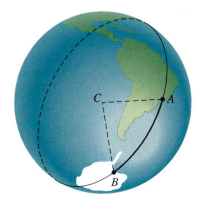

38 Nautical miles Refer to Exercise 37. If angle ACB has measure 1′, then the distance between A and B is a nautical mile. Approximate the number of land (statute) miles in a nautical mile.

39 Measuring angles using distance Refer to Exercise 37. If two points A and B are 500 miles apart, express angle ACB in radians and in degrees.

40 A hexagon is inscribed in a circle. If the difference between the area of the circle and the area of the hexagon is 24 m², use the formula for the area of a sector to approximate the radius r of the circle.

41 Window area A rectangular window measures 54 inches by 24 inches. There is a 17-inch wiper blade attached by a 5-inch arm at the center of the base of the window, as shown in the figure. If the arm rotates 120°, approximate the percentage of the window's area that is wiped by the blade.

Exercise 41

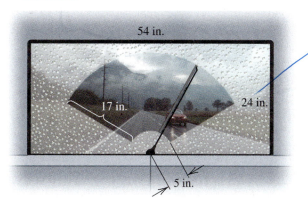

42 A tornado's core A simple model of the core of a tornado is a right circular cylinder that rotates about its axis. If a tornado has a core diameter of 200 feet and maximum wind speed of 180 mi/hr (or 264 ft/sec) at the perimeter of the core, approximate the number of revolutions the core makes each minute.

43 Earth's rotation Earth rotates about its axis once every 23 hours, 56 minutes, and 4 seconds. Approximate the number of radians Earth rotates in one second.

44 Earth's rotation Refer to Exercise 43. The equatorial radius of Earth is approximately 3963.3 miles. Find the linear speed of a point on the equator as a result of Earth's rotation.

Exer. 45–46: A wheel of the given radius is rotating at the indicated rate.
(a) Find the angular speed (in radians per minute).
(b) Find the linear speed of a point on the circumference (in ft/min).

45 radius 5 in.,   40 rpm   **46** radius 9 in.,   2400 rpm

47 Rotation of compact discs (CDs) The drive motor of a particular CD player is controlled to rotate at a speed of 200 rpm when reading a track 5.7 centimeters from the center of the CD. The speed of the drive motor must vary so that the reading of the data occurs at a constant rate.

(a) Find the angular speed (in radians per minute) of the drive motor when it is reading a track 5.7 centimeters from the center of the CD.

(b) Find the linear speed (in cm/sec) of a point on the CD that is 5.7 centimeters from the center of the CD.

(c) Find the angular speed (in rpm) of the drive motor when it is reading a track 3 centimeters from the center of the CD.

(d) Find a function S that gives the drive motor speed in rpm for any radius r in centimeters, where $2.3 \le r \le 5.9$. What type of variation exists between the drive motor speed and the radius of the track being read? Check your answer by graphing S and finding the speeds for $r = 3$ and $r = 5.7$.

48 **Tire revolutions** A typical tire for a compact car is 22 inches in diameter. If the car is traveling at a speed of 60 mi/hr, find the number of revolutions the tire makes per minute.

49 **Cargo winch** A large winch of diameter 3 feet is used to hoist cargo, as shown in the figure.

(a) Find the distance the cargo is lifted if the winch rotates through an angle of radian measure $7\pi/4$.

(b) Find the angle (in radians) through which the winch must rotate in order to lift the cargo d feet.

Exercise 49

50 **Pendulum's swing** A pendulum in a grandfather clock is 4 feet long and swings back and forth along a 6-inch arc. Approximate the angle (in degrees) through which the pendulum passes during one swing.

51 **Pizza values** A vender sells two sizes of pizza by the slice. The *small* slice is $\frac{1}{6}$ of a circular 18-inch-diameter pizza, and it sells for $2.00. The *large* slice is $\frac{1}{8}$ of a circular 26-inch-diameter pizza, and it sells for $3.00. Which slice provides more pizza per dollar?

52 **Bicycle mechanics** The sprocket assembly for a bicycle is shown in the figure. If the sprocket of radius r_1 rotates through an angle of θ_1 radians, find the corresponding angle of rotation for the sprocket of radius r_2.

Exercise 52

53 **Bicycle mechanics** Refer to Exercise 52. An expert cyclist can attain a speed of 40 mi/hr. If the sprocket assembly has $r_1 = 5$ in., $r_2 = 2$ in., and the wheel has a diameter of 28 inches, approximately how many revolutions per minute of the front sprocket wheel will produce a speed of 40 mi/hr? (*Hint:* First change 40 mi/hr to in./sec.)

54 **Magnetic pole drift** The geographic and magnetic north poles have different locations. Currently, the magnetic north pole is drifting westward through 0.0017 radian per year, where the angle of drift has its vertex at the center of Earth. If this movement continues, approximately how many years will it take for the magnetic north pole to drift a total of 5°?

6.2

Trigonometric Functions of Angles

We shall introduce the trigonometric functions in the manner in which they originated historically—as ratios of sides of a right triangle. A triangle is a **right triangle** if one of its angles is a right angle. If θ is any acute angle, we may consider a right triangle having θ as one of its angles, as in Figure 1,

Figure 1

Figure 2

*We will refer to these six trigonometric functions as **the** trigonometric functions. Here are some other, less common trigonometric functions that we will not use in this text:*

$$\text{vers } \theta = 1 - \cos \theta$$

$$\text{covers } \theta = 1 - \sin \theta$$

$$\text{exsec } \theta = \sec \theta - 1$$

$$\text{hav } \theta = \tfrac{1}{2} \text{ vers } \theta$$

Figure 3

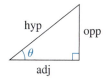

where the symbol ⌐ specifies the 90° angle. Six ratios can be obtained using the lengths a, b, and c of the sides of the triangle:

$$\frac{b}{c}, \quad \frac{a}{c}, \quad \frac{b}{a}, \quad \frac{a}{b}, \quad \frac{c}{a}, \quad \frac{c}{b}$$

We can show that these ratios depend only on θ, and not on the size of the triangle, as indicated in Figure 2. Since the two triangles have equal angles, they are similar, and therefore ratios of corresponding sides are proportional. For example,

$$\frac{b}{c} = \frac{b'}{c'}, \quad \frac{a}{c} = \frac{a'}{c'}, \quad \frac{b}{a} = \frac{b'}{a'}.$$

Thus, for each θ, the six ratios are uniquely determined and hence are functions of θ. They are called the **trigonometric functions*** and are designated as the **sine, cosine, tangent, cotangent, secant,** and **cosecant** functions, abbreviated **sin, cos, tan, cot, sec,** and **csc,** respectively. The symbol sin (θ), or sin θ, is used for the ratio b/c, which the sine function associates with θ. Values of the other five functions are denoted in similar fashion. To summarize, if θ is the acute angle of the right triangle in Figure 1, then, by definition,

$$\sin \theta = \frac{b}{c} \qquad \cos \theta = \frac{a}{c} \qquad \tan \theta = \frac{b}{a}$$

$$\csc \theta = \frac{c}{b} \qquad \sec \theta = \frac{c}{a} \qquad \cot \theta = \frac{a}{b}.$$

The domain of each of the six trigonometric functions is the set of all acute angles. Later in this section we will extend the domains to larger sets of angles, and in the next section, to real numbers.

If θ is the angle in Figure 1, we refer to the sides of the triangle of lengths a, b, and c as the **adjacent side, opposite side,** and **hypotenuse,** respectively. We shall use **adj, opp,** and **hyp** to denote the lengths of the sides. We may then represent the triangle as in Figure 3. With this notation, the trigonometric functions may be expressed as follows.

Definition of the Trigonometric Functions of an Acute Angle of a Right Triangle			
	$\sin \theta = \dfrac{\text{opp}}{\text{hyp}}$	$\cos \theta = \dfrac{\text{adj}}{\text{hyp}}$	$\tan \theta = \dfrac{\text{opp}}{\text{adj}}$
	$\csc \theta = \dfrac{\text{hyp}}{\text{opp}}$	$\sec \theta = \dfrac{\text{hyp}}{\text{adj}}$	$\cot \theta = \dfrac{\text{adj}}{\text{opp}}$

A mnemonic device for remembering the top row in the definition is

SOH CAH TOA,

where SOH is an abbreviation for Sin θ = Opp/Hyp, *and so forth.*

The formulas in the preceding definition can be applied to any right triangle without attaching the labels a, b, c to the sides. Since the lengths of the sides of a triangle are positive real numbers, *the values of the six trigonometric functions are positive for every acute angle θ.* Moreover, the hypotenuse is always greater than the adjacent or opposite side, and hence sin $\theta < 1$, cos $\theta < 1$, csc $\theta > 1$, and sec $\theta > 1$ for every acute angle θ.

Note that since

$$\sin \theta = \frac{\text{opp}}{\text{hyp}} \quad \text{and} \quad \csc \theta = \frac{\text{hyp}}{\text{opp}},$$

$\sin \theta$ and $\csc \theta$ are reciprocals of each other, giving us the two identities in the left-hand column of the next box. Similarly, $\cos \theta$ and $\sec \theta$ are reciprocals of each other, as are $\tan \theta$ and $\cot \theta$.

Reciprocal Identities

$$\sin \theta = \frac{1}{\csc \theta} \qquad \cos \theta = \frac{1}{\sec \theta} \qquad \tan \theta = \frac{1}{\cot \theta}$$

$$\csc \theta = \frac{1}{\sin \theta} \qquad \sec \theta = \frac{1}{\cos \theta} \qquad \cot \theta = \frac{1}{\tan \theta}$$

Several other important identities involving the trigonometric functions will be discussed at the end of this section.

 EXAMPLE 1   Finding trigonometric function values

If θ is an acute angle and $\cos \theta = \frac{3}{4}$, find the values of the trigonometric functions of θ.

Figure 4

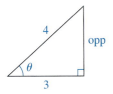

SOLUTION   We begin by sketching a right triangle having an acute angle θ with adj = 3 and hyp = 4, as shown in Figure 4, and proceed as follows:

$$3^2 + (\text{opp})^2 = 4^2 \qquad \text{Pythagorean theorem}$$
$$(\text{opp})^2 = 16 - 9 = 7 \qquad \text{isolate } (\text{opp})^2$$
$$\text{opp} = \sqrt{7} \qquad \text{take the square root}$$

Applying the definition of the trigonometric functions of an acute angle of a right triangle, we obtain the following:

$$\sin \theta = \frac{\text{opp}}{\text{hyp}} = \frac{\sqrt{7}}{4} \qquad \cos \theta = \frac{\text{adj}}{\text{hyp}} = \frac{3}{4} \qquad \tan \theta = \frac{\text{opp}}{\text{adj}} = \frac{\sqrt{7}}{3}$$

$$\csc \theta = \frac{\text{hyp}}{\text{opp}} = \frac{4}{\sqrt{7}} \qquad \sec \theta = \frac{\text{hyp}}{\text{adj}} = \frac{4}{3} \qquad \cot \theta = \frac{\text{adj}}{\text{opp}} = \frac{3}{\sqrt{7}}$$

In Example 1 we could have rationalized the denominators for $\csc \theta$ and $\cot \theta$, writing

$$\csc \theta = \frac{4\sqrt{7}}{7} \quad \text{and} \quad \cot \theta = \frac{3\sqrt{7}}{7}.$$

However, in most examples and exercises we will leave expressions in unrationalized form. An exception to this practice is the special trigonometric function values corresponding to 60°, 30°, and 45°, which are obtained in the following example.

EXAMPLE 2 Finding trigonometric function values of 60°, 30°, and 45°

Find the values of the trigonometric functions that correspond to θ:

(a) $\theta = 60°$ (b) $\theta = 30°$ (c) $\theta = 45°$

Figure 5

SOLUTION   Consider an equilateral triangle with sides of length 2. The median from one vertex to the opposite side bisects the angle at that vertex, as illustrated by the dashes in Figure 5. By the Pythagorean theorem, the side opposite 60° in the shaded right triangle has length $\sqrt{3}$. Using the formulas for the trigonometric functions of an acute angle of a right triangle, we obtain the values corresponding to 60° and 30° as follows:

(a) $\sin 60° = \dfrac{\sqrt{3}}{2}$ $\cos 60° = \dfrac{1}{2}$ $\tan 60° = \dfrac{\sqrt{3}}{1} = \sqrt{3}$

$\csc 60° = \dfrac{2}{\sqrt{3}} = \dfrac{2\sqrt{3}}{3}$ $\sec 60° = \dfrac{2}{1} = 2$ $\cot 60° = \dfrac{1}{\sqrt{3}} = \dfrac{\sqrt{3}}{3}$

(b) $\sin 30° = \dfrac{1}{2}$ $\cos 30° = \dfrac{\sqrt{3}}{2}$ $\tan 30° = \dfrac{1}{\sqrt{3}} = \dfrac{\sqrt{3}}{3}$

$\csc 30° = \dfrac{2}{1} = 2$ $\sec 30° = \dfrac{2}{\sqrt{3}} = \dfrac{2\sqrt{3}}{3}$ $\cot 30° = \dfrac{\sqrt{3}}{1} = \sqrt{3}$

Figure 6

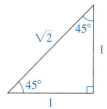

(c)   To find the values for $\theta = 45°$, we may consider an isosceles right triangle whose two equal sides have length 1, as illustrated in Figure 6. By the Pythagorean theorem, the length of the hypotenuse is $\sqrt{2}$. Hence, the values corresponding to 45° are as follows:

$$\sin 45° = \frac{1}{\sqrt{2}} = \frac{\sqrt{2}}{2} = \cos 45° \qquad \tan 45° = \frac{1}{1} = 1$$

$$\csc 45° = \frac{\sqrt{2}}{1} = \sqrt{2} = \sec 45° \qquad \cot 45° = \frac{1}{1} = 1$$

For reference, we list the values found in Example 2, together with the radian measures of the angles, in the following table. Two reasons for stressing these values are that they are exact and that they occur frequently in work involving trigonometry. Because of the importance of these special values, it is a good idea either to memorize the table or to learn to find the values quickly by using triangles, as in Example 2.

Special Values of the Trigonometric Functions

θ (radians)	θ (degrees)	$\sin\theta$	$\cos\theta$	$\tan\theta$	$\cot\theta$	$\sec\theta$	$\csc\theta$
$\dfrac{\pi}{6}$	$30°$	$\dfrac{1}{2}$	$\dfrac{\sqrt{3}}{2}$	$\dfrac{\sqrt{3}}{3}$	$\sqrt{3}$	$\dfrac{2\sqrt{3}}{3}$	2
$\dfrac{\pi}{4}$	$45°$	$\dfrac{\sqrt{2}}{2}$	$\dfrac{\sqrt{2}}{2}$	1	1	$\sqrt{2}$	$\sqrt{2}$
$\dfrac{\pi}{3}$	$60°$	$\dfrac{\sqrt{3}}{2}$	$\dfrac{1}{2}$	$\sqrt{3}$	$\dfrac{\sqrt{3}}{3}$	2	$\dfrac{2\sqrt{3}}{3}$

The next example illustrates a practical use for trigonometric functions of acute angles. Additional applications involving right triangles will be considered in Section 6.7.

EXAMPLE 3   Finding the height of a flagpole

A surveyor observes that at a point A, located on level ground a distance 25.0 feet from the base B of a flagpole, the angle between the ground and the top of the pole is 30°. Approximate the height h of the pole to the nearest tenth of a foot.

Figure 7

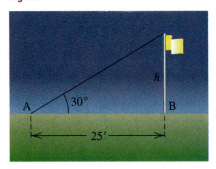

SOLUTION   Referring to Figure 7, we see that we want to relate the opposite side and the adjacent side, h and 25, respectively, to the 30° angle. This suggests that we use a trigonometric function involving those two sides— namely, tan or cot. It is usually easier to solve the problem if we select the function for which the variable is in the numerator. Hence, we have

$$\tan 30° = \frac{h}{25} \quad \text{or, equivalently,} \quad h = 25\tan 30°.$$

We use the value of tan 30° from Example 2 to find h:

$$h = 25\left(\frac{\sqrt{3}}{3}\right) \approx 14.4 \text{ ft}$$

It is possible to approximate, to any degree of accuracy, the values of the trigonometric functions for any acute angle. Calculators have keys labeled SIN, COS, and TAN that can be used to approximate values of these functions. The values of csc, sec, and cot may then be found by means of the reciprocal key. *Before using a calculator to find function values that correspond to the radian measure of an acute angle, be sure that the calculator is in radian mode. For values corresponding to degree measure, select degree mode.*

Figure 8

In degree mode

Figure 9

In radian mode

As an illustration (see Figure 8), to find sin 30° on a typical calculator, we place the calculator in degree mode and use the $\boxed{\text{SIN}}$ key to obtain sin 30° = 0.5, which is the exact value. Using the same procedure for 60°, we obtain a decimal approximation to $\sqrt{3}/2$, such as

$$\sin 60° \approx 0.8660.$$

Most calculators give eight- to ten-decimal-place accuracy for such function values; throughout the text, however, we will usually round off values to four decimal places.

To find a value such as cos 1.3 (see Figure 9), where 1.3 is the radian measure of an acute angle, we place the calculator in radian mode and use the $\boxed{\text{COS}}$ key, obtaining

$$\cos 1.3 \approx 0.2675.$$

For sec 1.3, we could find cos 1.3 and then use the reciprocal key, usually labeled $\boxed{1/x}$ or $\boxed{x^{-1}}$ (as shown in Figure 9), to obtain

$$\sec 1.3 = \frac{1}{\cos 1.3} \approx 3.7383.$$

The formulas listed in the box on the next page are, without doubt, the most important identities in trigonometry, because they can be used to simplify and unify many different aspects of the subject. Since the formulas are part of the foundation for work in trigonometry, they are called the *fundamental identities*.

Three of the fundamental identities involve squares, such as $(\sin \theta)^2$ and $(\cos \theta)^2$. In general, if n is an integer different from -1, then a power such as $(\cos \theta)^n$ is written $\cos^n \theta$. The symbols $\sin^{-1} \theta$ and $\cos^{-1} \theta$ are reserved for inverse trigonometric functions, which we will discuss in Section 6.4 and treat thoroughly in the next chapter. With this agreement on notation, we have, for example,

$$\cos^2 \theta = (\cos \theta)^2 = (\cos \theta)(\cos \theta)$$

$$\tan^3 \theta = (\tan \theta)^3 = (\tan \theta)(\tan \theta)(\tan \theta)$$

$$\sec^4 \theta = (\sec \theta)^4 = (\sec \theta)(\sec \theta)(\sec \theta)(\sec \theta).$$

Let us next list all the fundamental identities and then discuss the proofs. These identities are true for every acute angle θ, and θ may take on various forms. For example, using the first Pythagorean identity with $\theta = 4\alpha$, we know that

$$\sin^2 4\alpha + \cos^2 4\alpha = 1.$$

We shall see later that these identities are also true for other angles and for real numbers.

The Fundamental Identities

(1) The reciprocal identities:

$$\csc \theta = \frac{1}{\sin \theta} \qquad \sec \theta = \frac{1}{\cos \theta} \qquad \cot \theta = \frac{1}{\tan \theta}$$

(2) The tangent and cotangent identities:

$$\tan \theta = \frac{\sin \theta}{\cos \theta} \qquad \cot \theta = \frac{\cos \theta}{\sin \theta}$$

(3) The Pythagorean identities:

$$\sin^2 \theta + \cos^2 \theta = 1 \qquad 1 + \tan^2 \theta = \sec^2 \theta \qquad 1 + \cot^2 \theta = \csc^2 \theta$$

PROOFS

(1) The reciprocal identities were established earlier in this section.

Figure 10

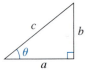

(2) To prove the tangent identity, we refer to the right triangle in Figure 10 and use definitions of trigonometric functions as follows:

$$\tan \theta = \frac{b}{a} = \frac{b/c}{a/c} = \frac{\sin \theta}{\cos \theta}$$

To verify the cotangent identity, we use a reciprocal identity and the tangent identity:

$$\cot \theta = \frac{1}{\tan \theta} = \frac{1}{\sin \theta / \cos \theta} = \frac{\cos \theta}{\sin \theta}$$

(3) The Pythagorean identities are so named because of the first step in the following proof. Referring to Figure 10, we obtain

$$b^2 + a^2 = c^2 \qquad \text{Pythagorean theorem}$$

$$\left(\frac{b}{c}\right)^2 + \left(\frac{a}{c}\right)^2 = \left(\frac{c}{c}\right)^2 \qquad \text{divide by } c^2$$

$$(\sin \theta)^2 + (\cos \theta)^2 = 1 \qquad \text{definitions of } \sin \theta \text{ and } \cos \theta$$

$$\sin^2 \theta + \cos^2 \theta = 1. \qquad \text{equivalent notation}$$

We may use this identity to verify the second Pythagorean identity as follows:

$$\frac{\sin^2 \theta + \cos^2 \theta}{\cos^2 \theta} = \frac{1}{\cos^2 \theta} \qquad \text{divide by } \cos^2 \theta$$

$$\frac{\sin^2 \theta}{\cos^2 \theta} + \frac{\cos^2 \theta}{\cos^2 \theta} = \frac{1}{\cos^2 \theta} \qquad \text{equivalent equation}$$

$$\left(\frac{\sin \theta}{\cos \theta}\right)^2 + \left(\frac{\cos \theta}{\cos \theta}\right)^2 = \left(\frac{1}{\cos \theta}\right)^2 \qquad \text{law of exponents}$$

$$\tan^2 \theta + 1 = \sec^2 \theta \qquad \text{tangent and reciprocal identities}$$

To prove the third Pythagorean identity, $1 + \cot^2 \theta = \csc^2 \theta$, we could divide both sides of the identity $\sin^2 \theta + \cos^2 \theta = 1$ by $\sin^2 \theta$. ◢

We can use the fundamental identities to express each trigonometric function in terms of any other trigonometric function. Two illustrations are given in the next example.

EXAMPLE 4   Using fundamental identities

Let θ be an acute angle.

(a) Express $\sin \theta$ in terms of $\cos \theta$.

(b) Express $\tan \theta$ in terms of $\sin \theta$.

SOLUTION

(a) We may proceed as follows:

$$\sin^2 \theta + \cos^2 \theta = 1 \qquad \text{Pythagorean identity}$$

$$\sin^2 \theta = 1 - \cos^2 \theta \qquad \text{isolate } \sin^2 \theta$$

$$\sin \theta = \pm\sqrt{1 - \cos^2 \theta} \qquad \text{take the square root}$$

$$\sin \theta = \sqrt{1 - \cos^2 \theta} \qquad \sin \theta > 0 \text{ for acute angles}$$

Later in this section (Example 12) we will consider a simplification involving a *non*-acute angle θ.

(b) If we begin with the fundamental identity

$$\tan \theta = \frac{\sin \theta}{\cos \theta},$$

then all that remains is to express $\cos \theta$ in terms of $\sin \theta$. We can do this by solving $\sin^2 \theta + \cos^2 \theta = 1$ for $\cos \theta$, obtaining

$$\cos \theta = \sqrt{1 - \sin^2 \theta} \quad \text{for} \quad 0 < \theta < \frac{\pi}{2}.$$

(continued)

Hence,

$$\tan \theta = \frac{\sin \theta}{\cos \theta} = \frac{\sin \theta}{\sqrt{1 - \sin^2 \theta}} \quad \text{for} \quad 0 < \theta < \frac{\pi}{2}.$$

Fundamental identities are often used to simplify expressions involving trigonometric functions, as illustrated in the next example.

EXAMPLE 5 **Showing that an equation is an identity**

Show that the following equation is an identity by transforming the left-hand side into the right-hand side:

$$(\sec \theta + \tan \theta)(1 - \sin \theta) = \cos \theta$$

SOLUTION We begin with the left-hand side and proceed as follows:

$$(\sec \theta + \tan \theta)(1 - \sin \theta) = \left(\frac{1}{\cos \theta} + \frac{\sin \theta}{\cos \theta} \right)(1 - \sin \theta) \qquad \text{reciprocal and tangent identities}$$

$$= \left(\frac{1 + \sin \theta}{\cos \theta} \right)(1 - \sin \theta) \qquad \text{add fractions}$$

$$= \frac{1 - \sin^2 \theta}{\cos \theta} \qquad \text{multiply}$$

$$= \frac{\cos^2 \theta}{\cos \theta} \qquad \sin^2 \theta + \cos^2 \theta = 1$$

$$= \cos \theta \qquad \text{cancel } \cos \theta$$

There are other ways to simplify the expression on the left-hand side in Example 5. We could first multiply the two factors and then simplify and combine terms. The method we employed—changing all expressions to expressions that involve only sines and cosines—is often useful. However, that technique does not always lead to the shortest possible simplification.

Hereafter, we shall use the phrase *verify an identity* instead of *show that an equation is an identity*. When verifying an identity, we often use fundamental identities and algebraic manipulations to simplify expressions, as we did in the preceding example. As with the fundamental identities, we understand that an identity that contains fractions is valid for all values of the variables such that no denominator is zero.

EXAMPLE 6   Verifying an identity

Verify the following identity by transforming the left-hand side into the right-hand side:

$$\frac{\tan \theta + \cos \theta}{\sin \theta} = \sec \theta + \cot \theta$$

SOLUTION   We may transform the left-hand side into the right-hand side as follows:

$$\frac{\tan \theta + \cos \theta}{\sin \theta} = \frac{\tan \theta}{\sin \theta} + \frac{\cos \theta}{\sin \theta} \qquad \text{divide numerator by } \sin \theta$$

$$= \frac{\left(\dfrac{\sin \theta}{\cos \theta}\right)}{\sin \theta} + \cot \theta \qquad \text{tangent and cotangent identities}$$

$$= \frac{\sin \theta}{\cos \theta} \cdot \frac{1}{\sin \theta} + \cot \theta \qquad \text{rule for quotients}$$

$$= \frac{1}{\cos \theta} + \cot \theta \qquad \text{cancel } \sin \theta$$

$$= \sec \theta + \cot \theta \qquad \text{reciprocal identity} \qquad \blacksquare$$

Figure 11

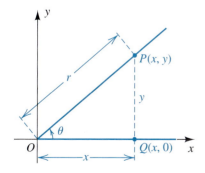

In Section 7.1 we will verify many other identities using methods similar to those used in Examples 5 and 6.

Since many applied problems involve angles that are not acute, it is necessary to extend the definition of the trigonometric functions. We make this extension by using the standard position of an angle θ on a rectangular coordinate system. If θ is acute, we have the situation illustrated in Figure 11, where we have chosen a point $P(x, y)$ on the terminal side of θ and where $d(O, P) = r = \sqrt{x^2 + y^2}$. Referring to triangle OQP, we have

$$\sin \theta = \frac{\text{opp}}{\text{hyp}} = \frac{y}{r}, \quad \cos \theta = \frac{\text{adj}}{\text{hyp}} = \frac{x}{r}, \quad \text{and} \quad \tan \theta = \frac{\text{opp}}{\text{adj}} = \frac{y}{x}.$$

We now wish to consider angles of the types illustrated in Figure 12 on the next page (or any *other* angle, either positive, negative, or zero). Note that in Figure 12 the value of x or y may be negative. In each case, side QP (opp in Figure 12) has length $|y|$, side OQ (adj in Figure 12) has length $|x|$, and the hypotenuse OP has length r. We shall define the six trigonometric functions so that their values agree with those given previously whenever the angle is acute. It is understood that if a zero denominator occurs, then the corresponding function value is undefined.

Figure 12

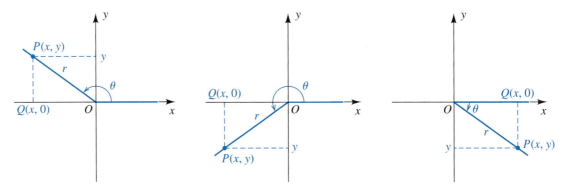

Definition of the Trigonometric Functions of Any Angle	Let θ be an angle in standard position on a rectangular coordinate system, and let $P(x, y)$ be any point other than the origin O on the terminal side of θ. If $d(O, P) = r = \sqrt{x^2 + y^2}$, then

$$\sin \theta = \frac{y}{r} \qquad \cos \theta = \frac{x}{r} \qquad \tan \theta = \frac{y}{x} \ \ (\text{if } x \neq 0)$$

$$\csc \theta = \frac{r}{y} \ \ (\text{if } y \neq 0) \quad \sec \theta = \frac{r}{x} \ \ (\text{if } x \neq 0) \quad \cot \theta = \frac{x}{y} \ \ (\text{if } y \neq 0).$$

We can show, using similar triangles, that the formulas in this definition do not depend on the point $P(x, y)$ that is chosen on the terminal side of θ. The fundamental identities, which were established for acute angles, are also true for trigonometric functions of any angle.

The domains of the sine and cosine functions consist of all angles θ. However, $\tan \theta$ and $\sec \theta$ are undefined if $x = 0$ (that is, if the terminal side of θ is on the y-axis). Thus, the domains of the tangent and the secant functions consist of all angles *except* those of radian measure $(\pi/2) + \pi n$ for any integer n. Some special cases are $\pm\pi/2$, $\pm 3\pi/2$, and $\pm 5\pi/2$. The corresponding degree measures are $\pm 90°$, $\pm 270°$, and $\pm 450°$.

The domains of the cotangent and cosecant functions consist of all angles except those that have $y = 0$ (that is, all angles except those having terminal sides on the x-axis). These are the angles of radian measure πn (or degree measure $180° \cdot n$) for any integer n.

Our discussion of domains is summarized in the following table, where n denotes any integer.

Function		Domain
sine,	cosine	every angle θ
tangent,	secant	every angle θ except $\theta = \dfrac{\pi}{2} + \pi n = 90° + 180° \cdot n$
cotangent,	cosecant	every angle θ except $\theta = \pi n = 180° \cdot n$

For any point $P(x, y)$ in the preceding definition, $|x| \le r$ and $|y| \le r$ or, equivalently, $|x/r| \le 1$ and $|y/r| \le 1$. Thus,

$$|\sin \theta| \le 1, \quad |\cos \theta| \le 1, \quad |\csc \theta| \ge 1, \quad \text{and} \quad |\sec \theta| \ge 1$$

for every θ in the domains of these functions.

EXAMPLE 7 **Finding trigonometric function values of an angle in standard position**

If θ is an angle in standard position on a rectangular coordinate system and if $P(-15, 8)$ is on the terminal side of θ, find the values of the six trigonometric functions of θ.

Figure 13

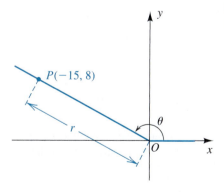

SOLUTION   The point $P(-15, 8)$ is shown in Figure 13. Applying the definition of the trigonometric functions of any angle with $x = -15$, $y = 8$, and

$$r = \sqrt{x^2 + y^2} = \sqrt{(-15)^2 + 8^2} = \sqrt{289} = 17,$$

we obtain the following:

$$\sin \theta = \frac{y}{r} = \frac{8}{17} \qquad \cos \theta = \frac{x}{r} = -\frac{15}{17} \qquad \tan \theta = \frac{y}{x} = -\frac{8}{15}$$

$$\csc \theta = \frac{r}{y} = \frac{17}{8} \qquad \sec \theta = \frac{r}{x} = -\frac{17}{15} \qquad \cot \theta = \frac{x}{y} = -\frac{15}{8}$$

EXAMPLE 8 **Finding trigonometric function values of an angle in standard position**

An angle θ is in standard position, and its terminal side lies in quadrant III on the line $y = 3x$. Find the values of the trigonometric functions of θ.

Figure 14

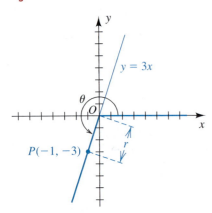

SOLUTION The graph of $y = 3x$ is sketched in Figure 14, together with the initial and terminal sides of θ. Since the terminal side of θ is in quadrant III, we begin by choosing a convenient negative value of x, say $x = -1$. Substituting in $y = 3x$ gives us $y = 3(-1) = -3$, and hence $P(-1, -3)$ is on the terminal side. Applying the definition of the trigonometric functions of any angle with

$$x = -1, \quad y = -3, \quad \text{and} \quad r = \sqrt{x^2 + y^2} = \sqrt{(-1)^2 + (-3)^2} = \sqrt{10}$$

gives us

$$\sin \theta = -\frac{3}{\sqrt{10}} \qquad \cos \theta = -\frac{1}{\sqrt{10}} \qquad \tan \theta = \frac{-3}{-1} = 3$$

$$\csc \theta = -\frac{\sqrt{10}}{3} \qquad \sec \theta = -\frac{\sqrt{10}}{1} \qquad \cot \theta = \frac{-1}{-3} = \frac{1}{3}. \qquad ◪$$

The definition of the trigonometric functions of any angle may be applied if θ is a quadrantal angle. The procedure is illustrated by the next example.

EXAMPLE 9 Finding trigonometric function values of a quadrantal angle

If $\theta = 3\pi/2$, find the values of the trigonometric functions of θ.

SOLUTION Note that $3\pi/2 = 270°$. If θ is placed in standard position, the terminal side of θ coincides with the negative y-axis, as shown in Figure 15. To apply the definition of the trigonometric functions of any angle, we may choose *any* point P on the terminal side of θ. For simplicity, we use $P(0, -1)$. In this case, $x = 0, y = -1, r = 1$, and hence

Figure 15

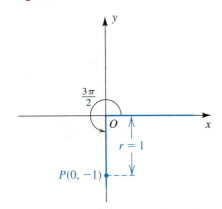

$$\sin \frac{3\pi}{2} = \frac{-1}{1} = -1 \qquad \cos \frac{3\pi}{2} = \frac{0}{1} = 0$$

$$\csc \frac{3\pi}{2} = \frac{1}{-1} = -1 \qquad \cot \frac{3\pi}{2} = \frac{0}{-1} = 0.$$

The tangent and secant functions are undefined, since the meaningless expressions $\tan \theta = (-1)/0$ and $\sec \theta = 1/0$ occur when we substitute in the appropriate formulas. ◪

Let us determine the signs associated with values of the trigonometric functions. If θ is in quadrant II and $P(x, y)$ is a point on the terminal side, then x is negative and y is positive. Hence, $\sin \theta = y/r$ and $\csc \theta = r/y$ are positive, and the other four trigonometric functions, which all involve x, are negative. Checking the remaining quadrants in a similar fashion, we obtain the following table.

Signs of the Trigonometric Functions

Quadrant containing θ	Positive functions	Negative functions
I	all	none
II	sin, csc	cos, sec, tan, cot
III	tan, cot	sin, csc, cos, sec
IV	cos, sec	sin, csc, tan, cot

Figure 16

Positive trigonometric functions

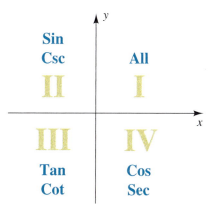

A mnemonic device for remembering the quadrants in which the trigonometric functions are positive is "A Smart Trig Class," which corresponds to All Sin Tan Cos.

The diagram in Figure 16 may be useful for remembering quadrants in which trigonometric functions are *positive*. If a function is not listed (such as cos in quadrant II), then that function is negative. We finish this section with three examples that require using the information in the preceding table.

EXAMPLE 10   Finding the quadrant containing an angle

Find the quadrant containing θ if both $\cos \theta > 0$ and $\sin \theta < 0$.

SOLUTION   Referring to the table of signs or Figure 16, we see that $\cos \theta > 0$ (cosine is positive) if θ is in quadrant I or IV and that $\sin \theta < 0$ (sine is negative) if θ is in quadrant III or IV. Hence, for both conditions to be satisfied, θ must be in quadrant IV.

EXAMPLE 11   Finding values of trigonometric functions from prescribed conditions

If $\sin \theta = \frac{3}{5}$ and $\tan \theta < 0$, use fundamental identities to find the values of the other five trigonometric functions.

SOLUTION   Since $\sin \theta = \frac{3}{5} > 0$ (positive) and $\tan \theta < 0$ (negative), θ is in quadrant II. Using the relationship $\sin^2 \theta + \cos^2 \theta = 1$ and the fact that $\cos \theta$ is negative in quadrant II, we have

$$\cos \theta = -\sqrt{1 - \sin^2 \theta} = -\sqrt{1 - \left(\tfrac{3}{5}\right)^2} = -\sqrt{\tfrac{16}{25}} = -\tfrac{4}{5}.$$

Next we use the tangent identity to obtain

$$\tan \theta = \frac{\sin \theta}{\cos \theta} = \frac{3/5}{-4/5} = -\frac{3}{4}.$$

Finally, using the reciprocal identities gives us

$$\csc \theta = \frac{1}{\sin \theta} = \frac{1}{3/5} = \frac{5}{3}$$

$$\sec \theta = \frac{1}{\cos \theta} = \frac{1}{-4/5} = -\frac{5}{4}$$

$$\cot \theta = \frac{1}{\tan \theta} = \frac{1}{-3/4} = -\frac{4}{3}.$$

EXAMPLE 12 Using fundamental identities

Rewrite $\sqrt{\cos^2 \theta + \sin^2 \theta + \cot^2 \theta}$ in nonradical form without using absolute values for $\pi < \theta < 2\pi$.

SOLUTION

$$\sqrt{\cos^2 \theta + \sin^2 \theta + \cot^2 \theta} = \sqrt{1 + \cot^2 \theta} \qquad \cos^2 \theta + \sin^2 \theta = 1$$
$$= \sqrt{\csc^2 \theta} \qquad 1 + \cot^2 \theta = \csc^2 \theta$$
$$= \left| \csc \theta \right| \qquad \sqrt{x^2} = |x|$$

Since $\pi < \theta < 2\pi$, we know that θ is in quadrant III or IV. Thus, $\csc \theta$ is *negative*, and by the definition of absolute value, we have

$$\left| \csc \theta \right| = -\csc \theta.$$

6.2 Exercises

Exer. 1–2: Use common sense to match the variables and the values. (The triangles are drawn to scale, and the angles are measured in radians.)

1

2

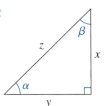

(a) α	(A) 7
(b) β	(B) 0.28
(c) x	(C) 24
(d) y	(D) 1.29
(e) z	(E) 25

(a) α	(A) 23.35
(b) β	(B) 16
(c) x	(C) 17
(d) y	(D) 0.82
(e) z	(E) 0.76

Exer. 3–10: Find the values of the six trigonometric functions for the angle θ.

3

4

5

6

7

8

9

10

Exer. 11–16: Find the exact values of x and y.

11

12

13
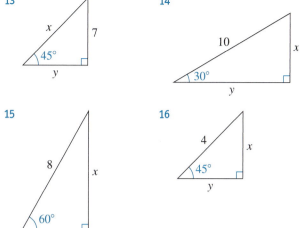
14

15

16

Exer. 17–22: Find the exact values of the trigonometric functions for the acute angle θ.

17 $\sin \theta = \frac{3}{5}$

18 $\cos \theta = \frac{8}{17}$

19 $\tan \theta = \frac{5}{12}$

20 $\cot \theta = \frac{7}{24}$

21 $\sec \theta = \frac{6}{5}$

22 $\csc \theta = 4$

23 Height of a tree A forester, 200 feet from the base of a redwood tree, observes that the angle between the ground and the top of the tree is 60°. Estimate the height of the tree.

24 Distance to Mt. Fuji The peak of Mt. Fuji in Japan is approximately 12,400 feet high. A trigonometry student, several miles away, notes that the angle between level ground and the peak is 30°. Estimate the distance from the student to the point on level ground directly beneath the peak.

25 Stonehenge blocks Stonehenge in Salisbury Plains, England, was constructed using solid stone blocks weighing over 99,000 pounds each. Lifting a single stone required 550 people, who pulled the stone up a ramp inclined at an angle of 9°. Approximate the distance that a stone was moved in order to raise it to a height of 30 feet.

26 Advertising sign height The highest advertising sign in the world is a large letter I situated at the top of the 73-story First Interstate World Center building in Los Angeles. At a distance of 200 feet from a point directly below the sign, the angle between the ground and the top of the sign is 78.87°. Approximate the height of the top of the sign.

27 Telescope resolution Two stars that are very close may appear to be one. The ability of a telescope to separate their images is called its resolution. The smaller the resolution, the better a telescope's ability to separate images in the sky. In a refracting telescope, resolution θ (see the figure) can be improved by using a lens with a larger diameter D. The relationship between θ in degrees and D in meters is given by $\sin \theta = 1.22\lambda/D$, where λ is the wavelength of light in meters. The largest refracting telescope in the world is at the University of Chicago. At a wavelength of $\lambda = 550 \times 10^{-9}$ meter, its resolution is 0.000 037 69°. Approximate the diameter of the lens.

Exercise 27

28 Moon phases The phases of the moon can be described using the phase angle θ, determined by the sun, the moon, and Earth, as shown in the figure. Because the moon orbits Earth, θ changes during the course of a month. The area of the region A of the moon, which appears illuminated to an observer on Earth, is given by $A = \frac{1}{2}\pi R^2(1 + \cos \theta)$, where $R = 1080$ mi is the radius of the moon. Approximate A for the following positions of the moon:

(a) $\theta = 0°$ (full moon)

(b) $\theta = 180°$ (new moon)

(c) $\theta = 90°$ (first quarter)

(d) $\theta = 103°$

Exercise 28

Exer. 29–34: Approximate to four decimal places, when appropriate.

29 (a) $\sin 42°$ (b) $\cos 77°$

(c) $\csc 123°$ (d) $\sec (-190°)$

30 (a) $\tan 282°$ (b) $\cot (-81°)$

(c) $\sec 202°$ (d) $\sin 97°$

31 (a) $\cot (\pi/13)$ (b) $\csc 1.32$

(c) $\cos (-8.54)$ (d) $\tan (3\pi/7)$

32 (a) $\sin (-0.11)$ (b) $\sec \frac{31}{27}$

(c) $\tan \left(-\frac{3}{13}\right)$ (d) $\cos 2.4\pi$

33 (a) $\sin 30°$ (b) $\sin 30$

(c) $\cos \pi°$ (d) $\cos \pi$

34 (a) $\sin 45°$ (b) $\sin 45$

(c) $\cos (3\pi/2)°$ (d) $\cos (3\pi/2)$

Exer. 35–38: Use the Pythagorean identities to write the expression as an integer.

35 (a) $\tan^2 4\beta - \sec^2 4\beta$ (b) $4 \tan^2 \beta - 4 \sec^2 \beta$

36 (a) $\csc^2 3\alpha - \cot^2 3\alpha$ (b) $3 \csc^2 \alpha - 3 \cot^2 \alpha$

37 (a) $5 \sin^2 \theta + 5 \cos^2 \theta$

(b) $5 \sin^2 (\theta/4) + 5 \cos^2 (\theta/4)$

38 (a) $7 \sec^2 \gamma - 7 \tan^2 \gamma$

(b) $7 \sec^2 (\gamma/3) - 7 \tan^2 (\gamma/3)$

Exer. 39–42: Simplify the expression.

39 $\dfrac{\sin^3 \theta + \cos^3 \theta}{\sin \theta + \cos \theta}$ 40 $\dfrac{\cot^2 \alpha - 4}{\cot^2 \alpha - \cot \alpha - 6}$

41 $\dfrac{2 - \tan \theta}{2 \csc \theta - \sec \theta}$ 42 $\dfrac{\csc \theta + 1}{(1/\sin^2 \theta) + \csc \theta}$

Exer. 43–48: Use fundamental identities to write the first expression in terms of the second, for any acute angle θ.

43 $\cot \theta, \sin \theta$ 44 $\tan \theta, \cos \theta$

45 $\sec \theta, \sin \theta$ 46 $\csc \theta, \cos \theta$

47 $\sin \theta, \sec \theta$ 48 $\cos \theta, \cot \theta$

Exer. 49–70: Verify the identity by transforming the left-hand side into the right-hand side.

49 $\cos \theta \sec \theta = 1$ 50 $\tan \theta \cot \theta = 1$

51 $\sin \theta \sec \theta = \tan \theta$ 52 $\sin \theta \cot \theta = \cos \theta$

53 $\dfrac{\csc \theta}{\sec \theta} = \cot \theta$ 54 $\cot \theta \sec \theta = \csc \theta$

55 $(1 + \cos 2\theta)(1 - \cos 2\theta) = \sin^2 2\theta$

56 $\cos^2 2\theta - \sin^2 2\theta = 2 \cos^2 2\theta - 1$

57 $\cos^2 \theta (\sec^2 \theta - 1) = \sin^2 \theta$

58 $(\tan \theta + \cot \theta) \tan \theta = \sec^2 \theta$

59 $\dfrac{\sin (\theta/2)}{\csc (\theta/2)} + \dfrac{\cos (\theta/2)}{\sec (\theta/2)} = 1$

60 $1 - 2 \sin^2 (\theta/2) = 2 \cos^2 (\theta/2) - 1$

61 $(1 + \sin \theta)(1 - \sin \theta) = \dfrac{1}{\sec^2 \theta}$

62 $(1 - \sin^2 \theta)(1 + \tan^2 \theta) = 1$

63 $\sec \theta - \cos \theta = \tan \theta \sin \theta$

64 $\dfrac{\sin \theta + \cos \theta}{\cos \theta} = 1 + \tan \theta$

65 $(\cot \theta + \csc \theta)(\tan \theta - \sin \theta) = \sec \theta - \cos \theta$

66 $\cot \theta + \tan \theta = \csc \theta \sec \theta$

67 $\sec^2 3\theta \csc^2 3\theta = \sec^2 3\theta + \csc^2 3\theta$

68 $\dfrac{1 + \cos^2 3\theta}{\sin^2 3\theta} = 2 \csc^2 3\theta - 1$

69 $\log \csc \theta = -\log \sin \theta$

70 $\log \tan \theta = \log \sin \theta - \log \cos \theta$

Exer. 71–74: Find the exact values of the six trigonometric functions of θ if θ is in standard position and P is on the terminal side.

71 $P(4, -3)$ 72 $P(-8, -15)$

73 $P(-2, -5)$ 74 $P(-1, 2)$

Exer. 75–80: Find the exact values of the six trigonometric functions of θ if θ is in standard position and the terminal side of θ is in the specified quadrant and satisfies the given condition.

75 II;   on the line $y = -4x$

76 IV;   on the line $3y + 5x = 0$

77 I;   on a line having slope $\frac{4}{3}$

78 III; bisects the quadrant

79 III; parallel to the line $2y - 7x + 2 = 0$

80 II;   parallel to the line through $A(1, 4)$ and $B(3, -2)$

Exer. 81–82: Find the exact values of the six trigonometric functions of each angle, whenever possible.

81 (a) 90°   (b) 0°   (c) $7\pi/2$ (d) 3π

82 (a) 180°   (b) −90°   (c) 2π (d) $5\pi/2$

Exer. 83–84: Find the quadrant containing θ if the given conditions are true.

83 (a) $\cos \theta > 0$ and $\sin \theta < 0$

    (b) $\sin \theta < 0$ and $\cot \theta > 0$

    (c) $\csc \theta > 0$ and $\sec \theta < 0$

    (d) $\sec \theta < 0$ and $\tan \theta > 0$

84 (a) $\tan \theta < 0$ and $\cos \theta > 0$

    (b) $\sec \theta > 0$ and $\tan \theta < 0$

    (c) $\csc \theta > 0$ and $\cot \theta < 0$

    (d) $\cos \theta < 0$ and $\csc \theta < 0$

Exer. 85–92: Use fundamental identities to find the values of the trigonometric functions for the given conditions.

85 $\tan \theta = -\frac{3}{4}$ and $\sin \theta > 0$ 86 $\cot \theta = \frac{3}{4}$ and $\cos \theta < 0$

87 $\sin \theta = -\frac{5}{13}$ and $\sec \theta > 0$ 88 $\cos \theta = \frac{1}{2}$ and $\sin \theta < 0$

89 $\cos \theta = -\frac{1}{3}$ and $\sin \theta < 0$ 90 $\csc \theta = 5$ and $\cot \theta < 0$

91 $\sec \theta = -4$ and $\csc \theta > 0$ 92 $\sin \theta = \frac{2}{5}$ and $\cos \theta < 0$

Exer. 93–98: Rewrite the expression in nonradical form without using absolute values for the indicated values of θ.

93 $\sqrt{\sec^2 \theta - 1}$; $\pi/2 < \theta < \pi$

94 $\sqrt{1 + \cot^2 \theta}$; $0 < \theta < \pi$

95 $\sqrt{1 + \tan^2 \theta}$; $3\pi/2 < \theta < 2\pi$

96 $\sqrt{\csc^2 \theta - 1}$; $3\pi/2 < \theta < 2\pi$

97 $\sqrt{\sin^2 (\theta/2)}$; $2\pi < \theta < 4\pi$

98 $\sqrt{\cos^2 (\theta/2)}$; $0 < \theta < \pi$

6.3

Trigonometric Functions of Real Numbers

The domain of each trigonometric function we have discussed is a set of angles. In calculus and in many applications, domains of functions consist of real numbers. To regard the domain of a trigonometric function as a subset of $\mathbb{R}$, we may use the following definition.

Definition of the Trigonometric Functions of Real Numbers	The **value of a trigonometric function at a real number** t is its value at an angle of t radians, provided that value exists.

Using this definition, we may interpret a notation such as sin 2 as *either* the sine of the real number 2 *or* the sine of an angle of 2 radians. As in Section 6.2, if degree measure is used, we shall write sin 2°. With this understanding,

$$\sin 2 \neq \sin 2°.$$

Figure 1

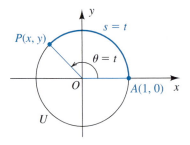

To find the values of trigonometric functions of real numbers with a calculator, we use the radian mode.

We may interpret trigonometric functions of real numbers geometrically by using a unit circle U—that is, a circle of radius 1, with center at the origin O of a rectangular coordinate plane. The circle U is the graph of the equation $x^2 + y^2 = 1$. Let t be a real number such that $0 < t < 2\pi$, and let θ denote the angle (in standard position) of radian measure t. One possibility is illustrated in Figure 1, where $P(x, y)$ is the point of intersection of the terminal side of θ and the unit circle U and where s is the length of the circular arc from $A(1, 0)$ to $P(x, y)$. Using the formula $s = r\theta$ for the length of a circular arc, with $r = 1$ and $\theta = t$, we see that

$$s = r\theta = 1(t) = t.$$

Figure 2

$\theta = t, t > 0$

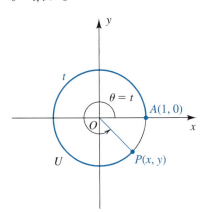

Thus, t *may be regarded either as the radian measure of the angle θ or as the length of the circular arc AP on U.*

Next consider *any* nonnegative real number t. If we regard the angle θ of radian measure t as having been generated by rotating the line segment OA about O in the counterclockwise direction, then t is the distance along U that A travels before reaching its final position $P(x, y)$. In Figure 2 we have illustrated a case for $t < 2\pi$; however, if $t > 2\pi$, then A may travel around U several times in a counterclockwise direction before reaching $P(x, y)$.

If $t < 0$, then the rotation of OA is in the *clockwise* direction, and the distance A travels before reaching $P(x, y)$ is $|t|$, as illustrated in Figure 3.

Figure 3

$\theta = t, t < 0$

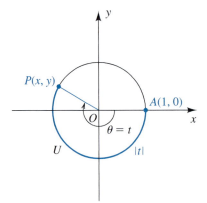

The preceding discussion indicates how *we may associate with each real number t a unique point $P(x, y)$ on U.* We shall call $P(x, y)$ the **point on the unit circle U that corresponds to t.** The coordinates (x, y) of P may be used to find the six trigonometric functions of t. Thus, by the definition of the trigonometric functions of real numbers together with the definition of the trigonometric functions of any angle (given in Section 6.2), we see that

$$\sin t = \sin \theta = \frac{y}{r} = \frac{y}{1} = y.$$

Using the same procedure for the remaining five trigonometric functions gives us the following formulas.

Definition of the Trigonometric Functions in Terms of a Unit Circle	If t is a real number and $P(x, y)$ is the point on the unit circle U that corresponds to t, then

$$\sin t = y \qquad\qquad \cos t = x \qquad\qquad \tan t = \frac{y}{x} \quad (\text{if } x \neq 0)$$

$$\csc t = \frac{1}{y} \quad (\text{if } y \neq 0) \qquad \sec t = \frac{1}{x} \quad (\text{if } x \neq 0) \qquad \cot t = \frac{x}{y} \quad (\text{if } y \neq 0).$$

The formulas in this definition express function values in terms of coordinates of a point P on a unit circle. For this reason, the trigonometric functions are sometimes referred to as the **circular functions.**

EXAMPLE 1   Finding values of the trigonometric functions

A point $P(x, y)$ on the unit circle U corresponding to a real number t is shown in Figure 4, for $\pi < t < 3\pi/2$. Find the values of the trigonometric functions at t.

SOLUTION   Referring to Figure 4, we see that the coordinates of the point $P(x, y)$ are

$$x = -\tfrac{3}{5}, \qquad y = -\tfrac{4}{5}.$$

Using the definition of the trigonometric functions in terms of a unit circle gives us

$$\sin t = y = -\frac{4}{5} \qquad\qquad \cos t = x = -\frac{3}{5} \qquad\qquad \tan t = \frac{y}{x} = \frac{-\frac{4}{5}}{-\frac{3}{5}} = \frac{4}{3}$$

$$\csc t = \frac{1}{y} = \frac{1}{-\frac{4}{5}} = -\frac{5}{4} \qquad \sec t = \frac{1}{x} = \frac{1}{-\frac{3}{5}} = -\frac{5}{3} \qquad \cot t = \frac{x}{y} = \frac{-\frac{3}{5}}{-\frac{4}{5}} = \frac{3}{4}.$$

Figure 4

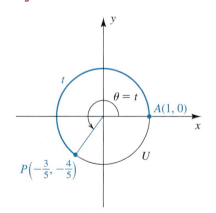

EXAMPLE 2   Finding a point on U relative to a given point

Let $P(t)$ denote the point on the unit circle U that corresponds to t for $0 \le t < 2\pi$. If $P(t) = \left(\frac{4}{5}, \frac{3}{5}\right)$, find

(a) $P(t + \pi)$ (b) $P(t - \pi)$ (c) $P(-t)$

SOLUTION

(a) The point $P(t)$ on U is plotted in Figure 5(a), where we have also shown the arc AP of length t. To find $P(t + \pi)$, we travel a distance π in the *counterclockwise* direction along U from $P(t)$, as indicated by the blue arc in the figure. Since π is one-half the circumference of U, this gives us the point $P(t + \pi) = \left(-\frac{4}{5}, -\frac{3}{5}\right)$ diametrically opposite $P(t)$.

Figure 5

(a)

(b)

(c)

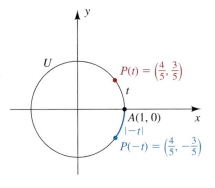

(b) To find $P(t - \pi)$, we travel a distance π in the *clockwise* direction along U from $P(t)$, as indicated in Figure 5(b). This gives us $P(t - \pi) = \left(-\frac{4}{5}, -\frac{3}{5}\right)$. Note that $P(t + \pi) = P(t - \pi)$.

(c) To find $P(-t)$, we travel along U a distance $\left|-t\right|$ in the *clockwise* direction from $A(1, 0)$, as indicated in Figure 5(c). This is equivalent to reflecting $P(t)$ through the x-axis. Thus, we merely change the sign of the y-coordinate of $P(t) = \left(\frac{4}{5}, \frac{3}{5}\right)$ to obtain $P(-t) = \left(\frac{4}{5}, -\frac{3}{5}\right)$. ◢

EXAMPLE 3 **Finding special values of the trigonometric functions**

Find the values of the trigonometric functions at t:

(a) $t = 0$ **(b)** $t = \dfrac{\pi}{4}$ **(c)** $t = \dfrac{\pi}{2}$

SOLUTION

(a) The point P on the unit circle U that corresponds to $t = 0$ has coordinates $(1, 0)$, as shown in Figure 6(a). Thus, we let $x = 1$ and $y = 0$ in the definition of the trigonometric functions in terms of a unit circle, obtaining

$$\sin 0 = y = 0 \qquad\qquad \cos 0 = x = 1$$

$$\tan 0 = \frac{y}{x} = \frac{0}{1} = 0 \qquad \sec 0 = \frac{1}{x} = \frac{1}{1} = 1.$$

Note that $\csc 0$ and $\cot 0$ are undefined, since $y = 0$ is a denominator.

Figure 6

(a)

Figure 6
(b)

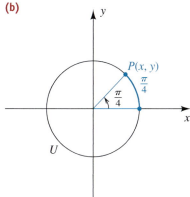

(b) If $t = \pi/4$, then the angle of radian measure $\pi/4$ shown in Figure 6(b) bisects the first quadrant and the point $P(x, y)$ lies on the line $y = x$. Since $P(x, y)$ is on the unit circle $x^2 + y^2 = 1$ and since $y = x$, we obtain

$$x^2 + x^2 = 1, \quad \text{or} \quad 2x^2 = 1.$$

Solving for x and noting that $x > 0$ gives us

$$x = \frac{1}{\sqrt{2}} = \frac{\sqrt{2}}{2}.$$

Thus, P is the point $\left(\sqrt{2}/2, \sqrt{2}/2\right)$. Letting $x = \sqrt{2}/2$ and $y = \sqrt{2}/2$ in the definition of the trigonometric functions in terms of a unit circle gives us

(c)

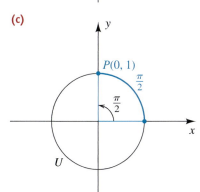

$$\sin \frac{\pi}{4} = \frac{\sqrt{2}}{2} \qquad \cos \frac{\pi}{4} = \frac{\sqrt{2}}{2} \qquad \tan \frac{\pi}{4} = \frac{\sqrt{2}/2}{\sqrt{2}/2} = 1$$

$$\csc \frac{\pi}{4} = \frac{2}{\sqrt{2}} = \sqrt{2} \qquad \sec \frac{\pi}{4} = \frac{2}{\sqrt{2}} = \sqrt{2} \qquad \cot \frac{\pi}{4} = \frac{\sqrt{2}/2}{\sqrt{2}/2} = 1.$$

(c) The point P on U that corresponds to $t = \pi/2$ has coordinates $(0, 1)$, as shown in Figure 6(c). Thus, we let $x = 0$ and $y = 1$ in the definition of the trigonometric functions in terms of a unit circle, obtaining

$$\sin \frac{\pi}{2} = 1 \qquad \cos \frac{\pi}{2} = 0 \qquad \csc \frac{\pi}{2} = \frac{1}{1} = 1 \qquad \cot \frac{\pi}{2} = \frac{0}{1} = 0.$$

The tangent and secant functions are undefined, since $x = 0$ is a denominator in each case.

A summary of the trigonometric functions of special angles appears in Appendix IV.

We shall use the unit circle formulation of the trigonometric functions to help obtain their graphs. If t is a real number and $P(x, y)$ is the point on the unit circle U that corresponds to t, then by the definition of the trigonometric functions in terms of a unit circle,

$$x = \cos t \quad \text{and} \quad y = \sin t.$$

Figure 7

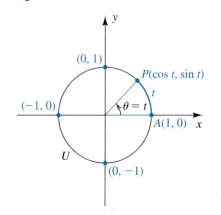

Thus, as shown in Figure 7, we may denote $P(x, y)$ by

$$P(\cos t, \sin t).$$

If $t > 0$, the real number t may be interpreted either as the radian measure of the angle θ or as the length of arc AP.

If we let t increase from 0 to 2π radians, the point $P(\cos t, \sin t)$ travels around the unit circle U one time in the counterclockwise direction. By observing the variation of the x- and y-coordinates of P, we obtain the next table. The notation $0 \to \pi/2$ in the first row of the table means that t increases from 0 to $\pi/2$, and the notation $(1, 0) \to (0, 1)$ denotes the corresponding variation of $P(\cos t, \sin t)$ as it travels along U from $(1, 0)$ to $(0, 1)$. If t increases from

0 to $\pi/2$, then sin t increases from 0 to 1, which we denote by $0 \rightarrow 1$. Moreover, sin t takes on every value between 0 and 1. If t increases from $\pi/2$ to π, then sin t decreases from 1 to 0, which is denoted by $1 \rightarrow 0$. Other entries in the table may be interpreted in similar fashion.

t	$P(\cos t, \sin t)$	$\cos t$	$\sin t$
$0 \rightarrow \dfrac{\pi}{2}$	$(1, 0) \rightarrow (0, 1)$	$1 \rightarrow 0$	$0 \rightarrow 1$
$\dfrac{\pi}{2} \rightarrow \pi$	$(0, 1) \rightarrow (-1, 0)$	$0 \rightarrow -1$	$1 \rightarrow 0$
$\pi \rightarrow \dfrac{3\pi}{2}$	$(-1, 0) \rightarrow (0, -1)$	$-1 \rightarrow 0$	$0 \rightarrow -1$
$\dfrac{3\pi}{2} \rightarrow 2\pi$	$(0, -1) \rightarrow (1, 0)$	$0 \rightarrow 1$	$-1 \rightarrow 0$

If t increases from 2π to 4π, the point $P(\cos t, \sin t)$ in Figure 7 traces the unit circle U again and the patterns for sin t and cos t are repeated—that is,

$$\sin (t + 2\pi) = \sin t \quad \text{and} \quad \cos (t + 2\pi) = \cos t$$

for every t in the interval $[0, 2\pi]$. The same is true if t increases from 4π to 6π, from 6π to 8π, and so on. In general, we have the following theorem.

Theorem on Repeated Function Values for sin and cos	If n is any integer, then $\qquad \sin (t + 2\pi n) = \sin t \quad \text{and} \quad \cos (t + 2\pi n) = \cos t.$

The repetitive variation of the sine and cosine functions is *periodic* in the sense of the following definition.

Definition of Periodic Function	A function f is **periodic** if there exists a positive real number k such that $\qquad f(t + k) = f(t)$ for every t in the domain of f. The least such positive real number k, if it exists, is the **period** of f.

You already have a common-sense grasp of the concept of the period of a function. For example, if you were asked on a Monday "What day of the week will it be in 15 days?" your response would be "Tuesday" due to your understanding that the days of the week repeat every 7 days and 15 is one day more

x	$y = \sin x$
0	0
$\dfrac{\pi}{4}$	$\dfrac{\sqrt{2}}{2} \approx 0.7$
$\dfrac{\pi}{2}$	1
$\dfrac{3\pi}{4}$	$\dfrac{\sqrt{2}}{2} \approx 0.7$
π	0
$\dfrac{5\pi}{4}$	$-\dfrac{\sqrt{2}}{2} \approx -0.7$
$\dfrac{3\pi}{2}$	-1
$\dfrac{7\pi}{4}$	$-\dfrac{\sqrt{2}}{2} \approx -0.7$
2π	0

than two complete periods of 7 days. From the discussion preceding the previous theorem, we see that the period of the sine and cosine functions is 2π.

We may now readily obtain the graphs of the sine and cosine functions. Since we wish to sketch these graphs on an xy-plane, let us replace the variable t by x and consider the equations

$$y = \sin x \quad \text{and} \quad y = \cos x.$$

We may think of x as the radian measure of any angle; however, in calculus, x is usually regarded as a real number. These are equivalent points of view, since the sine (or cosine) of an angle of x radians is the same as the sine (or cosine) of the real number x. The variable y denotes the function value that corresponds to x.

The table in the margin lists coordinates of several points on the graph of $y = \sin x$ for $0 \le x \le 2\pi$. Additional points can be determined using results on special angles, such as

$$\sin (\pi/6) = 1/2 \quad \text{and} \quad \sin (\pi/3) = \sqrt{3}/2 \approx 0.8660.$$

To sketch the graph for $0 \le x \le 2\pi$, we plot the points given by the table and remember that $\sin x$ increases on $[0, \pi/2]$, decreases on $[\pi/2, \pi]$ and $[\pi, 3\pi/2]$, and increases on $[3\pi/2, 2\pi]$. This gives us the sketch in Figure 8. Since the sine function is periodic, the pattern shown in Figure 8 is repeated to the right and to the left, in intervals of length 2π. This gives us the sketch in Figure 9.

Figure 8

Figure 9

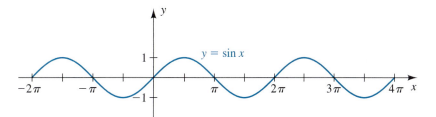

x	$y = \cos x$
0	1
$\dfrac{\pi}{4}$	$\dfrac{\sqrt{2}}{2} \approx 0.7$
$\dfrac{\pi}{2}$	0
$\dfrac{3\pi}{4}$	$-\dfrac{\sqrt{2}}{2} \approx -0.7$
π	-1
$\dfrac{5\pi}{4}$	$-\dfrac{\sqrt{2}}{2} \approx -0.7$
$\dfrac{3\pi}{2}$	0
$\dfrac{7\pi}{4}$	$\dfrac{\sqrt{2}}{2} \approx 0.7$
2π	1

We can use the same procedure to sketch the graph of $y = \cos x$. The table in the margin lists coordinates of several points on the graph for $0 \le x \le 2\pi$. Plotting these points leads to the part of the graph shown in Figure 10. Repeating this pattern to the right and to the left, in intervals of length 2π, we obtain the sketch in Figure 11.

Figure 10

Figure 11

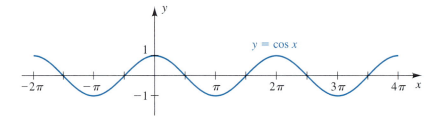

The part of the graph of the sine or cosine function corresponding to $0 \le x \le 2\pi$ is one **cycle.** We sometimes refer to a cycle as a **sine wave** or a **cosine wave.**

The range of the sine and cosine functions consists of all real numbers in the closed interval $[-1, 1]$. Since $\csc x = 1/\sin x$ and $\sec x = 1/\cos x$, it follows that the range of the cosecant and secant functions consists of all real numbers having absolute value greater than or equal to 1.

As we shall see, the range of the tangent and cotangent functions consists of all real numbers.

Before discussing graphs of the other trigonometric functions, let us establish formulas that involve functions of $-t$ for any t. Since a minus sign is involved, we call them *formulas for negatives.*

Formulas for Negatives	
	$\sin(-t) = -\sin t \qquad \cos(-t) = \cos t \qquad \tan(-t) = -\tan t$
	$\csc(-t) = -\csc t \qquad \sec(-t) = \sec t \qquad \cot(-t) = -\cot t$

Figure 12

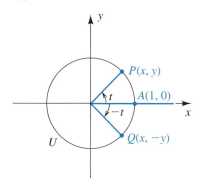

PROOFS   Consider the unit circle U in Figure 12. As t increases from 0 to 2π, the point $P(x, y)$ traces the unit circle U once in the counterclockwise direction and the point $Q(x, -y)$, corresponding to $-t$, traces U once in the clockwise direction. Applying the definition of the trigonometric functions of any angle (with $r = 1$), we have

$$\sin(-t) = -y = -\sin t$$

$$\cos(-t) = x = \cos t$$

$$\tan(-t) = \frac{-y}{x} = -\frac{y}{x} = -\tan t.$$

The proofs of the remaining three formulas are similar.

In the following illustration, formulas for negatives are used to find an exact value for each trigonometric function.

ILLUSTRATION   **Use of Formulas for Negatives**

■ $\sin(-45°) = -\sin 45° = -\dfrac{\sqrt{2}}{2}$

■ $\cos(-30°) = \cos 30° = \dfrac{\sqrt{3}}{2}$

■ $\tan\left(-\dfrac{\pi}{3}\right) = -\tan\left(\dfrac{\pi}{3}\right) = -\sqrt{3}$

■ $\csc(-30°) = -\csc 30° = -2$

■ $\sec(-60°) = \sec 60° = 2$

■ $\cot\left(-\dfrac{\pi}{4}\right) = -\cot\left(\dfrac{\pi}{4}\right) = -1$

We shall next use formulas for negatives to verify a trigonometric identity.

EXAMPLE 4   **Using formulas for negatives to verify an identity**

Verify the following identity by transforming the left-hand side into the right-hand side:

$$\sin(-x)\tan(-x) + \cos(-x) = \sec x$$

SOLUTION   We may proceed as follows:

$$\sin(-x)\tan(-x) + \cos(-x) = (-\sin x)(-\tan x) + \cos x \quad \text{formulas for negatives}$$

$$= \sin x \frac{\sin x}{\cos x} + \cos x \quad \text{tangent identity}$$

$$= \frac{\sin^2 x}{\cos x} + \cos x \quad \text{multiply}$$

$$= \frac{\sin^2 x + \cos^2 x}{\cos x} \quad \text{add terms}$$

$$= \frac{1}{\cos x} \quad \text{Pythagorean identity}$$

$$= \sec x \quad \text{reciprocal identity}$$

We may use the formulas for negatives to prove the following theorem.

Theorem on Even and Odd Trigonometric Functions	**(1)** The cosine and secant functions are even.
	(2) The sine, tangent, cotangent, and cosecant functions are odd.

PROOFS   We shall prove the theorem for the cosine and sine functions. If $f(x) = \cos x$, then

$$f(-x) = \cos(-x) = \cos x = f(x),$$

which means that the cosine function is even.

If $f(x) = \sin x$, then

$$f(-x) = \sin(-x) = -\sin x = -f(x).$$

Thus, the sine function is odd.

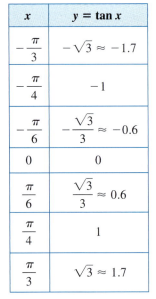

x	$y = \tan x$
$-\dfrac{\pi}{3}$	$-\sqrt{3} \approx -1.7$
$-\dfrac{\pi}{4}$	-1
$-\dfrac{\pi}{6}$	$-\dfrac{\sqrt{3}}{3} \approx -0.6$
0	0
$\dfrac{\pi}{6}$	$\dfrac{\sqrt{3}}{3} \approx 0.6$
$\dfrac{\pi}{4}$	1
$\dfrac{\pi}{3}$	$\sqrt{3} \approx 1.7$

Since the sine function is odd, its graph is symmetric with respect to the origin (see Figure 9). Since the cosine function is even, its graph is symmetric with respect to the y-axis (see Figure 11).

By the preceding theorem, the tangent function is odd, and hence the graph of $y = \tan x$ is symmetric with respect to the origin. The table in the margin lists some points on the graph if $-\pi/2 < x < \pi/2$. The corresponding points are plotted in Figure 13 on the next page. The values of $\tan x$ near $x = \pi/2$ require special attention. If we consider $\tan x = \sin x/\cos x$, then as x increases toward $\pi/2$, the numerator $\sin x$ approaches 1 and the

Figure 13

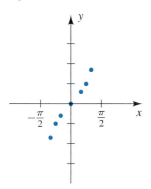

denominator cos x approaches 0. Consequently, tan x takes on large positive values. Following are some approximations of tan x for x close to $\pi/2 \approx 1.5708$:

$$\tan 1.57000 \approx \quad 1{,}255.8$$
$$\tan 1.57030 \approx \quad 2{,}014.8$$
$$\tan 1.57060 \approx \quad 5{,}093.5$$
$$\tan 1.57070 \approx \quad 10{,}381.3$$
$$\tan 1.57079 \approx 158{,}057.9$$

Notice how rapidly tan x increases as x approaches $\pi/2$. We say that tan x *increases without bound as x approaches $\pi/2$ through values less than $\pi/2$.* Similarly, if x approaches $-\pi/2$ through values *greater* than $-\pi/2$, then tan x *decreases without bound.* We may denote this variation using the notation introduced for rational functions in Section 4.5:

$$\text{as} \qquad x \to \frac{\pi^-}{2}, \quad \tan x \to \infty$$

$$\text{as} \quad x \to -\frac{\pi^+}{2}, \quad \tan x \to -\infty$$

This variation of tan x in the open interval $(-\pi/2, \pi/2)$ is illustrated in Figure 14. This portion of the graph is called one **branch** of the tangent. The lines $x = -\pi/2$ and $x = \pi/2$ are vertical asymptotes for the graph. The same pattern is repeated in the open intervals $(-3\pi/2, -\pi/2)$, $(\pi/2, 3\pi/2)$, and $(3\pi/2, 5\pi/2)$ and in similar intervals of length π, as shown in the figure. Thus, *the tangent function is periodic with period π.*

Figure 14 $y = \tan x$

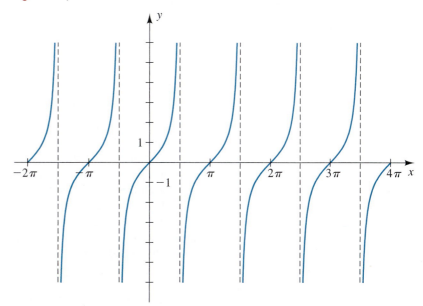

We may use the graphs of $y = \sin x$, $y = \cos x$, and $y = \tan x$ to help sketch the graphs of the remaining three trigonometric functions. For example, since $\csc x = 1/\sin x$, we may find the y-coordinate of a point on the graph of the cosecant function by taking the reciprocal of the corresponding y-coordinate on the sine graph for every value of x except $x = \pi n$ for any integer n. (If $x = \pi n$, $\sin x = 0$, and hence $1/\sin x$ is undefined.) As an aid to sketching the graph of the cosecant function, it is convenient to sketch the graph of the sine function (shown in red in Figure 15) and then take reciprocals to obtain points on the cosecant graph.

Figure 15 $y = \csc x$, $y = \sin x$

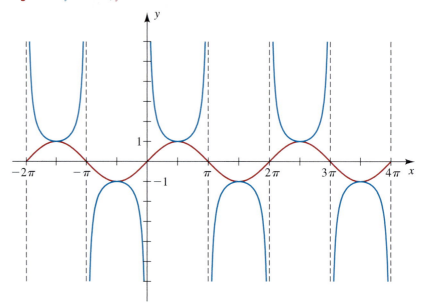

Notice the manner in which the cosecant function increases or decreases without bound as x approaches πn for any integer n. The graph has vertical asymptotes $x = \pi n$, as indicated in the figure. There is one **upper branch** of the cosecant on the interval $(0, \pi)$ and one **lower branch** on the interval $(\pi, 2\pi)$—together they compose one *cycle* of the cosecant.

Since $\sec x = 1/\cos x$ and $\cot x = 1/\tan x$, we may obtain the graphs of the secant and cotangent functions by taking reciprocals of y-coordinates of points on the graphs of the cosine and tangent functions, as illustrated in Figures 16 and 17 on the next page.

A graphical summary of the six trigonometric functions and their inverses (discussed in Section 7.6) appears in Appendix III.

Figure 16 $y = \sec x,\ y = \cos x$

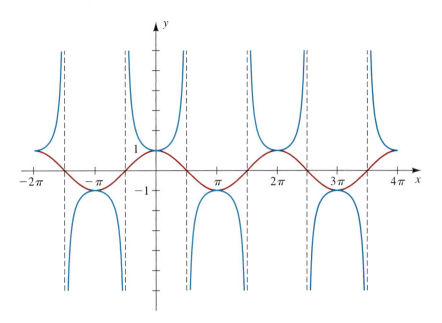

Figure 17 $y = \cot x,\ y = \tan x$

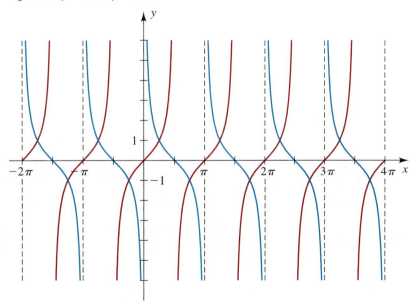

We have considered many properties of the six trigonometric functions of x, where x is a real number or the radian measure of an angle. The following chart contains a summary of important features of these functions (n denotes an arbitrary integer).

Summary of Features of the Trigonometric Functions and Their Graphs

Feature	$y = \sin x$	$y = \cos x$	$y = \tan x$	$y = \cot x$	$y = \sec x$	$y = \csc x$
Domain	$\mathbb{R}$	$\mathbb{R}$	$x \neq \dfrac{\pi}{2} + \pi n$	$x \neq \pi n$	$x \neq \dfrac{\pi}{2} + \pi n$	$x \neq \pi n$
Vertical asymptotes	none	none	$x = \dfrac{\pi}{2} + \pi n$	$x = \pi n$	$x = \dfrac{\pi}{2} + \pi n$	$x = \pi n$
Range	$[-1, 1]$	$[-1, 1]$	$\mathbb{R}$	$\mathbb{R}$	$(-\infty, -1] \cup [1, \infty)$	$(-\infty, -1] \cup [1, \infty)$
x-intercepts	πn	$\dfrac{\pi}{2} + \pi n$	πn	$\dfrac{\pi}{2} + \pi n$	none	none
y-intercept	0	1	0	none	1	none
Period	2π	2π	π	π	2π	2π
Even or odd	odd	even	odd	odd	even	odd
Symmetry	origin	y-axis	origin	origin	y-axis	origin

Figure 18
$y = \csc x, \ y = \sin x$

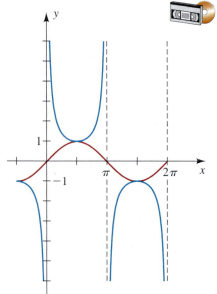

EXAMPLE 5   **Investigating the variation of csc x**

Investigate the variation of $\csc x$ as

$$x \to \pi^{-}, \quad x \to \pi^{+}, \quad x \to \frac{\pi}{2}^{-}, \quad \text{and} \quad x \to \frac{\pi}{6}^{+}.$$

SOLUTION   Referring to the graph of $y = \csc x$ in Figure 18 and using our knowledge of the special values of the sine and cosecant functions, we obtain the following:

as $x \to \pi^{-}$,	$\sin x \to 0$ (through positive values)	and $\csc x \to \infty$
as $x \to \pi^{+}$,	$\sin x \to 0$ (through negative values)	and $\csc x \to -\infty$
as $x \to \dfrac{\pi}{2}^{-}$,	$\sin x \to 1$	and $\csc x \to 1$
as $x \to \dfrac{\pi}{6}^{+}$,	$\sin x \to \dfrac{1}{2}$	and $\csc x \to 2$

EXAMPLE 6 Solving equations and inequalities that involve a trigonometric function

Find all values of x in the interval $[-2\pi, 2\pi]$ such that

(a) $\cos x = \frac{1}{2}$ **(b)** $\cos x > \frac{1}{2}$ **(c)** $\cos x < \frac{1}{2}$

Figure 19

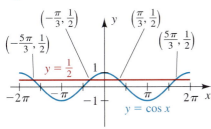

SOLUTION This problem can be easily solved by referring to the graphs of $y = \cos x$ and $y = \frac{1}{2}$, sketched on the same xy-plane in Figure 19 for $-2\pi \le x \le 2\pi$.

(a) The values of x such that $\cos x = \frac{1}{2}$ are the x-coordinates of the points at which the graphs intersect. Recall that $x = \pi/3$ satisfies the equation. By symmetry, $x = -\pi/3$ is another solution of $\cos x = \frac{1}{2}$. Since the cosine function has period 2π, the other values of x in $[-2\pi, 2\pi]$ such that $\cos x = \frac{1}{2}$ are

$$-\frac{\pi}{3} + 2\pi = \frac{5\pi}{3} \quad \text{and} \quad \frac{\pi}{3} - 2\pi = -\frac{5\pi}{3}.$$

(b) The values of x such that $\cos x > \frac{1}{2}$ can be found by determining where the graph of $y = \cos x$ in Figure 19 lies *above* the line $y = \frac{1}{2}$. This gives us the x-intervals

$$\left[-2\pi, -\frac{5\pi}{3}\right), \quad \left(-\frac{\pi}{3}, \frac{\pi}{3}\right), \quad \text{and} \quad \left(\frac{5\pi}{3}, 2\pi\right].$$

(c) To solve $\cos x < \frac{1}{2}$, we again refer to Figure 19 and note where the graph of $y = \cos x$ lies *below* the line $y = \frac{1}{2}$. This gives us the x-intervals

$$\left(-\frac{5\pi}{3}, -\frac{\pi}{3}\right) \quad \text{and} \quad \left(\frac{\pi}{3}, \frac{5\pi}{3}\right).$$

Another method of solving $\cos x < \frac{1}{2}$ is to note that the solutions are the open subintervals of $[-2\pi, 2\pi]$ that are *not* included in the intervals obtained in part (b). ◢

We have now discussed two different approaches to the trigonometric functions. The development in terms of angles and ratios, introduced in Section 6.2, has many applications in the sciences and engineering. The definition in terms of a unit circle, considered in this section, emphasizes the fact that the trigonometric functions have domains consisting of real numbers. Such functions are the building blocks for calculus. In addition, the unit circle approach is useful for discussing graphs and deriving trigonometric identities. You should work to become proficient in the use of both formulations of the trigonometric functions, since each will reinforce the other and thus facilitate your mastery of more advanced aspects of trigonometry.

6.3 *Exercises*

Exer. 1–4: A point $P(x, y)$ is shown on the unit circle U corresponding to a real number t. Find the values of the trigonometric functions at t.

1

2

3

4

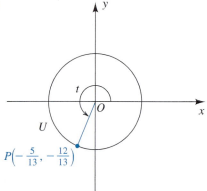

Exer. 5–8: Let $P(t)$ be the point on the unit circle U that corresponds to t. If $P(t)$ has the given rectangular coordinates, find

(a) $P(t + \pi)$ (b) $P(t - \pi)$ (c) $P(-t)$ (d) $P(-t - \pi)$

5 $\left(\frac{3}{5}, \frac{4}{5}\right)$ **6** $\left(-\frac{8}{17}, \frac{15}{17}\right)$

7 $\left(-\frac{12}{13}, -\frac{5}{13}\right)$ **8** $\left(\frac{7}{25}, -\frac{24}{25}\right)$

Exer. 9–16: Let P be the point on the unit circle U that corresponds to t. Find the coordinates of P and the exact values of the trigonometric functions of t, whenever possible.

9 (a) 2π (b) -3π

10 (a) $-\pi$ (b) 6π

11 (a) $3\pi/2$ (b) $-7\pi/2$

12 (a) $5\pi/2$ (b) $-\pi/2$

13 (a) $9\pi/4$ (b) $-5\pi/4$

14 (a) $3\pi/4$ (b) $-7\pi/4$

15 (a) $5\pi/4$ (b) $-\pi/4$

16 (a) $7\pi/4$ (b) $-3\pi/4$

Exer. 17–20: Use a formula for negatives to find the exact value.

17 (a) $\sin(-90°)$ (b) $\cos\left(-\frac{3\pi}{4}\right)$ (c) $\tan(-45°)$

18 (a) $\sin\left(-\frac{3\pi}{2}\right)$ (b) $\cos(-225°)$ (c) $\tan(-\pi)$

19 (a) $\cot\left(-\frac{3\pi}{4}\right)$ (b) $\sec(-180°)$ (c) $\csc\left(-\frac{3\pi}{2}\right)$

20 (a) $\cot(-225°)$ (b) $\sec\left(-\dfrac{\pi}{4}\right)$ (c) $\csc(-45°)$

Exer. 21–26: Verify the identity by transforming the left-hand side into the right-hand side.

21 $\sin(-x)\sec(-x) = -\tan x$

22 $\csc(-x)\cos(-x) = -\cot x$

23 $\dfrac{\cot(-x)}{\csc(-x)} = \cos x$ 24 $\dfrac{\sec(-x)}{\tan(-x)} = -\csc x$

25 $\dfrac{1}{\cos(-x)} - \tan(-x)\sin(-x) = \cos x$

26 $\cot(-x)\cos(-x) + \sin(-x) = -\csc x$

Exer. 27–38: Complete the statement by referring to a graph of a trigonometric function.

27 (a) As $x \to 0^+$, $\sin x \to$ ____

(b) As $x \to (-\pi/2)^-$, $\sin x \to$ ____

28 (a) As $x \to \pi^+$, $\sin x \to$ ____

(b) As $x \to (\pi/6)^-$, $\sin x \to$ ____

29 (a) As $x \to (\pi/4)^+$, $\cos x \to$ ____

(b) As $x \to \pi^-$, $\cos x \to$ ____

30 (a) As $x \to 0^+$, $\cos x \to$ ____

(b) As $x \to (-\pi/3)^-$, $\cos x \to$ ____

31 (a) As $x \to (\pi/4)^+$, $\tan x \to$ ____

(b) As $x \to (\pi/2)^+$, $\tan x \to$ ____

32 (a) As $x \to 0^+$, $\tan x \to$ ____

(b) As $x \to (-\pi/2)^-$, $\tan x \to$ ____

33 (a) As $x \to (-\pi/4)^-$, $\cot x \to$ ____

(b) As $x \to 0^+$, $\cot x \to$ ____

34 (a) As $x \to (\pi/6)^+$, $\cot x \to$ ____

(b) As $x \to \pi^-$, $\cot x \to$ ____

35 (a) As $x \to (\pi/2)^-$, $\sec x \to$ ____

(b) As $x \to (\pi/4)^+$, $\sec x \to$ ____

36 (a) As $x \to (\pi/2)^+$, $\sec x \to$ ____

(b) As $x \to 0^-$, $\sec x \to$ ____

37 (a) As $x \to 0^-$, $\csc x \to$ ____

(b) As $x \to (\pi/2)^+$, $\csc x \to$ ____

38 (a) As $x \to \pi^+$, $\csc x \to$ ____

(b) As $x \to (\pi/4)^-$, $\csc x \to$ ____

Exer. 39–46: Refer to the graph of $y = \sin x$ or $y = \cos x$ to find the exact values of x in the interval $[0, 4\pi]$ that satisfy the equation.

39 $\sin x = -1$ 40 $\sin x = 1$

41 $\sin x = \frac{1}{2}$ 42 $\sin x = -\sqrt{2}/2$

43 $\cos x = 1$ 44 $\cos x = -1$

45 $\cos x = \sqrt{2}/2$ 46 $\cos x = -\frac{1}{2}$

Exer. 47–50: Refer to the graph of $y = \tan x$ to find the exact values of x in the interval $(-\pi/2, 3\pi/2)$ that satisfy the equation.

47 $\tan x = 1$ 48 $\tan x = \sqrt{3}$

49 $\tan x = 0$ 50 $\tan x = -1/\sqrt{3}$

Exer. 51–54: Refer to the graph of the equation on the specified interval. Find all values of x such that for the real number a, (a) $y = a$, (b) $y > a$, and (c) $y < a$.

51 $y = \sin x$; $[-2\pi, 2\pi]$; $a = \frac{1}{2}$

52 $y = \cos x$; $[0, 4\pi]$; $a = \sqrt{3}/2$

53 $y = \cos x$; $[-2\pi, 2\pi]$; $a = -\frac{1}{2}$

54 $y = \sin x$; $[0, 4\pi]$; $a = -\sqrt{2}/2$

Exer. 55–62: Use the graph of a trigonometric function to sketch the graph of the equation without plotting points.

55 $y = 2 + \sin x$ 56 $y = 3 + \cos x$

57 $y = \cos x - 2$ 58 $y = \sin x - 1$

59 $y = 1 + \tan x$ 60 $y = \cot x - 1$

61 $y = \sec x - 2$ 62 $y = 1 + \csc x$

Exer. 63–66: Find the intervals between -2π and 2π on which the given function is (a) increasing or (b) decreasing.

63 secant   64 cosecant

65 tangent   66 cotangent

67 Practice sketching the graph of the sine function, taking different units of length on the horizontal and vertical axes. Practice sketching graphs of the cosine and tangent functions in the same manner. Continue this practice until you reach the stage at which, if you were awakened from a sound sleep in the middle of the night and asked to sketch one of these graphs, you could do so in less than thirty seconds.

68 Work Exercise 67 for the cosecant, secant, and cotangent functions.

Exer. 69–72: Use the figure to approximate the following to one decimal place.

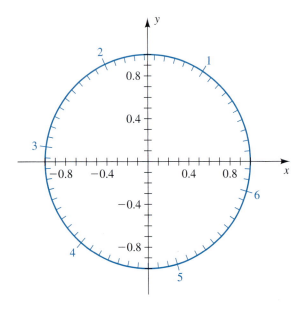

69 (a) sin 4   (b) sin (−1.2)

   (c) All numbers t between 0 and 2π such that sin $t = 0.5$

70 (a) sin 2   (b) sin (−2.3)

   (c) All numbers t between 0 and 2π such that sin $t = -0.2$

71 (a) cos 4   (b) cos (−1.2)

   (c) All numbers t between 0 and 2π such that cos $t = -0.6$

72 (a) cos 2   (b) cos (−2.3)

   (c) All numbers t between 0 and 2π such that cos $t = 0.2$

73 **Temperature-humidity relationship** On March 17, 1981, in Tucson, Arizona, the temperature in degrees Fahrenheit could be described by the equation

$$T(t) = -12 \cos\left(\frac{\pi}{12}t\right) + 60,$$

while the relative humidity in percent could be expressed by

$$H(t) = 20 \cos\left(\frac{\pi}{12}t\right) + 60,$$

where t is in hours and $t = 0$ corresponds to 6 A.M.

(a) Construct a table that lists the temperature and relative humidity every three hours, beginning at midnight.

(b) Determine the times when the maximums and minimums occurred for T and H.

(c) Discuss the relationship between the temperature and relative humidity on this day.

74 **Robotic arm movement** Trigonometric functions are used extensively in the design of industrial robots. Suppose that a robot's shoulder joint is motorized so that the angle θ increases at a constant rate of $\pi/12$ radian per second from an initial angle of $\theta = 0$. Assume that the elbow joint is always kept straight and that the arm has a constant length of 153 centimeters, as shown in the figure.

(a) Assume that $h = 50$ cm when $\theta = 0$. Construct a table that lists the angle θ and the height h of the robotic hand every second while $0 \le \theta \le \pi/2$.

(b) Determine whether or not a constant increase in the angle θ produces a constant increase in the height of the hand.

(c) Find the total distance that the hand moves.

Exercise 74

6.4

Values of the Trigonometric Functions

In previous sections we calculated special values of the trigonometric functions by using the definition of the trigonometric functions in terms of either an angle or a unit circle. In practice we most often use a calculator to approximate function values.

We will next show how the value of any trigonometric function at an angle of θ degrees or at a real number t can be found from its value in the θ-interval $(0°, 90°)$ or the t-interval $(0, \pi/2)$, respectively. This technique is sometimes necessary when a calculator is used to find all angles or real numbers that correspond to a given function value.

We shall make use of the following concept.

Definition of Reference Angle	Let θ be a nonquadrantal angle in standard position. The **reference angle** for θ is the acute angle θ_R that the terminal side of θ makes with the x-axis.

Figure 1 illustrates the reference angle θ_R for a nonquadrantal angle θ, with $0° < \theta < 360°$ or $0 < \theta < 2\pi$, in each of the four quadrants.

Figure 1 Reference angles

(a) Quadrant I

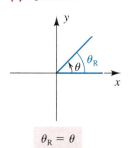

$$\theta_R = \theta$$

(b) Quadrant II

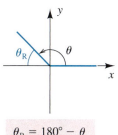

$$\theta_R = 180° - \theta$$
$$= \pi - \theta$$

(c) Quadrant III

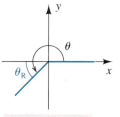

$$\theta_R = \theta - 180°$$
$$= \theta - \pi$$

(d) Quadrant IV

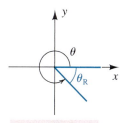

$$\theta_R = 360° - \theta$$
$$= 2\pi - \theta$$

The formulas below the axes in Figure 1 may be used to find the degree or radian measure of θ_R when θ is in degrees or radians, respectively. *For a nonquadrantal angle greater than 360° or less than 0°, first find the coterminal angle θ with $0° < \theta < 360°$ or $0 < \theta < 2\pi$, and then use the formulas in Figure 1.*

EXAMPLE 1 **Finding reference angles**

Find the reference angle θ_R for θ, and sketch θ and θ_R in standard position on the same coordinate plane.

(a) $\theta = 315°$ (b) $\theta = -240°$ (c) $\theta = \dfrac{5\pi}{6}$ (d) $\theta = 4$

Figure 2

(a)

(b)

(c)

(d)

SOLUTION

(a) The angle $\theta = 315°$ is in quadrant IV, and hence, as in Figure 1(d),

$$\theta_R = 360° - 315° = 45°.$$

The angles θ and θ_R are sketched in Figure 2(a).

(b) The angle between $0°$ and $360°$ that is coterminal with $\theta = -240°$ is

$$-240° + 360° = 120°,$$

which is in quadrant II. Using the formula in Figure 1(b) gives

$$\theta_R = 180° - 120° = 60°.$$

The angles θ and θ_R are sketched in Figure 2(b).

(c) Since the angle $\theta = 5\pi/6$ is in quadrant II, we have

$$\theta_R = \pi - \frac{5\pi}{6} = \frac{\pi}{6},$$

as shown in Figure 2(c).

(d) Since $\pi < 4 < 3\pi/2$, the angle $\theta = 4$ is in quadrant III. Using the formula in Figure 1(c), we obtain

$$\theta_R = 4 - \pi.$$

The angles are sketched in Figure 2(d).

We shall next show how reference angles can be used to find values of the trigonometric functions.

If θ is a nonquadrantal angle with reference angle θ_R, then we have $0° < \theta_R < 90°$ or $0 < \theta_R < \pi/2$. Let $P(x, y)$ be a point on the terminal side of θ, and consider the point $Q(x, 0)$ on the x-axis. Figure 3 illustrates a

Figure 3

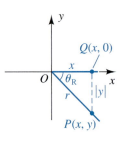

typical situation for θ in each quadrant. In each case, the lengths of the sides of triangle OQP are

$$d(O, Q) = |x|, \quad d(Q, P) = |y|, \quad \text{and} \quad d(O, P) = \sqrt{x^2 + y^2} = r.$$

We may apply the definition of the trigonometric functions of any angle and also use triangle OQP to obtain the following formulas:

$$\left| \sin \theta \right| = \left| \frac{y}{r} \right| = \frac{|y|}{|r|} = \frac{|y|}{r} = \sin \theta_R$$

$$\left| \cos \theta \right| = \left| \frac{x}{r} \right| = \frac{|x|}{|r|} = \frac{|x|}{r} = \cos \theta_R$$

$$\left| \tan \theta \right| = \left| \frac{y}{x} \right| = \frac{|y|}{|x|} = \tan \theta_R$$

These formulas lead to the next theorem. If θ is a quadrantal angle, the definition of the trigonometric functions of any angle should be used to find values.

<table>
<tr><td>Theorem on Reference Angles</td><td>If θ is a nonquadrantal angle in standard position, then to find the value of a trigonometric function at θ, find its value for the reference angle θ_R and prefix the appropriate sign.</td></tr>
</table>

The "appropriate sign" referred to in the theorem can be determined from the table of signs of the trigonometric functions given on page 371.

EXAMPLE 2   Using reference angles

Use reference angles to find the exact values of $\sin \theta$, $\cos \theta$, and $\tan \theta$ if

(a) $\theta = \dfrac{5\pi}{6}$ (b) $\theta = 315°$

SOLUTION

(a) The angle $\theta = 5\pi/6$ and its reference angle $\theta_R = \pi/6$ are sketched in Figure 4. Since θ is in quadrant II, $\sin \theta$ is positive and both $\cos \theta$ and $\tan \theta$ are negative. Hence, by the theorem on reference angles and known results about special angles, we obtain the following values:

Figure 4

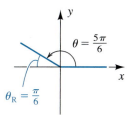

$$\sin \frac{5\pi}{6} = + \sin \frac{\pi}{6} = \frac{1}{2}$$

$$\cos \frac{5\pi}{6} = - \cos \frac{\pi}{6} = -\frac{\sqrt{3}}{2}$$

$$\tan \frac{5\pi}{6} = - \tan \frac{\pi}{6} = -\frac{\sqrt{3}}{3}$$

(continued)

Figure 5

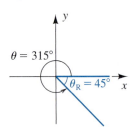

(b) The angle $\theta = 315°$ and its reference angle $\theta_R = 45°$ are sketched in Figure 5. Since θ is in quadrant IV, $\sin \theta < 0$, $\cos \theta > 0$, and $\tan \theta < 0$. Hence, by the theorem on reference angles, we obtain

$$\sin 315° = - \sin 45° = -\frac{\sqrt{2}}{2}$$

$$\cos 315° = + \cos 45° = \frac{\sqrt{2}}{2}$$

$$\tan 315° = - \tan 45° = -1.$$

If we use a calculator to approximate function values, reference angles are usually unnecessary. As an illustration, to find $\sin 210°$, we place the calculator in degree mode and obtain $\sin 210° = -0.5$, which is the exact value. Using the same procedure for $240°$, we obtain a decimal representation:

$$\sin 240° \approx -0.8660$$

A calculator should not be used to find the *exact* value of $\sin 240°$. In this case, we find the reference angle $60°$ of $240°$ and use the theorem on reference angles, together with known results about special angles, to obtain

$$\sin 240° = -\sin 60° = -\frac{\sqrt{3}}{2}.$$

Let us next consider the problem of solving an equation of the following type:

Problem: If θ is an acute angle and $\sin \theta = 0.6635$, approximate θ.

Most calculators have a key labeled $\boxed{\text{SIN}^{-1}}$ that can be used to help solve the equation. With some calculators, it may be necessary to use another key or a keystroke sequence such as $\boxed{\text{INV}}\boxed{\text{SIN}}$ (refer to the user manual for your calculator). We shall use the following notation when finding θ, where $0 \le k \le 1$:

$$\text{if} \quad \sin \theta = k, \quad \text{then} \quad \theta = \sin^{-1} k$$

This notation is similar to that used for the inverse function f^{-1} of a function f in Section 5.1, where we saw that under certain conditions,

$$\text{if} \quad f(x) = y, \quad \text{then} \quad x = f^{-1}(y).$$

For the problem $\sin \theta = 0.6635$, f is the sine function, $x = \theta$, and $y = 0.6635$. The notation $\sin^{-1}$ is based on the *inverse trigonometric functions* discussed in Section 7.6. At this stage of our work, *we shall regard $\sin^{-1}$ simply as an entry made on a calculator using a* $\boxed{\text{SIN}^{-1}}$ *key.* Thus, for the stated problem, we obtain

$$\theta = \sin^{-1}(0.6635) \approx 41.57° \approx 0.7255.$$

As indicated, when finding an angle, we will usually round off degree measure to the nearest $0.01°$ and radian measure to four decimal places.

Similarly, given $\cos \theta = k$ or $\tan \theta = k$, where θ is acute, we write

$$\theta = \cos^{-1} k \qquad \text{or} \qquad \theta = \tan^{-1} k$$

to indicate the use of a $\boxed{\text{COS}^{-1}}$ or $\boxed{\text{TAN}^{-1}}$ key on a calculator.

Given $\csc \theta$, $\sec \theta$, or $\cot \theta$, we use a reciprocal relationship to find θ, as indicated in the following illustration.

ILLUSTRATION   **Finding Acute Angle Solutions of Equations with a Calculator**

Equation	Calculator solution (degree and radian)
◼ $\sin \theta = 0.5$	$\theta = \sin^{-1}(0.5) = 30° \quad \approx 0.5236$
◼ $\cos \theta = 0.5$	$\theta = \cos^{-1}(0.5) = 60° \quad \approx 1.0472$
◼ $\tan \theta = 0.5$	$\theta = \tan^{-1}(0.5) \approx 26.57° \approx 0.4636$
◼ $\csc \theta = 2$	$\theta = \sin^{-1}\left(\frac{1}{2}\right) = 30° \quad \approx 0.5236$
◼ $\sec \theta = 2$	$\theta = \cos^{-1}\left(\frac{1}{2}\right) = 60° \quad \approx 1.0472$
◼ $\cot \theta = 2$	$\theta = \tan^{-1}\left(\frac{1}{2}\right) \approx 26.57° \approx 0.4636$

The same technique may be employed if θ is *any* angle or real number. Thus, using the $\boxed{\text{SIN}^{-1}}$ key, we obtain, in degree or radian mode,

$$\theta = \sin^{-1}(0.6635) \approx 41.57° \approx 0.7255,$$

which is the reference angle for θ. If $\sin \theta$ is *negative,* then a calculator gives us the *negative* of the reference angle. For example,

$$\sin^{-1}(-0.6635) \approx -41.57° \approx -0.7255.$$

Similarly, given $\cos \theta$ or $\tan \theta$, we find θ with a calculator by using $\boxed{\text{COS}^{-1}}$ or $\boxed{\text{TAN}^{-1}}$, respectively. The interval containing θ is listed in the next chart. It is important to note that if $\cos \theta$ is negative, then θ is *not* the negative of the reference angle, but instead is in the interval $\pi/2 < \theta \leq \pi$, or $90° < \theta \leq 180°$. The reasons for using these intervals are explained in Section 7.6. We may use reciprocal relationships to solve similar equations involving $\csc \theta$, $\sec \theta$, and $\cot \theta$.

Equation	Values of k	Calculator solution	Interval containing θ if a calculator is used
$\sin \theta = k$	$-1 \leq k \leq 1$	$\theta = \sin^{-1} k$	$-\dfrac{\pi}{2} \leq \theta \leq \dfrac{\pi}{2}$, or $\quad -90° \leq \theta \leq 90°$
$\cos \theta = k$	$-1 \leq k \leq 1$	$\theta = \cos^{-1} k$	$0 \leq \theta \leq \pi$, or $\quad 0° \leq \theta \leq 180°$
$\tan \theta = k$	any k	$\theta = \tan^{-1} k$	$-\dfrac{\pi}{2} < \theta < \dfrac{\pi}{2}$, or $\quad -90° < \theta < 90°$

The following illustration contains some specific examples for both degree and radian modes.

ILLUSTRATION

Finding Angles with a Calculator

Equation	**Calculator solution (degree and radian)**	
■ $\sin \theta = -0.5$	$\theta = \sin^{-1}(-0.5) = -30°$	≈ -0.5236
■ $\cos \theta = -0.5$	$\theta = \cos^{-1}(-0.5) = 120°$	≈ 2.0944
■ $\tan \theta = -0.5$	$\theta = \tan^{-1}(-0.5) \approx -26.57°$	≈ -0.4636

When using a calculator to find θ, be sure to keep the restrictions on θ in mind. If other values are desired, then reference angles or other methods may be employed, as illustrated in the next examples.

Figure 6

$\theta = 180° - \theta_R$
$\approx 155.2°$

EXAMPLE 3 Approximating an angle with a calculator

If $\tan \theta = -0.4623$ and $0° \leq \theta < 360°$, find θ to the nearest $0.1°$.

SOLUTION As pointed out in the preceding discussion, if we use a calculator (in degree mode) to find θ when $\tan \theta$ is negative, then the degree measure will be in the interval $(-90°, 0°)$. In particular, we obtain the following:

$$\theta = \tan^{-1}(-0.4623) \approx -24.8°$$

Since we wish to find values of θ between $0°$ and $360°$, we use the (approximate) reference angle $\theta_R \approx 24.8°$. There are two possible values of θ such that $\tan \theta$ is negative—one in quadrant II, the other in quadrant IV. If θ is in quadrant II and $0° \leq \theta < 360°$, we have the situation shown in Figure 6, and

$$\theta = 180° - \theta_R \approx 180° - 24.8° = 155.2°.$$

If θ is in quadrant IV and $0° \leq \theta < 360°$, then, as in Figure 7,

$$\theta = 360° - \theta_R \approx 360° - 24.8° = 335.2°.$$

Figure 7

$\theta = 360° - \theta_R$
$\approx 335.2°$

Figure 8

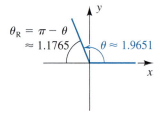

$\theta_R = \pi - \theta$
≈ 1.1765
$\theta \approx 1.9651$

EXAMPLE 4 Approximating an angle with a calculator

If $\cos \theta = -0.3842$ and $0 \leq \theta < 2\pi$, find θ to the nearest 0.0001 radian.

SOLUTION If we use a calculator (in radian mode) to find θ when $\cos \theta$ is negative, then the radian measure will be in the interval $[0, \pi]$. In particular, we obtain the following (shown in Figure 8):

$$\theta = \cos^{-1}(-0.3842) \approx 1.965\,137\,489$$

Since we wish to find values of θ between 0 and 2π, we use the (approximate) reference angle

$$\theta_R = \pi - \theta \approx 1.176\,455\,165.$$

There are two possible values of θ such that $\cos \theta$ is negative—the one we found in quadrant II and the other in quadrant III. If θ is in quadrant III, then

$$\theta = \pi + \theta_R \approx 4.318\,047\,819,$$

as shown in Figure 9.

Figure 9

$\theta = \pi + \theta_R$
≈ 4.3180

6.4   *Exercises*

Exer. 1–6: Find the reference angle θ_R if θ has the given measure.

1 (a) $240°$ (b) $340°$ (c) $-202°$ (d) $-660°$

2 (a) $165°$ (b) $275°$ (c) $-110°$ (d) $400°$

3 (a) $3\pi/4$ (b) $4\pi/3$ (c) $-\pi/6$ (d) $9\pi/4$

4 (a) $7\pi/4$ (b) $2\pi/3$ (c) $-3\pi/4$ (d) $-23\pi/6$

5 (a) 3 (b) -2 (c) 5.5 (d) 100

6 (a) 6 (b) -4 (c) 4.5 (d) 80

Exer. 7–18: Find the exact value.

7 (a) $\sin(2\pi/3)$ (b) $\sin(-5\pi/4)$

8 (a) $\sin 210°$ (b) $\sin(-315°)$

9 (a) $\cos 150°$ (b) $\cos(-60°)$

10 (a) $\cos(5\pi/4)$ (b) $\cos(-11\pi/6)$

11 (a) $\tan(5\pi/6)$ (b) $\tan(-\pi/3)$

12 (a) $\tan 330°$ (b) $\tan(-225°)$

13 (a) $\cot 120°$ (b) $\cot(-150°)$

14 (a) $\cot(3\pi/4)$ (b) $\cot(-2\pi/3)$

15 (a) $\sec(2\pi/3)$ (b) $\sec(-\pi/6)$

16 (a) $\sec 135°$ (b) $\sec(-210°)$

17 (a) $\csc 240°$ (b) $\csc(-330°)$

18 (a) $\csc(3\pi/4)$ (b) $\csc(-2\pi/3)$

Exer. 19–24: Approximate to three decimal places.

19 (a) $\sin 73°20'$ (b) $\cos 0.68$

20 (a) $\cos 38°30'$ (b) $\sin 1.48$

21 (a) $\tan 21°10'$ (b) $\cot 1.13$

22 (a) $\cot 9°10'$ (b) $\tan 0.75$

23 (a) $\sec 67°50'$ (b) $\csc 0.32$

24 (a) $\csc 43°40'$ (b) $\sec 0.26$

Exer. 25–32: Approximate the acute angle θ to the nearest (a) 0.01° and (b) 1′.

25 $\cos \theta = 0.8620$

26 $\sin \theta = 0.6612$

27 $\tan \theta = 3.7$

28 $\cos \theta = 0.8$

29 $\sin \theta = 0.4217$

30 $\tan \theta = 4.91$

31 $\sec \theta = 4.246$

32 $\csc \theta = 11$

Exer. 33–34: Approximate to four decimal places.

33 (a) $\sin 98°10'$ (b) $\cos 623.7°$ (c) $\tan 3$

(d) $\cot 231°40'$ (e) $\sec 1175.1°$ (f) $\csc 0.82$

34 (a) $\sin 496.4°$ (b) $\cos 0.65$ (c) $\tan 105°40'$

(d) $\cot 1030.2°$ (e) $\sec 1.46$ (f) $\csc 320°50'$

Exer. 35–36: Approximate, to the nearest 0.1°, all angles θ in the interval $[0°, 360°)$ that satisfy the equation.

35 (a) $\sin \theta = -0.5640$ (b) $\cos \theta = 0.7490$

(c) $\tan \theta = 2.798$ (d) $\cot \theta = -0.9601$

(e) $\sec \theta = -1.116$ (f) $\csc \theta = 1.485$

36 (a) $\sin \theta = 0.8225$ (b) $\cos \theta = -0.6604$

(c) $\tan \theta = -1.5214$ (d) $\cot \theta = 1.3752$

(e) $\sec \theta = 1.4291$ (f) $\csc \theta = -2.3179$

Exer. 37–38: Approximate, to the nearest 0.01 radian, all angles θ in the interval $[0, 2\pi)$ that satisfy the equation.

37 (a) $\sin \theta = 0.4195$ (b) $\cos \theta = -0.1207$

(c) $\tan \theta = -3.2504$ (d) $\cot \theta = 2.6815$

(e) $\sec \theta = 1.7452$ (f) $\csc \theta = -4.8521$

38 (a) $\sin \theta = -0.0135$ (b) $\cos \theta = 0.9235$

(c) $\tan \theta = 0.42$ (d) $\cot \theta = -2.731$

(e) $\sec \theta = -3.51$ (f) $\csc \theta = 1.258$

39 Thickness of the ozone layer The thickness of the ozone layer can be estimated using the formula

$$\ln I_0 - \ln I = kx \sec \theta,$$

where I_0 is the intensity of a particular wavelength of light from the sun before it reaches the atmosphere, I is the intensity of the same wavelength after passing through a layer of ozone x centimeters thick, k is the absorption constant of ozone for that wavelength, and θ is the acute angle that the sunlight makes with the vertical. Suppose that for a wavelength of 3055×10^{-8} centimeter with $k \approx 1.88$, I_0/I is measured as 1.72 and $\theta = 12°$. Approximate the thickness of the ozone layer to the nearest 0.01 centimeter.

40 Ozone calculations Refer to Exercise 39. If the ozone layer is estimated to be 0.31 centimeter thick and, for a wavelength of 3055×10^{-8} centimeter, I_0/I is measured as 2.05, approximate the angle the sun made with the vertical at the time of the measurement.

41 Solar radiation The amount of sunshine illuminating a wall of a building can greatly affect the energy efficiency of the building. The solar radiation striking a vertical wall that faces east is given by the formula

$$R = R_0 \cos \theta \sin \phi,$$

where R_0 is the maximum solar radiation possible, θ is the angle that the sun makes with the horizontal, and ϕ is the direction of the sun in the sky, with $\phi = 90°$ when the sun is in the east and $\phi = 0°$ when the sun is in the south.

(a) When does the maximum solar radiation R_0 strike the wall?

(b) What percentage of R_0 is striking the wall when θ is equal to $60°$ and the sun is in the southeast?

42 Meteorological calculations In the mid-latitudes it is sometimes possible to estimate the distance between consecutive regions of low pressure. If ϕ is the latitude (in degrees), R is Earth's radius (in kilometers), and v is the horizontal wind velocity (in km/hr), then the distance d (in kilometers) from one low pressure area to the next can be estimated using the formula

$$d = 2\pi \left(\frac{vR}{0.52 \cos \phi} \right)^{1/3}.$$

(a) At a latitude of $48°$, Earth's radius is approximately 6369 kilometers. Approximate d if the wind speed is 45 km/hr.

(b) If v and R are constant, how does d vary as the latitude increases?

43 Robot's arm Points on the terminal sides of angles play an important part in the design of arms for robots. Suppose a robot has a straight arm 18 inches long that can rotate about the origin in a coordinate plane. If the robot's hand is located at (18, 0) and then rotates through an angle of $60°$, what is the new location of the hand?

44 Robot's arm Suppose the robot's arm in Exercise 43 can change its length in addition to rotating about the origin. If the hand is initially at (12, 12), approximately how many degrees should the arm be rotated and how much should its length be changed to move the hand to $(-16, 10)$?

6.5
Trigonometric Graphs

In this section we consider graphs of the equations

$$y = a \sin (bx + c) \quad \text{and} \quad y = a \cos (bx + c)$$

for real numbers a, b, and c. Our goal is to sketch such graphs without plotting many points. To do so we shall use facts about the graphs of the sine and cosine functions discussed in Section 6.3.

Let us begin by considering the special case $c = 0$ and $b = 1$—that is,

$$y = a \sin x \quad \text{and} \quad y = a \cos x.$$

We can find y-coordinates of points on the graphs by multiplying y-coordinates of points on the graphs of $y = \sin x$ and $y = \cos x$ by a. To illustrate, if $y = 2 \sin x$, we multiply the y-coordinate of each point on the graph of

$y = \sin x$ by 2. This gives us Figure 1, where for comparison we also show the graph of $y = \sin x$. The procedure is the same as that for vertically stretching the graph of a function, discussed in Section 3.5.

As another illustration, if $y = \frac{1}{2} \sin x$, we multiply y-coordinates of points on the graph of $y = \sin x$ by $\frac{1}{2}$. This multiplication vertically compresses the graph of $y = \sin x$ by a factor of 2, as illustrated in Figure 2.

Figure 1

Figure 2

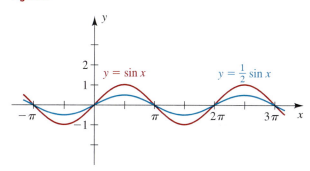

The following example illustrates a graph of $y = a \sin x$ with a negative.

EXAMPLE 1 Sketching the graph of an equation involving sin *x*

Sketch the graph of the equation $y = -2 \sin x$.

SOLUTION The graph of $y = -2 \sin x$ sketched in Figure 3 can be obtained by first sketching the graph of $y = \sin x$ (shown in the figure) and then multiplying y-coordinates by -2. An alternative method is to reflect the graph of $y = 2 \sin x$ (see Figure 1) through the x-axis.

Figure 3

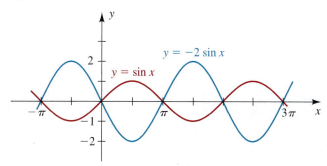

For any $a \neq 0$, the graph of $y = a \sin x$ has the general appearance of one of the graphs illustrated in Figures 1, 2, and 3. The amount of stretching of the graph of $y = \sin x$ and whether the graph is reflected are determined by the absolute value of a and the sign of a, respectively. The largest y-coordinate $|a|$ is the **amplitude of the graph** or, equivalently, the **amplitude of the function** f given by $f(x) = a \sin x$. In Figures 1 and 3 the amplitude is 2. In Figure 2 the amplitude is $\frac{1}{2}$. Similar remarks and techniques apply if $y = a \cos x$.

EXAMPLE 2 Sketching the graph of an equation involving cos x

Find the amplitude and sketch the graph of $y = 3 \cos x$.

SOLUTION By the preceding discussion, the amplitude is 3. As indicated in Figure 4, we first sketch the graph of $y = \cos x$ and then multiply y-coordinates by 3.

Figure 4

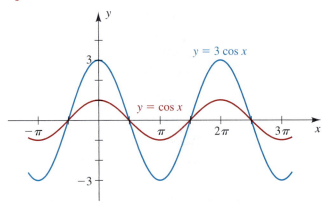

Let us next consider $y = a \sin bx$ and $y = a \cos bx$ for nonzero real numbers a and b. As before, the amplitude is $|a|$. If $b > 0$, then exactly one cycle occurs as bx increases from 0 to 2π or, equivalently, as x increases from 0 to $2\pi/b$. If $b < 0$, then $-b > 0$ and one cycle occurs as x increases from 0 to $2\pi/(-b)$. Thus, the period of the function f given by $f(x) = a \sin bx$ or $f(x) = a \cos bx$ is $2\pi/|b|$. For convenience, we shall also refer to $2\pi/|b|$ as the period of the *graph of f*. The next theorem summarizes our discussion.

Theorem on Amplitudes and Periods	If $y = a \sin bx$ or $y = a \cos bx$ for nonzero real numbers a and b, then the graph has amplitude $|a|$ and period $\dfrac{2\pi}{|b|}$.

We can also relate the role of b to the discussion of horizontally compressing and stretching a graph in Section 3.5. If $|b| > 1$, the graph of $y = \sin bx$ or $y = \cos bx$ can be considered to be compressed horizontally by a factor b. If $0 < |b| < 1$, the graphs are stretched horizontally by a factor $1/b$. This concept is illustrated in the next two examples.

EXAMPLE 3   **Finding an amplitude and a period**

Find the amplitude and the period and sketch the graph of $y = 3 \sin 2x$.

SOLUTION   Using the theorem on amplitudes and periods with $a = 3$ and $b = 2$, we obtain the following:

$$\text{amplitude: } |a| = |3| = 3$$

$$\text{period: } \quad \frac{2\pi}{|b|} = \frac{2\pi}{|2|} = \frac{2\pi}{2} = \pi$$

Figure 5

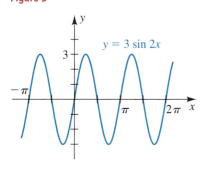

Thus, there is exactly one sine wave of amplitude 3 on the x-interval $[0, \pi]$. Sketching this wave and then extending the graph to the right and left gives us Figure 5. ▰

EXAMPLE 4   **Finding an amplitude and a period**

Find the amplitude and the period and sketch the graph of $y = 2 \sin \frac{1}{2}x$.

SOLUTION   Using the theorem on amplitudes and periods with $a = 2$ and $b = \frac{1}{2}$, we obtain the following:

$$\text{amplitude: } |a| = |2| = 2$$

$$\text{period: } \quad \frac{2\pi}{|b|} = \frac{2\pi}{\left|\frac{1}{2}\right|} = \frac{2\pi}{\frac{1}{2}} = 4\pi$$

Figure 6

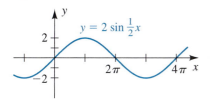

Thus, there is one sine wave of amplitude 2 on the interval $[0, 4\pi]$. Sketching this wave and extending it left and right gives us the graph in Figure 6. ▰

If $y = a \sin bx$ and if b is a large positive number, then the period $2\pi/b$ is small and the sine waves are close together, with b sine waves on the interval $[0, 2\pi]$. For example, in Figure 5, $b = 2$ and we have two sine waves on $[0, 2\pi]$. If b is a small positive number, then the period $2\pi/b$ is large and the waves are far apart. To illustrate, if $y = \sin \frac{1}{10}x$, then one-tenth of a sine wave occurs on $[0, 2\pi]$ and an interval 20π units long is required for one complete cycle. (See also Figure 6—for $y = 2 \sin \frac{1}{2}x$, one-half of a sine wave occurs on $[0, 2\pi]$.)

If $b < 0$, we can use the fact that $\sin (-x) = -\sin x$ to obtain the graph of $y = a \sin bx$. To illustrate, the graph of $y = \sin (-2x)$ is the same as the graph of $y = -\sin 2x$.

EXAMPLE 5 Finding an amplitude and a period

Find the amplitude and the period and sketch the graph of the equation $y = 2 \sin(-3x)$.

Figure 7

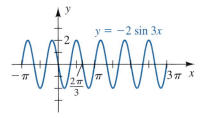

SOLUTION Since the sine function is odd, $\sin(-3x) = -\sin 3x$, and we may write the equation as $y = -2 \sin 3x$. The amplitude is $|-2| = 2$, and the period is $2\pi/3$. Thus, there is one cycle on an interval of length $2\pi/3$. The negative sign indicates a reflection through the x-axis. If we consider the interval $[0, 2\pi/3]$ and sketch a sine wave of amplitude 2 (reflected through the x-axis), the shape of the graph is apparent. The part of the graph in the interval $[0, 2\pi/3]$ is repeated periodically, as illustrated in Figure 7.

EXAMPLE 6 Finding an amplitude and a period

Find the amplitude and the period and sketch the graph of $y = 4 \cos \pi x$.

SOLUTION The amplitude is $|4| = 4$, and the period is $2\pi/\pi = 2$. Thus, there is exactly one cosine wave of amplitude 4 on the interval $[0, 2]$. Since the period does not contain the number π, it makes sense to use integer ticks on the x-axis. Sketching this wave and extending it left and right gives us the graph in Figure 8.

Figure 8

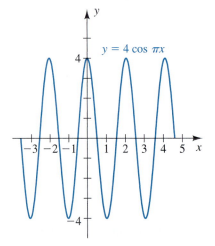

As discussed in Section 3.5, if f is a function and c is a positive real number, then the graph of $y = f(x) + c$ can be obtained by shifting the graph of $y = f(x)$ vertically upward a distance c. For the graph of $y = f(x) - c$, we shift the graph of $y = f(x)$ vertically downward a distance of c. In the next example we use this technique for a trigonometric graph.

Figure 9

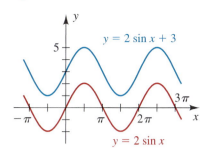

EXAMPLE 7 **Vertically shifting a trigonometric graph**

Sketch the graph of $y = 2 \sin x + 3$.

SOLUTION   It is important to note that $y \neq 2 \sin (x + 3)$. The graph of $y = 2 \sin x$ is sketched in red in Figure 9. If we shift this graph vertically upward a distance 3, we obtain the graph of $y = 2 \sin x + 3$.   ◨

Let us next consider the graph of

$$y = a \sin (bx + c).$$

As before, the amplitude is $|a|$, and the period is $2\pi/|b|$. One cycle occurs if $bx + c$ increases from 0 to 2π. Hence, we can find an interval containing exactly one sine wave by solving the following inequality for x:

$$0 \le bx + c \le 2\pi$$

$$-c \le bx \qquad \le 2\pi - c \qquad \text{subtract } c$$

$$-\frac{c}{b} \le \ x \qquad \le \frac{2\pi}{b} - \frac{c}{b} \qquad \text{divide by } b$$

The number $-c/b$ is the **phase shift** associated with the graph. The graph of $y = a \sin (bx + c)$ may be obtained by shifting the graph of $y = a \sin bx$ to the left if the phase shift is negative or to the right if the phase shift is positive.

Analogous results are true for $y = a \cos (bx + c)$. The next theorem summarizes our discussion.

Theorem on Amplitudes, Periods, and Phase Shifts

If $y = a \sin (bx + c)$ or $y = a \cos (bx + c)$ for nonzero real numbers a and b, then

(1) the amplitude is $|a|$, the period is $\dfrac{2\pi}{|b|}$, and the phase shift is $-\dfrac{c}{b}$;

(2) an interval containing exactly one cycle can be found by solving the inequality

$$0 \le bx + c \le 2\pi.$$

We will sometimes write $y = a \sin (bx + c)$ in the equivalent form $y = a \sin \left[b\left(x + \dfrac{c}{b} \right) \right]$.

EXAMPLE 8 **Finding an amplitude, a period, and a phase shift**

Find the amplitude, the period, and the phase shift and sketch the graph of

$$y = 3 \sin \left(2x + \frac{\pi}{2} \right).$$

SOLUTION The equation is of the form $y = a \sin (bx + c)$ with $a = 3$, $b = 2$, and $c = \pi/2$. Thus, the amplitude is $|a| = 3$, and the period is $2\pi/|b| = 2\pi/2 = \pi$.

By part (2) of the theorem on amplitudes, periods, and phase shifts, the phase shift and an interval containing one sine wave can be found by solving the following inequality:

$$0 \le 2x + \frac{\pi}{2} \le 2\pi$$

$$-\frac{\pi}{2} \le 2x \qquad \le \frac{3\pi}{2} \qquad \text{subtract } \frac{\pi}{2}$$

$$-\frac{\pi}{4} \le x \qquad \le \frac{3\pi}{4} \qquad \text{divide by 2}$$

Thus, the phase shift is $-\pi/4$, and one sine wave of amplitude 3 occurs on the interval $[-\pi/4, 3\pi/4]$. Sketching that wave and then repeating it to the right and left gives us the graph in Figure 10.

Figure 10

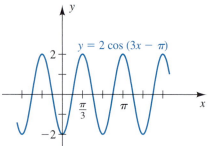

$$y = 3 \sin\left(2x + \frac{\pi}{2}\right)$$

EXAMPLE 9 Finding an amplitude, a period, and a phase shift

Find the amplitude, the period, and the phase shift and sketch the graph of $y = 2 \cos (3x - \pi)$.

SOLUTION The equation has the form $y = a \cos (bx + c)$ with $a = 2$, $b = 3$, and $c = -\pi$. Thus, the amplitude is $|a| = 2$, and the period is $2\pi/|b| = 2\pi/3$.

By part (2) of the theorem on amplitudes, periods, and phase shifts, the phase shift and an interval containing one cycle can be found by solving the following inequality:

$$0 \le 3x - \pi \le 2\pi$$

$$\pi \le 3x \qquad \le 3\pi \qquad \text{add } \pi$$

$$\frac{\pi}{3} \le x \qquad \le \pi \qquad \text{divide by 3}$$

Hence, the phase shift is $\pi/3$, and one cosine-type cycle (from maximum to maximum) of amplitude 2 occurs on the interval $[\pi/3, \pi]$. Sketching that part of the graph and then repeating it to the right and left gives us the sketch in Figure 11.

If we solve the inequality

Figure 11

$$y = 2 \cos (3x - \pi)$$

$$-\frac{\pi}{2} \le 3x - \pi \le \frac{3\pi}{2} \qquad \text{instead of} \qquad 0 \le 3x - \pi \le 2\pi,$$

we obtain the interval $\pi/6 \le x \le 5\pi/6$, which gives us a cycle between x-intercepts rather than a cycle between maximums.

EXAMPLE 10   **Finding an equation for a sine wave**

Express the equation for the sine wave shown in Figure 12 in the form

$$y = a \sin (bx + c)$$

for $a > 0$, $b > 0$, and the least positive real number c.

Figure 12

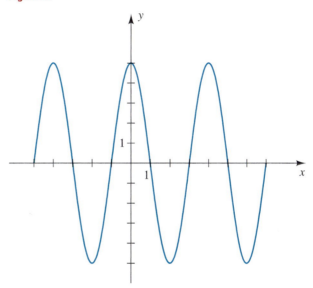

SOLUTION   The largest and smallest y-coordinates of points on the graph are 5 and -5, respectively. Hence, the amplitude is $a = 5$.

Since one sine wave occurs on the interval $[-1, 3]$, the period has value $3 - (-1) = 4$. Hence, by the theorem on amplitudes, periods, and phase shifts (with $b > 0$),

$$\frac{2\pi}{b} = 4 \qquad \text{or, equivalently,} \qquad b = \frac{\pi}{2}.$$

The phase shift is $-c/b = -c/(\pi/2)$. Since c is to be positive, the phase shift must be *negative;* that is, the graph in Figure 12 must be obtained by shifting the graph of $y = 5 \sin [(\pi/2)x]$ to the *left*. Since we want c to be as small as possible, we choose the phase shift -1. Hence,

$$-\frac{c}{\pi/2} = -1 \qquad \text{or, equivalently,} \qquad c = \frac{\pi}{2}.$$

Thus, the desired equation is

$$y = 5 \sin \left(\frac{\pi}{2}x + \frac{\pi}{2} \right).$$

(continued)

There are many other equations for the graph. For example, we could use the phase shifts $-5, -9, -13$, and so on, but these would not give us the *least* positive value for c. Two other equations for the graph are

$$y = 5 \sin \left(\frac{\pi}{2}x - \frac{3\pi}{2} \right) \quad \text{and} \quad y = -5 \sin \left(\frac{\pi}{2}x + \frac{3\pi}{2} \right).$$

However, neither of these equations satisfies the given criteria for a, b, and c, since in the first, $c < 0$, and in the second, $a < 0$ and c does not have its least positive value.

As an alternative solution, we could write

$$y = a \sin (bx + c) \quad \text{as} \quad y = a \sin \left[b \left(x + \frac{c}{b} \right) \right].$$

As before, we find $a = 5$ and $b = \pi/2$. Now since the graph has an x-intercept at $x = -1$, we can consider this graph to be a horizontal shift of the graph of $y = 5 \sin [(\pi/2)x]$ to the left by 1 unit—that is, replace x with $x + 1$. Thus, an equation is

$$y = 5 \sin \left[\frac{\pi}{2}(x + 1) \right], \quad \text{or} \quad y = 5 \sin \left(\frac{\pi}{2}x + \frac{\pi}{2} \right). \qquad \blacksquare$$

Many phenomena that occur in nature vary in a cyclic or rhythmic manner. It is sometimes possible to represent such behavior by means of trigonometric functions, as illustrated in the next two examples.

EXAMPLE 11   Analyzing the process of breathing

The rhythmic process of breathing consists of alternating periods of inhaling and exhaling. One complete cycle normally takes place every 5 seconds. If $F(t)$ denotes the air flow rate at time t (in liters per second) and if the maximum flow rate is 0.6 liter per second, find a formula of the form $F(t) = a \sin bt$ that fits this information.

SOLUTION   If $F(t) = a \sin bt$ for some $b > 0$, then the period of F is $2\pi/b$. In this application the period is 5 seconds, and hence

$$\frac{2\pi}{b} = 5, \quad \text{or} \quad b = \frac{2\pi}{5}.$$

Since the maximum flow rate corresponds to the amplitude a of F, we let $a = 0.6$. This gives us the formula

$$F(t) = 0.6 \sin \left(\frac{2\pi}{5}t \right). \qquad \blacksquare$$

EXAMPLE 12   Approximating the number of hours of daylight in a day

The number of hours of daylight $D(t)$ at a particular time of the year can be approximated by

$$D(t) = \frac{K}{2} \sin\left[\frac{2\pi}{365}(t - 79)\right] + 12$$

for t in days and $t = 0$ corresponding to January 1. The constant K determines the total variation in day length and depends on the latitude of the locale.

(a) For Boston, $K \approx 6$. Sketch the graph of D for $0 \le t \le 365$.

(b) When is the day length the longest? the shortest?

SOLUTION

(a) If $K = 6$, then $K/2 = 3$, and we may write $D(t)$ in the form

$$D(t) = f(t) + 12,$$

where

$$f(t) = 3 \sin\left[\frac{2\pi}{365}(t - 79)\right].$$

We shall sketch the graph of f and then apply a vertical shift through a distance 12.

As in part (2) of the theorem on amplitudes, periods, and phase shifts, we can obtain a t-interval containing exactly one cycle by solving the following inequality:

$$0 \le \frac{2\pi}{365}(t - 79) \le 2\pi$$

$$0 \le \quad t - 79 \le 365 \quad \text{multiply by } \frac{365}{2\pi}$$

$$79 \le \quad t \quad \le 444 \quad \text{add 79}$$

Hence, one sine wave occurs on the interval $[79, 444]$. Dividing this interval into four equal parts, we obtain the following table of values, which indicates the familiar sine wave pattern of amplitude 3.

t	79	170.25	261.5	352.75	444
$f(t)$	0	3	0	-3	0

If $t = 0$,

$$f(0) = 3 \sin\left[\frac{2\pi}{365}(-79)\right] \approx 3 \sin(-1.36) \approx -2.9.$$

Since the period of f is 365, this implies that $f(365) \approx -2.9$.

The graph of f for the interval $[0, 444]$ is sketched in Figure 13, with different scales on the axes and t rounded off to the nearest day.

(continued)

Figure 13

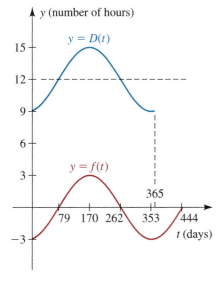

Applying a vertical shift of 12 units gives us the graph of D for $0 \le t \le 365$ shown in Figure 13.

(b) The longest day—that is, the largest value of $D(t)$—occurs 170 days after January 1. Except for leap year, this corresponds to June 20. The shortest day occurs 353 days after January 1, or December 20. ◢

6.5 Exercises

1 Find the amplitude and the period and sketch the graph of the equation:

(a) $y = 4 \sin x$ (b) $y = \sin 4x$

(c) $y = \frac{1}{4} \sin x$ (d) $y = \sin \frac{1}{4}x$

(e) $y = 2 \sin \frac{1}{4}x$ (f) $y = \frac{1}{2} \sin 4x$

(g) $y = -4 \sin x$ (h) $y = \sin(-4x)$

2 For equations analogous to those in (a)–(h) of Exercise 1 but involving the cosine, find the amplitude and the period and sketch the graph.

3 Find the amplitude and the period and sketch the graph of the equation:

(a) $y = 3 \cos x$ (b) $y = \cos 3x$

(c) $y = \frac{1}{3} \cos x$ (d) $y = \cos \frac{1}{3}x$

(e) $y = 2 \cos \frac{1}{3}x$ (f) $y = \frac{1}{2} \cos 3x$

(g) $y = -3 \cos x$ (h) $y = \cos(-3x)$

4 For equations analogous to those in (a)–(h) of Exercise 3 but involving the sine, find the amplitude and the period and sketch the graph.

Exer. 5–40: Find the amplitude, the period, and the phase shift and sketch the graph of the equation.

5 $y = \sin\left(x - \frac{\pi}{2}\right)$ 6 $y = \sin\left(x + \frac{\pi}{4}\right)$

7 $y = 3 \sin\left(x + \frac{\pi}{6}\right)$ 8 $y = 2 \sin\left(x - \frac{\pi}{3}\right)$

9 $y = \cos\left(x + \frac{\pi}{2}\right)$ 10 $y = \cos\left(x - \frac{\pi}{3}\right)$

11 $y = 4 \cos\left(x - \frac{\pi}{4}\right)$ 12 $y = 3 \cos\left(x + \frac{\pi}{6}\right)$

13 $y = \sin(2x - \pi) + 1$ 14 $y = -\sin(3x + \pi) - 1$

15 $y = -\cos(3x + \pi) - 2$ 16 $y = \cos(2x - \pi) + 2$

17 $y = -2 \sin(3x - \pi)$ 18 $y = 3 \cos(3x - \pi)$

19 $y = \sin\left(\frac{1}{2}x - \frac{\pi}{3}\right)$ 20 $y = \sin\left(\frac{1}{2}x + \frac{\pi}{4}\right)$

21 $y = 6 \sin \pi x$ 22 $y = 3 \cos \frac{\pi}{2}x$

23 $y = 2 \cos \frac{\pi}{2}x$ 24 $y = 4 \sin 3\pi x$

25 $y = \frac{1}{2} \sin 2\pi x$ 26 $y = \frac{1}{2} \cos \frac{\pi}{2}x$

27 $y = 5 \sin\left(3x - \frac{\pi}{2}\right)$ 28 $y = -4 \cos\left(2x + \frac{\pi}{3}\right)$

29 $y = 3 \cos\left(\frac{1}{2}x - \frac{\pi}{4}\right)$ 30 $y = -2 \sin\left(\frac{1}{2}x + \frac{\pi}{2}\right)$

31 $y = -5 \cos\left(\frac{1}{3}x + \frac{\pi}{6}\right)$ 32 $y = 4 \sin\left(\frac{1}{3}x - \frac{\pi}{3}\right)$

33 $y = 3 \cos(\pi x + 4\pi)$ 34 $y = -2 \sin(2\pi x + \pi)$

35 $y = -\sqrt{2} \sin\left(\frac{\pi}{2}x - \frac{\pi}{4}\right)$

36 $y = \sqrt{3} \cos\left(\frac{\pi}{4}x - \frac{\pi}{2}\right)$

37 $y = -2 \sin(2x - \pi) + 3$ 38 $y = 3 \cos(x + 3\pi) - 2$

39 $y = 5 \cos(2x + 2\pi) + 2$ 40 $y = -4 \sin(3x - \pi) - 3$

Exer. 41–44: The graph of an equation is shown in the figure. (a) Find the amplitude, period, and phase shift. (b) Write the equation in the form $y = a \sin (bx + c)$ for $a > 0$, $b > 0$, and the least positive real number c.

41

42

43

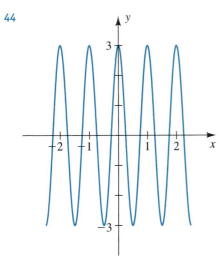

44

44 graph continues below:

45 **Electroencephalography** Shown in the figure is an electroencephalogram of human brain waves during deep sleep. If we use $W = a \sin (bt + c)$ to represent these waves, what is the value of b?

Exercise 45

0   1   2 (sec)

46 **Intensity of daylight** On a certain spring day with 12 hours of daylight, the light intensity I takes on its largest value of 510 calories/cm² at midday. If $t = 0$ corresponds to sunrise, find a formula $I = a \sin bt$ that fits this information.

47 **Heart action** The pumping action of the heart consists of the systolic phase, in which blood rushes from the left ventricle into the aorta, and the diastolic phase, during which the heart muscle relaxes. The function whose graph is shown in the figure is sometimes used to model one complete cycle of this process. For a particular individual, the systolic phase lasts $\frac{1}{4}$ second and has a maximum flow rate of 8 liters per minute. Find a and b.

Exercise 47

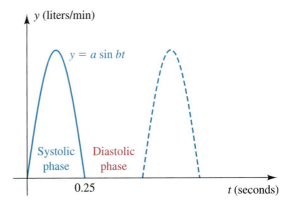

48 **Biorhythms** The popular biorhythm theory uses the graphs of three simple sine functions to make predictions about an individual's physical, emotional, and intellectual potential for a particular day. The graphs are given by $y = a \sin bt$

(continued)

for t in days, with $t = 0$ corresponding to birth and $a = 1$ denoting 100% potential.

(a) Find the value of b for the physical cycle, which has a period of 23 days; for the emotional cycle (period 28 days); and for the intellectual cycle (period 33 days).

(b) Evaluate the biorhythm cycles for a person who has just become 21 years of age and is exactly 7670 days old.

49 Tidal components The height of the tide at a particular point on shore can be predicted by using seven trigonometric functions (called tidal components) of the form

$$f(t) = a \cos (bt + c).$$

The principal lunar component may be approximated by

$$f(t) = a \cos \left(\frac{\pi}{6} t - \frac{11\pi}{12} \right),$$

where t is in hours and $t = 0$ corresponds to midnight. Sketch the graph of f if $a = 0.5$ m.

50 Tidal components Refer to Exercise 49. The principal solar diurnal component may be approximated by

$$f(t) = a \cos \left(\frac{\pi}{12} t - \frac{7\pi}{12} \right).$$

Sketch the graph of f if $a = 0.2$ m.

51 Hours of daylight in Fairbanks If the formula for $D(t)$ in Example 12 is used for Fairbanks, Alaska, then $K \approx 12$. Sketch the graph of D in this case for $0 \le t \le 365$.

52 Low temperature in Fairbanks Based on years of weather data, the expected low temperature T (in °F) in Fairbanks, Alaska, can be approximated by

$$T = 36 \sin \left[\frac{2\pi}{365} (t - 101) \right] + 14,$$

where t is in days and $t = 0$ corresponds to January 1.

(a) Sketch the graph of T for $0 \le t \le 365$.

(b) Predict when the coldest day of the year will occur.

Exer. 53–56: Scientists sometimes use the formula

$$f(t) = a \sin (bt + c) + d$$

to simulate temperature variations during the day, with time t in hours, temperature $f(t)$ in °C, and $t = 0$ corresponding to midnight. Assume that $f(t)$ is decreasing at midnight.

(a) **Determine values of a, b, c, and d that fit the information.**

(b) **Sketch the graph of f for $0 \le t \le 24$.**

53 The high temperature is 10°C, and the low temperature of -10°C occurs at 4 A.M.

54 The temperature at midnight is 15°C, and the high and low temperatures are 20°C and 10°C.

55 The temperature varies between 10°C and 30°C, and the average temperature of 20°C first occurs at 9 A.M.

56 The high temperature of 28°C occurs at 2 P.M., and the average temperature of 20°C occurs 6 hours later.

6.6

Additional Trigonometric Graphs

Methods we developed in Section 6.5 for the sine and cosine can be applied to the other four trigonometric functions; however, there are several differences. Since the tangent, cotangent, secant, and cosecant functions have no largest values, the notion of amplitude has no meaning. Moreover, we do not refer to cycles. For some tangent and cotangent graphs, we begin by sketching the portion between successive vertical asymptotes and then repeat that pattern to the right and to the left.

The graph of $y = a \tan x$ for $a > 0$ can be obtained by stretching or compressing the graph of $y = \tan x$. If $a < 0$, then we also use a reflection about the x-axis. Since the tangent function has period π, it is sufficient to sketch the branch between the two successive vertical asymptotes $x = -\pi/2$ and $x = \pi/2$. The same pattern occurs to the right and to the left, as in the next example.

EXAMPLE 1 **Sketching the graph of an equation involving tan x**

Sketch the graph of the equation:

(a) $y = 2 \tan x$ (b) $y = \frac{1}{2} \tan x$

SOLUTION   We begin by sketching the graph of one branch of $y = \tan x$, as shown in red in Figures 1 and 2, between the vertical asymptotes $x = -\pi/2$ and $x = \pi/2$.

(a) For $y = 2 \tan x$, we multiply the y-coordinate of each point by 2 and then extend the resulting branch to the right and left, as shown in Figure 1.

Figure 1 $y = 2 \tan x$

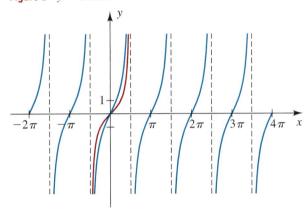

(b) For $y = \frac{1}{2} \tan x$, we multiply the y-coordinates by $\frac{1}{2}$, obtaining the sketch in Figure 2.

Figure 2 $y = \frac{1}{2} \tan x$

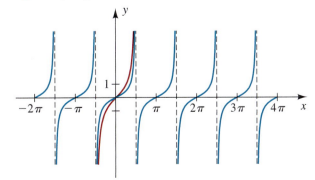

The method used in Example 1 can be applied to other functions. Thus, to sketch the graph of $y = 3 \sec x$, we could first sketch the graph of one branch of $y = \sec x$ and then multiply the y-coordinate of each point by 3.

The next theorem is an analogue of the theorem on amplitudes, periods, and phase shifts stated in Section 6.5 for the sine and cosine functions.

Theorem on the Graph of $y = a \tan (bx + c)$

If $y = a \tan (bx + c)$ for nonzero real numbers a and b, then

(1) the period is $\dfrac{\pi}{|b|}$ and the phase shift is $-\dfrac{c}{b}$;

(2) successive vertical asymptotes for the graph of one branch may be found by solving the inequality

$$-\frac{\pi}{2} < bx + c < \frac{\pi}{2}.$$

EXAMPLE 2 **Sketching the graph of an equation of the form $y = a \tan (bx + c)$**

Find the period and sketch the graph of $y = \dfrac{1}{2} \tan \left(x + \dfrac{\pi}{4} \right)$.

SOLUTION   The equation has the form given in the preceding theorem with $a = \frac{1}{2}$, $b = 1$, and $c = \pi/4$. Hence, by part (1), the period is given by $\pi/|b| = \pi/1 = \pi$.

As in part (2), to find successive vertical asymptotes we solve the following inequality:

$$-\frac{\pi}{2} \le x + \frac{\pi}{4} \le \frac{\pi}{2}$$

$$-\frac{3\pi}{4} \le x \qquad \le \frac{\pi}{4} \qquad \text{subtract } \frac{\pi}{4}$$

Because $a = \frac{1}{2}$, the graph of the equation on the interval $[-3\pi/4, \pi/4]$ has the shape of the graph of $y = \frac{1}{2} \tan x$ (see Figure 2). Sketching that branch and extending it to the right and left gives us Figure 3.

Note that since $c = \pi/4$ and $b = 1$, the phase shift is $-c/b = -\pi/4$. Hence, the graph can also be obtained by shifting the graph of $y = \frac{1}{2} \tan x$ in Figure 2 to the left a distance $\pi/4$.   ◢

If $y = a \cot (bx + c)$, we have a situation similar to that stated in the previous theorem. The only difference is part (2). Since successive vertical asymptotes for the graph of $y = \cot x$ are $x = 0$ and $x = \pi$ (see Figure 17 in Section 6.3), we obtain successive vertical asymptotes for the graph of one branch of $y = a \cot (bx + c)$ by solving the inequality

$$0 < bx + c < \pi.$$

Figure 3
$$y = \frac{1}{2} \tan \left(x + \frac{\pi}{4} \right)$$

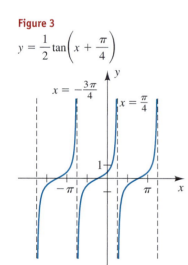

EXAMPLE 3   Sketching the graph of an equation
of the form $y = a \cot (bx + c)$

Find the period and sketch the graph of $y = \cot\left(2x - \dfrac{\pi}{2}\right)$.

Figure 4

$y = \cot\left(2x - \dfrac{\pi}{2}\right)$

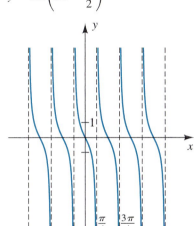

SOLUTION   Using the usual notation, we see that $a = 1$, $b = 2$, and $c = -\pi/2$. The period is $\pi/|b| = \pi/2$. Hence, the graph repeats itself in intervals of length $\pi/2$.

As in the discussion preceding this example, to find two successive vertical asymptotes for the graph of one branch we solve the inequality:

$$0 \leq 2x - \frac{\pi}{2} \leq \pi$$

$$\frac{\pi}{2} \leq 2x \qquad \leq \frac{3\pi}{2} \quad \text{add } \frac{\pi}{2}$$

$$\frac{\pi}{4} \leq \; x \qquad \leq \frac{3\pi}{4} \quad \text{divide by 2}$$

Since a is positive, we sketch a cotangent-shaped branch on the interval $[\pi/4, 3\pi/4]$ and then repeat it to the right and left in intervals of length $\pi/2$, as shown in Figure 4.

Graphs involving the secant and cosecant functions can be obtained by using methods similar to those for the tangent and cotangent or by taking reciprocals of corresponding graphs of the cosine and sine functions.

EXAMPLE 4   Sketching the graph of an equation
of the form $y = a \sec (bx + c)$

Sketch the graph of the equation:

(a) $y = \sec\left(x - \dfrac{\pi}{4}\right)$ **(b)** $y = 2 \sec\left(x - \dfrac{\pi}{4}\right)$

SOLUTION

(a) The graph of $y = \sec x$ is sketched (without asymptotes) in red in Figure 5 on the next page. The graph of $y = \cos x$ is sketched in black; notice that the asymptotes of $y = \sec x$ correspond to the zeros of $y = \cos x$. We can obtain the graph of $y = \sec\left(x - \dfrac{\pi}{4}\right)$ by shifting the graph of $y = \sec x$ to the right a distance $\pi/4$, as shown in blue in Figure 5.

(b) We can sketch this graph by multiplying the y-coordinates of the graph in part (a) by 2. This gives us Figure 6 on the next page.

(continued)

Figure 5 $y = \sec\left(x - \dfrac{\pi}{4}\right)$

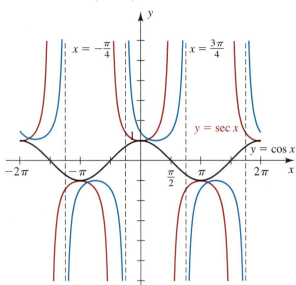

Figure 6 $y = 2\sec\left(x - \dfrac{\pi}{4}\right)$

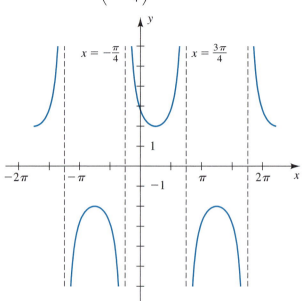

EXAMPLE 5 **Sketching the graph of an equation of the form $y = a \csc(bx + c)$**

Sketch the graph of $y = \csc(2x + \pi)$.

SOLUTION Since $\csc\theta = 1/\sin\theta$, we may write the given equation as

$$y = \frac{1}{\sin(2x + \pi)}.$$

Thus, we may obtain the graph of $y = \csc(2x + \pi)$ by finding the graph of $y = \sin(2x + \pi)$ and then taking the reciprocal of the y-coordinate of each point. Using $a = 1$, $b = 2$, and $c = \pi$, we see that the amplitude of $y = \sin(2x + \pi)$ is 1 and the period is $2\pi/|b| = 2\pi/2 = \pi$. To find an interval containing one cycle, we solve the inequality

$$0 \le 2x + \pi \le 2\pi$$
$$-\pi \le 2x \qquad \le \pi$$
$$-\frac{\pi}{2} \le \quad x \qquad \le \frac{\pi}{2}.$$

Figure 7

$y = \csc(2x + \pi)$

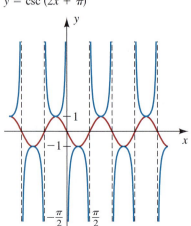

This leads to the graph in red in Figure 7. Taking reciprocals gives us the graph of $y = \csc(2x + \pi)$ shown in blue in the figure. Note that the zeros of the sine curve correspond to the asymptotes of the cosecant graph.

The next example involves the absolute value of a trigonometric function.

EXAMPLE 6 Sketching the graph of an equation involving an absolute value

Sketch the graph of $y = |\cos x| + 1$.

SOLUTION We shall sketch the graph in three stages. First, we sketch the graph of $y = \cos x$, as in Figure 8(a).

Next, we obtain the graph of $y = |\cos x|$ by reflecting the negative y-coordinates in Figure 8(a) through the x-axis. This gives us Figure 8(b).

Finally, we vertically shift the graph in (b) upward 1 unit to obtain Figure 8(c).

Figure 8

(a)

(b)

(c)

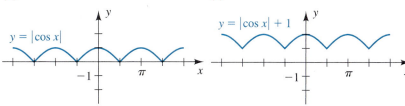

We have used three separate graphs for clarity. In practice, we could sketch the graphs successively on one coordinate plane. ◨

Figure 9

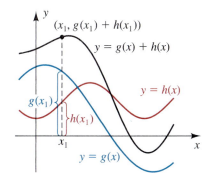

Mathematical applications often involve a function f that is a sum of two or more other functions. To illustrate, suppose

$$f(x) = g(x) + h(x),$$

where f, g, and h have the same domain D. A technique known as **addition of y-coordinates** is sometimes used to sketch the graph of f. The method is illustrated in Figure 9, where for each x_1, the y-coordinate $f(x_1)$ of a point on the graph of f is the *sum* $g(x_1) + h(x_1)$ of the y-coordinates of points on the graphs of g and h. The graph of f is obtained by *graphically adding* a sufficient number of such y-coordinates.

It is sometimes useful to compare the graph of a sum of functions with the individual functions, as illustrated in the next example.

Figure 10

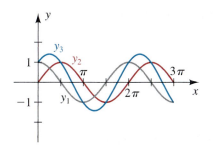

EXAMPLE 7 Sketching the graph of a sum of two trigonometric functions

Sketch the graph of $y_1 = \cos x$, $y_2 = \sin x$, and $y_3 = \cos x + \sin x$ on the same coordinate plane for $0 \le x \le 3\pi$.

SOLUTION Note that the graph of y_3 in Figure 10 intersects the graph of y_1 when $y_2 = 0$, and the graph of y_2 when $y_1 = 0$. The x-intercepts for y_3 correspond to the solutions of $y_2 = -y_1$. Finally, we see that the maximum and minimum values of y_3 occur when $y_1 = y_2$ (that is, when $x = \pi/4, 5\pi/4$, and $9\pi/4$). These y-values are

$$\sqrt{2}/2 + \sqrt{2}/2 = \sqrt{2} \qquad \text{and} \qquad -\sqrt{2}/2 + \left(-\sqrt{2}/2\right) = -\sqrt{2}. \quad ◨$$

The graph of an equation of the form

$$y = f(x) \sin(ax + b) \qquad \text{or} \qquad y = f(x) \cos(ax + b),$$

where f is a function and a and b are real numbers, is called a **damped sine wave** or **damped cosine wave,** respectively, and $f(x)$ is called the **damping factor.** The next example illustrates a method for graphing such equations.

EXAMPLE 8   **Sketching the graph of a damped sine wave**

Sketch the graph of f if $f(x) = 2^{-x} \sin x$.

SOLUTION   We first examine the absolute value of f:

$$
\begin{aligned}
|f(x)| &= |2^{-x} \sin x| &&\text{absolute value of both sides} \\
&= |2^{-x}||\sin x| &&|ab| = |a||b| \\
&\le |2^{-x}| \cdot 1 &&|\sin x| \le 1 \\
|f(x)| &\le 2^{-x} &&|2^{-x}| = 2^{-x} \text{ since } 2^{-x} > 0 \\
-2^{-x} \le f(x) &\le 2^{-x} &&|x| \le a \Longleftrightarrow -a \le x \le a
\end{aligned}
$$

The last inequality implies that the graph of f lies between the graphs of the equations $y = -2^{-x}$ and $y = 2^{-x}$. The graph of f will coincide with one of these graphs if $|\sin x| = 1$—that is, if $x = (\pi/2) + \pi n$ for some integer n.

Since $2^{-x} > 0$, the x-intercepts on the graph of f occur at $\sin x = 0$—that is, at $x = \pi n$. Because there are an infinite number of x-intercepts, this is an example of a function that intersects its horizontal asymptote an infinite number of times. With this information, we obtain the sketch shown in Figure 11.

Figure 11

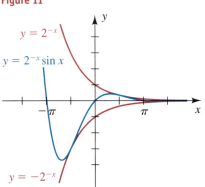

The damping factor in Example 8 is 2^{-x}. By using different damping factors, we can obtain other compressed or expanded variations of sine waves. The analysis of such graphs is important in physics and engineering.

6.6   *Exercises*

Exer. 1–52: Find the period and sketch the graph of the equation. Show the asymptotes.

1 $y = 4 \tan x$

2 $y = \frac{1}{4} \tan x$

3 $y = 3 \cot x$

4 $y = \frac{1}{3} \cot x$

5 $y = 2 \csc x$

6 $y = \frac{1}{2} \csc x$

7 $y = 3 \sec x$

8 $y = \frac{1}{4} \sec x$

9 $y = \tan\left(x - \dfrac{\pi}{4}\right)$

10 $y = \tan\left(x + \dfrac{\pi}{2}\right)$

11 $y = \tan 2x$

12 $y = \tan \frac{1}{2}x$

13 $y = \tan \frac{1}{4}x$

14 $y = \tan 4x$

15 $y = 2 \tan\left(2x + \dfrac{\pi}{2}\right)$

16 $y = \dfrac{1}{3} \tan\left(2x - \dfrac{\pi}{4}\right)$

17 $y = -\dfrac{1}{4} \tan \left(\dfrac{1}{2} x + \dfrac{\pi}{3} \right)$

18 $y = -3 \tan \left(\dfrac{1}{3} x - \dfrac{\pi}{3} \right)$

19 $y = \cot \left(x - \dfrac{\pi}{2} \right)$ 20 $y = \cot \left(x + \dfrac{\pi}{4} \right)$

21 $y = \cot 2x$ 22 $y = \cot \tfrac{1}{2} x$

23 $y = \cot \tfrac{1}{3} x$ 24 $y = \cot 3x$

25 $y = 2 \cot \left(2x + \dfrac{\pi}{2} \right)$ 26 $y = -\tfrac{1}{3} \cot (3x - \pi)$

27 $y = -\dfrac{1}{2} \cot \left(\dfrac{1}{2} x + \dfrac{\pi}{4} \right)$ 28 $y = 4 \cot \left(\dfrac{1}{3} x - \dfrac{\pi}{6} \right)$

29 $y = \sec \left(x - \dfrac{\pi}{2} \right)$ 30 $y = \sec \left(x - \dfrac{3\pi}{4} \right)$

31 $y = \sec 2x$ 32 $y = \sec \tfrac{1}{2} x$

33 $y = \sec \tfrac{1}{3} x$ 34 $y = \sec 3x$

35 $y = 2 \sec \left(2x - \dfrac{\pi}{2} \right)$ 36 $y = \dfrac{1}{2} \sec \left(2x - \dfrac{\pi}{2} \right)$

37 $y = -\dfrac{1}{3} \sec \left(\dfrac{1}{2} x + \dfrac{\pi}{4} \right)$

38 $y = -3 \sec \left(\dfrac{1}{3} x + \dfrac{\pi}{3} \right)$

39 $y = \csc \left(x - \dfrac{\pi}{2} \right)$ 40 $y = \csc \left(x + \dfrac{3\pi}{4} \right)$

41 $y = \csc 2x$ 42 $y = \csc \tfrac{1}{2} x$

43 $y = \csc \tfrac{1}{3} x$ 44 $y = \csc 3x$

45 $y = 2 \csc \left(2x + \dfrac{\pi}{2} \right)$ 46 $y = -\tfrac{1}{2} \csc (2x - \pi)$

47 $y = -\dfrac{1}{4} \csc \left(\dfrac{1}{2} x + \dfrac{\pi}{2} \right)$ 48 $y = 4 \csc \left(\dfrac{1}{2} x - \dfrac{\pi}{4} \right)$

49 $y = \tan \dfrac{\pi}{2} x$ 50 $y = \cot \pi x$

51 $y = \csc 2\pi x$ 52 $y = \sec \dfrac{\pi}{8} x$

53 Find an equation using the cotangent function that has the same graph as $y = \tan x$.

54 Find an equation using the cosecant function that has the same graph as $y = \sec x$.

Exer. 55–60: Use the graph of a trigonometric function to aid in sketching the graph of the equation without plotting points.

55 $y = |\sin x|$ 56 $y = |\cos x|$

57 $y = |\sin x| + 2$ 58 $y = |\cos x| - 3$

59 $y = -|\cos x| + 1$ 60 $y = -|\sin x| - 2$

Exer. 61–66: Sketch the graph of the equation.

61 $y = x + \cos x$ 62 $y = x - \sin x$

63 $y = 2^{-x} \cos x$ 64 $y = e^x \sin x$

65 $y = |x| \sin x$ 66 $y = |x| \cos x$

67 Radio signal intensity Radio stations often have more than one broadcasting tower because federal guidelines do not usually permit a radio station to broadcast its signal in all directions with equal power. Since radio waves can travel over long distances, it is important to control their directional patterns so that radio stations do not interfere with one another. Suppose that a radio station has two broadcasting towers located along a north-south line, as shown in the figure. If the radio station is broadcasting at a wavelength λ and the

distance between the two radio towers is equal to $\frac{1}{2}\lambda$, then the intensity I of the signal in the direction θ is given by

$$I = \tfrac{1}{2}I_0[1 + \cos(\pi \sin \theta)],$$

where I_0 is the maximum intensity. Approximate I in terms of I_0 for each θ.

(a) $\theta = 0$ (b) $\theta = \pi/3$ (c) $\theta = \pi/7$

Exercise 67

68 **Radio signal intensity** Refer to Exercise 67. Determine the directions in which I has maximum or minimum values.

69 **Earth's magnetic field** The strength of Earth's magnetic field varies with the depth below the surface. The strength at depth z and time t can sometimes be approximated using the damped sine wave

$$S = A_0 e^{-\alpha z} \sin(kt - \alpha z),$$

where A_0, α, and k are constants.

(a) What is the damping factor?

(b) Find the phase shift at depth z_0.

(c) At what depth is the amplitude of the wave one-half the amplitude of the surface strength?

6.7

Applied Problems

Trigonometry was developed to help solve problems involving angles and lengths of sides of triangles. Problems of that type are no longer the most important applications; however, questions about triangles still arise in physical situations. When considering such questions in this section, we shall restrict our discussion to right triangles. Triangles that do not contain a right angle will be considered in Chapter 8.

We shall often use the following notation. The vertices of a triangle will be denoted by A, B, and C; the angles at A, B, and C will be denoted by α, β, and γ, respectively; and the lengths of the sides opposite these angles by a, b, and c, respectively. The triangle itself will be referred to as *triangle ABC* (or denoted $\triangle ABC$). If a triangle is a right triangle and if one of the acute angles and a side are known or if two sides are given, then we may find the remaining parts by using the formulas in Section 6.2 that express the trigonometric functions as ratios of sides of a triangle. We can refer to the process of finding the remaining parts as **solving the triangle.**

In all examples *it is assumed that you know how to find trigonometric function values and angles by using either a calculator or results about special angles.*

EXAMPLE 1   Solving a right triangle

Solve $\triangle ABC$, given $\gamma = 90°$, $\alpha = 34°$, and $b = 10.5$.

SOLUTION   Since the sum of the three interior angles in a triangle is $180°$, we have $\alpha + \beta + \gamma = 180°$. Solving for the unknown angle β gives us

$$\beta = 180° - \alpha - \gamma = 180° - 34° - 90° = 56°.$$

Figure 1

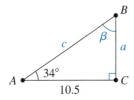

Referring to Figure 1, we obtain

$$\tan 34° = \frac{a}{10.5} \qquad\qquad \tan \alpha = \frac{\text{opp}}{\text{adj}}$$
$$a = (10.5)\tan 34° \approx 7.1. \quad \text{solve for } a; \text{ approximate}$$

To find side c, we can use either the cosine or the secant function, as follows in (1) or (2), respectively:

$$\textbf{(1)} \quad \cos 34° = \frac{10.5}{c} \qquad\qquad \cos \alpha = \frac{\text{adj}}{\text{hyp}}$$
$$c = \frac{10.5}{\cos 34°} \approx 12.7 \qquad \text{solve for } c; \text{ approximate}$$

$$\textbf{(2)} \quad \sec 34° = \frac{c}{10.5} \qquad\qquad \sec \alpha = \frac{\text{hyp}}{\text{adj}}$$
$$c = (10.5)\sec 34° \approx 12.7 \quad \text{solve for } c; \text{ approximate}$$

As illustrated in Example 1, when working with triangles, we usually round off answers. One reason for doing so is that in most applications the lengths of sides of triangles and measures of angles are found by mechanical devices and hence are only approximations to the exact values. Consequently, a number such as 10.5 in Example 1 is assumed to have been rounded off to the nearest tenth. We cannot expect more accuracy in the calculated values for the remaining sides, and therefore they should also be rounded off to the nearest tenth.

In finding angles, answers should be rounded off as indicated in the following table.

Number of significant figures for sides	Round off degree measure of angles to the nearest
2	1°
3	0.1°, or 10′
4	0.01°, or 1′

Justification of this table requires a careful analysis of problems that involve approximate data.

EXAMPLE 2 Solving a right triangle

Solve $\triangle ABC$, given $\gamma = 90°$, $a = 12.3$, and $b = 31.6$.

SOLUTION Referring to the triangle illustrated in Figure 2 gives us

$$\tan \alpha = \frac{12.3}{31.6}.$$

Figure 2

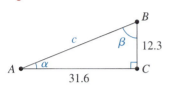

Since the sides are given with three significant figures, the rule stated in the preceding table tells us that α should be rounded off to the nearest 0.1°, or the nearest multiple of 10′. Using the degree mode on a calculator, we have

$$\alpha = \tan^{-1} \frac{12.3}{31.6} \approx 21.3° \qquad \text{or, equivalently,} \qquad \alpha \approx 21°20'.$$

Since α and β are complementary angles,

$$\beta = 90° - \alpha \approx 90° - 21.3° = 68.7°.$$

The only remaining part to find is c. We could use several relationships involving c to determine its value. Among these are

$$\cos \alpha = \frac{31.6}{c}, \quad \sec \beta = \frac{c}{12.3}, \qquad \text{and} \qquad a^2 + b^2 = c^2.$$

Whenever possible, it is best to use a relationship that involves only given information, since it doesn't depend on any previously calculated value. Hence, with $a = 12.3$ and $b = 31.6$, we have

$$c = \sqrt{a^2 + b^2} = \sqrt{(12.3)^2 + (31.6)^2} = \sqrt{1149.85} \approx 33.9.$$

Figure 3

As illustrated in Figure 3, if an observer at point X sights an object, then the angle that the line of sight makes with the horizontal line l is the **angle of elevation** of the object, if the object is above the horizontal line, or the **angle of depression** of the object, if the object is below the horizontal line. We use this terminology in the next two examples.

EXAMPLE 3 Using an angle of elevation

From a point on level ground 135 feet from the base of a tower, the angle of elevation of the top of the tower is 57°20′. Approximate the height of the tower.

SOLUTION   If we let d denote the height of the tower, then the given facts are represented by the triangle in Figure 4. Referring to the figure, we obtain

$$\tan 57°20' = \frac{d}{135} \qquad\qquad \tan 57°20' = \frac{\text{opp}}{\text{adj}}$$

$$d = 135 \tan 57°20' \approx 211. \qquad \text{solve for } d; \text{ approximate}$$

The tower is approximately 211 feet high.

Figure 4

EXAMPLE 4   Using angles of depression

From the top of a building that overlooks an ocean, an observer watches a boat sailing directly toward the building. If the observer is 100 feet above sea level and if the angle of depression of the boat changes from 25° to 40° during the period of observation, approximate the distance that the boat travels.

SOLUTION   As in Figure 5, let A and B be the positions of the boat that correspond to the 25° and 40° angles, respectively. Suppose that the observer is at point D and that C is the point 100 feet directly below. Let d denote the distance the boat travels, and let k denote the distance from B to C. If α and β

Figure 5

(continued)

denote angles DAC and DBC, respectively, then it follows from geometry (alternate interior angles) that $\alpha = 25°$ and $\beta = 40°$.

From triangle BCD:

$$\cot \beta = \cot 40° = \frac{k}{100} \qquad \cot \beta = \frac{\text{adj}}{\text{opp}}$$

$$k = 100 \cot 40° \qquad \text{solve for } k$$

From triangle DAC:

$$\cot \alpha = \cot 25° = \frac{d + k}{100} \qquad \cot \alpha = \frac{\text{adj}}{\text{opp}}$$

$$d + k = 100 \cot 25° \qquad \text{multiply by lcd}$$

$$d = 100 \cot 25° - k \qquad \text{solve for } d$$

$$= 100 \cot 25° - 100 \cot 40° \qquad k = 100 \cot 40°$$

$$= 100(\cot 25° - \cot 40°) \qquad \text{factor out } 100$$

$$\approx 100(2.145 - 1.192) \approx 95 \qquad \text{approximate}$$

Hence, the boat travels approximately 95 feet.

In certain navigation or surveying problems, the **direction,** or **bearing,** from a point P to a point Q is specified by stating the acute angle that segment PQ makes with the north-south line through P. We also state whether Q is north or south and east or west of P. Figure 6 illustrates four possibilities. The bearing from P to Q_1 is 25° east of north and is denoted by N25°E. We also refer to the **direction** N25°E, meaning the direction from P to Q_1. The bearings from P to Q_2, to Q_3, and to Q_4 are represented in a similar manner in the figure. Note that when this notation is used for bearings or directions, N or S always appears to the *left* of the angle and W or E to the *right*.

Figure 6

Figure 7

In air navigation, directions and bearings are specified by measuring from the north in a *clockwise* direction. In this case, a positive measure is assigned to the angle instead of the negative measure to which we are accustomed for clockwise rotations. Referring to Figure 7, we see that the direction of *PQ* is 40° and the direction of *PR* is 300°.

EXAMPLE 5   Using bearings

Two ships leave port at the same time, one ship sailing in the direction N23°E at a speed of 11 mi/hr and the second ship sailing in the direction S67°E at 15 mi/hr. Approximate the bearing from the second ship to the first, one hour later.

SOLUTION   The sketch in Figure 8 indicates the positions of the first and second ships at points *A* and *B,* respectively, after one hour. Point *C* represents the port. We wish to find the bearing from *B* to *A*. Note that

$$\angle ACB = 180° - 23° - 67° = 90°,$$

and hence triangle *ACB* is a right triangle. Thus,

$$\tan \beta = \frac{11}{15} \qquad\qquad \tan \beta = \frac{\text{opp}}{\text{adj}}$$

$$\beta = \tan^{-1} \tfrac{11}{15} \approx 36°. \qquad \text{solve for } \beta; \text{ approximate}$$

We have rounded β to the nearest degree because the sides of the triangles are given with two significant figures.

Referring to Figure 9, we obtain the following:

$$\angle CBD = 90° - \angle BCD = 90° - 67° = 23°$$

$$\angle ABD = \angle ABC + \angle CBD \approx 36° + 23° = 59°$$

$$\theta = 90° - \angle ABD \approx 90° - 59° = 31°$$

Thus, the bearing from *B* to *A* is approximately N31°W.

Trigonometric functions are useful in the investigation of vibratory or oscillatory motion, such as the motion of a particle in a vibrating guitar string or a spring that has been compressed or elongated and then released to oscillate back and forth. The fundamental type of particle displacement in these illustrations is *harmonic motion*.

Figure 8

Figure 9

Definition of Simple Harmonic Motion	A point moving on a coordinate line is in **simple harmonic motion** if its distance *d* from the origin at time *t* is given by either
	$$d = a \cos \omega t \qquad \text{or} \qquad d = a \sin \omega t,$$ where *a* and ω are constants, with $\omega > 0$.

In the preceding definition, the **amplitude** of the motion is the maximum displacement $|a|$ of the point from the origin. The **period** is the time $2\pi/\omega$ required for one complete oscillation. The reciprocal of the period, $\omega/(2\pi)$, is the number of oscillations per unit of time and is called the **frequency.**

A physical interpretation of simple harmonic motion can be obtained by considering a spring with an attached weight that is oscillating vertically relative to a coordinate line, as illustrated in Figure 10. The number d represents the coordinate of a fixed point Q in the weight, and we assume that the amplitude a of the motion is constant. In this case no frictional force is retarding the motion. If friction is present, then the amplitude decreases with time, and the motion is said to be *damped.*

EXAMPLE 6 Describing harmonic motion

Figure 10

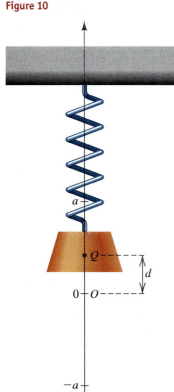

Suppose that the oscillation of the weight shown in Figure 10 is given by

$$d = 10 \cos\left(\frac{\pi}{6}t\right),$$

with t measured in seconds and d in centimeters. Discuss the motion of the weight.

SOLUTION By definition, the motion is simple harmonic with amplitude $a = 10$ cm. Since $\omega = \pi/6$, we obtain the following:

$$\text{period} = \frac{2\pi}{\omega} = \frac{2\pi}{\pi/6} = 12$$

Thus, in 12 seconds the weight makes one complete oscillation. The frequency is $\frac{1}{12}$, which means that one-twelfth of an oscillation takes place each second. The following table indicates the position of Q at various times.

t	0	1	2	3	4	5	6
$\dfrac{\pi}{6}t$	0	$\dfrac{\pi}{6}$	$\dfrac{\pi}{3}$	$\dfrac{\pi}{2}$	$\dfrac{2\pi}{3}$	$\dfrac{5\pi}{6}$	π
$\cos\left(\dfrac{\pi}{6}t\right)$	1	$\dfrac{\sqrt{3}}{2}$	$\dfrac{1}{2}$	0	$-\dfrac{1}{2}$	$-\dfrac{\sqrt{3}}{2}$	-1
d	10	$5\sqrt{3} \approx 8.7$	5	0	-5	$-5\sqrt{3} \approx -8.7$	-10

The initial position of Q is 10 centimeters above the origin O. It moves downward, gaining speed until it reaches O. Note that Q travels approximately $10 - 8.7 = 1.3$ cm during the first second, $8.7 - 5 = 3.7$ cm during the next second, and $5 - 0 = 5$ cm during the third second. It then slows down until it reaches a point 10 centimeters below O at the end of 6 seconds. The direction of motion is then reversed, and the weight moves upward, gaining speed until it reaches O. Once it reaches O, it slows down until it returns to its original position at the end of 12 seconds. The direction of motion is then reversed again, and the same pattern is repeated indefinitely. ◢

6.7 Exercises

Exer. 1–8: Given the indicated parts of triangle *ABC* with $\gamma = 90°$, find the exact values of the remaining parts.

1 $\alpha = 30°$, $b = 20$ 2 $\beta = 45°$, $b = 35$

3 $\beta = 45°$, $c = 30$ 4 $\alpha = 60°$, $c = 6$

5 $a = 5$, $b = 5$ 6 $a = 4\sqrt{3}$, $c = 8$

7 $b = 5\sqrt{3}$, $c = 10\sqrt{3}$ 8 $b = 7\sqrt{2}$, $c = 14$

Exer. 9–16: Given the indicated parts of triangle *ABC* with $\gamma = 90°$, approximate the remaining parts.

9 $\alpha = 37°$, $b = 24$ 10 $\beta = 64°20'$, $a = 20.1$

11 $\beta = 71°51'$, $b = 240.0$ 12 $\alpha = 31°10'$, $a = 510$

13 $a = 25$, $b = 45$ 14 $a = 31$, $b = 9.0$

15 $c = 5.8$, $b = 2.1$ 16 $a = 0.42$, $c = 0.68$

Exer. 17–24: Given the indicated parts of triangle *ABC* with $\gamma = 90°$, express the third part in terms of the first two.

17 α, c; b 18 β, c; b

19 β, b; a 20 α, b; a

21 α, a; c 22 β, a; c

23 a, c; b 24 a, b; c

25 **Height of a kite** A person flying a kite holds the string 4 feet above ground level. The string of the kite is taut and makes an angle of 60° with the horizontal (see the figure). Approximate the height of the kite above level ground if 500 feet of string is payed out.

Exercise 25

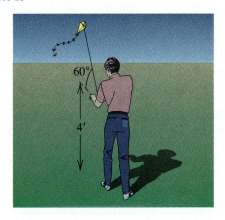

26 **Surveying** From a point 15 meters above level ground, a surveyor measures the angle of depression of an object on the ground at 68°. Approximate the distance from the object to the point on the ground directly beneath the surveyor.

27 **Airplane landing** A pilot, flying at an altitude of 5000 feet, wishes to approach the numbers on a runway at an angle of 10°. Approximate, to the nearest 100 feet, the distance from the airplane to the numbers at the beginning of the descent.

28 **Radio antenna** A guy wire is attached to the top of a radio antenna and to a point on horizontal ground that is 40.0 meters from the base of the antenna. If the wire makes an angle of 58°20' with the ground, approximate the length of the wire.

29 **Surveying** To find the distance d between two points P and Q on opposite shores of a lake, a surveyor locates a point R that is 50.0 meters from P such that RP is perpendicular to PQ, as shown in the figure. Next, using a transit, the surveyor measures angle PRQ as 72°40'. Find d.

Exercise 29

30 **Meteorological calculations** To measure the height h of a cloud cover, a meteorology student directs a spotlight vertically upward from the ground. From a point P on level ground that is d meters from the spotlight, the angle of elevation θ of the light image on the clouds is then measured (see the figure on the next page).

(a) Express h in terms of d and θ.

(b) Approximate h if $d = 1000$ m and $\theta = 59°$.

Exercise 30

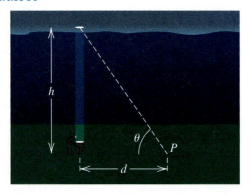

31 Altitude of a rocket A rocket is fired at sea level and climbs at a constant angle of 75° through a distance of 10,000 feet. Approximate its altitude to the nearest foot.

32 Airplane takeoff An airplane takes off at a 10° angle and travels at the rate of 250 ft/sec. Approximately how long does it take the airplane to reach an altitude of 15,000 feet?

33 Designing a drawbridge A drawbridge is 150 feet long when stretched across a river. As shown in the figure, the two sections of the bridge can be rotated upward through an angle of 35°.

(a) If the water level is 15 feet below the closed bridge, find the distance d between the end of a section and the water level when the bridge is fully open.

(b) Approximately how far apart are the ends of the two sections when the bridge is fully opened, as shown in the figure?

Exercise 33

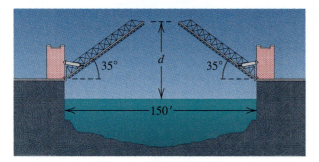

34 Designing a water slide Shown in the figure is part of a design for a water slide. Find the total length of the slide to the nearest foot.

Exercise 34

35 Sun's elevation Approximate the angle of elevation α of the sun if a person 5.0 feet tall casts a shadow 4.0 feet long on level ground (see the figure).

Exercise 35

36 Constructing a ramp A builder wishes to construct a ramp 24 feet long that rises to a height of 5.0 feet above level ground. Approximate the angle that the ramp should make with the horizontal.

37 Video game Shown in the figure is the screen for a simple video arcade game in which ducks move from A to B at the rate of 7 cm/sec. Bullets fired from point O travel 25 cm/sec. If a player shoots as soon as a duck appears at A, at which angle φ should the gun be aimed in order to score a direct hit?

Exercise 37

Venus occurs when Earth is at its minimum distance D_e from the sun and Venus is at its maximum distance D_v from the sun. If $D_e = 91{,}500{,}000$ mi and $D_v = 68{,}000{,}000$ mi, approximate the maximum elongation θ_{max} of Venus. Assume that the orbit of Venus is circular.

Exercise 40

38 Conveyor belt A conveyor belt 9 meters long can be hydraulically rotated up to an angle of 40° to unload cargo from airplanes (see the figure).

(a) Find, to the nearest degree, the angle through which the conveyor belt should be rotated up to reach a door that is 4 meters above the platform supporting the belt.

(b) Approximate the maximum height above the platform that the belt can reach.

Exercise 38

9 m

39 Tallest structure The tallest man-made structure in the world is a television transmitting tower located near Fargo, North Dakota. From a distance of 1 mile on level ground, its angle of elevation is $21°20'24''$. Determine its height to the nearest foot.

40 Elongation of Venus The *elongation* of the planet Venus is defined to be the angle θ determined by the sun, Earth, and Venus, as shown in the figure. Maximum elongation of

41 The Pentagon's ground area The Pentagon is the largest office building in the world in terms of ground area. The perimeter of the building has the shape of a regular pentagon with each side of length 921 feet. Find the area enclosed by the perimeter of the building.

42 A regular octagon is inscribed in a circle of radius 12.0 centimeters. Approximate the perimeter of the octagon.

43 A rectangular box has dimensions $8'' \times 6'' \times 4''$. Approximate, to the nearest tenth of a degree, the angle θ formed by a diagonal of the base and the diagonal of the box, as shown in the figure.

Exercise 43

44 Volume of a conical cup A conical paper cup has a radius of 2 inches. Approximate, to the nearest degree, the angle β (see the figure) so that the cone will have a volume of 20 in³.

Exercise 44

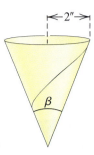

45 Height of a tower From a point P on level ground, the angle of elevation of the top of a tower is $26°50'$. From a point 25.0 meters closer to the tower and on the same line with P and the base of the tower, the angle of elevation of the top is $53°30'$. Approximate the height of the tower.

46 Ladder calculations A ladder 20 feet long leans against the side of a building, and the angle between the ladder and the building is $22°$.

(a) Approximate the distance from the bottom of the ladder to the building.

(b) If the distance from the bottom of the ladder to the building is increased by 3.0 feet, approximately how far does the top of the ladder move down the building?

47 Ascent of a hot-air balloon As a hot-air balloon rises vertically, its angle of elevation from a point P on level ground 110 kilometers from the point Q directly underneath the balloon changes from $19°20'$ to $31°50'$ (see the figure). Approximately how far does the balloon rise during this period?

Exercise 47

48 Height of a building From a point A that is 8.20 meters above level ground, the angle of elevation of the top of a building is $31°20'$ and the angle of depression of the base of the building is $12°50'$. Approximate the height of the building.

49 Radius of Earth A spacelab circles Earth at an altitude of 380 miles. When an astronaut views the horizon of Earth, the angle θ shown in the figure is $65.8°$. Use this information to estimate the radius of Earth.

Exercise 49

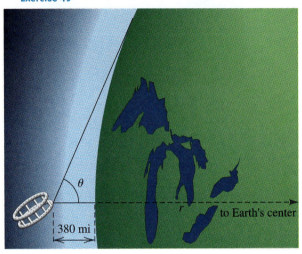

50 Length of an antenna A CB antenna is located on the top of a garage that is 16 feet tall. From a point on level ground that is 100 feet from a point directly below the antenna, the antenna subtends an angle of $12°$, as shown in the figure. Approximate the length of the antenna.

Exercise 50

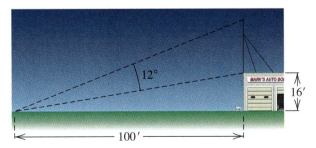

51 Speed of an airplane An airplane flying at an altitude of 10,000 feet passes directly over a fixed object on the ground. One minute later, the angle of depression of the object is $42°$. Approximate the speed of the airplane to the nearest mile per hour.

52 Height of a mountain A motorist, traveling along a level highway at a speed of 60 km/hr directly toward a mountain, observes that between 1:00 P.M. and 1:10 P.M. the angle of elevation of the top of the mountain changes from 10° to 70°. Approximate the height of the mountain.

53 Communications satellite Shown in the left part of the figure is a communications satellite with an equatorial orbit— that is, a nearly circular orbit in the plane determined by Earth's equator. If the satellite circles Earth at an altitude of $a = 22,300$ mi, its speed is the same as the rotational speed of Earth; to an observer on the equator, the satellite appears to be stationary—that is, its orbit is synchronous.

(a) Using $R = 4000$ mi for the radius of Earth, determine the percentage of the equator that is within signal range of such a satellite.

(b) As shown in the right part of the figure, three satellites are equally spaced in equatorial synchronous orbits. Use the value of θ obtained in part (a) to explain why all points on the equator are within signal range of at least one of the three satellites.

Exercise 53

54 Communications satellite Refer to Exercise 53. Shown in the figure is the area served by a communications satellite circling a planet of radius R at an altitude a. The portion of the planet's surface within range of the satellite is a spherical cap of depth d and surface area $A = 2\pi Rd$.

(a) Express d in terms of R and θ.

(b) Estimate the percentage of the planet's surface that is within signal range of a single satellite in equatorial synchronous orbit.

Exercise 54

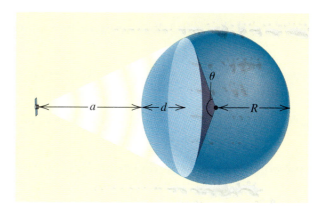

55 Height of a kite Generalize Exercise 25 to the case where the angle is α, the number of feet of string payed out is d, and the end of the string is held c feet above the ground. Express the height h of the kite in terms of α, d, and c.

56 Surveying Generalize Exercise 26 to the case where the point is d meters above level ground and the angle of depression is α. Express the distance x in terms of d and α.

57 Height of a tower Generalize Exercise 45 to the case where the first angle is α, the second angle is β, and the distance between the two points is d. Express the height h of the tower in terms of d, α, and β.

58 Generalize Exercise 42 to the case of an n-sided polygon inscribed in a circle of radius r. Express the perimeter P in terms of n and r.

59 Ascent of a hot-air balloon Generalize Exercise 47 to the case where the distance from P to Q is d kilometers and the angle of elevation changes from α to β.

60 Height of a building Generalize Exercise 48 to the case where point A is d meters above ground and the angles of elevation and depression are α and β, respectively. Express the height h of the building in terms of d, α, and β.

Exer. 61–62: Find the bearing from P to each of the points A, B, C, and D.

61

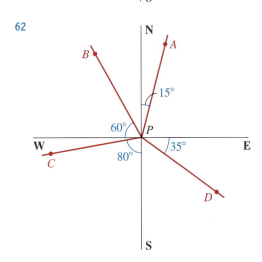

62

63 **Ship's bearings** A ship leaves port at 1:00 P.M. and sails in the direction N34°W at a rate of 24 mi/hr. Another ship leaves port at 1:30 P.M. and sails in the direction N56°E at a rate of 18 mi/hr.

   (a) Approximately how far apart are the ships at 3:00 P.M.?

   (b) What is the bearing, to the nearest degree, from the first ship to the second?

64 **Pinpointing a forest fire** From an observation point A, a forest ranger sights a fire in the direction S35°50′W (see the figure). From a point B, 5 miles due west of A, another ranger sights the same fire in the direction S54°10′E. Ap-

proximate, to the nearest tenth of a mile, the distance of the fire from A.

Exercise 64

65 **Airplane flight** An airplane flying at a speed of 360 mi/hr flies from a point A in the direction 137° for 30 minutes and then flies in the direction 227° for 45 minutes. Approximate, to the nearest mile, the distance from the airplane to A.

66 **Airplane flight plan** An airplane flying at a speed of 400 mi/hr flies from a point A in the direction 153° for 1 hour and then flies in the direction 63° for 1 hour.

   (a) In what direction does the plane need to fly in order to get back to point A?

   (b) How long will it take to get back to point A?

Exer. 67–70: The formula specifies the position of a point P that is moving harmonically on a vertical axis, where t is in seconds and d is in centimeters. Determine the amplitude, period, and frequency, and describe the motion of the point during one complete oscillation (starting at $t = 0$).

67 $d = 10 \sin 6\pi t$ 68 $d = \dfrac{1}{3} \cos \dfrac{\pi}{4} t$

69 $d = 4 \cos \dfrac{3\pi}{2} t$ 70 $d = 6 \sin \dfrac{2\pi}{3} t$

71 A point P in simple harmonic motion has a period of 3 seconds and an amplitude of 5 centimeters. Express the motion of P by means of an equation of the form $d = a \cos \omega t$.

72 A point P in simple harmonic motion has a frequency of $\frac{1}{2}$ oscillation per minute and an amplitude of 4 feet. Express the motion of P by means of an equation of the form $d = a \sin \omega t$.

73 Tsunamis A tsunami is a tidal wave caused by an earthquake beneath the sea. These waves can be more than 100 feet in height and can travel at great speeds. Engineers sometimes represent such waves by trigonometric expressions of the form $y = a \cos bt$ and use these representations to estimate the effectiveness of sea walls. Suppose that a wave has height $h = 50$ ft and period 30 minutes and is traveling at the rate of 180 ft/sec.

Exercise 73

(a) Let (x, y) be a point on the wave represented in the figure. Express y as a function of t if $y = 25$ ft when $t = 0$.

(b) The wave length L is the distance between two successive crests of the wave. Approximate L in feet.

74 Some Hawaiian tsunamis For an interval of 45 minutes, the tsunamis near Hawaii caused by the Chilean earthquake of 1960 could be modeled by the equation $y = 8 \sin \dfrac{\pi}{6} t$, where y is in feet and t is in minutes.

(a) Find the amplitude and period of the waves.

(b) If the distance from one crest of the wave to the next was 21 kilometers, what was the velocity of the wave? (Tidal waves can have velocities of more than 700 km/hr in deep sea water.)

CHAPTER 6 REVIEW EXERCISES

1 Find the radian measure that corresponds to each degree measure: $330°$, $405°$, $-150°$, $240°$, $36°$.

2 Find the degree measure that corresponds to each radian measure: $\dfrac{9\pi}{2}$, $-\dfrac{2\pi}{3}$, $\dfrac{7\pi}{4}$, 5π, $\dfrac{\pi}{5}$.

3 A central angle θ is subtended by an arc 20 centimeters long on a circle of radius 2 meters.

(a) Find the radian measure of θ.

(b) Find the area of the sector determined by θ.

4 (a) Find the length of the arc that subtends an angle of measure $70°$ on a circle of diameter 15 centimeters.

(b) Find the area of the sector in part (a).

5 Angular speed of phonograph records Two types of phonograph records, LP albums and singles, have diameters of 12 inches and 7 inches, respectively. The album rotates at a rate of $33\frac{1}{3}$ rpm, and the single rotates at 45 rpm. Find the angular speed (in radians per minute) of the album and of the single.

6 Linear speed on phonograph records Using the information in Exercise 5, find the linear speed (in ft/min) of a point on the circumference of the album and of the single.

Exer. 7–8: Find the exact values of x and y.

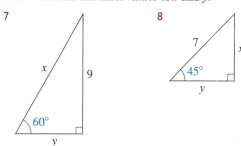

Exer. 9–10: Use fundamental identities to write the first expression in terms of the second, for any acute angle θ.

9 $\tan \theta$, $\sec \theta$

10 $\cot \theta$, $\csc \theta$

Exer. 11–20: Verify the identity by transforming the left-hand side into the right-hand side.

11 $\sin \theta (\csc \theta - \sin \theta) = \cos^2 \theta$

12 $\cos \theta (\tan \theta + \cot \theta) = \csc \theta$

13 $(\cos^2 \theta - 1)(\tan^2 \theta + 1) = 1 - \sec^2 \theta$

14 $\dfrac{\sec \theta - \cos \theta}{\tan \theta} = \dfrac{\tan \theta}{\sec \theta}$

15 $\dfrac{1 + \tan^2 \theta}{\tan^2 \theta} = \csc^2 \theta$

16 $\dfrac{\sec \theta + \csc \theta}{\sec \theta - \csc \theta} = \dfrac{\sin \theta + \cos \theta}{\sin \theta - \cos \theta}$

17 $\dfrac{\cot \theta - 1}{1 - \tan \theta} = \cot \theta$

18 $\dfrac{1 + \sec \theta}{\tan \theta + \sin \theta} = \csc \theta$

19 $\dfrac{\tan (-\theta) + \cot (-\theta)}{\tan \theta} = -\csc^2 \theta$

20 $-\dfrac{1}{\csc (-\theta)} - \dfrac{\cot (-\theta)}{\sec (-\theta)} = \csc \theta$

21 If θ is an acute angle of a right triangle and if the adjacent side and hypotenuse have lengths 4 and 7, respectively, find the values of the trigonometric functions of θ.

22 Whenever possible, find the exact values of the trigonometric functions of θ if θ is in standard position and satisfies the stated condition.

(a) The point $(30, -40)$ is on the terminal side of θ.

(b) The terminal side of θ is in quadrant II and is parallel to the line $2x + 3y + 6 = 0$.

(c) The terminal side of θ is on the negative y-axis.

23 Find the quadrant containing θ if θ is in standard position.

(a) $\sec \theta < 0$ and $\sin \theta > 0$

(b) $\cot \theta > 0$ and $\csc \theta < 0$

(c) $\cos \theta > 0$ and $\tan \theta < 0$

24 Find the exact values of the remaining trigonometric functions if

(a) $\sin \theta = -\frac{4}{5}$ and $\cos \theta = \frac{3}{5}$

(b) $\csc \theta = \dfrac{\sqrt{13}}{2}$ and $\cot \theta = -\dfrac{3}{2}$

Exer. 25–26: $P(t)$ denotes the point on the unit circle U that corresponds to the real number t.

25 Find the rectangular coordinates of $P(7\pi)$, $P(-5\pi/2)$, $P(9\pi/2)$, $P(-3\pi/4)$, $P(18\pi)$, and $P(\pi/6)$.

26 If $P(t)$ has coordinates $\left(-\frac{3}{5}, -\frac{4}{5}\right)$, find the coordinates of $P(t + 3\pi)$, $P(t - \pi)$, $P(-t)$, and $P(2\pi - t)$.

27 (a) Find the reference angle for each radian measure:
$$\frac{5\pi}{4}, \ -\frac{5\pi}{6}, \ -\frac{9\pi}{8}.$$

(b) Find the reference angle for each degree measure: $245°$, $137°$, $892°$.

28 Without the use of a calculator, find the exact values of the trigonometric functions corresponding to each real number, whenever possible.

(a) $\dfrac{9\pi}{2}$ (b) $-\dfrac{5\pi}{4}$ (c) 0 (d) $\dfrac{11\pi}{6}$

29 Find the exact value.

(a) $\cos 225°$ (b) $\tan 150°$ (c) $\sin \left(-\dfrac{\pi}{6}\right)$

(d) $\sec \dfrac{4\pi}{3}$ (e) $\cot \dfrac{7\pi}{4}$ (f) $\csc 300°$

30 If $\sin \theta = -0.7604$ and $\sec \theta$ is positive, approximate θ to the nearest $0.1°$ for $0° \le \theta < 360°$.

31 If $\tan \theta = 2.7381$, approximate θ to the nearest 0.0001 radian for $0 \le \theta < 2\pi$.

32 If $\sec \theta = 1.6403$, approximate θ to the nearest $0.01°$ for $0° \le \theta < 360°$.

Exer. 33–40: Find the amplitude and period and sketch the graph of the equation.

33 $y = 5 \cos x$

34 $y = \frac{2}{3} \sin x$

35 $y = \frac{1}{3} \sin 3x$

36 $y = -\frac{1}{2} \cos \frac{1}{3}x$

37 $y = -3 \cos \frac{1}{2}x$

38 $y = 4 \sin 2x$

39 $y = 2 \sin \pi x$

40 $y = 4 \cos \dfrac{\pi}{2}x - 2$

Exer. 41–44: The graph of an equation is shown in the figure. (a) Find the amplitude and period. (b) Express the equation in the form $y = a \sin bx$ or in the form $y = a \cos bx$.

41

$(1.5, -1.43)$

42

$\left(\frac{3\pi}{4}, -3.27\right)$

43

44

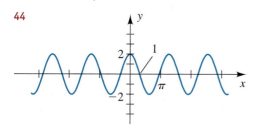

Exer. 45–56: Sketch the graph of the equation.

45 $y = 2 \sin\left(x - \dfrac{2\pi}{3}\right)$

46 $y = -3 \sin\left(\dfrac{1}{2}x - \dfrac{\pi}{4}\right)$

47 $y = -4 \cos\left(x + \dfrac{\pi}{6}\right)$

48 $y = 5 \cos\left(2x + \dfrac{\pi}{2}\right)$

49 $y = 2 \tan\left(\dfrac{1}{2}x - \pi\right)$

50 $y = -3 \tan\left(2x + \dfrac{\pi}{3}\right)$

51 $y = -4 \cot\left(2x - \dfrac{\pi}{2}\right)$

52 $y = 2 \cot\left(\dfrac{1}{2}x + \dfrac{\pi}{4}\right)$

53 $y = \sec\left(\dfrac{1}{2}x + \pi\right)$

54 $y = \sec\left(2x - \dfrac{\pi}{2}\right)$

55 $y = \csc\left(2x - \dfrac{\pi}{4}\right)$

56 $y = \csc\left(\dfrac{1}{2}x + \dfrac{\pi}{4}\right)$

Exer. 57–60: Given the indicated parts of triangle ABC with $\gamma = 90°$, approximate the remaining parts.

57 $\beta = 60°, \quad b = 40$

58 $\alpha = 54°40', \quad b = 220$

59 $a = 62, \quad b = 25$

60 $a = 9.0, \quad c = 41$

61 **Airplane propeller** The length of the largest airplane propeller ever used was 22 feet 7.5 inches. The plane was powered by four engines that turned the propeller at 545 revolutions per minute.

   (a) What was the angular speed of the propeller in radians per second?

   (b) Approximately how fast (in mi/hr) did the tip of the propeller travel along the circle it generated?

62 **The Eiffel Tower** When the top of the Eiffel Tower is viewed at a distance of 200 feet from the base, the angle of elevation is 79.2°. Estimate the height of the tower.

63 **Lasers and velocities** Lasers are used to accurately measure velocities of objects. Laser light produces an oscillating electromagnetic field E with a constant frequency f that can be described by

$$E = E_0 \cos(2\pi ft).$$

If a laser beam is pointed at an object moving toward the laser, light will be reflected toward the laser at a slightly higher frequency, in much the same way as a train whistle sounds higher when it is moving toward you. If Δf is this change in frequency and v is the object's velocity, then the equation

$$\Delta f = \dfrac{2fv}{c}$$

can be used to determine v, where $c = 186{,}000$ mi/sec is the velocity of the light. Approximate the velocity v of an object if $\Delta f = 10^8$ and $f = 10^{14}$.

64 The Great Pyramid The Great Pyramid of Egypt is 147 meters high, with a square base of side 230 meters (see the figure). Approximate, to the nearest degree, the angle φ formed when an observer stands at the midpoint of one of the sides and views the apex of the pyramid.

Exercise 64

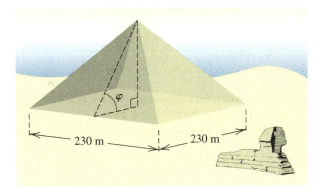

65 Venus When viewed from Earth over a period of time, the planet Venus appears to move back and forth along a line segment with the sun at its midpoint (see the figure). If *ES* is approximately 92,900,000 miles, then the maximum apparent distance of Venus from the sun occurs when angle *SEV* is approximately 47°. Assume that the orbit of Venus is circular and estimate the distance of Venus from the sun.

Exercise 65

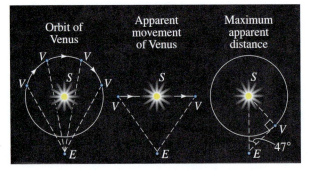

66 Constructing a conical cup A conical paper cup is constructed by removing a sector from a circle of radius

5 inches and attaching edge *OA* to *OB* (see the figure). Find angle *AOB* so that the cup has a depth of 4 inches.

Exercise 66

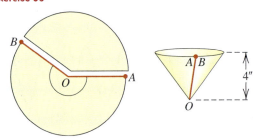

67 Length of a tunnel A tunnel for a new highway is to be cut through a mountain that is 260 feet high. At a distance of 200 feet from the base of the mountain, the angle of elevation is 36° (see the figure). From a distance of 150 feet on the other side, the angle of elevation is 47°. Approximate the length of the tunnel to the nearest foot.

Exercise 67

68 Height of a skyscraper When a certain skyscraper is viewed from the top of a building 50 feet tall, the angle of elevation is 59° (see the figure). When viewed from the street next to the shorter building, the angle of elevation is 62°.

(a) Approximately how far apart are the two structures?

(b) Approximate the height of the skyscraper to the nearest tenth of a foot.

Exercise 68

69 Height of a mountain When a mountaintop is viewed from the point P shown in the figure, the angle of elevation is α. From a point Q, which is d miles closer to the mountain, the angle of elevation increases to β.

(a) Show that the height h of the mountain is given by

$$h = \frac{d}{\cot \alpha - \cot \beta}.$$

(b) If $d = 2$ mi, $\alpha = 15°$, and $\beta = 20°$, approximate the height of the mountain.

Exercise 69

70 Height of a building An observer of height h stands on an incline at a distance d from the base of a building of height T, as shown in the figure. The angle of elevation from the observer to the top of the building is θ, and the incline makes an angle of α with the horizontal.

(a) Express T in terms of h, d, α, and θ.

(b) If $h = 6$ ft, $d = 50$ ft, $\alpha = 15°$, and $\theta = 31.4°$, estimate the height of the building.

Exercise 70

71 Illuminance A spotlight with intensity 5000 candles is located 15 feet above a stage. If the spotlight is rotated through an angle θ as shown in the figure, the illuminance E (in foot-candles) in the lighted area of the stage is given by

$$E = \frac{5000 \cos \theta}{s^2},$$

where s is the distance (in feet) that the light must travel.

(a) Find the illuminance if the spotlight is rotated through an angle of 30°.

(b) The maximum illuminance occurs when $\theta = 0°$. For what value of θ is the illuminance one-half the maximum value?

Exercise 71

72 Height of a mountain If a mountaintop is viewed from a point P due south of the mountain, the angle of elevation is α (see the figure). If viewed from a point Q that is d miles east of P, the angle of elevation is β.

(a) Show that the height h of the mountain is given by

$$h = \frac{d \sin \alpha \sin \beta}{\sqrt{\sin^2 \alpha - \sin^2 \beta}}.$$

(b) If $\alpha = 30°$, $\beta = 20°$, and $d = 10$ mi, approximate h to the nearest hundredth of a mile.

Exercise 72

(b) The volume V of the pyramid equals one-third the area of the base times the altitude. Express V in terms of a and θ.

Exercise 74

73 Mounting a projection unit The manufacturer of a computerized projection system recommends that a projection unit be mounted on the ceiling as shown in the figure. The distance from the end of the mounting bracket to the center of the screen is 85.5 inches, and the angle of depression is 30°.

(a) If the thickness of the screen is disregarded, how far from the wall should the bracket be mounted?

(b) If the bracket is 18 inches long and the screen is 6 feet high, determine the distance from the ceiling to the top edge of the screen.

Exercise 73

74 Pyramid relationships A pyramid has a square base and congruent triangular faces. Let θ be the angle that the altitude a of a triangular face makes with the altitude y of the pyramid, and let x be the length of a side (see the figure).

(a) Express the total surface area S of the four faces in terms of a and θ.

75 Surveying a bluff A surveyor, using a transit, sights the edge B of a bluff, as shown in the left part of the figure (not drawn to scale). Because of the curvature of Earth, the true elevation h of the bluff is larger than that measured by the surveyor. A cross-sectional schematic view of Earth is shown in the right part of the figure.

(a) If s is the length of arc PQ and R is the distance from P to the center C of Earth, express h in terms of R and s.

(b) If $R = 4000$ mi and $s = 50$ mi, estimate the elevation of the bluff in feet.

Exercise 75

76 Earthquake response To simulate the response of a structure to an earthquake, an engineer must choose a shape for the initial displacement of the beams in the building. When the beam has length L feet and the maximum displacement is a feet, the equation

$$y = a - a \cos \frac{\pi}{2L} x$$

has been used by engineers to estimate the displacement y (see the figure). If $a = 1$ and $L = 10$, sketch the graph of the equation for $0 \le x \le 10$.

Exercise 76

77 Circadian rhythms The variation in body temperature is an example of a circadian rhythm, a cycle of a biological process that repeats itself approximately every 24 hours. Body temperature is highest about 5 P.M. and lowest at 5 A.M. Let y denote the body temperature (in °F), and let $t = 0$ correspond to midnight. If the low and high body temperatures are 98.3° and 98.9°, respectively, find an equation having the form $y = 98.6 + a \sin(bt + c)$ that fits this information.

78 Temperature variation in Ottawa The annual variation in temperature T (in °C) in Ottawa, Canada, may be approximated by

$$T(t) = 15.8 \sin\left[\frac{\pi}{6}(t - 3)\right] + 5,$$

where t is the time in months and $t = 0$ corresponds to January 1.

(a) Sketch the graph of T for $0 \le t \le 12$.

(b) Find the highest temperature of the year and the date on which it occurs.

79 Water demand A reservoir supplies water to a community. During the summer months, the demand $D(t)$ for water (in ft³/day) is given by

$$D(t) = 2000 \sin\frac{\pi}{90}t + 4000,$$

where t is time in days and $t = 0$ corresponds to the beginning of summer.

(a) Sketch the graph of D for $0 \le t \le 90$.

(b) When is the demand for water the greatest?

80 Bobbing cork A cork bobs up and down in a lake. The distance from the bottom of the lake to the center of the cork at time $t \ge 0$ is given by $s(t) = 12 + \cos \pi t$, where $s(t)$ is in feet and t is in seconds.

(a) Describe the motion of the cork for $0 \le t \le 2$.

(b) During what time intervals is the cork rising?

CHAPTER 6 DISCUSSION EXERCISES

1 Determine the number of solutions of the equation

$$\cos x + \cos 2x + \cos 3x = \pi.$$

2 Racetrack coordinates Shown in the figure is a circular racetrack of diameter 2 kilometers. All races begin at S and proceed in a counterclockwise direction. Approximate, to four decimal places, the coordinates of the point at which the following races end relative to a rectangular coordinate system with origin at the center of the track and S on the positive x-axis.

Exercise 2

(a) A drag race of length 2 kilometers

(b) An endurance race of length 500 kilometers

3 Racetrack coordinates Work Exercise 2 for the track shown in the figure, if the origin of the rectangular coordinate system is at the center of the track and S is on the negative y-axis.

Exercise 3

4 Outboard motor propeller A 90-horsepower outboard motor at full throttle will rotate its propeller at 5000 revolutions per minute.

(a) Find the angular speed ω of the propeller in radians per second.

(b) The center of a 10-inch-diameter propeller is located 18 inches below the surface of the water. Express the depth $D(t) = a \cos(\omega t + c) + d$ of a point on the edge of a propeller blade as a function of time t, where t is in seconds. Assume that the point is initially at a depth of 23 inches.

5 Discuss the relationships among periodic functions, one-to-one functions, and inverse functions. With these relationships in mind, discuss what must happen for the trigonometric functions to have inverses.

7

Analytic Trigonometry

7.1 Verifying Trigonometric Identities

7.2 Trigonometric Equations

7.3 The Addition and Subtraction Formulas

7.4 Multiple-Angle Formulas

7.5 Product-to-Sum and Sum-to-Product Formulas

7.6 The Inverse Trigonometric Functions

In advanced mathematics, the natural sciences, and engineering, it is sometimes necessary to simplify complicated trigonometric expressions and to solve equations that involve trigonometric functions. These topics are discussed in the first two sections of this chapter. We then derive many useful formulas with respect to sums, differences, and multiples; for reference they are listed on the inside back cover of the text. In addition to formal manipulations, we also consider numerous applications of these formulas. The last section contains the definitions and properties of the inverse trigonometric functions.

7.1

Verifying Trigonometric Identities

A **trigonometric expression** contains symbols involving trigonometric functions.

ILLUSTRATION **Trigonometric Expressions**

■ $\quad x + \sin x$ ■ $\quad \dfrac{\sqrt{\theta} + 2^{\sin \theta}}{\cot \theta}$ ■ $\quad \dfrac{\cos (3t + 1)}{t^2 + \tan^2 (2 - t^2)}$

We assume that the domain of each variable in a trigonometric expression is the set of real numbers or angles for which the expression is meaningful. To provide manipulative practice in simplifying complicated trigonometric expressions, we shall use the fundamental identities (see page 364) and algebraic manipulations, as we did in Examples 5 and 6 of Section 6.2. In the first three examples our method consists of transforming the left-hand side of a given identity into the right-hand side, or vice versa.

EXAMPLE 1   **Verifying an identity**

Verify the identity $\sec \alpha - \cos \alpha = \sin \alpha \tan \alpha$.

SOLUTION   We transform the left-hand side into the right-hand side:

$$\sec \alpha - \cos \alpha = \frac{1}{\cos \alpha} - \cos \alpha \qquad \text{\color{red}{reciprocal identity}}$$

$$= \frac{1 - \cos^2 \alpha}{\cos \alpha} \qquad \text{\color{red}{add expressions}}$$

$$= \frac{\sin^2 \alpha}{\cos \alpha} \qquad \text{\color{red}{$\sin^2 \alpha + \cos^2 \alpha = 1$}}$$

$$= \sin \alpha \left(\frac{\sin \alpha}{\cos \alpha} \right) \qquad \text{\color{red}{equivalent expression}}$$

$$= \sin \alpha \tan \alpha \qquad \text{\color{red}{tangent identity}}$$

EXAMPLE 2   **Verifying an identity**

Verify the identity $\sec \theta = \sin \theta \, (\tan \theta + \cot \theta)$.

SOLUTION   Since the expression on the right-hand side is more complicated than that on the left-hand side, we transform the right-hand side into the left-hand side:

$$\sin \theta \, (\tan \theta + \cot \theta) = \sin \theta \left(\frac{\sin \theta}{\cos \theta} + \frac{\cos \theta}{\sin \theta} \right) \qquad \text{tangent and cotangent identities}$$

$$= \sin \theta \left(\frac{\sin^2 \theta + \cos^2 \theta}{\cos \theta \sin \theta} \right) \qquad \text{add fractions}$$

$$= \sin \theta \left(\frac{1}{\cos \theta \sin \theta} \right) \qquad \text{Pythagorean identity}$$

$$= \frac{1}{\cos \theta} \qquad \text{cancel } \sin \theta$$

$$= \sec \theta \qquad \text{reciprocal identity} \qquad \blacksquare$$

EXAMPLE 3 Verifying an identity

Verify the identity $\dfrac{\cos x}{1 - \sin x} = \dfrac{1 + \sin x}{\cos x}$.

SOLUTION Since the denominator of the left-hand side is a binomial and the denominator of the right-hand side is a monomial, we change the form of the fraction on the left-hand side by multiplying the numerator and denominator by the conjugate of the denominator and then use one of the Pythagorean identities:

$$\frac{\cos x}{1 - \sin x} = \frac{\cos x}{1 - \sin x} \cdot \frac{1 + \sin x}{1 + \sin x} \qquad \text{multiply numerator and denominator by } 1 + \sin x$$

$$= \frac{\cos x \, (1 + \sin x)}{1 - \sin^2 x} \qquad \text{property of quotients}$$

$$= \frac{\cos x \, (1 + \sin x)}{\cos^2 x} \qquad \sin^2 x + \cos^2 x = 1$$

$$= \frac{1 + \sin x}{\cos x} \qquad \text{cancel } \cos x \qquad \blacksquare$$

Another technique for showing that an equation $p = q$ is an identity is to begin by transforming the left-hand side p into another expression s, making sure that each step is *reversible*—that is, making sure it is possible to transform s back into p by reversing the procedure used in each step. In this case, the equation $p = s$ is an identity. Next, as a *separate* exercise, we show that the right-hand side q can also be transformed into the expression s by means of reversible steps and, therefore, that $q = s$ is an identity. It then follows that $p = q$ is an identity. This method is illustrated in the next example.

EXAMPLE 4 Verifying an identity

Verify the identity $(\tan \theta - \sec \theta)^2 = \dfrac{1 - \sin \theta}{1 + \sin \theta}$.

SOLUTION We shall verify the identity by showing that each side of the equation can be transformed into the same expression. First we work only with the left-hand side:

Work with the left-hand side.

$$(\tan \theta - \sec \theta)^2 = \tan^2 \theta - 2 \tan \theta \sec \theta + \sec^2 \theta \qquad \text{square expression}$$

$$= \left(\frac{\sin \theta}{\cos \theta}\right)^2 - 2\left(\frac{\sin \theta}{\cos \theta}\right)\left(\frac{1}{\cos \theta}\right) + \left(\frac{1}{\cos \theta}\right)^2$$

tangent and reciprocal identities

$$= \frac{\sin^2 \theta}{\cos^2 \theta} - \frac{2 \sin \theta}{\cos^2 \theta} + \frac{1}{\cos^2 \theta} \qquad \text{equivalent expression}$$

$$= \frac{\sin^2 \theta - 2 \sin \theta + 1}{\cos^2 \theta} \qquad \text{add fractions}$$

At this point it may not be obvious how we can obtain the right-hand side of the given equation from the last expression. Thus, we next work with only the right-hand side and try to obtain the last expression. Multiplying numerator and denominator by the conjugate of the denominator gives us the following:

equivalent expressions

Work with the right-hand side.

$$\frac{1 - \sin \theta}{1 + \sin \theta} = \frac{1 - \sin \theta}{1 + \sin \theta} \cdot \frac{1 - \sin \theta}{1 - \sin \theta} \qquad \begin{array}{l}\text{multiply numerator and}\\ \text{denominator by } 1 - \sin \theta\end{array}$$

$$= \frac{1 - 2 \sin \theta + \sin^2 \theta}{1 - \sin^2 \theta} \qquad \text{property of quotients}$$

$$= \frac{1 - 2 \sin \theta + \sin^2 \theta}{\cos^2 \theta} \qquad \sin^2 \theta + \cos^2 \theta = 1$$

The last expression is the same as that obtained from $(\tan \theta - \sec \theta)^2$. Since all steps are reversible, the given equation is an identity.

In calculus it is sometimes convenient to change the form of certain algebraic expressions by making a **trigonometric substitution,** as illustrated in the following example.

EXAMPLE 5 **Making a trigonometric substitution**

Express $\sqrt{a^2 - x^2}$ in terms of a trigonometric function of θ, without radicals, by making the substitution $x = a \sin \theta$ for $-\pi/2 \le \theta \le \pi/2$ and $a > 0$.

SOLUTION We proceed as follows:

$$\sqrt{a^2 - x^2} = \sqrt{a^2 - (a \sin \theta)^2} \qquad \text{let } x = a \sin \theta$$

$$= \sqrt{a^2 - a^2 \sin^2 \theta} \qquad \text{law of exponents}$$

$$= \sqrt{a^2(1 - \sin^2 \theta)} \qquad \text{factor out } a^2$$

$$= \sqrt{a^2 \cos^2 \theta} \qquad \sin^2 \theta + \cos^2 \theta = 1$$

$$= \sqrt{(a \cos \theta)^2} \qquad c^2 d^2 = (cd)^2$$

$$= |a \cos \theta| \qquad \sqrt{c^2} = |c|$$

$$= |a||\cos \theta| \qquad |cd| = |c||d|$$

$$= a \cos \theta \qquad \text{see below}$$

The last equality is true because (1) if $a > 0$, then $|a| = a$, and (2) if $-\pi/2 \le \theta \le \pi/2$, then $\cos\theta \ge 0$ and hence $|\cos\theta| = \cos\theta$.

We may also use a geometric solution. If $x = a\sin\theta$, then $\sin\theta = x/a$, and the triangle in Figure 1 illustrates the problem for $0 < \theta < \pi/2$. The third side of the triangle, $\sqrt{a^2 - x^2}$, can be found by using the Pythagorean theorem. From the figure we can see that

$$\cos\theta = \frac{\sqrt{a^2 - x^2}}{a} \qquad \text{or, equivalently,} \qquad \sqrt{a^2 - x^2} = a\cos\theta.$$

Figure 1

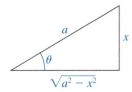

7.1   Exercises

Exer. 1–50: Verify the identity.

1 $\csc\theta - \sin\theta = \cot\theta\cos\theta$

2 $\sin x + \cos x\cot x = \csc x$

3 $\dfrac{\sec^2 2u - 1}{\sec^2 2u} = \sin^2 2u$

4 $\tan t + 2\cos t\csc t = \sec t\csc t + \cot t$

5 $\dfrac{\csc^2\theta}{1 + \tan^2\theta} = \cot^2\theta$

6 $(\tan u + \cot u)(\cos u + \sin u) = \csc u + \sec u$

7 $\dfrac{1 + \cos 3t}{\sin 3t} + \dfrac{\sin 3t}{1 + \cos 3t} = 2\csc 3t$

8 $\tan^2\alpha - \sin^2\alpha = \tan^2\alpha\sin^2\alpha$

9 $\dfrac{1}{1 - \cos\gamma} + \dfrac{1}{1 + \cos\gamma} = 2\csc^2\gamma$

10 $\dfrac{1 + \csc 3\beta}{\sec 3\beta} - \cot 3\beta = \cos 3\beta$

11 $(\sec u - \tan u)(\csc u + 1) = \cot u$

12 $\dfrac{\cot\theta - \tan\theta}{\sin\theta + \cos\theta} = \csc\theta - \sec\theta$

13 $\csc^4 t - \cot^4 t = \csc^2 t + \cot^2 t$

14 $\cos^4 2\theta + \sin^2 2\theta = \cos^2 2\theta + \sin^4 2\theta$

15 $\dfrac{\cos\beta}{1 - \sin\beta} = \sec\beta + \tan\beta$

16 $\dfrac{1}{\csc y - \cot y} = \csc y + \cot y$

17 $\dfrac{\tan^2 x}{\sec x + 1} = \dfrac{1 - \cos x}{\cos x}$

18 $\dfrac{\cot x}{\csc x + 1} = \dfrac{\csc x - 1}{\cot x}$

19 $\dfrac{\cot 4u - 1}{\cot 4u + 1} = \dfrac{1 - \tan 4u}{1 + \tan 4u}$

20 $\dfrac{1 + \sec 4x}{\sin 4x + \tan 4x} = \csc 4x$

21 $\sin^4 r - \cos^4 r = \sin^2 r - \cos^2 r$

22 $\sin^4\theta + 2\sin^2\theta\cos^2\theta + \cos^4\theta = 1$

23 $\tan^4 k - \sec^4 k = 1 - 2\sec^2 k$

24 $\sec^4 u - \sec^2 u = \tan^2 u + \tan^4 u$

25 $(\sec t + \tan t)^2 = \dfrac{1 + \sin t}{1 - \sin t}$

26 $\sec^2\gamma + \tan^2\gamma = (1 - \sin^4\gamma)\sec^4\gamma$

27 $(\sin^2\theta + \cos^2\theta)^3 = 1$

28 $\dfrac{\sin t}{1 - \cos t} = \csc t + \cot t$ **29** $\dfrac{1 + \csc\beta}{\cot\beta + \cos\beta} = \sec\beta$

30 $\dfrac{\cos^3 x - \sin^3 x}{\cos x - \sin x} = 1 + \sin x\cos x$

31 $(\csc t - \cot t)^4(\csc t + \cot t)^4 = 1$

32 $(a\cos t - b\sin t)^2 + (a\sin t + b\cos t)^2 = a^2 + b^2$

33 $\dfrac{\sin \alpha \cos \beta + \cos \alpha \sin \beta}{\cos \alpha \cos \beta - \sin \alpha \sin \beta} = \dfrac{\tan \alpha + \tan \beta}{1 - \tan \alpha \tan \beta}$

34 $\dfrac{\tan u - \tan v}{1 + \tan u \tan v} = \dfrac{\cot v - \cot u}{\cot u \cot v + 1}$

35 $\dfrac{\tan \alpha}{1 + \sec \alpha} + \dfrac{1 + \sec \alpha}{\tan \alpha} = 2 \csc \alpha$

36 $\dfrac{\csc x}{1 + \csc x} - \dfrac{\csc x}{1 - \csc x} = 2 \sec^2 x$

37 $\dfrac{1}{\tan \beta + \cot \beta} = \sin \beta \cos \beta$

38 $\dfrac{\cot y - \tan y}{\sin y \cos y} = \csc^2 y - \sec^2 y$

39 $\sec \theta + \csc \theta - \cos \theta - \sin \theta = \sin \theta \tan \theta + \cos \theta \cot \theta$

40 $\sin^3 t + \cos^3 t = (1 - \sin t \cos t)(\sin t + \cos t)$

41 $(1 - \tan^2 \phi)^2 = \sec^4 \phi - 4 \tan^2 \phi$

42 $\cos^4 w + 1 - \sin^4 w = 2 \cos^2 w$

43 $\dfrac{\cot (-t) + \tan (-t)}{\cot t} = -\sec^2 t$

44 $\dfrac{\csc (-t) - \sin (-t)}{\sin (-t)} = \cot^2 t$

45 $\log 10^{\tan t} = \tan t$ 　　　**46** $10^{\log |\sin t|} = |\sin t|$

47 $\ln \cot x = -\ln \tan x$ 　　**48** $\ln \sec \theta = -\ln \cos \theta$

49 $\ln |\sec \theta + \tan \theta| = -\ln |\sec \theta - \tan \theta|$

50 $\ln |\csc x - \cot x| = -\ln |\csc x + \cot x|$

Exer. 51–60: Show that the equation is *not* an identity. (*Hint:* Find one number for which the equation is false.)

51 $\cos t = \sqrt{1 - \sin^2 t}$

52 $\sqrt{\sin^2 t + \cos^2 t} = \sin t + \cos t$

53 $\sqrt{\sin^2 t} = \sin t$ 　　　**54** $\sec t = \sqrt{\tan^2 t + 1}$

55 $(\sin \theta + \cos \theta)^2 = \sin^2 \theta + \cos^2 \theta$

56 $\log \left(\dfrac{1}{\sin t}\right) = \dfrac{1}{\log \sin t}$

57 $\cos (-t) = -\cos t$ 　　　**58** $\sin (t + \pi) = \sin t$

59 $\cos (\sec t) = 1$ 　　　　**60** $\cot (\tan \theta) = 1$

Exer. 61–64: Either show that the equation *is* an identity or show that the equation *is not* an identity.

61 $(\sec x + \tan x)^2 = 2 \tan x (\tan x + \sec x)$

62 $\dfrac{\tan^2 x}{\sec x - 1} = \sec x$

63 $\cos x(\tan x + \cot x) = \csc x$

64 $\csc^2 x + \sec^2 x = \csc^2 x \sec^2 x$

Exer. 65–68: Refer to Example 5. Make the trigonometric substitution $x = a \sin \theta$ for $-\pi/2 < \theta < \pi/2$ and $a > 0$. Use fundamental identities to simplify the resulting expression.

65 $(a^2 - x^2)^{3/2}$ 　　　　　**66** $\dfrac{\sqrt{a^2 - x^2}}{x}$

67 $\dfrac{x^2}{\sqrt{a^2 - x^2}}$ 　　　　**68** $\dfrac{1}{x\sqrt{a^2 - x^2}}$

Exer. 69–72: Make the trigonometric substitution $x = a \tan \theta$ for $-\pi/2 < \theta < \pi/2$ and $a > 0$. Simplify the resulting expression.

69 $\sqrt{a^2 + x^2}$ 　　　　　**70** $\dfrac{1}{\sqrt{a^2 + x^2}}$

71 $\dfrac{1}{x^2 + a^2}$ 　　　　　**72** $\dfrac{(x^2 + a^2)^{3/2}}{x}$

Exer. 73–76: Make the trigonometric substitution $x = a \sec \theta$ for $0 < \theta < \pi/2$ and $a > 0$. Simplify the resulting expression.

73 $\sqrt{x^2 - a^2}$ 　　　　　**74** $\dfrac{1}{x^2\sqrt{x^2 - a^2}}$

75 $x^3\sqrt{x^2 - a^2}$ 　　　　**76** $\dfrac{\sqrt{x^2 - a^2}}{x^2}$

7.2

Trigonometric Equations

A **trigonometric equation** is an equation that contains trigonometric expressions. Each identity considered in the preceding section is an example of a trigonometric equation with every number (or angle) in the domain of the variable a solution of the equation. If a trigonometric equation is not an identity, we often find solutions by using techniques similar to those used for algebraic equations. The main difference is that we first solve the trigonometric equation for sin x, cos θ, and so on, and then find values of x or θ that satisfy the equation. Solutions may be expressed either as real numbers or as angles. Throughout our work we shall use the following rule: *If degree measure is not specified, then solutions of a trigonometric equation should be expressed in radian measure (or as real numbers). If solutions in degree measure are desired, an appropriate statement will be included in the example or exercise.*

EXAMPLE 1 **Solving a trigonometric equation involving the sine function**

Find the solutions of the equation sin $\theta = \frac{1}{2}$ if

(a) θ is in the interval $[0, 2\pi)$

(b) θ is any real number

SOLUTION

(a) If sin $\theta = \frac{1}{2}$, then the reference angle for θ is $\theta_R = \pi/6$. If we regard θ as an angle in standard position, then, since sin $\theta > 0$, the terminal side is in either quadrant I or quadrant II, as illustrated in Figure 1. Thus, there are two solutions for $0 \le \theta < 2\pi$:

$$\theta = \frac{\pi}{6} \quad \text{and} \quad \theta = \pi - \frac{\pi}{6} = \frac{5\pi}{6}$$

(b) Since the sine function has period 2π, we may obtain all solutions by adding multiples of 2π to $\pi/6$ and $5\pi/6$. This gives us

$$\theta = \frac{\pi}{6} + 2\pi n \quad \text{and} \quad \theta = \frac{5\pi}{6} + 2\pi n \quad \text{for every integer } n.$$

Figure 1

Figure 2

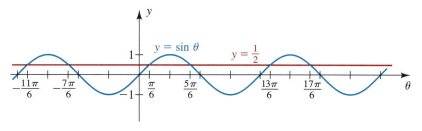

An alternative (graphical) solution involves determining where the graph of $y = \sin \theta$ intersects the horizontal line $y = \frac{1}{2}$, as illustrated in Figure 2.

EXAMPLE 2 **Solving a trigonometric equation involving the tangent function**

Find the solutions of the equation $\tan u = -1$.

SOLUTION Since the tangent function has period π, it is sufficient to find one real number u such that $\tan u = -1$ and then add multiples of π.

A portion of the graph of $y = \tan u$ is sketched in Figure 3. Since $\tan(3\pi/4) = -1$, one solution is $3\pi/4$; hence,

$$\text{if} \quad \tan u = -1, \quad \text{then} \quad u = \frac{3\pi}{4} + \pi n \quad \text{for every integer } n.$$

Figure 3 $y = \tan u$

Figure 4

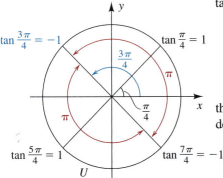

We could also have chosen $-\pi/4$ (or some other number u such that $\tan u = -1$) for the initial solution and written

$$u = -\frac{\pi}{4} + \pi n \quad \text{for every integer } n.$$

An alternative solution involves a unit circle. Using $\tan 3\pi/4 = -1$ and the fact that the period of the tangent is π, we can see from Figure 4 that the desired solutions are

$$u = \frac{3\pi}{4} + \pi n \quad \text{for every integer } n.$$

EXAMPLE 3 **Solving a trigonometric equation involving multiple angles**

(a) Solve the equation $\cos 2x = 0$, and express the solutions both in radians and in degrees.

(b) Find the solutions that are in the interval $[0, 2\pi)$ and, equivalently, $[0°, 360°)$.

SOLUTION

(a) We proceed as follows, where n denotes any integer:

$$\cos 2x = 0 \qquad \text{given}$$

$$\cos \theta = 0 \qquad \text{let } \theta = 2x$$

$$\theta = \frac{\pi}{2} + \pi n \qquad \text{refer to graph of } y = \cos x \text{ (page 382)}$$

$$2x = \frac{\pi}{2} + \pi n \qquad \theta = 2x$$

$$x = \frac{\pi}{4} + \frac{\pi}{2} n \qquad \text{divide by 2}$$

In degrees, we have $x = 45° + 90°n$.

(b) We may find particular solutions of the equation by substituting integers for n in either of the formulas for x obtained in part (a). Several such solutions are listed in the following table.

n	$\dfrac{\pi}{4} + \dfrac{\pi}{2}n$	$45° + 90°n$
-1	$\dfrac{\pi}{4} + \dfrac{\pi}{2}(-1) = -\dfrac{\pi}{4}$	$45° + 90°(-1) = -45°$
0	$\dfrac{\pi}{4} + \dfrac{\pi}{2}(0) = \dfrac{\pi}{4}$	$45° + 90°(0) = 45°$
1	$\dfrac{\pi}{4} + \dfrac{\pi}{2}(1) = \dfrac{3\pi}{4}$	$45° + 90°(1) = 135°$
2	$\dfrac{\pi}{4} + \dfrac{\pi}{2}(2) = \dfrac{5\pi}{4}$	$45° + 90°(2) = 225°$
3	$\dfrac{\pi}{4} + \dfrac{\pi}{2}(3) = \dfrac{7\pi}{4}$	$45° + 90°(3) = 315°$
4	$\dfrac{\pi}{4} + \dfrac{\pi}{2}(4) = \dfrac{9\pi}{4}$	$45° + 90°(4) = 405°$

Note that the solutions in the interval $[0, 2\pi)$ or, equivalently, $[0°, 360°)$ are given by $n = 0$, $n = 1$, $n = 2$, and $n = 3$. These solutions are

$$\frac{\pi}{4}, \frac{3\pi}{4}, \frac{5\pi}{4}, \frac{7\pi}{4} \qquad \text{or, equivalently,} \qquad 45°, 135°, 225°, 315°.$$

EXAMPLE 4 **Solving a trigonometric equation by factoring**

Solve the equation $\sin \theta \tan \theta = \sin \theta$.

SOLUTION

$$\sin \theta \tan \theta = \sin \theta \quad \text{given}$$

$$\sin \theta \tan \theta - \sin \theta = 0 \quad \text{make one side 0}$$

$$\sin \theta \, (\tan \theta - 1) = 0 \quad \text{factor out } \sin \theta$$

$$\sin \theta = 0, \quad \tan \theta - 1 = 0 \quad \text{zero factor theorem}$$

$$\sin \theta = 0, \quad \tan \theta = 1 \quad \text{solve for } \sin \theta \text{ and } \tan \theta$$

The solutions of the equation $\sin \theta = 0$ are $0, \pm \pi, \pm 2\pi, \ldots$. Thus,

$$\text{if} \quad \sin \theta = 0, \quad \text{then} \quad \theta = \pi n \quad \text{for every integer } n.$$

The tangent function has period π, and hence we find the solutions of the equation $\tan \theta = 1$ that are in the interval $(-\pi/2, \pi/2)$ and then add multiples of π. Since the only solution of $\tan \theta = 1$ in $(-\pi/2, \pi/2)$ is $\pi/4$, we see that

$$\text{if} \quad \tan \theta = 1, \quad \text{then} \quad \theta = \frac{\pi}{4} + \pi n \quad \text{for every integer } n.$$

Thus, the solutions of the given equation are

$$\pi n \quad \text{and} \quad \frac{\pi}{4} + \pi n \quad \text{for every integer } n.$$

Some *particular* solutions, obtained by letting $n = 0$, $n = 1$, $n = 2$, and $n = -1$, are

$$0, \quad \frac{\pi}{4}, \quad \pi, \quad \frac{5\pi}{4}, \quad 2\pi, \quad \frac{9\pi}{4}, \quad -\pi, \quad \text{and} \quad -\frac{3\pi}{4}. \qquad \text{◥}$$

In Example 4 it would have been incorrect to begin by dividing both sides by $\sin \theta$, since we would have lost the solutions of $\sin \theta = 0$.

EXAMPLE 5 **Solving a trigonometric equation by factoring**

Solve the equation $2 \sin^2 t - \cos t - 1 = 0$, and express the solutions both in radians and in degrees.

SOLUTION It appears that we have a quadratic equation in either $\sin t$ or $\cos t$. We do not have a simple substitution for $\cos t$ in terms of $\sin t$, but we do have one for $\sin^2 t$ in terms of $\cos^2 t$ ($\sin^2 t = 1 - \cos^2 t$), so we shall first express the equation in terms of $\cos t$ alone and then solve by factoring.

This is a quadratic equation in $\cos t$, so you could use the quadratic formula at this point. If you do so, remember to solve for <u>cos t</u>, not <u>t</u>.

$$2 \sin^2 t - \cos t - 1 = 0 \quad \text{given}$$

$$2(1 - \cos^2 t) - \cos t - 1 = 0 \quad \sin^2 t + \cos^2 t = 1$$

$$-2 \cos^2 t - \cos t + 1 = 0 \quad \text{simplify}$$

$\longrightarrow \quad 2 \cos^2 t + \cos t - 1 = 0 \quad$ multiply by -1

$$(2 \cos t - 1)(\cos t + 1) = 0 \quad \text{factor}$$

$$2 \cos t - 1 = 0, \quad \cos t + 1 = 0 \quad \text{zero factor theorem}$$

$$\cos t = \tfrac{1}{2}, \quad \cos t = -1 \quad \text{solve for } \cos t$$

Since the cosine function has period 2π, we may find all solutions of these equations by adding multiples of 2π to the solutions that are in the interval $[0, 2\pi)$.

If $\cos t = \frac{1}{2}$, the reference angle is $\pi/3$ (or $60°$). Since $\cos t$ is positive, the angle of radian measure t is in either quadrant I or quadrant IV. Hence, in the interval $[0, 2\pi)$, we see that

$$\text{if} \quad \cos t = \frac{1}{2}, \quad \text{then} \quad t = \frac{\pi}{3} \quad \text{or} \quad t = 2\pi - \frac{\pi}{3} = \frac{5\pi}{3}.$$

Referring to the graph of the cosine function, we see that

$$\text{if} \quad \cos t = -1, \quad \text{then} \quad t = \pi.$$

Thus, the solutions of the given equation are the following, where n is any integer:

$$\frac{\pi}{3} + 2\pi n, \quad \frac{5\pi}{3} + 2\pi n, \quad \text{and} \quad \pi + 2\pi n$$

In degree measure, we have

$$60° + 360°n, \quad 300° + 360°n, \quad \text{and} \quad 180° + 360°n.$$

EXAMPLE 6 **Solving a trigonometric equation by factoring**

Find the solutions of $4 \sin^2 x \tan x - \tan x = 0$ that are in the interval $[0, 2\pi)$.

SOLUTION
$$4 \sin^2 x \tan x - \tan x = 0 \qquad \textcolor{brown}{\text{given}}$$
$$\tan x \,(4 \sin^2 x - 1) = 0 \qquad \textcolor{brown}{\text{factor out } \tan x}$$
$$\tan x = 0, \quad 4 \sin^2 x - 1 = 0 \qquad \textcolor{brown}{\text{zero factor theorem}}$$
$$\tan x = 0, \qquad \sin^2 x = \tfrac{1}{4} \qquad \textcolor{brown}{\text{solve for } \tan x, \, \sin^2 x}$$
$$\tan x = 0, \qquad \sin x = \pm\tfrac{1}{2} \qquad \textcolor{brown}{\text{solve for } \sin x}$$

Figure 5

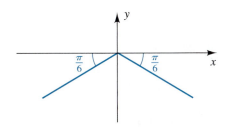

The reference angle $\pi/6$ for the third and fourth quadrants is shown in Figure 5. These angles, $7\pi/6$ and $11\pi/6$, are the solutions of the equation $\sin x = -\frac{1}{2}$ for $0 \le x < 2\pi$. The solutions of all three equations are listed in the following table.

Equation	Solutions in $[0, 2\pi)$	Refer to
$\tan x = 0$	$0, \pi$	Figure 3
$\sin x = \dfrac{1}{2}$	$\dfrac{\pi}{6}, \dfrac{5\pi}{6}$	Example 1
$\sin x = -\dfrac{1}{2}$	$\dfrac{7\pi}{6}, \dfrac{11\pi}{6}$	Figure 5 (use reference angle)

Thus, the given equation has the six solutions listed in the second column of the table.

EXAMPLE 7 Solving a trigonometric equation involving multiple angles

Find the solutions of $\csc^4 2u - 4 = 0$.

SOLUTION

$$\csc^4 2u - 4 = 0 \qquad \text{given}$$
$$(\csc^2 2u - 2)(\csc^2 2u + 2) = 0 \qquad \text{difference of two squares}$$
$$\csc^2 2u - 2 = 0, \quad \csc^2 2u + 2 = 0 \qquad \text{zero factor theorem}$$
$$\csc^2 2u = 2, \qquad \csc^2 2u = -2 \qquad \text{solve for } \csc^2 2u$$
$$\csc 2u = \pm\sqrt{2}, \qquad \csc 2u = \pm\sqrt{-2} \qquad \text{take square roots}$$

The second equation has no solution because $\sqrt{-2}$ is not a real number. The first equation is equivalent to

$$\sin 2u = \pm\frac{1}{\sqrt{2}} = \pm\frac{\sqrt{2}}{2}.$$

Since the reference angle for $2u$ is $\pi/4$, we obtain the following table, in which n denotes any integer.

Equation	Solution for $2u$	Solution for u
$\sin 2u = \dfrac{\sqrt{2}}{2}$	$2u = \dfrac{\pi}{4} + 2\pi n$	$u = \dfrac{\pi}{8} + \pi n$
	$2u = \dfrac{3\pi}{4} + 2\pi n$	$u = \dfrac{3\pi}{8} + \pi n$
$\sin 2u = -\dfrac{\sqrt{2}}{2}$	$2u = \dfrac{5\pi}{4} + 2\pi n$	$u = \dfrac{5\pi}{8} + \pi n$
	$2u = \dfrac{7\pi}{4} + 2\pi n$	$u = \dfrac{7\pi}{8} + \pi n$

The solutions of the given equation are listed in the last column. Note that *all* of these solutions can be written in the one form

$$u = \frac{\pi}{8} + \frac{\pi}{4}n.$$

The next example illustrates the use of a calculator in solving a trigonometric equation.

EXAMPLE 8 Approximating the solutions of a trigonometric equation

Approximate, to the nearest degree, the solutions of the following equation in the interval $[0°, 360°)$:

$$5 \sin \theta \tan \theta - 10 \tan \theta + 3 \sin \theta - 6 = 0$$

SOLUTION

$$5 \sin \theta \tan \theta - 10 \tan \theta + 3 \sin \theta - 6 = 0 \quad \text{given}$$

$$(5 \sin \theta \tan \theta - 10 \tan \theta) + (3 \sin \theta - 6) = 0 \quad \text{group terms}$$

$$5 \tan \theta (\sin \theta - 2) + 3(\sin \theta - 2) = 0 \quad \text{factor each group}$$

$$(5 \tan \theta + 3)(\sin \theta - 2) = 0 \quad \text{factor out } (\sin \theta - 2)$$

$$5 \tan \theta + 3 = 0, \qquad \sin \theta - 2 = 0 \quad \text{zero factor theorem}$$

$$\tan \theta = -\tfrac{3}{5}, \qquad \sin \theta = 2 \quad \text{solve for } \tan \theta \text{ and } \sin \theta$$

The equation $\sin \theta = 2$ has no solution, since $-1 \le \sin \theta \le 1$ for every θ. For $\tan \theta = -\tfrac{3}{5}$, we use a calculator in degree mode, obtaining

$$\theta = \tan^{-1}\left(-\tfrac{3}{5}\right) \approx -31°.$$

Hence, the reference angle is $\theta_R \approx 31°$. Since θ is in either quadrant II or quadrant IV, we obtain the following solutions:

$$\theta = 180° - \theta_R \approx 180° - 31° = 149°$$

$$\theta = 360° - \theta_R \approx 360° - 31° = 329°$$

EXAMPLE 9   Investigating the number of hours of daylight

In Boston, the number of hours of daylight $D(t)$ at a particular time of the year may be approximated by

$$D(t) = 3 \sin\left[\frac{2\pi}{365}(t - 79)\right] + 12,$$

with t in days and $t = 0$ corresponding to January 1. How many days of the year have more than 10.5 hours of daylight?

SOLUTION   The graph of D was discussed in Example 12 of Section 6.5 and is resketched in Figure 6. As illustrated in the figure, if we can find two numbers a and b with $D(a) = 10.5$, $D(b) = 10.5$, and $0 < a < b < 365$, then there will be more than 10.5 hours of daylight in the tth day of the year if $a < t < b$.

Let us solve the equation $D(t) = 10.5$ as follows:

$$3 \sin\left[\frac{2\pi}{365}(t - 79)\right] + 12 = 10.5 \qquad \text{let } D(t) = 10.5$$

$$3 \sin\left[\frac{2\pi}{365}(t - 79)\right] = -1.5 \qquad \text{subtract 12}$$

$$\sin\left[\frac{2\pi}{365}(t - 79)\right] = -0.5 = -\frac{1}{2} \qquad \text{divide by 3}$$

Figure 6

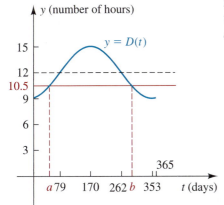

(continued)

If $\sin \theta = -\frac{1}{2}$, then the reference angle is $\pi/6$ and the angle θ is in either quadrant III or quadrant IV. Thus, we can find the numbers a and b by solving the equations

$$\frac{2\pi}{365}(t - 79) = \frac{7\pi}{6} \quad \text{and} \quad \frac{2\pi}{365}(t - 79) = \frac{11\pi}{6}.$$

From the first of these equations we obtain

$$t - 79 = \frac{7\pi}{6} \cdot \frac{365}{2\pi} = \frac{2555}{12} \approx 213,$$

and hence $\qquad t \approx 213 + 79, \qquad \text{or} \qquad t \approx 292.$

Similarly, the second equation gives us $t \approx 414$. Since the period of the function D is 365 days (see Figure 6), we obtain

$$t \approx 414 - 365, \qquad \text{or} \qquad t \approx 49.$$

Thus, there will be at least 10.5 hours of daylight from $t = 49$ to $t = 292$—that is, for 243 days of the year. ◤

EXAMPLE 10 Finding the minimum current in an electrical circuit

The current I (in amperes) in an alternating current circuit at time t (in seconds) is given by

$$I = 30 \sin\left(50\pi t - \frac{7\pi}{3}\right).$$

Find the smallest exact value of t for which $I = 15$.

SOLUTION Letting $I = 15$ in the given formula, we obtain

$$15 = 30 \sin\left(50\pi t - \frac{7\pi}{3}\right) \quad \text{or, equivalently,} \quad \sin\left(50\pi t - \frac{7\pi}{3}\right) = \frac{1}{2}.$$

Thus, the reference angle is $\pi/6$, and consequently

$$50\pi t - \frac{7\pi}{3} = \frac{\pi}{6} + 2\pi n \qquad \text{or} \qquad 50\pi t - \frac{7\pi}{3} = \frac{5\pi}{6} + 2\pi n,$$

where n is any integer. Solving for t gives us

$$t = \frac{\frac{15}{6} + 2n}{50} \qquad \text{or} \qquad t = \frac{\frac{19}{6} + 2n}{50}.$$

The smallest positive value of t will occur when one of the numerators of these two fractions has its least positive value. Since $\frac{15}{6} = 2.5$, $\frac{19}{6} \approx 3.17$, and $2(-1) = -2$, we see that the smallest positive value of t occurs when $n = -1$ in the first fraction—that is, when

$$t = \frac{\frac{15}{6} + 2(-1)}{50} = \frac{1}{100}.$$

◤

7.2   Exercises

Exer. 1–36: Find all solutions of the equation.

1 $\sin x = -\dfrac{\sqrt{2}}{2}$ 2 $\cos t = -1$

3 $\tan \theta = \sqrt{3}$ 4 $\cot \alpha = -\dfrac{1}{\sqrt{3}}$

5 $\sec \beta = 2$ 6 $\csc \gamma = \sqrt{2}$

7 $\sin x = \dfrac{\pi}{2}$ 8 $\cos x = -\dfrac{\pi}{3}$

9 $\cos \theta = \dfrac{1}{\sec \theta}$ 10 $\csc \theta \sin \theta = 1$

11 $2 \cos 2\theta - \sqrt{3} = 0$ 12 $2 \sin 3\theta + \sqrt{2} = 0$

13 $\sqrt{3} \tan \dfrac{1}{3}t = 1$ 14 $\cos \dfrac{1}{4}x = -\dfrac{\sqrt{2}}{2}$

15 $\sin\left(\theta + \dfrac{\pi}{4}\right) = \dfrac{1}{2}$ 16 $\cos\left(x - \dfrac{\pi}{3}\right) = -1$

17 $\sin\left(2x - \dfrac{\pi}{3}\right) = \dfrac{1}{2}$ 18 $\cos\left(4x - \dfrac{\pi}{4}\right) = \dfrac{\sqrt{2}}{2}$

19 $2 \cos t + 1 = 0$ 20 $\cot \theta + 1 = 0$

21 $\tan^2 x = 1$ 22 $4 \cos \theta - 2 = 0$

23 $(\cos \theta - 1)(\sin \theta + 1) = 0$

24 $2 \cos x = \sqrt{3}$

25 $\sec^2 \alpha - 4 = 0$ 26 $3 - \tan^2 \beta = 0$

27 $\sqrt{3} + 2 \sin \beta = 0$ 28 $4 \sin^2 x - 3 = 0$

29 $\cot^2 x - 3 = 0$ 30 $(\sin t - 1) \cos t = 0$

31 $(2 \sin \theta + 1)(2 \cos \theta + 3) = 0$

32 $(2 \sin u - 1)\left(\cos u - \sqrt{2}\right) = 0$

33 $\sin 2x \, (\csc 2x - 2) = 0$ 34 $\tan \alpha + \tan^2 \alpha = 0$

35 $\cos (\ln x) = 0$ 36 $\ln (\sin x) = 0$

Exer. 37–60: Find the solutions of the equation that are in the interval $[0, 2\pi)$.

37 $\cos\left(2x - \dfrac{\pi}{4}\right) = 0$ 38 $\sin\left(3x - \dfrac{\pi}{4}\right) = 1$

39 $2 - 8 \cos^2 t = 0$ 40 $\cot^2 \theta - \cot \theta = 0$

41 $2 \sin^2 u = 1 - \sin u$

42 $2 \cos^2 t + 3 \cos t + 1 = 0$

43 $\tan^2 x \sin x = \sin x$ 44 $\sec \beta \csc \beta = 2 \csc \beta$

45 $2 \cos^2 \gamma + \cos \gamma = 0$ 46 $\sin x - \cos x = 0$

47 $\sin^2 \theta + \sin \theta - 6 = 0$ 48 $2 \sin^2 u + \sin u - 6 = 0$

49 $1 - \sin t = \sqrt{3} \cos t$ 50 $\cos \theta - \sin \theta = 1$

51 $\cos \alpha + \sin \alpha = 1$ 52 $\sqrt{3} \sin t + \cos t = 1$

53 $2 \tan t - \sec^2 t = 0$ 54 $\tan \theta + \sec \theta = 1$

55 $\cot \alpha + \tan \alpha = \csc \alpha \sec \alpha$

56 $\sin x + \cos x \cot x = \csc x$

57 $2 \sin^3 x + \sin^2 x - 2 \sin x - 1 = 0$

58 $\sec^5 \theta = 4 \sec \theta$

59 $2 \tan t \csc t + 2 \csc t + \tan t + 1 = 0$

60 $2 \sin v \csc v - \csc v = 4 \sin v - 2$

Exer. 61–66: Approximate, to the nearest $10'$, the solutions of the equation in the interval $[0°, 360°)$.

61 $\sin^2 t - 4 \sin t + 1 = 0$

62 $\cos^2 t - 4 \cos t + 2 = 0$

63 $\tan^2 \theta + 3 \tan \theta + 2 = 0$

64 $2 \tan^2 x - 3 \tan x - 1 = 0$

65 $12 \sin^2 u - 5 \sin u - 2 = 0$

66 $5 \cos^2 \alpha + 3 \cos \alpha - 2 = 0$

67 Tidal waves A tidal wave of height 50 feet and period 30 minutes is approaching a sea wall that is 12.5 feet above sea level (see the figure). From a particular point on shore, the distance y from sea level to the top of the wave is given by

$$y = 25 \cos \dfrac{\pi}{15}t,$$

with t in minutes. For approximately how many minutes of each 30-minute period is the top of the wave above the level of the top of the sea wall?

Exercise 67

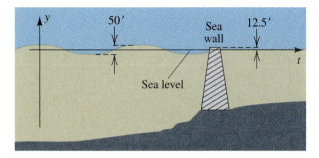

68 Temperature in Fairbanks The expected low temperature T (in °F) in Fairbanks, Alaska, may be approximated by

$$T = 36 \sin\left[\frac{2\pi}{365}(t - 101)\right] + 14,$$

where t is in days, with $t = 0$ corresponding to January 1. For how many days during the year is the low temperature expected to be below $-4°F$?

69 Intensity of sunlight On a clear day with D hours of daylight, the intensity of sunlight I (in calories/cm²) may be approximated by

$$I = I_M \sin^3 \frac{\pi t}{D} \quad \text{for} \quad 0 \le t \le D,$$

where $t = 0$ corresponds to sunrise and I_M is the maximum intensity. If $D = 12$, approximately how many hours after sunrise is $I = \frac{1}{2}I_M$?

70 Intensity of sunlight Refer to Exercise 69. On cloudy days, a better approximation of the sun intensity I is given by

$$I = I_M \sin^2 \frac{\pi t}{D}.$$

If $D = 12$, how many hours after sunrise is $I = \frac{1}{2}I_M$?

71 Protection from sunlight Refer to Exercises 69 and 70. A dermatologist recommends protection from the sun when the intensity I exceeds 75% of the maximum intensity. If $D = 12$ hours, approximate the number of hours for which protection is required on

(a) a clear day   (b) a cloudy day

72 Highway engineering In the study of frost penetration problems in highway engineering, the temperature T at time t hours and depth x feet is given by

$$T = T_0 e^{-\lambda x} \sin(\omega t - \lambda x),$$

where T_0, ω, and λ are constants and the period of T is 24 hours.

(a) Find a formula for the temperature at the surface.

(b) At what times is the surface temperature a minimum?

(c) If $\lambda = 2.5$, find the times when the temperature is a minimum at a depth of 1 foot.

73 Rabbit population Many animal populations, such as that of rabbits, fluctuate over ten-year cycles. Suppose that the number of rabbits at time t (in years) is given by

$$N(t) = 1000 \cos \frac{\pi}{5} t + 4000.$$

(a) Sketch the graph of N for $0 \le t \le 10$.

(b) For what values of t in part (a) does the rabbit population exceed 4500?

74 River flow rate The flow rate (or water discharge rate) at the mouth of the Orinoco River in South America may be approximated by

$$F(t) = 26{,}000 \sin\left[\frac{\pi}{6}(t - 5.5)\right] + 34{,}000,$$

where t is the time in months and $F(t)$ is the flow rate in m³/sec. For approximately how many months each year does the flow rate exceed 55,000 m³/sec?

75 Shown in the figure is a graph of $y = \frac{1}{2}x + \sin x$ for $-2\pi \le x \le 2\pi$. Using calculus, it can be shown that the x-coordinates of the turning points A, B, C, and D on the graph are solutions of the equation $\frac{1}{2} + \cos x = 0$. Determine the coordinates of these points.

Exercise 75

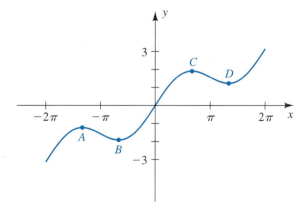

76 Shown in the figure is the graph of the equation

$$y = e^{-x/2} \sin 2x.$$

The x-coordinates of the turning points on the graph are solutions of $4 \cos 2x - \sin 2x = 0$. Approximate the x-coordinates of these points for $x > 0$.

Exercise 76

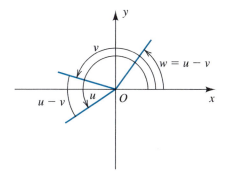

Exer. 77–78: If $I(t)$ is the current (in amperes) in an alternating current circuit at time t (in seconds), find the smallest exact value of t for which $I(t) = k$.

77 $I(t) = 20 \sin (60\pi t - 6\pi); \quad k = -10$

78 $I(t) = 40 \sin (100\pi t - 4\pi); \quad k = 20$

79 Weight at various latitudes The weight W of a person on the surface of Earth is directly proportional to the force of gravity g (in m/sec²). Because of rotation, Earth is flattened at the poles, and as a result weight will vary at different latitudes. If θ is the latitude, then g can be approximated by $g = 9.8066(1 - 0.00264 \cos 2\theta)$.

(a) At what latitude is $g = 9.8$?

(b) If a person weighs 150 pounds at the equator ($\theta = 0°$), at what latitude will the person weigh 150.5 pounds?

7.3

The Addition and Subtraction Formulas

In this section we derive formulas that involve trigonometric functions of $u + v$ or $u - v$ for any real numbers or angles u and v. These formulas are known as *addition* and *subtraction formulas,* respectively, or as *sum* and *difference identities.* The first formula that we will consider may be stated as follows.

Subtraction Formula for Cosine	$\cos (u - v) = \cos u \cos v + \sin u \sin v$

Figure 1

PROOF   Let u and v be any real numbers, and consider angles of radian measure u and v. Let $w = u - v$. Figure 1 illustrates one possibility with the angles in standard position. For convenience we have assumed that both u and v are positive and that $0 \le u - v < v$.

As in Figure 2, let $P(u_1, u_2)$, $Q(v_1, v_2)$, and $R(w_1, w_2)$ be the points on the terminal sides of the indicated angles that are each a distance 1 from the origin. In this case P, Q, and R are on the unit circle U with center at the origin. From the definition of trigonometric functions in terms of a unit circle,

$$\cos u = u_1 \qquad \cos v = v_1 \qquad \cos (u - v) = w_1$$
$$\sin u = u_2 \qquad \sin v = v_2 \qquad \sin (u - v) = w_2. \qquad (*)$$

(continued)

Figure 2

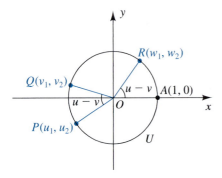

We next observe that the distance between $A(1, 0)$ and R must equal the distance between Q and P, because angles AOR and QOP have the same measure, $u - v$. Using the distance formula yields

$$d(A, R) = d(Q, P)$$
$$\sqrt{(w_1 - 1)^2 + (w_2 - 0)^2} = \sqrt{(u_1 - v_1)^2 + (u_2 - v_2)^2}.$$

Squaring both sides and simplifying the expressions under the radicals gives us

$$w_1^2 - 2w_1 + 1 + w_2^2 = u_1^2 - 2u_1v_1 + v_1^2 + u_2^2 - 2u_2v_2 + v_2^2.$$

Since the points (u_1, u_2), (v_1, v_2), and (w_1, w_2) are on the unit circle U and since an equation for U is $x^2 + y^2 = 1$, we may substitute 1 for each of $u_1^2 + u_2^2$, $v_1^2 + v_2^2$, and $w_1^2 + w_2^2$. Doing this and simplifying, we obtain

$$2 - 2w_1 = 2 - 2u_1v_1 - 2u_2v_2,$$

which reduces to

$$w_1 = u_1v_1 + u_2v_2.$$

Substituting from the formulas stated in (∗) gives us

$$\cos (u - v) = \cos u \cos v + \sin u \sin v,$$

which is what we wished to prove. It is possible to extend our discussion to all values of u and v.

The next example demonstrates the use of the subtraction formula in finding the *exact* value of cos 15°. Of course, if only an approximation were desired, we could use a calculator.

EXAMPLE 1   Using a subtraction formula

Find the exact value of cos 15° by using the fact that $15° = 60° - 45°$.

SOLUTION   We use the subtraction formula for cosine with $u = 60°$ and $v = 45°$:

$$\cos 15° = \cos (60° - 45°)$$
$$= \cos 60° \cos 45° + \sin 60° \sin 45°$$
$$= \frac{1}{2} \frac{\sqrt{2}}{2} + \frac{\sqrt{3}}{2} \frac{\sqrt{2}}{2}$$
$$= \frac{\sqrt{2} + \sqrt{6}}{4}$$

It is relatively easy to obtain a formula for cos $(u + v)$. We begin by writing $u + v$ as $u - (-v)$ and then use the subtraction formula for cosine:

$$\cos (u + v) = \cos [u - (-v)]$$
$$= \cos u \cos (-v) + \sin u \sin (-v)$$

Using the formulas for negatives, $\cos(-v) = \cos v$ and $\sin(-v) = -\sin v$, gives us the following addition formula for cosine.

Addition Formula for Cosine	$\cos(u + v) = \cos u \cos v - \sin u \sin v$

EXAMPLE 2   **Using an addition formula**

Find the exact value of $\cos \dfrac{7\pi}{12}$ by using the fact that $\dfrac{7\pi}{12} = \dfrac{\pi}{3} + \dfrac{\pi}{4}$.

SOLUTION   We apply the addition formula for cosine:

$$\cos \frac{7\pi}{12} = \cos\left(\frac{\pi}{3} + \frac{\pi}{4}\right)$$

$$= \cos \frac{\pi}{3} \cos \frac{\pi}{4} - \sin \frac{\pi}{3} \sin \frac{\pi}{4}$$

$$= \frac{1}{2}\frac{\sqrt{2}}{2} - \frac{\sqrt{3}}{2}\frac{\sqrt{2}}{2}$$

$$= \frac{\sqrt{2} - \sqrt{6}}{4}$$

Figure 3

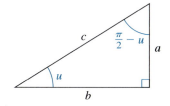

We refer to the sine and cosine functions as **cofunctions** of each other. Similarly, the tangent and cotangent functions are cofunctions, as are the secant and cosecant. If u is the radian measure of an acute angle, then the angle with radian measure $\pi/2 - u$ is complementary to u, and we may consider the right triangle shown in Figure 3. Using ratios, we see that

$$\sin u = \frac{a}{c} = \cos\left(\frac{\pi}{2} - u\right)$$

$$\cos u = \frac{b}{c} = \sin\left(\frac{\pi}{2} - u\right)$$

$$\tan u = \frac{a}{b} = \cot\left(\frac{\pi}{2} - u\right).$$

These three formulas and their analogues for $\sec u$, $\csc u$, and $\cot u$ state that *the function value of u equals the cofunction of the complementary angle $\pi/2 - u$.*

In the following formulas we use subtraction formulas to extend these relationships to any real number u, provided the function values are defined.

Cofunction Formulas	If u is a real number or the radian measure of an angle, then

$$(1)\ \cos\left(\frac{\pi}{2} - u\right) = \sin u \qquad\qquad (2)\ \sin\left(\frac{\pi}{2} - u\right) = \cos u$$

$$(3)\ \tan\left(\frac{\pi}{2} - u\right) = \cot u \qquad\qquad (4)\ \cot\left(\frac{\pi}{2} - u\right) = \tan u$$

$$(5)\ \sec\left(\frac{\pi}{2} - u\right) = \csc u \qquad\qquad (6)\ \csc\left(\frac{\pi}{2} - u\right) = \sec u$$

PROOFS Using the subtraction formula for cosine, we have

$$\cos\left(\frac{\pi}{2} - u\right) = \cos\frac{\pi}{2}\cos u + \sin\frac{\pi}{2}\sin u$$

$$= (0)\cos u + (1)\sin u = \sin u.$$

This gives us formula 1.

If we substitute $\pi/2 - v$ for u in the first formula, we obtain

$$\cos\left[\frac{\pi}{2} - \left(\frac{\pi}{2} - v\right)\right] = \sin\left(\frac{\pi}{2} - v\right),$$

or $$\cos v = \sin\left(\frac{\pi}{2} - v\right).$$

Since the symbol v is arbitrary, this equation is equivalent to the second cofunction formula:

$$\sin\left(\frac{\pi}{2} - u\right) = \cos u$$

Using the tangent identity, cofunction formulas 1 and 2, and the cotangent identity, we obtain a proof for the third formula:

$$\tan\left(\frac{\pi}{2} - u\right) = \frac{\sin\left(\frac{\pi}{2} - u\right)}{\cos\left(\frac{\pi}{2} - u\right)} = \frac{\cos u}{\sin u} = \cot u$$

The proofs of the remaining three formulas are similar.

An easy way to remember the cofunction formulas is to refer to the triangle in Figure 3.

We may now prove the following identities.

Addition and Subtraction Formulas for Sine and Tangent	**(1)** $\sin(u + v) = \sin u \cos v + \cos u \sin v$
	(2) $\sin(u - v) = \sin u \cos v - \cos u \sin v$
	(3) $\tan(u + v) = \dfrac{\tan u + \tan v}{1 - \tan u \tan v}$
	(4) $\tan(u - v) = \dfrac{\tan u - \tan v}{1 + \tan u \tan v}$

PROOFS   We shall prove formulas 1 and 3. Using the cofunction formulas and the subtraction formula for cosine, we can verify formula 1:

$$\sin(u + v) = \cos\left[\frac{\pi}{2} - (u + v)\right]$$

$$= \cos\left[\left(\frac{\pi}{2} - u\right) - v\right]$$

$$= \cos\left(\frac{\pi}{2} - u\right)\cos v + \sin\left(\frac{\pi}{2} - u\right)\sin v$$

$$= \sin u \cos v + \cos u \sin v$$

To verify formula 3, we begin as follows:

$$\tan(u + v) = \frac{\sin(u + v)}{\cos(u + v)}$$

$$= \frac{\sin u \cos v + \cos u \sin v}{\cos u \cos v - \sin u \sin v}$$

Dividing by $\cos u \cos v$ *will give us an expression involving tangents; dividing by* $\sin u \sin v$ *would give us an expression involving cotangents.*

If $\cos u \cos v \neq 0$, then we may divide the numerator and the denominator by $\cos u \cos v$, obtaining

$$\tan(u + v) = \frac{\left(\dfrac{\sin u}{\cos u}\right)\left(\dfrac{\cos v}{\cos v}\right) + \left(\dfrac{\cos u}{\cos u}\right)\left(\dfrac{\sin v}{\cos v}\right)}{\left(\dfrac{\cos u}{\cos u}\right)\left(\dfrac{\cos v}{\cos v}\right) - \left(\dfrac{\sin u}{\cos u}\right)\left(\dfrac{\sin v}{\cos v}\right)}$$

$$= \frac{\tan u + \tan v}{1 - \tan u \tan v}.$$

If $\cos u \cos v = 0$, then either $\cos u = 0$ or $\cos v = 0$. In this case, either $\tan u$ or $\tan v$ is undefined and the formula is invalid. Proofs of formulas 2 and 4 are left as exercises.   ◢

EXAMPLE 3 Using addition formulas to find
the quadrant containing an angle

Suppose $\sin \alpha = \frac{4}{5}$ and $\cos \beta = -\frac{12}{13}$, where α is in quadrant I and β is in quadrant II.

(a) Find the exact values of $\sin (\alpha + \beta)$ and $\tan (\alpha + \beta)$.

(b) Find the quadrant containing $\alpha + \beta$.

Figure 4

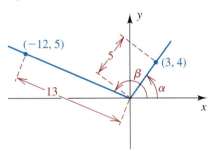

SOLUTION Angles α and β are illustrated in Figure 4. There is no loss of generality in regarding α and β as positive angles between 0 and 2π, as we have done in the figure. Since $\sin \alpha = \frac{4}{5}$, we may choose the point $(3, 4)$ on the terminal side of α. Similarly, since $\cos \beta = -\frac{12}{13}$, the point $(-12, 5)$ is on the terminal side of β. Referring to Figure 4 and using the definition of the trigonometric functions of any angle, we have

$$\cos \alpha = \tfrac{3}{5}, \quad \tan \alpha = \tfrac{4}{3}, \quad \sin \beta = \tfrac{5}{13}, \quad \tan \beta = -\tfrac{5}{12}.$$

(a) Addition formulas give us

$$\sin (\alpha + \beta) = \sin \alpha \cos \beta + \cos \alpha \sin \beta = \left(\tfrac{4}{5}\right)\left(-\tfrac{12}{13}\right) + \left(\tfrac{3}{5}\right)\left(\tfrac{5}{13}\right) = -\tfrac{33}{65}$$

$$\tan (\alpha + \beta) = \frac{\tan \alpha + \tan \beta}{1 - \tan \alpha \tan \beta} = \frac{\tfrac{4}{3} + \left(-\tfrac{5}{12}\right)}{1 - \left(\tfrac{4}{3}\right)\left(-\tfrac{5}{12}\right)} \cdot \frac{36}{36} = \frac{33}{56}.$$

(b) Since $\sin (\alpha + \beta)$ is negative and $\tan (\alpha + \beta)$ is positive, the angle $\alpha + \beta$ must be in quadrant III.

The next example illustrates a type of simplification of the difference quotient (introduced in Section 3.4) with the sine function. The resulting form is useful in calculus.

EXAMPLE 4 A formula used in calculus

If $f(x) = \sin x$ and $h \neq 0$, show that

$$\frac{f(x + h) - f(x)}{h} = \sin x \left(\frac{\cos h - 1}{h}\right) + \cos x \left(\frac{\sin h}{h}\right).$$

SOLUTION We use the definition of f and the addition formula for sine:

$$\frac{f(x + h) - f(x)}{h} = \frac{\sin (x + h) - \sin x}{h}$$

$$= \frac{\sin x \cos h + \cos x \sin h - \sin x}{h}$$

$$= \frac{\sin x (\cos h - 1) + \cos x \sin h}{h}$$

$$= \sin x \left(\frac{\cos h - 1}{h}\right) + \cos x \left(\frac{\sin h}{h}\right)$$

Addition formulas may also be used to derive **reduction formulas.** Reduction formulas may be used to change expressions such as

$$\sin\left(\theta + \frac{\pi}{2}n\right) \quad \text{and} \quad \cos\left(\theta + \frac{\pi}{2}n\right) \quad \text{for any integer } n$$

to expressions involving only $\sin\theta$ or $\cos\theta$. Similar formulas are true for the other trigonometric functions. Instead of deriving general reduction formulas, we shall illustrate two special cases in the next example.

EXAMPLE 5 Obtaining reduction formulas

Express in terms of a trigonometric function of θ alone:

(a) $\sin\left(\theta - \frac{3\pi}{2}\right)$ **(b)** $\cos(\theta + \pi)$

SOLUTION Using subtraction and addition formulas, we obtain the following:

(a) $\sin\left(\theta - \frac{3\pi}{2}\right) = \sin\theta\cos\frac{3\pi}{2} - \cos\theta\sin\frac{3\pi}{2}$

$$= \sin\theta\cdot(0) - \cos\theta\cdot(-1) = \cos\theta$$

(b) $\cos(\theta + \pi) = \cos\theta\cos\pi - \sin\theta\sin\pi$

$$= \cos\theta\cdot(-1) - \sin\theta\cdot(0) = -\cos\theta$$

EXAMPLE 6 Combining a sum involving the sine and cosine functions

Let a and b be real numbers with $a > 0$. Show that for every x,

$$a\cos Bx + b\sin Bx = A\cos(Bx - C),$$

Since $\cos u = \sin\left(\frac{\pi}{2} - u\right)$, we could also write the sum in terms of a sine function.

where $A = \sqrt{a^2 + b^2}$ and $\tan C = \dfrac{b}{a}$ with $-\dfrac{\pi}{2} < C < \dfrac{\pi}{2}$.

SOLUTION Given $a\cos Bx + b\sin Bx$, let us consider $\tan C = b/a$ with $-\pi/2 < C < \pi/2$. Thus, $b = a\tan C$, and we may write

$$a\cos Bx + b\sin Bx = a\cos Bx + (a\tan C)\sin Bx$$

$$= a\cos Bx + a\frac{\sin C}{\cos C}\sin Bx$$

$$= \frac{a}{\cos C}(\cos C\cos Bx + \sin C\sin Bx)$$

$$= (a\sec C)\cos(Bx - C).$$

We shall complete the proof by showing that $a\sec C = \sqrt{a^2 + b^2}$. Since $-\pi/2 < C < \pi/2$, it follows that $\sec C$ is positive, and hence

$$a\sec C = a\sqrt{1 + \tan^2 C}.$$

(continued)

Using $\tan C = b/a$ and $a > 0$, we obtain

$$a \sec C = a \sqrt{1 + \frac{b^2}{a^2}} = \sqrt{a^2 \left(1 + \frac{b^2}{a^2}\right)} = \sqrt{a^2 + b^2}.$$

EXAMPLE 7 An application of Example 6

If $f(x) = \cos x + \sin x$, use the formulas given in Example 6 to express $f(x)$ in the form $A \cos (Bx - C)$, and then sketch the graph of f.

SOLUTION Letting $a = 1$, $b = 1$, and $B = 1$ in the formulas from Example 6, we have

$$A = \sqrt{a^2 + b^2} = \sqrt{1 + 1} = \sqrt{2} \quad \text{and} \quad \tan C = \frac{b}{a} = \frac{1}{1} = 1.$$

Since $\tan C = 1$ and $-\pi/2 < C < \pi/2$, we have $C = \pi/4$. Substituting for a, b, A, B, and C in the formula

$$a \cos Bx + b \sin Bx = A \cos (Bx - C)$$

gives us

$$f(x) = \cos x + \sin x = \sqrt{2} \cos \left(x - \frac{\pi}{4}\right).$$

Figure 5

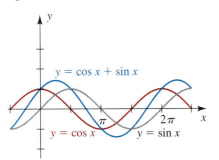

$y = \cos x + \sin x$

$y = \cos x$ $y = \sin x$

Comparing the last formula with the equation $y = a \cos (bx + c)$, which we discussed in Section 6.5, we see that the amplitude of the graph is $\sqrt{2}$, the period is 2π, and the phase shift is $\pi/4$. The graph of f is sketched in Figure 5, where we have also shown the graphs of $y = \sin x$ and $y = \cos x$. Our sketch agrees with that obtained in Chapter 6. (See Figure 10 in Section 6.6.)

7.3 Exercises

Exer. 1–4: Express as a cofunction of a complementary angle.

1 (a) $\sin 46°37'$ (b) $\cos 73°12'$

(c) $\tan \dfrac{\pi}{6}$ (d) $\sec 17.28°$

2 (a) $\tan 24°12'$ (b) $\sin 89°41'$

(c) $\cos \dfrac{\pi}{3}$ (d) $\cot 61.87°$

3 (a) $\cos \dfrac{7\pi}{20}$ (b) $\sin \dfrac{1}{4}$

(c) $\tan 1$ (d) $\csc 0.53$

4 (a) $\sin \dfrac{\pi}{12}$ (b) $\cos 0.64$

(c) $\tan \sqrt{2}$ (d) $\sec 1.2$

Exer. 5–10: Find the exact values.

5 (a) $\cos \dfrac{\pi}{4} + \cos \dfrac{\pi}{6}$

(b) $\cos \dfrac{5\pi}{12}$ $\left(\text{use } \dfrac{5\pi}{12} = \dfrac{\pi}{4} + \dfrac{\pi}{6}\right)$

6 (a) $\sin \dfrac{2\pi}{3} + \sin \dfrac{\pi}{4}$

(b) $\sin \dfrac{11\pi}{12}$ $\left(\text{use } \dfrac{11\pi}{12} = \dfrac{2\pi}{3} + \dfrac{\pi}{4}\right)$

7 (a) $\tan 60° + \tan 225°$

(b) $\tan 285°$ (use $285° = 60° + 225°$)

8 (a) $\cos 135° - \cos 60°$

(b) $\cos 75°$ (use $75° = 135° - 60°$)

9 (a) $\sin \dfrac{3\pi}{4} - \sin \dfrac{\pi}{6}$

(b) $\sin \dfrac{7\pi}{12}$ $\left(\text{use } \dfrac{7\pi}{12} = \dfrac{3\pi}{4} - \dfrac{\pi}{6} \right)$

10 (a) $\tan \dfrac{3\pi}{4} - \tan \dfrac{\pi}{6}$

(b) $\tan \dfrac{7\pi}{12}$ $\left(\text{use } \dfrac{7\pi}{12} = \dfrac{3\pi}{4} - \dfrac{\pi}{6} \right)$

Exer. 11–16: Express as a trigonometric function of one angle.

11 $\cos 48° \cos 23° + \sin 48° \sin 23°$

12 $\cos 13° \cos 50° - \sin 13° \sin 50°$

13 $\cos 10° \sin 5° - \sin 10° \cos 5°$

14 $\sin 57° \cos 4° + \cos 57° \sin 4°$

15 $\cos 3 \sin (-2) - \cos 2 \sin 3$

16 $\sin (-5) \cos 2 + \cos 5 \sin (-2)$

17 If α and β are acute angles such that $\cos \alpha = \frac{4}{5}$ and $\tan \beta = \frac{8}{15}$, find

(a) $\sin (\alpha + \beta)$ **(b)** $\cos (\alpha + \beta)$

(c) the quadrant containing $\alpha + \beta$

18 If α and β are acute angles such that $\csc \alpha = \frac{13}{12}$ and $\cot \beta = \frac{4}{3}$, find

(a) $\sin (\alpha + \beta)$ **(b)** $\tan (\alpha + \beta)$

(c) the quadrant containing $\alpha + \beta$

19 If $\sin \alpha = -\frac{4}{5}$ and $\sec \beta = \frac{5}{3}$ for a third-quadrant angle α and a first-quadrant angle β, find

(a) $\sin (\alpha + \beta)$ **(b)** $\tan (\alpha + \beta)$

(c) the quadrant containing $\alpha + \beta$

20 If $\tan \alpha = -\frac{7}{24}$ and $\cot \beta = \frac{3}{4}$ for a second-quadrant angle α and a third-quadrant angle β, find

(a) $\sin (\alpha + \beta)$ **(b)** $\cos (\alpha + \beta)$ **(c)** $\tan (\alpha + \beta)$

(d) $\sin (\alpha - \beta)$ **(e)** $\cos (\alpha - \beta)$ **(f)** $\tan (\alpha - \beta)$

21 If α and β are third-quadrant angles such that $\cos \alpha = -\frac{2}{5}$ and $\cos \beta = -\frac{3}{5}$, find

(a) $\sin (\alpha - \beta)$ **(b)** $\cos (\alpha - \beta)$

(c) the quadrant containing $\alpha - \beta$

22 If α and β are second-quadrant angles such that $\sin \alpha = \frac{2}{3}$ and $\cos \beta = -\frac{1}{3}$, find

(a) $\sin (\alpha + \beta)$ **(b)** $\tan (\alpha + \beta)$

(c) the quadrant containing $\alpha + \beta$

Exer. 23–34: Verify the reduction formula.

23 $\sin (\theta + \pi) = -\sin \theta$

24 $\sin \left(x + \dfrac{\pi}{2} \right) = \cos x$

25 $\sin \left(x - \dfrac{5\pi}{2} \right) = -\cos x$

26 $\sin \left(\theta - \dfrac{3\pi}{2} \right) = \cos \theta$

27 $\cos (\theta - \pi) = -\cos \theta$

28 $\cos \left(x + \dfrac{\pi}{2} \right) = -\sin x$

29 $\cos \left(x + \dfrac{3\pi}{2} \right) = \sin x$

30 $\cos \left(\theta - \dfrac{5\pi}{2} \right) = \sin \theta$

31 $\tan \left(x - \dfrac{\pi}{2} \right) = -\cot x$

32 $\tan (\pi - \theta) = -\tan \theta$

33 $\tan \left(\theta + \dfrac{\pi}{2} \right) = -\cot \theta$

34 $\tan (x + \pi) = \tan x$

Exer. 35–44: Verify the identity.

35 $\sin \left(\theta + \dfrac{\pi}{4} \right) = \dfrac{\sqrt{2}}{2} (\sin \theta + \cos \theta)$

36 $\cos \left(\theta + \dfrac{\pi}{4} \right) = \dfrac{\sqrt{2}}{2} (\cos \theta - \sin \theta)$

37 $\tan \left(u + \dfrac{\pi}{4} \right) = \dfrac{1 + \tan u}{1 - \tan u}$

38 $\tan \left(x - \dfrac{\pi}{4} \right) = \dfrac{\tan x - 1}{\tan x + 1}$

39 $\cos (u + v) + \cos (u - v) = 2 \cos u \cos v$

40 $\sin (u + v) + \sin (u - v) = 2 \sin u \cos v$

41 $\sin (u + v) \cdot \sin (u - v) = \sin^2 u - \sin^2 v$

42 $\cos (u + v) \cdot \cos (u - v) = \cos^2 u - \sin^2 v$

43 $\dfrac{1}{\cot \alpha - \cot \beta} = \dfrac{\sin \alpha \sin \beta}{\sin (\beta - \alpha)}$

44 $\dfrac{1}{\tan \alpha + \tan \beta} = \dfrac{\cos \alpha \cos \beta}{\sin (\alpha + \beta)}$

45 Express $\sin (u + v + w)$ in terms of trigonometric functions of u, v, and w. (*Hint:* Write

$$\sin (u + v + w) \quad \text{as} \quad \sin [(u + v) + w]$$

and use addition formulas.)

46 Express $\tan (u + v + w)$ in terms of trigonometric functions of u, v, and w.

47 Derive the formula $\cot (u + v) = \dfrac{\cot u \cot v - 1}{\cot u + \cot v}$.

48 If α and β are complementary angles, show that

$$\sin^2 \alpha + \sin^2 \beta = 1.$$

49 Derive the subtraction formula for the sine function.

50 Derive the subtraction formula for the tangent function.

51 If $f(x) = \cos x$, show that

$$\frac{f(x + h) - f(x)}{h} = \cos x \left(\frac{\cos h - 1}{h} \right) - \sin x \left(\frac{\sin h}{h} \right).$$

52 If $f(x) = \tan x$, show that

$$\frac{f(x + h) - f(x)}{h} = \sec^2 x \left(\frac{\sin h}{h} \right) \frac{1}{\cos h - \sin h \tan x}.$$

Exer. 53–54: (a) Compare the decimal approximations of both sides of equation (1). (b) Find the acute angle α such that equation (2) is an identity. (c) How does equation (1) relate to equation (2)?

53 (1) $\sin 63° - \sin 57° = \sin 3°$

(2) $\sin (\alpha + \beta) - \sin (\alpha - \beta) = \sin \beta$

54 (1) $\sin 35° + \sin 25° = \cos 5°$

(2) $\sin (\alpha + \beta) + \sin (\alpha - \beta) = \cos \beta$

Exer. 55–60: Use an addition or subtraction formula to find the solutions of the equation that are in the interval $[0, \pi)$.

55 $\sin 4t \cos t = \sin t \cos 4t$

56 $\cos 5t \cos 3t = \tfrac{1}{2} + \sin (-5t) \sin 3t$

57 $\cos 5t \cos 2t = -\sin 5t \sin 2t$

58 $\sin 3t \cos t + \cos 3t \sin t = -\tfrac{1}{2}$

59 $\tan 2t + \tan t = 1 - \tan 2t \tan t$

60 $\tan t - \tan 4t = 1 + \tan 4t \tan t$

Exer. 61–64: (a) Use the formula from Example 6 to express f in terms of the cosine function. (b) Determine the amplitude, period, and phase shift of f. (c) Sketch the graph of f.

61 $f(x) = \sqrt{3} \cos 2x + \sin 2x$

62 $f(x) = \cos 4x + \sqrt{3} \sin 4x$

63 $f(x) = 2 \cos 3x - 2 \sin 3x$

64 $f(x) = 5 \cos 10x - 5 \sin 10x$

Exer. 65–66: For certain applications in electrical engineering, the sum of several voltage signals or radio waves of the same frequency is expressed in the compact form $y = A \cos (Bt - C)$. Express the given signal in this form.

65 $y = 50 \sin 60\pi t + 40 \cos 60\pi t$

66 $y = 10 \sin \left(120\pi t - \dfrac{\pi}{2} \right) + 5 \sin 120\pi t$

67 Motion of a mass If a mass that is attached to a spring is raised y_0 feet and released with an initial vertical velocity of v_0 ft/sec, then the subsequent position y of the mass is given by

$$y = y_0 \cos \omega t + \frac{v_0}{\omega} \sin \omega t,$$

where t is time in seconds and ω is a positive constant.

(a) If $\omega = 1$, $y_0 = 2$ ft, and $v_0 = 3$ ft/sec, express y in the form $A \cos (Bt - C)$, and find the amplitude and period of the resulting motion.

(b) Determine the times when $y = 0$—that is, the times when the mass passes through the equilibrium position.

68 Motion of a mass Refer to Exercise 67. If $y_0 = 1$ and $\omega = 2$, find the initial velocities that result in an amplitude of 4 feet.

69 Pressure on the eardrum If a tuning fork is struck and then held a certain distance from the eardrum, the pressure $p_1(t)$ on the outside of the eardrum at time t may be represented by $p_1(t) = A \sin \omega t$, where A and ω are positive constants. If a second identical tuning fork is struck with a possibly different force and held a different distance from the eardrum (see the figure), its effect may be represented by the equation $p_2(t) = B \sin(\omega t + \tau)$, where B is a positive constant and $0 \le \tau \le 2\pi$. The total pressure $p(t)$ on the eardrum is given by

$$p(t) = A \sin \omega t + B \sin(\omega t + \tau).$$

(a) Show that $p(t) = a \cos \omega t + b \sin \omega t$, where

$$a = B \sin \tau \quad \text{and} \quad b = A + B \cos \tau.$$

(b) Show that the amplitude C of p is given by

$$C^2 = A^2 + B^2 + 2AB \cos \tau.$$

Exercise 69

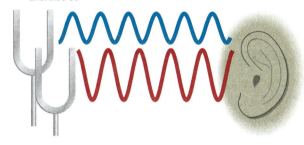

70 Destructive interference Refer to Exercise 69. Destructive interference occurs if the amplitude of the resulting sound wave is less than A. Suppose that the two tuning forks are struck with the same force—that is, $A = B$.

(a) When total destructive interference occurs, the amplitude of p is zero and no sound is heard. Find the least positive value of τ for which this occurs.

(b) Determine the τ-interval (a, b) for which destructive interference occurs and a has its least positive value.

71 Constructive interference Refer to Exercise 69. When two tuning forks are struck, constructive interference occurs if the amplitude C of the resulting sound wave is larger than either A or B (see the figure).

(a) Show that $C \le A + B$.

(b) Find the values of τ such that $C = A + B$.

(c) If $A \ge B$, determine a condition under which constructive interference will occur.

Exercise 71

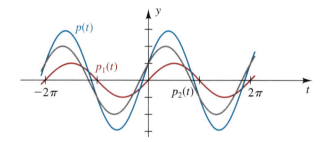

7.4

Multiple-Angle Formulas

We refer to the formulas considered in this section as **multiple-angle formulas.** In particular, the following identities are **double-angle formulas,** because they contain the expression $2u$.

Double-Angle Formulas	
	(1) $\sin 2u = 2 \sin u \cos u$
	(2) (a) $\cos 2u = \cos^2 u - \sin^2 u$
	(b) $\cos 2u = 1 - 2 \sin^2 u$
	(c) $\cos 2u = 2 \cos^2 u - 1$
	(3) $\tan 2u = \dfrac{2 \tan u}{1 - \tan^2 u}$

PROOFS Each of these formulas may be proved by letting $v = u$ in the appropriate addition formulas. If we use the formula for $\sin(u + v)$, then

$$\sin 2u = \sin(u + u)$$
$$= \sin u \cos u + \cos u \sin u$$
$$= 2 \sin u \cos u.$$

Using the formula for $\cos(u + v)$, we have

$$\cos 2u = \cos(u + u)$$
$$= \cos u \cos u - \sin u \sin u$$
$$= \cos^2 u - \sin^2 u.$$

To obtain the other two forms for $\cos 2u$ in 2(b) and 2(c), we use the fundamental identity $\sin^2 u + \cos^2 u = 1$. Thus,

$$\cos 2u = \cos^2 u - \sin^2 u$$
$$= (1 - \sin^2 u) - \sin^2 u$$
$$= 1 - 2 \sin^2 u.$$

Similarly, if we substitute for $\sin^2 u$ instead of $\cos^2 u$, we obtain

$$\cos 2u = \cos^2 u - (1 - \cos^2 u)$$
$$= 2 \cos^2 u - 1.$$

Formula 3 for $\tan 2u$ may be obtained by letting $v = u$ in the formula for $\tan(u + v)$.

EXAMPLE 1 Using double-angle formulas

If $\sin \alpha = \frac{4}{5}$ and α is an acute angle, find the exact values of $\sin 2\alpha$ and $\cos 2\alpha$.

SOLUTION If we regard α as an acute angle of a right triangle, as shown in Figure 1, we obtain $\cos \alpha = \frac{3}{5}$. We next substitute in double-angle formulas:

$$\sin 2\alpha = 2 \sin \alpha \cos \alpha = 2\left(\tfrac{4}{5}\right)\left(\tfrac{3}{5}\right) = \tfrac{24}{25}$$

$$\cos 2\alpha = \cos^2 \alpha - \sin^2 \alpha = \left(\tfrac{3}{5}\right)^2 - \left(\tfrac{4}{5}\right)^2 = \tfrac{9}{25} - \tfrac{16}{25} = -\tfrac{7}{25}$$

The next example demonstrates how to change a multiple-angle expression to a single-angle expression.

EXAMPLE 2 Changing the form of $\cos 3\theta$

Express $\cos 3\theta$ in terms of $\cos \theta$.

Figure 1

SOLUTION

$$\cos 3\theta = \cos (2\theta + \theta) \qquad \textcolor{red}{3\theta = 2\theta + \theta}$$
$$= \cos 2\theta \cos \theta - \sin 2\theta \sin \theta \qquad \textcolor{red}{\text{addition formula}}$$
$$= (2 \cos^2 \theta - 1) \cos \theta - (2 \sin \theta \cos \theta) \sin \theta \qquad \textcolor{red}{\text{double-angle formulas}}$$
$$= 2 \cos^3 \theta - \cos \theta - 2 \cos \theta \sin^2 \theta \qquad \textcolor{red}{\text{multiply}}$$
$$= 2 \cos^3 \theta - \cos \theta - 2 \cos \theta (1 - \cos^2 \theta) \qquad \textcolor{red}{\sin^2 \theta + \cos^2 \theta = 1}$$
$$= 4 \cos^3 \theta - 3 \cos \theta \qquad \textcolor{red}{\text{simplify}} \qquad \blacksquare$$

We call each of the next three formulas a **half-angle identity,** because the number u is one-half the number $2u$.

Half-Angle Identities

(1) $\sin^2 u = \dfrac{1 - \cos 2u}{2}$ $\qquad$ **(2)** $\cos^2 u = \dfrac{1 + \cos 2u}{2}$

$$\textbf{(3)} \ \tan^2 u = \frac{1 - \cos 2u}{1 + \cos 2u}$$

PROOFS The first identity may be verified as follows:

$$\cos 2u = 1 - 2 \sin^2 u \qquad \textcolor{red}{\text{double-angle formula 2(b)}}$$
$$2 \sin^2 u = 1 - \cos 2u \qquad \textcolor{red}{\text{isolate } 2 \sin^2 u}$$
$$\sin^2 u = \frac{1 - \cos 2u}{2} \qquad \textcolor{red}{\text{divide by 2}}$$

The second identity can be derived in similar fashion by starting with

$$\cos 2u = 2 \cos^2 u - 1.$$

The third identity can be obtained from identities 1 and 2 by noting that

$$\tan^2 u = (\tan u)^2 = \left(\frac{\sin u}{\cos u} \right)^2 = \frac{\sin^2 u}{\cos^2 u}. \qquad \blacksquare$$

Half-angle identities may be used to express even powers of trigonometric functions in terms of functions with exponent 1, as illustrated in the next two examples.

EXAMPLE 3 **Using half-angle identities to verify an identity**

Verify the identity $\sin^2 x \cos^2 x = \frac{1}{8}(1 - \cos 4x)$.

SOLUTION

$$\sin^2 x \cos^2 x = \left(\frac{1 - \cos 2x}{2}\right)\left(\frac{1 + \cos 2x}{2}\right) \quad \text{half-angle identities}$$

$$= \tfrac{1}{4}(1 - \cos^2 2x) \quad \text{multiply}$$

$$= \tfrac{1}{4}(\sin^2 2x) \quad \sin^2 2x + \cos^2 2x = 1$$

$$= \frac{1}{4}\left(\frac{1 - \cos 4x}{2}\right) \quad \begin{array}{l}\text{half-angle identity with}\\ u = 2x\end{array}$$

$$= \tfrac{1}{8}(1 - \cos 4x) \quad \text{multiply}$$

EXAMPLE 4 **Using half-angle identities to reduce a power of cos t**

Express $\cos^4 t$ in terms of values of the cosine function with exponent 1.

SOLUTION

$$\cos^4 t = (\cos^2 t)^2 \quad \text{law of exponents}$$

$$= \left(\frac{1 + \cos 2t}{2}\right)^2 \quad \text{half-angle identity}$$

$$= \tfrac{1}{4}(1 + 2 \cos 2t + \cos^2 2t) \quad \text{square}$$

$$= \frac{1}{4}\left(1 + 2 \cos 2t + \frac{1 + \cos 4t}{2}\right) \quad \text{half-angle identity with } u = 2t$$

$$= \tfrac{3}{8} + \tfrac{1}{2} \cos 2t + \tfrac{1}{8} \cos 4t \quad \text{simplify}$$

Substituting $v/2$ for u in the three half-angle identities gives us

$$\sin^2 \frac{v}{2} = \frac{1 - \cos v}{2} \qquad \cos^2 \frac{v}{2} = \frac{1 + \cos v}{2} \qquad \tan^2 \frac{v}{2} = \frac{1 - \cos v}{1 + \cos v}.$$

Taking the square roots of both sides of each of these equations, we obtain the following, which we call the *half-angle formulas* in order to distinguish them from the half-angle identities.

Half-Angle Formulas

(1) $\sin \dfrac{v}{2} = \pm\sqrt{\dfrac{1 - \cos v}{2}}$ **(2)** $\cos \dfrac{v}{2} = \pm\sqrt{\dfrac{1 + \cos v}{2}}$

(3) $\tan \dfrac{v}{2} = \pm\sqrt{\dfrac{1 - \cos v}{1 + \cos v}}$

When using a half-angle formula, we choose either the $+$ or the $-$, depending on the quadrant containing the angle of radian measure $v/2$. Thus, for $\sin (v/2)$ we use $+$ if $v/2$ is an angle in quadrant I or II or $-$ if $v/2$ is in quadrant III or IV. For $\cos (v/2)$ we use $+$ if $v/2$ is in quadrant I or IV, and so on.

EXAMPLE 5   **Using half-angle formulas for the sine and cosine**

Find exact values for

(a) $\sin 22.5°$ **(b)** $\cos 112.5°$

SOLUTION

(a) We choose the positive sign because $22.5°$ is in quadrant I, and hence $\sin 22.5° > 0$.

$$\sin 22.5° = +\sqrt{\frac{1 - \cos 45°}{2}} \qquad \text{half-angle formula for sine with } v = 45°$$

$$= \sqrt{\frac{1 - \sqrt{2}/2}{2}} \qquad \cos 45° = \frac{\sqrt{2}}{2}$$

$$= \frac{\sqrt{2 - \sqrt{2}}}{2} \qquad \text{multiply radicand by } \frac{2}{2} \text{ and simplify}$$

(b) Similarly, we choose the negative sign because $112.5°$ is in quadrant II, and so $\cos 112.5° < 0$.

$$\cos 112.5° = -\sqrt{\frac{1 + \cos 225°}{2}} \qquad \begin{array}{l}\text{half-angle formula for} \\ \text{cosine with } v = 225°\end{array}$$

$$= -\sqrt{\frac{1 - \sqrt{2}/2}{2}} \qquad \cos 225° = -\frac{\sqrt{2}}{2}$$

$$= -\frac{\sqrt{2 - \sqrt{2}}}{2} \qquad \text{multiply radicand by } \frac{2}{2} \text{ and simplify} \quad \blacksquare$$

We can obtain an alternative form for the half-angle formula for $\tan (v/2)$. Multiplying the numerator and denominator of the radicand in the third half-angle formula by $1 - \cos v$ gives us

$$\tan \frac{v}{2} = \pm\sqrt{\frac{1 - \cos v}{1 + \cos v} \cdot \frac{1 - \cos v}{1 - \cos v}}$$

$$= \pm\sqrt{\frac{(1 - \cos v)^2}{1 - \cos^2 v}}$$

$$= \pm\sqrt{\frac{(1 - \cos v)^2}{\sin^2 v}} = \pm\frac{1 - \cos v}{\sin v}.$$

We can eliminate the $\pm$ sign in the preceding formula. First note that the numerator $1 - \cos v$ is never negative. We can show that $\tan (v/2)$ and $\sin v$ always have the same sign. For example, if $0 < v < \pi$, then $0 < v/2 < \pi/2$, and consequently both $\sin v$ and $\tan (v/2)$ are positive. If $\pi < v < 2\pi$, then $\pi/2 < v/2 < \pi$, and hence both $\sin v$ and $\tan (v/2)$ are negative, which gives us the first of the next two identities. The second identity for tan $(v/2)$ may be obtained by multiplying the numerator and denominator of the radicand in the third half-angle formula by $1 + \cos v$.

Half-Angle Formulas for the Tangent	**(1)** $\tan \dfrac{v}{2} = \dfrac{1 - \cos v}{\sin v}$	**(2)** $\tan \dfrac{v}{2} = \dfrac{\sin v}{1 + \cos v}$

Figure 2

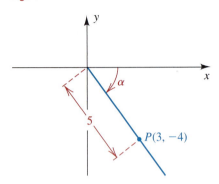

EXAMPLE 6 Using a half-angle formula for the tangent

If $\tan \alpha = -\dfrac{4}{3}$ and α is in quadrant IV, find $\tan \dfrac{\alpha}{2}$.

SOLUTION   If we choose the point $(3, -4)$ on the terminal side of α, as illustrated in Figure 2, then $\sin \alpha = -\frac{4}{5}$ and $\cos \alpha = \frac{3}{5}$. Applying the first half-angle formula for the tangent, we obtain

$$\tan \frac{\alpha}{2} = \frac{1 - \cos \alpha}{\sin \alpha} = \frac{1 - \frac{3}{5}}{-\frac{4}{5}} = -\frac{1}{2}.$$

EXAMPLE 7 Finding the *x*-intercepts of a graph

Figure 3

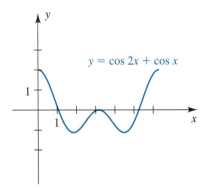

A graph of the equation $y = \cos 2x + \cos x$ for $0 \le x \le 2\pi$ is sketched in Figure 3. The *x*-intercepts appear to be approximately 1.1, 3.1, and 5.2. Find their exact values and three-decimal-place approximations.

SOLUTION   To find the *x*-intercepts, we proceed as follows:

$$\cos 2x + \cos x = 0 \qquad \text{let } y = 0$$
$$(2 \cos^2 x - 1) + \cos x = 0 \qquad \text{double-angle formula 2(c)}$$
$$2 \cos^2 x + \cos x - 1 = 0 \qquad \text{equivalent equation}$$
$$(2 \cos x - 1)(\cos x + 1) = 0 \qquad \text{factor}$$
$$2 \cos x - 1 = 0, \quad \cos x + 1 = 0 \qquad \text{zero factor theorem}$$
$$\cos x = \tfrac{1}{2}, \qquad \cos x = -1 \qquad \text{solve for } \cos x$$

The solutions of the last two equations in the interval $[0, 2\pi]$ give us the following exact and approximate *x*-intercepts:

$$\frac{\pi}{3} \approx 1.047, \quad \frac{5\pi}{3} \approx 5.236, \quad \pi \approx 3.142$$

Figure 4

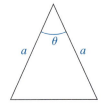

EXAMPLE 8 Deriving a formula for the area of an isosceles triangle

An **isosceles triangle** has two equal sides of length a, and the angle between them is θ (see Figure 4). Express the area A of the triangle in terms of a and θ.

Figure 5

SOLUTION From Figure 5 we see that the altitude from point P bisects θ and that $A = \frac{1}{2}(2k)h = kh$. Thus, we have the following, where $\theta/2$ is an acute angle:

$$\sin \frac{\theta}{2} = \frac{k}{a} \qquad\qquad \cos \frac{\theta}{2} = \frac{h}{a} \qquad \text{see Figure 5}$$

$$k = a \sin \frac{\theta}{2} \qquad\qquad h = a \cos \frac{\theta}{2} \qquad \text{solve for } k \text{ and } h$$

We next find the area:

$$A = a^2 \sin \frac{\theta}{2} \cos \frac{\theta}{2} \qquad \text{substitute in } A = kh \qquad (*)$$

$$= a^2 \sqrt{\frac{1 - \cos \theta}{2}} \sqrt{\frac{1 + \cos \theta}{2}} \qquad \text{half-angle formulas with } \theta/2 \text{ in quadrant I}$$

$$= a^2 \sqrt{\frac{1 - \cos^2 \theta}{4}} \qquad \text{law of radicals}$$

$$= a^2 \sqrt{\frac{\sin^2 \theta}{4}} \qquad \sin^2 \theta + \cos^2 \theta = 1$$

$$= \tfrac{1}{2} a^2 |\sin \theta| \qquad \text{take the square root}$$

$$= \tfrac{1}{2} a^2 \sin \theta \qquad \sin \theta > 0 \text{ for } 0° < \theta < 180°$$

Another method for simplifying $(*)$ is to write the double-angle formula for the sine, $\sin 2u = 2 \sin u \cos u$, as

$$\sin u \cos u = \tfrac{1}{2} \sin 2u \qquad (**)$$

and proceed as follows:

$$A = a^2 \sin \frac{\theta}{2} \cos \frac{\theta}{2} \qquad \text{substitute in } A = kh$$

$$= a^2 \cdot \frac{1}{2} \sin \left(2 \cdot \frac{\theta}{2} \right) \qquad \text{let } u = \frac{\theta}{2} \text{ in } (**)$$

$$= \tfrac{1}{2} a^2 \sin \theta \qquad \text{simplify} \qquad \blacksquare$$

7.4 Exercises

Exer. 1–4: Find the exact values of sin 2θ, cos 2θ, and tan 2θ for the given values of θ.

1 $\cos \theta = \frac{3}{5};$ $\quad 0° < \theta < 90°$

2 $\cot \theta = \frac{4}{3};$ $\quad 180° < \theta < 270°$

3 $\sec \theta = -3;$ $\quad 90° < \theta < 180°$

4 $\sin \theta = -\frac{4}{5};$ $\quad 270° < \theta < 360°$

Exer. 5–8: Find the exact values of sin $(\theta/2)$, cos $(\theta/2)$, and tan $(\theta/2)$ for the given conditions.

5 $\sec \theta = \frac{5}{4};$ $\quad 0° < \theta < 90°$

6 $\csc \theta = -\frac{5}{3};$ $\quad -90° < \theta < 0°$

7 $\tan \theta = 1;$ $\quad -180° < \theta < -90°$

8 $\sec \theta = -4;$ $\quad 180° < \theta < 270°$

Exer. 9–10: Use half-angle formulas to find the exact values.

9 (a) $\cos 67°30'$ (b) $\sin 15°$ (c) $\tan \dfrac{3\pi}{8}$

10 (a) $\cos 165°$ (b) $\sin 157°30'$ (c) $\tan \dfrac{\pi}{8}$

Exer. 11–30: Verify the identity.

11 $\sin 10\theta = 2 \sin 5\theta \cos 5\theta$

12 $\cos^2 3x - \sin^2 3x = \cos 6x$

13 $4 \sin \dfrac{x}{2} \cos \dfrac{x}{2} = 2 \sin x$

14 $\dfrac{\sin^2 2\alpha}{\sin^2 \alpha} = 4 - 4 \sin^2 \alpha$

15 $(\sin t + \cos t)^2 = 1 + \sin 2t$

16 $\csc 2u = \tfrac{1}{2} \csc u \sec u$

17 $\sin 3u = \sin u\,(3 - 4 \sin^2 u)$

18 $\sin 4t = 4 \sin t \cos t\,(1 - 2 \sin^2 t)$

19 $\cos 4\theta = 8 \cos^4 \theta - 8 \cos^2 \theta + 1$

20 $\cos 6t = 32 \cos^6 t - 48 \cos^4 t + 18 \cos^2 t - 1$

21 $\sin^4 t = \tfrac{3}{8} - \tfrac{1}{2} \cos 2t + \tfrac{1}{8} \cos 4t$

22 $\cos^4 x - \sin^4 x = \cos 2x$

23 $\sec 2\theta = \dfrac{\sec^2 \theta}{2 - \sec^2 \theta}$ 24 $\cot 2u = \dfrac{\cot^2 u - 1}{2 \cot u}$

25 $2 \sin^2 2t + \cos 4t = 1$ 26 $\tan \theta + \cot \theta = 2 \csc 2\theta$

27 $\tan 3u = \dfrac{\tan u\,(3 - \tan^2 u)}{1 - 3 \tan^2 u}$

28 $\dfrac{1 + \sin 2v + \cos 2v}{1 + \sin 2v - \cos 2v} = \cot v$

29 $\tan \dfrac{\theta}{2} = \csc \theta - \cot \theta$

30 $\tan^2 \dfrac{\theta}{2} = 1 - 2 \cot \theta \csc \theta + 2 \cot^2 \theta$

Exer. 31–34: Express in terms of the cosine function with exponent 1.

31 $\cos^4 \dfrac{\theta}{2}$ 32 $\cos^4 2x$

33 $\sin^4 2x$ 34 $\sin^4 \dfrac{\theta}{2}$

Exer. 35–42: Find the solutions of the equation that are in the interval $[0, 2\pi)$.

35 $\sin 2t + \sin t = 0$ 36 $\cos t - \sin 2t = 0$

37 $\cos u + \cos 2u = 0$ 38 $\cos 2\theta - \tan \theta = 1$

39 $\tan 2x = \tan x$ 40 $\tan 2t - 2 \cos t = 0$

41 $\sin \tfrac{1}{2}u + \cos u = 1$ 42 $2 - \cos^2 x = 4 \sin^2 \tfrac{1}{2}x$

43 If $a > 0$, $b > 0$, and $0 < u < \pi/2$, show that

$$a \sin u + b \cos u = \sqrt{a^2 + b^2} \, \sin (u + v)$$

for $0 < v < \pi/2$, with

$$\sin v = \dfrac{b}{\sqrt{a^2 + b^2}} \quad \text{and} \quad \cos v = \dfrac{a}{\sqrt{a^2 + b^2}}.$$

44 Use Exercise 43 to express $8 \sin u + 15 \cos u$ in the form $c \sin (u + v)$.

45 A graph of $y = \cos 2x + 2 \cos x$ for $0 \le x \le 2\pi$ is shown in the figure.

(a) Approximate the x-intercepts to two decimal places.

(b) The x-coordinates of the turning points P, Q, and R on the graph are solutions of $\sin 2x + \sin x = 0$. Find the coordinates of these points.

Exercise 45

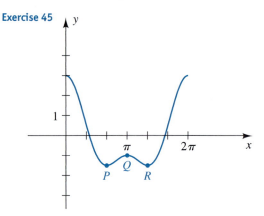

46 A graph of $y = \cos x - \sin 2x$ for $-2\pi \le x \le 2\pi$ is shown in the figure.

(a) Find the x-intercepts.

(b) The x-coordinates of the eight turning points on the graph are solutions of $\sin x + 2 \cos 2x = 0$. Approximate these x-coordinates to two decimal places.

Exercise 46

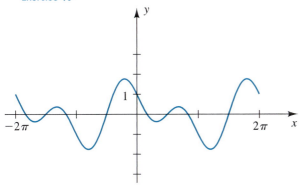

47 A graph of $y = \cos 3x - 3 \cos x$ for $-2\pi \le x \le 2\pi$ is shown in the figure.

(a) Find the x-intercepts. (*Hint:* Use the formula for $\cos 3\theta$ given in Example 2.)

(b) The x-coordinates of the 13 turning points on the graph are solutions of $\sin 3x - \sin x = 0$. Find these x-coordinates. (*Hint:* Use the formula for $\sin 3u$ in Exercise 17.)

Exercise 47

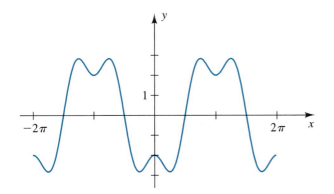

48 A graph of $y = \sin 4x - 4 \sin x$ for $-2\pi \le x \le 2\pi$ is shown in the figure. Find the x-intercepts. (*Hint:* Use the formula for $\sin 4t$ in Exercise 18.)

Exercise 48

49 Planning a railroad route Shown in the figure is a proposed railroad route through three towns located at points A, B, and C. At B, the track will turn toward C at an angle θ.

(a) Show that the total distance d from A to C is given by $d = 20 \tan \frac{1}{2}\theta + 40$.

(b) Because of mountains between A and C, the turning point B must be at least 20 miles from A. Is there a route that avoids the mountains and measures exactly 50 miles?

Exercise 49

50 Projectile's range If a projectile is fired from ground level with an initial velocity of v ft/sec and at an angle of θ degrees with the horizontal, the range R of the projectile is given by

$$R = \frac{v^2}{16} \sin \theta \cos \theta.$$

If $v = 80$ ft/sec, approximate the angles that result in a range of 150 feet.

51 Constructing a rain gutter Shown in the figure is a design for a rain gutter.

(a) Express the volume V as a function of θ. (*Hint:* See Example 8.)

(b) Approximate the acute angle θ that results in a volume of 2 ft³.

Exercise 51

52 Designing curbing A highway engineer is designing curbing for a street at an intersection where two highways meet at an angle ϕ, as shown in the figure. The curbing between points A and B is to be constructed using a circle that is tangent to the highway at these two points.

(a) Show that the relationship between the radius R of the circle and the distance d in the figure is given by the equation $d = R \tan(\phi/2)$.

(b) If $\phi = 45°$ and $d = 20$ ft, approximate R and the length of the curbing.

Exercise 52

53 Arterial bifurcation A common form of cardiovascular branching is bifurcation, in which an artery splits into two smaller blood vessels. The bifurcation angle θ is the angle formed by the two smaller arteries. In the figure, the line through A and D bisects θ and is perpendicular to the line through B and C.

(a) Show that the length l of the artery from A to B is given by $l = a + \dfrac{b}{2} \tan \dfrac{\theta}{4}$.

(b) Estimate the length l from the three measurements $a = 10$ mm, $b = 6$ mm, and $\theta = 156°$.

Exercise 53

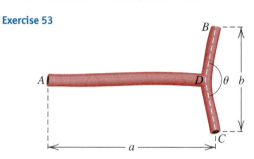

54 Heat production in an AC circuit By definition, the average value of $f(t) = c + a \cos bt$ for one or more complete cycles is c (see the figure).

(a) Use a double-angle formula to find the average value of $f(t) = \sin^2 \omega t$ for $0 \le t \le 2\pi/\omega$, with t in seconds.

(b) In an electrical circuit with an alternating current $I = I_0 \sin \omega t$, the rate r (in calories/sec) at which heat is produced in an R-ohm resistor is given by $r = RI^2$. Find the average rate at which heat is produced for one complete cycle.

Exercise 54

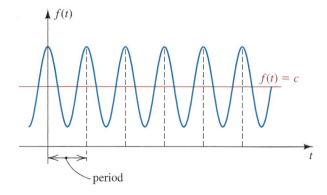

7.5

Product-to-Sum and Sum-to-Product Formulas

The following formulas may be used to change the form of certain trigonometric expressions from products to sums. We refer to these as **product-to-sum formulas** even though two of the formulas express a product as a difference, because any difference $x - y$ of two real numbers is also a sum $x + (-y)$. These formulas are frequently used in calculus as an aid in a process called *integration*.

Product-to-Sum Formulas	
	(1) $\sin u \cos v = \frac{1}{2}[\sin (u + v) + \sin (u - v)]$
	(2) $\cos u \sin v = \frac{1}{2}[\sin (u + v) - \sin (u - v)]$
	(3) $\cos u \cos v = \frac{1}{2}[\cos (u + v) + \cos (u - v)]$
	(4) $\sin u \sin v = \frac{1}{2}[\cos (u - v) - \cos (u + v)]$

PROOFS   Let us add the left-hand and right-hand sides of the addition and subtraction formulas for the sine function, as follows:

$$\sin (u + v) = \quad \sin u \cos v + \cos u \sin v$$
$$\underline{\sin (u - v) = \quad \sin u \cos v - \cos u \sin v}$$
$$\sin (u + v) + \sin (u - v) = 2 \sin u \cos v$$

Dividing both sides of the last equation by 2 gives us formula 1.

Formula 2 is obtained by *subtracting* the left- and right-hand sides of the addition and subtraction formulas for the sine function. Formulas 3 and 4 are developed in a similar fashion, using the addition and subtraction formulas for the cosine function.   ◢

EXAMPLE 1   Using product-to-sum formulas

Express as a sum:

(a) $\sin 4\theta \cos 3\theta$ **(b)** $\sin 3x \sin x$

SOLUTION

(a) We use product-to-sum formula 1 with $u = 4\theta$ and $v = 3\theta$.

$$\sin 4\theta \cos 3\theta = \frac{1}{2}[\sin (4\theta + 3\theta) + \sin (4\theta - 3\theta)]$$
$$= \frac{1}{2}(\sin 7\theta + \sin \theta)$$

We can also obtain this relationship by using product-to-sum formula 2.

(b) We use product-to-sum formula 4 with $u = 3x$ and $v = x$:

$$\sin 3x \sin x = \frac{1}{2}[\cos (3x - x) - \cos (3x + x)]$$
$$= \frac{1}{2}(\cos 2x - \cos 4x)$$ ◢

We may use the product-to-sum formulas to express a sum or difference as a product. To obtain forms that can be applied more easily, we shall change the notation as follows. If we let

$$u + v = a \quad \text{and} \quad u - v = b,$$

then $(u + v) + (u - v) = a + b$, which simplifies to

$$u = \frac{a + b}{2}.$$

Similarly, since $(u + v) - (u - v) = a - b$, we obtain

$$v = \frac{a - b}{2}.$$

We now substitute for $u + v$ and $u - v$ on the right-hand sides of the product-to-sum formulas and for u and v on the left-hand sides. If we then multiply by 2, we obtain the following sum-to-product formulas.

Sum-to-Product Formulas

$$\textbf{(1)} \ \sin a + \sin b = 2 \sin \frac{a + b}{2} \cos \frac{a - b}{2}$$

$$\textbf{(2)} \ \sin a - \sin b = 2 \cos \frac{a + b}{2} \sin \frac{a - b}{2}$$

$$\textbf{(3)} \ \cos a + \cos b = 2 \cos \frac{a + b}{2} \cos \frac{a - b}{2}$$

$$\textbf{(4)} \ \cos a - \cos b = -2 \sin \frac{a + b}{2} \sin \frac{a - b}{2}$$

EXAMPLE 2 **Using a sum-to-product formula**

Express $\sin 5x - \sin 3x$ as a product.

SOLUTION We use sum-to-product formula 2 with $a = 5x$ and $b = 3x$:

$$\sin 5x - \sin 3x = 2 \cos \frac{5x + 3x}{2} \sin \frac{5x - 3x}{2}$$

$$= 2 \cos 4x \sin x \qquad \blacksquare$$

EXAMPLE 3 **Using sum-to-product formulas to verify an identity**

Verify the identity $\dfrac{\sin 3t + \sin 5t}{\cos 3t - \cos 5t} = \cot t.$

SOLUTION   We first use a sum-to-product formula for the numerator and one for the denominator:

$$\frac{\sin 3t + \sin 5t}{\cos 3t - \cos 5t} = \frac{2 \sin \dfrac{3t + 5t}{2} \cos \dfrac{3t - 5t}{2}}{-2 \sin \dfrac{3t + 5t}{2} \sin \dfrac{3t - 5t}{2}}$$ sum-to-product formulas 1 and 4

$$= \frac{2 \sin 4t \cos (-t)}{-2 \sin 4t \sin (-t)}$$ simplify

$$= \frac{\cos (-t)}{-\sin (-t)}$$ cancel $2 \sin 4t$

$$= \frac{\cos t}{\sin t}$$ formulas for negatives

$$= \cot t$$ cotangent identity   ◢

EXAMPLE 4   **Using a sum-to-product formula to solve an equation**

Find the solutions of $\sin 5x + \sin x = 0$.

SOLUTION   Changing a sum to a product allows us to use the zero factor theorem to solve the equation.

$$\sin 5x + \sin x = 0$$ given

$$2 \sin \frac{5x + x}{2} \cos \frac{5x - x}{2} = 0$$ sum-to-product formula 1

$$\sin 3x \cos 2x = 0$$ simplify and divide by 2

$$\sin 3x = 0, \quad \cos 2x = 0$$ zero factor theorem

The solutions of the last two equations are

$$3x = \pi n \quad \text{and} \quad 2x = \frac{\pi}{2} + \pi n \quad \text{for every integer } n.$$

Dividing by 3 and 2, respectively, we obtain

$$\frac{\pi}{3} n \quad \text{and} \quad \frac{\pi}{4} + \frac{\pi}{2} n \quad \text{for every integer } n.$$ ◢

EXAMPLE 5   **Finding the x-intercepts of a graph**

A graph of the equation $y = \cos x - \cos 3x - \sin 2x$ is shown in Figure 1. Find the 13 x-intercepts that are in the interval $[-2\pi, 2\pi]$.

Figure 1

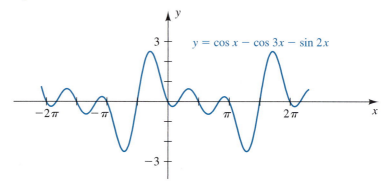

SOLUTION To find the *x*-intercepts, we proceed as follows:

$$\cos x - \cos 3x - \sin 2x = 0 \qquad \text{let } y = 0$$

$$(\cos x - \cos 3x) - \sin 2x = 0 \qquad \text{group the first two terms}$$

$$-2 \sin \frac{x + 3x}{2} \sin \frac{x - 3x}{2} - \sin 2x = 0 \qquad \text{sum-to-product formula 4}$$

$$-2 \sin 2x \sin (-x) - \sin 2x = 0 \qquad \text{simplify}$$

$$2 \sin 2x \sin x - \sin 2x = 0 \qquad \text{formula for negatives}$$

$$\sin 2x \, (2 \sin x - 1) = 0 \qquad \text{factor out } \sin 2x$$

$$\sin 2x = 0, \quad 2 \sin x - 1 = 0 \qquad \text{zero factor theorem}$$

$$\sin 2x = 0, \qquad \sin x = \tfrac{1}{2} \qquad \text{solve for } \sin x$$

The equation $\sin 2x = 0$ has solutions $2x = \pi n$, or, dividing by 2,

$$x = \frac{\pi}{2} n \quad \text{for every integer } n.$$

If we let $n = 0, \pm 1, \pm 2, \pm 3$, and ± 4, we obtain nine *x*-intercepts in $[-2\pi, 2\pi]$:

$$0, \quad \pm \frac{\pi}{2}, \quad \pm \pi, \quad \pm \frac{3\pi}{2}, \quad \pm 2\pi$$

The solutions of the equation $\sin x = \tfrac{1}{2}$ are

$$\frac{\pi}{6} + 2\pi n \quad \text{and} \quad \frac{5\pi}{6} + 2\pi n \quad \text{for every integer } n.$$

The four solutions in $[-2\pi, 2\pi]$ are obtained by letting $n = 0$ and $n = -1$:

$$\frac{\pi}{6}, \quad \frac{5\pi}{6}, \quad -\frac{11\pi}{6}, \quad -\frac{7\pi}{6}$$

7.5 Exercises

Exer. 1–8: Express as a sum or difference.

1 $\sin 7t \sin 3t$

2 $\sin (-4x) \cos 8x$

3 $\cos 6u \cos (-4u)$

4 $\cos 4t \sin 6t$

5 $2 \sin 9\theta \cos 3\theta$

6 $2 \sin 7\theta \sin 5\theta$

7 $3 \cos x \sin 2x$

8 $5 \cos u \cos 5u$

Exer. 9–16: Express as a product.

9 $\sin 6\theta + \sin 2\theta$

10 $\sin 4\theta - \sin 8\theta$

11 $\cos 5x - \cos 3x$

12 $\cos 5t + \cos 6t$

13 $\sin 3t - \sin 7t$

14 $\cos \theta - \cos 5\theta$

15 $\cos x + \cos 2x$

16 $\sin 8t + \sin 2t$

Exer. 17–24: Verify the identity.

17 $\dfrac{\sin 4t + \sin 6t}{\cos 4t - \cos 6t} = \cot t$

18 $\dfrac{\sin \theta + \sin 3\theta}{\cos \theta + \cos 3\theta} = \tan 2\theta$

19 $\dfrac{\sin u + \sin v}{\cos u + \cos v} = \tan \dfrac{1}{2}(u + v)$

20 $\dfrac{\sin u - \sin v}{\cos u - \cos v} = -\cot \dfrac{1}{2}(u + v)$

21 $\dfrac{\sin u - \sin v}{\sin u + \sin v} = \dfrac{\tan \frac{1}{2}(u - v)}{\tan \frac{1}{2}(u + v)}$

22 $\dfrac{\cos u - \cos v}{\cos u + \cos v} = -\tan \dfrac{1}{2}(u + v) \tan \dfrac{1}{2}(u - v)$

23 $4 \cos x \cos 2x \sin 3x = \sin 2x + \sin 4x + \sin 6x$

24 $\dfrac{\cos t + \cos 4t + \cos 7t}{\sin t + \sin 4t + \sin 7t} = \cot 4t$

Exer. 25–26: Express as a sum.

25 $(\sin ax)(\cos bx)$

26 $(\cos au)(\cos bu)$

Exer. 27–34: Use sum-to-product formulas to find the solutions of the equation.

27 $\sin 5t + \sin 3t = 0$

28 $\sin t + \sin 3t = \sin 2t$

29 $\cos x = \cos 3x$

30 $\cos 4x - \cos 3x = 0$

31 $\cos 3x + \cos 5x = \cos x$

32 $\cos 3x = -\cos 6x$

33 $\sin 2x - \sin 5x = 0$

34 $\sin 5x - \sin x = 2 \cos 3x$

Exer. 35–36: Shown in the figure is a graph of the function f for $0 \le x \le 2\pi$. Use a sum-to-product formula to help find the x-intercepts.

35 $f(x) = \cos x + \cos 3x$

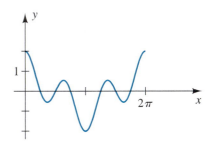

36 $f(x) = \sin 4x - \sin x$

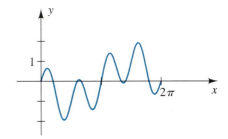

37 Refer to Exercise 47 of Section 7.4. The graph of the equation $y = \cos 3x - 3 \cos x$ has 13 turning points for $-2\pi \le x \le 2\pi$. The x-coordinates of these points are solutions of the equation $\sin 3x - \sin x = 0$. Use a sum-to-product formula to find these x-coordinates.

38 Refer to Exercise 48 of Section 7.4. The x-coordinates of the turning points on the graph of $y = \sin 4x - 4 \sin x$ are solutions of $\cos 4x - \cos x = 0$. Use a sum-to-product formula to find these x-coordinates for $-2\pi \le x \le 2\pi$.

39 Vibration of a violin string Mathematical analysis of a vibrating violin string of length l involves functions such that

$$f(x) = \sin \left(\frac{\pi n}{l} x \right) \cos \left(\frac{k \pi n}{l} t \right),$$

where n is an integer, k is a constant, and t is time. Express f as a sum of two sine functions.

40 Pressure on the eardrum If two tuning forks are struck simultaneously with the same force and are then held at the same distance from the eardrum, the pressure on the outside of the eardrum at time t is given by

$$p(t) = a \cos \omega_1 t + a \cos \omega_2 t,$$

where a, ω_1, and ω_2 are constants. If ω_1 and ω_2 are almost equal, a tone is produced that alternates between loudness and virtual silence. This phenomenon is known as beats.

(a) Use a sum-to-product formula to express $p(t)$ as a product.

(b) Show that $p(t)$ may be considered as a cosine wave with approximate period $2\pi/\omega_1$ and variable amplitude $f(t) = 2a \cos \frac{1}{2}(\omega_1 - \omega_2)t$. Find the maximum amplitude.

(c) Shown in the figure is a graph of the equation

$$p(t) = \cos 4.5t + \cos 3.5t.$$

Near-silence occurs at points A and B, where the variable amplitude $f(t)$ in part (b) is zero. Find the coordinates of these points, and determine how frequently near-silence occurs.

(d) Use the graph to show that the function p in part (c) has period 4π. Conclude that the maximum amplitude of 2 occurs every 4π units of time.

Exercise 40

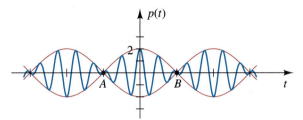

7.6

The Inverse Trigonometric Functions

Recall from Section 5.1 that to define the inverse function f^{-1} of a function f, it is essential that f be one-to-one; that is, if $a \neq b$ in the domain of f, then $f(a) \neq f(b)$. The inverse function f^{-1} *reverses* the correspondence given by f; that is,

$$u = f(v) \qquad \text{if and only if} \qquad v = f^{-1}(u).$$

The following general relationships involving f and f^{-1} were discussed in Section 5.1.

Relationships Between f^{-1} and f	
	(1) $y = f^{-1}(x)$ if and only if $x = f(y)$, where x is in the domain of f^{-1} and y is in the domain of f
	(2) domain of f^{-1} = range of f
	(3) range of f^{-1} = domain of f
	(4) $f(f^{-1}(x)) = x$ for every x in the domain of f^{-1}
	(5) $f^{-1}(f(y)) = y$ for every y in the domain of f
	(6) The point (a, b) is on the graph of f if and only if the point (b, a) is on the graph of f^{-1}.
	(7) The graphs of f^{-1} and f are reflections of each other through the line $y = x$.

We shall use relationship 1 to define each of the inverse trigonometric functions.

The sine function is not one-to-one, since different numbers, such as $\pi/6$, $5\pi/6$, and $-7\pi/6$, yield the same function value $\left(\frac{1}{2}\right)$. If we restrict the domain to $[-\pi/2, \pi/2]$, then, as illustrated by the blue portion of the graph of $y = \sin x$ in Figure 1, we obtain a one-to-one (increasing) function that takes on every value of the sine function once and only once. We use this *new* function with domain $[-\pi/2, \pi/2]$ and range $[-1, 1]$ to define the *inverse sine function*.

Figure 1

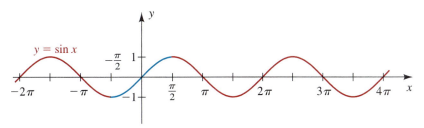

Definition of the Inverse Sine Function	The **inverse sine function,** denoted by $\sin^{-1}$, is defined by $$y = \sin^{-1} x \quad \text{if and only if} \quad x = \sin y$$ for $-1 \le x \le 1$ and $-\dfrac{\pi}{2} \le y \le \dfrac{\pi}{2}$.

The domain of the inverse sine function is $[-1, 1]$, and the range is $[-\pi/2, \pi/2]$.

The notation $y = \sin^{-1} x$ is sometimes read "y is the inverse sine of x." The equation $x = \sin y$ in the definition allows us to regard y as an angle, so $y = \sin^{-1} x$ may also be read "y is the angle whose sine is x" (with $-\pi/2 \le y \le \pi/2$).

The inverse sine function is also called the **arcsine function,** and arcsin x may be used in place of $\sin^{-1} x$. If $t = \arcsin x$, then $\sin t = x$, and t may be interpreted as an *arc length* on the unit circle U with center at the origin. We will use both notations—$\sin^{-1}$ and arcsin—throughout our work.

Several values of the inverse sine function are listed in the next chart.

Note on notation:

While $(\sin x)^{-1} = \dfrac{1}{\sin x} = \csc x,$

none *of these equal* $\sin^{-1} x.$

🔻 *Warning!* 🔻

It is *essential* to choose the value y in the range $[-\pi/2, \pi/2]$ of $\sin^{-1}$. Thus, even though $\sin (5\pi/6) = \frac{1}{2}$, the number $y = 5\pi/6$ is not the inverse function value $\sin^{-1} \frac{1}{2}$.

Equation	Equivalent statement		Solution
$y = \sin^{-1}\left(\dfrac{1}{2}\right)$	$\sin y = \dfrac{1}{2}$	and $-\dfrac{\pi}{2} \le y \le \dfrac{\pi}{2}$	$y = \dfrac{\pi}{6}$
$y = \sin^{-1}\left(-\dfrac{1}{2}\right)$	$\sin y = -\dfrac{1}{2}$	and $-\dfrac{\pi}{2} \le y \le \dfrac{\pi}{2}$	$y = -\dfrac{\pi}{6}$
$y = \sin^{-1}(1)$	$\sin y = 1$	and $-\dfrac{\pi}{2} \le y \le \dfrac{\pi}{2}$	$y = \dfrac{\pi}{2}$
$y = \arcsin(0)$	$\sin y = 0$	and $-\dfrac{\pi}{2} \le y \le \dfrac{\pi}{2}$	$y = 0$
$y = \arcsin\left(-\dfrac{\sqrt{3}}{2}\right)$	$\sin y = -\dfrac{\sqrt{3}}{2}$	and $-\dfrac{\pi}{2} \le y \le \dfrac{\pi}{2}$	$y = -\dfrac{\pi}{3}$

Figure 2

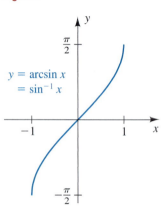

$y = \arcsin x$
$= \sin^{-1} x$

We have now justified the method of solving an equation of the form $\sin \theta = k$ as discussed in Chapter 6. We see that the calculator key $\boxed{\text{SIN}^{-1}}$ used to obtain $\theta = \sin^{-1} k$ gives us the value of the inverse sine function.

Relationship 7 for the graphs of f and f^{-1} tells us that we can sketch the graph of $y = \sin^{-1} x$ by reflecting the blue portion of Figure 1 through the line $y = x$. We can also use the equation $x = \sin y$ with the restriction $-\pi/2 \le y \le \pi/2$ to find points on the graph. This gives us Figure 2.

Relationship 4, $f(f^{-1}(x)) = x$, and relationship 5, $f^{-1}(f(y)) = y$, which hold for any inverse function f^{-1}, give us the following properties.

Properties of sin⁻¹

(1) $\sin(\sin^{-1} x) = \sin(\arcsin x) = x$ if $-1 \le x \le 1$

(2) $\sin^{-1}(\sin y) = \arcsin(\sin y) = y$ if $-\dfrac{\pi}{2} \le y \le \dfrac{\pi}{2}$

EXAMPLE 1   Using properties of sin⁻¹

Find the exact value:

(a) $\sin\left(\sin^{-1}\dfrac{1}{2}\right)$ (b) $\sin^{-1}\left(\sin\dfrac{\pi}{4}\right)$ (c) $\sin^{-1}\left(\sin\dfrac{2\pi}{3}\right)$

SOLUTION

(a) The *difficult* way to find the value of this expression is to first find the angle $\sin^{-1}\frac{1}{2}$, namely $\pi/6$, and then evaluate $\sin(\pi/6)$, obtaining $\frac{1}{2}$. The *easy* way is to use property 1 of $\sin^{-1}$:

$$\text{since} \quad -1 \le \tfrac{1}{2} \le 1, \quad \sin\left(\sin^{-1}\tfrac{1}{2}\right) = \tfrac{1}{2}$$

(b) Since $-\pi/2 \le \pi/4 \le \pi/2$, we can use property 2 of $\sin^{-1}$ to obtain

$$\sin^{-1}\left(\sin\frac{\pi}{4}\right) = \frac{\pi}{4}.$$

(c) Be careful! Since $2\pi/3$ is *not* between $-\pi/2$ and $\pi/2$, we cannot use property 2 of $\sin^{-1}$. Instead, we first evaluate the inner expression, $\sin(2\pi/3)$, and then use the definition of $\sin^{-1}$, as follows:

$$\sin^{-1}\left(\sin\frac{2\pi}{3}\right) = \sin^{-1}\left(\frac{\sqrt{3}}{2}\right) = \frac{\pi}{3}$$

EXAMPLE 2 **Finding a value of $\sin^{-1}$**

Find the exact value of y if $y = \sin^{-1}\left(\tan\frac{3\pi}{4}\right)$.

SOLUTION   We first evaluate the inner expression—$\tan(3\pi/4)$—and then find the inverse sine of that number:

$$y = \sin^{-1}\left(\tan\frac{3\pi}{4}\right) = \sin^{-1}(-1)$$

In words, we have "y is the angle whose sine is -1." It may be helpful to recall the arcsine values by associating them with the angles corresponding to the blue portion of the unit circle shown in Figure 3. From the figure we see that $-\pi/2$ is the angle whose sine is -1. It follows that $y = -\pi/2$, and hence

$$y = \sin^{-1}\left(\tan\frac{3\pi}{4}\right) = -\frac{\pi}{2}.$$

The other trigonometric functions may also be used to introduce inverse trigonometric functions. The procedure is first to determine a convenient subset of the domain in order to obtain a one-to-one function. If the domain of the cosine function is restricted to the interval $[0, \pi]$, as illustrated by the blue portion of the graph of $y = \cos x$ in Figure 4, we obtain a one-to-one (decreasing) function that takes on every value of the cosine function once and only once. Then, we use this *new* function with domain $[0, \pi]$ and range $[-1, 1]$ to define the *inverse cosine function*.

Figure 3

Figure 4

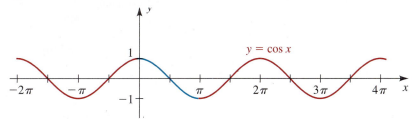

Definition of the Inverse Cosine Function	The **inverse cosine function,** denoted by $\cos^{-1}$, is defined by $$y = \cos^{-1} x \quad \text{if and only if} \quad x = \cos y$$ for $-1 \le x \le 1$ and $0 \le y \le \pi$.

The domain of the inverse cosine function is $[-1, 1]$, and the range is $[0, \pi]$. Note that the range of $\cos^{-1}$ is *not* the same as the range of $\sin^{-1}$ but their domains are equal.

The notation $y = \cos^{-1} x$ may be read "y is the inverse cosine of x" or "y is the angle whose cosine is x" (with $0 \le y \le \pi$).

The inverse cosine function is also called the **arccosine function,** and the notation arccos x is used interchangeably with $\cos^{-1} x$.

Several values of the inverse cosine function are listed in the next chart.

◤ Warning! ◤

It is *essential* to choose the value y in the range $[0, \pi]$ of $\cos^{-1}$.

Figure 5

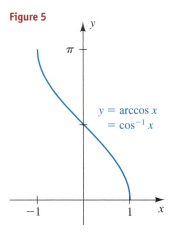

$y = \text{arccos } x = \cos^{-1} x$

Equation	Equivalent statement	Solution
$y = \cos^{-1}\left(\dfrac{1}{2}\right)$	$\cos y = \dfrac{1}{2}$ and $0 \le y \le \pi$	$y = \dfrac{\pi}{3}$
$y = \cos^{-1}\left(-\dfrac{1}{2}\right)$	$\cos y = -\dfrac{1}{2}$ and $0 \le y \le \pi$	$y = \dfrac{2\pi}{3}$
$y = \cos^{-1}(1)$	$\cos y = 1$ and $0 \le y \le \pi$	$y = 0$
$y = \text{arccos }(0)$	$\cos y = 0$ and $0 \le y \le \pi$	$y = \dfrac{\pi}{2}$
$y = \text{arccos}\left(-\dfrac{\sqrt{3}}{2}\right)$	$\cos y = -\dfrac{\sqrt{3}}{2}$ and $0 \le y \le \pi$	$y = \dfrac{5\pi}{6}$

Figure 6

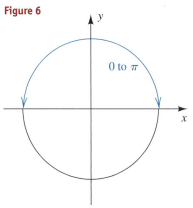

0 to π

We can sketch the graph of $y = \cos^{-1} x$ by reflecting the blue portion of Figure 4 through the line $y = x$. This gives us the sketch in Figure 5. We could also use the equation $x = \cos y$, with $0 \le y \le \pi$, to find points on the graph. As indicated by the graph, *the values of the inverse cosine function are never negative.*

As in Example 2 and Figure 3 for the arcsine, it may be helpful to associate the arccosine values with the angles corresponding to the blue arc in Figure 6.

Using relationships 4 and 5 for general inverse functions f and f^{-1}, we obtain the following properties.

Properties of cos⁻¹	**(1)** $\cos (\cos^{-1} x) = \cos (\arccos x) = x$ if $-1 \leq x \leq 1$
	(2) $\cos^{-1} (\cos y) = \arccos (\cos y) = y$ if $0 \leq y \leq \pi$

EXAMPLE 3 **Using properties of cos⁻¹**

Find the exact value:

(a) $\cos [\cos^{-1} (-0.5)]$ **(b)** $\cos^{-1} (\cos 3.14)$ **(c)** $\cos^{-1} \left[\sin \left(-\dfrac{\pi}{6} \right) \right]$

SOLUTION   For parts (a) and (b), we may use properties 1 and 2 of cos⁻¹, respectively.

(a) Since $-1 \leq -0.5 \leq 1$, $\cos [\cos^{-1} (-0.5)] = -0.5$.

(b) Since $0 \leq 3.14 \leq \pi$, $\cos^{-1} (\cos 3.14) = 3.14$.

(c) We first find $\sin (-\pi/6)$ and then use the definition of cos⁻¹, as follows:

$$\cos^{-1} \left[\sin \left(-\frac{\pi}{6} \right) \right] = \cos^{-1} \left(-\frac{1}{2} \right) = \frac{2\pi}{3}$$

EXAMPLE 4 **Finding a trigonometric function value**

Find the exact value of $\sin \left[\arccos \left(-\frac{2}{3} \right) \right]$.

SOLUTION   If we let $\theta = \arccos \left(-\frac{2}{3} \right)$, then, using the definition of the inverse cosine function, we have

$$\cos \theta = -\frac{2}{3} \quad \text{and} \quad 0 \leq \theta \leq \pi.$$

Figure 7

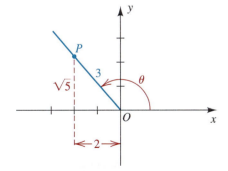

Hence, θ is in quadrant II, as illustrated in Figure 7. If we choose the point P on the terminal side with x-coordinate -2, the hypotenuse of the triangle in the figure must have length 3, since $\cos \theta = -\frac{2}{3}$. Thus, by the Pythagorean theorem, the y-coordinate of P is

$$\sqrt{3^2 - 2^2} = \sqrt{9 - 4} = \sqrt{5},$$

and therefore

$$\sin \left[\arccos \left(-\frac{2}{3} \right) \right] = \sin \theta = \frac{\sqrt{5}}{3}.$$

If we restrict the domain of the tangent function of the branch defined on the open interval $(-\pi/2, \pi/2)$, we obtain a one-to-one (increasing) function (see Figure 3 in Section 7.2). We use this *new* function to define the *inverse tangent function*.

Definition of the Inverse Tangent Function	The **inverse tangent function,** or **arctangent function,** denoted by $\tan^{-1}$ or arctan, is defined by $$y = \tan^{-1} x = \arctan x \quad \text{if and only if} \quad x = \tan y$$ for any real number x and for $-\dfrac{\pi}{2} < y < \dfrac{\pi}{2}$.

Figure 8

$y = \arctan x$
$= \tan^{-1} x$

The domain of the arctangent function is $\mathbb{R}$, and the range is the open interval $(-\pi/2,\ \pi/2)$.

We can obtain the graph of $y = \tan^{-1} x$ in Figure 8 by sketching the graph of $x = \tan y$ for $-\pi/2 < y < \pi/2$. Note that the two *vertical* asymptotes, $x = \pm\pi/2$, of the tangent function correspond to the two *horizontal* asymptotes, $y = \pm\pi/2$, of the arctangent function.

As with $\sin^{-1}$ and $\cos^{-1}$, we have the following properties for $\tan^{-1}$.

Properties of $\tan^{-1}$	**(1)** $\tan\,(\tan^{-1} x) = \tan\,(\arctan x) = x \quad$ for every x **(2)** $\tan^{-1}\,(\tan y) = \arctan\,(\tan y) = y \quad$ if $\quad -\dfrac{\pi}{2} < y < \dfrac{\pi}{2}$

EXAMPLE 5 Using properties of $\tan^{-1}$

Find the exact value:

(a) $\tan\,(\tan^{-1} 1000)$ **(b)** $\tan^{-1}\left(\tan \dfrac{\pi}{4}\right)$ **(c)** $\arctan\,(\tan \pi)$

SOLUTION

(a) By property 1 of $\tan^{-1}$,

$$\tan\,(\tan^{-1} 1000) = 1000.$$

(b) Since $-\pi/2 < \pi/4 < \pi/2$, we have, by property 2 of $\tan^{-1}$,

$$\tan^{-1}\left(\tan \frac{\pi}{4}\right) = \frac{\pi}{4}.$$

(c) Since $\pi > \pi/2$, we cannot use the second property of $\tan^{-1}$. Thus, we first find $\tan \pi$ and then evaluate, as follows:

$$\arctan\,(\tan \pi) = \arctan 0 = 0$$

EXAMPLE 6   Finding a trigonometric function value

Find the exact value of $\sec\left(\arctan\frac{2}{3}\right)$.

Figure 9

SOLUTION   If we let $y = \arctan\frac{2}{3}$, then $\tan y = \frac{2}{3}$. We wish to find $\sec y$. Since $-\pi/2 < \arctan x < \pi/2$ for every x and $\tan y > 0$, it follows that $0 < y < \pi/2$. Thus, we may regard y as the radian measure of an angle of a right triangle such that $\tan y = \frac{2}{3}$, as illustrated in Figure 9. By the Pythagorean theorem, the hypotenuse is $\sqrt{3^2 + 2^2} = \sqrt{13}$. Referring to the triangle, we obtain

$$\sec\left(\arctan\frac{2}{3}\right) = \sec y = \frac{\sqrt{13}}{3}.$$

EXAMPLE 7   Finding a trigonometric function value

Find the exact value of $\sin\left(\arctan\frac{1}{2} - \arccos\frac{4}{5}\right)$.

Figure 10

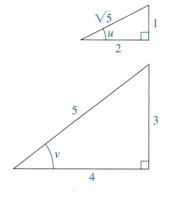

SOLUTION   If we let

$$u = \arctan\frac{1}{2} \qquad \text{and} \qquad v = \arccos\frac{4}{5},$$

then

$$\tan u = \frac{1}{2} \qquad \text{and} \qquad \cos v = \frac{4}{5}.$$

We wish to find $\sin(u - v)$. Since u and v are in the interval $(0, \pi/2)$, they can be considered as the radian measures of positive acute angles, and we may refer to the right triangles in Figure 10. This gives us

$$\sin u = \frac{1}{\sqrt{5}}, \quad \cos u = \frac{2}{\sqrt{5}}, \quad \sin v = \frac{3}{5}, \quad \text{and} \quad \cos v = \frac{4}{5}.$$

By the subtraction formula for sine,

$$\begin{aligned}
\sin(u - v) &= \sin u \cos v - \cos u \sin v \\
&= \frac{1}{\sqrt{5}}\frac{4}{5} - \frac{2}{\sqrt{5}}\frac{3}{5} \\
&= \frac{-2}{5\sqrt{5}}, \quad \text{or} \quad \frac{-2\sqrt{5}}{25}.
\end{aligned}$$

EXAMPLE 8   Changing an expression involving $\sin^{-1} x$ to an algebraic expression

If $-1 \le x \le 1$, rewrite $\cos(\sin^{-1} x)$ as an algebraic expression in x.

SOLUTION   Let

$$y = \sin^{-1} x \qquad \text{or, equivalently,} \qquad \sin y = x.$$

(continued)

We wish to express $\cos y$ in terms of x. Since $-\pi/2 \le y \le \pi/2$, it follows that $\cos y \ge 0$, and hence (from $\sin^2 y + \cos^2 y = 1$)

$$\cos y = \sqrt{1 - \sin^2 y} = \sqrt{1 - x^2}.$$

Consequently, $\qquad \cos(\sin^{-1} x) = \sqrt{1 - x^2}.$

Figure 11

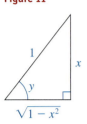

Note that $\sin y = \dfrac{x}{1} = x.$

The last identity is also evident geometrically if $0 < x < 1$. In this case $0 < y < \pi/2$, and we may regard y as the radian measure of an angle of a right triangle such that $\sin y = x$, as illustrated in Figure 11. (The side of length $\sqrt{1 - x^2}$ is found by the Pythagorean theorem.) Referring to the triangle, we have

$$\cos(\sin^{-1} x) = \cos y = \frac{\sqrt{1 - x^2}}{1} = \sqrt{1 - x^2}.$$

Most of the trigonometric equations we considered in Section 7.2 had solutions that were rational multiples of π, such as $\pi/3$, $3\pi/4$, π, and so on. If solutions of trigonometric equations are not of that type, we can sometimes use inverse functions to express them in exact form, as illustrated in the next example.

EXAMPLE 9 Using inverse trigonometric functions to solve an equation

Figure 12
(a)

(b)

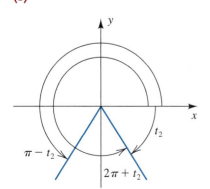

Find the solutions of $5 \sin^2 t + 3 \sin t - 1 = 0$ in $[0, 2\pi)$.

SOLUTION The equation may be regarded as a quadratic equation in $\sin t$. Applying the quadratic formula gives us

$$\sin t = \frac{-3 \pm \sqrt{3^2 - 4(5)(-1)}}{2(5)} = \frac{-3 \pm \sqrt{29}}{10}.$$

Using the definition of the inverse sine function, we obtain the following solutions:

$$t_1 = \sin^{-1} \tfrac{1}{10}\left(-3 + \sqrt{29}\right) \approx 0.2408$$
$$t_2 = \sin^{-1} \tfrac{1}{10}\left(-3 - \sqrt{29}\right) \approx -0.9946$$

Since the range of arcsin is $[-\pi/2, \pi/2]$, we know that t_1 is in $[0, \pi/2]$ and t_2 is in $[-\pi/2, 0]$. Using t_1 as a reference angle, we also have $\pi - t_1$ as a solution in quadrant II, as shown in Figure 12(a). We can add 2π to t_2 to obtain a solution in quadrant IV, as shown in Figure 12(b). The solution in quadrant III is $\pi - t_2$, not $\pi + t_2$, because t_2 is negative.

Hence, with t_1 and t_2 as previously defined, the four exact solutions are

$$t_1, \quad \pi - t_1, \quad \pi - t_2, \quad \text{and} \quad 2\pi + t_2,$$

and the four approximate solutions are

$$0.2408, \quad 2.9008, \quad 4.1361, \quad \text{and} \quad 5.2886.$$

The next example illustrates one of many identities that are true for inverse trigonometric functions.

EXAMPLE 10   **Verifying an identity involving inverse trigonometric functions**

Verify the identity $\sin^{-1} x + \cos^{-1} x = \dfrac{\pi}{2}$ for $-1 \le x \le 1$.

SOLUTION   Let

$$\alpha = \sin^{-1} x \qquad \text{and} \qquad \beta = \cos^{-1} x.$$

We wish to show that $\alpha + \beta = \pi/2$. From the definitions of $\sin^{-1}$ and $\cos^{-1}$,

$$\sin \alpha = x \quad \text{for} \quad -\frac{\pi}{2} \le \alpha \le \frac{\pi}{2}$$

and

$$\cos \beta = x \quad \text{for} \quad 0 \le \beta \le \pi.$$

Adding the two inequalities on the right, we see that

$$-\frac{\pi}{2} \le \alpha + \beta \le \frac{3\pi}{2}.$$

Note also that

$$\cos \alpha = \sqrt{1 - \sin^2 \alpha} = \sqrt{1 - x^2}$$

and

$$\sin \beta = \sqrt{1 - \cos^2 \beta} = \sqrt{1 - x^2}.$$

Using the addition formula for sine, we obtain

$$\begin{aligned}
\sin (\alpha + \beta) &= \sin \alpha \cos \beta + \cos \alpha \sin \beta \\
&= x \cdot x + \sqrt{1 - x^2}\,\sqrt{1 - x^2} \\
&= x^2 + (1 - x^2) = 1.
\end{aligned}$$

Since $\alpha + \beta$ is in the interval $[-\pi/2, 3\pi/2]$, the equation $\sin (\alpha + \beta) = 1$ has only one solution, $\alpha + \beta = \pi/2$, which is what we wished to show.

We may interpret the identity geometrically if $0 < x < 1$. If we construct a right triangle with one side of length x and hypotenuse of length 1, as illustrated in Figure 13, then angle β at B is an angle whose cosine is x; that is, $\beta = \cos^{-1} x$. Similarly, angle α at A is an angle whose sine is x; that is, $\alpha = \sin^{-1} x$. Since the acute angles of a right triangle are complementary, $\alpha + \beta = \pi/2$ or, equivalently,

$$\sin^{-1} x + \cos^{-1} x = \frac{\pi}{2}.$$

Figure 13

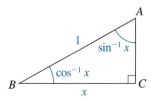

Each of the remaining inverse trigonometric functions is defined in the same manner as the first three—by choosing a domain D in which the corresponding trigonometric function is one-to-one and then using the usual technique (where y is in D):

$$y = \cot^{-1} x \quad \text{if and only if} \quad x = \cot y$$
$$y = \sec^{-1} x \quad \text{if and only if} \quad x = \sec y$$
$$y = \csc^{-1} x \quad \text{if and only if} \quad x = \csc y$$

The function $\sec^{-1}$ is used in calculus; however, $\cot^{-1}$ and $\csc^{-1}$ are seldom used. Because of their limited use in applications, we will not consider examples or exercises pertaining to these functions. We will merely summarize typical domains, ranges, and graphs in the following chart. A similar summary for the six trigonometric functions and their inverses appears in Appendix III.

Summary of Features of $\cot^{-1}$, $\sec^{-1}$, and $\csc^{-1}$

Feature	$y = \cot^{-1} x$	$y = \sec^{-1} x$	$y = \csc^{-1} x$
Domain	$\mathbb{R}$	$\lvert x \rvert \geq 1$	$\lvert x \rvert \geq 1$
Range	$(0, \pi)$	$\left[0, \dfrac{\pi}{2}\right) \cup \left[\pi, \dfrac{3\pi}{2}\right)$	$\left(-\pi, -\dfrac{\pi}{2}\right] \cup \left(0, \dfrac{\pi}{2}\right]$
Graph			

7.6 Exercises

Exer. 1–22: Find the exact value of the expression whenever it is defined.

1 (a) $\sin^{-1}\left(-\dfrac{\sqrt{2}}{2}\right)$ (b) $\cos^{-1}\left(-\dfrac{1}{2}\right)$ 2 (a) $\sin^{-1}\left(-\dfrac{1}{2}\right)$ (b) $\cos^{-1}\left(-\dfrac{\sqrt{2}}{2}\right)$

(c) $\tan^{-1}\left(-\sqrt{3}\right)$ (c) $\tan^{-1}(-1)$

3 (a) $\arcsin \dfrac{\sqrt{3}}{2}$ (b) $\arccos \dfrac{\sqrt{2}}{2}$ (c) $\arctan \dfrac{1}{\sqrt{3}}$

4 (a) $\arcsin 0$ (b) $\arccos(-1)$ (c) $\arctan 0$

5 (a) $\sin^{-1} \dfrac{\pi}{3}$ (b) $\cos^{-1} \dfrac{\pi}{2}$ (c) $\tan^{-1} 1$

6 (a) $\arcsin \dfrac{\pi}{2}$ (b) $\arccos \dfrac{\pi}{3}$ (c) $\arctan\left(-\dfrac{\sqrt{3}}{3}\right)$

7 (a) $\sin\left[\arcsin\left(-\dfrac{3}{10}\right)\right]$ (b) $\cos\left(\arccos \dfrac{1}{2}\right)$

   (c) $\tan(\arctan 14)$

8 (a) $\sin\left(\sin^{-1} \dfrac{2}{3}\right)$ (b) $\cos\left[\cos^{-1}\left(-\dfrac{1}{5}\right)\right]$

   (c) $\tan[\tan^{-1}(-9)]$

9 (a) $\sin^{-1}\left(\sin \dfrac{\pi}{3}\right)$ (b) $\cos^{-1}\left[\cos\left(\dfrac{5\pi}{6}\right)\right]$

   (c) $\tan^{-1}\left[\tan\left(-\dfrac{\pi}{6}\right)\right]$

10 (a) $\arcsin\left[\sin\left(-\dfrac{\pi}{2}\right)\right]$ (b) $\arccos(\cos 0)$

   (c) $\arctan\left(\tan \dfrac{\pi}{4}\right)$

11 (a) $\arcsin\left(\sin \dfrac{5\pi}{4}\right)$ (b) $\arccos\left(\cos \dfrac{5\pi}{4}\right)$

   (c) $\arctan\left(\tan \dfrac{7\pi}{4}\right)$

12 (a) $\sin^{-1}\left(\sin \dfrac{2\pi}{3}\right)$ (b) $\cos^{-1}\left(\cos \dfrac{4\pi}{3}\right)$

   (c) $\tan^{-1}\left(\tan \dfrac{7\pi}{6}\right)$

13 (a) $\sin\left[\cos^{-1}\left(-\dfrac{1}{2}\right)\right]$ (b) $\cos(\tan^{-1} 1)$

   (c) $\tan[\sin^{-1}(-1)]$

14 (a) $\sin\left(\tan^{-1} \sqrt{3}\right)$ (b) $\cos(\sin^{-1} 1)$

   (c) $\tan(\cos^{-1} 0)$

15 (a) $\cot\left(\sin^{-1} \dfrac{2}{3}\right)$ (b) $\sec\left[\tan^{-1}\left(-\dfrac{3}{5}\right)\right]$

   (c) $\csc\left[\cos^{-1}\left(-\dfrac{1}{4}\right)\right]$

16 (a) $\cot\left[\sin^{-1}\left(-\dfrac{2}{5}\right)\right]$ (b) $\sec\left(\tan^{-1} \dfrac{7}{4}\right)$

   (c) $\csc\left(\cos^{-1} \dfrac{1}{5}\right)$

17 (a) $\sin\left(\arcsin \dfrac{1}{2} + \arccos 0\right)$

   (b) $\cos\left[\arctan\left(-\dfrac{3}{4}\right) - \arcsin \dfrac{4}{5}\right]$

   (c) $\tan\left(\arctan \dfrac{4}{3} + \arccos \dfrac{8}{17}\right)$

18 (a) $\sin\left[\sin^{-1} \dfrac{5}{13} - \cos^{-1}\left(-\dfrac{3}{5}\right)\right]$

   (b) $\cos\left(\sin^{-1} \dfrac{4}{5} + \tan^{-1} \dfrac{3}{4}\right)$

   (c) $\tan\left[\cos^{-1} \dfrac{1}{2} - \sin^{-1}\left(-\dfrac{1}{2}\right)\right]$

19 (a) $\sin\left[2\arccos\left(-\dfrac{3}{5}\right)\right]$ (b) $\cos\left(2\sin^{-1} \dfrac{15}{17}\right)$

   (c) $\tan\left(2\tan^{-1} \dfrac{3}{4}\right)$

20 (a) $\sin\left(2\tan^{-1} \dfrac{5}{12}\right)$ (b) $\cos\left(2\arccos \dfrac{9}{41}\right)$

   (c) $\tan\left[2\arcsin\left(-\dfrac{8}{17}\right)\right]$

21 (a) $\sin\left[\dfrac{1}{2}\sin^{-1}\left(-\dfrac{7}{25}\right)\right]$ (b) $\cos\left(\dfrac{1}{2}\tan^{-1} \dfrac{8}{15}\right)$

   (c) $\tan\left(\dfrac{1}{2}\cos^{-1} \dfrac{3}{5}\right)$

22 (a) $\sin\left[\dfrac{1}{2}\cos^{-1}\left(-\dfrac{3}{5}\right)\right]$ (b) $\cos\left(\dfrac{1}{2}\sin^{-1} \dfrac{12}{13}\right)$

   (c) $\tan\left(\dfrac{1}{2}\tan^{-1} \dfrac{40}{9}\right)$

Exer. 23–30: Write the expression as an algebraic expression in x for $x > 0$.

23 $\sin(\tan^{-1} x)$ 24 $\tan(\arccos x)$

25 $\sec\left(\sin^{-1} \dfrac{x}{\sqrt{x^2 + 4}}\right)$ 26 $\cot\left(\sin^{-1} \dfrac{\sqrt{x^2 - 9}}{x}\right)$

27 $\sin(2\sin^{-1} x)$ 28 $\cos(2\tan^{-1} x)$

29 $\cos\left(\dfrac{1}{2}\arccos x\right)$ 30 $\tan\left(\dfrac{1}{2}\cos^{-1} \dfrac{1}{x}\right)$

Exer. 31–32: Complete the statements.

31 (a) As $x \to -1^+$, $\sin^{-1} x \to$ _____

(b) As $x \to 1^-$, $\cos^{-1} x \to$ _____

(c) As $x \to \infty$, $\tan^{-1} x \to$ _____

32 (a) As $x \to 1^-$, $\sin^{-1} x \to$ _____

(b) As $x \to -1^+$, $\cos^{-1} x \to$ _____

(c) As $x \to -\infty$, $\tan^{-1} x \to$ _____

Exer. 33–42: Sketch the graph of the equation.

33 $y = \sin^{-1} 2x$

34 $y = \frac{1}{2} \sin^{-1} x$

35 $y = \sin^{-1} (x + 1)$

36 $y = \sin^{-1} (x - 2) + \dfrac{\pi}{2}$

37 $y = \cos^{-1} \frac{1}{2} x$

38 $y = 2 \cos^{-1} x$

39 $y = 2 + \tan^{-1} x$

40 $y = \tan^{-1} 2x$

41 $y = \sin (\arccos x)$

42 $y = \sin (\sin^{-1} x)$

Exer. 43–46: The given equation has the form $y = f(x)$. (a) Find the domain of f. (b) Find the range of f. (c) Solve for x in terms of y.

43 $y = \frac{1}{2} \sin^{-1} (x - 3)$

44 $y = 3 \tan^{-1} (2x + 1)$

45 $y = 4 \cos^{-1} \frac{2}{3} x$

46 $y = 2 \sin^{-1} (3x - 4)$

Exer. 47–50: Solve the equation for x in terms of y if x is restricted to the given interval.

47 $y = -3 - \sin x;$ $\left[-\dfrac{\pi}{2}, \dfrac{\pi}{2} \right]$

48 $y = 2 + 3 \sin x;$ $\left[-\dfrac{\pi}{2}, \dfrac{\pi}{2} \right]$

49 $y = 15 - 2 \cos x;$ $[0, \pi]$

50 $y = 6 - 3 \cos x;$ $[0, \pi]$

Exer. 51–52: Solve the equation for x in terms of y if $0 < x < \pi$ and $0 < y < \pi$.

51 $\dfrac{\sin x}{3} = \dfrac{\sin y}{4}$

52 $\dfrac{4}{\sin x} = \dfrac{7}{\sin y}$

Exer. 53–64: Use inverse trigonometric functions to find the solutions of the equation that are in the given interval, and approximate the solutions to four decimal places.

53 $\cos^2 x + 2 \cos x - 1 = 0;$ $[0, 2\pi)$

54 $\sin^2 x - \sin x - 1 = 0;$ $[0, 2\pi)$

55 $2 \tan^2 t + 9 \tan t + 3 = 0;$ $\left(-\dfrac{\pi}{2}, \dfrac{\pi}{2} \right)$

56 $3 \sin^2 t + 7 \sin t + 3 = 0;$ $\left[-\dfrac{\pi}{2}, \dfrac{\pi}{2} \right]$

57 $15 \cos^4 x - 14 \cos^2 x + 3 = 0;$ $[0, \pi]$

58 $3 \tan^4 \theta - 19 \tan^2 \theta + 2 = 0;$ $\left(-\dfrac{\pi}{2}, \dfrac{\pi}{2} \right)$

59 $6 \sin^3 \theta + 18 \sin^2 \theta - 5 \sin \theta - 15 = 0;$ $\left(-\dfrac{\pi}{2}, \dfrac{\pi}{2} \right)$

60 $6 \sin 2x - 8 \cos x + 9 \sin x - 6 = 0;$ $\left(-\dfrac{\pi}{2}, \dfrac{\pi}{2} \right)$

61 $(\cos x)(15 \cos x + 4) = 3;$ $[0, 2\pi)$

62 $6 \sin^2 x = \sin x + 2;$ $[0, 2\pi)$

63 $3 \cos 2x - 7 \cos x + 5 = 0;$ $[0, 2\pi)$

64 $\sin 2x = -1.5 \cos x;$ $[0, 2\pi)$

Exer. 65–66: If an earthquake has a total horizontal displacement of S meters along its fault line, then the horizontal movement M of a point on the surface of Earth d kilometers from the fault line can be estimated using the formula

$$M = \dfrac{S}{2} \left(1 - \dfrac{2}{\pi} \tan^{-1} \dfrac{d}{D} \right),$$

where D is the depth (in kilometers) below the surface of the focal point of the earthquake.

65 **Earthquake movement** For the San Francisco earthquake of 1906, S was 4 meters and D was 3.5 kilometers. Approximate M for the stated values of d.

(a) 1 kilometer   (b) 4 kilometers

(c) 10 kilometers

66 **Earthquake movement** Approximate the depth D of the focal point of an earthquake with $S = 3$ m if a point on the surface of Earth 5 kilometers from the fault line moved 0.6 meter horizontally.

67 A golfer's drive A golfer, centered in a 30-yard-wide straight fairway, hits a ball 280 yards. Approximate the largest angle the drive can have from the center of the fairway if the ball is to stay in the fairway (see the figure).

Exercise 67

68 Placing a wooden brace A 14-foot piece of lumber is to be placed as a brace, as shown in the figure. Assuming all the lumber is 2 inches by 4 inches, find α and β.

Exercise 68

69 Tracking a sailboat As shown in the figure, a sailboat is following a straight-line course l. (Assume that the shoreline is parallel to the north-south line.) The shortest distance from a tracking station T to the course is d miles. As the boat sails, the tracking station records its distance k from T and its direction θ with respect to T. Angle α specifies the direction of the sailboat.

(a) Express α in terms of d, k, and θ.

(b) Estimate α to the nearest degree if $d = 50$ mi, $k = 210$ mi, and $\theta = 53.4°$.

Exercise 69

70 Calculating viewing angles An art critic whose eye level is 6 feet above the floor views a painting that is 10 feet in height and is mounted 4 feet above the floor, as shown in the figure.

(a) If the critic is standing x feet from the wall, express the viewing angle θ in terms of x.

(b) Use the addition formula for tangent to show that
$$\theta = \tan^{-1}\left(\frac{10x}{x^2 - 16}\right).$$

(c) For what value of x is $\theta = 45°$?

Exercise 70

Exer. 71–76: Verify the identity.

71 $\sin^{-1} x = \tan^{-1} \dfrac{x}{\sqrt{1 - x^2}}$

72 $\arccos x + \arccos \sqrt{1 - x^2} = \dfrac{\pi}{2}, 0 \le x \le 1$

73 $\arcsin (-x) = -\arcsin x$

74 $\arccos (-x) = \pi - \arccos x$

75 $\arctan x + \arctan \dfrac{1}{x} = \dfrac{\pi}{2}, x > 0$

76 $2 \cos^{-1} x = \cos^{-1} (2x^2 - 1), 0 \le x \le 1$

CHAPTER 7 REVIEW EXERCISES

Exer. 1–22: Verify the identity.

1 $(\cot^2 x + 1)(1 - \cos^2 x) = 1$

2 $\cos \theta + \sin \theta \tan \theta = \sec \theta$

3 $\dfrac{(\sec^2 \theta - 1) \cot \theta}{\tan \theta \sin \theta + \cos \theta} = \sin \theta$

4 $(\tan x + \cot x)^2 = \sec^2 x \csc^2 x$

5 $\dfrac{1}{1 + \sin t} = (\sec t - \tan t) \sec t$

6 $\dfrac{\sin (\alpha - \beta)}{\cos (\alpha + \beta)} = \dfrac{\tan \alpha - \tan \beta}{1 - \tan \alpha \tan \beta}$

7 $\tan 2u = \dfrac{2 \cot u}{\csc^2 u - 2}$ **8** $\cos^2 \dfrac{v}{2} = \dfrac{1 + \sec v}{2 \sec v}$

9 $\dfrac{\tan^3 \phi - \cot^3 \phi}{\tan^2 \phi + \csc^2 \phi} = \tan \phi - \cot \phi$

10 $\dfrac{\sin u + \sin v}{\csc u + \csc v} = \dfrac{1 - \sin u \sin v}{-1 + \csc u \csc v}$

11 $\left(\dfrac{\sin^2 x}{\tan^4 x}\right)^3 \left(\dfrac{\csc^3 x}{\cot^6 x}\right)^2 = 1$

12 $\dfrac{\cos \gamma}{1 - \tan \gamma} + \dfrac{\sin \gamma}{1 - \cot \gamma} = \cos \gamma + \sin \gamma$

13 $\dfrac{\cos (-t)}{\sec (-t) + \tan (-t)} = 1 + \sin t$

14 $\dfrac{\cot (-t) + \csc (-t)}{\sin (-t)} = \dfrac{1}{1 - \cos t}$

15 $\sqrt{\dfrac{1 - \cos t}{1 + \cos t}} = \dfrac{1 - \cos t}{|\sin t|}$

16 $\sqrt{\dfrac{1 - \sin \theta}{1 + \sin \theta}} = \dfrac{|\cos \theta|}{1 + \sin \theta}$ **17** $\cos \left(x - \dfrac{5\pi}{2}\right) = \sin x$

18 $\tan \left(x + \dfrac{3\pi}{4}\right) = \dfrac{\tan x - 1}{1 + \tan x}$

19 $\frac{1}{4} \sin 4\beta = \sin \beta \cos^3 \beta - \cos \beta \sin^3 \beta$

20 $\tan \frac{1}{2} \theta = \csc \theta - \cot \theta$

21 $\sin 8\theta = 8 \sin \theta \cos \theta (1 - 2 \sin^2 \theta)(1 - 8 \sin^2 \theta \cos^2 \theta)$

22 $\arctan x = \dfrac{1}{2} \arctan \dfrac{2x}{1 - x^2}, \ -1 < x < 1$

Exer. 23–40: Find the solutions of the equation that are in the interval $[0, 2\pi)$.

23 $2 \cos^3 \theta - \cos \theta = 0$ **24** $2 \cos \alpha + \tan \alpha = \sec \alpha$

25 $\sin \theta = \tan \theta$ **26** $\csc^5 \theta - 4 \csc \theta = 0$

27 $2 \cos^3 t + \cos^2 t - 2 \cos t - 1 = 0$

28 $\cos x \cot^2 x = \cos x$ **29** $\sin \beta + 2 \cos^2 \beta = 1$

30 $\cos 2x + 3 \cos x + 2 = 0$

31 $2 \sec u \sin u + 2 = 4 \sin u + \sec u$

32 $\tan 2x \cos 2x = \sin 2x$

33 $2 \cos 3x \cos 2x = 1 - 2 \sin 3x \sin 2x$

34 $\sin x \cos 2x + \cos x \sin 2x = 0$

35 $\cos \pi x + \sin \pi x = 0$ **36** $\sin 2u = \sin u$

37 $2 \cos^2 \frac{1}{2} \theta - 3 \cos \theta = 0$ **38** $\sec 2x \csc 2x = 2 \csc 2x$

39 $\sin 5x = \sin 3x$ **40** $\cos 3x = -\cos 2x$

Exer. 41–44: Find the exact value.

41 $\cos 75°$ **42** $\tan 285°$

43 $\sin 195°$ **44** $\csc \dfrac{\pi}{8}$

Exer. 45–56: If θ and ϕ are acute angles such that $\csc \theta = \frac{5}{3}$ and $\cos \phi = \frac{8}{17}$, find the exact value.

45 $\sin (\theta + \phi)$ **46** $\cos (\theta + \phi)$

47 $\tan (\phi + \theta)$ **48** $\tan (\theta - \phi)$

49 $\sin (\phi - \theta)$ **50** $\sin (\theta - \phi)$

51 $\sin 2\phi$ **52** $\cos 2\phi$

53 $\tan 2\theta$ **54** $\sin \frac{1}{2} \theta$

55 $\tan \frac{1}{2} \theta$ **56** $\cos \frac{1}{2} \phi$

57 Express as a sum or difference:

(a) $\sin 7t \sin 4t$ (b) $\cos \frac{1}{4}u \cos \left(-\frac{1}{6}u\right)$

(c) $6 \cos 5x \sin 3x$ (d) $4 \sin 3\theta \cos 7\theta$

58 Express as a product:

(a) $\sin 8u + \sin 2u$ (b) $\cos 3\theta - \cos 8\theta$

(c) $\sin \frac{1}{4}t - \sin \frac{1}{5}t$ (d) $3 \cos 2x + 3 \cos 6x$

Exer. 59–70: Find the exact value of the expression whenever it is defined.

59 $\cos^{-1}\left(\dfrac{\sqrt{3}}{2}\right)$ **60** $\arcsin\left(\dfrac{\sqrt{2}}{2}\right)$

61 $\arctan \sqrt{3}$ **62** $\arccos\left(\tan \dfrac{3\pi}{4}\right)$

63 $\arcsin\left(\sin \dfrac{5\pi}{4}\right)$ **64** $\cos^{-1}\left(\cos \dfrac{5\pi}{4}\right)$

65 $\sin\left[\arccos\left(-\dfrac{\sqrt{3}}{2}\right)\right]$ **66** $\tan\left(\tan^{-1} 2\right)$

67 $\sec\left(\sin^{-1} \frac{3}{2}\right)$ **68** $\cos^{-1}\left(\sin 0\right)$

69 $\cos\left(\sin^{-1} \frac{15}{17} - \sin^{-1} \frac{8}{17}\right)$ **70** $\cos\left(2 \sin^{-1} \frac{4}{5}\right)$

Exer. 71–74: Sketch the graph of the equation.

71 $y = \cos^{-1} 3x$ **72** $y = 4 \sin^{-1} x$

73 $y = 1 - \sin^{-1} x$ **74** $y = \sin\left(\frac{1}{2} \cos^{-1} x\right)$

75 Express $\cos\left(\alpha + \beta + \gamma\right)$ in terms of trigonometric functions of α, β, and γ.

76 **Force of a foot** When an individual is walking, the magnitude F of the vertical force of one foot on the ground (see the figure) can be described by

$$F = A(\cos bt - a \cos 3bt),$$

where t is time in seconds, $A > 0$, $b > 0$, and $0 < a < 1$.

Exercise 76

(a) Show that $F = 0$ when $t = -\pi/(2b)$ and $t = \pi/(2b)$. (The time $t = -\pi/(2b)$ corresponds to the moment when the foot first touches the ground and the weight of the body is being supported by the other foot.)

(b) The maximum force occurs when

$$3a \sin 3bt = \sin bt.$$

If $a = \frac{1}{3}$, find the solutions of this equation for the interval $-\pi/(2b) < t < \pi/(2b)$.

(c) If $a = \frac{1}{3}$, express the maximum force in terms of A.

77 Shown in the figure is a graph of the equation

$$y = \sin x - \frac{1}{2} \sin 2x + \frac{1}{3} \sin 3x.$$

The x-coordinates of the turning points are solutions of the equation $\cos x - \cos 2x + \cos 3x = 0$. Use a sum-to-product formula to find these x-coordinates.

Exercise 77

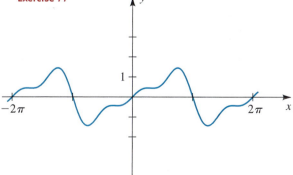

78 Visual distinction The human eye can distinguish between two distant points P and Q provided the angle of resolution θ is not too small. Suppose P and Q are x units apart and are d units from the eye, as illustrated in the figure.

(a) Express x in terms of d and θ.

(b) For a person with normal vision, the smallest distinguishable angle of resolution is about 0.0005 radian. If a pen 6 inches long is viewed by such an individual at a distance of d feet, for what values of d will the end points of the pen be distinguishable?

Exercise 78

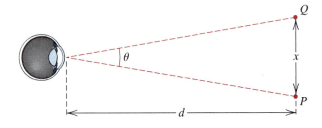

79 Satellites A satellite S circles a planet at a distance d miles from the planet's surface. The portion of the planet's surface that is visible from the satellite is determined by the angle θ indicated in the figure.

Exercise 79

(a) Assuming that the planet is spherical in shape, express d in terms of θ and the radius r of the planet.

(b) Approximate θ for a satellite 300 miles from the surface of Earth, using $r = 4000$ mi.

80 Urban canyons Because of the tall buildings and relatively narrow streets in some inner cities, the amount of sunlight illuminating these "canyons" is greatly reduced. If h is the average height of the buildings and w is the width of the street, the narrowness N of the street is defined by $N = h/w$. The angle θ of the horizon is defined by $\tan \theta = N$. (The value $\theta = 63°$ may result in an 85% loss of illumination.) Approximate the angle of the horizon for the following values of h and w.

(a) $h = 400$ ft, $w = 80$ ft

(b) $h = 55$ m, $w = 30$ m

CHAPTER 7 DISCUSSION EXERCISES

1 Verify the following identity:

$$\frac{\tan x}{1 - \cot x} + \frac{\cot x}{1 - \tan x} = 1 + \sec x \csc x$$

(*Hint:* At some point, consider a special factoring.)

2 Refer to Example 5 of Section 7.1. Suppose $0 \le \theta < 2\pi$, and rewrite the conclusion using a piecewise-defined function.

3 How many solutions does the following equation have on $[0, 2\pi)$? Find the largest one.

$$3 \cos 45x + 4 \sin 45x = 5$$

4 Shown in the figure is a function called a *sawtooth function*.

(a) Define an inverse sawtooth function (**arcsaw**), including its domain and range.

(b) Find arcsaw (1.7) and arcsaw (−0.8).

(c) Formulate two properties of arcsaw (similar to the sin (sin⁻¹) property).

(d) Graph the arcsaw function.

Exercise 4

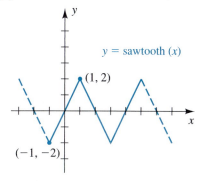

$y = \text{sawtooth}(x)$

$(1, 2)$

$(-1, -2)$

5 There are several interesting exact relationships between π and inverse trigonometric functions such as

$$\frac{\pi}{4} = 4 \tan^{-1}\left(\frac{1}{5}\right) - \tan^{-1}\left(\frac{1}{239}\right).$$

Use trigonometric identities to prove that this relationship is true. Two other relationships are

$$\frac{\pi}{4} = \tan^{-1}\left(\frac{1}{2}\right) + \tan^{-1}\left(\frac{1}{5}\right) + \tan^{-1}\left(\frac{1}{8}\right)$$

and $\quad \pi = \tan^{-1} 1 + \tan^{-1} 2 + \tan^{-1} 3.$

8

Applications of Trigonometry

8.1 The Law of Sines

8.2 The Law of Cosines

8.3 Vectors

8.4 The Dot Product

8.5 Trigonometric Form for Complex Numbers

8.6 De Moivre's Theorem and *n*th Roots of Complex Numbers

In the first two sections of this chapter we consider methods of solving oblique triangles using the law of sines and the law of cosines. The next two sections contain an introduction to vectors—a topic that has many applications in engineering, the natural sciences, and advanced mathematics. We then introduce the trigonometric form for complex numbers and use it to find all n solutions of equations of the form $w^n = z$, where n is any positive integer and w and z are complex numbers.

8.1

The Law of Sines

Figure 1

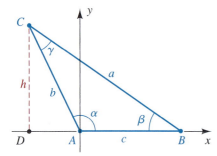

An **oblique triangle** is a triangle that does not contain a right angle. We shall use the letters A, B, C, a, b, c, α, β, and γ for parts of triangles, as we did in Chapter 6. Given triangle ABC, let us place angle α in standard position so that B is on the positive x-axis. The case for α obtuse is illustrated in Figure 1; however, the following discussion is also valid if α is acute.

Consider the line through C parallel to the y-axis and intersecting the x-axis at point D. If we let $d(C, D) = h$, then the y-coordinate of C is h. From the definition of the trigonometric functions of any angle,

$$\sin \alpha = \frac{h}{b}, \quad \text{so} \quad h = b \sin \alpha.$$

Referring to right triangle BDC, we see that

$$\sin \beta = \frac{h}{a}, \quad \text{so} \quad h = a \sin \beta.$$

Equating the two expressions for h gives us

$$b \sin \alpha = a \sin \beta,$$

which we may write as
$$\frac{\sin \alpha}{a} = \frac{\sin \beta}{b}.$$

If we place α in standard position with C on the positive x-axis, then by the same reasoning,

$$\frac{\sin \alpha}{a} = \frac{\sin \gamma}{c}.$$

The last two equalities give us the following result.

The Law of Sines	If ABC is an oblique triangle labeled in the usual manner (as in Figure 1), then
	$$\frac{\sin \alpha}{a} = \frac{\sin \beta}{b} = \frac{\sin \gamma}{c}.$$

Note that the law of sines consists of the following three formulas:

$$(1) \ \frac{\sin \alpha}{a} = \frac{\sin \beta}{b} \qquad (2) \ \frac{\sin \alpha}{a} = \frac{\sin \gamma}{c} \qquad (3) \ \frac{\sin \beta}{b} = \frac{\sin \gamma}{c}$$

To apply any one of these formulas to a specific triangle, we must know the values of three of the four variables. If we substitute these three values into the appropriate formula, we can then solve for the value of the fourth variable. It follows that the law of sines can be used to find the remaining parts of an oblique triangle whenever we know either of the following (the three letters in parentheses are used to denote the known parts, with S representing a side and A an angle):

(1) two sides and an angle *opposite* one of them (SSA)

(2) two angles and any side (AAS or ASA)

In the next section we will discuss the law of cosines and show how it can be used to find the remaining parts of an oblique triangle when given the following:

(1) two sides and the angle *between* them (SAS)

(2) three sides (SSS)

The law of sines cannot be applied directly to the last two cases.

The law of sines can also be written in the form

$$\frac{a}{\sin \alpha} = \frac{b}{\sin \beta} = \frac{c}{\sin \gamma}.$$

Instead of memorizing the three formulas associated with the law of sines, it may be more convenient to remember the following statement, which takes all of them into account.

The Law of Sines (General Form)	In any triangle, the ratio of the sine of an angle to the side opposite that angle is equal to the ratio of the sine of another angle to the side opposite that angle.

In examples and exercises involving triangles, we shall assume that known lengths of sides and angles have been obtained by measurement and hence are approximations to exact values. Unless directed otherwise, when finding parts of triangles we will round off answers according to the following rule: *If known sides or angles are stated to a certain accuracy, then unknown sides or angles should be calculated to the same accuracy.* To illustrate, if known sides are stated to the nearest 0.1, then unknown sides should be calculated to the nearest 0.1. If known angles are stated to the nearest $10'$, then unknown angles should be calculated to the nearest $10'$. Similar remarks hold for accuracy to the nearest 0.01, 0.1°, and so on.

Figure 2

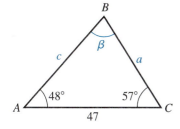

EXAMPLE 1   Using the law of sines (ASA)

Solve $\triangle ABC$, given $\alpha = 48°$, $\gamma = 57°$, and $b = 47$.

SOLUTION   The triangle is sketched in Figure 2. Since the sum of the angles of a triangle is 180°,

$$\beta = 180° - 57° - 48° = 75°.$$

(continued)

Since side b and all three angles are known, we can find a by using a form of the law of sines that involves a, α, b, and β:

$$\frac{a}{\sin \alpha} = \frac{b}{\sin \beta} \qquad \text{law of sines}$$

$$a = \frac{b \sin \alpha}{\sin \beta} \qquad \text{solve for } a$$

$$= \frac{47 \sin 48°}{\sin 75°} \qquad \text{substitute for } b, \alpha, \text{ and } \beta$$

$$\approx 36 \qquad \text{approximate to the nearest integer}$$

To find c, we merely replace $\dfrac{a}{\sin \alpha}$ with $\dfrac{c}{\sin \gamma}$ in the preceding solution for a, obtaining

$$c = \frac{b \sin \gamma}{\sin \beta} = \frac{47 \sin 57°}{\sin 75°} \approx 41.$$

Data such as those in Example 1 lead to exactly one triangle ABC. However, if two sides and an angle *opposite* one of them are given, a unique triangle is not always determined. To illustrate, suppose that a and b are to be lengths of sides of triangle ABC and that a given angle α is to be opposite the side of length a. Let us examine the case for α acute. Place α in standard position and consider the line segment AC of length b on the terminal side of α, as shown in Figure 3. The third vertex, B, should be somewhere on the x-axis. Since the length a of the side opposite α is given, we may find B by striking off a circular arc of length a with center at C. The four possible outcomes are illustrated in Figure 4 (without the coordinate axes).

Figure 3

Figure 4

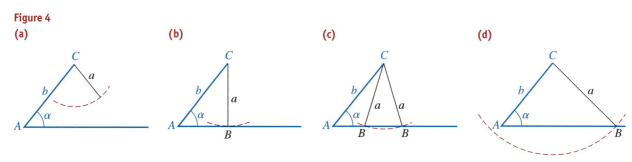

The four possibilities in the figure may be described as follows:

(a) The arc does not intersect the x-axis, and no triangle is formed.

(b) The arc is tangent to the x-axis, and a right triangle is formed.

(c) The arc intersects the positive x-axis in two distinct points, and two triangles are formed.

(d) The arc intersects both the positive and the nonpositive parts of the x-axis, and one triangle is formed.

Figure 5

(a) $a < b$

(b) $a > b$

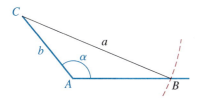

The particular case that occurs in a given problem will become evident when the solution is attempted. For example, if we solve the equation

$$\frac{\sin \alpha}{a} = \frac{\sin \beta}{b}$$

and obtain $\sin \beta > 1$, then no triangle exists and we have case (a). If we obtain $\sin \beta = 1$, then $\beta = 90°$ and hence (b) occurs. If $\sin \beta < 1$, then there are two possible choices for the angle β. By checking both possibilities, we may determine whether (c) or (d) occurs.

If the measure of α is greater than $90°$, then a triangle exists if and only if $a > b$ (see Figure 5). Since we may have more than one possibility when two sides and an angle opposite one of them are given, this situation is sometimes called the **ambiguous case.**

EXAMPLE 2   Using the law of sines (SSA)

Solve $\triangle ABC$, given $\alpha = 67°$, $a = 100$, and $c = 125$.

 SOLUTION   Since we know α, a, and c, we can find γ by using a form of the law of sines that involves a, α, c, and γ:

$$\frac{\sin \gamma}{c} = \frac{\sin \alpha}{a} \qquad \text{law of sines}$$

$$\sin \gamma = \frac{c \sin \alpha}{a} \qquad \text{solve for } \sin \gamma$$

$$= \frac{125 \sin 67°}{100} \qquad \text{substitute for } c, \alpha, \text{ and } a$$

$$\approx 1.1506 \qquad \text{approximate}$$

Since $\sin \gamma$ *cannot* be greater than 1, no triangle can be constructed with the given parts.   ◢

EXAMPLE 3   Using the law of sines (SSA)

Solve $\triangle ABC$, given $a = 12.4$, $b = 8.7$, and $\beta = 36.7°$.

SOLUTION   To find α, we proceed as follows:

$$\frac{\sin \alpha}{a} = \frac{\sin \beta}{b} \qquad \text{law of sines}$$

$$\sin \alpha = \frac{a \sin \beta}{b} \qquad \text{solve for } \sin \alpha$$

$$= \frac{12.4 \sin 36.7°}{8.7} \qquad \text{substitute for } a, \beta, \text{ and } b$$

$$\approx 0.8518 \qquad \text{approximate}$$

(continued)

There are two possible angles α between $0°$ and $180°$ such that $\sin \alpha$ is approximately 0.8518. The reference angle α_R is

$$\alpha_R \approx \sin^{-1}(0.8518) \approx 58.4°.$$

Consequently, the two possibilities for α are

$$\alpha_1 \approx 58.4° \quad \text{and} \quad \alpha_2 = 180° - \alpha_1 \approx 121.6°.$$

Figure 6

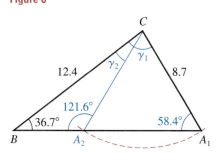

The angle $\alpha_1 \approx 58.4°$ gives us triangle A_1BC in Figure 6, and $\alpha_2 \approx 121.6°$ gives us triangle A_2BC.

If we let γ_1 and γ_2 denote the third angles of the triangles A_1BC and A_2BC corresponding to the angles α_1 and α_2, respectively, then

$$\gamma_1 = 180° - \alpha_1 - \beta \approx 180° - 58.4° - 36.7° \approx 84.9°$$
$$\gamma_2 = 180° - \alpha_2 - \beta \approx 180° - 121.6° - 36.7° \approx 21.7°.$$

If $c_1 = \overline{BA_1}$ is the side opposite γ_1 in triangle A_1BC, then

$$\frac{c_1}{\sin \gamma_1} = \frac{a}{\sin \alpha_1} \qquad \text{law of sines}$$

$$c_1 = \frac{a \sin \gamma_1}{\sin \alpha_1} \qquad \text{solve for } c_1$$

$$\approx \frac{12.4 \sin 84.9°}{\sin 58.4°} \approx 14.5. \qquad \text{substitute and approximate}$$

Thus, the remaining parts of triangle A_1BC are

$$\alpha_1 \approx 58.4°, \quad \gamma_1 \approx 84.9°, \quad \text{and} \quad c_1 \approx 14.5.$$

Similarly, if $c_2 = \overline{BA_2}$ is the side opposite γ_2 in $\triangle A_2BC$, then

$$c_2 = \frac{a \sin \gamma_2}{\sin \alpha_2} \approx \frac{12.4 \sin 21.7°}{\sin 121.6°} \approx 5.4,$$

and the remaining parts of triangle A_2BC are

$$\alpha_2 \approx 121.6°, \quad \gamma_2 \approx 21.7°, \quad \text{and} \quad c_2 \approx 5.4. \qquad \blacksquare$$

Figure 7

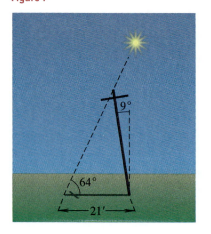

EXAMPLE 4 Using an angle of elevation

When the angle of elevation of the sun is $64°$, a telephone pole that is tilted at an angle of $9°$ directly away from the sun casts a shadow 21 feet long on level ground. Approximate the length of the pole.

Figure 8

SOLUTION   The problem is illustrated in Figure 7. Triangle ABC in Figure 8 also displays the given facts. Note that in Figure 8 we have calculated the following angles:

$$\beta = 90° - 9° = 81°$$

$$\gamma = 180° - 64° - 81° = 35°$$

To find the length of the pole—that is, side a of triangle ABC—we proceed as follows:

$$\frac{a}{\sin 64°} = \frac{21}{\sin 35°} \qquad \text{law of sines}$$

$$a = \frac{21 \sin 64°}{\sin 35°} \approx 33 \qquad \text{solve for } a \text{ and approximate}$$

Thus, the telephone pole is approximately 33 feet in length.

EXAMPLE 5   Using bearings

Figure 9

A point P on level ground is 3.0 kilometers due north of a point Q. A runner proceeds in the direction N25°E from Q to a point R, and then from R to P in the direction S70°W. Approximate the distance run.

SOLUTION   The notation used to specify directions was introduced in Section 6.7. The arrows in Figure 9 show the path of the runner, together with a north-south (dashed) line from R to another point S.

Since the lines through PQ and RS are parallel, it follows from geometry that the alternate interior angles PQR and QRS both have measure 25°. Hence,

$$\angle PRQ = \angle PRS - \angle QRS = 70° - 25° = 45°.$$

These observations give us triangle PQR in Figure 10 with

$$\angle QPR = 180° - 25° - 45° = 110°.$$

We apply the law of sines to find both q and p:

$$\frac{q}{\sin 25°} = \frac{3.0}{\sin 45°} \qquad \text{and} \qquad \frac{p}{\sin 110°} = \frac{3.0}{\sin 45°}$$

$$q = \frac{3.0 \sin 25°}{\sin 45°} \approx 1.8 \quad \text{and} \qquad p = \frac{3.0 \sin 110°}{\sin 45°} \approx 4.0$$

Figure 10

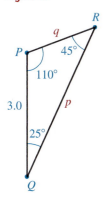

The distance run, $p + q$, is approximately $4.0 + 1.8 = 5.8$ km.

EXAMPLE 6   Locating a school of fish

A commercial fishing boat uses sonar equipment to detect a school of fish 2 miles east of the boat and traveling in the direction of N51°W at a rate of 8 mi/hr (see Figure 11 on the next page).

Figure 11

(a) If the boat travels at 20 mi/hr, approximate, to the nearest 0.1°, the direction it should head to intercept the school of fish.

(b) Find, to the nearest minute, the time it will take the boat to reach the fish.

SOLUTION

Figure 12

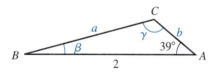

(a) The problem is illustrated by the triangle in Figure 12, with the school of fish at A, the boat at B, and the point of interception at C. Note that angle $\alpha = 90° - 51° = 39°$. To obtain β, we begin as follows:

$$\frac{\sin \beta}{b} = \frac{\sin 39°}{a} \qquad \text{law of sines}$$

$$\sin \beta = \frac{b}{a} \sin 39° \qquad \text{solve for } \sin \beta \qquad (*)$$

We next find b/a, letting t denote the amount of time required for the boat and fish to meet at C:

$$a = 20t, \qquad b = 8t \qquad \text{(distance)} = \text{(rate)(time)}$$

$$\frac{b}{a} = \frac{8t}{20t} = \frac{2}{5} \qquad \text{divide } b \text{ by } a$$

$$\sin \beta = \tfrac{2}{5} \sin 39° \qquad \text{substitute for } b/a \text{ in } (*)$$

$$\beta = \sin^{-1}\left(\tfrac{2}{5} \sin 39°\right) \approx 14.6° \qquad \text{approximate}$$

Since $90° - 14.6° = 75.4°$, the boat should travel in the (approximate) direction N75.4°E.

(b) We can find t using the relationship $a = 20t$. Let us first find the distance a from B to C. Since the only known side is 2, we need to find the angle γ opposite the side of length 2 in order to use the law of sines. We begin by noting that

$$\gamma \approx 180° - 39° - 14.6° = 126.4°.$$

To find side a, we have

$$\frac{a}{\sin \alpha} = \frac{c}{\sin \gamma} \qquad \text{law of sines}$$

$$a = \frac{c \sin \alpha}{\sin \gamma} \qquad \text{solve for } a$$

$$\approx \frac{2 \sin 39°}{\sin 126.4°} \approx 1.56 \text{ mi.} \qquad \text{substitute and approximate}$$

Using $a = 20t$, we find the time t for the boat to reach C:

$$t = \frac{a}{20} \approx \frac{1.56}{20} \approx 0.08 \text{ hr} \approx 5 \text{ min}$$

8.1 Exercises

Exer. 1–16: Solve $\triangle ABC$.

1 $\alpha = 41°$, $\quad \gamma = 77°$, $\quad a = 10.5$

2 $\beta = 20°$, $\quad \gamma = 31°$, $\quad b = 210$

3 $\alpha = 27°40'$, $\quad \beta = 52°10'$, $\quad a = 32.4$

4 $\beta = 50°50'$, $\quad \gamma = 70°30'$, $\quad c = 537$

5 $\alpha = 42°10'$, $\quad \gamma = 61°20'$, $\quad b = 19.7$

6 $\alpha = 103.45°$, $\quad \gamma = 27.19°$, $\quad b = 38.84$

7 $\gamma = 81°$, $\quad c = 11$, $\quad b = 12$

8 $\alpha = 32.32°$, $\quad c = 574.3$, $\quad a = 263.6$

9 $\gamma = 53°20'$, $\quad a = 140$, $\quad c = 115$

10 $\alpha = 27°30'$, $\quad c = 52.8$, $\quad a = 28.1$

11 $\gamma = 47.74°$, $\quad a = 131.08$, $\quad c = 97.84$

12 $\alpha = 42.17°$, $\quad a = 5.01$, $\quad b = 6.12$

13 $\alpha = 65°10'$, $\quad a = 21.3$, $\quad b = 18.9$

14 $\beta = 113°10'$, $\quad b = 248$, $\quad c = 195$

15 $\beta = 121.624°$, $\quad b = 0.283$, $\quad c = 0.178$

16 $\gamma = 73.01°$, $\quad a = 17.31$, $\quad c = 20.24$

17 Surveying To find the distance between two points A and B that lie on opposite banks of a river, a surveyor lays off a line segment AC of length 240 yards along one bank and de-termines that the measures of $\angle BAC$ and $\angle ACB$ are 63°20' and 54°10', respectively (see the figure). Approximate the distance between A and B.

Exercise 17

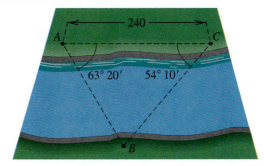

18 Surveying To determine the distance between two points A and B, a surveyor chooses a point C that is 375 yards from A and 530 yards from B. If $\angle BAC$ has measure 49°30', ap-proximate the distance between A and B.

19 Cable car route As shown in the figure on the next page, a cable car carries passengers from a point A, which is 1.2 miles from a point B at the base of a mountain, to a point P at the top of the mountain. The angles of elevation of P from A and B are 21° and 65°, respectively.

(a) Approximate the distance between A and P.

(b) Approximate the height of the mountain.

Exercise 19

20 Length of a shadow A straight road makes an angle of 15° with the horizontal. When the angle of elevation of the sun is 57°, a vertical pole at the side of the road casts a shadow 75 feet long directly down the road, as shown in the figure. Approximate the length of the pole.

Exercise 20

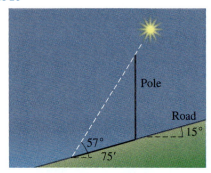

21 Height of a hot-air balloon The angles of elevation of a balloon from two points A and B on level ground are 24°10′ and 47°40′, respectively. As shown in the figure, points A and B are 8.4 miles apart, and the balloon is between the points, in the same vertical plane. Approximate the height of the balloon above the ground.

Exercise 21

22 Installing a solar panel Shown in the figure is a solar panel 10 feet in width, which is to be attached to a roof that makes an angle of 25° with the horizontal. Approximate the length d

of the brace that is needed for the panel to make an angle of 45° with the horizontal.

Exercise 22

23 Distance to an airplane A straight road makes an angle of 22° with the horizontal. From a certain point P on the road, the angle of elevation of an airplane at point A is 57°. At the same instant, from another point Q, 100 meters farther up the road, the angle of elevation is 63°. As indicated in the figure, the points P, Q, and A lie in the same vertical plane. Approximate the distance from P to the airplane.

Exercise 23

24 Surveying A surveyor notes that the direction from point A to point B is S63°W and the direction from A to point C is S38°W. The distance from A to B is 239 yards, and the distance from B to C is 374 yards. Approximate the distance from A to C.

25 Sighting a forest fire A forest ranger at an observation point A sights a fire in the direction N27°10′E. Another ranger at an observation point B, 6.0 miles due east of A, sights the same fire at N52°40′W. Approximate the distance from each of the observation points to the fire.

26 Leaning tower of Pisa The leaning tower of Pisa was originally perpendicular to the ground and 179 feet tall. Because of sinking into the earth, it now leans at a certain angle θ from the perpendicular, as shown in the figure. When the top of the tower is viewed from a point 150 feet from the center of its base, the angle of elevation is 53.3°.

(a) Approximate the angle θ.

(b) Approximate the distance d that the center of the top of the tower has moved from the perpendicular.

Exercise 26

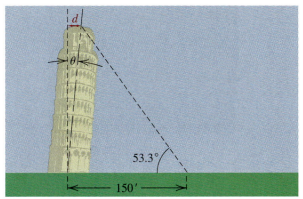

27 Height of a cathedral A cathedral is located on a hill, as shown in the figure. When the top of the spire is viewed from the base of the hill, the angle of elevation is 48°. When it is viewed at a distance of 200 feet from the base of the hill, the angle of elevation is 41°. The hill rises at an angle of 32°. Approximate the height of the cathedral.

Exercise 27

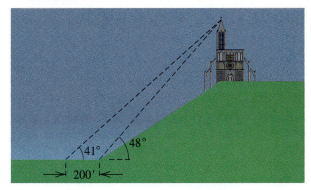

28 Sighting from a helicopter A helicopter hovers at an altitude that is 1000 feet above a mountain peak of altitude 5210 feet, as shown in the figure. A second, taller peak is viewed from both the mountaintop and the helicopter. From the helicopter, the angle of depression is 43°, and from the mountaintop, the angle of elevation is 18°.

(a) Approximate the distance from peak to peak.

(b) Approximate the altitude of the taller peak.

Exercise 28

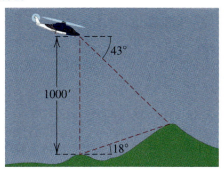

29 The volume V of the right triangular prism shown in the figure is $\frac{1}{3}Bh$, where B is the area of the base and h is the height of the prism.

(a) Approximate h.　　(b) Approximate V.

Exercise 29

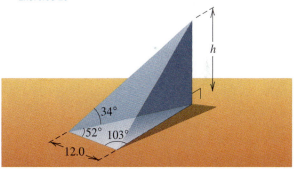

30 Design for a jet fighter Shown in the figure on the next page is a plan for the top of a wing of a jet fighter.

(a) Approximate angle ϕ.

(b) If the fuselage is 4.80 feet wide, approximate the wing span CC'.

(c) Approximate the area of triangle ABC.

Exercise 30

tions. An offshore oil well at point R in the figure is viewed from points P and Q, and $\angle QPR$ and $\angle RQP$ are found to be $55°50'$ and $65°22'$, respectively. If points P and Q have coordinates $(1487.7, 3452.8)$ and $(3145.8, 5127.5)$, respectively, approximate the coordinates of R.

Exercise 31

31 Software for surveyors Computer software for surveyors makes use of coordinate systems to locate geographic posi-

---

$$\underline{\textbf{8.2}}$$

The Law of Cosines

In the preceding section we stated that the law of sines cannot be applied directly to find the remaining parts of an oblique triangle given either of the following:

(1) two sides and the angle *between* them (SAS)

(2) three sides (SSS)

For these cases we may apply the *law of cosines,* which follows.

The Law of Cosines	If ABC is a triangle labeled in the usual manner (as in Figure 1), then

$$\textbf{(1)} \quad a^2 = b^2 + c^2 - 2bc \cos \alpha$$
$$\textbf{(2)} \quad b^2 = a^2 + c^2 - 2ac \cos \beta$$
$$\textbf{(3)} \quad c^2 = a^2 + b^2 - 2ab \cos \gamma$$

Figure 1

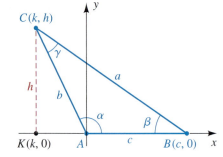

PROOF Let us prove the first formula. Given triangle ABC, place α in standard position, as illustrated in Figure 1. We have pictured α as obtuse; however, our discussion is also valid if α is acute. Consider the dashed line through C, parallel to the y-axis and intersecting the x-axis at the point $K(k, 0)$. If we let $d(C, K) = h$, then C has coordinates (k, h). By the definition of the trigonometric functions of any angle,

$$\cos \alpha = \frac{k}{b} \quad \text{and} \quad \sin \alpha = \frac{h}{b}.$$

Solving for k and h gives us

$$k = b \cos \alpha \quad \text{and} \quad h = b \sin \alpha.$$

Since the segment AB has length c, the coordinates of B are $(c, 0)$, and we obtain the following:

$$
\begin{aligned}
a^2 = [d(B, C)]^2 &= (k - c)^2 + (h - 0)^2 && \text{distance formula} \\
&= (b \cos \alpha - c)^2 + (b \sin \alpha)^2 && \text{substitute for } k \text{ and } h \\
&= b^2 \cos^2 \alpha - 2bc \cos \alpha + c^2 + b^2 \sin^2 \alpha && \text{square} \\
&= b^2(\cos^2 \alpha + \sin^2 \alpha) + c^2 - 2bc \cos \alpha && \text{factor the first and last terms} \\
&= b^2 + c^2 - 2bc \cos \alpha && \text{Pythagorean identity}
\end{aligned}
$$

Our result is the first formula stated in the law of cosines. The second and third formulas may be obtained by placing β and γ, respectively, in standard position on a coordinate system. ◢

Note that if $\alpha = 90°$ in Figure 1, then $\cos \alpha = 0$ and the law of cosines reduces to $a^2 = b^2 + c^2$. This shows that the Pythagorean theorem is a special case of the law of cosines.

Instead of memorizing each of the three formulas of the law of cosines, it is more convenient to remember the following statement, which takes all of them into account.

The Law of Cosines (General Form)	The square of the length of any side of a triangle equals the sum of the squares of the lengths of the other two sides minus twice the product of the lengths of the other two sides and the cosine of the angle between them.

Given two sides and the included angle of a triangle, we can use the law of cosines to find the third side. We may then use the law of sines to find another angle of the triangle. Whenever this procedure is followed, it is best to find the angle opposite the shortest side, since that angle is always acute. In this way, we avoid the possibility of obtaining two solutions when solving a trigonometric equation involving that angle, as illustrated in the following example.

EXAMPLE 1 Using the law of cosines (SAS)

Solve $\triangle ABC$, given $a = 5.0$, $c = 8.0$, and $\beta = 77°$.

SOLUTION   The triangle is sketched in Figure 2. Since β is the angle *between* sides a and c, we begin by approximating b (the side opposite β) as follows:

$$
\begin{aligned}
b^2 &= a^2 + c^2 - 2ac \cos \beta && \text{law of cosines} \\
&= (5.0)^2 + (8.0)^2 - 2(5.0)(8.0) \cos 77° && \text{substitute for } a, c, \text{ and } \beta \\
&= 89 - 80 \cos 77° \approx 71.0 && \text{simplify and approximate} \\
b &\approx \sqrt{71.0} \approx 8.4 && \text{take the square root}
\end{aligned}
$$

(continued)

Figure 2

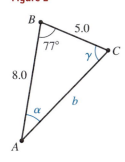

Let us find another angle of the triangle using the law of sines. In accordance with the remarks preceding this example, we will apply the law of sines and find α, since it is the angle opposite the shortest side a.

$$\frac{\sin \alpha}{a} = \frac{\sin \beta}{b} \qquad \text{law of sines}$$

$$\sin \alpha = \frac{a \sin \beta}{b} \qquad \text{solve for } \sin \alpha$$

$$\approx \frac{5.0 \sin 77°}{\sqrt{71.0}} \approx 0.5782 \qquad \text{substitute and approximate}$$

Since α is acute,

$$\alpha = \sin^{-1}(0.5782) \approx 35.3° \approx 35°.$$

Finally, since $\alpha + \beta + \gamma = 180°$, we have

$$\gamma = 180° - \alpha - \beta \approx 180° - 35° - 77° = 68°. \qquad \blacksquare$$

Given the three sides of a triangle, we can use the law of cosines to find *any* of the three angles. We shall always find the largest angle first—that is, *the angle opposite the longest side*—since this practice will guarantee that the remaining angles are acute. We may then find another angle of the triangle by using either the law of sines or the law of cosines. Note that when an angle is found by means of the law of cosines, there is no ambiguous case, since we always obtain a unique angle between $0°$ and $180°$.

EXAMPLE 2 Using the law of cosines (SSS)

If triangle ABC has sides $a = 90$, $b = 70$, and $c = 40$, approximate angles α, β, and γ to the nearest degree.

SOLUTION In accordance with the remarks preceding this example, we first find the angle opposite the longest side a. Thus, we choose the form of the law of cosines that involves α and proceed as follows:

$$a^2 = b^2 + c^2 - 2bc \cos \alpha \qquad \text{law of cosines}$$

$$\cos \alpha = \frac{b^2 + c^2 - a^2}{2bc} \qquad \text{solve for } \cos \alpha$$

$$= \frac{70^2 + 40^2 - 90^2}{2(70)(40)} = -\frac{2}{7} \qquad \text{substitute and simplify}$$

$$\alpha = \cos^{-1}\left(-\tfrac{2}{7}\right) \approx 106.6° \approx 107° \qquad \text{approximate } \alpha$$

We may now use either the law of sines or the law of cosines to find β. Let's use the law of cosines in this case:

$$b^2 = a^2 + c^2 - 2ac \cos \beta \qquad \text{law of cosines}$$

$$\cos \beta = \frac{a^2 + c^2 - b^2}{2ac} \qquad \text{solve for } \cos \beta$$

$$= \frac{90^2 + 40^2 - 70^2}{2(90)(40)} = \frac{2}{3} \qquad \text{substitute and simplify}$$

$$\beta = \cos^{-1}\left(\tfrac{2}{3}\right) \approx 48.2° \approx 48° \qquad \text{approximate } \beta$$

At this point in the solution, we could find γ by using the relationship $\alpha + \beta + \gamma = 180°$. But if either α or β was incorrectly calculated, then γ would be incorrect. Alternatively, we can approximate γ and then check that the sum of the three angles is $180°$. Thus,

$$\cos \gamma = \frac{a^2 + b^2 - c^2}{2ab}, \qquad \text{so} \qquad \gamma = \cos^{-1}\frac{90^2 + 70^2 - 40^2}{2(90)(70)} \approx 25°.$$

Note that $\alpha + \beta + \gamma = 107° + 48° + 25° = 180°$. ◢

EXAMPLE 3   Approximating the diagonals of a parallelogram

A parallelogram has sides of lengths 30 centimeters and 70 centimeters and one angle of measure $65°$. Approximate the length of each diagonal to the nearest centimeter.

Figure 3

SOLUTION   The parallelogram $ABCD$ and its diagonals AC and BD are shown in Figure 3. Using triangle ABC with $\angle ABC = 65°$, we may approximate AC as follows:

$$(AC)^2 = 30^2 + 70^2 - 2(30)(70) \cos 65° \qquad \text{law of cosines}$$

$$\approx 900 + 4900 - 1775 = 4025 \qquad \text{approximate}$$

$$AC \approx \sqrt{4025} \approx 63 \text{ cm} \qquad \text{take the square root}$$

Similarly, using triangle BAD and $\angle BAD = 180° - 65° = 115°$, we may approximate BD as follows:

$$(BD)^2 = 30^2 + 70^2 - 2(30)(70) \cos 115° \approx 7575 \qquad \text{law of cosines}$$

$$BD \approx \sqrt{7575} \approx 87 \text{ cm} \qquad \text{take the square root} \quad ◢$$

EXAMPLE 4   Finding the length of a cable

A vertical pole 40 feet tall stands on a hillside that makes an angle of $17°$ with the horizontal. Approximate the minimal length of cable that will reach from the top of the pole to a point 72 feet downhill from the base of the pole.

Figure 4

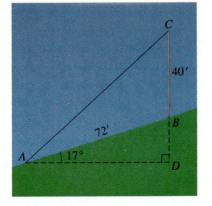

SOLUTION   The sketch in Figure 4 depicts the given data. We wish to find AC. Referring to the figure, we see that

$$\angle ABD = 90° - 17° = 73° \qquad \text{and} \qquad \angle ABC = 180° - 73° = 107°.$$

Using triangle ABC, we may approximate AC as follows:

$$(AC)^2 = 72^2 + 40^2 - 2(72)(40) \cos 107° \approx 8468 \qquad \text{law of cosines}$$

$$AC \approx \sqrt{8468} \approx 92 \text{ ft} \qquad \text{take the square root} \quad ◢$$

The law of cosines can be used to derive a formula for the area of a triangle. Let us first prove a preliminary result.

Given triangle ABC, place angle α in standard position (see Figure 1). As shown in the proof of the law of cosines, the altitude h from vertex C is $h = b \sin \alpha$. Since the area $\mathcal{A}$ of the triangle is given by $\mathcal{A} = \frac{1}{2}ch$, we see that

$$\mathcal{A} = \tfrac{1}{2}bc \sin \alpha.$$

Our argument is independent of the specific angle that is placed in standard position. By taking β and γ in standard position, we obtain the formulas

$$\mathcal{A} = \tfrac{1}{2}ac \sin \beta \quad \text{and} \quad \mathcal{A} = \tfrac{1}{2}ab \sin \gamma.$$

All three formulas are covered in the following statement.

Area of a Triangle	The area of a triangle equals one-half the product of the lengths of any two sides and the sine of the angle between them.

The next two examples illustrate uses of this result.

EXAMPLE 5   Approximating the area of a triangle

Approximate the area of triangle ABC if $a = 2.20$ cm, $b = 1.30$ cm, and $\gamma = 43.2°$.

Figure 5

$b = 1.3$ cm, Area ≈ 0.98 cm²
$\gamma = 43.2°$
$a = 2.2$ cm

SOLUTION   Since γ is the angle between sides a and b as shown in Figure 5, we may use the preceding result directly, as follows:

$$\mathcal{A} = \tfrac{1}{2}ab \sin \gamma \qquad \text{area of a triangle formula}$$
$$= \tfrac{1}{2}(2.20)(1.30) \sin 43.2° \approx 0.98 \text{ cm}^2 \qquad \text{substitute and approximate}$$

EXAMPLE 6   Approximating the area of a triangle

Approximate the area of triangle ABC if $a = 5.0$ cm, $b = 3.0$ cm, and $\alpha = 37°$.

SOLUTION   To apply the formula for the area of a triangle, we must find the angle γ between known sides a and b. Since we are given a, b, and α, let us first find β as follows:

$$\frac{\sin \beta}{b} = \frac{\sin \alpha}{a} \qquad \text{law of sines}$$

$$\sin \beta = \frac{b \sin \alpha}{a} \qquad \text{solve for } \sin \beta$$

$$= \frac{3.0 \sin 37°}{5.0} \qquad \text{substitute for } b, \alpha, \text{ and } a$$

$$\beta_R = \sin^{-1}\left(\frac{3.0 \sin 37°}{5.0}\right) \approx 21° \qquad \text{reference angle for } \beta$$

$$\beta \approx 21° \quad \text{or} \quad \beta \approx 159° \qquad \beta_R \text{ or } 180° - \beta_R$$

We reject $\beta \approx 159°$, because then $\alpha + \beta = 196° \geq 180°$. Hence, $\beta \approx 21°$ and

$$\gamma = 180° - \alpha - \beta \approx 180° - 37° - 21° = 122°.$$

Finally, we approximate the area of the triangle as follows:

$$\mathcal{A} = \tfrac{1}{2}ab \sin \gamma \qquad \text{area of a triangle formula}$$

$$\approx \tfrac{1}{2}(5.0)(3.0) \sin 122° \approx 6.4 \text{ cm}^2 \qquad \text{substitute and approximate}$$

We will use the preceding result for the area of a triangle to derive *Heron's formula,* which expresses the area of a triangle in terms of the lengths of its sides.

Heron's Formula	The area $\mathcal{A}$ of a triangle with sides a, b, and c is given by $$\mathcal{A} = \sqrt{s(s - a)(s - b)(s - c)},$$ where s is one-half the perimeter; that is, $s = \tfrac{1}{2}(a + b + c)$.

PROOF   The following equations are equivalent:

$$\mathcal{A} = \tfrac{1}{2}bc \sin \alpha$$

$$= \sqrt{\tfrac{1}{4}b^2 c^2 \sin^2 \alpha}$$

$$= \sqrt{\tfrac{1}{4}b^2 c^2 (1 - \cos^2 \alpha)}$$

$$= \sqrt{\tfrac{1}{2}bc(1 + \cos \alpha) \cdot \tfrac{1}{2}bc(1 - \cos \alpha)}$$

We shall obtain Heron's formula by replacing the expressions under the final radical sign by expressions involving only a, b, and c. We solve formula 1 of the law of cosines for $\cos \alpha$ and then substitute, as follows:

$$\frac{1}{2}bc(1 + \cos \alpha) = \frac{1}{2}bc\left(1 + \frac{b^2 + c^2 - a^2}{2bc}\right)$$

$$= \frac{1}{2}bc\left(\frac{2bc + b^2 + c^2 - a^2}{2bc}\right)$$

$$= \frac{2bc + b^2 + c^2 - a^2}{4}$$

$$= \frac{(b + c)^2 - a^2}{4}$$

$$= \frac{(b + c) + a}{2} \cdot \frac{(b + c) - a}{2}$$

(continued)

We use the same type of manipulations on the second expression under the radical sign:

$$\frac{1}{2}bc(1 - \cos\alpha) = \frac{a - b + c}{2} \cdot \frac{a + b - c}{2}$$

If we now substitute for the expressions under the radical sign, we obtain

$$\mathcal{A} = \sqrt{\frac{b + c + a}{2} \cdot \frac{b + c - a}{2} \cdot \frac{a - b + c}{2} \cdot \frac{a + b - c}{2}}.$$

Letting $s = \frac{1}{2}(a + b + c)$, we see that

$$s - a = \frac{b + c - a}{2}, \quad s - b = \frac{a - b + c}{2}, \quad s - c = \frac{a + b - c}{2}.$$

Substitution in the above formula for $\mathcal{A}$ gives us Heron's formula. ◢

EXAMPLE 7 **Using Heron's formula**

A triangular field has sides of lengths 125 yards, 160 yards, and 225 yards. Approximate the number of acres in the field. (One acre is equivalent to 4840 square yards.)

SOLUTION We first find one-half the perimeter of the field with $a = 125$, $b = 160$, and $c = 225$, as well as the values of $s - a$, $s - b$, and $s - c$:

$$s = \tfrac{1}{2}(125 + 160 + 225) = \tfrac{1}{2}(510) = 255$$
$$s - a = 255 - 125 = 130$$
$$s - b = 255 - 160 = 95$$
$$s - c = 255 - 225 = 30$$

Substituting in Heron's formula gives us

$$\mathcal{A} = \sqrt{(255)(130)(95)(30)} \approx 9720 \text{ yd}^2.$$

Since there are 4840 square yards in one acre, the number of acres is $\frac{9720}{4840}$, or approximately 2. ◢

8.2 Exercises

Exer. 1–2: Use common sense to match the variables and the values. (The triangles are drawn to scale, and the angles are measured in radians.)

1

(a) α (A) 12.60

(b) β (B) 1.10

(c) γ (C) 10

(d) x (D) 0.79

(e) y (E) 13.45

(f) z (F) 1.26

2

(a) α (A) 3

(b) β (B) 0.87

(c) γ (C) 8.24

(d) x (D) 1.92

(e) y (E) 6.72

(f) z (F) 0.35

Exer. 3–4: Given the indicated parts of △*ABC*, what angle (*α*, *β*, or *γ*) or side (*a*, *b*, or *c*) would you find next, and what would you use to find it?

3 (a)

(b)

(c)

(d)

(e)

(f)

4 (a)

(b)

(c)

(d)

(e)

(f)

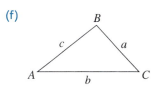

Exer. 5–14: Solve △*ABC*.

5 $\alpha = 60°$, $b = 20$, $c = 30$

6 $\gamma = 45°$, $b = 10.0$, $a = 15.0$

7 $\beta = 150°$, $a = 150$, $c = 30$

8 $\beta = 73°50'$, $c = 14.0$, $a = 87.0$

9 $\gamma = 115°10'$, $a = 1.10$, $b = 2.10$

10 $\alpha = 23°40'$, $c = 4.30$, $b = 70.0$

11 $a = 2.0$, $b = 3.0$, $c = 4.0$

12 $a = 10$, $b = 15$, $c = 12$

13 $a = 25.0$, $b = 80.0$, $c = 60.0$

14 $a = 20.0$, $b = 20.0$, $c = 10.0$

15 Dimensions of a triangular plot The angle at one corner of a triangular plot of ground is $73°40'$, and the sides that meet at this corner are 175 feet and 150 feet long. Approximate the length of the third side.

16 Surveying To find the distance between two points A and B, a surveyor chooses a point C that is 420 yards from A and 540 yards from B. If angle ACB has measure $63°10'$, approximate the distance between A and B.

17 Distance between automobiles Two automobiles leave a city at the same time and travel along straight highways that differ in direction by $84°$. If their speeds are 60 mi/hr and 45 mi/hr, respectively, approximately how far apart are the cars at the end of 20 minutes?

18 Angles of a triangular plot A triangular plot of land has sides of lengths 420 feet, 350 feet, and 180 feet. Approximate the smallest angle between the sides.

19 Distance between ships A ship leaves port at 1:00 P.M. and travels S35°E at the rate of 24 mi/hr. Another ship leaves the same port at 1:30 P.M. and travels S20°W at 18 mi/hr. Approximately how far apart are the ships at 3:00 P.M.?

20 Flight distance An airplane flies 165 miles from point A in the direction $130°$ and then travels in the direction $245°$ for 80 miles. Approximately how far is the airplane from A?

21 Jogger's course A jogger runs at a constant speed of one mile every 8 minutes in the direction S40°E for 20 minutes and then in the direction N20°E for the next 16 minutes. Approximate, to the nearest tenth of a mile, the straight-line distance from the endpoint to the starting point of the jogger's course.

22 Surveying Two points P and Q on level ground are on opposite sides of a building. To find the distance between the points, a surveyor chooses a point R that is 300 feet from P and 438 feet from Q and then determines that angle PRQ has measure $37°40'$ (see the figure). Approximate the distance between P and Q.

Exercise 22

23 Motorboat's course A motorboat traveled along a triangular course having sides of lengths 2 kilometers, 4 kilometers, and 3 kilometers, respectively. The first side was traversed in the direction N20°W and the second in a direction Sθ°W, where $\theta°$ is the degree measure of an acute angle. Approximate, to the nearest minute, the direction in which the third side was traversed.

24 Angle of a box The rectangular box shown in the figure has dimensions $8'' \times 6'' \times 4''$. Approximate the angle θ formed by a diagonal of the base and a diagonal of the $6'' \times 4''$ side.

Exercise 24

25 Distances in a baseball diamond A baseball diamond has four bases (forming a square) that are 90 feet apart; the pitcher's mound is 60.5 feet from home plate. Approximate the distance from the pitcher's mound to each of the other three bases.

26 A rhombus has sides of length 100 centimeters, and the angle at one of the vertices is $70°$. Approximate the lengths of the diagonals to the nearest tenth of a centimeter.

27 Reconnaissance A reconnaissance airplane P, flying at 10,000 feet above a point R on the surface of the water, spots a submarine S at an angle of depression of $37°$ and a tanker T at an angle of depression of $21°$, as shown in the figure. In addition, $\angle SPT$ is found to be $110°$. Approximate the distance between the submarine and the tanker.

Exercise 27

28 Correcting a ship's course A cruise ship sets a course N47°E from an island to a port on the mainland, which is 150 miles away. After moving through strong currents, the ship is off course at a position P that is N33°E and 80 miles from the island, as illustrated in the figure.

(a) Approximately how far is the ship from the port?

(b) In what direction should the ship head to correct its course?

Exercise 28

29 Seismology Seismologists investigate the structure of Earth's interior by analyzing seismic waves caused by earthquakes. If the interior of Earth is assumed to be homogeneous, then these waves will travel in straight lines at a constant velocity v. The figure shows a cross-sectional view of Earth, with the epicenter at E and an observation station at S. Use the law of cosines to show that the time t for a wave to travel through Earth's interior from E to S is given by

$$t = \frac{2R}{v} \sin \frac{\theta}{2},$$

where R is the radius of Earth and θ is the indicated angle with vertex at the center of Earth.

Exercise 29

30 Calculating distances The distance across the river shown in the figure can be found without measuring angles. Two points B and C on the opposite shore are selected, and line segments AB and AC are extended as shown. Points D and E are chosen as indicated, and distances BC, BD, BE, CD,

and CE are then measured. Suppose that $BC = 184$ ft, $BD = 102$ ft, $BE = 218$ ft, $CD = 236$ ft, and $CE = 80$ ft.

(a) Approximate the distances AB and AC.

(b) Approximate the shortest distance across the river from point A.

Exercise 30

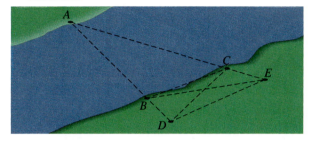

31 Penrose tiles Penrose tiles are formed from a rhombus $ABCD$ having sides of length 1 and an interior angle of 72°. First a point P is located that lies on the diagonal AC and is a distance 1 from vertex C, and then segments PB and PD are drawn to the other vertices of the diagonal, as shown in the figure. The two tiles formed are called a dart and a kite. Three-dimensional counterparts of these tiles have been applied in molecular chemistry.

(a) Find the degree measures of $\angle BPC$, $\angle APB$, and $\angle ABP$.

(b) Approximate, to the nearest 0.01, the length of segment BP.

(c) Approximate, to the nearest 0.01, the area of a kite and the area of a dart.

Exercise 31

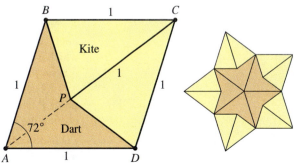

32 Automotive design The rear hatchback door of an automobile is 42 inches long. A strut with a fully extended length of 24 inches is to be attached to the door and the body of the car so that when the door is opened completely, the strut is vertical and the rear clearance is 32 inches, as shown in the figure. Approximate the lengths of segments TQ and TP.

Exercise 32

Exer. 33–40: Approximate the area of triangle ABC.

33 $\alpha = 60°$, $b = 20$, $c = 30$

34 $\gamma = 45°$, $b = 10.0$, $a = 15.0$

35 $\alpha = 40.3°$, $\beta = 62.9°$, $b = 5.63$

36 $\alpha = 35.7°$, $\gamma = 105.2°$, $b = 17.2$

37 $\alpha = 80.1°$, $a = 8.0$, $b = 3.4$

38 $\gamma = 32.1°$, $a = 14.6$, $c = 15.8$

39 $a = 25.0$, $b = 80.0$, $c = 60.0$

40 $a = 20.0$, $b = 20.0$, $c = 10.0$

Exer. 41–42: A triangular field has sides of lengths a, b, and c (in yards). Approximate the number of acres in the field (1 acre = 4840 yd²).

41 $a = 115$, $b = 140$, $c = 200$

42 $a = 320$, $b = 350$, $c = 500$

Exer. 43–44: Approximate the area of a parallelogram that has sides of lengths a and b (in feet) if one angle at a vertex has measure θ.

43 $a = 12.0$, $b = 16.0$, $\theta = 40°$

44 $a = 40.3$, $b = 52.6$, $\theta = 100°$

8.3

Vectors

Quantities such as area, volume, length, temperature, and time have magnitude only and can be completely characterized by a single real number (with an appropriate unit of measurement such as in², ft³, cm, deg, or sec). A quantity of this type is a **scalar quantity,** and the corresponding real number is a **scalar.** A concept such as velocity or force has both magnitude and direction and is often represented by a **directed line segment**—that is, a line segment to which a direction has been assigned. Another name for a directed line segment is a **vector.**

As shown in Figure 1, we use $\overrightarrow{PQ}$ to denote the vector with **initial point** P and **terminal point** Q, and we indicate the direction of the vector by placing the arrowhead at Q. The **magnitude** of $\overrightarrow{PQ}$ is the length of the segment PQ and is denoted by $\|\overrightarrow{PQ}\|$. As in the figure, we use boldface letters such as **u** and **v** to denote vectors whose endpoints are not specified. In handwritten work, a notation such as $\vec{u}$ or $\vec{v}$ is often used.

Vectors that have the same magnitude and direction are said to be **equivalent.** In mathematics, a vector is determined only by its magnitude and direc-

Figure 1
Equal vectors

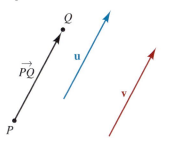

tion, not by its location. Thus, we regard equivalent vectors, such as those in Figure 1, as **equal** and write

$$\mathbf{u} = \overrightarrow{PQ}, \quad \mathbf{v} = \overrightarrow{PQ}, \quad \text{and} \quad \mathbf{u} = \mathbf{v}.$$

Thus, *a vector may be translated from one location to another, provided neither the magnitude nor the direction is changed.*

We can represent many physical concepts by vectors. To illustrate, suppose an airplane is descending at a constant speed of 100 mi/hr and the line of flight makes an angle of 20° with the horizontal. Both of these facts are represented by the vector **v** of magnitude 100 in Figure 2. The vector **v** is a **velocity vector.**

Figure 3
Force vector

Figure 2   Velocity vector

A vector that represents a pull or push of some type is a **force vector.** The force exerted when a person holds a 5-pound weight is illustrated by the vector **F** of magnitude 5 in Figure 3. This force has the same magnitude as the force exerted on the weight by gravity, but it acts in the opposite direction. As a result, there is no movement upward or downward.

We sometimes use $\overrightarrow{AB}$ to represent the path of a point (or particle) as it moves along the line segment from A to B. We then refer to $\overrightarrow{AB}$ as a **displacement** of the point (or particle). As in Figure 4, a displacement $\overrightarrow{AB}$ followed by a displacement $\overrightarrow{BC}$ leads to the same point as the single displacement $\overrightarrow{AC}$. By definition, the vector AC is the **sum** of $\overrightarrow{AB}$ and $\overrightarrow{BC}$, and we write

Figure 4
Sum of vectors

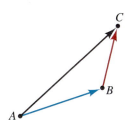

$$\overrightarrow{AC} = \overrightarrow{AB} + \overrightarrow{BC}.$$

Since vectors may be translated from one location to another, *any* two vectors may be added by placing the initial point of the second vector on the terminal point of the first and then drawing the line segment from the initial point of the first to the terminal point of the second, as in Figure 4. We refer to this method of vector addition as using the **triangle law.**

Another way to find the sum is to choose vector PQ and vector PR that are equal to $\overrightarrow{AB}$ and $\overrightarrow{BC}$, respectively, and have the same initial point P, as shown in Figure 5. If we construct parallelogram $RPQS$, then, since $\overrightarrow{PR} = \overrightarrow{QS}$,

Figure 5
Resultant force

Figure 5
Resultant force

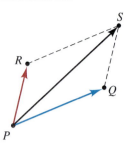

it follows that $\overrightarrow{PS} = \overrightarrow{PQ} + \overrightarrow{PR}$. If $\overrightarrow{PQ}$ and $\overrightarrow{PR}$ are two forces acting at P, then $\overrightarrow{PS}$ is the **resultant force**—that is, the single force that produces the same effect as the two combined forces. We refer to this method of vector addition as using the **parallelogram law.**

If m is a scalar and $\mathbf{v}$ is a vector, then $m\mathbf{v}$ is defined as a vector whose magnitude is $|m|$ times $\|\mathbf{v}\|$ (the magnitude of $\mathbf{v}$) and whose direction is either the same as that of $\mathbf{v}$ (if $m > 0$) or opposite that of $\mathbf{v}$ (if $m < 0$). Illustrations are given in Figure 6. We refer to $m\mathbf{v}$ as a **scalar multiple** of $\mathbf{v}$.

Figure 6 Scalar multiples

Figure 7

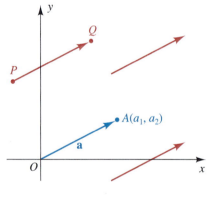

Throughout the remainder of this section we shall restrict our discussion to vectors that lie in an *xy*-plane. If $\overrightarrow{PQ}$ is such a vector, then, as indicated in Figure 7, there are many vectors that are equivalent to $\overrightarrow{PQ}$; however, there is exactly *one* equivalent vector $\mathbf{a} = \overrightarrow{OA}$ with initial point at the origin. In this sense, *each vector determines a unique ordered pair of real numbers,* the coordinates (a_1, a_2) of the terminal point A. Conversely, every ordered pair (a_1, a_2) determines the vector OA, where A has coordinates (a_1, a_2). Thus, *there is a one-to-one correspondence between vectors in an xy-plane and ordered pairs of real numbers.* This correspondence allows us to interpret a vector as both a directed line segment *and* an ordered pair of real numbers. To avoid confusion with the notation for open intervals or points, we use the symbol $\langle a_1, a_2 \rangle$ (referred to as *wedge notation*) for an ordered pair that represents a vector, and we denote it by a boldface letter—for example, $\mathbf{a} = \langle a_1, a_2 \rangle$. The numbers a_1 and a_2 are the **components** of the vector $\langle a_1, a_2 \rangle$. If A is the point (a_1, a_2), as in Figure 7, we call $\overrightarrow{OA}$ the **position vector** for $\langle a_1, a_2 \rangle$ or for the *point A.*

Figure 8
Magnitude $\|\mathbf{a}\|$

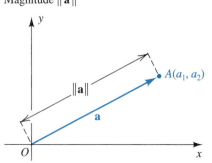

The preceding discussion shows that vectors have two different natures, one geometric and the other algebraic. Often we do not distinguish between the two. It should always be clear from our discussion whether we are referring to ordered pairs or directed line segments.

The *magnitude* of the vector $\mathbf{a} = \langle a_1, a_2 \rangle$ is, by definition, the length of its position vector OA, as illustrated in Figure 8.

Definition of the Magnitude of a Vector	The **magnitude** of the vector $\mathbf{a} = \langle a_1, a_2 \rangle$, denoted by $\|\mathbf{a}\|$, is given by $$\|\mathbf{a}\| = \|\langle a_1, a_2 \rangle\| = \sqrt{a_1^2 + a_2^2}.$$

Figure 9

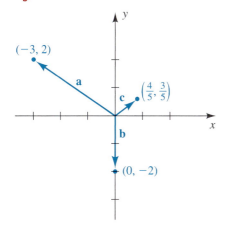

EXAMPLE 1 Finding the magnitude of a vector

Sketch the vectors

$$\mathbf{a} = \langle -3, 2 \rangle, \quad \mathbf{b} = \langle 0, -2 \rangle, \quad \mathbf{c} = \left\langle \tfrac{4}{5}, \tfrac{3}{5} \right\rangle$$

on a coordinate plane, and find the magnitude of each vector.

SOLUTION The vectors are sketched in Figure 9. By the definition of the magnitude of a vector,

$$\|\mathbf{a}\| = \|\langle -3, 2 \rangle\| = \sqrt{(-3)^2 + 2^2} = \sqrt{13}$$
$$\|\mathbf{b}\| = \|\langle 0, -2 \rangle\| = \sqrt{0^2 + (-2)^2} = \sqrt{4} = 2$$
$$\|\mathbf{c}\| = \left\|\left\langle \tfrac{4}{5}, \tfrac{3}{5} \right\rangle\right\| = \sqrt{\left(\tfrac{4}{5}\right)^2 + \left(\tfrac{3}{5}\right)^2} = \sqrt{\tfrac{16}{25} + \tfrac{9}{25}} = \sqrt{\tfrac{25}{25}} = 1.$$

Consider the vector OA and the vector OB corresponding to $\mathbf{a} = \langle a_1, a_2 \rangle$ and $\mathbf{b} = \langle b_1, b_2 \rangle$, respectively, as illustrated in Figure 10. If $\overrightarrow{OC}$ corresponds to $\mathbf{c} = \langle a_1 + b_1, a_2 + b_2 \rangle$, we can show, using slopes, that the points O, A, C, and B are vertices of a parallelogram; that is,

$$\overrightarrow{OA} + \overrightarrow{OB} = \overrightarrow{OC}.$$

Figure 10

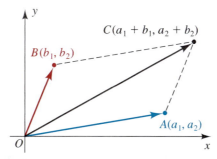

Expressing this equation in terms of ordered pairs leads to the following.

Definition of Addition of Vectors	$$\langle a_1, a_2 \rangle + \langle b_1, b_2 \rangle = \langle a_1 + b_1, a_2 + b_2 \rangle$$

Note that to add two vectors, we add corresponding components.

ILLUSTRATION **Addition of Vectors**

■ $\langle 3, -4 \rangle + \langle 2, 7 \rangle = \langle 3 + 2, -4 + 7 \rangle = \langle 5, 3 \rangle$

■ $\langle 5, 1 \rangle + \langle -5, 1 \rangle = \langle 5 + (-5), 1 + 1 \rangle = \langle 0, 2 \rangle$

It can also be shown that if m is a scalar and $\overrightarrow{OA}$ corresponds to $\mathbf{a} = \langle a_1, a_2 \rangle$, then the ordered pair determined by $m\overrightarrow{OA}$ is (ma_1, ma_2), as illustrated in Figure 11 for $m > 1$. This leads to the next definition.

Figure 11

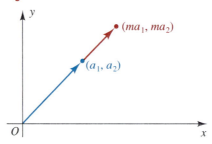

Definition of a Scalar Multiple of a Vector	$m\langle a_1, a_2 \rangle = \langle ma_1, ma_2 \rangle$

Thus, to find a scalar multiple of a vector, we multiply each component by the scalar.

ILLUSTRATION **Scalar Multiple of a Vector**

■ $2\langle -3, 4 \rangle = \langle 2(-3), 2(4) \rangle = \langle -6, 8 \rangle$

■ $-2\langle -3, 4 \rangle = \langle (-2)(-3), (-2)(4) \rangle = \langle 6, -8 \rangle$

■ $1\langle 5, 2 \rangle = \langle 1 \cdot 5, 1 \cdot 2 \rangle = \langle 5, 2 \rangle$

EXAMPLE 2 **Finding a scalar multiple of a vector**

If $\mathbf{a} = \langle 2, 1 \rangle$, find $3\mathbf{a}$ and $-2\mathbf{a}$, and sketch each vector in a coordinate plane.

SOLUTION   Using the definition of scalar multiples of vectors, we find

$$3\mathbf{a} = 3\langle 2, 1 \rangle = \langle 3 \cdot 2, 3 \cdot 1 \rangle = \langle 6, 3 \rangle$$
$$-2\mathbf{a} = -2\langle 2, 1 \rangle = \langle (-2) \cdot 2, (-2) \cdot 1 \rangle = \langle -4, -2 \rangle.$$

The vectors are sketched in Figure 12 on the next page.

Figure 12

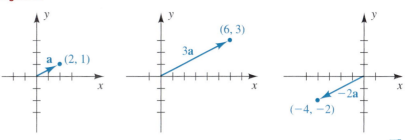

The **zero vector 0** and the **negative** $-\mathbf{a}$ of a vector $\mathbf{a} = \langle a_1, a_2 \rangle$ are defined as follows.

Definition of 0 and −a	$\mathbf{0} = \langle 0, 0 \rangle$ and $-\mathbf{a} = -\langle a_1, a_2 \rangle = \langle -a_1, -a_2 \rangle$

ILLUSTRATION **The Zero Vector and the Negative of a Vector**

- $\langle 3, 5 \rangle + \mathbf{0} = \langle 3, 5 \rangle + \langle 0, 0 \rangle = \langle 3 + 0, 5 + 0 \rangle = \langle 3, 5 \rangle$
- $-\langle 3, -5 \rangle = \langle -3, -(-5) \rangle = \langle -3, 5 \rangle$
- $\langle 3, -5 \rangle + \langle -3, 5 \rangle = \langle 3 + (-3), -5 + 5 \rangle = \langle 0, 0 \rangle = \mathbf{0}$
- $0\langle 2, 3 \rangle = \langle 0 \cdot 2, 0 \cdot 3 \rangle = \langle 0, 0 \rangle = \mathbf{0}$
- $5 \cdot \mathbf{0} = 5\langle 0, 0 \rangle = \langle 5 \cdot 0, 5 \cdot 0 \rangle = \langle 0, 0 \rangle = \mathbf{0}$

We next state properties of addition and scalar multiples of vectors for any vectors $\mathbf{a}$, $\mathbf{b}$, $\mathbf{c}$ and scalars m, n. You should have little difficulty in remembering these properties, since they resemble familiar properties of real numbers.

Properties of Addition and Scalar Multiples of Vectors	**(1)** $\mathbf{a} + \mathbf{b} = \mathbf{b} + \mathbf{a}$	**(5)** $m(\mathbf{a} + \mathbf{b}) = m\mathbf{a} + m\mathbf{b}$
	(2) $\mathbf{a} + (\mathbf{b} + \mathbf{c}) = (\mathbf{a} + \mathbf{b}) + \mathbf{c}$	**(6)** $(m + n)\mathbf{a} = m\mathbf{a} + n\mathbf{a}$
	(3) $\mathbf{a} + \mathbf{0} = \mathbf{a}$	**(7)** $(mn)\mathbf{a} = m(n\mathbf{a}) = n(m\mathbf{a})$
	(4) $\mathbf{a} + (-\mathbf{a}) = \mathbf{0}$	**(8)** $1\mathbf{a} = \mathbf{a}$
		(9) $0\mathbf{a} = \mathbf{0} = m\mathbf{0}$

PROOFS Let $\mathbf{a} = \langle a_1, a_2 \rangle$ and $\mathbf{b} = \langle b_1, b_2 \rangle$. To prove property 1, we note that

$$\mathbf{a} + \mathbf{b} = \langle a_1 + b_1, a_2 + b_2 \rangle = \langle b_1 + a_1, b_2 + a_2 \rangle = \mathbf{b} + \mathbf{a}.$$

(continued)

The proof of property 5 is as follows:

$$
\begin{aligned}
m(\mathbf{a} + \mathbf{b}) &= m\langle a_1 + b_1, a_2 + b_2 \rangle && \text{definition of addition} \\
&= \langle m(a_1 + b_1), m(a_2 + b_2) \rangle && \text{definition of scalar multiple} \\
&= \langle ma_1 + mb_1, ma_2 + mb_2 \rangle && \text{distributive property} \\
&= \langle ma_1, ma_2 \rangle + \langle mb_1, mb_2 \rangle && \text{definition of addition} \\
&= m\mathbf{a} + m\mathbf{b} && \text{definition of scalar multiple}
\end{aligned}
$$

Proofs of the remaining properties are similar and are left as exercises.

Vector subtraction (denoted by $-$) is defined by $\mathbf{a} - \mathbf{b} = \mathbf{a} + (-\mathbf{b})$. If we use the ordered pair notation for $\mathbf{a}$ and $\mathbf{b}$, then $-\mathbf{b} = \langle -b_1, -b_2 \rangle$, and we obtain the following.

Definition of Subtraction of Vectors	$\mathbf{a} - \mathbf{b} = \langle a_1, a_2 \rangle - \langle b_1, b_2 \rangle = \langle a_1 - b_1, a_2 - b_2 \rangle$

Thus, to find $\mathbf{a} - \mathbf{b}$, we merely subtract the components of $\mathbf{b}$ from the corresponding components of $\mathbf{a}$.

ILLUSTRATION **Subtraction of Vectors** If $\mathbf{a} = \langle 5, -4 \rangle$ and $\mathbf{b} = \langle -3, 2 \rangle$

■ $\mathbf{a} - \mathbf{b} = \langle 5, -4 \rangle - \langle -3, 2 \rangle$
$$= \langle 5 - (-3), -4 - 2 \rangle = \langle 8, -6 \rangle$$

■ $2\mathbf{a} - 3\mathbf{b} = 2\langle 5, -4 \rangle - 3\langle -3, 2 \rangle$
$$= \langle 10, -8 \rangle - \langle -9, 6 \rangle = \langle 10 - (-9), -8 - 6 \rangle = \langle 19, -14 \rangle$$

Figure 13

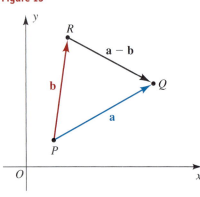

If $\mathbf{a}$ and $\mathbf{b}$ are arbitrary vectors, then

$$\mathbf{b} + (\mathbf{a} - \mathbf{b}) = \mathbf{a};$$

that is, $\mathbf{a} - \mathbf{b}$ *is the vector that, when added to* $\mathbf{b}$, *gives us* $\mathbf{a}$. If we represent $\mathbf{a}$ and $\mathbf{b}$ by vector PQ and vector PR *with the same initial point,* as in Figure 13, then $\overrightarrow{RQ}$ represents $\mathbf{a} - \mathbf{b}$.

The special vectors $\mathbf{i}$ and $\mathbf{j}$ are defined as follows.

Definition of i and j	$\mathbf{i} = \langle 1, 0 \rangle, \qquad \mathbf{j} = \langle 0, 1 \rangle$

A **unit vector** is a vector of magnitude 1. The vectors $\mathbf{i}$ and $\mathbf{j}$ are unit vectors, as is the vector $\mathbf{c} = \left\langle \frac{4}{5}, \frac{3}{5} \right\rangle$ in Example 1.

The vectors **i** and **j** can be used to obtain an alternative way of denoting vectors. Specifically, if $\mathbf{a} = \langle a_1, a_2 \rangle$, then

$$\mathbf{a} = \langle a_1, 0 \rangle + \langle 0, a_2 \rangle = a_1 \langle 1, 0 \rangle + a_2 \langle 0, 1 \rangle.$$

This result gives us the following.

i, j Form for Vectors	$\mathbf{a} = \langle a_1, a_2 \rangle = a_1\mathbf{i} + a_2\mathbf{j}$

ILLUSTRATION **i, j Form**

■ $\langle 5, 2 \rangle = 5\mathbf{i} + 2\mathbf{j}$

■ $\langle -3, 4 \rangle = -3\mathbf{i} + 4\mathbf{j}$

■ $\langle 0, -6 \rangle = 0\mathbf{i} + (-6)\mathbf{j} = -6\mathbf{j}$

Vectors corresponding to **i**, **j**, and an arbitrary vector **a** are illustrated in Figure 14. Since **i** and **j** are unit vectors, $a_1\mathbf{i}$ and $a_2\mathbf{j}$ may be represented by horizontal and vertical vectors of magnitudes $|a_1|$ and $|a_2|$, respectively, as illustrated in Figure 15. For this reason we call a_1 the **horizontal component** and a_2 the **vertical component** of the vector **a**.

Figure 14 $\mathbf{a} = \langle a_1, a_2 \rangle$

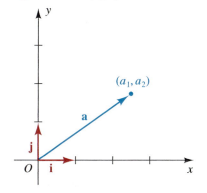

Figure 15 $\mathbf{a} = a_1\mathbf{i} + a_2\mathbf{j}$

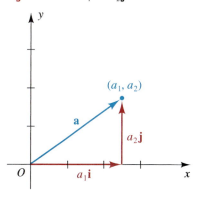

The vector sum $a_1\mathbf{i} + a_2\mathbf{j}$ is a **linear combination** of **i** and **j**. Rules for addition, subtraction, and multiplication by a scalar m may be written as follows, with $\mathbf{b} = \langle b_1, b_2 \rangle = b_1\mathbf{i} + b_2\mathbf{j}$:

$$(a_1\mathbf{i} + a_2\mathbf{j}) + (b_1\mathbf{i} + b_2\mathbf{j}) = (a_1 + b_1)\mathbf{i} + (a_2 + b_2)\mathbf{j}$$
$$(a_1\mathbf{i} + a_2\mathbf{j}) - (b_1\mathbf{i} + b_2\mathbf{j}) = (a_1 - b_1)\mathbf{i} + (a_2 - b_2)\mathbf{j}$$
$$m(a_1\mathbf{i} + a_2\mathbf{j}) = (ma_1)\mathbf{i} + (ma_2)\mathbf{j}$$

These formulas show that we may regard linear combinations of **i** and **j** as algebraic sums.

EXAMPLE 3   Expressing a vector as a linear combination of i and j

If $\mathbf{a} = 5\mathbf{i} + \mathbf{j}$ and $\mathbf{b} = 4\mathbf{i} - 7\mathbf{j}$, express $3\mathbf{a} - 2\mathbf{b}$ as a linear combination of $\mathbf{i}$ and $\mathbf{j}$.

SOLUTION
$$3\mathbf{a} - 2\mathbf{b} = 3(5\mathbf{i} + \mathbf{j}) - 2(4\mathbf{i} - 7\mathbf{j})$$
$$= (15\mathbf{i} + 3\mathbf{j}) - (8\mathbf{i} - 14\mathbf{j})$$
$$= 7\mathbf{i} + 17\mathbf{j}$$

Figure 16

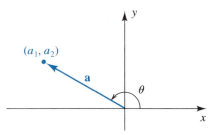

Let θ be an angle in standard position, measured from the positive x-axis to the vector $\mathbf{a} = \langle a_1, a_2 \rangle = a_1\mathbf{i} + a_2\mathbf{j}$, as illustrated in Figure 16. Since

$$\cos \theta = \frac{a_1}{\|\mathbf{a}\|} \quad \text{and} \quad \sin \theta = \frac{a_2}{\|\mathbf{a}\|},$$

we obtain the following formulas.

Formulas for Horizontal and Vertical Components of a $= \langle a_1, a_2 \rangle$	If the vector $\mathbf{a}$ and the angle θ are defined as above, then $a_1 = \|\mathbf{a}\| \cos \theta \quad$ and $\quad a_2 = \|\mathbf{a}\| \sin \theta.$

Using these formulas, we have
$$\mathbf{a} = \langle a_1, a_2 \rangle = \langle \|\mathbf{a}\| \cos \theta, \|\mathbf{a}\| \sin \theta \rangle$$
$$= \|\mathbf{a}\| \cos \theta \, \mathbf{i} + \|\mathbf{a}\| \sin \theta \, \mathbf{j}$$
$$= \|\mathbf{a}\|(\cos \theta \, \mathbf{i} + \sin \theta \, \mathbf{j}).$$

EXAMPLE 4   Expressing wind velocity as a vector

If the wind is blowing at 12 mi/hr in the direction N40°W, express its velocity as a vector $\mathbf{v}$.

SOLUTION   The vector $\mathbf{v}$ and the angle $\theta = 90° + 40° = 130°$ are illustrated in Figure 17. Using the formulas for horizontal and vertical components with $\mathbf{v} = \langle v_1, v_2 \rangle$ gives us

$$v_1 = \|\mathbf{v}\| \cos \theta = 12 \cos 130°, \qquad v_2 = \|\mathbf{v}\| \sin \theta = 12 \sin 130°.$$

Hence,
$$\mathbf{v} = v_1\mathbf{i} + v_2\mathbf{j}$$
$$= (12 \cos 130°)\mathbf{i} + (12 \sin 130°)\mathbf{j}$$
$$\approx (-7.7)\mathbf{i} + (9.2)\mathbf{j}.$$

Figure 17

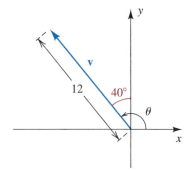

EXAMPLE 5   Finding a vector of specified direction and magnitude

Find a vector $\mathbf{b}$ in the opposite direction of $\mathbf{a} = \langle 5, -12 \rangle$ that has magnitude 6.

Figure 18

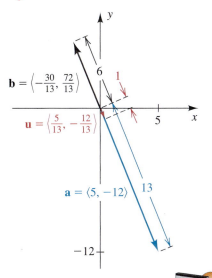

$$b = \left\langle -\frac{30}{13}, \frac{72}{13} \right\rangle$$

$$u = \left\langle \frac{5}{13}, -\frac{12}{13} \right\rangle$$

$$a = \langle 5, -12 \rangle$$

SOLUTION   The magnitude of **a** is given by

$$\| a \| = \sqrt{5^2 + (-12)^2} = \sqrt{25 + 144} = \sqrt{169} = 13.$$

A unit vector **u** in the direction of **a** can be found by multiplying **a** by $1/\| a \|$. Thus,

$$u = \frac{1}{\| a \|} a = \frac{1}{13}\langle 5, -12 \rangle = \left\langle \frac{5}{13}, -\frac{12}{13} \right\rangle.$$

Multiplying **u** by 6 gives us a vector of magnitude 6 in the direction of **a**, so we'll multiply **u** by -6 to obtain the desired vector **b**, as shown in Figure 18:

$$b = -6u = -6\left\langle \frac{5}{13}, -\frac{12}{13} \right\rangle = \left\langle -\frac{30}{13}, \frac{72}{13} \right\rangle$$

EXAMPLE 6   Finding a resultant vector

Two forces $\overrightarrow{PQ}$ and $\overrightarrow{PR}$ of magnitudes 5.0 kilograms and 8.0 kilograms, respectively, act at a point P. The direction of $\overrightarrow{PQ}$ is N20°E, and the direction of $\overrightarrow{PR}$ is N65°E. Approximate the magnitude and direction of the resultant $\overrightarrow{PS}$.

Figure 19

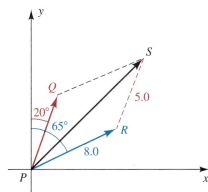

SOLUTION   The forces are represented geometrically in Figure 19. Note that the angles from the positive x-axis to $\overrightarrow{PQ}$ and $\overrightarrow{PR}$ have measures 70° and 25°, respectively. Using the formulas for horizontal and vertical components, we obtain the following:

$$\overrightarrow{PQ} = (5 \cos 70°)\mathbf{i} + (5 \sin 70°)\mathbf{j}$$
$$\overrightarrow{PR} = (8 \cos 25°)\mathbf{i} + (8 \sin 25°)\mathbf{j}$$

Since $\overrightarrow{PS} = \overrightarrow{PQ} + \overrightarrow{PR}$,

$$\overrightarrow{PS} = (5 \cos 70° + 8 \cos 25°)\mathbf{i} + (5 \sin 70° + 8 \sin 25°)\mathbf{j}$$
$$\approx 8.9606\mathbf{i} + 8.0794\mathbf{j} \approx (9.0)\mathbf{i} + (8.1)\mathbf{j}.$$

Consequently,

$$\| \overrightarrow{PS} \| \approx \sqrt{(9.0)^2 + (8.1)^2} \approx 12.1.$$

We can also find $\| \overrightarrow{PS} \|$ by using the law of cosines (see Example 3 of Section 8.2). Since $\angle QPR = 45°$, it follows that $\angle PRS = 135°$, and hence

$$\| \overrightarrow{PS} \|^2 = (8.0)^2 + (5.0)^2 - 2(8.0)(5.0) \cos 135° \approx 145.6$$

and
$$\| \overrightarrow{PS} \| \approx \sqrt{145.6} \approx 12.1.$$

If θ is the angle from the positive x-axis to the resultant PS, then using the (approximate) coordinates (8.9606, 8.0794) of S, we obtain the following:

$$\tan \theta \approx \frac{8.0794}{8.9606} \approx 0.9017$$
$$\theta \approx \tan^{-1}(0.9017) \approx 42°$$

Hence, the direction of $\overrightarrow{PS}$ is approximately N(90° − 42°)E = N48°E.

8.3 Exercises

Exer. 1–6: Find $a + b$, $a - b$, $4a + 5b$, $4a - 5b$, and $\|a\|$.

1 $a = \langle 2, -3 \rangle$, $b = \langle 1, 4 \rangle$

2 $a = \langle -2, 6 \rangle$, $b = \langle 2, 3 \rangle$

3 $a = -\langle 7, -2 \rangle$, $b = 4\langle -2, 1 \rangle$

4 $a = 2\langle 5, -4 \rangle$, $b = -\langle 6, 0 \rangle$

5 $a = i + 2j$, $b = 3i - 5j$

6 $a = -3i + j$, $b = -3i + j$

Exer. 7–10: Sketch vectors corresponding to a, b, $a + b$, $2a$, and $-3b$.

7 $a = 3i + 2j$, $b = -i + 5j$

8 $a = -5i + 2j$, $b = i - 3j$

9 $a = \langle -4, 6 \rangle$, $b = \langle -2, 3 \rangle$

10 $a = \langle 2, 0 \rangle$, $b = \langle -2, 0 \rangle$

Exer. 11–16: Use components to express the sum or difference as a scalar multiple of one of the vectors a, b, c, d, e, or f shown in the figure.

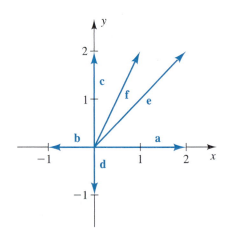

11 $a + b$

12 $c - d$

13 $b + e$

14 $f - b$

15 $b + d$

16 $e + c$

Exer. 17–26: If $a = \langle a_1, a_2 \rangle$, $b = \langle b_1, b_2 \rangle$, $c = \langle c_1, c_2 \rangle$, and m and n are real numbers, prove the stated property.

17 $a + (b + c) = (a + b) + c$

18 $a + 0 = a$

19 $a + (-a) = 0$

20 $(m + n)a = ma + na$

21 $(mn)a = m(na) = n(ma)$

22 $1a = a$

23 $0a = 0 = m0$

24 $(-m)a = -ma$

25 $-(a + b) = -a - b$

26 $m(a - b) = ma - mb$

27 If $v = \langle a, b \rangle$, prove that the magnitude of $2v$ is twice the magnitude of v.

28 If $v = \langle a, b \rangle$ and k is any real number, prove that the magnitude of kv is $|k|$ times the magnitude of v.

Exer. 29–36: Find the magnitude of the vector a and the smallest positive angle θ from the positive x-axis to the vector OP that corresponds to a.

29 $a = \langle 3, -3 \rangle$

30 $a = \langle -2, -2\sqrt{3} \rangle$

31 $a = \langle -5, 0 \rangle$

32 $a = \langle 0, 10 \rangle$

33 $a = -4i + 5j$

34 $a = 10i - 10j$

35 $a = -18j$

36 $a = 2i - 3j$

Exer. 37–40: The vectors a and b represent two forces acting at the same point, and θ is the smallest positive angle between a and b. Approximate the magnitude of the resultant force.

37 $\|a\| = 40$ lb, $\|b\| = 70$ lb, $\theta = 45°$

38 $\|a\| = 5.5$ lb, $\|b\| = 6.2$ lb, $\theta = 60°$

39 $\|a\| = 2.0$ lb, $\|b\| = 8.0$ lb, $\theta = 120°$

40 $\|a\| = 30$ lb, $\|b\| = 50$ lb, $\theta = 150°$

Exer. 41–44: The magnitudes and directions of two forces acting at a point P are given in (a) and (b). Approximate the magnitude and direction of the resultant vector.

41 (a) 90 lb, N75°W   (b) 60 lb, S5°E

42 (a) 20 lb, S17°W   (b) 50 lb, N82°W

43 (a) 6.0 lb, 110°   (b) 2.0 lb, 215°

44 (a) 70 lb, 320°   (b) 40 lb, 30°

Exer. 45–48: Approximate the horizontal and vertical components of the vector that is described.

45 **Releasing a football** A quarterback releases a football with a speed of 50 ft/sec at an angle of 35° with the horizontal.

46 **Pulling a sled** A child pulls a sled through the snow by exerting a force of 20 pounds at an angle of 40° with the horizontal.

47 **Biceps muscle** The biceps muscle, in supporting the forearm and a weight held in the hand, exerts a force of 20 pounds. As shown in the figure, the muscle makes an angle of 108° with the forearm.

Exercise 47

48 **Jet's approach** A jet airplane approaches a runway at an angle of 7.5° with the horizontal, traveling at a speed of 160 mi/hr.

Exer. 49–52: Find a unit vector that has (a) the same direction as the vector a and (b) the opposite direction of the vector a.

49 $\mathbf{a} = -8\mathbf{i} + 15\mathbf{j}$ 50 $\mathbf{a} = 5\mathbf{i} - 3\mathbf{j}$

51 $\mathbf{a} = \langle 2, -5 \rangle$ 52 $\mathbf{a} = \langle 0, 6 \rangle$

53 Find a vector that has the same direction as $\langle -6, 3 \rangle$ and

  (a) twice the magnitude

  (b) one-half the magnitude

54 Find a vector that has the opposite direction of $8\mathbf{i} - 5\mathbf{j}$ and

  (a) three times the magnitude

  (b) one-third the magnitude

55 Find a vector of magnitude 6 that has the opposite direction of $\mathbf{a} = 4\mathbf{i} - 7\mathbf{j}$.

56 Find a vector of magnitude 4 that has the opposite direction of $\mathbf{a} = \langle 2, -5 \rangle$.

Exer. 57–60: If forces F_1, F_2, ... , F_n act at a point P, the net (or resultant) force F is the sum $F_1 + F_2 + \cdots + F_n$. If $F = 0$, the forces are said to be in equilibrium. The given forces act at the origin O of an xy-plane.

(a) Find the net force F.
(b) Find an additional force G such that equilibrium occurs.

57 $\mathbf{F_1} = \langle 4, 3 \rangle$, $\mathbf{F_2} = \langle -2, -3 \rangle$, $\mathbf{F_3} = \langle 5, 2 \rangle$

58 $\mathbf{F_1} = \langle -3, -1 \rangle$, $\mathbf{F_2} = \langle 0, -3 \rangle$, $\mathbf{F_3} = \langle 3, 4 \rangle$

59

60

61 Tugboat force Two tugboats are towing a large ship into port, as shown in the figure. The larger tug exerts a force of 4000 pounds on its cable, and the smaller tug exerts a force of 3200 pounds on its cable. If the ship is to travel on a straight line l, approximate the angle θ that the larger tug must make with l.

Exercise 61

62 Gravity simulation Shown in the figure is a simple apparatus that may be used to simulate gravity conditions on other planets. A rope is attached to an astronaut who maneuvers on an inclined plane that makes an angle of θ degrees with the horizontal.

(a) If the astronaut weighs 160 pounds, find the x- and y-components of the downward force (see the figure for axes).

(b) The y-component in part (a) is the weight of the astronaut relative to the inclined plane. The astronaut would weigh 27 pounds on the moon and 60 pounds on Mars. Approximate the angles θ (to the nearest 0.01°) so that the inclined-plane apparatus will simulate walking on these surfaces.

63 Airplane course and ground speed An airplane with an airspeed of 200 mi/hr is flying in the direction 50°, and a 40 mi/hr wind is blowing directly from the west. As shown in the figure, these facts may be represented by vectors **p** and **w** of magnitudes 200 and 40, respectively. The direction of the resultant **p** + **w** gives the true course of the airplane relative to the ground, and the magnitude ∥**p** + **w**∥ is the ground speed of the airplane. Approximate the true course and ground speed.

Exercise 63

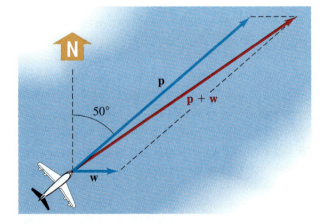

64 Airplane course and ground speed Refer to Exercise 63. An airplane is flying in the direction 140° with an airspeed of 500 mi/hr, and a 30 mi/hr wind is blowing in the direction 65°. Approximate the true course and ground speed of the airplane.

65 **Airplane course and ground speed** An airplane pilot wishes to maintain a true course in the direction 250° with a ground speed of 400 mi/hr when the wind is blowing directly north at 50 mi/hr. Approximate the required airspeed and compass heading.

66 **Wind direction and speed** An airplane is flying in the direction 20° with an airspeed of 300 mi/hr. Its ground speed and true course are 350 mi/hr and 30°, respectively. Approximate the direction and speed of the wind.

67 **Rowboat navigation** The current in a river flows directly from the west at a rate of 1.5 ft/sec. A person who rows a boat at a rate of 4 ft/sec in still water wishes to row directly north across the river. Approximate, to the nearest degree, the direction in which the person should row.

68 **Motorboat navigation** For a motorboat moving at a speed of 30 mi/hr to travel directly north across a river, it must aim at a point that has the bearing N15°E. If the current is flowing directly west, approximate the rate at which it flows.

69 **Flow of ground water** Ground-water contaminants can enter a community's drinking water by migrating through porous rock into the aquifer. If underground water flows with a velocity v_1 through an interface between one type of rock and a second type of rock, its velocity changes to v_2, and both the direction and the speed of the flow can be obtained using the formula

$$\frac{\|v_1\|}{\|v_2\|} = \frac{\tan \theta_1}{\tan \theta_2},$$

where the angles θ_1 and θ_2 are as shown in the figure. For sandstone, $\|v_1\| = 8.2$ cm/day; for limestone, $\|v_2\| = 3.8$ cm/day. If $\theta_1 = 30°$, approximate the vectors v_1 and v_2 in **i**, **j** form.

Exercise 69

70 **Flow of ground water** Refer to Exercise 69. Contaminated ground water is flowing through silty sand with the direction of flow θ_1 and speed (in cm/day) given by the vector $v_1 = 20\mathbf{i} - 82\mathbf{j}$. When the flow enters a region of clean sand, its rate increases to 725 cm/day. Find the new direction of flow by approximating θ_2.

71 **Robotic movement** Vectors are useful for describing movement of robots.

(a) The robot's arm illustrated in the first figure can rotate at the joint connections P and Q. The upper arm, represented by **a**, is 15 inches long, and the forearm (including the hand), represented by **b**, is 17 inches long. Approximate the coordinates of the point R in the hand by using **a** + **b**.

Exercise 71(a)

(b) If the upper arm is rotated 85° and the forearm is rotated an additional 35°, as illustrated in the second figure, approximate the new coordinates of R by using **c** + **d**.

Exercise 71(b)

72 Robotic movement Refer to Exercise 71.

(a) Suppose the wrist joint of the robot's arm is allowed to rotate at the joint connection S and the arm is located as shown in the first figure. The upper arm has a length of 15 inches; the forearm, without the hand, has a length of 10 inches; and the hand has a length of 7 inches. Approximate the coordinates of R by using $\mathbf{a} + \mathbf{b} + \mathbf{c}$.

Exercise 72(a)

(b) Suppose the robot's upper arm is rotated 75°, and then the forearm is rotated −80°, and finally the hand is rotated an additional 40°, as shown in the second figure. Approximate the new coordinates of R by using $\mathbf{d} + \mathbf{e} + \mathbf{f}$.

Exercise 72(b)

73 Stonehenge forces Refer to Exercise 25 in Section 6.2. In the construction of Stonehenge, groups of 550 people were used to pull 99,000-pound blocks of stone up ramps inclined at 9°. Ignoring friction, determine the force that each person had to contribute in order to move the stone up the ramp.

Exercise 73

8.4

The Dot Product

The *dot product* of two vectors has many applications. We begin with an algebraic definition.

Definition of the Dot Product	Let $\mathbf{a} = \langle a_1, a_2 \rangle = a_1\mathbf{i} + a_2\mathbf{j}$ and $\mathbf{b} = \langle b_1, b_2 \rangle = b_1\mathbf{i} + b_2\mathbf{j}$. The **dot product** of $\mathbf{a}$ and $\mathbf{b}$, denoted $\mathbf{a} \cdot \mathbf{b}$, is $$\mathbf{a} \cdot \mathbf{b} = \langle a_1, a_2 \rangle \cdot \langle b_1, b_2 \rangle = a_1b_1 + a_2b_2.$$

The symbol $\mathbf{a} \cdot \mathbf{b}$ is read "$\mathbf{a}$ dot $\mathbf{b}$." We also refer to the dot product as the **scalar product** or the **inner product**. Note that $\mathbf{a} \cdot \mathbf{b}$ *is a real number and not a vector,* as illustrated in the following example.

EXAMPLE 1   Finding the dot product of two vectors

Find $\mathbf{a} \cdot \mathbf{b}$.

(a) $\mathbf{a} = \langle -5, 3 \rangle, \quad \mathbf{b} = \langle 2, 6 \rangle$ (b) $\mathbf{a} = 4\mathbf{i} + 6\mathbf{j}, \quad \mathbf{b} = 3\mathbf{i} - 7\mathbf{j}$

SOLUTION

(a) $\langle -5, 3 \rangle \cdot \langle 2, 6 \rangle = (-5)(2) + (3)(6) = -10 + 18 = 8$

(b) $(4\mathbf{i} + 6\mathbf{j}) \cdot (3\mathbf{i} - 7\mathbf{j}) = (4)(3) + (6)(-7) = 12 - 42 = -30$

Properties of the Dot Product	If $\mathbf{a}$, $\mathbf{b}$, and $\mathbf{c}$ are vectors and m is a real number, then
	$\textbf{(1) } \mathbf{a} \cdot \mathbf{a} = \|\mathbf{a}\|^2$
	$\textbf{(2) } \mathbf{a} \cdot \mathbf{b} = \mathbf{b} \cdot \mathbf{a}$
	$\textbf{(3) } \mathbf{a} \cdot (\mathbf{b} + \mathbf{c}) = \mathbf{a} \cdot \mathbf{b} + \mathbf{a} \cdot \mathbf{c}$
	$\textbf{(4) } (m\mathbf{a}) \cdot \mathbf{b} = m(\mathbf{a} \cdot \mathbf{b}) = \mathbf{a} \cdot (m\mathbf{b})$
	$\textbf{(5) } \mathbf{0} \cdot \mathbf{a} = 0$

PROOF   The proof of each property follows from the definition of the dot product and the properties of real numbers. Thus, if $\mathbf{a} = \langle a_1, a_2 \rangle$, $\mathbf{b} = \langle b_1, b_2 \rangle$, and $\mathbf{c} = \langle c_1, c_2 \rangle$, then

$$\mathbf{a} \cdot (\mathbf{b} + \mathbf{c}) = \langle a_1, a_2 \rangle \cdot \langle b_1 + c_1, b_2 + c_2 \rangle \quad \text{\color{red}definition of addition}$$
$$= a_1(b_1 + c_1) + a_2(b_2 + c_2) \quad \text{\color{red}definition of dot product}$$
$$= (a_1 b_1 + a_2 b_2) + (a_1 c_1 + a_2 c_2) \quad \text{\color{red}real number properties}$$
$$= \mathbf{a} \cdot \mathbf{b} + \mathbf{a} \cdot \mathbf{c}, \quad \text{\color{red}definition of dot product}$$

which proves property 3. The proofs of the remaining properties are left as exercises.

Figure 1

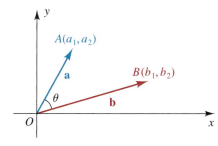

Any two nonzero vectors $\mathbf{a} = \langle a_1, a_2 \rangle$ and $\mathbf{b} = \langle b_1, b_2 \rangle$ may be represented in a coordinate plane by directed line segments from the origin O to the points $A(a_1, a_2)$ and $B(b_1, b_2)$, respectively. The **angle θ between $\mathbf{a}$ and $\mathbf{b}$** is, by definition, $\angle AOB$ (see Figure 1). Note that $0 \leq \theta \leq \pi$ and that $\theta = 0$ if $\mathbf{a}$ and $\mathbf{b}$ have the same direction or $\theta = \pi$ if $\mathbf{a}$ and $\mathbf{b}$ have opposite directions.

Definition of Parallel and Orthogonal Vectors	Let θ be the angle between two nonzero vectors $\mathbf{a}$ and $\mathbf{b}$.
	(1) $\mathbf{a}$ and $\mathbf{b}$ are **parallel** if $\theta = 0$ or $\theta = \pi$.
	(2) $\mathbf{a}$ and $\mathbf{b}$ are **orthogonal** if $\theta = \dfrac{\pi}{2}$.

The vectors **a** and **b** in Figure 1 are parallel if and only if they lie on the same line that passes through the origin. In this case, **b** = m**a** for some real number m. The vectors are orthogonal if and only if they lie on mutually perpendicular lines that pass through the origin. We assume that the zero vector **0** is parallel and orthogonal to *every* vector **a**.

The next theorem shows the close relationship between the angle between two vectors and their dot product.

Theorem on the Dot Product	If θ is the angle between two nonzero vectors **a** and **b**, then $$\mathbf{a} \cdot \mathbf{b} = \|\mathbf{a}\| \|\mathbf{b}\| \cos \theta.$$

PROOF If **a** and **b** are not parallel, we have a situation similar to that illustrated in Figure 1. We may then apply the law of cosines to triangle AOB. Since the lengths of the three sides of the triangle are $\|\mathbf{a}\|$, $\|\mathbf{b}\|$, and $d(A, B)$,

$$[d(A, B)]^2 = \|\mathbf{a}\|^2 + \|\mathbf{b}\|^2 - 2\|\mathbf{a}\| \|\mathbf{b}\| \cos \theta.$$

Using the distance formula and the definition of the magnitude of a vector, we obtain

$$(b_1 - a_1)^2 + (b_2 - a_2)^2 = (a_1^2 + a_2^2) + (b_1^2 + b_2^2) - 2\|\mathbf{a}\| \|\mathbf{b}\| \cos \theta,$$

which reduces to

$$-2a_1 b_1 - 2a_2 b_2 = -2\|\mathbf{a}\| \|\mathbf{b}\| \cos \theta.$$

Dividing both sides of the last equation by -2 gives us

$$a_1 b_1 + a_2 b_2 = \|\mathbf{a}\| \|\mathbf{b}\| \cos \theta,$$

which is equivalent to what we wished to prove, since the left-hand side is **a** · **b**.

If **a** and **b** are parallel, then either $\theta = 0$ or $\theta = \pi$, and therefore **b** = m**a** for some real number m with $m > 0$ if $\theta = 0$ and $m < 0$ if $\theta = \pi$. We can show, using properties of the dot product, that $\mathbf{a} \cdot (m\mathbf{a}) = \|\mathbf{a}\| \|m\mathbf{a}\| \cos \theta$, and hence the theorem is true for all nonzero vectors **a** and **b**. ◢

Theorem on the Cosine of the Angle Between Vectors	If θ is the angle between two nonzero vectors **a** and **b**, then $$\cos \theta = \frac{\mathbf{a} \cdot \mathbf{b}}{\|\mathbf{a}\| \|\mathbf{b}\|}.$$

EXAMPLE 2 Finding the angle between two vectors

Find the angle between **a** = $\langle 4, -3 \rangle$ and **b** = $\langle 1, 2 \rangle$.

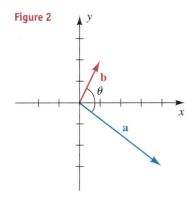

Figure 2

SOLUTION   The vectors are sketched in Figure 2. We apply the preceding theorem:

$$\cos \theta = \frac{\mathbf{a} \cdot \mathbf{b}}{\|\mathbf{a}\| \|\mathbf{b}\|} = \frac{(4)(1) + (-3)(2)}{\sqrt{16 + 9} \sqrt{1 + 4}} = \frac{-2}{5\sqrt{5}}, \quad \text{or} \quad \frac{-2\sqrt{5}}{25}$$

Hence,

$$\theta = \arccos \left(\frac{-2\sqrt{5}}{25} \right) \approx 100.3°.$$

EXAMPLE 3   Showing that two vectors are parallel

Let $\mathbf{a} = \frac{1}{2}\mathbf{i} - 3\mathbf{j}$ and $\mathbf{b} = -2\mathbf{i} + 12\mathbf{j}$.

(a) Show that **a** and **b** are parallel.

(b) Find the scalar m such that $\mathbf{b} = m\mathbf{a}$.

SOLUTION

(a) By definition, the vectors **a** and **b** are parallel if and only if the angle θ between them is either 0 or π. Since

$$\cos \theta = \frac{\mathbf{a} \cdot \mathbf{b}}{\|\mathbf{a}\| \|\mathbf{b}\|} = \frac{\left(\frac{1}{2}\right)(-2) + (-3)(12)}{\sqrt{\frac{1}{4} + 9} \sqrt{4 + 144}} = \frac{-37}{37} = -1,$$

we conclude that

$$\theta = \arccos (-1) = \pi.$$

(b) Since **a** and **b** are parallel, there *is* a scalar m such that $\mathbf{b} = m\mathbf{a}$; that is,

$$-2\mathbf{i} + 12\mathbf{j} = m\left(\tfrac{1}{2}\mathbf{i} - 3\mathbf{j}\right) = \tfrac{1}{2}m\mathbf{i} - 3m\mathbf{j}.$$

Equating the coefficients of **i** and **j** gives us

$$-2 = \tfrac{1}{2}m \quad \text{and} \quad 12 = -3m.$$

Thus, $m = -4$; that is, $\mathbf{b} = -4\mathbf{a}$. Note that **a** and **b** have opposite directions, since $m < 0$.

Using the formula $\mathbf{a} \cdot \mathbf{b} = \|\mathbf{a}\| \|\mathbf{b}\| \cos \theta$, together with the fact that two vectors are orthogonal if and only if the angle between them is $\pi/2$ (or one of the vectors is **0**), gives us the following result.

Theorem on Orthogonal Vectors	Two vectors **a** and **b** are orthogonal if and only if $\mathbf{a} \cdot \mathbf{b} = 0$.

EXAMPLE 4   Showing that two vectors are orthogonal

Show that the pair of vectors is orthogonal:

(a) **i**, **j**   (b) $2\mathbf{i} + 3\mathbf{j}$, $6\mathbf{i} - 4\mathbf{j}$

SOLUTION   We may use the theorem on orthogonal vectors to prove orthogonality by showing that the dot product of each pair is zero:

(a) $\mathbf{i} \cdot \mathbf{j} = \langle 1, 0 \rangle \cdot \langle 0, 1 \rangle = (1)(0) + (0)(1) = 0 + 0 = 0$

(b) $(2\mathbf{i} + 3\mathbf{j}) \cdot (6\mathbf{i} - 4\mathbf{j}) = (2)(6) + (3)(-4) = 12 - 12 = 0$

Definition of comp$_b$ a

Let θ be the angle between two nonzero vectors **a** and **b**. The **component of a along b**, denoted by comp$_b$ **a**, is given by

$$\text{comp}_b\, \mathbf{a} = \|\mathbf{a}\|\, \cos \theta.$$

The geometric significance of the preceding definition with θ acute or obtuse is illustrated in Figure 3, where the x- and y-axes are not shown.

Figure 3   comp$_b$ **a** = $\|\mathbf{a}\|$ cos θ

(a)   (b)

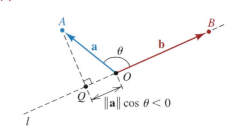

If angle θ is acute, then, as in Figure 3(a), we can form a right triangle by constructing a line segment AQ perpendicular to the line l through O and B. Note that $\overrightarrow{OQ}$ has the same direction as $\overrightarrow{OB}$. Referring to part (a) of the figure, we see that

$$\cos \theta = \frac{d(O, Q)}{\|\mathbf{a}\|} \quad \text{or, equivalently,} \quad \|\mathbf{a}\| \cos \theta = d(O, Q).$$

If θ is obtuse, then, as in Figure 3(b), we again construct AQ perpendicular to l. In this case, the direction of $\overrightarrow{OQ}$ is opposite that of $\overrightarrow{OB}$, and since $\cos \theta$ is negative,

$$\cos \theta = \frac{-d(O, Q)}{\|\mathbf{a}\|} \quad \text{or, equivalently,} \quad \|\mathbf{a}\| \cos \theta = -d(O, Q).$$

special cases for the component of a along b
$\left\{\begin{array}{l}\end{array}\right.$
(1) If $\theta = \pi/2$, then **a** is orthogonal to **b** and comp$_b$ **a** = 0.

(2) If $\theta = 0$, then **a** has the same direction as **b** and comp$_b$ **a** = $\|\mathbf{a}\|$.

(3) If $\theta = \pi$, then **a** and **b** have opposite directions and comp$_b$ **a** = $-\|\mathbf{a}\|$.

The preceding discussion shows that the component of **a** along **b** may be found by *projecting* the endpoint of **a** onto the line *l* containing **b**. For this reason, $\|\mathbf{a}\| \cos \theta$ is sometimes called the **projection of a on b** and is denoted by $\text{proj}_\mathbf{b}\ \mathbf{a}$. The following formula shows how to compute this projection *without* knowing the angle θ.

Formula for comp$_\mathbf{b}$ a	If **a** and **b** are nonzero vectors, then $$\text{comp}_\mathbf{b}\ \mathbf{a} = \frac{\mathbf{a} \cdot \mathbf{b}}{\|\mathbf{b}\|}.$$

PROOF   If θ is the angle between **a** and **b**, then, from the theorem on the dot product,

$$\mathbf{a} \cdot \mathbf{b} = \|\mathbf{a}\| \|\mathbf{b}\| \cos \theta.$$

Dividing both sides of this equation by $\|\mathbf{b}\|$ gives us

$$\frac{\mathbf{a} \cdot \mathbf{b}}{\|\mathbf{b}\|} = \|\mathbf{a}\| \cos \theta = \text{comp}_\mathbf{b}\ \mathbf{a}.$$

Figure 4

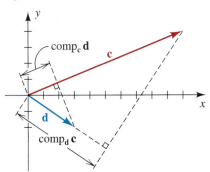

EXAMPLE 5   **Finding the components of one vector along another**

If $\mathbf{c} = 10\mathbf{i} + 4\mathbf{j}$ and $\mathbf{d} = 3\mathbf{i} - 2\mathbf{j}$, find comp$_\mathbf{d}$ **c** and comp$_\mathbf{c}$ **d**, and illustrate these numbers graphically.

SOLUTION   The vectors **c** and **d** and the desired components are illustrated in Figure 4. We use the formula for comp$_\mathbf{b}$ **a**, as follows:

$$\text{comp}_\mathbf{d}\ \mathbf{c} = \frac{\mathbf{c} \cdot \mathbf{d}}{\|\mathbf{d}\|} = \frac{(10)(3) + (4)(-2)}{\sqrt{3^2 + (-2)^2}} = \frac{22}{\sqrt{13}} \approx 6.10$$

$$\text{comp}_\mathbf{c}\ \mathbf{d} = \frac{\mathbf{d} \cdot \mathbf{c}}{\|\mathbf{c}\|} = \frac{(3)(10) + (-2)(4)}{\sqrt{10^2 + 4^2}} = \frac{22}{\sqrt{116}} \approx 2.04$$

We shall conclude this section with a physical application of the dot product. First let us briefly discuss the scientific concept of *work*.

A **force** may be thought of as the physical entity that is used to describe a push or pull on an object. For example, a force is needed to push or pull an object along a horizontal plane, to lift an object off the ground, or to move a charged particle through an electromagnetic field. Forces are often measured in pounds. If an object weighs 10 pounds, then, by definition, the force required to lift it (or hold it off the ground) is 10 pounds. A force of this type is a **constant force,** since its magnitude does not change while it is applied to the given object.

If a constant force F is applied to an object, moving it a distance d in the direction of the force, then, by definition, the **work** W done is

$$W = Fd.$$

If F is measured in pounds and d in feet, then the units for W are foot-pounds (ft-lb). In the cgs (centimeter-gram-second) system a **dyne** is used as the unit of force. If F is expressed in dynes and d in centimeters, then the unit for W is the dyne-centimeter, or **erg.** In the mks (meter-kilogram-second) system the **newton** is used as the unit of force. If F is in newtons and d is in meters, then the unit for W is the newton-meter, or **joule.**

EXAMPLE 6   Finding the work done by a constant force

Find the work done in pushing an automobile along a level road from a point A to another point B, 40 feet from A, while exerting a constant force of 90 pounds.

SOLUTION   The problem is illustrated in Figure 5, where we have pictured the road as part of a line l. Since the constant force is $F = 90$ lb and the distance the automobile moves is $d = 40$ feet, the work done is

$$W = (90)(40) = 3600 \text{ ft-lb.}$$

Figure 5

The formula $W = Fd$ is very restrictive, since it can be used only if the force is applied along the line of motion. More generally, suppose that a vector **a** represents a force and that its point of application moves along a vector **b**. This is illustrated in Figure 6, where the force **a** is used to pull an object along a level path from O to B, and $\mathbf{b} = \overrightarrow{OB}$.

Figure 6

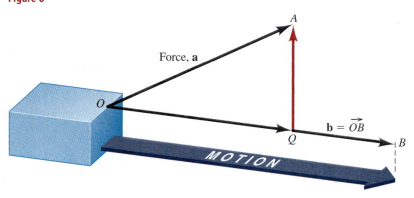

The vector **a** is the sum of the vectors OQ and QA, where $\overrightarrow{QA}$ is orthogonal to **b**. Since $\overrightarrow{QA}$ does not contribute to the horizontal movement, we may assume that the motion from O to B is caused by $\overrightarrow{OQ}$ alone. Applying $W = Fd$, we know that the work is the product of $\|\overrightarrow{OQ}\|$ and $\|\mathbf{b}\|$. Since the magnitude $\|\overrightarrow{OQ}\| = \text{comp}_b\,\mathbf{a}$, we obtain

$$W = (\text{comp}_b\,\mathbf{a})\|\mathbf{b}\| = (\|\mathbf{a}\|\cos\theta)\|\mathbf{b}\| = \mathbf{a}\cdot\mathbf{b},$$

where θ represents $\angle AOQ$. This leads to the following definition.

Definition of Work	The **work** W done by a constant force **a** as its point of application moves along a vector **b** is $W = \mathbf{a}\cdot\mathbf{b}$.

Figure 7

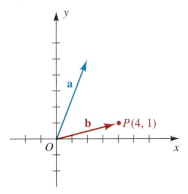

EXAMPLE 7   Finding the work done by a constant force

The magnitude and direction of a constant force are given by $\mathbf{a} = 2\mathbf{i} + 5\mathbf{j}$. Find the work done if the point of application of the force moves from the origin to the point $P(4, 1)$.

SOLUTION   The force **a** and the vector $\mathbf{b} = \overrightarrow{OP}$ are sketched in Figure 7. Since $\mathbf{b} = \langle 4, 1 \rangle = 4\mathbf{i} + \mathbf{j}$, we have, from the preceding definition,

$$W = \mathbf{a}\cdot\mathbf{b} = (2\mathbf{i} + 5\mathbf{j})\cdot(4\mathbf{i} + \mathbf{j})$$
$$= (2)(4) + (5)(1) = 13.$$

If, for example, the unit of length is feet and the magnitude of the force is measured in pounds, then the work done is 13 ft-lb.

EXAMPLE 8   Finding the work done against gravity

A small cart weighing 100 pounds is pushed up an incline that makes an angle of 30° with the horizontal, as shown in Figure 8. Find the work done against gravity in pushing the cart a distance of 80 feet.

Figure 8

Figure 9

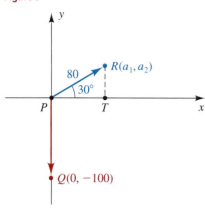

SOLUTION Let us introduce an xy-coordinate system, as shown in Figure 9. The vector PQ represents the force of gravity acting vertically downward with a magnitude of 100 pounds. The corresponding vector $\mathbf{F}$ is $0\mathbf{i} - 100\mathbf{j}$. The point of application of this force moves along the vector PR of magnitude 80. If $\overrightarrow{PR}$ corresponds to $\mathbf{a} = a_1\mathbf{i} + a_2\mathbf{j}$, then, referring to triangle PTR, we see that

$$a_1 = 80 \cos 30° = 40\sqrt{3}$$
$$a_2 = 80 \sin 30° = 40,$$

and hence

$$\mathbf{a} = 40\sqrt{3}\,\mathbf{i} + 40\mathbf{j}.$$

Applying the definition, we find that the work done *by* gravity is

$$\mathbf{F} \cdot \mathbf{a} = (0\mathbf{i} - 100\mathbf{j}) \cdot \left(40\sqrt{3}\,\mathbf{i} + 40\mathbf{j}\right) = 0 - 4000 = -4000 \text{ ft-lb.}$$

The work done *against* gravity is

$$-\mathbf{F} \cdot \mathbf{a} = 4000 \text{ ft-lb.}$$

8.4 Exercises

Exer. 1–8: Find (a) the dot product of the two vectors and (b) the angle between the two vectors.

1 $\langle -2, 5 \rangle$, $\langle 3, 6 \rangle$

2 $\langle 4, -7 \rangle$, $\langle -2, 3 \rangle$

3 $4\mathbf{i} - \mathbf{j}$, $-3\mathbf{i} + 2\mathbf{j}$

4 $8\mathbf{i} - 3\mathbf{j}$, $2\mathbf{i} - 7\mathbf{j}$

5 $9\mathbf{i}$, $5\mathbf{i} + 4\mathbf{j}$

6 $6\mathbf{j}$, $-4\mathbf{i}$

7 $\langle 10, 7 \rangle$, $\left\langle -2, -\frac{7}{5} \right\rangle$

8 $\langle -3, 6 \rangle$, $\langle -1, 2 \rangle$

Exer. 9–12: Show that the vectors are orthogonal.

9 $\langle 4, -1 \rangle$, $\langle 2, 8 \rangle$

10 $\langle 3, 6 \rangle$, $\langle 4, -2 \rangle$

11 $-4\mathbf{j}$, $-7\mathbf{i}$

12 $8\mathbf{i} - 4\mathbf{j}$, $-6\mathbf{i} - 12\mathbf{j}$

Exer. 13–16: Show that the vectors are parallel, and determine whether they have the same direction or opposite directions.

13 $\mathbf{a} = 3\mathbf{i} - 5\mathbf{j}$, $\mathbf{b} = -\frac{12}{7}\mathbf{i} + \frac{20}{7}\mathbf{j}$

14 $\mathbf{a} = -\frac{5}{2}\mathbf{i} + 6\mathbf{j}$, $\mathbf{b} = -10\mathbf{i} + 24\mathbf{j}$

15 $\mathbf{a} = \left\langle \frac{2}{3}, \frac{1}{2} \right\rangle$, $\mathbf{b} = \langle 8, 6 \rangle$

16 $\mathbf{a} = \langle 6, 18 \rangle$, $\mathbf{b} = \langle -4, -12 \rangle$

Exer. 17–20: Determine m such that the two vectors are orthogonal.

17 $3\mathbf{i} - 2\mathbf{j}$, $4\mathbf{i} + 5m\mathbf{j}$

18 $4m\mathbf{i} + \mathbf{j}$, $9m\mathbf{i} - 25\mathbf{j}$

19 $9\mathbf{i} - 16m\mathbf{j}$, $\mathbf{i} + 4m\mathbf{j}$

20 $5m\mathbf{i} + 3\mathbf{j}$, $2\mathbf{i} + 7\mathbf{j}$

Exer. 21–28: Given that $\mathbf{a} = \langle 2, -3 \rangle$, $\mathbf{b} = \langle 3, 4 \rangle$, and $\mathbf{c} = \langle -1, 5 \rangle$, find the number.

21 (a) $\mathbf{a} \cdot (\mathbf{b} + \mathbf{c})$ (b) $\mathbf{a} \cdot \mathbf{b} + \mathbf{a} \cdot \mathbf{c}$

22 (a) $\mathbf{b} \cdot (\mathbf{a} - \mathbf{c})$ (b) $\mathbf{b} \cdot \mathbf{a} - \mathbf{b} \cdot \mathbf{c}$

23 $(2\mathbf{a} + \mathbf{b}) \cdot (3\mathbf{c})$

24 $(\mathbf{a} - \mathbf{b}) \cdot (\mathbf{b} + \mathbf{c})$

25 $\text{comp}_{\mathbf{c}}\, \mathbf{b}$

26 $\text{comp}_{\mathbf{b}}\, \mathbf{c}$

27 $\text{comp}_{\mathbf{b}}\, (\mathbf{a} + \mathbf{c})$

28 $\text{comp}_{\mathbf{c}}\, \mathbf{c}$

Exer. 29–32: If $\mathbf{c}$ represents a constant force, find the work done if the point of application of $\mathbf{c}$ moves along the line segment from P to Q.

29 $\mathbf{c} = 3\mathbf{i} + 4\mathbf{j}$; $P(0, 0)$, $Q(5, -2)$

30 $\mathbf{c} = -10\mathbf{i} + 12\mathbf{j}$; $P(0, 0)$, $Q(4, 7)$

31 $\mathbf{c} = 6\mathbf{i} + 4\mathbf{j}$; $P(2, -1)$, $Q(4, 3)$
    $\left(\textit{Hint:}\text{ Find a vector }\mathbf{b} = \langle b_1, b_2 \rangle\text{ such that }\mathbf{b} = \overrightarrow{PQ}.\right)$

32 $c = -i + 7j$; $P(-2, 5)$, $Q(6, 1)$

33 A constant force of magnitude 4 has the same direction as **j**. Find the work done if its point of application moves from $P(0, 0)$ to $Q(8, 3)$.

34 A constant force of magnitude 10 has the same direction as $-i$. Find the work done if its point of application moves from $P(0, 1)$ to $Q(1, 0)$.

Exer. 35–40: Prove the property if a and b are vectors and *m* is a real number.

35 $a \cdot a = \|a\|^2$

36 $a \cdot b = b \cdot a$

37 $(ma) \cdot b = m(a \cdot b)$

38 $m(a \cdot b) = a \cdot (mb)$

39 $0 \cdot a = 0$

40 $(a + b) \cdot (a - b) = a \cdot a - b \cdot b$

41 Pulling a wagon A child pulls a wagon along level ground by exerting a force of 20 pounds on a handle that makes an angle of 30° with the horizontal, as shown in the figure. Find the work done in pulling the wagon 100 feet.

Exercise 41

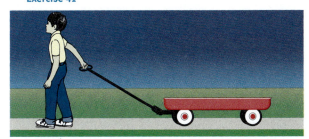

42 Pulling a wagon Refer to Exercise 41. Find the work done if the wagon is pulled, with the same force, 100 feet up an incline that makes an angle of 30° with the horizontal, as shown in the figure.

Exercise 42

43 The sun's rays The sun has a radius of 432,000 miles, and its center is 93,000,000 miles from the center of Earth. Let **v** and **w** be the vectors illustrated in the figure.

(a) Express **v** and **w** in **i**, **j** form.

(b) Approximate the angle between **v** and **w**.

Exercise 43

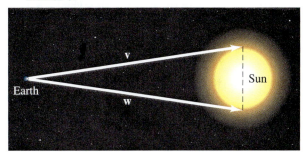

44 July sunlight The intensity I of sunlight (in watts/m²) can be calculated using the formula $I = ke^{-c/\sin \phi}$, where k and c are positive constants and ϕ is the angle between the sun's rays and the horizon. The amount of sunlight striking a vertical wall facing the sun is equal to the component of the sun's rays along the horizontal. If, during July, $\phi = 30°$, $k = 978$, and $c = 0.136$, approximate the total amount of sunlight striking a vertical wall that has an area of 160 m².

Exer. 45–46: Vectors are used extensively in computer graphics to perform shading. When light strikes a flat surface, it is reflected, and that area should not be shaded. Suppose that an incoming ray of light is represented by a vector L and that N is a vector orthogonal to the flat surface, as shown in the figure. The ray of reflected light can be represented by the vector R and is calculated using the formula R = 2(N · L)N − L. Compute R for the vectors L and N.

45 Reflected light $L = \langle -\frac{4}{5}, \frac{3}{5} \rangle$, $N = \langle 0, 1 \rangle$

46 Reflected light $L = \langle \frac{12}{13}, -\frac{5}{13} \rangle$, $N = \langle \frac{1}{2}\sqrt{2}, \frac{1}{2}\sqrt{2} \rangle$

Exercises 45–46

Exer. 47–48: Vectors are used in computer graphics to calculate the lengths of shadows over flat surfaces. The length of an object can sometimes be represented by a vector a. If a single light source is shining down on the object, then the length of its shadow on the ground will be equal to the absolute value of the component of the vector a along the direction of the ground, as shown in the figure. Compute the length of the shadow for the specified vector a if the ground is level.

47 Shadow on level ground $\mathbf{a} = \langle 2.6, 4.5 \rangle$

48 Shadow on level ground $\mathbf{a} = \langle -3.1, 7.9 \rangle$

Exercises 47–48

Exer. 49–50: Refer to Exercises 47 and 48. An object represented by a vector a is held over a flat surface inclined at an angle θ, as shown in the figure. If a light is shining directly downward, approximate the length of the shadow to two decimal places for the specified values of the vector a and θ.

49 Shadow on inclined plane $\mathbf{a} = \langle 25.7, -3.9 \rangle$, $\theta = 12°$

50 Shadow on inclined plane $\mathbf{a} = \langle -13.8, 19.4 \rangle$, $\theta = -17°$

Exercises 49–50

51 **Determining horsepower** The amount of horsepower P produced by an engine can be determined by using the formula $P = \frac{1}{550}(\mathbf{F} \cdot \mathbf{v})$, where $\mathbf{F}$ is the force (in pounds) exerted by the engine and $\mathbf{v}$ is the velocity (in ft/sec) of an object moved by the engine. An engine pulls with a force of 2200 pounds on a cable that makes an angle θ with the horizontal, moving a cart horizontally, as shown in the figure. Find the horsepower of the engine if the speed of the cart is 8 ft/sec when $\theta = 30°$.

Exercise 51

8.5

Trigonometric Form for Complex Numbers

In Section 1.1 we represented real numbers geometrically by using points on a coordinate line. We can obtain geometric representations for complex numbers by using points in a coordinate plane. Specifically, each complex number $a + bi$ determines a unique ordered pair (a, b). The corresponding point $P(a, b)$ in a coordinate plane is the **geometric representation** of $a + bi$. To emphasize that we are assigning complex numbers to points in a plane, we may label the point $P(a, b)$ as $a + bi$. A coordinate plane with a complex number assigned to each point is referred to as a **complex** (or Argand) **plane** instead of an *xy*-plane. The *x*-axis is the **real axis** and the *y*-axis is the **imaginary axis.** In Figure 1 (on the next page) we have represented several complex numbers geometrically. Note that to obtain the point corresponding to the conjugate $a - bi$ of any complex number $a + bi$, we simply reflect through the real axis.

Figure 1

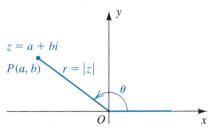

The absolute value of a real number a (denoted $|a|$) is the distance between the origin and the point on the x-axis that corresponds to a. Thus, it is natural to interpret the absolute value of a complex number as the distance between the origin of a complex plane and the point (a, b) that corresponds to $a + bi$.

Definition of the Absolute Value of a Complex Number	If $z = a + bi$ is a complex number, then its **absolute value,** denoted by $\|a + bi\|$, is $$\sqrt{a^2 + b^2}.$$

 EXAMPLE 1   **Finding the absolute value of a complex number**

Find

(a) $|2 - 6i|$ **(b)** $|3i|$

SOLUTION   We use the previous definition:

(a) $|2 - 6i| = \sqrt{2^2 + (-6)^2} = \sqrt{40} = 2\sqrt{10} \approx 6.3$

(b) $|3i| = |0 + 3i| = \sqrt{0^2 + 3^2} = \sqrt{9} = 3$

Figure 2

$z = a + bi = r(\cos \theta + i \sin \theta)$

The points corresponding to all complex numbers that have a fixed absolute value k are on a circle of radius k with center at the origin in the complex plane. For example, the points corresponding to the complex numbers z with $|z| = 1$ are on a unit circle.

Let us consider a nonzero complex number $z = a + bi$ and its geometric representation $P(a, b)$, as illustrated in Figure 2. Let θ be any angle in standard position whose terminal side lies on the segment OP, and let $r = |z| = \sqrt{a^2 + b^2}$. Since $\cos \theta = a/r$ and $\sin \theta = b/r$, we see that $a = r \cos \theta$ and $b = r \sin \theta$. Substituting for a and b in $z = a + bi$, we obtain

$$z = a + bi = (r \cos \theta) + (r \sin \theta)i = r(\cos \theta + i \sin \theta).$$

This expression is called the **trigonometric** (or **polar**) **form for the complex number $a + bi$.** A common abbreviation is

$$r(\cos \theta + i \sin \theta) = r \operatorname{cis} \theta.$$

The trigonometric form for $z = a + bi$ is not unique, since there are an unlimited number of different choices for the angle θ. When the trigonometric form is used, the absolute value r of z is sometimes referred to as the **modulus** of z and an angle θ associated with z as an **argument** (or **amplitude**) of z. We may summarize our discussion as follows.

Trigonometric (or Polar) Form for a Complex Number	Let $z = a + bi$. If $r = \lvert z \rvert = \sqrt{a^2 + b^2}$ and if θ is an argument of z, then $$z = r(\cos \theta + i \sin \theta) = r \operatorname{cis} \theta.$$

Euler's formula,

$$\cos \theta + i \sin \theta = e^{i\theta},$$

gives us yet another form for the complex number $z = a + bi$, commonly called the **exponential form;** that is,

$$z = r(\cos \theta + i \sin \theta) = re^{i\theta}.$$

See Exercise 6 of the Discussion Exercises at the end of the chapter for some related problems.

EXAMPLE 2 **Expressing a complex number in trigonometric form**

Express the complex number in trigonometric form with $0 \le \theta < 2\pi$:

(a) $-4 + 4i$ **(b)** $2\sqrt{3} - 2i$ **(c)** $2 + 7i$ **(d)** $-2 + 7i$

SOLUTION   We begin by representing each complex number geometrically and labeling its modulus r and argument θ, as in Figure 3.

Figure 3

(a)

(b)

(c)

(d)

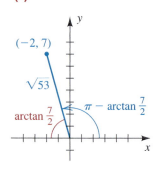

We next substitute for r and θ in the trigonometric form:

(a) $-4 + 4i = 4\sqrt{2}\left(\cos\dfrac{3\pi}{4} + i\sin\dfrac{3\pi}{4}\right) = 4\sqrt{2}\,\text{cis}\,\dfrac{3\pi}{4}$

(b) $2\sqrt{3} - 2i = 4\left(\cos\dfrac{11\pi}{6} + i\sin\dfrac{11\pi}{6}\right) = 4\,\text{cis}\,\dfrac{11\pi}{6}$

(c) $2 + 7i = \sqrt{53}\left[\cos\left(\arctan\tfrac{7}{2}\right) + i\sin\left(\arctan\tfrac{7}{2}\right)\right] = \sqrt{53}\,\text{cis}\left(\arctan\tfrac{7}{2}\right)$

(d) $-2 + 7i = \sqrt{53}\left[\cos\left(\pi - \arctan\tfrac{7}{2}\right) + i\sin\left(\pi - \arctan\tfrac{7}{2}\right)\right]$
$ = \sqrt{53}\,\text{cis}\left(\pi - \arctan\tfrac{7}{2}\right)$

If we allow arbitrary values for θ, there are many other trigonometric forms for the complex numbers in Example 2. Thus, for $-4 + 4i$ in part (a) we could use

$$\theta = \dfrac{3\pi}{4} + 2\pi n \quad \text{for any integer } n.$$

If, for example, we let $n = 1$ and $n = -1$, we obtain

$$4\sqrt{2}\,\text{cis}\,\dfrac{11\pi}{4} \quad \text{and} \quad 4\sqrt{2}\,\text{cis}\left(-\dfrac{5\pi}{4}\right),$$

respectively. In general, arguments for the same complex number always differ by a multiple of 2π.

If complex numbers are expressed in trigonometric form, then multiplication and division may be performed as indicated in the next theorem.

Theorem on Products and Quotients of Complex Numbers	If trigonometric forms for two complex numbers z_1 and z_2 are $z_1 = r_1(\cos\theta_1 + i\sin\theta_1)$ and $z_2 = r_2(\cos\theta_2 + i\sin\theta_2),$ then **(1)** $z_1 z_2 = r_1 r_2[\cos(\theta_1 + \theta_2) + i\sin(\theta_1 + \theta_2)]$ **(2)** $\dfrac{z_1}{z_2} = \dfrac{r_1}{r_2}[\cos(\theta_1 - \theta_2) + i\sin(\theta_1 - \theta_2)], z_2 \neq 0$

PROOF We may prove (1) as follows:

$$z_1 z_2 = r_1(\cos\theta_1 + i\sin\theta_1) \cdot r_2(\cos\theta_2 + i\sin\theta_2)$$
$$= r_1 r_2[(\cos\theta_1\cos\theta_2 - \sin\theta_1\sin\theta_2)$$
$$+ i(\sin\theta_1\cos\theta_2 + \cos\theta_1\sin\theta_2)]$$

Applying the addition formulas for $\cos(\theta_1 + \theta_2)$ and $\sin(\theta_1 + \theta_2)$ gives us (1). We leave the proof of (2) as an exercise.

Part (1) of the preceding theorem states that *the modulus of a product of two complex numbers is the product of their moduli, and an argument is the sum of their arguments.* An analogous statement can be made for (2).

EXAMPLE 3 Using trigonometric forms to find products and quotients

If $z_1 = 2\sqrt{3} - 2i$ and $z_2 = -1 + \sqrt{3}i$, use trigonometric forms to find
(a) z_1z_2 and (b) z_1/z_2. Check by using algebraic methods.

SOLUTION The complex number $2\sqrt{3} - 2i$ is represented geometrically in Figure 3(b). If we use $\theta = -\pi/6$ in the trigonometric form, then

$$z_1 = 2\sqrt{3} - 2i = 4\left[\cos\left(-\frac{\pi}{6}\right) + i\sin\left(-\frac{\pi}{6}\right)\right].$$

The complex number $z_2 = -1 + \sqrt{3}i$ is represented geometrically in Figure 4. A trigonometric form is

$$z_2 = -1 + \sqrt{3}i = 2\left(\cos\frac{2\pi}{3} + i\sin\frac{2\pi}{3}\right).$$

(a) We apply part (1) of the theorem on products and quotients of complex numbers:

$$z_1z_2 = 4 \cdot 2\left[\cos\left(-\frac{\pi}{6} + \frac{2\pi}{3}\right) + i\sin\left(-\frac{\pi}{6} + \frac{2\pi}{3}\right)\right]$$

$$= 8\left(\cos\frac{\pi}{2} + i\sin\frac{\pi}{2}\right) = 8(0 + i) = 8i$$

Figure 5 gives a geometric interpretation of the product z_1z_2.
Using algebraic methods to check our result, we have

$$z_1z_2 = (2\sqrt{3} - 2i)(-1 + \sqrt{3}i)$$

$$= (-2\sqrt{3} + 2\sqrt{3}) + (2 + 6)i = 0 + 8i = 8i.$$

(b) We apply part (2) of the theorem:

$$\frac{z_1}{z_2} = \frac{4}{2}\left[\cos\left(-\frac{\pi}{6} - \frac{2\pi}{3}\right) + i\sin\left(-\frac{\pi}{6} - \frac{2\pi}{3}\right)\right]$$

$$= 2\left[\cos\left(-\frac{5\pi}{6}\right) + i\sin\left(-\frac{5\pi}{6}\right)\right]$$

$$= 2\left[-\frac{\sqrt{3}}{2} + i\left(-\frac{1}{2}\right)\right] = -\sqrt{3} - i$$

Figure 6 gives a geometric interpretation of the quotient z_1/z_2.

Figure 4

Figure 5

Figure 6

Using algebraic methods to check our result, we multiply the numerator and denominator by the conjugate of the denominator to obtain

$$\frac{z_1}{z_2} = \frac{2\sqrt{3} - 2i}{-1 + \sqrt{3}\,i} \cdot \frac{-1 - \sqrt{3}\,i}{-1 - \sqrt{3}\,i}$$

$$= \frac{\left(-2\sqrt{3} - 2\sqrt{3}\right) + (2 - 6)i}{(-1)^2 + \left(\sqrt{3}\right)^2}$$

$$= \frac{-4\sqrt{3} - 4i}{4} = -\sqrt{3} - i.$$

8.5 *Exercises*

Exer. 1–10: Find the absolute value.

1 $|3 - 4i|$ 2 $|5 + 8i|$

3 $|-6 - 7i|$ 4 $|1 - i|$

5 $|8i|$ 6 $|i^7|$

7 $|i^{500}|$ 8 $|-15i|$

9 $|0|$ 10 $|-15|$

Exer. 11–20: Represent the complex number geometrically.

11 $4 + 2i$ 12 $-5 + 3i$

13 $3 - 5i$ 14 $-2 - 6i$

15 $-(3 - 6i)$ 16 $(1 + 2i)^2$

17 $2i(2 + 3i)$ 18 $(-3i)(2 - i)$

19 $(1 + i)^2$ 20 $4(-1 + 2i)$

Exer. 21–46: Express the complex number in trigonometric form with $0 \le \theta < 2\pi$.

21 $1 - i$ 22 $\sqrt{3} + i$

23 $-4\sqrt{3} + 4i$ 24 $-2 - 2i$

25 $2\sqrt{3} + 2i$ 26 $3 - 3\sqrt{3}\,i$

27 $-4 - 4i$ 28 $-10 + 10i$

29 $-20i$ 30 $-6i$

31 12 32 15

33 -7 34 -5

35 $6i$ 36 $4i$

37 $-5 - 5\sqrt{3}\,i$ 38 $\sqrt{3} - i$

39 $2 + i$ 40 $3 + 2i$

41 $-3 + i$ 42 $-4 + 2i$

43 $-5 - 3i$ 44 $-2 - 7i$

45 $4 - 3i$ 46 $1 - 3i$

Exer. 47–56: Express in the form $a + bi$, where a and b are real numbers.

47 $4\left(\cos\dfrac{\pi}{4} + i\,\sin\dfrac{\pi}{4}\right)$ 48 $8\left(\cos\dfrac{7\pi}{4} + i\,\sin\dfrac{7\pi}{4}\right)$

49 $6\left(\cos\dfrac{2\pi}{3} + i\,\sin\dfrac{2\pi}{3}\right)$ 50 $12\left(\cos\dfrac{4\pi}{3} + i\,\sin\dfrac{4\pi}{3}\right)$

51 $5(\cos\pi + i\,\sin\pi)$ 52 $3\left(\cos\dfrac{3\pi}{2} + i\,\sin\dfrac{3\pi}{2}\right)$

53 $\sqrt{34}\ \text{cis}\ \left(\tan^{-1}\frac{3}{5}\right)$ 54 $\sqrt{53}\ \text{cis}\ \left[\tan^{-1}\left(-\frac{2}{7}\right)\right]$

55 $\sqrt{5}\ \text{cis}\ \left[\tan^{-1}\left(-\frac{1}{2}\right)\right]$ 56 $\sqrt{10}\ \text{cis}\ (\tan^{-1}3)$

Exer. 57–64: Use trigonometric forms to find $z_1 z_2$ and z_1/z_2.

57 $z_1 = -1 + i$, $z_2 = 1 + i$

58 $z_1 = \sqrt{3} - i$, $z_2 = -\sqrt{3} - i$

59 $z_1 = -2 - 2\sqrt{3}\,i$, $z_2 = 5i$

60 $z_1 = -5 + 5i$, $z_2 = -3i$

61 $z_1 = -10$, $z_2 = -4$

62 $z_1 = 2i$, $z_2 = -3i$

63 $z_1 = 4$, $z_2 = 2 - i$

64 $z_1 = -3$, $z_2 = 5 + 2i$

65 Prove (2) of the theorem on products and quotients of complex numbers.

66 (a) Extend (1) of the theorem on products and quotients of complex numbers to three complex numbers.

(b) Generalize (1) of the theorem to n complex numbers.

Exer. 67–70: The trigonometric form of complex numbers is often used by electrical engineers to describe the current I, voltage V, and impedance Z in electrical circuits with alternating current. Impedance is the opposition to the flow of current in a circuit. Most common electrical devices operate on 115-volt, alternating current. The relationship among these three quantities is $I = V/Z$. Approximate the unknown quantity, and express the answer in rectangular form to two decimal places.

67 Finding voltage $I = 10$ cis $35°$, $Z = 3$ cis $20°$

68 Finding voltage $I = 12$ cis $5°$, $Z = 100$ cis $90°$

69 Finding impedance $I = 8$ cis $5°$, $V = 115$ cis $45°$

70 Finding current $Z = 78$ cis $61°$, $V = 163$ cis $17°$

71 Modulus of impedance The modulus of the impedance Z represents the total opposition to the flow of electricity in a circuit and is measured in ohms. If $Z = 14 - 13i$, compute $|Z|$.

72 Resistance and reactance The absolute value of the real part of Z represents the resistance in an electrical circuit; the absolute value of the complex part represents the reactance. Both quantities are measured in ohms. If $V = 220$ cis $34°$ and $I = 5$ cis $90°$, approximate the resistance and the reactance.

73 Actual voltage The real part of V represents the actual voltage delivered to an electrical appliance in volts. Approximate this voltage when $I = 4$ cis $90°$ and $Z = 18$ cis $(-78°)$.

74 Actual current The real part of I represents the actual current delivered to an electrical appliance in amps. Approximate this current when $V = 163$ cis $43°$ and $Z = 100$ cis $17°$.

8.6

De Moivre's Theorem and nth Roots of Complex Numbers

If z is a complex number and n is a positive integer, then a complex number w is an **nth root** of z if $w^n = z$. We will show that every nonzero complex number has n different nth roots. Since $\mathbb{R}$ is contained in $\mathbb{C}$, it will also follow that every nonzero real number has n different nth (complex) roots. If a is a positive real number and $n = 2$, then we already know that the roots are $\sqrt{a}$ and $-\sqrt{a}$.

If, in the theorem on products and quotients of complex numbers, we let both z_1 and z_2 equal the complex number $z = r(\cos \theta + i \sin \theta)$, we obtain

$$z^2 = r \cdot r[\cos (\theta + \theta) + i \sin (\theta + \theta)]$$
$$= r^2(\cos 2\theta + i \sin 2\theta).$$

Applying the same theorem to z^2 and z gives us

$$z^2 \cdot z = (r^2 \cdot r)[\cos (2\theta + \theta) + i \sin (2\theta + \theta)],$$

or

$$z^3 = r^3(\cos 3\theta + i \sin 3\theta).$$

Applying the theorem to z^3 and z, we obtain

$$z^4 = r^4(\cos 4\theta + i \sin 4\theta).$$

In general, we have the following result, named after the French mathematician Abraham De Moivre (1667–1754).

De Moivre's Theorem	For every integer n, $$[r(\cos \theta + i \sin \theta)]^n = r^n(\cos n\theta + i \sin n\theta).$$

We will use only positive integers for n in examples and exercises involving De Moivre's theorem. However, for completeness, the theorem holds for $n = 0$ and n negative if we use the respective real number exponent definitions—that is, $z^0 = 1$ and $z^{-n} = 1/z^n$, where z is a nonzero complex number and n is a positive integer.

EXAMPLE 1 Using De Moivre's theorem

Use De Moivre's theorem to change $(1 + i)^{20}$ to the form $a + bi$, where a and b are real numbers.

SOLUTION It would be tedious to change $(1 + i)^{20}$ using algebraic methods. Let us therefore introduce a trigonometric form for $1 + i$. Referring to Figure 1, we see that

$$1 + i = \sqrt{2}\left(\cos\frac{\pi}{4} + i\sin\frac{\pi}{4}\right).$$

We now apply De Moivre's theorem:

$$(1 + i)^{20} = (2^{1/2})^{20}\left[\cos\left(20\cdot\frac{\pi}{4}\right) + i\sin\left(20\cdot\frac{\pi}{4}\right)\right]$$

$$= 2^{10}(\cos 5\pi + i\sin 5\pi) = 2^{10}(-1 + 0i) = -1024$$

The number -1024 is of the form $a + bi$ with $a = -1024$ and $b = 0$. ◻

If a nonzero complex number z has an nth root w, then $w^n = z$. If trigonometric forms for w and z are

$$w = s(\cos\alpha + i\sin\alpha) \qquad \text{and} \qquad z = r(\cos\theta + i\sin\theta), \qquad (*)$$

then applying De Moivre's theorem to $w^n = z$ yields

$$s^n(\cos n\alpha + i\sin n\alpha) = r(\cos\theta + i\sin\theta).$$

If two complex numbers are equal, then so are their absolute values. Consequently, $s^n = r$, and since s and r are nonnegative, $s = \sqrt[n]{r}$. Substituting s^n for r in the last displayed equation and dividing both sides by s^n, we obtain

$$\cos n\alpha + i\sin n\alpha = \cos\theta + i\sin\theta.$$

Since the arguments of equal complex numbers differ by a multiple of 2π, there is an integer k such that $n\alpha = \theta + 2\pi k$. Dividing both sides of the last equation by n, we see that

$$\alpha = \frac{\theta + 2\pi k}{n} \quad \text{for some integer } k.$$

Substituting in the trigonometric form for w (see $(*)$) gives us the formula

$$w = \sqrt[n]{r}\left[\cos\left(\frac{\theta + 2\pi k}{n}\right) + i\sin\left(\frac{\theta + 2\pi k}{n}\right)\right].$$

If we substitute $k = 0, 1, \ldots, n - 1$ successively, we obtain n different nth roots of z. No other value of k will produce a new nth root. For example, if

Figure 1

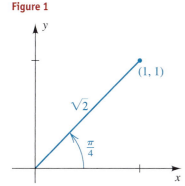

$k = n$, we obtain the angle $(\theta + 2\pi n)/n$, or $(\theta/n) + 2\pi$, which gives us the same nth root as $k = 0$. Similarly, $k = n + 1$ yields the same nth root as $k = 1$, and so on. The same is true for negative values of k. We have proved the following theorem.

Theorem on nth Roots

If $z = r(\cos\theta + i\sin\theta)$ is any nonzero complex number and if n is any positive integer, then z has exactly n different nth roots $w_0, w_1, w_2, \ldots, w_{n-1}$. These roots, for θ in radians, are

$$w_k = \sqrt[n]{r}\left[\cos\left(\frac{\theta + 2\pi k}{n}\right) + i\sin\left(\frac{\theta + 2\pi k}{n}\right)\right]$$

or, equivalently, for θ in degrees,

$$w_k = \sqrt[n]{r}\left[\cos\left(\frac{\theta + 360°k}{n}\right) + i\sin\left(\frac{\theta + 360°k}{n}\right)\right],$$

where $k = 0, 1, \ldots, n - 1$.

The nth roots of z in this theorem all have absolute value $\sqrt[n]{r}$, and hence their geometric representations lie on a circle of radius $\sqrt[n]{r}$ with center at O. Moreover, they are equispaced on this circle, since the difference in the arguments of successive nth roots is $2\pi/n$ (or $360°/n$).

EXAMPLE 2   **Finding the fourth roots of a complex number**

(a) Find the four fourth roots of $-8 - 8\sqrt{3}i$.

(b) Represent the roots geometrically.

SOLUTION

(a) The geometric representation of $-8 - 8\sqrt{3}i$ is shown in Figure 2. Introducing trigonometric form, we have

$$-8 - 8\sqrt{3}i = 16(\cos 240° + i\sin 240°).$$

Using the theorem on nth roots with $n = 4$ and noting that $\sqrt[4]{16} = 2$, we find that the fourth roots are

$$w_k = 2\left[\cos\left(\frac{240° + 360°k}{4}\right) + i\sin\left(\frac{240° + 360°k}{4}\right)\right]$$

for $k = 0, 1, 2, 3$. This formula may be written

$$w_k = 2[\cos(60° + 90°k) + i\sin(60° + 90°k)].$$

Figure 2

Figure 3

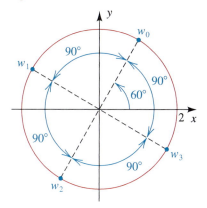

Substituting 0, 1, 2, and 3 for k in $(60° + 90°k)$ gives us the four fourth roots:

$$w_0 = 2(\cos 60° + i \sin 60°) = 1 + \sqrt{3}\,i$$
$$w_1 = 2(\cos 150° + i \sin 150°) = -\sqrt{3} + i$$
$$w_2 = 2(\cos 240° + i \sin 240°) = -1 - \sqrt{3}\,i$$
$$w_3 = 2(\cos 330° + i \sin 330°) = \sqrt{3} - i$$

(b) By the comments preceding this example, all roots lie on a circle of radius $\sqrt[4]{16} = 2$ with center at O. The first root, w_0, has an argument of 60°, and successive roots are spaced apart $360°/4 = 90°$, as shown in Figure 3. ◢

EXAMPLE 3 **Finding the sixth roots of a real number**

(a) Find the six sixth roots of -1.

(b) Represent the roots geometrically.

SOLUTION

(a) Writing $-1 = 1(\cos \pi + i \sin \pi)$ and using the theorem on nth roots with $n = 6$, we find that the sixth roots of -1 are given by

$$w_k = 1\left[\cos\left(\frac{\pi + 2\pi k}{6}\right) + i \sin\left(\frac{\pi + 2\pi k}{6}\right)\right]$$

for $k = 0, 1, 2, 3, 4, 5$. Substituting 0, 1, 2, 3, 4, 5 for k, we obtain the six sixth roots of -1:

$$w_0 = \cos\frac{\pi}{6} + i \sin\frac{\pi}{6} = \frac{\sqrt{3}}{2} + \frac{1}{2}i$$

$$w_1 = \cos\frac{\pi}{2} + i \sin\frac{\pi}{2} = i$$

$$w_2 = \cos\frac{5\pi}{6} + i \sin\frac{5\pi}{6} = -\frac{\sqrt{3}}{2} + \frac{1}{2}i$$

$$w_3 = \cos\frac{7\pi}{6} + i \sin\frac{7\pi}{6} = -\frac{\sqrt{3}}{2} - \frac{1}{2}i$$

$$w_4 = \cos\frac{3\pi}{2} + i \sin\frac{3\pi}{2} = -i$$

$$w_5 = \cos\frac{11\pi}{6} + i \sin\frac{11\pi}{6} = \frac{\sqrt{3}}{2} - \frac{1}{2}i$$

Figure 4

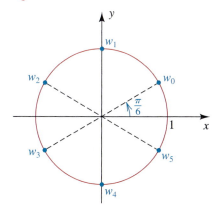

(b) Since $\sqrt[6]{1} = 1$, the points that represent the roots of -1 all lie on the unit circle shown in Figure 4. Moreover, they are equispaced on this circle by $\pi/3$ radians, or 60°. ◢

The special case in which $z = 1$ is of particular interest. The n distinct nth roots of 1 are called the **nth roots of unity.** In particular, if $n = 3$, we call these roots the **cube roots of unity.**

EXAMPLE 4 Finding the cube roots of unity

Find the three cube roots of unity.

SOLUTION Writing $1 = 1(\cos 0 + i \sin 0)$ and using the theorem on nth roots with $n = 3$, we obtain

$$w_k = 1\left[\cos \frac{2\pi k}{3} + i \sin \frac{2\pi k}{3}\right]$$

for $k = 0, 1, 2$. Substituting for k gives us the three roots:

$$w_0 = \cos 0 + i \sin 0 = 1$$

$$w_1 = \cos \frac{2\pi}{3} + i \sin \frac{2\pi}{3} = -\frac{1}{2} + \frac{\sqrt{3}}{2} i$$

$$w_2 = \cos \frac{4\pi}{3} + i \sin \frac{4\pi}{3} = -\frac{1}{2} - \frac{\sqrt{3}}{2} i$$

Note that finding the nth roots of a complex number c, as we did in Examples 2–4, is equivalent to finding all the solutions of the equation

$$x^n = c, \quad \text{or} \quad x^n - c = 0.$$

We will use this concept in Exercises 23–30.

8.6 Exercises

Exer. 1–12: Use De Moivre's theorem to change the given complex number to the form $a + bi$, where a and b are real numbers.

1 $(3 + 3i)^5$

2 $(1 + i)^{12}$

3 $(1 - i)^{10}$

4 $(-1 + i)^8$

5 $\left(1 - \sqrt{3}\,i\right)^3$

6 $\left(1 - \sqrt{3}\,i\right)^5$

7 $\left(-\frac{\sqrt{2}}{2} + \frac{\sqrt{2}}{2} i\right)^{15}$

8 $\left(\frac{\sqrt{2}}{2} + \frac{\sqrt{2}}{2} i\right)^{25}$

9 $\left(-\frac{\sqrt{3}}{2} - \frac{1}{2} i\right)^{20}$

10 $\left(-\frac{\sqrt{3}}{2} - \frac{1}{2} i\right)^{50}$

11 $\left(\sqrt{3} + i\right)^7$

12 $(-2 - 2i)^{10}$

13 Find the two square roots of $1 + \sqrt{3}\,i$.

14 Find the two square roots of $-9i$.

15 Find the four fourth roots of $-1 - \sqrt{3}\,i$.

16 Find the four fourth roots of $-8 + 8\sqrt{3}\,i$.

17 Find the three cube roots of $-27i$.

18 Find the three cube roots of $64i$.

Exer. 19–22: Find the indicated roots, and represent them geometrically.

19 The six sixth roots of unity

20 The eight eighth roots of unity

21 The five fifth roots of $1 + i$

22 The five fifth roots of $-\sqrt{3} - i$

Exer. 23–30: Find the solutions of the equation.

23 $x^4 - 16 = 0$

24 $x^6 - 64 = 0$

25 $x^6 + 64 = 0$

26 $x^5 + 1 = 0$

27 $x^3 + 8i = 0$

28 $x^3 - 64i = 0$

29 $x^5 - 243 = 0$

30 $x^4 + 81 = 0$

31 Use Euler's formula to prove De Moivre's theorem.

CHAPTER 8 REVIEW EXERCISES

Exer. 1–4: Find the exact values of the remaining parts of triangle *ABC*.

1 $\alpha = 60°$, $b = 6$, $c = 7$

2 $\gamma = 30°$, $a = 2\sqrt{3}$, $c = 2$

3 $\alpha = 60°$, $\beta = 45°$, $b = 100$

4 $a = 2$, $b = 3$, $c = 4$

Exer. 5–8: Approximate the remaining parts of triangle *ABC*.

5 $\beta = 67°$, $\gamma = 75°$, $b = 12$

6 $\alpha = 23°30'$, $c = 125$, $a = 152$

7 $\beta = 115°$, $a = 4.6$, $c = 7.3$

8 $a = 37$, $b = 55$, $c = 43$

Exer. 9–10: Approximate the area of triangle *ABC* to the nearest 0.1 square unit.

9 $\alpha = 75°$, $b = 20$, $c = 30$

10 $a = 4$, $b = 7$, $c = 10$

11 If $\mathbf{a} = \langle -4, 5 \rangle$ and $\mathbf{b} = \langle 2, -8 \rangle$, sketch vectors corresponding to

(a) $\mathbf{a} + \mathbf{b}$ (b) $\mathbf{a} - \mathbf{b}$ (c) $2\mathbf{a}$ (d) $-\frac{1}{2}\mathbf{b}$

12 If $\mathbf{a} = 2\mathbf{i} + 5\mathbf{j}$ and $\mathbf{b} = 4\mathbf{i} - \mathbf{j}$, find the vector or number corresponding to

(a) $4\mathbf{a} + \mathbf{b}$ (b) $2\mathbf{a} - 3\mathbf{b}$

(c) $\|\mathbf{a} - \mathbf{b}\|$ (d) $\|\mathbf{a}\| - \|\mathbf{b}\|$

13 **A ship's course** A ship is sailing at a speed of 14 mi/hr in the direction S50°E. Express its velocity **v** as a vector.

14 The magnitudes and directions of two forces are 72 lb, S60°E and 46 lb, N74°E, respectively. Approximate the magnitude and direction of the resultant force.

15 Find a vector that has the opposite direction of $\mathbf{a} = 8\mathbf{i} - 6\mathbf{j}$ and twice the magnitude.

16 Find a vector of magnitude 4 that has the same direction as $\mathbf{a} = \langle -3, 7 \rangle$.

17 If $\mathbf{a} = \langle a_1, a_2 \rangle$, $\mathbf{r} = \langle x, y \rangle$, and $c > 0$, describe the set of all points $P(x, y)$ such that $\|\mathbf{r} - \mathbf{a}\| = c$.

18 If **a** and **b** are vectors with the same initial point and angle θ between them, prove that

$$\|\mathbf{a} - \mathbf{b}\|^2 = \|\mathbf{a}\|^2 + \|\mathbf{b}\|^2 - 2\|\mathbf{a}\|\|\mathbf{b}\| \cos \theta.$$

19 **Wind speed and direction** An airplane is flying in the direction 80° with an airspeed of 400 mi/hr. Its ground speed and true course are 390 mi/hr and 90°, respectively. Approximate the direction and speed of the wind.

20 If $\mathbf{a} = \langle 2, -3 \rangle$ and $\mathbf{b} = \langle -1, -4 \rangle$, find each of the following:

(a) $\mathbf{a} \cdot \mathbf{b}$ (b) the angle between **a** and **b**

(c) $\text{comp}_{\mathbf{a}}\, \mathbf{b}$

21 If $\mathbf{a} = 6\mathbf{i} - 2\mathbf{j}$ and $\mathbf{b} = \mathbf{i} + 3\mathbf{j}$, find each of the following:

(a) $(2\mathbf{a} - 3\mathbf{b}) \cdot \mathbf{a}$

(b) the angle between **a** and $\mathbf{a} + \mathbf{b}$

(c) $\text{comp}_{\mathbf{a}}\, (\mathbf{a} + \mathbf{b})$

22 A constant force has the magnitude and direction of the vector $\mathbf{a} = 7\mathbf{i} + 4\mathbf{j}$. Find the work done when the point of application of **a** moves along the *x*-axis from $P(-5, 0)$ to $Q(3, 0)$.

Exer. 23–28: Express the complex number in trigonometric form with $0 \leq \theta < 2\pi$.

23 $-10 + 10i$ **24** $2 - 2\sqrt{3}\, i$

25 -17 **26** $-12i$

27 $-5\sqrt{3} - 5i$ **28** $4 + 5i$

Exer. 29–30: Express in the form $a + bi$, where a and b are real numbers.

29 $20\left(\cos\dfrac{11\pi}{6} + i\sin\dfrac{11\pi}{6}\right)$ **30** $13\text{ cis }\left(\tan^{-1}\frac{5}{12}\right)$

Exer. 31–32: Use trigonometric forms to find z_1z_2 and z_1/z_2.

31 $z_1 = -3\sqrt{3} - 3i$, $z_2 = 2\sqrt{3} + 2i$

32 $z_1 = 2\sqrt{2} + 2\sqrt{2}i$, $z_2 = -1 - i$

Exer. 33–36: Use De Moivre's theorem to change the given complex number to the form $a + bi$, where a and b are real numbers.

33 $\left(-\sqrt{3} + i\right)^9$

34 $\left(\dfrac{\sqrt{2}}{2} - \dfrac{\sqrt{2}}{2}i\right)^{30}$

35 $(3 - 3i)^5$

36 $\left(2 + 2\sqrt{3}i\right)^{10}$

37 Find the three cube roots of -27.

38 Let $z = 1 - \sqrt{3}i$.

(a) Find z^{24}.   (b) Find the three cube roots of z.

39 Find the solutions of the equation $x^5 - 32 = 0$.

40 Skateboard racecourse A course for a skateboard race consists of a 200-meter downhill run and a 150-meter level portion. The angle of elevation of the starting point of the race from the finish line is 27.4°. What angle does the hill make with the horizontal?

41 Distances to planets The distances between Earth and nearby planets can be approximated using the phase angle α, as shown in the figure. Suppose that the distance between Earth and the sun is 93,000,000 miles and the distance between Venus and the sun is 67,000,000 miles. Approximate the distance between Earth and Venus to the nearest million miles when $\alpha = 34°$.

Exercise 41

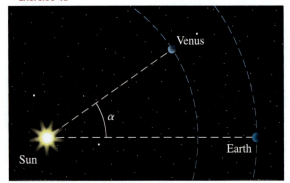

42 Height of a skyscraper If a skyscraper is viewed from the top of a 50-foot building, the angle of elevation is 59°. If it is viewed from street level, the angle of elevation is 62° (see the figure).

(a) Use the law of sines to approximate the shortest distance between the tops of the two buildings.

(b) Approximate the height of the skyscraper.

Exercise 42

43 Distances between cities The beach communities of San Clemente and Long Beach are 41 miles apart, along a fairly straight stretch of coastline. Shown in the figure is the triangle formed by the two cities and the town of Avalon at the southeast corner of Santa Catalina Island. Angles ALS and ASL are found to be 66.4° and 47.2°, respectively.

(a) Approximate the distance from Avalon to each of the two cities.

(b) Approximate the shortest distance from Avalon to the coast.

Exercise 43

44 Surveying A surveyor wishes to find the distance between two inaccessible points A and B. As shown in the figure, two points C and D are selected from which it is possible to view both A and B. The distance CD and the angles ACD, ACB, BDC, and BDA are then measured. If $CD = 120$ ft, $\angle ACD = 115°$, $\angle ACB = 92°$, $\angle BDC = 125°$, and $\angle BDA = 100°$, approximate the distance AB.

Exercise 44

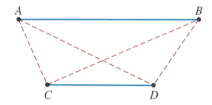

45 Radio contact Two girls with two-way radios are at the intersection of two country roads that meet at a $105°$ angle (see the figure). One begins walking in a northerly direction along one road at a rate of 5 mi/hr; at the same time the other walks east along the other road at the same rate. If each radio has a range of 10 miles, how long will the girls maintain contact?

Exercise 45

10 mi

105°

46 Robotic design Shown in the figure is a design for a robotic arm with two moving parts. The dimensions are chosen to emulate a human arm. The upper arm AC and lower arm CP rotate through angles θ_1 and θ_2, respectively, to hold an object at point $P(x, y)$.

(a) Show that $\angle ACP = 180° - (\theta_2 - \theta_1)$.

(b) Find $d(A, P)$, and then use part (a) and the law of cosines to show that

$$1 + \cos(\theta_2 - \theta_1) = \frac{x^2 + (y - 26)^2}{578}.$$

(c) If $x = 25$, $y = 4$, and $\theta_1 = 135°$, approximate θ_2.

Exercise 46

47 Rescue efforts A child is trapped 45 feet down an abandoned mine shaft that slants at an angle of $78°$ from the horizontal. A rescue tunnel is to be dug 50 feet from the shaft opening (see the figure).

(a) At what angle θ should the tunnel be dug?

(b) If the tunnel can be dug at a rate of 3 ft/hr, how many hours will it take to reach the child?

Exercise 47

48 Design for a jet fighter Shown in the figure is a plan for the top of a wing of a jet fighter.

(a) Approximate angle ϕ.

(b) Approximate the area of quadrilateral $ABCD$.

(c) If the fuselage is 5.8 feet wide, approximate the wing span CC'.

Exercise 48

CHAPTER 8 DISCUSSION EXERCISES

1 Mollweide's formula The following equation, called *Mollweide's formula*, is sometimes used to check solutions to triangles because it involves all the angles and sides:

$$\frac{a+b}{c} = \frac{\cos\frac{1}{2}(\alpha - \beta)}{\sin\frac{1}{2}\gamma}$$

(a) Use the law of sines to show that

$$\frac{a+b}{c} = \frac{\sin\alpha + \sin\beta}{\sin\gamma}.$$

(b) Use a sum-to-product formula and a double-angle formula to verify Mollweide's formula.

2 Use the trigonometric form of a complex number to show that $z^{-n} = 1/z^n$, where n is a positive integer.

3 Discuss the algebraic and geometric similarities of the cube roots of any positive real number a.

4 Suppose that two vectors $\mathbf{v}$ and $\mathbf{w}$ have the same initial point, that the angle between them is θ, and that $\mathbf{v} \neq m\mathbf{w}$ (m is a real number).

(a) What is the geometric interpretation of $\mathbf{v} - \mathbf{w}$?

(b) How could you find $\|\mathbf{v} - \mathbf{w}\|$?

5 A vector approach to the laws of sines and cosines

(a) From the figure we see that $\mathbf{c} = \mathbf{b} + \mathbf{a}$. Use horizontal and vertical components to write $\mathbf{c}$ in terms of $\mathbf{i}$ and $\mathbf{j}$.

Exercise 5

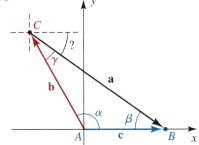

(b) Now find the magnitude of $\mathbf{c}$, using the answer to part (a), and simplify to the point where you have proved the law of cosines.

(c) If $\mathbf{c}$ lies on the x-axis, then its $\mathbf{j}$-component is zero. Use this fact to prove the law of sines.

6 Euler's formula and other results The following are some interesting and unexpected results involving complex numbers and topics that have been previously discussed.

(a) Leonhard Euler (1707–1783) gave us the following formula:

$$e^{i\theta} = \cos\theta + i\sin\theta$$

If we let $\theta = \pi$, we obtain $e^{i\pi} = -1$ or, equivalently,

$$e^{i\pi} + 1 = 0,$$

an equation relating five of the most important numbers in mathematics. Find $e^{2\pi i}$.

(b) We define the logarithm of a complex number $z \neq 0$ as follows:

$$\text{LN } z = \ln |z| + i(\theta + 2\pi n),$$

where ln is the natural logarithm function, θ is an argument of z, and n is an integer. The **principal value** of LN z is the value that corresponds to $n = 0$ and $-\pi < \theta \le \pi$. Find the principal values of LN (-1) and LN i.

(c) We define the complex power w of a complex number $z \neq 0$ as follows:

$$z^w = e^{w \text{ LN } z}$$

We use principal values of LN z to find principal values of z^w. Find principal values of $\sqrt{i}$ and i^i.

7 An interesting identity? Suppose α, β, and γ are angles in an oblique triangle. Prove or disprove the following statement: The sum of the tangents of α, β, and γ is equal to the product of the tangents of α, β, and γ.

8 Forces of hanging wires A 5-pound ornament hangs from two wires as shown in the figure. Show that the magnitudes of the tensions (forces) in the wires are given by

$$\|\mathbf{T}_1\| = \frac{5 \cos \beta}{\sin (\alpha + \beta)} \quad \text{and} \quad \|\mathbf{T}_2\| = \frac{5 \cos \alpha}{\sin (\alpha + \beta)}.$$

Exercise 8

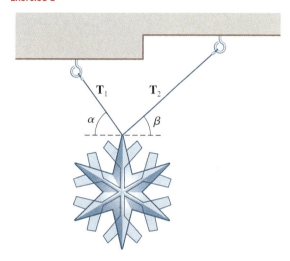

9

Systems of Equations and Inequalities

9.1 Systems of Equations

9.2 Systems of Linear Equations in Two Variables

9.3 Systems of Inequalities

9.4 Linear Programming

9.5 Systems of Linear Equations in More Than Two Variables

9.6 The Algebra of Matrices

9.7 The Inverse of a Matrix

9.8 Determinants

9.9 Properties of Determinants

9.10 Partial Fractions

Applications of mathematics sometimes require working simultaneously with more than one equation in several variables—that is, with a system of equations. In this chapter we develop methods for finding solutions that are common to all the equations in a system. Of particular importance are the techniques involving matrices, because they are well suited for computer programs and can be readily applied to systems containing any number of linear equations in any number of variables. We shall also consider systems of inequalities and linear programming—topics that are of major importance in business applications and statistics. The last part of the chapter provides an introduction to the algebra of matrices and determinants.

9.1

Systems of Equations

Figure 1

Figure 2

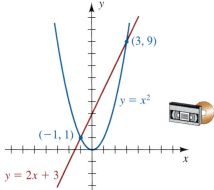

Consider the graphs of the two functions f and g, illustrated in Figure 1. In applications it is often necessary to find points such as $P(a, b)$ and $Q(c, d)$ at which the graphs intersect. Since $P(a, b)$ is on each graph, the pair (a, b) is a **solution** of *both* of the equations $y = f(x)$ and $y = g(x)$; that is,

$$b = f(a) \quad \text{and} \quad b = g(a).$$

We say that (a, b) is a solution of the **system of equations** (or simply **system**)

$$\begin{cases} y = f(x) \\ y = g(x) \end{cases}$$

where the brace is used to indicate that the equations are to be treated simultaneously. Similarly, the pair (c, d) is a solution of the system. To **solve** a system of equations means to find all the solutions.

As a special case, consider the system

$$\begin{cases} y = x^2 \\ y = 2x + 3 \end{cases}$$

The graphs of the equations are the parabola and line sketched in Figure 2. The following table shows that the points $(-1, 1)$ and $(3, 9)$ are on both graphs.

(x, y)	$y = x^2$	$y = 2x + 3$
$(-1, 1)$	$1 = (-1)^2$, or $1 = 1$	$1 = 2(-1) + 3$, or $1 = 1$
$(3, 9)$	$9 = 3^2$, or $9 = 9$	$9 = 2(3) + 3$, or $9 = 9$

Hence, $(-1, 1)$ and $(3, 9)$ are solutions of the system.

The preceding discussion does not give us a strategy for actually finding the solutions. The next two examples illustrate how to find the solutions of the system using only algebraic methods.

EXAMPLE 1 **Solving a system of two equations**

Solve the system

$$\begin{cases} y = x^2 \\ y = 2x + 3 \end{cases}$$

SOLUTION If (x, y) is a solution of the system, then the variable y in the equation $y = 2x + 3$ must satisfy the condition $y = x^2$. Hence, we *substitute* x^2 for y in $y = 2x + 3$:

$$x^2 = 2x + 3 \qquad \text{substitute } y = x^2 \text{ in } y = 2x + 3$$
$$x^2 - 2x - 3 = 0 \qquad \text{subtract } 2x + 3$$
$$(x + 1)(x - 3) = 0 \qquad \text{factor}$$
$$x + 1 = 0, \quad x - 3 = 0 \qquad \text{zero factor theorem}$$
$$x = -1, \quad x = 3 \qquad \text{solve for } x$$

This gives us the x-values for the solutions (x, y) of the system. To find the corresponding y-values, we may use either $y = x^2$ or $y = 2x + 3$. Using $y = x^2$, we find that

$$\text{if} \quad x = -1, \quad \text{then} \quad y = (-1)^2 = 1$$

and

$$\text{if} \quad x = 3, \quad \text{then} \quad y = 3^2 = 9.$$

Hence, the solutions of the system are $(-1, 1)$ and $(3, 9)$.

We could also have found the solutions by substituting $y = 2x + 3$ in the *first* equation, $y = x^2$, obtaining

$$2x + 3 = x^2.$$

The remainder of the solution is the same.

Given the system in Example 1, we *could* have solved one of the equations for x in terms of y and then substituted in the other equation, obtaining an equation in y alone. Solving the latter equation would give us the y-values for the solutions of the system. The x-values could then be found using one of the given equations. In general, we may use the following guidelines, where u and v denote any two variables (*possibly x and y*). This technique is called the **method of substitution.**

Guidelines for the Method of Substitution for Two Equations in Two Variables	1 Solve one of the equations for one variable u in terms of the other variable v.
	2 Substitute the expression for u found in guideline 1 in the other equation, obtaining an equation in v alone.
	3 Find the solutions of the equation in v obtained in guideline 2.
	4 Substitute the v-values found in guideline 3 in the equation of guideline 1 to find the corresponding u-values.
	5 Check each pair (u, v) found in guideline 4 in the given system.

EXAMPLE 2   Using the method of substitution

Solve the following system and then sketch the graph of each equation, showing the points of intersection:

$$\begin{cases} x + y^2 = 6 \\ x + 2y = 3 \end{cases}$$

SOLUTION   We must first decide which equation to solve and which variable to solve for. Let's examine the possibilities.

Solve the first equation for y: $\quad y = \pm\sqrt{6 - x}$

Solve the first equation for x: $\quad x = 6 - y^2$

Solve the second equation for y: $\quad y = (3 - x)/2$

Solve the second equation for x: $\quad x = 3 - 2y$

(continued)

Guideline 1 Looking ahead to guideline 2, we note that solving either equation for x will result in a simple substitution. Thus, we will use $x = 3 - 2y$ and follow the guidelines with $u = x$ and $v = y$.

Guideline 2 Substitute the expression for x found in guideline 1 in the first equation of the system:

$$(3 - 2y) + y^2 = 6 \quad \text{substitute } x = 3 - 2y \text{ in } x + y^2 = 6$$
$$y^2 - 2y - 3 = 0 \quad \text{simplify}$$

Guideline 3 Solve the equation in guideline 2 for y:

$$(y - 3)(y + 1) = 0 \quad \text{factor } y^2 - 2y - 3$$
$$y - 3 = 0, \quad y + 1 = 0 \quad \text{zero factor theorem}$$
$$y = 3, \quad\quad y = -1 \quad \text{solve for } y$$

These are the only possible y-values for the solutions of the system.

Guideline 4 Use the equation $x = 3 - 2y$ from guideline 1 to find the corresponding x-values:

$$\text{if} \quad y = 3, \quad \text{then} \quad x = 3 - 2(3) = 3 - 6 = -3$$
$$\text{if} \quad y = -1, \quad \text{then} \quad x = 3 - 2(-1) = 3 + 2 = 5$$

Thus, possible solutions are $(-3, 3)$ and $(5, -1)$.

Guideline 5 Substituting $x = -3$ and $y = 3$ in $x + y^2 = 6$, the first equation of the system, yields $-3 + 9 = 6$, a true statement. Substituting $x = -3$ and $y = 3$ in $x + 2y = 3$, the second equation of the system, yields $-3 + 6 = 3$, also a true statement. Hence, $(-3, 3)$ is a solution of the system. In a similar manner, we may check that $(5, -1)$ is also a solution.

The graphs of the two equations (a parabola and a line, respectively) are sketched in Figure 3, showing the two points of intersection.

Figure 3

In future examples we will not list the specific guidelines that are used in finding solutions of systems.

In solving certain systems using the method of substitution, it is convenient to let u or v in the guidelines denote an *expression* involving another variable. This technique is illustrated in the next example with $u = x^2$.

EXAMPLE 3 **Using the method of substitution**

Solve the following system and then sketch the graph of each equation, showing the points of intersection:

$$\begin{cases} x^2 + y^2 = 25 \\ x^2 + y = 19 \end{cases}$$

SOLUTION We proceed as follows:

$$x^2 = 19 - y \qquad \text{solve } x^2 + y = 19 \text{ for } x^2$$
$$(19 - y) + y^2 = 25 \qquad \text{substitute } x^2 = 19 - y \text{ in } x^2 + y^2 = 25$$
$$y^2 - y - 6 = 0 \qquad \text{simplify}$$
$$(y - 3)(y + 2) = 0 \qquad \text{factor}$$
$$y - 3 = 0, \quad y + 2 = 0 \qquad \text{zero factor theorem}$$
$$y = 3, \qquad y = -2 \qquad \text{solve for } y$$

These are the only possible y-values for the solutions of the system. To find the corresponding x-values, we use $x^2 = 19 - y$:

If $y = 3$, then $x^2 = 19 - 3 = 16$ and $x = \pm 4$

If $y = -2$, then $x^2 = 19 - (-2) = 21$ and $x = \pm\sqrt{21}$

Thus, the only possible solutions of the system are

$$(4, 3), \quad (-4, 3), \quad \left(\sqrt{21}, -2\right), \quad \text{and} \quad \left(-\sqrt{21}, -2\right).$$

We can check by substitution in the given equations that all four pairs are solutions.

The graph of $x^2 + y^2 = 25$ is a circle of radius 5 with center at the origin, and the graph of $y = 19 - x^2$ is a parabola with a vertical axis. The graphs are sketched in Figure 4. The points of intersection correspond to the solutions of the system.

There are, of course, other ways to find the solutions. We could solve the first equation for x^2, $x^2 = 25 - y^2$, and then substitute in the second, obtaining $25 - y^2 + y = 19$. Another method is to solve the second equation for y, $y = 19 - x^2$, and substitute in the first. ◢

Figure 4

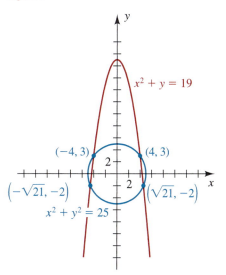

We can also consider equations in three variables x, y, and z, such as

$$x^2y + xz + 3^y = 4z^3.$$

Such an equation has a **solution** (a, b, c) if substitution of a, b, and c, for x, y, and z, respectively, yields a true statement. We refer to (a, b, c) as an **ordered triple** of real numbers. Systems of equations are **equivalent systems** provided they have the same solutions. A system of equations in three variables and the solutions of the system are defined as in the two-variable case. Similarly, we can consider systems of *any* number of equations in *any* number of variables.

The method of substitution can be extended to these more complicated systems. For example, given three equations in three variables, suppose that it is possible to solve one of the equations for one variable in terms of the remaining two variables. By substituting that expression in each of the other equations, we obtain a system of two equations in two variables. The solutions of the two-variable system can then be used to find the solutions of the original system.

EXAMPLE 4   Solving a system of three equations

Solve the system

$$\begin{cases} x - y + z = 2 \\ xyz = 0 \\ 2y + z = 1 \end{cases}$$

SOLUTION   We proceed as follows:

$$z = 1 - 2y \quad \text{solve } 2y + z = 1 \text{ for } z$$

$$\begin{cases} x - y + (1 - 2y) = 2 \\ xy(1 - 2y) = 0 \end{cases} \quad \begin{array}{l} \text{substitute } z = 1 - 2y \text{ in the} \\ \text{first two equations} \end{array}$$

$$\begin{cases} x - 3y - 1 = 0 \\ xy(1 - 2y) = 0 \end{cases} \quad \text{equivalent system}$$

We now find the solutions of the last system:

$$x = 3y + 1 \quad \text{solve } x - 3y - 1 = 0 \text{ for } x$$

$$(3y + 1)y(1 - 2y) = 0 \quad \begin{array}{l} \text{substitute } x = 3y + 1 \text{ in} \\ xy(1 - 2y) = 0 \end{array}$$

$$3y + 1 = 0, \quad y = 0, \quad 1 - 2y = 0 \quad \text{zero factor theorem}$$

$$y = -\tfrac{1}{3}, \quad y = 0, \qquad y = \tfrac{1}{2} \quad \text{solve for } y$$

These are the only possible y-values for the solutions of the system.

To obtain the corresponding x-values, we substitute for y in the equation $x = 3y + 1$, obtaining

$$x = 0, \qquad x = 1, \qquad \text{and} \qquad x = \tfrac{5}{2}.$$

Using $z = 1 - 2y$ gives us the corresponding z-values

$$z = \tfrac{5}{3}, \qquad z = 1, \qquad \text{and} \qquad z = 0.$$

Thus, the solutions (x, y, z) of the original system must be among the ordered triples

$$\left(0, -\tfrac{1}{3}, \tfrac{5}{3}\right), \qquad (1, 0, 1), \qquad \text{and} \qquad \left(\tfrac{5}{2}, \tfrac{1}{2}, 0\right).$$

Checking each shows that the three ordered triples are solutions of the system.

EXAMPLE 5   An application of a system of equations

Is it possible to construct an aquarium with a glass top and two square ends that holds 16 ft^3 of water and requires 40 ft^2 of glass? (Disregard the thickness of the glass.)

SOLUTION   We begin by sketching a typical aquarium and labeling it as in Figure 5, with x and y in feet. Referring to the figure and using formulas for volume and area, we see that

$$\text{volume of the aquarium} = x^2y \qquad \text{length} \times \text{width} \times \text{height}$$

$$\text{square feet of glass required} = 2x^2 + 4xy. \qquad \text{2 ends, 2 sides, top, and bottom}$$

Figure 5

Since the volume is to be 16 ft^3 and the area of the glass required is 40 ft^2, we obtain the following system of equations:

$$\begin{cases} x^2y = 16 \\ 2x^2 + 4xy = 40 \end{cases}$$

We find the solutions as follows:

$$y = \frac{16}{x^2} \qquad \text{solve } x^2y = 16 \text{ for } y$$

$$2x^2 + 4x\left(\frac{16}{x^2}\right) = 40 \qquad \text{substitute } y = \frac{16}{x^2} \text{ in } 2x^2 + 4xy = 40$$

$$x^2 + \frac{32}{x} = 20 \qquad \text{cancel } x, \text{ and divide by 2}$$

$$x^3 + 32 = 20x \qquad \text{multiply by } x \ (x \neq 0)$$

$$x^3 - 20x + 32 = 0 \qquad \text{subtract } 20x$$

We next look for rational solutions of the last equation. Dividing the polynomial $x^3 - 20x + 32$ synthetically by $x - 2$ gives us

$$\begin{array}{r|rrrr} 2 & 1 & 0 & -20 & 32 \\ & & 2 & 4 & -32 \\ \hline & 1 & 2 & -16 & 0 \end{array}$$

Thus, one solution of $x^3 - 20x + 32 = 0$ is 2, and the remaining two solutions are zeros of the quotient $x^2 + 2x - 16$—that is, roots of the depressed equation

$$x^2 + 2x - 16 = 0.$$

(continued)

By the quadratic formula,

$$x = \frac{-2 \pm \sqrt{2^2 - 4(1)(-16)}}{2(1)} = \frac{-2 \pm 2\sqrt{17}}{2} = -1 \pm \sqrt{17}.$$

Since x is positive, we may discard $x = -1 - \sqrt{17}$. Hence, the only possible values of x are

$$x = 2 \quad \text{and} \quad x = -1 + \sqrt{17} \approx 3.12.$$

The corresponding y-values can be found by substituting for x in the equation $y = 16/x^2$. Letting $x = 2$ gives us $y = \frac{16}{4} = 4$. Using these values, we obtain the dimensions 2 feet by 2 feet by 4 feet for the aquarium.

Letting $x = -1 + \sqrt{17}$, we obtain $y = 16/(-1 + \sqrt{17})^2$, which simplifies to $y = \frac{1}{8}(9 + \sqrt{17}) \approx 1.64$. Thus, approximate dimensions for another aquarium are 3.12 feet by 3.12 feet by 1.64 feet. ◢

9.1 Exercises

Exer. 1–30: Use the method of substitution to solve the system.

1. $\begin{cases} y = x^2 - 4 \\ y = 2x - 1 \end{cases}$

2. $\begin{cases} y = x^2 + 1 \\ x + y = 3 \end{cases}$

3. $\begin{cases} y^2 = 1 - x \\ x + 2y = 1 \end{cases}$

4. $\begin{cases} y^2 = x \\ x + 2y + 3 = 0 \end{cases}$

5. $\begin{cases} 2y = x^2 \\ y = 4x^3 \end{cases}$

6. $\begin{cases} x - y^3 = 1 \\ 2x = 9y^2 + 2 \end{cases}$

7. $\begin{cases} x + 2y = -1 \\ 2x - 3y = 12 \end{cases}$

8. $\begin{cases} 3x - 4y + 20 = 0 \\ 3x + 2y + 8 = 0 \end{cases}$

9. $\begin{cases} 2x - 3y = 1 \\ -6x + 9y = 4 \end{cases}$

10. $\begin{cases} 4x - 5y = 2 \\ 8x - 10y = -5 \end{cases}$

11. $\begin{cases} x + 3y = 5 \\ x^2 + y^2 = 25 \end{cases}$

12. $\begin{cases} 3x - 4y = 25 \\ x^2 + y^2 = 25 \end{cases}$

13. $\begin{cases} x^2 + y^2 = 8 \\ y - x = 4 \end{cases}$

14. $\begin{cases} x^2 + y^2 = 25 \\ 3x + 4y = -25 \end{cases}$

15. $\begin{cases} x^2 + y^2 = 9 \\ y - 3x = 2 \end{cases}$

16. $\begin{cases} x^2 + y^2 = 16 \\ y + 2x = -1 \end{cases}$

17. $\begin{cases} x^2 + y^2 = 16 \\ 2y - x = 4 \end{cases}$

18. $\begin{cases} x^2 + y^2 = 1 \\ y + 2x = -3 \end{cases}$

19. $\begin{cases} (x - 1)^2 + (y + 2)^2 = 10 \\ x + y = 1 \end{cases}$

20. $\begin{cases} xy = 2 \\ 3x - y + 5 = 0 \end{cases}$

21. $\begin{cases} y = 20/x^2 \\ y = 9 - x^2 \end{cases}$

22. $\begin{cases} x = y^2 - 4y + 5 \\ x - y = 1 \end{cases}$

23. $\begin{cases} y^2 - 4x^2 = 4 \\ 9y^2 + 16x^2 = 140 \end{cases}$

24. $\begin{cases} 25y^2 - 16x^2 = 400 \\ 9y^2 - 4x^2 = 36 \end{cases}$

25. $\begin{cases} x^2 - y^2 = 4 \\ x^2 + y^2 = 12 \end{cases}$

26. $\begin{cases} 6x^3 - y^3 = 1 \\ 3x^3 + 4y^3 = 5 \end{cases}$

27. $\begin{cases} x + 2y - z = -1 \\ 2x - y + z = 9 \\ x + 3y + 3z = 6 \end{cases}$

28. $\begin{cases} 2x - 3y - z^2 = 0 \\ x - y - z^2 = -1 \\ x^2 - xy = 0 \end{cases}$

29. $\begin{cases} x^2 + z^2 = 5 \\ 2x + y = 1 \\ y + z = 1 \end{cases}$

30. $\begin{cases} x + 2z = 1 \\ 2y - z = 4 \\ xyz = 0 \end{cases}$

31. Find the values of b such that the system represented in the graph on the next page has
   (a) one solution   (b) two solutions   (c) no solution

Exercise 31

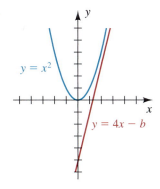

$y = x^2$

$y = 4x - b$

Interpret (a)–(c) graphically.

32 Find the values of b such that the system

$$\begin{cases} x^2 + y^2 = 4 \\ y = x + b \end{cases}$$

has

(a) one solution   (b) two solutions

(c) no solution

Interpret (a)–(c) graphically.

33 Is there a real number x such that $x = 2^{-x}$? Decide by displaying graphically the system

$$\begin{cases} y = x \\ y = 2^{-x} \end{cases}$$

34 Is there a real number x such that $x = \log x$? Decide by displaying graphically the system

$$\begin{cases} y = x \\ y = \log x \end{cases}$$

35 Shown in the figure is the graph of $x = y^2$ and a line of slope m that passes through the point $(4, 2)$. Find the value of m such that the line intersects the graph only at $(4, 2)$ and interpret graphically.

Exercise 35

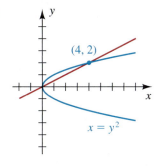

$(4, 2)$

$x = y^2$

36 Shown in the figure is the graph of $y = x^2$ and a line of slope m that passes through the point $(1, 1)$. Find the value of m such that the line intersects the graph only at $(1, 1)$, and interpret graphically.

Exercise 36

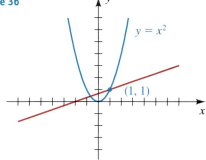

$y = x^2$

$(1, 1)$

Exer. 37–38: Find an exponential function of the form $f(x) = ba^x + c$ for the graph.

37

$(1, 7)$

$\left(-1, \frac{5}{3}\right)$ $(0, 3)$

38

$(-1, 9)$

$(0, 3)$ $\left(1, \frac{3}{2}\right)$

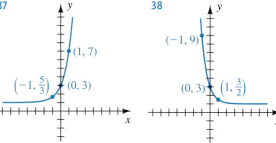

39 The perimeter of a rectangle is 40 inches, and its area is 96 in². Find its length and width.

40 Constructing tubing Sections of cylindrical tubing are to be made from thin rectangular sheets that have an area of 200 in² (see the figure). Is it possible to construct a tube that has a volume of 200 in³? If so, find r and h.

Exercise 40

r

200 in² h h

41 Fish population In fishery science, spawner-recruit functions are used to predict the number of adult fish R in next year's breeding population from an estimate S of the number of fish presently spawning.

(a) For a certain species of fish, $R = aS/(S + b)$. Estimate a and b from the data in the following table.

Year	1998	1999	2000
Number spawning	40,000	60,000	72,000

(b) Predict the breeding population for the year 2001.

42 Fish population Refer to Exercise 41. Ricker's spawner-recruit function is given by

$$R = aSe^{-bS}$$

for positive constants a and b. This relationship predicts low recruitment from very high stocks and has been found to be appropriate for many species, such as arctic cod. Rework Exercise 41 using Ricker's spawner-recruit function.

43 Competition for food A *competition model* is a collection of equations that specifies how two or more species interact in competition for the food resources of an ecosystem. Let x and y denote the numbers (in hundreds) of two competing species, and suppose that the respective rates of growth R_1 and R_2 are given by

$$R_1 = 0.01x(50 - x - y),$$
$$R_2 = 0.02y(100 - y - 0.5x).$$

Determine the population levels (x, y) at which both rates of growth are zero. (Such population levels are called *stationary points*.)

44 Fencing a region A rancher has 2420 feet of fence to enclose a rectangular region that lies along a straight river. If no fence is used along the river (see the figure), is it possible to enclose 10 acres of land? Recall that 1 acre = 43,560 ft².

Exercise 44

45 Constructing an aquarium Refer to Example 5. Is it possible to construct a small aquarium with an *open* top and two square ends that holds 2 ft³ of water and requires 8 ft² of glass? If so, approximate the dimensions. (Disregard the thickness of the glass.)

46 Isoperimetric problem The isoperimetric problem is to prove that of all plane geometric figures with the same perimeter (isoperimetric figures), the circle has the greatest area. Show that no rectangle has both the same area and the same perimeter as any circle.

47 Moiré pattern A moiré pattern is formed when two geometrically regular patterns are superimposed. Shown in the figure is a pattern obtained from the family of circles $x^2 + y^2 = n^2$ and the family of horizontal lines $y = m$ for integers m and n.

(a) Show that the points of intersection of the circle $x^2 + y^2 = n^2$ and the line $y = n - 1$ lie on a parabola.

(b) Work part (a) using the line $y = n - 2$.

Exercise 47

y

$y = m$

$x^2 + y^2 = n^2$

x

48 Dimensions of a pill A spherical pill has diameter 1 centimeter. A second pill in the shape of a right circular cylinder is to be manufactured with the same volume and twice the surface area of the spherical pill.

(a) If r is the radius and h is the height of the cylindrical pill, show that $6r^2h = 1$ and $r^2 + rh = 1$. Conclude that $6r^3 - 6r + 1 = 0$.

(b) The positive solutions of $6r^3 - 6r + 1 = 0$ are approximately 0.172 and 0.903. Find the corresponding heights, and interpret these results.

49 Hammer throw A hammer thrower is working on his form in a small practice area. The hammer spins, generating a circle with a radius of 5 feet, and when released, it hits a tall screen that is 50 feet from the center of the throwing area. Let coordinate axes be introduced as shown in the figure (not to scale).

(a) If the hammer is released at $(-4, -3)$ and travels in the tangent direction, where will it hit the screen?

(b) If the hammer is to hit at $(0, -50)$, where on the circle should it be released?

Exercise 49

50 Path of a tossed ball A person throws a ball from the edge of a hill, at an angle of 45° with the horizontal, as illustrated in the figure. The ball lands 50 feet down the hill, which has slope $-\frac{3}{4}$. Using calculus, it can be shown that the path of the ball is given by $y = ax^2 + x + c$ for some constants a and c.

(a) Disregarding the height of the person, find an equation for the path.

(b) What is the maximum height of the ball *off the ground?*

Exercise 50

9.2

Systems of Linear Equations in Two Variables

An equation $ax + by = c$ (or, equivalently, $ax + by - c = 0$), with a and b not both zero, is a **linear equation in two variables** x and y. Similarly, the equation $ax + by + cz = d$ is a linear equation in three variables x, y, and z. We may also consider linear equations in four, five, or *any* number of variables. The most common systems of equations are those in which every equation is linear. In this section we shall consider only systems of two linear equations in two variables. Systems involving more than two variables are discussed in a later section.

Two systems of equations are equivalent if they have the same solutions. To find the solutions of a system, we may manipulate the equations until we obtain an equivalent system of simple equations for which the solutions can be found readily. Some manipulations (or *transformations*) that lead to equivalent systems are stated in the next theorem.

Theorem on Equivalent Systems	Given a system of equations, an equivalent system results if
	(1) two equations are interchanged.
	(2) an equation is multiplied or divided by a nonzero constant.
	(3) a constant multiple of one equation is added to another equation.

A *constant multiple* of an equation is obtained by multiplying *each* term of the equation by the same nonzero constant k. When applying part (3) of the theorem, we often use the phrase *add to one equation k times any other equation*. To *add* two equations means to add corresponding sides of the equations.

The next example illustrates how the theorem on equivalent systems may be used to solve a system of linear equations.

EXAMPLE 1 **Using the theorem on equivalent systems**

Solve the system

$$\begin{cases} x + 3y = -1 \\ 2x - y = 5 \end{cases}$$

SOLUTION We often multiply one of the equations by a constant that will give us the additive inverse of the coefficient of one of the variables in the other equation. Doing so enables us to add the two equations and obtain an equation in only one variable, as follows:

$$\begin{cases} x + 3y = -1 \\ 6x - 3y = 15 \end{cases} \quad \text{multiply the second equation by 3}$$

$$\begin{cases} x + 3y = -1 \\ 7x = 14 \end{cases} \quad \text{add the first equation to the second}$$

We see from the last system that $7x = 14$, and hence $x = \frac{14}{7} = 2$. To find the corresponding y-value, we substitute 2 for x in $x + 3y = -1$, obtaining $y = -1$. Thus, $(2, -1)$ is the only solution of the system.

There are many other ways to use the theorem on equivalent systems to find the solution. Another approach is to proceed as follows:

Figure 1

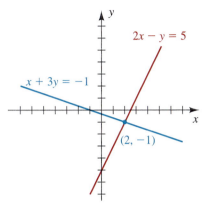

$$\begin{cases} x + 3y = -1 \\ 2x - y = 5 \end{cases} \quad \text{given}$$

$$\begin{cases} -2x - 6y = 2 \\ 2x - y = 5 \end{cases} \quad \text{multiply the first equation by } -2$$

$$\begin{cases} -2x - 6y = 2 \\ \quad\quad - 7y = 7 \end{cases} \quad \text{add the first equation to the second}$$

We see from the last system that $-7y = 7$, or $y = -1$. To find the corresponding x-value, we could substitute -1 for y in $x + 3y = -1$, obtaining $x = 2$. Hence, $(2, -1)$ is the solution.

The graphs of the two equations are lines that intersect at the point $(2, -1)$, as shown in Figure 1.

The technique used in Example 1 is called the **method of elimination,** since it involves the elimination of a variable from one of the equations. The method of elimination usually leads to solutions in fewer steps than does the method of substitution discussed in the preceding section.

EXAMPLE 2 **A system of linear equations with an infinite number of solutions**

Solve the system

$$\begin{cases} 3x + y = 6 \\ 6x + 2y = 12 \end{cases}$$

SOLUTION Multiplying the second equation by $\frac{1}{2}$ gives us

$$\begin{cases} 3x + y = 6 \\ 3x + y = 6 \end{cases}$$

Figure 2

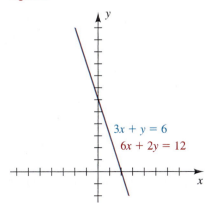

Thus, (a, b) is a solution if and only if $3a + b = 6$—that is, $b = 6 - 3a$. It follows that the solutions consist of ordered pairs of the form $(a, 6 - 3a)$, where a is any real number. If we wish to find particular solutions, we may substitute various values for a. A few solutions are $(0, 6), (1, 3), (3, -3), (-2, 12)$, and $\left(\sqrt{2}, 6 - 3\sqrt{2}\right)$.

It is incorrect to say that the solution is "all reals." It is correct to say that the solution is the set of all ordered pairs such that $3x + y = 6$, which can be written

$$\{(x, y) \,|\, 3x + y = 6\}.$$

The graph of each equation is the same line, as shown in Figure 2.

Figure 3

EXAMPLE 3   A system of linear equations with no solutions

Solve the system

$$\begin{cases} 3x + y = 6 \\ 6x + 2y = 20 \end{cases}$$

SOLUTION   If we add to the second equation -2 times the first equation, $-6x - 2y = -12$, we obtain the equivalent system

$$\begin{cases} 3x + y = 6 \\ 0 = 8 \end{cases}$$

The last equation can be written $0x + 0y = 8$, which is false for every ordered pair (x, y). Thus, the system has no solution.

The graphs of the two equations in the given system are lines that have the same slope and hence are parallel (see Figure 3). The conclusion that the system has no solution corresponds to the fact that these lines do not intersect.

The preceding three examples illustrate typical outcomes of solving a system of two linear equations in two variables: there is either exactly one solution, an infinite number of solutions, or no solution. A system is **consistent** if it has at least one solution. A system with an infinite number of solutions is **dependent and consistent.** A system is **inconsistent** if it has no solution.

Since the graph of any linear equation $ax + by = c$ is a line, *exactly one* of the three cases listed in the following table holds for any system of two such equations.

Characteristics of a System of Two Linear Equations in Two Variables

Graphs	Number of solutions	Classification
Nonparallel lines	One solution	Consistent system
Identical lines	Infinite number of solutions	Dependent and consistent system
Parallel lines	No solution	Inconsistent system

In practice, there should be little difficulty determining which of the three cases occurs. The case of the unique solution will become apparent when suitable transformations are applied to the system, as illustrated in Example 1. The case of an infinite number of solutions is similar to that of Example 2, where one of the equations can be transformed into the other. The case of no solution is indicated by a contradiction, such as the statement $0 = 8$, which appeared in Example 3.

In the process of solving a system, suppose we obtain for x a rational number such as $-\frac{41}{29}$. Substituting $-\frac{41}{29}$ for x to find the value of y is cumbersome. It is easier to select a different multiplier for each of the original equations that will enable us to eliminate x and solve for y. This technique is illustrated in the next example.

EXAMPLE 4   Solving a system

Solve the system

$$\begin{cases} 4x + 7y = 11 \\ 3x - 2y = -9 \end{cases}$$

SOLUTION   We select multipliers to eliminate y. (The least common multiple of 7 and 2 is 14.)

$$\begin{cases} 8x + 14y = 22 & \text{multiply the first equation by 2} \\ 21x - 14y = -63 & \text{multiply the second equation by 7} \end{cases}$$

Adding the first equation to the second gives us

$$29x = -41, \quad \text{so} \quad x = -\tfrac{41}{29}.$$

Next, we return to the original system and select multipliers to eliminate x. (The least common multiple of 4 and 3 is 12.)

$$\begin{cases} 4x + 7y = 11 \\ 3x - 2y = -9 \end{cases} \quad \text{original system}$$

$$\begin{cases} 12x + 21y = 33 & \text{multiply the first equation by 3} \\ -12x + 8y = 36 & \text{multiply the second equation by } -4 \end{cases}$$

Adding the equations gives us

$$29y = 69, \quad \text{so} \quad y = \tfrac{69}{29}.$$

Hence, the solution is $\left(-\tfrac{41}{29}, \tfrac{69}{29}\right)$.

✔ **Check** $(x, y) = \left(-\tfrac{41}{29}, \tfrac{69}{29}\right)$

We substitute the values of x and y into the original equations.

$$4x + 7y = 4\left(-\tfrac{41}{29}\right) + 7\left(\tfrac{69}{29}\right) = -\tfrac{164}{29} + \tfrac{483}{29} = \tfrac{319}{29} = 11 \qquad \text{first equation checks}$$

$$3x - 2y = 3\left(-\tfrac{41}{29}\right) - 2\left(\tfrac{69}{29}\right) = -\tfrac{123}{29} - \tfrac{138}{29} = -\tfrac{261}{29} = -9 \qquad \text{so does the second}$$

◢

Certain applied problems can be solved by introducing systems of two linear equations, as illustrated in the next two examples.

EXAMPLE 5   An application of a system of linear equations

A produce company has a 100-acre farm on which it grows lettuce and cabbage. Each acre of cabbage requires 600 hours of labor, and each acre of lettuce needs 400 hours of labor. If 45,000 hours are available and if all land and labor resources are to be used, find the number of acres of each crop that should be planted.

SOLUTION   Let us introduce variables to denote the unknown quantities as follows:

$$x = \text{number of acres of cabbage}$$
$$y = \text{number of acres of lettuce}$$

Thus, the number of hours of labor required for each crop can be expressed as follows:

$$600x = \text{number of hours required for cabbage}$$
$$400y = \text{number of hours required for lettuce}$$

Using the facts that the total number of acres is 100 and the total number of hours available is 45,000 leads to the following system:

$$\begin{cases} x + y = 100 \\ 600x + 400y = 45{,}000 \end{cases}$$

We next use the method of elimination:

$$\begin{cases} x + y = 100 \\ 6x + 4y = 450 \end{cases} \qquad \textcolor{red}{\text{divide the second equation by 100}}$$

$$\begin{cases} -6x - 6y = -600 \\ 6x + 4y = 450 \end{cases} \qquad \textcolor{red}{\text{multiply the first equation by } -6}$$

$$\begin{cases} -6x - 6y = -600 \\ -2y = -150 \end{cases} \qquad \textcolor{red}{\text{add the first equation to the second}}$$

We see from the last equation that $-2y = -150$, or $y = 75$. Substituting 75 for y in $x + y = 100$ gives us $x = 25$. Hence, the company should plant 25 acres of cabbage and 75 acres of lettuce.

✔ **Check**   Planting 25 acres of cabbage and 75 acres of lettuce requires $(25)(600) + (75)(400) = 45{,}000$ hours of labor. Thus, all 100 acres of land and 45,000 hours of labor are used.   ◢

EXAMPLE 6   **Finding the speed of the current in a river**

A motorboat, operating at full throttle, made a trip 4 miles upstream (against a constant current) in 15 minutes. The return trip (with the same current and at full throttle) took 12 minutes. Find the speed of the current and the equivalent speed of the boat in still water.

SOLUTION   We begin by introducing variables to denote the unknown quantities. Thus, let

$$x = \text{speed of boat (in mi/hr)}$$
$$y = \text{speed of current (in mi/hr)}.$$

We plan to use the formula $d = rt$, where d denotes the distance traveled, r the rate, and t the time. Since the current slows the boat as it travels upstream but adds to its speed as it travels downstream, we obtain

$$\text{upstream rate} = x - y \quad (\text{in mi/hr})$$
$$\text{downstream rate} = x + y \quad (\text{in mi/hr}).$$

The time (in hours) traveled in each direction is

$$\text{upstream time} = \frac{15}{60} = \frac{1}{4} \text{ hr}$$
$$\text{downstream time} = \frac{12}{60} = \frac{1}{5} \text{ hr}.$$

The distance is 4 miles for each trip. Substituting in $d = rt$ gives us the system

$$\begin{cases} 4 = (x - y)\left(\frac{1}{4}\right) \\ 4 = (x + y)\left(\frac{1}{5}\right) \end{cases}$$

Applying the theorem on equivalent systems, we obtain

$$\begin{cases} x - y = 16 \\ x + y = 20 \end{cases} \quad \text{multiply the first equation by 4 and the second by 5}$$

$$\begin{cases} x - y = 16 \\ 2x \quad\;\; = 36 \end{cases} \quad \text{add the first equation to the second}$$

We see from the last equation that $2x = 36$, or $x = 18$. Substituting 18 for x in $x + y = 20$ gives us $y = 2$. Hence, the speed of the boat in still water is 18 mi/hr, and the speed of the current is 2 mi/hr.

✔ **Check** The upstream rate is $18 - 2 = 16$ mi/hr, and the downstream rate is $18 + 2 = 20$ mi/hr. An upstream 4-mile trip would take $\frac{4}{16} = \frac{1}{4}$ hr = 15 min, and a downstream 4-mile trip would take $\frac{4}{20} = \frac{1}{5}$ hr = 12 min. ◢

9.2 Exercises

Exer. 1–20: Solve the system.

1. $\begin{cases} 2x + 3y = 2 \\ x - 2y = 8 \end{cases}$

2. $\begin{cases} 4x + 5y = 13 \\ 3x + y = -4 \end{cases}$

3. $\begin{cases} 2x + 5y = 16 \\ 3x - 7y = 24 \end{cases}$

4. $\begin{cases} 7x - 8y = 9 \\ 4x + 3y = -10 \end{cases}$

5. $\begin{cases} 3r + 4s = 3 \\ r - 2s = -4 \end{cases}$

6. $\begin{cases} 9u + 2v = 0 \\ 3u - 5v = 17 \end{cases}$

7. $\begin{cases} 5x - 6y = 4 \\ 3x + 7y = 8 \end{cases}$

8. $\begin{cases} 2x + 8y = 7 \\ 3x - 5y = 4 \end{cases}$

9. $\begin{cases} \frac{1}{3}c + \frac{1}{2}d = 5 \\ c - \frac{2}{3}d = -1 \end{cases}$

10. $\begin{cases} \frac{1}{2}t - \frac{1}{5}v = \frac{3}{2} \\ \frac{2}{3}t + \frac{1}{4}v = \frac{5}{12} \end{cases}$

11. $\begin{cases} \sqrt{3}x - \sqrt{2}y = 2\sqrt{3} \\ 2\sqrt{2}x + \sqrt{3}y = \sqrt{2} \end{cases}$

12. $\begin{cases} 0.11x - 0.03y = 0.25 \\ 0.12x + 0.05y = 0.70 \end{cases}$

13. $\begin{cases} 2x - 3y = 5 \\ -6x + 9y = 12 \end{cases}$

14. $\begin{cases} 3p - q = 7 \\ -12p + 4q = 3 \end{cases}$

15. $\begin{cases} 3m - 4n = 2 \\ -6m + 8n = -4 \end{cases}$

16. $\begin{cases} x - 5y = 2 \\ 3x - 15y = 6 \end{cases}$

17 $\begin{cases} 2y - 5x = 0 \\ 3y + 4x = 0 \end{cases}$

18 $\begin{cases} 3x + 7y = 9 \\ y = 5 \end{cases}$

19 $\begin{cases} \dfrac{2}{x} + \dfrac{3}{y} = -2 \\ \dfrac{4}{x} - \dfrac{5}{y} = 1 \end{cases}$ $\left(\text{Hint: Let } u = \dfrac{1}{x} \text{ and } v = \dfrac{1}{y}. \right)$

20 $\begin{cases} \dfrac{3}{x-1} + \dfrac{4}{y+2} = 2 \\ \dfrac{6}{x-1} - \dfrac{7}{y+2} = -3 \end{cases}$

21 Ticket sales The price of admission to a high school play was $3.00 for students and $4.50 for nonstudents. If 450 tickets were sold for a total of $1555.50, how many of each kind were purchased?

22 Air travel An airline that flies from Los Angeles to Albuquerque with a stopover in Phoenix charges a fare of $45 to Phoenix and a fare of $60 from Los Angeles to Albuquerque. A total of 185 passengers boarded the plane in Los Angeles, and fares totaled $10,500. How many passengers got off the plane in Phoenix?

23 Crayon dimensions A crayon 8 centimeters in length and 1 centimeter in diameter will be made from 5 cm³ of colored wax. The crayon is to have the shape of a cylinder surmounted by a small conical tip (see the figure). Find the length x of the cylinder and the height y of the cone.

Exercise 23

24 Rowing a boat A man rows a boat 500 feet upstream against a constant current in 10 minutes. He then rows 300 feet downstream (with the same current) in 5 minutes. Find the speed of the current and the equivalent rate at which he can row in still water.

25 Table top dimensions A large table for a conference room is to be constructed in the shape of a rectangle with two semicircles at the ends (see the figure). The table is to have a perimeter of 40 feet, and the area of the rectangular por-

tion is to be twice the sum of the areas of the two ends. Find the length l and the width w of the rectangular portion.

Exercise 25

26 Investment income A woman has $15,000 to invest in two funds that pay simple interest at the rates of 6% and 8% per year. Interest on the 6% fund is tax-exempt; however, income tax must be paid on interest on the 8% fund. Being in a high tax bracket, the woman does not wish to invest the entire sum in the 8% account. Is there a way of investing the money so that she will receive $1000 in interest at the end of one year?

27 Bobcat population A bobcat population is classified by age into kittens (less than 1 year old) and adults (at least 1 year old). All adult females, including those born the prior year, have a litter each June, with an average litter size of 3 kittens. The springtime population of bobcats in a certain area is estimated to be 6000, and the male-female ratio is one. Estimate the number of adults and kittens in the population.

28 Flow rates A 300-gallon water storage tank is filled by a single inlet pipe, and two identical outlet pipes can be used to supply water to the surrounding fields (see the figure). It takes 5 hours to fill an empty tank when both outlet pipes are open. When one outlet pipe is closed, it takes 3 hours to fill the tank. Find the flow rates (in gallons per hour) in and out of the pipes.

Exercise 28

29 Mixing a silver alloy A silversmith has two alloys, one containing 35% silver and the other 60% silver. How much of each should be melted and combined to obtain 100 grams of an alloy containing 50% silver?

30 Mixing nuts A merchant wishes to mix peanuts costing $3 per pound with cashews costing $8 per pound to obtain 60 pounds of a mixture costing $5 per pound. How many pounds of each variety should be mixed?

31 Air travel An airplane, flying with a tail wind, travels 1200 miles in 2 hours. The return trip, against the wind, takes $2\frac{1}{2}$ hours. Find the cruising speed of the plane and the speed of the wind (assume that both rates are constant).

32 Filling orders A stationery company sells two types of notepads to college bookstores, the first wholesaling for 50¢ and the second for 70¢. The company receives an order for 500 notepads, together with a check for $286. If the order fails to specify the number of each type, how should the company fill the order?

33 Acceleration As a ball rolls down an inclined plane, its velocity $v(t)$ (in cm/sec) at time t (in seconds) is given by $v(t) = v_0 + at$ for initial velocity v_0 and acceleration a (in cm/sec^2). If $v(2) = 16$ and $v(5) = 25$, find v_0 and a.

34 Vertical projection If an object is projected vertically upward from an altitude of s_0 feet with an initial velocity of v_0 ft/sec, then its distance $s(t)$ above the ground after t seconds is

$$s(t) = -16t^2 + v_0 t + s_0.$$

If $s(1) = 84$ and $s(2) = 116$, what are v_0 and s_0?

35 Planning production A small furniture company manufactures sofas and recliners. Each sofa requires 8 hours of labor and $60 in materials, while a recliner can be built for $35 in 6 hours. The company has 340 hours of labor available each week and can afford to buy $2250 worth of materials. How many recliners and sofas can be produced if all labor hours and all materials must be used?

36 Livestock diet A rancher is preparing an oat-cornmeal mixture for livestock. Each ounce of oats provides 4 grams of protein and 18 grams of carbohydrates, and an ounce of cornmeal provides 3 grams of protein and 24 grams of carbohydrates. How many ounces of each can be used to meet the nutritional goals of 200 grams of protein and 1320 grams of carbohydrates per feeding?

37 Services swap A plumber and an electrician are each doing repairs on their offices and agree to swap services. The number of hours spent on each of the projects is shown in the following table.

	Plumber's office	Electrician's office
Plumber's hours	6	4
Electrician's hours	5	6

They would prefer to call the matter even, but because of tax laws, they must charge for all work performed. They agree to select hourly wage rates so that the bill on each project will match the income that each person would ordinarily receive for a comparable job.

(a) If x and y denote the hourly wages of the plumber and electrician, respectively, show that

$$6x + 5y = 10x \quad \text{and} \quad 4x + 6y = 11y.$$

Describe the solutions to this system.

(b) If the plumber ordinarily makes $20 per hour, what should the electrician charge?

38 Find equations for the altitudes of the triangle with vertices $A(-3, 2)$, $B(5, 4)$, and $C(3, -8)$, and find the point at which the altitudes intersect.

39 Warming trend in Paris As a result of urbanization, the temperatures in Paris have increased. In 1891 the average daily minimum and maximum temperatures were 5.8°C and 15.1°C, respectively. Between 1891 and 1968, these average temperatures rose 0.019°C/yr and 0.011°C/yr, respectively. Assuming the increases were linear, find the year when the difference between the minimum and maximum temperatures was 9°C, and determine the corresponding average maximum temperature.

40 Long distance telephone rates A telephone company charges customers a certain amount for the first minute of a long distance call and another amount for each additional minute. A customer makes two calls to the same city— a 36-minute call for $2.93 and a 13-minute call for $1.09.

(a) Determine the cost for the first minute and the cost for each additional minute.

(b) If there is a federal tax rate of 3.2% and a state tax rate of 7.2% on all long distance calls, find, to the nearest minute, the longest call to the same city whose cost will not exceed $5.00.

41 VCR taping An avid tennis watcher wants to record 6 hours of a major tournament on a single tape. Her tape can hold 5 hours and 20 minutes at the LP speed and 8 hours at the slower SLP speed. The LP speed produces a better quality picture, so she wishes to maximize the time recorded at the LP speed. Find the amount of time to be recorded at each speed.

42 Price and demand Suppose consumers will buy 1,000,000 T-shirts if the selling price is $10, but for each $1 increase in price, they will buy 100,000 fewer T-shirts. Moreover, suppose vendors will order 2,000,000 T-shirts if the selling price is $15, and for every $1 increase in price, they will order an additional 150,000.

(a) Express the number Q of T-shirts consumers will buy if the selling price is p dollars.

(b) Express the number K of T-shirts vendors will order if the selling price is p dollars.

(c) Determine the market price—that is, the price when $Q = K$.

Exer. 43–46: Solve the system for a and b. (*Hint:* Treat terms such as e^{3x}, $\cos x$, and $\sin x$ as "constant coefficients.")

43 $\begin{cases} ae^{3x} + be^{-3x} = 0 \\ a(3e^{3x}) + b(-3e^{-3x}) = e^{3x} \end{cases}$

44 $\begin{cases} ae^{-x} + be^{4x} = 0 \\ -ae^{-x} + b(4e^{4x}) = 2 \end{cases}$

45 $\begin{cases} a\cos x + b\sin x = 0 \\ -a\sin x + b\cos x = \tan x \end{cases}$

46 $\begin{cases} a\cos x + b\sin x = 0 \\ -a\sin x + b\cos x = \sin x \end{cases}$

9.3

Systems of Inequalities

In Chapter 2 we restricted our discussion of inequalities to inequalities in one variable. We shall now consider inequalities in two variables x and y, such as those shown in the following illustration.

ILLUSTRATION **Inequalities in x and y**

■ $y^2 < x + 4$ ■ $3x - 4y > 12$ ■ $x^2 + y^2 \leq 16$

A **solution** of an inequality in x and y is an ordered pair (a, b) that yields a true statement if a and b are substituted for x and y, respectively. To **solve** an inequality in x and y means to find all the solutions. The **graph** of such an inequality is the set of all points (a, b) in an xy-plane that correspond to the solutions. Two inequalities are **equivalent** if they have the same solutions.

Given an inequality in x and y, if we replace the inequality symbol with an equal sign, we obtain an equation whose graph usually separates the xy-plane into two regions. We shall consider only equations having the property that if R is one such region and if a **test point** (p, q) in R yields a solution of the inequality, then *every* point in R yields a solution. The following guidelines may then be used to sketch the graph of the inequality.

<table>
<tr>
<td>

**Guidelines for
Sketching the Graph of
an Inequality in x and y**

</td>
<td>

1 Replace the inequality symbol with an equal sign, and graph the result-ing equation. Use dashes if the inequality symbol is $<$ or $>$ to indicate that no point on the graph yields a solution. Use a solid line or curve for $\leq$ or $\geq$ to indicate that solutions of the equation are also solutions of the inequality.

2 If R is a region of the xy-plane determined by the graph in guideline 1 and if a test point (p, q) in R yields a solution of the inequality, then every point in R yields a solution. Shade R to indicate this fact. If (p, q) is not a solution, then *no* point in R yields a solution and R is left unshaded.

</td>
</tr>
</table>

The use of these guidelines is demonstrated in the next example.

Figure 1

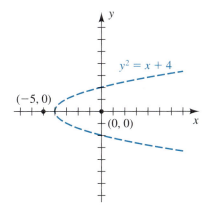

EXAMPLE 1   Sketching the graph of an inequality

Find the solutions and sketch the graph of the inequality $y^2 < x + 4$.

SOLUTION

Guideline 1   We replace $<$ with $=$, obtaining $y^2 = x + 4$. The graph of this equation is a parabola, symmetric with respect to the x-axis and having x-intercept -4 and y-intercepts ± 2. Since the inequality symbol is $<$, we sketch the parabola using dashes, as in Figure 1.

Guideline 2   The graph in guideline 1 separates the xy-plane into two regions, one to the *left* of the parabola and the other to the *right*. Let us choose test points $(-5, 0)$ and $(0, 0)$ in the regions (see Figure 1) and substitute for x and y in $y^2 < x + 4$ as follows:

Figure 2

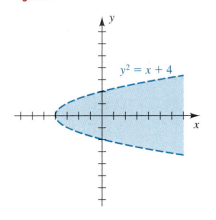

Test point $(-5, 0)$ LS: $0^2 = 0$
$\qquad\qquad\qquad\quad$ RS: $-5 + 4 = -1$

Since $0 < -1$ is a *false* statement, $(-5, 0)$ is *not* a solution of the inequality. Hence, *no* point to the left of the parabola is a solution, and we leave that re-gion unshaded.

Test point $(0, 0)$ LS: $0^2 = 0$
$\qquad\qquad\qquad$ RS: $0 + 4 = 4$

Since $0 < 4$ is a *true* statement, $(0, 0)$ *is* a solution of the inequality. Hence, *all* points to the right of the parabola are solutions, so we shade this region, as in Figure 2.   ◢

A **linear inequality** is an inequality that can be written in one of the fol-lowing forms, where a, b, and c are real numbers:

$$ax + by < c, \quad ax + by > c, \quad ax + by \leq c, \quad ax + by \geq c$$

Figure 3

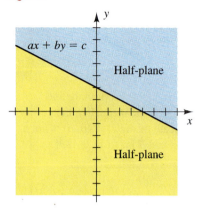

The line $ax + by = c$ separates the xy-plane into two **half-planes,** as illustrated in Figure 3. The solutions of the inequality consist of all points in *one* of these half-planes, where the line is included for $\leq$ or $\geq$ and is not included for $<$ or $>$. *For a linear inequality, only one test point* (p, q) *is required,* because if (p, q) is a solution, then the half-plane with (p, q) in it contains all the solutions, whereas if (p, q) is *not* a solution, then the *other* half-plane contains the solutions.

EXAMPLE 2 **Sketching the graph of a linear inequality**

Sketch the graph of the inequality $3x - 4y > 12$.

SOLUTION Replacing $>$ with $=$ gives us the line $3x - 4y = 12$, sketched with dashes in Figure 4. This line separates the xy-plane into two half-planes, one *above* the line and the other *below* the line. It is convenient to choose the test point $(0, 0)$ above the line and substitute in $3x - 4y > 12$, as follows:

Figure 4

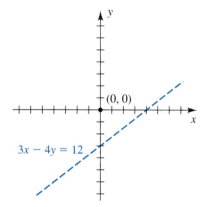

Test point $(0, 0)$ LS: $3 \cdot 0 - 4 \cdot 0 = 0 - 0 = 0$
                              RS: 12

Since $0 > 12$ is a false statement, $(0, 0)$ is not a solution. Thus, no point above the line is a solution, and the solutions of $3x - 4y > 12$ are given by the points in the half-plane *below* the line. The graph is sketched in Figure 5.

Figure 5

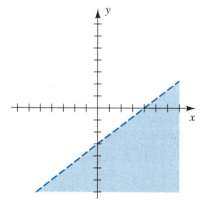

As we did with equations, we sometimes work simultaneously with several inequalities in two variables—that is, with a **system of inequalities.** The **solutions** of a system of inequalities are the solutions common to all inequalities in the system. The **graph** of a system of inequalities consists of the points corresponding to the solutions. The following examples illustrate a method for solving systems of inequalities.

Figure 6

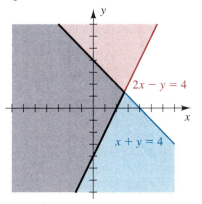

EXAMPLE 3 Solving a system of linear inequalities

Sketch the graph of the system

$$\begin{cases} x + y \le 4 \\ 2x - y \le 4 \end{cases}$$

SOLUTION We replace each $\le$ with $=$ and then sketch the resulting lines, as shown in Figure 6. Using the test point $(0, 0)$, we see that the solutions of the system correspond to the points *below* (and on) the line $x + y = 4$ and *above* (and on) the line $2x - y = 4$. Shading these half-planes with different colors, as in Figure 6, we have as the graph of the system the points that are in *both* regions, indicated by the purple portion of the figure.

EXAMPLE 4 Solving a system of linear inequalities

Sketch the graph of the system

$$\begin{cases} x + y \le 4 \\ 2x - y \le 4 \\ x \ge 0 \\ y \ge 0 \end{cases}$$

Figure 7

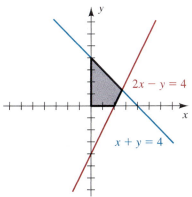

SOLUTION The first two inequalities are the same as those considered in Example 3, and hence the points on the graph of the system must lie within the purple region shown in Figure 6. In addition, the third and fourth inequalities in the system tell us that the points must lie in the first quadrant or on its boundaries. This gives us the region shown in Figure 7.

EXAMPLE 5 Solving a system of inequalities containing absolute values

Sketch the graph of the system

$$\begin{cases} |x| \le 2 \\ |y| > 1 \end{cases}$$

SOLUTION Using properties of absolute values (listed on page 108), we see that (x, y) is a solution of the system if and only if *both* of the following conditions are true:

(1) $-2 \le x \le 2$

(2) $y < -1$ or $y > 1$

Thus, a point (x, y) on the graph of the system must lie between (or on) the vertical lines $x = \pm 2$ and also either below the horizontal line $y = -1$ or above the line $y = 1$. The graph is sketched in Figure 8.

Figure 8

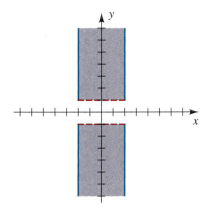

EXAMPLE 6 Solving a system of inequalities

Sketch the graph of the system

$$\begin{cases} x^2 + y^2 \le 16 \\ x + y \ge 2 \end{cases}$$

Figure 9

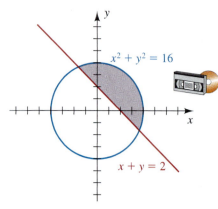

SOLUTION   The graphs of $x^2 + y^2 = 16$ and $x + y = 2$ are the circle and line, respectively, shown in Figure 9. Using the test point $(0, 0)$, we see that the points that yield solutions of the system must lie inside (or on) the circle and also above (or on) the line. This gives us the region sketched in Figure 9.

EXAMPLE 7   Finding a system of inequalities from a graph

Find a system of inequalities for the shaded region shown in Figure 10.

SOLUTION   An equation of the circle is $x^2 + y^2 = 5^2$. Since the *interior* of the solid circle is shaded, the shaded region (including the circle) can be described by $x^2 + y^2 \leq 25$. The *exterior* of the circle could be described by $x^2 + y^2 > 25$.

Because the shaded region is *below* the dashed line with equation $y = \frac{3}{4}x$, it is described by the inequality $y < \frac{3}{4}x$. Lastly, since the shaded region is *above* the solid horizontal line $y = -3$, we use $y \geq -3$. Thus, a system is

$$\begin{cases} x^2 + y^2 \leq 25 \\ \quad\;\; y < \frac{3}{4}x \\ \quad\;\; y \geq -3 \end{cases}$$

Figure 10

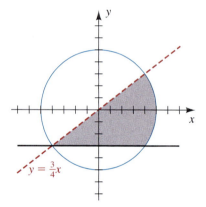

EXAMPLE 8   An application of a system of inequalities

The manager of a baseball team wishes to buy bats and balls costing $12 and $3 each, respectively. At least five bats and ten balls are required, and the total cost is not to exceed $180. Find a system of inequalities that describes all possibilities, and sketch the graph.

SOLUTION   We begin by letting x denote the number of bats and y the number of balls. Since the cost of a bat is $12 and the cost of a ball is $3, we see that

$$12x = \text{cost of } x \text{ bats}$$
$$3y = \text{cost of } y \text{ balls.}$$

Since the total cost is not to exceed $180, we must have

$$12x + 3y \leq 180$$

or, equivalently,

$$y \leq -4x + 60.$$

Since at least five bats and ten balls are required, we also have

$$x \geq 5 \quad \text{and} \quad y \geq 10.$$

The graph of $y \leq -4x + 60$ is the half-plane that lies *below* (or on) the line $y = -4x + 60$ shown in Figure 11.

Figure 11

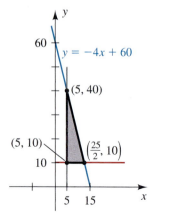

The graph of $x \geq 5$ is the region to the right of (or on) the vertical line $x = 5$, and the graph of $y \geq 10$ is the region above (or on) the horizontal line $y = 10$.

The graph of the system—that is, the points common to the three half-planes—is the triangular region sketched in Figure 11.

9.3 Exercises

Exer. 1–10: Sketch the graph of the inequality.

1 $3x - 2y < 6$

2 $4x + 3y < 12$

3 $2x + 3y \geq 2y + 1$

4 $2x - y > 3$

5 $y + 2 < x^2$

6 $y^2 - x \leq 0$

7 $x^2 + 1 \leq y$

8 $y - x^3 < 1$

9 $yx^2 \geq 1$

10 $x^2 + 4 \geq y$

Exer. 11–26: Sketch the graph of the system of inequalities.

11 $\begin{cases} 3x + y < 3 \\ 4 - y < 2x \end{cases}$

12 $\begin{cases} y + 2 < 2x \\ y - x > 4 \end{cases}$

13 $\begin{cases} y - x < 0 \\ 2x + 5y < 10 \end{cases}$

14 $\begin{cases} 2y - x \leq 4 \\ 3y + 2x < 6 \end{cases}$

15 $\begin{cases} 3x + y \leq 6 \\ y - 2x \geq 1 \\ x \geq -2 \\ y \leq 4 \end{cases}$

16 $\begin{cases} 3x - 4y \geq 12 \\ x - 2y \leq 2 \\ x \geq 9 \\ y \leq 5 \end{cases}$

17 $\begin{cases} x + 2y \leq 8 \\ 0 \leq x \leq 4 \\ 0 \leq y \leq 3 \end{cases}$

18 $\begin{cases} 2x + 3y \geq 6 \\ 0 \leq x \leq 5 \\ 0 \leq y \leq 4 \end{cases}$

19 $\begin{cases} |x| \geq 2 \\ |y| < 3 \end{cases}$

20 $\begin{cases} |x| \geq 4 \\ |y| \geq 3 \end{cases}$

21 $\begin{cases} |x + 2| \leq 1 \\ |y - 3| < 5 \end{cases}$

22 $\begin{cases} |x - 2| \leq 5 \\ |y - 4| > 2 \end{cases}$

23 $\begin{cases} x^2 + y^2 \leq 4 \\ x + y \geq 1 \end{cases}$

24 $\begin{cases} x^2 + y^2 > 1 \\ x^2 + y^2 < 4 \end{cases}$

25 $\begin{cases} x^2 \leq 1 - y \\ x \geq 1 + y \end{cases}$

26 $\begin{cases} x - y^2 < 0 \\ x + y^2 > 0 \end{cases}$

Exer. 27–34: Find a system of inequalities whose graph is shown.

27

28

29

32

30

33

31

34

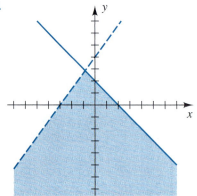

35 Inventory levels A store sells two brands of television sets. Customer demand indicates that it is necessary to stock at least twice as many sets of brand A as of brand B. It is also necessary to have on hand at least 10 sets of brand B. There is room for not more than 100 sets in the store. Find and graph a system of inequalities that describes all possibilities for stocking the two brands.

36 Ticket prices An auditorium contains 600 seats. For an upcoming event, tickets will be priced at $8 for some seats and $5 for others. At least 225 tickets are to be priced at $5, and total sales of at least $3000 are desired. Find and graph a system of inequalities that describes all possibilities for pricing the two types of tickets.

37 Investment strategy A woman with $15,000 to invest decides to place at least $2000 in a high-risk, high-yield investment and at least three times that amount in a low-risk, low-yield investment. Find and graph a system of inequalities that describes all possibilities for placing the money in the two investments.

38 Inventory levels The manager of a college bookstore stocks two types of notebooks, the first wholesaling for 55¢ and the second for 85¢. The maximum amount to be spent is $600, and an inventory of at least 300 of the 85¢ variety and 400 of the 55¢ variety is desired. Find and graph a system of inequalities that describes all possibilities for stocking the two types of notebooks.

39 Dimensions of a can An aerosol can is to be constructed in the shape of a circular cylinder with a small cone on the top. The total height of the can including the conical top is to be no more than 9 inches, and the cylinder must contain at least 75% of the total volume. In addition, the height of the conical top must be at least 1 inch. Find and graph a system of inequalities that describes all possibilities for the relationship between the height y of the cylinder and the height x of the cone.

40 Dimensions of a window A stained-glass window is to be constructed in the form of a rectangle surmounted by a semicircle (see the figure). The total height h of the window can be no more than 6 feet, and the area of the rectangular part must be at least twice the area of the semicircle. In addition, the diameter d of the semicircle must be at least 2 feet. Find and graph a system of inequalities that describes all possibilities for the base and height of the rectangular part.

Exercise 40

41 Locating a power plant A nuclear power plant will be constructed to serve the power needs of cities A and B. City B is 100 miles due east of A. The state has promised that the plant will be at least 60 miles from each city. It is not possible, however, to locate the plant south of either city because of rough terrain, and the plant must be within 100 miles of both A and B. Assuming A is at the origin, find and graph a system of inequalities that describes all possible locations for the plant.

42 Allocating space A man has a rectangular back yard that is 50 feet wide and 60 feet deep. He plans to construct a pool area and a patio area, as shown in the figure, where $y \geq 10$. He can spend at most $10,500 on the project. The patio area must be at least as large as the pool area. The pool area will cost $5 per square foot, and the patio will cost $3 per square foot. Find and graph a system of inequalities that describes all possibilities for the width of the patio and pool areas.

Exercise 42

43 Forest growth Temperature and precipitation have a significant effect on plant life. If either the average annual temperature or the amount of precipitation is too low, trees and forests cannot grow. Instead, only grasslands and deserts will exist. The relationship between average annual temperature T (in °F) and average annual precipitation P (in inches) is a linear inequality. In order for forests to grow in a region, T and P must satisfy the inequality $29T - 39P < 450$, where $33 \leq T \leq 80$ and $13 \leq P \leq 45$.

(a) Determine whether forests can grow in Winnipeg, where $T = 37°F$ and $P = 21.2$ in.

(b) Graph the inequality, with T on the horizontal axis and P on the vertical axis.

(c) Identify the region on the graph that represents where forests can grow.

44 Grassland growth Refer to Exercise 43. If the average annual precipitation P (in inches) is too low or the average annual temperature T (in °F) is too high, forests and grasslands become deserts. The conditions necessary for grasslands to grow are given by a linear inequality. T and P must satisfy $22P - 3T > 33$, where $33 \leq T \leq 80$ and $13 \leq P \leq 45$.

(a) Determine whether grasslands can grow in Phoenix, where $T = 70°F$ and $P = 7.8$ in.

(b) Graph the inequality for forests and the inequality for grasslands on the same coordinate axes.

(c) Determine the region on the graph that represents where grasslands can exist but forests cannot.

9.4

Linear Programming

Figure 1

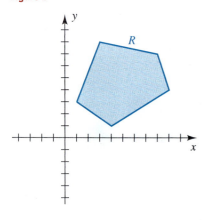

If a system of inequalities contains only linear inequalities of the form

$$ax + by \leq c \quad \text{or} \quad ax + by \geq c,$$

where a, b, and c are real numbers, then the graph of the system may be a region R in the xy-plane bounded by a polygon—possibly of the type illustrated in Figure 1 (for a specific illustration, see Example 4 and Figure 7 of Section 9.3). For problems in **linear programming,** we consider such systems together with an expression of the form

$$C = Ax + By + K,$$

where A, B, and K are real numbers and (x, y) is a point in R (that is, a solution of the system). Since for each (x, y) we obtain a specific value for C, we call C a *function of two variables x and y*. In linear programming, C is called an **objective function,** and the inequalities in the system are referred to as the **constraints** on C. The solutions of the system—that is, the pairs (x, y) corresponding to the points in R—are called the **feasible solutions** for the problem.

In typical business applications, the value of C may represent cost, profit, loss, or a physical resource, and the goal is to find a specific point (x, y) in R at which C takes on its maximum or minimum value. The methods of linear programming greatly simplify the task of finding this point. Specifically, it can be shown that *the maximum and minimum values of C occur at a vertex of R.* This fact is used in the next example.

Figure 2

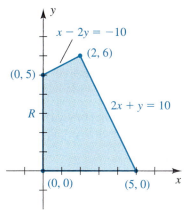

EXAMPLE 1   Finding the maximum and minimum values of an objective function

Find the maximum and minimum values of the objective function given by $C = 7x + 3y$ subject to the following constraints:

$$\begin{cases} x - 2y \geq -10 \\ 2x + y \leq 10 \\ x \geq 0 \\ y \geq 0 \end{cases}$$

SOLUTION   The graph of the system of inequalities determined by the constraints is the region R bounded by the quadrilateral sketched in Figure 2. From the preceding discussion, the maximum and minimum values of C must occur at a vertex of R. The values at the vertices are given in the following table.

Vertex	Value of $C = 7x + 3y$
$(0, 0)$	$7(0) + 3(0) = 0$
$(0, 5)$	$7(0) + 3(5) = 15$
$(5, 0)$	$7(5) + 3(0) = 35$
$(2, 6)$	$7(2) + 3(6) = 32$

Hence, the minimum value $C = 0$ occurs if $x = 0$ and $y = 0$. The maximum value $C = 35$ occurs if $x = 5$ and $y = 0$.

Figure 3

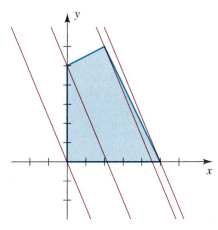

In the preceding example, we say that the maximum value of C on R occurs at the vertex $(5, 0)$. To verify this fact, let us solve $C = 7x + 3y$ for y, obtaining

$$y = -\frac{7}{3}x + \frac{C}{3}.$$

For each C, the graph of this equation is a line of slope $-\frac{7}{3}$ and y-intercept $C/3$, as illustrated in Figure 3. To find the maximum value of C, we simply determine which of these lines that intersect the region has the largest y-intercept $C/3$. Referring to Figure 3, we see that the required line passes through $(5, 0)$. Similarly, for the minimum value of C, we determine the line having equation $y = (-7/3)x + (C/3)$ that intersects the region and has the *smallest* y-intercept. This is the line through $(0, 0)$.

We shall call a problem that can be expressed in the form of Example 1 a **linear programming problem.** To solve such problems, we may use the following guidelines.

Guidelines for Solving a Linear Programming Problem	1 Sketch the region R determined by the system of constraints.
	2 Find the vertices of R.
	3 Calculate the value of the objective function C at each vertex of R.
	4 Select the maximum or minimum value(s) of C in guideline 3.

In the next example we encounter a linear programming problem in which the minimum value of the objective function occurs at more than one point.

Figure 4

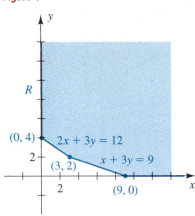

EXAMPLE 2   Solving a linear programming problem

Find the minimum value of the objective function $C = 2x + 6y$ subject to the following constraints:

$$\begin{cases} 2x + 3y \geq 12 \\ x + 3y \geq 9 \\ x \geq 0 \\ y \geq 0 \end{cases}$$

SOLUTION   We shall follow the guidelines.

Guideline 1   The graph of the system of inequalities determined by the constraints is the unbounded region R sketched in Figure 4.

Guideline 2   The vertices of R are $(0, 4)$, $(3, 2)$, and $(9, 0)$, as shown in the figure.

Guideline 3   The value of C at each vertex of R is given in the following table.

Vertex	Value of $C = 2x + 6y$
$(0, 4)$	$2(0) + 6(4) = 24$
$(3, 2)$	$2(3) + 6(2) = 18$
$(9, 0)$	$2(9) + 6(0) = 18$

Guideline 4   The table in guideline 3 shows that the minimum value of C, 18, occurs at *two* vertices, $(3, 2)$ and $(9, 0)$. Moreover, if (x, y) is any point on the line segment joining these points, then (x, y) is a solution of the equation $x + 3y = 9$, and hence

$$C = 2x + 6y = 2(x + 3y) = 2(9) = 18.$$

Thus, the minimum value $C = 18$ occurs at *every* point on this line segment.

In the next two examples we consider applications of linear programming. For such problems it is necessary to use given information and data to formulate the system of constraints and the objective function. Once this has been accomplished, we may apply the guidelines as we did in the solution to Example 2.

EXAMPLE 3 **Maximizing profit**

A firm manufactures two products, X and Y. For each product, it is necessary to use three different machines, A, B, and C. To manufacture one unit of product X, machine A must be used for 3 hours, machine B for 1 hour, and machine C for 1 hour. To manufacture one unit of product Y requires 2 hours on A, 2 hours on B, and 1 hour on C. The profit on product X is $500 per unit, and the profit on product Y is $350 per unit. Machine A is available for a total of 24 hours per day; however, B can be used for only 16 hours and C for 9 hours. Assuming the machines are available when needed (subject to the noted total hour restrictions), determine the number of units of each product that should be manufactured each day in order to maximize the profit.

SOLUTION The following table summarizes the data given in the statement of the problem.

Machine	Hours required for one unit of X	Hours required for one unit of Y	Hours available
A	3	2	24
B	1	2	16
C	1	1	9

Let us introduce the following variables:

$$x = \text{number of units of X manufactured each day}$$
$$y = \text{number of units of Y manufactured each day}$$

Using the first row of the table, we note that each unit of X requires 3 hours on machine A, and hence x units require $3x$ hours. Similarly, since each unit of Y requires 2 hours on A, y units require $2y$ hours. Hence, the total number of hours per day that machine A must be used is $3x + 2y$. This, together with the fact that A can be used for at most 24 hours per day, gives us the first constraint in the following system of inequalities—that is, $3x + 2y \leq 24$. The second and third constraints are obtained by using the same type of reasoning for rows 2 and 3 of the table. The last two constraints, $x \geq 0$ and $y \geq 0$, are true because x and y cannot be negative.

$$\begin{cases} 3x + 2y \leq 24 \\ x + 2y \leq 16 \\ x + y \leq 9 \\ x \geq 0 \\ y \geq 0 \end{cases}$$

Figure 5

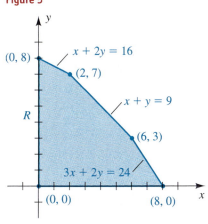

The graph of this system is the region R in Figure 5.

(continued)

Since the production of each unit of product X yields a profit of $500 and each unit of product Y yields a profit of $350, the profit P obtained by producing x units of X together with y units of Y is

$$P = 500x + 350y.$$

This is the objective function for the problem. The maximum value of P must occur at one of the vertices of R in Figure 5. The values of P at these vertices are given in the following table.

Vertex	Value of $P = 500x + 350y$
(0, 0)	$500(0) + 350(0) = 0$
(0, 8)	$500(0) + 350(8) = 2800$
(8, 0)	$500(8) + 350(0) = 4000$
(2, 7)	$500(2) + 350(7) = 3450$
(6, 3)	$500(6) + 350(3) = 4050$

We see from the table that a maximum profit of $4050 occurs for a daily production of 6 units of product X and 3 units of product Y.

Example 3 illustrates maximization of profit. The next example demonstrates how linear programming can be used to minimize the cost in a certain situation.

EXAMPLE 4   **Minimizing cost**

A distributor of compact disk players has two warehouses, W_1 and W_2. There are 80 units stored at W_1 and 70 units at W_2. Two customers, A and B, order 35 units and 60 units, respectively. The shipping cost from each warehouse to A and B is determined according to the following table. How should the order be filled to minimize the total shipping cost?

Warehouse	Customer	Shipping cost per unit
W_1	A	$ 8
W_1	B	12
W_2	A	10
W_2	B	13

SOLUTION   Let us begin by introducing the following variables:

$$x = \text{number of units sent to A from } W_1$$
$$y = \text{number of units sent to B from } W_1$$

Since A ordered 35 units and B ordered 60 units, we must have

$$35 - x = \text{number of units sent to A from } W_2$$
$$60 - y = \text{number of units sent to B from } W_2.$$

Our goal is to determine values for x and y that make the total shipping cost minimal.

The number of units shipped from W_1 cannot exceed 80, and the number shipped from W_2 cannot exceed 70. Expressing these facts in terms of inequalities gives us

$$\begin{cases} x + y \le 80 \\ (35 - x) + (60 - y) \le 70 \end{cases}$$

Simplifying, we obtain the first two constraints in the following system. The last two constraints are true because the largest values of x and y are 35 and 60, respectively.

$$\begin{cases} x + y \le 80 \\ x + y \ge 25 \\ 0 \le x \le 35 \\ 0 \le y \le 60 \end{cases}$$

The graph of this system is the region R shown in Figure 6.

Let C denote the total cost (in dollars) of shipping the disk players to customers A and B. We see from the table of shipping costs that the following are true:

$$\text{cost of shipping 35 units to A} = 8x + 10(35 - x)$$
$$\text{cost of shipping 60 units to B} = 12y + 13(60 - y)$$

Hence, the total cost is

$$C = 8x + 10(35 - x) + 12y + 13(60 - y).$$

Simplifying gives us the following objective function:

$$C = 1130 - 2x - y$$

To determine the minimum value of C on R, we need check only the vertices shown in Figure 6, as in the following table.

Figure 6

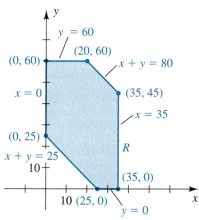

Vertex	Value of $C = 1130 - 2x - y$
$(0, 25)$	$1130 - 2(0) - 25 = 1105$
$(0, 60)$	$1130 - 2(0) - 60 = 1070$
$(20, 60)$	$1130 - 2(20) - 60 = 1030$
$(35, 45)$	$1130 - 2(35) - 45 = 1015$
$(35, 0)$	$1130 - 2(35) - 0 = 1060$
$(25, 0)$	$1130 - 2(25) - 0 = 1080$

(continued)

We see from the table that the minimal shipping cost, $1015, occurs if $x = 35$ and $y = 45$. This means that the distributor should ship all of the disk players to A from W_1 and none from W_2. In addition, the distributor should ship 45 units to B from W_1 and 15 units to B from W_2. (Note that the *maximum* shipping cost will occur if $x = 0$ and $y = 25$—that is, if all 35 units are shipped to A from W_2 and if B receives 25 units from W_1 and 35 units from W_2.) ◢

The examples in this section are elementary linear programming problems in two variables that can be solved by basic methods. The much more complicated problems in many variables that occur in practice may be solved by employing matrix techniques (discussed later) that are adapted for solution by computers.

9.4 Exercises

Exer. 1–2: Find the maximum and minimum values of the objective function C on the region in the figure.

1 $C = 3x + 2y + 5$

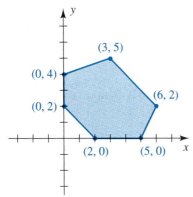

2 $C = 2x + 7y + 3$

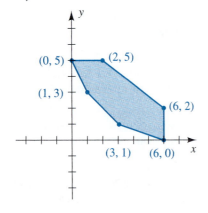

Exer. 3–4: Sketch the region R determined by the given constraints, and label its vertices. Find the maximum value of C on R.

3 $C = 3x + y$; $x \ge 0, y \ge 0,$
$3x - 4y \ge -12,$ $3x + 2y \le 24,$ $3x - y \le 15$

4 $C = 4x - 2y$;
$x - 2y \ge -8,$ $7x - 2y \le 28,$ $x + y \ge 4$

Exer. 5–6: Sketch the region R determined by the given constraints, and label its vertices. Find the minimum value of C on R.

5 $C = 3x + 6y$; $x \ge 0, y \ge 0,$
$2x + 3y \ge 12,$ $2x + 5y \ge 16$

6 $C = 6x + y$; $y \ge 0,$
$3x + y \ge 3,$ $x + 5y \le 15,$ $2x + y \le 12$

Exer. 7–8: Sketch the region R determined by the given constraints, and label its vertices. Describe the set of points for which C is a maximum on R.

7 $C = 2x + 4y$; $x \ge 0, y \ge 0,$
$x - 2y \ge -8,$ $\frac{1}{2}x + y \le 6,$ $3x + 2y \le 24$

8 $C = 6x + 3y$; $x \ge 2, y \ge 1,$
$2x + 3y \le 19,$ $x + 0.5y \le 6.5$

9 Production scheduling A manufacturer of tennis rackets makes a profit of $15 on each oversized racket and $8 on each standard racket. To meet dealer demand, daily production of standard rackets should be between 30 and 80, and production of oversized rackets should be between 10 and 30. To maintain high quality, the total number of rackets produced should not exceed 80 per day. How many of each type should be manufactured daily to maximize the profit?

10 **Production scheduling** A manufacturer of CB radios makes a profit of $25 on a deluxe model and $30 on a standard model. The company wishes to produce at least 80 deluxe models and at least 100 standard models per day. To maintain high quality, the daily production should not exceed 200 radios. How many of each type should be produced daily in order to maximize the profit?

11 **Minimizing cost** Two substances, S and T, each contain two types of ingredients, I and G. One pound of S contains 2 ounces of I and 4 ounces of G. One pound of T contains 2 ounces of I and 6 ounces of G. A manufacturer plans to combine quantities of the two substances to obtain a mixture that contains at least 9 ounces of I and 20 ounces of G. If the cost of S is $3 per pound and the cost of T is $4 per pound, how much of each substance should be used to keep the cost to a minimum?

12 **Maximizing gross profit** A stationery company makes two types of notebooks: a deluxe notebook with subject dividers, which sells for $1.25, and a regular notebook, which sells for $0.90. The production cost is $1.00 for each deluxe notebook and $0.75 for each regular notebook. The company has the facilities to manufacture between 2000 and 3000 deluxe and between 3000 and 6000 regular notebooks, but not more than 7000 altogether. How many notebooks of each type should be manufactured to maximize the difference between the selling prices and the production costs?

13 **Minimizing shipping costs** Refer to Example 4 of this section. If the shipping costs are $12 per unit from W_1 to A, $10 per unit from W_2 to A, $16 per unit from W_1 to B, and $12 per unit from W_2 to B, determine how the order should be filled to minimize shipping cost.

14 **Minimizing cost** A coffee company purchases mixed lots of coffee beans and then grades them into premium, regular, and unusable beans. The company needs at least 280 tons of premium-grade and 200 tons of regular-grade coffee beans. The company can purchase ungraded coffee from two suppliers in any amount desired. Samples from the two suppliers contain the following percentages of premium, regular, and unusable beans:

Supplier	Premium	Regular	Unusable
A	20%	50%	30%
B	40%	20%	40%

If supplier A charges $125 per ton and B charges $200 per ton, how much should the company purchase from each supplier to fulfill its needs at minimum cost?

15 **Planning crop acreage** A farmer, in the business of growing fodder for livestock, has 90 acres available for planting alfalfa and corn. The cost of seed per acre is $4 for alfalfa and $6 for corn. The total cost of labor will amount to $20 per acre for alfalfa and $10 per acre for corn. The expected income is $110 per acre from alfalfa and $150 per acre from corn. If the farmer does not wish to spend more than $480 for seed and $1400 for labor, how many acres of each crop should be planted to obtain the maximum profit?

16 **Machinery scheduling** A small firm manufactures bookshelves and desks for microcomputers. For each product it is necessary to use a table saw and a power router. To manufacture each bookshelf, the saw must be used for $\frac{1}{2}$ hour and the router for 1 hour. A desk requires the use of each machine for 2 hours. The profits are $20 per bookshelf and $50 per desk. If the saw can be used for 8 hours per day and the router for 12 hours per day, how many bookshelves and desks should be manufactured each day to maximize the profit?

17 **Minimizing a mixture's cost** Three substances, X, Y, and Z, each contain four ingredients, A, B, C, and D. The percentage of each ingredient and the cost in cents per ounce of each substance are given in the following table.

Substance	Ingredients				Cost per ounce
	A	B	C	D	
X	20%	10%	25%	45%	25¢
Y	20%	40%	15%	25%	35¢
Z	10%	20%	25%	45%	50¢

If the cost is to be minimal, how many ounces of each substance should be combined to obtain a mixture of 20 ounces containing at least 14% A, 16% B, and 20% C? What combination would make the cost greatest?

18 **Maximizing profit** A man plans to operate a stand at a one-day fair at which he will sell bags of peanuts and bags of candy. He has $400 available to purchase his stock, which will cost 40¢ per bag of peanuts and 80¢ per bag of candy. He intends to sell the peanuts at $1.00 and the

candy at $1.60 per bag. His stand can accommodate up to 500 bags of peanuts and 400 bags of candy. From past experience he knows that he will sell no more than a total of 700 bags. Find the number of bags of each that he should have available in order to maximize his profit. What is the maximum profit?

19 **Maximizing passenger capacity** A small community wishes to purchase used vans and small buses for its public transportation system. The community can spend no more than $100,000 for the vehicles and no more than $500 per month for maintenance. The vans sell for $10,000 each and average $100 per month in maintenance costs. The corresponding cost estimates for each bus are $20,000 and $75 per month. If each van can carry 15 passengers and each bus can accommodate 25 riders, determine the number of vans and buses that should be purchased to maximize the passenger capacity of the system.

20 **Minimizing fuel cost** Refer to Exercise 19. The monthly fuel cost (based on 5000 miles of service) is $550 for each van and $850 for each bus. Find the number of vans and buses that should be purchased to minimize the monthly fuel costs if the passenger capacity of the system must be at least 75.

21 **Stocking a fish farm** A fish farmer will purchase no more than 5000 young trout and bass from the hatchery and will feed them a special diet for the next year. The cost of food per fish will be $0.50 for trout and $0.75 for bass, and the total cost is not to exceed $3000. At the end of the year, a typical trout will weigh 3 pounds, and a bass will weigh 4 pounds. How many fish of each type should be stocked in the pond in order to maximize the total number of pounds of fish at the end of the year?

22 **Dietary planning** A hospital dietician wishes to prepare a corn-squash vegetable dish that will provide at least 3 grams of protein and cost no more than 36¢ per serving. An ounce of creamed corn provides $\frac{1}{2}$ gram of protein and costs 4¢. An ounce of squash supplies $\frac{1}{4}$ gram of protein and costs 3¢. For taste, there must be at least 2 ounces of corn and at least as much squash as corn. It is important to keep the total number of ounces in a serving as small as possible. Find the combination of corn and squash that will minimize the amount of ingredients used per serving.

23 **Planning storage units** A contractor has a large building that she wishes to convert into a series of rental storage spaces. She will construct basic 8 ft × 10 ft units and deluxe 12 ft × 10 ft units that contain extra shelves and a clothes closet. Market considerations dictate that there be at least twice as many basic units as deluxe units and that the basic units rent for $40 per month and the deluxe units for $75 per month. At most 7200 ft² is available for the storage spaces, and no more than $30,000 can be spent on construction. If each basic unit will cost $300 to make and each deluxe unit will cost $600, how many units of each type should be constructed to maximize monthly revenue?

24 **A moose's diet** A moose feeding primarily on tree leaves and aquatic plants is capable of digesting no more than 33 kilograms of these foods daily. Although the aquatic plants are lower in energy content, the animal must eat at least 17 kilograms to satisfy its sodium requirement. A kilogram of leaves provides four times as much energy as a kilogram of aquatic plants. Find the combination of foods that maximizes the daily energy intake.

9.5

Systems of Linear Equations in More Than Two Variables

For systems of linear equations containing more than two variables, we can use either the method of substitution explained in Section 9.1 or the method of elimination developed in Section 9.2. The method of elimination is the shorter and more straightforward technique for finding solutions. In addition, it leads to the matrix technique, discussed in this section.

EXAMPLE 1 **Using the method of elimination to solve a system of linear equations**

Solve the system

$$\begin{cases} x - 2y + 3z = 4 \\ 2x + y - 4z = 3 \\ -3x + 4y - z = -2 \end{cases}$$

SOLUTION

$$\begin{cases} x - 2y + 3z = 4 \\ \quad\quad 5y - 10z = -5 \\ -3x + 4y - z = -2 \end{cases}$$

add -2 times the first equation to the second equation

$$\begin{cases} x - 2y + 3z = 4 \\ \quad\quad 5y - 10z = -5 \\ \quad -2y + 8z = 10 \end{cases}$$

add 3 times the first equation to the third equation

$$\begin{cases} x - 2y + 3z = 4 \\ \quad\quad y - 2z = -1 \\ \quad\quad y - 4z = -5 \end{cases}$$

multiply the second equation by $\frac{1}{5}$ and the third equation by $-\frac{1}{2}$

$$\begin{cases} x - 2y + 3z = 4 \\ \quad\quad y - 2z = -1 \\ \quad\quad -2z = -4 \end{cases}$$

add -1 times the second equation to the third equation

$$\begin{cases} x - 2y + 3z = 4 \\ \quad\quad y - 2z = -1 \\ \quad\quad\quad z = 2 \end{cases}$$

multiply the third equation by $-\frac{1}{2}$

The solutions of the last system are easy to find by **back substitution.** From the third equation, we see that $z = 2$. Substituting 2 for z in the second equation, $y - 2z = -1$, we get $y = 3$. Finally, we find the x-value by substituting $y = 3$ and $z = 2$ in the first equation, $x - 2y + 3z = 4$, obtaining $x = 4$. Thus, there is one solution, $(4, 3, 2)$. ◢

Any system of three linear equations in three variables has either a *unique solution,* an *infinite number of solutions,* or *no solution.* As for two equations in two variables, the terminology used to describe these systems is *consistent, dependent and consistent,* or *inconsistent,* respectively.

If we analyze the method of solution in Example 1, we see that the symbols used for the variables are immaterial. The *coefficients* of the variables are what we must consider. Thus, if different symbols such as r, s, and t are used for the variables, we obtain the system

$$\begin{cases} r - 2s + 3t = 4 \\ 2r + s - 4t = 3 \\ -3r + 4s - t = -2 \end{cases}$$

The method of elimination can then proceed exactly as in the example. Since this is true, it is possible to simplify the process. Specifically, we introduce a scheme for keeping track of the coefficients in such a way that we do not have to write down the variables. Referring to the preceding system, we first check that variables appear in the same order in each equation and that terms not involving variables are to the right of the equal signs. We then list the numbers that are involved in the equations as follows:

$$\begin{bmatrix} 1 & -2 & 3 & 4 \\ 2 & 1 & -4 & 3 \\ -3 & 4 & -1 & -2 \end{bmatrix}$$

An array of numbers of this type is called a **matrix.** The **rows** of the matrix are the numbers that appear next to each other *horizontally:*

$$\begin{array}{rrrrl} 1 & -2 & 3 & 4 & \text{first row, } R_1 \\ 2 & 1 & -4 & 3 & \text{second row, } R_2 \\ -3 & 4 & -1 & -2 & \text{third row, } R_3 \end{array}$$

The **columns** of the matrix are the numbers that appear next to each other *vertically:*

first column, C_1	second column, C_2	third column, C_3	fourth column, C_4
1	-2	3	4
2	1	-4	3
-3	4	-1	-2

The matrix obtained from a system of linear equations in the preceding manner is the **matrix of the system.** If we delete the last column of this matrix, the remaining array of numbers is the **coefficient matrix.** Since the matrix of the system can be obtained from the coefficient matrix by adjoining one column, we call it the **augmented coefficient matrix** or simply the **augmented matrix.** Later, when we use matrices to find the solutions of a system of linear equations, we shall introduce a vertical line segment in the augmented matrix to indicate where the equal signs would appear in the corresponding system of equations, as in the next illustration.

ILLUSTRATION **Coefficient Matrix and Augmented Matrix**

	system	coefficient matrix	augmented matrix

$$\begin{cases} x - 2y + 3z = 4 \\ 2x + y - 4z = 3 \\ -3x + 4y - z = -2 \end{cases} \qquad \begin{bmatrix} 1 & -2 & 3 \\ 2 & 1 & -4 \\ -3 & 4 & -1 \end{bmatrix} \qquad \left[\begin{array}{rrr|r} 1 & -2 & 3 & 4 \\ 2 & 1 & -4 & 3 \\ -3 & 4 & -1 & -2 \end{array}\right]$$

Before discussing a matrix method of solving a system of linear equations, let us state a general definition of a matrix. We shall use a **double subscript notation,** denoting the number that appears in row i and column j by a_{ij}. The **row subscript** of a_{ij} is i, and the **column subscript** is j.

| **Definition of a Matrix** | Let m and n be positive integers. An **$m \times n$ matrix** is an array of the following form, where each a_{ij} is a real number: |

$$\begin{bmatrix} a_{11} & a_{12} & a_{13} & \cdots & a_{1n} \\ a_{21} & a_{22} & a_{23} & \cdots & a_{2n} \\ a_{31} & a_{32} & a_{33} & \cdots & a_{3n} \\ \cdot & \cdot & \cdot & & \cdot \\ \cdot & \cdot & \cdot & & \cdot \\ \cdot & \cdot & \cdot & & \cdot \\ a_{m1} & a_{m2} & a_{m3} & \cdots & a_{mn} \end{bmatrix}$$

The notation $m \times n$ in the definition is read "m by n." We often say that the matrix *is $m \times n$* and call $m \times n$ the **size** of the matrix. It is possible to consider matrices in which the symbols a_{ij} represent complex numbers, polynomials, or other mathematical objects. The rows and columns of a matrix are defined as before. Thus, the matrix in the definition has m rows and n columns. Note that a_{23} is in row 2 and column 3 and a_{32} is in row 3 and column 2. Each a_{ij} is an **element of the matrix.** If $m = n$, the matrix is a **square matrix of order n** and the elements $a_{11}, a_{22}, a_{33}, \ldots, a_{nn}$ are the **main diagonal elements.**

ILLUSTRATION $m \times n$ Matrices

$$2 \times 3 \qquad 2 \times 2 \qquad 1 \times 3 \qquad 3 \times 2 \qquad 3 \times 1$$

$$\blacksquare \begin{bmatrix} -5 & 3 & 1 \\ 7 & 0 & -2 \end{bmatrix} \quad \blacksquare \begin{bmatrix} 5 & -1 \\ 2 & 3 \end{bmatrix} \quad \blacksquare \begin{bmatrix} 3 & 1 & -2 \end{bmatrix} \quad \blacksquare \begin{bmatrix} 2 & -1 \\ 0 & 1 \\ 8 & 3 \end{bmatrix} \quad \blacksquare \begin{bmatrix} -4 \\ 0 \\ 5 \end{bmatrix}$$

To find the solutions of a system of linear equations, we begin with the augmented matrix. If a variable does not appear in an equation, we assume that the coefficient is zero. We then work with the rows of the matrix *just as though they were equations.* The only items missing are the symbols for the variables, the addition or subtraction signs used between terms, and the equal signs. We simply keep in mind that the numbers in the first column are the coefficients of the first variable, the numbers in the second column are the coefficients of the second variable, and so on. The rules for transforming a matrix are formulated so that they always produce a matrix of an equivalent system of equations.

The next theorem is a restatement, in terms of matrices, of the theorem on equivalent systems in Section 9.2. In part (2) of the theorem, the terminology *a row is multiplied by a nonzero constant* means that each element in the row is multiplied by the constant. To *add* two rows of a matrix, as in part (3), we add corresponding elements in each row.

Theorem on Matrix Row Transformations	Given a matrix of a system of linear equations, a matrix of an equivalent system results if **(1)** two rows are interchanged. **(2)** a row is multiplied or divided by a nonzero constant. **(3)** a constant multiple of one row is added to another row.

We refer to 1–3 as the **elementary row transformations** of a matrix. If a matrix is obtained from another matrix by one or more elementary row transformations, the two matrices are said to be **equivalent** or, more precisely, **row equivalent.** We shall use the symbols in the following chart to denote elementary row transformations of a matrix, where the arrow $\rightarrow$ may be read "replaces." Thus, for the transformation $k\mathrm{R}_i \rightarrow \mathrm{R}_i$, the constant multiple $k\mathrm{R}_i$ *replaces* R_i. Similarly, for $k\mathrm{R}_i + \mathrm{R}_j \rightarrow \mathrm{R}_j$, the sum $k\mathrm{R}_i + \mathrm{R}_j$ *replaces* R_j. For convenience, we shall write $(-1)\mathrm{R}_i$ as $-\mathrm{R}_i$.

Elementary Row Transformations of a Matrix

Symbol	Meaning
$\mathrm{R}_i \leftrightarrow \mathrm{R}_j$	Interchange rows i and j
$k\mathrm{R}_i \rightarrow \mathrm{R}_i$	Multiply row i by k
$k\mathrm{R}_i + \mathrm{R}_j \rightarrow \mathrm{R}_j$	Add k times row i to row j

We shall next rework Example 1 using matrices. You should compare the two solutions, since analogous steps are used in each case.

EXAMPLE 2 **Using matrices to solve a system of linear equations**

Solve the system

$$\begin{cases} x - 2y + 3z = 4 \\ 2x + y - 4z = 3 \\ -3x + 4y - z = -2 \end{cases}$$

SOLUTION We begin with the matrix of the system—that is, with the augmented matrix:

$$\begin{bmatrix} 1 & -2 & 3 & 4 \\ 2 & 1 & -4 & 3 \\ -3 & 4 & -1 & -2 \end{bmatrix}$$

We next apply elementary row transformations to obtain another (simpler) matrix of an equivalent system of equations. These transformations correspond to the manipulations used for equations in Example 1. We will place appropriate symbols between equivalent matrices.

$$\begin{bmatrix} 1 & -2 & 3 & | & 4 \\ 2 & 1 & -4 & | & 3 \\ -3 & 4 & -1 & | & -2 \end{bmatrix} \begin{array}{l} \\ -2R_1 + R_2 \rightarrow R_2 \\ 3R_1 + R_3 \rightarrow R_3 \end{array} \begin{bmatrix} 1 & -2 & 3 & | & 4 \\ 0 & 5 & -10 & | & -5 \\ 0 & -2 & 8 & | & 10 \end{bmatrix}$$ add $-2R_1$ to R_2
add $3R_1$ to R_3

$$\begin{array}{l} \frac{1}{5}R_2 \rightarrow R_2 \\ -\frac{1}{2}R_3 \rightarrow R_3 \end{array} \begin{bmatrix} 1 & -2 & 3 & | & 4 \\ 0 & 1 & -2 & | & -1 \\ 0 & 1 & -4 & | & -5 \end{bmatrix}$$ multiply R_2 by $\frac{1}{5}$
multiply R_3 by $-\frac{1}{2}$

$$-R_2 + R_3 \rightarrow R_3 \begin{bmatrix} 1 & -2 & 3 & | & 4 \\ 0 & 1 & -2 & | & -1 \\ 0 & 0 & -2 & | & -4 \end{bmatrix}$$ add $-R_2$ to R_3

$$-\frac{1}{2}R_3 \rightarrow R_3 \begin{bmatrix} 1 & -2 & 3 & | & 4 \\ 0 & 1 & -2 & | & -1 \\ 0 & 0 & 1 & | & 2 \end{bmatrix}$$ multiply R_3 by $-\frac{1}{2}$

We use the last matrix to return to the system of equations

$$\begin{bmatrix} 1 & -2 & 3 & | & 4 \\ 0 & 1 & -2 & | & -1 \\ 0 & 0 & 1 & | & 2 \end{bmatrix} \iff \begin{cases} x - 2y + 3z = 4 \\ y - 2z = -1 \\ z = 2 \end{cases}$$

which is equivalent to the original system. The solution $x = 4$, $y = 3$, $z = 2$ may now be found by back substitution, as in Example 1. ◢

The final matrix in the solution of Example 2 is in **echelon form.** In general, a matrix is in echelon form if it satisfies the following conditions.

Echelon Form of a Matrix

(1) The first nonzero number in each row, reading from left to right, is 1.

(2) The column containing the first nonzero number in any row is to the left of the column containing the first nonzero number in the row below.

(3) Rows consisting entirely of zeros may appear at the bottom of the matrix.

The following is an illustration of matrices in echelon form. The symbols a_{ij} represent real numbers.

ILLUSTRATION **Echelon Form**

■ $$\begin{bmatrix} 1 & a_{12} & a_{13} & a_{14} \\ 0 & 1 & a_{23} & a_{24} \\ 0 & 0 & 1 & a_{34} \end{bmatrix}$$

■ $$\begin{bmatrix} 1 & a_{12} & a_{13} & a_{14} & a_{15} & a_{16} & a_{17} \\ 0 & 1 & a_{23} & a_{24} & a_{25} & a_{26} & a_{27} \\ 0 & 0 & 0 & 1 & a_{35} & a_{36} & a_{37} \\ 0 & 0 & 0 & 0 & 0 & 1 & a_{47} \\ 0 & 0 & 0 & 0 & 0 & 0 & 0 \\ 0 & 0 & 0 & 0 & 0 & 0 & 0 \end{bmatrix}$$

The following guidelines may be used to find echelon forms.

Guidelines for Finding the Echelon Form of a Matrix	*1* Locate the *first* column that contains nonzero elements, and apply elementary row transformations to get the number 1 into the first row of that column.
	2 Apply elementary row transformations of the type $kR_1 + R_j \rightarrow R_j$ for $j > 1$ to get 0 underneath the number 1 obtained in guideline 1 in each of the remaining rows.
	3 *Disregard the first row.* Locate the next column that contains nonzero elements, and apply elementary row transformations to get the number 1 into the *second* row of that column.
	4 Apply elementary row transformations of the type $kR_2 + R_j \rightarrow R_j$ for $j > 2$ to get 0 underneath the number 1 obtained in guideline 3 in each of the remaining rows.
	5 *Disregard the first and second rows.* Locate the next column that contains nonzero elements, and repeat the procedure.
	6 Continue the process until the echelon form is reached.

Not all echelon forms contain rows consisting of only zeros (see Example 2).

We can use elementary row operations to transform the matrix of any system of linear equations to echelon form. The echelon form can then be used to produce a system of equations that is equivalent to the original system. The solutions of the given system may be found by back substitution. The next example illustrates this technique for a system of four linear equations.

EXAMPLE 3   Using an echelon form to solve a system of linear equations

Solve the system

$$\begin{cases} -2x + 3y + 4z = -1 \\ x - 2z + 2w = 1 \\ y + z - w = 0 \\ 3x + y - 2z - w = 3 \end{cases}$$

SOLUTION   We have arranged the equations so that the same variables appear in vertical columns. We begin with the augmented matrix and then obtain an echelon form as described in the guidelines.

$$\begin{bmatrix} -2 & 3 & 4 & 0 & | & -1 \\ 1 & 0 & -2 & 2 & | & 1 \\ 0 & 1 & 1 & -1 & | & 0 \\ 3 & 1 & -2 & -1 & | & 3 \end{bmatrix} \xrightarrow{R_1 \leftrightarrow R_2} \begin{bmatrix} 1 & 0 & -2 & 2 & | & 1 \\ -2 & 3 & 4 & 0 & | & -1 \\ 0 & 1 & 1 & -1 & | & 0 \\ 3 & 1 & -2 & -1 & | & 3 \end{bmatrix}$$

$$\begin{matrix} \\ 2R_1 + R_2 \to R_2 \\ \\ -3R_1 + R_4 \to R_4 \end{matrix} \begin{bmatrix} 1 & 0 & -2 & 2 & | & 1 \\ 0 & 3 & 0 & 4 & | & 1 \\ 0 & 1 & 1 & -1 & | & 0 \\ 0 & 1 & 4 & -7 & | & 0 \end{bmatrix}$$

$$R_2 \leftrightarrow R_3 \begin{bmatrix} 1 & 0 & -2 & 2 & | & 1 \\ 0 & 1 & 1 & -1 & | & 0 \\ 0 & 3 & 0 & 4 & | & 1 \\ 0 & 1 & 4 & -7 & | & 0 \end{bmatrix}$$

$$\begin{matrix} \\ \\ -3R_2 + R_3 \to R_3 \\ -R_2 + R_4 \to R_4 \end{matrix} \begin{bmatrix} 1 & 0 & -2 & 2 & | & 1 \\ 0 & 1 & 1 & -1 & | & 0 \\ 0 & 0 & -3 & 7 & | & 1 \\ 0 & 0 & 3 & -6 & | & 0 \end{bmatrix}$$

$$\begin{matrix} \\ \\ \\ R_3 + R_4 \to R_4 \end{matrix} \begin{bmatrix} 1 & 0 & -2 & 2 & | & 1 \\ 0 & 1 & 1 & -1 & | & 0 \\ 0 & 0 & -3 & 7 & | & 1 \\ 0 & 0 & 0 & 1 & | & 1 \end{bmatrix}$$

$$\begin{matrix} \\ \\ -\frac{1}{3}R_3 \to R_3 \\ \\ \end{matrix} \begin{bmatrix} 1 & 0 & -2 & 2 & | & 1 \\ 0 & 1 & 1 & -1 & | & 0 \\ 0 & 0 & 1 & -\frac{7}{3} & | & -\frac{1}{3} \\ 0 & 0 & 0 & 1 & | & 1 \end{bmatrix}$$

The final matrix is in echelon form and corresponds to the following system of equations:

$$\begin{cases} x & - 2z + 2w = 1 \\ y + z - w = 0 \\ z - \frac{7}{3}w = -\frac{1}{3} \\ w = 1 \end{cases}$$

We now use back substitution to find the solution. From the last equation we see that $w = 1$. Substituting in the third equation, $z - \frac{7}{3}w = -\frac{1}{3}$, we get

$$z - \frac{7}{3}(1) = -\frac{1}{3}, \quad \text{or} \quad z = \frac{6}{3} = 2.$$

(continued)

Substituting $w = 1$ and $z = 2$ in the second equation, $y + z - w = 0$, we obtain

$$y + 2 - 1 = 0, \quad \text{or} \quad y = -1.$$

Finally, from the first equation, $x - 2z + 2w = 1$, we have

$$x - 2(2) + 2(1) = 1, \quad \text{or} \quad x = 3.$$

Hence, the system has one solution, $x = 3$, $y = -1$, $z = 2$, and $w = 1$. ◢

After obtaining an echelon form, it is often convenient to apply additional elementary row operations of the type $kR_i + R_j \rightarrow R_j$ so that 0 also appears *above* the first 1 in each row. We refer to the resulting matrix as being in **reduced echelon form.** The following is an illustration of matrices in reduced echelon form. (Compare them with the echelon forms on page 603.)

ILLUSTRATION   **Reduced Echelon Form**

$$
\blacksquare \quad
\begin{bmatrix}
1 & 0 & 0 & a_{14} \\
0 & 1 & 0 & a_{24} \\
0 & 0 & 1 & a_{34}
\end{bmatrix}
\qquad
\blacksquare \quad
\begin{bmatrix}
1 & 0 & a_{13} & 0 & a_{15} & 0 & a_{17} \\
0 & 1 & a_{23} & 0 & a_{25} & 0 & a_{27} \\
0 & 0 & 0 & 1 & a_{35} & 0 & a_{37} \\
0 & 0 & 0 & 0 & 0 & 1 & a_{47} \\
0 & 0 & 0 & 0 & 0 & 0 & 0 \\
0 & 0 & 0 & 0 & 0 & 0 & 0
\end{bmatrix}
$$

EXAMPLE 4   **Using a reduced echelon form to solve a system of linear equations**

Solve the system in Example 3 using reduced echelon form.

SOLUTION   We begin with the echelon form obtained in Example 3 and apply additional row operations as follows:

$$
\begin{bmatrix}
1 & 0 & -2 & 2 & \bigm| & 1 \\
0 & 1 & 1 & -1 & \bigm| & 0 \\
0 & 0 & 1 & -\frac{7}{3} & \bigm| & -\frac{1}{3} \\
0 & 0 & 0 & 1 & \bigm| & 1
\end{bmatrix}
\begin{array}{l}
-2R_4 + R_1 \rightarrow R_1 \\
R_4 + R_2 \rightarrow R_2 \\
\frac{7}{3}R_4 + R_3 \rightarrow R_3
\end{array}
\begin{bmatrix}
1 & 0 & -2 & 0 & \bigm| & -1 \\
0 & 1 & 1 & 0 & \bigm| & 1 \\
0 & 0 & 1 & 0 & \bigm| & 2 \\
0 & 0 & 0 & 1 & \bigm| & 1
\end{bmatrix}
$$

$$
\begin{array}{l}
2R_3 + R_1 \rightarrow R_1 \\
-R_3 + R_2 \rightarrow R_2
\end{array}
\begin{bmatrix}
1 & 0 & 0 & 0 & \bigm| & 3 \\
0 & 1 & 0 & 0 & \bigm| & -1 \\
0 & 0 & 1 & 0 & \bigm| & 2 \\
0 & 0 & 0 & 1 & \bigm| & 1
\end{bmatrix}
$$

The system of equations corresponding to the reduced echelon form gives us the solution *without* using back substitution:

$$x = 3, \quad y = -1, \quad z = 2, \quad w = 1$$ ◢

Sometimes it is necessary to consider systems in which the number of equations is not the same as the number of variables. The same matrix techniques are applicable, as illustrated in the next example.

EXAMPLE 5   Solving a system of two linear equations in three variables

Solve the system

$$\begin{cases} 2x + 3y + 4z = 1 \\ 3x + 4y + 5z = 3 \end{cases}$$

SOLUTION   We shall begin with the augmented matrix and then find a reduced echelon form. There are many different ways of getting the number 1 into the first position of the first row. For example, the elementary row transformation $\frac{1}{2}R_1 \rightarrow R_1$ or $-\frac{1}{3}R_2 + R_1 \rightarrow R_1$ would accomplish this in one step. Another way, which does not involve fractions, is demonstrated in the following steps:

$$\begin{bmatrix} 2 & 3 & 4 & | & 1 \\ 3 & 4 & 5 & | & 3 \end{bmatrix} \xrightarrow{R_1 \leftrightarrow R_2} \begin{bmatrix} 3 & 4 & 5 & | & 3 \\ 2 & 3 & 4 & | & 1 \end{bmatrix}$$

$$\xrightarrow{-R_2 + R_1 \rightarrow R_1} \begin{bmatrix} 1 & 1 & 1 & | & 2 \\ 2 & 3 & 4 & | & 1 \end{bmatrix}$$

$$\xrightarrow{-2R_1 + R_2 \rightarrow R_2} \begin{bmatrix} 1 & 1 & 1 & | & 2 \\ 0 & 1 & 2 & | & -3 \end{bmatrix}$$

$$\xrightarrow{-R_2 + R_1 \rightarrow R_1} \begin{bmatrix} 1 & 0 & -1 & | & 5 \\ 0 & 1 & 2 & | & -3 \end{bmatrix}$$

The reduced echelon form is the matrix of the system

$$\begin{cases} x \quad\ - z = \ \ 5 \\ \quad\ \ y + 2z = -3 \end{cases}$$

or, equivalently,

$$\begin{cases} x = \quad z + 5 \\ y = -2z - 3 \end{cases}$$

There are an infinite number of solutions to this system; they can be found by assigning z any value c and then using the last two equations to express x and y in terms of c. This gives us

$$x = c + 5, \qquad y = -2c - 3, \qquad z = c.$$

Thus, the solutions of the system consist of all ordered triples of the form

$$(c + 5, -2c - 3, c)$$

for any real number c. The solutions may be checked by substituting $c + 5$ for x, $-2c - 3$ for y, and c for z in the two original equations.

We can obtain any number of solutions for the system by substituting specific real numbers for c. For example, if $c = 0$, we obtain $(5, -3, 0)$; if $c = 2$, we have $(7, -7, 2)$; and so on.

(continued)

There are other ways to specify the general solution. For example, starting with $x = z + 5$ and $y = -2z - 3$, we could let $z = d - 5$ for any real number d. In this case,

$$x = z + 5 = (d - 5) + 5 = d$$
$$y = -2z - 3 = -2(d - 5) - 3 = -2d + 7,$$

and the solutions of the system have the form

$$(d, -2d + 7, d - 5).$$

These triples produce the same solutions as $(c + 5, -2c - 3, c)$. For example, if $d = 5$, we get $(5, -3, 0)$; if $d = 7$, we obtain $(7, -7, 2)$; and so on.

A system of linear equations is **homogeneous** if all the terms that do not contain variables—that is, the *constant terms*—are zero. A system of homogeneous equations always has the **trivial solution** obtained by substituting zero for each variable. Nontrivial solutions sometimes exist. The procedure for finding solutions is the same as that used for nonhomogeneous systems.

EXAMPLE 6   **Solving a homogeneous system of linear equations**

Solve the homogeneous system

$$\begin{cases} x - y + 4z = 0 \\ 2x + y - z = 0 \\ -x - y + 2z = 0 \end{cases}$$

SOLUTION   We begin with the augmented matrix and find a reduced echelon form:

$$\begin{bmatrix} 1 & -1 & 4 & \bigm| & 0 \\ 2 & 1 & -1 & \bigm| & 0 \\ -1 & -1 & 2 & \bigm| & 0 \end{bmatrix} \begin{matrix} \\ -2R_1 + R_2 \to R_2 \\ R_1 + R_3 \to R_3 \end{matrix} \begin{bmatrix} 1 & -1 & 4 & \bigm| & 0 \\ 0 & 3 & -9 & \bigm| & 0 \\ 0 & -2 & 6 & \bigm| & 0 \end{bmatrix}$$

$$\begin{matrix} \\ \frac{1}{3}R_2 \to R_2 \\ -\frac{1}{2}R_3 \to R_3 \end{matrix} \begin{bmatrix} 1 & -1 & 4 & \bigm| & 0 \\ 0 & 1 & -3 & \bigm| & 0 \\ 0 & 1 & -3 & \bigm| & 0 \end{bmatrix}$$

$$\begin{matrix} R_2 + R_1 \to R_1 \\ \\ -R_2 + R_3 \to R_3 \end{matrix} \begin{bmatrix} 1 & 0 & 1 & \bigm| & 0 \\ 0 & 1 & -3 & \bigm| & 0 \\ 0 & 0 & 0 & \bigm| & 0 \end{bmatrix}$$

The reduced echelon form corresponds to the system

$$\begin{cases} x \quad + z = 0 \\ y - 3z = 0 \end{cases}$$

or, equivalently,

$$\begin{cases} x = -z \\ y = 3z \end{cases}$$

Assigning any value c to z, we obtain $x = -c$ and $y = 3c$. The solutions consist of all ordered triples of the form $(-c, 3c, c)$ for any real number c.

EXAMPLE 7   **A homogeneous system with only the trivial solution**

Solve the system

$$\begin{cases} x + y + z = 0 \\ x - y + z = 0 \\ x - y - z = 0 \end{cases}$$

SOLUTION   We begin with the augmented matrix and find a reduced echelon form:

$$\begin{bmatrix} 1 & 1 & 1 & | & 0 \\ 1 & -1 & 1 & | & 0 \\ 1 & -1 & -1 & | & 0 \end{bmatrix} \begin{array}{c} \\ -\mathbf{R}_1 + \mathbf{R}_2 \to \mathbf{R}_2 \\ -\mathbf{R}_1 + \mathbf{R}_3 \to \mathbf{R}_3 \end{array} \begin{bmatrix} 1 & 1 & 1 & | & 0 \\ 0 & -2 & 0 & | & 0 \\ 0 & -2 & -2 & | & 0 \end{bmatrix}$$

$$\begin{array}{c} \\ -\tfrac{1}{2}\mathbf{R}_2 \to \mathbf{R}_2 \\ -\tfrac{1}{2}\mathbf{R}_3 \to \mathbf{R}_3 \end{array} \begin{bmatrix} 1 & 1 & 1 & | & 0 \\ 0 & 1 & 0 & | & 0 \\ 0 & 1 & 1 & | & 0 \end{bmatrix}$$

$$\begin{array}{c} -\mathbf{R}_2 + \mathbf{R}_1 \to \mathbf{R}_1 \\ \\ -\mathbf{R}_2 + \mathbf{R}_3 \to \mathbf{R}_3 \end{array} \begin{bmatrix} 1 & 0 & 1 & | & 0 \\ 0 & 1 & 0 & | & 0 \\ 0 & 0 & 1 & | & 0 \end{bmatrix}$$

$$\begin{array}{c} -\mathbf{R}_3 + \mathbf{R}_1 \to \mathbf{R}_1 \\ \\ \\ \end{array} \begin{bmatrix} 1 & 0 & 0 & | & 0 \\ 0 & 1 & 0 & | & 0 \\ 0 & 0 & 1 & | & 0 \end{bmatrix}$$

The reduced echelon form is the matrix of the system

$$x = 0, \qquad y = 0, \qquad z = 0.$$

Thus, the only solution for the given system is the trivial one, $(0, 0, 0)$.

The next two examples illustrate applied problems.

 EXAMPLE 8   **Using a system of equations to determine maximum profit**

A manufacturer of electrical equipment has the following information about the weekly profit from the production and sale of an electric motor.

Production level x	25	50	100
Profit $P(x)$ (dollars)	5250	7500	4500

(a) Determine a, b, and c so that the graph of $P(x) = ax^2 + bx + c$ fits this information.

(b) According to the quadratic function P in part (a), how many motors should be produced each week for maximum profit? What is the maximum weekly profit?

SOLUTION

(a) We see from the table that the graph of $P(x) = ax^2 + bx + c$ contains the points $(25, 5250)$, $(50, 7500)$, and $(100, 4500)$. This gives us the system of equations

$$\begin{cases} 5250 = 625a + 25b + c \\ 7500 = 2500a + 50b + c \\ 4500 = 10{,}000a + 100b + c \end{cases}$$

It is easy to solve any of the equations for c, so we'll start solving the system by solving the first equation for c,

$$c = 5250 - 625a - 25b,$$

and then substituting that expression for c in the other two equations:

$$\begin{cases} 7500 = 2500a + 50b + (5250 - 625a - 25b) \\ 4500 = 10{,}000a + 100b + (5250 - 625a - 25b) \end{cases}$$

Note that we have reduced the system of three equations and three variables to two equations and two variables. Simplifying the system gives us

$$\begin{cases} 1875a + 25b = 2250 \\ 9375a + 75b = -750 \end{cases}$$

At this point we could divide the equations by 25, but we see that 75 is just 3 times 25, so we'll use the method of elimination to eliminate b:

*Note that we have used **both** the method of substitution and the method of elimination in solving this system of equations.*

$$\begin{cases} -5625a - 75b = -6750 \qquad \text{\color{red}multiply the first} \\ 9375a + 75b = -750 \qquad \text{\color{red}equation by } -3 \end{cases}$$

Adding the equations gives us $3750a = -7500$, so $a = -2$. We can verify that the solution is $a = -2$, $b = 240$, $c = 500$.

(b) From part (a),

$$P(x) = -2x^2 + 240x + 500.$$

Since $a = -2 < 0$, the graph of the quadratic function P is a parabola that opens downward. By the formula on page 189, the x-coordinate of the vertex (the highest point on the parabola) is

$$x = \frac{-b}{2a} = \frac{-240}{2(-2)} = \frac{-240}{-4} = 60.$$

Hence, for the maximum profit, the manufacturer should produce and sell 60 motors per week. The maximum weekly profit is

$$P(60) = -2(60)^2 + 240(60) + 500 = \$7700.$$

EXAMPLE 9   **Solving a mixture problem**

A merchant wishes to mix two grades of peanuts costing \$3 and \$4 per pound, respectively, with cashews costing \$8 per pound, to obtain 140 pounds of a mixture costing \$6 per pound. If the merchant also wants the amount of lower-grade peanuts to be twice that of the higher-grade peanuts, how many pounds of each variety should be mixed?

SOLUTION   Let us introduce three variables, as follows:

$$x = \text{number of pounds of peanuts at \$3 per pound}$$
$$y = \text{number of pounds of peanuts at \$4 per pound}$$
$$z = \text{number of pounds of cashews at \$8 per pound}$$

We refer to the statement of the problem and obtain the following system:

$$\begin{cases} x + y + z = 140 & \text{weight equation} \\ 3x + 4y + 8z = 6(140) & \text{value equation} \\ x \qquad\qquad = 2y & \text{constraint} \end{cases}$$

You may verify that the solution of this system is $x = 40$, $y = 20$, $z = 80$. Thus, the merchant should use 40 pounds of the \$3/lb peanuts, 20 pounds of the \$4/lb peanuts, and 80 pounds of cashews.

Sometimes we can combine row transformations to simplify our work. For example, consider the augmented matrix

$$\begin{bmatrix} 11 & 3 & 8 & 9 \\ 7 & -2 & 2 & 1 \\ 0 & 87 & 80 & 94 \end{bmatrix}.$$

To obtain a 1 in the first column, it appears we have to multiply row 1 by $\frac{1}{11}$ or row 2 by $\frac{1}{7}$. However, we can multiply row 1 by 2 and row 2 by -3 and then add those two rows to obtain

$$2(11) + (-3)(7) = 22 + (-21) = 1$$

in column one, as shown in the next matrix:

$$2R_1 - 3R_2 \rightarrow R_1 \begin{bmatrix} 1 & 12 & 10 & 15 \\ 7 & -2 & 2 & 1 \\ 0 & 87 & 80 & 94 \end{bmatrix}.$$

We can then proceed to find the reduced echelon form without the cumbersome use of fractions. This process is called using a **linear combination of rows.**

Exercises

Exer. 1–22: Use matrices to solve the system.

1. $\begin{cases} x - 2y - 3z = -1 \\ 2x + y + z = 6 \\ x + 3y - 2z = 13 \end{cases}$

2. $\begin{cases} x + 3y - z = -3 \\ 3x - y + 2z = 1 \\ 2x - y + z = -1 \end{cases}$

3. $\begin{cases} 5x + 2y - z = -7 \\ x - 2y + 2z = 0 \\ 3y + z = 17 \end{cases}$

4. $\begin{cases} 4x - y + 3z = 6 \\ -8x + 3y - 5z = -6 \\ 5x - 4y = -9 \end{cases}$

5. $\begin{cases} 2x + 6y - 4z = 1 \\ x + 3y - 2z = 4 \\ 2x + y - 3z = -7 \end{cases}$

6. $\begin{cases} x + 3y - 3z = -5 \\ 2x - y + z = -3 \\ -6x + 3y - 3z = 4 \end{cases}$

7. $\begin{cases} 2x - 3y + 2z = -3 \\ -3x + 2y + z = 1 \\ 4x + y - 3z = 4 \end{cases}$

8. $\begin{cases} 2x - 3y + z = 2 \\ 3x + 2y - z = -5 \\ 5x - 2y + z = 0 \end{cases}$

9. $\begin{cases} x + 3y + z = 0 \\ x + y - z = 0 \\ x - 2y - 4z = 0 \end{cases}$

10. $\begin{cases} 2x - y + z = 0 \\ x - y - 2z = 0 \\ 2x - 3y - z = 0 \end{cases}$

11. $\begin{cases} 2x + y + z = 0 \\ x - 2y - 2z = 0 \\ x + y + z = 0 \end{cases}$

12. $\begin{cases} x + y - 2z = 0 \\ x - y - 4z = 0 \\ y + z = 0 \end{cases}$

13. $\begin{cases} 3x - 2y + 5z = 7 \\ x + 4y - z = -2 \end{cases}$

14. $\begin{cases} 2x - y + 4z = 8 \\ -3x + y - 2z = 5 \end{cases}$

15. $\begin{cases} 4x - 2y + z = 5 \\ 3x + y - 4z = 0 \end{cases}$

16. $\begin{cases} 5x + 2y - z = 10 \\ y + z = -3 \end{cases}$

17. $\begin{cases} 5x + 2z = 1 \\ y - 3z = 2 \\ 2x + y = 3 \end{cases}$

18. $\begin{cases} 2x - 3y = 12 \\ 3y + z = -2 \\ 5x - 3z = 3 \end{cases}$

19. $\begin{cases} 4x - 3y = 1 \\ 2x + y = -7 \\ -x + y = -1 \end{cases}$

20. $\begin{cases} 2x + 3y = -2 \\ x + y = 1 \\ x - 2y = 13 \end{cases}$

21. $\begin{cases} 2x + 3y = 5 \\ x - 3y = 4 \\ x + y = -2 \end{cases}$

22. $\begin{cases} 4x - y = 2 \\ 2x + 2y = 1 \\ 4x - 5y = 3 \end{cases}$

23 **Mixing acid solutions** Three solutions contain a certain acid. The first contains 10% acid, the second 30%, and the third 50%. A chemist wishes to use all three solutions to obtain a 50-liter mixture containing 32% acid. If the chemist wants to use twice as much of the 50% solution as of the 30% solution, how many liters of each solution should be used?

24 **Filling a pool** A swimming pool can be filled by three pipes, A, B, and C. Pipe A alone can fill the pool in 8 hours. If pipes A and C are used together, the pool can be filled in 6 hours; if B and C are used together, it takes 10 hours. How long does it take to fill the pool if all three pipes are used?

25 **Production capability** A company has three machines, A, B, and C, that are each capable of producing a certain item. However, because of a lack of skilled operators, only two of the machines can be used simultaneously. The following table indicates production over a three-day period, using various combinations of the machines. How long would it take each machine, if used alone, to produce 1000 items?

Machines used	Hours used	Items produced
A and B	6	4500
A and C	8	3600
B and C	7	4900

26 **Electrical resistance** In electrical circuits, the formula $1/R = (1/R_1) + (1/R_2)$ is used to find the total resistance R if two resistors R_1 and R_2 are connected in parallel. Given three resistors, A, B, and C, suppose that the total resistance is 48 ohms if A and B are connected in parallel, 80 ohms if B and C are connected in parallel, and 60 ohms if A and C are connected in parallel. Find the resistances of A, B, and C.

27 **Mixing fertilizers** A supplier of lawn products has three types of grass fertilizer, G_1, G_2, and G_3, having nitrogen contents of 30%, 20%, and 15%, respectively. The supplier plans to mix them, obtaining 600 pounds of fertilizer with a

25% nitrogen content. The mixture is to contain 100 pounds more of type G_3 than of type G_2. How much of each type should be used?

28 Particle acceleration If a particle moves along a coordinate line with a constant acceleration a (in cm/sec^2), then at time t (in seconds) its distance $s(t)$ (in centimeters) from the origin is

$$s(t) = \tfrac{1}{2}at^2 + v_0t + s_0$$

for velocity v_0 and distance s_0 from the origin at $t = 0$. If the distances of the particle from the origin at $t = \tfrac{1}{2}, t = 1$, and $t = \tfrac{3}{2}$ are 7, 11, and 17, respectively, find a, v_0, and s_0.

29 Electrical currents Shown in the figure is a schematic of an electrical circuit containing three resistors, a 6-volt battery, and a 12-volt battery. It can be shown, using Kirchhoff's laws, that the three currents I_1, I_2, and I_3 are solutions of the following system of equations:

$$\begin{cases} I_1 - I_2 + I_3 = 0 \\ R_1I_1 + R_2I_2 = 6 \\ R_2I_2 + R_3I_3 = 12 \end{cases}$$

Find the three currents if

(a) $R_1 = R_2 = R_3 = 3$ ohms

(b) $R_1 = 4$ ohms, $R_2 = 1$ ohm, and $R_3 = 4$ ohms

Exercise 29

30 Bird population A stable population of 35,000 birds lives on three islands. Each year 10% of the population on island A migrates to island B, 20% of the population on island B migrates to island C, and 5% of the population on island C migrates to island A. Find the number of birds on each island if the population count on each island does not vary from year to year.

31 Blending coffees A shop specializes in preparing blends of gourmet coffees. From Colombian, Brazilian, and Kenyan coffees, the owner wishes to prepare 1-pound bags that will sell for $8.50. The cost per pound of these coffees is $10, $6, and $8, respectively. The amount of Colombian is to be three times the amount of Brazilian. Find the amount of each type of coffee in the blend.

32 Weights of chains There are three chains, weighing 450, 610, and 950 ounces, each consisting of links of three different sizes. Each chain has 10 small links. The chains also have 20, 30, and 40 medium links and 30, 40, and 70 large links, respectively. Find the weights of the small, medium, and large links.

33 Traffic flow Shown in the figure is a system of four one-way streets leading into the center of a city. The numbers in the figure denote the average number of vehicles per hour that travel in the directions shown. A total of 300 vehicles enter the area and 300 vehicles leave the area every hour. Signals at intersections A, B, C, and D are to be timed in order to avoid congestion, and this timing will determine traffic flow rates x_1, x_2, x_3, and x_4.

Exercise 33

(a) If the number of vehicles entering an intersection per hour must equal the number leaving the intersection per hour, describe the traffic flow rates at each intersection with a system of equations.

(b) If the signal at intersection C is timed so that x_3 is equal to 100, find x_1, x_2, and x_4.

(c) Make use of the system in part (a) to explain why $75 \leq x_3 \leq 150$.

34 If $f(x) = ax^3 + bx + c$, determine a, b, and c such that the graph of f passes through the points $P(-3, -12)$, $Q(-1, 22)$, and $R(2, 13)$.

Exer. 35–36: Find an equation of the circle of the form $x^2 + y^2 + ax + by + c = 0$ that passes through the given points.

35 $P(2, 1)$, $Q(-1, -4)$, $R(3, 0)$

36 $P(-5, 5)$, $Q(-2, -4)$, $R(2, 4)$

Exer. 37–38: Find an equation of the cubic polynomial $f(x) = ax^3 + bx^2 + cx + d$ that passes through the given points.

37 $P(0, -6)$, $Q(1, -11)$, $R(-1, -5)$, $S(2, -14)$

38 $P(0, 4)$, $Q(1, 2)$, $R(-1, 10)$, $S(2, -2)$

9.6

The Algebra of Matrices

Matrices were introduced in Section 9.5 as an aid to finding solutions of systems of equations. In this section we discuss some of the properties of matrices. These properties are important in advanced fields of mathematics and in applications.

In the following definition, the symbol (a_{ij}) denotes an $m \times n$ matrix A of the type displayed in the definition on page 601. We use similar notations for the matrices B and C.

Definition of Equality and Addition of Matrices	Let $A = (a_{ij})$, $B = (b_{ij})$, and $C = (c_{ij})$ be $m \times n$ matrices. **(1)** $A = B$ if and only if $a_{ij} = b_{ij}$ for every i and j. **(2)** $C = A + B$ if and only if $c_{ij} = a_{ij} + b_{ij}$ for every i and j.

Note that two matrices are equal if and only if they have the same size and corresponding elements are equal.

ILLUSTRATION   **Equality of Matrices**

■ $\begin{bmatrix} 1 & 0 & 5 \\ \sqrt[3]{8} & 3^2 & -2 \end{bmatrix} = \begin{bmatrix} (-1)^2 & 0 & \sqrt{25} \\ 2 & 9 & -2 \end{bmatrix}$

Using the parentheses notation for matrices, we may write the definition of addition of two $m \times n$ matrices as

$$(a_{ij}) + (b_{ij}) = (a_{ij} + b_{ij}).$$

Thus, to add two matrices, we add the elements in corresponding positions in each matrix. *Two matrices can be added only if they have the same size.*

ILLUSTRATION **Addition of Matrices**

■
$$\begin{bmatrix} 4 & -5 \\ 0 & 4 \\ -6 & 1 \end{bmatrix} + \begin{bmatrix} 3 & 2 \\ 7 & -4 \\ -2 & 1 \end{bmatrix} = \begin{bmatrix} 4+3 & -5+2 \\ 0+7 & 4+(-4) \\ -6+(-2) & 1+1 \end{bmatrix} = \begin{bmatrix} 7 & -3 \\ 7 & 0 \\ -8 & 2 \end{bmatrix}$$

■
$$\begin{bmatrix} 2 & 3 \\ -4 & 1 \end{bmatrix} + \begin{bmatrix} -2 & -3 \\ 4 & -1 \end{bmatrix} = \begin{bmatrix} 0 & 0 \\ 0 & 0 \end{bmatrix}$$

■
$$\begin{bmatrix} 1 & 3 & -2 \\ 0 & -5 & 4 \end{bmatrix} + \begin{bmatrix} 0 & 0 & 0 \\ 0 & 0 & 0 \end{bmatrix} = \begin{bmatrix} 1 & 3 & -2 \\ 0 & -5 & 4 \end{bmatrix}$$

The **$m \times n$ zero matrix,** denoted by O, is the matrix with m rows and n columns in which every element is 0.

ILLUSTRATION **Zero Matrices**

■
$$\begin{bmatrix} 0 & 0 \\ 0 & 0 \end{bmatrix}$$
 ■
$$\begin{bmatrix} 0 & 0 \\ 0 & 0 \\ 0 & 0 \end{bmatrix}$$
 ■
$$\begin{bmatrix} 0 & 0 & 0 & 0 \\ 0 & 0 & 0 & 0 \end{bmatrix}$$

The **additive inverse** $-A$ of the matrix $A = (a_{ij})$ is the matrix $(-a_{ij})$ obtained by changing the sign of each nonzero element of A.

ILLUSTRATION **Additive Inverse**

■
$$-\begin{bmatrix} 2 & -3 & 4 \\ -1 & 0 & 5 \end{bmatrix} = \begin{bmatrix} -2 & 3 & -4 \\ 1 & 0 & -5 \end{bmatrix}$$

The proof of the next theorem follows from the definition of addition of matrices.

Theorem on Matrix Properties	If A, B, and C are $m \times n$ matrices and if O is the $m \times n$ zero matrix, then
	(1) $A + B = B + A$
	(2) $A + (B + C) = (A + B) + C$
	(3) $A + O = A$
	(4) $A + (-A) = O$

Subtraction of two $m \times n$ matrices is defined by

$$A - B = A + (-B).$$

Using the parentheses notation, we have

$$(a_{ij}) - (b_{ij}) = (a_{ij}) + (-b_{ij})$$
$$= (a_{ij} - b_{ij}).$$

Thus, to subtract two matrices, we subtract the elements in corresponding positions.

ILLUSTRATION **Subtraction of Matrices**

■ $\begin{bmatrix} 4 & -5 \\ 0 & 4 \\ -6 & 1 \end{bmatrix} - \begin{bmatrix} 3 & 2 \\ 7 & -4 \\ -2 & 1 \end{bmatrix} = \begin{bmatrix} 4-3 & -5-2 \\ 0-7 & 4-(-4) \\ -6-(-2) & 1-1 \end{bmatrix} = \begin{bmatrix} 1 & -7 \\ -7 & 8 \\ -4 & 0 \end{bmatrix}$

Definition of the Product of a Real Number and a Matrix

The **product** of a real number c and an $m \times n$ matrix $A = (a_{ij})$ is

$$cA = (ca_{ij}).$$

Note that to find cA, we multiply each element of A by c.

ILLUSTRATION **Product of a Real Number and a Matrix**

■ $3 \begin{bmatrix} 4 & -1 \\ 2 & 3 \end{bmatrix} = \begin{bmatrix} 3 \cdot 4 & 3 \cdot (-1) \\ 3 \cdot 2 & 3 \cdot 3 \end{bmatrix} = \begin{bmatrix} 12 & -3 \\ 6 & 9 \end{bmatrix}$

We can prove the following.

Theorem on Matrix Properties

If A and B are $m \times n$ matrices and if c and d are real numbers, then

(1) $c(A + B) = cA + cB$

(2) $(c + d)A = cA + dA$

(3) $(cd)A = c(dA)$

The next definition, of the product AB of two matrices, may seem unusual, but it has many uses in mathematics and applications. For multiplication, unlike addition, A and B may have different sizes; however, *the number of columns of A must be the same as the number of rows of B.* Thus, if A is $m \times n$, then B must be $n \times p$ for some p. As we shall see, the size of AB is then $m \times p$. If $C = AB$, then a method for finding the element c_{ij} in row i and column j of C is given in the following guidelines.

Guidelines for Finding c_{ij} in the Product $C = AB$ if A is $m \times n$ and B is $n \times p$

1 Single out the ith row, R_i, of A and the jth column, C_j, of B:

$$\begin{bmatrix} a_{11} & a_{12} & \cdots & a_{1n} \\ \cdot & \cdot & & \cdot \\ \cdot & \cdot & & \cdot \\ a_{i1} & a_{i2} & \cdots & a_{in} \\ \cdot & \cdot & & \cdot \\ \cdot & \cdot & & \cdot \\ a_{m1} & a_{m2} & \cdots & a_{mn} \end{bmatrix} \begin{bmatrix} b_{11} & \cdots & b_{1j} & \cdots & b_{1p} \\ b_{21} & \cdots & b_{2j} & \cdots & b_{2p} \\ \cdot & & \cdot & & \cdot \\ \cdot & & \cdot & & \cdot \\ b_{n1} & \cdots & b_{nj} & \cdots & b_{np} \end{bmatrix}$$

2 *Simultaneously* move to the right along R_i and down C_j, multiplying pairs of elements, to obtain

$$a_{i1}b_{1j}, \ a_{i2}b_{2j}, \ a_{i3}b_{3j}, \ \ldots, \ a_{in}b_{nj}.$$

3 Add the products of the pairs in guideline 2 to obtain c_{ij}:

$$c_{ij} = a_{i1}b_{1j} + a_{i2}b_{2j} + a_{i3}b_{3j} + \cdots + a_{in}b_{nj}$$

Using the guidelines, we see that the element c_{11} in the first row and the first column of $C = AB$ is

$$c_{11} = a_{11}b_{11} + a_{12}b_{21} + a_{13}b_{31} + \cdots + a_{1n}b_{n1}.$$

The element c_{mp} in the last row and the last column of $C = AB$ is

$$c_{mp} = a_{m1}b_{1p} + a_{m2}b_{2p} + a_{m3}b_{3p} + \cdots + a_{mn}b_{np}.$$

The preceding discussion is summarized in the next definition.

Definition of the Product of Two Matrices

Let $A = (a_{ij})$ be an $m \times n$ matrix and let $B = (b_{ij})$ be an $n \times p$ matrix. The **product** AB is the $m \times p$ matrix $C = (c_{ij})$ such that

$$c_{ij} = a_{i1}b_{1j} + a_{i2}b_{2j} + a_{i3}b_{3j} + \cdots + a_{in}b_{nj}$$

for $i = 1, 2, 3, \ldots, m$ and $j = 1, 2, 3, \ldots, p$.

The following diagram may help you remember the relationship between sizes of matrices when working with a product AB.

The next illustration contains some special cases.

ILLUSTRATION **Sizes of Matrices in Products**

Size of A	Size of B	Size of AB
■ 2×3	3×5	2×5
■ 4×2	2×3	4×3
■ 3×1	1×3	3×3
■ 1×3	3×1	1×1
■ 5×3	3×5	5×5
■ 5×3	5×3	AB is not defined

In the following example we find the product of two specific matrices.

EXAMPLE 1 **Finding the product of two matrices**

Find the product AB if

$$A = \begin{bmatrix} 1 & 2 & -3 \\ 4 & 0 & -2 \end{bmatrix} \quad \text{and} \quad B = \begin{bmatrix} 5 & -4 & 2 & 0 \\ -1 & 6 & 3 & 1 \\ 7 & 0 & 5 & 8 \end{bmatrix}.$$

SOLUTION   The matrix A is 2×3, and the matrix B is 3×4. Hence, the product $C = AB$ is defined and is 2×4. We next use the guidelines to find the elements $c_{11}, c_{12}, \ldots, c_{24}$ of the product. For instance, to find the element c_{23} we single out the second row, R_2, of A and the third column, C_3, of B, as illustrated below, and then use guidelines 2 and 3 to obtain

$$c_{23} = 4 \cdot 2 + 0 \cdot 3 + (-2) \cdot 5 = -2.$$

$$\begin{bmatrix} 1 & 2 & -3 \\ 4 & 0 & -2 \end{bmatrix} \begin{bmatrix} 5 & -4 & \boxed{2} & 0 \\ -1 & 6 & \boxed{3} & 1 \\ 7 & 0 & \boxed{5} & 8 \end{bmatrix} = \begin{bmatrix} \blacksquare & \blacksquare & \blacksquare & \blacksquare \\ \blacksquare & \blacksquare & -2 & \blacksquare \end{bmatrix}$$

Similarly, to find the element c_{12} in row 1 and column 2 of the product, we proceed as follows:

$$c_{12} = 1 \cdot (-4) + 2 \cdot 6 + (-3) \cdot 0 = 8$$

$$\begin{bmatrix} 1 & 2 & -3 \\ 4 & 0 & -2 \end{bmatrix} \begin{bmatrix} 5 & -4 & 2 & 0 \\ -1 & 6 & 3 & 1 \\ 7 & 0 & 5 & 8 \end{bmatrix} = \begin{bmatrix} \blacksquare & 8 & \blacksquare & \blacksquare \\ \blacksquare & \blacksquare & -2 & \blacksquare \end{bmatrix}$$

The remaining elements of the product are calculated as follows, where we have indicated the row of A and the column of B that are used when guideline 1 is applied.

Row of A	Column of B	Element of C
R_1	C_1	$c_{11} = 1 \cdot 5 \quad + 2 \cdot (-1) + (-3) \cdot 7 = -18$
R_1	C_3	$c_{13} = 1 \cdot 2 \quad + 2 \cdot 3 \quad + (-3) \cdot 5 = -7$
R_1	C_4	$c_{14} = 1 \cdot 0 \quad + 2 \cdot 1 \quad + (-3) \cdot 8 = -22$
R_2	C_1	$c_{21} = 4 \cdot 5 \quad + 0 \cdot (-1) + (-2) \cdot 7 = 6$
R_2	C_2	$c_{22} = 4 \cdot (-4) + 0 \cdot 6 \quad + (-2) \cdot 0 = -16$
R_2	C_4	$c_{24} = 4 \cdot 0 \quad + 0 \cdot 1 \quad + (-2) \cdot 8 = -16$

Hence,

$$AB = \begin{bmatrix} 1 & 2 & -3 \\ 4 & 0 & -2 \end{bmatrix} \begin{bmatrix} 5 & -4 & 2 & 0 \\ -1 & 6 & 3 & 1 \\ 7 & 0 & 5 & 8 \end{bmatrix}$$

$$= \begin{bmatrix} -18 & 8 & -7 & -22 \\ 6 & -16 & -2 & -16 \end{bmatrix}.$$

A matrix is a **row matrix** if it has only one row. A **column matrix** has only one column. The following illustration contains some products involving row and column matrices. You should check each entry in the products.

ILLUSTRATION **Products Involving Row and Column Matrices**

■ $\begin{bmatrix} -2 & 4 \\ 0 & -1 \\ 5 & 3 \end{bmatrix} \begin{bmatrix} -2 \\ 1 \end{bmatrix} = \begin{bmatrix} 8 \\ -1 \\ -7 \end{bmatrix}$ ■ $\begin{bmatrix} 3 & -1 & 2 \end{bmatrix} \begin{bmatrix} -2 & 4 \\ 0 & -1 \\ 5 & 3 \end{bmatrix} = \begin{bmatrix} 4 & 19 \end{bmatrix}$

■ $\begin{bmatrix} -2 \\ 3 \end{bmatrix} \begin{bmatrix} 1 & 5 \end{bmatrix} = \begin{bmatrix} -2 & -10 \\ 3 & 15 \end{bmatrix}$ ■ $\begin{bmatrix} 1 & 5 \end{bmatrix} \begin{bmatrix} -2 \\ 3 \end{bmatrix} = \begin{bmatrix} 13 \end{bmatrix}$

The product operation for matrices is not commutative. For example, if A is 2×3 and B is 3×4, then AB may be found, since the number of columns of A is the same as the number of rows of B. However, BA is undefined, since the number of columns of B is different from the number of rows of A. Even if AB and BA are both defined, it is often true that these products are different. This is illustrated in the next example, along with the fact that the product of two nonzero matrices may equal a zero matrix.

EXAMPLE 2 **Matrix multiplication is not commutative**

If $A = \begin{bmatrix} 2 & 2 \\ -1 & -1 \end{bmatrix}$ and $B = \begin{bmatrix} 1 & 2 \\ 1 & 2 \end{bmatrix}$, show that $AB \neq BA$.

SOLUTION Using the definition of the product of two matrices, we obtain the following:

$$AB = \begin{bmatrix} 2 & 2 \\ -1 & -1 \end{bmatrix} \begin{bmatrix} 1 & 2 \\ 1 & 2 \end{bmatrix} = \begin{bmatrix} 4 & 8 \\ -2 & -4 \end{bmatrix}$$

$$BA = \begin{bmatrix} 1 & 2 \\ 1 & 2 \end{bmatrix} \begin{bmatrix} 2 & 2 \\ -1 & -1 \end{bmatrix} = \begin{bmatrix} 0 & 0 \\ 0 & 0 \end{bmatrix}$$

Hence, $AB \neq BA$. Note that the last equality shows that *the product of two nonzero matrices can equal a zero matrix.* ◢

Although matrix multiplication is not commutative, it is associative. Thus, if A is $m \times n$, B is $n \times p$, and C is $p \times q$, then

$$A(BC) = (AB)C.$$

The distributive properties also hold if the matrices involved have the proper number of rows and columns. If A_1 and A_2 are $m \times n$ matrices and if B_1 and B_2 are $n \times p$ matrices, then

$$A_1(B_1 + B_2) = A_1B_1 + A_1B_2$$
$$(A_1 + A_2)B_1 = A_1B_1 + A_2B_1.$$

As a special case, if all matrices are square, of order n, then both the associative and the distributive property are true.

We conclude this section with an application of the product of two matrices.

EXAMPLE 3   An application of a matrix product

(a) Three investors, I_1, I_2, and I_3, each own a certain number of shares of four stocks, S_1, S_2, S_3, and S_4, according to matrix A. Matrix B contains the present value V of each share of each stock. Find AB, and interpret the meaning of its elements.

number of shares of stock   share value

$$\text{investors} \begin{cases} I_1 \\ I_2 \\ I_3 \end{cases} \begin{matrix} S_1 & S_2 & S_3 & S_4 \\ \begin{bmatrix} 50 & 100 & 30 & 25 \\ 100 & 150 & 10 & 30 \\ 100 & 50 & 40 & 100 \end{bmatrix} \end{matrix} = A, \qquad \text{stocks} \begin{cases} S_1 \\ S_2 \\ S_3 \\ S_4 \end{cases} \begin{matrix} V \\ \begin{bmatrix} 20.37 \\ 16.21 \\ 90.80 \\ 42.75 \end{bmatrix} \end{matrix} = B$$

(b) Matrix C contains the change in the value of each stock for the last week. Find AC, and interpret the meaning of its elements.

$$\text{stocks} \begin{cases} S_1 \\ S_2 \\ S_3 \\ S_4 \end{cases} \begin{bmatrix} +1.03 \\ -0.22 \\ -1.35 \\ +0.15 \end{bmatrix} = C$$

SOLUTION

(a) Since A is a 3×4 matrix and B is a 4×1 matrix, the product AB is a 3×1 matrix:

$$AB = \begin{bmatrix} 50 & 100 & 30 & 25 \\ 100 & 150 & 10 & 30 \\ 100 & 50 & 40 & 100 \end{bmatrix} \begin{bmatrix} 20.37 \\ 16.21 \\ 90.80 \\ 42.75 \end{bmatrix} = \begin{bmatrix} 6432.25 \\ 6659.00 \\ 10{,}754.50 \end{bmatrix}$$

The first element in the product AB, 6432.25, was obtained from the computation

$$50(20.37) + 100(16.21) + 30(90.80) + 25(42.75)$$

and represents the total value that investor I_1 has in all four stocks. Similarly, the second and third elements represent the total value for investors I_2 and I_3, respectively.

(b)

$$AC = \begin{bmatrix} 50 & 100 & 30 & 25 \\ 100 & 150 & 10 & 30 \\ 100 & 50 & 40 & 100 \end{bmatrix} \begin{bmatrix} +1.03 \\ -0.22 \\ -1.35 \\ +0.15 \end{bmatrix} = \begin{bmatrix} -7.25 \\ 61.00 \\ 53.00 \end{bmatrix}$$

The first element in the product AC, -7.25, indicates that the total value that investor I_1 has in all four stocks went down \$7.25 in the last week. The second and third elements indicate that the total value that investors I_2 and I_3 have in all four stocks went up \$61.00 and \$53.00, respectively.

9.6   *Exercises*

Exer. 1–8: Find, if possible, $A + B$, $A - B$, $2A$, and $-3B$.

1 $A = \begin{bmatrix} 5 & -2 \\ 1 & 3 \end{bmatrix}$, $B = \begin{bmatrix} 4 & 1 \\ -3 & 2 \end{bmatrix}$

2 $A = \begin{bmatrix} 3 & 0 \\ -1 & 2 \end{bmatrix}$, $B = \begin{bmatrix} 3 & -4 \\ 1 & 1 \end{bmatrix}$

3 $A = \begin{bmatrix} 6 & -1 \\ 2 & 0 \\ -3 & 4 \end{bmatrix}$, $B = \begin{bmatrix} 3 & 1 \\ -1 & 5 \\ 6 & 0 \end{bmatrix}$

4 $A = \begin{bmatrix} 0 & -2 & 7 \\ 5 & 4 & -3 \end{bmatrix}$, $B = \begin{bmatrix} 8 & 4 & 0 \\ 0 & 1 & 4 \end{bmatrix}$

5 $A = \begin{bmatrix} 4 & -3 & 2 \end{bmatrix}$, $B = \begin{bmatrix} 7 & 0 & -5 \end{bmatrix}$

6 $A = \begin{bmatrix} 7 \\ -16 \end{bmatrix}$, $B = \begin{bmatrix} -11 \\ 9 \end{bmatrix}$

7 $A = \begin{bmatrix} 3 & -2 & 2 \\ 0 & 1 & -4 \\ -3 & 2 & -1 \end{bmatrix}$, $B = \begin{bmatrix} 4 & 0 \\ 2 & -1 \\ -1 & 3 \end{bmatrix}$

8 $A = \begin{bmatrix} 2 & 1 \end{bmatrix}$, $B = \begin{bmatrix} 3 & -1 & 5 \end{bmatrix}$

Exer. 9–10: Find the given element of the matrix product $C = AB$ in the listed exercise.

9 c_{21}; Exercise 15

10 c_{23}; Exercise 16

Exer. 11–22: Find, if possible, AB and BA.

11 $A = \begin{bmatrix} 2 & 6 \\ 3 & -4 \end{bmatrix}$, $\qquad$ $B = \begin{bmatrix} 5 & -2 \\ 1 & 7 \end{bmatrix}$

12 $A = \begin{bmatrix} 4 & -2 \\ -2 & 1 \end{bmatrix}$, $\qquad$ $B = \begin{bmatrix} 2 & 1 \\ 4 & 2 \end{bmatrix}$

13 $A = \begin{bmatrix} 3 & 0 & -1 \\ 0 & 4 & 2 \\ 5 & -3 & 1 \end{bmatrix}$, $\qquad$ $B = \begin{bmatrix} 1 & -5 & 0 \\ 4 & 1 & -2 \\ 0 & -1 & 3 \end{bmatrix}$

14 $A = \begin{bmatrix} 5 & 0 & 0 \\ 0 & -3 & 0 \\ 0 & 0 & 2 \end{bmatrix}$, $\qquad$ $B = \begin{bmatrix} 3 & 0 & 0 \\ 0 & 4 & 0 \\ 0 & 0 & -2 \end{bmatrix}$

15 $A = \begin{bmatrix} 4 & -3 & 1 \\ -5 & 2 & 2 \end{bmatrix}$, $\qquad$ $B = \begin{bmatrix} 2 & 1 \\ 0 & 1 \\ -4 & 7 \end{bmatrix}$

16 $A = \begin{bmatrix} 2 & 1 & -1 & 0 \\ 3 & -2 & 0 & 5 \\ -2 & 1 & 4 & 2 \end{bmatrix}$, $\qquad$ $B = \begin{bmatrix} 5 & -3 & 1 \\ 1 & 2 & 0 \\ -1 & 0 & 4 \\ 0 & -2 & 3 \end{bmatrix}$

17 $A = \begin{bmatrix} 1 & 2 & 3 \\ 4 & 5 & 6 \\ 7 & 8 & 9 \end{bmatrix}$, $\qquad$ $B = \begin{bmatrix} 1 & 0 & 0 \\ 0 & 1 & 0 \\ 0 & 0 & 1 \end{bmatrix}$

18 $A = \begin{bmatrix} 1 & 2 & 3 \\ 2 & 3 & 1 \\ 3 & 1 & 2 \end{bmatrix}$, $\qquad$ $B = \begin{bmatrix} 2 & 0 & 0 \\ 0 & 2 & 0 \\ 0 & 0 & 2 \end{bmatrix}$

19 $A = \begin{bmatrix} -3 & 7 & 2 \end{bmatrix}$, $\qquad$ $B = \begin{bmatrix} 1 \\ 4 \\ -5 \end{bmatrix}$

20 $A = \begin{bmatrix} 4 & 8 \end{bmatrix}$, $\qquad$ $B = \begin{bmatrix} -3 \\ 2 \end{bmatrix}$

21 $A = \begin{bmatrix} 2 & 0 & 1 \\ -1 & 2 & 0 \end{bmatrix}$, $\qquad$ $B = \begin{bmatrix} 1 & -1 & 2 \\ 3 & 1 & 0 \\ 0 & 2 & 1 \end{bmatrix}$

22 $A = \begin{bmatrix} 3 & -1 & 4 \end{bmatrix}$, $\qquad$ $B = \begin{bmatrix} -2 \\ 5 \end{bmatrix}$

Exer. 23–26: Find AB.

23 $A = \begin{bmatrix} 4 & -2 \\ 0 & 3 \\ -7 & 5 \end{bmatrix}$, $\qquad$ $B = \begin{bmatrix} 3 \\ 4 \end{bmatrix}$

24 $A = \begin{bmatrix} 4 \\ -3 \\ 2 \end{bmatrix}$, $\qquad$ $B = \begin{bmatrix} 5 & 1 \end{bmatrix}$

25 $A = \begin{bmatrix} 2 & 1 & 0 & -3 \\ -7 & 0 & -2 & 4 \end{bmatrix}$, $\qquad$ $B = \begin{bmatrix} 4 & -2 & 0 \\ 1 & 1 & -2 \\ 0 & 0 & 5 \\ -3 & -1 & 0 \end{bmatrix}$

26 $A = \begin{bmatrix} 1 & 2 & -3 \\ 4 & -5 & 6 \end{bmatrix}$, $\qquad$ $B = \begin{bmatrix} 1 & -1 & 0 & 2 \\ -2 & 3 & 1 & 0 \\ 0 & 4 & 0 & -3 \end{bmatrix}$

Exer. 27–30: Let

$$A = \begin{bmatrix} 1 & 2 \\ 0 & -3 \end{bmatrix}, \quad B = \begin{bmatrix} 2 & -1 \\ 3 & 1 \end{bmatrix}, \quad C = \begin{bmatrix} 3 & 1 \\ -2 & 0 \end{bmatrix}.$$

Verify the statement.

27 $(A + B)(A - B) \neq A^2 - B^2$, where $A^2 = AA$ and $B^2 = BB$.

28 $(A + B)(A + B) \neq A^2 + 2AB + B^2$

29 $A(B + C) = AB + AC$

30 $A(BC) = (AB)C$

Exer 31–34: Verify the identity for

$$A = \begin{bmatrix} a & b \\ c & d \end{bmatrix}, \quad B = \begin{bmatrix} p & q \\ r & s \end{bmatrix}, \quad C = \begin{bmatrix} w & x \\ y & z \end{bmatrix},$$

and real numbers m and n.

31 $m(A + B) = mA + mB$ $\qquad$ 32 $(m + n)A = mA + nA$

33 $A(B + C) = AB + AC$ $\qquad$ 34 $A(BC) = (AB)C$

35 **Value of inventory** A store stocks these sizes of towels, each available in five colors: small, priced at \$1.39 each; medium, priced at \$2.99 each; and large, priced at \$4.99 each. The store's current inventory is as follows:

Towel size	Colors				
	White	Tan	Beige	Pink	Yellow
Small	400	400	300	250	100
Medium	550	450	500	200	100
Large	500	500	600	300	200

(a) Organize these data into an inventory matrix A and a price matrix B so that the product $C = AB$ is defined.

(b) Find C.

(c) Interpret the meaning of element c_{51} in C.

36 Building costs A housing contractor has orders for 4 one-bedroom units, 10 two-bedroom units, and 6 three-bedroom units. The labor and material costs (in thousands of dollars) are given in the following table.

	1-Bedroom	2-Bedroom	3-Bedroom
Labor	34	40	43
Materials	50	60	67

(a) Organize these data into an order matrix A and a cost matrix B so that the product $C = AB$ is defined.

(b) Find C.

(c) Interpret the meaning of each element in C.

9.7

The Inverse of a Matrix

Throughout this section and the next two sections we shall restrict our discussion to *square* matrices. The symbol I_n will denote the square matrix of order n that has 1 in each position on the main diagonal and 0 elsewhere. We call I_n the **identity matrix of order n.**

ILLUSTRATION **Identity Matrices**

■ $I_2 = \begin{bmatrix} 1 & 0 \\ 0 & 1 \end{bmatrix}$ ■ $I_3 = \begin{bmatrix} 1 & 0 & 0 \\ 0 & 1 & 0 \\ 0 & 0 & 1 \end{bmatrix}$

We can show that if A is any square matrix of order n, then

$$AI_n = A = I_n A.$$

ILLUSTRATION $AI_2 = A = I_2 A$

■ $\begin{bmatrix} a_{11} & a_{12} \\ a_{21} & a_{22} \end{bmatrix} \begin{bmatrix} 1 & 0 \\ 0 & 1 \end{bmatrix} = \begin{bmatrix} a_{11} & a_{12} \\ a_{21} & a_{22} \end{bmatrix} = \begin{bmatrix} 1 & 0 \\ 0 & 1 \end{bmatrix} \begin{bmatrix} a_{11} & a_{12} \\ a_{21} & a_{22} \end{bmatrix}$

Recall that when we are working with a nonzero real number b, the unique number b^{-1} (the multiplicative inverse of b) may be multiplied times b to obtain the multiplicative identity (the number 1)—that is,

$$b \cdot b^{-1} = 1.$$

We have a similar situation with matrices.

Definition of the Inverse of a Matrix

Let A be a square matrix of order n. If there exists a matrix B such that

$$AB = I_n = BA,$$

then B is called the **inverse** of A and is denoted A^{-1} (read "A inverse").

If a square matrix A has an inverse, then we say that A is **invertible.** If a matrix is not square, then it cannot have an inverse. For matrices (unlike real numbers), the symbol $1/A$ does not represent the inverse A^{-1}.

If A is invertible, we can calculate A^{-1} using elementary row operations. If $A = (a_{ij})$ is $n \times n$, we begin with the $n \times 2n$ matrix formed by *adjoining* I_n to A:

$$\left[\begin{array}{cccc|cccc} a_{11} & a_{12} & \cdots & a_{1n} & 1 & 0 & \cdots & 0 \\ a_{21} & a_{22} & \cdots & a_{2n} & 0 & 1 & \cdots & 0 \\ \vdots & \vdots & & \vdots & \vdots & \vdots & & \vdots \\ a_{n1} & a_{n2} & \cdots & a_{nn} & 0 & 0 & \cdots & 1 \end{array}\right]$$

We next apply a succession of elementary row transformations, as we did in Section 9.5 to find reduced echelon forms, until we arrive at a matrix of the form

$$\left[\begin{array}{cccc|cccc} 1 & 0 & \cdots & 0 & b_{11} & b_{12} & \cdots & b_{1n} \\ 0 & 1 & \cdots & 0 & b_{21} & b_{22} & \cdots & b_{2n} \\ \vdots & \vdots & & \vdots & \vdots & \vdots & & \vdots \\ 0 & 0 & \cdots & 1 & b_{n1} & b_{n2} & \cdots & b_{nn} \end{array}\right]$$

in which the identity matrix I_n appears to the left of the vertical rule. It can be shown that the $n \times n$ matrix (b_{ij}) is the inverse of A—that is, $B = A^{-1}$.

EXAMPLE 1 **Finding the inverse of a 2 × 2 matrix**

Find A^{-1} if $A = \begin{bmatrix} 3 & 5 \\ 1 & 4 \end{bmatrix}$.

SOLUTION   We begin with the matrix

$$\left[\begin{array}{cc|cc} 3 & 5 & 1 & 0 \\ 1 & 4 & 0 & 1 \end{array}\right].$$

Next we perform elementary row transformations until the identity matrix I_2 appears on the left of the vertical rule, as follows:

$$\left[\begin{array}{cc|cc} 3 & 5 & 1 & 0 \\ 1 & 4 & 0 & 1 \end{array}\right] \xrightarrow{\mathbf{R_1 \leftrightarrow R_2}} \left[\begin{array}{cc|cc} 1 & 4 & 0 & 1 \\ 3 & 5 & 1 & 0 \end{array}\right]$$

$$\xrightarrow{\mathbf{-3R_1 + R_2 \to R_2}} \left[\begin{array}{cc|cc} 1 & 4 & 0 & 1 \\ 0 & -7 & 1 & -3 \end{array}\right]$$

$$\xrightarrow{-\frac{1}{7}\mathbf{R_2 \to R_2}} \left[\begin{array}{cc|cc} 1 & 4 & 0 & 1 \\ 0 & 1 & -\frac{1}{7} & \frac{3}{7} \end{array}\right]$$

$$\xrightarrow{\mathbf{-4R_2 + R_1 \to R_1}} \left[\begin{array}{cc|cc} 1 & 0 & \frac{4}{7} & -\frac{5}{7} \\ 0 & 1 & -\frac{1}{7} & \frac{3}{7} \end{array}\right]$$

By the previous discussion,

$$A^{-1} = \begin{bmatrix} \frac{4}{7} & -\frac{5}{7} \\ -\frac{1}{7} & \frac{3}{7} \end{bmatrix} = \frac{1}{7} \begin{bmatrix} 4 & -5 \\ -1 & 3 \end{bmatrix}.$$

Let us verify that $AA^{-1} = I_2 = A^{-1}A$:

$$\begin{bmatrix} 3 & 5 \\ 1 & 4 \end{bmatrix} \begin{bmatrix} \frac{4}{7} & -\frac{5}{7} \\ -\frac{1}{7} & \frac{3}{7} \end{bmatrix} = \begin{bmatrix} 1 & 0 \\ 0 & 1 \end{bmatrix} = \begin{bmatrix} \frac{4}{7} & -\frac{5}{7} \\ -\frac{1}{7} & \frac{3}{7} \end{bmatrix} \begin{bmatrix} 3 & 5 \\ 1 & 4 \end{bmatrix}$$

EXAMPLE 2   **Finding the inverse of a 3 × 3 matrix**

Find A^{-1} if $A = \begin{bmatrix} -1 & 3 & 1 \\ 2 & 5 & 0 \\ 3 & 1 & -2 \end{bmatrix}$.

SOLUTION

$$\left[\begin{array}{ccc|ccc} -1 & 3 & 1 & 1 & 0 & 0 \\ 2 & 5 & 0 & 0 & 1 & 0 \\ 3 & 1 & -2 & 0 & 0 & 1 \end{array}\right] \xrightarrow{-\mathbf{R}_1 \to \mathbf{R}_1} \left[\begin{array}{ccc|ccc} 1 & -3 & -1 & -1 & 0 & 0 \\ 2 & 5 & 0 & 0 & 1 & 0 \\ 3 & 1 & -2 & 0 & 0 & 1 \end{array}\right]$$

$$\begin{array}{c} \\ -2\mathbf{R}_1 + \mathbf{R}_2 \to \mathbf{R}_2 \\ -3\mathbf{R}_1 + \mathbf{R}_3 \to \mathbf{R}_3 \end{array} \left[\begin{array}{ccc|ccc} 1 & -3 & -1 & -1 & 0 & 0 \\ 0 & 11 & 2 & 2 & 1 & 0 \\ 0 & 10 & 1 & 3 & 0 & 1 \end{array}\right]$$

$$\begin{array}{c} \\ -\mathbf{R}_3 + \mathbf{R}_2 \to \mathbf{R}_2 \\ \\ \end{array} \left[\begin{array}{ccc|ccc} 1 & -3 & -1 & -1 & 0 & 0 \\ 0 & 1 & 1 & -1 & 1 & -1 \\ 0 & 10 & 1 & 3 & 0 & 1 \end{array}\right]$$

$$\begin{array}{c} 3\mathbf{R}_2 + \mathbf{R}_1 \to \mathbf{R}_1 \\ \\ -10\mathbf{R}_2 + \mathbf{R}_3 \to \mathbf{R}_3 \end{array} \left[\begin{array}{ccc|ccc} 1 & 0 & 2 & -4 & 3 & -3 \\ 0 & 1 & 1 & -1 & 1 & -1 \\ 0 & 0 & -9 & 13 & -10 & 11 \end{array}\right]$$

$$\begin{array}{c} \\ \\ -\frac{1}{9}\mathbf{R}_3 \to \mathbf{R}_3 \end{array} \left[\begin{array}{ccc|ccc} 1 & 0 & 2 & -4 & 3 & -3 \\ 0 & 1 & 1 & -1 & 1 & -1 \\ 0 & 0 & 1 & -\frac{13}{9} & \frac{10}{9} & -\frac{11}{9} \end{array}\right]$$

$$\begin{array}{c} -2\mathbf{R}_3 + \mathbf{R}_1 \to \mathbf{R}_1 \\ -\mathbf{R}_3 + \mathbf{R}_2 \to \mathbf{R}_2 \\ \\ \end{array} \left[\begin{array}{ccc|ccc} 1 & 0 & 0 & -\frac{10}{9} & \frac{7}{9} & -\frac{5}{9} \\ 0 & 1 & 0 & \frac{4}{9} & -\frac{1}{9} & \frac{2}{9} \\ 0 & 0 & 1 & -\frac{13}{9} & \frac{10}{9} & -\frac{11}{9} \end{array}\right]$$

Consequently,

$$A^{-1} = \begin{bmatrix} -\frac{10}{9} & \frac{7}{9} & -\frac{5}{9} \\ \frac{4}{9} & -\frac{1}{9} & \frac{2}{9} \\ -\frac{13}{9} & \frac{10}{9} & -\frac{11}{9} \end{bmatrix} = \frac{1}{9} \begin{bmatrix} -10 & 7 & -5 \\ 4 & -1 & 2 \\ -13 & 10 & -11 \end{bmatrix}.$$

You may verify that $AA^{-1} = I_3 = A^{-1}A$.

Not all square matrices are invertible. In fact, if the procedure used in Examples 1 and 2 does not lead to an identity matrix to the left of the vertical rule, then the matrix A has no inverse—that is, A is not invertible.

We may apply inverses of matrices to solutions of systems of linear equations. Consider the case of two linear equations in two unknowns:

$$\begin{cases} a_{11}x + a_{12}y = k_1 \\ a_{21}x + a_{22}y = k_2 \end{cases}$$

This system can be expressed in terms of matrices as

$$\begin{bmatrix} a_{11}x + a_{12}y \\ a_{21}x + a_{22}y \end{bmatrix} = \begin{bmatrix} k_1 \\ k_2 \end{bmatrix}.$$

If we let

$$A = \begin{bmatrix} a_{11} & a_{12} \\ a_{21} & a_{22} \end{bmatrix}, \qquad X = \begin{bmatrix} x \\ y \end{bmatrix}, \qquad \text{and} \qquad B = \begin{bmatrix} k_1 \\ k_2 \end{bmatrix},$$

then a *matrix form* for the system is

$$AX = B.$$

If A^{-1} exists, then multiplying both sides of the last equation by A^{-1} gives us $A^{-1}AX = A^{-1}B$. Since $A^{-1}A = I_2$ and $I_2X = X$, this leads to

$$X = A^{-1}B,$$

from which the solution (x, y) may be found. This technique (which we refer to as the *inverse method*) may be extended to systems of n linear equations in n unknowns.

EXAMPLE 3   Solving a system of linear equations using the inverse method

Solve the system of equations:

$$\begin{cases} -x + 3y + z = 1 \\ 2x + 5y = 3 \\ 3x + y - 2z = -2 \end{cases}$$

SOLUTION   If we let

$$A = \begin{bmatrix} -1 & 3 & 1 \\ 2 & 5 & 0 \\ 3 & 1 & -2 \end{bmatrix}, \qquad X = \begin{bmatrix} x \\ y \\ z \end{bmatrix}, \qquad \text{and} \qquad B = \begin{bmatrix} 1 \\ 3 \\ -2 \end{bmatrix},$$

then $AX = B$. This implies that $X = A^{-1}B$. The matrix A^{-1} was found in Example 2. Hence,

$$\begin{bmatrix} x \\ y \\ z \end{bmatrix} = \frac{1}{9}\begin{bmatrix} -10 & 7 & -5 \\ 4 & -1 & 2 \\ -13 & 10 & -11 \end{bmatrix}\begin{bmatrix} 1 \\ 3 \\ -2 \end{bmatrix} = \frac{1}{9}\begin{bmatrix} 21 \\ -3 \\ 39 \end{bmatrix} = \begin{bmatrix} \frac{7}{3} \\ -\frac{1}{3} \\ \frac{13}{3} \end{bmatrix}.$$

Thus, $x = \frac{7}{3}$, $y = -\frac{1}{3}$, $z = \frac{13}{3}$, and the ordered triple $\left(\frac{7}{3}, -\frac{1}{3}, \frac{13}{3}\right)$ is the solution of the given system.

If we are solving a system of linear equations without the aid of any computational device, then the inverse method of solution in Example 3 is beneficial only if A^{-1} is known (or can be easily computed) or if many systems with the same coefficient matrix are to be considered.

If we are using a computational device and if the coefficient matrix is not invertible, then the inverse method cannot be used, and the preferred method of solution is the matrix method discussed in Section 9.5. There are other important uses for the inverse of a matrix that arise in more advanced fields of mathematics and in applications of such fields.

9.7   *Exercises*

Exer. 1–2: Show that B is the inverse of A.

1 $A = \begin{bmatrix} 5 & 7 \\ 2 & 3 \end{bmatrix}$, $B = \begin{bmatrix} 3 & -7 \\ -2 & 5 \end{bmatrix}$

2 $A = \begin{bmatrix} 8 & -5 \\ -3 & 2 \end{bmatrix}$, $B = \begin{bmatrix} 2 & 5 \\ 3 & 8 \end{bmatrix}$

Exer. 3–12: Find the inverse of the matrix if it exists.

3 $\begin{bmatrix} 2 & -4 \\ 1 & 3 \end{bmatrix}$

4 $\begin{bmatrix} 3 & 2 \\ 4 & 5 \end{bmatrix}$

5 $\begin{bmatrix} 2 & 4 \\ 4 & 8 \end{bmatrix}$

6 $\begin{bmatrix} 3 & -1 \\ 6 & -2 \end{bmatrix}$

7 $\begin{bmatrix} 3 & -1 & 0 \\ 2 & 2 & 0 \\ 0 & 0 & 4 \end{bmatrix}$

8 $\begin{bmatrix} 3 & 0 & 2 \\ 0 & 1 & 0 \\ -4 & 0 & 2 \end{bmatrix}$

9 $\begin{bmatrix} -2 & 2 & 3 \\ 1 & -1 & 0 \\ 0 & 1 & 4 \end{bmatrix}$

10 $\begin{bmatrix} 1 & 2 & 3 \\ -2 & 1 & 0 \\ 3 & -1 & 1 \end{bmatrix}$

11 $\begin{bmatrix} 2 & 0 & 0 \\ 0 & 4 & 0 \\ 0 & 0 & 6 \end{bmatrix}$

12 $\begin{bmatrix} 1 & 1 & 1 \\ 2 & 2 & 2 \\ 3 & 3 & 3 \end{bmatrix}$

13 State conditions on a and b that guarantee that the matrix $\begin{bmatrix} a & 0 \\ 0 & b \end{bmatrix}$ has an inverse, and find a formula for the inverse if it exists.

14 If $abc \neq 0$, find the inverse of $\begin{bmatrix} a & 0 & 0 \\ 0 & b & 0 \\ 0 & 0 & c \end{bmatrix}$.

15 If $A = \begin{bmatrix} a_{11} & a_{12} & a_{13} \\ a_{21} & a_{22} & a_{23} \\ a_{31} & a_{32} & a_{33} \end{bmatrix}$, show that $AI_3 = A = I_3A$.

16 Show that $AI_4 = A = I_4A$ for every square matrix A of order 4.

Exer. 17–20: Solve the system using the inverse method. Refer to Exercises 3–4 and 9–10.

17 $\begin{cases} 2x - 4y = c \\ x + 3y = d \end{cases}$

(a) $\begin{bmatrix} c \\ d \end{bmatrix} = \begin{bmatrix} 3 \\ 1 \end{bmatrix}$ (b) $\begin{bmatrix} c \\ d \end{bmatrix} = \begin{bmatrix} -2 \\ 5 \end{bmatrix}$

18 $\begin{cases} 3x + 2y = c \\ 4x + 5y = d \end{cases}$

(a) $\begin{bmatrix} c \\ d \end{bmatrix} = \begin{bmatrix} -1 \\ 1 \end{bmatrix}$ (b) $\begin{bmatrix} c \\ d \end{bmatrix} = \begin{bmatrix} 4 \\ 3 \end{bmatrix}$

19 $\begin{cases} -2x + 2y + 3z = c \\ x - y \quad\quad = d \\ \quad\quad y + 4z = e \end{cases}$

(a) $\begin{bmatrix} c \\ d \\ e \end{bmatrix} = \begin{bmatrix} 1 \\ 3 \\ -2 \end{bmatrix}$ (b) $\begin{bmatrix} c \\ d \\ e \end{bmatrix} = \begin{bmatrix} -1 \\ 0 \\ 4 \end{bmatrix}$

20 $\begin{cases} x + 2y + 3z = c \\ -2x + y \quad\quad = d \\ 3x - y + z = e \end{cases}$

(a) $\begin{bmatrix} c \\ d \\ e \end{bmatrix} = \begin{bmatrix} -1 \\ 4 \\ 2 \end{bmatrix}$ (b) $\begin{bmatrix} c \\ d \\ e \end{bmatrix} = \begin{bmatrix} -3 \\ -2 \\ 1 \end{bmatrix}$

9.8

Determinants

Associated with each square matrix A is a number called the **determinant of A,** denoted by $|A|$. This notation should not be confused with the symbol for the absolute value of a real number. To avoid any misunderstanding, the expression "det A" is sometimes used in place of $|A|$. We shall define $|A|$ by beginning with the case in which A has order 1 and then increasing the order one at a time. As we shall see in Section 9.9, these definitions arise in a natural way when systems of linear equations are solved.

If A is a square matrix of order 1, then A has only one element. Thus, $A = [a_{11}]$ and we define $|A| = a_{11}$. If A is a square matrix of order 2, then

$$A = \begin{bmatrix} a_{11} & a_{12} \\ a_{21} & a_{22} \end{bmatrix},$$

and the determinant of A is defined by

$$|A| = a_{11}a_{22} - a_{21}a_{12}.$$

Another notation for $|A|$ is obtained by replacing the brackets used for A with vertical bars, as follows.

Definition of the Determinant of a 2 × 2 Matrix A	$|A| = \begin{vmatrix} a_{11} & a_{12} \\ a_{21} & a_{22} \end{vmatrix} = a_{11}a_{22} - a_{21}a_{12}$

EXAMPLE 1 **Finding the determinant of a 2 × 2 matrix**

Find $|A|$ if $A = \begin{bmatrix} 2 & -1 \\ 4 & -3 \end{bmatrix}$.

SOLUTION By definition,

$$|A| = \begin{vmatrix} 2 & -1 \\ 4 & -3 \end{vmatrix} = (2)(-3) - (4)(-1) = -6 + 4 = -2.$$

To assist in finding determinants for square matrices of order $n > 1$, we introduce the following terminology.

Definition of Minors and Cofactors	Let $A = (a_{ij})$ be a square matrix of order $n > 1$. **(1)** The **minor** M_{ij} of the element a_{ij} is the determinant of the matrix of order $n - 1$ obtained by deleting row i and column j. **(2)** The **cofactor** A_{ij} of the element a_{ij} is $A_{ij} = (-1)^{i+j}M_{ij}$.

To determine the minor of an element, we delete the row and column in which the element appears and then find the determinant of the resulting square matrix. This process is demonstrated in the following illustration, where deletions of rows and columns in a 3 × 3 matrix are indicated with horizontal and vertical line segments, respectively.

To obtain the cofactor of a_{ij} of a square matrix $A = (a_{ij})$, we find the minor and multiply it by 1 or -1, depending on whether the sum of i and j is even or odd, respectively, as demonstrated in the illustration.

ILLUSTRATION **Minors and Cofactors**

Matrix	**Minor**	**Cofactor**
■ $\begin{bmatrix} a_{11} & a_{12} & a_{13} \\ a_{21} & a_{22} & a_{23} \\ a_{31} & a_{32} & a_{33} \end{bmatrix}$	$M_{11} = \begin{vmatrix} a_{22} & a_{23} \\ a_{32} & a_{33} \end{vmatrix}$ $= a_{22}a_{33} - a_{32}a_{23}$	$A_{11} = (-1)^{1+1}M_{11} = M_{11}$
■ $\begin{bmatrix} a_{11} & a_{12} & a_{13} \\ a_{21} & a_{22} & a_{23} \\ a_{31} & a_{32} & a_{33} \end{bmatrix}$	$M_{12} = \begin{vmatrix} a_{21} & a_{23} \\ a_{31} & a_{33} \end{vmatrix}$ $= a_{21}a_{33} - a_{31}a_{23}$	$A_{12} = (-1)^{1+2}M_{12} = -M_{12}$
■ $\begin{bmatrix} a_{11} & a_{12} & a_{13} \\ a_{21} & a_{22} & a_{23} \\ a_{31} & a_{32} & a_{33} \end{bmatrix}$	$M_{23} = \begin{vmatrix} a_{11} & a_{12} \\ a_{31} & a_{32} \end{vmatrix}$ $= a_{11}a_{32} - a_{31}a_{12}$	$A_{23} = (-1)^{2+3}M_{23} = -M_{23}$

For the matrix in the preceding illustration, there are six other minors—$M_{13}, M_{21}, M_{22}, M_{31}, M_{32}$, and M_{33}—that can be obtained in similar fashion.

Another way to remember the sign $(-1)^{i+j}$ associated with the cofactor A_{ij} is to consider the following checkerboard style of plus and minus signs:

$$\begin{bmatrix} + & - & + & - & \cdots \\ - & + & - & + & \cdots \\ + & - & + & - & \cdots \\ - & + & - & + & \cdots \\ \vdots & \vdots & \vdots & \vdots & \end{bmatrix}$$

EXAMPLE 2 **Finding minors and cofactors**

If $A = \begin{bmatrix} 1 & -3 & 3 \\ 4 & 2 & 0 \\ -2 & -7 & 5 \end{bmatrix}$, find $M_{11}, M_{21}, M_{22}, A_{11}, A_{21}$, and A_{22}.

SOLUTION   Deleting appropriate rows and columns of A, we obtain

$$M_{11} = \begin{vmatrix} 2 & 0 \\ -7 & 5 \end{vmatrix} = (2)(5) - (-7)(0) = 10$$

$$M_{21} = \begin{vmatrix} -3 & 3 \\ -7 & 5 \end{vmatrix} = (-3)(5) - (-7)(3) = 6$$

$$M_{22} = \begin{vmatrix} 1 & 3 \\ -2 & 5 \end{vmatrix} = (1)(5) - (-2)(3) = 11.$$

To obtain the cofactors, we prefix the corresponding minors with the proper signs. Thus, using the definition of cofactor, we have

$$A_{11} = (-1)^{1+1}M_{11} = (1)(10) = 10$$
$$A_{21} = (-1)^{2+1}M_{21} = (-1)(6) = -6$$
$$A_{22} = (-1)^{2+2}M_{22} = (1)(11) = 11.$$

We can also use the checkerboard style of plus and minus signs to determine the proper signs. ◤

The determinant $|A|$ of a square matrix of order 3 is defined as follows.

Definition of the Determinant of a 3 × 3 Matrix A	$|A| = \begin{vmatrix} a_{11} & a_{12} & a_{13} \\ a_{21} & a_{22} & a_{23} \\ a_{31} & a_{32} & a_{33} \end{vmatrix} = a_{11}A_{11} + a_{12}A_{12} + a_{13}A_{13}$

Since cofactors $A_{11} = (-1)^{1+1}M_{11} = M_{11}$, $A_{12} = (-1)^{1+2}M_{12} = -M_{12}$, and $A_{13} = (-1)^{1+3}M_{13} = M_{13}$, the preceding definition may also be written

$$|A| = a_{11}M_{11} - a_{12}M_{12} + a_{13}M_{13}.$$

If we express M_{11}, M_{12}, and M_{13} using elements of A and rearrange terms, we obtain the following formula for $|A|$:

$$|A| = a_{11}a_{22}a_{33} - a_{11}a_{23}a_{32} - a_{12}a_{21}a_{33} + a_{12}a_{23}a_{31} + a_{13}a_{21}a_{32} - a_{13}a_{22}a_{31}$$

The definition of $|A|$ for a square matrix A of order 3 displays a pattern of multiplying each element in row 1 by its cofactor and then adding to find $|A|$. This process is referred to as *expanding $|A|$ by the first row*. By actually carrying out the computations, we can show that $|A|$ *can be expanded in similar fashion by using any row or column*. As an illustration, the expansion by the second column is

$$|A| = a_{12}A_{12} + a_{22}A_{22} + a_{32}A_{32}$$

$$= a_{12}\left(-\begin{vmatrix} a_{21} & a_{23} \\ a_{31} & a_{33} \end{vmatrix}\right) + a_{22}\left(+\begin{vmatrix} a_{11} & a_{13} \\ a_{31} & a_{33} \end{vmatrix}\right) + a_{32}\left(-\begin{vmatrix} a_{11} & a_{13} \\ a_{21} & a_{23} \end{vmatrix}\right).$$

Applying the definition to the determinants in parentheses, multiplying as indicated, and rearranging the terms in the sum, we could arrive at the formula for $|A|$ in terms of the elements of A. Similarly, the expansion by the third row is

$$|A| = a_{31}A_{31} + a_{32}A_{32} + a_{33}A_{33}.$$

Once again we can show that this result agrees with previous expansions.

EXAMPLE 3   Finding the determinant of a 3 × 3 matrix

Find $|A|$ if $A = \begin{bmatrix} -1 & 3 & 1 \\ 2 & 5 & 0 \\ 3 & 1 & -2 \end{bmatrix}$.

SOLUTION   Since the second row contains a zero, we shall expand $|A|$ by that row, because then we need to evaluate only two cofactors. Thus,

$$|A| = (2)A_{21} + (5)A_{22} + (0)A_{23}.$$

Using the definition of cofactors, we have

$$A_{21} = (-1)^{2+1}M_{21} = -\begin{vmatrix} 3 & 1 \\ 1 & -2 \end{vmatrix} = -[(3)(-2) - (1)(1)] = 7$$

$$A_{22} = (-1)^{2+2}M_{22} = \begin{vmatrix} -1 & 1 \\ 3 & -2 \end{vmatrix} = [(-1)(-2) - (3)(1)] = -1.$$

Consequently,

$$|A| = (2)(7) + (5)(-1) + (0)A_{23} = 14 - 5 + 0 = 9.$$

The following definition of the determinant of a matrix of arbitrary order n is patterned after that used for the determinant of a matrix of order 3.

Definition of the Determinant of an $n \times n$ Matrix A	The **determinant** $\|A\|$ **of a matrix** A **of order** n is the cofactor expansion by the first row: $$\|A\| = a_{11}A_{11} + a_{12}A_{12} + \cdots + a_{1n}A_{1n}$$ In terms of minors, $$\|A\| = a_{11}M_{11} - a_{12}M_{12} + \cdots + a_{1n}(-1)^{1+n}M_{1n}.$$

The number $|A|$ may be found by using *any* row or column, as stated in the following theorem.

Theorem on Expansion of Determinants	If A is a square matrix of order $n > 1$, then the determinant $\|A\|$ may be found by multiplying the elements of any row (or column) by their respective cofactors and adding the resulting products.

This theorem is useful if many zeros appear in a row or column, as illustrated in the following example.

EXAMPLE 4 Finding the determinant of a 4 × 4 matrix

Find $|A|$ if $A = \begin{bmatrix} 1 & 0 & 2 & 5 \\ -2 & 1 & 5 & 0 \\ 0 & 0 & -3 & 0 \\ 0 & -1 & 0 & 3 \end{bmatrix}$.

SOLUTION Note that all but one of the elements in the third row are zero. Hence, if we expand $|A|$ by the third row, there will be at most one nonzero term. Specifically,

$$|A| = (0)A_{31} + (0)A_{32} + (-3)A_{33} + (0)A_{34} = -3A_{33}$$

with

$$A_{33} = (-1)^{3+3}M_{33} = M_{33} = \begin{vmatrix} 1 & 0 & 5 \\ -2 & 1 & 0 \\ 0 & -1 & 3 \end{vmatrix}.$$

We expand M_{33} by column 1:

$$M_{33} = (1)\begin{vmatrix} 1 & 0 \\ -1 & 3 \end{vmatrix} - (-2)\begin{vmatrix} 0 & 5 \\ -1 & 3 \end{vmatrix} + (0)\begin{vmatrix} 0 & 5 \\ 1 & 0 \end{vmatrix}$$

$$= (1)(3) - (-2)(5) + (0)(-5) = 3 + 10 + 0 = 13$$

Thus,

$$|A| = -3A_{33} = (-3)(13) = -39.$$

In general, if all but one element in some row (or column) of A are zero and if the determinant $|A|$ is expanded by that row (or column), then all terms drop out except the product of that element with its cofactor. If *all* elements in a row (or column) are zero, we have the following.

Theorem on a Row of Zeros	If every element in a row (or column) of a square matrix A is zero, then $|A| = 0$.

PROOF If every element in a row (or column) of a square matrix A is zero, then the expansion by that row (or column) is a sum of terms that are zero (since each term is zero times its respective cofactor). Hence, this sum is equal to zero, and we conclude that $|A| = 0$.

In the previous section we found that if we could not obtain the identity matrix on the left side of the adjoined matrix, then the original matrix was not invertible. If we obtain a row of zeros in this process, we certainly cannot obtain the identity matrix. Combining this fact with the previous theorem leads to the following theorem.

Theorem on Matrix Invertibility	If A is a square matrix, then A is invertible if and only if $|A| \neq 0$.

9.8 Exercises

Exer. 1–4: Find all the minors and cofactors of the elements in the matrix.

1 $\begin{bmatrix} 7 & -1 \\ 5 & 0 \end{bmatrix}$

2 $\begin{bmatrix} -6 & 4 \\ 3 & 2 \end{bmatrix}$

3 $\begin{bmatrix} 2 & 4 & -1 \\ 0 & 3 & 2 \\ -5 & 7 & 0 \end{bmatrix}$

4 $\begin{bmatrix} 5 & -2 & 1 \\ 4 & 7 & 0 \\ -3 & 4 & -1 \end{bmatrix}$

Exer. 5–8: Find the determinant of the matrix in the given exercise.

5 Exercise 1

6 Exercise 2

7 Exercise 3

8 Exercise 4

Exer. 9–20: Find the determinant of the matrix.

9 $\begin{bmatrix} -5 & 4 \\ -3 & 2 \end{bmatrix}$

10 $\begin{bmatrix} 6 & 4 \\ -3 & 2 \end{bmatrix}$

11 $\begin{bmatrix} a & -a \\ b & -b \end{bmatrix}$

12 $\begin{bmatrix} c & d \\ -d & c \end{bmatrix}$

13 $\begin{bmatrix} 3 & 1 & -2 \\ 4 & 2 & 5 \\ -6 & 3 & -1 \end{bmatrix}$

14 $\begin{bmatrix} 2 & -5 & 1 \\ -3 & 1 & 6 \\ 4 & -2 & 3 \end{bmatrix}$

15 $\begin{bmatrix} -5 & 4 & 1 \\ 3 & -2 & 7 \\ 2 & 0 & 6 \end{bmatrix}$

16 $\begin{bmatrix} 2 & 7 & -3 \\ 1 & 0 & 4 \\ 4 & -1 & -2 \end{bmatrix}$

17 $\begin{bmatrix} 3 & -1 & 2 & 0 \\ 4 & 0 & -3 & 5 \\ 0 & 6 & 0 & 0 \\ 1 & 3 & -4 & 2 \end{bmatrix}$

18 $\begin{bmatrix} 2 & 5 & 1 & 0 \\ -4 & 0 & -3 & 0 \\ 3 & -2 & 1 & 6 \\ -1 & 4 & 2 & 0 \end{bmatrix}$

19 $\begin{bmatrix} 0 & b & 0 & 0 \\ 0 & 0 & c & 0 \\ a & 0 & 0 & 0 \\ 0 & 0 & 0 & d \end{bmatrix}$

20 $\begin{bmatrix} a & u & v & w \\ 0 & b & x & y \\ 0 & 0 & c & z \\ 0 & 0 & 0 & d \end{bmatrix}$

Exer. 21–28: Verify the identity by expanding each determinant.

21 $\begin{vmatrix} a & b \\ c & d \end{vmatrix} = - \begin{vmatrix} c & d \\ a & b \end{vmatrix}$

22 $\begin{vmatrix} a & b \\ c & d \end{vmatrix} = - \begin{vmatrix} b & a \\ d & c \end{vmatrix}$

23 $\begin{vmatrix} a & kb \\ c & kd \end{vmatrix} = k \begin{vmatrix} a & b \\ c & d \end{vmatrix}$

24 $\begin{vmatrix} a & b \\ kc & kd \end{vmatrix} = k \begin{vmatrix} a & b \\ c & d \end{vmatrix}$

25 $\begin{vmatrix} a & b \\ c & d \end{vmatrix} = \begin{vmatrix} a & b \\ ka + c & kb + d \end{vmatrix}$

26 $\begin{vmatrix} a & b \\ c & d \end{vmatrix} = \begin{vmatrix} a & ka + b \\ c & kc + d \end{vmatrix}$

27 $\begin{vmatrix} a & b \\ c & d \end{vmatrix} + \begin{vmatrix} a & e \\ c & f \end{vmatrix} = \begin{vmatrix} a & b + e \\ c & d + f \end{vmatrix}$

28 $\begin{vmatrix} a & b \\ c & d \end{vmatrix} + \begin{vmatrix} a & b \\ e & f \end{vmatrix} = \begin{vmatrix} a & b \\ c + e & d + f \end{vmatrix}$

29 Let $A = (a_{ij})$ be a square matrix of order n such that $a_{ij} = 0$ if $i < j$. Show that $|A| = a_{11}a_{22} \cdots a_{nn}$.

30 If $A = (a_{ij})$ is any 2×2 matrix such that $|A| \neq 0$, show that A has an inverse, and find a general formula for A^{-1}.

Exer. 31–34: Let $I = I_2$ be the identity matrix of order 2, and let $f(x) = |A - xI|$. Find (a) the polynomial $f(x)$ and (b) the zeros of $f(x)$. (In the study of matrices, $f(x)$ is the *characteristic polynomial of A*, and the zeros of $f(x)$ are the *characteristic values (eigenvalues) of A*.)

31 $A = \begin{bmatrix} 1 & 2 \\ 3 & 2 \end{bmatrix}$

32 $A = \begin{bmatrix} 3 & 1 \\ 2 & 2 \end{bmatrix}$

33 $A = \begin{bmatrix} -3 & -2 \\ 2 & 2 \end{bmatrix}$

34 $A = \begin{bmatrix} 2 & -4 \\ -3 & 5 \end{bmatrix}$

Exer. 35–38: Let $I = I_3$ and let $f(x) = |A - xI|$. Find (a) the polynomial $f(x)$ and (b) the zeros of $f(x)$.

35 $A = \begin{bmatrix} 1 & 0 & 0 \\ 1 & 0 & -2 \\ -1 & 1 & -3 \end{bmatrix}$

36 $A = \begin{bmatrix} 2 & 1 & 0 \\ -1 & 0 & 0 \\ 1 & 3 & 2 \end{bmatrix}$

37 $A = \begin{bmatrix} 0 & 2 & -2 \\ -1 & 3 & 1 \\ -3 & 3 & 1 \end{bmatrix}$

38 $A = \begin{bmatrix} 3 & 2 & 2 \\ 1 & 0 & 2 \\ -1 & -1 & 0 \end{bmatrix}$

Exer. 39–42: Express the determinant in the form
$ai + bj + ck$ **for real numbers** a, b, **and** c.

39 $\begin{vmatrix} i & j & k \\ 2 & -1 & 6 \\ -3 & 5 & 1 \end{vmatrix}$
40 $\begin{vmatrix} i & j & k \\ 1 & -2 & 3 \\ 2 & 1 & -4 \end{vmatrix}$
41 $\begin{vmatrix} i & j & k \\ 5 & -6 & -1 \\ 3 & 0 & 1 \end{vmatrix}$
42 $\begin{vmatrix} i & j & k \\ 4 & -6 & 2 \\ -2 & 3 & -1 \end{vmatrix}$

9.9

Properties of Determinants

Evaluating a determinant by using the expansion theorem stated in Section 9.8 is inefficient for matrices of high order. For example, if a determinant of a matrix of order 10 is expanded by any row, a sum of 10 terms is obtained, and each term contains the determinant of a matrix of order 9, which is a cofactor of the original matrix. If any of the latter determinants is expanded by a row (or column), a sum of 9 terms is obtained, each containing the determinant of a matrix of order 8. Hence, at this stage there are 90 determinants of matrices of order 8 to evaluate. The process could be continued until only determinants of matrices of order 2 remain. You may verify that there are 1,814,400 such matrices of order 2! Unless many elements of the original matrix are zero, it is an enormous task to carry out all of the computations.

In this section we discuss rules that simplify the process of evaluating determinants. The main use for these rules is to introduce zeros into the determinant. They may also be used to change the determinant to **echelon form**—that is, to a form in which the elements below the main diagonal elements are all zero (see Section 9.5). The transformations on rows stated in the next theorem are the same as the elementary row transformations of a matrix introduced in Section 9.5. However, for determinants we may also use similar transformations on columns.

Theorem on Row and Column Transformations of a Determinant	Let A be a square matrix of order n.
	(1) If a matrix B is obtained from A by interchanging two rows (or columns), then $\lvert B \rvert = -\lvert A \rvert$.
	(2) If B is obtained from A by multiplying every element of one row (or column) of A by a real number k, then $\lvert B \rvert = k\lvert A \rvert$.
	(3) If B is obtained from A by adding k times any row (or column) of A to another row (or column) for a real number k, then $\lvert B \rvert = \lvert A \rvert$—that is, the determinants of B and A are equal.

When using the theorem, we refer to the rows (or columns) of the *determinant* in the obvious way. For example, property 3 may be phrased as follows: *Adding k times any row (or column) to another row (or column) of a determinant does not affect the value of the determinant.*

Row transformations of determinants will be specified by means of the symbols $R_i \leftrightarrow R_j$, $kR_i \rightarrow R_i$, and $kR_i + R_j \rightarrow R_j$, which were introduced in Section 9.5. Analogous symbols are used for column transformations. For example, $kC_i + C_j \rightarrow C_j$ means "add k times the ith column to the jth column."

Property 2 of the theorem on row and column transformations is useful for finding factors of determinants. To illustrate, for a determinant of a matrix of order 3, we have the following:

$$\begin{vmatrix} a_{11} & a_{12} & a_{13} \\ ka_{21} & ka_{22} & ka_{23} \\ a_{31} & a_{32} & a_{33} \end{vmatrix} = k \begin{vmatrix} a_{11} & a_{12} & a_{13} \\ a_{21} & a_{22} & a_{23} \\ a_{31} & a_{32} & a_{33} \end{vmatrix}$$

Similar formulas hold if k is a common factor of the elements of any other row or column. When referring to this manipulation, we often use the phrase *k is a common factor of the row (or column)*.

The following are illustrations of the preceding theorem, with the reason for each equality stated at the right.

ILLUSTRATION **Transformation of Determinants**

$\blacksquare$ $\begin{vmatrix} 2 & 0 & 1 \\ 6 & 4 & 3 \\ 0 & 3 & 5 \end{vmatrix} = - \begin{vmatrix} 6 & 4 & 3 \\ 2 & 0 & 1 \\ 0 & 3 & 5 \end{vmatrix}$ $R_1 \leftrightarrow R_2$ (property 1)

$\blacksquare$ $\begin{vmatrix} 2 & 0 & 1 \\ 6 & 4 & 3 \\ 0 & 3 & 5 \end{vmatrix} = 2 \begin{vmatrix} 1 & 0 & 1 \\ 3 & 4 & 3 \\ 0 & 3 & 5 \end{vmatrix}$ 2 is a common factor of column 1 (property 2)

$\blacksquare$ $\begin{vmatrix} 1 & -3 & 4 \\ 2 & -1 & 0 \\ 3 & 1 & 6 \end{vmatrix} = \begin{vmatrix} -5 & -3 & 4 \\ 0 & -1 & 0 \\ 5 & 1 & 6 \end{vmatrix}$ $2C_2 + C_1 \rightarrow C_1$ (property 3)

$\blacksquare$ $\begin{vmatrix} 1 & -3 & 4 \\ 2 & -1 & 0 \\ 3 & 1 & 6 \end{vmatrix} = \begin{vmatrix} 1 & -3 & 4 \\ 0 & 5 & -8 \\ 0 & 10 & -6 \end{vmatrix}$ $-2R_1 + R_2 \rightarrow R_2$
$-3R_1 + R_3 \rightarrow R_3$
(property 3 applied twice)

Theorem on Identical Rows	If two rows (or columns) of a square matrix A are identical, then $|A| = 0$.

PROOF   If B is the matrix obtained from A by interchanging the two identical rows (or columns), then B and A are the same and, consequently, $|B| = |A|$. However, by property 1 of the theorem on row and column transformations of a determinant, $|B| = -|A|$, and hence $-|A| = |A|$. Thus, $2|A| = 0$, and therefore $|A| = 0$. $\blacksquare$

EXAMPLE 1 Using row and column transformations

Find $|A|$ if $A = \begin{bmatrix} 2 & 3 & 0 & 4 \\ 0 & 5 & -1 & 6 \\ 1 & 0 & -2 & 3 \\ -3 & 2 & 0 & -5 \end{bmatrix}$.

SOLUTION We plan to use property 3 of the theorem on row and column transformations of a determinant to introduce three zeros in some row or column. It is convenient to work with an element of the matrix that equals 1, since this enables us to avoid the use of fractions. If 1 is not an element of the original matrix, it is always possible to introduce the number 1 by using property 2 or 3 of the theorem. In this example, 1 appears in row 3, and we proceed as follows, with the reason for each equality stated at the right.

$$\begin{vmatrix} 2 & 3 & 0 & 4 \\ 0 & 5 & -1 & 6 \\ 1 & 0 & -2 & 3 \\ -3 & 2 & 0 & -5 \end{vmatrix} = \begin{vmatrix} 0 & 3 & 4 & -2 \\ 0 & 5 & -1 & 6 \\ 1 & 0 & -2 & 3 \\ 0 & 2 & -6 & 4 \end{vmatrix} \qquad \begin{aligned} -2R_3 + R_1 &\to R_1 \\[3em] 3R_3 + R_4 &\to R_4 \end{aligned}$$

$$= (1) \cdot (-1)^{3+1} \begin{vmatrix} 3 & 4 & -2 \\ 5 & -1 & 6 \\ 2 & -6 & 4 \end{vmatrix} \qquad \begin{aligned} \text{expand by the} \\ \text{first column} \end{aligned}$$

$$= \begin{vmatrix} 23 & 4 & 22 \\ 0 & -1 & 0 \\ -28 & -6 & -32 \end{vmatrix} \qquad \begin{aligned} 5C_2 + C_1 &\to C_1 \\[1.5em] 6C_2 + C_3 &\to C_3 \end{aligned}$$

$$= (-1) \cdot (-1)^{2+2} \begin{vmatrix} 23 & 22 \\ -28 & -32 \end{vmatrix} \qquad \begin{aligned} \text{expand by the} \\ \text{second row} \end{aligned}$$

$$= (-1)[(23)(-32) - (-28)(22)] \qquad \begin{aligned} \text{definition of} \\ \text{determinant} \end{aligned}$$

$$= 120$$

The next two examples illustrate the use of property 2 of the theorem on row and column transformations of a determinant.

EXAMPLE 2 Removing common factors from rows

Find $|A|$ if $A = \begin{bmatrix} 14 & -6 & 4 \\ 4 & -5 & 12 \\ -21 & 9 & -6 \end{bmatrix}$.

SOLUTION

$$|A| = 2 \begin{vmatrix} 7 & -3 & 2 \\ 4 & -5 & 12 \\ -21 & 9 & -6 \end{vmatrix} \qquad \text{2 is a common factor of row 1}$$

$$= (2)(-3) \begin{vmatrix} 7 & -3 & 2 \\ 4 & -5 & 12 \\ 7 & -3 & 2 \end{vmatrix} \qquad \text{-3 is a common factor of row 3}$$

$$= 0 \qquad \text{two rows are identical}$$

EXAMPLE 3 **Removing a common factor from a column**

Without expanding, show that $a - b$ is a factor of $|A|$ if

$$A = \begin{bmatrix} 1 & 1 & 1 \\ a & b & c \\ a^2 & b^2 & c^2 \end{bmatrix}.$$

SOLUTION

$$\begin{vmatrix} 1 & 1 & 1 \\ a & b & c \\ a^2 & b^2 & c^2 \end{vmatrix} = \begin{vmatrix} 0 & 1 & 1 \\ a-b & b & c \\ a^2-b^2 & b^2 & c^2 \end{vmatrix} \qquad -C_2 + C_1 \rightarrow C_1$$

$$= (a-b) \begin{vmatrix} 0 & 1 & 1 \\ 1 & b & c \\ a+b & b^2 & c^2 \end{vmatrix} \qquad \begin{array}{l} a - b \text{ is a common factor} \\ \text{of column 1} \end{array}$$

Hence, $|A|$ is equal to $a - b$ times the last determinant, and so $a - b$ is a factor of $|A|$.

Determinants arise in the study of solutions of systems of linear equations. To illustrate, let us consider two linear equations in two variables x and y:

$$\begin{cases} a_{11}x + a_{12}y = k_1 \\ a_{21}x + a_{22}y = k_2 \end{cases}$$

where at least one nonzero coefficient appears in each equation. We may assume that $a_{11} \neq 0$, for otherwise $a_{12} \neq 0$ and we could then regard y as the first variable instead of x. We shall use elementary row transformations to obtain the matrix of an equivalent system with $a_{21} = 0$, as follows:

$$\begin{bmatrix} a_{11} & a_{12} & \bigm| & k_1 \\ a_{21} & a_{22} & \bigm| & k_2 \end{bmatrix} \; -\frac{a_{21}}{a_{11}}\mathbf{R_1} + \mathbf{R_2} \rightarrow \mathbf{R_2} \begin{bmatrix} a_{11} & a_{12} & \bigm| & k_1 \\ 0 & a_{22} - \left(\dfrac{a_{21}a_{12}}{a_{11}}\right) & \bigm| & k_2 - \left(\dfrac{a_{21}k_1}{a_{11}}\right) \end{bmatrix}$$

$$a_{11}\mathbf{R_2} \rightarrow \mathbf{R_2} \begin{bmatrix} a_{11} & a_{12} & \bigm| & k_1 \\ 0 & (a_{11}a_{22} - a_{21}a_{12}) & \bigm| & (a_{11}k_2 - a_{21}k_1) \end{bmatrix}$$

Thus, the given system is equivalent to

$$\begin{cases} a_{11}x + a_{12}y = k_1 \\ (a_{11}a_{22} - a_{21}a_{12})y = a_{11}k_2 - a_{21}k_1 \end{cases}$$

which may also be written

$$\begin{cases} a_{11}x + a_{12}y = k_1 \\ \begin{vmatrix} a_{11} & a_{12} \\ a_{21} & a_{22} \end{vmatrix} y = \begin{vmatrix} a_{11} & k_1 \\ a_{21} & k_2 \end{vmatrix} \end{cases}$$

If $\begin{vmatrix} a_{11} & a_{12} \\ a_{21} & a_{22} \end{vmatrix} \neq 0$, we can solve the second equation for y, obtaining

$$y = \frac{\begin{vmatrix} a_{11} & k_1 \\ a_{21} & k_2 \end{vmatrix}}{\begin{vmatrix} a_{11} & a_{12} \\ a_{21} & a_{22} \end{vmatrix}}.$$

The corresponding value for x may be found by substituting for y in the first equation, which leads to

The proof of this statement is left as Discussion Exercise 7 at the end of the chapter.

$$x = \frac{\begin{vmatrix} k_1 & a_{12} \\ k_2 & a_{22} \end{vmatrix}}{\begin{vmatrix} a_{11} & a_{12} \\ a_{21} & a_{22} \end{vmatrix}}. \qquad (*)$$

This proves that *if the determinant of the coefficient matrix of a system of two linear equations in two variables is not zero, then the system has a unique solution.* The last two formulas for x and y as quotients of determinants constitute **Cramer's rule** for two variables.

There is an easy way to remember Cramer's rule. Let

$$D = \begin{bmatrix} a_{11} & a_{12} \\ a_{21} & a_{22} \end{bmatrix}$$

be the coefficient matrix of the system, and let D_x denote the matrix obtained from D by replacing the coefficients a_{11}, a_{21} of x by the numbers k_1, k_2, respectively. Similarly, let D_y denote the matrix obtained from D by replacing the coefficients a_{12}, a_{22} of y by the numbers k_1, k_2, respectively. Thus,

$$D_x = \begin{bmatrix} k_1 & a_{12} \\ k_2 & a_{22} \end{bmatrix}, \quad D_y = \begin{bmatrix} a_{11} & k_1 \\ a_{21} & k_2 \end{bmatrix}.$$

If $|D| \neq 0$, the solution (x, y) is given by the following formulas.

Cramer's Rule for Two Variables

$$x = \frac{|D_x|}{|D|}, \quad y = \frac{|D_y|}{|D|}$$

 EXAMPLE 4 **Using Cramer's rule to solve a system of two linear equations**

Use Cramer's rule to solve the system

$$\begin{cases} 2x - 3y = -4 \\ 5x + 7y = 1 \end{cases}$$

SOLUTION The determinant of the coefficient matrix is

$$|D| = \begin{vmatrix} 2 & -3 \\ 5 & 7 \end{vmatrix} = 29.$$

Using the notation introduced previously, we have

$$|D_x| = \begin{vmatrix} -4 & -3 \\ 1 & 7 \end{vmatrix} = -25, \quad |D_y| = \begin{vmatrix} 2 & -4 \\ 5 & 1 \end{vmatrix} = 22.$$

Hence, $\qquad x = \dfrac{|D_x|}{|D|} = \dfrac{-25}{29}, \quad y = \dfrac{|D_y|}{|D|} = \dfrac{22}{29}.$

Thus, the system has the unique solution $\left(-\frac{25}{29}, \frac{22}{29}\right)$.

Cramer's rule can be extended to systems of n linear equations in n variables $x_1, x_2, \ldots, x_n$, where the ith equation has the form

$$a_{i1}x_1 + a_{i2}x_2 + \cdots + a_{in}x_n = k_i.$$

To solve such a system, let D denote the coefficient matrix and let D_{x_j} denote the matrix obtained by replacing the coefficients of x_j in D by the numbers $k_1, \ldots, k_n$ that appear in the column to the right of the equal signs in the system. If $|D| \neq 0$, then the system has the following unique solution.

Cramer's Rule (General Form)

$$x_1 = \frac{|D_{x_1}|}{|D|}, \quad x_2 = \frac{|D_{x_2}|}{|D|}, \quad \ldots, \quad x_n = \frac{|D_{x_n}|}{|D|}$$

EXAMPLE 5 **Using Cramer's rule to solve a system of three linear equations**

Use Cramer's rule to solve the system

$$\begin{cases} x - 2z = 3 \\ - y + 3z = 1 \\ 2x + 5z = 0 \end{cases}$$

SOLUTION We shall merely list the various determinants. You should check the results.

$$|D| = \begin{vmatrix} 1 & 0 & -2 \\ 0 & -1 & 3 \\ 2 & 0 & 5 \end{vmatrix} = -9, \quad |D_x| = \begin{vmatrix} 3 & 0 & -2 \\ 1 & -1 & 3 \\ 0 & 0 & 5 \end{vmatrix} = -15$$

$$|D_y| = \begin{vmatrix} 1 & 3 & -2 \\ 0 & 1 & 3 \\ 2 & 0 & 5 \end{vmatrix} = 27, \quad |D_z| = \begin{vmatrix} 1 & 0 & 3 \\ 0 & -1 & 1 \\ 2 & 0 & 0 \end{vmatrix} = 6$$

By Cramer's rule, the solution is

$$x = \frac{|D_x|}{|D|} = \frac{-15}{-9} = \frac{5}{3}, \quad y = \frac{|D_y|}{|D|} = \frac{27}{-9} = -3, \quad z = \frac{|D_z|}{|D|} = \frac{6}{-9} = -\frac{2}{3}.$$

Cramer's rule is an inefficient method to apply if the system has a large number of equations, since many determinants of matrices of high order must be evaluated. Note also that Cramer's rule cannot be used directly if $|D| = 0$ or if the number of equations is not the same as the number of variables. For numerical calculations, the inverse method and the matrix method are superior to Cramer's rule; however, the Cramer's rule formulation is theoretically useful.

9.9 Exercises

Exer. 1–14: Without expanding, explain why the statement is true.

1 $\begin{vmatrix} 1 & 0 & 1 \\ 0 & 1 & 1 \\ 1 & 1 & 0 \end{vmatrix} = - \begin{vmatrix} 1 & 0 & 1 \\ 1 & 1 & 0 \\ 0 & 1 & 1 \end{vmatrix}$

2 $\begin{vmatrix} 1 & 0 & 1 \\ 0 & 1 & 1 \\ 1 & 1 & 0 \end{vmatrix} = - \begin{vmatrix} 1 & 1 & 0 \\ 0 & 1 & 1 \\ 1 & 0 & 1 \end{vmatrix}$

3 $\begin{vmatrix} 1 & 0 & 1 \\ 2 & 1 & 0 \\ 1 & 1 & 2 \end{vmatrix} = \begin{vmatrix} 1 & 0 & 1 \\ 2 & 1 & 0 \\ 0 & 1 & 1 \end{vmatrix}$

4 $\begin{vmatrix} 1 & 1 & 2 \\ 1 & 0 & 1 \\ 2 & 1 & 1 \end{vmatrix} = \begin{vmatrix} 0 & 1 & 1 \\ 1 & 0 & 1 \\ 2 & 1 & 1 \end{vmatrix}$

5 $\begin{vmatrix} 2 & 4 & 2 \\ 1 & 2 & 4 \\ 2 & 6 & 4 \end{vmatrix} = 4 \begin{vmatrix} 1 & 2 & 1 \\ 1 & 2 & 4 \\ 1 & 3 & 2 \end{vmatrix}$

6 $\begin{vmatrix} 2 & 1 & 6 \\ 4 & 3 & 3 \\ 2 & 1 & 3 \end{vmatrix} = 6 \begin{vmatrix} 1 & 1 & 2 \\ 2 & 3 & 1 \\ 1 & 1 & 1 \end{vmatrix}$

7 $\begin{vmatrix} 1 & -1 & 2 \\ 1 & 2 & -1 \\ 1 & -1 & 2 \end{vmatrix} = 0$

8 $\begin{vmatrix} 1 & -1 & 1 \\ 0 & 1 & 0 \\ -1 & 1 & -1 \end{vmatrix} = 0$

9 $\begin{vmatrix} 1 & 5 \\ -3 & 2 \end{vmatrix} = -\begin{vmatrix} 1 & 5 \\ 3 & -2 \end{vmatrix}$

10 $\begin{vmatrix} 2 & -2 \\ 1 & 1 \end{vmatrix} = -\begin{vmatrix} -2 & 2 \\ 1 & 1 \end{vmatrix}$

11 $\begin{vmatrix} 0 & 0 & 1 \\ 1 & 0 & 0 \\ 0 & 0 & 2 \end{vmatrix} = 0$ 12 $\begin{vmatrix} 1 & 0 & 1 \\ 0 & 0 & 0 \\ 1 & 1 & 0 \end{vmatrix} = 0$

13 $\begin{vmatrix} 1 & -1 & -2 \\ -1 & 2 & 1 \\ 0 & 1 & 1 \end{vmatrix} = \begin{vmatrix} 1 & -1 & 0 \\ -1 & 2 & -1 \\ 0 & 1 & 1 \end{vmatrix}$

14 $\begin{vmatrix} a & 0 & 0 \\ 0 & b & 0 \\ 0 & 0 & c \end{vmatrix} = -\begin{vmatrix} 0 & 0 & a \\ 0 & b & 0 \\ c & 0 & 0 \end{vmatrix}$

Exer. 15–24: Find the determinant of the matrix after introducing zeros, as in Example 1.

15 $\begin{bmatrix} 3 & 1 & 0 \\ -2 & 0 & 1 \\ 1 & 3 & -1 \end{bmatrix}$ 16 $\begin{bmatrix} -3 & 0 & 4 \\ 1 & 2 & 0 \\ 4 & 1 & -1 \end{bmatrix}$

17 $\begin{bmatrix} 5 & 4 & 3 \\ -3 & 2 & 1 \\ 0 & 7 & -2 \end{bmatrix}$ 18 $\begin{bmatrix} 0 & 2 & -6 \\ 5 & 1 & -3 \\ 6 & -2 & 5 \end{bmatrix}$

19 $\begin{bmatrix} 2 & 2 & -3 \\ 3 & 6 & 9 \\ -2 & 5 & 4 \end{bmatrix}$ 20 $\begin{bmatrix} 3 & 8 & 5 \\ 5 & 3 & -6 \\ 2 & 4 & -2 \end{bmatrix}$

21 $\begin{bmatrix} 3 & 1 & -2 & 2 \\ 2 & 0 & 1 & 4 \\ 0 & 1 & 3 & 5 \\ -1 & 2 & 0 & -3 \end{bmatrix}$ 22 $\begin{bmatrix} 3 & 2 & 0 & 4 \\ -2 & 0 & 5 & 0 \\ 4 & -3 & 1 & 6 \\ 2 & -1 & 2 & 0 \end{bmatrix}$

23 $\begin{bmatrix} 2 & -2 & 0 & 0 & -3 \\ 3 & 0 & 3 & 2 & -1 \\ 0 & 1 & -2 & 0 & 2 \\ -1 & 2 & 0 & 3 & 0 \\ 0 & 4 & 1 & 0 & 0 \end{bmatrix}$

24 $\begin{bmatrix} 2 & 0 & -1 & 0 & 2 \\ 1 & 3 & 0 & 0 & 1 \\ 0 & 4 & 3 & 0 & -1 \\ -1 & 2 & 0 & -2 & 0 \\ 0 & 1 & 5 & 0 & -4 \end{bmatrix}$

25 Show that
$$\begin{vmatrix} 1 & 1 & 1 \\ a & b & c \\ a^2 & b^2 & c^2 \end{vmatrix} = (a - b)(b - c)(c - a).$$
(*Hint:* See Example 3.)

26 Show that
$$\begin{vmatrix} 1 & 1 & 1 \\ a & b & c \\ a^3 & b^3 & c^3 \end{vmatrix} = (a - b)(b - c)(c - a)(a + b + c).$$

27 If
$$A = \begin{bmatrix} a_{11} & a_{12} & a_{13} & a_{14} \\ 0 & a_{22} & a_{23} & a_{24} \\ 0 & 0 & a_{33} & a_{34} \\ 0 & 0 & 0 & a_{44} \end{bmatrix},$$
show that $|A| = a_{11}a_{22}a_{33}a_{44}$.

28 If
$$A = \begin{bmatrix} a & b & 0 & 0 \\ c & d & 0 & 0 \\ 0 & 0 & e & f \\ 0 & 0 & g & h \end{bmatrix},$$
show that
$$|A| = \begin{vmatrix} a & b \\ c & d \end{vmatrix} \begin{vmatrix} e & f \\ g & h \end{vmatrix}.$$

29 If $A = (a_{ij})$ and $B = (b_{ij})$ are arbitrary square matrices of order 2, show that $|AB| = |A||B|$.

30 If $A = (a_{ij})$ is a square matrix of order n and k is any real number, show that $|kA| = k^n|A|$. (*Hint:* Use property 2 of the theorem on row and column transformations of a determinant.)

31 Use properties of determinants to show that the following is an equation of a line through the points (x_1, y_1) and (x_2, y_2):

$$\begin{vmatrix} x & y & 1 \\ x_1 & y_1 & 1 \\ x_2 & y_2 & 1 \end{vmatrix} = 0$$

32 Use properties of determinants to show that the following is an equation of a circle through three noncollinear points (x_1, y_1), (x_2, y_2), and (x_3, y_3):

$$\begin{vmatrix} x^2 + y^2 & x & y & 1 \\ x_1^2 + y_1^2 & x_1 & y_1 & 1 \\ x_2^2 + y_2^2 & x_2 & y_2 & 1 \\ x_3^2 + y_3^2 & x_3 & y_3 & 1 \end{vmatrix} = 0$$

Exer. 33–42: Use Cramer's rule, whenever applicable, to solve the system.

33 $\begin{cases} 2x + 3y = 2 \\ x - 2y = 8 \end{cases}$

34 $\begin{cases} 4x + 5y = 13 \\ 3x + y = -4 \end{cases}$

35 $\begin{cases} 2x + 5y = 16 \\ 3x - 7y = 24 \end{cases}$

36 $\begin{cases} 7x - 8y = 9 \\ 4x + 3y = -10 \end{cases}$

37 $\begin{cases} 2x - 3y = 5 \\ -6x + 9y = 12 \end{cases}$

38 $\begin{cases} 3p - q = 7 \\ -12p + 4q = 3 \end{cases}$

39 $\begin{cases} x - 2y - 3z = -1 \\ 2x + y + z = 6 \\ x + 3y - 2z = 13 \end{cases}$

40 $\begin{cases} x + 3y - z = -3 \\ 3x - y + 2z = 1 \\ 2x - y + z = -1 \end{cases}$

41 $\begin{cases} 5x + 2y - z = -7 \\ x - 2y + 2z = 0 \\ 3y + z = 17 \end{cases}$

42 $\begin{cases} 4x - y + 3z = 6 \\ -8x + 3y - 5z = -6 \\ 5x - 4y = -9 \end{cases}$

43 Use Cramer's rule to solve the system for x.

$$\begin{cases} ax + by + cz = d \\ ex + fz = g \\ hx + iy = j \end{cases}$$

9.10

Partial Fractions

In this section we show how systems of equations can be used to help decompose rational expressions into sums of simpler expressions. This technique is useful in advanced mathematics courses.

We may verify that

$$\frac{2}{x^2 - 1} = \frac{1}{x - 1} + \frac{-1}{x + 1}$$

by adding the fractions $1/(x - 1)$ and $-1/(x + 1)$ to obtain $2/(x^2 - 1)$. The expression on the right-hand side of this equation is called the *partial fraction decomposition* of $2/(x^2 - 1)$.

It is theoretically possible to write *any* rational expression as a sum of rational expressions whose denominators involve powers of polynomials of degree not greater than two. Specifically, if $f(x)$ and $g(x)$ are polynomials *and the degree of $f(x)$ is less than the degree of $g(x)$*, it can be proved that

$$\frac{f(x)}{g(x)} = F_1 + F_2 + \cdots + F_r$$

such that each F_k has one of the forms

$$\frac{A}{(px + q)^m} \quad \text{or} \quad \frac{Ax + B}{(ax^2 + bx + c)^n},$$

where A and B are real numbers, m and n are nonnegative integers, and the quadratic polynomial $ax^2 + bx + c$ is irreducible over $\mathbb{R}$ (that is, has no real zero). The sum $F_1 + F_2 + \cdots + F_r$ is the **partial fraction decomposition** of $f(x)/g(x)$, and each F_k is a **partial fraction.**

For the partial fraction decomposition of $f(x)/g(x)$ to be found, *it is essential that $f(x)$ have lower degree than $g(x)$.* If this is not the case, we can use long division to obtain such an expression. For example, given

$$\frac{x^3 - 6x^2 + 5x - 3}{x^2 - 1},$$

we obtain

$$\frac{x^3 - 6x^2 + 5x - 3}{x^2 - 1} = x - 6 + \frac{6x - 9}{x^2 - 1}.$$

We then find the partial fraction decomposition of $(6x - 9)/(x^2 - 1)$.

The following guidelines can be used to obtain decompositions.

Guidelines for Finding Partial Fraction Decompositions of $f(x)/g(x)$	

1 If the degree of the numerator $f(x)$ is not lower than the degree of the denominator $g(x)$, use long division to obtain the proper form.

2 Factor the denominator $g(x)$ into a product of linear factors $px + q$ or irreducible quadratic factors $ax^2 + bx + c$, and collect repeated factors so that $g(x)$ is a product of *different* factors of the form $(px + q)^m$ or $(ax^2 + bx + c)^n$ for a nonnegative integer m or n.

3 Apply the following rules to the factors found in guideline 2.
Rule A: For each factor of the form $(px + q)^m$ with $m \geq 1$, the partial fraction decomposition contains a sum of m partial fractions of the form

$$\frac{A_1}{px + q} + \frac{A_2}{(px + q)^2} + \cdots + \frac{A_m}{(px + q)^m},$$

where each numerator A_k is a real number.
Rule B: For each factor of the form $(ax^2 + bx + c)^n$ with $n \geq 1$ and $ax^2 + bx + c$ irreducible, the partial fraction decomposition contains a sum of n partial fractions of the form

$$\frac{A_1 x + B_1}{ax^2 + bx + c} + \frac{A_2 x + B_2}{(ax^2 + bx + c)^2} + \cdots + \frac{A_n x + B_n}{(ax^2 + bx + c)^n},$$

where each A_k and each B_k is a real number.

4 Find the numbers A_k and B_k in guideline 3.

We shall apply the preceding guidelines in the following examples. For the sake of convenience, we will use the variables *A, B, C,* and so on, rather than the subscripted variables A_k and B_k given in the guidelines.

EXAMPLE 1 **A partial fraction decomposition in which each denominator is linear**

Find the partial fraction decomposition of

$$\frac{4x^2 + 13x - 9}{x^3 + 2x^2 - 3x}.$$

SOLUTION

Guideline 1   The degree of the numerator, 2, is less than the degree of the denominator, 3, so long division is not required.

Guideline 2   We factor the denominator:

$$x^3 + 2x^2 - 3x = x(x^2 + 2x - 3) = x(x + 3)(x - 1)$$

Guideline 3   Each factor of the denominator has the form stated in Rule A with $m = 1$. Thus, to the factor x there corresponds a partial fraction of the form A/x. Similarly, to the factors $x + 3$ and $x - 1$ there correspond partial fractions of the form $B/(x + 3)$ and $C/(x - 1)$, respectively. The partial fraction decomposition has the form

$$\frac{4x^2 + 13x - 9}{x^3 + 2x^2 - 3x} = \frac{A}{x} + \frac{B}{x + 3} + \frac{C}{x - 1}.$$

Guideline 4   We find the values of A, B, and C in guideline 3. Multiplying both sides of the partial fraction decomposition by the least common denominator, $x(x + 3)(x - 1)$, gives us

$$
\begin{aligned}
4x^2 + 13x - 9 &= A(x + 3)(x - 1) + Bx(x - 1) + Cx(x + 3) \\
&= A(x^2 + 2x - 3) + B(x^2 - x) + C(x^2 + 3x) \\
&= (A + B + C)x^2 + (2A - B + 3C)x - 3A.
\end{aligned}
$$

Equating the coefficients of like powers of x on each side of the last equation, we obtain the system of equations

$$
\begin{cases}
A + B + C = 4 \\
2A - B + 3C = 13 \\
-3A = -9
\end{cases}
$$

Using the methods of Section 9.5 yields the solution $A = 3$, $B = -1$, and $C = 2$. Hence, the partial fraction decomposition is

$$\frac{4x^2 + 13x - 9}{x(x + 3)(x - 1)} = \frac{3}{x} + \frac{-1}{x + 3} + \frac{2}{x - 1}.$$

There is an alternative way to find A, B, and C if all factors of the denominator are linear and nonrepeated, as in this example. Instead of equating coefficients and using a system of equations, we begin with the equation

$$4x^2 + 13x - 9 = A(x + 3)(x - 1) + Bx(x - 1) + Cx(x + 3).$$

We next substitute values for x that make the factors, x, $x - 1$, and $x + 3$, equal to zero. If we let $x = 0$ and simplify, we obtain

$$-9 = -3A, \quad \text{or} \quad A = 3.$$

Letting $x = 1$ in the equation leads to $8 = 4C$, or $C = 2$. Finally, if $x = -3$, then we have $-12 = 12B$, or $B = -1$. ◢

EXAMPLE 2 **A partial fraction decomposition containing a repeated linear factor**

Find the partial fraction decomposition of

$$\frac{x^2 + 10x - 36}{x(x - 3)^2}.$$

SOLUTION

Guideline 1 The degree of the numerator, 2, is less than the degree of the denominator, 3, so long division is not required.

Guideline 2 The denominator, $x(x - 3)^2$, is already in factored form.

Guideline 3 By Rule A with $m = 1$, there is a partial fraction of the form A/x corresponding to the factor x. Next, applying Rule A with $m = 2$, we find that the factor $(x - 3)^2$ determines a sum of two partial fractions of the form $B/(x - 3)$ and $C/(x - 3)^2$. Thus, the partial fraction decomposition has the form

$$\frac{x^2 + 10x - 36}{x(x - 3)^2} = \frac{A}{x} + \frac{B}{x - 3} + \frac{C}{(x - 3)^2}.$$

Guideline 4 To find A, B, and C, we begin by multiplying both sides of the partial fraction decomposition in guideline 3 by the lcd, $x(x - 3)^2$:

$$\begin{aligned} x^2 + 10x - 36 &= A(x - 3)^2 + Bx(x - 3) + Cx \\ &= A(x^2 - 6x + 9) + B(x^2 - 3x) + Cx \\ &= (A + B)x^2 + (-6A - 3B + C)x + 9A \end{aligned}$$

We next equate the coefficients of like powers of x, obtaining the system

$$\begin{cases} A + B & = 1 \\ -6A - 3B + C = 10 \\ 9A & = -36 \end{cases}$$

This system of equations has the solution $A = -4$, $B = 5$, and $C = 1$. The partial fraction decomposition is therefore

$$\frac{x^2 + 10x - 36}{x(x - 3)^2} = \frac{-4}{x} + \frac{5}{x - 3} + \frac{1}{(x - 3)^2}.$$

As in Example 1, we could also obtain A and C by beginning with the equation

$$x^2 + 10x - 36 = A(x - 3)^2 + Bx(x - 3) + Cx$$

(continued)

and then substituting values for x that make the factors, $x - 3$ and x, equal to zero. Thus, letting $x = 3$, we obtain $3 = 3C$, or $C = 1$. Letting $x = 0$ gives us $-36 = 9A$, or $A = -4$. The value of B may then be found by using one of the equations in the system.

EXAMPLE 3 **A partial fraction decomposition containing an irreducible quadratic factor**

Find the partial fraction decomposition of

$$\frac{4x^3 - x^2 + 15x - 29}{2x^3 - x^2 + 8x - 4}.$$

SOLUTION

Guideline 1 The degree of the numerator, 3, is *equal* to the degree of the denominator. Thus, long division is required, and we obtain

$$\frac{4x^3 - x^2 + 15x - 29}{2x^3 - x^2 + 8x - 4} = 2 + \frac{x^2 - x - 21}{2x^3 - x^2 + 8x - 4}.$$

Guideline 2 The denominator may be factored by grouping, as follows:

$$2x^3 - x^2 + 8x - 4 = x^2(2x - 1) + 4(2x - 1) = (x^2 + 4)(2x - 1)$$

Guideline 3 Applying Rule B to the irreducible quadratic factor $x^2 + 4$ in guideline 2, we see that one partial fraction has the form $(Ax + B)/(x^2 + 4)$. By Rule A, there is also a partial fraction $C/(2x - 1)$ corresponding to $2x - 1$. Consequently,

$$\frac{x^2 - x - 21}{2x^3 - x^2 + 8x - 4} = \frac{Ax + B}{x^2 + 4} + \frac{C}{2x - 1}.$$

Guideline 4 Multiplying both sides of the partial fraction decomposition in guideline 3 by the lcd, $(x^2 + 4)(2x - 1)$, we obtain

$$
\begin{aligned}
x^2 - x - 21 &= (Ax + B)(2x - 1) + C(x^2 + 4) \\
&= 2Ax^2 - Ax + 2Bx - B + Cx^2 + 4C \\
&= (2A + C)x^2 + (-A + 2B)x - B + 4C.
\end{aligned}
$$

This leads to the system

$$
\begin{cases}
2A \quad\quad\;\; + \;\; C = \quad 1 \\
-A + 2B \quad\quad\;\;\; = \; -1 \\
\quad\;\; - \;\; B + 4C = -21
\end{cases}
$$

This system has the solution $A = 3$, $B = 1$, and $C = -5$. Thus, the partial fraction decomposition in guideline 3 is

$$\frac{x^2 - x - 21}{2x^3 - x^2 + 8x - 4} = \frac{3x + 1}{x^2 + 4} + \frac{-5}{2x - 1},$$

and therefore the decomposition of the given expression (see guideline 1) is

$$\frac{4x^3 - x^2 + 15x - 29}{2x^3 - x^2 + 8x - 4} = 2 + \frac{3x + 1}{x^2 + 4} + \frac{-5}{2x - 1}.$$

EXAMPLE 4 **A partial fraction decomposition containing a repeated quadratic factor**

Find the partial fraction decomposition of

$$\frac{5x^3 - 3x^2 + 7x - 3}{(x^2 + 1)^2}.$$

SOLUTION

Guideline 1 The degree of the numerator, 3, is less than the degree of the denominator, 4, so long division is not required.

Guideline 2 The denominator, $(x^2 + 1)^2$, is already in factored form.

Guideline 3 We apply Rule B with $n = 2$ to $(x^2 + 1)^2$, to obtain the partial fraction decomposition

$$\frac{5x^3 - 3x^2 + 7x - 3}{(x^2 + 1)^2} = \frac{Ax + B}{x^2 + 1} + \frac{Cx + D}{(x^2 + 1)^2}.$$

Guideline 4 Multiplying both sides of the decomposition in guideline 3 by $(x^2 + 1)^2$ gives us

$$5x^3 - 3x^2 + 7x - 3 = (Ax + B)(x^2 + 1) + Cx + D$$
$$= Ax^3 + Bx^2 + (A + C)x + (B + D).$$

Comparing the coefficients of x^3 and x^2, we obtain $A = 5$ and $B = -3$. From the coefficients of x, we see that $A + C = 7$. Thus, $C = 7 - A = 7 - 5 = 2$. Finally, comparing the constant terms gives us the equation $B + D = -3$, and so $D = -3 - B = -3 - (-3) = 0$. Therefore, the partial fraction decomposition is

$$\frac{5x^3 - 3x^2 + 7x - 3}{(x^2 + 1)^2} = \frac{5x - 3}{x^2 + 1} + \frac{2x}{(x^2 + 1)^2}.$$

9.10 *Exercises*

Exer. 1–28: Find the partial fraction decomposition.

1 $\dfrac{8x - 1}{(x - 2)(x + 3)}$

2 $\dfrac{x - 29}{(x - 4)(x + 1)}$

3 $\dfrac{x + 34}{x^2 - 4x - 12}$

4 $\dfrac{5x - 12}{x^2 - 4x}$

5 $\dfrac{4x^2 - 15x - 1}{(x - 1)(x + 2)(x - 3)}$

6 $\dfrac{x^2 + 19x + 20}{x(x + 2)(x - 5)}$

7 $\dfrac{4x^2 - 5x - 15}{x^3 - 4x^2 - 5x}$

8 $\dfrac{37 - 11x}{(x + 1)(x^2 - 5x + 6)}$

$$\frac{+\ 3}{\ -\ 1)^2}$$

10 $\dfrac{5x^2 - 4}{x^2(x + 2)}$

21 $\dfrac{4x^3 - x^2 + 4x + 2}{(x^2 + 1)^2}$

22 $\dfrac{3x^3 + 13x - 1}{(x^2 + 4)^2}$

11 $\dfrac{19x^2 + 50x - 25}{3x^3 - 5x^2}$

12 $\dfrac{10 - x}{x^2 + 10x + 25}$

23 $\dfrac{2x^4 - 2x^3 + 6x^2 - 5x + 1}{x^3 - x^2 + x - 1}$

24 $\dfrac{x^3}{x^3 - 3x^2 + 9x - 27}$

13 $\dfrac{x^2 - 6}{(x + 2)^2(2x - 1)}$

14 $\dfrac{2x^2 + x}{(x - 1)^2(x + 1)^2}$

25 $\dfrac{3x^2 - 16}{x^2 - 4x}$

26 $\dfrac{2x^2 + 7x}{x^2 + 6x + 9}$

15 $\dfrac{3x^3 + 11x^2 + 16x + 5}{x(x + 1)^3}$

16 $\dfrac{4x^3 + 3x^2 + 5x - 2}{x^3(x + 2)}$

27 $\dfrac{4x^3 + 4x^2 - 4x + 2}{2x^2 - x - 1}$

17 $\dfrac{x^2 + x - 6}{(x^2 + 1)(x - 1)}$

18 $\dfrac{x^2 - x - 21}{(x^2 + 4)(2x - 1)}$

28 $\dfrac{x^5 - 5x^4 + 7x^3 - x^2 - 4x + 12}{x^3 - 3x^2}$

19 $\dfrac{9x^2 - 3x + 8}{x^3 + 2x}$

20 $\dfrac{2x^3 + 2x^2 + 4x - 3}{x^4 + x^2}$

CHAPTER 9 REVIEW EXERCISES

Exer. 1–16: Solve the system.

1 $\begin{cases} 2x - 3y = 4 \\ 5x + 4y = 1 \end{cases}$

2 $\begin{cases} x - 3y = 4 \\ -2x + 6y = 2 \end{cases}$

11 $\begin{cases} 4x - 3y - z = 0 \\ x - y - z = 0 \\ 3x - y + 3z = 0 \end{cases}$

12 $\begin{cases} 2x + y - z = 0 \\ x - 2y + z = 0 \\ 3x + 3y + 2z = 0 \end{cases}$

3 $\begin{cases} y + 4 = x^2 \\ 2x + y = -1 \end{cases}$

4 $\begin{cases} x^2 + y^2 = 25 \\ x - y = 7 \end{cases}$

13 $\begin{cases} 4x + 2y - z = 1 \\ 3x + 2y + 4z = 2 \end{cases}$

14 $\begin{cases} 2x + y = 6 \\ x - 3y = 17 \\ 3x + 2y = 7 \end{cases}$

5 $\begin{cases} 9x^2 + 16y^2 = 140 \\ x^2 - 4y^2 = 4 \end{cases}$

6 $\begin{cases} 2x = y^2 + 3z \\ x = y^2 + z - 1 \\ x^2 = xz \end{cases}$

15 $\begin{cases} \dfrac{4}{x} + \dfrac{1}{y} + \dfrac{2}{z} = 4 \\ \dfrac{2}{x} + \dfrac{3}{y} - \dfrac{1}{z} = 1 \\ \dfrac{1}{x} + \dfrac{1}{y} + \dfrac{1}{z} = 4 \end{cases}$

7 $\begin{cases} \dfrac{1}{x} + \dfrac{3}{y} = 7 \\ \dfrac{4}{x} - \dfrac{2}{y} = 1 \end{cases}$

8 $\begin{cases} 2^x + 3^{y+1} = 10 \\ 2^{x+1} - 3^y = 5 \end{cases}$

9 $\begin{cases} 3x + y - 2z = -1 \\ 2x - 3y + z = 4 \\ 4x + 5y - z = -2 \end{cases}$

10 $\begin{cases} x + 3y = 0 \\ y - 5z = 3 \\ 2x + z = -1 \end{cases}$

16 $\begin{cases} 2x - y + 3z - w = -3 \\ 3x + 2y - z + w = 13 \\ x - 3y + z - 2w = -4 \\ -x + y + 4z + 3w = 0 \end{cases}$

Exer. 17–20: Sketch the graph of the system.

17 $\begin{cases} x^2 + y^2 < 16 \\ y - x^2 > 0 \end{cases}$

18 $\begin{cases} y - x \le 0 \\ y + x \ge 2 \\ x \le 5 \end{cases}$

19 $\begin{cases} x - 2y \le 2 \\ y - 3x \le 4 \\ 2x + y \le 4 \end{cases}$

20 $\begin{cases} x^2 - y < 0 \\ y - 2x < 5 \\ xy < 0 \end{cases}$

Exer. 21–30: Express as a single matrix.

21 $\begin{bmatrix} 2 & -1 & 0 \\ 3 & 0 & -2 \end{bmatrix} \begin{bmatrix} 2 & -1 & 3 \\ 0 & 3 & 0 \\ 1 & 4 & 2 \end{bmatrix}$

22 $\begin{bmatrix} 4 & 2 \\ 5 & -3 \end{bmatrix} \begin{bmatrix} 3 \\ 7 \end{bmatrix}$

23 $\begin{bmatrix} 2 & 0 \\ 1 & 4 \\ -2 & 3 \end{bmatrix} \begin{bmatrix} 0 & 2 & -3 \\ 4 & 5 & 1 \end{bmatrix}$

24 $\begin{bmatrix} 0 & -2 & 3 \\ 4 & 1 & 2 \end{bmatrix} \begin{bmatrix} 2 & 0 \\ 3 & 8 \\ 2 & -7 \end{bmatrix}$

25 $2\begin{bmatrix} 0 & -1 & -4 \\ 3 & 2 & 1 \end{bmatrix} - 3\begin{bmatrix} 4 & -2 & 1 \\ 0 & 5 & -1 \end{bmatrix}$

26 $\begin{bmatrix} 1 & 3 \\ 2 & 4 \end{bmatrix} \begin{bmatrix} a & 0 \\ 0 & a \end{bmatrix}$

27 $\begin{bmatrix} a & 0 \\ 0 & b \end{bmatrix} \begin{bmatrix} 1 & 3 \\ 2 & 4 \end{bmatrix}$

28 $\begin{bmatrix} 3 & 2 \\ 0 & 0 \end{bmatrix} \begin{bmatrix} -2 & 0 \\ 3 & 0 \end{bmatrix}$

29 $\begin{bmatrix} 1 & 2 \\ 3 & 4 \end{bmatrix} \left\{ \begin{bmatrix} 2 & -4 \\ 3 & 7 \end{bmatrix} + \begin{bmatrix} 1 & 5 \\ -2 & -3 \end{bmatrix} \right\}$

30 $\begin{bmatrix} 3 & 2 & 5 \\ -3 & 4 & 7 \\ 6 & 5 & 1 \end{bmatrix} \begin{bmatrix} 3 & 2 & 5 \\ -3 & 4 & 7 \\ 6 & 5 & 1 \end{bmatrix}^{-1}$

Exer. 31–34: Find the inverse of the matrix.

31 $\begin{bmatrix} 5 & -4 \\ -3 & 2 \end{bmatrix}$

32 $\begin{bmatrix} 2 & -1 & 0 \\ 1 & 4 & 2 \\ 3 & -2 & 1 \end{bmatrix}$

33 $\begin{bmatrix} 1 & 0 & 0 \\ 0 & 4 & 7 \\ 0 & 1 & 2 \end{bmatrix}$

34 $\begin{bmatrix} 2 & 0 & 5 \\ 0 & 3 & -1 \\ 3 & 4 & 0 \end{bmatrix}$

35 Use the result of Exercise 31 to solve the system

$$\begin{cases} 5x - 4y = 30 \\ -3x + 2y = -16 \end{cases}$$

36 Use the result of Exercise 32 to solve the system

$$\begin{cases} 2x - y = -5 \\ x + 4y + 2z = 15 \\ 3x - 2y + z = -7 \end{cases}$$

Exer. 37–46: Find the determinant of the matrix.

37 $[-6]$

38 $\begin{bmatrix} 3 & 4 \\ -6 & -5 \end{bmatrix}$

39 $\begin{bmatrix} 3 & -4 \\ 6 & 8 \end{bmatrix}$

40 $\begin{bmatrix} 0 & 4 & -3 \\ 2 & 0 & 4 \\ -5 & 1 & 0 \end{bmatrix}$

41 $\begin{bmatrix} 2 & -3 & 5 \\ -4 & 1 & 3 \\ 3 & 2 & -1 \end{bmatrix}$

42 $\begin{bmatrix} 3 & 1 & -2 \\ -5 & 2 & -4 \\ 7 & 3 & -6 \end{bmatrix}$

43 $\begin{bmatrix} 5 & 0 & 0 & 0 \\ 6 & -3 & 0 & 0 \\ 1 & 4 & -4 & 0 \\ 7 & 2 & 3 & 2 \end{bmatrix}$

44 $\begin{bmatrix} 1 & 2 & 0 & 3 & 1 \\ -2 & -1 & 4 & 1 & 2 \\ 3 & 0 & -1 & 0 & -1 \\ 2 & -3 & 2 & -4 & 2 \\ -1 & 1 & 0 & 1 & 3 \end{bmatrix}$

45 $\begin{bmatrix} 2 & 0 & 1 & 0 & -1 \\ 0 & 1 & 0 & 1 & 2 \\ 2 & -2 & 1 & -2 & 0 \\ 0 & 0 & -2 & 0 & 1 \\ 1 & -1 & 0 & -1 & 0 \end{bmatrix}$

46 $\begin{bmatrix} 1 & 2 & 0 & 0 & 0 \\ 3 & 4 & 0 & 0 & 0 \\ 0 & 0 & 1 & 2 & 3 \\ 0 & 0 & 2 & -1 & 1 \\ 0 & 0 & 1 & 3 & -1 \end{bmatrix}$

Exer. 47–48: Solve the equation $|A - xI| = 0$.

47 $A = \begin{bmatrix} 2 & 3 \\ 1 & -4 \end{bmatrix}, \qquad I = I_2$

48 $A = \begin{bmatrix} 2 & -1 & 3 \\ 0 & 4 & 0 \\ 1 & 0 & -2 \end{bmatrix}, \quad I = I_3$

Exer. 49–50: Without expanding, explain why the statement is true.

49 $\begin{vmatrix} 2 & 4 & -6 \\ 1 & 4 & 3 \\ 2 & 2 & 0 \end{vmatrix} = 12 \begin{vmatrix} 1 & 1 & -1 \\ 1 & 2 & 1 \\ 2 & 1 & 0 \end{vmatrix}$

50 $\begin{vmatrix} a & b & c \\ d & e & f \\ g & h & k \end{vmatrix} = \begin{vmatrix} d & e & f \\ g & h & k \\ a & b & c \end{vmatrix}$

51 Find the determinant of the $n \times n$ matrix (a_{ij}) in which $a_{ij} = 0$ for $i \neq j$.

52 Without expanding, show that

$$\begin{vmatrix} 1 & a & b+c \\ 1 & b & a+c \\ 1 & c & a+b \end{vmatrix} = 0.$$

Exer. 53–54: Use Cramer's rule to solve the system.

53 $\begin{cases} 5x - 6y = 4 \\ 3x + 7y = 8 \end{cases}$

54 $\begin{cases} 2x - 3y + 2z = -3 \\ -3x + 2y + z = 1 \\ 4x + y - 3z = 4 \end{cases}$

Exer. 55–58: Find the partial fraction decomposition.

55 $\dfrac{4x^2 + 54x + 134}{(x + 3)(x^2 + 4x - 5)}$

56 $\dfrac{2x^2 + 7x + 9}{x^2 + 2x + 1}$

57 $\dfrac{x^2 + 14x - 13}{x^3 + 5x^2 + 4x + 20}$

58 $\dfrac{x^3 + 2x^2 + 2x + 16}{x^4 + 7x^2 + 10}$

59 Watering a field A rotating sprinkler head with a range of 50 feet is to be placed in the center of a rectangular field (see the figure). If the area of the field is 4000 ft² and the water is to just reach the corners, find the dimensions of the field.

Exercise 59

60 Find equations of the two lines that are tangent to the circle $x^2 + y^2 = 1$ and pass through the point $(0, 3)$. (*Hint:* Let $y = mx + 3$, and determine conditions on m that will ensure that the system has only one solution.)

61 Payroll accounting An accountant must pay taxes and payroll bonuses to employees from the company's profits of $50,000. The total tax is 40% of the amount left after bonuses are paid, and the total paid in bonuses is 10% of the amount left after taxes. Find the total tax and the total bonus amount.

62 Track dimensions A circular track is to have a 10-foot-wide running lane around the outside (see the figure). The inside distance around the track is to be 90% of the outside distance. Find the dimensions of the track.

Exercise 62

63 **Flow rates** Three inlet pipes, A, B, and C, can be used to fill a 1000-ft^3 water storage tank. When all three pipes are in operation, the tank can be filled in 10 hours. When only pipes A and B are used, the time increases to 20 hours. With pipes A and C, the tank can be filled in 12.5 hours. Find the individual flow rates (in ft^3/hr) for each of the three pipes.

64. **Warehouse shipping charges** To fill an order for 150 office desks, a furniture distributor must ship the desks from two warehouses. The shipping cost per desk is $24 from the western warehouse and $35 from the eastern warehouse. If the total shipping charge is $4205, how many desks are shipped from each location?

65 **Express-mail rates** An express-mail company charges $15 for overnight delivery of a letter, provided the dimensions of the standard envelope satisfy the following three conditions: (a) the length, the larger of the two dimensions, must be at most 12 inches; (b) the width must be at most 8 inches; (c) the width must be at least one-half the length. Find and graph a system of inequalities that describes all the possibilities for dimensions of a standard envelope.

66 **Activities of a deer** A deer spends the day in three basic activities: resting, searching for food, and grazing. At least 6 hours each day must be spent resting, and the number of hours spent searching for food will be at least two times the number of hours spent grazing. Using x as the number of hours spent searching for food and y as the number of hours spent grazing, find and graph the system of inequalities that describes the possible divisions of the day.

67 **Production scheduling** A company manufactures a power lawn mower and a power edger. These two products are of such high quality that the company can sell all the products it makes, but production capability is limited in the areas of machining, welding, and assembly. Each week the company has 600 hours available for machining, 300 hours for welding, and 550 hours for assembly. The number of hours required for the production of a single item is shown in the following table. The profits from the sale of a mower and an edger are $100 and $80, respectively. How many mowers and edgers should be made each week to maximize the profit?

Product	Machining	Welding	Assembly
Mower	6	2	5
Edger	4	3	5

68 **Maximizing investment income** A retired couple wishes to invest $150,000, diversifying the investment in three areas: a high-risk stock that has an expected annual rate of return (or interest) of 15%, a low-risk stock that has an expected annual return of 10%, and government-issued bonds that pay annual interest of 8% and involve no risk. To protect the value of the investment, the couple wishes to place at least twice as much in the low-risk stock as in the high-risk stock and use the remainder to buy bonds. How should the money be invested to maximize the expected annual return?

CHAPTER 9 DISCUSSION EXERCISES

1 (a) It is easy to see that the system

$$\begin{cases} x + 2y = 4 \\ x + 2y = 5 \end{cases}$$

has no solution. Let $x + by = 5$ be the second equation, and solve the system for $b = 1.99$ and $b = 1.999$. Note that a small change in b produces a large change in x and y. Such a system is known as an *ill-conditioned*

system (a precise definition is given in most numerical analysis texts).

(b) Solve this system for x and y in terms of b, and explain why a small change in b (for b near 2) produces a large change in x and y.

(c) If b gets very large, what happens to the solution of the system?

2 **Bird migration trends** Refer to Exercise 30 of Section 9.5. Suppose the initial bird populations on islands A, B, and C are 12,000, 9000, and 14,000, respectively.

(a) Represent the initial populations with a 1×3 matrix D. Represent the proportions of the populations that migrate to each island with a 3×3 matrix E. (*Hint:* The first row of E is 0.90, 0.10, and 0.00—indicating that 90% of the birds on A stay on A, 10% of the birds on A migrate to B, and no birds on A migrate to C.)

(b) Find the product $F = DE$, and interpret the meaning of the elements of F.

(c) Multiply F times E, and continue to multiply the result by E until a pattern becomes apparent. What is your conclusion?

3 Explain why a nonsquare matrix A cannot have an inverse.

4 **Distributing money** A college president has received budgets from the athletic director (AD), dean of students (DS), and student senate president (SP), in which they propose to allocate department funds to the three basic areas of student scholarships, activities, and services, as shown in the table.

	Scholarships	Activities	Services
AD	50%	40%	10%
DS	30%	20%	50%
SP	20%	40%	40%

The Board of Regents has requested that the overall distribution of funding to these three areas be in the following proportions: scholarships, 34%; activities, 33%; and services, 33%. Determine what percentage of the total funds the president should allocate to each department so that the percentages spent in these three areas conform to the Board of Regents' requirements.

5 If $x^4 + ax^2 + bx + c = 0$ has roots $x = -1, 2$, and 3, find a, b, c, and the fourth root of the equation.

6 Find, if possible, an equation of

(a) a line

(b) a circle

(c) a parabola with vertical axis

(d) a cubic

(e) an exponential

that passes through the points $P(-1, 3)$, $Q(0, 4)$, and $R(3, 2)$.

7 Prove ($*$) on page 638.

10

Sequences, Series, and Probability

10.1 Infinite Sequences and Summation Notation

10.2 Arithmetic Sequences

10.3 Geometric Sequences

10.4 Mathematical Induction

10.5 The Binomial Theorem

10.6 Permutations

10.7 Distinguishable Permutations and Combinations

10.8 Probability

Sequences and summation notation, discussed in the first section, are very important in advanced mathematics and applications. Of special interest are arithmetic and geometric sequences, considered in Sections 10.2 and 10.3. We then discuss the method of mathematical induction, a process that is often used to prove that each statement in an infinite sequence of statements is true. As an application, we use it to prove the binomial theorem in Section 10.5. The last part of the chapter deals with counting processes that occur frequently in mathematics and everyday life. These include the concepts of permutations, combinations, and probability.

10.1

Infinite Sequences and Summation Notation

An arbitrary *infinite sequence* may be denoted as follows:

Infinite Sequence Notation	$a_1, a_2, a_3, \ldots, a_n, \ldots$

For convenience, we often refer to infinite sequences as *sequences.* We may regard an infinite sequence as a collection of real numbers that is in one-to-one correspondence with the positive integers. Each number a_k is a **term** of the sequence. The sequence is *ordered* in the sense that there is a **first term** a_1, a *second term* a_2, a *forty-fifth term* a_{45}, and, if n denotes an arbitrary positive integer, an **nth term** a_n. Infinite sequences are often defined by stating a formula for the nth term.

Infinite sequences occur frequently in mathematics. For example, the sequence

$$0.6, \quad 0.66, \quad 0.666, \quad 0.6666, \quad 0.66666, \quad \ldots$$

may be used to represent the rational number $\frac{2}{3}$. In this case the nth term gets closer and closer to $\frac{2}{3}$ as n increases.

We may regard an infinite sequence as a function. Recall from Section 3.4 that a function f is a correspondence that assigns to each number x in the domain D exactly one number $f(x)$ in the range R. If we restrict the domain to the positive integers $1, 2, 3, \ldots$, we obtain an infinite sequence, as in the following definition.

Definition of Infinite Sequence	An **infinite sequence** is a function whose domain is the set of positive integers.

In our work, the range of an infinite sequence will be a set of real numbers.

If a function f is an infinite sequence, then to each positive integer n there corresponds a real number $f(n)$. These numbers in the range of f may be represented by writing

$$f(1), f(2), f(3), \ldots, f(n), \ldots.$$

To obtain the subscript form of a sequence, as shown at the beginning of this section, we let $a_n = f(n)$ for every positive integer n.

If we regard a sequence as a function f, then we may consider its graph in an xy-plane. Since the domain of f is the set of positive integers, the only points on the graph are

$$(1, a_1), (2, a_2), (3, a_3), \ldots, (n, a_n), \ldots,$$

where a_n is the nth term of the sequence as shown in Figure 1. We sometimes use the graph of a sequence to illustrate the behavior of the nth term a_n as n increases without bound.

Figure 1   Graph of a sequence

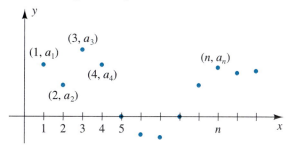

From the definition of equality of functions we see that a sequence

$$a_1, a_2, a_3, \ldots, a_n, \ldots$$

is **equal** to a sequence

$$b_1, b_2, b_3, \ldots, b_n, \ldots$$

if and only if $a_k = b_k$ for every positive integer k.

Another notation for a sequence with nth term a_n is $\{a_n\}$. For example, the sequence $\{2^n\}$ has nth term $a_n = 2^n$. Using sequence notation, we write this sequence as follows:

$$2^1, 2^2, 2^3, \ldots, 2^n, \ldots$$

By definition, the sequence $\{2^n\}$ is the function f with $f(n) = 2^n$ for every positive integer n.

EXAMPLE 1   Finding terms of a sequence

List the first four terms and the tenth term of each sequence:

(a) $\left\{ \dfrac{n}{n+1} \right\}$ **(b)** $\{2 + (0.1)^n\}$ **(c)** $\left\{ (-1)^{n+1} \dfrac{n^2}{3n-1} \right\}$ **(d)** $\{4\}$

SOLUTION   To find the first four terms, we substitute, successively, $n = 1$, 2, 3, and 4 in the formula for a_n. The tenth term is found by substituting 10 for n. Doing this and simplifying gives us the following:

(continued)

Sequence	nth term a_n	First four terms	Tenth term
(a) $\left\{\dfrac{n}{n+1}\right\}$	$\dfrac{n}{n+1}$	$\dfrac{1}{2}, \dfrac{2}{3}, \dfrac{3}{4}, \dfrac{4}{5}$	$\dfrac{10}{11}$
(b) $\{2 + (0.1)^n\}$	$2 + (0.1)^n$	2.1, 2.01, 2.001, 2.0001	2.000 000 000 1
(c) $\left\{(-1)^{n+1}\dfrac{n^2}{3n-1}\right\}$	$(-1)^{n+1}\dfrac{n^2}{3n-1}$	$\dfrac{1}{2}, -\dfrac{4}{5}, \dfrac{9}{8}, -\dfrac{16}{11}$	$-\dfrac{100}{29}$
(d) $\{4\}$	4	4, 4, 4, 4	4

EXAMPLE 2   **Graphing a sequence**

Graph the sequence in Example 1(a)—that is,

$$\left\{\frac{n}{n+1}\right\}.$$

SOLUTION   The domain values are

$$1, 2, 3, \ldots, n, \ldots.$$

The range values are

$$\frac{1}{1+1}, \quad \frac{2}{2+1}, \quad \frac{3}{3+1}, \quad \ldots, \quad \frac{n}{n+1}, \quad \ldots$$

or, equivalently,

$$\frac{1}{2}, \quad \frac{2}{3}, \quad \frac{3}{4}, \quad \ldots, \quad \frac{n}{n+1}, \quad \ldots.$$

A plot of the ordered pairs $(n, n/(n+1))$ is shown in Figure 2.

Figure 2

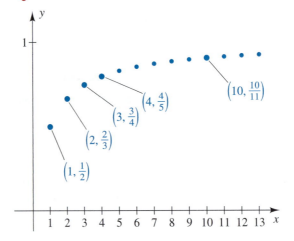

For some sequences we state the first term a_1, together with a rule for obtaining any term a_{k+1} from the preceding term a_k whenever $k \geq 1$. We call such a statement a **recursive definition,** and the sequence is said to be defined **recursively.**

EXAMPLE 3   Finding terms of a recursively defined sequence

Find the first four terms and the nth term of the infinite sequence defined recursively as follows:

$$a_1 = 3, \quad a_{k+1} = 2a_k \quad \text{for} \quad k \geq 1$$

SOLUTION   The first four terms are

$$a_1 = 3 \qquad\qquad\qquad\qquad\qquad\qquad \text{given}$$
$$a_2 = 2a_1 = 2 \cdot 3 = 6 \qquad\qquad\qquad k = 1$$
$$a_3 = 2a_2 = 2 \cdot 2 \cdot 3 = 2^2 \cdot 3 = 12 \qquad k = 2$$
$$a_4 = 2a_3 = 2 \cdot 2 \cdot 2 \cdot 3 = 2^3 \cdot 3 = 24. \quad k = 3$$

We have written the terms as products to gain some insight into the nature of the nth term. Continuing, we obtain $a_5 = 2^4 \cdot 3$, $a_6 = 2^5 \cdot 3$, and, in general,

$$a_n = 2^{n-1} \cdot 3$$

for every positive integer n.

If only the first few terms of an infinite sequence are known, then it is impossible to predict additional terms. For example, if we were given $3, 6, 9, \ldots$ and asked to find the fourth term, we could not proceed without further information. The infinite sequence with nth term

$$a_n = 3n + (1 - n)^3(2 - n)^2(3 - n)$$

has for its first four terms 3, 6, 9, and 120. It is possible to describe sequences in which the first three terms are 3, 6, and 9 and the fourth term is *any* given number. This shows that when we work with an infinite sequence it is essential to have either specific information about the nth term or a general scheme for obtaining each term from the preceding one. (See Exercise 1 of the Chapter 10 Discussion Exercises for a related problem.)

We sometimes need to find the sum of many terms of an infinite sequence. To express such sums easily, we use **summation notation.** Given an infinite sequence

$$a_1, a_2, a_3, \ldots, a_n, \ldots,$$

the symbol $\sum_{k=1}^{m} a_k$ represents the sum of the first m terms, as follows.

Summation Notation	$$\sum_{k=1}^{m} a_k = a_1 + a_2 + a_3 + \cdots + a_m$$

The Greek capital letter sigma, Σ, indicates a sum, and the symbol a_k represents the kth term. The letter k is the **index of summation,** or the **summation variable,** and the numbers 1 and m indicate the smallest and largest values of the summation variable, respectively.

EXAMPLE 4 **Evaluating a sum**

Find the sum $\displaystyle\sum_{k=1}^{4} k^2(k-3)$.

SOLUTION In this case, $a_k = k^2(k-3)$. To find the sum, we merely substitute, in succession, the integers 1, 2, 3, and 4 for k and add the resulting terms:

$$\sum_{k=1}^{4} k^2(k-3) = 1^2(1-3) + 2^2(2-3) + 3^2(3-3) + 4^2(4-3)$$

$$= (-2) + (-4) + 0 + 16 = 10$$

The letter we use for the summation variable is immaterial. To illustrate, if j is the summation variable, then

$$\sum_{j=1}^{m} a_j = a_1 + a_2 + a_3 + \cdots + a_m,$$

which is the same sum as $\sum_{k=1}^{m} a_k$. As a specific example, the sum in Example 4 can be written

$$\sum_{j=1}^{4} j^2(j-3).$$

If n is a positive integer, then the sum of the first n terms of an infinite sequence will be denoted by S_n. For example, given the infinite sequence $a_1, a_2, a_3, \ldots, a_n, \ldots,$

$$S_1 = a_1$$
$$S_2 = a_1 + a_2$$
$$S_3 = a_1 + a_2 + a_3$$
$$S_4 = a_1 + a_2 + a_3 + a_4$$

and, in general,

$$S_n = \sum_{k=1}^{n} a_k = a_1 + a_2 + \cdots + a_n.$$

Note that we can also write

$$S_1 = a_1$$
$$S_2 = S_1 + a_2$$
$$S_3 = S_2 + a_3$$
$$S_4 = S_3 + a_4$$

and, for every $n > 1$,

$$S_n = S_{n-1} + a_n.$$

The real number S_n is called the **nth partial sum** of the infinite sequence $a_1, a_2, a_3, \ldots, a_n, \ldots$, and the sequence

$$S_1, S_2, S_3, \ldots, S_n, \ldots$$

is called a **sequence of partial sums.** Sequences of partial sums are important in the study of *infinite series,* a topic in calculus. We shall discuss some special types of infinite series in Section 10.3.

EXAMPLE 5   **Finding the terms of a sequence of partial sums**

Find the first four terms and the nth term of the sequence of partial sums associated with the sequence $1, 2, 3, \ldots, n, \ldots$ of positive integers.

SOLUTION   If we let $a_n = n$, then the first four terms of the sequence of partial sums are

$$S_1 = a_1 = 1$$
$$S_2 = S_1 + a_2 = 1 + 2 = 3$$
$$S_3 = S_2 + a_3 = 3 + 3 = 6$$
$$S_4 = S_3 + a_4 = 6 + 4 = 10.$$

The nth partial sum S_n (that is, the sum of $1, 2, 3, \ldots, n$) can be written in either of the following forms:

$$S_n = 1 + 2 \quad + 3 \quad + \cdots + (n - 2) + (n - 1) + n$$
$$S_n = n + (n - 1) + (n - 2) + \cdots + 3 \quad + 2 \quad + 1$$

Adding corresponding terms on each side of these equations gives us

$$2S_n = \underbrace{(n + 1) + (n + 1) + (n + 1) + \cdots + (n + 1) + (n + 1) + (n + 1)}_{n \text{ times}}.$$

Since the expression $(n + 1)$ appears n times on the right-hand side of the last equation, we see that

$$2S_n = n(n + 1) \qquad \text{or, equivalently,} \qquad S_n = \frac{n(n + 1)}{2}.$$

If a_k is the same for every positive integer k—say $a_k = c$ for a real number c—then

$$\sum_{k=1}^{n} a_k = a_1 + a_2 + a_3 + \cdots + a_n$$
$$= c + c + c + \cdots + c = nc.$$

We have proved property 1 of the following theorem.

Theorem on the Sum of a Constant	$(1)\ \displaystyle\sum_{k=1}^{n} c = nc$ $\qquad$ $(2)\ \displaystyle\sum_{k=m}^{n} c = (n - m + 1)c$

To prove property 2, we may write

$$\sum_{k=m}^{n} c = \sum_{k=1}^{n} c - \sum_{k=1}^{m-1} c \qquad \text{subtract the first } m - 1 \text{ terms from the sum of } n \text{ terms}$$

$$= nc - (m - 1)c \qquad \text{use property 1 for each sum}$$

$$= [n - (m - 1)]c \qquad \text{factor out } c$$

$$= (n - m + 1)c. \qquad \text{simplify}$$

ILLUSTRATION   **Sum of a Constant**

■ $\displaystyle\sum_{k=1}^{4} 7 = 4 \cdot 7 = 28$

■ $\displaystyle\sum_{k=1}^{10} \pi = 10 \cdot \pi = 10\pi$

■ $\displaystyle\sum_{k=3}^{8} 9 = (8 - 3 + 1)(9) = 6(9) = 54$

■ $\displaystyle\sum_{k=10}^{20} 5 = (20 - 10 + 1)(5) = 11(5) = 55$

As shown in property 2 of the preceding theorem, the domain of the summation variable does not have to begin at 1. For example,

$$\sum_{k=4}^{8} a_k = a_4 + a_5 + a_6 + a_7 + a_8.$$

As another variation, if the first term of an infinite sequence is a_0, as in

$$a_0, a_1, a_2, \ldots, a_n, \ldots,$$

then we may consider sums of the form

$$\sum_{k=0}^{n} a_k = a_0 + a_1 + a_2 + \cdots + a_n,$$

which is the sum of the first $n + 1$ terms of the sequence.

If the summation variable does not appear in the term a_k, then the *entire term* may be considered a constant. For example,

$$\sum_{j=1}^{n} a_k = n \cdot a_k,$$

since j does not appear in the term a_k.

Summation notation can be used to denote polynomials. Thus, if

$$f(x) = a_0 + a_1 x + a_2 x^2 + \cdots + a_n x^n,$$

then

$$f(x) = \sum_{k=0}^{n} a_k x^k.$$

The following theorem concerning sums has many uses.

Theorem on Sums	If $a_1, a_2, \ldots, a_n, \ldots$ and $b_1, b_2, \ldots, b_n, \ldots$ are infinite sequences, then for every positive integer n,

$$\textbf{(1)} \quad \sum_{k=1}^{n} (a_k + b_k) = \sum_{k=1}^{n} a_k + \sum_{k=1}^{n} b_k$$

$$\textbf{(2)} \quad \sum_{k=1}^{n} (a_k - b_k) = \sum_{k=1}^{n} a_k - \sum_{k=1}^{n} b_k$$

$$\textbf{(3)} \quad \sum_{k=1}^{n} c a_k = c \left(\sum_{k=1}^{n} a_k \right) \text{ for every real number } c$$

PROOFS To prove formula 1, we first write

$$\sum_{k=1}^{n} (a_k + b_k) = (a_1 + b_1) + (a_2 + b_2) + (a_3 + b_3) + \cdots + (a_n + b_n).$$

Using commutative and associative properties of real numbers many times, we may rearrange the terms on the right-hand side to produce

$$\sum_{k=1}^{n} (a_k + b_k) = (a_1 + a_2 + a_3 + \cdots + a_n) + (b_1 + b_2 + b_3 + \cdots + b_n)$$

$$= \sum_{k=1}^{n} a_k + \sum_{k=1}^{n} b_k.$$

For a proof of formula 3, we have

$$\sum_{k=1}^{n} (c a_k) = c a_1 + c a_2 + c a_3 + \cdots + c a_n$$

$$= c(a_1 + a_2 + a_3 + \cdots + a_n)$$

$$= c \left(\sum_{k=1}^{n} a_k \right).$$

The proof of formula 2 is left as an exercise.

10.1 *Exercises*

Exer. 1–16: Find the first four terms and the eighth term of the sequence.

1 $\{12 - 3n\}$

2 $\left\{\dfrac{3}{5n - 2}\right\}$

3 $\left\{\dfrac{3n - 2}{n^2 + 1}\right\}$

4 $\left\{10 + \dfrac{1}{n}\right\}$

5 $\{9\}$

6 $\{\sqrt{2}\}$

7 $\{2 + (-0.1)^n\}$

8 $\{4 + (0.1)^n\}$

9 $\left\{(-1)^{n-1}\dfrac{n + 7}{2n}\right\}$

10 $\left\{(-1)^n \dfrac{6 - 2n}{\sqrt{n + 1}}\right\}$

11 $\{1 + (-1)^{n+1}\}$

12 $\{(-1)^{n+1} + (0.1)^{n-1}\}$

13 $\left\{\dfrac{2^n}{n^2 + 2}\right\}$

14 $\{(n - 1)(n - 2)(n - 3)\}$

15 a_n is the number of decimal places in $(0.1)^n$.

16 a_n is the number of positive integers less than n^3.

Exer. 17–20: Graph the sequence.

17 $\left\{\dfrac{1}{\sqrt{n}}\right\}$

18 $\left\{\dfrac{1}{n}\right\}$

19 $\{(-1)^{n+1}n^2\}$

20 $\{(-1)^n(2n + 1)\}$

Exer. 21–28: Find the first five terms of the recursively defined infinite sequence.

21 $a_1 = 2$, $\quad a_{k+1} = 3a_k - 5$

22 $a_1 = 5$, $\quad a_{k+1} = 7 - 2a_k$

23 $a_1 = -3$, $\quad a_{k+1} = a_k^2$

24 $a_1 = 128$, $\quad a_{k+1} = \frac{1}{4}a_k$

25 $a_1 = 5$, $\quad a_{k+1} = ka_k$

26 $a_1 = 3$, $\quad a_{k+1} = 1/a_k$

27 $a_1 = 2$, $\quad a_{k+1} = (a_k)^k$

28 $a_1 = 2$, $\quad a_{k+1} = (a_k)^{1/k}$

Exer. 29–32: Find the first four terms of the sequence of partial sums for the given sequence.

29 $\{3 + \frac{1}{2}n\}$

30 $\{1/n^2\}$

31 $\{(-1)^n n^{-1/2}\}$

32 $\{(-1)^n(1/2)^n\}$

Exer. 33–48: Find the sum.

33 $\displaystyle\sum_{k=1}^{5} (2k - 7)$

34 $\displaystyle\sum_{k=1}^{6} (10 - 3k)$

35 $\displaystyle\sum_{k=1}^{4} (k^2 - 5)$

36 $\displaystyle\sum_{k=1}^{10} [1 + (-1)^k]$

37 $\displaystyle\sum_{k=0}^{5} k(k - 2)$

38 $\displaystyle\sum_{k=0}^{4} (k - 1)(k - 3)$

39 $\displaystyle\sum_{k=3}^{6} \dfrac{k - 5}{k - 1}$

40 $\displaystyle\sum_{k=1}^{6} \dfrac{3}{k + 1}$

41 $\displaystyle\sum_{k=1}^{5} (-3)^{k-1}$

42 $\displaystyle\sum_{k=0}^{4} 3(2^k)$

43 $\displaystyle\sum_{k=1}^{100} 100$

44 $\displaystyle\sum_{k=1}^{1000} 5$

45 $\displaystyle\sum_{k=253}^{571} \dfrac{1}{3}$

46 $\displaystyle\sum_{k=137}^{428} 2.1$

47 $\displaystyle\sum_{j=1}^{7} \frac{1}{2}k^2$

48 $\displaystyle\sum_{k=0}^{5} (3j + 2)$

49 Prove formula 2 of the theorem on sums.

50 Extend formula 1 of the theorem on sums to
$$\sum_{k=1}^{n} (a_k + b_k + c_k).$$

51 Consider the sequence defined recursively by $a_1 = 5$, $a_{k+1} = \sqrt{a_k}$ for $k \geq 1$. Describe what happens to the terms of the sequence as k increases.

52 Approximations to π may be obtained from the sequence

$$x_1 = 3, \quad x_{k+1} = x_k - \tan x_k.$$

Use the $\boxed{\text{TAN}}$ key for tan.

(a) Find the first five terms of this sequence.

(b) What happens to the terms of the sequence when $x_1 = 6$?

53 Bode's sequence Bode's sequence, defined by

$$a_1 = 0.4, \quad a_k = 0.1(3 \cdot 2^{k-2} + 4) \quad \text{for} \quad k \geq 2,$$

can be used to approximate distances of planets from the sun. These distances are measured in astronomical units, with 1 AU = 93,000,000 mi. For example, the third term corresponds to Earth and the fifth term to the minor planet Ceres. Approximate the first five terms of the sequence.

54 Growth of bacteria The number of bacteria in a certain culture is initially 500, and the culture doubles in size every day.

(a) Find the number of bacteria present after one day, two days, and three days.

(b) Find a formula for the number of bacteria present after n days.

55 The Fibonacci sequence The Fibonacci sequence is defined recursively by

$$a_1 = 1, \quad a_2 = 1, \quad a_{k+1} = a_k + a_{k-1} \quad \text{for} \quad k \geq 2.$$

(a) Find the first ten terms of the sequence.

(b) The terms of the sequence $r_k = a_{k+1}/a_k$ give progressively better approximations to τ, the golden ratio. Approximate the first ten terms of this sequence.

56 The Fibonacci sequence The Fibonacci sequence can be defined by the formula

$$a_n = \frac{1}{\sqrt{5}}\left(\frac{1 + \sqrt{5}}{2}\right)^n - \frac{1}{\sqrt{5}}\left(\frac{1 - \sqrt{5}}{2}\right)^n.$$

Find the first eight terms, and show that they agree with those found using the definition in Exercise 55.

57 Chlorine levels Chlorine is often added to swimming pools to control microorganisms. If the level of chlorine rises above 3 ppm (parts per million), swimmers will experience burning eyes and skin discomfort. If the level drops below 1 ppm, there is a possibility that the water will turn green because of a large algae count. Chlorine must be added to pool water at regular intervals. If no chlorine is added to a

pool during a 24-hour period, approximately 20% of the chlorine will dissipate into the atmosphere and 80% will remain in the water.

(a) Determine a recursive sequence a_n that expresses the amount of chlorine present after n days if the pool has a_0 ppm of chlorine initially and no chlorine is added.

(b) If a pool has 7 ppm of chlorine initially, construct a table to determine the first day on which the chlorine level will drop below 3 ppm.

58 Chlorine levels Refer to Exercise 57. Suppose a pool has 2 ppm of chlorine initially, and 0.5 ppm of chlorine is added to the pool at the end of each day.

(a) Find a recursive sequence a_n that expresses the amount of chlorine present after n days.

(b) Determine the amount of chlorine in the pool after 15 days and after a long period of time.

(c) Estimate the amount of chlorine that needs to be added daily in order to stabilize the pool's chlorine level at 1.5 ppm.

59 Golf club costs A golf club company sells driver heads as follows:

Number of heads	1–4	5–9	10+
Cost per head	$89.95	$87.95	$85.95

Find a piecewise-defined function C that specifies the total cost for n heads. Sketch a graph of C.

60 DVD player costs An electronics wholesaler sells DVD players at $20 each for the first 4 units. All units after the first 4 sell for $17 each. Find a piecewise-defined function C that specifies the total cost for n players. Sketch a graph of C.

Exer. 61–62: Some calculators use an algorithm similar to the following to approximate $\sqrt{N}$ for a positive real number N: Let $x_1 = N/2$ and find successive approximations $x_2, x_3, \ldots$ by using

$$x_2 = \frac{1}{2}\left(x_1 + \frac{N}{x_1}\right), \quad x_3 = \frac{1}{2}\left(x_2 + \frac{N}{x_2}\right), \quad \cdots$$

until the desired accuracy is obtained. Use this method to approximate the radical to six-decimal-place accuracy.

61 $\sqrt{5}$ **62** $\sqrt{18}$

63 The equation $\frac{1}{3}\sqrt[3]{x} - x + 2 = 0$ has a root near 2. To approximate this root, rewrite the equation as $x = \frac{1}{3}\sqrt[3]{x} + 2$. Let $x_1 = 2$ and find successive approximations $x_2, x_3, \ldots$ by using the formulas

$$x_2 = \tfrac{1}{3}\sqrt[3]{x_1} + 2, \quad x_3 = \tfrac{1}{3}\sqrt[3]{x_2} + 2, \quad \ldots$$

until four-decimal-place accuracy is obtained.

64 The equation $2x + \dfrac{1}{x^4 + x + 2} = 0$ has a root near 0. Use

a procedure similar to that in Exercise 63 to approximate this root to four-decimal-place accuracy.

Exer. 65–66: (a) Show that f takes on both positive and negative values on the interval [1, 2]. (b) Use the method of Exercise 63, with $x_1 = 1.5$, to approximate a zero of f to two-decimal-place accuracy.

65 $f(x) = x - 2 + \log x$

66 $f(x) = \log x - 10^{-x}$ (*Hint:* Solve for x in $\log x$.)

10.2

Arithmetic Sequences

In this section and the next we consider two special types of sequences: arithmetic and geometric. The first type may be defined as follows.

Definition of Arithmetic Sequence	A sequence $a_1, a_2, \ldots, a_n, \ldots$ is an **arithmetic sequence** if there is a real number d such that for every positive integer k, $$a_{k+1} = a_k + d.$$ The number $d = a_{k+1} - a_k$ is called the **common difference** of the sequence.

Note that the common difference d is the difference of *any* two successive terms of an arithmetic sequence.

ILLUSTRATION **Arithmetic Sequence and Common Difference**

■ $-3, 2, 7, 12, \ldots, 5n - 8, \ldots$ common difference $= 2 - (-3)$
$= 5$

■ $17, 10, 3, -4, \ldots, 24 - 7n, \ldots$ common difference $= 10 - 17$
$= -7$

EXAMPLE 1   **Showing that a sequence is arithmetic**

Show that the sequence

$$1, 4, 7, 10, \ldots, 3n - 2, \ldots$$

is arithmetic, and find the common difference.

SOLUTION   If $a_n = 3n - 2$, then for every positive integer k,

$$a_{k+1} - a_k = [3(k + 1) - 2] - (3k - 2)$$
$$= 3k + 3 - 2 - 3k + 2 = 3.$$

Hence, the given sequence is arithmetic with common difference 3.

Given an arithmetic sequence, we know that

$$a_{k+1} = a_k + d$$

for every positive integer k. This gives us a recursive formula for obtaining successive terms. Beginning with any real number a_1, we can obtain an arithmetic sequence with common difference d simply by adding d to a_1, then to $a_1 + d$, and so on, obtaining

$$a_1, \quad a_1 + d, \quad a_1 + 2d, \quad a_1 + 3d, \quad a_1 + 4d, \quad \ldots.$$

The nth term a_n of this sequence is given by the next formula.

The nth Term of an Arithmetic Sequence	$$a_n = a_1 + (n - 1)d$$

EXAMPLE 2 **Finding a specific term of an arithmetic sequence**

The first three terms of an arithmetic sequence are 20, 16.5, and 13. Find the fifteenth term.

SOLUTION The common difference is

$$a_2 - a_1 = 16.5 - 20 = -3.5.$$

Substituting $n = 15$, $a_1 = 20$, and $d = -3.5$ in the formula for the nth term of an arithmetic sequence, $a_n = a_1 + (n - 1)d$, gives us

$$a_{15} = 20 + (15 - 1)(-3.5) = 20 - 49 = -29. \qquad \blacksquare$$

EXAMPLE 3 **Finding a specific term of an arithmetic sequence**

If the fourth term of an arithmetic sequence is 5 and the ninth term is 20, find the sixth term.

SOLUTION We are given $a_4 = 5$ and $a_9 = 20$ and wish to find a_6. The following are equivalent systems of equations in the variables a_1 and d:

$$\begin{cases} a_4 = a_1 + (4 - 1)d & \text{let } n = 4 \text{ in } a_n = a_1 + (n-1)d \\ a_9 = a_1 + (9 - 1)d & \text{let } n = 9 \text{ in } a_n = a_1 + (n-1)d \end{cases}$$

$$\begin{cases} 5 = a_1 + 3d & a_4 = 5 \\ 20 = a_1 + 8d & a_9 = 20 \end{cases}$$

Subtracting the first equation of the system from the second equation gives us $15 = 5d$, or $d = 3$. Substituting 3 for d in the first equation, $5 = a_1 + 3d$, yields $a_1 = 5 - 3d = 5 - 3(3) = -4$. Hence, to find the sixth term we have

$$a_6 = a_1 + (6 - 1)d \qquad \text{let } n = 6 \text{ in } a_n = a_1 + (n-1)d$$
$$= (-4) + (5)(3) = 11. \quad a_1 = -4 \text{ and } d = 3 \qquad \blacksquare$$

The next theorem contains a formula for the nth partial sum S_n of an arithmetic sequence.

Theorem: Formulas for S_n	If $a_1, a_2, a_3, \ldots, a_n, \ldots$ is an arithmetic sequence with common difference d, then the nth partial sum S_n (that is, the sum of the first n terms) is given by either
	$$S_n = \frac{n}{2}[2a_1 + (n-1)d] \qquad \text{or} \qquad S_n = \frac{n}{2}(a_1 + a_n).$$

PROOF We may write

$$S_n = a_1 + a_2 \quad + a_3 \quad + \cdots + a_n$$
$$= a_1 + (a_1 + d) + (a_1 + 2d) + \cdots + [a_1 + (n-1)d].$$

Employing the commutative and associative properties of real numbers many times, we obtain

$$S_n = (a_1 + a_1 + a_1 + \cdots + a_1) + [d + 2d + \cdots + (n-1)d],$$

with a_1 appearing n times within the first pair of parentheses. Thus,

$$S_n = na_1 + d[1 + 2 + \cdots + (n-1)].$$

The expression within brackets is the sum of the first $n - 1$ positive integers. Using the formula for the sum of the first n positive integers, $S_n = n(n + 1)/2$, from Example 5 of Section 10.1, but with $n - 1$ in place of n and n in place of $n + 1$, we have

$$1 + 2 + \cdots + (n-1) = \frac{(n-1)n}{2}.$$

Substituting in the last equation for S_n and factoring out $n/2$ gives us

$$S_n = na_1 + d\frac{(n-1)n}{2} = \frac{n}{2}[2a_1 + (n-1)d].$$

Since $a_n = a_1 + (n-1)d$, the last equation is equivalent to

$$S_n = \frac{n}{2}(a_1 + a_n).$$

EXAMPLE 4 **Finding a sum of even integers**

Find the sum of all the even integers from 2 through 100.

SOLUTION This problem is equivalent to finding the sum of the first 50 terms of the arithmetic sequence

$$2, 4, 6, \ldots, 2n, \ldots.$$

Substituting $n = 50$, $a_1 = 2$, and $a_{50} = 100$ in $S_n = (n/2)(a_1 + a_n)$ produces

$$S_{50} = \tfrac{50}{2}(2 + 100) = 2550.$$

Alternatively, we may use $S_n = \dfrac{n}{2}[2a_1 + (n - 1)d]$ with $d = 2$:

$$S_{50} = \tfrac{50}{2}[2(2) + (50 - 1)2] = 25[4 + 98] = 2550$$

The **arithmetic mean** of two numbers a and b is defined as $(a + b)/2$. This is the **average** of a and b. Note that the three numbers

$$a, \quad \frac{a + b}{2}, \quad \text{and} \quad b$$

constitute a (finite) arithmetic sequence with a common difference of $d = \frac{1}{2}(b - a)$. This concept may be generalized as follows: If $c_1, c_2, \ldots, c_k$ are real numbers such that

$$a, c_1, c_2, \ldots, c_k, b$$

is a (finite) arithmetic sequence, then $c_1, c_2, \ldots, c_k$ are k **arithmetic means** between the numbers a and b. The process of determining these numbers is referred to as *inserting k arithmetic means between a and b*.

EXAMPLE 5   Inserting arithmetic means

Insert three arithmetic means between 2 and 9.

SOLUTION   We wish to find three real numbers $c_1, c_2,$ and c_3 such that the following is a (finite) arithmetic sequence:

$$2, c_1, c_2, c_3, 9$$

We may regard this sequence as an arithmetic sequence with first term $a_1 = 2$ and fifth term $a_5 = 9$. To find the common difference d, we may proceed as follows:

$$
\begin{aligned}
a_5 &= a_1 + (5 - 1)d \qquad &&\text{let } n = 5 \text{ in } a_n = a_1 + (n - 1)d \\
9 &= 2 + 4d &&a_5 = 9 \text{ and } a_1 = 2 \\
d &= \tfrac{7}{4} &&\text{solve for } d
\end{aligned}
$$

Thus, the arithmetic means are

$$
\begin{aligned}
c_1 &= a_1 + d = 2 + \tfrac{7}{4} = \tfrac{15}{4} \\
c_2 &= c_1 + d = \tfrac{15}{4} + \tfrac{7}{4} = \tfrac{22}{4} = \tfrac{11}{2} \\
c_3 &= c_2 + d = \tfrac{22}{4} + \tfrac{7}{4} = \tfrac{29}{4}.
\end{aligned}
$$

EXAMPLE 6   An application of an arithmetic sequence

A carpenter wishes to construct a ladder with nine rungs whose lengths decrease uniformly from 24 inches at the base to 18 inches at the top. Determine the lengths of the seven intermediate rungs.

Figure 1

$a_1 = 18$
a_2
a_3
a_4
a_5
a_6
a_7
a_8
$a_9 = 24$

SOLUTION   The ladder is sketched in Figure 1. The lengths of the rungs are to form an arithmetic sequence $a_1, a_2, \ldots, a_9$ with $a_1 = 18$ and $a_9 = 24$. Hence, we need to insert seven arithmetic means between 18 and 24. Using $a_n = a_1 + (n - 1)d$ with $n = 9$, $a_1 = 18$, and $a_9 = 24$ gives us

$$24 = 18 + 8d \quad \text{or, equivalently,} \quad 8d = 6.$$

Hence, $d = \frac{6}{8} = 0.75$, and the intermediate rungs have lengths (in inches)

$$18.75, \quad 19.5, \quad 20.25, \quad 21, \quad 21.75, \quad 22.5, \quad \text{and} \quad 23.25.$$

It is sometimes desirable to express a sum in terms of summation notation, as illustrated in the next example.

EXAMPLE 7   Expressing a sum in summation notation

Express in terms of summation notation:

$$\tfrac{1}{4} + \tfrac{2}{9} + \tfrac{3}{14} + \tfrac{4}{19} + \tfrac{5}{24} + \tfrac{6}{29}$$

SOLUTION   The six terms of the sum do not form an arithmetic sequence; however, the numerators and denominators of the fractions, *considered separately,* are each an arithmetic sequence. Specifically, we have the following:

Numerators: 1, 2, 3, 4, 5, 6   common difference 1

Denominators: 4, 9, 14, 19, 24, 29   common difference 5

Using the formula for the *n*th term of an arithmetic sequence twice, we obtain the following *n*th term for each sequence:

$$a_n = a_1 + (n - 1)d = 1 + (n - 1)1 = n$$
$$a_n = a_1 + (n - 1)d = 4 + (n - 1)5 = 5n - 1$$

Hence, the *n*th term of the given sum is $n/(5n - 1)$, and we may write

$$\frac{1}{4} + \frac{2}{9} + \frac{3}{14} + \frac{4}{19} + \frac{5}{24} + \frac{6}{29} = \sum_{n=1}^{6} \frac{n}{5n - 1}.$$

10.2 *Exercises*

Exer. 1–2: Show that the given sequence is arithmetic, and find the common difference.

1 $-6, -2, 2, \ldots, 4n - 10, \ldots$

2 $53, 48, 43, \ldots, 58 - 5n, \ldots$

Exer. 3–10: Find the *n*th term, the fifth term, and the tenth term of the arithmetic sequence.

3 $2, 6, 10, 14, \ldots$

4 $16, 13, 10, 7, \ldots$

5 $3, 2.7, 2.4, 2.1, \ldots$

6 $-6, -4.5, -3, -1.5, \ldots$

7 $-7, -3.9, -0.8, 2.3, \ldots$

8 $x - 8, x - 3, x + 2, x + 7, \ldots$

9 $\ln 3, \ln 9, \ln 27, \ln 81, \ldots$

10 $\log 1000, \log 100, \log 10, \log 1, \ldots$

Exer. 11–12: Find the common difference for the arithmetic sequence with the specified terms.

11 $a_2 = 21, a_6 = -11$

12 $a_4 = 14, a_{11} = 35$

Exer. 13–18: Find the specified term of the arithmetic sequence that has the two given terms.

13 $a_{12};\quad a_1 = 9.1,\qquad a_2 = 7.5$

14 $a_{11};\quad a_1 = 2 + \sqrt{2},\quad a_2 = 3$

15 $a_1;\qquad a_6 = 2.7,\qquad a_7 = 5.2$

16 $a_1;\qquad a_8 = 47,\qquad a_9 = 53$

17 $a_{15};\quad a_3 = 7,\qquad a_{20} = 43$

18 $a_{10};\quad a_2 = 1,\qquad a_{18} = 49$

Exer. 19–22: Find the sum S_n of the arithmetic sequence that satisfies the stated conditions.

19 $a_1 = 40,\quad d = -3,\quad n = 30$

20 $a_1 = 5,\qquad d = 0.1,\quad n = 40$

21 $a_1 = -9,\quad a_{10} = 15,\quad n = 10$

22 $a_7 = \frac{7}{3},\qquad d = -\frac{2}{3},\quad n = 15$

Exer. 23–28: Find the sum.

23 $\displaystyle\sum_{k=1}^{20} (3k - 5)$

24 $\displaystyle\sum_{k=1}^{12} (7 - 4k)$

25 $\displaystyle\sum_{k=1}^{18} \left(\tfrac{1}{2}k + 7\right)$

26 $\displaystyle\sum_{k=1}^{10} \left(\tfrac{1}{4}k + 3\right)$

27 $\displaystyle\sum_{k=126}^{592} (5k + 2j)$

28 $\displaystyle\sum_{k=88}^{371} (3j - 2k)$

Exer. 29–34: Express the sum in terms of summation notation. (Answers are not unique.)

29 $4 + 11 + 18 + 25 + 32$

30 $3 + 8 + 13 + 18 + 23$

31 $4 + 11 + 18 + \cdots + 466$

32 $3 + 8 + 13 + \cdots + 463$

33 $\frac{3}{7} + \frac{6}{11} + \frac{9}{15} + \frac{12}{19} + \frac{15}{23} + \frac{18}{27}$

34 $\frac{5}{13} + \frac{10}{11} + \frac{15}{9} + \frac{20}{7}$

Exer. 35–36: Express the sum in terms of summation notation and find the sum.

35 $8 + 19 + 30 + \cdots + 16{,}805$

36 $2 + 11 + 20 + \cdots + 16{,}058$

Exer. 37–40: Find the number of terms in the arithmetic sequence with the given conditions.

37 $a_1 = -2,\quad d = \tfrac{1}{4},\qquad S = 21$

38 $a_1 = -1,\quad d = \tfrac{1}{5},\qquad S = 21$

39 $a_1 = -\tfrac{29}{6},\quad d = \tfrac{1}{3},\qquad S = -36$

40 $a_6 = -3,\quad d = 0.2,\quad S = -33$

41 Insert five arithmetic means between 2 and 10.

42 Insert three arithmetic means between 3 and -5.

43 (a) Find the number of integers between 32 and 395 that are divisible by 6.

    (b) Find their sum.

44 (a) Find the number of negative integers greater than -500 that are divisible by 33.

    (b) Find their sum.

45 Log pile A pile of logs has 24 logs in the bottom layer, 23 in the second layer, 22 in the third, and so on. The top layer contains 10 logs. Find the total number of logs in the pile.

46 Stadium seating The first ten rows of seating in a certain section of a stadium have 30 seats, 32 seats, 34 seats, and so on. The eleventh through the twentieth rows each contain 50 seats. Find the total number of seats in the section.

47 Constructing a grain bin A grain bin is to be constructed in the shape of a frustum of a cone (see the figure). The bin is to be 10 feet tall with 11 metal rings positioned uniformly around it, from the 4-foot opening at the bottom to the

24-foot opening at the top. Find the total length of metal needed to make the rings.

Exercise 47

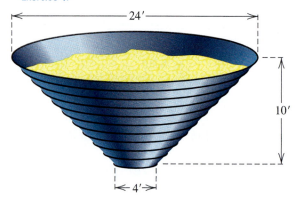

24'

10'

4'→

48 Coasting downhill A bicycle rider coasts downhill, traveling 4 feet the first second. In each succeeding second, the rider travels 5 feet farther than in the preceding second. If the rider reaches the bottom of the hill in 11 seconds, find the total distance traveled.

49 Prize money A contest will have five cash prizes totaling $5000, and there will be a $100 difference between successive prizes. Find the first prize.

50 Sales bonuses A company is to distribute $46,000 in bonuses to its top ten salespeople. The tenth salesperson on the list will receive $1000, and the difference in bonus money between successively ranked salespeople is to be constant. Find the bonus for each salesperson.

51 Distance an object falls Assuming air resistance is negligible, a small object that is dropped from a hot air balloon falls 16 feet during the first second, 48 feet during the second second, 80 feet during the third second, 112 feet during the fourth second, and so on. Find an expression for the distance the object falls in n seconds.

52 If f is a linear function, show that the sequence with nth term $a_n = f(n)$ is an arithmetic sequence.

53 Genetic sequence The sequence defined recursively by $x_{k+1} = x_k/(1 + x_k)$ occurs in genetics in the study of the elimination of a deficient gene from a population. Show that the sequence whose nth term is $1/x_n$ is arithmetic.

54 Dimensions of a maze Find the total length of the red-line curve in the figure if the width of the maze formed by the curve is 16 inches and all halls in the maze have width 1 inch. What is the length if the width of the maze is 32 inches?

Exercise 54

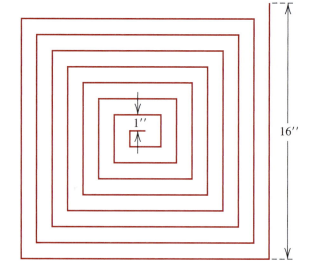

1''

16''

Exer. 55–56: Depreciation methods are sometimes used by businesses and individuals to estimate the value of an asset over a life span of n years. In the sum-of-year's-digits method, for each year $k = 1, 2, 3, \ldots, n$, the value of an asset is decreased by the fraction $A_k = \dfrac{n - k + 1}{T_n}$ of its initial cost, where $T_n = 1 + 2 + 3 + \cdots + n$.

55 (a) If $n = 8$, find $A_1, A_2, A_3, \ldots, A_8$.

(b) Show that the sequence in (a) is arithmetic, and find S_8.

(c) If the initial value of an asset is $1000, how much has been depreciated after 4 years?

56 (a) If n is any positive integer, find $A_1, A_2, A_3, \ldots, A_n$.

(b) Show that the sequence in (a) is arithmetic, and find S_n.

10.3
Geometric Sequences

The second special type of sequence that we will discuss—the geometric sequence—occurs frequently in applications.

Definition of Geometric Sequence	A sequence $a_1, a_2, \ldots, a_n, \ldots$ is a **geometric sequence** if $a_1 \neq 0$ and if there is a real number $r \neq 0$ such that for every positive integer k, $$a_{k+1} = a_k r.$$ The number $r = \dfrac{a_{k+1}}{a_k}$ is called the **common ratio** of the sequence.

Note that the common ratio $r = a_{k+1}/a_k$ is the ratio of *any* two successive terms of a geometric sequence.

ILLUSTRATION **Geometric Sequence and Common Ratio**

■ $6, -12, 24, -48, \ldots, (-2)^{n-1}(6), \ldots$ common ratio $= \frac{-12}{6} = -2$

■ $9, 3, 1, \frac{1}{3}, \ldots, (3)^{3-n}, \ldots$ common ratio $= \frac{3}{9} = \frac{1}{3}$

The formula $a_{k+1} = a_k r$ provides a recursive method for obtaining terms of a geometric sequence. Beginning with any nonzero real number a_1, we multiply by the number r successively, obtaining

$$a_1, \quad a_1 r, \quad a_1 r^2, \quad a_1 r^3, \quad \ldots.$$

The nth term a_n of this sequence is given by the following formula.

Formula for the nth Term of a Geometric Sequence	$$a_n = a_1 r^{n-1}$$

EXAMPLE 1 **Finding terms of a geometric sequence**

A geometric sequence has first term 3 and common ratio $-\frac{1}{2}$. Find the first five terms and the tenth term.

SOLUTION If we multiply $a_1 = 3$ successively by $r = -\frac{1}{2}$, then the first five terms are
$$3, \quad -\tfrac{3}{2}, \quad \tfrac{3}{4}, \quad -\tfrac{3}{8}, \quad \tfrac{3}{16}.$$

Using the formula $a_n = a_1 r^{n-1}$ with $n = 10$, we find that the tenth term is
$$a_{10} = a_1 r^9 = 3\left(-\tfrac{1}{2}\right)^9 = -\tfrac{3}{512}.$$

EXAMPLE 2 Finding a specific term of a geometric sequence

The third term of a geometric sequence is 5, and the sixth term is -40. Find the eighth term.

SOLUTION We are given $a_3 = 5$ and $a_6 = -40$ and wish to find a_8. The following are equivalent systems of equations in the variables a_1 and r:

$$\begin{cases} a_3 = a_1 r^{3-1} & \text{let } n = 3 \text{ in } a_n = a_1 r^{n-1} \\ a_6 = a_1 r^{6-1} & \text{let } n = 6 \text{ in } a_n = a_1 r^{n-1} \end{cases}$$

$$\begin{cases} 5 = a_1 r^2 & a_3 = 5 \\ -40 = a_1 r^5 & a_6 = -40 \end{cases}$$

Solving the first equation of the system for a_1 gives us $a_1 = 5/r^2$. Substituting this expression in the second equation yields

$$-40 = \frac{5}{r^2} \cdot r^5.$$

Simplifying, we get $r^3 = -8$, and hence $r = -2$. We next use $a_1 = 5/r^2$ to obtain

$$a_1 = \frac{5}{(-2)^2} = \frac{5}{4}.$$

Finally, using $a_n = a_1 r^{n-1}$ with $n = 8$ gives us

$$a_8 = a_1 r^7 = \left(\tfrac{5}{4}\right)(-2)^7 = -160.$$

The next theorem contains a formula for the nth partial sum S_n of a geometric sequence.

Theorem: Formula for S_n	The nth partial sum S_n of a geometric sequence with first term a_1 and common ratio $r \neq 1$ is $$S_n = a_1 \frac{1 - r^n}{1 - r}.$$

PROOF By definition, the nth partial sum S_n of a geometric sequence is

$$S_n = a_1 + a_1 r + a_1 r^2 + \cdots + a_1 r^{n-2} + a_1 r^{n-1}. \qquad \textbf{(1)}$$

Multiplying both sides of (1) by r, we obtain

$$r S_n = a_1 r + a_1 r^2 + a_1 r^3 + \cdots + a_1 r^{n-1} + a_1 r^n. \qquad \textbf{(2)}$$

If we subtract equation (2) from equation (1), all but two terms on the right-hand side cancel and we obtain the following:

$$S_n - rS_n = a_1 - a_1 r^n \quad \text{subtract (2) from (1)}$$

$$S_n(1 - r) = a_1(1 - r^n) \quad \text{factor both sides}$$

$$S_n = a_1 \frac{1 - r^n}{1 - r} \quad \text{divide by } (1 - r)$$

EXAMPLE 3   Finding a sum of terms of a geometric sequence

If the sequence $1, 0.3, 0.09, 0.027, \ldots$ is a geometric sequence, find the sum of the first five terms.

SOLUTION   If we let $a_1 = 1$, $r = 0.3$, and $n = 5$ in the formula for S_n stated in the preceding theorem, we obtain

$$S_5 = a_1 \frac{1 - r^5}{1 - r} = (1) \frac{1 - (0.3)^5}{1 - 0.3} = 1.4251.$$

EXAMPLE 4   The rapid growth of terms of a geometric sequence

A man wishes to save money by setting aside 1 cent the first day, 2 cents the second day, 4 cents the third day, and so on.

(a) If he continues to double the amount set aside each day, how much must he set aside on the fifteenth day?

(b) Assuming he does not run out of money, what is the total amount saved at the end of the 30 days?

SOLUTION

(a) The amount (in cents) set aside on successive days forms a geometric sequence

$$1, 2, 4, 8, \ldots,$$

with first term 1 and common ratio 2. We find the amount to be set aside on the fifteenth day by using $a_n = a_1 r^{n-1}$ with $a_1 = 1$ and $n = 15$:

$$a_{15} = a_1 r^{14} = 1 \cdot 2^{14} = 16,384$$

Thus, $163.84 should be set aside on the fifteenth day.

(b) To find the total amount saved after 30 days, we use the formula for S_n with $n = 30$, obtaining (in cents)

$$S_{30} = (1) \frac{1 - 2^{30}}{1 - 2} = 1,073,741,823.$$

Thus, the total amount saved is $10,737,418.23.

The terminology used with geometric sequences is analogous to that used with arithmetic sequences. If a and b are positive real numbers, then a positive number c is called the **geometric mean** of a and b if a, c, b is a geometric sequence. If the common ratio is r, then

$$r = \frac{c}{a} = \frac{b}{c}, \qquad \text{or} \qquad c^2 = ab.$$

Taking the square root of both sides of the last equation, we see that *the geometric mean of the positive numbers a and b is* $\sqrt{ab}$. As a generalization, k positive real numbers $c_1, c_2, \ldots, c_k$ are k **geometric means** between a and b if $a, c_1, c_2, \ldots, c_k, b$ is a geometric sequence. The process of determining these numbers is referred to as *inserting k geometric means between a and b*.

ILLUSTRATION   **Geometric Means**

	Numbers	Geometric mean
■	20, 45	$\sqrt{20 \cdot 45} = \sqrt{900} = 30$
■	3, 4	$\sqrt{3 \cdot 4} = \sqrt{12} \approx 3.46$

Given the geometric series with first term a_1 and common ratio $r \neq 1$, we may write the formula for S_n of the preceding theorem in the form

$$S_n = \frac{a_1}{1 - r} - \frac{a_1}{1 - r} r^n.$$

If $|r| < 1$, then r^n *approaches* 0 *as n increases* without bound. Thus, S_n approaches $a_1/(1 - r)$ as n increases without bound. Using the notation we developed for rational functions in Section 4.5, we have

$$S_n \to \frac{a_1}{1 - r} \quad \text{as} \quad n \to \infty.$$

The number $a_1/(1 - r)$ is called the *sum S* of the **infinite geometric series**

$$a_1 + a_1 r + a_1 r^2 + \cdots + a_1 r^{n-1} + \cdots.$$

This gives us the following result.

Theorem on the Sum of an Infinite Geometric Series	If $|r| < 1$, then the infinite geometric series $$a_1 + a_1 r + a_1 r^2 + \cdots + a_1 r^{n-1} + \cdots$$ has the sum $$S = \frac{a_1}{1 - r}.$$

The preceding theorem implies that as we add more terms of the indicated infinite geometric series, the sum gets closer to $a_1/(1 - r)$. The next example illustrates how the theorem can be used to show that every real number represented by a repeating decimal is rational.

EXAMPLE 5 **Expressing an infinite repeating decimal as a rational number**

Find a rational number that corresponds to $5.4\overline{27}$.

SOLUTION   We can write the decimal expression $5.4272727\ldots$ as

$$5.4 + 0.027 + 0.00027 + 0.0000027 + \cdots.$$

Beginning with the second term, 0.027, the above sum has the form given in the theorem on the sum of an infinite geometric series, with $a_1 = 0.027$ and $r = 0.01$. Hence, the sum S of this infinite geometric series is

$$S = \frac{a_1}{1 - r} = \frac{0.027}{1 - 0.01} = \frac{0.027}{0.990} = \frac{27}{990} = \frac{3}{110}.$$

Thus, the desired number is

$$5.4 + \tfrac{3}{110} = \tfrac{594}{110} + \tfrac{3}{110} = \tfrac{597}{110}.$$

A check by division shows that $\tfrac{597}{110}$ corresponds to $5.4\overline{27}$.

In general, given any infinite sequence, $a_1, a_2, \ldots, a_n, \ldots$, the expression

$$a_1 + a_2 + \cdots + a_n + \cdots$$

is called an **infinite series** or simply a **series.** We denote this series by

$$\sum_{n=1}^{\infty} a_n.$$

Each number a_k is a **term** of the series, and a_n is the **nth term.** Since only *finite* sums may be added algebraically, it is necessary to define what is meant by an *infinite sum.* Consider the sequence of partial sums

$$S_1, S_2, \ldots, S_n, \ldots.$$

If there is a number S such that $S_n \to S$ as $n \to \infty$, then, as in our discussion of infinite geometric series, S is the **sum** of the infinite series and we write

$$S = a_1 + a_2 + \cdots + a_n + \cdots.$$

In the previous example we found that the infinite repeating decimal $5.4272727\ldots$ corresponds to the rational number $\tfrac{597}{110}$. Since $\tfrac{597}{110}$ is the sum of an infinite series determined by the decimal, we may write

$$\tfrac{597}{110} = 5.4 + 0.027 + 0.00027 + 0.0000027 + \cdots.$$

If the terms of an infinite sequence are alternately positive and negative, as in the expression

$$a_1 + (-a_2) + a_3 + (-a_4) + \cdots + [(-1)^{n+1}a_n] + \cdots$$

for positive real numbers a_k, then the expression is an **alternating infinite series** and we write it in the form

$$a_1 - a_2 + a_3 - a_4 + \cdots + (-1)^{n+1}a_n + \cdots.$$

The most common types of alternating infinite series are infinite geometric series in which the common ratio r is negative.

EXAMPLE 6   Finding the sum of an infinite geometric series

Find the sum S of the alternating infinite geometric series

$$\sum_{n=1}^{\infty} 3\left(-\tfrac{2}{3}\right)^{n-1} = 3 - 2 + \tfrac{4}{3} - \tfrac{8}{9} + \cdots + 3\left(-\tfrac{2}{3}\right)^{n-1} + \cdots.$$

SOLUTION   Using the formula for S in the theorem on the sum of an infinite geometric series, with $a_1 = 3$ and $r = -\tfrac{2}{3}$, we obtain

$$S = \frac{a_1}{1 - r} = \frac{3}{1 - \left(-\tfrac{2}{3}\right)} = \frac{3}{\tfrac{5}{3}} = \frac{9}{5}.$$ ◢

Figure 1

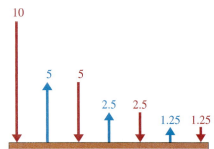

EXAMPLE 7   An application of an infinite geometric series

A rubber ball is dropped from a height of 10 meters. Suppose it rebounds one-half the distance after each fall, as illustrated by the arrows in Figure 1. Find the total distance the ball travels.

SOLUTION   The sum of the distances the ball travels downward and the sum of the distances it travels on the rebounds form two infinite geometric series:

$$\textit{Downward series:} \quad 10 + 5 + 2.5 + 1.25 + 0.625 + \cdots$$
$$\textit{Upward series:} \quad 5 + 2.5 + 1.25 + 0.625 + \cdots$$

We assume that the total distance S the ball travels can be found by adding the sums of these infinite series. This gives us

$$S = 10 + 2[5 + 2.5 + 1.25 + 0.625 + \cdots]$$
$$= 10 + 2\left[5 + 5\left(\tfrac{1}{2}\right) + 5\left(\tfrac{1}{2}\right)^2 + 5\left(\tfrac{1}{2}\right)^3 + \cdots\right].$$

Using the formula $S = a_1/(1 - r)$ with $a_1 = 5$ and $r = \tfrac{1}{2}$, we obtain

$$S = 10 + 2\left(\frac{5}{1 - \tfrac{1}{2}}\right) = 10 + 2(10) = 30 \text{ m}.$$

A related problem: Does this ball ever come to rest? See Discussion Exercise 5 at the end of this chapter.   ◢

10.3 *Exercises*

Exer. 1–2: Show that the given sequence is geometric, and find the common ratio.

1 $5, -\frac{5}{4}, \frac{5}{16}, \ldots, 5\left(-\frac{1}{4}\right)^{n-1}, \ldots$

2 $\frac{1}{7}, \frac{3}{7}, \frac{9}{7}, \ldots, \frac{1}{7}(3)^{n-1}, \ldots$

Exer. 3–14: Find the *n*th term, the fifth term, and the eighth term of the geometric sequence.

3 $8, 4, 2, 1, \ldots$

4 $4, 1.2, 0.36, 0.108, \ldots$

5 $300, -30, 3, -0.3, \ldots$

6 $1, -\sqrt{3}, 3, -3\sqrt{3}, \ldots$

7 $5, 25, 125, 625, \ldots$

8 $2, 6, 18, 54, \ldots$

9 $4, -6, 9, -13.5, \ldots$

10 $162, -54, 18, -6, \ldots$

11 $1, -x^2, x^4, -x^6, \ldots$

12 $1, -\frac{x}{3}, \frac{x^2}{9}, -\frac{x^3}{27}, \ldots$

13 $2, 2^{x+1}, 2^{2x+1}, 2^{3x+1}, \ldots$

14 $10, 10^{2x-1}, 10^{4x-3}, 10^{6x-5}, \ldots$

Exer. 15–16: Find all possible values of *r* for a geometric sequence with the two given terms.

15 $a_4 = 3, a_6 = 9$

16 $a_3 = 4, a_7 = \frac{1}{4}$

17 Find the sixth term of the geometric sequence whose first two terms are 4 and 6.

18 Find the seventh term of the geometric sequence whose second and third terms are 2 and $-\sqrt{2}$.

19 Given a geometric sequence with $a_4 = 4$ and $a_7 = 12$, find r and a_{10}.

20 Given a geometric sequence with $a_2 = 3$ and $a_5 = -81$, find r and a_9.

Exer. 21–26: Find the sum.

21 $\sum_{k=1}^{10} 3^k$

22 $\sum_{k=1}^{9} \left(-\sqrt{5}\right)^k$

23 $\sum_{k=0}^{9} \left(-\frac{1}{2}\right)^{k+1}$

24 $\sum_{k=1}^{7} (3^{-k})$

25 $\sum_{k=16}^{26} (2^{k-14} + 5j)$

26 $\sum_{k=8}^{14} (3^{k-7} + 2j^2)$

Exer. 27–30: Express the sum in terms of summation notation. (Answers are not unique.)

27 $2 + 4 + 8 + 16 + 32 + 64 + 128$

28 $2 - 4 + 8 - 16 + 32 - 64$

29 $\frac{1}{4} - \frac{1}{12} + \frac{1}{36} - \frac{1}{108}$

30 $3 + \frac{3}{5} + \frac{3}{25} + \frac{3}{125} + \frac{3}{625}$

Exer. 31–38: Find the sum of the infinite geometric series if it exists.

31 $1 - \frac{1}{2} + \frac{1}{4} - \frac{1}{8} + \cdots$

32 $2 + \frac{2}{3} + \frac{2}{9} + \frac{2}{27} + \cdots$

33 $1.5 + 0.015 + 0.00015 + \cdots$

34 $1 - 0.1 + 0.01 - 0.001 + \cdots$

35 $\sqrt{2} - 2 + \sqrt{8} - 4 + \cdots$

36 $1 + \frac{3}{2} + \frac{9}{4} + \frac{27}{8} + \cdots$

37 $256 + 192 + 144 + 108 + \cdots$

38 $250 - 100 + 40 - 16 + \cdots$

Exer. 39–46: Find the rational number represented by the repeating decimal.

39 $0.\overline{23}$

40 $0.0\overline{71}$

41 $2.4\overline{17}$

42 $10.\overline{5}$

43 $5.\overline{146}$

44 $3.2\overline{394}$

45 $1.\overline{6124}$

46 $123.61\overline{83}$

47 Find the geometric mean of 12 and 48.

48 Find the geometric mean of 20 and 25.

49 Insert two geometric means between 4 and 500.

50 Insert three geometric means between 2 and 512.

51 **Using a vacuum pump** A vacuum pump removes one-half of the air in a container with each stroke. After 10 strokes, what percentage of the original amount of air remains in the container?

52 **Calculating depreciation** The yearly depreciation of a certain machine is 25% of its value at the beginning of the year. If the original cost of the machine is $20,000, what is its value after 6 years?

53 Growth of bacteria A certain culture initially contains 10,000 bacteria and increases by 20% every hour.

(a) Find a formula for the number $N(t)$ of bacteria present after t hours.

(b) How many bacteria are in the culture at the end of 10 hours?

54 Interest on savings An amount of money P is deposited in a savings account that pays interest at a rate of r percent per year compounded quarterly; the principal and accumulated interest are left in the account. Find a formula for the total amount in the account after n years.

55 Rebounding ball A rubber ball is dropped from a height of 60 feet. If it rebounds approximately two-thirds the distance after each fall, use an infinite geometric series to approximate the total distance the ball travels.

56 Motion of a pendulum The bob of a pendulum swings through an arc 24 centimeters long on its first swing. If each successive swing is approximately five-sixths the length of the preceding swing, use an infinite geometric series to approximate the total distance the bob travels.

57 Multiplier effect A manufacturing company that has just located in a small community will pay two million dollars per year in salaries. It has been estimated that 60% of these salaries will be spent in the local area, and 60% of the money spent will again change hands within the community. This process, called the *multiplier effect,* will be repeated ad infinitum. Find the total amount of local spending that will be generated by company salaries.

58 Pest eradication In a pest eradication program, N sterilized male flies are released into the general population each day. It is estimated that 90% of these flies will survive a given day.

(a) Show that the number of sterilized flies in the population n days after the program has begun is

$$N + (0.9)N + (0.9)^2N + \cdots + (0.9)^{n-1}N.$$

(b) If the *long-range* goal of the program is to keep 20,000 sterilized males in the population, how many flies should be released each day?

59 Drug dosage A certain drug has a half-life of about 2 hours in the bloodstream. The drug is formulated to be administered in doses of D milligrams every 4 hours, but D is yet to be determined.

(a) Show that the number of milligrams of drug in the bloodstream after the nth dose has been administered is

$$D + \tfrac{1}{4}D + \cdots + \left(\tfrac{1}{4}\right)^{n-1}D$$

and that this sum is approximately $\tfrac{4}{3}D$ for large values of n.

(b) A level of more than 500 milligrams of the drug in the bloodstream is considered to be dangerous. Find the largest possible dose that can be given repeatedly over a long period of time.

60 Genealogy Shown in the figure is a family tree displaying the current generation (you) and 3 prior generations, with a total of 12 grandparents. If you were to trace your family history back 10 generations, how many grandparents would you find?

Exercise 60

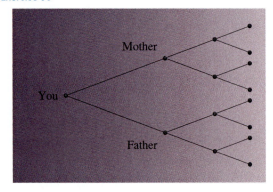

61 The first figure shows some terms of a sequence of squares $S_1, S_2, \ldots, S_k, \ldots$. Let a_k, A_k, and P_k denote the side, area, and perimeter, respectively, of the square S_k. The square S_{k+1} is constructed from S_k by connecting four points on S_k, with each point a distance of $\tfrac{1}{4}a_k$ from a vertex, as shown in the second figure.

(a) Find the relationship between a_{k+1} and a_k.

(b) Find a_n, A_n, and P_n.

(c) Calculate $\displaystyle\sum_{n=1}^{\infty} P_n$.

Exercise 61

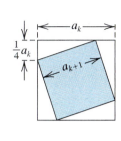

62 The figure shows several terms of a sequence consisting of alternating circles and squares. Each circle is inscribed in a square, and each square (excluding the largest) is inscribed in a circle. Let S_n denote the area of the nth square and C_n the area of the nth circle.

(a) Find the relationships between S_n and C_n and between C_n and S_{n+1}.

(b) What portion of the largest square is shaded in the figure?

Exercise 62

63 **Sierpinski sieve** The Sierpinski sieve, designed in 1915, is an example of a fractal. It can be constructed by starting with a solid black equilateral triangle. This triangle is divided into four congruent equilateral triangles, and the middle triangle is removed. On the next step, each of the three remaining equilateral triangles is divided into four congruent equilateral triangles, and the middle triangle in each of these triangles is removed, as shown in the first figure. On the third step, nine triangles are removed. If the process is continued indefinitely, the Sierpinski sieve results (see the second figure).

Exercise 63

(a) Find a geometric sequence a_k that gives the number of triangles removed on the kth step.

(b) Calculate the number of triangles removed on the fifteenth step.

(c) Suppose the initial triangle has an area of 1 unit. Find a geometric sequence b_k that gives the area removed on the kth step.

(d) Determine the area removed on the seventh step.

64 **Sierpinski sieve** Refer to Exercise 63.

(a) Write a geometric series that calculates the total number of triangles removed after n steps.

(b) Determine the total number of triangles removed after 12 steps.

(c) Write a geometric series that calculates the total area removed after n steps.

(d) Determine the total area removed after 12 steps.

65 **Annuity** If a deposit of $100 is made on the first day of each month into an account that pays 6% interest per year compounded monthly, determine the amount in the account after 18 years.

66 **Annuity** Refer to Exercise 65. Show that if the monthly deposit is P dollars and the rate is $r\%$ per year compounded monthly, then the amount A in the account after n months is given by

$$A = P\left(\frac{12}{r} + 1\right)\left[\left(1 + \frac{r}{12}\right)^n - 1\right].$$

67 **Annuity** Use Exercise 66 to find A when $P = \$100$, $r = 8\%$, and $n = 60$.

68 **Annuity** Refer to Exercise 66. If $r = 10\%$, approximately how many years are required to accumulate $100,000 if the monthly deposit P is

(a) $100   (b) $200

Exer. 69–70: The *double-declining balance method* is a method of depreciation in which, for each year $k = 1, 2, 3, \ldots, n$, the value of an asset is decreased by the fraction

$$A_k = \frac{2}{n}\left(1 - \frac{2}{n}\right)^{k-1} \text{ of its initial cost.}$$

69 (a) If $n = 5$, find $A_1, A_2, \ldots, A_5$.

(b) Show that the sequence in (a) is geometric, and find S_5.

(c) If the initial value of an asset is $25,000, how much of its value has been depreciated after 2 years?

70 (a) If n is any positive integer, find $A_1, A_2, \ldots, A_n$.

(b) Show that the sequence in (a) is geometric, and find S_n.

10.4

Mathematical Induction

If n is a positive integer and we let P_n denote the mathematical statement $(xy)^n = x^n y^n$, we obtain the following *infinite sequence of statements:*

Statement P_1: $\quad (xy)^1 = x^1 y^1$

Statement P_2: $\quad (xy)^2 = x^2 y^2$

Statement P_3: $\quad (xy)^3 = x^3 y^3$

$$\vdots \qquad\qquad \vdots$$

Statement P_n: $\quad (xy)^n = x^n y^n$

$$\vdots \qquad\qquad \vdots$$

It is easy to show that P_1, P_2, and P_3 are *true* statements. However, it is impossible to check the validity of P_n for *every* positive integer n. Showing that P_n is true for every n requires the following principle.

Principle of Mathematical Induction	If with each positive integer n there is associated a statement P_n, then all the statements P_n are true, provided the following two conditions are satisfied. **(1)** P_1 is true. **(2)** Whenever k is a positive integer such that P_k is true, then P_{k+1} is also true.

To help us understand this principle, we consider an infinite sequence of statements labeled

$$P_1, P_2, P_3, \ldots, P_n, \ldots$$

that satisfy conditions (1) and (2). By (1), statement P_1 is true. Since condition (2) holds, whenever a statement P_k is true the *next* statement P_{k+1} is also true. Hence, since P_1 is true, P_2 is also true, by (2). However, if P_2 is true, then, by (2), we see that the next statement P_3 is true. Once again, if P_3 is true, then, by (2), P_4 is also true. If we continue in this manner, we can argue that if n is any *particular* integer, then P_n is true, since we can use condition (2) one step at a time, eventually reaching P_n. Although this type of reasoning does not actually *prove* the principle of mathematical induction, it certainly makes it plausible. The principle is proved in advanced algebra using postulates for the positive integers.

When applying the principle of mathematical induction, we always follow two steps.

Steps in Applying the Principle of Mathematical Induction	*1* Show that P_1 is true.
	2 *Assume* that P_k is true, and then prove that P_{k+1} is true.

Step 2 often causes confusion. Note that we do not *prove* that P_k is true (except for $k = 1$). Instead, we show that *if P_k happens to be true*, then the statement P_{k+1} is also true. We refer to the assumption that P_k is true as the **induction hypothesis.**

EXAMPLE 1   Using the principle of mathematical induction

Use mathematical induction to prove that for every positive integer n, the sum of the first n positive integers is

$$\frac{n(n + 1)}{2}.$$

SOLUTION   If n is any positive integer, let P_n denote the statement

$$1 + 2 + 3 + \cdots + n = \frac{n(n + 1)}{2}.$$

The following are some special cases of P_n.
   If $n = 1$, then P_1 is

$$1 = \frac{1(1 + 1)}{2}; \quad \text{that is,} \quad 1 = 1.$$

   If $n = 2$, then P_2 is

$$1 + 2 = \frac{2(2 + 1)}{2}; \quad \text{that is,} \quad 3 = 3.$$

   If $n = 3$, then P_3 is

$$1 + 2 + 3 = \frac{3(3 + 1)}{2}; \quad \text{that is,} \quad 6 = 6.$$

Although it is instructive to check the validity of P_n for several values of n as we have done, it is unnecessary to do so. We need only apply the two-step process outlined prior to this example. Thus, we proceed as follows:

Step 1   If we substitute $n = 1$ in P_n, then the left-hand side contains only the number 1 and the right-hand side is $\dfrac{1(1 + 1)}{2}$, which also equals 1. Hence, P_1 is true.

(continued)

Step 2 Assume that P_k is true. Thus, the induction hypotheses is

$$1 + 2 + 3 + \cdots + k = \frac{k(k + 1)}{2}.$$

Our goal is to show that P_{k+1} is true—that is, that

$$1 + 2 + 3 + \cdots + k + (k + 1) = \frac{(k + 1)[(k + 1) + 1]}{2}.$$

We may prove that the last formula is true by rewriting the left-hand side and using the induction hypothesis as follows:

$$1 + 2 + 3 + \cdots + k + (k + 1) = (1 + 2 + 3 + \cdots + k) + (k + 1) \qquad \text{group the first } k \text{ terms}$$

$$= \frac{k(k + 1)}{2} + (k + 1) \qquad \text{induction hypothesis}$$

$$= \frac{k(k + 1) + 2(k + 1)}{2} \qquad \text{add terms}$$

$$= \frac{(k + 1)(k + 2)}{2} \qquad \text{factor out } k + 1$$

$$= \frac{(k + 1)[(k + 1) + 1]}{2} \qquad \text{change form of } k + 2$$

This shows that P_{k+1} is true, and therefore the proof by mathematical induction is complete. ◢

EXAMPLE 2 **Using the principle of mathematical induction**

Prove that for each positive integer n,

$$1^2 + 3^2 + \cdots + (2n - 1)^2 = \frac{n(2n - 1)(2n + 1)}{3}.$$

SOLUTION For each positive integer n, let P_n denote the given statement. Note that this is a formula for the sum of the squares of the first n odd positive integers. We again follow the two-step procedure.

Step 1 Substituting 1 for n in P_n, we obtain

$$1^2 = \frac{(1)(2 - 1)(2 + 1)}{3} = \frac{3}{3} = 1.$$

This shows that P_1 is true.

Step 2 Assume that P_k is true. Thus, the induction hypothesis is

$$1^2 + 3^2 + \cdots + (2k - 1)^2 = \frac{k(2k - 1)(2k + 1)}{3}.$$

We wish to show that P_{k+1} is true—that is, that

$$1^2 + 3^2 + \cdots + [2(k+1) - 1]^2 = \frac{(k+1)[2(k+1) - 1][2(k+1) + 1]}{3}.$$

This equation simplifies to

$$1^2 + 3^2 + \cdots + (2k+1)^2 = \frac{(k+1)(2k+1)(2k+3)}{3}.$$

Remember that the next to last term on the left-hand side of the equation (the kth term) is $(2k - 1)^2$. In a manner similar to that used in the solution of Example 1, we may prove the formula for P_{k+1} by rewriting the left-hand side and using the induction hypothesis as follows:

$$1^2 + 3^2 + \cdots + (2k+1)^2 = [1^2 + 3^2 + \cdots + (2k-1)^2] + (2k+1)^2 \quad \text{group the first } k \text{ terms}$$

$$= \frac{k(2k-1)(2k+1)}{3} + (2k+1)^2 \quad \text{induction hypothesis}$$

$$= \frac{k(2k-1)(2k+1) + 3(2k+1)^2}{3} \quad \text{add terms}$$

$$= \frac{(2k+1)[k(2k-1) + 3(2k+1)]}{3} \quad \text{factor out } 2k+1$$

$$= \frac{(2k+1)(2k^2 + 5k + 3)}{3} \quad \text{simplify}$$

$$= \frac{(k+1)(2k+1)(2k+3)}{3} \quad \text{factor and change order}$$

This shows that P_{k+1} is true, and hence P_n is true for every n. ◢

EXAMPLE 3 **Using the principle of mathematical induction**

Prove that 2 is a factor of $n^2 + 5n$ for every positive integer n.

SOLUTION For each positive integer n, let P_n denote the following statement:

$$2 \text{ is a factor of } n^2 + 5n$$

We shall follow the two-step procedure.

Step 1 If $n = 1$, then

$$n^2 + 5n = 1^2 + 5 \cdot 1 = 6 = 2 \cdot 3.$$

Thus, 2 is a factor of $n^2 + 5n$ for $n = 1$; that is, P_1 is true.

Step 2 Assume that P_k is true. Thus, the induction hypothesis is

$$2 \text{ is a factor of } k^2 + 5k$$

or, equivalently, $$k^2 + 5k = 2p$$

for some integer p. *(continued)*

We wish to show that P_{k+1} is true—that is, that

$$2 \text{ is a factor of } (k+1)^2 + 5(k+1).$$

We may do this as follows:

$$
\begin{aligned}
(k+1)^2 + 5(k+1) &= k^2 + 2k + 1 + 5k + 5 && \text{multiply} \\
&= (k^2 + 5k) + (2k + 6) && \text{rearrange terms} \\
&= 2p + 2(k+3) && \text{induction hypothesis,} \\
& && \text{factor } 2k + 6 \\
&= 2(p + k + 3) && \text{factor out 2}
\end{aligned}
$$

Since 2 is a factor of the last expression, P_{k+1} is true, and hence P_n is true for every n. ◢

Let j be a positive integer, and suppose that with each integer $n \geq j$ there is associated a statement P_n. For example, if $j = 6$, then the statements are numbered $P_6, P_7, P_8, \ldots$. The principle of mathematical induction may be extended to cover this situation. To prove that the statements P_n are true for $n \geq j$, we use the following two steps, in the same manner as we did for $n \geq 1$.

Steps in Applying the Extended Principle of Mathematical Induction for $P_k, k \geq j$	1 Show that P_j is true. 2 Assume that P_k is true with $k \geq j$, and then prove that P_{k+1} is true.

EXAMPLE 4   **Using the extended principle of mathematical induction**

Let a be a nonzero real number such that $a > -1$. Prove that

$$(1 + a)^n > 1 + na$$

for every integer $n \geq 2$.

SOLUTION   For each positive integer n, let P_n denote the inequality $(1 + a)^n > 1 + na$. Note that P_1 is *false*, since $(1 + a)^1 = 1 + (1)(a)$. However, we can show that P_n is true for $n \geq 2$ by using the extended principle with $j = 2$.

Step 1   We first note that $(1 + a)^2 = 1 + 2a + a^2$. Since $a \neq 0$, we have $a^2 > 0$, and so $1 + 2a + a^2 > 1 + 2a$ or, equivalently, $(1 + a)^2 > 1 + 2a$. Hence, P_2 is true.

Step 2   Assume that P_k is true. Thus, the induction hypothesis is

$$(1 + a)^k > 1 + ka.$$

We wish to show that P_{k+1} is true—that is, that

$$(1 + a)^{k+1} > 1 + (k+1)a.$$

To prove the last inequality, we first observe the following:

$$(1 + a)^{k+1} = (1 + a)^k(1 + a)^1 \quad \text{law of exponents}$$
$$> (1 + ka)(1 + a) \quad \text{induction hypothesis and } 1 + a > 0$$

We next note that

$$(1 + ka)(1 + a) = 1 + ka + a + ka^2 \quad \text{multiply}$$
$$= 1 + (ka + a) + ka^2 \quad \text{group terms}$$
$$= 1 + (k + 1)a + ka^2 \quad \text{factor out } a$$
$$> 1 + (k + 1)a. \quad \text{since } ka^2 > 0$$

The last two inequalities give us

$$(1 + a)^{k+1} > 1 + (k + 1)a.$$

Thus, P_{k+1} is true, and the proof by mathematical induction is complete. ✓

We have looked at several examples of proving statements by using the principle of mathematical induction. You may be wondering "Where do these statements come from?" These statements can often be "discovered" by observing patterns, combining results from several areas of mathematics, or recognizing certain types or categories of relationships. Two such statements are given in Exercises 37 and 38 in this section.

10.4   *Exercises*

Exer. 1–26: Prove that the statement is true for every positive integer *n*.

1 $2 + 4 + 6 + \cdots + 2n = n(n + 1)$

2 $1 + 4 + 7 + \cdots + (3n - 2) = \dfrac{n(3n - 1)}{2}$

3 $1 + 3 + 5 + \cdots + (2n - 1) = n^2$

4 $3 + 9 + 15 + \cdots + (6n - 3) = 3n^2$

5 $2 + 7 + 12 + \cdots + (5n - 3) = \frac{1}{2}n(5n - 1)$

6 $2 + 6 + 18 + \cdots + 2 \cdot 3^{n-1} = 3^n - 1$

7 $1 + 2 \cdot 2 + 3 \cdot 2^2 + \cdots + n \cdot 2^{n-1} = 1 + (n - 1) \cdot 2^n$

8 $(-1)^1 + (-1)^2 + (-1)^3 + \cdots + (-1)^n = \dfrac{(-1)^n - 1}{2}$

9 $1^2 + 2^2 + 3^2 + \cdots + n^2 = \dfrac{n(n + 1)(2n + 1)}{6}$

10 $1^3 + 2^3 + 3^3 + \cdots + n^3 = \left[\dfrac{n(n + 1)}{2}\right]^2$

11 $\dfrac{1}{1 \cdot 2} + \dfrac{1}{2 \cdot 3} + \dfrac{1}{3 \cdot 4} + \cdots + \dfrac{1}{n(n + 1)} = \dfrac{n}{n + 1}$

12 $\dfrac{1}{1 \cdot 2 \cdot 3} + \dfrac{1}{2 \cdot 3 \cdot 4} + \dfrac{1}{3 \cdot 4 \cdot 5} + \cdots +$
$\dfrac{1}{n(n + 1)(n + 2)} = \dfrac{n(n + 3)}{4(n + 1)(n + 2)}$

13 $3 + 3^2 + 3^3 + \cdots + 3^n = \frac{3}{2}(3^n - 1)$

14 $1^3 + 3^3 + 5^3 + \cdots + (2n - 1)^3 = n^2(2n^2 - 1)$

15 $n < 2^n$ 16 $1 + 2n \le 3^n$

17 $1 + 2 + 3 + \cdots + n < \frac{1}{8}(2n + 1)^2$

18 If $0 < a < b$, then $\left(\dfrac{a}{b}\right)^{n+1} < \left(\dfrac{a}{b}\right)^n$.

19 3 is a factor of $n^3 - n + 3$.

20 2 is a factor of $n^2 + n$. 21 4 is a factor of $5^n - 1$.

22 9 is a factor of $10^{n+1} + 3 \cdot 10^n + 5$.

23 If a is greater than 1, then $a^n > 1$.

24 If $r \neq 1$, then

$$a + ar + ar^2 + \cdots + ar^{n-1} = \frac{a(1 - r^n)}{1 - r}.$$

25 $a - b$ is a factor of $a^n - b^n$.

(*Hint:* $a^{k+1} - b^{k+1} = a^k(a - b) + (a^k - b^k)b$.)

26 $a + b$ is a factor of $a^{2n-1} + b^{2n-1}$.

Exer. 27–32: Find the smallest positive integer j for which the statement is true. Use the extended principle of mathematical induction to prove that the formula is true for every integer greater than j.

27 $n + 12 \leq n^2$

28 $n^2 + 18 \leq n^3$

29 $5 + \log_2 n \leq n$

30 $n^2 \leq 2^n$

31 $2n + 2 \leq 2^n$

32 $n \log_2 n + 20 \leq n^2$

Exer. 33–36: Express the sum in terms of n.

33 $\displaystyle\sum_{k=1}^{n} (k^2 + 3k + 5)$

(*Hint:* Use the theorem on sums to write the sum as

$$\sum_{k=1}^{n} k^2 + 3\sum_{k=1}^{n} k + \sum_{k=1}^{n} 5.$$

Next use Exercise 9 above, Example 5 of Section 10.1, and the theorem on the sum of a constant.)

34 $\displaystyle\sum_{k=1}^{n} (3k^2 - 2k + 1)$

35 $\displaystyle\sum_{k=1}^{n} (2k - 3)^2$

36 $\displaystyle\sum_{k=1}^{n} (k^3 + 2k^2 - k + 4)$ (*Hint:* Use Exercise 10.)

Exer. 37–38: (a) Evaluate the given formula for the stated values of n, and solve the resulting system of equations for $a, b, c,$ and d. (This method can sometimes be used to obtain formulas for sums.) (b) Compare the result in part (a) with the indicated exercise, and explain why this method does not prove that the formula is true for every n.

37 $1^2 + 2^2 + 3^2 + \cdots + n^2 = an^3 + bn^2 + cn$;

$n = 1, 2, 3$ (Exercise 9)

38 $1^3 + 2^3 + 3^3 + \cdots + n^3 = an^4 + bn^3 + cn^2 + dn$;

$n = 1, 2, 3, 4$ (Exercise 10)

Exer. 39–42: Prove that the statement is true for every positive integer n.

39 $\sin(\theta + n\pi) = (-1)^n \sin\theta$

40 $\cos(\theta + n\pi) = (-1)^n \cos\theta$

41 Prove De Moivre's theorem:

$$[r(\cos\theta + i\sin\theta)]^n = r^n(\cos n\theta + i\sin n\theta)$$

for every positive integer n.

42 Prove that for every positive integer $n \geq 3$, the sum of the interior angles of an n-sided polygon is given by the expression $(n - 2) \cdot 180°$.

10.5
The Binomial Theorem

A **binomial** is a sum $a + b$, where a and b represent numbers. If n is a positive integer, then a general formula for *expanding* $(a + b)^n$ (that is, for expressing it as a sum) is given by the **binomial theorem.** In this section we shall use mathematical induction to establish this general formula. The following special cases can be obtained by multiplication:

$$(a + b)^2 = a^2 + 2ab + b^2$$
$$(a + b)^3 = a^3 + 3a^2b + 3ab^2 + b^3$$
$$(a + b)^4 = a^4 + 4a^3b + 6a^2b^2 + 4ab^3 + b^4$$
$$(a + b)^5 = a^5 + 5a^4b + 10a^3b^2 + 10a^2b^3 + 5ab^4 + b^5$$

These expansions of $(a + b)^n$ for $n = 2, 3, 4,$ and 5 have the following properties.

(1) There are $n + 1$ terms, the first being a^n and the last b^n.

(2) As we proceed from any term to the next, the power of a decreases by 1 and the power of b increases by 1. For each term, the sum of the exponents of a and b is n.

(3) Each term has the form $(c)a^{n-k}b^k$, where the coefficient c is an integer and $k = 0, 1, 2, \ldots, n$.

(4) The following formula is true for each of the first n terms of the expansion:

$$\frac{(\text{coefficient of term}) \cdot (\text{exponent of } a)}{\text{number of term}} = \text{coefficient of next term}$$

The following table illustrates property 4 for the expansion of $(a + b)^5$.

Term	Number of term	Coefficient of term	Exponent of a	Coefficient of next term
a^5	1	1	5	$\dfrac{1 \cdot 5}{1} = 5$
$5a^4b$	2	5	4	$\dfrac{5 \cdot 4}{2} = 10$
$10a^3b^2$	3	10	3	$\dfrac{10 \cdot 3}{3} = 10$
$10a^2b^3$	4	10	2	$\dfrac{10 \cdot 2}{4} = 5$
$5ab^4$	5	5	1	$\dfrac{5 \cdot 1}{5} = 1$

Let us next consider $(a + b)^n$ for an arbitrary positive integer n. The first term is a^n, which has coefficient 1. If we assume that property 4 is true, we obtain the successive coefficients listed in the next table.

Term	Number of term	Coefficient of term	Exponent of a	Coefficient of next term
a^n	1	1	n	$\dfrac{1 \cdot n}{1} = n$
$\dfrac{n}{1}a^{n-1}b$	2	$\dfrac{n}{1}$	$n - 1$	$\dfrac{n(n-1)}{2 \cdot 1}$
$\dfrac{n(n-1)}{2 \cdot 1}a^{n-2}b^2$	3	$\dfrac{n(n-1)}{2 \cdot 1}$	$n - 2$	$\dfrac{n(n-1)(n-2)}{3 \cdot 2 \cdot 1}$
$\dfrac{n(n-1)(n-2)}{3 \cdot 2 \cdot 1}a^{n-3}b^3$	4	$\dfrac{n(n-1)(n-2)}{3 \cdot 2 \cdot 1}$	$n - 3$	$\dfrac{n(n-1)(n-2)(n-3)}{4 \cdot 3 \cdot 2 \cdot 1}$

The pattern that appears in the fifth column leads to the following formula for the coefficient of the general term.

Coefficient of the $(k + 1)$st Term in the Expansion of $(a + b)^n$	$\dfrac{n \cdot (n-1) \cdot (n-2) \cdot (n-3) \cdots \cdots (n-k+1)}{k \cdot (k-1) \cdots \cdots 3 \cdot 2 \cdot 1}, \quad k = 1, 2, \ldots, n$

The $(k + 1)$st coefficient can be written in a compact form by using **factorial notation.** If n is any nonnegative integer, then the symbol $n!$ (*n factorial*) is defined as follows.

Definition of $n!$	(1) $n! = n(n - 1)(n - 2) \cdot \cdots \cdot 1$ if $n > 0$ (2) $0! = 1$

Thus, if $n > 0$, then $n!$ is the product of the first n positive integers. The definition $0! = 1$ is used so that certain formulas involving factorials are true for all *nonnegative* integers.

ILLUSTRATION   *n* Factorial

- $1! = 1$
- $2! = 2 \cdot 1 = 2$
- $3! = 3 \cdot 2 \cdot 1 = 6$
- $4! = 4 \cdot 3 \cdot 2 \cdot 1 = 24$

- $5! = 5 \cdot 4 \cdot 3 \cdot 2 \cdot 1 = 120$
- $6! = 6 \cdot 5 \cdot 4 \cdot 3 \cdot 2 \cdot 1 = 720$
- $7! = 7 \cdot 6 \cdot 5 \cdot 4 \cdot 3 \cdot 2 \cdot 1 = 5040$
- $8! = 8 \cdot 7 \cdot 6 \cdot 5 \cdot 4 \cdot 3 \cdot 2 \cdot 1 = 40{,}320$

Notice the rapid growth of $n!$ as n increases.

We sometimes wish to simplify quotients where both the numerator and the denominator contain factorials, as shown in the next illustration.

ILLUSTRATION   **Simplifying Quotients of Factorials**

- $\dfrac{7!}{5!} = \dfrac{7 \cdot 6 \cdot 5!}{5!} = 7 \cdot 6 = 42$

- $\dfrac{10!}{6!} = \dfrac{10 \cdot 9 \cdot 8 \cdot 7 \cdot 6!}{6!} = 10 \cdot 9 \cdot 8 \cdot 7 = 5040$

As in the preceding illustration, if n and k are positive integers and $k < n$, then

$$\frac{n!}{(n - k)!} = \frac{n \cdot (n - 1) \cdot (n - 2) \cdot \cdots \cdot (n - k + 1) \cdot [(n - k)!]}{(n - k)!}$$

$$= n \cdot (n - 1) \cdot (n - 2) \cdot \cdots \cdot (n - k + 1),$$

which is the numerator of the coefficient of the $(k + 1)$st term of $(a + b)^n$. Dividing by the denominator $k!$ gives us the following alternative form for the $(k + 1)$st coefficient:

$$\frac{n \cdot (n - 1) \cdot (n - 2) \cdot \cdots \cdot (n - k + 1)}{k!} = \frac{n!}{k!\,(n - k)!}$$

These numbers are called **binomial coefficients** and are often denoted by either the symbol $\dbinom{n}{k}$ or the symbol $C(n, k)$. Thus, we have the following.

Coefficient of the $(k + 1)$st Term in the Expansion of $(a + b)^n$ (Alternative Form)	$\dbinom{n}{k} = C(n, k) = \dfrac{n!}{k!\,(n - k)!}, \quad k = 0, 1, 2, \ldots, n$

The symbols $\dbinom{n}{k}$ and $C(n, k)$ are sometimes read "n choose k."

EXAMPLE 1 Evaluating $\dbinom{n}{k}$

Find $\dbinom{5}{0}$, $\dbinom{5}{1}$, $\dbinom{5}{2}$, $\dbinom{5}{3}$, $\dbinom{5}{4}$, and $\dbinom{5}{5}$.

SOLUTION   These six numbers are the coefficients in the expansion of $(a + b)^5$, which we tabulated earlier in this section. By definition,

$$\binom{5}{0} = \frac{5!}{0!\,(5 - 0)!} = \frac{5!}{0!\,5!} = \frac{5!}{1 \cdot 5!} = 1$$

$$\binom{5}{1} = \frac{5!}{1!\,(5 - 1)!} = \frac{5!}{1!\,4!} = \frac{5!}{1 \cdot 4!} = \frac{5 \cdot 4!}{4!} = 5$$

$$\binom{5}{2} = \frac{5!}{2!\,(5 - 2)!} = \frac{5!}{2!\,3!} = \frac{5 \cdot 4 \cdot 3!}{2 \cdot 3!} = \frac{20}{2} = 10$$

$$\binom{5}{3} = \frac{5!}{3!\,(5 - 3)!} = \frac{5!}{3!\,2!} = \frac{5 \cdot 4 \cdot 3!}{3! \cdot 2} = \frac{20}{2} = 10$$

$$\binom{5}{4} = \frac{5!}{4!\,(5 - 4)!} = \frac{5!}{4!\,1!} = \frac{5!}{4! \cdot 1} = \frac{5 \cdot 4!}{4!} = 5$$

$$\binom{5}{5} = \frac{5!}{5!\,(5 - 5)!} = \frac{5!}{5!\,0!} = \frac{5!}{5! \cdot 1} = 1.$$

EXAMPLE 2 Simplifying a quotient of factorials

Rewrite $(3n + 3)!/(3n)!$ as an expression that does not contain factorials.

SOLUTION   By the definition of $n!$, we can write $(3n + 3)!$ as

$$(3n + 3)(3n + 2)(3n + 1)\underbrace{(3n)(3n - 1)(3n - 2) \cdots (3)(2)(1)}_{(3n)!}.$$

Thus,

$$\frac{(3n + 3)!}{(3n)!} = \frac{(3n + 3)(3n + 2)(3n + 1)(3n)!}{(3n)!} \qquad \text{definition of } n!$$

$$= (3n + 3)(3n + 2)(3n + 1). \qquad \text{cancel } (3n)! \neq 0$$

The binomial theorem may be stated as follows.

The Binomial Theorem

$$(a + b)^n = a^n + \binom{n}{1}a^{n-1}b + \binom{n}{2}a^{n-2}b^2 + \cdots + \binom{n}{k}a^{n-k}b^k + \cdots + \binom{n}{n-1}ab^{n-1} + b^n$$

Using summation notation, we may write the binomial theorem

$$(a + b)^n = \sum_{k=0}^{n} \binom{n}{k}a^{n-k}b^k.$$

Note that there are $n + 1$ terms (not n terms) in the expansion of $(a + b)^n$, and so

$$\binom{n}{k}a^{n-k}b^k \text{ is a formula for the } (k + 1)\text{st term of the expansion.}$$

An alternative statement of the binomial theorem is as follows. (A proof is given at the end of this section.)

**The Binomial Theorem
(Alternative Form)**

$$(a + b)^n = a^n + na^{n-1}b + \frac{n(n - 1)}{2!}a^{n-2}b^2 + \cdots + \frac{n(n - 1)(n - 2) \cdots (n - k + 1)}{k!}a^{n-k}b^k + \cdots + nab^{n-1} + b^n$$

The following examples may be solved either by using the general formulas for the binomial theorem or by repeated use of property 4, stated at the beginning of this section.

EXAMPLE 3 **Finding a binomial expansion**

Find the binomial expansion of $(2x + 3y^2)^4$.

SOLUTION We use the binomial theorem with $a = 2x$, $b = 3y^2$, and $n = 4$:

$$(2x + 3y^2)^4 = (2x)^4 + \binom{4}{1}(2x)^3(3y^2)^1 + \binom{4}{2}(2x)^2(3y^2)^2 + \binom{4}{3}(2x)^1(3y^2)^3 + (3y^2)^4$$

$$= 16x^4 + 4(8x^3)(3y^2) + 6(4x^2)(9y^4) + 4(2x)(27y^6) + 81y^8$$

$$= 16x^4 + 96x^3y^2 + 216x^2y^4 + 216xy^6 + 81y^8$$

Examining the terms of the expansion from left to right, we see that the exponents on x decrease by 1 and that the exponents on y increase by 2. It is a good idea to check for exponent patterns after simplifying a binomial expansion.

The next example illustrates that if either a or b is negative, then the terms of the expansion are alternately positive and negative.

EXAMPLE 4 Finding a binomial expansion

Expand $\left(\dfrac{1}{x} - 2\sqrt{x}\right)^5$.

SOLUTION The binomial coefficients for $(a + b)^5$ were calculated in Example 1. Thus, if we let $a = 1/x$, $b = -2\sqrt{x}$, and $n = 5$ in the binomial theorem, we obtain

$$\left(\frac{1}{x} - 2\sqrt{x}\right)^5 = \left(\frac{1}{x}\right)^5 + 5\left(\frac{1}{x}\right)^4(-2\sqrt{x})^1 + 10\left(\frac{1}{x}\right)^3(-2\sqrt{x})^2$$

$$+ 10\left(\frac{1}{x}\right)^2(-2\sqrt{x})^3 + 5\left(\frac{1}{x}\right)^1(-2\sqrt{x})^4 + (-2\sqrt{x})^5,$$

which can be written as

$$\left(\frac{1}{x} - 2\sqrt{x}\right)^5 = \frac{1}{x^5} - \frac{10}{x^{7/2}} + \frac{40}{x^2} - \frac{80}{x^{1/2}} + 80x - 32x^{5/2}.$$

To find a specific term in the expansion of $(a + b)^n$, it is convenient to first find the exponent k that is to be assigned to b. Notice that, by the binomial theorem, *the exponent of b is always one less than the number of the term.* Once k is found, we know that the exponent of a is $n - k$ and the coefficient is $\dbinom{n}{k}$.

EXAMPLE 5 Finding a specific term of a binomial expansion

Find the fifth term in the expansion of $\left(x^3 + \sqrt{y}\right)^{13}$.

SOLUTION Let $a = x^3$ and $b = \sqrt{y}$. The exponent of b in the fifth term is $k = 5 - 1 = 4$, and hence the exponent of a is $n - k = 13 - 4 = 9$. From the discussion of the preceding paragraph we obtain

$(k + 1)\text{st term} = \dbinom{n}{k}a^{n-k}b^k$

$$\binom{13}{4}(x^3)^9\left(\sqrt{y}\right)^4 = \frac{13!}{4!\,(13 - 4)!}x^{27}y^2 = \frac{13 \cdot 12 \cdot 11 \cdot 10}{4!}x^{27}y^2 = 715x^{27}y^2.$$

EXAMPLE 6 Finding a specific term of a binomial expansion

Find the term involving q^{10} in the binomial expansion of $\left(\frac{1}{3}p + q^2\right)^{12}$.

SOLUTION From the statement of the binomial theorem with $a = \frac{1}{3}p$, $b = q^2$, and $n = 12$, each term in the expansion has the form

$$\binom{n}{k}a^{n-k}b^k = \binom{12}{k}\left(\frac{1}{3}p\right)^{12-k}(q^2)^k.$$

(continued)

Since $(q^2)^k = q^{2k}$, we must let $k = 5$ to obtain the term involving q^{10}. Doing so gives us

$$\binom{12}{5}\left(\frac{1}{3}p\right)^{12-5}(q^2)^5 = \frac{12!}{5!\,(12-5)!}\left(\frac{1}{3}\right)^7 p^7 q^{10} = \frac{88}{243}\,p^7 q^{10}.$$

There is an interesting triangular array of numbers, called **Pascal's triangle,** that can be used to obtain binomial coefficients. The numbers are arranged as follows:

$$
\begin{array}{ccccccccccc}
 & & & & & 1 & & & & & \\
 & & & & 1 & & 1 & & & & \\
 & & & 1 & & 2 & & 1 & & & \\
 & & 1 & & 3 & & 3 & & 1 & & \\
 & 1 & & 4 & & 6 & & 4 & & 1 & \\
1 & & 5 & & 10 & & 10 & & 5 & & 1 \\
\end{array}
$$

$$
\begin{array}{ccccccccccccc}
1 & & 6 & & 15 & & 20 & & 15 & & 6 & & 1 \\
\end{array}
$$

The numbers in the second row are the coefficients in the expansion of $(a + b)^1$; those in the third row are the coefficients determined by $(a + b)^2$; those in the fourth row are obtained from $(a + b)^3$; and so on. Each number in the array that is different from 1 can be found by adding the two numbers in the previous row that appear above and immediately to the left and right of the number, as illustrated in the solution of the next example.

EXAMPLE 7   Using Pascal's triangle

Find the eighth row of Pascal's triangle, and use it to expand $(a + b)^7$.

SOLUTION   Let us rewrite the seventh row and then use the process described above. In the following display the arrows indicate which two numbers in row seven are added to obtain the numbers in row eight.

The eighth row gives us the coefficients in the expansion of $(a + b)^7$:

$$(a + b)^7 = a^7 + 7a^6b + 21a^5b^2 + 35a^4b^3 + 35a^3b^4 + 21a^2b^5 + 7ab^6 + b^7$$

Pascal's triangle is useful for expanding small powers of $a + b$; however, for expanding large powers or finding a specific term, as in Examples 5 and 6, the general formula given by the binomial theorem is more useful.

We shall conclude this section by giving a proof of the binomial theorem using mathematical induction.

PROOF OF THE BINOMIAL THEOREM   For each positive integer n, let P_n denote the statement given in the alternative form of the binomial theorem.

Step 1   If $n = 1$, the statement reduces to $(a + b)^1 = a^1 + b^1$. Consequently, P_1 is true.

Step 2   Assume that P_k is true. Thus, the induction hypothesis is

$$(a + b)^k = a^k + ka^{k-1}b + \frac{k(k - 1)}{2!}a^{k-2}b^2 + \cdots + \frac{k(k - 1)(k - 2) \cdots (k - r + 2)}{(r - 1)!}a^{k-r+1}b^{r-1}$$

$$+ \frac{k(k - 1)(k - 2) \cdots (k - r + 1)}{r!}a^{k-r}b^r + \cdots + kab^{k-1} + b^k.$$

We have shown both the rth term and the $(r + 1)$st term in the above expansion.

To prove that P_{k+1} is true, we first write

$$(a + b)^{k+1} = (a + b)^k(a + b).$$

Using the induction hypothesis to substitute for $(a + b)^k$ and then multiplying that expression by $a + b$, we obtain

$$(a + b)^{k+1} = \left[a^{k+1} + ka^kb + \frac{k(k - 1)}{2!}a^{k-1}b^2 + \cdots + \frac{k(k - 1) \cdots (k - r + 1)}{r!}a^{k-r+1}b^r + \cdots + ab^k \right]$$

$$+ \left[a^kb + ka^{k-1}b^2 + \cdots + \frac{k(k - 1) \cdots (k - r + 2)}{(r - 1)!}a^{k-r+1}b^r + \cdots + kab^k + b^{k+1} \right],$$

where the terms in the first pair of brackets result from multiplying the right-hand side of the induction hypothesis by a and the terms in the second pair of brackets result from multiplying by b. We next rearrange and combine terms:

$$(a + b)^{k+1} = a^{k+1} + (k + 1)a^kb + \left(\frac{k(k - 1)}{2!} + k \right)a^{k-1}b^2 + \cdots$$

$$+ \left(\frac{k(k - 1) \cdots (k - r + 1)}{r!} + \frac{k(k - 1) \cdots (k - r + 2)}{(r - 1)!} \right)a^{k-r+1}b^r$$

$$+ \cdots + (1 + k)ab^k + b^{k+1}$$

If the coefficients are simplified, we obtain statement P_n with $k + 1$ substituted for n. Thus, P_{k+1} is true, and therefore P_n holds for every positive integer n, which completes the proof.   ◢

10.5 *Exercises*

Exer. 1–12: Evaluate the expression.

1 $2!\,6!$ 2 $3!\,4!$

3 $7!\,0!$ 4 $5!\,0!$

5 $\dfrac{8!}{5!}$ 6 $\dfrac{6!}{3!}$

7 $\dbinom{5}{5}$ 8 $\dbinom{7}{0}$

9 $\dbinom{7}{5}$ 10 $\dbinom{8}{4}$

11 $\dbinom{13}{4}$ 12 $\dbinom{52}{2}$

Exer. 13–16: Rewrite as an expression that does not contain factorials.

13 $\dfrac{n!}{(n-2)!}$ 14 $\dfrac{(n+1)!}{(n-1)!}$

15 $\dfrac{(2n+2)!}{(2n)!}$ 16 $\dfrac{(3n+1)!}{(3n-1)!}$

Exer. 17–30: Use the binomial theorem to expand and simplify.

17 $(4x-y)^3$ 18 $(x^2+2y)^3$

19 $(x+y)^6$ 20 $(x+y)^4$

21 $(x-y)^7$ 22 $(x-y)^5$

23 $(3t-5s)^4$ 24 $(2t-s)^5$

25 $\left(\tfrac{1}{3}x+y^2\right)^5$ 26 $\left(\tfrac{1}{2}x+y^3\right)^4$

27 $\left(\dfrac{1}{x^2}+3x\right)^6$ 28 $\left(\dfrac{1}{x^3}-2x\right)^5$

29 $\left(\sqrt{x}-\dfrac{1}{\sqrt{x}}\right)^5$ 30 $\left(\sqrt{x}+\dfrac{1}{\sqrt{x}}\right)^5$

Exer. 31–46: Without expanding completely, find the indicated term(s) in the expansion of the expression.

31 $(3c^{2/5}+c^{4/5})^{25}$;   first three terms

32 $(x^3+5x^{-2})^{20}$;   first three terms

33 $(4z^{-1}-3z)^{15}$;   last three terms

34 $(s-2t^3)^{12}$;   last three terms

35 $\left(\dfrac{3}{c}+\dfrac{c^2}{4}\right)^7$;   sixth term

36 $\left(3x^2-\sqrt{y}\right)^9$;   fifth term

37 $\left(\tfrac{1}{3}u+4v\right)^8$;   seventh term

38 $(3x^2-y^3)^{10}$;   fourth term

39 $(x^{1/2}+y^{1/2})^8$;   middle term

40 $(rs^2+t)^7$;   two middle terms

41 $(2y+x^2)^8$;   term that contains x^{10}

42 $(x^2-2y^3)^5$;   term that contains y^6

43 $(3y^3-2x^2)^4$;   term that contains y^9

44 $\left(\sqrt{c}+\sqrt{d}\right)^8$;   term that contains c^2

45 $\left(3x-\dfrac{1}{4x}\right)^6$;   term that does not contain x

46 $(xy-3y^{-3})^8$;   term that does not contain y

47 Approximate $(1.2)^{10}$ by using the first three terms in the expansion of $(1+0.2)^{10}$, and compare your answer with that obtained using a calculator.

48 Approximate $(0.9)^4$ by using the first three terms in the expansion of $(1-0.1)^4$, and compare your answer with that obtained using a calculator.

Exer. 49–50: Simplify the expression using the binomial theorem.

49 $\dfrac{(x + h)^4 - x^4}{h}$

50 $\dfrac{(x + h)^5 - x^5}{h}$

51 Show that $\dbinom{n}{1} = \dbinom{n}{n - 1}$ for $n \geq 1$.

52 Show that $\dbinom{n}{0} = \dbinom{n}{n}$ for $n \geq 0$.

10.6

Permutations

Suppose that four teams are involved in a tournament in which first, second, third, and fourth places will be determined. For identification purposes, we label the teams A, B, C, and D. Let us find the number of different ways that first and second place can be decided. It is convenient to use a **tree diagram,** as in Figure 1. After the word START, the four possibilities for first place are listed. From each of these an arrow points to a possible second-place finisher. The final standings list the possible outcomes, from left to right. They are found by following the different paths (*branches* of the tree) that lead from the word START to the second-place team. The total number of outcomes is 12, which is the product of the number of choices (4) for first place and the number of choices (3) for second place (after first has been determined).

Let us now find the total number of ways that first, second, third, and fourth positions can be filled. To sketch a tree diagram, we may begin by drawing arrows from the word START to each possible first-place finisher A, B, C, or D. Next we draw arrows from those to possible second-place finishers, as was done in Figure 1. From each second-place position we then draw arrows indicating the possible third-place positions. Finally, we draw arrows to the fourth-place team. If we consider only the case in which team A finishes in first place, we have the diagram shown in Figure 2.

Figure 1

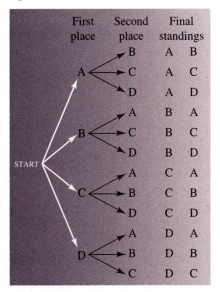

Figure 2

Note that there are six possible final standings in which team A occupies first place. In a complete tree diagram there would also be three other branches of this type corresponding to first-place finishes for B, C, and D. A complete diagram would display the following 24 possibilities for the final standings:

A first	ABCD, ABDC, ACBD, ACDB, ADBC, ADCB,
B first	BACD, BADC, BCAD, BCDA, BDAC, BDCA,
C first	CABD, CADB, CBAD, CBDA, CDAB, CDBA,
D first	DABC, DACB, DBAC, DBCA, DCAB, DCBA.

Note that the number of possibilities (24) is the product of the number of ways (4) that first place may occur, the number of ways (3) that second place may occur (after first place has been determined), the number of possible outcomes (2) for third place (after the first two places have been decided), and the number of ways (1) that fourth place can occur (after the first three places have been taken).

The preceding discussion illustrates the following general rule, which we accept as a basic axiom of counting.

Fundamental Counting Principle	Let $E_1, E_2, \ldots, E_k$ be a sequence of k events. If, for each i, the event E_i can occur in m_i ways, then the total number of ways all the events may take place is the product $m_1 m_2 \cdots m_k$.

Returning to our first illustration, we let E_1 represent the determination of the first-place team, so that $m_1 = 4$. If E_2 denotes the determination of the second-place team, then $m_2 = 3$. Hence, the number of outcomes for the sequence E_1, E_2 is $4 \cdot 3 = 12$, which is the same as that found by means of the tree diagram. If we proceed to E_3, the determination of the third-place team, then $m_3 = 2$, and hence $m_1 m_2 m_3 = 24$. Finally, if E_1, E_2, and E_3 have occurred, there is only one possible outcome for E_4. Thus, $m_4 = 1$, and $m_1 m_2 m_3 m_4 = 24$.

Instead of teams, let us now regard a, b, c, and d merely as symbols and consider the various *orderings,* or *arrangements,* that may be assigned to these symbols, taking them either two at a time, three at a time, or four at a time. By abstracting in this way we may apply our methods to other similar situations. The arrangements we have discussed are **arrangements without repetitions,** since a symbol may not be used more than once in an arrangement. In Example 1 we shall consider arrangements in which repetitions *are* allowed.

Previously we defined ordered pairs and ordered triples. Similarly, an *ordered 4-tuple* is a set containing four elements x_1, x_2, x_3, x_4 in which an ordering has been specified, so that one of the elements may be referred to as the *first element,* another as the *second element,* and so on. The symbol (x_1, x_2, x_3, x_4) is used for the ordered 4-tuple having first element x_1, second element x_2, third element x_3, and fourth element x_4. In general, for any positive integer r, we speak of the **ordered r-tuple**

$$(x_1, x_2, \ldots, x_r)$$

as a set of r elements in which x_1 is designated as the first element, x_2 as the second element, and so on.

EXAMPLE 1   Determining the number of r-tuples

Using only the letters $a, b, c,$ and $d,$ determine how many of the following can be obtained:

(a) ordered triples   **(b)** ordered 4-tuples   **(c)** ordered r-tuples

SOLUTION

(a) We must determine the number of symbols of the form (x_1, x_2, x_3) that can be obtained using only the letters $a, b, c,$ and $d.$ This is not the same as listing first, second, and third place as in our previous illustration, since we have not ruled out the possibility of repetitions. For example, (a, b, a), (a, a, b), and (a, a, a) are different ordered triples. If, for $i = 1, 2, 3,$ we let E_i represent the determination of x_i in the ordered triple (x_1, x_2, x_3), then, since repetitions are allowed, there are four possibilities—$a, b, c,$ and d—for each of $E_1, E_2,$ and E_3. Hence, by the fundamental counting principle, the total number of ordered triples is $4 \cdot 4 \cdot 4$, or 64.

(b) The number of possible ordered 4-tuples of the form (x_1, x_2, x_3, x_4) is $4 \cdot 4 \cdot 4 \cdot 4$, or 256.

(c) The number of ordered r-tuples is the product $4 \cdot 4 \cdot 4 \cdots\cdots 4$, with 4 appearing as a factor r times. That product equals 4^r.

 ### EXAMPLE 2   Choosing class officers

A class consists of 60 girls and 40 boys. In how many ways can a president, vice-president, treasurer, and secretary be chosen if the treasurer must be a girl, the secretary must be a boy, and a student may not hold more than one office?

SOLUTION   If an event is specialized in some way (for example, the treasurer *must* be a girl), then that event should be considered before any nonspecialized events. Thus, we let E_1 represent the choice of treasurer and E_2 the choice of secretary. Next we let E_3 and E_4 denote the choices for president and vice-president, respectively. As in the fundamental counting principle, we let m_i denote the number of different ways E_i can occur for $i = 1, 2, 3,$ and 4. It follows that $m_1 = 60$, $m_2 = 40$, $m_3 = 60 + 40 - 2 = 98$, and $m_4 = 97$. By the fundamental counting principle, the total number of possibilities is

$$m_1 m_2 m_3 m_4 = 60 \cdot 40 \cdot 98 \cdot 97 = 22{,}814{,}400.$$

When working with sets, we are usually not concerned about the order or arrangement of the elements. In the remainder of this section, however, the arrangement of the elements will be our main concern.

Definition of Permutation	Let S be a set of n elements and let $1 \leq r \leq n$. A **permutation** of r elements of S is an arrangement, without repetitions, of r elements.

We also use the phrase **permutation of n elements taken r at a time.** The symbol $P(n, r)$ will denote the number of different permutations of r elements that can be obtained from a set of n elements. As a special case, $P(n, n)$ denotes the number of arrangements of n elements of S—that is, the number of ways of arranging *all* the elements of S.

In our first discussion involving the four teams A, B, C, and D, we had $P(4, 2) = 12$, since there are 12 different ways of arranging the four teams in groups of two. We also showed that the number of ways to arrange all the elements A, B, C, and D is 24. In permutation notation we would write this result as $P(4, 4) = 24$.

The next theorem gives us a general formula for $P(n, r)$.

Theorem on the Number of Different Permutations	Let S be a set of n elements and let $1 \leq r \leq n$. The number of different permutations of r elements of S is $$P(n, r) = n(n - 1)(n - 2) \cdots (n - r + 1).$$

PROOF The problem of determining $P(n, r)$ is equivalent to determining the number of different r-tuples $(x_1, x_2, \ldots, x_r)$ such that each x_i is an element of S and no element of S appears twice in the same r-tuple. We may find this number by means of the fundamental counting principle. For each $i = 1, 2, \ldots, r$, let E_i represent the determination of the element x_i and let m_i be the number of different ways of choosing x_i. We wish to apply the sequence $E_1, E_2, \ldots, E_r$. We have n possible choices for x_1, and consequently $m_1 = n$. Since repetitions are not allowed, we have $n - 1$ choices for x_2, so $m_2 = n - 1$. Continuing in this manner, we successively obtain $m_3 = n - 2$, $m_4 = n - 3$, and ultimately $m_r = n - (r - 1)$ or, equivalently, $m_r = n - r + 1$. Hence, using the fundamental counting principle, we obtain the formula for $P(n, r)$. ◢

Note that *the formula for $P(n, r)$ in the previous theorem contains exactly r factors on the right-hand side,* as shown in the following illustration.

ILLUSTRATION **Number of Different Permutations**

- $P(n, 1) = n$ ■ $P(n, 3) = n(n - 1)(n - 2)$

- $P(n, 2) = n(n - 1)$ ■ $P(n, 4) = n(n - 1)(n - 2)(n - 3)$

EXAMPLE 3 Evaluating $P(n, r)$

Find $P(5, 2)$, $P(6, 4)$, and $P(5, 5)$.

SOLUTION We will use the formula for $P(n, r)$ in the preceding theorem. In each case, we first calculate the value of $(n - r + 1)$.

$$5 - 2 + 1 = \underline{4}, \quad \text{so} \quad P(5, 2) = 5 \cdot \underline{4} = 20$$
$$6 - 4 + 1 = \underline{3}, \quad \text{so} \quad P(6, 4) = 6 \cdot 5 \cdot 4 \cdot \underline{3} = 360$$
$$5 - 5 + 1 = \underline{1}, \quad \text{so} \quad P(5, 5) = 5 \cdot 4 \cdot 3 \cdot 2 \cdot \underline{1} = 120$$

EXAMPLE 4 Arranging the batting order for a baseball team

A baseball team consists of nine players. Find the number of ways of arranging the first four positions in the batting order if the pitcher is excluded.

SOLUTION We wish to find the number of permutations of 8 objects taken 4 at a time. Using the formula for $P(n, r)$ with $n = 8$ and $r = 4$, we have $n - r + 1 = 5$, and it follows that

$$P(8, 4) = 8 \cdot 7 \cdot 6 \cdot 5 = 1680.$$

The next result gives us a form for $P(n, r)$ that involves the factorial symbol.

Factorial Form for $P(n, r)$	If n is a positive integer and $1 \le r \le n$, then $$P(n, r) = \frac{n!}{(n - r)!}.$$

PROOF If we let $r = n$ in the formula for $P(n, r)$ in the theorem on permutations, we obtain the number of different arrangements of *all* the elements of a set consisting of n elements. In this case,

$$n - r + 1 = n - n + 1 = 1$$

and

$$P(n, n) = n(n - 1)(n - 2) \cdots 3 \cdot 2 \cdot 1 = n!.$$

Consequently, $P(n, n)$ is the product of the first n positive integers. This result is also given by the factorial form, for if $r = n$, then

$$P(n, n) = \frac{n!}{(n - n)!} = \frac{n!}{0!} = \frac{n!}{1} = n!.$$

If $1 \le r < n$, then

$$\frac{n!}{(n - r)!} = \frac{n(n - 1)(n - 2) \cdots (n - r + 1) \cdot [(n - r)!]}{(n - r)!}$$
$$= n(n - 1)(n - 2) \cdots (n - r + 1).$$

This agrees with the formula for $P(n, r)$ in the theorem on permutations.

EXAMPLE 5   **Evaluating $P(n, r)$ using factorials**

Use the factorial form for $P(n, r)$ to find $P(5, 2)$, $P(6, 4)$, and $P(5, 5)$.

SOLUTION

$$P(5, 2) = \frac{5!}{(5 - 2)!} = \frac{5!}{3!} = \frac{5 \cdot 4 \cdot 3!}{3!} = 5 \cdot 4 = 20$$

$$P(6, 4) = \frac{6!}{(6 - 4)!} = \frac{6!}{2!} = \frac{6 \cdot 5 \cdot 4 \cdot 3 \cdot 2!}{2!} = 6 \cdot 5 \cdot 4 \cdot 3 = 360$$

$$P(5, 5) = \frac{5!}{(5 - 5)!} = \frac{5!}{0!} = \frac{5!}{1} = 5 \cdot 4 \cdot 3 \cdot 2 \cdot 1 = 120$$

10.6   *Exercises*

Exer. 1–8: Find the number.

1 $P(7, 3)$

2 $P(8, 5)$

3 $P(9, 6)$

4 $P(5, 3)$

5 $P(5, 5)$

6 $P(4, 4)$

7 $P(6, 1)$

8 $P(5, 1)$

Exer. 9–12: Simplify the permutation.

9 $P(n, 0)$

10 $P(n, 1)$

11 $P(n, n - 1)$

12 $P(n, 2)$

13 How many three-digit numbers can be formed from the digits 1, 2, 3, 4, and 5 if repetitions

(a) are not allowed?   (b) are allowed?

14 Work Exercise 13 for four-digit numbers.

15 How many numbers can be formed from the digits 1, 2, 3, and 4 if repetitions are not allowed? (*Note:* 42 and 231 are examples of such numbers.)

16 Determine the number of positive integers less than 10,000 that can be formed from the digits 1, 2, 3, and 4 if repetitions are allowed.

17 Basketball standings If eight basketball teams are in a tournament, find the number of different ways that first, second, and third place can be decided, assuming ties are not allowed.

18 Basketball standings Work Exercise 17 for 12 teams.

19 Wardrobe mix 'n' match A girl has four skirts and six blouses. How many different skirt-blouse combinations can she wear?

20 Wardrobe mix 'n' match Refer to Exercise 19. If the girl also has three sweaters, how many different skirt-blouse-sweater combinations can she wear?

21 License plate numbers In a certain state, automobile license plates start with one letter of the alphabet, followed by five digits (0, 1, 2, ..., 9). Find how many different license plates are possible if

(a) the first digit following the letter cannot be 0

(b) the first letter cannot be O or I and the first digit cannot be 0

22 Tossing dice Two dice are tossed, one after the other. In how many different ways can they fall? List the number of different ways the sum of the dots can equal

(a) 3   (b) 5   (c) 7   (d) 9   (e) 11

23 Seating arrangement A row of six seats in a classroom is to be filled by selecting individuals from a group of ten students.

(a) In how many different ways can the seats be occupied?

(b) If there are six boys and four girls in the group and if boys and girls are to be alternated, find the number of different seating arrangements.

24 Scheduling courses A student in a certain college may take mathematics at 8, 10, 11, or 2 o'clock; English at 9, 10, 1, or 2; and history at 8, 11, 2, or 3. Find the number of different ways in which the student can schedule the three courses.

25 True-or-false test In how many different ways can a test consisting of ten true-or-false questions be completed?

26 Multiple-choice test A test consists of six multiple-choice questions, and there are five choices for each question. In how many different ways can the test be completed?

27 Seating arrangement In how many different ways can eight people be seated in a row?

28 Book arrangement In how many different ways can ten books be arranged on a shelf?

29 Semaphore With six different flags, how many different signals can be sent by placing three flags, one above the other, on a flag pole?

30 Selecting books In how many different ways can five books be selected from a twelve-volume set of books?

31 Radio call letters How many four-letter radio station call letters can be formed if the first letter must be K or W and repetitions

(a) are not allowed?   (b) are allowed?

32 Fraternity designations There are 24 letters in the Greek alphabet. How many fraternities may be specified by choosing three Greek letters if repetitions

(a) are not allowed?   (b) are allowed?

33 Phone numbers How many seven-digit phone numbers can be formed from the digits $0, 1, 2, 3, \ldots, 9$ if the first digit may not be 0?

34 Baseball batting order After selecting nine players for a baseball game, the manager of the team arranges the batting order so that the pitcher bats last and the best hitter bats third. In how many different ways can the remainder of the batting order be arranged?

35 ATM access code A customer remembers that 2, 4, 7, and 9 are the digits of a four-digit access code for an automatic bank-teller machine. Unfortunately, the customer has forgotten the order of the digits. Find the largest possible number of trials necessary to obtain the correct code.

36 ATM access code Work Exercise 35 if the digits are 2, 4, and 7 and one of these digits is repeated in the four-digit code.

37 Selecting theater seats Three married couples have purchased tickets for a play. Spouses are to be seated next to each other, and the six seats are in a row. In how many ways can the six people be seated?

38 Horserace results Ten horses are entered in a race. If the possibility of a tie for any place is ignored, in how many ways can the first-, second-, and third-place winners be determined?

39 Lunch possibilities Owners of a restaurant advertise that they offer 1,114,095 different lunches based on the fact that they have 16 "free fixins" to go along with any of their 17 menu items (sandwiches, hot dogs, and salads). How did they arrive at that number?

40 Shuffling cards

(a) In how many ways can a standard deck of 52 cards be shuffled?

(b) In how many ways can the cards be shuffled so that the four aces appear on the top of the deck?

41 Numerical palindromes A palindrome is an integer, such as 45654, that reads the same backward and forward.

(a) How many five-digit palindromes are there?

(b) How many n-digit palindromes are there?

42 Color arrangements Each of the six squares shown in the figure is to be filled with any one of ten possible colors. How many ways are there of coloring the strip shown in the figure so that no two adjacent squares have the same color?

Exercise 42

43 The graph of

$$y = \frac{x! \, e^x}{x^x \sqrt{2\pi x}}$$

has a horizontal asymptote of $y = 1$. Use this fact to find an approximation for $n!$ if n is a large positive integer.

44 (a) What happens if a calculator is used to find $P(150, 50)$? Explain.

(b) Approximate r if $P(150, 50) = 10^r$ by using the following formula from advanced mathematics:

$$\log n! \approx \frac{n \ln n - n}{\ln 10}$$

10.7

Distinguishable Permutations and Combinations

Certain problems involve finding different arrangements of objects, some of which are indistinguishable. For example, suppose we are given five disks of the same size, of which three are black, one is white, and one is red. Let us find the number of ways they can be arranged in a row so that different color arrangements are obtained. If the disks were all different colors, then the number of arrangements would be 5!, or 120. However, since some of the disks have the same appearance, we cannot obtain 120 different arrangements. To clarify this point, let us write

$$\text{B} \quad \text{B} \quad \text{B} \quad \text{W} \quad \text{R}$$

for the arrangement having black disks in the first three positions in the row, the white disk in the fourth position, and the red disk in the fifth position. The first three disks can be arranged in 3!, or 6, different ways, but these arrangements cannot be distinguished from one another because the first three disks look alike. We say that those 3! permutations are **nondistinguishable.** Similarly, given any other arrangement, say

$$\text{B} \quad \text{R} \quad \text{B} \quad \text{W} \quad \text{B,}$$

there are 3! different ways of arranging the three black disks, but again each such arrangement is nondistinguishable from the others. Let us call two arrangements of objects **distinguishable permutations** if one arrangement cannot be obtained from the other by rearranging like objects. Thus, B B B W R and B R B W B are distinguishable permutations of the five disks. Let k denote the number of distinguishable permutations. Since to each such arrangement there correspond 3! *nondistinguishable* permutations, we must have $3!\,k = 5!$, the number of permutations of five *different* objects. Hence, $k = 5!/3! = 5 \cdot 4 = 20$. By the same type of reasoning we can obtain the following extension of this discussion.

First Theorem on Distinguishable Permutations	If r objects in a collection of n objects are alike and if the remaining objects are different from each other and from the r objects, then the number of distinguishable permutations of the n objects is $$\frac{n!}{r!}.$$

We can generalize this theorem to the case in which there are several sub-collections of nondistinguishable objects. For example, consider eight disks, of which four are black, three are white, and one is red. In this case, with each arrangement, such as

$$\text{B} \quad \text{W} \quad \text{B} \quad \text{W} \quad \text{B} \quad \text{W} \quad \text{B} \quad \text{R,}$$

there are 4! arrangements of the black disks and 3! arrangements of the white disks that have no effect on the color arrangement. Hence, 4! 3! possible arrangements of the disks will not produce distinguishable permutations. If we let k denote the number of *distinguishable* permutations, then $4!\,3!\,k = 8!$, since 8! is the number of permutations we would obtain if the disks were all different. Thus, the number of distinguishable permutations is

$$k = \frac{8!}{4!\,3!} = \frac{8 \cdot 7 \cdot 6 \cdot 5}{3!} \cdot \frac{4!}{4!} = 280.$$

The following general result can be proved.

Second Theorem on Distinguishable Permutations	If, in a collection of n objects, n_1 are alike of one kind, n_2 are alike of another kind, $\ldots$, n_k are alike of a further kind, and $$n = n_1 + n_2 + \cdots + n_k,$$ then the number of distinguishable permutations of the n objects is $$\frac{n!}{n_1!\,n_2! \cdots n_k!}.$$

 EXAMPLE 1 Finding a number of distinguishable permutations

Find the number of distinguishable permutations of the letters in the word *Mississippi*.

SOLUTION In this example we are given a collection of eleven objects in which four are of one kind (the letter s), four are of another kind (i), two are of a third kind (p), and one is of a fourth kind (M). Hence, by the preceding theorem, we have $11 = 4 + 4 + 2 + 1$ and the number of distinguishable permutations is

$$\frac{11!}{4!\,4!\,2!\,1!} = 34{,}650.$$

When we work with permutations, our concern is with the orderings or arrangements of elements. Let us now ignore the order or arrangement of elements and consider the following question: Given a set containing n distinct elements, in how many ways can a subset of r elements be chosen with $r \le n$? Before answering, let us state a definition.

Definition of Combination	Let S be a set of n elements and let $1 \le r \le n$. A **combination** of r elements of S is a subset of S that contains r distinct elements.

If S contains n elements, we also use the phrase **combination of n elements taken r at a time.** The symbol $C(n, r)$ will denote the number of combinations of r elements that can be obtained from a set of n elements.

Theorem on the Number of Combinations	The number of combinations of r elements that can be obtained from a set of n elements is $$C(n, r) = \frac{n!}{(n - r)!\, r!}, \quad 1 \le r \le n.$$

The formula for $C(n, r)$ is identical to the formula for the binomial coefficient $\binom{n}{r}$ in Section 10.5.

PROOF If S contains n elements, then, to find $C(n, r)$, we must find the total number of subsets of the form

$$\{x_1, x_2, \ldots, x_r\}$$

such that the x_i are *different* elements of S. Since the r elements $x_1, x_2, \ldots, x_r$ can be arranged in $r!$ different ways, each such subset produces $r!$ different r-tuples. Thus, the total number of different r-tuples is $r!\, C(n, r)$. However, in the previous section we found that the total number of r-tuples is

$$P(n, r) = \frac{n!}{(n - r)!}.$$

Hence, $$r!\, C(n, r) = \frac{n!}{(n - r)!}.$$

Dividing both sides of the last equation by $r!$ gives us the formula for $C(n, r)$.

From the proof, note that

$$P(n, r) = r!\, C(n, r),$$

which means that there are *more permutations than combinations* when we choose a subset of r elements from a set of n elements. To remember this relationship, consider a presidency, say Bush-Quayle. There is only one group or combination of these two people, but when a president–vice-president ordering is associated with these two people, there are two permutations, and Bush-Quayle is clearly different from Quayle-Bush.

As you read the examples and work the exercises, keep the following in mind.

If the order of selection *is* important, use a *permutation*.

If the order of selection *is not* important, use a *combination*.

EXAMPLE 2 **Choosing a baseball squad**

A little league baseball squad has six outfielders, seven infielders, five pitchers, and two catchers. Each outfielder can play any of the three outfield positions, and each infielder can play any of the four infield positions. In how many ways can a team of nine players be chosen?

Remember—if the order of selection can be ignored, use a combination.

SOLUTION   The number of ways of choosing three outfielders from the six candidates is

$$C(6, 3) = \frac{6!}{(6-3)!\,3!} = \frac{6!}{3!\,3!} = \frac{6 \cdot 5 \cdot 4 \cdot 3!}{3 \cdot 2 \cdot 1 \cdot 3!} = \frac{6 \cdot 5 \cdot 4}{3 \cdot 2 \cdot 1} = 20.$$

The number of ways of choosing the four infielders is

$$C(7, 4) = \frac{7!}{(7-4)!\,4!} = \frac{7!}{3!\,4!} = \frac{7 \cdot 6 \cdot 5 \cdot 4!}{3 \cdot 2 \cdot 1 \cdot 4!} = \frac{7 \cdot 6 \cdot 5}{3 \cdot 2 \cdot 1} = 35.$$

There are five ways of choosing a pitcher and two choices for the catcher. It follows from the fundamental counting principle that the total number of ways to choose a team is

$$20 \cdot 35 \cdot 5 \cdot 2 = 7000.$$

EXAMPLE 3   Being dealt a full house

In one type of poker, a five-card hand is dealt from a standard 52-card deck.

(a) How many hands are possible?

(b) A *full house* is a hand that consists of three cards of one denomination and two cards of another denomination. (The 13 denominations are 2's, 3's, 4's, 5's, 6's, 7's, 8's, 9's, 10's, J's, Q's, K's, and A's.) How many hands are full houses?

Figure 1

The order of selection is not important, so use combinations. →

The order of selection is important, so → *use a permutation.*

SOLUTION

(a) The order in which the five cards are dealt is not important, so we use a combination:

$$C(52, 5) = \frac{52!}{(52-5)!\,5!} = \frac{52 \cdot 51 \cdot 50 \cdot 49 \cdot 48 \cdot 47!}{47! \cdot 5 \cdot 4 \cdot 3 \cdot 2 \cdot 1} = 2{,}598{,}960$$

(b) We first determine how many ways we can be dealt a specific full house—say 3 aces and 2 kings (see Figure 1). There are four cards of each denomination and the order of selection can be ignored, so we use combinations:

$$\text{number of ways to get 3 A's} = C(4, 3)$$
$$\text{number of ways to get 2 K's} = C(4, 2)$$

Now we must pick the two denominations. Since 3 A's and 2 K's is a different full house than 3 K's and 2 A's, the order of selecting the denominations is important, and so we use a permutation:

$$\text{number of ways to select two denominations} = P(13, 2)$$

By the fundamental counting principle, the number of full houses is

$$C(4, 3) \cdot C(4, 2) \cdot P(13, 2) = 4 \cdot 6 \cdot 156 = 3744.$$

Note that if $r = n$, the formula for $C(n, r)$ becomes

$$C(n, n) = \frac{n!}{(n-n)!\,n!} = \frac{n!}{0!\,n!} = \frac{n!}{1 \cdot n!} = 1.$$

It is convenient to assign a meaning to $C(n, r)$ if $r = 0$. If the formula is to be true in this case, then we must have

$$C(n, 0) = \frac{n!}{(n - 0)!\,0!} = \frac{n!}{n!\,0!} = \frac{n!}{n! \cdot 1} = 1.$$

Hence, we *define* $C(n, 0) = 1$, which is the same as $C(n, n)$. Finally, for consistency, we also *define* $C(0, 0) = 1$. Thus, $C(n, r)$ has meaning for all nonnegative integers n and r with $r \leq n$.

EXAMPLE 4 **Finding the number of subsets of a set**

Let S be a set of n elements. Find the number of distinct subsets of S.

SOLUTION Let r be any nonnegative integer such that $r \leq n$. From our previous work, the number of subsets of S that consist of r elements is $C(n, r)$, or $\binom{n}{r}$. Hence, to find the total number of subsets, it suffices to find the sum

$$\binom{n}{0} + \binom{n}{1} + \binom{n}{2} + \binom{n}{3} + \cdots + \binom{n}{n}. \tag{$*$}$$

Recalling the formula for the binomial theorem,

$$(a + b)^n = \sum_{k=0}^{n} \binom{n}{k} a^{n-k} b^k,$$

we can see that the indicated sum $(*)$ is precisely the binomial expansion of $(1 + 1)^n$. Thus, there are 2^n subsets of a set of n elements. In particular, a set of 3 elements has 2^3, or 8, different subsets. A set of 4 elements has 2^4, or 16, subsets. A set of 10 elements has 2^{10}, or 1024, subsets. ◢

Pascal's triangle, introduced in Section 10.5, can easily be remembered by the following combination form:

$$\binom{0}{0}$$

$$\binom{1}{0} \quad \binom{1}{1}$$

$$\binom{2}{0} \quad \binom{2}{1} \quad \binom{2}{2}$$

$$\binom{3}{0} \quad \binom{3}{1} \quad \binom{3}{2} \quad \binom{3}{3}$$

$$\binom{4}{0} \quad \binom{4}{1} \quad \binom{4}{2} \quad \binom{4}{3} \quad \binom{4}{4}$$

Combining this information with that in Example 4, we conclude that the third coefficient in the expansion of $(a + b)^4$, $\binom{4}{2}$, is exactly the same as the number of two-element subsets of a set that contains four elements. We leave it as an exercise to find a generalization of the last statement (see Discussion Exercise 4 at the end of the chapter).

10.7 *Exercises*

Exer. 1–8: Find the number.

1 $C(7, 3)$

2 $C(8, 4)$

3 $C(9, 8)$

4 $C(6, 2)$

5 $C(n, n - 1)$

6 $C(n, 1)$

7 $C(7, 0)$

8 $C(5, 5)$

Exer. 9–10: Find the number of possible color arrangements for the 12 given disks, arranged in a row.

9 5 black, 3 red, 2 white, 2 green

10 3 black, 3 red, 3 white, 3 green

11 Find the number of distinguishable permutations of the letters in the word *bookkeeper*.

12 Find the number of distinguishable permutations of the letters in the word *moon*. List all the permutations.

13 Choosing basketball teams Ten people wish to play in a basketball game. In how many different ways can two teams of five players each be formed?

14 Selecting test questions A student may answer any six of ten questions on an examination.

(a) In how many ways can six questions be selected?

(b) How many selections are possible if the first two questions must be answered?

Exer. 15–16: Consider any eight points such that no three are collinear.

15 How many lines are determined?

16 How many triangles are determined?

17 Book arrangement A student has five mathematics books, four history books, and eight fiction books. In how many different ways can they be arranged on a shelf if books in the same category are kept next to one another?

18 Selecting a basketball team A basketball squad consists of twelve players.

(a) Disregarding positions, in how many ways can a team of five be selected?

(b) If the center of a team must be selected from two specific individuals on the squad and the other four members of the team from the remaining ten players, find the number of different teams possible.

19 Selecting a football team A football squad consists of three centers, ten linemen who can play either guard or tackle, three quarterbacks, six halfbacks, four ends, and four fullbacks. A team must have one center, two guards, two tackles, two ends, two halfbacks, a quarterback, and a fullback. In how many different ways can a team be selected from the squad?

20 Arranging keys on a ring In how many different ways can seven keys be arranged on a key ring if the keys can slide completely around the ring?

21 Committee selection A committee of 3 men and 2 women is to be chosen from a group of 12 men and 8 women. Determine the number of different ways of selecting the committee.

22 Birth order Let the letters G and B denote a girl birth and a boy birth, respectively. For a family of three boys and three girls, one possible birth order is G G G B B B. How many birth orders are possible for these six children?

Exer. 23–24: Shown in each figure is a street map and a possible path from point A to point B. How many possible paths are there from A to B if moves are restricted to the right or up? (*Hint:* If R denotes a move one unit right and U denotes a move one unit up, then the path in Exercise 23 can be specified by **R U U R R R U R**.)

23

24

25 Lotto selections To win a state lottery game, a player must correctly select six numbers from the numbers 1 through 49.

(a) Find the total number of selections possible.

(b) Work part (a) if a player selects only even numbers.

26 Office assignments A mathematics department has ten faculty members but only nine offices, so one office must be shared by two individuals. In how many different ways can the offices be assigned?

27 Tennis tournament In a round-robin tennis tournament, every player meets every other player exactly once. How many players can participate in a tournament of 45 matches?

28 True-or-false test A true-or-false test has 20 questions.

(a) In how many different ways can the test be completed?

(b) In how many different ways can a student answer 10 questions correctly?

29 Basketball championship series The winner of the seven-game NBA championship series is the team that wins four games. In how many different ways can the series be extended to seven games?

30 A geometric design is determined by joining every pair of vertices of an octagon (see the figure).

(a) How many triangles in the design have their three vertices on the octagon?

(b) How many quadrilaterals in the design have their four vertices on the octagon?

Exercise 30

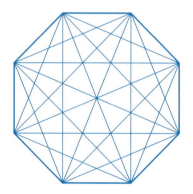

31 Ice cream selections An ice cream parlor stocks 31 different flavors and advertises that it serves almost 4500 different triple scoop cones, with each scoop being a different flavor. How was this number obtained?

32 Choices of hamburger condiments A fast food restaurant advertises that it offers any combination of 8 condiments on a hamburger, thus giving a customer 256 choices. How was this number obtained?

33 Scholarship selection A committee is going to select 30 students from a pool of 1000 to receive scholarships. How may ways could the students be selected if each scholarship is worth

(a) the same amount?

(b) a different amount?

34 Track rankings Twelve sprinters are running a heat; those with the best four times will advance to the finals.

(a) In how many ways can this group of four be selected?

(b) If the four best times will be seeded (ranked) in the finals, in how many ways can this group of four be selected and seeded?

35 Poker hands Refer to Example 3. How many hands will have exactly three kings?

36 Bridge hands How many 13-card hands dealt from a standard deck will have exactly seven spades?

Exer. 37–38: (a) Calculate the sum S_n for $n = 1, 2, 3, \ldots, 10$, where if $n < r$, then $\binom{n}{r} = 0$. (b) Predict a general formula for S_n.

37 $\binom{n}{1} + \binom{n}{3} + \binom{n}{5} + \binom{n}{7} + \cdots$

38 $(1)\binom{n}{1} - (2)\binom{n}{2} + (3)\binom{n}{3} - (4)\binom{n}{4} + (5)\binom{n}{5} - \cdots$

10.8
Probability

If two dice are tossed, what are the chances of rolling a 7? If a person is dealt five cards from a standard deck of 52 playing cards, what is the likelihood of obtaining three aces? In the seventeenth century, similar questions about games of chance led to the study of *probability*. Since that time, the theory of probability has grown extensively. It is now used to predict outcomes of a large variety of situations that arise in the natural and social sciences.

Any chance process, such as flipping a coin, rolling a die, being dealt a card from a deck, determining if a manufactured item is defective, or finding the blood pressure of an individual, is an **experiment.** A result of an experiment is an **outcome.** We will restrict our discussion to experiments for which outcomes are **equally likely** unless stated otherwise. This means, for example, that if a coin is flipped, we assume that the possibility of obtaining a head is the same as that of obtaining a tail. Similarly, if a die is tossed, we assume that the die is *fair*—that is, there is an equal chance of obtaining either a 1, 2, 3, 4, 5, or 6. The set S of all possible outcomes of an experiment is the **sample space** of the experiment. Thus, if the experiment consists of flipping a coin and we let H or T denote the outcome of obtaining a head or tail, respectively, then the sample space S may be denoted by

$$S = \{H, T\}.$$

If a fair die is tossed as an experiment, then the set S of all possible outcomes (the sample space) is

$$S = \{1, 2, 3, 4, 5, 6\}.$$

The following definition expresses, in mathematical terms, the notion of obtaining *particular* outcomes of an experiment.

Definition of Event	Let S be the sample space of an experiment. An **event** associated with the experiment is any subset E of S.

Let us consider the experiment of tossing a single die, so that the sample space is $S = \{1, 2, 3, 4, 5, 6\}$. If $E = \{4\}$, then the event E associated with the experiment consists of the outcome of obtaining a 4 on the toss. Different events may be associated with the same experiment. For example, if we let $E = \{1, 3, 5\}$, then this event consists of obtaining an odd number on a toss of the die.

As another illustration, suppose the experiment consists of flipping two coins, one after the other. If we let HH denote the outcome in which two heads appear, HT that of a head on the first coin and a tail on the second, and so on, then the sample space S of the experiment may be denoted by

$$S = \{HH, HT, TH, TT\}.$$

If we let $\qquad\qquad E = \{HT, TH\},$

then the event E consists of the appearance of a head on one of the coins and a tail on the other.

Next we shall define what is meant by the *probability* of an event. Throughout our discussion we will assume that the sample space S of an experiment contains only a finite number of elements. If E is an event, *the symbols $n(E)$ and $n(S)$ will denote the number of elements in E and S, respectively.* Keep in mind that E and S consist of outcomes that are equally likely.

Definition of the Probability of an Event	Let S be the sample space of an experiment and E an event. The **probability** $P(E)$ of E is given by $$P(E) = \frac{n(E)}{n(S)}.$$

Since E is a subset of S, we see that

$$0 \le n(E) \le n(S).$$

Dividing by $n(S)$, we obtain

$$\frac{0}{n(S)} \le \frac{n(E)}{n(S)} \le \frac{n(S)}{n(S)} \qquad \text{or, equivalently,} \qquad 0 \le P(E) \le 1.$$

Note that $P(E) = 0$ if E contains no elements, and $P(E) = 1$ if $E = S$.

The next example provides three illustrations of the preceding definition if E contains exactly one element.

EXAMPLE 1 Finding the probability of an event

(a) If a coin is flipped, find the probability that a head will turn up.

(b) If a fair die is tossed, find the probability of obtaining a 4.

(c) If two coins are flipped, find the probability that both coins turn up heads.

SOLUTION For each experiment we shall list sets S and E and then use the definition of probability of an event to find $P(E)$.

(a) $S = \{H, T\}, \qquad\qquad E = \{H\}, \qquad P(E) = \dfrac{n(E)}{n(S)} = \dfrac{1}{2}$

(b) $S = \{1, 2, 3, 4, 5, 6\},$ $\quad E = \{4\},$ $\quad P(E) = \dfrac{n(E)}{n(S)} = \dfrac{1}{6}$

(c) $S = \{HH, HT, TH, TT\},$ $\quad E = \{HH\},$ $\quad P(E) = \dfrac{n(E)}{n(S)} = \dfrac{1}{4}$

In part (a) of Example 1 we found that the probability of obtaining a head on a flip of a coin is $\frac{1}{2}$. We take this to mean that if a coin is flipped many times, the number of times that a head turns up should be approximately one-half the total number of flips. Thus, for 100 flips, a head should turn up approximately 50 times. It is unlikely that this number will be *exactly* 50. A probability of $\frac{1}{2}$ implies that if we let the number of flips increase, then the number of times a head turns up *approaches* one-half the total number of flips. Similar remarks can be made for parts (b) and (c) of Example 1.

In the next two examples we consider experiments in which an event contains more than one element.

EXAMPLE 2 Finding probabilities when two dice are tossed

If two dice are tossed, what is the probability of rolling a sum of

(a) 7? **(b)** 9?

SOLUTION Let us refer to one die as *the first die* and the other as *the second die*. We shall use ordered pairs to represent outcomes as follows: (2, 4) denotes the outcome of obtaining a 2 on the first die and a 4 on the second; (5, 3) represents a 5 on the first die and a 3 on the second; and so on. Since there are six different possibilities for the first number of the ordered pair and, with each of these, six possibilities for the second number, the total number of ordered pairs is $6 \times 6 = 36$. Hence, if S is the sample space, then $n(S) = 36$.

(a) The event E corresponding to rolling a sum of 7 is given by

$$E = \{(1, 6), (2, 5), (3, 4), (4, 3), (5, 2), (6, 1)\},$$

and consequently $\qquad P(E) = \dfrac{n(E)}{n(S)} = \dfrac{6}{36} = \dfrac{1}{6}.$

(b) If E is the event corresponding to rolling a sum of 9, then

$$E = \{(3, 6), (4, 5), (5, 4), (6, 3)\}$$

and $\qquad P(E) = \dfrac{n(E)}{n(S)} = \dfrac{4}{36} = \dfrac{1}{9}.$

In the next example (and in the exercises), when it is stated that one or more cards are drawn from a deck, we mean that each card is removed from a standard 52-card deck and is *not* replaced before the next card is drawn.

EXAMPLE 3 Finding the probability of drawing a certain hand of cards

Suppose five cards are drawn from a deck of cards. Find the probability that all five cards are hearts.

SOLUTION The sample space S of the experiment is the set of all possible five-card hands that can be formed from the 52 cards in the deck. It follows from our work in the preceding section that $n(S) = C(52, 5)$.

Since there are 13 cards in the heart suit, the number of different ways of obtaining a hand that contains five hearts is $C(13, 5)$. Hence, if E represents this event, then

$$P(E) = \frac{n(E)}{n(S)} = \frac{C(13, 5)}{C(52, 5)} = \frac{\dfrac{13!}{5!8!}}{\dfrac{52!}{5!47!}} = \frac{1287}{2{,}598{,}960} \approx 0.0005 = \frac{5}{10{,}000} = \frac{1}{2000}.$$

This result implies that if the experiment is performed many times, a five-card heart hand should be drawn approximately once every 2000 times. ◢

Suppose S is the sample space of an experiment and E_1 and E_2 are two events associated with the experiment. If E_1 and E_2 have no elements in common, they are called *disjoint sets* and we write $E_1 \cap E_2 = \varnothing$ (the *empty set*). In this case, if one event occurs, the other cannot occur; they are **mutually exclusive events**. Thus, if $E = E_1 \cup E_2$, then

$$n(E) = n(E_1 \cup E_2) = n(E_1) + n(E_2).$$

Hence,

$$P(E) = \frac{n(E_1) + n(E_2)}{n(S)} = \frac{n(E_1)}{n(S)} + \frac{n(E_2)}{n(S)},$$

or

$$P(E) = P(E_1) + P(E_2).$$

The probability of E is therefore the sum of the probabilities of E_1 and E_2. We have proved the following.

Theorem on Mutually Exclusive Events	If E_1 and E_2 are mutually exclusive events and $E = E_1 \cup E_2$, then $$P(E) = P(E_1 \cup E_2) = P(E_1) + P(E_2).$$

The preceding theorem can be extended to any number of events E_1, $E_2, \ldots, E_k$ that are mutually exclusive in the sense that if $i \neq j$, then $E_i \cap E_j = \varnothing$. The conclusion of the theorem is then

$$P(E) = P(E_1 \cup E_2 \cup \cdots \cup E_k) = P(E_1) + P(E_2) + \cdots + P(E_k).$$

EXAMPLE 4 **Finding probabilities when two dice are tossed**

If two dice are tossed, find the probability of rolling a sum of either 7 or 9.

SOLUTION Let E_1 denote the event of rolling 7 and E_2 that of rolling 9. Since E_1 and E_2 cannot occur simultaneously, they are mutually exclusive events. We wish to find the probability of the event $E = E_1 \cup E_2$. From Example 2 we know that $P(E_1) = \frac{6}{36}$ and $P(E_2) = \frac{4}{36}$. Hence, by the last theorem,

$$P(E) = P(E_1) + P(E_2)$$
$$= \tfrac{6}{36} + \tfrac{4}{36} = \tfrac{10}{36} = 0.2\overline{7}.$$

If E_1 and E_2 are events that possibly have elements in common, then the following can be proved.

Theorem on the Probability of the Occurrence of Either of Two Events	If E_1 and E_2 are any two events, then $$P(E_1 \cup E_2) = P(E_1) + P(E_2) - P(E_1 \cap E_2).$$

Note that if E_1 and E_2 are mutually exclusive, then $E_1 \cap E_2 = \varnothing$ and $P(E_1 \cap E_2) = 0$. Hence, the last theorem includes, as a special case, the theorem on mutually exclusive events.

 EXAMPLE 5 **Finding the probability of selecting a certain card from a deck**

If a single card is selected from a deck, find the probability that the card is either a jack or a spade.

SOLUTION   Let E_1 denote the event that the card is a jack and E_2 the event that it is a spade. The events E_1 and E_2 are *not* mutually exclusive, since there is one card—the jack of spades—in both events, and hence $P(E_1 \cap E_2) = \frac{1}{52}$. By the preceding theorem, the probability that the card is either a jack or a spade is

$$P(E_1 \cup E_2) = P(E_1) + P(E_2) - P(E_1 \cap E_2)$$
$$= \tfrac{4}{52} + \tfrac{13}{52} - \tfrac{1}{52} = \tfrac{16}{52} \approx 0.31.$$

In solving probability problems, it is often helpful to categorize the outcomes of a sample space S into an event E and the set E' of elements of S that are not in E. We call E' the **complement** of E. Note that

$$E \cup E' = S \qquad \text{and} \qquad n(E) + n(E') = n(S).$$

Dividing both sides of the last equation by $n(S)$ gives us

$$\frac{n(E)}{n(S)} + \frac{n(E')}{n(S)} = 1.$$

Hence,

$$P(E) + P(E') = 1, \qquad \text{or} \qquad P(E) = 1 - P(E').$$

We shall use the last formula in the next example.

EXAMPLE 6 **Finding the probability of drawing a certain hand of cards**

If 13 cards are drawn from a deck, what is the probability that at least 2 of the cards are hearts?

SOLUTION If $P(k)$ denotes the probability of getting k hearts, then the probability of getting *at least* two hearts is

$$P(2) + P(3) + P(4) + \cdots + P(13).$$

Since the only remaining probabilities are $P(0)$ and $P(1)$, the desired probability is equal to

$$1 - [P(0) + P(1)].$$

To calculate $P(k)$ for any k, we may regard the deck as being split into two groups: hearts and non-hearts. For $P(0)$ we note that of the 13 hearts in the deck, we get none; and of the 39 non-hearts, we get 13. Since the number of ways to choose 13 cards from a 52-card deck is $C(52, 13)$, we see that

$$P(0) = \frac{n(0)}{n(S)} = \frac{C(13, 0) \cdot C(39, 13)}{C(52, 13)} \approx 0.0128.$$

The probability $P(1)$ corresponds to getting 1 of the hearts and 12 of the 39 non-hearts. Thus,

$$P(1) = \frac{n(1)}{n(S)} = \frac{C(13, 1) \cdot C(39, 12)}{C(52, 13)} \approx 0.0801.$$

Hence, the desired probability is

$$1 - [P(0) + P(1)] \approx 1 - [0.0128 + 0.0801] = 0.9071. \qquad \blacksquare$$

The words *probability* and *odds* are often used interchangeably. While knowing one allows us to calculate the other, they are quite different.

Definition of the Odds of an Event	Let S be the sample space of an experiment, E an event, and E' its complement. The **odds** $O(E)$ in favor of the event E occurring are given by $$n(E) \quad \text{to} \quad n(E').$$

The odds $n(E)$ to $n(E')$ are sometimes denoted by $n(E):n(E')$.

We can think of the odds in favor of an event E as the number of ways E occurs compared to the number of ways E doesn't occur. Similarly, the odds *against* E occurring are given by $n(E')$ to $n(E)$.

EXAMPLE 7 **Finding odds when two dice are tossed**

If two dice are tossed and E is the event of rolling a sum of 7, what are the odds

(a) in favor of E? **(b)** against E?

SOLUTION   From Example 2, we have $n(E) = 6$ and $n(S) = 36$, so

$$n(E') = n(S) - n(E) = 36 - 6 = 30.$$

(a) The odds in favor of rolling a sum of 7 are $n(E)$ to $n(E')$ or

6   to   30   or, equivalently,   1   to   5.

(b) The odds against rolling a sum of 7 are $n(E')$ to $n(E)$ or

30   to   6   or, equivalently,   5   to   1.

EXAMPLE 8   **Finding probabilities and odds**

(a) If $P(E) = 0.75$, find $O(E)$.

(b) If $O(E)$ are 6 to 5, find $P(E)$.

SOLUTION

(a) Since $P(E) = 0.75 = \frac{3}{4}$ and $P(E) = n(E)/n(S)$, we can let

$$n(E) = 3 \quad \text{and} \quad n(S) = 4.$$

Thus, $n(E') = n(S) - n(E) = 4 - 3 = 1$, and $O(E)$ are given by

$n(E)$ to $n(E')$,   or   3   to   1.

(b) Since $O(E)$ are 6 to 5 and $O(E)$ are $n(E)$ to $n(E')$, we can let

$$n(E) = 6 \quad \text{and} \quad n(E') = 5.$$

Thus, $n(S) = n(E) + n(E') = 6 + 5 = 11$, and

$$P(E) = \frac{n(E)}{n(S)} = \frac{6}{11}.$$

Two events E_1 and E_2 are said to be **independent events** if the occurrence of one does not influence the occurrence of the other.

Theorem on Independent Events	If E_1 and E_2 are independent events, then $$P(E_1 \cap E_2) = P(E_1) \cdot P(E_2).$$

In words, the theorem states that if E_1 and E_2 are independent events, the probability that *both* E_1 and E_2 occur simultaneously is the product of their probabilities. Note that if two events E_1 and E_2 are mutually exclusive, then $P(E_1 \cap E_2) = 0$ and they cannot be independent. (We assume that both E_1 and E_2 are not empty.)

EXAMPLE 9   **An application of probability to an electrical system**

Figure 1

An electrical system has open-close switches s_1, s_2, and s_3, as shown in Figure 1. The switches operate independently of one another, and current will flow from A to B either if s_1 is closed or if *both* s_2 and s_3 are closed.

(a) If S_k denotes the event that s_k is closed, where $k = 1, 2, 3$, express, in terms of $P(S_1)$, $P(S_2)$, and $P(S_3)$, the probability p that current will flow from A to B.

(b) Find p if $P(S_k) = \frac{1}{2}$ for each k.

SOLUTION

(a) The probability p that either S_1 or *both* S_2 and S_3 occur is

$$p = P(S_1 \cup (S_2 \cap S_3)).$$

Using the theorem on the probability of the occurrence of either of two events S_1 or $S_2 \cap S_3$, we obtain

$$p = P(S_1) + P(S_2 \cap S_3) - P(S_1 \cap (S_2 \cap S_3)).$$

Applying the theorem on independent events twice gives us

$$p = P(S_1) + P(S_2) \cdot P(S_3) - P(S_1) \cdot P(S_2 \cap S_3).$$

Finally, using the theorem on independent events one more time, we see that

$$p = P(S_1) + P(S_2) \cdot P(S_3) - P(S_1) \cdot P(S_2) \cdot P(S_3).$$

(b) If $P(S_k) = \frac{1}{2}$ for each k, then from part (a) the probability that current will flow from A to B is

$$p = \tfrac{1}{2} + \tfrac{1}{2} \cdot \tfrac{1}{2} - \tfrac{1}{2} \cdot \tfrac{1}{2} \cdot \tfrac{1}{2} = \tfrac{5}{8} = 0.625.$$

Figure 2

First card

$P(F) = \dfrac{12}{52}$

$P(F') = \dfrac{40}{52}$

EXAMPLE 10   **Using a tree diagram to find a probability**

If two cards are drawn from a deck, what is the probability that at least one of the cards will be a face card?

SOLUTION   Let F denote the event of drawing a face card. There are 12 face cards in a 52-card deck, so $P(F) = \frac{12}{52}$. We can depict this probability, as well as the probability of its complement, with the *tree diagram* shown in Figure 2.

The probabilities for the second card depend on what the first card was. To cover all possibilities for the second card, we attach branches with similar probabilities to the end of each branch of the first tree diagram, as shown in Figure 3.

Figure 3

The Products column lists the probabilities for all two-card possibilities; for example, the probability that both cards will be face cards is $\frac{132}{2652}$. The vertical sums must equal 1—calculating these is a good way to check your computations. To answer the question, we can add the first three probabilities in the Products column or subtract the fourth probability from 1. Using the latter approach, we have

$$1 - \frac{1560}{2652} = \frac{1092}{2652} = \frac{7}{17} \approx 41\%.$$

It is often of interest to know what amount of return we can expect on an investment in a game of chance. The following definition will help us answer questions that fall in this category.

Definition of Expected Value	Suppose a variable can have payoff amounts $a_1, a_2, \ldots, a_n$ with corresponding probabilities $p_1, p_2, \ldots, p_n$. The **expected value** EV of the variable is given by
	$$\text{EV} = a_1 p_1 + a_2 p_2 + \cdots + a_n p_n = \sum_{k=1}^{n} a_k p_k.$$

EXAMPLE 11 Expected value of a single pull-tab

States that run lotteries often offer games in which a certain number of pull-tabs are printed, some being redeemable for money and the rest worthless. Suppose that in a particular game there are 4000 pull-tabs, 432 of which are redeemable according to the following table.

Number of pull-tabs	Value
4	$100
8	50
20	20
400	2

Find the expected value of a pull-tab that sells for $1.

SOLUTION   The payoff amounts $100, $50, $20, and $2 have probabilities $\frac{4}{4000}, \frac{8}{4000}, \frac{20}{4000}$, and $\frac{400}{4000}$, respectively. The remaining 3568 pull-tabs have a payoff amount of $0. By the preceding definition, the expected value of a single pull-tab is

$$\text{EV} = 100 \cdot \tfrac{4}{4000} + 50 \cdot \tfrac{8}{4000} + 20 \cdot \tfrac{20}{4000} + 2 \cdot \tfrac{400}{4000} + 0 \cdot \tfrac{3568}{4000}$$
$$= \tfrac{2000}{4000} = \$0.50.$$

Thus, after subtracting the $1 cost of the pull-tab, we can expect to *lose* $0.50 on each pull-tab we buy. Note that we cannot lose $0.50 on any individual pull-tab, but we can expect to lose this amount on each pull-tab in the long run. This game yields a terribly poor return for the buyer and a healthy profit for the seller.   ◢

   The expected value of $0.50 obtained in Example 11 may be considered to be the amount we would expect to pay to play the game if the game were *fair*—that is, if we would not expect to win or lose any money after playing the game many times.
   In this section we have merely introduced several basic concepts about probability. The interested person is referred to entire books and courses devoted to this branch of mathematics.

10.8   *Exercises*

Exer. 1–2: A single card is drawn from a deck. Find the probability and the odds that the card is as specified.

1 (a) greater than 9   (b) a king or a queen

   (c) a king, a queen, or a jack

2 (a) a heart

   (b) a heart or a diamond

   (c) a heart, a diamond, or a club

Exer. 3–4: A single die is tossed. Find the probability and the odds that the die is as specified.

3 (a) a 4   (b) a 6   (c) a 4 or a 6

4 (a) an even number   (b) a number divisible by 5

   (c) an even number or a number divisible by 5

Exer. 5–6: An urn contains five red balls, six green balls, and four white balls. If a single ball is drawn, find the probability and the odds that the ball is as specified.

5 (a) red   (b) green   (c) red or white

6 (a) white   (b) green or white   (c) not green

Exer. 7–8: Two dice are tossed. Find the probability and the odds that the sum is as specified.

7 (a) 11   (b) 8   (c) 11 or 8

8 (a) greater than 9   (b) an odd number

Exer. 9–10: Three dice are tossed. Find the probability of the specified event.

9 A sum of 5

10 A 6 turns up on exactly one die

11 If three coins are flipped, find the probability that exactly two heads turn up.

12 If four coins are flipped, find the probability of obtaining two heads and two tails.

13 If $P(E) = \frac{5}{7}$, find $O(E)$ and $O(E')$.

14 If $P(E) = 0.4$, find $O(E)$ and $O(E')$.

15 If $O(E)$ are 9 to 5, find $O(E')$ and $P(E)$.

16 If $O(E')$ are 7 to 3, find $O(E)$ and $P(E)$.

Exer. 17–18: For the given value of $P(E)$, approximate $O(E)$ in terms of "X to 1."

17 $P(E) \approx 0.659$ 18 $P(E) \approx 0.822$

Exer. 19–24: Suppose five cards are drawn from a deck. Find the probability of obtaining the indicated cards.

19 Four of a kind (such as four aces or four kings)

20 Three aces and two kings

21 Four diamonds and one spade

22 Five face cards

23 A flush (five cards of the same suit)

24 A royal flush (an ace, king, queen, jack, and 10 of the same suit)

25 If a single die is tossed, find the probability of obtaining an odd number or a prime number.

26 A single card is drawn from a deck. Find the probability that the card is either red or a face card.

27 If the probability of a baseball player's getting a hit in one time at bat is 0.326, find the probability that the player gets no hits in 4 times at bat.

28 If the probability of a basketball player's making a free throw is 0.9, find the probability that the player makes at least 1 of 2 free throws.

Exer. 29–30: The outcomes 1, 2, . . . , 6 of an experiment and their probabilities are listed in the table.

Outcome	1	2	3	4	5	6
Probability	0.25	0.10	0.15	0.20	0.25	0.05

For the indicated events, find (a) $P(E_2)$, (b) $P(E_1 \cap E_2)$, (c) $P(E_1 \cup E_2)$, and (d) $P(E_2 \cup E_3')$.

29 $E_1 = \{1, 2\};$ $E_2 = \{2, 3, 4\};$ $E_3 = \{4, 6\}$

30 $E_1 = \{1, 2, 3, 6\};$ $E_2 = \{3, 4\};$ $E_3 = \{4, 5, 6\}$

Exer. 31–32: A box contains 10 red chips, 20 blue chips, and 30 green chips. If 5 chips are drawn from the box, find the probability of drawing the indicated chips.

31 (a) all blue   (b) at least 1 green

   (c) at most 1 red

32 (a) exactly 4 green   (b) at least 2 red

   (c) at most 2 blue

33 True-or-false test A true-or-false test consists of eight questions. If a student guesses the answer for each question, find the probability that

   (a) eight answers are correct

   (b) seven answers are correct and one is incorrect

   (c) six answers are correct and two are incorrect

   (d) at least six answers are correct

34 Committee selection A 6-member committee is to be chosen by drawing names of individuals from a hat. If the hat contains the names of 8 men and 14 women, find the probability that the committee will consist of 3 men and 3 women.

Exer. 35–36: Five cards are drawn from a deck. Find the probability of the specified event.

35 Obtaining at least one ace

36 Obtaining at least one heart

37 Card and die experiment Each suit in a deck is made up of an ace (A), nine numbered cards (2, 3, . . . , 10), and three face cards (J, Q, K). An experiment consists of drawing a single card from a deck followed by rolling a single die.

(a) Describe the sample space S of the experiment, and find $n(S)$.

(b) Let E_1 be the event consisting of the outcomes in which a numbered card is drawn and the number of dots on the die is the same as the number on the card. Find $n(E_1)$, $n(E_1')$, and $P(E_1)$.

(c) Let E_2 be the event in which the card drawn is a face card, and let E_3 be the event in which the number of dots on the die is even. Are E_2 and E_3 mutually exclusive? Are they independent? Find $P(E_2)$, $P(E_3)$, $P(E_2 \cap E_3)$, and $P(E_2 \cup E_3)$.

(d) Are E_1 and E_2 mutually exclusive? Are they independent? Find $P(E_1 \cap E_2)$ and $P(E_1 \cup E_2)$.

38 Letter and number experiment An experiment consists of selecting a letter from the alphabet and one of the digits 0, 1, . . . , 9.

(a) Describe the sample space S of the experiment, and find $n(S)$.

(b) Suppose the letters of the alphabet are assigned numbers as follows: $A = 1, B = 2, . . . , Z = 26$. Let E_1 be the event in which the units digit of the number assigned to the letter of the alphabet is the same as the digit selected. Find $n(E_1)$, $n(E_1')$, and $P(E_1)$.

(c) Let E_2 be the event that the letter is one of the five vowels and E_3 the event that the digit is a prime number. Are E_2 and E_3 mutually exclusive? Are they independent? Find $P(E_2)$, $P(E_3)$, $P(E_2 \cap E_3)$, and $P(E_2 \cup E_3)$.

(d) Let E_4 be the event that the numerical value of the letter is even. Are E_2 and E_4 mutually exclusive? Are they independent? Find $P(E_2 \cap E_4)$ and $P(E_2 \cup E_4)$.

39 Tossing dice If two dice are tossed, find the probability that the sum is greater than 5.

40 Tossing dice If three dice are tossed, find the probability that the sum is less than 16.

41 Family makeup Assuming that girl-boy births are equally probable, find the probability that a family with five children has

(a) all boys   (b) at least one girl

42 Slot machine A standard slot machine contains three reels, and each reel contains 20 symbols. If the first reel has five bells, the middle reel four bells, and the last reel two bells, find the probability of obtaining three bells in a row.

43 ESP experiment In a simple experiment designed to test ESP, four cards (jack, queen, king, and ace) are shuffled and then placed face down on a table. The subject then attempts to identify each of the four cards, giving a different name to each of the cards. If the individual is guessing, find the probability of correctly identifying

(a) all four cards   (b) exactly two of the four cards

44 Tossing dice Three dice are tossed.

(a) Find the probability that all dice show the same number of dots.

(b) Find the probability that the numbers of dots on the dice are all different.

(c) Work parts (a) and (b) for n dice.

45 Trick dice For a normal die, the sum of the dots on opposite faces is 7. Shown in the figure is a pair of trick dice in which the *same* number of dots appears on opposite faces. Find the probability of rolling a sum of

(a) 7   (b) 8

Exercise 45

46 Carnival game In a common carnival game, three balls are rolled down an incline into slots numbered 1 through 9, as shown in the figure. Because the slots are so narrow, players have no control over where the balls collect. A prize is given if the sum of the three numbers is less than 7. Find the probability of winning a prize.

Exercise 46

47 Smoking deaths During 1990, smoking caused 418,890 deaths in the United States. Of these deaths, cardiovascular disease accounted for 179,820, cancer for 151,322, and respiratory diseases such as emphysema for 84,475.

(a) Find the probability that a smoking-related death was the result of either cardiovascular disease or cancer.

(b) Determine the probability that a smoking-related death was not the result of respiratory diseases.

48 Starting work times In a survey about what time people go to work, it was found that 8.2 million people go to work between midnight and 6 A.M., 60.4 million between 6 A.M. and 9 A.M., and 18.3 million between 9 A.M. and midnight.

(a) Find the probability that a person goes to work between 6 A.M. and midnight.

(b) Determine the probability that a person goes to work between midnight and 6 A.M.

49 Arsenic exposure and cancer In a certain county, 2% of the people have cancer. Of those with cancer, 70% have been exposed to high levels of arsenic. Of those without cancer, 10% have been exposed. What percentage of the people who have been exposed to high levels of arsenic have cancer? (*Hint*: Use a tree diagram.)

50 Computers and defective chips A computer manufacturer buys 30% of its chips from supplier A and the rest from supplier B. Two percent of the chips from supplier A are defective, as are 4% of the chips from supplier B. Approximately what percentage of the defective chips are from supplier B?

51 Probability demonstration Shown in the figure is a small version of a probability demonstration device. A small ball is dropped into the top of the maze and tumbles to the bottom. Each time the ball strikes an obstacle, there is a 50% chance that the ball will move to the left. Find the probability that the ball ends up in the slot

(a) on the far left   (b) in the middle

Exercise 51

52 Roulette In the American version of roulette, a ball is spun around a wheel and has an equal chance of landing in any one of 38 slots numbered 0, 00, 1, 2, ..., 36. Shown in the figure is a standard betting layout for roulette, where the color of the oval corresponds to the color of the slot on the wheel. Find the probability that the ball lands

(a) in a black slot

(b) in a black slot twice in succession

Exercise 52

53 Selecting lottery numbers In one version of a popular lottery game, a player selects six of the numbers from 1 to 54. The agency in charge of the lottery also selects six numbers. What is the probability that the player will match the six numbers if two 50¢ tickets are purchased? (This jackpot is worth at least $2 million in prize money and grows according to the number of tickets sold.)

54 Lottery Refer to Exercise 53. The player can win about $1000 for matching five of the six numbers and about $40 for matching four of the six numbers. Find the probability that the player will win some amount of prize money on the purchase of one ticket.

55 Quality control In a quality control procedure to test for defective light bulbs, two light bulbs are randomly selected from a large sample without replacement. If either light bulb is defective, the entire lot is rejected. Suppose a sample of 200 light bulbs contains 5 defective light bulbs. Find the probability that the sample will be rejected. (*Hint:* First calculate the probability that neither bulb is defective.)

56 Life expectancy A man is 54 years old and a woman is 34 years old. The probability that the man will be alive in 10 years is 0.74, whereas the probability that the woman will be alive 10 years from now is 0.94. Assume that their life expectancies are unrelated.

(a) Find the probability that they will both be alive 10 years from now.

(b) Determine the probability that neither one will be alive 10 years from now.

(c) Determine the probability that at least one of the two will be alive 10 years from now.

57 Shooting craps In the game of *craps*, there are two ways a player can win a *pass line* bet. The player wins immediately if two dice are rolled and their sum is 7 or 11. If their sum is 4, 5, 6, 8, 9, or 10, the player can still win a pass line bet if this same number (called the *point*) is rolled again before a 7 is rolled. Find the probability that the player wins

(a) a pass line bet on the first roll

(b) a pass line bet with a 4 on the first roll

(c) on any pass line bet

58 Crapless craps Refer to Exercise 57. In the game of craps, a player loses a pass line bet if a sum of 2, 3, or 12 is obtained on the first roll (referred to as "craps"). In another version of the game, called *crapless craps,* the player does not lose by rolling craps and does not win by rolling an 11 on the first roll. Instead, the player wins if the first roll is a 7 or if the point (2–12, excluding 7) is repeated before a 7 is rolled. Find the probability that the player wins on a pass line bet in crapless craps.

59 Birthday probability
(a) Show that the probability p that n people all have different birthdays is given by

$$p = \frac{365!}{365^n(365 - n)!}.$$

(b) If a room contains 32 people, approximate the probability that two or more people have the *same* birthday. (First approximate $\ln p$ by using the following formula from advanced mathematics:

$$\ln n! \approx n \ln n - n.)$$

60 Birthday probability Refer to Exercise 59. Find the smallest number of people in a room such that the probability that everyone has a different birthday is less than $\frac{1}{2}$. *Hint:* Rewrite the formula for p in part (a) of the previous exercise as

$$\frac{365}{365} \cdot \frac{364}{365} \cdot \frac{363}{365} \cdot \dots \cdot \frac{365 - n + 1}{365}.$$

61 A bet in craps Refer to Exercise 57. A player receives $2 for winning a $1 pass line bet. Approximate the expected value of a $1 bet.

62 A bet in roulette Refer to Exercise 52. If a player bets $1 that the ball will land in a black slot, he or she will receive $2 if it does. Approximate the expected value of a $1 bet.

63 Contest prize winning A contest offers the following cash prizes:

Number of prizes	1	10	100	1000
Prize values	$1,000,000	$100,000	$10,000	$1000

If the sponsor expects 20 million contestants, find the expected value for a single contestant.

64 Tournament prize winnings A bowling tournament is handicapped so that all 80 bowlers are equally matched. The tournament prizes are listed in the table.

Place	1st	2nd	3rd	4th	5th–10th
Prize	$1000	$500	$300	$200	$100

Find the expected winnings for one contestant.

CHAPTER 10 REVIEW EXERCISES

Exer. 1–4: Find the first four terms and the seventh term of the sequence that has the given nth term.

1 $\left\{\dfrac{5n}{3 - 2n^2}\right\}$

2 $\{(-1)^{n+1} - (0.1)^n\}$

3 $\left\{1 + \left(-\tfrac{1}{2}\right)^{n-1}\right\}$

4 $\left\{\dfrac{2^n}{(n + 1)(n + 2)(n + 3)}\right\}$

Exer. 5–8: Find the first five terms of the recursively defined infinite sequence.

5 $a_1 = 10, \quad a_{k+1} = 1 + (1/a_k)$

6 $a_1 = 2, \quad a_{k+1} = a_k!$

7 $a_1 = 9, \quad a_{k+1} = \sqrt{a_k}$

8 $a_1 = 1, \quad a_{k+1} = (1 + a_k)^{-1}$

Exer. 9–12: Evaluate.

9 $\displaystyle\sum_{k=1}^{5} (k^2 + 4)$

10 $\displaystyle\sum_{k=2}^{6} \dfrac{2k - 8}{k - 1}$

11 $\displaystyle\sum_{k=7}^{100} 10$

12 $\displaystyle\sum_{k=1}^{4} (2^k - 10)$

Exer. 13–24: Express the sum in terms of summation notation. (Answers are not unique.)

13 $3 + 6 + 9 + 12 + 15$

14 $4 + 2 + 1 + \tfrac{1}{2} + \tfrac{1}{4} + \tfrac{1}{8}$

15 $\dfrac{1}{1 \cdot 2} + \dfrac{1}{2 \cdot 3} + \dfrac{1}{3 \cdot 4} + \cdots + \dfrac{1}{99 \cdot 100}$

16 $\dfrac{1}{1 \cdot 2 \cdot 3} + \dfrac{1}{2 \cdot 3 \cdot 4} + \dfrac{1}{3 \cdot 4 \cdot 5} + \cdots + \dfrac{1}{98 \cdot 99 \cdot 100}$

17 $\tfrac{1}{2} + \tfrac{2}{5} + \tfrac{3}{8} + \tfrac{4}{11}$

18 $\tfrac{1}{4} + \tfrac{2}{9} + \tfrac{3}{14} + \tfrac{4}{19}$

19 $100 - 95 + 90 - 85 + 80$

20 $1 - \tfrac{1}{2} + \tfrac{1}{3} - \tfrac{1}{4} + \tfrac{1}{5} - \tfrac{1}{6} + \tfrac{1}{7}$

21 $a_0 + a_1x^4 + a_2x^8 + \cdots + a_{25}x^{100}$

22 $a_0 + a_1x^3 + a_2x^6 + \cdots + a_{20}x^{60}$

23 $1 - \dfrac{x^2}{2} + \dfrac{x^4}{4} - \dfrac{x^6}{6} + \cdots + (-1)^n\dfrac{x^{2n}}{2n}$

24 $1 + x + \dfrac{x^2}{2} + \dfrac{x^3}{3} + \cdots + \dfrac{x^n}{n}$

25 Find the tenth term and the sum of the first ten terms of the arithmetic sequence whose first two terms are $4 + \sqrt{3}$ and 3.

26 Find the sum of the first eight terms of the arithmetic sequence in which the fourth term is 9 and the common difference is -5.

27 The fifth and thirteenth terms of an arithmetic sequence are 5 and 77, respectively. Find the first term and the tenth term.

28 Find the number of terms in the arithmetic sequence with $a_1 = 1, d = 5$, and $S = 342$.

29 Insert four arithmetic means between 20 and -10.

30 Find the tenth term of the geometric sequence whose first two terms are $\tfrac{1}{8}$ and $\tfrac{1}{4}$.

31 If a geometric sequence has 3 and -0.3 as its third and fourth terms, respectively, find the eighth term.

32 Given a geometric sequence with $a_3 = 16$ and $a_7 = 625$, find a_8.

33 Find the geometric mean of 4 and 8.

34 In a certain geometric sequence, the eighth term is 100 and the common ratio is $-\tfrac{3}{2}$. Find the first term.

35 Given an arithmetic sequence such that $S_{12} = 402$ and $a_{12} = 50$, find a_1 and d.

36 Given a geometric sequence such that $a_5 = \tfrac{1}{16}$ and $r = \tfrac{3}{2}$, find a_1 and S_5.

Exer. 37–40: Evaluate.

37 $\displaystyle\sum_{k=1}^{15} (5k - 2)$

38 $\displaystyle\sum_{k=1}^{10} \left(6 - \tfrac{1}{2}k\right)$

39 $\displaystyle\sum_{k=1}^{10} \left(2^k - \tfrac{1}{2}\right)$

40 $\displaystyle\sum_{k=1}^{8} \left(\tfrac{1}{2} - 2^k\right)$

41 Find the sum of the infinite geometric series

$$1 - \tfrac{2}{5} + \tfrac{4}{25} - \tfrac{8}{125} + \cdots.$$

42 Find the rational number whose decimal representation is $6.\overline{274}$.

Exer. 43–47: Prove that the statement is true for every positive integer n.

43 $2 + 5 + 8 + \cdots + (3n - 1) = \dfrac{n(3n + 1)}{2}$

44 $2^2 + 4^2 + 6^2 + \cdots + (2n)^2 = \dfrac{2n(2n + 1)(n + 1)}{3}$

45 $\dfrac{1}{1 \cdot 3} + \dfrac{1}{3 \cdot 5} + \dfrac{1}{5 \cdot 7} + \cdots + \dfrac{1}{(2n - 1)(2n + 1)} =$

$$\dfrac{n}{2n + 1}$$

46 $1 \cdot 2 + 2 \cdot 3 + 3 \cdot 4 + \cdots + n(n + 1) =$

$$\dfrac{n(n + 1)(n + 2)}{3}$$

47 3 is a factor of $n^3 + 2n$.

48 Prove that $n^2 + 3 < 2^n$ for every positive integer $n \geq 5$.

Exer. 49–50: Find the smallest positive integer j for which the statement is true. Use the extended principle of mathematical induction to prove that the formula is true for every integer greater than j.

49 $2^n \leq n!$

50 $10^n \leq n^n$

Exer. 51–52: Use the binomial theorem to expand and simplify the expression.

51 $(x^2 - 3y)^6$

52 $(2x + y^3)^4$

Exer. 53–56: Without expanding completely, find the indicated term(s) in the expansion of the expression.

53 $(x^{2/5} + 2x^{-3/5})^{20}$; first three terms

54 $\left(y^3 - \tfrac{1}{2}c^2\right)^9$; sixth term

55 $(4x^2 - y)^7$; term that contains x^{10}

56 $(2c^3 + 5c^{-2})^{10}$; term that does not contain c

57 Building blocks Ten-foot lengths of 2×2 lumber are to be cut into five pieces to form children's building blocks; the lengths of the five blocks are to form an arithmetic sequence.

  (a) Show that the difference d in lengths must be less than 1 foot.

(b) If the smallest block is to have a length of 6 inches, find the lengths of the other four pieces.

58 Constructing a ladder A ladder is to be constructed with 16 rungs whose lengths decrease uniformly from 20 inches at the base to 16 inches at the top. Find the total length of material needed for the rungs.

59 Shown in the first figure is a broken-line curve obtained by taking two adjacent sides of a square, each of length s_n, decreasing the length of the side by a factor f with $0 < f < 1$, and forming two sides of a smaller square, each of length $s_{n+1} = f \cdot s_n$. The process is then repeated ad infinitum. If $s_1 = 1$ in the second figure, express the length of the resulting (infinite) broken-line curve in terms of f.

Exercise 59

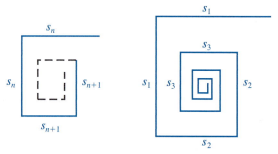

60 The commutative and associative laws of addition guarantee that the sum of integers 1 through 10 is independent of the order in which the numbers are added. In how many different ways can these integers be summed?

61 Selecting cards

  (a) In how many ways can 13 cards be selected from a deck?

  (b) In how many ways can 13 cards be selected to obtain five spades, three hearts, three clubs, and two diamonds?

62 How many four-digit numbers can be formed from the digits 1, 2, 3, 4, 5, and 6 if repetitions

  (a) are not allowed?   **(b)** are allowed?

63 Selecting test questions

  (a) If a student must answer 8 of 12 questions on an examination, how many different selections of questions are possible?

  (b) How many selections are possible if the first three questions must be answered?

64 Color arrangements If six black, five red, four white, and two green disks are to be arranged in a row, what is the number of possible color arrangements?

65 If $O(E)$ are 8 to 5, find $O(E')$ and $P(E)$.

66 Coin toss Find the probability that the coins will match if

(a) two boys each toss a coin

(b) three boys each toss a coin

67 Dealing cards If four cards are dealt from a deck, find the probability that

(a) all four cards will be the same color

(b) the cards dealt will alternate red-black-red-black

68 Raffle probabilities If 1000 tickets are sold for a raffle, find the probability of winning if an individual purchases

(a) 1 ticket   (b) 10 tickets   (c) 50 tickets

69 Coin toss If four coins are flipped, find the probability and the odds of obtaining one head and three tails.

70 True-or-false quiz A quiz consists of six true-or-false questions; at least four correct answers are required for a passing grade. If a student guesses at each answer, what is the probability of

(a) passing?   (b) failing?

71 Die and card probabilities If a single die is tossed and then a card is drawn from a deck, what is the probability of obtaining

(a) a 6 on the die and the king of hearts?

(b) a 6 on the die or the king of hearts?

72 Population demographics In a town of 5000 people, 1000 are over 60 years old and 2000 are female. It is known that 40% of the females are over 60. What is the probability that a randomly chosen individual from the town is either female or over 60?

73 Backgammon moves In the game of backgammon, players are allowed to move their counters the same number of spaces as the sum of the dots on two dice. However, if a double is rolled (that is, both dice show the same number of dots), then players may move their counters twice the sum of the dots. What is the probability that a player will be able to move his or her counters at least 10 spaces on a given roll?

74 Games in a series Two equally matched baseball teams are playing a series of games. The first team to win four games wins the series. Find the expected number of games in the series.

CHAPTER 10 DISCUSSION EXERCISES

1 A test question lists the first four terms of a sequence as 2, 4, 6, and 8 and asks for the fifth term. Show that the fifth term can be *any* real number a by finding the nth term of a sequence that has for its first five terms 2, 4, 6, 8, and a.

2 Decide whether ☐ should be replaced by $\leq$ or $\geq$ in

$$ n \boxed{} (\ln n)^3 $$

for the statement to be true when $n \geq j$, where j is the smallest positive integer for which the statement is true. Find j.

3 Determine the largest factorial that your calculator can compute. Some typical values are 69! and 449!. Speculate as to why these numbers are the maximum values that your calculator can compute.

4 Find a relationship between the coefficients in the expansion of $(a + b)^n$ and the number of distinct subsets of an n-element set.

5 Rebounding ball When a ball is dropped from a height of h feet, it reaches the ground in $\sqrt{h}/4$ seconds. The ball rebounds to a height of d feet in $\sqrt{d}/4$ seconds. If a rubber ball is dropped from a height of 10 feet and rebounds to one-half of its height after each fall, for approximately how many seconds does the ball travel?

6 Slot tournament A slot tournament will be held over a 30-day month, eight hours each day, with 36 contestants each hour. The prize structure is as follows:

(continued)

Place	1st	2nd	3rd	4th	5th
Prize $	4000	2000	1500	1000	800

Place	6th	7th	8th	9th	10th
Prize $	600	500	400	300	200

Place	11th–50th	51st–100th	101st–300th	301st–500th
Prize $	100	75	50	25

There is also a daily prize awarded as follows: $250 for first, $100 for second, and $50 for third. How much would you expect to pay for an entry fee if the tournament is to be fair?

7 **Prize money** Suppose that the tenth prize of a $1600 tournament will be $100 and each place should be worth approximately 10% more than the next place. Discuss the realistic distribution of prize values if they are rounded to the nearest penny, dollar, five dollars, and ten dollars.

8 **Pizza toppings** A pizza parlor sponsored an advertisement claiming that it gave you a total of 1,048,576 possible ways to order 2 pizzas, with up to 5 toppings on each. Discuss how the company computed the number of possible ways to order, and determine how many toppings are available.

9 **Powerball** Powerball is a popular lottery game played in many states. The player selects five integers from 1 to 49 and one integer from 1 to 42. These numbers correspond to five white balls and one red Powerball drawn by the Multi-State Lottery Association. To win the jackpot (which has reached $150 million), the player must match all six numbers. The prizes for all paying matches are listed in the table.

Match	Prize
5 white and red	jackpot
5 white	$100,000
4 white and red	$5000
4 white	$100
3 white and red	$100
3 white	$7
2 white and red	$7
1 white and red	$4
red only	$3

(a) What is the probability of winning the jackpot?

(b) What is the probability of winning any prize?

(c) What is the expected value of the game without the jackpot?

(d) How much does the jackpot need to be worth for this lottery to be considered a fair game?

10 **Probability and odds confusion** Analyze the following statement: "There is a 20% chance that a male applicant will be admitted, but the odds are three times more favorable for a female applicant." What is the probability that a female applicant will be admitted?

11 Let $a = 0$ and $b = 1$ in

$$(a + b)^n = \sum_{k=0}^{n}\binom{n}{k}a^{n-k}b^k$$

and discuss the result.

12 Investigate the partial sums of

$$\sum_{n=0}^{\infty}(-1)^n\frac{3^{3/2}}{2^{3n+2}}\left(\frac{2}{3n + 1} + \frac{1}{3n + 2}\right)$$

and discuss them.

13 (a) Examine the following identities for tan nx in terms of tan x:

$$\tan 2x = \frac{2\tan x}{1 - \tan^2 x}$$

$$\tan 3x = \frac{3\tan x - \tan^3 x}{1 - 3\tan^2 x}$$

$$\tan 4x = \frac{4\tan x - 4\tan^3 x}{1 - 6\tan^2 x + \tan^4 x}$$

By using a pattern formed by the three identities, predict an identity for tan $5x$ in terms of tan x.

(b) Listed below are identities for cos $2x$ and sin $2x$:

$$\cos 2x = 1\cos^2 x \qquad -1\sin^2 x$$
$$\sin 2x = \qquad 2\cos x \sin x$$

Write similar identities for cos $3x$ and sin $3x$ and then cos $4x$ and sin $4x$. Use a pattern to predict identities for cos $5x$ and sin $5x$.

11

Topics from Analytic Geometry

11.1 Parabolas

11.2 Ellipses

11.3 Hyperbolas

11.4 Plane Curves and Parametric Equations

11.5 Polar Coordinates

11.6 Polar Equations of Conics

Plane geometry includes the study of figures—such as lines, circles, and triangles—that lie in a plane. Theorems are proved by reasoning deductively from certain postulates. In analytic geometry, plane geometric figures are investigated by introducing coordinate systems and then using equations and formulas. If the study of analytic geometry were to be summarized by means of one statement, perhaps the following would be appropriate: Given an equation, find its graph, and conversely, given a graph, find its equation. In this chapter we shall apply coordinate methods to several basic plane figures.

11.1

Parabolas

The *conic sections,* also called *conics,* can be obtained by intersecting a double-napped right circular cone with a plane. By varying the position of the plane, we obtain a *circle,* an *ellipse,* a *parabola,* or a *hyperbola,* as illustrated in Figure 1.

Figure 1
(a) Circle   **(b)** Ellipse   **(c)** Parabola   **(d)** Hyperbola

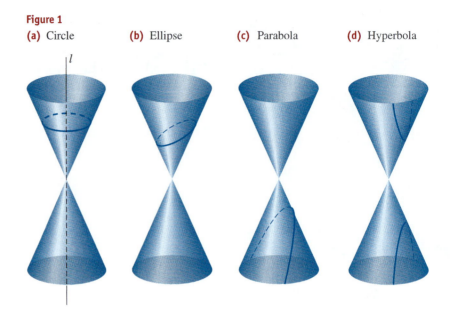

Degenerate conics are obtained if the plane intersects the cone in only one point or along either one or two lines that lie on the cone. Conic sections were studied extensively by the ancient Greeks, who discovered properties that enable us to state their definitions in terms of points and lines, as we do in our discussion.

From our work in Section 3.6, if $a \neq 0$, the graph of $y = ax^2 + bx + c$ is a *parabola* with a vertical axis. We shall next state a general definition of a parabola and derive equations for parabolas that have either a vertical axis or a horizontal axis.

Definition of a Parabola	A **parabola** is the set of all points in a plane equidistant from a fixed point F (the **focus**) and a fixed line l (the **directrix**) that lie in the plane.

We shall assume that F is not on l, for this would result in a line. If P is a point in the plane and P' is the point on l determined by a line through P that

Figure 2

Directrix

Figure 3

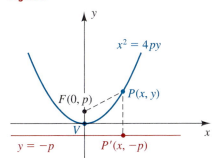

is perpendicular to l (see Figure 2), then, by the preceding definition, P is on the parabola if and only if the distances $d(P, F)$ and $d(P, P')$ are equal. The **axis** of the parabola is the line through F that is perpendicular to the directrix. The **vertex** of the parabola is the point V on the axis halfway from F to l. The vertex is the point on the parabola that is closest to the directrix.

To obtain a simple equation for a parabola, place the y-axis along the axis of the parabola, with the origin at the vertex V, as shown in Figure 3. In this case, the focus F has coordinates $(0, p)$ for some real number $p \neq 0$, and the equation of the directrix is $y = -p$. (The figure shows the case $p > 0$.) By the distance formula, a point $P(x, y)$ is on the graph of the parabola if and only if $d(P, F) = d(P, P')$—that is, if

$$\sqrt{(x - 0)^2 + (y - p)^2} = \sqrt{(x - x)^2 + (y + p)^2}.$$

We square both sides and simplify:

$$x^2 + (y - p)^2 = (y + p)^2$$

$$x^2 + y^2 - 2py + p^2 = y^2 + 2py + p^2$$

$$x^2 = 4py$$

An equivalent equation for the parabola is

$$y = \frac{1}{4p} x^2.$$

We have shown that the coordinates of every point (x, y) on the parabola satisfy $x^2 = 4py$. Conversely, if (x, y) is a solution of $x^2 = 4py$, then by reversing the previous steps we see that the point (x, y) is on the parabola.

If $p > 0$, the parabola opens upward, as in Figure 3. If $p < 0$, the parabola opens downward. The graph is symmetric with respect to the y-axis, since substitution of $-x$ for x does not change the equation $x^2 = 4py$.

If we interchange the roles of x and y, we obtain

$$y^2 = 4px \qquad \text{or, equivalently,} \qquad x = \frac{1}{4p} y^2.$$

This is an equation of a parabola with vertex at the origin, focus $F(p, 0)$, and opening right if $p > 0$ or left if $p < 0$. The equation of the directrix is $x = -p$.

For convenience we often refer to "the parabola $x^2 = 4py$" (or $y^2 = 4px$) instead of "the parabola with equation $x^2 = 4py$" (or $y^2 = 4px$).

The next chart summarizes our discussion.

Parabolas with Vertex $V(0, 0)$

Equation, focus, directrix	Graph for $p > 0$	Graph for $p < 0$
$x^2 = 4py$ or $y = \dfrac{1}{4p}x^2$ Focus: $F(0, p)$ Directrix: $y = -p$		
$y^2 = 4px$ or $x = \dfrac{1}{4p}y^2$ Focus: $F(p, 0)$ Directrix: $x = -p$		

We see from the chart that for any nonzero real number a, the graph of the **standard equation** $y = ax^2$ or $x = ay^2$ is a parabola with vertex $V(0, 0)$. Moreover, $a = 1/(4p)$ or, equivalently, $p = 1/(4a)$, where $|p|$ is the distance between the focus F and vertex V. To find the directrix l, recall that l is also a distance $|p|$ from V.

EXAMPLE 1 **Finding the focus and directrix of a parabola**

Find the focus and directrix of the parabola $y = -\frac{1}{6}x^2$, and sketch its graph.

SOLUTION The equation has the form $y = ax^2$, with $a = -\frac{1}{6}$. As in the preceding chart, $a = 1/(4p)$, and hence

$$p = \frac{1}{4a} = \frac{1}{4\left(-\frac{1}{6}\right)} = \frac{1}{-\frac{4}{6}} = -\frac{3}{2}.$$

Figure 4

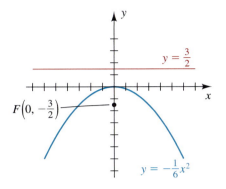

Thus, the parabola opens downward and has focus $F\left(0, -\frac{3}{2}\right)$, as illustrated in Figure 4. The directrix is the horizontal line $y = \frac{3}{2}$, which is a distance $\frac{3}{2}$ above V, as shown in the figure.

EXAMPLE 2 Finding an equation of a parabola satisfying prescribed conditions

(a) Find an equation of a parabola that has vertex at the origin, opens right, and passes through the point $P(7, -3)$.

(b) Find the focus of the parabola.

SOLUTION

(a) The parabola is sketched in Figure 5. An equation of a parabola with vertex at the origin that opens right has the form $x = ay^2$ for some number a. If $P(7, -3)$ is on the graph, then we can substitute 7 for x and -3 for y to find a:

$$7 = a(-3)^2, \text{ or } a = \tfrac{7}{9}$$

Hence, an equation for the parabola is $x = \frac{7}{9}y^2$.

(b) The focus is a distance p to the right of the vertex. Since $a = \frac{7}{9}$, we have

$$p = \frac{1}{4a} = \frac{1}{4\left(\frac{7}{9}\right)} = \frac{9}{28}.$$

Thus, the focus has coordinates $\left(\frac{9}{28}, 0\right)$.

Figure 5

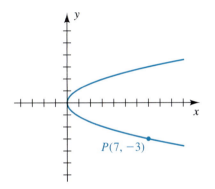

If we take a standard equation of a parabola (of the form $x^2 = 4py$) and replace x with $x - h$ and y with $y - k$, then

$$x^2 = 4py \quad \text{becomes} \quad (x - h)^2 = 4p(y - k). \tag{$*$}$$

From our discussion of translations in Section 3.5, we recognize that the graph of the second equation can be obtained from the graph of the first equation by shifting it h units to the right and k units up—thereby moving the vertex from $(0, 0)$ to (h, k). Squaring the left-hand side of ($*$) and simplifying leads to an equation of the form $y = ax^2 + bx + c$, where a, b, and c are real numbers.

Similarly, if we begin with $(y - k)^2 = 4p(x - h)$, it may be written in the form $x = ay^2 + by + c$. In the following chart $V(h, k)$ has been placed in the first quadrant, but the information given in the leftmost column holds true regardless of the position of V.

Parabolas with Vertex $V(h, k)$

Equation, focus, directrix	Graph for $p > 0$	Graph for $p < 0$
$(x - h)^2 = 4p(y - k)$ or $y = ax^2 + bx + c$, where $p = \dfrac{1}{4a}$ Focus: $F(h, k + p)$ Directrix: $y = k - p$		
$(y - k)^2 = 4p(x - h)$ or $x = ay^2 + by + c$, where $p = \dfrac{1}{4a}$ Focus: $F(h + p, k)$ Directrix: $x = h - p$		

EXAMPLE 3   Sketching a parabola with a horizontal axis

Discuss and sketch the graph of $2x = y^2 + 8y + 22$.

SOLUTION   The equation can be written in the form shown in the second row of the preceding chart, $x = ay^2 + by + c$, so we see from the chart that the graph is a parabola with a horizontal axis. We first write the given equation as

$$y^2 + 8y + \underline{} = 2x - 22 + \underline{}$$

and then complete the square by adding $\left[\frac{1}{2}(8)\right]^2 = 16$ to both sides:

$$y^2 + 8y + 16 = 2x - 6$$

$$(y + 4)^2 = 2(x - 3)$$

Figure 6

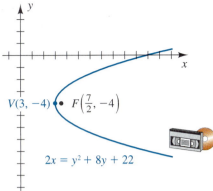

$2x = y^2 + 8y + 22$

Referring to the last chart, we see that $h = 3$, $k = -4$, and $4p = 2$ or, equivalently, $p = \frac{1}{2}$. This gives us the following.

The vertex $V(h, k)$ is $V(3, -4)$.

The focus is $F(h + p, k) = F\left(3 + \frac{1}{2}, -4\right)$, or $F\left(\frac{7}{2}, -4\right)$.

The directrix is $x = h - p = 3 - \frac{1}{2}$, or $x = \frac{5}{2}$.

The parabola is sketched in Figure 6.

EXAMPLE 4 Finding an equation of a parabola given its vertex and directrix

A parabola has vertex $V(-4, 2)$ and directrix $y = 5$. Express the equation of the parabola in the form $y = ax^2 + bx + c$.

SOLUTION The vertex and directrix are shown in Figure 7. The dashes indicate a possible position for the parabola. The last chart shows that an equation of the parabola is

$$(x - h)^2 = 4p(y - k),$$

with $h = -4$ and $k = 2$ and with p equal to *negative* 3, since V is 3 units *below* the directrix. This gives us

$$(x + 4)^2 = -12(y - 2).$$

The last equation can be expressed in the form $y = ax^2 + bx + c$, as follows:

$$x^2 + 8x + 16 = -12y + 24$$
$$12y = -x^2 - 8x + 8$$
$$y = -\tfrac{1}{12}x^2 - \tfrac{2}{3}x + \tfrac{2}{3}$$

Figure 7

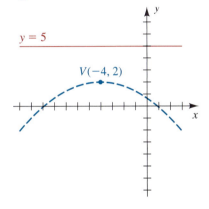

$y = 5$

$V(-4, 2)$

An important property is associated with a tangent line to a parabola. (A *tangent line* to a parabola is a line that has exactly one point in common with the parabola but does not cut through the parabola.) Suppose l is the tangent line at a point $P(x_1, y_1)$ on the graph of $y^2 = 4px$, and let F be the focus. As in Figure 8, let α denote the angle between l and the line segment FP, and let β denote the angle between l and the indicated horizontal half-line with endpoint P. It can be shown that $\alpha = \beta$. This *reflective property* has many applications. For example, the shape of the mirror in a searchlight is obtained by revolving a parabola about its axis. The resulting three-dimensional surface is said to be *generated* by the parabola and is called a **paraboloid.** The **focus** of the paraboloid is the same as the focus of the generating parabola. If a light source is placed at F, then, by a law of physics (the angle of reflection equals the angle of incidence), a beam of light will be reflected along a line parallel to the axis (see Figure 9(a)). The same principle is used in the construction of mirrors for telescopes or solar ovens—a beam of light coming toward the parabolic mirror and parallel to the axis will be reflected into the focus (see

Figure 8

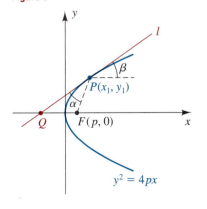

$y^2 = 4px$

Figure 9

(a) Searchlight mirror

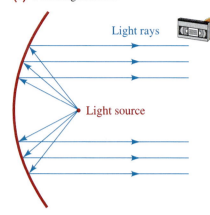

Light rays

Light source

(b) Telescope mirror

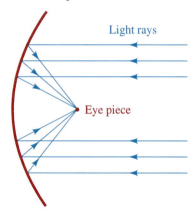

Light rays

Eye piece

Figure 9(b)). Antennas for radar systems, radio telescopes, and field micro-phones used at football games also make use of this property.

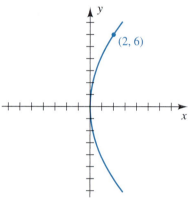

EXAMPLE 5   Locating the focus of a satellite TV antenna

The interior of a satellite TV antenna is a dish having the shape of a (finite) pa-raboloid that has diameter 12 feet and is 2 feet deep, as shown in Figure 10. Find the distance from the center of the dish to the focus.

Figure 10   **Figure 11**

2'   12'

$(2, 6)$

SOLUTION   The generating parabola is sketched on an xy-plane in Fig-ure 11, where we have taken the vertex of the parabola at the origin and its axis along the x-axis. An equation of the parabola is $y^2 = 4px$, where p is the re-quired distance from the center of the dish to the focus. Since the point $(2, 6)$ is on the parabola, we obtain

$$6^2 = 4p \cdot 2, \quad \text{or} \quad p = \tfrac{36}{8} = 4.5 \text{ ft.}$$

11.1   *Exercises*

Exer. 1–12: Find the vertex, focus, and directrix of the parabola. Sketch its graph, showing the focus and the directrix.

1 $8y = x^2$

2 $20x = y^2$

3 $2y^2 = -3x$

4 $x^2 = -3y$

5 $(x + 2)^2 = -8(y - 1)$

6 $(x - 3)^2 = \tfrac{1}{2}(y + 1)$

7 $(y - 2)^2 = \tfrac{1}{4}(x - 3)$

8 $(y + 1)^2 = -12(x + 2)$

9 $y = x^2 - 4x + 2$

10 $y^2 + 14y + 4x + 45 = 0$

11 $x^2 + 20y = 10$

12 $y^2 - 4y - 2x - 4 = 0$

Exer. 13–18: Find an equation for the parabola shown in the figure.

13

14

15

16

17

18

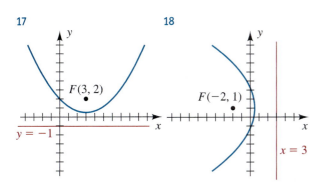

Exer. 19–30: Find an equation of the parabola that satisfies the given conditions.

19 Focus $F(2, 0)$,   directrix $x = -2$

20 Focus $F(0, -4)$,   directrix $y = 4$

21 Focus $F(6, 4)$,   directrix $y = -2$

22 Focus $F(-3, -2)$,   directrix $y = 1$

23 Vertex $V(3, -5)$,   directrix $x = 2$

24 Vertex $V(-2, 3)$,   directrix $y = 5$

25 Vertex $V(-1, 0)$ focus $F(-4, 0)$

26 Vertex $V(1, -2)$,   focus $F(1, 0)$

27 Vertex at the origin, symmetric to the y-axis, and passing through the point $(2, -3)$

28 Vertex at the origin, symmetric to the y-axis, and passing through the point $(6, 3)$

29 Vertex $V(-3, 5)$, axis parallel to the x-axis, and passing through the point $(5, 9)$

30 Vertex $V(3, -2)$, axis parallel to the x-axis, and y-intercept 1

Exer. 31–34: Find an equation for the set of points in an xy-plane that are equidistant from the point P and the line l.

31 $P(0, 5)$; $l: y = -3$ **32** $P(7, 0)$; $l: x = 1$

33 $P(-6, 3)$; $l: x = -2$ **34** $P(5, -2)$; $l: y = 4$

Exer. 35–38: Find an equation for the indicated half of the parabola.

35 Lower half of $(y + 1)^2 = x + 3$

36 Upper half of $(y - 2)^2 = x - 4$

37 Right half of $(x + 1)^2 = y - 4$

38 Left half of $(x + 3)^2 = y + 2$

Exer. 39–40: Find an equation for the parabola that has a vertical axis and passes through the given points.

39 $P(2, 5)$, $Q(-2, -3)$, $R(1, 6)$

40 $P(3, -1)$, $Q(1, -7)$, $R(-2, 14)$

Exer. 41–42: Find an equation for the parabola that has a horizontal axis and passes through the given points.

41 $P(-1, 1)$, $Q(11, -2)$, $R(5, -1)$

42 $P(2, 1)$, $Q(6, 2)$, $R(12, -1)$

43 **Telescope mirror** A mirror for a reflecting telescope has the shape of a (finite) paraboloid of diameter 8 inches and depth 1 inch. How far from the center of the mirror will the incoming light collect?

Exercise 43

44 **Antenna dish** A satellite antenna dish has the shape of a paraboloid that is 10 feet across at the open end and is 3 feet deep. At what distance from the center of the dish should the receiver be placed to receive the greatest intensity of sound waves?

45 **Searchlight reflector** A searchlight reflector has the shape of a paraboloid, with the light source at the focus. If the reflector is 3 feet across at the opening and 1 foot deep, where is the focus?

46 **Flashlight mirror** A flashlight mirror has the shape of a paraboloid of diameter 4 inches and depth $\frac{3}{4}$ inch, as shown in the figure. Where should the bulb be placed so that the emitted light rays are parallel to the axis of the paraboloid?

Exercise 46

47 **Receiving dish** A sound receiving dish used at outdoor sporting events is constructed in the shape of a paraboloid, with its focus 5 inches from the vertex. Determine the width of the dish if the depth is to be 2 feet.

48 **Receiving dish** Work Exercise 47 if the receiver is 9 inches from the vertex.

49 **Parabolic reflector**

(a) The focal length of the (finite) paraboloid in the figure is the distance p between its vertex and focus. Express p in terms of r and h.

(b) A reflector is to be constructed with a focal length of 10 feet and a depth of 5 feet. Find the radius of the reflector.

Exercise 49

50 **Confocal parabolas** The parabola $y^2 = 4p(x + p)$ has its focus at the origin and axis along the x-axis. By assigning different values to p, we obtain a family of confocal parabolas, as shown in the figure. Such families occur in the study of electricity and magnetism. Show that there are exactly two parabolas in the family that pass through a given point $P(x_1, y_1)$ if $y_1 \neq 0$.

Exercise 50

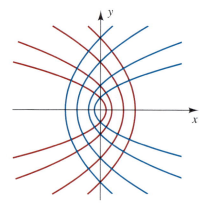

51 **Jodrell Bank radio telescope** A radio telescope has the shape of a paraboloid of revolution, with focal length p and diameter of base $2a$. From calculus, the surface area S available for collecting radio waves is

$$S = \frac{8\pi p^2}{3}\left[\left(1 + \frac{a^2}{4p^2}\right)^{3/2} - 1\right].$$

One of the largest radio telescopes, located in Jodrell Bank, Cheshire, England, has diameter 250 feet and focal length 50 feet. Approximate S to the nearest square foot.

52 **Satellite path** A satellite will travel in a parabolic path near a planet if its velocity v in meters per second satisfies the equation $v = \sqrt{2k/r}$, where r is the distance in meters between the satellite and the center of the planet and k is a positive constant. The planet will be located at the focus of the parabola, and the satellite will pass by the planet once. Suppose a satellite is designed to follow a parabolic path and travel within 58,000 miles of Mars, as shown in the figure.

(a) Determine an equation of the form $x = ay^2$ that describes its flight path.

Exercise 52

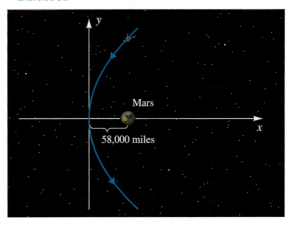

(b) For Mars, $k = 4.28 \times 10^{13}$. Approximate the maximum velocity of the satellite.

(c) Find the velocity of the satellite when its y-coordinate is 100,000 miles.

11.2

Ellipses

An ellipse may be defined as follows. (*Foci* is the plural of *focus*.)

Definition of an Ellipse	An **ellipse** is the set of all points in a plane, the sum of whose distances from two fixed points (the **foci**) in the plane is a positive constant.

We can construct an ellipse on paper as follows: Insert two pushpins in the paper at any points F and F', and fasten the ends of a piece of string to the pins. After looping the string around a pencil and drawing it tight, as at point P in Figure 1, move the pencil, keeping the string tight. The sum of the distances $d(P, F)$ and $d(P, F')$ is the length of the string and hence is constant; thus, the pencil will trace out an ellipse with foci at F and F'. The midpoint of the segment $F'F$ is called the **center** of the ellipse. By changing the positions of F and F' while keeping the length of the string fixed, we can vary the shape of the ellipse considerably. If F and F' are far apart so that $d(F, F')$ is almost the same as the length of the string, the ellipse is flat. If $d(F, F')$ is close to zero, the ellipse is almost circular. If $F = F'$, we obtain a circle with center F.

Figure 1

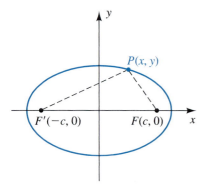

To obtain a simple equation for an ellipse, choose the x-axis as the line through the two foci F and F', with the center of the ellipse at the origin. If F has coordinates $(c, 0)$ with $c > 0$, then, as in Figure 2, F' has coordinates $(-c, 0)$. Hence, the distance between F and F' is $2c$. The constant sum of the distances of P from F and F' will be denoted by $2a$. To obtain points that are not on the x-axis, we must have $2a > 2c$—that is, $a > c$. By definition, $P(x, y)$ is on the ellipse if and only if the following equivalent equations are true:

$$d(P, F) + d(P, F') = 2a$$

$$\sqrt{(x - c)^2 + (y - 0)^2} + \sqrt{(x + c)^2 + (y - 0)^2} = 2a$$

$$\sqrt{(x - c)^2 + y^2} = 2a - \sqrt{(x + c)^2 + y^2}$$

Squaring both sides of the last equation gives us

$$x^2 - 2cx + c^2 + y^2 = 4a^2 - 4a\sqrt{(x + c)^2 + y^2} + x^2 + 2cx + c^2 + y^2,$$

or

$$a\sqrt{(x + c)^2 + y^2} = a^2 + cx.$$

Figure 2

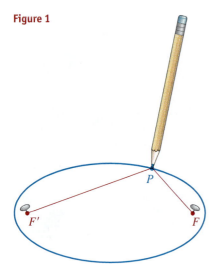

Squaring both sides again yields

$$a^2(x^2 + 2cx + c^2 + y^2) = a^4 + 2a^2cx + c^2x^2,$$

or $\qquad\qquad x^2(a^2 - c^2) + a^2y^2 = a^2(a^2 - c^2).$

Dividing both sides by $a^2(a^2 - c^2)$, we obtain

$$\frac{x^2}{a^2} + \frac{y^2}{a^2 - c^2} = 1.$$

Recalling that $a > c$ and therefore $a^2 - c^2 > 0$, we let

$$b = \sqrt{a^2 - c^2}, \qquad \text{or} \qquad b^2 = a^2 - c^2.$$

This substitution gives us the equation

$$\frac{x^2}{a^2} + \frac{y^2}{b^2} = 1.$$

Since $c > 0$ and $b^2 = a^2 - c^2$, it follows that $a^2 > b^2$ and hence $a > b$.

We have shown that the coordinates of every point (x, y) on the ellipse in Figure 3 satisfy the equation $(x^2/a^2) + (y^2/b^2) = 1$. Conversely, if (x, y) is a solution of this equation, then by reversing the preceding steps we see that the point (x, y) is on the ellipse.

Figure 3

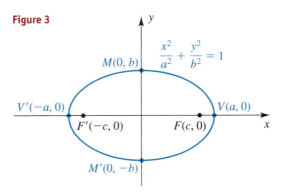

We may find the x-intercepts of the ellipse by letting $y = 0$ in the equation. Doing so gives us $x^2/a^2 = 1$, or $x^2 = a^2$. Consequently, the x-intercepts are a and $-a$. The corresponding points $V(a, 0)$ and $V'(-a, 0)$ on the graph are called the **vertices** of the ellipse (see Figure 3). The line segment $V'V$ is called the **major axis**. Similarly, letting $x = 0$ in the equation, we obtain $y^2/b^2 = 1$, or $y^2 = b^2$. Hence, the y-intercepts are b and $-b$. The segment between $M'(0, -b)$ and $M(0, b)$ is called the **minor axis** of the ellipse. The major axis is always longer than the minor axis, since $a > b$.

Applying tests for symmetry, we see that the ellipse is symmetric with respect to the x-axis, the y-axis, and the origin.

Similarly, if we take the foci on the y-axis, we obtain the equation

Figure 4

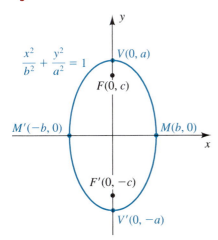

$$\frac{x^2}{b^2} + \frac{y^2}{a^2} = 1.$$

In this case, the vertices of the ellipse are $(0, \pm a)$ and the endpoints of the minor axis are $(\pm b, 0)$, as shown in Figure 4.

The preceding discussion may be summarized as follows.

Standard Equations of an Ellipse with Center at the Origin	The graph of

The graph of

$$\frac{x^2}{a^2} + \frac{y^2}{b^2} = 1 \qquad \text{or} \qquad \frac{x^2}{b^2} + \frac{y^2}{a^2} = 1,$$

where $a > b > 0$, is an ellipse with center at the origin. The length of the major axis is $2a$, and the length of the minor axis is $2b$. The foci are a distance c from the origin, where $c^2 = a^2 - b^2$.

To help you remember the relationship for the foci, think of the right triangle formed by a ladder of length a leaning against a building, as shown in Figure 5. By the Pythagorean Theorem, $b^2 + c^2 = a^2$. In this position, the ends of the ladder are at a focus and an endpoint of the minor axis. If the ladder falls, the ends of the ladder will be at the center of the ellipse and an endpoint of the major axis.

Figure 5

We have shown that an equation of an ellipse with center at the origin and foci on a coordinate axis can always be written in the form

$$\frac{x^2}{p} + \frac{y^2}{q} = 1, \quad \text{or} \quad qx^2 + py^2 = pq,$$

with p and q positive and $p \neq q$. If $p > q$, the major axis is on the x-axis; if $p < q$, the major axis is on the y-axis. It is unnecessary to memorize these facts, because in any given problem the major axis can be determined by examining the x- and y-intercepts.

EXAMPLE 1 Sketching an ellipse with center at the origin

Sketch the graph of $2x^2 + 9y^2 = 18$, and find the foci.

SOLUTION   The graph is an ellipse with center at the origin and foci on a coordinate axis. To find the x-intercepts, we let $y = 0$, obtaining

$$2x^2 = 18, \quad \text{or} \quad x = \pm 3.$$

To find the y-intercepts, we let $x = 0$, obtaining

$$9y^2 = 18, \quad \text{or} \quad y = \pm\sqrt{2}.$$

We are now able to sketch the ellipse (Figure 6). Since $\sqrt{2} \approx 1.4 < 3$, the major axis is on the x-axis.

To find the foci, we let $a = 3$ and $b = \sqrt{2}$ and calculate

$$c^2 = a^2 - b^2 = 3^2 - \left(\sqrt{2}\right)^2 = 7.$$

Thus, $c = \sqrt{7}$, and the foci are $\left(\pm\sqrt{7}, 0\right)$.   ◢

Figure 6

Figure 7

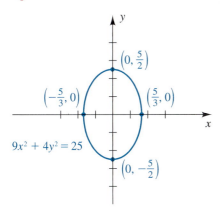

$\left(0, \frac{5}{2}\right)$

$\left(-\frac{5}{3}, 0\right)$ $\left(\frac{5}{3}, 0\right)$

$9x^2 + 4y^2 = 25$

$\left(0, -\frac{5}{2}\right)$

EXAMPLE 2   **Sketching an ellipse with center at the origin**

Sketch the graph of $9x^2 + 4y^2 = 25$, and find the foci.

SOLUTION   As in Example 1, the graph is an ellipse with center at the origin and foci on a coordinate axis. To find the x-intercepts, we let $y = 0$, obtaining

$$9x^2 = 25, \quad \text{or} \quad x = \pm\tfrac{5}{3}.$$

To find the y-intercepts, we let $x = 0$, obtaining

$$4y^2 = 25, \quad \text{or} \quad y = \pm\tfrac{5}{2}.$$

These calculations give us the sketch in Figure 7. Since $\frac{5}{3} < \frac{5}{2}$, the major axis is on the y-axis.

To find the foci, we let $a = \frac{5}{2}$ and $b = \frac{5}{3}$ and calculate

$$c^2 = a^2 - b^2 = \left(\tfrac{5}{2}\right)^2 - \left(\tfrac{5}{3}\right)^2 = \tfrac{125}{36}.$$

Thus, $c = \sqrt{125/36} = 5\sqrt{5}/6 \approx 1.86$, and the foci are $\left(0, \pm 5\sqrt{5}/6\right)$.

EXAMPLE 3   **Finding an equation of an ellipse given its vertices and foci**

Find an equation of the ellipse with vertices $(\pm 4, 0)$ and foci $(\pm 2, 0)$.

SOLUTION   Since the foci are on the x-axis and are equidistant from the origin, the major axis is on the x-axis and the ellipse has center $(0, 0)$. Thus, a general equation of an ellipse is

$$\frac{x^2}{a^2} + \frac{y^2}{b^2} = 1.$$

Since the vertices are $(\pm 4, 0)$, we conclude that $a = 4$. Since the foci are $(\pm 2, 0)$, we have $c = 2$. Hence,

$$b^2 = a^2 - c^2 = 4^2 - 2^2 = 12,$$

and an equation of the ellipse is

$$\frac{x^2}{16} + \frac{y^2}{12} = 1.$$

In certain applications it is necessary to work with only one-half of an ellipse. The next example indicates how to find equations in such cases.

EXAMPLE 4   **Finding equations for half-ellipses**

Find equations for the upper half, lower half, left half, and right half of the ellipse $9x^2 + 4y^2 = 25$.

Figure 8

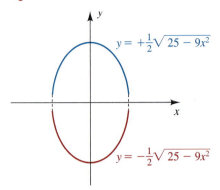

SOLUTION The graph of the entire ellipse was sketched in Figure 7. To find equations for the upper and lower halves, we solve for y in terms of x, as follows:

$$9x^2 + 4y^2 = 25 \qquad \text{given}$$

$$y^2 = \frac{25 - 9x^2}{4} \qquad \text{solve for } y^2$$

$$y = \pm \sqrt{\frac{25 - 9x^2}{4}} = \pm \frac{1}{2}\sqrt{25 - 9x^2} \quad \text{take the square root}$$

Since $\sqrt{25 - 9x^2} \geq 0$, it follows that equations for the upper and lower halves are $y = \frac{1}{2}\sqrt{25 - 9x^2}$ and $y = -\frac{1}{2}\sqrt{25 - 9x^2}$, respectively, as shown in Figure 8.

To find equations for the left and right halves, we use a procedure similar to that above and solve for x in terms of y, obtaining

$$x = \pm \sqrt{\frac{25 - 4y^2}{9}} = \pm \frac{1}{3}\sqrt{25 - 4y^2}.$$

The left half of the ellipse has the equation $x = -\frac{1}{3}\sqrt{25 - 4y^2}$, and the right half is given by $x = \frac{1}{3}\sqrt{25 - 4y^2}$, as shown in Figure 9.

Figure 9

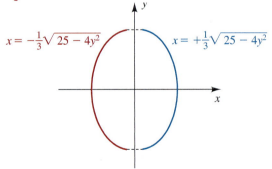

If we take a standard equation of an ellipse $(x^2/a^2 + y^2/b^2 = 1)$ and replace x with $x - h$ and y with $y - k$, then

$$\frac{x^2}{a^2} + \frac{y^2}{b^2} = 1 \quad \text{becomes} \quad \frac{(x - h)^2}{a^2} + \frac{(y - k)^2}{b^2} = 1. \qquad (*)$$

The graph of $(*)$ is an ellipse with center (h, k). Squaring terms in $(*)$ and simplifying gives us an equation of the form

$$Ax^2 + Cy^2 + Dx + Ey + F = 0,$$

where the coefficients are real numbers and both A and C are positive. Conversely, if we start with such an equation, then by completing squares we can obtain a form that helps give us the center of the ellipse and the lengths of the major and minor axes. This technique is illustrated in the next example.

EXAMPLE 5   Sketching an ellipse with center (h, k)

Discuss and sketch the graph of the equation

$$16x^2 + 9y^2 + 64x - 18y - 71 = 0.$$

SOLUTION   We begin by grouping the terms containing x and those containing y:

$$(16x^2 + 64x) + (9y^2 - 18y) = 71$$

Next, we factor out the coefficients of x^2 and y^2 as follows:

$$16(x^2 + 4x + \underline{}) + 9(y^2 - 2y + \underline{}) = 71$$

We now complete the squares for the expressions within parentheses:

$$16(x^2 + 4x + 4) + 9(y^2 - 2y + 1) = 71 + \underline{16 \cdot 4} + \underline{9 \cdot 1}$$

By adding 4 to the expression within the first parentheses we have added 64 to the left-hand side of the equation, and hence we must compensate by adding 64 to the right-hand side. Similarly, by adding 1 to the expression within the second parentheses we have added 9 to the left-hand side, and consequently we must also add 9 to the right-hand side. The last equation may be written

$$16(x + 2)^2 + 9(y - 1)^2 = 144.$$

Dividing by 144 to obtain 1 on the right-hand side gives us

$$\frac{(x + 2)^2}{9} + \frac{(y - 1)^2}{16} = 1.$$

The graph of the last equation is an ellipse with center $C(-2, 1)$ and major axis on the vertical line $x = -2$ (since $9 < 16$). Using $a = 4$ and $b = 3$ gives us the ellipse in Figure 10.

Figure 10

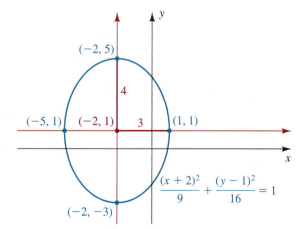

(continued)

To find the foci, we first calculate

$$c^2 = a^2 - b^2 = 4^2 - 3^2 = 7.$$

The distance from the center of the ellipse to the foci is $c = \sqrt{7}$. Since the center is $(-2, 1)$, the foci are $\left(-2, 1 \pm \sqrt{7}\right)$. ◼

Ellipses can be very flat or almost circular. To obtain information about the *roundness* of an ellipse, we sometimes use the term *eccentricity,* which is defined as follows, with a, b, and c having the same meanings as before.

Definition of Eccentricity	The **eccentricity** e of an ellipse is $$e = \frac{\text{distance from center to focus}}{\text{distance from center to vertex}} = \frac{c}{a} = \frac{\sqrt{a^2 - b^2}}{a}.$$

Consider the ellipse $(x^2/a^2) + (y^2/b^2) = 1$, and suppose that the length $2a$ of the major axis is fixed and the length $2b$ of the minor axis is variable (note that $0 < b < a$). Since b^2 is positive, $a^2 - b^2 < a^2$ and hence $\sqrt{a^2 - b^2} < a$. Dividing both sides of the last inequality by a gives us $\sqrt{a^2 - b^2}/a < 1$, or $0 < e < 1$. If b is close to 0 (c is close to a), then $\sqrt{a^2 - b^2} \approx a$, $e \approx 1$, and the ellipse is very flat. This case is illustrated in Figure 11(a), with $a = 2$, $b = 0.3$, and $e \approx 0.99$. If b is close to a (c is close to 0), then $\sqrt{a^2 - b^2} \approx 0$, $e \approx 0$, and the ellipse is almost circular. This case is illustrated in Figure 11(b), with $a = 2$, $b = 1.9999$, and $e \approx 0.01$.

Figure 11
(a) Eccentricity almost 1   **(b)** Eccentricity almost 0

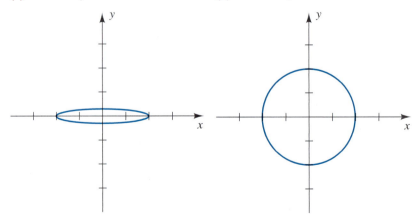

In Figure 11(a), the foci are close to the vertices.

In Figure 11(b), the foci are close to the origin.

Note that the ellipse in Figure 5 has eccentricity $\frac{5}{13} \approx 0.38$ and appears to be nearly circular.

After many years of analyzing an enormous amount of empirical data, the German astronomer Johannes Kepler (1571–1630) formulated three laws that describe the motion of planets about the sun. Kepler's first law states that the orbit of each planet in the solar system is an ellipse with the sun at one focus. Most of these orbits are almost circular, so their corresponding eccentricities are close to 0. To illustrate, for Earth, $e \approx 0.017$; for Mars, $e \approx 0.093$; and for Uranus, $e \approx 0.046$. The orbits of Mercury and Pluto are less circular, with eccentricities of 0.206 and 0.249, respectively.

Many comets have elliptical orbits with the sun at a focus. In this case the eccentricity e is close to 1, and the ellipse is very flat. In the next example we use the **astronomical unit** (AU)—that is, the average distance from Earth to the sun—to specify large distances (1 AU $\approx$ 93,000,000 mi).

EXAMPLE 6   Approximating a distance in an elliptical path

Halley's comet has an elliptical orbit with eccentricity $e = 0.967$. The closest that Halley's comet comes to the sun is 0.587 AU. Approximate the maximum distance of the comet from the sun, to the nearest 0.1 AU.

SOLUTION   Figure 12 illustrates the orbit of the comet, where c is the distance from the center of the ellipse to a focus (the sun) and $2a$ is the length of the major axis.

Figure 12

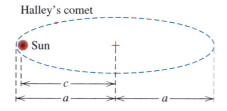

Since $a - c$ is the minimum distance between the sun and the comet, we have (in AU)

$$a - c = 0.587, \quad \text{or} \quad a = c + 0.587.$$

Since $e = c/a = 0.967$, we obtain the following:

$$
\begin{aligned}
c &= 0.967a & &\text{multiply by } a \\
&= 0.967(c + 0.587) & &\text{substitute for } a \\
&\approx 0.967c + 0.568 & &\text{multiply} \\
c - 0.967c &\approx 0.568 & &\text{subtract } 0.967c \\
c(1 - 0.967) &\approx 0.568 & &\text{factor out } c \\
c &\approx \frac{0.568}{0.033} \approx 17.2 & &\text{solve for } c
\end{aligned}
$$

Since $a = c + 0.587$, we obtain

$$a \approx 17.2 + 0.587 \approx 17.8,$$

and the maximum distance between the sun and the comet is

$$a + c \approx 17.8 + 17.2 = 35.0 \text{ AU.}$$

Figure 13

Figure 14

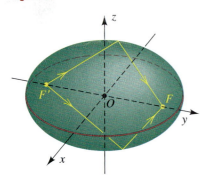

An ellipse has a *reflective property* analogous to that of the parabola discussed at the end of the previous section. To illustrate, let l denote the tangent line at a point P on an ellipse with foci F and F', as shown in Figure 13. If α is the acute angle between $F'P$ and l and if β is the acute angle between FP and l, it can be shown that $\alpha = \beta$. Thus, if a ray of light or sound emanates from one focus, it is reflected to the other focus. This property is used in the design of certain types of optical equipment.

If the ellipse with center O and foci F' and F on the x-axis is revolved about the x-axis, as illustrated in Figure 14, we obtain a three-dimensional surface called an **ellipsoid**. The upper half or lower half is a **hemi-ellipsoid**, as is the right half or left half. Sound waves or other impulses that are emitted from the focus F' will be reflected off the ellipsoid into the focus F. This property is used in the design of *whispering galleries*—structures with ellipsoidal ceilings, in which a person who whispers at one focus can be heard at the other focus. Examples of whispering galleries may be found in the Rotunda of the Capitol Building in Washington, D.C., and in the Mormon Tabernacle in Salt Lake City.

The reflective property of ellipsoids (and hemi-ellipsoids) is used in modern medicine in a device called a *lithotripter,* which disintegrates kidney stones by means of high-energy underwater shock waves. After taking extremely accurate measurements, the operator positions the patient so that the stone is at a focus. Ultra–high frequency shock waves are then produced at the other focus, and reflected waves break up the kidney stone. Recovery time with this technique is usually 3–4 days, instead of the 2–3 weeks with conventional surgery. Moreover, the mortality rate is less than 0.01%, as compared to 2–3% for traditional surgery (see Exercises 51–52).

11.2 Exercises

Exer. 1–14: Find the vertices and foci of the ellipse. Sketch its graph, showing the foci.

1 $\dfrac{x^2}{9} + \dfrac{y^2}{4} = 1$

2 $\dfrac{x^2}{25} + \dfrac{y^2}{16} = 1$

3 $\dfrac{x^2}{15} + \dfrac{y^2}{16} = 1$

4 $\dfrac{x^2}{45} + \dfrac{y^2}{49} = 1$

5 $4x^2 + y^2 = 16$

6 $y^2 + 9x^2 = 9$

7 $4x^2 + 25y^2 = 1$

8 $10y^2 + x^2 = 5$

9 $\dfrac{(x-3)^2}{16} + \dfrac{(y+4)^2}{9} = 1$

10 $\dfrac{(x+2)^2}{25} + \dfrac{(y-3)^2}{4} = 1$

11 $4x^2 + 9y^2 - 32x - 36y + 64 = 0$

12 $x^2 + 2y^2 + 2x - 20y + 43 = 0$

13 $25x^2 + 4y^2 - 250x - 16y + 541 = 0$

14 $4x^2 + y^2 = 2y$

Exer. 15–18: Find an equation for the ellipse shown in the figure.

15

16

17

18

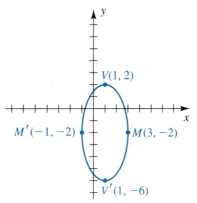

Exer. 19–30: Find an equation for the ellipse that has its center at the origin and satisfies the given conditions.

19 Vertices $V(\pm8, 0)$,   foci $F(\pm5, 0)$

20 Vertices $V(0, \pm7)$,   foci $F(0, \pm2)$

21 Vertices $V(0, \pm5)$,   minor axis of length 3

22 Foci $F(\pm3, 0)$,   minor axis of length 2

23 Vertices $V(0, \pm6)$,   passing through $(3, 2)$

24 Passing through $(2, 3)$ and $(6, 1)$

25 Eccentricity $\frac{3}{4}$,   vertices $V(0, \pm4)$

26 Eccentricity $\frac{1}{2}$,   vertices on the x-axis, passing through $(1, 3)$

27 x-intercepts ±2, y-intercepts $\pm\frac{1}{3}$

28 x-intercepts $\pm\frac{1}{2}$, y-intercepts ±4

29 Horizontal major axis of length 8, minor axis of length 5

30 Vertical major axis of length 7, minor axis of length 6

Exer. 31–32: Find the points of intersection of the graphs of the equations. Sketch both graphs on the same coordinate plane, and show the points of intersection.

31 $\begin{cases} x^2 + 4y^2 = 20 \\ x + 2y = 6 \end{cases}$ **32** $\begin{cases} x^2 + 4y^2 = 36 \\ x^2 + y^2 = 12 \end{cases}$

Exer. 33–36: Find an equation for the set of points in an xy-plane such that the sum of the distances from F and F' is k.

33 $F(3, 0)$, $F'(-3, 0)$; $k = 10$

34 $F(12, 0)$, $F'(-12, 0)$; $k = 26$

35 $F(0, 15)$, $F'(0, -15)$; $k = 34$

36 $F(0, 8)$, $F'(0, -8)$; $k = 20$

Exer. 37–38: Find an equation for the ellipse with foci F and F' that passes through P. Sketch the ellipse.

37   **38**

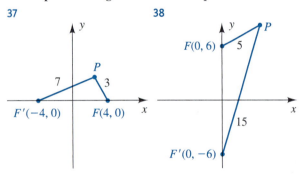

Exer. 39–46: Determine whether the graph of the equation is the upper, lower, left, or right half of an ellipse, and find an equation for the ellipse.

39 $y = 11\sqrt{1 - \dfrac{x^2}{49}}$ **40** $y = -6\sqrt{1 - \dfrac{x^2}{25}}$

41 $x = -\dfrac{1}{3}\sqrt{9 - y^2}$ **42** $x = \dfrac{4}{5}\sqrt{25 - y^2}$

43 $x = 1 + 2\sqrt{1 - \dfrac{(y + 2)^2}{9}}$

44 $x = -2 - 5\sqrt{1 - \dfrac{(y - 1)^2}{16}}$

45 $y = 2 - 7\sqrt{1 - \dfrac{(x + 1)^2}{9}}$

46 $y = -1 + \sqrt{1 - \dfrac{(x - 3)^2}{16}}$

47 Dimensions of an arch An arch of a bridge is semi-elliptical, with major axis horizontal. The base of the arch is 30 feet across, and the highest part of the arch is 10 feet above the horizontal roadway, as shown in the figure. Find the height of the arch 6 feet from the center of the base.

Exercise 47

48 Designing a bridge A bridge is to be constructed across a river that is 200 feet wide. The arch of the bridge is to be semielliptical and must be constructed so that a ship less than 50 feet wide and 30 feet high can pass safely through the arch, as shown in the figure.

(a) Find an equation for the arch.

(b) Approximate the height of the arch in the middle of the bridge.

Exercise 48

49 Earth's orbit Assume that the length of the major axis of Earth's orbit is 186,000,000 miles and that the eccentricity is 0.017. Approximate, to the nearest 1000 miles, the maximum and minimum distances between Earth and the sun.

50 Mercury's orbit The planet Mercury travels in an elliptical orbit that has eccentricity 0.206 and major axis of length 0.774 AU. Find the maximum and minimum distances between Mercury and the sun.

51 Elliptical reflector The basic shape of an elliptical reflector is a hemi-ellipsoid of height h and diameter k, as shown in the figure. Waves emitted from focus F will reflect off the surface into focus F'.

(a) Express the distances $d(V, F)$ and $d(V, F')$ in terms of h and k.

(b) An elliptical reflector of height 17 centimeters is to be constructed so that waves emitted from F are reflected to a point F' that is 32 centimeters from V. Find the diameter of the reflector and the location of F.

Exercise 51

52 Lithotripter operation A lithotripter of height 15 centimeters and diameter 18 centimeters is to be constructed (see the figure). High-energy underwater shock waves will be emitted from the focus F that is closest to the vertex V.

(a) Find the distance from V to F.

(b) How far from V (in the vertical direction) should a kidney stone be located?

Exercise 52

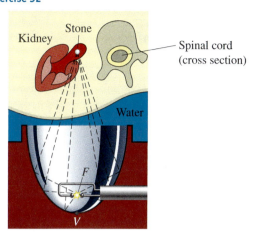

53 Whispering gallery The ceiling of a whispering gallery has the shape of the hemi-ellipsoid shown in Figure 14, with the highest point of the ceiling 15 feet above the elliptical floor and the vertices of the floor 50 feet apart. If two people are standing at the foci F' and F, how far from the vertices are their feet?

54 An ellipse has a vertex at the origin and foci $F_1(p, 0)$ and $F_2(p + 2c, 0)$, as shown in the figure. If the focus at F_1 is fixed and (x, y) is on the ellipse, show that y^2 approaches $4px$ as $c \to \infty$. (Thus, as $c \to \infty$, the ellipse takes on the shape of a parabola.)

Exercise 54

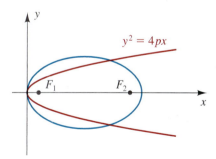

11.3

Hyperbolas

The definition of a hyperbola is similar to that of an ellipse. The only change is that instead of using the *sum* of distances from two fixed points, we use the *difference*.

Definition of a Hyperbola	A **hyperbola** is the set of all points in a plane, the difference of whose distances from two fixed points (the **foci**) in the plane is a positive constant.

Figure 1

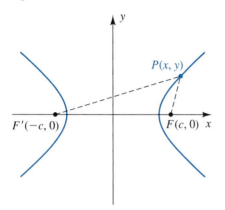

To find a simple equation for a hyperbola, we choose a coordinate system with foci at $F(c, 0)$ and $F'(-c, 0)$ and denote the (constant) distance by $2a$. The midpoint of the segment $F'F$ (the origin) is called the **center** of the hyperbola. Referring to Figure 1, we see that a point $P(x, y)$ is on the hyperbola if and only if either of the following is true:

$$(1) \ d(P, F') - d(P, F) = 2a \quad \text{or} \quad (2) \ d(P, F) - d(P, F') = 2a$$

If P is not on the x-axis, then from Figure 1 we see that

$$d(P, F) < d(F', F) + d(P, F'),$$

because the length of one side of a triangle is always less than the sum of the lengths of the other two sides. Similarly,

$$d(P, F') < d(F', F) + d(P, F).$$

Equivalent forms for the previous two inequalities are

$$d(P, F) - d(P, F') < d(F', F) \quad \text{and} \quad d(P, F') - d(P, F) < d(F', F).$$

Since the differences on the left-hand sides of these inequalities both equal $2a$ and since $d(F', F) = 2c$, the last two inequalities imply that $2a < 2c$, or $a < c$. (Recall that for ellipses we had $a > c$.)

Next, equations (1) and (2) may be replaced by the single equation

$$|d(P, F) - d(P, F')| = 2a.$$

Using the distance formula to find $d(P, F)$ and $d(P, F')$, we obtain an equation of the hyperbola:

$$\left| \sqrt{(x - c)^2 + (y - 0)^2} - \sqrt{(x + c)^2 + (y - 0)^2} \right| = 2a$$

Employing the type of simplification procedure that we used to derive an equation for an ellipse, we can rewrite the preceding equation as

$$\frac{x^2}{a^2} - \frac{y^2}{c^2 - a^2} = 1.$$

Finally, if we let

$$b^2 = c^2 - a^2 \quad \text{with} \quad b > 0$$

in the preceding equation, we obtain

$$\frac{x^2}{a^2} - \frac{y^2}{b^2} = 1.$$

We have shown that the coordinates of every point (x, y) on the hyperbola in Figure 1 satisfy the equation $(x^2/a^2) - (y^2/b^2) = 1$. Conversely, if (x, y) is a solution of this equation, then by reversing steps we see that the point (x, y) is on the hyperbola.

Applying tests for symmetry, we see that the hyperbola is symmetric with respect to both axes and the origin. We may find the x-intercepts of the hyperbola by letting $y = 0$ in the equation. Doing so gives us $x^2/a^2 = 1$, or $x^2 = a^2$, and consequently the x-intercepts are a and $-a$. The corresponding points $V(a, 0)$ and $V'(-a, 0)$ on the graph are called the **vertices** of the hyperbola (see Figure 2). The line segment $V'V$ is called the **transverse axis.** The graph has no y-intercept, since the equation $-y^2/b^2 = 1$ has the *complex* solutions $y = \pm bi$. The points $W(0, b)$ and $W'(0, -b)$ are endpoints of the **conjugate axis** $W'W$. The points W and W' are not on the hyperbola; however, as we shall see, they are useful for describing the graph.

Figure 2

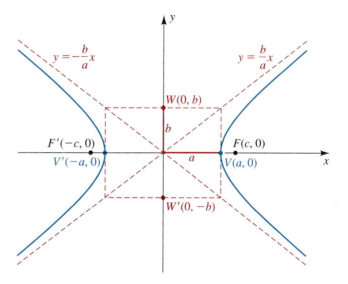

Solving the equation $(x^2/a^2) - (y^2/b^2) = 1$ for y gives us

$$y = \pm \frac{b}{a} \sqrt{x^2 - a^2}.$$

If $x^2 - a^2 < 0$ or, equivalently, $-a < x < a$, then there are no points (x, y) on the graph. There *are* points $P(x, y)$ on the graph if $x \geq a$ or $x \leq -a$.

It can be shown that *the lines* $y = \pm (b/a)x$ *are asymptotes for the hyperbola.* These asymptotes serve as excellent guides for sketching the graph. A convenient way to sketch the asymptotes is to first plot the vertices $V(a, 0)$, $V'(-a, 0)$ and the points $W(0, b)$, $W'(0, -b)$ (see Figure 2). If vertical and horizontal lines are drawn through these endpoints of the transverse and conjugate axes, respectively, then the diagonals of the resulting **auxiliary rectangle** have slopes b/a and $-b/a$. Hence, by extending these diagonals we obtain the asymptotes $y = \pm(b/a)x$. The hyperbola is then sketched as in Figure 2, using the asymptotes as guides. The two parts that make up the hyperbola are called the **right branch** and the **left branch** of the hyperbola.

Similarly, if we take the foci on the y-axis, we obtain the equation

$$\frac{y^2}{a^2} - \frac{x^2}{b^2} = 1.$$

In this case, the vertices of the hyperbola are $(0, \pm a)$ and the endpoints of the conjugate axis are $(\pm b, 0)$, as shown in Figure 3. The asymptotes are $y = \pm(a/b)x$ (*not* $y = \pm(b/a)x$, as in the previous case), and we now refer to the two parts that make up the hyperbola as the **upper branch** and the **lower branch** of the hyperbola.

Figure 3

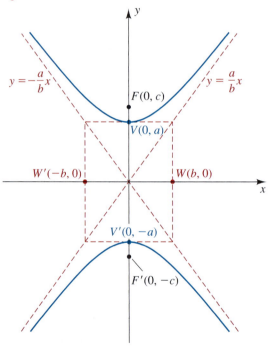

The preceding discussion may be summarized as follows.

Standard Equations of a Hyperbola with Center at the Origin	The graph of

$$\frac{x^2}{a^2} - \frac{y^2}{b^2} = 1 \qquad \text{or} \qquad \frac{y^2}{a^2} - \frac{x^2}{b^2} = 1$$

is a hyperbola with center at the origin. The length of the transverse axis is $2a$, and the length of the conjugate axis is $2b$. The foci are a distance c from the origin, where $c^2 = a^2 + b^2$.

We have shown that an equation of a hyperbola with center at the origin and foci on a coordinate axis can always be written in the form

$$\frac{x^2}{p} + \frac{y^2}{q} = 1, \quad \text{or} \quad qx^2 + py^2 = pq,$$

where p and q have opposite signs. The vertices are on the x-axis if p is positive or on the y-axis if q is positive.

EXAMPLE 1 Sketching a hyperbola with center at the origin

Sketch the graph of $9x^2 - 4y^2 = 36$. Find the foci and equations of the asymptotes.

Figure 4

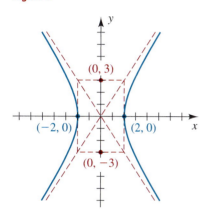

SOLUTION From the remarks preceding this example, the graph is a hyperbola with center at the origin. To express the given equation in a standard form, we divide both sides by 36 and simplify, obtaining

$$\frac{x^2}{4} - \frac{y^2}{9} = 1.$$

Comparing $(x^2/4) - (y^2/9) = 1$ to $(x^2/a^2) - (y^2/b^2) = 1$, we see that $a^2 = 4$ and $b^2 = 9$; that is, $a = 2$ and $b = 3$. The hyperbola has its vertices on the x-axis, since there are x-intercepts and no y-intercepts. The vertices $(\pm 2, 0)$ and the endpoints $(0, \pm 3)$ of the conjugate axis determine the auxiliary rectangle whose diagonals (extended) give us the asymptotes. The graph of the equation is sketched in Figure 4.

To find the foci, we calculate

$$c^2 = a^2 + b^2 = 4 + 9 = 13.$$

Thus, $c = \sqrt{13}$, and the foci are $(\pm\sqrt{13}, 0)$.

The equations of the asymptotes, $y = \pm\frac{3}{2}x$, can be found by referring to the graph or to the equations $y = \pm(b/a)x$. ◣

The preceding example indicates that for hyperbolas it is not always true that $a > b$, as is the case for ellipses. In fact, we may have $a < b$, $a > b$, or $a = b$.

EXAMPLE 2 Sketching a hyperbola with center at the origin

Sketch the graph of $4y^2 - 2x^2 = 1$. Find the foci and equations of the asymptotes.

SOLUTION To express the given equation in a standard form, we write

$$\frac{y^2}{\frac{1}{4}} - \frac{x^2}{\frac{1}{2}} = 1.$$

Thus,

$$a^2 = \tfrac{1}{4}, \qquad b^2 = \tfrac{1}{2}, \qquad \text{and} \qquad c^2 = a^2 + b^2 = \tfrac{3}{4},$$

(continued)

Figure 5

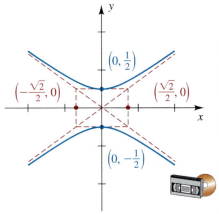

and consequently

$$a = \frac{1}{2}, \qquad b = \frac{1}{\sqrt{2}} = \frac{\sqrt{2}}{2}, \qquad \text{and} \qquad c = \frac{\sqrt{3}}{2}.$$

The hyperbola has its vertices on the y-axis, since there are y-intercepts and no x-intercepts. The vertices are $\left(0, \pm\frac{1}{2}\right)$, the endpoints of the conjugate axes are $\left(\pm\sqrt{2}/2, 0\right)$, and the foci are $\left(0, \pm\sqrt{3}/2\right)$. The graph is sketched in Figure 5.

To find the equations of the asymptotes, we refer to the figure or use $y = \pm(a/b)x$, obtaining $y = \pm\left(\sqrt{2}/2\right)x$. ◢

EXAMPLE 3   **Finding an equation of a hyperbola satisfying prescribed conditions**

A hyperbola has vertices $(\pm3, 0)$ and passes through the point $P(5, 2)$. Find its equation, foci, and asymptotes.

Figure 6

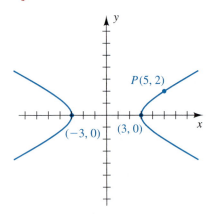

SOLUTION   We begin by sketching a hyperbola with vertices $(\pm3, 0)$ that passes through the point $P(5, 2)$, as in Figure 6.

An equation of the hyperbola has the form

$$\frac{x^2}{3^2} - \frac{y^2}{b^2} = 1.$$

Since $P(5, 2)$ is on the hyperbola, the x- and y-coordinates satisfy this equation; that is,

$$\frac{5^2}{3^2} - \frac{2^2}{b^2} = 1.$$

Solving for b^2 gives us $b^2 = \frac{9}{4}$, and hence an equation for the hyperbola is

$$\frac{x^2}{9} - \frac{y^2}{\frac{9}{4}} = 1$$

or, equivalently,

$$x^2 - 4y^2 = 9.$$

To find the foci, we first calculate

$$c^2 = a^2 + b^2 = 9 + \frac{9}{4} = \frac{45}{4}.$$

Hence, $c = \sqrt{\frac{45}{4}} = \frac{3}{2}\sqrt{5} \approx 3.35$, and the foci are $\left(\pm\frac{3}{2}\sqrt{5}, 0\right)$.

The general equations of the asymptotes are $y = \pm(b/a)x$. Substituting $a = 3$ and $b = \frac{3}{2}$ gives us $y = \pm\frac{1}{2}x$. ◢

The next example indicates how to find equations for certain parts of a hyperbola.

EXAMPLE 4   Finding equations of portions of a hyperbola

The hyperbola $9x^2 - 4y^2 = 36$ was discussed in Example 1. Solve the equation as indicated, and describe the resulting graph.

(a) For x in terms of y **(b)** For y in terms of x

SOLUTION

(a) We solve for x in terms of y as follows:

$$9x^2 - 4y^2 = 36 \qquad \text{given}$$
$$x^2 = \frac{36 + 4y^2}{9} \qquad \text{solve for } x^2$$
$$x = \pm\tfrac{2}{3}\sqrt{9 + y^2} \qquad \text{factor out 4, and take the square root}$$

The graph of the equation $x = \tfrac{2}{3}\sqrt{9 + y^2}$ is the right branch of the hyperbola sketched in Figure 4 (and repeated in Figure 7), and the graph of $x = -\tfrac{2}{3}\sqrt{9 + y^2}$ is the left branch.

(b) We solve for y in terms of x as follows:

$$9x^2 - 4y^2 = 36 \qquad \text{given}$$
$$y^2 = \frac{9x^2 - 36}{4} \qquad \text{solve for } y^2$$
$$y = \pm\tfrac{3}{2}\sqrt{x^2 - 4} \qquad \text{factor out 9, and take the square root}$$

The graph of $y = \tfrac{3}{2}\sqrt{x^2 - 4}$ is the upper half of the right and left branches, and the graph of $y = -\tfrac{3}{2}\sqrt{x^2 - 4}$ is the lower half of these branches.

As was the case for ellipses, we may use translations to help sketch hyperbolas that have centers at some point $(h, k) \neq (0, 0)$. The following example illustrates this technique.

EXAMPLE 5   Sketching a hyperbola with center (h, k)

Discuss and sketch the graph of the equation

$$9x^2 - 4y^2 - 54x - 16y + 29 = 0.$$

SOLUTION   We arrange our work using a procedure similar to that used for ellipses in Example 5 of the previous section:

$$(9x^2 - 54x) + (-4y^2 - 16y) = -29 \qquad \text{group terms}$$
$$9(x^2 - 6x + \underline{}) - 4(y^2 + 4y + \underline{}) = -29 \qquad \text{factor out 9 and } -4$$
$$9(x^2 - 6x + 9) - 4(y^2 + 4y + 4) = -29 + \underline{9 \cdot 9} - \underline{4 \cdot 4}$$
$$\text{complete the squares}$$
$$9(x - 3)^2 - 4(y + 2)^2 = 36 \qquad \text{factor, and simplify}$$
$$\frac{(x - 3)^2}{4} - \frac{(y + 2)^2}{9} = 1 \qquad \text{divide by 36}$$

(continued)

Figure 7

(0, 3)

(−2, 0)   (2, 0)

(0, −3)

Figure 8

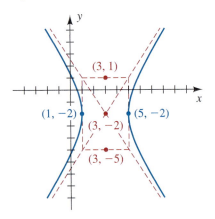

The last equation indicates that the hyperbola has center $C(3, -2)$ with vertices and foci on the horizontal line $y = -2$, because the term containing x is positive. We also know that

$$a^2 = 4, \qquad b^2 = 9, \qquad \text{and} \qquad c^2 = a^2 + b^2 = 13.$$

Hence,

$$a = 2, \qquad b = 3, \qquad \text{and} \qquad c = \sqrt{13}.$$

As illustrated in Figure 8, the vertices are $(3 \pm 2, -2)$—that is, $(5, -2)$ and $(1, -2)$. The endpoints of the conjugate axis are $(3, -2 \pm 3)$—that is, $(3, 1)$ and $(3, -5)$. The foci are $\left(3 \pm \sqrt{13}, -2\right)$, and equations of the asymptotes are

$$y + 2 = \pm\tfrac{3}{2}(x - 3). \qquad \qquad ◪$$

The results of Sections 11.1 through 11.3 indicate that the graph of every equation of the form

$$Ax^2 + Cy^2 + Dx + Ey + F = 0$$

is a conic, except for certain degenerate cases in which a point, one or two lines, or no graph is obtained. Although we have considered only special examples, our methods can be applied to any such equation. If A and C are equal and not 0, then the graph, when it exists, is a circle or, in exceptional cases, a point. If A and C are unequal but have the same sign, an equation is obtained whose graph, when it exists, is an ellipse (or a point). If A and C have opposite signs, an equation of a hyperbola is obtained or possibly, in the degenerate case, two intersecting straight lines. If either A or C (but not both) is 0, the graph is a parabola or, in certain cases, a pair of parallel lines.

We shall conclude this section with an application involving hyperbolas.

EXAMPLE 6 Locating a ship

Coast Guard station A is 200 miles directly east of another station B. A ship is sailing on a line parallel to and 50 miles north of the line through A and B. Radio signals are sent out from A and B at the rate of 980 ft/μsec (microsecond). If, at 1:00 P.M., the signal from B reaches the ship 400 microseconds after the signal from A, locate the position of the ship at that time.

Figure 9
(a)

(b)

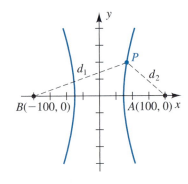

SOLUTION Let us introduce a coordinate system, as shown in Figure 9(a), with the stations at points A and B on the x-axis and the ship at P on the line $y = 50$. Since at 1:00 P.M. it takes 400 microseconds longer for the signal to arrive from B than from A, the difference $d_1 - d_2$ in the indicated distances at that time is

$$d_1 - d_2 = (980)(400) = 392{,}000 \text{ ft.}$$

Dividing by 5280 (ft/mi) gives us

$$d_1 - d_2 = \frac{392{,}000}{5280} = 74.\overline{24} \text{ mi.}$$

At 1:00 P.M., point P is on the right branch of a hyperbola whose equation is $(x^2/a^2) - (y^2/b^2) = 1$ (see Figure 9(b)), consisting of all points whose difference in distances from the foci B and A is $d_1 - d_2$. In our derivation of the equation $(x^2/a^2) - (y^2/b^2) = 1$, we let $d_1 - d_2 = 2a$; it follows that in the present situation

$$a = \frac{74.\overline{24}}{2} = 37.\overline{12} \qquad \text{and} \qquad a^2 \approx 1378.$$

Since the distance c from the origin to either focus is 100,

$$b^2 = c^2 - a^2 \approx 10{,}000 - 1378, \qquad \text{or} \qquad b^2 \approx 8622.$$

Hence, an (approximate) equation for the hyperbola that has foci A and B and passes through P is

$$\frac{x^2}{1378} - \frac{y^2}{8622} = 1.$$

If we let $y = 50$ (the y-coordinate of P), we obtain

$$\frac{x^2}{1378} - \frac{2500}{8622} = 1.$$

Solving for x gives us $x \approx 42.16$. Rounding off to the nearest mile, we find that the coordinates of P are approximately $(42, 50)$. ◢

Figure 10

An extension of the method used in Example 6 is the basis for the navigational system LORAN (for Long Range Navigation). This system involves two pairs of radio transmitters, such as those located at T, T' and S, S' in Figure 10. Suppose that signals sent out by the transmitters at T and T' reach a radio receiver in a ship located at some point P. The difference in the times of arrival of the signals can be used to determine the difference in the distances of P from T and T'. Thus, P lies on one branch of a hyperbola with foci at T and T'. Repeating this process for the other pair of transmitters, we see that P also lies on one branch of a hyperbola with foci at S and S'. The intersection of these two branches determines the position of P.

A hyperbola has a *reflective property* analogous to that of the ellipse discussed in the previous section. To illustrate, let l denote the tangent line at a point P on a hyperbola with foci F and F', as shown in Figure 11. If α is the acute angle between $F'P$ and l and if β is the acute angle between FP and l, it can be shown that $\alpha = \beta$. If a ray of light is directed along the line l_1 toward F, it will be reflected back at P along the line l_2 toward F'. This property is used in the design of telescopes of the Cassegrain type (see Exercise 64).

Figure 11

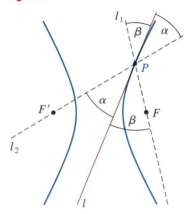

11.3   Exercises

Exer. 1–16: Find the vertices, the foci, and the equations of the asymptotes of the hyperbola. Sketch its graph, showing the asymptotes and the foci.

1 $\dfrac{x^2}{9} - \dfrac{y^2}{4} = 1$

2 $\dfrac{y^2}{49} - \dfrac{x^2}{16} = 1$

3 $\dfrac{y^2}{9} - \dfrac{x^2}{4} = 1$

4 $\dfrac{x^2}{49} - \dfrac{y^2}{16} = 1$

5 $x^2 - \dfrac{y^2}{24} = 1$

6 $y^2 - \dfrac{x^2}{15} = 1$

7 $y^2 - 4x^2 = 16$

8 $x^2 - 2y^2 = 8$

9 $16x^2 - 36y^2 = 1$

10 $y^2 - 16x^2 = 1$

11 $\dfrac{(y + 2)^2}{9} - \dfrac{(x + 2)^2}{4} = 1$

12 $\dfrac{(x - 3)^2}{25} - \dfrac{(y - 1)^2}{4} = 1$

13 $144x^2 - 25y^2 + 864x - 100y - 2404 = 0$

14 $y^2 - 4x^2 - 12y - 16x + 16 = 0$

15 $4y^2 - x^2 + 40y - 4x + 60 = 0$

16 $25x^2 - 9y^2 + 100x - 54y + 10 = 0$

Exer. 17–20: Find an equation for the hyperbola shown in the figure.

17

18

19

20

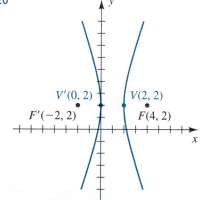

Exer. 21–32: Find an equation for the hyperbola that has its center at the origin and satisfies the given conditions.

21 Foci $F(0, \pm 4)$,   vertices $V(0, \pm 1)$

22 Foci $F(\pm 8, 0)$,   vertices $V(\pm 5, 0)$

23 Foci $F(\pm 5, 0)$,   vertices $V(\pm 3, 0)$

24 Foci $F(0, \pm 3)$,   vertices $V(0, \pm 2)$

25 Foci $F(0, \pm 5)$,   conjugate axis of length 4

26 Vertices $V(\pm 4, 0)$,   passing through $(8, 2)$

27 Vertices $V(\pm 3, 0)$,   asymptotes $y = \pm 2x$

28 Foci $F(0, \pm 10)$,   asymptotes $y = \pm \frac{1}{3}x$

29 x-intercepts ± 5,   asymptotes $y = \pm 2x$

30 y-intercepts ± 2,   asymptotes $y = \pm \frac{1}{4}x$

31 Vertical transverse axis of length 10, conjugate axis of length 14

32 Horizontal transverse axis of length 6, conjugate axis of length 2

Exer. 33–42: Identify the graph of the equation as a parabola (with vertical or horizontal axis), circle, ellipse, or hyperbola.

33 $\frac{1}{3}(x + 2) = y^2$

34 $y^2 = \frac{14}{3} - x^2$

35 $x^2 + 6x - y^2 = 7$

36 $x^2 + 4x + 4y^2 - 24y = -36$

37 $-x^2 = y^2 - 25$

38 $x = 2x^2 - y + 4$

39 $4x^2 - 16x + 9y^2 + 36y = -16$

40 $x + 4 = y^2 + y$

41 $x^2 + 3x = 3y - 6$

42 $9x^2 - y^2 = 10 - 2y$

Exer. 43–44: Find the points of intersection of the graphs of the equations. Sketch both graphs on the same coordinate plane, and show the points of intersection.

43 $\begin{cases} y^2 - 4x^2 = 16 \\ y - x = 4 \end{cases}$

44 $\begin{cases} x^2 - y^2 = 4 \\ y^2 - 3x = 0 \end{cases}$

Exer. 45–48: Find an equation for the set of points in an *xy*-plane such that the difference of the distances from *F* and *F'* is *k*.

45 $F(13, 0)$, $F'(-13, 0)$; $k = 24$

46 $F(5, 0)$, $F'(-5, 0)$; $k = 8$

47 $F(0, 10)$, $F'(0, -10)$; $k = 16$

48 $F(0, 17)$, $F'(0, -17)$; $k = 30$

Exer. 49–50: Find an equation for the hyperbola with foci *F* and *F'* that passes through *P*. Sketch the hyperbola.

49 50

Exer. 51–58: Describe the part of a hyperbola given by the equation.

51 $x = \frac{5}{4}\sqrt{y^2 + 16}$

52 $x = -\frac{5}{4}\sqrt{y^2 + 16}$

53 $y = \frac{3}{7}\sqrt{x^2 + 49}$

54 $y = -\frac{3}{7}\sqrt{x^2 + 49}$

55 $y = -\frac{9}{4}\sqrt{x^2 - 16}$

56 $y = \frac{9}{4}\sqrt{x^2 - 16}$

57 $x = -\frac{2}{3}\sqrt{y^2 - 36}$

58 $x = \frac{2}{3}\sqrt{y^2 - 36}$

59 The graphs of the equations

$$\frac{x^2}{a^2} - \frac{y^2}{b^2} = 1 \quad \text{and} \quad \frac{x^2}{a^2} - \frac{y^2}{b^2} = -1$$

are called *conjugate hyperbolas*. Sketch the graphs of both equations on the same coordinate plane, with $a = 5$ and $b = 3$, and describe the relationship between the two graphs.

60 Find an equation of the hyperbola with foci $(h \pm c, k)$ and vertices $(h \pm a, k)$, where

$$0 < a < c \quad \text{and} \quad c^2 = a^2 + b^2.$$

61 **Alpha particles** In 1911, the physicist Ernest Rutherford (1871–1937) discovered that if alpha particles are shot toward the nucleus of an atom, they are eventually repulsed away from the nucleus along hyperbolic paths. The figure illustrates the path of a particle that starts toward the origin along the line $y = \frac{1}{2}x$ and comes within 3 units of the nucleus. Find an equation of the path.

Exercise 61

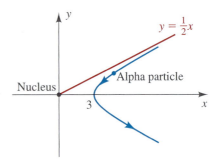

62 **Airplane maneuver** An airplane is flying along the hyperbolic path illustrated in the figure. If an equation of the path is $2y^2 - x^2 = 8$, determine how close the airplane comes to a town located at $(3, 0)$. (*Hint:* Let *S* denote the square of the distance from a point (x, y) on the path to $(3, 0)$, and find the minimum value of *S*.)

Exercise 62

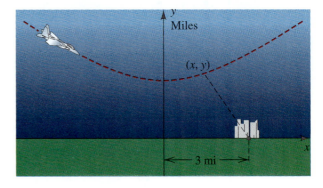

63 Locating a ship A ship is traveling a course that is 100 miles from, and parallel to, a straight shoreline. The ship sends out a distress signal that is received by two Coast Guard stations A and B, located 200 miles apart, as shown in the figure. By measuring the difference in signal reception times, it is determined that the ship is 160 miles closer to B than to A. Where is the ship?

Exercise 63

64 Design of a telescope The Cassegrain telescope design (dating back to 1672) makes use of the reflective properties of both the parabola and the hyperbola. Shown in the figure is a (split) parabolic mirror, with focus at F_1 and axis along the line l, and a hyperbolic mirror, with one focus also at F_1 and transverse axis along l. Where do incoming light waves parallel to the common axis finally collect?

Exercise 64

65 Comet's path Comets can travel in elliptical, parabolic, or hyperbolic paths around the sun. If a comet travels in a parabolic or hyperbolic path, it will pass by the sun once and never return. Suppose that a comet's coordinates in miles can be described by the equation

$$\frac{x^2}{26 \times 10^{14}} - \frac{y^2}{18 \times 10^{14}} = 1 \quad \text{for} \quad x > 0,$$

where the sun is located at a focus, as shown in the figure.

(a) Approximate the coordinates of the sun.

(b) For the comet to maintain a hyperbolic trajectory, the minimum velocity v of the comet, in meters per second, must satisfy $v > \sqrt{2k/r}$, where r is the distance between the comet and the center of the sun in meters and $k = 1.325 \times 10^{20}$ is a constant. Determine v when r is minimum.

Exercise 65

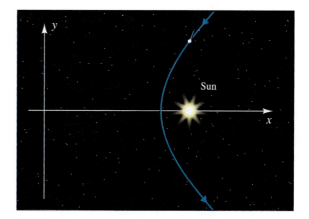

11.4

Plane Curves and Parametric Equations

If f is a function, the graph of the equation $y = f(x)$ is often called a *plane curve*. However, this definition is restrictive, because it excludes many useful graphs. The following definition is more general.

Definition of Plane Curve	A **plane curve** is a set C of ordered pairs $(f(t), g(t))$, where f and g are functions defined on an interval I.

For simplicity, we often refer to a plane curve as a **curve.** The **graph** of C in the preceding definition consists of all points $P(t) = (f(t), g(t))$ in an xy-plane, for t in I. We shall use the term *curve* interchangeably with *graph of a curve.* We sometimes regard the point $P(t)$ as tracing the curve C as t varies through the interval I.

The graphs of several curves are sketched in Figure 1, where I is a closed interval $[a, b]$—that is, $a \le t \le b$. In part (a) of the figure, $P(a) \ne P(b)$, and $P(a)$ and $P(b)$ are called the **endpoints** of C. The curve in (a) intersects itself; that is, two different values of t produce the same point. If $P(a) = P(b)$, as in Figure 1(b), then C is a **closed curve.** If $P(a) = P(b)$ and C does not intersect itself at any other point, as in Figure 1(c), then C is a **simple closed curve.**

Figure 1

(a) Curve

(b) Closed curve

(c) Simple closed curve

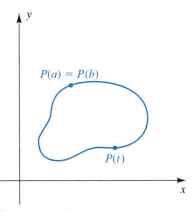

A convenient way to represent curves is given in the next definition.

Definition of Parametric Equations	Let C be the curve consisting of all ordered pairs $(f(t), g(t))$, where f and g are defined on an interval I. The equations $$x = f(t), \quad y = g(t),$$ for t in I, are **parametric equations** for C with **parameter** t.

The curve C in this definition is referred to as a **parametrized curve,** and the parametric equations are a **parametrization** for C. We often use the notation

$$x = f(t), \quad y = g(t); \quad t \text{ in } I$$

to indicate the domain I of f and g. We can refer to these equations as the **x-equation** and the **y-equation.**

Sometimes it may be possible to eliminate the parameter and obtain a familiar equation in x and y for C. In simple cases we can sketch a graph of a parametrized curve by plotting points and connecting them in order of increasing t, as illustrated in the next example.

EXAMPLE 1   Sketching the graph of a parametrized curve

Sketch the graph of the curve C that has the parametrization

$$x = 2t, \quad y = t^2 - 1; \quad -1 \le t \le 2.$$

SOLUTION   We use the parametric equations to tabulate coordinates of points $P(x, y)$ on C, as follows.

t	-1	$-\frac{1}{2}$	0	$\frac{1}{2}$	1	$\frac{3}{2}$	2
x	-2	-1	0	1	2	3	4
y	0	$-\frac{3}{4}$	-1	$-\frac{3}{4}$	0	$\frac{5}{4}$	3

Figure 2
$x = 2t, y = t^2 - 1; -1 \le t \le 2$

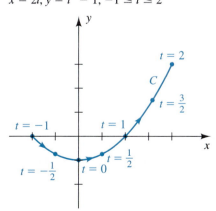

Plotting points leads to the sketch in Figure 2. The arrowheads on the graph indicate the direction in which $P(x, y)$ traces the curve as t *increases* from -1 to 2.

We may obtain a more familiar description of the graph by eliminating the parameter. Solving the x-equation for t, we obtain $t = \frac{1}{2}x$. Substituting this expression for t in the y-equation gives us

$$y = \left(\tfrac{1}{2}x\right)^2 - 1.$$

The graph of this equation in x and y is a parabola symmetric with respect to the y-axis with vertex $(0, -1)$. However, since $x = 2t$ and $-1 \le t \le 2$, we see that $-2 \le x \le 4$ for points (x, y) on C, and hence C is that part of the parabola between the points $(-2, 0)$ and $(4, 3)$ shown in Figure 2.   ◢

As indicated by the arrowheads in Figure 2, the point $P(x, y)$ traces the curve C from *left to right* as t increases. The parametric equations

$$x = -2t, \quad y = t^2 - 1; \quad -2 \le t \le 1$$

give us the same graph; however, as t increases, $P(x, y)$ traces the curve from *right to left*. For other parametrizations, the point $P(x, y)$ may oscillate back and forth as t increases.

The **orientation** of a parametrized curve C is the direction determined by *increasing* values of the parameter. We often indicate an orientation by placing arrowheads on C, as in Figure 2. If $P(x, y)$ moves back and forth as t increases, we may place arrows *alongside* of C.

As we have observed, a curve may have different orientations, depending on the parametrization. To illustrate, the curve C in Example 1 is given parametrically by any of the following:

$$x = 2t, \quad y = t^2 - 1; \qquad -1 \le t \le 2$$
$$x = t, \quad y = \tfrac{1}{4}t^2 - 1; \qquad -2 \le t \le 4$$
$$x = t^3, \quad y = \tfrac{1}{4}t^6 - 1; \quad \sqrt[3]{-2} \le t \le \sqrt[3]{4}$$

The next example demonstrates that it is sometimes useful to eliminate the parameter *before* plotting points.

EXAMPLE 2   Describing the motion of a point

A point moves in a plane such that its position $P(x, y)$ at time t is given by

$$x = a \cos t, \quad y = a \sin t; \quad t \text{ in } \mathbb{R},$$

where $a > 0$. Describe the motion of the point.

SOLUTION   When x and y contain trigonometric functions of t, we can often eliminate the parameter t by isolating the trigonometric functions, squaring both sides of the equations, and then using one of the Pythagorean identities, as follows:

$$x = a \cos t, \quad y = a \sin t \quad \text{given}$$
$$\frac{x}{a} = \cos t, \quad \frac{y}{a} = \sin t \quad \text{isolate } \cos t \text{ and } \sin t$$
$$\frac{x^2}{a^2} = \cos^2 t, \quad \frac{y^2}{a^2} = \sin^2 t \quad \text{square both sides}$$
$$\frac{x^2}{a^2} + \frac{y^2}{a^2} = 1 \quad \cos^2 t + \sin^2 t = 1$$
$$x^2 + y^2 = a^2 \quad \text{multiply by } a^2$$

Figure 3

$x = a \cos t, y = a \sin t; t \text{ in } \mathbb{R}$

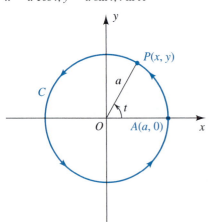

This shows that the point $P(x, y)$ moves on the circle C of radius a with center at the origin (see Figure 3). The point is at $A(a, 0)$ when $t = 0$, at $(0, a)$ when $t = \pi/2$, at $(-a, 0)$ when $t = \pi$, at $(0, -a)$ when $t = 3\pi/2$, and back at $A(a, 0)$ when $t = 2\pi$. Thus, P moves around C in a counterclockwise direction, making one revolution every 2π units of time. The orientation of C is indicated by the arrowheads in Figure 3.

Note that in this example we may interpret t geometrically as the radian measure of the angle generated by the line segment OP.

EXAMPLE 3   Sketching the graph of a parametrized curve

Sketch the graph of the curve C that has the parametrization

$$x = -2 + t^2, \quad y = 1 + 2t^2; \quad t \text{ in } \mathbb{R},$$

and indicate the orientation.

SOLUTION   To eliminate the parameter, we use the x-equation to obtain $t^2 = x + 2$ and then substitute for t^2 in the y-equation. Thus,

$$y = 1 + 2(x + 2).$$

The graph of the last equation is the line of slope 2 through the point $(-2, 1)$, as indicated by the dashes in Figure 4(a). However, since $t^2 \geq 0$, we see from the parametric equations for C that

$$x = -2 + t^2 \geq -2 \quad \text{and} \quad y = 1 + 2t^2 \geq 1.$$

Thus, the graph of C is that part of the line to the right of $(-2, 1)$ (the point corresponding to $t = 0$), as shown in Figure 4(b). The orientation is indicated by the arrows alongside of C. As t increases in the interval $(-\infty, 0]$, $P(x, y)$ moves down the curve toward the point $(-2, 1)$. As t increases in $[0, \infty)$, $P(x, y)$ moves up the curve away from $(-2, 1)$.

Figure 4

(a)

(b)

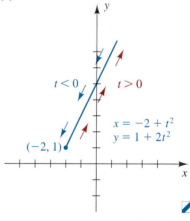

If a curve C is described by an equation $y = f(x)$ for some function f, then an easy way to obtain parametric equations for C is to let

$$x = t, \quad y = f(t),$$

where t is in the domain of f. For example, if $y = x^3$, then parametric equations are

$$x = t, \quad y = t^3; \quad t \text{ in } \mathbb{R}.$$

We can use many different substitutions for x, provided that as t varies through some interval, x takes on every value in the domain of f. Thus, the graph of $y = x^3$ is also given by

$$x = t^{1/3}, \quad y = t; \quad t \text{ in } \mathbb{R}.$$

Note, however, that the parametric equations

$$x = \sin t, \quad y = \sin^3 t; \quad t \text{ in } \mathbb{R}$$

give only that part of the graph of $y = x^3$ between the points $(-1, -1)$ and $(1, 1)$.

EXAMPLE 4   Finding parametric equations for a line

Find three parametrizations for the line of slope m through the point (x_1, y_1).

SOLUTION   By the point-slope form, an equation for the line is

$$y - y_1 = m(x - x_1). \tag{*}$$

If we let $x = t$, then $y - y_1 = m(t - x_1)$ and we obtain the parametrization

$$x = t, \quad y = y_1 + m(t - x_1); \quad t \text{ in } \mathbb{R}.$$

We obtain another parametrization for the line if we let $x - x_1 = t$ in (*). In this case $y - y_1 = mt$, and we have

$$x = x_1 + t, \quad y = y_1 + mt; \quad t \text{ in } \mathbb{R}.$$

As a third illustration, if we let $x - x_1 = \tan t$ in (*), then

$$x = x_1 + \tan t, \quad y = y_1 + m \tan t; \quad -\frac{\pi}{2} < t < \frac{\pi}{2}.$$

There are many other parametrizations for the line.   ◢

In the next example, we use parametric equations to model the path of a projectile (object). These equations are developed by means of methods in physics and calculus. We assume that the object is moving near the surface of Earth under the influence of gravity alone; that is, air resistance and other forces that could affect acceleration are negligible. We also assume that the ground is level and the curvature of Earth is not a factor in determining the path of the object.

EXAMPLE 5   The path of a projectile

The path of a projectile at time t can be modeled using the parametric equations

$$x(t) = (s \cos \alpha)t, \quad y(t) = -\tfrac{1}{2}gt^2 + (s \sin \alpha)t + h; \quad t \geq 0, \tag{1}$$

where, at $t = 0$, s is the speed of the projectile in ft/sec, α is the angle the path makes with the horizontal, and h is the height in feet. The acceleration due to gravity is $g = 32$ ft/sec². Suppose that the projectile is fired at a speed of 1024 ft/sec at an angle of 30° from the horizontal from a height of 2304 feet (see Figure 5).

(a) Find parametric equations for the projectile.

(b) Find the range r of the projectile—that is, the horizontal distance it travels before hitting the ground.

(c) Find an equation in x and y for the projectile.

(d) Find the point and time at which the projectile reaches its maximum altitude.

Figure 5

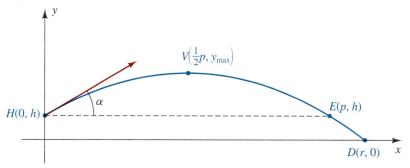

SOLUTION

(a) Substituting 1024 for s, 30° for α, 32 for g, and 2304 for h in the parametric equations in (1) gives

$$x = (1024 \cos 30°)t, \quad y = -\tfrac{1}{2}(32)t^2 + (1024 \sin 30°)t + 2304; \quad t \geq 0.$$

Simplifying yields

$$x = 512\sqrt{3}\,t, \quad y = -16t^2 + 512t + 2304; \quad t \geq 0. \tag{2}$$

(b) To find the range r of the projectile, we must find the point D in Figure 5 at which the projectile hits the ground. Since the y-coordinate of D is 0, we let $y = 0$ in the y-equation of (2) and solve for t:

$$
\begin{array}{ll}
y = -16t^2 + 512t + 2304 & \text{given in (2)} \\
0 = -16t^2 + 512t + 2304 & \text{let } y = 0 \\
0 = t^2 - 32t - 144 & \text{divide by } -16 \\
0 = (t - 36)(t + 4) & \text{factor}
\end{array}
$$

Since $t \geq 0$, we must have $t = 36$ sec. We can now use the x-equation of (2) to obtain the range:

$$x = 512\sqrt{3}\,t = 512\sqrt{3}(36) = 18{,}432\sqrt{3} \approx 31{,}925 \text{ ft}$$

(continued)

(c) To eliminate the parameter t, we solve the x-equation in (2) for t and substitute this expression for t in the y-equation in (2):

$$x = 512\sqrt{3}\,t \quad \text{implies} \quad t = \frac{x}{512\sqrt{3}} \qquad\qquad \text{solve } x\text{-equation in (2) for } t$$

$$y = -16t^2 + 512t + 2304 \qquad\qquad\qquad \text{y-equation in (2)}$$

$$y = -16\left(\frac{x}{512\sqrt{3}}\right)^2 + 512\left(\frac{x}{512\sqrt{3}}\right) + 2304 \quad \text{let } t = \frac{x}{512\sqrt{3}}$$

$$y = -\frac{1}{49{,}152}x^2 + \frac{1}{\sqrt{3}}x + 2304 \qquad\qquad \text{simplify} \qquad\qquad (3)$$

The last equation is of the form $y = ax^2 + bx + c$, showing that the path of the projectile is parabolic.

(d) The y-coordinate of point E in Figure 5 is 2304, so we can find the value of t at E by solving the equation $y = 2304$:

$$y = -16t^2 + 512t + 2304 \qquad \text{given in (2)}$$
$$2304 = -16t^2 + 512t + 2304 \qquad \text{let } y = 2304$$
$$0 = -16t^2 + 512t \qquad \text{subtract 2304}$$
$$0 = -16t(t - 32) \qquad \text{factor}$$

So if $y = 2304$, $t = 0$ or $t = 32$. Since the path is parabolic, the x-coordinate of V is one-half of the x-coordinate p of E. Also, the value of t at V is one-half the value of t at E, so $t = \frac{1}{2}(32) = 16$ at V. We can find the x- and y-values at V by substituting 16 for t in (2):

$$x = 512\sqrt{3}\,t = 512\sqrt{3}(16) = 8192\sqrt{3} \approx 14{,}189 \text{ ft}$$

and

$$y = -16t^2 + 512t + 2304 = -16(16)^2 + 512(16) + 2304 = 6400 \text{ ft}$$

Thus, the projectile reaches its maximum altitude when $t = 16$ at approximately $(14{,}189, 6400)$.

An alternative way of finding the maximum altitude is to use the theorem for locating the vertex of a parabola to find the x-value ($x = -b/(2a)$) of the highest point on the graph of equation (3) and then use the equations in (2) to find t and y. ◢

See Discussion Exercises 7 and 8 at the end of the chapter for related problems concerning Example 5.

EXAMPLE 6   Finding parametric equations for a cycloid

The curve traced by a fixed point P on the circumference of a circle as the circle rolls along a line in a plane is called a **cycloid.** Find parametric equations for a cycloid.

SOLUTION   Suppose the circle has radius a and that it rolls along (and above) the x-axis in the positive direction. If one position of P is the origin, then Figure 6 depicts part of the curve and a possible position of the circle. The V-shaped part of the curve at $x = 2\pi a$ is called a **cusp.**

Figure 6

Figure 7

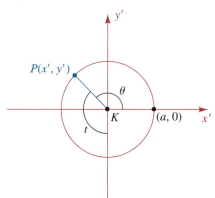

Let K denote the center of the circle and T the point of tangency with the x-axis. We introduce, as a parameter t, the radian measure of angle TKP. The distance the circle has rolled is $d(O, T) = at$ (formula for the length of a circular arc). Consequently, the coordinates of K are $(x, y) = (at, a)$. If we consider an $x'y'$-coordinate system with origin at $K(at, a)$ and if $P(x', y')$ denotes the point P relative to this system, then, by adding x' and y' to the x- and y-coordinates of K, we obtain

$$x = at + x', \qquad y = a + y'.$$

If, as in Figure 7, θ denotes an angle in standard position on the $x'y'$-plane, then $\theta + t = 3\pi/2$ or, equivalently, $\theta = (3\pi/2) - t$. Hence,

$$x' = a \cos \theta = a \cos \left(\frac{3\pi}{2} - t \right) = -a \sin t$$

$$y' = a \sin \theta = a \sin \left(\frac{3\pi}{2} - t \right) = -a \cos t,$$

and substitution in $x = at + x'$, $y = a + y'$ gives us parametric equations for the cycloid:

$$x = a(t - \sin t), \quad y = a(1 - \cos t); \quad t \text{ in } \mathbb{R}$$

If $a < 0$, then the graph of $x = a(t - \sin t)$, $y = a(1 - \cos t)$ is the inverted cycloid that results if the circle of Example 6 rolls *below* the x-axis.

Figure 8

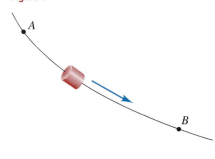

This curve has a number of important physical properties. To illustrate, suppose a thin wire passes through two fixed points A and B, as shown in Figure 8, and that the shape of the wire can be changed by bending it in any manner. Suppose further that a bead is allowed to slide along the wire and the only force acting on the bead is gravity. We now ask which of all the possible paths will allow the bead to slide from A to B in the least amount of time. It is natural to believe that the desired path is the straight line segment from A to B; however, this is not the correct answer. The path that requires the least time coincides with the graph of an inverted cycloid with A at the origin. Because the velocity of the bead increases more rapidly along the cycloid than along the line through A and B, the bead reaches B more rapidly, even though the distance is greater.

There is another interesting property of this **curve of least descent**. Suppose that A is the origin and B is the point with x-coordinate $\pi|a|$—that is, the lowest point on the cycloid in the first arc to the right of A. If the bead is released at *any* point between A and B, it can be shown that the *time* required for it to reach B is always the *same*.

Variations of the cycloid occur in applications. For example, if a motorcycle wheel rolls along a straight road, then the curve traced by a fixed point on one of the spokes is a cycloidlike curve. In this case the curve does not have cusps, nor does it intersect the road (the x-axis) as does the graph of a cycloid. If the wheel of a train rolls along a railroad track, then the curve traced by a fixed point on the circumference of the wheel (which extends below the track) contains loops at regular intervals. Other cycloids are defined in Exercises 37 and 38.

11.4 Exercises

Exer. 1–22: Find an equation in x and y whose graph contains the points on the curve C. Sketch the graph of C, and indicate the orientation.

1   $x = t - 2$, $y = 2t + 3$; $0 \le t \le 5$

2   $x = 1 - 2t$, $y = 1 + t$; $-1 \le t \le 4$

3   $x = t^2 + 1$, $y = t^2 - 1$; $-2 \le t \le 2$

4   $x = t^3 + 1$, $y = t^3 - 1$; $-2 \le t \le 2$

5   $x = 4t^2 - 5$, $y = 2t + 3$; t in $\mathbb{R}$

6   $x = t^3$, $y = t^2$; t in $\mathbb{R}$

7   $x = e^t$, $y = e^{-2t}$; t in $\mathbb{R}$

8   $x = \sqrt{t}$, $y = 3t + 4$; $t \ge 0$

9   $x = 2 \sin t$, $y = 3 \cos t$; $0 \le t \le 2\pi$

10   $x = \cos t - 2$, $y = \sin t + 3$; $0 \le t \le 2\pi$

11   $x = \sec t$, $y = \tan t$; $-\pi/2 < t < \pi/2$

12   $x = \cos 2t$, $y = \sin t$; $-\pi \le t \le \pi$

13   $x = t^2$, $y = 2 \ln t$; $t > 0$

14 $x = \cos^3 t,$ $y = \sin^3 t;$ $0 \le t \le 2\pi$

15 $x = \sin t,$ $y = \csc t;$ $0 < t \le \pi/2$

16 $x = e^t,$ $y = e^{-t};$ t in $\mathbb{R}$

17 $x = t,$ $y = \sqrt{t^2 - 1};$ $|t| \ge 1$

18 $x = -2\sqrt{1 - t^2},$ $y = t;$ $|t| \le 1$

19 $x = t,$ $y = \sqrt{t^2 - 2t + 1};$ $0 \le t \le 4$

20 $x = 2t,$ $y = 8t^3;$ $-1 \le t \le 1$

21 $x = (t + 1)^3,$ $y = (t + 2)^2;$ $0 \le t \le 2$

22 $x = \tan t,$ $y = 1;$ $-\pi/2 < t < \pi/2$

23 (a) Describe the graph of a curve C that has the parametrization

$$x = 3 + 2\sin t, \quad y = -2 + 2\cos t; \quad 0 \le t \le 2\pi.$$

(b) Change the parametrization to

$$x = 3 - 2\sin t, \quad y = -2 + 2\cos t; \quad 0 \le t \le 2\pi$$

and describe how this changes the graph from part (a).

(c) Change the parametrization to

$$x = 3 - 2\sin t, \quad y = -2 - 2\cos t; \quad 0 \le t \le 2\pi$$

and describe how this changes the graph from part (a).

24 (a) Describe the graph of a curve C that has the parametrization

$$x = -2 + 3\sin t, \quad y = 3 - 3\cos t; \quad 0 \le t \le 2\pi.$$

(b) Change the parametrization to

$$x = -2 - 3\sin t, \quad y = 3 + 3\cos t; \quad 0 \le t \le 2\pi$$

and describe how this changes the graph from part (a).

(c) Change the parametrization to

$$x = -2 + 3\sin t, \quad y = 3 + 3\cos t; \quad 0 \le t \le 2\pi$$

and describe how this changes the graph from part (a).

Exer. 25–26: Curves C_1, C_2, C_3, and C_4 are given parametrically, for t in $\mathbb{R}$. Sketch their graphs, and indicate orientations.

25 $C_1: x = t^2,$ $y = t$
$C_2: x = t^4,$ $y = t^2$
$C_3: x = \sin^2 t,$ $y = \sin t$
$C_4: x = e^{2t},$ $y = -e^t$

26 $C_1: x = t,$ $y = 1 - t$
$C_2: x = 1 - t^2,$ $y = t^2$
$C_3: x = \cos^2 t,$ $y = \sin^2 t$
$C_4: x = \ln t - t,$ $y = 1 + t - \ln t; t > 0$

Exer. 27–28: The parametric equations specify the position of a moving point $P(x, y)$ at time t. Sketch the graph, and indicate the motion of P as t increases.

27 (a) $x = \cos t,$ $y = \sin t;$ $0 \le t \le \pi$

(b) $x = \sin t,$ $y = \cos t;$ $0 \le t \le \pi$

(c) $x = t,$ $y = \sqrt{1 - t^2};$ $-1 \le t \le 1$

28 (a) $x = t^2,$ $y = 1 - t^2;$ $0 \le t \le 1$

(b) $x = 1 - \ln t,$ $y = \ln t;$ $1 \le t \le e$

(c) $x = \cos^2 t,$ $y = \sin^2 t;$ $0 \le t \le 2\pi$

29 Show that

$$x = a\cos t + h, \quad y = b\sin t + k; \quad 0 \le t \le 2\pi$$

are parametric equations of an ellipse with center (h, k) and axes of lengths $2a$ and $2b$.

30 Show that

$$x = a\sec t + h, \quad y = b\tan t + k;$$
$$-\pi/2 < t < 3\pi/2 \text{ and } t \ne \pi/2$$

are parametric equations of a hyperbola with center (h, k), transverse axis of length $2a$, and conjugate axis of length $2b$. Determine the values of t for each branch.

Exer. 31–32: (a) Find three parametrizations that give the same graph as the given equation. (b) Find three parametrizations that give only a portion of the graph of the given equation.

31 $y = x^2$ **32** $y = \ln x$

Exer. 33–36: Refer to the equations in (1) of Example 5. Find the range and maximum altitude for the given values.

33 $s = 256\sqrt{3},$ $\alpha = 60°,$ $h = 400$

34 $s = 512\sqrt{2},$ $\alpha = 45°,$ $h = 1088$

35 $s = 704,$ $\alpha = 45°,$ $h = 0$

36 $s = 2448,$ $\alpha = 30°,$ $h = 0$

37 A circle C of radius b rolls on the outside of the circle $x^2 + y^2 = a^2$, and $b < a$. Let P be a fixed point on C, and let the initial position of P be $A(a, 0)$, as shown in the figure. If the parameter t is the angle from the positive x-axis to the line segment from O to the center of C, show that parametric equations for the curve traced by P (an *epicycloid*) are

$$x = (a + b) \cos t - b \cos \left(\frac{a + b}{b} t \right),$$

$$y = (a + b) \sin t - b \sin \left(\frac{a + b}{b} t \right); \quad 0 \le t \le 2\pi.$$

Exercise 37

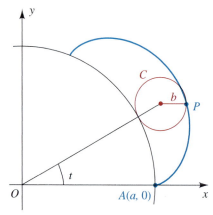

38 If the circle C of Exercise 37 rolls on the inside of the second circle (see the figure), then the curve traced by P is a *hypocycloid*.

(a) Show that parametric equations for this curve are

$$x = (a - b) \cos t + b \cos \left(\frac{a - b}{b} t \right),$$

$$y = (a - b) \sin t - b \sin \left(\frac{a - b}{b} t \right); \quad 0 \le t \le 2\pi.$$

(b) If $b = \frac{1}{4}a$, show that $x = a \cos^3 t$, $y = a \sin^3 t$, and sketch the graph.

Exercise 38

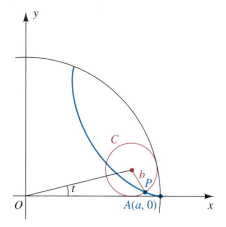

39 If $b = \frac{1}{3}a$ in Exercise 37, find parametric equations for the epicycloid and sketch the graph.

40 The radius of circle B is one-third that of circle A. How many revolutions will circle B make as it rolls around circle A until it reaches its starting point? (*Hint:* Use Exercise 39.)

11.5
Polar Coordinates

In a rectangular coordinate system, the ordered pair (a, b) denotes the point whose directed distances from the x- and y-axes are b and a, respectively. Another method for representing points is to use *polar coordinates*. We begin with a fixed point O (the **origin,** or **pole**) and a directed half-line (the **polar axis**) with endpoint O. Next we consider any point P in the plane different from O. If, as illustrated in Figure 1, $r = d(O, P)$ and θ denotes the measure of any angle determined by the polar axis and OP, then r and θ are **polar coordinates** of P and the symbols (r, θ) or $P(r, \theta)$ are used to denote P. As usual, θ is considered positive if the angle is generated by a counterclockwise rotation of the polar axis and negative if the rotation is clockwise. Either radian or degree measure may be used for θ.

Figure 1

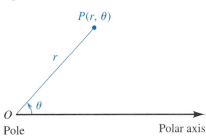

The polar coordinates of a point are not unique. For example, $(3, \pi/4)$, $(3, 9\pi/4)$, and $(3, -7\pi/4)$ all represent the same point (see Figure 2). We shall also allow r to be negative. In this case, instead of measuring $|r|$ units along the terminal side of the angle θ, we measure along the half-line with endpoint O that has direction *opposite* that of the terminal side. The points corresponding to the pairs $(-3, 5\pi/4)$ and $(-3, -3\pi/4)$ are also plotted in Figure 2.

Figure 2

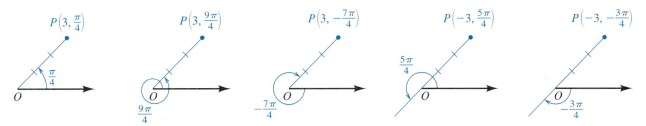

We agree that the pole O has polar coordinates $(0, \theta)$ for *any* θ. An assignment of ordered pairs of the form (r, θ) to points in a plane is a **polar coordinate system,** and the plane is an **$r\theta$-plane.**

Let us next superimpose an *xy*-plane on an $r\theta$-plane so that the positive *x*-axis coincides with the polar axis. Any point P in the plane may then be assigned rectangular coordinates (x, y) or polar coordinates (r, θ). If $r > 0$, we have a situation similar to that illustrated in Figure 3(a); if $r < 0$, we have that shown in part (b) of the figure. In Figure 3(b), for later purposes, we have also plotted the point P', having polar coordinates $(|r|, \theta)$ and rectangular coordinates $(-x, -y)$.

Figure 3

(a) $r > 0$

(b) $r < 0$

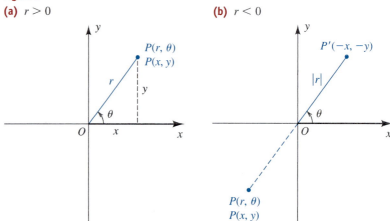

The following result specifies relationships between (x, y) and (r, θ), where it is assumed that the positive x-axis coincides with the polar axis.

Relationships Between Rectangular and Polar Coordinates	The rectangular coordinates (x, y) and polar coordinates (r, θ) of a point P are related as follows: (1) $x = r \cos \theta, \quad y = r \sin \theta$ (2) $r^2 = x^2 + y^2, \quad \tan \theta = \dfrac{y}{x} \quad \text{if } x \neq 0$

PROOFS

(1) Although we have pictured θ as an acute angle in Figure 3, the discussion that follows is valid for all angles.

   If $r > 0$, as in Figure 3(a), then $\cos \theta = x/r$ and $\sin \theta = y/r$, and hence

$$x = r \cos \theta, \qquad y = r \sin \theta.$$

If $r < 0$, then $|r| = -r$, and from Figure 3(b) we see that

$$\cos \theta = \frac{-x}{|r|} = \frac{-x}{-r} = \frac{x}{r}, \qquad \sin \theta = \frac{-y}{|r|} = \frac{-y}{-r} = \frac{y}{r}.$$

Multiplication by r gives us relationship 1, and therefore these formulas hold if r is either positive or negative.

   If $r = 0$, then the point is the pole, and we again see that the formulas in (1) are true.

(2) The formulas in relationship 2 follow readily from Figure 3(a). By the Pythagorean theorem, $x^2 + y^2 = r^2$. From the definition of the trigonometric functions of any angle, $\tan \theta = y/x$ (if $x \neq 0$). If $x = 0$, then $\theta = (\pi/2) + \pi n$ from some integer n. ◢

We may use the preceding result to change from one system of coordinates to the other.

Figure 4

EXAMPLE 1   Changing polar coordinates to rectangular coordinates

If $(r, \theta) = (4, 7\pi/6)$ are polar coordinates of a point P, find the rectangular coordinates of P.

SOLUTION   The point P is plotted in Figure 4. Substituting $r = 4$ and $\theta = 7\pi/6$ in relationship 1 of the preceding result, we obtain the following:

$$x = r \cos \theta = 4 \cos (7\pi/6) = 4\left(-\sqrt{3}/2\right) = -2\sqrt{3}$$
$$y = r \sin \theta = 4 \sin (7\pi/6) = 4(-1/2) = -2$$

Hence, the rectangular coordinates of P are $(x, y) = \left(-2\sqrt{3}, -2\right)$. ◢

EXAMPLE 2   Changing rectangular coordinates to polar coordinates

If $(x, y) = \left(-1, \sqrt{3}\right)$ are rectangular coordinates of a point P, find three different pairs of polar coordinates (r, θ) for P.

Figure 5
(a)

(b)

(c)

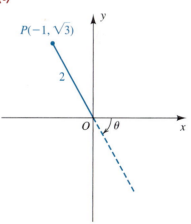

SOLUTION   Three possibilities for θ are illustrated in Figure 5(a)–(c). Using $x = -1$ and $y = \sqrt{3}$ in relationship 2 between rectangular and polar coordinates, we obtain

$$r^2 = x^2 + y^2 = (-1)^2 + \left(\sqrt{3}\right)^2 = 4,$$

(continued)

and since r is positive in Figure 5(a), $r = 2$. Using

$$\tan \theta = \frac{y}{x} = \frac{\sqrt{3}}{-1} = -\sqrt{3},$$

we see that the reference angle for θ is $\theta_R = \pi/3$, and hence

$$\theta = \pi - \frac{\pi}{3} = \frac{2\pi}{3}.$$

Thus, $(2, 2\pi/3)$ is one pair of polar coordinates for P.

Referring to Figure 5(b) and the values obtained for P in Figure 5(a), we get

$$r = 2 \qquad \text{and} \qquad \theta = \frac{2\pi}{3} + 2\pi = \frac{8\pi}{3}.$$

Hence, $(2, 8\pi/3)$ is another pair of polar coordinates for P.

In Figure 5(c), $\theta = -\pi/3$. In this case we use $r = -2$ to obtain $(-2, -\pi/3)$ as a third pair of polar coordinates for P. ◳

A **polar equation** is an equation in r and θ. A **solution** of a polar equation is an ordered pair (a, b) that leads to equality if a is substituted for r and b for θ. The **graph** of a polar equation is the set of all points (in an $r\theta$-plane) that correspond to the solutions.

The simplest polar equations are $r = a$ and $\theta = a$, where a is a nonzero real number. Since the solutions of the polar equation $r = a$ are of the form (a, θ) for *any* angle θ, it follows that the graph is a circle of radius $|a|$ with center at the pole. A graph for $a > 0$ is sketched in Figure 6. The same graph is obtained for $r = -a$.

The solutions of the polar equation $\theta = a$ are of the form (r, a) for *any* real number r. Since the coordinate a (the angle) is constant, the graph of $\theta = a$ is a line through the origin, as illustrated in Figure 7 for an acute angle a.

We may use the relationships between rectangular and polar coordinates to transform a polar equation to an equation in x and y, and vice versa. This procedure is illustrated in the next three examples.

Figure 6

Figure 7

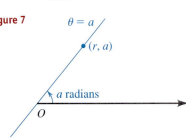

EXAMPLE 3　Finding a polar equation of a line

Find a polar equation of an arbitrary line.

SOLUTION　Every line in an xy-coordinate plane is the graph of a linear equation that can be written in the form $ax + by = c$. Using the formulas $x = r \cos \theta$ and $y = r \sin \theta$ gives us the following equivalent polar equations:

$$ar \cos \theta + br \sin \theta = c \qquad \text{substitute for } x \text{ and } y$$

$$r(a \cos \theta + b \sin \theta) = c \qquad \text{factor out } r$$

If $a \cos \theta + b \sin \theta \neq 0$, the last equation may be written as follows:

$$r = \frac{c}{a \cos \theta + b \sin \theta}$$

EXAMPLE 4   **Changing an equation in x and y to a polar equation**

Find a polar equation for the hyperbola $x^2 - y^2 = 16$.

SOLUTION   Using the formulas $x = r \cos \theta$ and $y = r \sin \theta$, we obtain the following polar equations:

$$\begin{array}{ll}
(r \cos \theta)^2 - (r \sin \theta)^2 = 16 & \text{substitute for } x \text{ and } y \\
r^2 \cos^2 \theta - r^2 \sin^2 \theta = 16 & \text{square the terms} \\
r^2(\cos^2 \theta - \sin^2 \theta) = 16 & \text{factor out } r^2 \\
r^2 \cos 2\theta = 16 & \text{double-angle formula} \\
r^2 = \dfrac{16}{\cos 2\theta} & \text{divide by } \cos 2\theta
\end{array}$$

The division by $\cos 2\theta$ is allowable because $\cos 2\theta \neq 0$. (Note that if $\cos 2\theta = 0$, then $r^2 \cos 2\theta \neq 16$.) We may also write the polar equation as $r^2 = 16 \sec 2\theta$.

EXAMPLE 5   **Changing a polar equation to an equation in x and y**

Find an equation in x and y that has the same graph as the polar equation $r = a \sin \theta$, with $a \neq 0$. Sketch the graph.

SOLUTION   A formula that relates $\sin \theta$ and y is given by $y = r \sin \theta$. To introduce the expression $r \sin \theta$ into the equation $r = a \sin \theta$, we multiply both sides by r, obtaining

$$r^2 = ar \sin \theta.$$

Next, if we substitute $x^2 + y^2$ for r^2 and y for $r \sin \theta$, the last equation becomes

$$x^2 + y^2 = ay,$$

or

$$x^2 + y^2 - ay = 0.$$

Completing the square in y gives us

$$x^2 + y^2 - ay + \left(\frac{a}{2}\right)^2 = \left(\frac{a}{2}\right)^2,$$

or

$$x^2 + \left(y - \frac{a}{2}\right)^2 = \left(\frac{a}{2}\right)^2.$$

In the xy-plane, the graph of the last equation is a circle with center $(0, a/2)$ and radius $|a|/2$, as illustrated in Figure 8 for the case $a > 0$ (the solid circle) and $a < 0$ (the dashed circle).

Figure 8

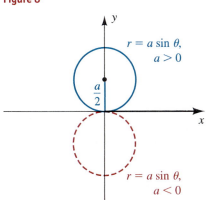

$r = a \sin \theta,$
$a > 0$

$\dfrac{a}{2}$

$r = a \sin \theta,$
$a < 0$

Figure 9

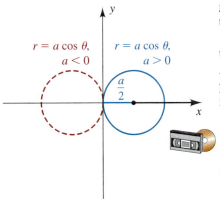

$r = a \cos \theta,$
$a < 0$

$r = a \cos \theta,$
$a > 0$

$\frac{a}{2}$

Using the same method as in the preceding example, we can show that the graph of $r = a \cos \theta$, with $a \neq 0$, is a circle of radius $a/2$ of the type illustrated in Figure 9.

In the following examples we obtain the graphs of polar equations by plotting points and examining the relationship between θ-intervals and r-intervals. As you proceed through this section, you should try to recognize forms of polar equations so that you will be able to sketch their graphs by plotting few, if any, points.

EXAMPLE 6 **Sketching the graph of a polar equation**

Sketch the graph of the polar equation $r = 4 \sin \theta$.

SOLUTION The proof that the graph of $r = 4 \sin \theta$ is a circle was given in Example 5. The following table displays some solutions of the equation. We have included a third row in the table that contains one-decimal-place approximations to r.

θ	0	$\dfrac{\pi}{6}$	$\dfrac{\pi}{4}$	$\dfrac{\pi}{3}$	$\dfrac{\pi}{2}$	$\dfrac{2\pi}{3}$	$\dfrac{3\pi}{4}$	$\dfrac{5\pi}{6}$	π
r	0	2	$2\sqrt{2}$	$2\sqrt{3}$	4	$2\sqrt{3}$	$2\sqrt{2}$	2	0
r **(approx.)**	0	2	2.8	3.5	4	3.5	2.8	2	0

Figure 10

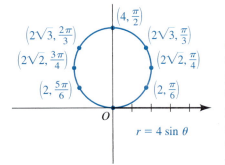

$\left(4, \frac{\pi}{2}\right)$

$\left(2\sqrt{3}, \frac{2\pi}{3}\right)$ $\left(2\sqrt{3}, \frac{\pi}{3}\right)$

$\left(2\sqrt{2}, \frac{3\pi}{4}\right)$ $\left(2\sqrt{2}, \frac{\pi}{4}\right)$

$\left(2, \frac{5\pi}{6}\right)$ $\left(2, \frac{\pi}{6}\right)$

O

$r = 4 \sin \theta$

As an aid to plotting points in the $r\theta$-plane shown in Figure 10, we have extended the polar axis in the negative direction and introduced a vertical line through the pole (this line is the graph of the equation $\theta = \pi/2$). Additional points obtained by letting θ vary from π to 2π lie on the same circle. For example, the solution $(-2, 7\pi/6)$ gives us the same point as $(2, \pi/6)$; the point corresponding to $(-2\sqrt{2}, 5\pi/4)$ is the same as that obtained from $(2\sqrt{2}, \pi/4)$; and so on. If we let θ increase through all real numbers, we obtain the same points again and again because of the periodicity of the sine function. ◢

EXAMPLE 7 **Sketching the graph of a polar equation**

Sketch the graph of the polar equation $r = 2 + 2 \cos \theta$.

SOLUTION Since the cosine function decreases from 1 to -1 as θ varies from 0 to π, it follows that r decreases from 4 to 0 in this θ-interval. The following table exhibits some solutions of $r = 2 + 2 \cos \theta$, together with one-decimal-place approximations to r.

θ	0	$\dfrac{\pi}{6}$	$\dfrac{\pi}{4}$	$\dfrac{\pi}{3}$	$\dfrac{\pi}{2}$	$\dfrac{2\pi}{3}$	$\dfrac{3\pi}{4}$	$\dfrac{5\pi}{6}$	π
r	4	$2 + \sqrt{3}$	$2 + \sqrt{2}$	3	2	1	$2 - \sqrt{2}$	$2 - \sqrt{3}$	0
r (approx.)	4	3.7	3.4	3	2	1	0.6	0.3	0

Plotting points in an $r\theta$-plane leads to the upper half of the graph sketched in Figure 11. (We have used polar coordinate graph paper, which displays lines through O at various angles and concentric circles with centers at the pole.)

Figure 11

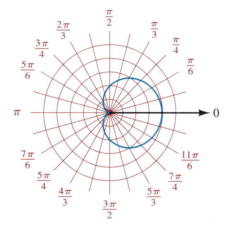

If θ increases from π to 2π, then $\cos \theta$ increases from -1 to 1, and consequently r increases from 0 to 4. Plotting points for $\pi \le \theta \le 2\pi$ gives us the lower half of the graph.

The same graph may be obtained by taking other intervals of length 2π for θ.

The heart-shaped graph in Example 7 is a **cardioid.** In general, the graph of any of the polar equations in Figure 12 on the next page, with $a \ne 0$, is a cardioid.

Figure 12

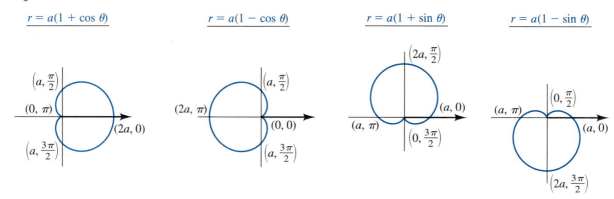

$r = a(1 + \cos \theta)$ $r = a(1 - \cos \theta)$ $r = a(1 + \sin \theta)$ $r = a(1 - \sin \theta)$

If a and b are not zero, then the graphs of the following polar equations are **limaçons:**

$$r = a + b \cos \theta \qquad r = a + b \sin \theta$$

Note that the special limaçons in which $|a| = |b|$ are cardioids.

Using the θ-interval $[0, 2\pi]$ (or $[-\pi, \pi]$) is usually sufficient to graph polar equations. For equations with more complex graphs, it is often helpful to graph by using subintervals of $[0, 2\pi]$ that are determined by the θ-values that make $r = 0$—that is, the **pole values.** We will demonstrate this technique in the next example.

EXAMPLE 8   **Sketching the graph of a polar equation**

Sketch the graph of the polar equation $r = 2 + 4 \cos \theta$.

Figure 13

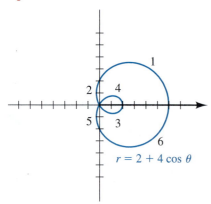

$r = 2 + 4 \cos \theta$

SOLUTION   We first find the pole values by solving the equation $r = 0$:

$$2 + 4 \cos \theta = 0 \qquad \text{let } r = 0$$
$$\cos \theta = -\tfrac{1}{2} \qquad \text{solve for } \cos \theta$$
$$\theta = \frac{2\pi}{3}, \frac{4\pi}{3} \qquad \text{solve for } \theta \text{ in } [0, 2\pi]$$

We next construct a table of θ-values from 0 to 2π, using subintervals determined by the quadrantal angles and the pole values. The row numbers on the left-hand side correspond to the numbers in Figure 13.

	θ	$\cos\theta$	$4\cos\theta$	$r = 2 + 4\cos\theta$
(1)	$0 \to \pi/2$	$1 \to 0$	$4 \to 0$	$6 \to 2$
(2)	$\pi/2 \to 2\pi/3$	$0 \to -1/2$	$0 \to -2$	$2 \to 0$
(3)	$2\pi/3 \to \pi$	$-1/2 \to -1/2$	$-2 \to -4$	$0 \to -2$
(4)	$\pi \to 4\pi/3$	$-1 \to -1/2$	$-4 \to -2$	$-2 \to 0$
(5)	$4\pi/3 \to 3\pi/2$	$-1/2 \to 0$	$-2 \to 0$	$0 \to 2$
(6)	$3\pi/2 \to 2\pi$	$0 \to 1$	$0 \to 4$	$2 \to 6$

You should verify the table entries with the figure, especially for rows 3 and 4 (in which the value of r is negative). The graph is called a limaçon with an inner loop.

The following chart summarizes the four categories of limaçons according to the ratio of a and b in the listed general equations.

Limaçons $a \pm b \cos\theta, a \pm b \sin\theta$ $(a > 0, b > 0)$

Name	Limaçon with an inner loop	Cardioid	Limaçon with a dimple	Convex limaçon
Condition	$\dfrac{a}{b} < 1$	$\dfrac{a}{b} = 1$	$1 < \dfrac{a}{b} < 2$	$\dfrac{a}{b} \geq 2$
Specific graph				
Specific equation	$r = 2 + 4\cos\theta$	$r = 4 + 4\cos\theta$	$r = 6 + 4\cos\theta$	$r = 8 + 4\cos\theta$

EXAMPLE 9 Sketching the graph of a polar equation

Sketch the graph of the polar equation $r = a \sin 2\theta$ for $a > 0$.

SOLUTION The following table contains θ-intervals and the corresponding values of r. The row numbers on the left-hand side correspond to the numbers in Figure 14 on the next page.

(continued)

Figure 14

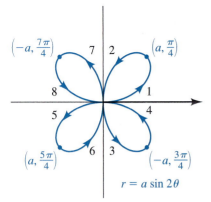

$r = a \sin 2\theta$

	θ	2θ	$\sin 2\theta$	$r = a \sin 2\theta$
(1)	$0 \to \pi/4$	$0 \to \pi/2$	$0 \to 1$	$0 \to a$
(2)	$\pi/4 \to \pi/2$	$\pi/2 \to \pi$	$1 \to 0$	$a \to 0$
(3)	$\pi/2 \to 3\pi/4$	$\pi \to 3\pi/2$	$0 \to -1$	$0 \to -a$
(4)	$3\pi/4 \to \pi$	$3\pi/2 \to 2\pi$	$-1 \to 0$	$-a \to 0$
(5)	$\pi \to 5\pi/4$	$2\pi \to 5\pi/2$	$0 \to 1$	$0 \to a$
(6)	$5\pi/4 \to 3\pi/2$	$5\pi/2 \to 3\pi$	$1 \to 0$	$a \to 0$
(7)	$3\pi/2 \to 7\pi/4$	$3\pi \to 7\pi/2$	$0 \to -1$	$0 \to -a$
(8)	$7\pi/4 \to 2\pi$	$7\pi/2 \to 4\pi$	$-1 \to 0$	$-a \to 0$

You should verify the table entries with the figure, especially for rows 3, 4, 7, and 8 (in which the value of r is negative).

The graph in Example 9 is a **four-leafed rose.** In general, a polar equation of the form

$$r = a \sin n\theta \qquad \text{or} \qquad r = a \cos n\theta$$

for any positive integer n greater than 1 and any nonzero real number a has a graph that consists of a number of loops through the origin. If n is even, there are $2n$ loops, and if n is odd, there are n loops.

The graph of the polar equation $r = a\theta$ for any nonzero real number a is a **spiral of Archimedes.** The case $a = 1$ is considered in the next example.

Figure 15

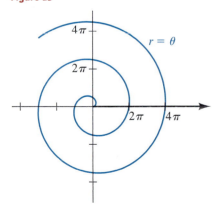

$r = \theta$

EXAMPLE 10   Sketching the graph of a spiral of Archimedes

Sketch the graph of the polar equation $r = \theta$ for $\theta \geq 0$.

SOLUTION   The graph consists of all points that have polar coordinates of the form (c, c) for every real number $c \geq 0$. Thus, the graph contains the points $(0, 0)$, $(\pi/2, \pi/2)$, (π, π), and so on. As θ increases, r increases at the same rate, and the spiral winds around the origin in a counterclockwise direction, intersecting the polar axis at $0, 2\pi, 4\pi, \ldots$, as illustrated in Figure 15.

If θ is allowed to be negative, then as θ decreases through negative values, the resulting spiral winds around the origin and is the symmetric image, with respect to the vertical axis, of the curve sketched in Figure 15.

If we superimpose an xy-plane on an $r\theta$-plane, then the graph of a polar equation may be symmetric with respect to the x-axis (the polar axis), the y-axis (the line $\theta = \pi/2$), or the origin (the pole). Some typical symmetries are illustrated in Figure 16. The next result summarizes these symmetries.

Figure 16 Symmetries of graphs of polar equations

(a) Polar axis

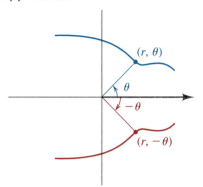

(b) Line $\theta = \pi/2$

(c) Pole

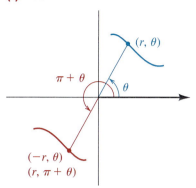

Tests for Symmetry	**(1)** The graph of $r = f(\theta)$ is symmetric with respect to the polar axis if substitution of $-\theta$ for θ leads to an equivalent equation.
	(2) The graph of $r = f(\theta)$ is symmetric with respect to the vertical line $\theta = \pi/2$ if substitution of either (a) $\pi - \theta$ for θ or (b) $-r$ for r and $-\theta$ for θ leads to an equivalent equation.
	(3) The graph of $r = f(\theta)$ is symmetric with respect to the pole if substitution of either (a) $\pi + \theta$ for θ or (b) $-r$ for r leads to an equivalent equation.

To illustrate, since $\cos(-\theta) = \cos\theta$, the graph of the polar equation $r = 2 + 4\cos\theta$ in Example 8 is symmetric with respect to the polar axis, by test 1. Since $\sin(\pi - \theta) = \sin\theta$, the graph in Example 6 is symmetric with respect to the line $\theta = \pi/2$, by test 2. The graph of the four-leafed rose in Example 9 is symmetric with respect to the polar axis, the line $\theta = \pi/2$, and the pole. Other tests for symmetry may be stated; however, those we have listed are among the easiest to apply.

Unlike the graph of an equation in x and y, the graph of a polar equation $r = f(\theta)$ can be symmetric with respect to the polar axis, the line $\theta = \pi/2$, or the pole *without* satisfying one of the preceding tests for symmetry. This is true because of the many different ways of specifying a point in polar coordinates.

Another difference between rectangular and polar coordinate systems is that the points of intersection of two graphs cannot always be found by solving the polar equations simultaneously. To illustrate, from Example 6, the graph of $r = 4\sin\theta$ is a circle of diameter 4 with center at $(2, \pi/2)$ (see Figure 17). Similarly, the graph of $r = 4\cos\theta$ is a circle of diameter 4 with center at $(2, 0)$ on the polar axis. Referring to Figure 17, we see that the coordinates of the point of intersection $P(2\sqrt{2}, \pi/4)$ in quadrant I satisfy both equations; however, the origin O, which is on each circle, *cannot* be found by

Figure 17

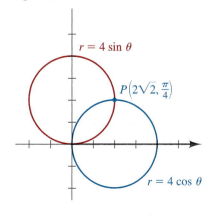

solving the equations simultaneously. Thus, in searching for points of inter-section of polar graphs, it is sometimes necessary to refer to the graphs themselves, *in addition* to solving the two equations simultaneously.

An alternative method is to use different (equivalent) equations for the graphs. See Discussion Exercise 12 at the end of the chapter.

11.5 *Exercises*

1 Which polar coordinates represent the same point as $(3, \pi/3)$?

(a) $(3, 7\pi/3)$ (b) $(3, -\pi/3)$

(c) $(-3, 4\pi/3)$ (d) $(3, -2\pi/3)$

(e) $(-3, -2\pi/3)$ (f) $(-3, -\pi/3)$

2 Which polar coordinates represent the same point as $(4, -\pi/2)$?

(a) $(4, 5\pi/2)$ (b) $(4, 7\pi/2)$

(c) $(-4, -\pi/2)$ (d) $(4, -5\pi/2)$

(e) $(-4, -3\pi/2)$ (f) $(-4, \pi/2)$

Exer. 3–8: Change the polar coordinates to rectangular coordinates.

3 (a) $(3, \pi/4)$ (b) $(-1, 2\pi/3)$

4 (a) $(5, 5\pi/6)$ (b) $(-6, 7\pi/3)$

5 (a) $(8, -2\pi/3)$ (b) $(-3, 5\pi/3)$

6 (a) $(4, -\pi/4)$ (b) $(-2, 7\pi/6)$

7 $\left(6, \arctan \frac{3}{4}\right)$

8 $\left(10, \arccos\left(-\frac{1}{3}\right)\right)$

Exer. 9–12: Change the rectangular coordinates to polar coordinates with $r > 0$ and $0 \le \theta \le 2\pi$.

9 (a) $(-1, 1)$ (b) $\left(-2\sqrt{3}, -2\right)$

10 (a) $\left(3\sqrt{3}, 3\right)$ (b) $(2, -2)$

11 (a) $\left(7, -7\sqrt{3}\right)$ (b) $(5, 5)$

12 (a) $\left(-2\sqrt{2}, -2\sqrt{2}\right)$ (b) $\left(-4, 4\sqrt{3}\right)$

Exer. 13–26: Find a polar equation that has the same graph as the equation in x and y.

13 $x = -3$ 14 $y = 2$

15 $x^2 + y^2 = 16$ 16 $x^2 + y^2 = 2$

17 $y^2 = 6x$ 18 $x^2 = 8y$

19 $x + y = 3$ 20 $2y = -x + 4$

21 $2y = -x$ 22 $y = 6x$

23 $y^2 - x^2 = 4$ 24 $xy = 8$

25 $(x - 1)^2 + y^2 = 1$

26 $(x + 2)^2 + (y - 3)^2 = 13$

Exer. 27–44: Find an equation in x and y that has the same graph as the polar equation. Use it to help sketch the graph in an $r\theta$-plane.

27 $r \cos \theta = 5$ 28 $r \sin \theta = -2$

29 $r - 6 \sin \theta = 0$ 30 $r = 2$

31 $\theta = \pi/4$

32 $r = 4 \sec \theta$

33 $r^2(4 \sin^2 \theta - 9 \cos^2 \theta) = 36$

34 $r^2(\cos^2 \theta + 4 \sin^2 \theta) = 16$

35 $r^2 \cos 2\theta = 1$

36 $r^2 \sin 2\theta = 4$

37 $r(\sin \theta - 2 \cos \theta) = 6$

38 $r(3 \cos \theta - 4 \sin \theta) = 12$

39 $r(\sin \theta + r \cos^2 \theta) = 1$

40 $r(r \sin^2 \theta - \cos \theta) = 3$

41 $r = 8 \sin \theta - 2 \cos \theta$

42 $r = 2 \cos \theta - 4 \sin \theta$

43 $r = \tan \theta$

44 $r = 6 \cot \theta$

Exer. 45–78: Sketch the graph of the polar equation.

45 $r = 5$

46 $r = -2$

47 $\theta = -\pi/6$

48 $\theta = \pi/4$

49 $r = 3 \cos \theta$

50 $r = -2 \sin \theta$

51 $r = 4 \cos \theta + 2 \sin \theta$

52 $r = 6 \cos \theta - 2 \sin \theta$

53 $r = 4(1 - \sin \theta)$

54 $r = 3(1 + \cos \theta)$

55 $r = -6(1 + \cos \theta)$

56 $r = 2(1 + \sin \theta)$

57 $r = 2 + 4 \sin \theta$

58 $r = 1 + 2 \cos \theta$

59 $r = \sqrt{3} - 2 \sin \theta$

60 $r = 2\sqrt{3} - 4 \cos \theta$

61 $r = 2 - \cos \theta$

62 $r = 5 + 3 \sin \theta$

63 $r = 4 \csc \theta$

64 $r = -3 \sec \theta$

65 $r = 8 \cos 3\theta$

66 $r = 2 \sin 4\theta$

67 $r = 3 \sin 2\theta$

68 $r = 8 \cos 5\theta$

69 $r^2 = 4 \cos 2\theta$ (lemniscate)

70 $r^2 = -16 \sin 2\theta$

71 $r = 2^\theta, \theta \geq 0$ (spiral)

72 $r = e^{2\theta}, \theta \geq 0$ (logarithmic spiral)

73 $r = 2\theta, \theta \geq 0$

74 $r\theta = 1, \theta > 0$ (spiral)

75 $r = 6 \sin^2 (\theta/2)$

76 $r = -4 \cos^2 (\theta/2)$

77 $r = 2 + 2 \sec \theta$ (conchoid)

78 $r = 1 - \csc \theta$

79 If $P_1(r_1, \theta_1)$ and $P_2(r_2, \theta_2)$ are points in an $r\theta$-plane, use the law of cosines to prove that
$$[d(P_1, P_2)]^2 = r_1^2 + r_2^2 - 2r_1r_2 \cos (\theta_2 - \theta_1).$$

80 Prove that the graph of each polar equation is a circle, and find its center and radius.

(a) $r = a \sin \theta, a \neq 0$

(b) $r = b \cos \theta, b \neq 0$

(c) $r = a \sin \theta + b \cos \theta, a \neq 0$ and $b \neq 0$

11.6

Polar Equations of Conics

The following theorem combines the definitions of parabola, ellipse, and hyperbola into a unified description of the conic sections. The constant e in the statement of the theorem is the **eccentricity** of the conic. The point F is a **focus** of the conic, and the line l is a **directrix.** Possible positions of F and l are illustrated in Figure 1.

Figure 1

l

Q

P

F
Focus

Directrix

Theorem on Conics	Let F be a fixed point and l a fixed line in a plane. The set of all points P in the plane, such that the ratio $d(P, F)/d(P, Q)$ is a positive constant e with $d(P, Q)$ the distance from P to l, is a conic section. The conic is a parabola if $e = 1$, an ellipse if $0 < e < 1$, and a hyperbola if $e > 1$.

PROOF   If $e = 1$, then $d(P, F) = d(P, Q)$, and, by definition, the resulting conic is a parabola with focus F and directrix l.

Suppose next that $0 < e < 1$. It is convenient to introduce a polar coordinate system in the plane with F as the pole and l perpendicular to the polar axis at the point $D(d, 0)$, with $d > 0$, as illustrated in Figure 2. If $P(r, \theta)$ is a point in the plane such that $d(P, F)/d(P, Q) = e < 1$, then P lies to the left of l. Let C be the projection of P on the polar axis. Since

$$d(P, F) = r \qquad \text{and} \qquad d(P, Q) = d - r \cos \theta,$$

it follows that P satisfies the condition in the theorem if and only if the following are true:

$$\frac{r}{d - r \cos \theta} = e$$

$$r = de - er \cos \theta$$

$$r(1 + e \cos \theta) = de$$

$$r = \frac{de}{1 + e \cos \theta}$$

Figure 2

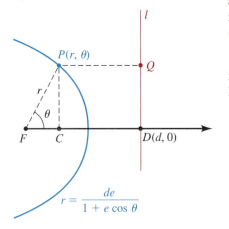

l

$P(r, \theta)$

Q

r

θ

F C $D(d, 0)$

$$r = \frac{de}{1 + e \cos \theta}$$

The same equations are obtained if $e = 1$; however, there is no point (r, θ) on the graph if $1 + \cos \theta = 0$.

An equation in x and y corresponding to $r = de - er \cos \theta$ is

$$\sqrt{x^2 + y^2} = de - ex.$$

Squaring both sides and rearranging terms leads to

$$(1 - e^2)x^2 + 2de^2x + y^2 = d^2e^2.$$

Completing the square and simplifying, we obtain

$$\left(x + \frac{de^2}{1 - e^2} \right)^2 + \frac{y^2}{1 - e^2} = \frac{d^2e^2}{(1 - e^2)^2}.$$

Finally, dividing both sides by $d^2e^2/(1 - e^2)^2$ gives us an equation of the form

$$\frac{(x - h)^2}{a^2} + \frac{y^2}{b^2} = 1,$$

with $h = -de^2/(1 - e^2)$. Consequently, the graph is an ellipse with center at the point $(h, 0)$ on the x-axis and with

$$a^2 = \frac{d^2e^2}{(1 - e^2)^2} \qquad \text{and} \qquad b^2 = \frac{d^2e^2}{1 - e^2}.$$

Since

$$c^2 = a^2 - b^2 = \frac{d^2e^4}{(1 - e^2)^2},$$

we obtain $c = de^2/(1 - e^2)$, and hence $|h| = c$. This proves that F is a focus of the ellipse. It also follows that $e = c/a$. A similar proof may be given for the case $e > 1$. ◢

We also can show that every conic that is not degenerate may be described by means of the statement in the theorem on conics. This gives us a formulation of conic sections that is equivalent to the one used previously. Since the theorem includes all three types of conics, it is sometimes regarded as a definition for the conic sections.

If we had chosen the focus F to the *right* of the directrix, as illustrated in Figure 3 (with $d > 0$), then the equation $r = de/(1 - e \cos \theta)$ would have resulted. (Note the minus sign in place of the plus sign.) Other sign changes occur if d is allowed to be negative.

If we had taken l *parallel* to the polar axis through one of the points $(d, \pi/2)$ or $(d, 3\pi/2)$, as illustrated in Figure 4, then the corresponding equations would have contained $\sin \theta$ instead of $\cos \theta$.

Figure 3

Figure 4

(a)

(b)

The following theorem summarizes our discussion.

Theorem on Polar Equations of Conics	A polar equation that has one of the four forms $$r = \frac{de}{1 \pm e \cos \theta} \quad \text{or} \quad r = \frac{de}{1 \pm e \sin \theta}$$ is a conic section. The conic is a parabola if $e = 1$, an ellipse if $0 < e < 1$, or a hyperbola if $e > 1$.

EXAMPLE 1 Sketching the graph of a polar equation of an ellipse

Describe and sketch the graph of the polar equation

$$r = \frac{10}{3 + 2 \cos \theta}.$$

SOLUTION   We first divide the numerator and denominator of the fraction by 3 to obtain the constant term 1 in the denominator:

$$r = \frac{\frac{10}{3}}{1 + \frac{2}{3} \cos \theta}$$

This equation has one of the forms in the preceding theorem, with $e = \frac{2}{3}$. Thus, the graph is an ellipse with focus F at the pole and major axis along the polar axis. We find the endpoints of the major axis by letting $\theta = 0$ and $\theta = \pi$. This gives us the points $V(2, 0)$ and $V'(10, \pi)$. Hence,

$$2a = d(V', V) = 12, \quad \text{or} \quad a = 6.$$

The center of the ellipse is the midpoint $(4, \pi)$ of the segment $V'V$. Using the fact that $e = c/a$, we obtain

$$c = ae = 6\left(\tfrac{2}{3}\right) = 4.$$

Hence, $b^2 = a^2 - c^2 = 6^2 - 4^2 = 36 - 16 = 20.$

Thus, $b = \sqrt{20}$. The graph is sketched in Figure 5. For reference, we have superimposed an xy-coordinate system on the polar system.   ◢

Figure 5

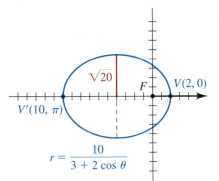

EXAMPLE 2 Sketching the graph of a polar equation of a hyperbola

Describe and sketch the graph of the polar equation

$$r = \frac{10}{2 + 3 \sin \theta}.$$

SOLUTION To express the equation in the proper form, we divide the numerator and denominator of the fraction by 2:

$$r = \frac{5}{1 + \frac{3}{2} \sin \theta}$$

Thus, $e = \frac{3}{2}$, and, by the theorem on polar equations of conics, the graph is a hyperbola with a focus at the pole. The expression $\sin \theta$ tells us that the transverse axis of the hyperbola is perpendicular to the polar axis. To find the vertices, we let $\theta = \pi/2$ and $\theta = 3\pi/2$ in the given equation. This gives us the points $V(2, \pi/2)$ and $V'(-10, 3\pi/2)$. Hence,

$$2a = d(V, V') = 8, \quad \text{or} \quad a = 4.$$

The points $(5, 0)$ and $(5, \pi)$ on the graph can be used to sketch the lower branch of the hyperbola. The upper branch is obtained by symmetry, as illustrated in Figure 6. If we desire more accuracy or additional information, we calculate

$$c = ae = 4(\tfrac{3}{2}) = 6$$

and

$$b^2 = c^2 - a^2 = 6^2 - 4^2 = 36 - 16 = 20.$$

Asymptotes may then be constructed in the usual way. ◢

Figure 6

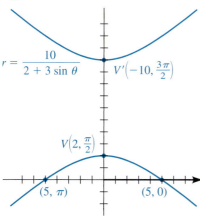

$$r = \frac{10}{2 + 3 \sin \theta}$$

$$V'\left(-10, \tfrac{3\pi}{2}\right)$$

$$V\left(2, \tfrac{\pi}{2}\right)$$

$$(5, \pi) \qquad (5, 0)$$

EXAMPLE 3 Sketching the graph of a polar equation of a parabola

Sketch the graph of the polar equation

$$r = \frac{15}{4 - 4 \cos \theta}.$$

SOLUTION To obtain the proper form, we divide the numerator and denominator by 4:

$$r = \frac{\frac{15}{4}}{1 - \cos \theta}$$

Consequently, $e = 1$, and, by the theorem on polar equations of conics, the graph is a parabola with focus at the pole. We may obtain a sketch by plotting the points that correspond to the quadrantal angles indicated in the following table.

(continued)

Figure 7

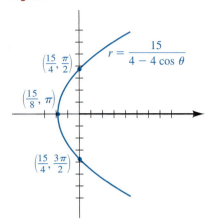

$$\left(\tfrac{15}{4}, \tfrac{\pi}{2}\right)$$

$$r = \frac{15}{4 - 4\cos\theta}$$

$$\left(\tfrac{15}{8}, \pi\right)$$

$$\left(\tfrac{15}{4}, \tfrac{3\pi}{2}\right)$$

θ	0	$\dfrac{\pi}{2}$	π	$\dfrac{3\pi}{2}$
r	undefined	$\tfrac{15}{4}$	$\tfrac{15}{8}$	$\tfrac{15}{4}$

Note that there is no point on the graph corresponding to $\theta = 0$, since the denominator $1 - \cos\theta$ is 0 for that value. Plotting the three points and using the fact that the graph is a parabola with focus at the pole gives us the sketch in Figure 7. ◢

If we desire only a rough sketch of a conic, then the technique employed in Example 3 is recommended. To use this method, we plot (if possible) points corresponding to $\theta = 0$, $\pi/2$, π, and $3\pi/2$. These points, together with the type of conic (obtained from the value of the eccentricity e), readily lead to the sketch.

EXAMPLE 4   **Expressing a polar equation of a conic in terms of x and y**

Find an equation in x and y that has the same graph as the polar equation

$$r = \frac{15}{4 - 4\cos\theta}.$$

SOLUTION

$r(4 - 4\cos\theta) = 15$	multiply by the lcd
$4r - 4r\cos\theta = 15$	distribute
$4\left(\pm\sqrt{x^2 + y^2}\right) - 4x = 15$	substitute for r and $r\cos\theta$
$4\left(\pm\sqrt{x^2 + y^2}\right) = 15 + 4x$	isolate the radical term
$16(x^2 + y^2) = 225 + 120x + 16x^2$	square both sides
$16y^2 = 225 + 120x$	simplify

We may write the last equation as $x = \frac{16}{120}y^2 - \frac{225}{120}$ or, simplified, $x = \frac{2}{15}y^2 - \frac{15}{8}$. We recognize this equation as that of a parabola with vertex $V\left(-\frac{15}{8}, 0\right)$ and opening to the right. Its graph on an xy-coordinate system would be the same as the graph in Figure 7. ◢

EXAMPLE 5   **Finding a polar equation of a conic satisfying prescribed conditions**

Find a polar equation of the conic with a focus at the pole, eccentricity $e = \frac{1}{2}$, and directrix $r = -3\sec\theta$.

SOLUTION   The equation $r = -3\sec\theta$ of the directrix may be written $r\cos\theta = -3$, which is equivalent to $x = -3$ in a rectangular coordinate system. This gives us the situation illustrated in Figure 3, with $d = 3$. Hence, a polar equation has the form

$$r = \frac{de}{1 - e\cos\theta}.$$

We now substitute $d = 3$ and $e = \frac{1}{2}$:

$$r = \frac{3\left(\frac{1}{2}\right)}{1 - \frac{1}{2}\cos\theta} \quad \text{or, equivalently,} \quad r = \frac{3}{2 - \cos\theta}$$

11.6   *Exercises*

Exer. 1–12: Find the eccentricity, and classify the conic. Sketch the graph, and label the vertices.

1 $r = \dfrac{12}{6 + 2\sin\theta}$

2 $r = \dfrac{12}{6 - 2\sin\theta}$

3 $r = \dfrac{12}{2 - 6\cos\theta}$

4 $r = \dfrac{12}{2 + 6\cos\theta}$

5 $r = \dfrac{3}{2 + 2\cos\theta}$

6 $r = \dfrac{3}{2 - 2\sin\theta}$

7 $r = \dfrac{4}{\cos\theta - 2}$

8 $r = \dfrac{4\sec\theta}{2\sec\theta - 1}$

9 $r = \dfrac{6\csc\theta}{2\csc\theta + 3}$

10 $r = \dfrac{8\csc\theta}{2\csc\theta - 5}$

11 $r = \dfrac{4\csc\theta}{1 + \csc\theta}$

12 $r = \csc\theta\,(\csc\theta - \cot\theta)$

Exer. 13–24: Find equations in x and y for the polar equations in Exercises 1–12.

Exer. 25–32: Find a polar equation of the conic with focus at the pole that has the given eccentricity and equation of directrix.

25 $e = \frac{1}{3}, \quad r = 2\sec\theta$

26 $e = 1, \quad r\cos\theta = 5$

27 $e = \frac{4}{3}, \quad r\cos\theta = -3$

28 $e = 3, \quad r = -4\sec\theta$

29 $e = 1, \quad r\sin\theta = -2$

30 $e = 4, \quad r = -3\csc\theta$

31 $e = \frac{2}{5}, \quad r = 4\csc\theta$

32 $e = \frac{3}{4}, \quad r\sin\theta = 5$

Exer. 33–34: Find a polar equation of the parabola with focus at the pole and the given vertex.

33 $V\left(4, \dfrac{\pi}{2}\right)$

34 $V(5, 0)$

Exer. 35–36: An ellipse has a focus at the pole with the given center C and vertex V. Find (a) the eccentricity and (b) a polar equation for the ellipse.

35 $C\left(3, \dfrac{\pi}{2}\right), V\left(1, \dfrac{3\pi}{2}\right)$

36 $C(2, \pi), V(1, 0)$

37 Kepler's first law Kepler's first law asserts that planets travel in elliptical orbits with the sun at one focus. To find an equation of an orbit, place the pole O at the center of the sun and the polar axis along the major axis of the ellipse (see the figure).

(a) Show that an equation of the orbit is

$$r = \frac{(1 - e^2)a}{1 - e\cos\theta},$$

where e is the eccentricity and $2a$ is the length of the major axis.

(b) The perihelion distance r_{per} and aphelion distance r_{aph} are defined as the minimum and maximum distances, respectively, of a planet from the sun. Show that

$$r_{\text{per}} = a(1 - e) \quad \text{and} \quad r_{\text{aph}} = a(1 + e).$$

Exercise 37

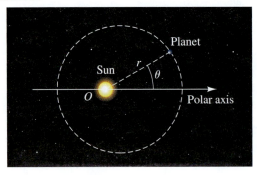

38 Kepler's first law Refer to Exercise 37. The planet Pluto travels in an elliptical orbit of eccentricity 0.249. If the perihelion distance is 29.62 AU, find a polar equation for the orbit and estimate the aphelion distance.

39 Earth's orbit The closest Earth gets to the sun is about 91,405,950 miles, and the farthest Earth gets from the sun is about 94,505,420 miles. Referring to the formulas in Exercise 37, find formulas for a and e in terms of r_{per} and r_{aph}.

CHAPTER 11 REVIEW EXERCISES

Exer. 1–16: Find the vertices and foci of the conic, and sketch its graph.

1 $y^2 = 64x$

2 $y = 8x^2 + 32x + 33$

3 $9y^2 = 144 - 16x^2$

4 $9y^2 = 144 + 16x^2$

5 $x^2 - y^2 - 4 = 0$

6 $25x^2 + 36y^2 = 1$

7 $25y = 100 - x^2$

8 $3x^2 + 4y^2 - 18x + 8y + 19 = 0$

9 $x^2 - 9y^2 + 8x + 90y - 210 = 0$

10 $x = 2y^2 + 8y + 3$

11 $4x^2 + 9y^2 + 24x - 36y + 36 = 0$

12 $4x^2 - y^2 - 40x - 8y + 88 = 0$

13 $y^2 - 8x + 8y + 32 = 0$

14 $4x^2 + y^2 - 24x + 4y + 36 = 0$

15 $x^2 - 9y^2 + 8x + 7 = 0$

16 $y^2 - 2x^2 + 6y + 8x - 3 = 0$

Exer. 17–18: Find the standard equation of a parabola with a vertical axis that satisfies the given conditions.

17 x-intercepts -10 and -4, y-intercept 80

18 x-intercepts -11 and 3, passing through $(2, 39)$

Exer. 19–28: Find an equation for the conic that satisfies the given conditions.

19 Hyperbola, with vertices $V(0, \pm 7)$ and endpoints of conjugate axis $(\pm 3, 0)$

20 Parabola, with focus $F(-4, 0)$ and directrix $x = 4$

21 Parabola, with focus $F(0, -10)$ and directrix $y = 10$

22 Parabola, with vertex at the origin, symmetric to the x-axis, and passing through the point $(5, -1)$

23 Ellipse, with vertices $V(0, \pm 10)$ and foci $F(0, \pm 5)$

24 Hyperbola, with foci $F(\pm10, 0)$ and vertices $V(\pm5, 0)$

25 Hyperbola, with vertices $V(0, \pm6)$ and asymptotes $y = \pm9x$

26 Ellipse, with foci $F(\pm2, 0)$ and passing through the point $(2, \sqrt{2})$

27 Ellipse, with eccentricity $\frac{2}{3}$ and endpoints of minor axis $(\pm5, 0)$

28 Ellipse, with eccentricity $\frac{3}{4}$ and foci $F(\pm12, 0)$

29 (a) Determine A so that the point $(2, -3)$ is on the conic $Ax^2 + 2y^2 = 4$.

(b) Is the conic an ellipse or a hyperbola?

30 If a square with sides parallel to the coordinate axes is inscribed in the ellipse $(x^2/a^2) + (y^2/b^2) = 1$, express the area A of the square in terms of a and b.

31 Find the standard equation of the circle that has center at the focus of the parabola $y = \frac{1}{8}x^2$ and passes through the origin.

32 **Focal length and angular velocity** A cylindrical container, partially filled with mercury, is rotated about its axis so that the angular speed of each cross section is ω radians/second. From physics, the function f, whose graph generates the inside surface of the mercury (see the figure), is given by

$$f(x) = \frac{1}{64}\omega^2 x^2 + k,$$

where k is a constant. Determine the angular speed ω that will result in a focal length of 2 feet.

Exercise 32

$y = f(x)$

Exer. 33–37: **Find an equation in x and y whose graph contains the points on the curve C. Sketch the graph of C, and indicate the orientation.**

33 $x = 3 + 4t$, $\quad y = t - 1$; $\quad -2 \le t \le 2$

34 $x = \sqrt{-t}$, $\quad y = t^2 - 4$; $\quad t \le 0$

35 $x = \cos^2 t - 2$, $\quad y = \sin t + 1$; $\quad 0 \le t \le 2\pi$

36 $x = \sqrt{t}$, $\quad y = 2^{-t}$; $\quad t \ge 0$

37 $x = \dfrac{1}{t} + 1$, $\quad y = \dfrac{2}{t} - t$; $\quad 0 < t \le 4$

38 Curves C_1, C_2, C_3, and C_4 are given parametrically for t in $\mathbb{R}$. Sketch their graphs, and discuss their similarities and differences.

$C_1: x = t$, $\quad y = \sqrt{16 - t^2}$
$C_2: x = -\sqrt{16 - t}$, $\quad y = -\sqrt{t}$
$C_3: x = 4\cos t$, $\quad y = 4\sin t$
$C_4: x = e^t$, $\quad y = -\sqrt{16 - e^{2t}}$

39 Refer to the equations in (1) of Example 5 in Section 11.4. Find the range and maximum altitude for $s = 1024$, $\alpha = 30°$, and $h = 5120$.

40 List two polar coordinate points that represent the same point as $(2, \pi/4)$.

41 Change $(5, 7\pi/4)$ to rectangular coordinates.

42 Change $(2\sqrt{3}, -2)$ to polar coordinates with $r > 0$ and $0 \le \theta < 2\pi$.

Exer. 43–46: **Find a polar equation that has the same graph as the equation in x and y.**

43 $y^2 = 4x$

44 $x^2 + y^2 - 3x + 4y = 0$

45 $2x - 3y = 8$

46 $x^2 + y^2 = 2xy$

Exer. 47–52: **Find an equation in x and y that has the same graph as the polar equation.**

47 $r^2 = \tan\theta$

48 $r = 2\cos\theta + 3\sin\theta$

49 $r^2 = 4\sin 2\theta$

50 $\theta = \sqrt{3}$

51 $r = 5\sec\theta + 3r\sec\theta$

52 $r^2\sin\theta = 6\csc\theta + r\cot\theta$

Exer. 53–64: Sketch the graph of the polar equation.

53 $r = -4 \sin \theta$

54 $r = 8 \sec \theta$

55 $r = 3 \sin 5\theta$

56 $r = 6 - 3 \cos \theta$

57 $r = 3 - 3 \sin \theta$

58 $r = 2 + 4 \cos \theta$

59 $r^2 = 9 \sin 2\theta$

60 $2r = \theta$

61 $r = \dfrac{8}{1 - 3 \sin \theta}$

62 $r = 6 - r \cos \theta$

63 $r = \dfrac{6}{3 + 2 \cos \theta}$

64 $r = \dfrac{-6 \csc \theta}{1 - 2 \csc \theta}$

CHAPTER 11 DISCUSSION EXERCISES

1 On a parabola, the line segment through the focus, perpendicular to the axis, and intercepted by the parabola is called the *focal chord* or *latus rectum*. The length of the focal chord is called the *focal width*. Find a formula for the focal width w in terms of the focal length $|p|$.

2 On the graph of a hyperbola with center at the origin O, draw a circle with center at the origin and radius $r = d(O, F)$, where F denotes a focus of the hyperbola. What relationship do you observe?

3 A point $P(x, y)$ is on an ellipse if and only if

$$d(P, F) + d(P, F') = 2a.$$

If $b^2 = a^2 - c^2$, derive the general equation of an ellipse — that is,

$$\frac{x^2}{a^2} + \frac{y^2}{b^2} = 1.$$

4 A point $P(x, y)$ is on a hyperbola if and only if

$$|d(P, F) - d(P, F')| = 2a.$$

If $c^2 = a^2 + b^2$, derive the general equation of a hyperbola — that is,

$$\frac{x^2}{a^2} - \frac{y^2}{b^2} = 1.$$

5 A point $P(x, y)$ is the same distance from $(4, 0)$ as it is from the circle $x^2 + y^2 = 4$, as illustrated in the figure. Show that the collection of all such points forms a branch of a hyperbola, and sketch its graph.

Exercise 5

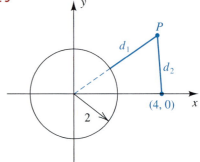

6 **Design of a telescope** Refer to Exercise 64 in Section 11.3. Suppose the upper branch of the hyperbola (shown) has equation $y = \dfrac{a}{b} \sqrt{x^2 + b^2}$ and an equation of the parabola is $y = dx^2$. Find d in terms of a and b.

Exercise 6

7 Maximizing a projectile's range As in Example 5 in Section 11.4, suppose a projectile is to be fired at a speed of 1024 ft/sec from a height of 2304 feet. Approximate the angle that maximizes the range.

8 Generalizations for a projectile's path If $h = 0$, the equations in (1) of Example 5 in Section 11.4 become

$$x(t) = (s \cos \alpha)t, \quad y(t) = -\tfrac{1}{2}gt^2 + (s \sin \alpha)t; \quad t \geq 0.$$

Show that each statement is true.

(a) The projectile strikes the ground when

$$t = \frac{2s \sin \alpha}{g}.$$

(b) The range r of the projectile is

$$r = \frac{s^2 \sin 2\alpha}{g}.$$

(c) The angle that maximizes the range r is 45°.

(d) The path of the projectile in rectangular coordinates is

$$y = -\frac{g}{2s^2 \cos^2 \alpha}x^2 + (\tan \alpha)x.$$

(e) The time at which the maximum height is reached is

$$t = \frac{s \sin \alpha}{g}.$$

(f) The maximum height reached is

$$y = \frac{s^2 \sin^2 \alpha}{2g}.$$

9 Investigating a Lissajous figure Find an equation in x and y for the curve given by

$$x = \sin 2t, \quad y = \cos t; \quad 0 \leq t \leq 2\pi.$$

10 Sketch the graphs of the equations $r = f(\theta) = 2 + 4 \cos \theta$, $r = f(\theta - \alpha)$, and $r = f(\theta + \alpha)$ for $\alpha = \pi/4$. Try as many values of α as necessary to generalize results concerning the graphs of $r = f(\theta - \alpha)$ and $r = f(\theta + \alpha)$, where $\alpha > 0$.

11 Generalized roses Examine the graph of $r = \sin n\theta$ for odd values of n and even values of n. Derive an expression for the *leaf angle* (the number of degrees between consecutive pole values). What other generalizations do you observe? How do the graphs change if sin is replaced by cos?

12 Figure 17 of Section 11.5 shows the circles $r = 4 \sin \theta$ and $r = 4 \cos \theta$. Solve this system of equations for (r, θ) solutions. Now find equations in x and y that have the same graphs as the polar equations. Solve this system for (x, y) solutions, convert them to (r, θ) solutions, and explain why your answer to the first system did not reveal the solution at the pole.

Appendixes

I   **Common Graphs and Their Equations**

II   **A Summary of Graph Transformations**

III   **Graphs of Trigonometric Functions and Their Inverses**

IV   **Values of the Trigonometric Functions of Special Angles on a Unit Circle**

APPENDIX I

Common Graphs and Their Equations

(Graphs of conics appear on the back endpaper of this text.)

Horizontal line; constant function

Vertical line

Identity function

Absolute value function

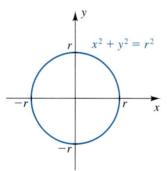

Circle with center $(0, 0)$ and radius r

Semicircles

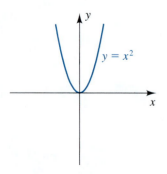

Parabola with vertical axis; squaring function

Parabola with horizontal axis

Square root function

Cube root function

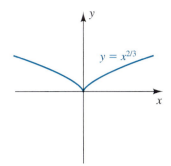

A graph with a cusp at the origin

Cubing function

Greatest integer function

Reciprocal function

A rational function

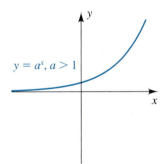

Exponential growth function
(includes natural exponential
function)

Exponential decay function

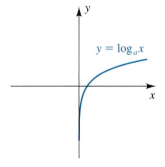

Logarithmic function
(includes common and
natural logarithmic
functions)

APPENDIX II
A Summary of Graph Transformations

The graph of $y = f(x)$ is shown in black in each figure. The domain of f is $[-1, 3]$ and the range of f is $[-4, 3]$.

$$y = g(x) = f(x) + 3$$

The graph of f is shifted vertically upward 3 units.
Domain of g: $[-1, 3]$ Range of g: $[-1, 6]$

$$y = h(x) = f(x) - 4$$

The graph of f is shifted vertically downward 4 units.
Domain of h: $[-1, 3]$ Range of h: $[-8, -1]$

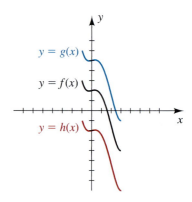

$$y = g(x) = f(x - 3)$$

The graph of f is shifted horizontally to the right 3 units.
Domain of g: $[2, 6]$ Range of g: $[-4, 3]$

$$y = h(x) = f(x + 6)$$

The graph of f is shifted horizontally to the left 6 units.
Domain of h: $[-7, -3]$ Range of h: $[-4, 3]$

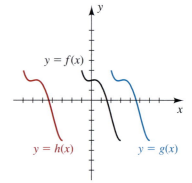

$$y = g(x) = 2f(x) \quad [2 > 1]$$

The graph of f is stretched vertically by a factor of 2.
Domain of g: $[-1, 3]$ Range of g: $[-8, 6]$

$$y = h(x) = \tfrac{1}{2}f(x) \quad \left[\tfrac{1}{2} < 1\right]$$

The graph of f is compressed vertically by a factor of 2.
Domain of h: $[-1, 3]$ Range of h: $\left[-2, \tfrac{3}{2}\right]$

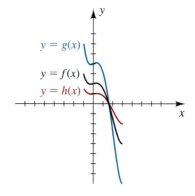

$$y = g(x) = f(2x) \quad [2 > 1]$$

The graph of f is compressed horizontally by a factor of 2.
Domain of g: $\left[-\frac{1}{2}, \frac{3}{2}\right]$ Range of g: $[-4, 3]$

$$y = h(x) = f\left(\tfrac{1}{2}x\right) \quad \left[\tfrac{1}{2} < 1\right]$$

The graph of f is stretched horizontally by a factor of 2.
Domain of h: $[-2, 6]$ Range of h: $[-4, 3]$

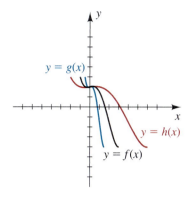

$$y = g(x) = -f(x)$$

The graph of f is reflected through the x-axis.
Domain of g: $[-1, 3]$ Range of g: $[-3, 4]$

$$y = h(x) = f(-x)$$

The graph of f is reflected through the y-axis.
Domain of h: $[-3, 1]$ Range of h: $[-4, 3]$

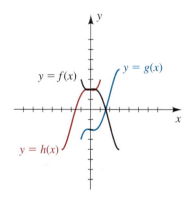

$$y = g(x) = |f(x)|$$

Reflect points on f with negative y-values through the x-axis.
Domain of g: $[-1, 3]$ Range of g: $[0, 4]$

$$y = h(x) = f(|x|)$$

Reflect points on f with positive x-values through the y-axis.
Domain of h: $[-3, 3]$ Range of h: $[-4, 3]$ at most.
In this case, the range is a
subset of $[-4, 3]$.

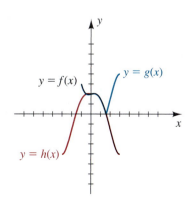

801

APPENDIX III

Graphs of Trigonometric Functions and Their Inverses

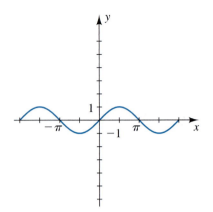

$y = \sin x$

Domain: $\mathbb{R}$

Range: $[-1, 1]$

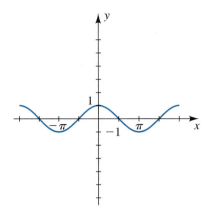

$y = \cos x$

Domain: $\mathbb{R}$

Range: $[-1, 1]$

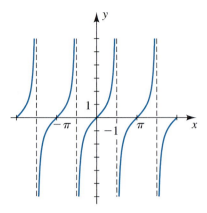

$y = \tan x$

Domain: $x \neq \dfrac{\pi}{2} + \pi n$

Range: $\mathbb{R}$

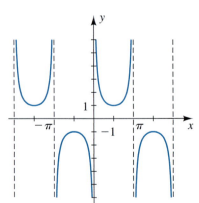

$y = \csc x$

Domain: $x \neq \pi n$

Range: $(-\infty, -1] \cup [1, \infty)$

$y = \sec x$

Domain: $x \neq \dfrac{\pi}{2} + \pi n$

Range: $(-\infty, -1] \cup [1, \infty)$

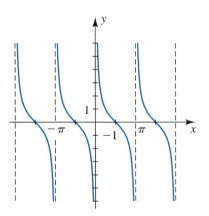

$y = \cot x$

Domain: $x \neq \pi n$

Range: $\mathbb{R}$

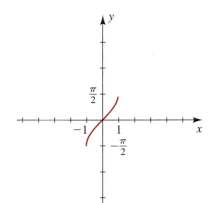

$y = \sin^{-1} x$

Domain: $[-1, 1]$

Range: $\left[-\dfrac{\pi}{2}, \dfrac{\pi}{2}\right]$

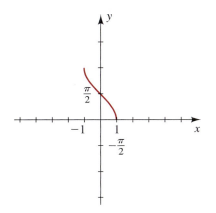

$y = \cos^{-1} x$

Domain: $[-1, 1]$

Range: $[0, \pi]$

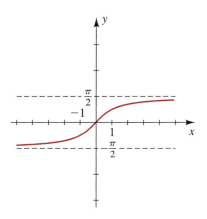

$y = \tan^{-1} x$

Domain: $\mathbb{R}$

Range: $\left(-\dfrac{\pi}{2}, \dfrac{\pi}{2}\right)$

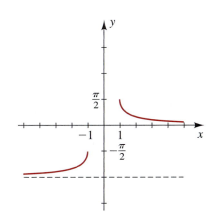

$y = \csc^{-1} x$

Domain: $(-\infty, -1] \cup [1, \infty)$

Range: $\left(-\pi, -\dfrac{\pi}{2}\right] \cup \left(0, \dfrac{\pi}{2}\right]$

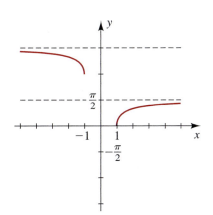

$y = \sec^{-1} x$

Domain: $(-\infty, -1] \cup [1, \infty)$

Range: $\left[0, \dfrac{\pi}{2}\right) \cup \left[\pi, \dfrac{3\pi}{2}\right)$

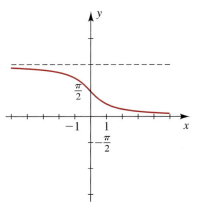

$y = \cot^{-1} x$

Domain: $\mathbb{R}$

Range: $(0, \pi)$

APPENDIX IV

Values of the Trigonometric Functions of Special Angles on a Unit Circle

$$P(x, y) = P(\cos t, \sin t)$$

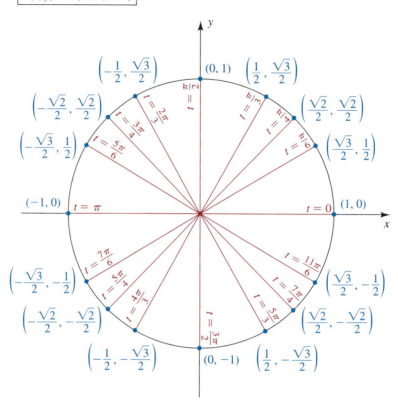

To find the values of the other trigonometric functions, use the following definitions:

$$\tan t = \frac{y}{x} \text{ (if } x \neq 0) \qquad \cot t = \frac{x}{y} \text{ (if } y \neq 0)$$

$$\sec t = \frac{1}{x} \text{ (if } x \neq 0) \qquad \csc t = \frac{1}{y} \text{ (if } y \neq 0)$$

Answers to Selected Exercises

A *Student's Solutions Manual* to accompany this textbook is available from your college bookstore. The guide contains detailed solutions to approximately one-half of the exercises, as well as strategies for solving other exercises in the text.

Chapter 1

EXERCISES 1.1

1 (a) Negative   (b) Positive   (c) Negative
(d) Positive
3 (a) $<$ (b) $>$ (c) $=$
5 (a) $>$ (b) $>$ (c) $>$
7 (a) $x < 0$ (b) $y \geq 0$ (c) $q \leq \pi$ (d) $2 < d < 4$
(e) $t \geq 5$ (f) $-z \leq 3$ (g) $\dfrac{p}{q} \leq 7$ (h) $\dfrac{1}{w} \geq 9$
(i) $|x| > 7$ 9 (a) 5   (b) 3   (c) 11
11 (a) -15 (b) -3 (c) 11
13 (a) $4 - \pi$ (b) $4 - \pi$ (c) $1.5 - \sqrt{2}$
15 (a) 4   (b) 12   (c) 12   (d) 8
17 (a) 10   (b) 9   (c) 9   (d) 19   19 $|7 - x| < 5$
21 $|-3 - x| \geq 8$ 23 $|x - 4| \leq 3$ 25 $-x - 3$
27 $2 - x$ 29 $b - a$ 31 $x^2 + 4$ 33 $\neq$ 35 $=$
37 $\neq$ 39 $=$ 41 (a) 8.4652   (b) 14.1428
43 (a) 6.557×10^{-1} (b) 6.708×10^1
45 Construct a right triangle with sides of lengths $\sqrt{2}$ and 1. The hypotenuse will have length $\sqrt{3}$. Next, construct a right triangle with sides of lengths $\sqrt{3}$ and $\sqrt{2}$. The hypotenuse will have length $\sqrt{5}$.
47 The large rectangle has area $a(b + c)$. The sum of the areas of the two small rectangles is $ab + ac$.
49 (a) 4.27×10^5 (b) 9.8×10^{-8} (c) 8.1×10^8
51 (a) 830,000   (b) 0.000 000 000 002 9
(c) 563,000,000
53 1.7×10^{-24} 55 5.87×10^{12} 57 1.678×10^{-24} g
59 4.1472×10^6 frames
61 (a) 201.6 lb   (b) 32.256 tons

EXERCISES 1.2

1 $\dfrac{16}{81}$ 3 $\dfrac{9}{8}$ 5 $\dfrac{-47}{3}$ 7 $\dfrac{1}{8}$ 9 $\dfrac{1}{25}$ 11 $8x^9$

13 $\dfrac{6}{x}$ 15 $-2a^{14}$ 17 $\dfrac{9}{2}$ 19 $\dfrac{12u^{11}}{v^2}$ 21 $\dfrac{4}{xy}$

23 $\dfrac{9y^6}{x^8}$ 25 $\dfrac{81}{64}y^6$ 27 $\dfrac{s^6}{4r^8}$ 29 $\dfrac{20y}{x^3}$ 31 $9x^{10}y^{14}$

33 $8a^2$ 35 $24x^{3/2}$ 37 $\dfrac{1}{9a^4}$ 39 $\dfrac{8}{x^{1/2}}$ 41 $4x^2y^4$

43 $\dfrac{3}{x^3y^2}$ 45 1   47 $x^{3/4}$ 49 $(a + b)^{2/3}$
51 $(x^2 + y^2)^{1/2}$ 53 (a) $4x\sqrt{x}$ (b) $8x\sqrt{x}$
55 (a) $8 - \sqrt[3]{y}$ (b) $\sqrt[3]{8 - y}$ 57 9   59 $-2\sqrt[5]{2}$
61 $\dfrac{1}{2}\sqrt[3]{4}$ 63 $\dfrac{3y^3}{x^2}$ 65 $\dfrac{2a^2}{b}$ 67 $\dfrac{1}{2y^2}\sqrt{6xy}$
69 $\dfrac{xy}{3}\sqrt[3]{6y}$ 71 $\dfrac{x}{3}\sqrt[4]{15x^2y^3}$ 73 $\dfrac{1}{2}\sqrt[3]{20x^4y^2}$
75 $\dfrac{3x^5}{y^2}$ 77 $\dfrac{2x}{y^2}\sqrt[5]{x^2y^4}$ 79 $-3tv^2$ 81 $|x^3|y^2$
83 $x^2|y - 1|^3$ 85 $\neq$; $(a^r)^2 = a^{2r} \neq a^{(r^2)}$
87 $\neq$; $(ab)^{xy} = a^{xy}b^{xy} \neq a^xb^y$
89 $=$; $\sqrt[n]{\dfrac{1}{c}} = \left(\dfrac{1}{c}\right)^{1/n} = \dfrac{1^{1/n}}{c^{1/n}} = \dfrac{1}{\sqrt[n]{c}}$
91 (a) 1.5518   (b) 8.5499
93 (a) 2.0351   (b) 3.9670   95 $232,825.78
97 2.82 m   99 The 120-kg lifter

101

Height	Weight	Height	Weight
64	137	72	168
65	141	73	172
66	145	74	176
67	148	75	180
68	152	76	184
69	156	77	188
70	160	78	192
71	164	79	196

EXERCISES 1.3

1 $12x^3 - 13x + 1$ 3 $x^3 - 2x^2 + 4$
5 $6x^2 + x - 35$ 7 $15x^2 + 31xy + 14y^2$
9 $6u^2 - 13u - 12$ 11 $6x^3 + 37x^2 + 30x - 25$
13 $3t^4 + 5t^3 - 15t^2 + 9t - 10$
15 $2x^6 + 2x^5 - 2x^4 + 8x^3 + 10x^2 - 10x - 10$
17 $4y^2 - 5x$ 19 $3v^2 - 2u^2 + uv^2$ 21 $4x^2 - 9y^2$
23 $x^4 - 4y^2$ 25 $x^4 + 5x^2 - 36$
27 $9x^2 + 12xy + 4y^2$ 29 $x^4 - 6x^2y^2 + 9y^4$
31 $x^4 - 8x^2 + 16$ 33 $x - y$ 35 $x - y$
37 $x^3 - 6x^2y + 12xy^2 - 8y^3$
39 $8x^3 + 36x^2y + 54xy^2 + 27y^3$
41 $a^2 + b^2 + c^2 + 2ab - 2ac - 2bc$
43 $4x^2 + y^2 + 9z^2 + 4xy - 12xz - 6yz$

45 $s(r + 4t)$ **47** $3a^2b(b - 2)$ **49** $3x^2y^2(y - 3x)$

51 $5x^3y^2(3y^3 - 5x + 2x^3y^2)$ **53** $(8x + 3)(x - 7)$

55 Irreducible   **57** $(3x - 4)(2x + 5)$

59 $(3x - 5)(4x - 3)$ **61** $(2x - 5)^2$ **63** $(5z + 3)^2$

65 $(5x + 2y)(9x + 4y)$ **67** $(6r + 5t)(6r - 5t)$

69 $(z^2 + 8w)(z^2 - 8w)$ **71** $x^2(x + 2)(x - 2)$

73 Irreducible   **75** $3(5x + 4y)(5x - 4y)$

77 $(4x + 3)(16x^2 - 12x + 9)$

79 $(4x - y^2)(16x^2 + 4xy^2 + y^4)$

81 $(7x + y^3)(49x^2 - 7xy^3 + y^6)$

83 $(5 - 3x)(25 + 15x + 9x^2)$

85 $(2x + y)(a - 3b)$ **87** $3(x + 3)(x - 3)(x + 1)$

89 $(x - 1)(x + 2)(x^2 + x + 1)$ **91** $(a^2 + b^2)(a - b)$

93 $(a + b)(a - b)(a^2 - ab + b^2)(a^2 + ab + b^2)$

95 $(x + 2 + 3y)(x + 2 - 3y)$

97 $(y + 4 + x)(y + 4 - x)$

99 $(y + 2)(y^2 - 2y + 4)(y - 1)(y^2 + y + 1)$

101 $(x^8 + 1)(x^4 + 1)(x^2 + 1)(x + 1)(x - 1)$

103 Area of I is $(x - y)x$, area of II is $(x - y)y$, and
$$A = x^2 - y^2 = (x - y)x + (x - y)y$$
$$= (x - y)(x + y).$$

105 **(a)** 1525.7; 1454.7

(b) As people age, they require fewer calories. Coefficients of w and h are positive because large people require more calories.

EXERCISES 1.4

1 $\dfrac{22}{75}$ **3** $\dfrac{7}{120}$ **5** $\dfrac{x + 3}{x - 4}$ **7** $\dfrac{y + 5}{y^2 + 5y + 25}$

9 $\dfrac{4 - r}{r^2}$ **11** $\dfrac{x}{x - 1}$ **13** $\dfrac{a}{(a^2 + 4)(5a + 2)}$

15 $\dfrac{-3}{x + 2}$ **17** $\dfrac{6s - 7}{(3s + 1)^2}$ **19** $\dfrac{5x^2 + 2}{x^3}$

21 $\dfrac{4(2t + 5)}{t + 2}$ **23** $\dfrac{2(2x + 3)}{3x - 4}$ **25** $\dfrac{2x - 1}{x}$

27 $\dfrac{p^2 + 2p + 4}{p - 3}$ **29** $\dfrac{11u^2 + 18u + 5}{u(3u + 1)}$ **31** $-\dfrac{x + 5}{(x + 2)^2}$

33 $a + b$ **35** $\dfrac{x^2 + xy + y^2}{x + y}$ **37** $x + y$

39 $\dfrac{2x^2 + 7x + 15}{x^2 + 10x + 7}$ **41** $-\dfrac{3}{(x - 1)(a - 1)}$

43 $2x + h - 3$ **45** $-\dfrac{3x^2 + 3xh + h^2}{x^3(x + h)^3}$

47 $\dfrac{-12}{(3x + 3h - 1)(3x - 1)}$ **49** $\dfrac{t + 10\sqrt{t} + 25}{t - 25}$

51 $(9x + 4y)(3\sqrt{x} + 2\sqrt{y})$ **53** $\dfrac{\sqrt[3]{a^2} + \sqrt[3]{ab} + \sqrt[3]{b^2}}{a - b}$

55 $\dfrac{1}{(a + b)(\sqrt{a} + \sqrt{b})}$ **57** $\dfrac{2}{\sqrt{2(x + h) + 1} + \sqrt{2x + 1}}$

59 $\dfrac{-1}{\sqrt{1 - x - h} + \sqrt{1 - x}}$ **61** $4x^{4/3} - x^{1/3} + 5x^{-2/3}$

63 $x^{-1} + 4x^{-3} + 4x^{-5}$ **65** $\dfrac{1 + x^5}{x^3}$ **67** $\dfrac{1 - x^2}{x^{1/2}}$

69 $(3x + 2)^3(36x^2 - 37x + 6)$

71 $\dfrac{(2x + 1)^2(8x^2 + x - 24)}{(x^2 - 4)^{1/2}}$ **73** $\dfrac{(3x + 1)^5(39x - 89)}{(2x - 5)^{1/2}}$

75 $\dfrac{27x^2 - 24x + 2}{(6x + 1)^4}$ **77** $\dfrac{4x(1 - x^2)}{(x^2 + 2)^4}$ **79** $\dfrac{x^2 + 12}{(x^2 + 4)^{4/3}}$

81 $\dfrac{6(3 - 2x)}{(4x^2 + 9)^{3/2}}$

CHAPTER 1 REVIEW EXERCISES

1 **(a)** $-\dfrac{5}{12}$ **(b)** $\dfrac{39}{20}$ **(c)** $-\dfrac{13}{56}$ **(d)** $\dfrac{5}{8}$

2 **(a)** $<$ **(b)** $>$ **(c)** $>$

3 **(a)** $x < 0$ **(b)** $\dfrac{1}{3} < a < \dfrac{1}{2}$ **(c)** $|x| \le 4$

4 **(a)** 7   **(b)** -1 **(c)** $\dfrac{1}{6}$ **5 (a)** 5   **(b)** 5   **(c)** 7

6 **(a)** No   **(b)** No   **(c)** Yes

7 **(a)** 9.37×10^{10} **(b)** 4.02×10^{-6}

8 **(a)** 68,000,000   **(b)** 0.000 73   **9** $-x - 3$

10 $-(x - 2)(x - 3)$ **11** $\dfrac{-71}{9}$ **12** $\dfrac{1}{8}$ **13** $18a^5b^5$

14 $\dfrac{3y}{r^2}$ **15** $\dfrac{xy^5}{9}$ **16** $\dfrac{b^3}{a^8}$ **17** $-\dfrac{p^8}{2q}$ **18** $c^{1/3}$

19 $\dfrac{x^3z}{y^{10}}$ **20** $\dfrac{16x^2}{z^4y^6}$ **21** $\dfrac{b^6}{a^2}$ **22** $\dfrac{27u^2v^{27}}{16w^{20}}$ **23** $s + r$

24 $u + v$ **25** s **26** $\dfrac{y - x^2}{x^2y}$ **27** $\dfrac{x^8}{y^2}$

28 $2xyz\sqrt[3]{x^2z}$ **29** $\dfrac{1}{2}\sqrt[3]{2}$ **30** $\dfrac{ab}{c}\sqrt{bc}$

31 $2x^2y\sqrt[3]{x}$ **32** $2ab\sqrt{ac}$ **33** $\dfrac{1 - \sqrt{t}}{t}$ **34** c^2d^4

35 $\dfrac{2x}{y^2}$ **36** $a + 2b$ **37** $\dfrac{1}{2\pi}\sqrt[3]{4\pi}$ **38** $\dfrac{1}{3y}\sqrt[3]{3x^2y^2}$

39 $\dfrac{1 - 2\sqrt{x} + x}{1 - x}$ **40** $\dfrac{\sqrt{a} - \sqrt{a - 2}}{2}$

41 $(9x + y)(3\sqrt{x} - \sqrt{y})$ **42** $\dfrac{x + 6\sqrt{x} + 9}{9 - x}$

43 $x^4 + x^3 - x^2 + x - 2$

44 $3z^4 - 4z^3 - 3z^2 + 4z + 1$ **45** $-x^2 + 18x + 7$

46 $8x^3 + 2x^2 - 43x + 35$

47 $3y^5 - 2y^4 - 8y^3 + 10y^2 - 3y - 12$

48 $15x^3 - 53x^2 - 102x - 40$ **49** $a^4 - b^4$

50 $3p^2q - 2q^2 + \dfrac{5}{3}p$ **51** $6a^2 + 11ab - 35b^2$

52 $16r^4 - 24r^2s + 9s^2$ **53** $169a^4 - 16b^2$

54 $a^6 - 2a^5 + a^4$ **55** $8a^3 + 12a^2b + 6ab^2 + b^3$

56 $c^6 - 3c^4d^2 + 3c^2d^4 - d^6$ **57** $81x^4 - 72x^2y^2 + 16y^4$

58 $a^2 + b^2 + c^2 + d^2 + 2(ab + ac + ad + bc + bd + cd)$

59 $10w(6x + 7)$ **60** $2r^2s^3(r + 2s)(r - 2s)$

61 $(14x + 9)(2x - 1)$ **62** $(4a^2 + 3b^2)^2$

63 $(y - 4z)(2w + 3x)$ **64** $(2c^2 + 3)(c - 6)$

65 $8(x + 2y)(x^2 - 2xy + 4y^2)$

66 $u^3v(v - u)(v^2 + uv + u^2)$

67 $(p^4 + q^4)(p^2 + q^2)(p + q)(p - q)$ **68** $x^2(x - 4)^2$

69 $(w^2 + 1)(w^4 - w^2 + 1)$ **70** $3(x + 2)$

71 Irreducible   **72** $(x - 7 + 7y)(x - 7 - 7y)$

73 $(x - 2)(x + 2)^2(x^2 - 2x + 4)$ **74** $4x^2(x^2 + 3x + 5)$

75 $\dfrac{3x - 5}{2x + 1}$ **76** $\dfrac{r^2 + rt + t^2}{r + t}$ **77** $\dfrac{3x + 2}{x(x - 2)}$

78 $\dfrac{27}{(4x - 5)(10x + 1)}$ **79** $\dfrac{5x^2 - 6x - 20}{x(x + 2)^2}$ **80** $\dfrac{x^3 + 1}{x^2 + 1}$

81 $\dfrac{-2x^2 - x - 3}{x(x + 1)(x + 3)}$ **82** $\dfrac{ab}{a + b}$ **83** $x + 5$

84 $\dfrac{1}{x + 3}$ **85** $(x^2 + 1)^{1/2}(x + 5)^3(7x^2 + 15x + 4)$

86 $\dfrac{2(5x^2 + x + 4)}{(6x + 1)^{2/3}(4 - x^2)^2}$ **87** 2.75×10^{13} cells

88 Between 2.94×10^9 and 3.78×10^9 beats

89 0.54 m^2 **90** 0.13 dyne-cm

CHAPTER 1 DISCUSSION EXERCISES

1 1 gallon ≈ 0.13368 ft^3; 586.85 ft^2

2 Either $a = 0$ or $b = 0$

3 Add and subtract $10x$; $x + 5 \pm \sqrt{10x}$ are the factors.

4 The first expression can be evaluated at $x = 1$.

5 They get close to the ratio of leading coefficients as x gets larger.

7 If x is the age and y is the height, show that the final value is $100x + y$.

8 $V_{\text{out}} = \frac{1}{3}V_{\text{in}}$

Chapter 2

EXERCISES 2.1

1 $\dfrac{5}{3}$ **3** 1 **5** $\dfrac{26}{7}$ **7** $\dfrac{35}{17}$ **9** $\dfrac{23}{18}$ **11** $-\dfrac{1}{40}$

13 $\dfrac{49}{4}$ **15** $\dfrac{4}{3}$ **17** $-\dfrac{24}{29}$ **19** $\dfrac{7}{31}$ **21** $-\dfrac{3}{61}$

23 $\dfrac{29}{4}$ **25** $\dfrac{31}{18}$ **27** No solution

29 All real numbers except $\dfrac{1}{2}$ **31** $\dfrac{5}{9}$ **33** $-\dfrac{2}{3}$

35 No solution   **37** 0 **39** All real numbers except ± 2

41 No solution   **43** No solution

45 $(4x - 3)^2 - 16x^2 = (16x^2 - 24x + 9) - 16x^2 = 9 - 24x$

47 $\dfrac{x^2 - 9}{x + 3} = \dfrac{(x + 3)(x - 3)}{x + 3} = x - 3$

49 $\dfrac{3x^2 + 8}{x} = \dfrac{3x^2}{x} + \dfrac{8}{x} = \dfrac{8}{x} + 3x$ **51** $-\dfrac{19}{3}$

53 (a) Yes (b) No, 5 is not a solution of the first equation.

55 Choose any a and b such that $b = -\dfrac{5}{3}a$.

57 $x + 1 = x + 2$ **59** $K = \dfrac{D - L}{E + T}$ **61** $Q = \dfrac{1}{M - 1}$

63 $P = \dfrac{I}{rt}$ **65** $h = \dfrac{2A}{b}$ **67** $m = \dfrac{Fd^2}{gM}$

69 $w = \dfrac{P - 2l}{2}$ **71** $b_1 = \dfrac{2A - hb_2}{h}$

73 $q = \dfrac{p(1 - S)}{S(1 - p)}$ **75** $q = \dfrac{fp}{p - f}$

EXERCISES 2.2

1 88   **3** $820   **5** (a) 125   (b) 21

7 180 mo (or 15 yr)   **9** Not possible   **11** 200 children

13 $\dfrac{14}{3}$ oz of 30% glucose solution and $\dfrac{7}{3}$ oz of water

15 194.6 g of British sterling silver and 5.4 g of copper

17 (a) After 64 sec   (b) 96 m and 128 m, respectively

19 6 mi/hr   **21** (a) $\dfrac{5}{9}$ mi/hr   (b) $2\dfrac{2}{9}$ mi

23 1237.5 ft

25 (a) 4050 ft^2 (b) 2592 ft^2 (c) 3600 ft^2

27 $\dfrac{19}{2} - \dfrac{3\pi}{8} \approx 8.32$ ft   **29** 55 ft   **31** 36 min

33 36 min   **35** 27

37 (a) 40.96°F   (b) 6909 ft   **39** 37°F

EXERCISES 2.3

1 $-\dfrac{3}{2}, \dfrac{4}{3}$ **3** $-\dfrac{6}{5}, \dfrac{2}{3}$ **5** $-\dfrac{9}{2}, \dfrac{3}{4}$ **7** $-\dfrac{2}{3}, \dfrac{1}{5}$

9 $-\dfrac{5}{2}$ **11** $-\dfrac{1}{2}$ **13** $-\dfrac{34}{5}$

15 (a) No, -4 is not a solution of $x = 4$.   (b) Yes

17 ± 13 **19** $\pm\dfrac{3}{5}$ **21** $3 \pm \sqrt{17}$ **23** $-2 \pm \dfrac{1}{2}\sqrt{11}$

25 (a) $\dfrac{81}{4}$ **(b)** 16   **(c)** ± 12 **(d)** ± 7

27 $-3 \pm \sqrt{2}$ **29** $\dfrac{3}{2} \pm \sqrt{5}$ **31** $-\dfrac{1}{2}, \dfrac{2}{3}$

33 $-2 \pm \sqrt{2}$ **35** $\dfrac{3}{4} \pm \dfrac{1}{4}\sqrt{41}$ **37** $\dfrac{4}{3} \pm \dfrac{1}{3}\sqrt{22}$

39 $\dfrac{5}{2} \pm \dfrac{1}{2}\sqrt{15}$ **41** $\dfrac{9}{2}$ **43** No real solutions

45 $(x + 6)(x - 5)$ **47** $(2x - 3)(6x + 1)$

49 (a) $x = \dfrac{y \pm \sqrt{2y^2 - 1}}{2}$ **(b)** $y = -2x \pm \sqrt{8x^2 + 1}$

51 $v = \sqrt{\dfrac{2K}{m}}$ **53** $r = \dfrac{-\pi h + \sqrt{\pi^2 h^2 + 2\pi A}}{2\pi}$

55 $r = r_0\sqrt{1 - (V/V_{\max})}$ **57** $\sqrt{150/\pi} \approx 6.9$ cm
59 (a) After 1 sec and after 3 sec   **(b)** After 4 sec
61 (a) 4320 m   **(b)** 96.86°C   **63** 2 ft
65 12 ft by 12 ft

67 $3 + \dfrac{1}{2}\sqrt{14} \approx 4.9$ mi or $3 - \dfrac{1}{2}\sqrt{14} \approx 1.1$ mi

69 (a) $d = 100\sqrt{20t^2 + 4t + 1}$ **(b)** 3:30 P.M.
71 14 in. by 27 in.   **73** 7 mi/hr   **75** 300 pairs
77 2 ft   **79** 15.89 sec
81 (a) $0; -4{,}500{,}000$ **(b)** 2.13×10^{-7}

EXERCISES 2.4

1 $2 + 4i$ **3** $18 - 3i$ **5** $41 - 11i$ **7** $17 - i$
9 $21 - 20i$ **11** $-24 - 7i$ **13** 25   **15** $-i$

17 i **19** $\dfrac{3}{10} - \dfrac{3}{5}i$ **21** $\dfrac{1}{2} - i$ **23** $\dfrac{34}{53} + \dfrac{40}{53}i$

25 $\dfrac{2}{5} + \dfrac{4}{5}i$ **27** $-142 - 65i$ **29** $-2 - 14i$

31 $-\dfrac{44}{113} + \dfrac{95}{113}i$ **33** $\dfrac{21}{2}i$ **35** $x = 4, y = -1$

37 $x = 3, y = -4$ **39** $3 \pm 2i$ **41** $-2 \pm 3i$

43 $\dfrac{5}{2} \pm \dfrac{1}{2}\sqrt{55}i$ **45** $-\dfrac{1}{8} \pm \dfrac{1}{8}\sqrt{47}i$

47 $-5, \dfrac{5}{2} \pm \dfrac{5}{2}\sqrt{3}i$ **49** $\pm 4, \pm 4i$ **51** $\pm 2i, \pm \dfrac{3}{2}i$

53 $0, -\dfrac{3}{2} \pm \dfrac{1}{2}\sqrt{7}i$

55 $\overline{z + w} = \overline{(a + bi) + (c + di)}$
$= \overline{(a + c) + (b + d)i} = (a + c) - (b + d)i$
$= (a - bi) + (c - di) = \bar{z} + \bar{w}$

57 $\overline{z \cdot w} = \overline{(a + bi) \cdot (c + di)}$
$= \overline{(ac - bd) + (ad + bc)i}$
$= (ac - bd) - (ad + bc)i$
$= ac - adi - bd - bci$
$= a(c - di) - bi(c - di)$
$= (a - bi) \cdot (c - di) = \bar{z} \cdot \bar{w}$
59 If $\bar{z} = z$, then $a - bi = a + bi$ and hence $-bi = bi$,
or $2bi = 0$. Thus, $b = 0$ and $z = a$ is real.
Conversely, if z is real, then $b = 0$ and hence
$\bar{z} = \overline{a + 0i} = a - 0i = a + 0i = z$.

EXERCISES 2.5

1 $-15, 7$ **3** $-\dfrac{2}{3}, 2$ **5** No solution   **7** $\pm\dfrac{2}{3}, 2$

9 $\pm\dfrac{1}{2}\sqrt{6}, -\dfrac{5}{2}, 0$ **11** $0, 25$ **13** $-\dfrac{57}{5}$ **15** $\dfrac{9}{5}$

17 $\pm\dfrac{1}{2}\sqrt{62}$ **19** 6   **21** 6   **23** 5, 7   **25** -3

27 -1 **29** $-\dfrac{5}{4}$ **31** 3   **33** 0, 4   **35** $\pm 3, \pm 4$

37 $\pm\dfrac{1}{10}\sqrt{70 \pm 10\sqrt{29}}$ **39** $\pm 2, \pm 3$ **41** $\dfrac{8}{27}, -8$

43 $\dfrac{25}{4}, \dfrac{16}{9}$ **45** $-\dfrac{4}{3}, -\dfrac{2}{3}$ **47** $\dfrac{5}{2}$ **49** 0, 4096

51 (a) 8   **(b)** ± 8 **(c)** No real solutions   **(d)** 625
(e) No real solutions

53 $l = \dfrac{gT^2}{4\pi^2}$ **55** $h = \dfrac{1}{\pi r}\sqrt{S^2 - \pi^2 r^4}$ **57** $h \approx 97\%$ of L

59 9.16 ft/sec   **61** \$4.00   **63** $2\sqrt[3]{\dfrac{432}{\pi}} \approx 10.3$ cm

65 53.4%
67 There are two possible routes, corresponding to
$x \approx 0.6743$ mi and $x \approx 2.2887$ mi.

EXERCISES 2.6

1 (a) $-2 < 2$ **(b)** $-11 < -7$ **(c)** $-\dfrac{7}{3} < -1$

(d) $1 < \dfrac{7}{3}$

3 $(-\infty, -2)$ **5** $[4, \infty)$

7 $(-2, 4]$ **9** $[3, 7]$

11 $[-2, 5)$

$-2 \quad 0 \quad 5$

13 $-5 < x \le 8$ **15** $-4 \le x \le -1$ **17** $x \ge 4$

19 $x < -5$ **21** $\left(\dfrac{16}{3}, \infty\right)$ **23** $\left(-\infty, -\dfrac{4}{3}\right]$

25 $(12, \infty)$ **27** $[-6, \infty)$ **29** $(1, 6)$ **31** $[9, 19]$

33 $\left(-\dfrac{26}{3}, \dfrac{16}{3}\right]$ **35** $(6, 12]$ **37** $\left(-\infty, \dfrac{8}{53}\right]$

39 $\left(-\infty, \dfrac{4}{5}\right)$ **41** $\left(-\dfrac{2}{3}, \infty\right)$ **43** $\left(\dfrac{4}{3}, \infty\right)$

45 All real numbers except 1 **47** $(-3, 3)$

49 $(-\infty, -5] \cup [5, \infty)$ **51** $(-3.01, -2.99)$

53 $(-\infty, -2.1] \cup [-1.9, \infty)$ **55** $\left(-\dfrac{9}{2}, -\dfrac{1}{2}\right)$

57 $\left[\dfrac{3}{5}, \dfrac{9}{5}\right]$ **59** $(-\infty, \infty)$ **61** $(-\infty, 3) \cup (3, \infty)$

63 $\left(-\infty, -\dfrac{8}{3}\right] \cup [4, \infty)$ **65** $\left(-\infty, \dfrac{7}{4}\right) \cup \left(\dfrac{13}{4}, \infty\right)$

67 $(-4, 4)$ **69** $(-2, 1) \cup (3, 6)$

71 (a) $-8, -2$ **(b)** $-8 < x < -2$
(c) $(-\infty, -8) \cup (-2, \infty)$

73 $|w - 148| \le 2$ **75** $5 < |T_1 - T_2| < 10$

77 $86 \le F \le 104$ **79** $R \ge 11$ **81** $4 \le p < 6$

83 $6\dfrac{2}{3}$ yr **85 (a)** 5 ft 8 in. **(b)** $65.52 \le h \le 66.48$

EXERCISES 2.7

1 $\left(-\dfrac{1}{3}, \dfrac{1}{2}\right)$ **3** $[-2, 1] \cup [4, \infty)$ **5** $(-2, 3)$

7 $(-\infty, -2) \cup (4, \infty)$ **9** $\left(-\infty, -\dfrac{5}{2}\right] \cup [1, \infty)$

11 $(2, 4)$ **13** $(-4, 4)$ **15** $\left(-\dfrac{3}{5}, \dfrac{3}{5}\right)$

17 $(-\infty, 0] \cup \left[\dfrac{9}{16}, \infty\right)$ **19** $(-\infty, -2] \cup [2, \infty)$

21 $\{-2\} \cup [2, \infty)$ **23** $(-\infty, -2) \cup (-2, -1) \cup \{0\}$

25 $(-2, 0) \cup (0, 1]$ **27** $(-2, 2] \cup (5, \infty)$

29 $(-\infty, -3) \cup (0, 3)$ **31** $\left(\dfrac{3}{2}, \dfrac{7}{3}\right)$

33 $(-\infty, -1) \cup \left(2, \dfrac{7}{2}\right]$ **35** $\left(-1, \dfrac{2}{3}\right] \cup [4, \infty)$

37 $\left(1, \dfrac{5}{3}\right) \cup [2, 5]$ **39** $(-1, 0) \cup (1, \infty)$

41 $[0, 2] \cup [3, 5]$ **43** $\dfrac{1}{2}$ sec **45** $0 \le v < 30$

47 $0 < S < 4000$ **49** height $> 25,600$ km

51 $70.5 \le V \le 81.4$

CHAPTER 2 REVIEW EXERCISES

1 $-\dfrac{5}{6}$ **2** 5 **3** -32 **4** No solution

5 Every $x > 0$ **6** $-4, \dfrac{3}{2}$ **7** $-\dfrac{2}{3} \pm \dfrac{1}{3}\sqrt{19}$

8 $\dfrac{5}{2} \pm \dfrac{1}{2}\sqrt{29}$ **9** $\dfrac{1}{2} \pm \dfrac{1}{2}\sqrt{21}$ **10** $\pm\dfrac{5}{2}, \pm\sqrt{2}$

11 $-27, 125$ **12** $\pm\dfrac{1}{2}\sqrt{7}, -\dfrac{2}{5}$ **13** $\dfrac{1}{5} \pm \dfrac{1}{5}\sqrt{14}\,i$

14 $-\dfrac{1}{6} \pm \dfrac{1}{6}\sqrt{71}\,i$ **15** $\pm\dfrac{1}{2}\sqrt{14}\,i, \pm\dfrac{2}{3}\sqrt{3}\,i$

16 $\pm\dfrac{1}{2}\sqrt{6 \pm 2\sqrt{5}}$ **17** $-\dfrac{3}{2}, 2$ **18** $-5, 4$

19 $\dfrac{1}{4}, \dfrac{1}{9}$ **20** $\dfrac{13}{4}$ **21** 2 **22** $-3, 1$ **23** 5

24 ± 8 **25** $2 \pm \sqrt{3}$ **26** $-5 \pm \sqrt{13}\,i$ **27** 3

28 $\left(\dfrac{2}{3}, \infty\right)$ **29** $\left(-\dfrac{11}{4}, \dfrac{9}{4}\right)$ **30** $\left[\dfrac{13}{23}, \infty\right)$

31 $\left(-\infty, -\dfrac{3}{10}\right)$ **32** $\left(-7, \dfrac{7}{2}\right)$

33 $(-\infty, 1) \cup (5, \infty)$ **34** $[0, 6]$

35 $\left(-\infty, \dfrac{11}{3}\right] \cup [7, \infty)$ **36** $(2, 4) \cup (8, 10)$

37 $\left(-\infty, -\dfrac{3}{2}\right) \cup \left(\dfrac{2}{5}, \infty\right)$ **38** $[-2, 5]$

39 $(-\infty, -2) \cup \{0\} \cup [3, \infty)$ **40** $(-3, -1) \cup (-1, 2]$

41 $\left(-\infty, -\dfrac{3}{2}\right) \cup (2, 9)$ **42** $(-\infty, -5) \cup [-1, 5)$

43 $(1, \infty)$ **44** $(0, 1) \cup (2, 3)$ **45** $C = \dfrac{2}{P + N - 1}$

46 $D = \dfrac{CB^3}{(A + E)^3}$ **47** $r = \sqrt[3]{\dfrac{3V}{4\pi}}$

48 $R = \sqrt[4]{\dfrac{8FVL}{\pi P}}$ **49** $h = R \pm \dfrac{1}{2}\sqrt{4R^2 - c^2}$

50 $r = \dfrac{-\pi hR + \sqrt{12\pi hV - 3\pi^2 h^2 R^2}}{2\pi h}$ **51** $15 + 2i$

52 $-28 + 6i$ **53** $-55 + 48i$ **54** $\dfrac{9}{85} + \dfrac{2}{85}i$

55 $-\dfrac{9}{53} - \dfrac{48}{53}i$ **56** $-2 - 5i$ **57** 56

58 $R_2 = \dfrac{10}{3}$ ohms   **59** 11.055%   **60** 60.3 g

61 6 oz of vegetables and 4 oz of meat

62 315.8 g of ethyl alcohol and 84.2 g of water

63 80 gal of 20% solution and 40 gal of 50% solution

64 75 mi   **65** 2   **66** 64 mi/hr

67 $\dfrac{640}{11} \approx 58.2$ mi/hr   **68** 1 hr 40 min   **69** 165 mi

70 $10 - 5\sqrt{3} \approx 1.34$ mi   **71** $3\sqrt{5} - 6 \approx 0.71$ micron

72 **(a)** $d = \sqrt{2900t^2 - 200t + 4}$

  (b) $t = \dfrac{5 + 2\sqrt{19{,}603}}{145} \approx 1.97$, or approximately

   11:58 A.M.

73 There are two arrangements: 40 ft $\times$ 25 ft and 50 ft $\times$ 20 ft.

74 **(a)** $2\sqrt{2}$ ft   **(b)** 2 ft   **75** 12 ft by 48 ft

76 10 ft by 4 ft   **77** After $7\dfrac{2}{3}$ yr   **78** $4 \le p \le 8$

79 Over \$100,000   **80** $T > 279.57$ K

81 $\dfrac{\pi}{5}\sqrt{10} \le T \le \dfrac{2\pi}{7}\sqrt{5}$

82 $v < \dfrac{626.4}{\sqrt{6472}} \approx 7.786$ km/sec   **83** $20 \le w \le 25$

84 36 to 38 trees/acre   **85** \$320 to \$340

CHAPTER 2 DISCUSSION EXERCISES

1 No   **2** $\dfrac{-b}{2a}$

3 **(a)** $\dfrac{ac + bd}{a^2 + b^2} + \dfrac{ad - bc}{a^2 + b^2}i$ **(b)** Yes

  (c) a and b cannot both be 0

5 $a > 0, D \le 0: x \in \mathbb{R}$;

  $a > 0, D > 0: (-\infty, x_1] \cup [x_2, \infty)$;

  $a < 0, D < 0: \{\ \}$;

  $a < 0, D = 0: x = \dfrac{-b}{2a}$;

  $a < 0, D > 0: [x_1, x_2]$

6 **(a)** 11,006 ft   **(b)** $h = \dfrac{1}{6}(2497D - 497G - 64{,}000)$

8 $1/10^{1000}$; $cx - 2/c$ must be nonnegative

Chapter 3

EXERCISES 3.1

1

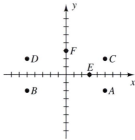

3 The line bisecting quadrants I and III

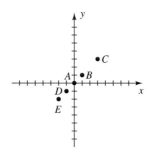

5 $A(3, 3)$, $B(-3, 3)$, $C(-3, -3)$, $D(3, -3)$, $E(3, 0)$, $F(0, 3)$

7 **(a)** The line parallel to the y-axis that intersects the x-axis at $(-2, 0)$

  (b) The line parallel to the x-axis that intersects the y-axis at $(0, 3)$

  (c) All points to the right of and on the y-axis

  (d) All points in quadrants I and III

  (e) All points below the x-axis

  (f) All points on the y-axis

9 **(a)** $\sqrt{29}$ **(b)** $\left(5, -\dfrac{1}{2}\right)$

11 **(a)** $\sqrt{13}$ **(b)** $\left(-\dfrac{7}{2}, -1\right)$

13 **(a)** 4   **(b)** $(5, -3)$

15 $d(A, C)^2 = d(A, B)^2 + d(B, C)^2$; area = 28

17 $d(A, B) = d(B, C) = d(C, D) = d(D, A)$ and $d(A, C)^2 = d(A, B)^2 + d(B, C)^2$

19 $(13, -28)$ **21** $d(A, C) = d(B, C) = \sqrt{145}$

23 $5x + 2y = 3$

25 $\sqrt{x^2 + y^2} = 5$; a circle of radius 5 with center at the origin

27 $\left(0, 3 + \sqrt{11}\right), \left(0, 3 - \sqrt{11}\right)$ **29** $(-2, -1)$

31 $a < \dfrac{2}{5}$ or $a > 4$

33 Let M be the midpoint of the hypotenuse. Show that
$$d(A, M) = d(B, M) = d(O, M) = \frac{1}{2}\sqrt{a^2 + b^2}.$$

EXERCISES 3.2

Exer. 1–20: x-intercept(s) is listed, followed by y-intercept(s).

1 1.5; -3

3 1; 1

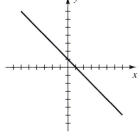

5 0; 0

7 $\pm\frac{1}{2}\sqrt{2}$; -1

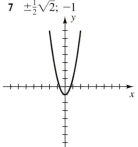

9 0; 0

11 3; $\pm\sqrt{3}$

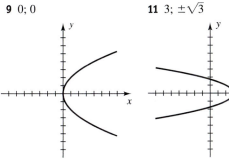

13 0; 0

15 2; -8

17 0; 0

19 16; -4

21 (a) 5, 7   **(b)** 9, 11   **(c)** 13

23

25

27

29

31

33

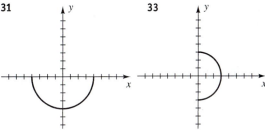

35 $(x - 2)^2 + (y + 3)^2 = 25$ **37** $\left(x - \dfrac{1}{4}\right)^2 + y^2 = 5$

39 $(x + 4)^2 + (y - 6)^2 = 41$

41 $(x + 3)^2 + (y - 6)^2 = 9$

43 $(x + 4)^2 + (y - 4)^2 = 16$

45 $(x - 1)^2 + (y - 2)^2 = 34$ **47** $C(2, -3); r = 7$

49 $C(0, -2); r = 11$ **51** $C(3, -1); r = \dfrac{1}{2}\sqrt{70}$

53 $C(-2, 1); r = 0$ (a point)

55 Not a circle, since r^2 cannot equal -2

57 $y = \sqrt{36 - x^2}; y = -\sqrt{36 - x^2}; x = \sqrt{36 - y^2};$
   $x = -\sqrt{36 - y^2}$

59 $y = -1 + \sqrt{49 - (x - 2)^2};$
   $y = -1 - \sqrt{49 - (x - 2)^2};$
   $x = 2 + \sqrt{49 - (y + 1)^2}; x = 2 - \sqrt{49 - (y + 1)^2}$

61 $(x + 3)^2 + (y - 2)^2 = 4^2$ **63** $y = -\sqrt{4^2 - x^2}$

65 (a) Inside   (b) On   (c) Outside

67 (a) 2   (b) $3 \pm \sqrt{5}$

69 $(x + 2)^2 + (y - 3)^2 = 25$ **71** $\sqrt{5}$

73 $(-\infty, -3) \cup (2, \infty)$ **75** $(-1, 0) \cup (0, 1)$

EXERCISES 3.3

1 $m = -\dfrac{3}{4}$ **3** $m = 0$

5 m is undefined

7 The slopes of opposite sides are equal.

9 The slopes of opposite sides are equal, and the slopes of two adjacent sides are negative reciprocals.

11 $(-12, 0)$

13

15

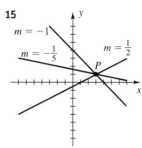

17 $y + 3 = \pm\dfrac{5}{4}(x - 2)$

19

21 (a) $x = 5$ (b) $y = -2$ **23** $4x + y = 17$

25 $3x + y = 12$ **27** $11x + 7y = 9$

29 $5x - 2y = 18$ **31** $5x + 2y = 29$

33 $y = \dfrac{3}{4}x - 3$ **35** $y = -\dfrac{1}{3}x + \dfrac{11}{3}$

37 $5x - 7y = -15$ **39** $y = -x$

41 $m = -\dfrac{2}{3}, b = 5$ **43** $m = \dfrac{4}{3}, b = -3$

   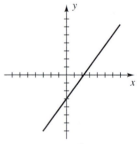

45 (a) $y = 3$ (b) $y = -\dfrac{1}{2}x$ (c) $y = -\dfrac{3}{2}x + 1$

   (d) $y + 2 = -(x - 3)$

47 $\dfrac{x}{3/2} + \dfrac{y}{-3} = 1$ **49** $(x - 3)^2 + (y + 2)^2 = 49$

51 Approximately 23 weeks

53 (a) 25.2 tons   (b) As large as 3.4 tons

55 (a) $y = \dfrac{5}{14}x$ (b) 58

57 (a) $W = \dfrac{20}{3}t + 10$ (b) 50 lb   (c) 9 yr

(d)

59 $H = -\dfrac{8}{3}T + \dfrac{7520}{3}$

61 (a) $T = 0.032t + 13.5$ (b) 16.22°C

63 (a) $E = 0.55R + 3600$ (b) $P = 0.45R - 3600$

(c) $8000

65 (a) Yes: the creature at $x = 3$ (b) No

67 34.95 mi/hr   **69** $a = 0.321; b = -0.9425$

EXERCISES 3.4

1 $-6, -4, -24$ **3** $-12, -22, -36$

5 (a) $5a - 2$ (b) $-5a - 2$ (c) $-5a + 2$
(d) $5a + 5h - 2$ (e) $5a + 5h - 4$ (f) 5

7 (a) $-a^2 + 4$ (b) $-a^2 + 4$ (c) $a^2 - 4$
(d) $-a^2 - 2ah - h^2 + 4$ (e) $-a^2 - h^2 + 8$
(f) $-2a - h$

9 (a) $a^2 - a + 3$ (b) $a^2 + a + 3$ (c) $-a^2 + a - 3$
(d) $a^2 + 2ah + h^2 - a - h + 3$
(e) $a^2 + h^2 - a - h + 6$ (f) $2a + h - 1$

11 (a) $\dfrac{4}{a^2}$ (b) $\dfrac{1}{4a^2}$ (c) $4a$ (d) $2a$

13 (a) $\dfrac{2a}{a^2 + 1}$ (b) $\dfrac{a^2 + 1}{2a}$ (c) $\dfrac{2\sqrt{a}}{a + 1}$

(d) $\dfrac{\sqrt{2a^3 + 2a}}{a^2 + 1}$

15 The graph is that of a function because it passes the vertical line test.

17 $D = [-4, 1] \cup [2, 4); R = [-3, 3)$

19 (a) $[-3, 4]$ (b) $[-2, 2]$ (c) 0   (d) $-1, \dfrac{1}{2}, 2$

(e) $\left(-1, \dfrac{1}{2}\right) \cup (2, 4]$

21 $\left[-\dfrac{7}{2}, \infty\right)$ **23** $[-3, 3]$

25 All real numbers except $-2, 0,$ and 2

27 $\left[\dfrac{3}{2}, 4\right) \cup (4, \infty)$ **29** $(2, \infty)$ **31** $[-2, 2]$

33 (a) $D = [-5, -3) \cup (-1, 1] \cup (2, 4];$
$R = \{-3\} \cup [-1, 4]$
(b) Increasing on $[-4, -3) \cup [3, 4];$
decreasing on $[-5, -4] \cup (2, 3];$
constant on $(-1, 1]$

35

37 (a)

(b) $D = (-\infty, \infty),$
$R = (-\infty, \infty)$

(c) Increasing on
$(-\infty, \infty)$

39 (a)

(b) $D = (-\infty, \infty),$
$R = (-\infty, 4]$

(c) Increasing on
$(-\infty, 0],$
decreasing on
$[0, \infty)$

41 (a)

(b) $D = [-4, \infty),$
$R = [0, \infty)$

(c) Increasing on
$[-4, \infty)$

43 (a)

(b) $D = (-\infty, \infty)$,
    $R = \{-2\}$
(c) Constant on
    $(-\infty, \infty)$

3 Odd **5** Even **7** Neither **9** Even **11** Odd

13

15

45 (a)

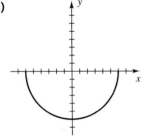

(b) $D = [-6, 6]$,
    $R = [-6, 0]$
(c) Decreasing on
    $[-6, 0]$,
    increasing on
    $[0, 6]$

17

19

47 $h + 1$

49 $2x + h$ **51** $\dfrac{1}{\sqrt{x - 3} + \sqrt{a - 3}}$

21

23

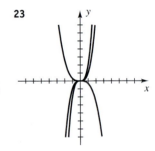

53 $f(x) = \dfrac{1}{6}x + \dfrac{3}{2}$ **55** Yes   **57** No   **59** Yes

61 No   **63** No   **65** $V(x) = 4x(15 - x)(10 - x)$

67 (a) $y(x) = \dfrac{500}{x}$ **(b)** $C(x) = 300x + \dfrac{100{,}000}{x} - 600$

69 $S(h) = 6h - 50$

71 (a) $y(t) = 2.5t + 33$

(b)

(7, 50.5)   (10, 58)   The yearly increase
                            in height
(6, 48)

25

(c) 58 in.

73 $d(t) = 2\sqrt{t^2 + 2500}$

75 (a) $y(h) = \sqrt{h^2 + 2hr}$ **(b)** 1280.6 mi

77 $d(x) = \sqrt{90{,}400 + x^2}$

27 $(-2, 4)$ **29** $(7, -3)$ **31** $(6, 2)$

33 The graph of f is shifted 2 units to the right and 3 units up.

35 The graph of f is reflected about the y-axis and shifted 2 units down.

37 The graph of f is compressed vertically by a factor of 2 and reflected about the x-axis.

EXERCISES 3.5

1 $f(-2) = 7, g(-2) = 6$

39 The graph of f is stretched horizontally by a factor of 3, stretched vertically by a factor of 2, and reflected about the x-axis.

41 (a)

(b)

(c)

(d)

(e)

(f)

(g)

(h)

(i)

(j)

(k)

(l)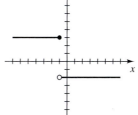

43 (a) $y = f(x + 9) + 1$ **(b)** $y = -f(x)$
(c) $y = -f(x + 7) - 1$

45 (a) $y = f(x + 4)$ **(b)** $y = f(x) + 1$ **(c)** $y = f(-x)$

47

49

51

53 (a)

(b)

(c)

(d)

(e)

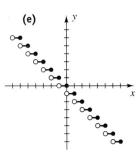

55 If $x > 0$, two different points on the graph have x-coordinate x.

57

59

61

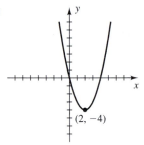

63 (a) $D = [-2, 6]$, $R = [-16, 8]$
　 (b) $D = [-4, 12]$, $R = [-4, 8]$
　 (c) $D = [1, 9]$, $R = [-3, 9]$
　 (d) $D = [-4, 4]$, $R = [-7, 5]$
　 (e) $D = [-6, 2]$, $R = [-4, 8]$
　 (f) $D = [-2, 6]$, $R = [-8, 4]$
　 (g) $D = [-6, 6]$, $R = [-4, 8]$
　 (h) $D = [-2, 6]$, $R = [0, 8]$

65 $T(x) = \begin{cases} 0.15x & \text{if } 0 \le x \le 20{,}000 \\ 0.20x - 1000 & \text{if } x > 20{,}000 \end{cases}$

67 $R(x) = \begin{cases} 1.20x & \text{if } 0 \le x \le 10{,}000 \\ 1.50x - 3000 & \text{if } 10{,}000 < x \le 15{,}000 \\ 1.80x - 7500 & \text{if } x > 15{,}000 \end{cases}$

69 (a) 194.80, 234.80

　 (b) $C_1(x) = \begin{cases} 119.80 & \text{if } 0 \le x \le 200 \\ 119.80 + 0.25(x - 200) & \text{if } x > 200 \end{cases}$

　　 $C_2(x) = 159.8 + 0.15x$ for $x \ge 0$

　 (c) I if $x \in [0, 900)$, II if $x > 900$

EXERCISES 3.6

1 $y = a(x + 3)^2 + 1$　　**3** $y = ax^2 - 3$
5 $f(x) = -(x + 2)^2 - 4$　　**7** $f(x) = 2(x - 3)^2 + 4$
9 $f(x) = -3(x + 1)^2 - 2$

11 $f(x) = -\dfrac{3}{4}(x - 6)^2 - 7$

13 (a) $0, 4$　　　　　　　　**(c)**
　 (b) Min: $f(2) = -4$

$(2, -4)$

15 (a) $-\dfrac{3}{4}, \dfrac{5}{3}$

(b) Max: $f\left(\dfrac{11}{24}\right) = \dfrac{841}{48}$

(c)

$\left(\frac{11}{24}, \frac{841}{48}\right)$

17 (a) $-\dfrac{4}{3}$

(b) Min: $f\left(-\dfrac{4}{3}\right) = 0$

(c)

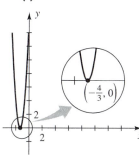

$\left(-\dfrac{4}{3}, 0\right)$

19 (a) None
(b) Min: $f(-2) = 5$
(c)

$(-2, 5)$

21 (a) $5 \pm \dfrac{1}{2}\sqrt{14} \approx 6.87, 3.13$ **(b)** Max: $f(5) = 7$

(c)

$(5, 7)$

23 $y = \dfrac{1}{8}(x - 4)^2 - 1$ **25** $y = -\dfrac{4}{9}(x + 2)^2 + 4$

27 $y = -\dfrac{1}{2}(x + 2)(x - 4)$

29 $y = 3(x - 0)^2 - 2$ **31** $y = -\dfrac{5}{9}(x - 3)^2 + 5$

33 $y = -\dfrac{1}{4}(x - 1)^2 + 4$ **35** 6.125   **37** 24.72 km

39 10.5 lb   **41 (a)** 424 ft   **(b)** 100 ft   **43** 20 and 20

45 (a) $y(x) = 250 - \dfrac{3}{4}x$ **(b)** $A(x) = x\left(250 - \dfrac{3}{4}x\right)$

(c) $166\dfrac{2}{3}$ ft by 125 ft

47 $y = -\dfrac{4}{27}\left(x - \dfrac{9}{2}\right)^2 + 3$

49 (a) $y = \dfrac{1}{500}x^2 + 10$ **(b)** 282 ft   **51** 2 ft

53 500 pairs

55 (a) $R(x) = 500x(30 - x)$
(b) \$15

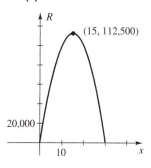

$(15, 112,500)$

20,000

10

57 (a) $f(x) = \begin{cases} \dfrac{4}{25}x + 80 & \text{if } -800 \le x < -500 \\[2mm] -\dfrac{1}{6250}x^2 + 40 & \text{if } -500 \le x \le 500 \\[2mm] -\dfrac{4}{25}x + 80 & \text{if } 500 < x \le 800 \end{cases}$

EXERCISES 3.7

1 (a) 15   **(b)** -3 **(c)** 54   **(d)** $\dfrac{2}{3}$

3 (a) $3x^2 + 1; \ 3 - x^2; \ 2x^4 + 3x^2 - 2; \ \dfrac{x^2 + 2}{2x^2 - 1}$

   (b) $\mathbb{R}$ **(c)** All real numbers except $\pm\dfrac{1}{2}\sqrt{2}$

5 (a) $2\sqrt{x + 5}; \ 0; \ x + 5; \ 1$ **(b)** $[-5, \infty)$ **(c)** $(-5, \infty)$

7 (a) $\dfrac{3x^2 + 6x}{(x - 4)(x + 5)}; \ \dfrac{x^2 + 14x}{(x - 4)(x + 5)}; \ \dfrac{2x^2}{(x - 4)(x + 5)};$

   $\dfrac{2(x + 5)}{x - 4}$

   (b) All real numbers except -5 and 4
   (c) All real numbers except -5, 0, and 4

9 (a) $-2x^2 - 1$ **(b)** $-4x^2 + 4x - 1$ **(c)** $4x - 3$
   (d) $-x^4$

11 (a) $6x + 9$ **(b)** $6x - 8$ **(c)** -3 **(d)** 10

13 (a) $75x^2 + 4$ **(b)** $15x^2 + 20$ **(c)** 304 **(d)** 155

15 (a) $8x^2 - 2x - 5$ **(b)** $4x^2 + 6x - 9$ **(c)** 31
(d) 45

17 (a) $8x^3 - 20x$ **(b)** $128x^3 - 20x$ **(c)** -24
(d) 3396

19 (a) 7 **(b)** -7 **(c)** 7 **(d)** -7

21 (a) $x + 2 - 3\sqrt{x + 2}; [-2, \infty)$
(b) $\sqrt{x^2 - 3x + 2}; (-\infty, 1] \cup [2, \infty)$

23 (a) $3x - 4; [0, \infty)$
(b) $\sqrt{3x^2 - 12}; (-\infty, -2] \cup [2, \infty)$

25 (a) $\sqrt{\sqrt{x + 5} - 2}; [-1, \infty)$
(b) $\sqrt{\sqrt{x - 2} + 5}; [2, \infty)$

27 (a) $\sqrt{3 - \sqrt{x^2 - 16}}; [-5, -4] \cup [4, 5]$
(b) $\sqrt{-x - 13}; (-\infty, -13]$

29 (a) $x; \mathbb{R}$ **(b)** $x; \mathbb{R}$

31 (a) $\dfrac{1}{x^6}$; all nonzero real numbers

(b) $\dfrac{1}{x^6}$; all nonzero real numbers

33 (a) $\dfrac{1}{5 - x}$; all real numbers except 4 and 5

(b) $\dfrac{-2x + 5}{-3x + 7}$; all real numbers except 2 and $\dfrac{7}{3}$

35 $-3 \pm \sqrt{2}$

37 (a) 5 **(b)** 6 **(c)** 6 **(d)** 5 **(e)** Not possible

39 $20\sqrt{x^2 + 1}$ **41** Odd **43** 31.26

45 $A(t) = 36\pi t^2$ **47** $r(t) = 9\sqrt[3]{t}$

49 $h(t) = 5\sqrt{t^2 + 8t}$

51 $d(t) = \sqrt{90{,}400 + (500 + 150t)^2}$

Exer. 53–60: Answers are not unique.

53 $u = x^2 + 3x, y = u^{1/3}$ **55** $u = x - 3, y = u^{-4}$

57 $u = x^4 - 2x^2 + 5, y = u^5$

59 $u = \sqrt{x + 4}, y = \dfrac{u - 2}{u + 2}$ **61** 5×10^{-13}

CHAPTER 3 REVIEW EXERCISES

1 The points in quadrants II and IV

2 $d(A, B)^2 + d(A, C)^2 = d(B, C)^2$; area $= 10$

3 (a) $\sqrt{265}$ **(b)** $\left(-\dfrac{13}{2}, 1\right)$ **(c)** $(-11, -23)$

4 $(0, 1), (0, 11)$ **5** $-2 < a < 1$

6 $(x - 7)^2 + (y + 4)^2 = 149$

7 $(x - 3)^2 + (y + 2)^2 = 169$

8 $x = -2 - \sqrt{9 - y^2}$ **9** $-\dfrac{11}{19}$

10 The slope of AD and BC is $\dfrac{2}{3}$.

11 (a) $18x + 6y = 7$ **(b)** $2x - 6y = 3$

12 $y = -\dfrac{8}{3}x + 8$ **13** $(x + 5)^2 + (y + 1)^2 = 81$

14 $x + y = -3$ **15** $5x - y = 23$

16 $2x - 3y = 5$ **17** $C(0, 6); r = \sqrt{5}$

18 $C(-3, 2); r = \dfrac{1}{2}\sqrt{13}$

19 (a) $\dfrac{1}{2}$ **(b)** $-\dfrac{1}{\sqrt{2}}$ **(c)** 0 **(d)** $-\dfrac{x}{\sqrt{3 - x}}$

(e) $-\dfrac{x}{\sqrt{x + 3}}$ **(f)** $\dfrac{x^2}{\sqrt{x^2 + 3}}$ **(g)** $\dfrac{x^2}{x + 3}$

20 Positive **21** Positive

22 (a) $\left[\dfrac{4}{3}, \infty\right); [0, \infty)$

(b) All real numbers except -3; $(0, \infty)$

23 $-2a - h + 1$ **24** $-\dfrac{1}{(a + h + 2)(a + 2)}$

25 $f(x) = \dfrac{5}{2}x - \dfrac{1}{2}$

26 (a) Odd **(b)** Neither **(c)** Even

Exer. 27–40: x-intercept(s) is listed, followed by y-intercept(s).

27 -5; none **28** None; 3.5

29 1.6; 4

30 4; $-\frac{4}{3}$

37 0, 8; 0

38 −3; ±3

31 0; 0

32 0; 0

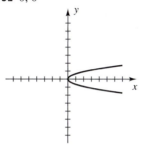

39 $3 \pm \sqrt{2}$; 7

40 −3, 1; 3

33 1; 1

34 1; −1

41 $\left(\sqrt{8}, \sqrt{8} \right)$

42 The graph of $y = -f(x - 2)$ is the graph of $y = f(x)$ shifted to the right 2 units and reflected about the x-axis.

43 (a)

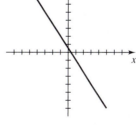

(b) $D = \mathbb{R}; R = \mathbb{R}$
(c) Decreasing on $(-\infty, \infty)$

35 ±4; ±4

36 None; 8

44 (a)

500

500

(b) $D = \mathbb{R};$ $R = \{1000\}$
(c) Constant on $(-\infty, \infty)$

45 (a)

(b) $D = \mathbb{R}$;
$R = [0, \infty)$
(c) Decreasing on
$(-\infty, -3]$,
increasing on
$[-3, \infty)$

46 (a)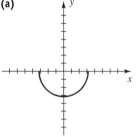

(b) $D = \left(-\sqrt{10}, \sqrt{10}\right)$;
$R = \left(-\sqrt{10}, 0\right)$
(c) Decreasing on
$\left[-\sqrt{10}, 0\right]$,
increasing on
$\left[0, \sqrt{10}\right]$

47 (a)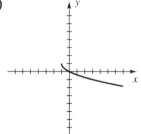

(b) $D = [-1, \infty)$;
$R = (-\infty, 1]$
(c) Decreasing on
$[-1, \infty)$

48 (a)

(b) $D = (-\infty, 2]$;
$R = [0, \infty)$
(c) Decreasing on
$(-\infty, 2]$

49 (a)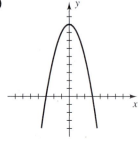

(b) $D = \mathbb{R}$;
$R = (-\infty, 9]$
(c) Increasing on
$(-\infty, 0]$,
decreasing on
$[0, \infty)$

50 (a)

(b) $D = \mathbb{R}$;
$R = [7, \infty)$
(c) Decreasing on
$(-\infty, -3]$,
increasing on
$[-3, \infty)$

51 (a)

(b) $D = \mathbb{R}$;
$R = [0, \infty)$
(c) Decreasing on
$(-\infty, 0]$,
increasing on
$[0, 2]$, constant
on $[2, \infty)$

52 (a)

(b) $D = \mathbb{R}$; $R = \{\ldots, -3, -1, 1, 3, \ldots\}$
(c) Constant on $[n, n + 1)$, where n is any integer

53 (a) **(b)**

(c)

(d)

(g)

(e)

(f)

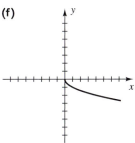

55 $2x - 5y = 10$ **56** $(x + 2)^2 + (y - 1)^2 = 25$

57 $y = \dfrac{1}{2}(x - 2)^2 - 4$ **58** $y = -|x - 2| - 1$

59 Min: $f(-3) = 4$ **60** Max: $f(5) = -7$

61 Max: $f(-1) = -37$ **62** Min: $f(4) = -108$

63 $f(x) = -2(x - 3)^2 + 4$ **64** $y = \dfrac{3}{2}(x - 3)^2 - 2$

65 **(a)** $[0, 2]$ **(b)** $(0, 2]$ **66** **(a)** -1 **(b)** $\sqrt{13}$

67 **(a)** $18x^2 + 9x - 1$ **(b)** $6x^2 - 15x + 5$

68 **(a)** $\sqrt{\dfrac{3 + 2x^2}{x^2}}$ **(b)** $\dfrac{1}{3x + 2}$

69 **(a)** $\sqrt{28 - x}$; $[3, 28]$

    (b) $\sqrt{\sqrt{25 - x^2} - 3}$; $[-4, 4]$

70 **(a)** $\dfrac{1}{x + 3}$; all real numbers except -3 and 0

    (b) $\dfrac{6x + 4}{x}$; all real numbers except $-\dfrac{2}{3}$ and 0

71 $u = x^2 - 5x$, $y = \sqrt[3]{u}$ **72** Between 36.1 ft and 60.1 ft

73 **(a)** 245 ft   **(b)** 2028

74 **(a)** $V = 6000t + 89{,}000$ **(b)** $2\dfrac{1}{3}$

75 **(a)** $F = \dfrac{9}{5}C + 32$ **(b)** $1.8°F$

76 **(a)** $C_1(x) = \dfrac{1}{16}x$ **(b)** $C_2(x) = \dfrac{5}{88}x + 50$ **(c)** 8800

77 **(a)** $y(x) = -\dfrac{4}{5}x + 20$ **(b)** $V(x) = 4x\left(-\dfrac{4}{5}x + 20\right)$

78 $C(r) = \dfrac{3\pi(r^3 + 16)}{10r}$

79 **(a)** $V = 10t$

    (b) $V = 200h^2$ for $0 \le h \le 6$;

       $V = 7200 + 3200(h - 6)$ for $6 < h \le 9$

    (c) $h = \sqrt{\dfrac{t}{20}}$ for $0 \le t \le 720$; $h = 6 + \dfrac{t - 720}{320}$ for

       $720 < t \le 1680$

80 **(a)** $r = \dfrac{1}{2}x$ **(b)** $y = \dfrac{5}{4\pi} - \dfrac{1}{48}x^3$

54 **(a)**

(b)

(c)

(d)

(e)

(f)

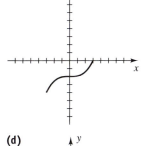

81 (a) $y(h) = \dfrac{bh}{a - b}$ **(b)** $V(h) = \dfrac{1}{3}\pi h(a^2 + ab + b^2)$

(c) $\dfrac{200}{7\pi} \approx 9.1$ ft

82 $\dfrac{18}{13}$ hr after 1:00 P.M., or about 2:23 P.M.

83 Radius of semicircle is $\dfrac{1}{8\pi}$ mi; length of rectangle is $\dfrac{1}{8}$ mi.

84 (a) 1 sec   **(b)** 4 ft

(c) On the moon, 6 sec and 24 ft

85 (a) $(87.5, 17.5)$ **(b)** 30.625 units

CHAPTER 3 DISCUSSION EXERCISES

2 (a) $g(x) = -\dfrac{1}{2}x + 3$ **(b)** $g(x) = -\dfrac{1}{2}x - 3$

(c) $g(x) = -\dfrac{1}{2}x + 7$ **(d)** $g(x) = -\dfrac{1}{2}x$

4 $2ax + ah + b$ **5** m_{PQ}; the slope of the tangent line at P

6 $R(x_3, y_3) = \left(\left(1 - \dfrac{m}{n}\right)x_1 + \dfrac{m}{n}x_2, \left(1 - \dfrac{m}{n}\right)y_1 + \dfrac{m}{n}y_2\right)$

7 $h = -ad^2$ **8** $f(x) = 40 - 20[\![-x/15]\!]$

9 $x = \dfrac{0.4996 + \sqrt{(-0.4996)^2 - 4(0.0833)(3.5491 - D)}}{2(0.0833)}$

10 $f(x) = \begin{cases} 0.132(x - 1)^2 + 0.7 & \text{if } 1 \le x \le 6 \\ -0.517x + 7.102 & \text{if } 6 < x \le 12 \end{cases}$

Chapter 4

EXERCISES 4.1

1 (a)   **(b)**

3 (a)   **(b)**

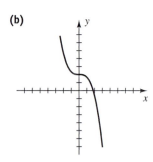

5 $f(3) = -2 < 0$, $f(4) = 10 > 0$

7 $f(2) = 5 > 0$, $f(3) = -5 < 0$

9 $f\left(-\dfrac{1}{2}\right) = \dfrac{19}{32} > 0$, $f(-1) = -1 < 0$

11 (a) C   **(b)** D   **(c)** B   **(d)** A

13 $f(x) > 0$ if $x > 2$,   **15** $f(x) > 0$ if $|x| < 2$,
$f(x) < 0$ if $x < 2$ $f(x) < 0$ if $|x| > 2$

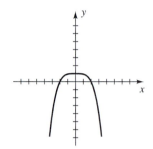

17 $f(x) > 0$ if $|x| > 2$,   **19** $f(x) > 0$ if $x < -2$ or
$f(x) < 0$ if $0 < |x| < 2$ $0 < x < 5$, $f(x) < 0$ if
$-2 < x < 0$ or $x > 5$

21 $f(x) > 0$ if $-2 < x < 3$ or $x > 4$, $f(x) < 0$ if $x < -2$ or $3 < x < 4$

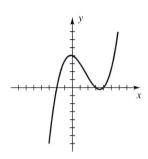

23 $f(x) > 0$ if $x > 2$, $f(x) < 0$ if $x < -2$ or $|x| < 2$

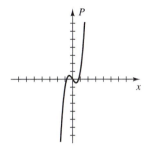

25 $f(x) > 0$ if $|x| > 2$ or $|x| < \sqrt{2}$, $f(x) < 0$ if $\sqrt{2} < |x| < 2$

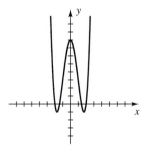

27 $f(x) > 0$ if $|x| > 2$, $f(x) < 0$ if $|x| < 2$, $x \neq 0$, $x \neq 1$

29

31 (a)

(b) $-abc$ **(c)** $(-\infty, a) \cup (b, c)$ **(d)** $[a, b] \cup [c, \infty)$

33 If n is even, then $(-x)^n = x^n$ and hence $f(-x) = f(x)$. Thus, f is an even function.

35 $-\dfrac{4}{3}$ **37** ± 4

39 $P(x) > 0$ on $\left(-\frac{1}{5}\sqrt{15}, 0\right)$ and $\left(\frac{1}{5}\sqrt{15}, \infty\right)$;

$P(x) < 0$ on $\left(-\infty, -\frac{1}{5}\sqrt{15}\right)$ and $\left(0, \frac{1}{5}\sqrt{15}\right)$

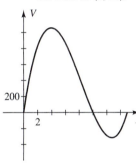

41 (b) $V(x) > 0$ on $(0, 10)$ and $(15, \infty)$; allowable values for x are in $(0, 10)$.

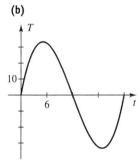

43 (a) $T > 0$ for $0 < t < 12$; $T < 0$ for $12 < t < 24$

(b)

(c) $T(6) = 32.4 > 32$, $T(7) = 29.75 < 32$

45 (a) $N(t) > 0$ for $0 < t < 5$

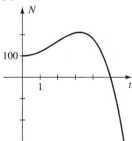

(b) The population becomes extinct after 5 years.

47 (a)

x	$f(x)$	$g(x)$	$h(x)$	$k(x)$
-60	25,920,000	25,902,001	25,937,999	26,135,880
-40	5,120,000	5,112,001	5,127,999	5,183,920
-20	320,000	318,001	321,999	327,960
20	320,000	318,001	321,999	312,040
40	5,120,000	5,112,001	5,127,999	5,056,080
60	25,920,000	25,902,001	25,937,999	25,704,120

(b) They become similar.   **(c)** $2x^4$

EXERCISES 4.2

1 $2x^2 - x + 3; 4x - 3$ **3** $\dfrac{3}{2}x; \dfrac{1}{2}x - 4$

5 $0; 7x + 2$ **7** $\dfrac{9}{2}; \dfrac{53}{2}$ **9** 26   **11** 7

13 $f(-3) = 0$ **15** $f(-2) = 0$ **17** $x^3 - 3x^2 - 10x$

19 $x^4 - 2x^3 - 9x^2 + 2x + 8$

21 $2x^2 + x + 6; 7$

23 $x^2 - 3x + 1; -8$

25 $3x^4 - 6x^3 + 12x^2 - 18x + 36; -65$

27 $4x^3 + 2x^2 - 4x - 2; 0$

29 73   **31** -0.0444

33 $8 + 7\sqrt{3}$

35 $f(-2) = 0$ **37** $f\left(\dfrac{1}{2}\right) = 0$

39 $3, 5$ **41** $f(c) > 0$ **43** -14

45 If $f(x) = x^n - y^n$ and n is even, then $f(-y) = 0$.

47 (a) $V = \pi x^2(6 - x)$

(b) $\left(\dfrac{1}{2}(5 + \sqrt{45}), \dfrac{1}{2}(7 - \sqrt{45})\right)$

49 (a) $A = 8x - 2x^3$ **(b)** $\sqrt{13} - 1 \approx 2.61$

EXERCISES 4.3

1 $-4x^3 + 16x^2 - 4x - 24$ **3** $3x^3 + 3x^2 - 36x$

5 $-2x^3 + 6x^2 - 8x + 24$

7 $x^4 + 2x^3 - 23x^2 - 24x + 144$

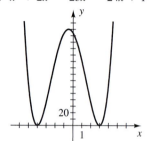

9 $3x^6 - 27x^5 + 81x^4 - 81x^3$

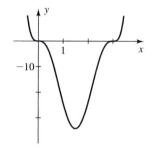

11 $f(x) = \dfrac{7}{9}(x + 1)\left(x - \dfrac{3}{2}\right)(x - 3)$

13 $f(x) = -1(x - 1)^2(x - 3)$

15 $-\dfrac{2}{3}$ (multiplicity 1); 0 (multiplicity 2);

$\dfrac{5}{2}$ (multiplicity 3)

17 $-\dfrac{3}{2}$ (multiplicity 2); 0 (multiplicity 3)

19 -4 (multiplicity 3); -3 (multiplicity 2);

3 (multiplicity 5)

21 $\pm 4i, \pm 3$ (each of multiplicity 1)

23 $f(x) = (x + 3)^2(x + 2)(x - 1)$

25 $f(x) = (x - 1)^5(x + 1)$

Exer. 27–34: The types of possible solutions are listed in the order positive, negative, nonreal complex.

27 3, 0, 0 or 1, 0, 2   **29** 0, 1, 2

31 2, 2, 0; 2, 0, 2; 0, 2, 2; 0, 0, 4

33 2, 3, 0; 2, 1, 2; 0, 3, 2; 0, 1, 4

35 Upper, 5; lower, -2 **37** Upper, 2; lower, -2

39 Upper, 3; lower, -3

41 $f(x) = -\dfrac{1}{4}(x + 1)^2(x - 1)(x - 2)^3$

43 (a) $f(x) = a(x + 3)^3(x + 1)(x - 2)^2$ **(b)** 108

45 No   **49** Yes: $1.5(x - 2)(x - 5.2)(x - 10.1)$

49 $f(t) = \dfrac{5}{3528}t(t - 5)(t - 19)(t - 24)$

EXERCISES 4.4

1 $x^2 - 6x + 13$ **3** $(x - 2)(x^2 + 4x + 29)$

5 $x(x + 1)(x^2 - 6x + 10)$

7 $(x^2 - 8x + 25)(x^2 + 4x + 5)$

9 $x(x^2 + 4)(x^2 - 2x + 2)$

Exer. 11–14: Show that none of the possible rational roots listed satisfy the equation.

11 $\pm 1, \pm 2, \pm 3, \pm 6$ **13** $\pm 1, \pm 2$ **15** $-2, -1, 4$

17 $-3, 2, \dfrac{5}{2}$ **19** $-7, \pm\sqrt{2}, 4$

21 $-3, -\dfrac{2}{3}, 0$ (multiplicity 2), $\dfrac{1}{2}$

23 $-\dfrac{3}{4}, -\dfrac{3}{4} \pm \dfrac{3}{4}\sqrt{7}i$

25 $f(x) = (3x + 2)(2x - 1)(x - 1)^2(x - 2)$

27 No. If i is a root, then $-i$ is also a root. Hence, the polynomial would have factors $x - 1, x + 1, x - i, x + i$ and therefore would be of degree greater than 3.

29 Since n is odd and nonreal complex zeros occur in conjugate pairs for polynomials with real coefficients, there must be at least one real zero.

31 (a) The two boxes correspond to $x = 5$ and $x = 5(2 - \sqrt{2})$.
   (b) The box corresponding to $x = 5$

33 (c) In feet: 5, 12, and 13   **35 (b)** 4 ft

EXERCISES 4.5

1 (a)

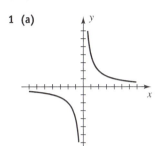

(b) $D =$ all nonzero real numbers; $R = D$
(c) Decreasing on $(-\infty, 0)$ and on $(0, \infty)$

3 VA: $x = 3$;
HA: $y = -2$;
hole: $\left(6, -\dfrac{22}{3}\right)$

5

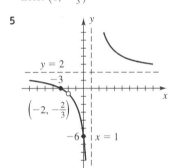

$f(x) = \dfrac{2(x + 3)(x + 2)}{(x - 1)(x + 2)}$

7

9

11

13

15

17

19

21

23

25

27

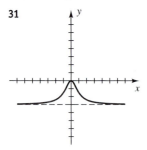

29

31

33 $y = x - 2$

35 $y = -\dfrac{1}{2}x$

$y = \left(-\dfrac{1}{2}\right)x$

37 $f(x) = \dfrac{2x - 3}{x + 1}$ for $x \neq -2$

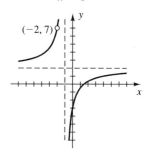

$(-2, 7)$

39 $f(x) = \dfrac{-1}{x + 1}$ for $x \neq 1$

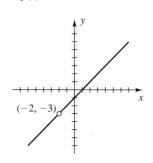

$\left(1, -\dfrac{1}{2}\right)$

41 $f(x) = x - 1$ for $x \neq -2$

$(-2, -3)$

43 $f(x) = \dfrac{x + 2}{x + 1}$ for $x \neq -2$

$(-2, 0)$

45 $f(x) = \dfrac{3 - x}{x - 4}$ **47** $f(x) = \dfrac{6x^2 - 6x - 12}{x^3 - 7x + 6}$

49 (a) $h = \dfrac{16}{(r + 0.5)^2} - 1$ (b) $V(r) = \pi r^2 h$

(c) Exclude $r \leq 0$ and $r \geq 3.5$.

51 (a) $V(t) = 50 + 5t, A(t) = 0.5t$ (b) $\dfrac{t}{10t + 100}$

(c) As $t \to \infty$, $c(t) \to 0.1$ lb of salt per gal.

53 (a) $0 < S < 4000$ (b) 4500   (c) 2000

(d) A 125% increase in S produces only a 12.5% increase in R.

55 (a) The graph of g is the horizontal line $y = 1$ with holes at $x = 0, \pm1, \pm2, \pm3$.
  (b) The graph of h is the graph of p with holes at $x = 0$, $\pm1, \pm2, \pm3$.

57 (a) $y = \dfrac{132 - 48x}{x - 4}$

(b)

x	y
2.8	2
3.0	12
3.2	27
3.4	52
3.6	102
3.8	252
4.0	undefined

(c)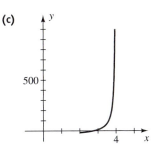

(d) $x = 4$
(e) Regardless of the number of additional credit hours obtained at 4.0, a cumulative GPA of 4.0 is not attainable.

EXERCISES 4.6

1 $u = kv;\ k = \dfrac{2}{5}$ **3** $r = k\dfrac{s}{t};\ k = -14$

5 $y = k\dfrac{x^2}{z^3};\ k = 27$ **7** $z = kx^2y^3;\ k = -\dfrac{2}{49}$

9 $y = k\dfrac{x}{z^2};\ k = 36$ **11** $y = k\dfrac{\sqrt{x}}{z^3};\ k = \dfrac{40}{3}$

13 (a) $P = kd$ **(b)** 59   **(c)** 295 lb/ft^2
  (d)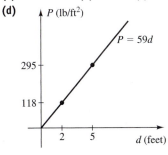

15 (a) $R = k\dfrac{l}{d^2}$ **(b)** $\dfrac{1}{40{,}000}$
  (c)
  (d) $\dfrac{50}{9}$ ohms

17 (a) $P = k\sqrt{l}$ **(b)** $\dfrac{3}{4}\sqrt{2}$ **(c)** $\dfrac{3}{2}\sqrt{3}$ sec

19 (a) $T = kd^{3/2}$ **(b)** $\dfrac{365}{(93)^{3/2}}$ **(c)** 223.2 days

21 (a) $V = k\sqrt{L}$ **(b)** $\dfrac{7}{2}\sqrt{2}$ **(c)** 60.6 mi/hr

23 (a) $W = kh^3$ **(b)** $\dfrac{25}{27}$ **(c)** 154 lb

25 (a) $F = kPr^4$ **(b)** About 2.05 times as hard
27 Increases 250%   **29** d is multiplied by 9.
31 $y = 1.2x$ **33** $y = -\dfrac{10.1}{x^2}$

CHAPTER 4 REVIEW EXERCISES

1 $f(x) > 0$ if $x > -2$,
  $f(x) < 0$ if $x < -2$

2 $f(x) > 0$ if $x < -\sqrt[6]{32}$
  or $x > \sqrt[6]{32}, f(x) < 0$
  if $-\sqrt[6]{32} < x < \sqrt[6]{32}$

3 $f(x) > 0$ if $-2 < x < 1$
or $1 < x < 3$, $f(x) < 0$
if $x < -2$ or $x > 3$

4 $f(x) > 0$ if $-1 < x < 0$
or $0 < x < 2$,
$f(x) < 0$ if $x < -1$
or $x > 2$

5 $f(x) > 0$ if $-4 < x < 0$,
or $x > 2$, $f(x) < 0$ if
$x < -4$ or $0 < x < 2$

6 $f(x) > 0$ if $-4 < x < -2$,
$0 < x < 2$, or $x > 4$,
$f(x) < 0$ if $x < -4$,
$-2 < x < 0$, or $2 < x < 4$

7 $f(0) = -9 < 100$ and $f(10) = 561 > 100$. By the
intermediate value theorem for polynomial functions, f
takes on every value between -9 and 561. Hence, there is
at least one real number a in $[0, 10]$ such that $f(a) = 100$.

8 Let $f(x) = x^5 - 3x^4 - 2x^3 - x + 1$. $f(0) = 1 > 0$ and
$f(1) = -4 < 0$. By the intermediate value theorem for
polynomial functions, f takes on every value between -4
and 1. Hence, there is at least one real number a in $[0, 1]$
such that $f(a) = 0$.

9 $3x^2 + 2$; $-21x^2 + 5x - 9$ **10** $4x - 1$; $2x - 1$

11 -132 **12** $f(3) = 0$

13 $6x^4 - 12x^3 + 24x^2 - 52x + 104$; -200

14 $2x^2 + \left(5 + 2\sqrt{2}\right)x + \left(2 + 5\sqrt{2}\right)$; $11 + 2\sqrt{2}$

15 $\dfrac{2}{41}(x^2 + 6x + 34)(x + 1)$

16 $\dfrac{1}{4}x(x^2 - 2x + 2)(x - 3)$

17 $x^7 + 6x^6 + 9x^5$

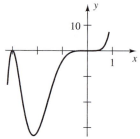

18 $(x - 2)^3(x + 3)(x - 1)$

19 1 (multiplicity 5); -3 (multiplicity 1)

20 $0, \pm i$ (all have multiplicity 2)

21 (a) Either 3 positive and 1 negative or 1 positive,
1 negative, and 2 nonreal complex

   (b) Upper bound, 3; lower bound, -1

22 (a) Either 2 positive and 3 negative; 2 positive, 1 negative,
and 2 nonreal complex; 3 negative and 2 nonreal
complex; or 1 negative and 4 nonreal complex

   (b) Upper bound, 2; lower bound, -3

23 Since there are only even powers,
$7x^6 + 2x^4 + 3x^2 + 10 \geq 10$ for every real number x.

24 $-3, -2, -2 \pm i$ **25** $-\dfrac{1}{2}, \dfrac{1}{4}, \dfrac{3}{2}$ **26** $\pm\sqrt{6}, \pm 1$

27 $f(x) = -\dfrac{1}{6}(x + 2)^3(x - 1)^2(x - 3)$

28 $f(x) = \dfrac{1}{16}(x + 3)^2 x^2(x - 3)^2$

29 VA: $x = 5$; HA: $y = \dfrac{4}{3}$; x-intercept: 1;

y-intercept: $\dfrac{4}{15}$; hole: $\left(-2, \dfrac{4}{7}\right)$

30

31

32

33

41 27

42

$y = \dfrac{288}{x^2}$

34

35

43 (a) $\dfrac{1}{15{,}000}$

(b) $y \approx 0.9754 < 1$ if $x = 6.1$, and
$y \approx 1.0006 > 1$ if $x = 6.2$

44 (a) $V = \dfrac{1}{4\pi} x(l^2 - x^2)$

(b) If $x > 0$, $V > 0$ when $0 < x < l$.

45 $t = 4$ (10:00 A.M.) and $t = 16 - 4\sqrt{6} \approx 6.2020$
(12:12 P.M.)

46 $\sqrt{5} < t < 4$

47 (a) $R = k$

(b) k is the maximum rate at which the liver can remove
alcohol from the bloodstream.

36

37

48 (a) $C(100) = \$2{,}000{,}000.00$ and $C(90) \approx \$163{,}636.36$

(b)

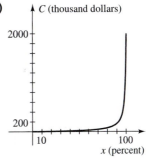

49 375   **50** 10,125 watts

38

39

40 $f(x) = \dfrac{3(x - 5)(x - 2)}{2(x + 3)(x - 2)}$ or $f(x) = \dfrac{3x^2 - 21x + 30}{2x^2 + 2x - 12}$

CHAPTER 4 DISCUSSION EXERCISES

2 Yes   **4** No   **5** $n + 1$ **7** $f(x) = \dfrac{(x^2 + 1)(x - 1)}{(x^2 + 1)(x - 2)}$

8 (a) No

(b) Yes, when $x = \dfrac{cd - af}{ae - bd}$, provided the denominator is
not zero

9 (a) \$1476

(b) Not valid for high confidence values

10 The second integer

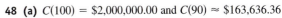

Chapter 5

EXERCISES 5.1

1 (a) 4   **(b)** Not possible
3 (a) Yes   **(b)** No   **(c)** Not a function
5 Yes   **7** No   **9** Yes   **11** No   **13** No   **15** Yes

Exer. 17–20: Show that $f(g(x)) = x = g(f(x))$.

17

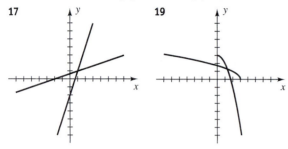

19

21 $(-\infty, 0) \cup (0, \infty); (-\infty, 1) \cup (1, \infty)$
23 $\left(-\infty, \frac{4}{3}\right) \cup \left(\frac{4}{3}, \infty\right); \left(-\infty, \frac{8}{3}\right) \cup \left(\frac{8}{3}, \infty\right)$
25 $f^{-1}(x) = \dfrac{x - 5}{3}$ **27** $f^{-1}(x) = \dfrac{2x + 1}{3x}$
29 $f^{-1}(x) = \dfrac{5x + 2}{2x - 3}$ **31** $f^{-1}(x) = -\sqrt{\dfrac{2 - x}{3}}$
33 $f^{-1}(x) = \sqrt[3]{\dfrac{x + 5}{2}}$ **35** $f^{-1}(x) = 3 - x^2, x \geq 0$
37 $f^{-1}(x) = (x - 1)^3$ **39** $f^{-1}(x) = x$
41 $f^{-1}(x) = 3 + \sqrt{x + 9}$ **43 (a)** 3   **(b)** -1 **(c)** 5
45 (a)

(b) $D = [-1, 2];$
$R = \left[\dfrac{1}{2}, 4\right]$
(c) $D_1 = \left[\dfrac{1}{2}, 4\right];$
$R_1 = [-1, 2]$

47 (a)

(b) $D = [-3, 3];$
$R = [-2, 2]$
(c) $D_1 = [-2, 2];$
$R_1 = [-3, 3]$

49 (a) Since f is one-to-one, an inverse exists;
$$f^{-1}(x) = \frac{x - b}{a}$$
(b) No; not one-to-one
51 (c) The graph of f is symmetric about the line $y = x$.
Thus, $f(x) = f^{-1}(x)$.
53 (a) 805 ft³/min
(b) $V^{-1}(x) = \dfrac{1}{35}x$. Given an air circulation of x cubic feet
per minute, $V^{-1}(x)$ computes the maximum number of
people that should be in the restaurant at one time.
(c) 67

EXERCISES 5.2

1 5   **3** $-1, 3$ **5** $-\dfrac{4}{99}$ **7** $\dfrac{18}{5}$ **9** 3

11 (a)

(b)

(c)

(d)

(e)

(f)

(g)

(h)

(i)

(j)

13

15

17

19

21

23

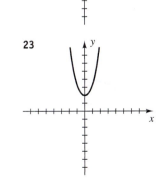

25 $f(x) = 2\left(\frac{5}{2}\right)^x$ **27** $f(x) = 2\left(\frac{2}{3}\right)^x - 3$

29 $f(x) = 8\left(\frac{1}{2}\right)^x$ **31** $f(x) = 180(1.5)^{-x} + 32$

33 (a) 90   **(b)** 59   **(c)** 35

35 (a) 1039; 3118; 5400   **37 (a)** 50 mg; 25 mg;

(b)
$$\frac{25}{2}\sqrt{2} \approx 17.7 \text{ mg}$$

(b)

39 $-\dfrac{1}{1600}$

41 (a) \$1010.00   **(b)** \$1061.52   **(c)** \$1126.83
   (d) \$10,892.55

43 (a) \$7800   **(b)** \$4790   **(c)** \$2942

45 \$161,657,351,965.80

47 (a) Examine the pattern formed by the value y in the
   year n.
   (b) Solve $s = (1 - a)^T y_0$ for a.

49 (a) \$925.75   **(b)** \$243,270

51 \$6346.40

53 (a) 180.1206   **(b)** 20.9758   **(c)** 7.3639

55 $y = 0.03(1.0588)^t$; 58¢

57 (a) \$746,648.43; \$1,192,971   **(b)** 12.44%

   (c) exponential; polynomial

EXERCISES 5.3

1 (a)

(b)

3 (a) 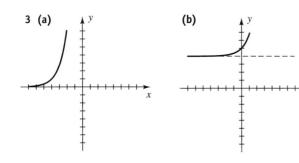 **(b)**

15 (a) 0 **(b)** 1 **(c)** Not possible **(d)** 2 **(e)** 8
(f) 3 **(g)** -2

17 (a) 3 **(b)** 5 **(c)** 2 **(d)** -4 **(e)** 2
(f) -3 **(g)** $3e^2$

19 4 **21** No solution **23** $-1, -2$ **25** 13

27 27 **29** $\pm \dfrac{1}{e}$ **31** 3 **33** 3

5 \$1510.59 **7** \$13,806.92 **9** 13% **11** 3, 4

13 -1 **15** $-\dfrac{3}{4}, 0$ **17** $\dfrac{4}{(e^x + e^{-x})^2}$ **19** 27.43 g

21 280.0 million **23** 13.5% **25** 5610

27 7.44 in. **29** 75.77 cm; 15.98 cm/yr

31 \$11.25 per hr **33 (a)** 7.19% **(b)** 7.25%

35

35 (a) **(b)**

(c) 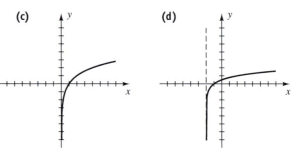 **(d)**

EXERCISES 5.4

1 (a) $\log_4 64 = 3$ **(b)** $\log_4 \dfrac{1}{64} = -3$

(c) $\log_t s = r$ **(d)** $\log_3 (4 - t) = x$

(e) $\log_5 \dfrac{a + b}{a} = 7t$ **(f)** $\log_{0.7} (5.3) = t$

3 (a) $2^5 = 32$ **(b)** $3^{-5} = \dfrac{1}{243}$ **(c)** $t^p = r$

(d) $3^5 = (x + 2)$ **(e)** $2^{3x+4} = m$ **(f)** $b^{3/2} = 512$

5 $t = 3 \log_a \dfrac{5}{2}$ **7** $t = \log_a \left(\dfrac{H - K}{C} \right)$

9 $t = \dfrac{1}{C} \log_a \left(\dfrac{A - D}{B} \right)$

11 (a) $\log 100{,}000 = 5$ **(b)** $\log 0.001 = -3$

(c) $\log (y + 1) = x$ **(d)** $\ln p = 7$

(e) $\ln (3 - x) = 2t$

13 (a) $10^{50} = x$ **(b)** $10^{20t} = x$ **(c)** $e^{0.1} = x$

(d) $e^{4+3x} = w$ **(e)** $e^{1/6} = z - 2$

(e) **(f)**

(g) **(h)**

(i)

(j)

(k)

(l)

37

$x = -10$

39

41

43 $f(x) = \log_3 x$

45 $f(x) = -F(x)$ **47** $f(x) = F(x - 2)$

49 $f(x) = F(x) + 1$

51 (a) 4240 **(b)** 8.85 **(c)** 0.0237 **(d)** 9.97
(e) 1.05 **(f)** 0.202

53 $f(x) = 1000e^{x \ln 1.05}$; 4.88% **55** $t = -1600 \log_2 \left(\dfrac{q}{q_0} \right)$

57 $t = -\dfrac{L}{R} \ln \left(\dfrac{I}{20} \right)$ **59 (a)** 2 **(b)** 4 **(c)** 5

61 (a) 10 **(b)** 30 **(c)** 40 **63** In the year 2079
65 (a) $W = 2.4e^{1.84h}$ **(b)** 37.92 kg
67 (a) 10,007 ft **(b)** 18,004 ft
69 (a) 305.9 kg **(b)** (1) 20 yr (2) 19.8 yr
71 10.1 mi **73** $2^{1/8} \approx 1.09$
75 (a) Pedestrians have faster average walking speeds in
large cities.
(b) 570,000
77 (a) 8.4877 **(b)** −0.0601
79 30%

EXERCISES 5.5

1 (a) $\log_4 x + \log_4 z$ **(b)** $\log_4 y - \log_4 x$

(c) $\dfrac{1}{3} \log_4 z$

3 $3 \log_a x + \log_a w - 2 \log_a y - 4 \log_a z$

5 $\dfrac{1}{3} \log z - \log x - \dfrac{1}{2} \log y$

7 $\dfrac{7}{4} \ln x - \dfrac{5}{4} \ln y - \dfrac{1}{4} \ln z$

9 (a) $\log_3 (5xy)$ **(b)** $\log_3 \dfrac{2z}{x}$ **(c)** $\log_3 y^5$

11 $\log_a \dfrac{x^2 \sqrt[3]{x - 2}}{(2x + 3)^5}$ **13** $\log \dfrac{y^{13/3}}{x^2}$ **15** $\ln x$ **17** $\dfrac{7}{2}$

19 $5\sqrt{5}$ **21** No solution **23** −7 **25** 1

27 −2 **29** $\dfrac{-1 + \sqrt{65}}{2}$ **31** $-1 + \sqrt{1 + e}$

33 $3 + \sqrt{10}$

35

37

39

41

43

45

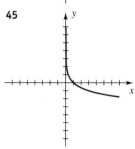

47 $f(x) = \log_2 x^2$ **49** $f(x) = \log_2 (8x)$ **51** $\approx +7$

53 $y = \dfrac{b}{x^k}$ **55**

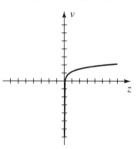

57 **(a)** 0   **(b)** $R(2x) = R(x) + a \log 2$ **59** 0.29 cm

EXERCISES 5.6

1 $\dfrac{\log 8}{\log 5} \approx 1.29$ **3** $4 - \dfrac{\log 5}{\log 3} \approx 2.54$ **5** 1.1133

7 -0.7325 **9** 2   **11** $\dfrac{\log (2/81)}{\log 24} \approx -1.16$

13 $\dfrac{\log (8/25)}{\log (4/5)} \approx 5.11$ **15** -3 **17** 5

19 $\dfrac{2}{3} \sqrt{\dfrac{101}{11}} \approx 2.02$ **21** 1, 2

23 $\dfrac{\log (4 + \sqrt{19})}{\log 4} \approx 1.53$ **25** 1 or 100   **27** 10^{100}

29 10,000   **31** ln 3   **33** 7

35 $x = \log \left(y \pm \sqrt{y^2 - 1}\right)$

37 $x = \dfrac{1}{2} \log \left(\dfrac{1 + y}{1 - y}\right)$ **39** $x = \ln \left(y + \sqrt{y^2 + 1}\right)$

41 $x = \dfrac{1}{2} \ln \left(\dfrac{y + 1}{y - 1}\right)$

43 y-intercept $= \log_2 3$
≈ 1.5850

45 x-intercept $= \log_4 3$
≈ 0.7925

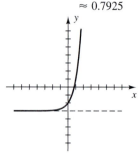

47 **(a)** 2.2   **(b)** 5   **(c)** 8.3

49 Basic if pH > 7, acidic if pH < 7

51 11.58 yr $\approx$ 11 yr 7 mo   **53** 86.4 m

55 **(a)**   **(b)** 6.58 min

57 **(a)** $t = \dfrac{\log (F/F_0)}{\log (1 - m)}$ **(b)** After 13,863 generations

59 **(a)** 4.28 ft   **(b)** 24.8 yr   **61** $\dfrac{\ln (25/6)}{\ln (200/35)} \approx 0.82$

63 The suspicion is correct.

65 The suspicion is incorrect.   **67** -0.5764

69 (4)   **71** 1998

CHAPTER 5 REVIEW EXERCISES

1 Yes

2

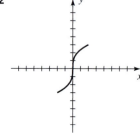

3 (a) $f^{-1}(x) = \dfrac{10 - x}{15}$ **4 (a)** $f^{-1}(x) = -\sqrt{\dfrac{9 - x}{2}}$

(b)   **(b)**

5 (a) 2   **(b)** 4   **(c)** 2   **(d)** 2   **(e)** $x > 2$
6 (a) 5   **(b)** 7   **(c)** 4
   (d) Not enough information is given.

7   **8**

9   **10**

11   **12**

13   **14**

15   **16**

17   **18**

19   **20**

21

22

51 **(a)** 1.89   **(b)** 78.3   **(c)** 0.472

52 **(a)** 0.924   **(b)** 0.00375   **(c)** 6.05

53 **(a)** $D = (-1, \infty), R = \mathbb{R}$
   (b) $y = 2^x - 1, D = \mathbb{R}, R = (-1, \infty)$

54 **(a)** $D = \mathbb{R}, R = (-2, \infty)$
   (b) $y = 3 - \log_2 (x + 2), D = (-2, \infty), R = \mathbb{R}$

55 **(a)** 2000
   (b) $2000(3^{1/6}) \approx 2401; 2000(3^{1/2}) = 3464; 6000$

56 \$1125.51

57 **(a)**   **(b)** 8 days

23 **(a)** −4   **(b)** 0   **(c)** 1   **(d)** 4   **(e)** 6   **(f)** 8
   (g) $\dfrac{1}{2}$

24 **(a)** $\dfrac{1}{3}$ **(b)** 0   **(c)** 1   **(d)** 5   **(e)** 1   **(f)** 25
   (g) $\dfrac{1}{3}$

25 0   **26** $-\dfrac{6}{5}$ **27** 9   **28** 9   **29** $\dfrac{33}{47}$ **30** 1

31 $-1 + \sqrt{3}$ **32** 99   **33** $5 - \dfrac{\log 6}{\log 2}$

34 $\pm \sqrt{\dfrac{\log 7}{\log 3}}$ **35** $\dfrac{\log (3/8)}{\log (32/9)}$ **36** 1   **37** $\dfrac{1}{4}, 1, 4$

38 No solution   **39** $\sqrt{5}$ **40** 2   **41** $0, \pm 1$

42 $\ln 2$ **43** **(a)** $-3, 2$ **(b)** 2

44 **(a)** 8   **(b)** ± 4

45 $4 \log x + \dfrac{2}{3} \log y - \dfrac{1}{3} \log z$

46 $-\log (xy^2)$ **47** $f(x) = 6\left(\dfrac{4}{3}\right)^x$

48

49 $x = \log \left(\dfrac{1 \pm \sqrt{1 - 4y^2}}{2y} \right)$

50 If $y < 0$, then $x = \log \left(\dfrac{1 - \sqrt{1 + 4y^2}}{2y} \right)$.

     If $y > 0$, then $x = \log \left(\dfrac{1 + \sqrt{1 + 4y^2}}{2y} \right)$.

58 $N = 1000 \left(\dfrac{3}{5} \right)^{t/3}$

59 **(a)** After 11.39 yr   **(b)** 6.30 yr   **60** 3.16%

61 $t = (\ln 100) \dfrac{L}{R} \approx 4.6 \dfrac{L}{R}$

62 **(a)** $I = I_0 10^{\alpha/10}$
   (b) Examine $I(\alpha + 1)$, where $I(\alpha)$ is the intensity corresponding to α decibels.

63 $t = -\dfrac{1}{k} \ln \left(\dfrac{a - L}{ab} \right)$ **64** $A = 10^{(R+5.1)/2.3} - 3000$

65 $\dfrac{A_1}{A_2} = \dfrac{10^{(R+5.1)/2.3} - 3000}{10^{(R+7.5)/2.3} - 34{,}000}$ **66** 26,615.9 mi^2

67 $h = \dfrac{\ln (29/p)}{0.000034}$ **68** $v = a \ln \left(\dfrac{m_1 + m_2}{m_1} \right)$

69 **(a)** $n = 10^{7.7 - 0.9R}$ **(b)** 12,589; 1585; 200

70 **(a)** $E = 10^{11.4 + 1.5R}$ **(b)** 10^{24} ergs   **71** 110 days

72 86.8 cm; 9.715 cm/yr   **73** $t = -\dfrac{L}{R} \ln \left(\dfrac{V - RI}{V} \right)$

74 **(a)** 26,749 yr   **(b)** 30%   **75** 21.00 yr

76 3196 yr

CHAPTER 5 DISCUSSION EXERCISES

1 **(a)** $f^{-1}(x) = \sqrt[3]{1-x} + 1$

2 $f^{-1}(x) = \dfrac{x}{\sqrt{81-x^2}}$. The vertical asymptotes are $x = \pm 9$.

The horizontal asymptotes of f are $y = \pm 9$.

3 **(a)** *Hint:* Take the natural logarithm of both sides first.

(b) Note that $f(e) = \dfrac{1}{e}$. Any horizontal line $y = k$, with

$0 < k < \dfrac{1}{e}$, will intersect the graph at points

$\left(x_1, \dfrac{\ln x_1}{x_1}\right)$ and $\left(x_2, \dfrac{\ln x_2}{x_2}\right)$, where $1 < x_1 < e$ and

$x_2 > e$.

4 7.16 yr

5 *Hint:* Check the restrictions for the logarithm laws.

6 **(a)** The difference is in the compounding.

(b) Closer to the graph of the second function

7 8.447177%; $1,025,156.25

8 **(a)** 3.5 earthquakes $= 1$ bomb, 425 bombs $= 1$ eruption

(b) 9.22; no

9 e^b, with $b = \dfrac{11 \ln 5 \cdot \ln 7}{\ln 35}$

Chapter 6

EXERCISES 6.1

Exer. 1–4: The answers are not unique.

1 **(a)** $480°, 840°, -240°, -600°$

(b) $495°, 855°, -225°, -585°$

(c) $330°, 690°, -390°, -750°$

3 **(a)** $260°, 980°, -100°, -460°$

(b) $\dfrac{17\pi}{6}, \dfrac{29\pi}{6}, -\dfrac{7\pi}{6}, -\dfrac{19\pi}{6}$

(c) $\dfrac{7\pi}{4}, \dfrac{15\pi}{4}, -\dfrac{9\pi}{4}, -\dfrac{17\pi}{4}$

5 **(a)** $84°42'26''$ **(b)** $57.5°$

7 **(a)** $131°8'23''$ **(b)** $43.58°$

9 **(a)** $\dfrac{5\pi}{6}$ **(b)** $-\dfrac{\pi}{3}$ **(c)** $\dfrac{5\pi}{4}$

11 **(a)** $\dfrac{5\pi}{2}$ **(b)** $\dfrac{2\pi}{5}$ **(c)** $\dfrac{5\pi}{9}$

13 **(a)** $120°$ **(b)** $330°$ **(c)** $135°$

15 **(a)** $-630°$ **(b)** $1260°$ **(c)** $20°$

17 $114°35'30''$ **19** $286°28'44''$ **21** $37.6833°$

23 $115.4408°$ **25** $63°10'8''$ **27** $310°37'17''$

29 2.5 cm

31 **(a)** $2\pi \approx 6.28$ cm **(b)** $8\pi \approx 25.13$ cm^2

33 **(a)** $1.75; \dfrac{315}{\pi} \approx 100.27°$ **(b)** 14 cm^2

35 **(a)** $\dfrac{20\pi}{9} \approx 6.98$ m **(b)** $\dfrac{80\pi}{9} \approx 27.93$ m^2

37 In miles: **(a)** 4189 **(b)** 3142 **(c)** 2094

(d) 698 **(e)** 70

39 $\dfrac{1}{8}$ radian $\approx 7°10'$ **41** 37.1%

43 7.29×10^{-5} rad/sec

45 **(a)** 80π rad/min **(b)** $\dfrac{100\pi}{3} \approx 104.72$ ft/min

47 **(a)** 400π rad/min **(b)** 38π cm/sec **(c)** 380 rpm

(d) $S(r) = \dfrac{1140}{r}$; inversely

49 **(a)** $\dfrac{21\pi}{8} \approx 8.25$ ft **(b)** $\dfrac{2}{3}d$

51 Large **53** 192.08 rev/min

EXERCISES 6.2

1 **(a)** B **(b)** D **(c)** A **(d)** C **(e)** E

Note: Answers are in the order *sin, cos, tan, cot, sec, csc* for any exercises that require the values of the six trigonometric functions.

3 $\dfrac{4}{5}, \dfrac{3}{5}, \dfrac{4}{3}, \dfrac{3}{4}, \dfrac{5}{3}, \dfrac{5}{4}$

5 $\dfrac{2}{5}, \dfrac{\sqrt{21}}{5}, \dfrac{2}{\sqrt{21}}, \dfrac{\sqrt{21}}{2}, \dfrac{5}{\sqrt{21}}, \dfrac{5}{2}$

7 $\dfrac{a}{\sqrt{a^2+b^2}}, \dfrac{b}{\sqrt{a^2+b^2}}, \dfrac{a}{b}, \dfrac{b}{a}, \dfrac{\sqrt{a^2+b^2}}{b}, \dfrac{\sqrt{a^2+b^2}}{a}$

9 $\dfrac{b}{c}, \dfrac{\sqrt{c^2-b^2}}{c}, \dfrac{b}{\sqrt{c^2-b^2}}, \dfrac{\sqrt{c^2-b^2}}{b}, \dfrac{c}{\sqrt{c^2-b^2}}, \dfrac{c}{b}$

11 $x = 8; y = 4\sqrt{3}$ **13** $x = 7\sqrt{2}; y = 7$

15 $x = 4\sqrt{3}; y = 4$

17 $\dfrac{3}{5}, \dfrac{4}{5}, \dfrac{3}{4}, \dfrac{4}{3}, \dfrac{5}{4}, \dfrac{5}{3}$ **19** $\dfrac{5}{13}, \dfrac{12}{13}, \dfrac{5}{12}, \dfrac{12}{5}, \dfrac{13}{12}, \dfrac{13}{5}$

21 $\dfrac{\sqrt{11}}{6}, \dfrac{5}{6}, \dfrac{\sqrt{11}}{5}, \dfrac{5}{\sqrt{11}}, \dfrac{6}{5}, \dfrac{6}{\sqrt{11}}$

23 $200\sqrt{3} \approx 346.4$ ft   **25** 192 ft   **27** 1.02 m

29 (a) 0.6691   **(b)** 0.2250   **(c)** 1.1924   **(d)** -1.0154

31 (a) 4.0572   **(b)** 1.0323   **(c)** -0.6335 **(d)** 4.3813

33 (a) 0.5   **(b)** -0.9880 **(c)** 0.9985   **(d)** -1

35 (a) -1 **(b)** -4

37 (a) 5   **(b)** 5

39 $1 - \sin\theta\cos\theta$ **41** $\sin\theta$

43 $\cot\theta = \dfrac{\sqrt{1 - \sin^2\theta}}{\sin\theta}$ **45** $\sec\theta = \dfrac{1}{\sqrt{1 - \sin^2\theta}}$

47 $\sin\theta = \dfrac{\sqrt{\sec^2\theta - 1}}{\sec\theta}$

Exer. 49–70: Typical verifications are given.

49 $\cos\theta\sec\theta = \cos\theta\,(1/\cos\theta) = 1$

51 $\sin\theta\sec\theta = \sin\theta\,(1/\cos\theta) = \sin\theta/\cos\theta = \tan\theta$

53 $\dfrac{\csc\theta}{\sec\theta} = \dfrac{1/\sin\theta}{1/\cos\theta} = \dfrac{\cos\theta}{\sin\theta} = \cot\theta$

55 $(1 + \cos 2\theta)(1 - \cos 2\theta) = 1 - \cos^2 2\theta = \sin^2 2\theta$

57 $\cos^2\theta\,(\sec^2\theta - 1) = \cos^2\theta\,(\tan^2\theta)$
$$= \cos^2\theta \cdot \dfrac{\sin^2\theta}{\cos^2\theta} = \sin^2\theta$$

59 $\dfrac{\sin(\theta/2)}{\csc(\theta/2)} + \dfrac{\cos(\theta/2)}{\sec(\theta/2)} = \dfrac{\sin(\theta/2)}{1/\sin(\theta/2)} + \dfrac{\cos(\theta/2)}{1/\cos(\theta/2)}$
$$= \sin^2(\theta/2) + \cos^2(\theta/2) = 1$$

61 $(1 + \sin\theta)(1 - \sin\theta) = 1 - \sin^2\theta = \cos^2\theta$
$$= \dfrac{1}{\sec^2\theta}$$

63 $\sec\theta - \cos\theta = \dfrac{1}{\cos\theta} - \cos\theta = \dfrac{1 - \cos^2\theta}{\cos\theta} = \dfrac{\sin^2\theta}{\cos\theta}$
$$= \dfrac{\sin\theta}{\cos\theta} \cdot \sin\theta = \tan\theta\sin\theta$$

65 $(\cot\theta + \csc\theta)(\tan\theta - \sin\theta)$
$= \cot\theta\tan\theta - \cot\theta\sin\theta + \csc\theta\tan\theta$
  $- \csc\theta\sin\theta$
$= \dfrac{1}{\tan\theta}\tan\theta - \dfrac{\cos\theta}{\sin\theta}\sin\theta + \dfrac{1}{\sin\theta}\dfrac{\sin\theta}{\cos\theta} - \dfrac{1}{\sin\theta}\sin\theta$
$= 1 - \cos\theta + \dfrac{1}{\cos\theta} - 1 = -\cos\theta + \sec\theta$
$\qquad\qquad\qquad = \sec\theta - \cos\theta$

67 $\sec^2 3\theta\csc^2 3\theta = (1 + \tan^2 3\theta)(1 + \cot^2 3\theta)$
$= 1 + \tan^2 3\theta + \cot^2 3\theta + 1$
$= \sec^2 3\theta + \csc^2 3\theta$

69 $\log\csc\theta = \log\left(\dfrac{1}{\sin\theta}\right) = \log 1 - \log\sin\theta$
$= 0 - \log\sin\theta = -\log\sin\theta$

71 $-\dfrac{3}{5}, \dfrac{4}{5}, -\dfrac{3}{4}, -\dfrac{4}{3}, \dfrac{5}{4}, -\dfrac{5}{3}$

73 $-\dfrac{5}{\sqrt{29}}, -\dfrac{2}{\sqrt{29}}, \dfrac{5}{2}, \dfrac{2}{5}, -\dfrac{\sqrt{29}}{2}, -\dfrac{\sqrt{29}}{5}$

75 $\dfrac{4}{\sqrt{17}}, \dfrac{1}{\sqrt{17}}, -4, -\dfrac{1}{4}, -\sqrt{17}, \dfrac{\sqrt{17}}{4}$

77 $\dfrac{4}{5}, \dfrac{3}{5}, \dfrac{4}{3}, \dfrac{3}{4}, \dfrac{5}{3}, \dfrac{5}{4}$

79 $-\dfrac{7}{\sqrt{53}}, -\dfrac{2}{\sqrt{53}}, \dfrac{7}{2}, \dfrac{2}{7}, -\dfrac{\sqrt{53}}{2}, -\dfrac{\sqrt{53}}{7}$

Note: U denotes *undefined.*

81 (a) 1, 0, U, 0, U, 1   **(b)** 0, 1, 0, U, 1, U
   (c) -1, 0, U, 0, U, -1 **(d)** 0, -1, 0, U, -1, U

83 (a) IV   **(b)** III   **(c)** II   **(d)** III

85 $\dfrac{3}{5}, -\dfrac{4}{5}, -\dfrac{3}{4}, -\dfrac{4}{3}, -\dfrac{5}{4}, \dfrac{5}{3}$

87 $\dfrac{5}{13}, \dfrac{12}{13}, -\dfrac{5}{12}, -\dfrac{12}{5}, \dfrac{13}{12}, -\dfrac{13}{5}$

89 $-\dfrac{\sqrt{8}}{3}, -\dfrac{1}{3}, \sqrt{8}, \dfrac{1}{\sqrt{8}}, -3, -\dfrac{3}{\sqrt{8}}$

91 $\dfrac{\sqrt{15}}{4}, -\dfrac{1}{4}, -\sqrt{15}, -\dfrac{1}{\sqrt{15}}, -4, \dfrac{4}{\sqrt{15}}$

93 $-\tan\theta$ **95** $\sec\theta$ **97** $-\sin\dfrac{\theta}{2}$

EXERCISES 6.3

1 $\dfrac{8}{17}, -\dfrac{15}{17}, -\dfrac{8}{15}, -\dfrac{15}{8}, -\dfrac{17}{15}, \dfrac{17}{8}$

3 $-\dfrac{7}{25}, \dfrac{24}{25}, -\dfrac{7}{24}, -\dfrac{24}{7}, \dfrac{25}{24}, -\dfrac{25}{7}$

5 (a) $\left(-\dfrac{3}{5}, -\dfrac{4}{5}\right)$ **(b)** $\left(-\dfrac{3}{5}, -\dfrac{4}{5}\right)$

   (c) $\left(\dfrac{3}{5}, -\dfrac{4}{5}\right)$ **(d)** $\left(-\dfrac{3}{5}, \dfrac{4}{5}\right)$

7 (a) $\left(\dfrac{12}{13}, \dfrac{5}{13}\right)$ **(b)** $\left(\dfrac{12}{13}, \dfrac{5}{13}\right)$

   (c) $\left(-\dfrac{12}{13}, \dfrac{5}{13}\right)$ **(d)** $\left(\dfrac{12}{13}, -\dfrac{5}{13}\right)$

Note: U denotes *undefined.*

9 (a) (1, 0); 0, 1, 0, U, 1, U
   (b) $(-1, 0)$; 0, -1, 0, U, -1, U

11 (a) $(0, -1)$; -1, 0, U, 0, U, -1
   (b) (0, 1); 1, 0, U, 0, U, 1

13 (a) $\left(\dfrac{\sqrt{2}}{2}, \dfrac{\sqrt{2}}{2}\right)$; $\dfrac{\sqrt{2}}{2}, \dfrac{\sqrt{2}}{2}, 1, 1, \sqrt{2}, \sqrt{2}$

(b) $\left(-\dfrac{\sqrt{2}}{2}, \dfrac{\sqrt{2}}{2}\right)$; $\dfrac{\sqrt{2}}{2}, -\dfrac{\sqrt{2}}{2}, -1, -1, -\sqrt{2}, \sqrt{2}$

15 (a) $\left(-\dfrac{\sqrt{2}}{2}, -\dfrac{\sqrt{2}}{2}\right)$; $-\dfrac{\sqrt{2}}{2}, -\dfrac{\sqrt{2}}{2}, 1, 1, -\sqrt{2}, -\sqrt{2}$

(b) $\left(\dfrac{\sqrt{2}}{2}, -\dfrac{\sqrt{2}}{2}\right)$; $-\dfrac{\sqrt{2}}{2}, \dfrac{\sqrt{2}}{2}, -1, -1, \sqrt{2}, -\sqrt{2}$

17 (a) -1 **(b)** $-\dfrac{\sqrt{2}}{2}$ **(c)** -1

19 (a) 1 **(b)** -1 **(c)** 1

Exer. 21–26: Typical verifications are given.

21 $\sin(-x)\sec(-x) = (-\sin x)\sec x$
$= (-\sin x)(1/\cos x)$
$= -\tan x$

23 $\dfrac{\cot(-x)}{\csc(-x)} = \dfrac{-\cot x}{-\csc x} = \dfrac{\cos x/\sin x}{1/\sin x} = \cos x$

25 $\dfrac{1}{\cos(-x)} - \tan(-x)\sin(-x)$

$= \dfrac{1}{\cos x} - (-\tan x)(-\sin x)$

$= \dfrac{1}{\cos x} - \dfrac{\sin x}{\cos x}\sin x$

$= \dfrac{1 - \sin^2 x}{\cos x} = \dfrac{\cos^2 x}{\cos x} = \cos x$

27 (a) 0 **(b)** -1 **29 (a)** $\dfrac{\sqrt{2}}{2}$ **(b)** -1

31 (a) 1 **(b)** $-\infty$ **33 (a)** -1 **(b)** ∞

35 (a) ∞ **(b)** $\sqrt{2}$ **37 (a)** $-\infty$ **(b)** 1

39 $\dfrac{3\pi}{2}, \dfrac{7\pi}{2}$ **41** $\dfrac{\pi}{6}, \dfrac{5\pi}{6}, \dfrac{13\pi}{6}, \dfrac{17\pi}{6}$ **43** $0, 2\pi, 4\pi$

45 $\dfrac{\pi}{4}, \dfrac{7\pi}{4}, \dfrac{9\pi}{4}, \dfrac{15\pi}{4}$ **47** $\dfrac{\pi}{4}, \dfrac{5\pi}{4}$ **49** $0, \pi$

51 (a) $-\dfrac{11\pi}{6}, -\dfrac{7\pi}{6}, \dfrac{\pi}{6}, \dfrac{5\pi}{6}$

(b) $-\dfrac{11\pi}{6} < x < -\dfrac{7\pi}{6}$ and $\dfrac{\pi}{6} < x < \dfrac{5\pi}{6}$

(c) $-2\pi \le x < -\dfrac{11\pi}{6}, -\dfrac{7\pi}{6} < x < \dfrac{\pi}{6}$, and

$\dfrac{5\pi}{6} < x \le 2\pi$

53 (a) $-\dfrac{4\pi}{3}, -\dfrac{2\pi}{3}, \dfrac{2\pi}{3}, \dfrac{4\pi}{3}$

(b) $-2\pi \le x < -\dfrac{4\pi}{3}, -\dfrac{2\pi}{3} < x < \dfrac{2\pi}{3}$, and

$\dfrac{4\pi}{3} < x \le 2\pi$

(c) $-\dfrac{4\pi}{3} < x < -\dfrac{2\pi}{3}$ and $\dfrac{2\pi}{3} < x < \dfrac{4\pi}{3}$

55 **57**

59 **61**

63 (a) $\left[-2\pi, -\dfrac{3\pi}{2}\right), \left(-\dfrac{3\pi}{2}, -\pi\right], \left[0, \dfrac{\pi}{2}\right), \left(\dfrac{\pi}{2}, \pi\right]$

(b) $\left[-\pi, -\dfrac{\pi}{2}\right), \left(-\dfrac{\pi}{2}, 0\right], \left[\pi, \dfrac{3\pi}{2}\right), \left(\dfrac{3\pi}{2}, 2\pi\right]$

65 (a) The tangent function increases on *all* intervals on
which it is defined. Between -2π and
2π, these intervals are $\left[-2\pi, -\dfrac{3\pi}{2}\right), \left(-\dfrac{3\pi}{2}, -\dfrac{\pi}{2}\right),$
$\left(-\dfrac{\pi}{2}, \dfrac{\pi}{2}\right), \left(\dfrac{\pi}{2}, \dfrac{3\pi}{2}\right),$ and $\left(\dfrac{3\pi}{2}, 2\pi\right].$

(b) The tangent function is *never* decreasing on any interval for which it is defined.

69 (a) -0.8 **(b)** -0.9 **(c)** $0.5, 2.6$
71 (a) -0.7 **(b)** 0.4 **(c)** $2.2, 4.1$

73 (a)

Time	T	H	Time	T	H
12 A.M.	60	60	12 P.M.	60	60
3 A.M.	52	74	3 P.M.	68	46
6 A.M.	48	80	6 P.M.	72	40
9 A.M.	52	74	9 P.M.	68	46

(b) Max: 72°F at 6:00 P.M., 80% at 6:00 A.M.;
min: 48°F at 6:00 A.M., 40% at 6:00 P.M.

EXERCISES 6.4

1 (a) 60° **(b)** 20° **(c)** 22° **(d)** 60°

3 (a) $\dfrac{\pi}{4}$ **(b)** $\dfrac{\pi}{3}$ **(c)** $\dfrac{\pi}{6}$ **(d)** $\dfrac{\pi}{4}$

5 (a) $\pi - 3 \approx 8.1°$ **(b)** $\pi - 2 \approx 65.4°$
(c) $2\pi - 5.5 \approx 44.9°$ **(d)** $32\pi - 100 \approx 30.4°$

7 (a) $\dfrac{\sqrt{3}}{2}$ **(b)** $\dfrac{\sqrt{2}}{2}$ **9 (a)** $-\dfrac{\sqrt{3}}{2}$ **(b)** $\dfrac{1}{2}$

11 (a) $-\dfrac{\sqrt{3}}{3}$ **(b)** $-\sqrt{3}$ **13 (a)** $-\dfrac{\sqrt{3}}{3}$ **(b)** $\sqrt{3}$

15 (a) -2 **(b)** $\dfrac{2}{\sqrt{3}}$ **17 (a)** $-\dfrac{2}{\sqrt{3}}$ **(b)** 2

19 (a) 0.958 **(b)** 0.778 **21 (a)** 0.387 **(b)** 0.472
23 (a) 2.650 **(b)** 3.179 **25 (a)** 30.46° **(b)** 30°27′
27 (a) 74.88° **(b)** 74°53′
29 (a) 24.94° **(b)** 24°57′
31 (a) 76.38° **(b)** 76°23′
33 (a) 0.9899 **(b)** -0.1097 **(c)** -0.1425
(d) 0.7907 **(e)** -11.2493 **(f)** 1.3677
35 (a) 214.3°, 325.7° **(b)** 41.5°, 318.5°
(c) 70.3°, 250.3° **(d)** 133.8°, 313.8°
(e) 153.6°, 206.4° **(f)** 42.3°, 137.7°
37 (a) 0.43, 2.71 **(b)** 1.69, 4.59 **(c)** 1.87, 5.01
(d) 0.36, 3.50 **(e)** 0.96, 5.32 **(f)** 3.35, 6.07
39 0.28 cm
41 (a) The maximum occurs when the sun is rising in
the east.
(b) $\dfrac{\sqrt{2}}{4} \approx 35\%$
43 $\left(9, 9\sqrt{3}\right)$

EXERCISES 6.5

1 (a) $4, 2\pi$

(b) $1, \dfrac{\pi}{2}$

(c) $\dfrac{1}{4}, 2\pi$

(d) $1, 8\pi$

(e) $2, 8\pi$

(f) $\dfrac{1}{2}, \dfrac{\pi}{2}$

(g) $4, 2\pi$

(h) $1, \dfrac{\pi}{2}$

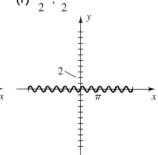

3 (a) $3, 2\pi$

(b) $1, \dfrac{2\pi}{3}$

5 $1, 2\pi, \dfrac{\pi}{2}$

7 $3, 2\pi, -\dfrac{\pi}{6}$

(c) $\dfrac{1}{3}, 2\pi$

(d) $1, 6\pi$

9 $1, 2\pi, -\dfrac{\pi}{2}$

11 $4, 2\pi, \dfrac{\pi}{4}$

(e) $2, 6\pi$

(f) $\dfrac{1}{2}, \dfrac{2\pi}{3}$

13 $1, \pi, \dfrac{\pi}{2}$

15 $1, \dfrac{2\pi}{3}, -\dfrac{\pi}{3}$

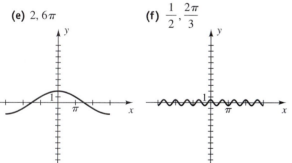

(g) $3, 2\pi$

(h) $1, \dfrac{2\pi}{3}$

17 $2, \dfrac{2\pi}{3}, \dfrac{\pi}{3}$

19 $1, 4\pi, \dfrac{2\pi}{3}$

21 6, 2, 0

23 2, 4, 0

37 2, π, $\dfrac{\pi}{2}$

39 5, π, $-\pi$

25 $\dfrac{1}{2}$, 1, 0

27 5, $\dfrac{2\pi}{3}$, $\dfrac{\pi}{6}$

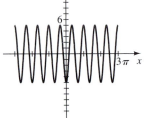

41 (a) 4, 2π, $-\pi$ **(b)** $y = 4 \sin (x + \pi)$

43 (a) 2, 4, -3 **(b)** $y = 2 \sin \left(\dfrac{\pi}{2} x + \dfrac{3\pi}{2} \right)$

45 4π **47** $a = 8$, $b = 4\pi$

49

51

29 3, 4π, $\dfrac{\pi}{2}$

31 5, 6π, $-\dfrac{\pi}{2}$

53 (a) $f(t) = 10 \sin \left[\dfrac{\pi}{12} (t - 10) \right] + 0$, with $a = 10$,

$b = \dfrac{\pi}{12}$, $c = -\dfrac{5\pi}{6}$, $d = 0$

(b)

33 3, 2, -4

35 $\sqrt{2}$, 4, $\dfrac{1}{2}$

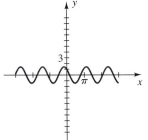

55 (a) $f(t) = 10 \sin \left[\dfrac{\pi}{12}(t - 9) \right] + 20$, with $a = 10$,

$b = \dfrac{\pi}{12}, c = -\dfrac{3\pi}{4}, d = 20$

(b)

EXERCISES 6.6

1 π

3 π

5 2π

7 2π

9 π

11 $\dfrac{\pi}{2}$

13 4π

15 $\dfrac{\pi}{2}$

17 2π

19 π

21 $\dfrac{\pi}{2}$

23 3π

25 $\dfrac{\pi}{2}$

27 2π

41 π

43 6π

29 2π

31 π

45 π

47 4π

33 6π

35 π

49 2

51 1

37 4π

39 2π

53 $y = -\cot\left(x + \dfrac{\pi}{2}\right)$

55

57

59

61

63

65

67 (a) I_0 **(b)** $0.044I_0$ **(c)** $0.603I_0$

69 (a) $A_0e^{-\alpha z}$ **(b)** $\dfrac{\alpha}{k}z_0$ **(c)** $\dfrac{\ln 2}{\alpha}$

EXERCISES 6.7

1 $\beta = 60°$, $a = \dfrac{20}{3}\sqrt{3}$, $c = \dfrac{40}{3}\sqrt{3}$

3 $\alpha = 45°$, $a = b = 15\sqrt{2}$

5 $\alpha = \beta = 45°$, $c = 5\sqrt{2}$

7 $\alpha = 60°$, $\beta = 30°$, $a = 15$

9 $\beta = 53°$, $a \approx 18$, $c \approx 30$

11 $\alpha = 18°9'$, $a \approx 78.7$, $c \approx 252.6$

13 $\alpha \approx 29°$, $\beta \approx 61°$, $c \approx 51$

15 $\alpha \approx 69°$, $\beta \approx 21°$, $a \approx 5.4$ **17** $b = c \cos \alpha$

19 $a = b \cot \beta$ **21** $c = a \csc \alpha$

23 $b = \sqrt{c^2 - a^2}$

25 $250\sqrt{3} + 4 \approx 437$ ft   **27** 28,800 ft   **29** 160 m

31 9659 ft   **33 (a)** 58 ft   **(b)** 27 ft   **35** 51°20′

37 16.3°   **39** 2063 ft   **41** 1,459,379 ft^2

43 21.8°   **45** 20.2 m   **47** 29.7 km   **49** 3944 mi

51 126 mi/hr

53 (a) 45%

   (b) Each satellite has a signal range of more than 120°.

55 $h = d \sin \alpha + c$ **57** $h = \dfrac{d}{\cot \alpha - \cot \beta}$

59 $h = d(\tan \beta - \tan \alpha)$

61 N70°E; N40°W; S15°W; S25°E

63 (a) 55 mi   **(b)** S63°E   **65** 324 mi

67 Amplitude, 10 cm; period, $\dfrac{1}{3}$ sec; frequency, 3 osc/sec.

   The point is at the origin at $t = 0$. It moves upward with decreasing speed, reaching the point with coordinate 10 at $t = \dfrac{1}{12}$. It then reverses direction and moves downward, gaining speed until it reaches the origin at $t = \dfrac{1}{6}$. It continues downward with decreasing speed, reaching the point with coordinate -10 at $t = \dfrac{1}{4}$. It then reverses direction and moves upward with increasing speed, returning to the origin at $t = \dfrac{1}{3}$.

69 Amplitude, 4 cm; period, $\dfrac{4}{3}$ sec; frequency, $\dfrac{3}{4}$ osc/sec.

   The motion is similar to that in Exercise 67; however, the point starts 4 units above the origin and moves downward, reaching the origin at $t = \dfrac{1}{3}$ and the point with coordinate -4 at $t = \dfrac{2}{3}$. It then reverses direction and moves upward, reaching the origin at $t = 1$ and its initial point at $t = \dfrac{4}{3}$.

71 $d = 5 \cos \dfrac{2\pi}{3} t$

73 (a) $y = 25 \cos \dfrac{\pi}{15} t$

   (b) 324,000 ft

CHAPTER 6 REVIEW EXERCISES

1 $\dfrac{11\pi}{6}, \dfrac{9\pi}{4}, -\dfrac{5\pi}{6}, \dfrac{4\pi}{3}, \dfrac{\pi}{5}$

2 $810°, -120°, 315°, 900°, 36°$

3 (a) 0.1   **(b)** 0.2 m^2

4 (a) $\dfrac{35\pi}{12}$ cm   **(b)** $\dfrac{175\pi}{16}$ cm^2

5 $\dfrac{200\pi}{3}, 90\pi$ **6** $\dfrac{100\pi}{3}, \dfrac{105\pi}{4}$

7 $x = 6\sqrt{3}; y = 3\sqrt{3}$ **8** $x = \dfrac{7}{2}\sqrt{2}; y = \dfrac{7}{2}\sqrt{2}$

9 $\tan \theta = \sqrt{\sec^2 \theta - 1}$ **10** $\cot \theta = \sqrt{\csc^2 \theta - 1}$

Exer. 11–20: Typical verifications are given.

11 $\sin \theta (\csc \theta - \sin \theta) = \sin \theta \csc \theta - \sin^2 \theta$
$$= 1 - \sin^2 \theta = \cos^2 \theta$$

12 $\cos \theta (\tan \theta + \cot \theta) = \cos \theta \cdot \dfrac{\sin \theta}{\cos \theta} + \cos \theta \cdot \dfrac{\cos \theta}{\sin \theta}$
$$= \sin \theta + \frac{\cos^2 \theta}{\sin \theta}$$
$$= \frac{\sin^2 \theta + \cos^2 \theta}{\sin \theta}$$
$$= \frac{1}{\sin \theta} = \csc \theta$$

13 $(\cos^2 \theta - 1)(\tan^2 \theta + 1) = (\cos^2 \theta - 1)(\sec^2 \theta)$
$$= \cos^2 \theta \sec^2 \theta - \sec^2 \theta$$
$$= 1 - \sec^2 \theta$$

14 $\dfrac{\sec \theta - \cos \theta}{\tan \theta} = \dfrac{\dfrac{1}{\cos \theta} - \cos \theta}{\dfrac{\sin \theta}{\cos \theta}} = \dfrac{\dfrac{1 - \cos^2 \theta}{\cos \theta}}{\dfrac{\sin \theta}{\cos \theta}} = \dfrac{\dfrac{\sin^2 \theta}{\cos \theta}}{\dfrac{\sin \theta}{\cos \theta}}$

$$= \frac{\dfrac{\sin \theta}{\cos \theta}}{\dfrac{1}{\cos \theta}} = \frac{\tan \theta}{\sec \theta}$$

15 $\dfrac{1 + \tan^2 \theta}{\tan^2 \theta} = \dfrac{1}{\tan^2 \theta} + \dfrac{\tan^2 \theta}{\tan^2 \theta} = \cot^2 \theta + 1 = \csc^2 \theta$

16 $\dfrac{\sec \theta + \csc \theta}{\sec \theta - \csc \theta} = \dfrac{\dfrac{1}{\cos \theta} + \dfrac{1}{\sin \theta}}{\dfrac{1}{\cos \theta} - \dfrac{1}{\sin \theta}} = \dfrac{\dfrac{\sin \theta + \cos \theta}{\cos \theta \sin \theta}}{\dfrac{\sin \theta - \cos \theta}{\cos \theta \sin \theta}}$

$$= \frac{\sin \theta + \cos \theta}{\sin \theta - \cos \theta}$$

17 $\dfrac{\cot \theta - 1}{1 - \tan \theta} = \dfrac{\dfrac{\cos \theta}{\sin \theta} - 1}{1 - \dfrac{\sin \theta}{\cos \theta}} = \dfrac{\dfrac{\cos \theta - \sin \theta}{\sin \theta}}{\dfrac{\cos \theta - \sin \theta}{\cos \theta}}$

$$= \frac{(\cos \theta - \sin \theta) \cos \theta}{(\cos \theta - \sin \theta) \sin \theta} = \frac{\cos \theta}{\sin \theta} = \cot \theta$$

18 $\dfrac{1 + \sec \theta}{\tan \theta + \sin \theta} = \dfrac{1 + \dfrac{1}{\cos \theta}}{\dfrac{\sin \theta}{\cos \theta} + \dfrac{\sin \theta \cos \theta}{\cos \theta}} = \dfrac{\dfrac{\cos \theta + 1}{\cos \theta}}{\dfrac{\sin \theta (1 + \cos \theta)}{\cos \theta}}$

$$= \frac{1}{\sin \theta} = \csc \theta$$

19 $\dfrac{\tan (-\theta) + \cot (-\theta)}{\tan \theta} = \dfrac{-\tan \theta - \cot \theta}{\tan \theta} = -\dfrac{\tan \theta}{\tan \theta} - \dfrac{\cot \theta}{\tan \theta}$
$$= -1 - \cot^2 \theta = -(1 + \cot^2 \theta)$$
$$= -\csc^2 \theta$$

20 $-\dfrac{1}{\csc (-\theta)} - \dfrac{\cot (-\theta)}{\sec (-\theta)} = -\dfrac{1}{-\csc \theta} - \dfrac{-\cot \theta}{\sec \theta}$
$$= \sin \theta + \frac{\cos \theta / \sin \theta}{1 / \cos \theta}$$
$$= \sin \theta + \frac{\cos^2 \theta}{\sin \theta}$$
$$= \frac{\sin^2 \theta + \cos^2 \theta}{\sin \theta}$$
$$= \frac{1}{\sin \theta} = \csc \theta$$

21 $\dfrac{\sqrt{33}}{7}, \dfrac{4}{7}, \dfrac{\sqrt{33}}{4}, \dfrac{4}{\sqrt{33}}, \dfrac{7}{4}, \dfrac{7}{\sqrt{33}}$

22 (a) $-\dfrac{4}{5}, \dfrac{3}{5}, -\dfrac{4}{3}, \dfrac{3}{4}, \dfrac{5}{3}, -\dfrac{5}{4}$

(b) $\dfrac{2}{\sqrt{13}}, -\dfrac{3}{\sqrt{13}}, -\dfrac{2}{3}, -\dfrac{3}{2}, -\dfrac{\sqrt{13}}{3}, \dfrac{\sqrt{13}}{2}$

(c) $-1, 0, U, 0, U, -1$

23 (a) II **(b)** III **(c)** IV

24 (a) $-\dfrac{4}{5}, \dfrac{3}{5}, -\dfrac{4}{3}, \dfrac{3}{4}, \dfrac{5}{3}, -\dfrac{5}{4}$

(b) $\dfrac{2}{\sqrt{13}}, -\dfrac{3}{\sqrt{13}}, -\dfrac{2}{3}, -\dfrac{3}{2}, -\dfrac{\sqrt{13}}{3}, \dfrac{\sqrt{13}}{2}$

25 $(-1, 0); (0, -1); (0, 1);$
$$\left(-\frac{\sqrt{2}}{2}, -\frac{\sqrt{2}}{2}\right); (1, 0); \left(\frac{\sqrt{3}}{2}, \frac{1}{2}\right)$$

26 $\left(\dfrac{3}{5}, \dfrac{4}{5}\right); \left(\dfrac{3}{5}, \dfrac{4}{5}\right); \left(-\dfrac{3}{5}, \dfrac{4}{5}\right); \left(-\dfrac{3}{5}, \dfrac{4}{5}\right)$

27 (a) $\dfrac{\pi}{4}, \dfrac{\pi}{6}, \dfrac{\pi}{8}$ **(b)** $65°, 43°, 8°$

28 (a) $1, 0, U, 0, U, 1$

(b) $\dfrac{\sqrt{2}}{2}, -\dfrac{\sqrt{2}}{2}, -1, -1, -\sqrt{2}, \sqrt{2}$

(c) $0, 1, 0, U, 1, U$

(d) $-\dfrac{1}{2}, \dfrac{\sqrt{3}}{2}, -\dfrac{\sqrt{3}}{3}, -\sqrt{3}, \dfrac{2}{\sqrt{3}}, -2$

29 (a) $-\dfrac{\sqrt{2}}{2}$ **(b)** $-\dfrac{\sqrt{3}}{3}$ **(c)** $-\dfrac{1}{2}$ **(d)** -2

(e) -1 **(f)** $-\dfrac{2}{\sqrt{3}}$

30 310.5° **31** 1.2206; 4.3622 **32** 52.44°; 307.56°

33 5, 2π

34 $\dfrac{2}{3}$, 2π

35 $\dfrac{1}{3}$, $\dfrac{2\pi}{3}$

36 $\dfrac{1}{2}$, 6π

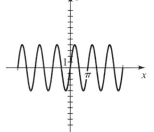

37 3, 4π

38 4, π

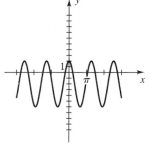

39 2, 2

40 4, 4

41 (a) 1.43, 2 **(b)** $y = 1.43 \sin \pi x$

42 (a) 3.27, 3π **(b)** $y = -3.27 \sin \dfrac{2}{3} x$

43 (a) 3, $\dfrac{4\pi}{3}$ **(b)** $y = -3 \cos \dfrac{3}{2} x$

44 (a) 2, 4 **(b)** $y = 2 \cos \dfrac{\pi}{2} x$

45

46

47

48

49

50

51

52

53

54

55

56

57 $\alpha = 30°, a \approx 23, c \approx 46$

58 $\beta = 35°20', a \approx 310, c \approx 380$

59 $\alpha \approx 68°, \beta \approx 22°, c \approx 67$

60 $\alpha \approx 13°, \beta \approx 77°, b = 40$

61 (a) $\dfrac{109\pi}{6}$ (b) 440.2 **62** 1048 ft

63 0.093 mi/sec **64** 52°

65 Approximately 67,900,000 mi

66 $\dfrac{6\pi}{5}$ radians = 216° **67** 250 ft

68 (a) 231.0 ft (b) 434.5 **69** (b) 2 mi

70 (a) $T = h + d(\cos \alpha \tan \theta - \sin \alpha)$ (b) 22.54 ft

71 (a) $\dfrac{25}{3}\sqrt{3} \approx 14.43$ ft-candles (b) 37.47°

72 (b) 4.69 **73** (a) 74.05 in. (b) 24.75 in.

74 (a) $S = 4a^2 \sin \theta$ (b) $V = \dfrac{4}{3}a^3 \sin^2 \theta \cos \theta$

75 (a) $h = R \sec \dfrac{s}{R} - R$ (b) $h \approx 1650$ ft

76

77 $y = 98.6 + (0.3) \sin\left(\dfrac{\pi}{12}t - \dfrac{11\pi}{12}\right)$

78 (a)

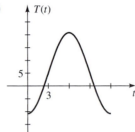

(b) 20.8°C on July 1

79 (a)

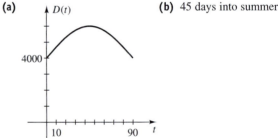

(b) 45 days into summer

80 (a) The cork is in simple harmonic motion.

(b) $1 \le t \le 2$

CHAPTER 6 DISCUSSION EXERCISES

1 None

2 (a) $x \approx -0.4161, y \approx 0.9093$

(b) $x \approx -0.8838, y \approx -0.4678$

3 (a) $x \approx 1.8415, y \approx -0.5403$

(b) $x \approx -1.2624, y \approx 0.9650$

4 (a) $\dfrac{500\pi}{3}$ rad/sec (b) $D(t) = 5 \cos\left(\dfrac{500\pi}{3}t\right) + 18$

Chapter 7

EXERCISES 7.1

Exer. 1–50: Typical verifications are given for Exercises 1, 5, 9,..., 49.

1 $\csc\theta - \sin\theta = \dfrac{1}{\sin\theta} - \sin\theta = \dfrac{1-\sin^2\theta}{\sin\theta} = \dfrac{\cos^2\theta}{\sin\theta}$

$\qquad = \dfrac{\cos\theta}{\sin\theta}\cos\theta = \cot\theta\cos\theta$

5 $\dfrac{\csc^2\theta}{1+\tan^2\theta} = \dfrac{\csc^2\theta}{\sec^2\theta} = \dfrac{1/\sin^2\theta}{1/\cos^2\theta} = \dfrac{\cos^2\theta}{\sin^2\theta}$

$\qquad = \left(\dfrac{\cos\theta}{\sin\theta}\right)^2 = \cot^2\theta$

9 $\dfrac{1}{1-\cos\gamma} + \dfrac{1}{1+\cos\gamma} = \dfrac{1+\cos\gamma+1-\cos\gamma}{1-\cos^2\gamma}$

$\qquad = \dfrac{2}{\sin^2\gamma} = 2\csc^2\gamma$

13 $\csc^4 t - \cot^4 t = (\csc^2 t + \cot^2 t)(\csc^2 t - \cot^2 t)$

$\qquad = (\csc^2 t + \cot^2 t)(1)$

$\qquad = \csc^2 t + \cot^2 t$

17 $\dfrac{\tan^2 x}{\sec x + 1} = \dfrac{\sec^2 x - 1}{\sec x + 1} = \dfrac{(\sec x + 1)(\sec x - 1)}{\sec x + 1}$

$\qquad = \sec x - 1 = \dfrac{1}{\cos x} - 1 = \dfrac{1-\cos x}{\cos x}$

21 $\sin^4 r - \cos^4 r = (\sin^2 r - \cos^2 r)(\sin^2 r + \cos^2 r)$

$\qquad = (\sin^2 r - \cos^2 r)(1)$

$\qquad = \sin^2 r - \cos^2 r$

25 $(\sec t + \tan t)^2 = \left(\dfrac{1}{\cos t} + \dfrac{\sin t}{\cos t}\right)^2 = \left(\dfrac{1+\sin t}{\cos t}\right)^2$

$\qquad = \dfrac{(1+\sin t)^2}{\cos^2 t} = \dfrac{(1+\sin t)^2}{1-\sin^2 t}$

$\qquad = \dfrac{(1+\sin t)^2}{(1+\sin t)(1-\sin t)} = \dfrac{1+\sin t}{1-\sin t}$

29 $\dfrac{1+\csc\beta}{\cot\beta+\cos\beta} = \dfrac{1+\dfrac{1}{\sin\beta}}{\dfrac{\cos\beta}{\sin\beta}+\cos\beta} = \dfrac{\dfrac{\sin\beta+1}{\sin\beta}}{\dfrac{\cos\beta+\cos\beta\sin\beta}{\sin\beta}}$

$\qquad = \dfrac{\sin\beta+1}{\cos\beta(1+\sin\beta)} = \dfrac{1}{\cos\beta} = \sec\beta$

33 $RS = \dfrac{\tan\alpha+\tan\beta}{1-\tan\alpha\tan\beta} = \dfrac{\dfrac{\sin\alpha}{\cos\alpha}+\dfrac{\sin\beta}{\cos\beta}}{1-\dfrac{\sin\alpha}{\cos\alpha}\cdot\dfrac{\sin\beta}{\cos\beta}}$

$\qquad = \dfrac{\dfrac{\sin\alpha\cos\beta+\cos\alpha\sin\beta}{\cos\alpha\cos\beta}}{\dfrac{\cos\alpha\cos\beta-\sin\alpha\sin\beta}{\cos\alpha\cos\beta}}$

$\qquad = \dfrac{\sin\alpha\cos\beta+\cos\alpha\sin\beta}{\cos\alpha\cos\beta-\sin\alpha\sin\beta}$

$\qquad = LS$

37 $\dfrac{1}{\tan\beta+\cot\beta} = \dfrac{1}{\dfrac{\sin\beta}{\cos\beta}+\dfrac{\cos\beta}{\sin\beta}} = \dfrac{1}{\dfrac{\sin^2\beta+\cos^2\beta}{\cos\beta\sin\beta}}$

$\qquad = \sin\beta\cos\beta$

41 $RS = \sec^4\phi - 4\tan^2\phi = (\sec^2\phi)^2 - 4\tan^2\phi$

$\qquad = (1+\tan^2\phi)^2 - 4\tan^2\phi$

$\qquad = 1 + 2\tan^2\phi + \tan^4\phi - 4\tan^2\phi$

$\qquad = 1 - 2\tan^2\phi + \tan^4\phi$

$\qquad = (1-\tan^2\phi)^2 = LS$

45 $\log 10^{\tan t} = \log_{10} 10^{\tan t} = \tan t$, since $\log_a a^x = x$.

49 $\ln|\sec\theta+\tan\theta| = \ln\left|\dfrac{(\sec\theta+\tan\theta)(\sec\theta-\tan\theta)}{\sec\theta-\tan\theta}\right|$

$\qquad = \ln\left|\dfrac{\sec^2\theta-\tan^2\theta}{\sec\theta-\tan\theta}\right|$

$\qquad = \ln\left|\dfrac{1}{\sec\theta-\tan\theta}\right|$

$\qquad = \ln|1| - \ln|\sec\theta-\tan\theta|$

$\qquad = -\ln|\sec\theta-\tan\theta|$

Exer. 51–62: A typical value of t or θ and the resulting nonequality are given.

51 π, $-1\neq 1$ **53** $\dfrac{3\pi}{2}$, $1\neq -1$ **55** $\dfrac{\pi}{4}$, $2\neq 1$

57 π, $-1\neq 1$ **59** $\dfrac{\pi}{4}$, $\cos\sqrt{2}\neq 1$

61 Not an identity   **63** Identity

65 $a^3\cos^3\theta$ **67** $a\tan\theta\sin\theta$ **69** $a\sec\theta$

71 $\dfrac{1}{a^2}\cos^2\theta$ **73** $a\tan\theta$ **75** $a^4\sec^3\theta\tan\theta$

EXERCISES 7.2

Exer. 1–34: n denotes any integer.

1 $\dfrac{5\pi}{4} + 2\pi n, \dfrac{7\pi}{4} + 2\pi n$ **3** $\dfrac{\pi}{3} + \pi n$

5 $\dfrac{\pi}{3} + 2\pi n, \dfrac{5\pi}{3} + 2\pi n$

7 No solution, since $\dfrac{\pi}{2} > 1$.

9 All θ except $\theta = \dfrac{\pi}{2} + \pi n$

11 $\dfrac{\pi}{12} + \pi n, \dfrac{11\pi}{12} + \pi n$ **13** $\dfrac{\pi}{2} + 3\pi n$

15 $-\dfrac{\pi}{12} + 2\pi n, \dfrac{7\pi}{12} + 2\pi n$

17 $\dfrac{\pi}{4} + \pi n, \dfrac{7\pi}{12} + \pi n$ **19** $\dfrac{2\pi}{3} + 2\pi n, \dfrac{4\pi}{3} + 2\pi n$

21 $\dfrac{\pi}{4} + \dfrac{\pi}{2} n$ **23** $2\pi n, \dfrac{3\pi}{2} + 2\pi n$

25 $\dfrac{\pi}{3} + \pi n, \dfrac{2\pi}{3} + \pi n$ **27** $\dfrac{4\pi}{3} + 2\pi n, \dfrac{5\pi}{3} + 2\pi n$

29 $\dfrac{\pi}{6} + \pi n, \dfrac{5\pi}{6} + \pi n$ **31** $\dfrac{7\pi}{6} + 2\pi n, \dfrac{11\pi}{6} + 2\pi n$

33 $\dfrac{\pi}{12} + \pi n, \dfrac{5\pi}{12} + \pi n$ **35** $e^{(\pi/2) + \pi n}$

37 $\dfrac{3\pi}{8}, \dfrac{7\pi}{8}, \dfrac{11\pi}{8}, \dfrac{15\pi}{8}$ **39** $\dfrac{\pi}{3}, \dfrac{2\pi}{3}, \dfrac{4\pi}{3}, \dfrac{5\pi}{3}$

41 $\dfrac{\pi}{6}, \dfrac{5\pi}{6}, \dfrac{3\pi}{2}$ **43** $0, \pi, \dfrac{\pi}{4}, \dfrac{3\pi}{4}, \dfrac{5\pi}{4}, \dfrac{7\pi}{4}$

45 $\dfrac{\pi}{2}, \dfrac{3\pi}{2}, \dfrac{2\pi}{3}, \dfrac{4\pi}{3}$ **47** No solution   **49** $\dfrac{11\pi}{6}, \dfrac{\pi}{2}$

51 $0, \dfrac{\pi}{2}$ **53** $\dfrac{\pi}{4}, \dfrac{5\pi}{4}$

55 All α in $[0, 2\pi)$ except $0, \dfrac{\pi}{2}, \pi$, and $\dfrac{3\pi}{2}$

57 $\dfrac{\pi}{2}, \dfrac{3\pi}{2}, \dfrac{7\pi}{6}, \dfrac{11\pi}{6}$ **59** $\dfrac{3\pi}{4}, \dfrac{7\pi}{4}$

61 $15°30', 164°30'$ **63** $135°, 315°, 116°30', 296°30'$
65 $41°50', 138°10', 194°30', 345°30'$ **67** 10
69 $t \approx 3.50$ and $t \approx 8.50$ **71 (a)** 3.29   **(b)** 4

73 (a)

(b) $0 \le t < \dfrac{5}{3}$ and $\dfrac{25}{3} < t \le 10$

75 $A\left(-\dfrac{4\pi}{3}, -\dfrac{2\pi}{3} + \dfrac{1}{2}\sqrt{3}\right), B\left(-\dfrac{2\pi}{3}, -\dfrac{\pi}{3} - \dfrac{1}{2}\sqrt{3}\right),$
$C\left(\dfrac{2\pi}{3}, \dfrac{\pi}{3} + \dfrac{1}{2}\sqrt{3}\right), D\left(\dfrac{4\pi}{3}, \dfrac{2\pi}{3} - \dfrac{1}{2}\sqrt{3}\right)$

77 $\dfrac{7}{360}$ **79 (a)** 37.6°   **(b)** 52.5°

EXERCISES 7.3

1 (a) $\cos 43°23'$ **(b)** $\sin 16°48'$ **(c)** $\cot \dfrac{\pi}{3}$
(d) $\csc 72.72°$

3 (a) $\sin \dfrac{3\pi}{20}$ **(b)** $\cos\left(\dfrac{2\pi - 1}{4}\right)$ **(c)** $\cot\left(\dfrac{\pi - 2}{2}\right)$
(d) $\sec\left(\dfrac{\pi}{2} - 0.53\right)$

5 (a) $\dfrac{\sqrt{2} + \sqrt{3}}{2}$ **(b)** $\dfrac{\sqrt{6} - \sqrt{2}}{4}$

7 (a) $\sqrt{3} + 1$ **(b)** $-2 - \sqrt{3}$

9 (a) $\dfrac{\sqrt{2} - 1}{2}$ **(b)** $\dfrac{\sqrt{6} + \sqrt{2}}{4}$

11 $\cos 25°$ **13** $\sin(-5°)$ **15** $\sin(-5)$

17 (a) $\dfrac{77}{85}$ **(b)** $\dfrac{36}{85}$ **(c)** I

19 (a) $-\dfrac{24}{25}$ **(b)** $-\dfrac{24}{7}$ **(c)** IV

21 (a) $\dfrac{3\sqrt{21} - 8}{25} \approx 0.23$ **(b)** $\dfrac{4\sqrt{21} + 6}{25} \approx 0.97$ **(c)** I

23 $\sin(\theta + \pi) = \sin \theta \cos \pi + \cos \theta \sin \pi$
$= \sin \theta(-1) + \cos \theta(0) = -\sin \theta$

25 $\sin\left(x - \dfrac{5\pi}{2}\right) = \sin x \cos \dfrac{5\pi}{2} - \cos x \sin \dfrac{5\pi}{2}$
$= -\cos x$

27 $\cos(\theta - \pi) = \cos \theta \cos \pi + \sin \theta \sin \pi = -\cos \theta$

29 $\cos\left(x + \dfrac{3\pi}{2}\right) = \cos x \cos \dfrac{3\pi}{2} - \sin x \sin \dfrac{3\pi}{2}$
$= \sin x$

31 $\tan\left(x - \dfrac{\pi}{2}\right) = \dfrac{\sin\left(x - \dfrac{\pi}{2}\right)}{\cos\left(x - \dfrac{\pi}{2}\right)}$

$= \dfrac{\sin x \cos\dfrac{\pi}{2} - \cos x \sin\dfrac{\pi}{2}}{\cos x \cos\dfrac{\pi}{2} + \sin x \sin\dfrac{\pi}{2}}$

$= \dfrac{-\cos x}{\sin x} = -\cot x$

33 $\tan\left(\theta + \dfrac{\pi}{2}\right) = \cot\left[\dfrac{\pi}{2} - \left(\theta + \dfrac{\pi}{2}\right)\right]$

$= \cot(-\theta) = -\cot\theta$

35 $\sin\left(\theta + \dfrac{\pi}{4}\right) = \sin\theta\cos\dfrac{\pi}{4} + \cos\theta\sin\dfrac{\pi}{4}$

$= \dfrac{\sqrt{2}}{2}\sin\theta + \dfrac{\sqrt{2}}{2}\cos\theta$

$= \dfrac{\sqrt{2}}{2}(\sin\theta + \cos\theta)$

37 $\tan\left(u + \dfrac{\pi}{4}\right) = \dfrac{\tan u + \tan\dfrac{\pi}{4}}{1 - \tan u \tan\dfrac{\pi}{4}} = \dfrac{1 + \tan u}{1 - \tan u}$

39 $\cos(u + v) + \cos(u - v)$

$= (\cos u \cos v - \sin u \sin v) +$
$(\cos u \cos v + \sin u \sin v)$
$= 2\cos u \cos v$

41 $\sin(u + v) \cdot \sin(u - v)$

$= (\sin u \cos v + \cos u \sin v) \cdot$
$(\sin u \cos v - \cos u \sin v)$
$= \sin^2 u \cos^2 v - \cos^2 u \sin^2 v$
$= \sin^2 u(1 - \sin^2 v) - (1 - \sin^2 u)\sin^2 v$
$= \sin^2 u - \sin^2 u \sin^2 v - \sin^2 v + \sin^2 u \sin^2 v$
$= \sin^2 u - \sin^2 v$

43 $\dfrac{1}{\cot\alpha - \cot\beta} = \dfrac{1}{\dfrac{\cos\alpha}{\sin\alpha} - \dfrac{\cos\beta}{\sin\beta}}$

$= \dfrac{1}{\dfrac{\cos\alpha\sin\beta - \cos\beta\sin\alpha}{\sin\alpha\sin\beta}}$

$= \dfrac{\sin\alpha\sin\beta}{\sin(\beta - \alpha)}$

45 $\sin u \cos v \cos w + \cos u \sin v \cos w +$
$\cos u \cos v \sin w - \sin u \sin v \sin w$

47 $\cot(u + v) = \dfrac{\cos(u + v)}{\sin(u + v)}$

$= \dfrac{(\cos u \cos v - \sin u \sin v)(1/\sin u \sin v)}{(\sin u \cos v + \cos u \sin v)(1/\sin u \sin v)}$

$= \dfrac{\cot u \cot v - 1}{\cot v + \cot u}$

49 $\sin(u - v) = \sin[u + (-v)]$

$= \sin u \cos(-v) + \cos u \sin(-v)$
$= \sin u \cos v - \cos u \sin v$

51 $\dfrac{f(x + h) - f(x)}{h} = \dfrac{\cos(x + h) - \cos x}{h}$

$= \dfrac{\cos x \cos h - \sin x \sin h - \cos x}{h}$

$= \dfrac{\cos x \cos h - \cos x}{h} - \dfrac{\sin x \sin h}{h}$

$= \cos x\left(\dfrac{\cos h - 1}{h}\right) - \sin x\left(\dfrac{\sin h}{h}\right)$

53 (a) Each side ≈ 0.0523 **(b)** $\alpha = 60°$
(c) $\alpha = 60°$, $\beta = 3°$

55 $0, \dfrac{\pi}{3}, \dfrac{2\pi}{3}$ **57** $\dfrac{\pi}{6}, \dfrac{\pi}{2}, \dfrac{5\pi}{6}$

59 $\dfrac{\pi}{12}, \dfrac{5\pi}{12}; \dfrac{3\pi}{4}$ is extraneous

61 (a) $f(x) = 2\cos\left(2x - \dfrac{\pi}{6}\right)$ **(b)** $2, \pi, \dfrac{\pi}{12}$

(c)

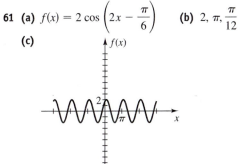

63 (a) $f(x) = 2\sqrt{2} \cos\left(3x + \dfrac{\pi}{4}\right)$ **(b)** $2\sqrt{2}, \dfrac{2\pi}{3}, -\dfrac{\pi}{12}$

(c)

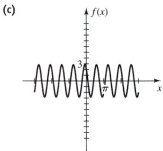

65 $y = 10\sqrt{41} \cos\left(60\pi t - \tan^{-1}\dfrac{5}{4}\right)$

$\approx 10\sqrt{41} \cos(60\pi t - 0.8961)$

67 (a) $y = \sqrt{13} \cos(t - C)$ with $\tan C = \dfrac{3}{2}$; $\sqrt{13}, 2\pi$

(b) $t = C + \dfrac{\pi}{2} + \pi n \approx 2.55 + \pi n$ for every nonnegative integer n

69 (a) $p(t) = A \sin \omega t + B \sin(\omega t + \tau)$
$= A \sin \omega t + B(\sin \omega t \cos \tau + \cos \omega t \sin \tau)$
$= (B \sin \tau) \cos \omega t + (A + B \cos \tau) \sin \omega t$
$= a \cos \omega t + b \sin \omega t$
with $a = B \sin \tau$ and $b = A + B \cos \tau$

(b) $C^2 = (B \sin \tau)^2 + (A + B \cos \tau)^2$
$= B^2 \sin^2 \tau + A^2 + 2AB \cos \tau + B^2 \cos^2 \tau$
$= A^2 + B^2(\sin^2 \tau + \cos^2 \tau) + 2AB \cos \tau$
$= A^2 + B^2 + 2AB \cos \tau$

71 (a) $C^2 = A^2 + B^2 + 2AB \cos \tau \le A^2 + B^2 + 2AB$,
since $\cos \tau \le 1$ and $A > 0$, $B > 0$. Thus,
$C^2 \le (A + B)^2$, and hence $C \le A + B$.

(b) $0, 2\pi$ **(c)** $\cos \tau > -B/(2A)$

EXERCISES 7.4

1 $\dfrac{24}{25}, -\dfrac{7}{25}, -\dfrac{24}{7}$ **3** $-\dfrac{4}{9}\sqrt{2}, -\dfrac{7}{9}, \dfrac{4}{7}\sqrt{2}$

5 $\dfrac{1}{10}\sqrt{10}, \dfrac{3}{10}\sqrt{10}, \dfrac{1}{3}$

7 $-\dfrac{1}{2}\sqrt{2 + \sqrt{2}}, \dfrac{1}{2}\sqrt{2 - \sqrt{2}}, -\sqrt{2} - 1$

9 (a) $\dfrac{1}{2}\sqrt{2 - \sqrt{2}}$ **(b)** $\dfrac{1}{2}\sqrt{2 - \sqrt{3}}$ **(c)** $\sqrt{2} + 1$

11 $\sin 10\theta = \sin(2 \cdot 5\theta) = 2 \sin 5\theta \cos 5\theta$

13 $4 \sin \dfrac{x}{2} \cos \dfrac{x}{2} = 2 \cdot 2 \sin \dfrac{x}{2} \cos \dfrac{x}{2} = 2 \sin\left(2 \cdot \dfrac{x}{2}\right)$
$= 2 \sin x$

15 $(\sin t + \cos t)^2 = \sin^2 t + 2 \sin t \cos t + \cos^2 t$
$= 1 + \sin 2t$

17 $\sin 3u = \sin(2u + u)$
$= \sin 2u \cos u + \cos 2u \sin u$
$= (2 \sin u \cos u) \cos u + (1 - 2 \sin^2 u) \sin u$
$= 2 \sin u \cos^2 u + \sin u - 2 \sin^3 u$
$= 2 \sin u(1 - \sin^2 u) + \sin u - 2 \sin^3 u$
$= 2 \sin u - 2 \sin^3 u + \sin u - 2 \sin^3 u$
$= 3 \sin u - 4 \sin^3 u = \sin u(3 - 4 \sin^2 u)$

19 $\cos 4\theta = \cos(2 \cdot 2\theta) = 2 \cos^2 2\theta - 1$
$= 2(2 \cos^2 \theta - 1)^2 - 1$
$= 2(4 \cos^4 \theta - 4 \cos^2 \theta + 1) - 1$
$= 8 \cos^4 \theta - 8 \cos^2 \theta + 1$

21 $\sin^4 t = (\sin^2 t)^2 = \left(\dfrac{1 - \cos 2t}{2}\right)^2$
$= \dfrac{1}{4}(1 - 2 \cos 2t + \cos^2 2t)$
$= \dfrac{1}{4} - \dfrac{1}{2} \cos 2t + \dfrac{1}{4}\left(\dfrac{1 + \cos 4t}{2}\right)$
$= \dfrac{1}{4} - \dfrac{1}{2} \cos 2t + \dfrac{1}{8} + \dfrac{1}{8} \cos 4t$
$= \dfrac{3}{8} - \dfrac{1}{2} \cos 2t + \dfrac{1}{8} \cos 4t$

23 $\sec 2\theta = \dfrac{1}{\cos 2\theta} = \dfrac{1}{2 \cos^2 \theta - 1} = \dfrac{1}{2\left(\dfrac{1}{\sec^2 \theta}\right) - 1}$
$= \dfrac{1}{\dfrac{2 - \sec^2 \theta}{\sec^2 \theta}} = \dfrac{\sec^2 \theta}{2 - \sec^2 \theta}$

25 $2 \sin^2 2t + \cos 4t = 2 \sin^2 2t + \cos(2 \cdot 2t)$
$= 2 \sin^2 2t + (1 - 2 \sin^2 2t) = 1$

27 $\tan 3u = \tan(2u + u) = \dfrac{\tan 2u + \tan u}{1 - \tan 2u \tan u}$
$= \dfrac{\dfrac{2 \tan u}{1 - \tan^2 u} + \tan u}{1 - \dfrac{2 \tan u}{1 - \tan^2 u} \cdot \tan u}$
$= \dfrac{\dfrac{2 \tan u + \tan u - \tan^3 u}{1 - \tan^2 u}}{\dfrac{1 - \tan^2 u - 2 \tan^2 u}{1 - \tan^2 u}}$
$= \dfrac{3 \tan u - \tan^3 u}{1 - 3 \tan^2 u} = \dfrac{\tan u(3 - \tan^2 u)}{1 - 3 \tan^2 u}$

29 $\tan\dfrac{\theta}{2} = \dfrac{1-\cos\theta}{\sin\theta} = \dfrac{1}{\sin\theta} - \dfrac{\cos\theta}{\sin\theta} = \csc\theta - \cot\theta$

31 $\dfrac{3}{8} + \dfrac{1}{2}\cos\theta + \dfrac{1}{8}\cos 2\theta$

33 $\dfrac{3}{8} - \dfrac{1}{2}\cos 4x + \dfrac{1}{8}\cos 8x$ **35** $0,\ \pi,\ \dfrac{2\pi}{3},\ \dfrac{4\pi}{3}$

37 $\dfrac{\pi}{3},\ \dfrac{5\pi}{3},\ \pi$ **39** $0,\ \pi$ **41** $0,\ \dfrac{\pi}{3},\ \dfrac{5\pi}{3}$

45 (a) 1.20, 5.09

(b) $P\!\left(\dfrac{2\pi}{3}, -1.5\right),\ Q(\pi, -1),\ R\!\left(\dfrac{4\pi}{3}, -1.5\right)$

47 (a) $-\dfrac{3\pi}{2},\ -\dfrac{\pi}{2},\ \dfrac{\pi}{2},\ \dfrac{3\pi}{2}$

(b) $0,\ \pm\pi,\ \pm 2\pi,\ \pm\dfrac{\pi}{4},\ \pm\dfrac{3\pi}{4},\ \pm\dfrac{5\pi}{4},\ \pm\dfrac{7\pi}{4}$

49 (b) Yes, point B is 25 miles from A.

51 (a) $V = \dfrac{5}{2}\sin\theta$ **(b)** 53.13°   **53 (b)** 12.43 mm

EXERCISES 7.5

1 $\dfrac{1}{2}\cos 4t - \dfrac{1}{2}\cos 10t$ **3** $\dfrac{1}{2}\cos 2u + \dfrac{1}{2}\cos 10u$

5 $\sin 12\theta + \sin 6\theta$ **7** $\dfrac{3}{2}\sin 3x + \dfrac{3}{2}\sin x$

9 $2\sin 4\theta\cos 2\theta$ **11** $-2\sin 4x\sin x$

13 $-2\cos 5t\sin 2t$ **15** $2\cos\dfrac{3}{2}x\cos\dfrac{1}{2}x$

17 $\dfrac{\sin 4t + \sin 6t}{\cos 4t - \cos 6t} = \dfrac{2\sin 5t\cos t}{2\sin 5t\sin t} = \cot t$

19 $\dfrac{\sin u + \sin v}{\cos u + \cos v} = \dfrac{2\sin\frac{1}{2}(u+v)\cos\frac{1}{2}(u-v)}{2\cos\frac{1}{2}(u+v)\cos\frac{1}{2}(u-v)}$

$= \tan\dfrac{1}{2}(u+v)$

21 $\dfrac{\sin u - \sin v}{\sin u + \sin v} = \dfrac{2\cos\frac{1}{2}(u+v)\sin\frac{1}{2}(u-v)}{2\sin\frac{1}{2}(u+v)\cos\frac{1}{2}(u-v)}$

$= \cot\dfrac{1}{2}(u+v)\tan\dfrac{1}{2}(u-v)$

$= \dfrac{\tan\frac{1}{2}(u-v)}{\tan\frac{1}{2}(u+v)}$

23 $4\cos x\cos 2x\sin 3x = 2\cos 2x\,(2\sin 3x\cos x)$
$= 2\cos 2x\,(\sin 4x + \sin 2x)$
$= (2\cos 2x\sin 4x) + (2\cos 2x\sin 2x)$
$= [\sin 6x - \sin(-2x)] + (\sin 4x - \sin 0)$
$= \sin 2x + \sin 4x + \sin 6x$

25 $\dfrac{1}{2}\sin[(a+b)x] + \dfrac{1}{2}\sin[(a-b)x]$ **27** $\dfrac{\pi}{4}n$

29 $\dfrac{\pi}{2}n$ **31** $\dfrac{\pi}{2} + \pi n,\ \dfrac{\pi}{12} + \dfrac{\pi}{2}n,\ \dfrac{5\pi}{12} + \dfrac{\pi}{2}n$

33 $\dfrac{\pi}{7} + \dfrac{2\pi}{7}n,\ \dfrac{2\pi}{3}n$ **35** $\dfrac{\pi}{4},\ \dfrac{3\pi}{4},\ \dfrac{5\pi}{4},\ \dfrac{7\pi}{4},\ \dfrac{\pi}{2},\ \dfrac{3\pi}{2}$

37 $0,\ \pm\pi,\ \pm 2\pi,\ \pm\dfrac{\pi}{4},\ \pm\dfrac{3\pi}{4},\ \pm\dfrac{5\pi}{4},\ \pm\dfrac{7\pi}{4}$

39 $f(x) = \dfrac{1}{2}\sin\dfrac{\pi n}{l}(x+kt) + \dfrac{1}{2}\sin\dfrac{\pi n}{l}(x-kt)$

EXERCISES 7.6

1 (a) $-\dfrac{\pi}{4}$ **(b)** $\dfrac{2\pi}{3}$ **(c)** $-\dfrac{\pi}{3}$

3 (a) $\dfrac{\pi}{3}$ **(b)** $\dfrac{\pi}{4}$ **(c)** $\dfrac{\pi}{6}$

5 (a) Not defined   **(b)** Not defined   **(c)** $\dfrac{\pi}{4}$

7 (a) $-\dfrac{3}{10}$ **(b)** $\dfrac{1}{2}$ **(c)** 14

9 (a) $\dfrac{\pi}{3}$ **(b)** $\dfrac{5\pi}{6}$ **(c)** $-\dfrac{\pi}{6}$

11 (a) $-\dfrac{\pi}{4}$ **(b)** $\dfrac{3\pi}{4}$ **(c)** $-\dfrac{\pi}{4}$

13 (a) $\dfrac{\sqrt{3}}{2}$ **(b)** $\dfrac{\sqrt{2}}{2}$ **(c)** Not defined

15 (a) $\dfrac{\sqrt{5}}{2}$ **(b)** $\dfrac{\sqrt{34}}{5}$ **(c)** $\dfrac{4}{\sqrt{15}}$

17 (a) $\dfrac{\sqrt{3}}{2}$ **(b)** 0 **(c)** $-\dfrac{77}{36}$

19 (a) $-\dfrac{24}{25}$ **(b)** $-\dfrac{161}{289}$ **(c)** $\dfrac{24}{7}$

21 (a) $-\dfrac{1}{10}\sqrt{2}$ **(b)** $\dfrac{4}{17}\sqrt{17}$ **(c)** $\dfrac{1}{2}$

23 $\dfrac{x}{\sqrt{x^2+1}}$ **25** $\dfrac{\sqrt{x^2+4}}{2}$ **27** $2x\sqrt{1-x^2}$

29 $\sqrt{\dfrac{1+x}{2}}$ **31 (a)** $-\dfrac{\pi}{2}$ **(b)** 0 **(c)** $\dfrac{\pi}{2}$

33

35

37

39

41

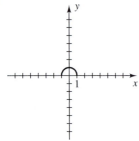

43 (a) $2 \leq x \leq 4$

(b) $-\dfrac{\pi}{4} \leq y \leq \dfrac{\pi}{4}$

(c) $x = \sin 2y + 3$

45 (a) $-\dfrac{3}{2} \leq x \leq \dfrac{3}{2}$ **(b)** $0 \leq y \leq 4\pi$

(c) $x = \dfrac{3}{2} \cos \dfrac{1}{4} y$

47 $x = \sin^{-1}(-y - 3)$ **49** $x = \cos^{-1}\left[\dfrac{1}{2}(15 - y)\right]$

51 $x = x_R$ or $x = \pi - x_R$, where $x_R = \sin^{-1}\left(\dfrac{3}{4} \sin y\right)$

53 $\cos^{-1}\left(-1 + \sqrt{2}\right) \approx 1.1437$,
$2\pi - \cos^{-1}\left(-1 + \sqrt{2}\right) \approx 5.1395$

55 $\tan^{-1}\dfrac{1}{4}\left(-9 + \sqrt{57}\right) \approx -0.3478$,

$\tan^{-1}\dfrac{1}{4}\left(-9 - \sqrt{57}\right) \approx -1.3337$

57 $\cos^{-1}\dfrac{1}{5}\sqrt{15} \approx 0.6847$, $\cos^{-1}\left(-\dfrac{1}{5}\sqrt{15}\right) \approx 2.4569$,

$\cos^{-1}\dfrac{1}{3}\sqrt{3} \approx 0.9553$, $\cos^{-1}\left(-\dfrac{1}{3}\sqrt{3}\right) \approx 2.1863$

59 $\sin^{-1}\left(\pm\dfrac{1}{6}\sqrt{30}\right) \approx \pm 1.1503$

61 $\cos^{-1}\left(-\dfrac{3}{5}\right) \approx 2.2143$, $\cos^{-1}\dfrac{1}{3} \approx 1.2310$,

$2\pi - \cos^{-1}\left(-\dfrac{3}{5}\right) \approx 4.0689$, $2\pi - \cos^{-1}\dfrac{1}{3} \approx 5.0522$

63 $\cos^{-1}\dfrac{2}{3} \approx 0.8411$, $2\pi - \cos^{-1}\dfrac{2}{3} \approx 5.4421$,

$\dfrac{\pi}{3} \approx 1.0472$, $\dfrac{5\pi}{3} \approx 5.2360$

65 (a) 1.65 m **(b)** 0.92 m **(c)** 0.43 m **67** 3.07°

69 (a) $\alpha = \theta - \sin^{-1}\dfrac{d}{k}$ **(b)** 40°

71 Let $\alpha = \sin^{-1} x$ and $\beta = \tan^{-1}\dfrac{x}{\sqrt{1 + x^2}}$ with

$-\dfrac{\pi}{2} < \alpha < \dfrac{\pi}{2}$ and $-\dfrac{\pi}{2} < \beta < \dfrac{\pi}{2}$. Thus, $\sin \alpha = x$
and $\sin \beta = x$. Since the sine function is one-to-one on

$\left(-\dfrac{\pi}{2}, \dfrac{\pi}{2}\right)$, we have $\alpha = \beta$.

73 Let $\alpha = \arcsin(-x)$ and $\beta = \arcsin x$ with

$-\dfrac{\pi}{2} \leq \alpha \leq \dfrac{\pi}{2}$ and $-\dfrac{\pi}{2} \leq \beta \leq \dfrac{\pi}{2}$. Thus,
$\sin \alpha = -x$ and $\sin \beta = x$. Consequently,
$\sin \alpha = -\sin \beta = \sin(-\beta)$. Since the sine function
is one-to-one on $\left[-\dfrac{\pi}{2}, \dfrac{\pi}{2}\right]$, we have $\alpha = -\beta$.

75 Let $\alpha = \arctan x$ and $\beta = \arctan(1/x)$. Since $x > 0$, we
have $0 < \alpha < \dfrac{\pi}{2}$ and $0 < \beta < \dfrac{\pi}{2}$, and hence
$0 < \alpha + \beta < \pi$. Thus,

$\tan(\alpha + \beta) = \dfrac{\tan \alpha + \tan \beta}{1 - \tan \alpha \tan \beta} = \dfrac{x + (1/x)}{1 - x \cdot (1/x)} = $

$\dfrac{x + (1/x)}{0}$. Since the denominator is 0, $\tan(\alpha + \beta)$ is

undefined and hence $\alpha + \beta = \dfrac{\pi}{2}$.

CHAPTER 7 REVIEW EXERCISES

1 $(\cot^2 x + 1)(1 - \cos^2 x) = (\csc^2 x)(\sin^2 x) = 1$

2 $\cos \theta + \sin \theta \tan \theta = \cos \theta + \sin \theta \cdot \dfrac{\sin \theta}{\cos \theta}$

$= \dfrac{\cos^2 \theta + \sin^2 \theta}{\cos \theta} = \dfrac{1}{\cos \theta} = \sec \theta$

3 $\dfrac{(\sec^2 \theta - 1) \cot \theta}{\tan \theta \sin \theta + \cos \theta} = \dfrac{(\tan^2 \theta) \cot \theta}{\dfrac{\sin \theta}{\cos \theta} \cdot \sin \theta + \cos \theta}$

$\qquad = \dfrac{\tan \theta}{\dfrac{\sin^2 \theta + \cos^2 \theta}{\cos \theta}} = \dfrac{\sin \theta / \cos \theta}{1/\cos \theta}$

$\qquad = \sin \theta$

4 $(\tan x + \cot x)^2 = \left(\dfrac{\sin x}{\cos x} + \dfrac{\cos x}{\sin x}\right)^2$

$\qquad = \left(\dfrac{\sin^2 x + \cos^2 x}{\cos x \sin x}\right)^2$

$\qquad = \dfrac{1}{\cos^2 x \sin^2 x} = \sec^2 x \csc^2 x$

5 $\dfrac{1}{1 + \sin t} \cdot \dfrac{1 - \sin t}{1 - \sin t} = \dfrac{1 - \sin t}{1 - \sin^2 t} = \dfrac{1 - \sin t}{\cos^2 t}$

$\qquad = \dfrac{1 - \sin t}{\cos t} \cdot \dfrac{1}{\cos t}$

$\qquad = \left(\dfrac{1}{\cos t} - \dfrac{\sin t}{\cos t}\right) \cdot \sec t$

$\qquad = (\sec t - \tan t) \sec t$

6 $\dfrac{\sin (\alpha - \beta)}{\cos (\alpha + \beta)} = \dfrac{(\sin \alpha \cos \beta - \cos \alpha \sin \beta)/\cos \alpha \cos \beta}{(\cos \alpha \cos \beta - \sin \alpha \sin \beta)/\cos \alpha \cos \beta}$

$\qquad = \dfrac{\tan \alpha - \tan \beta}{1 - \tan \alpha \tan \beta}$

7 $\tan 2u = \dfrac{2 \tan u}{1 - \tan^2 u} = \dfrac{2 \cdot \dfrac{1}{\cot u}}{1 - \dfrac{1}{\cot^2 u}} = \dfrac{\dfrac{2}{\cot u}}{\dfrac{\cot^2 u - 1}{\cot^2 u}}$

$\qquad = \dfrac{2 \cot u}{\cot^2 u - 1} = \dfrac{2 \cot u}{(\csc^2 u - 1) - 1} = \dfrac{2 \cot u}{\csc^2 u - 2}$

8 $\cos^2 \dfrac{v}{2} = \dfrac{1 + \cos v}{2} = \dfrac{1 + \dfrac{1}{\sec v}}{2} = \dfrac{\dfrac{\sec v + 1}{\sec v}}{2}$

$\qquad = \dfrac{1 + \sec v}{2 \sec v}$

9 $\dfrac{\tan^3 \phi - \cot^3 \phi}{\tan^2 \phi + \csc^2 \phi}$

$\qquad = \dfrac{(\tan \phi - \cot \phi)[(\tan^2 \phi + \tan \phi \cot \phi + \cot^2 \phi)]}{[\tan^2 \phi + (1 + \cot^2 \phi)]}$

$\qquad = \tan \phi - \cot \phi$

10 $LS = \dfrac{\sin u + \sin v}{\csc u + \csc v} = \dfrac{\sin u + \sin v}{\dfrac{1}{\sin u} + \dfrac{1}{\sin v}} = \dfrac{\sin u + \sin v}{\dfrac{\sin v + \sin u}{\sin u \sin v}}$

$\qquad = \sin u \sin v$

$RS = \dfrac{1 - \sin u \sin v}{-1 + \csc u \csc v} = \dfrac{1 - \sin u \sin v}{-1 + \dfrac{1}{\sin u \sin v}}$

$\qquad = \dfrac{1 - \sin u \sin v}{\dfrac{1 - \sin u \sin v}{\sin u \sin v}}$

$\qquad = \sin u \sin v$

Since the LS and RS equal the same expression and the steps are reversible, the identity is verified.

11 $\left(\dfrac{\sin^2 x}{\tan^4 x}\right)^3 \left(\dfrac{\csc^3 x}{\cot^6 x}\right)^2 = \left(\dfrac{\sin^6 x}{\tan^{12} x}\right) \left(\dfrac{\csc^6 x}{\cot^{12} x}\right) = \dfrac{(\sin x \csc x)^6}{(\tan x \cot x)^{12}}$

$\qquad = \dfrac{(1)^6}{(1)^{12}} = 1$

12 $\dfrac{\cos \gamma}{1 - \tan \gamma} + \dfrac{\sin \gamma}{1 - \cot \gamma} = \dfrac{\cos \gamma}{\dfrac{\cos \gamma - \sin \gamma}{\cos \gamma}} + \dfrac{\sin \gamma}{\dfrac{\sin \gamma - \cos \gamma}{\sin \gamma}}$

$\qquad = \dfrac{\cos^2 \gamma}{\cos \gamma - \sin \gamma} + \dfrac{\sin^2 \gamma}{\sin \gamma - \cos \gamma}$

$\qquad = \dfrac{\cos^2 \gamma - \sin^2 \gamma}{\cos \gamma - \sin \gamma}$

$\qquad = \dfrac{(\cos \gamma + \sin \gamma)(\cos \gamma - \sin \gamma)}{\cos \gamma - \sin \gamma}$

$\qquad = \cos \gamma + \sin \gamma$

13 $\dfrac{\cos (-t)}{\sec (-t) + \tan (-t)} = \dfrac{\cos t}{\sec t - \tan t} = \dfrac{\cos t}{\dfrac{1}{\cos t} - \dfrac{\sin t}{\cos t}}$

$\qquad = \dfrac{\cos t}{\dfrac{1 - \sin t}{\cos t}} = \dfrac{\cos^2 t}{1 - \sin t} = \dfrac{1 - \sin^2 t}{1 - \sin t}$

$\qquad = \dfrac{(1 - \sin t)(1 + \sin t)}{1 - \sin t} = 1 + \sin t$

14 $\dfrac{\cot (-t) + \csc (-t)}{\sin (-t)} = \dfrac{-\cot t - \csc t}{-\sin t} = \dfrac{\dfrac{\cos t}{\sin t} + \dfrac{1}{\sin t}}{\sin t}$

$\qquad = \dfrac{\cos t + 1}{\sin^2 t} = \dfrac{\cos t + 1}{1 - \cos^2 t}$

$\qquad = \dfrac{\cos t + 1}{(1 - \cos t)(1 + \cos t)} = \dfrac{1}{1 - \cos t}$

15 $\sqrt{\dfrac{1 - \cos t}{1 + \cos t}} = \sqrt{\dfrac{(1 - \cos t)}{(1 + \cos t)} \cdot \dfrac{(1 - \cos t)}{(1 - \cos t)}}$

$\qquad = \sqrt{\dfrac{(1 - \cos t)^2}{1 - \cos^2 t}}$

$\qquad = \sqrt{\dfrac{(1 - \cos t)^2}{\sin^2 t}} = \dfrac{|1 - \cos t|}{|\sin t|} = \dfrac{1 - \cos t}{|\sin t|},$

since $(1 - \cos t) \ge 0$.

16 $\sqrt{\dfrac{1 - \sin \theta}{1 + \sin \theta}} = \sqrt{\dfrac{(1 - \sin \theta)}{(1 + \sin \theta)} \cdot \dfrac{(1 + \sin \theta)}{(1 + \sin \theta)}}$

$\qquad = \sqrt{\dfrac{1 - \sin^2 \theta}{(1 + \sin \theta)^2}}$

$\qquad = \sqrt{\dfrac{\cos^2 \theta}{(1 + \sin \theta)^2}}$

$\qquad = \dfrac{|\cos \theta|}{|1 + \sin \theta|} = \dfrac{|\cos \theta|}{1 + \sin \theta},$

since $(1 + \sin \theta) \geq 0$.

17 $\cos\left(x - \dfrac{5\pi}{2}\right) = \cos x \cos \dfrac{5\pi}{2} + \sin x \sin \dfrac{5\pi}{2} = \sin x$

18 $\tan\left(x + \dfrac{3\pi}{4}\right) = \dfrac{\tan x + \tan \dfrac{3\pi}{4}}{1 - \tan x \tan \dfrac{3\pi}{4}} = \dfrac{\tan x - 1}{1 + \tan x}$

19 $\dfrac{1}{4} \sin 4\beta = \dfrac{1}{4} \sin (2 \cdot 2\beta) = \dfrac{1}{4}(2 \sin 2\beta \cos 2\beta)$

$\qquad = \dfrac{1}{2}(2 \sin \beta \cos \beta)(\cos^2 \beta - \sin^2 \beta)$

$\qquad = \sin \beta \cos^3 \beta - \cos \beta \sin^3 \beta$

20 $\tan \dfrac{1}{2}\theta = \dfrac{1 - \cos \theta}{\sin \theta} = \dfrac{1}{\sin \theta} - \dfrac{\cos \theta}{\sin \theta} = \csc \theta - \cot \theta$

21 $\sin 8\theta = 2 \sin 4\theta \cos 4\theta$

$\qquad = 2(2 \sin 2\theta \cos 2\theta)(1 - 2 \sin^2 2\theta)$

$\qquad = 8 \sin \theta \cos \theta (1 - 2 \sin^2 \theta)[1 - 2(2 \sin \theta \cos \theta)^2]$

$\qquad = 8 \sin \theta \cos \theta (1 - 2 \sin^2 \theta)(1 - 8 \sin^2 \theta \cos^2 \theta)$

22 Let $\alpha = \arctan x$ and $\beta = \arctan \dfrac{2x}{1 - x^2}$. Because

$-1 < x < 1, -\dfrac{\pi}{4} < \alpha < \dfrac{\pi}{4}$. Thus, $\tan \alpha = x$ and

$\tan \beta = \dfrac{2x}{1 - x^2} = \dfrac{2 \tan \alpha}{1 - \tan^2 \alpha} = \tan 2\alpha$. Since the tangent

function is one-to-one on $\left(-\dfrac{\pi}{2}, \dfrac{\pi}{2}\right)$, we have $\beta = 2\alpha$ or,

equivalently, $\alpha = \dfrac{1}{2}\beta$.

23 $\dfrac{\pi}{2}, \dfrac{3\pi}{2}, \dfrac{\pi}{4}, \dfrac{7\pi}{4}, \dfrac{3\pi}{4}, \dfrac{5\pi}{4}$ **24** $\dfrac{7\pi}{6}, \dfrac{11\pi}{6}$ **25** $0, \pi$

26 $\dfrac{\pi}{4}, \dfrac{3\pi}{4}, \dfrac{5\pi}{4}, \dfrac{7\pi}{4}$ **27** $0, \pi, \dfrac{2\pi}{3}, \dfrac{4\pi}{3}$

28 $\dfrac{\pi}{2}, \dfrac{3\pi}{2}, \dfrac{\pi}{4}, \dfrac{5\pi}{4}, \dfrac{3\pi}{4}, \dfrac{7\pi}{4}$ **29** $\dfrac{7\pi}{6}, \dfrac{11\pi}{6}, \dfrac{\pi}{2}$

30 $\dfrac{2\pi}{3}, \dfrac{4\pi}{3}, \pi$ **31** $\dfrac{\pi}{6}, \dfrac{5\pi}{6}, \dfrac{\pi}{3}, \dfrac{5\pi}{3}$

32 All x in $[0, 2\pi)$ except $\dfrac{\pi}{4}, \dfrac{3\pi}{4}, \dfrac{5\pi}{4}, \dfrac{7\pi}{4}$

33 $\dfrac{\pi}{3}, \dfrac{5\pi}{3}$ **34** $0, \dfrac{\pi}{3}, \dfrac{2\pi}{3}, \pi, \dfrac{4\pi}{3}, \dfrac{5\pi}{3}$

35 $\dfrac{3}{4}, \dfrac{7}{4}, \dfrac{11}{4}, \dfrac{15}{4}, \dfrac{19}{4}, \dfrac{23}{4}$ **36** $0, \pi, \dfrac{\pi}{3}, \dfrac{5\pi}{3}$

37 $\dfrac{\pi}{3}, \dfrac{5\pi}{3}$ **38** $\dfrac{\pi}{6}, \dfrac{5\pi}{6}, \dfrac{7\pi}{6}, \dfrac{11\pi}{6}$

39 $0, \dfrac{\pi}{8}, \dfrac{3\pi}{8}, \dfrac{5\pi}{8}, \dfrac{7\pi}{8}, \pi, \dfrac{9\pi}{8}, \dfrac{11\pi}{8}, \dfrac{13\pi}{8}, \dfrac{15\pi}{8}$

40 $\dfrac{\pi}{5}, \dfrac{3\pi}{5}, \pi, \dfrac{7\pi}{5}, \dfrac{9\pi}{5}$ **41** $\dfrac{\sqrt{6} - \sqrt{2}}{4}$

42 $-2 - \sqrt{3}$ **43** $\dfrac{\sqrt{2} - \sqrt{6}}{4}$ **44** $\dfrac{2}{\sqrt{2 - \sqrt{2}}}$

45 $\dfrac{84}{85}$ **46** $-\dfrac{13}{85}$ **47** $-\dfrac{84}{13}$ **48** $-\dfrac{36}{77}$

49 $\dfrac{36}{85}$ **50** $-\dfrac{36}{85}$ **51** $\dfrac{240}{289}$ **52** $-\dfrac{161}{289}$

53 $\dfrac{24}{7}$ **54** $\dfrac{1}{10}\sqrt{10}$ **55** $\dfrac{1}{3}$ **56** $\dfrac{5}{34}\sqrt{34}$

57 (a) $\dfrac{1}{2} \cos 3t - \dfrac{1}{2} \cos 11t$

(b) $\dfrac{1}{2} \cos \dfrac{1}{12}u + \dfrac{1}{2} \cos \dfrac{5}{12}u$

(c) $3 \sin 8x - 3 \sin 2x$ **(d)** $2 \sin 10\theta - 2 \sin 4\theta$

58 (a) $2 \sin 5u \cos 3u$ **(b)** $2 \sin \dfrac{11}{2}\theta \sin \dfrac{5}{2}\theta$

(c) $2 \cos \dfrac{9}{40}t \sin \dfrac{1}{40}t$ **(d)** $6 \cos 4x \cos 2x$

59 $\dfrac{\pi}{6}$ **60** $\dfrac{\pi}{4}$ **61** $\dfrac{\pi}{3}$ **62** π **63** $-\dfrac{\pi}{4}$

64 $\dfrac{3\pi}{4}$ **65** $\dfrac{1}{2}$ **66** 2 **67** Not defined **68** $\dfrac{\pi}{2}$

69 $\dfrac{240}{289}$ **70** $-\dfrac{7}{25}$

71

72

73

74

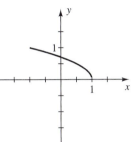

75 $\cos (\alpha + \beta + \gamma) = \cos [(\alpha + \beta) + \gamma]$
$= \cos (\alpha + \beta) \cos \gamma - \sin (\alpha + \beta) \sin \gamma$
$= (\cos \alpha \cos \beta - \sin \alpha \sin \beta) \cos \gamma -$
$(\sin \alpha \cos \beta + \cos \alpha \sin \beta) \sin \gamma$
$= \cos \alpha \cos \beta \cos \gamma - \sin \alpha \sin \beta \cos \gamma -$
$\sin \alpha \cos \beta \sin \gamma - \cos \alpha \sin \beta \sin \gamma$

76 (b) $t = 0, \pm \dfrac{\pi}{4b}$ **(c)** $\dfrac{2}{3}\sqrt{2A}$

77 $\pm \dfrac{\pi}{4}, \pm \dfrac{3\pi}{4}, \pm \dfrac{5\pi}{4}, \pm \dfrac{7\pi}{4}, \pm \dfrac{\pi}{3}, \pm \dfrac{5\pi}{3}$

78 (a) $x = 2d \tan \dfrac{1}{2}\theta$ **(b)** $d \le 1000$ ft

79 (a) $d = r\left(\sec \dfrac{1}{2}\theta - 1\right)$ **(b)** $43°$

80 (a) $78.7°$ **(b)** $61.4°$

CHAPTER 7 DISCUSSION EXERCISES

1 *Hint:* Factor $\sin^3 x - \cos^3 x$ as the difference of cubes.

2 $\sqrt{a^2 - x^2}$
$= \begin{cases} a \cos \theta & \text{if } 0 \le \theta \le \pi/2 \text{ or } 3\pi/2 \le \theta < 2\pi \\ -a \cos \theta & \text{if } \pi/2 < \theta < 3\pi/2 \end{cases}$

3 45; approximately 6.164

4 (a) The **inverse sawtooth function,** denoted by saw^{-1}, is defined by $y = \text{saw}^{-1} x$ iff $x = \text{saw } y$ for $-2 \le x \le 2$ and $-1 \le y \le 1$.
(b) $0.85; -0.4$
(c) $\text{saw } (\text{saw}^{-1} x) = x$ if $-2 \le x \le 2$;
$\text{saw}^{-1} (\text{saw } y) = y$ if $-1 \le y \le 1$
(d)

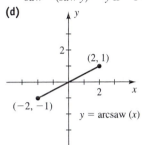

5 *Hint:* Write the equation in the form $\dfrac{\pi}{4} + \alpha = 4\theta$, and take the tangent of both sides.

Chapter 8

EXERCISES 8.1

1 $\beta = 62°, b \approx 14.1, c \approx 15.6$
3 $\gamma = 100°10', b \approx 55.1, c \approx 68.7$
5 $\beta = 78°30', a \approx 13.6, c \approx 17.8$
7 No triangle exists.
9 $\alpha \approx 77°30', \beta \approx 49°10', b \approx 108$;
   $\alpha \approx 102°30', \beta \approx 24°10', b \approx 59$
11 $\alpha \approx 82.54°, \beta \approx 49.72°, b \approx 100.85$;
   $\alpha \approx 97.46°, \beta \approx 34.80°, b \approx 75.45$
13 $\beta \approx 53°40', \gamma \approx 61°10', c \approx 20.6$
15 $\alpha \approx 25.993°, \gamma \approx 32.383°, a \approx 0.146$ **17** 219 yd
19 (a) 1.6 mi   **(b)** 0.6 mi   **21** 2.7 mi   **23** 628 m
25 3.7 mi from A and 5.4 mi from B   **27** 350 ft
29 (a) 18.7   **(b)** 814   **31** (3949.9, 2994.2)

EXERCISES 8.2

1 (a) B   **(b)** F   **(c)** D   **(d)** E
(e) A   **(f)** C
3 (a) α, law of sines   **(b)** a, law of cosines
(c) Any angle, law of cosines
(d) Not enough information given
(e) $\gamma, \alpha + \beta + \gamma = 180°$
(f) c, law of sines; or $\gamma, \alpha + \beta + \gamma = 180°$
5 $a \approx 26, \beta \approx 41°, \gamma \approx 79°$
7 $b \approx 180, \alpha \approx 25°, \gamma \approx 5°$
9 $c \approx 2.75, \alpha \approx 21°10', \beta \approx 43°40'$

11 $\alpha \approx 29°$, $\beta \approx 47°$, $\gamma \approx 104°$

13 $\alpha \approx 12°30'$, $\beta \approx 136°30'$, $\gamma \approx 31°00'$ **15** 196 ft

17 24 mi   **19** 39 mi   **21** 2.3 mi   **23** N55°31′E

25 63.7 ft from first and third base; 66.8 ft from second base

27 37,039 ft $\approx$ 7 mi

29 *Hint:* Use the formula $\sin \dfrac{\theta}{2} = \sqrt{\dfrac{1 - \cos \theta}{2}}$.

31 (a) 72°, 108°, 36°   **(b)** 0.62   **(c)** 0.59, 0.36

Exer. 33–40: The answer is in square units.

33 260   **35** 11.21   **37** 13.1   **39** 517.0

41 1.62 acres   **43** 123.4 ft^2

EXERCISES 8.3

1 $\langle 3, 1 \rangle, \langle 1, -7 \rangle, \langle 13, 8 \rangle, \langle 3, -32 \rangle, \sqrt{13}$

3 $\langle -15, 6 \rangle, \langle 1, -2 \rangle, \langle -68, 28 \rangle, \langle 12, -12 \rangle, \sqrt{53}$

5 $4\mathbf{i} - 3\mathbf{j}, -2\mathbf{i} + 7\mathbf{j}, 19\mathbf{i} - 17\mathbf{j}, -11\mathbf{i} + 33\mathbf{j}, \sqrt{5}$

7 Terminal points are   **9** Terminal points are
(3, 2), (−1, 5), (2, 7),   (−4, 6), (−2, 3),
(6, 4), (3, −15).   (−6, 9), (−8, 12),
                               (6, −9).

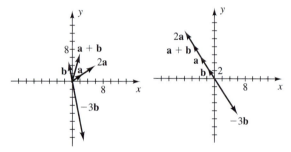

11 $-\mathbf{b}$ **13** $\mathbf{f}$ **15** $-\dfrac{1}{2}\mathbf{e}$

17 $\mathbf{a} + (\mathbf{b} + \mathbf{c}) = \langle a_1, a_2 \rangle + (\langle b_1, b_2 \rangle + \langle c_1, c_2 \rangle)$
$= \langle a_1, a_2 \rangle + \langle b_1 + c_1, b_2 + c_2 \rangle$
$= \langle a_1 + b_1 + c_1, a_2 + b_2 + c_2 \rangle$
$= \langle a_1 + b_1, a_2 + b_2 \rangle + \langle c_1, c_2 \rangle$
$= (\langle a_1, a_2 \rangle + \langle b_1, b_2 \rangle) + \langle c_1, c_2 \rangle$
$= (\mathbf{a} + \mathbf{b}) + \mathbf{c}$

19 $\mathbf{a} + (-\mathbf{a}) = \langle a_1, a_2 \rangle + (-\langle a_1, a_2 \rangle)$
$= \langle a_1, a_2 \rangle + \langle -a_1, -a_2 \rangle$
$= \langle a_1 - a_1, a_2 - a_2 \rangle$
$= \langle 0, 0 \rangle = \mathbf{0}$

21 $(mn)\mathbf{a} = (mn)\langle a_1, a_2 \rangle$
$= \langle (mn)a_1, (mn)a_2 \rangle$
$= \langle mna_1, mna_2 \rangle$
$= m\langle na_1, na_2 \rangle$ or $n\langle ma_1, ma_2 \rangle$
$= m(n\langle a_1, a_2 \rangle)$ or $n(m\langle a_1, a_2 \rangle)$
$= m(n\mathbf{a})$ or $n(m\mathbf{a})$

23 $0\mathbf{a} = 0\langle a_1, a_2 \rangle = \langle 0a_1, 0a_2 \rangle = \langle 0, 0 \rangle = \mathbf{0}$.
Also, $m\mathbf{0} = m\langle 0, 0 \rangle = \langle m0, m0 \rangle = \langle 0, 0 \rangle = \mathbf{0}$.

25 $-(\mathbf{a} + \mathbf{b}) = -(\langle a_1, a_2 \rangle + \langle b_1, b_2 \rangle)$
$= -(\langle a_1 + b_1, a_2 + b_2 \rangle)$
$= \langle -(a_1 + b_1), -(a_2 + b_2) \rangle$
$= \langle -a_1 - b_1, -a_2 - b_2 \rangle$
$= \langle -a_1, -a_2 \rangle + \langle -b_1, -b_2 \rangle$
$= -\mathbf{a} + (-\mathbf{b}) = -\mathbf{a} - \mathbf{b}$

27 $\|2\mathbf{v}\| = \|2\langle a, b \rangle\| = \|\langle 2a, 2b \rangle\| = \sqrt{(2a)^2 + (2b)^2}$
$= \sqrt{4a^2 + 4b^2} = 2\sqrt{a^2 + b^2} = 2\|\langle a, b \rangle\|$
$= 2\|\mathbf{v}\|$

29 $3\sqrt{2}; \dfrac{7\pi}{4}$ **31** 5; π **33** $\sqrt{41}$; $\tan^{-1}\left(-\dfrac{5}{4}\right) + \pi$

35 $18; \dfrac{3\pi}{2}$ **37** 102 lb   **39** 7.2 lb

41 89 lb; S66°W   **43** 5.8 lb; 129°

45 40.96; 28.68   **47** −6.18; 19.02

49 (a) $-\dfrac{8}{17}\mathbf{i} + \dfrac{15}{17}\mathbf{j}$ **(b)** $\dfrac{8}{17}\mathbf{i} - \dfrac{15}{17}\mathbf{j}$

51 (a) $\left\langle \dfrac{2}{\sqrt{29}}, -\dfrac{5}{\sqrt{29}} \right\rangle$ **(b)** $\left\langle -\dfrac{2}{\sqrt{29}}, \dfrac{5}{\sqrt{29}} \right\rangle$

53 (a) $\langle -12, 6 \rangle$ **(b)** $\left\langle -3, \dfrac{3}{2} \right\rangle$

55 $-\dfrac{24}{\sqrt{65}}\mathbf{i} + \dfrac{42}{\sqrt{65}}\mathbf{j}$

57 (a) $\mathbf{F} = \langle 7, 2 \rangle$ **(b)** $\mathbf{G} = -\mathbf{F} = \langle -7, -2 \rangle$

59 (a) $\mathbf{F} \approx \langle -5.86, 1.13 \rangle$
(b) $\mathbf{G} = -\mathbf{F} \approx \langle 5.86, -1.13 \rangle$

61 $\sin^{-1}(0.4) \approx 23.6°$ **63** 56°; 232 mi/hr

65 420 mi/hr; 244°   **67** N22°W

69 $\mathbf{v}_1 \approx 4.1\mathbf{i} - 7.10\mathbf{j}$; $\mathbf{v}_2 \approx 0.98\mathbf{i} - 3.67\mathbf{j}$

71 (a) (24.51, 20.57)   **(b)** (−24.57, 18.10)

73 28.2 lb/person

EXERCISES 8.4

1 (a) 24   **(b)** $\cos^{-1}\left(\dfrac{24}{\sqrt{29}\sqrt{45}}\right) \approx 48°22'$

3 (a) −14   **(b)** $\cos^{-1}\left(\dfrac{-14}{\sqrt{17}\sqrt{13}}\right) \approx 160°21'$

5 (a) 45   **(b)** $\cos^{-1}\left(\dfrac{45}{\sqrt{81}\sqrt{41}}\right) \approx 38°40'$

7 (a) $-\dfrac{149}{5}$ **(b)** $\cos^{-1}\left(\dfrac{-149/5}{\sqrt{149}\sqrt{149/25}}\right) = 180°$

9 $\langle 4, -1 \rangle \cdot \langle 2, 8 \rangle = 0$ **11** $(-4\mathbf{j}) \cdot (-7\mathbf{i}) = 0$

13 Opposite   **15** Same   **17** $\dfrac{6}{5}$ **19** $\pm\dfrac{3}{8}$

21 (a) −23   **(b)** −23   **23** −51

25 $17/\sqrt{26} \approx 3.33$ **27** 2.2 **29** 7

31 28 **33** 12

35 $\mathbf{a} \cdot \mathbf{a} = \langle a_1, a_2 \rangle \cdot \langle a_1, a_2 \rangle = a_1^2 + a_2^2$
$= \left(\sqrt{a_1^2 + a_2^2}\right)^2 = \|\mathbf{a}\|^2$

37 $(m\mathbf{a}) \cdot \mathbf{b} = (m\langle a_1, a_2 \rangle) \cdot \langle b_1, b_2 \rangle$
$= \langle ma_1, ma_2 \rangle \cdot \langle b_1, b_2 \rangle$
$= ma_1 b_1 + ma_2 b_2$
$= m(a_1 b_1 + a_2 b_2) = m(\mathbf{a} \cdot \mathbf{b})$

39 $\mathbf{0} \cdot \mathbf{a} = \langle 0, 0 \rangle \cdot \langle a_1, a_2 \rangle = 0(a_1) + 0(a_2)$
$= 0 + 0 = 0$

41 $1000\sqrt{3} \approx 1732$ ft-lb

43 (a) $\mathbf{v} = (93 \times 10^6)\mathbf{i} + (0.432 \times 10^6)\mathbf{j}$;
$\mathbf{w} = (93 \times 10^6)\mathbf{i} - (0.432 \times 10^6)\mathbf{j}$
(b) $0.53°$

45 $\left\langle \dfrac{4}{5}, \dfrac{3}{5} \right\rangle$ **47** 2.6 **49** 24.33

51 $16\sqrt{3} \approx 27.7$ horsepower

EXERCISES 8.5

1 5 **3** $\sqrt{85}$ **5** 8 **7** 1 **9** 0

Note: Point P is the point corresponding to the geometric representation.

11 $P(4, 2)$ **13** $P(3, -5)$ **15** $P(-3, 6)$
17 $P(-6, 4)$ **19** $P(0, 2)$

21 $\sqrt{2} \text{ cis } \dfrac{7\pi}{4}$ **23** $8 \text{ cis } \dfrac{5\pi}{6}$ **25** $4 \text{ cis } \dfrac{\pi}{6}$

27 $4\sqrt{2} \text{ cis } \dfrac{5\pi}{4}$ **29** $20 \text{ cis } \dfrac{3\pi}{2}$ **31** $12 \text{ cis } 0$

33 $7 \text{ cis } \pi$ **35** $6 \text{ cis } \dfrac{\pi}{2}$ **37** $10 \text{ cis } \dfrac{4\pi}{3}$

39 $\sqrt{5} \text{ cis } \left(\tan^{-1} \dfrac{1}{2} \right)$

41 $\sqrt{10} \text{ cis } \left[\tan^{-1} \left(-\dfrac{1}{3} \right) + \pi \right]$

43 $\sqrt{34} \text{ cis } \left(\tan^{-1} \dfrac{3}{5} + \pi \right)$

45 $5 \text{ cis } \left[\tan^{-1} \left(-\dfrac{3}{4} \right) + 2\pi \right]$

47 $2\sqrt{2} + 2\sqrt{2}i$ **49** $-3 + 3\sqrt{3}i$ **51** -5

53 $5 + 3i$ **55** $2 - i$ **57** $-2, i$

59 $10\sqrt{3} - 10i, -\dfrac{2}{5}\sqrt{3} + \dfrac{2}{5}i$ **61** $40, \dfrac{5}{2}$

63 $8 - 4i, \dfrac{8}{5} + \dfrac{4}{5}i$ **67** $17.21 + 24.57i$

69 $11.01 + 9.24i$ **71** $\sqrt{365}$ ohms **73** 70.43 volts

EXERCISES 8.6

1 $-972 - 972i$ **3** $-32i$ **5** -8

7 $-\dfrac{1}{2}\sqrt{2} - \dfrac{1}{2}\sqrt{2}i$ **9** $-\dfrac{1}{2} - \dfrac{1}{2}\sqrt{3}i$

11 $-64\sqrt{3} - 64i$ **13** $\pm\left(\dfrac{1}{2}\sqrt{6} + \dfrac{1}{2}\sqrt{2}i \right)$

15 $\pm\left(\dfrac{\sqrt[4]{2}}{2} + \dfrac{\sqrt[4]{18}}{2}i \right), \pm\left(\dfrac{\sqrt[4]{18}}{2} - \dfrac{\sqrt[4]{2}}{2}i \right)$

17 $3i, \pm\dfrac{3}{2}\sqrt{3} - \dfrac{3}{2}i$

19 $\pm 1, \dfrac{1}{2} \pm \dfrac{1}{2}\sqrt{3}i,$ **21** $\sqrt[10]{2} \text{ cis } \theta$ with $\theta = 9°,$
$-\dfrac{1}{2} \pm \dfrac{1}{2}\sqrt{3}i$ $81°, 153°, 225°, 297°$

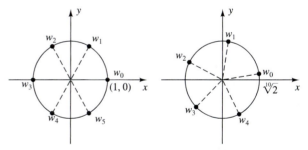

23 $\pm 2, \pm 2i$ **25** $\pm 2i, \sqrt{3} \pm i, -\sqrt{3} \pm i$

27 $2i, \pm\sqrt{3} - i$

29 $3 \text{ cis } \theta$ with $\theta = 0°, 72°, 144°, 216°, 288°$

31 $[r(\cos \theta + i \sin \theta)]^n = [r(e^{i\theta})]^n$
$= r^n (e^{i\theta})^n$
$= r^n e^{i(n\theta)}$
$= r^n(\cos n\theta + i \sin n\theta)$

CHAPTER 8 REVIEW EXERCISES

1 $a = \sqrt{43}, \beta = \cos^{-1}\left(\dfrac{4}{43}\sqrt{43} \right), \gamma = \cos^{-1}\left(\dfrac{5}{86}\sqrt{43} \right)$

2 $\alpha = 60°, \beta = 90°, b = 4; \alpha = 120°, \beta = 30°, b = 2$

3 $\gamma = 75°, a = 50\sqrt{6}, c = 50\left(1 + \sqrt{3}\right)$

4 $\alpha = \cos^{-1}\left(\dfrac{7}{8}\right), \beta = \cos^{-1}\left(\dfrac{11}{16}\right), \gamma = \cos^{-1}\left(-\dfrac{1}{4}\right)$

5 $\alpha = 38°, a \approx 8.0, c \approx 13$

6 $\gamma \approx 19°10', \beta \approx 137°20', b \approx 258$

7 $\alpha \approx 24°, \gamma \approx 41°, b \approx 10.1$

8 $\alpha \approx 42°, \beta \approx 87°, \gamma \approx 51°$ **9** 290   **10** 10.9

11 Terminal points are $(-2, -3), (-6, 13),$ $(-8, 10), (-1, 4).$

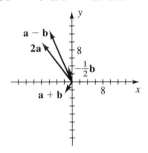

12 **(a)** $12\mathbf{i} + 19\mathbf{j}$ **(b)** $-8\mathbf{i} + 13\mathbf{j}$ **(c)** $\sqrt{40} \approx 6.32$
   (d) $\sqrt{29} - \sqrt{17} \approx 1.26$

13 $\langle 14 \cos 40°, -14 \sin 40° \rangle$ **14** 109 lb; S78°E

15 $-16\mathbf{i} + 12\mathbf{j}$

16 $\left\langle -\dfrac{12}{\sqrt{58}}, \dfrac{28}{\sqrt{58}} \right\rangle$

17 Circle with center (a_1, a_2) and radius c

18 The vectors $\mathbf{a}, \mathbf{b},$ and $\mathbf{a} - \mathbf{b}$ form a triangle with the vector $\mathbf{a} - \mathbf{b}$ opposite angle θ. The conclusion is a direct application of the law of cosines with sides $\|\mathbf{a}\|, \|\mathbf{b}\|,$ and $\|\mathbf{a} - \mathbf{b}\|.$

19 183°; 70 mi/hr

20 **(a)** 10   **(b)** $\cos^{-1}\left(\dfrac{10}{\sqrt{13}\sqrt{17}}\right) \approx 47°44'$ **(c)** $\dfrac{10}{\sqrt{13}}$

21 **(a)** 80   **(b)** $\cos^{-1}\left(\dfrac{40}{\sqrt{40}\sqrt{50}}\right) \approx 26°34'$ **(c)** $\sqrt{40}$

22 56

23 $10\sqrt{2} \operatorname{cis} \dfrac{3\pi}{4}$ **24** $4 \operatorname{cis} \dfrac{5\pi}{3}$ **25** $17 \operatorname{cis} \pi$

26 $12 \operatorname{cis} \dfrac{3\pi}{2}$ **27** $10 \operatorname{cis} \dfrac{7\pi}{6}$

28 $\sqrt{41} \operatorname{cis} \left(\tan^{-1} \dfrac{5}{4}\right)$ **29** $10\sqrt{3} - 10i$

30 $12 + 5i$ **31** $-12 - 12\sqrt{3}i, -\dfrac{3}{2}$

32 $-4\sqrt{2}i, -2\sqrt{2}$ **33** $-512i$ **34** i

35 $-972 + 972i$ **36** $-2^{19} - 2^{19}\sqrt{3}i$

37 $-3, \dfrac{3}{2} \pm \dfrac{3}{2}\sqrt{3}i$

38 **(a)** 2^{24} **(b)** $\sqrt[3]{2} \operatorname{cis} \theta$ with $\theta = 100°, 220°, 340°$

39 $2 \operatorname{cis} \theta$ with $\theta = 0°, 72°, 144°, 216°, 288°$

40 47.6°   **41** 53,000,000 mi

42 **(a)** 449 ft   **(b)** 434 ft

43 **(a)** 33 mi, 41 mi   **(b)** 30 mi   **44** 204

45 1 hour and 16 minutes   **46 (c)** 158°

47 **(a)** 47°   **(b)** 20

48 **(a)** 72°   **(b)** 181.6 ft^2 **(c)** 37.6 ft

CHAPTER 8 DISCUSSION EXERCISES

4 **(b)** *Hint:* Law of cosines

5 **(a)** $(\|\mathbf{b}\| \cos \alpha + \|\mathbf{a}\| \cos \beta)\mathbf{i} +$ $(\|\mathbf{b}\| \sin \alpha - \|\mathbf{a}\| \sin \beta)\mathbf{j}$

6 **(a)** 1   **(b)** $\pi i; \dfrac{\pi}{2}i$ **(c)** $\dfrac{\sqrt{2}}{2} + \dfrac{\sqrt{2}}{2}i; e^{-\pi/2} \approx 0.2079$

7 The statement is true.

Chapter 9

EXERCISES 9.1

1 $(3, 5), (-1, -3)$ **3** $(1, 0), (-3, 2)$

5 $(0, 0), \left(\dfrac{1}{8}, \dfrac{1}{128}\right)$ **7** $(3, -2)$ **9** No solution

11 $(-4, 3), (5, 0)$ **13** $(-2, 2)$

15 $\left(-\dfrac{3}{5} + \dfrac{1}{10}\sqrt{86}, \dfrac{1}{5} + \dfrac{3}{10}\sqrt{86}\right),$ $\left(-\dfrac{3}{5} - \dfrac{1}{10}\sqrt{86}, \dfrac{1}{5} - \dfrac{3}{10}\sqrt{86}\right)$

17 $(-4, 0), \left(\dfrac{12}{5}, \dfrac{16}{5}\right)$ **19** $(0, 1), (4, -3)$

21 $(\pm 2, 5), \left(\pm\sqrt{5}, 4\right)$

23 $\left(\sqrt{2}, \pm 2\sqrt{3}\right), \left(-\sqrt{2}, \pm 2\sqrt{3}\right)$

25 $\left(2\sqrt{2}, \pm 2\right), \left(-2\sqrt{2}, \pm 2\right)$ **27** $(3, -1, 2)$

29 $(1, -1, 2), (-1, 3, -2)$

31 **(a)** $b = 4$; tangent
   (b) $b < 4$; intersect twice
   (c) $b > 4$; no intersection

33 Yes; a solution occurs between 0 and 1.

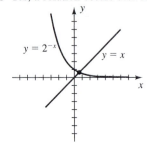

35 $\frac{1}{4}$; tangent   **37** $f(x) = 2(3)^x + 1$

39 12 in. $\times$ 8 in.

41 **(a)** $a = 120{,}000, b = 40{,}000$ **(b)** 77,143

43 $(0, 0), (0, 100), (50, 0)$; the fourth solution $(-100, 150)$ is not meaningful.

45 Yes; 1 ft $\times$ 1 ft $\times$ 2 ft or

$$\frac{\sqrt{13} - 1}{2} \text{ ft} \times \frac{\sqrt{13} - 1}{2} \text{ ft} \times \frac{8}{(\sqrt{13} - 1)^2} \text{ ft}$$
$$\approx 1.30 \text{ ft} \times 1.30 \text{ ft} \times 1.18 \text{ ft}$$

47 The points are on the parabola **(a)** $y = \frac{1}{2}x^2 - \frac{1}{2}$ and

(b) $y = \frac{1}{4}x^2 - 1$.

49 **(a)** $(31.25, -50)$

(b) $\left(-\frac{3}{2}\sqrt{11}, -\frac{1}{2}\right) \approx (-4.975, -0.5)$

EXERCISES 9.2

1 $(4, -2)$ **3** $(8, 0)$ **5** $\left(-1, \frac{3}{2}\right)$ **7** $\left(\frac{76}{53}, \frac{28}{53}\right)$

9 $\left(\frac{51}{13}, \frac{96}{13}\right)$ **11** $\left(\frac{8}{7}, -\frac{3}{7}\sqrt{6}\right)$ **13** No solution

15 All ordered pairs (m, n) such that $3m - 4n = 2$

17 $(0, 0)$ **19** $\left(-\frac{22}{7}, -\frac{11}{5}\right)$

21 313 students, 137 nonstudents

23 $x = \left(\frac{30}{\pi}\right) - 4 \approx 5.55$ cm, $y = 12 - \left(\frac{30}{\pi}\right) \approx 2.45$ cm

25 $l = 10$ ft, $w = \frac{20}{\pi}$ ft   **27** 2400 adults, 3600 kittens

29 40 g of 35% alloy, 60 g of 60% alloy

31 540 mi/hr, 60 mi/hr   **33** $v_0 = 10, a = 3$

35 20 sofas, 30 recliners

37 **(a)** $\left(c, \frac{4}{5}c\right)$ for an arbitrary $c > 0$ **(b)** $16 per hour

39 1928; 15.5°C   **41** LP: 4 hr, SLP: 2 hr

43 $a = \frac{1}{6}, b = -\frac{1}{6}e^{6x}$ **45** $a = \cos x - \sec x, b = \sin x$

EXERCISES 9.3

1

3

$y = -2x + 1$

$3x - 2y = 6$

5

$y = x^2 - 2$

7

$y = x^2 + 1$

9

$y = \frac{1}{x^2}$

11

$3x + y = 3$
$4 - y = 2x$
5
-5

13

$2x + 5y = 10$
$y - x = 0$

15

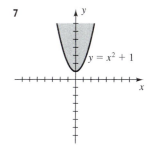
$y - 2x = 1$
$y = 4$
$x = -2$
$3x + y = 6$

17

19

21

23

25

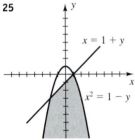

27 $0 \leq x < 3, y < -x + 4, y \geq x - 4$

29 $x^2 + y^2 \leq 9, y > -2x + 4$

31 $y < x, y \leq -x + 4, (x - 2)^2 + (y - 2)^2 \leq 8$

33 $y > \frac{1}{8}x + \frac{1}{2}, y \leq x + 4, y \leq -\frac{3}{4}x + 4$

35 If x and y denote the numbers of sets of brand A and brand B, respectively, then a system is $x \geq 20$, $y \geq 10, x \geq 2y, x + y \leq 100$.

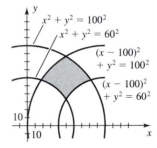

37 If x and y denote the amounts placed in the high-risk and low-risk investment, respectively, then a system is $x \geq 2000, y \geq 3x,$ $x + y \leq 15,000.$

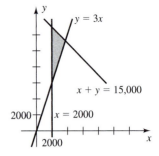

39 $x + y \leq 9, y \geq x, x \geq 1$

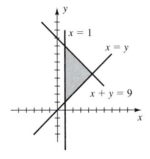

41 If the plant is located at (x, y), then a system is $60^2 \leq x^2 + y^2 \leq 100^2, 60^2 \leq (x - 100)^2 + y^2 \leq 100^2,$ $y \geq 0.$

43 (a) Yes
(b)

(c) Region above the line

EXERCISES 9.4

1 Maximum of 27 at (6, 2); minimum of 9 at (0, 2)

3 Maximum of 21 at (6, 3)   **5** Minimum of 21 at (3, 2)

7 C has the maximum value 24 for any point on the line segment from (2, 5) to (6, 3).

9 50 standard and 30 oversized

11 3.5 lb of S and 1 lb of T

13 Send 25 from W_1 to A and 0 from W_1 to B.
Send 10 from W_2 to A and 60 from W_2 to B.

15 None of alfalfa and 80 acres of corn

17 Minimum cost: 16 oz X, 4 oz Y, 0 oz Z;
maximum cost: 0 oz X, 8 oz Y, 12 oz Z

19 2 vans and 4 buses   **21** 3000 trout and 2000 bass

23 60 small units and 20 deluxe units

EXERCISES 9.5

1 $(2, 3, -1)$ **3** $(-2, 4, 5)$ **5** No solution

7 $\left(\dfrac{2}{3}, \dfrac{31}{21}, \dfrac{1}{21}\right)$

Exer. 9–16: There are other forms for the answers; c is any real number.

9 $(2c, -c, c)$ **11** $(0, -c, c)$

13 $\left(\dfrac{12}{7} - \dfrac{9}{7}c, \dfrac{4}{7}c - \dfrac{13}{14}, c\right)$

15 $\left(\dfrac{7}{10}c + \dfrac{1}{2}, \dfrac{19}{10}c - \dfrac{3}{2}, c\right)$

17 $\left(\dfrac{1}{11}, \dfrac{31}{11}, \dfrac{3}{11}\right)$ **19** $(-2, -3)$ **21** No solution

23 17 of 10%, 11 of 30%, 22 of 50%

25 4 hr for A, 2 hr for B, 5 hr for C

27 380 lb of G_1, 60 lb of G_2, 160 lb of G_3

29 (a) $I_1 = 0, I_2 = 2, I_3 = 2$ (b) $I_1 = \dfrac{3}{4}, I_2 = 3, I_3 = \dfrac{9}{4}$

31 $\dfrac{3}{8}$ lb Colombian, $\dfrac{1}{8}$ lb Brazilian, $\dfrac{1}{2}$ lb Kenyan

33 (a) A: $x_1 + x_4 = 75$, B: $x_1 + x_2 = 150$,
C: $x_2 + x_3 = 225$, D: $x_3 + x_4 = 150$
(b) $x_1 = 25, x_2 = 125, x_4 = 50$
(c) $x_3 = 150 - x_4 \le 150$;
$x_3 = 225 - x_2 = 225 - (150 - x_1) = 75 + x_1 \ge 75$

35 $x^2 + y^2 - x + 3y - 6 = 0$

37 $f(x) = x^3 - 2x^2 - 4x - 6$

EXERCISES 9.6

1 $\begin{bmatrix} 9 & -1 \\ -2 & 5 \end{bmatrix}, \begin{bmatrix} 1 & -3 \\ 4 & 1 \end{bmatrix}, \begin{bmatrix} 10 & -4 \\ 2 & 6 \end{bmatrix}, \begin{bmatrix} -12 & -3 \\ 9 & -6 \end{bmatrix}$

3 $\begin{bmatrix} 9 & 0 \\ 1 & 5 \\ 3 & 4 \end{bmatrix}, \begin{bmatrix} 3 & -2 \\ 3 & -5 \\ -9 & 4 \end{bmatrix}, \begin{bmatrix} 12 & -2 \\ 4 & 0 \\ -6 & 8 \end{bmatrix}, \begin{bmatrix} -9 & -3 \\ 3 & -15 \\ -18 & 0 \end{bmatrix}$

5 $\begin{bmatrix} 11 & -3 & -3 \end{bmatrix}, \begin{bmatrix} -3 & -3 & 7 \end{bmatrix}$,
$\begin{bmatrix} 8 & -6 & 4 \end{bmatrix}, \begin{bmatrix} -21 & 0 & 15 \end{bmatrix}$

7 Not possible, not possible,
$\begin{bmatrix} 6 & -4 & 4 \\ 0 & 2 & -8 \\ -6 & 4 & -2 \end{bmatrix}, \begin{bmatrix} -12 & 0 \\ -6 & 3 \\ 3 & -9 \end{bmatrix}$

9 -18 **11** $\begin{bmatrix} 16 & 38 \\ 11 & -34 \end{bmatrix}, \begin{bmatrix} 4 & 38 \\ 23 & -22 \end{bmatrix}$

13 $\begin{bmatrix} 3 & -14 & -3 \\ 16 & 2 & -2 \\ -7 & -29 & 9 \end{bmatrix}, \begin{bmatrix} 3 & -20 & -11 \\ 2 & 10 & -4 \\ 15 & -13 & 1 \end{bmatrix}$

15 $\begin{bmatrix} 4 & 8 \\ -18 & 11 \end{bmatrix}, \begin{bmatrix} 3 & -4 & 4 \\ -5 & 2 & 2 \\ -51 & 26 & 10 \end{bmatrix}$

17 $\begin{bmatrix} 1 & 2 & 3 \\ 4 & 5 & 6 \\ 7 & 8 & 9 \end{bmatrix}, \begin{bmatrix} 1 & 2 & 3 \\ 4 & 5 & 6 \\ 7 & 8 & 9 \end{bmatrix}$

19 [15], $\begin{bmatrix} -3 & 7 & 2 \\ -12 & 28 & 8 \\ 15 & -35 & -10 \end{bmatrix}$

21 $\begin{bmatrix} 2 & 0 & 5 \\ 5 & 3 & -2 \end{bmatrix}$, not possible   **23** $\begin{bmatrix} 4 \\ 12 \\ -1 \end{bmatrix}$

25 $\begin{bmatrix} 18 & 0 & -2 \\ -40 & 10 & -10 \end{bmatrix}$

35 (a) $A = \begin{bmatrix} 400 & 550 & 500 \\ 400 & 450 & 500 \\ 300 & 500 & 600 \\ 250 & 200 & 300 \\ 100 & 100 & 200 \end{bmatrix}$, $B = \begin{bmatrix} \$1.39 \\ \$2.99 \\ \$4.99 \end{bmatrix}$

(b) $\begin{bmatrix} \$4695.50 \\ \$4396.50 \\ \$4906.00 \\ \$2442.50 \\ \$1436.00 \end{bmatrix}$

(c) The \$1436.00 represents the amount the store would receive if all the yellow towels were sold.

EXERCISES 9.7

1 Show that $AB = I_2$ and $BA = I_2$.

3 $\frac{1}{10} \begin{bmatrix} 3 & 4 \\ -1 & 2 \end{bmatrix}$ **5** Does not exist

7 $\frac{1}{8} \begin{bmatrix} 2 & 1 & 0 \\ -2 & 3 & 0 \\ 0 & 0 & 2 \end{bmatrix}$ **9** $\frac{1}{3} \begin{bmatrix} -4 & -5 & 3 \\ -4 & -8 & 3 \\ 1 & 2 & 0 \end{bmatrix}$

11 $\begin{bmatrix} \frac{1}{2} & 0 & 0 \\ 0 & \frac{1}{4} & 0 \\ 0 & 0 & \frac{1}{6} \end{bmatrix}$ **13** $ab \neq 0$; $\begin{bmatrix} \frac{1}{a} & 0 \\ 0 & \frac{1}{b} \end{bmatrix}$

17 (a) $\left(\frac{13}{10}, -\frac{1}{10} \right)$ **(b)** $\left(\frac{7}{5}, \frac{6}{5} \right)$

19 (a) $\left(-\frac{25}{3}, -\frac{34}{3}, \frac{7}{3} \right)$ **(b)** $\left(\frac{16}{3}, \frac{16}{3}, -\frac{1}{3} \right)$

EXERCISES 9.8

1 $M_{11} = 0 = A_{11}$; $M_{12} = 5$; $A_{12} = -5$;
$M_{21} = -1$; $A_{21} = 1$; $M_{22} = 7 = A_{22}$

3 $M_{11} = -14 = A_{11}$; $M_{12} = 10$; $A_{12} = -10$;
$M_{13} = 15 = A_{13}$; $M_{21} = 7$; $A_{21} = -7$;
$M_{22} = -5 = A_{22}$; $M_{23} = 34$; $A_{23} = -34$;
$M_{31} = 11 = A_{31}$; $M_{32} = 4$; $A_{32} = -4$;
$M_{33} = 6 = A_{33}$

5 5   **7** -83 **9** 2   **11** 0   **13** -125 **15** 48
17 -216 **19** $abcd$ **31 (a)** $x^2 - 3x - 4$ **(b)** $-1, 4$
33 (a) $x^2 + x - 2$ **(b)** $-2, 1$
35 (a) $-x^3 - 2x^2 + x + 2$ **(b)** $-2, -1, 1$
37 (a) $-x^3 + 4x^2 + 4x - 16$ **(b)** $-2, 2, 4$
39 $-31i - 20j + 7k$ **41** $-6i - 8j + 18k$

EXERCISES 9.9

1 $R_2 \leftrightarrow R_3$ **3** $-R_1 + R_3 \rightarrow R_3$
5 2 is a common factor of R_1 and R_3.
7 R_1 and R_3 are identical.
9 -1 is a common factor of R_2.
11 Every number in C_2 is 0.   **13** $2C_1 + C_3 \rightarrow C_3$
15 -10 **17** -142 **19** -183 **21** 44   **23** 359
33 $(4, -2)$ **35** $(8, 0)$
37 $|D| = 0$, so Cramer's rule cannot be used.
39 $(2, 3, -1)$ **41** $(-2, 4, 5)$
43 $x = \dfrac{cgi - dfi + bfj}{cei - afi + bfh}$

EXERCISES 9.10

1 $\dfrac{3}{x - 2} + \dfrac{5}{x + 3}$ **3** $\dfrac{5}{x - 6} - \dfrac{4}{x + 2}$

5 $\dfrac{2}{x - 1} + \dfrac{3}{x + 2} - \dfrac{1}{x - 3}$ **7** $\dfrac{3}{x} + \dfrac{2}{x - 5} - \dfrac{1}{x + 1}$

9 $\dfrac{2}{x - 1} + \dfrac{5}{(x - 1)^2}$ **11** $-\dfrac{7}{x} + \dfrac{5}{x^2} + \dfrac{40}{3x - 5}$

13 $\dfrac{24/25}{x + 2} + \dfrac{2/5}{(x + 2)^2} - \dfrac{23/25}{2x - 1}$

15 $\dfrac{5}{x} - \dfrac{2}{x + 1} + \dfrac{3}{(x + 1)^3}$ **17** $-\dfrac{2}{x - 1} + \dfrac{3x + 4}{x^2 + 1}$

19 $\dfrac{4}{x} + \dfrac{5x - 3}{x^2 + 2}$ **21** $\dfrac{4x - 1}{x^2 + 1} + \dfrac{3}{(x^2 + 1)^2}$

23 $2x + \dfrac{1}{x - 1} + \dfrac{3x}{x^2 + 1}$ **25** $3 + \dfrac{4}{x} + \dfrac{8}{x - 4}$

27 $2x + 3 + \dfrac{2}{x - 1} - \dfrac{3}{2x + 1}$

CHAPTER 9 REVIEW EXERCISES

1 $\left(\dfrac{19}{23}, -\dfrac{18}{23} \right)$ **2** No solution   **3** $(-3, 5), (1, -3)$

4 $(4, -3), (3, -4)$ **5** $\left(2\sqrt{3}, \pm\sqrt{2}\right), \left(-2\sqrt{3}, \pm\sqrt{2}\right)$

6 $(-1, \pm 1, -1), \left(0, \pm\dfrac{1}{2}\sqrt{6}, -\dfrac{1}{2}\right)$ **7** $\left(\dfrac{14}{17}, \dfrac{14}{27}\right)$

8 $\left(\log_2 \dfrac{25}{7}, \log_3 \dfrac{15}{7}\right)$ **9** $\left(\dfrac{6}{11}, -\dfrac{7}{11}, 1\right)$

10 $\left(-\dfrac{6}{29}, \dfrac{2}{29}, -\dfrac{17}{29}\right)$

11 $(-2c, -3c, c)$ for any real number c **12** $(0, 0, 0)$

13 $\left(5c - 1, -\dfrac{19}{2}c + \dfrac{5}{2}, c\right)$ for any real number c

14 $(5, -4)$ **15** $\left(-1, \dfrac{1}{2}, \dfrac{1}{3}\right)$ **16** $(3, -1, -2, 4)$

17

18

19

20

21 $\begin{bmatrix} 4 & -5 & 6 \\ 4 & -11 & 5 \end{bmatrix}$ **22** $\begin{bmatrix} 26 \\ -6 \end{bmatrix}$ **23** $\begin{bmatrix} 0 & 4 & -6 \\ 16 & 22 & 1 \\ 12 & 11 & 9 \end{bmatrix}$

24 $\begin{bmatrix} 0 & -37 \\ 15 & -6 \end{bmatrix}$ **25** $\begin{bmatrix} -12 & 4 & -11 \\ 6 & -11 & 5 \end{bmatrix}$

26 $\begin{bmatrix} a & 3a \\ 2a & 4a \end{bmatrix}$ **27** $\begin{bmatrix} a & 3a \\ 2b & 4b \end{bmatrix}$ **28** $\begin{bmatrix} 0 & 0 \\ 0 & 0 \end{bmatrix}$

29 $\begin{bmatrix} 5 & 9 \\ 13 & 19 \end{bmatrix}$ **30** $\begin{bmatrix} 1 & 0 & 0 \\ 0 & 1 & 0 \\ 0 & 0 & 1 \end{bmatrix}$ **31** $-\dfrac{1}{2}\begin{bmatrix} 2 & 4 \\ 3 & 5 \end{bmatrix}$

32 $\dfrac{1}{11}\begin{bmatrix} 8 & 1 & -2 \\ 5 & 2 & -4 \\ -14 & 1 & 9 \end{bmatrix}$ **33** $\begin{bmatrix} 1 & 0 & 0 \\ 0 & 2 & -7 \\ 0 & -1 & 4 \end{bmatrix}$

34 $\dfrac{1}{37}\begin{bmatrix} -4 & -20 & 15 \\ 3 & 15 & -2 \\ 9 & 8 & -6 \end{bmatrix}$ **35** $(2, -5)$ **36** $(-1, 3, 2)$

37 -6 **38** 9 **39** 48 **40** -86 **41** -84

42 0 **43** 120 **44** -76 **45** 0 **46** -50

47 $-1 \pm 2\sqrt{3}$ **48** $4, \pm\sqrt{7}$

49 2 is a common factor of R_1, 2 is a common factor of C_2, and 3 is a common factor of C_3.

50 Interchange R_1 with R_2 and then R_2 with R_3 to obtain the determinant on the right. The effect is to multiply by -1 twice.

51 $a_{11}a_{22}a_{33} \cdots a_{nn}$ **53** $\left(\dfrac{76}{53}, \dfrac{28}{53}\right)$ **54** $\left(\dfrac{2}{3}, \dfrac{31}{21}, \dfrac{1}{21}\right)$

55 $\dfrac{8}{x - 1} - \dfrac{3}{x + 5} - \dfrac{1}{x + 3}$ **56** $2 + \dfrac{3}{x + 1} + \dfrac{4}{(x + 1)^2}$

57 $-\dfrac{2}{x + 5} + \dfrac{3x - 1}{x^2 + 4}$ **58** $\dfrac{4}{x^2 + 2} + \dfrac{x - 2}{x^2 + 5}$

59 $40\sqrt{5}$ ft $\times 20\sqrt{5}$ ft   **60** $y = \pm 2\sqrt{2}x + 3$

61 Tax $= \$18,750$; bonus $= \$3,125$

62 Inside radius $= 90$ ft, outside radius $= 100$ ft

63 In ft^3/hr: A, 30; B, 20; C, 50

64 Western 95, eastern 55

65 If x and y denote the length and width, respectively, then a system is $x \le 12$, $y \le 8$, $y \ge \dfrac{1}{2}x$.

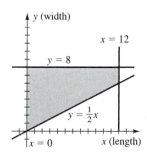

66 $x + y \le 18$, $x \ge 2y$, $x \ge 0$, $y \ge 0$

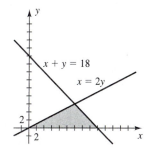

67 80 mowers and 30 edgers

68 High-risk $50,000; low-risk $100,000; bonds $0

CHAPTER 9 DISCUSSION EXERCISES

1 (a) $b = 1.99$, $x = 204$, $y = -100$;
 $b = 1.999$, $x = 2004$, $y = -1000$

 (b) $x = \dfrac{4b - 10}{b - 2}$, $y = \dfrac{1}{b - 2}$

 (c) It gets close to $(4, 0)$.

2 (a) $D = [12{,}000 \quad 9000 \quad 14{,}000]$;

$$E = \begin{bmatrix} 0.90 & 0.10 & 0.00 \\ 0.00 & 0.80 & 0.20 \\ 0.05 & 0.00 & 0.95 \end{bmatrix}$$

 (b) The elements of $F = [11{,}500 \quad 8400 \quad 15{,}100]$ represent the populations on islands A, B, and C, respectively, after one year.

 (c) The population stabilizes with 10,000 birds on A, 5000 birds on B, and 20,000 birds on C.

3 *Hint:* Assign a size to A, and examine the definition of an inverse.

4 AD: 35%, DS: $33\frac{1}{3}$%, SP: $31\frac{2}{3}$%

5 $a = -15$, $b = 10$, $c = 24$; the fourth root is -4

6 (a) Not possible **(b)** $x^2 + y^2 - 1.8x - 4.2y + 0.8 = 0$

 (c) $f(x) = -\dfrac{5}{12}x^2 + \dfrac{7}{12}x + 4$

 (d) $f(x) = ax^3 + \left(-2a - \dfrac{5}{12}\right)x^2 + \left(-3a + \dfrac{7}{12}\right)x + 4,$
  where a is any nonzero real number

 (e) Not possible

Chapter 10

EXERCISES 10.1

1 $9, 6, 3, 0; -12$

3 $\dfrac{1}{2}, \dfrac{4}{5}, \dfrac{7}{10}, \dfrac{10}{17}, \dfrac{22}{65}$

5 $9, 9, 9, 9; 9$

7 $1.9, 2.01, 1.999, 2.0001; 2.000\,000\,01$

9 $4, -\dfrac{9}{4}, \dfrac{5}{3}, -\dfrac{11}{8}; -\dfrac{15}{16}$ **11** $2, 0, 2, 0; 0$

13 $\dfrac{2}{3}, \dfrac{2}{3}, \dfrac{8}{11}, \dfrac{8}{9}; \dfrac{128}{33}$ **15** $1, 2, 3, 4; 8$

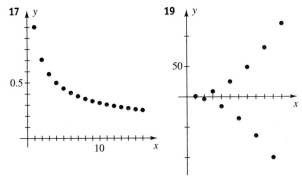

17

19

21 $2, 1, -2, -11, -38$ **23** $-3, 3^2, 3^4, 3^8, 3^{16}$

25 $5, 5, 10, 30, 120$ **27** $2, 2, 4, 4^3, 4^{12}$

29 $\dfrac{7}{2}, \dfrac{15}{2}, 12, 17$

31 $-1, -1 + \dfrac{1}{\sqrt{2}}, -1 + \dfrac{1}{\sqrt{2}} - \dfrac{1}{\sqrt{3}}, -\dfrac{1}{2} + \dfrac{1}{\sqrt{2}} - \dfrac{1}{\sqrt{3}}$

33 -5 **35** 10 **37** 25 **39** $-\dfrac{17}{15}$ **41** 61

43 $10{,}000$ **45** $\dfrac{319}{3}$ **47** $\dfrac{7}{2}k^2$

49 $\displaystyle\sum_{k=1}^{n} (a_k - b_k)$

$= (a_1 - b_1) + (a_2 - b_2) + \cdots + (a_n - b_n)$
$= (a_1 + a_2 + \cdots + a_n) + (-b_1 - b_2 - \cdots - b_n)$
$= (a_1 + a_2 + \cdots + a_n) - (b_1 + b_2 + \cdots + b_n)$
$= \displaystyle\sum_{k=1}^{n} a_k - \sum_{k=1}^{n} b_k$

51 As k increases, the terms approach 1.

53 $0.4, 0.7, 1, 1.6, 2.8$

55 (a) $1, 1, 2, 3, 5, 8, 13, 21, 34, 55$

 (b) $1, 2, 1.5, 1.\overline{6}, 1.6, 1.625, 1.6153846, 1.6190476,$
  $1.6176471, 1.6181818$

57 (a) $a_n = 0.8a_{n-1}$ **(b)** The fourth day

59 $C(n) = \begin{cases} 89.95n & \text{if } 1 \le n \le 4 \\ 87.95n & \text{if } 5 \le n \le 9 \\ 85.95n & \text{if } n \ge 10 \end{cases}$

61 2.236068   **63** 2.4493

65 (a) $f(1) = -1 < 0,\ f(2) \approx 0.30 > 0$ **(b)** 1.76

EXERCISES 10.2

1 Show that $a_{k+1} - a_k = 4$.   **3** $4n - 2$; 18; 38

5 $3.3 - 0.3n$; 1.8; 0.3   **7** $3.1n - 10.1$; 5.4; 20.9

9 $\ln 3^n$; $\ln 3^5$; $\ln 3^{10}$ **11** -8 **13** -8.5

15 -9.8 **17** $\dfrac{551}{17}$ **19** -105 **21** 30   **23** 530

25 $\dfrac{423}{2}$ **27** $934j + 838{,}265$

29 $\displaystyle\sum_{n=1}^{5} (7n - 3)$ or $\displaystyle\sum_{n=0}^{4} (4 + 7n)$

31 $\displaystyle\sum_{n=1}^{67} (7n - 3)$ or $\displaystyle\sum_{n=0}^{66} (4 + 7n)$

33 $\displaystyle\sum_{n=1}^{6} \dfrac{3n}{4n + 3}$ or $\displaystyle\sum_{n=0}^{5} \dfrac{3 + 3n}{7 + 4n}$

35 $\displaystyle\sum_{n=1}^{1528} (11n - 3) = 12{,}845{,}132$

37 24   **39** 12 or 18   **41** $\dfrac{10}{3}, \dfrac{14}{3}, 6, \dfrac{22}{3}, \dfrac{26}{3}$

43 (a) 60   **(b)** 12,780   **45** 255   **47** 154π ft

49 $1200   **51** $16n^2$

53 Show that the $(n + 1)$st term is 1 greater than the nth term.

55 (a) $\dfrac{8}{36}, \dfrac{7}{36}, \dfrac{6}{36}, \ldots, \dfrac{1}{36}$ **(b)** $d = -\dfrac{1}{36}$; 1   **(c)** $722.22

EXERCISES 10.3

1 Show that $\dfrac{a_{k+1}}{a_k} = -\dfrac{1}{4}$.   **3** $8\left(\dfrac{1}{2}\right)^{n-1} = 2^{4-n}$; $\dfrac{1}{2}$; $\dfrac{1}{16}$

5 $300(-0.1)^{n-1}$; 0.03; -0.00003 **7** 5^n; 3125; 390,625

9 $4(-1.5)^{n-1}$; 20.25; -68.34375

11 $(-1)^{n-1}x^{2n-2}$; x^8; $-x^{14}$ **13** $2^{(n-1)x+1}$; 2^{4x+1}; 2^{7x+1}

15 $\pm\sqrt{3}$ **17** $\dfrac{243}{8}$ **19** $\sqrt[3]{3}$; 36   **21** 88,572

23 $-\dfrac{341}{1024}$ **25** $8188 + 55j$ **27** $\displaystyle\sum_{n=1}^{7} 2^n$

29 $\displaystyle\sum_{n=1}^{4} (-1)^{n+1} \dfrac{1}{4}\left(\dfrac{1}{3}\right)^{n-1}$ **31** $\dfrac{2}{3}$ **33** $\dfrac{50}{33}$

35 Since $|r| = \sqrt{2} > 1$, the sum does not exist.

37 1024   **39** $\dfrac{23}{99}$ **41** $\dfrac{2393}{990}$ **43** $\dfrac{5141}{999}$ **45** $\dfrac{16{,}123}{9999}$

47 24   **49** 4, 20, 100, 500   **51** $\dfrac{25}{256}\% \approx 0.1\%$

53 (a) $N(t) = 10{,}000(1.2)^t$ **(b)** 61,917   **55** 300 ft

57 $3,000,000   **59 (b)** 375 mg

61 (a) $a_{k+1} = \dfrac{1}{4}\sqrt{10}a_k$

(b) $a_n = \left(\dfrac{1}{4}\sqrt{10}\right)^{n-1} a_1,\ A_n = \left(\dfrac{5}{8}\right)^{n-1} A_1,$

$P_n = \left(\dfrac{1}{4}\sqrt{10}\right)^{n-1} P_1$ **(c)** $\dfrac{16a_1}{4 - \sqrt{10}}$

63 (a) $a_k = 3^{k-1}$ **(b)** 4,782,969

(c) $b_k = \dfrac{3^{k-1}}{4^k} = \dfrac{1}{4}\left(\dfrac{3}{4}\right)^{k-1}$ **(d)** $\dfrac{729}{16{,}384} \approx 4.45\%$

65 $38,929.00   **67** $7396.67

69 (a) $\dfrac{2}{5}, \dfrac{6}{25}, \dfrac{18}{125}, \dfrac{54}{625}, \dfrac{162}{3125}$

(b) $r = \dfrac{3}{5}$; $\dfrac{2882}{3125} = 0.92224$ **(c)** $16,000

EXERCISES 10.4

Exer. 1–32: A typical proof is given for Exercises 1, 5, 9, ..., 29.

1 (1) P_1 is true, since $2(1) = 1(1 + 1) = 2$.

(2) Assume P_k is true:

$2 + 4 + 6 + \cdots + 2k = k(k + 1)$. Hence,

$2 + 4 + 6 + \cdots + 2k + 2(k + 1)$

$\qquad = k(k + 1) + 2(k + 1)$

$\qquad = (k + 1)(k + 2)$

$\qquad = (k + 1)(k + 1 + 1)$.

Thus, P_{k+1} is true, and the proof is complete.

5 (1) P_1 is true, since $5(1) - 3 = \dfrac{1}{2}(1)[5(1) - 1] = 2$.

(2) Assume P_k is true:

$2 + 7 + 12 + \cdots + (5k - 3) = \dfrac{1}{2}k(5k - 1)$.

Hence,

$2 + 7 + 12 + \cdots + (5k - 3) + 5(k + 1) - 3$

$\qquad = \dfrac{1}{2}k(5k - 1) + 5(k + 1) - 3$

$\qquad = \dfrac{5}{2}k^2 + \dfrac{9}{2}k + 2$

$\qquad = \dfrac{1}{2}(5k^2 + 9k + 4)$

$\qquad = \dfrac{1}{2}(k + 1)(5k + 4)$

$\qquad = \dfrac{1}{2}(k + 1)[5(k + 1) - 1]$.

Thus, P_{k+1} is true, and the proof is complete.

9 (1) P_1 is true, since $(1)^1 = \dfrac{1(1+1)[2(1)+1]}{6} = 1$.

(2) Assume P_k is true:

$$1^2 + 2^2 + 3^2 + \cdots + k^2 = \dfrac{k(k+1)(2k+1)}{6}.$$

Hence,

$$1^2 + 2^2 + 3^2 + \cdots + k^2 + (k+1)^2$$
$$= \dfrac{k(k+1)(2k+1)}{6} + (k+1)^2$$
$$= (k+1)\left[\dfrac{k(2k+1)}{6} + \dfrac{6(k+1)}{6}\right]$$
$$= \dfrac{(k+1)(2k^2 + 7k + 6)}{6}$$
$$= \dfrac{(k+1)(k+2)(2k+3)}{6}.$$

Thus, P_{k+1} is true, and the proof is complete.

13 (1) P_1 is true, since $3^1 = \dfrac{3}{2}(3^1 - 1) = 3$.

(2) Assume P_k is true:

$$3 + 3^2 + 3^3 + \cdots + 3^k = \dfrac{3}{2}(3^k - 1). \text{ Hence,}$$
$$3 + 3^2 + 3^3 + \cdots + 3^k + 3^{k+1}$$
$$= \dfrac{3}{2}(3^k - 1) + 3^{k+1}$$
$$= \dfrac{3}{2} \cdot 3^k - \dfrac{3}{2} + 3 \cdot 3^k$$
$$= \dfrac{9}{2} \cdot 3^k - \dfrac{3}{2}$$
$$= \dfrac{3}{2}(3 \cdot 3^k - 1)$$
$$= \dfrac{3}{2}(3^{k+1} - 1).$$

Thus, P_{k+1} is true, and the proof is complete.

17 (1) P_1 is true, since $1 < \dfrac{1}{8}[2(1)+1]^2 = \dfrac{9}{8}$.

(2) Assume P_k is true:

$$1 + 2 + 3 + \cdots + k < \dfrac{1}{8}(2k+1)^2. \text{ Hence,}$$
$$1 + 2 + 3 + \cdots + k + (k+1)$$
$$< \dfrac{1}{8}(2k+1)^2 + (k+1)$$
$$= \dfrac{1}{2}k^2 + \dfrac{3}{2}k + \dfrac{9}{8}$$
$$= \dfrac{1}{8}(4k^2 + 12k + 9)$$
$$= \dfrac{1}{8}(2k+3)^2$$
$$= \dfrac{1}{8}[2(k+1)+1]^2.$$

Thus, P_{k+1} is true, and the proof is complete.

21 (1) For $n = 1$, $5^n - 1 = 4$ and 4 is a factor of 4.

(2) Assume 4 is a factor of $5^k - 1$. The $(k+1)$st term is
$$5^{k+1} - 1 = 5 \cdot 5^k - 1$$
$$= 5 \cdot 5^k - 5 + 4$$
$$= 5(5^k - 1) + 4.$$
By the induction hypothesis, 4 is a factor of $5^k - 1$ and 4 is a factor of 4, so 4 is a factor of the $(k+1)$st term. Thus, P_{k+1} is true, and the proof is complete.

25 (1) For $n = 1$, $a - b$ is a factor of $a^1 - b^1$.

(2) Assume $a - b$ is a factor of $a^k - b^k$. Following the hint for the $(k+1)$st term,
$$a^{k+1} - b^{k+1} = a^k \cdot a - b \cdot a^k + b \cdot a^k - b^k \cdot b$$
$$= a^k(a - b) + (a^k - b^k)b.$$
Since $(a - b)$ is a factor of $a^k(a - b)$ and since by the induction hypothesis $a - b$ is a factor of $(a^k - b^k)$, it follows that $a - b$ is a factor of the $(k+1)$st term. Thus, P_{k+1} is true, and the proof is complete.

29 (1) P_8 is true, since $5 + \log_2 8 \le 8$.

(2) Assume P_k is true: $5 + \log_2 k \le k$. Hence,
$$5 + \log_2(k+1) < 5 + \log_2(k+k)$$
$$= 5 + \log_2 2k$$
$$= 5 + \log_2 2 + \log_2 k$$
$$= (5 + \log_2 k) + 1$$
$$\le k + 1.$$
Thus, P_{k+1} is true, and the proof is complete.

33 $\dfrac{n^3 + 6n^2 + 20n}{3}$　　**35** $\dfrac{4n^3 - 12n^2 + 11n}{3}$

37 (a) $a + b + c = 1$, $8a + 4b + 2c = 5$,

$$27a + 9b + 3c = 14; \; a = \dfrac{1}{3}, b = \dfrac{1}{2}, c = \dfrac{1}{6}$$

(b) The method used in part (a) shows that the formula is true for only $n = 1, 2, 3$.

39 (1) For $n = 1$,
$$\sin(\theta + 1\pi) = \sin\theta\cos\pi + \cos\theta\sin\pi$$
$$= -\sin\theta = (-1)^1\sin\theta.$$
(2) Assume P_k is true: $\sin(\theta + k\pi) = (-1)^k\sin\theta$.
Hence,
$$\sin[\theta + (k+1)\pi]$$
$$= \sin[(\theta + k\pi) + \pi]$$
$$= \sin(\theta + k\pi)\cos\pi + \cos(\theta + k\pi)\sin\pi$$
$$= [(-1)^k\sin\theta]\cdot(-1) + \cos(\theta + k\pi)\cdot(0)$$
$$= (-1)^{k+1}\sin\theta.$$
Thus, P_{k+1} is true, and the proof is complete.

41 (1) For $n = 1$,
$$[r(\cos\theta + i\sin\theta)]^1 = r^1[\cos(1\theta) + i\sin(1\theta)].$$
(2) Assume P_k is true:
$$[r(\cos\theta + i\sin\theta)]^k = r^k(\cos k\theta + i\sin k\theta).$$
Hence,
$$[r(\cos\theta + i\sin\theta)]^{k+1}$$
$$= [r(\cos\theta + i\sin\theta)]^k[r(\cos\theta + i\sin\theta)]$$
$$= r^k[\cos k\theta + i\sin k\theta][r(\cos\theta + i\sin\theta)]$$
$$= r^{k+1}[(\cos k\theta\cos\theta - \sin k\theta\sin\theta) +$$
$$i(\sin k\theta\cos\theta + \cos k\theta\sin\theta)]$$
$$= r^{k+1}[\cos(k+1)\theta + i\sin(k+1)\theta].$$
Thus, P_{k+1} is true, and the proof is complete.

EXERCISES 10.5

1 1440   **3** 5040   **5** 336   **7** 1   **9** 21
11 715   **13** $n(n-1)$ **15** $(2n+2)(2n+1)$
17 $64x^3 - 48x^2y + 12xy^2 - y^3$
19 $x^6 + 6x^5y + 15x^4y^2 + 20x^3y^3 + 15x^2y^4 + 6xy^5 + y^6$
21 $x^7 - 7x^6y + 21x^5y^2 - 35x^4y^3 + 35x^3y^4 - 21x^2y^5$
$$+ 7xy^6 - y^7$$
23 $81t^4 - 540t^3s + 1350t^2s^2 - 1500ts^3 + 625s^4$
25 $\dfrac{1}{243}x^5 + \dfrac{5}{81}x^4y^2 + \dfrac{10}{27}x^3y^4 + \dfrac{10}{9}x^2y^6 + \dfrac{5}{3}xy^8 + y^{10}$
27 $x^{-12} + 18x^{-9} + 135x^{-6} + 540x^{-3} + 1215 + 1458x^3$
$$+ 729x^6$$
29 $x^{5/2} - 5x^{3/2} + 10x^{1/2} - 10x^{-1/2} + 5x^{-3/2} - x^{-5/2}$
31 $3^{25}c^{10} + 25\cdot 3^{24}c^{52/5} + 300\cdot 3^{23}c^{54/5}$
33 $-1680\cdot 3^{13}z^{11} + 60\cdot 3^{14}z^{13} - 3^{15}z^{15}$ **35** $\dfrac{189}{1024}c^8$
37 $\dfrac{114{,}688}{9}u^2v^6$ **39** $70x^2y^2$ **41** $448y^3x^{10}$
43 $-216y^9x^2$ **45** $-\dfrac{135}{16}$ **47** 4.8, 6.19
49 $4x^3 + 6x^2h + 4xh^2 + h^3$

51 $\dbinom{n}{1} = \dfrac{n!}{(n-1)!\,1!} = n$ and
$$\binom{n}{n-1} = \dfrac{n!}{[n-(n-1)]!\,(n-1)!}$$
$$= \dfrac{n!}{1!\,(n-1)!} = n$$

EXERCISES 10.6

1 210   **3** 60,480   **5** 120   **7** 6   **9** 1   **11** $n!$
13 **(a)** 60   **(b)** 125   **15** 64   **17** $P(8, 3) = 336$
19 24   **21** **(a)** 2,340,000   **(b)** 2,160,000
23 **(a)** 151,200   **(b)** 5760   **25** 1024
27 $P(8, 8) = 40{,}320$ **29** $P(6, 3) = 120$
31 **(a)** 27,600   **(b)** 35,152   **33** 9,000,000
35 $P(4, 4) = 24$ **37** $3!\cdot 2^3 = 48$
39 $(2^{16} - 1)\cdot 17$
41 **(a)** 900
　(b) If n is even, $9\cdot 10^{(n/2)-1}$; if n is odd, $9\cdot 10^{(n-1)/2}$.
43 $n! \approx \dfrac{n^n\sqrt{2\pi n}}{e^n}$

EXERCISES 10.7

1 35   **3** 9   **5** n **7** 1   **9** $\dfrac{12!}{5!\,3!\,2!\,2!} = 166{,}320$
11 $\dfrac{10!}{3!\,2!\,2!\,1!\,1!\,1!} = 151{,}200$ **13** $C(10, 5) = 252$
15 $C(8, 2) = 28$ **17** $(5!\cdot 4!\cdot 8!)\cdot 3! = 696{,}729{,}600$
19 $3\cdot C(10, 2)\cdot C(8, 2)\cdot C(4, 2)\cdot C(6, 2)\cdot 3\cdot 4$
$$= 4{,}082{,}400$$
21 $C(12, 3)\cdot C(8, 2) = 6160$ **23** $C(8, 3) = 56$
25 **(a)** $C(49, 6) = 13{,}983{,}816$ **(b)** $C(24, 6) = 134{,}596$
27 $C(n, 2) = 45$ and hence $n = 10$ **29** $C(6, 3) = 20$
31 By finding $C(31, 3) = 4495$
33 **(a)** $C(1000, 30) \approx 2.43 \times 10^{57}$
　(b) $P(1000, 30) \approx 6.44 \times 10^{89}$
35 $C(4, 3)\cdot C(48, 2) = 4512$
37 **(a)** 1, 2, 4, 8, 16, 32, 64, 128, 256, 512
　(b) $S_n = 2^{n-1}$

EXERCISES 10.8

1 **(a)** $\dfrac{4}{52}$; 1 to 12   **(b)** $\dfrac{8}{52}$; 2 to 11   **(c)** $\dfrac{12}{52}$; 3 to 10
3 **(a)** $\dfrac{1}{6}$; 1 to 5   **(b)** $\dfrac{1}{6}$; 1 to 5   **(c)** $\dfrac{2}{6}$; 1 to 2
5 **(a)** $\dfrac{5}{15}$; 1 to 2   **(b)** $\dfrac{6}{15}$; 2 to 3   **(c)** $\dfrac{9}{15}$; 3 to 2

7 (a) $\frac{2}{36}$; 1 to 17 **(b)** $\frac{5}{36}$; 5 to 31 **(c)** $\frac{7}{36}$; 7 to 29

9 $\frac{6}{216}$ **11** $\frac{3}{8}$ **13** 5 to 2; 2 to 5 **15** 5 to 9; $\frac{9}{14}$

17 1.93 to 1 **19** $\frac{48 \cdot 13}{C(52, 5)} \approx 0.00024$

21 $\frac{C(13, 4) \cdot C(13, 1)}{C(52, 5)} \approx 0.00358$

23 $\frac{C(13, 5) \cdot 4}{C(52, 5)} \approx 0.00198$ **25** $\frac{4}{6}$

27 $(0.674)^4 \approx 0.2064$

29 (a) 0.45 **(b)** 0.10 **(c)** 0.70 **(d)** 0.95

31 (a) $\frac{C(20, 5) \cdot C(40, 0)}{C(60, 5)} \approx 0.0028$

(b) $1 - \frac{C(30, 0) \cdot C(30, 5)}{C(60, 5)} \approx 0.9739$

(c) $\frac{C(10, 0) \cdot C(50, 5)}{C(60, 5)} + \frac{C(10, 1) \cdot C(50, 4)}{C(60, 5)} \approx 0.8096$

33 (a) $\frac{C(8, 8)}{2^8} \approx 0.00391$ **(b)** $\frac{C(8, 7)}{2^8} = 0.03125$

(c) $\frac{C(8, 6)}{2^8} = 0.109375$

(d) $\frac{C(8, 6) + C(8, 7) + C(8, 8)}{2^8} \approx 0.14453$

35 $1 - \frac{C(48, 5)}{C(52, 5)} \approx 0.34116$

37 (a) A representative outcome is (nine of clubs, 3); 312

(b) 20; 292; $\frac{20}{312}$ **(c)** No; yes; $\frac{72}{312}$; $\frac{156}{312}$; $\frac{36}{312}$; $\frac{192}{312}$

(d) Yes; no; 0; $\frac{92}{312}$

39 $1 - \frac{10}{36} = \frac{26}{36}$ **41 (a)** $\frac{1}{32}$ **(b)** $1 - \frac{1}{32} = \frac{31}{32}$

43 (a) $\frac{C(4, 4)}{4!} = \frac{1}{24}$ **(b)** $\frac{C(4, 2)}{4!} = \frac{1}{4}$

45 (a) 0 **(b)** $\frac{1}{9}$

47 (a) $\frac{331,142}{418,890} \approx 0.791$ **(b)** $\frac{334,415}{418,890} \approx 0.798$

49 12.5% **51 (a)** $\frac{1}{16}$ **(b)** $\frac{C(4, 2)}{2^4} = \frac{6}{16}$

53 $\frac{2}{25,827,165}$ (about 1 chance in 13 million)

55 $\frac{1970}{39,800} \approx 0.0495$

57 (a) $\frac{8}{36}$ **(b)** $\frac{1}{36}$ **(c)** $\frac{244}{495} \approx 0.4929$

59 (b) 0.76 **61** \$0.99 **63** \$0.20

CHAPTER 10 REVIEW EXERCISES

1 $5, -2, -1, -\frac{20}{29}; -\frac{7}{19}$

2 $0.9, -1.01, 0.999, -1.0001; 0.999\,999\,9$

3 $2, \frac{1}{2}, \frac{5}{4}, \frac{7}{8}; \frac{65}{64}$ **4** $\frac{1}{12}, \frac{1}{15}, \frac{1}{15}, \frac{8}{105}; \frac{8}{45}$

5 $10, \frac{11}{10}, \frac{21}{11}, \frac{32}{21}, \frac{53}{32}$ **6** 2, 2, 2, 2, 2

7 $9, 3, \sqrt{3}, \sqrt[4]{3}, \sqrt[8]{3}$ **8** $1, \frac{1}{2}, \frac{2}{3}, \frac{3}{5}, \frac{5}{8}$

9 75 **10** $-\frac{37}{10}$ **11** 940 **12** -10 **13** $\sum_{n=1}^{5} 3n$

14 $\sum_{n=1}^{6} 2^{3-n}$ **15** $\sum_{n=1}^{99} \frac{1}{n(n+1)}$ **16** $\sum_{n=1}^{98} \frac{1}{n(n+1)(n+2)}$

17 $\sum_{n=1}^{4} \frac{n}{3n-1}$ **18** $\sum_{n=1}^{4} \frac{n}{5n-1}$

19 $\sum_{n=1}^{5} (-1)^{n+1}(105 - 5n)$ **20** $\sum_{n=1}^{7} (-1)^{n-1} \frac{1}{n}$

21 $\sum_{n=0}^{25} a_n x^{4n}$ **22** $\sum_{n=0}^{20} a_n x^{3n}$ **23** $1 + \sum_{k=1}^{n} (-1)^k \frac{x^{2k}}{2k}$

24 $1 + \sum_{k=1}^{n} \frac{x^k}{k}$ **25** $-5 - 8\sqrt{3}; -5 - 35\sqrt{3}$

26 52 **27** $-31; 50$ **28** 12

29 $20, 14, 8, 2, -4, -10$ **30** 64 **31** -0.00003

32 1562.5 or -1562.5 **33** $4\sqrt{2}$ **34** $-\frac{12,800}{2187}$

35 17; 3 **36** $\frac{1}{81}; \frac{211}{1296}$ **37** 570 **38** 32.5

39 2041 **40** -506 **41** $\frac{5}{7}$ **42** $\frac{6268}{999}$

43 (1) P_1 is true, since $3(1) - 1 = \dfrac{1[3(1) + 1]}{2} = 2.$

(2) Assume P_k is true:

$$2 + 5 + 8 + \cdots + (3k - 1) = \frac{k(3k + 1)}{2}.$$

Hence,

$$2 + 5 + 8 + \cdots + (3k - 1) + 3(k + 1) - 1$$
$$= \frac{k(3k + 1)}{2} + 3(k + 1) - 1$$
$$= \frac{3k^2 + k + 6k + 4}{2}$$
$$= \frac{3k^2 + 7k + 4}{2}$$
$$= \frac{(k + 1)(3k + 4)}{2}$$
$$= \frac{(k + 1)[3(k + 1) + 1]}{2}.$$

Thus, P_{k+1} is true, and the proof is complete.

44 (1) P_1 is true, since $[2(1)]^2 = \dfrac{[2(1)][2(1) + 1][1 + 1]}{3} = 4.$

(2) Assume P_k is true:

$$2^2 + 4^2 + 6^2 + \cdots + (2k)^2 = \frac{(2k)(2k + 1)(k + 1)}{3}.$$

Hence,

$$2^2 + 4^2 + 6^2 + \cdots + (2k)^2 + [2(k + 1)]^2$$
$$= \frac{(2k)(2k + 1)(k + 1)}{3} + [2(k + 1)]^2$$
$$= (k + 1)\left(\frac{4k^2 + 2k}{3} + \frac{12(k + 1)}{3}\right)$$
$$= \frac{(k + 1)(4k^2 + 14k + 12)}{3}$$
$$= \frac{2(k + 1)(2k + 3)(k + 2)}{3}.$$

Thus, P_{k+1} is true, and the proof is complete.

45 (1) P_1 is true, since $\dfrac{1}{[2(1) - 1][2(1) + 1]} = \dfrac{1}{2(1) + 1} = \dfrac{1}{3}.$

(2) Assume P_k is true:

$$\frac{1}{1 \cdot 3} + \frac{1}{3 \cdot 5} + \frac{1}{5 \cdot 7} + \cdots + \frac{1}{(2k - 1)(2k + 1)} = \frac{k}{2k + 1}.$$

Hence,

$$\frac{1}{1 \cdot 3} + \frac{1}{3 \cdot 5} + \frac{1}{5 \cdot 7} + \cdots + \frac{1}{(2k - 1)(2k + 1)}$$
$$+ \frac{1}{(2k + 1)(2k + 3)} = \frac{k}{2k + 1} + \frac{1}{(2k + 1)(2k + 3)}$$
$$= \frac{k(2k + 3) + 1}{(2k + 1)(2k + 3)}$$
$$= \frac{2k^2 + 3k + 1}{(2k + 1)(2k + 3)}$$
$$= \frac{(2k + 1)(k + 1)}{(2k + 1)(2k + 3)}$$
$$= \frac{k + 1}{2(k + 1) + 1}.$$

Thus, P_{k+1} is true, and the proof is complete.

46 (1) P_1 is true, since $1(1 + 1) = \dfrac{(1)(1 + 1)(1 + 2)}{3} = 2.$

(2) Assume P_k is true:

$$1 \cdot 2 + 2 \cdot 3 + 3 \cdot 4 + \cdots + k(k + 1)$$
$$= \frac{k(k + 1)(k + 2)}{3}.$$

Hence,

$$1 \cdot 2 + 2 \cdot 3 + 3 \cdot 4 + \cdots + k(k + 1)$$
$$+ (k + 1)(k + 2)$$
$$= \frac{k(k + 1)(k + 2)}{3} + (k + 1)(k + 2)$$
$$= (k + 1)(k + 2)\left(\frac{k}{3} + 1\right)$$
$$= \frac{(k + 1)(k + 2)(k + 3)}{3}.$$

Thus, P_{k+1} is true, and the proof is complete.

47 (1) For $n = 1$, $n^3 + 2n = 3$ and 3 is a factor of 3.

(2) Assume 3 is a factor of $k^3 + 2k$. The $(k + 1)$st term is
$$(k + 1)^3 + 2(k + 1) = k^3 + 3k^2 + 5k + 3$$
$$= (k^3 + 2k) + (3k^2 + 3k + 3)$$
$$= (k^3 + 2k) + 3(k^2 + k + 1).$$

By the induction hypothesis, 3 is a factor of $k^3 + 2k$ and 3 is a factor of $3(k^2 + k + 1)$, so 3 is a factor of the $(k + 1)$st term.

Thus, P_{k+1} is true, and the proof is complete.

48 (1) P_3 is true, since $5^2 + 3 < 2^5$.
   (2) Assume P_k is true: $k^2 + 3 < 2^k$. Hence,
$$(k + 1)^2 + 3 = k^2 + 2k + 4$$
$$= (k^2 + 3) + (k + 1)$$
$$< 2^k + (k + 1)$$
$$< 2^k + 2^k$$
$$= 2 \cdot 2^k = 2^{k+1}$$
   Thus, P_{k+1} is true, and the proof is complete.

49 (1) P_4 is true, since $2^4 \le 4!$.
   (2) Assume P_k is true: $2^k \le k!$. Hence,
$$2^{k+1} = 2 \cdot 2^k \le 2 \cdot k! < (k + 1) \cdot k! = (k + 1)!.$$
   Thus, P_{k+1} is true, and the proof is complete.

50 (1) P_{10} is true, since $10^{10} \le 10^{10}$.
   (2) Assume P_k is true: $10^k \le k^k$. Hence,
$$10^{k+1} = 10 \cdot 10^k \le 10 \cdot k^k < (k + 1) \cdot k^k$$
$$< (k + 1) \cdot (k + 1)^k = (k + 1)^{k+1}.$$
   Thus, P_{k+1} is true, and the proof is complete.

51 $x^{12} - 18x^{10}y + 135x^8y^2 - 540x^6y^3 + 1215x^4y^4$
$$- 1458x^2y^5 + 729y^6$$

52 $16x^4 + 32x^3y^3 + 24x^2y^6 + 8xy^9 + y^{12}$

53 $x^8 + 40x^7 + 760x^6$ **54** $-\dfrac{63}{16}y^{12}c^{10}$

55 $21{,}504x^{10}y^2$ **56** $52{,}500{,}000$

57 (a) $d = 1 - \dfrac{1}{2}a_1$ (b) In ft: $1\dfrac{1}{4}, 2, 2\dfrac{3}{4}, 3\dfrac{1}{2}$

58 24 ft   **59** $\dfrac{2}{1 - f}$ **60** $P(10, 10) = 3{,}628{,}800$

61 (a) $P(52, 13) \approx 3.954 \times 10^{21}$
   (b) $P(13, 5) \cdot P(13, 3) \cdot P(13, 3) \cdot P(13, 2)$
$$\approx 7.094 \times 10^{13}$$

62 (a) $P(6, 4) = 360$ (b) $6^4 = 1296$

63 (a) $C(12, 8) = 495$ (b) $C(9, 5) = 126$

64 $\dfrac{17!}{6! \, 5! \, 4! \, 2!} = 85{,}765{,}680$ **65** 5 to 8; $\dfrac{8}{13}$

66 (a) $\dfrac{2}{4}$ (b) $\dfrac{2}{8}$

67 (a) $\dfrac{P(26, 4) \cdot 2}{P(52, 4)} \approx 0.1104$ (b) $\dfrac{26^2 \cdot 25^2}{P(52, 4)} \approx 0.0650$

68 (a) $\dfrac{1}{1000}$ (b) $\dfrac{10}{1000}$ (c) $\dfrac{50}{1000}$

69 $\dfrac{C(4, 1)}{2^4} = \dfrac{4}{16}$; 1 to 3

70 (a) $\dfrac{C(6, 4) + C(6, 5) + C(6, 6)}{2^6} = \dfrac{22}{64}$
   (b) $1 - \dfrac{22}{64} = \dfrac{42}{64}$

71 (a) $\dfrac{1}{312}$ (b) $\dfrac{57}{312}$ **72** 0.44   **73** $\dfrac{8}{36}$

74 5.8125

CHAPTER 10 DISCUSSION EXERCISES

1 $a_n = 2n + \dfrac{1}{24}(n - 1)(n - 2)(n - 3)(n - 4)(a - 10)$
   (The answer is not unique.)

2 $\ge$; $j = 94$

3 Examine the number of digits in the exponent of the value in scientific notation.

4 The $(k + 1)$st coefficient $(k = 0, 1, 2, \ldots, n)$ of the expansion of $(a + b)^n$, namely $\dbinom{n}{k}$, is the same as the number of k-element subsets of an n-element set.

5 4.61   **6** $5.33

7 Penny amounts:
   $237.37 $215.63 $195.89 $177.95 $161.65
   $146.85 $133.40 $121.18 $110.08 $100.00
   Realistic ten dollar amounts:
   $240.00 $220.00 $200.00 $180.00 $160.00
   $140.00 $130.00 $120.00 $110.00 $100.00

8 11 toppings are available.

9 (a) $\dfrac{1}{80{,}089{,}128}$ (b) $\dfrac{2{,}303{,}805}{80{,}089{,}128}$ (about 1 in 35)
   (c) $\dfrac{16{,}663{,}144}{80{,}089{,}128} \approx 0.21$ (d) $63{,}425{,}984

10 0.43   **11** $0^0 = 1$ **12** The sum equals π.

13 (a) $\tan 5x = \dfrac{5 \tan x - 10 \tan^3 x + \tan^5 x}{1 - 10 \tan^2 x + 5 \tan^4 x}$
   (b) $\cos 5x = 1 \cos^5 x - 10 \cos^3 x \sin^2 x + 5 \cos x \sin^4 x$;
   $\sin 5x = 5 \cos^4 x \sin x - 10 \cos^2 x \sin^3 x + 1 \sin^5 x$

Chapter 11

EXERCISES 11.1

1 $V(0, 0)$; $F(0, 2)$; $y = -2$ **3** $V(0, 0)$; $F\left(-\dfrac{3}{8}, 0\right)$;

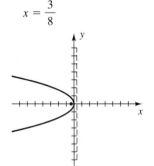

5 $V(-2, 1)$; $F(-2, -1)$;
$y = 3$

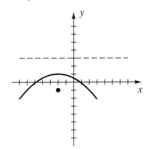

7 $V(3, 2)$; $F\left(\dfrac{49}{16}, 2\right)$;
$x = \dfrac{47}{16}$

45 $\dfrac{9}{16}$ ft from the center of the paraboloid

47 $2\sqrt{480} \approx 43.82$ in.

49 (a) $p = \dfrac{r^2}{4h}$ **(b)** $10\sqrt{2}$ ft   **51** $64{,}968 \text{ ft}^2$

EXERCISES 11.2

1 $V(\pm 3, 0)$; $F(\pm\sqrt{5}, 0)$

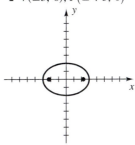

3 $V(0, \pm 4)$; $F(0, \pm 1)$

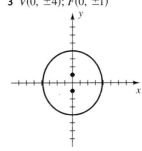

9 $V(2, -2)$; $F\left(2, -\dfrac{7}{4}\right)$;
$y = -\dfrac{9}{4}$

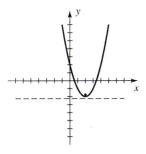

11 $V\left(0, \dfrac{1}{2}\right)$; $F\left(0, -\dfrac{9}{2}\right)$;
$y = \dfrac{11}{2}$

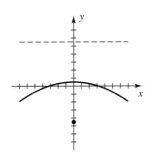

5 $V(0, \pm 4)$; $F(0, \pm 2\sqrt{3})$

7 $V\left(\pm\dfrac{1}{2}, 0\right)$;
$F\left(\pm\dfrac{1}{10}\sqrt{21}, 0\right)$

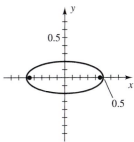

13 $y^2 = 20(x - 1)$

15 $(x + 2)^2 = -16(y - 3)$

17 $(x - 3)^2 = 6\left(y - \dfrac{1}{2}\right)$

19 $y^2 = 8x$

21 $(x - 6)^2 = 12(y - 1)$

23 $(y + 5)^2 = 4(x - 3)$

25 $y^2 = -12(x + 1)$

27 $3x^2 = -4y$

29 $(y - 5)^2 = 2(x + 3)$

31 $x^2 = 16(y - 1)$

33 $(y - 3)^2 = -8(x + 4)$

35 $y = -\sqrt{x + 3} - 1$

37 $x = \sqrt{y - 4} - 1$

39 $y = -x^2 + 2x + 5$

41 $x = y^2 - 3y + 1$

43 4 in.

9 $V(3 \pm 4, -4)$;
$F(3 \pm \sqrt{7}, -4)$

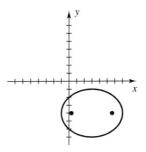

11 $V(4 \pm 3, 2)$;
$F(4 \pm \sqrt{5}, 2)$

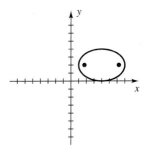

13 $V(5, 2 \pm 5)$;
$F\left(5, 2 \pm \sqrt{21}\right)$

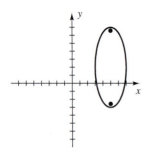

15 $\dfrac{x^2}{4} + \dfrac{y^2}{36} = 1$ **17** $\dfrac{(x+2)^2}{25} + \dfrac{(y-1)^2}{4} = 1$

19 $\dfrac{x^2}{64} + \dfrac{y^2}{39} = 1$ **21** $\dfrac{4x^2}{9} + \dfrac{y^2}{25} = 1$

23 $\dfrac{8x^2}{81} + \dfrac{y^2}{36} = 1$ **25** $\dfrac{x^2}{7} + \dfrac{y^2}{16} = 1$

27 $\dfrac{x^2}{4} + 9y^2 = 1$ **29** $\dfrac{x^2}{16} + \dfrac{4y^2}{25} = 1$

31 $(2, 2), (4, 1)$

33 $\dfrac{x^2}{25} + \dfrac{y^2}{16} = 1$ **35** $\dfrac{x^2}{64} + \dfrac{y^2}{289} = 1$

37 $\dfrac{x^2}{25} + \dfrac{y^2}{9} = 1$

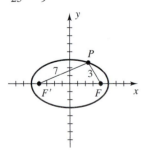

39 Upper half of $\dfrac{x^2}{49} + \dfrac{y^2}{121} = 1$

41 Left half of $x^2 + \dfrac{y^2}{9} = 1$

43 Right half of $\dfrac{(x-1)^2}{4} + \dfrac{(y+2)^2}{9} = 1$

45 Lower half of $\dfrac{(x+1)^2}{9} + \dfrac{(y-2)^2}{49} = 1$

47 $\sqrt{84} \approx 9.2$ ft   **49** 94,581,000; 91,419,000

51 (a) $d = h - \sqrt{h^2 - \dfrac{1}{4}k^2}$; $d' = h + \sqrt{h^2 - \dfrac{1}{4}k^2}$

(b) 16 cm; 2 cm from V

53 5 ft

EXERCISES 11.3

1 $V(\pm 3, 0)$; $F\left(\pm\sqrt{13}, 0\right)$;
$y = \pm\dfrac{2}{3}x$

3 $V(0, \pm 3)$; $F\left(0, \pm\sqrt{13}\right)$;
$y = \pm\dfrac{3}{2}x$

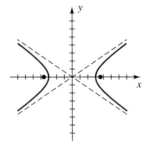

5 $V(\pm 1, 0)$; $F(\pm 5, 0)$;
$y = \pm\sqrt{24}x$

7 $V(0, \pm 4)$; $F\left(0, \pm 2\sqrt{5}\right)$;
$y = \pm 2x$

9 $V\left(\pm\dfrac{1}{4}, 0\right)$;

$F\left(\pm\dfrac{1}{12}\sqrt{13}, 0\right)$;

$y = \pm\dfrac{2}{3}x$

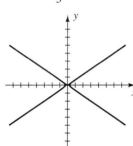

11 $V(-2, -2 \pm 3)$;

$F\left(-2, -2 \pm \sqrt{13}\right)$;

$(y + 2) = \pm\dfrac{3}{2}(x + 2)$

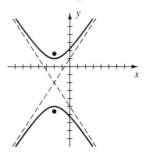

13 $V(-3 \pm 5, -2)$;

$F(-3 \pm 13, -2)$;

$(y + 2) = \pm\dfrac{12}{5}(x + 3)$

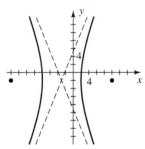

15 $V(-2, -5 \pm 3)$;

$F\left(-2, -5 \pm 3\sqrt{5}\right)$;

$(y + 5) = \pm\dfrac{1}{2}(x + 2)$

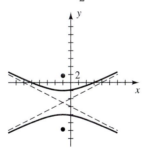

17 $\dfrac{x^2}{9} - \dfrac{y^2}{16} = 1$ **19** $(y + 3)^2 - \dfrac{(x + 2)^2}{3} = 1$

21 $y^2 - \dfrac{x^2}{15} = 1$ **23** $\dfrac{x^2}{9} - \dfrac{y^2}{16} = 1$ **25** $\dfrac{y^2}{21} - \dfrac{x^2}{4} = 1$

27 $\dfrac{x^2}{9} - \dfrac{y^2}{36} = 1$ **29** $\dfrac{x^2}{25} - \dfrac{y^2}{100} = 1$ **31** $\dfrac{y^2}{25} - \dfrac{x^2}{49} = 1$

33 Parabola with horizontal axis

35 Hyperbola **37** Circle **39** Ellipse

41 Parabola with vertical axis **43** $(0, 4)$, $\left(\dfrac{8}{3}, \dfrac{20}{3}\right)$

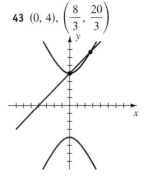

45 $\dfrac{x^2}{144} - \dfrac{y^2}{25} = 1$ **47** $\dfrac{y^2}{64} - \dfrac{x^2}{36} = 1$

49 $\dfrac{y^2}{16} - \dfrac{x^2}{9} = 1$

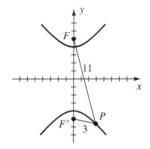

51 Right branch of $\dfrac{x^2}{25} - \dfrac{y^2}{16} = 1$

53 Upper branch of $\dfrac{y^2}{9} - \dfrac{x^2}{49} = 1$

55 Lower halves of the branches of $\dfrac{x^2}{16} - \dfrac{y^2}{81} = 1$

57 Left halves of the branches of $\dfrac{y^2}{36} - \dfrac{x^2}{16} = 1$

59 The graphs have the
same asymptotes.

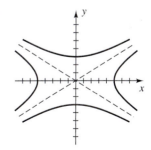

61 $x = \sqrt{9 + 4y^2}$

63 If a coordinate system similar to that in Example 6 is
introduced, then the ship's coordinates are
$\left(\dfrac{80}{3}\sqrt{34}, 100\right) \approx (155.5, 100)$.

65 (a) $(6.63 \times 10^7, 0)$ **(b)** $v > 103,600$ m/sec

EXERCISES 11.4

1 $y = 2x + 7$

3 $y = x - 2$

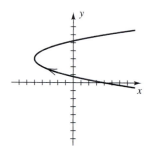

5 $(y - 3)^2 = x + 5$

7 $y = 1/x^2$

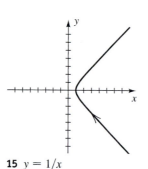

9 $\dfrac{x^2}{4} + \dfrac{y^2}{9} = 1$

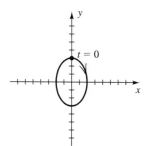

11 $x^2 - y^2 = 1$

13 $y = \ln x$

15 $y = 1/x$

17 $y = \sqrt{x^2 - 1}$

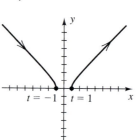

19 $y = |x - 1|$

21 $y = (x^{1/3} + 1)^2$

23 (a) The graph is a circle with center $(3, -2)$ and radius 2. Its orientation is clockwise, and it starts and ends at the point $(3, 0)$.
(b) The orientation changes to counterclockwise.
(c) The starting and ending point changes to $(3, -4)$.

25 C_1 C_2

C_3 C_4

27 (a)

(b)

(c)

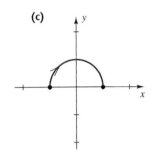

31 Answers are not unique.
  (a) (1) $x = t,$ $\quad y = t^2;$ $\qquad t \in \mathbb{R}$
  (2) $x = \tan t,$ $\quad y = \tan^2 t;$ $\qquad -\dfrac{\pi}{2} < t < \dfrac{\pi}{2}$
  (3) $x = t^3,$ $\quad y = t^6;$ $\qquad t \in \mathbb{R}$
  (b) (1) $x = e^t,$ $\quad y = e^{2t};$ $\qquad t \in \mathbb{R}$ (only gives $x > 0$)
  (2) $x = \sin t,$ $\quad y = \sin^2 t;$ $\qquad t \in \mathbb{R}$ (only gives $-1 \le x \le 1$)
  (3) $x = \tan^{-1} t,$ $\quad y = (\tan^{-1} t)^2;$ $\quad t \in \mathbb{R} \left(\text{only gives} \right.$

$$\left. -\dfrac{\pi}{2} < x < \dfrac{\pi}{2} \right)$$

33 $3200\sqrt{3};\ 2704$ **35** $15{,}488;\ 3872$
39 $x = 4b \cos t - b \cos 4t,$
$\quad y = 4b \sin t - b \sin 4t$

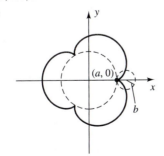

EXERCISES 11.5

1 (a), (c), (e)
3 (a) $\left(\dfrac{3}{2} \sqrt{2},\ \dfrac{3}{2} \sqrt{2} \right)$ **(b)** $\left(\dfrac{1}{2},\ -\dfrac{1}{2} \sqrt{3} \right)$
5 (a) $\left(-4,\ -4\sqrt{3} \right)$ **(b)** $\left(-\dfrac{3}{2},\ \dfrac{3}{2} \sqrt{3} \right)$
7 $\left(\dfrac{24}{5},\ \dfrac{18}{5} \right)$ **9 (a)** $\left(\sqrt{2},\ \dfrac{3\pi}{4} \right)$ **(b)** $\left(4,\ \dfrac{7\pi}{6} \right)$
11 (a) $\left(14,\ \dfrac{5\pi}{3} \right)$ **(b)** $\left(5\sqrt{2},\ \dfrac{\pi}{4} \right)$
13 $r = -3 \sec \theta$ **15** $r = 4$ **17** $r = 6 \cot \theta \csc \theta$
19 $r = \dfrac{3}{\cos \theta + \sin \theta}$ **21** $\theta = \tan^{-1} \left(-\dfrac{1}{2} \right)$
23 $r^2 = -4 \sec 2\theta$ **25** $r = 2 \cos \theta$
27 $x = 5$ **29** $x^2 + (y - 3)^2 = 9$

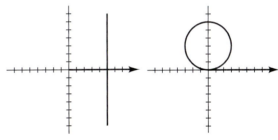

31 $y = x$ **33** $\dfrac{y^2}{9} - \dfrac{x^2}{4} = 1$

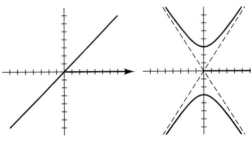

35 $x^2 - y^2 = 1$ **37** $y - 2x = 6$

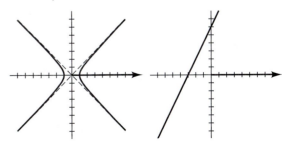

39 $y = -x^2 + 1$

57

59

41 $(x + 1)^2 + (y - 4)^2 = 17$ **43** $y^2 = \dfrac{x^4}{1 - x^2}$

61

63

45

47

65

67

49

51

69

71

53

55

73

75

77

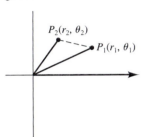

79 Let $P_1(r_1, \theta_1)$ and $P_2(r_2, \theta_2)$ be points in an $r\theta$-plane. Let $a = r_1$, $b = r_2$, $c = d(P_1, P_2)$, and $\gamma = \theta_2 - \theta_1$. Substituting into the law of cosines, $c^2 = a^2 + b^2 - 2ab \cos \gamma$, gives us the formula.

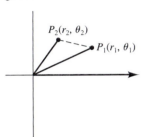

EXERCISES 11.6

1 $\dfrac{1}{3}$, ellipse

3 3, hyperbola

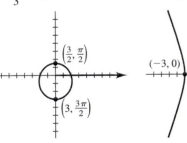

5 1, parabola

7 $\dfrac{1}{2}$, ellipse

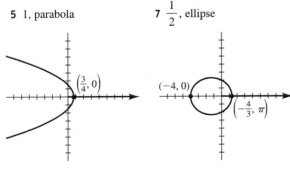

9 $\dfrac{3}{2}$, hyperbola

11 1, parabola

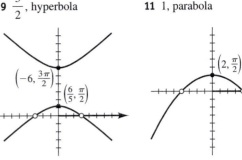

13 $9x^2 + 8y^2 + 12y - 36 = 0$

15 $8x^2 - y^2 + 36x + 36 = 0$

17 $4y^2 + 12x - 9 = 0$ **19** $3x^2 + 4y^2 + 8x - 16 = 0$

21 $4x^2 - 5y^2 + 36y - 36 = 0$; $x \neq \pm 3$

23 $x^2 + 8y - 16 = 0$; $x \neq \pm 4$ **25** $r = \dfrac{2}{3 + \cos \theta}$

27 $r = \dfrac{12}{3 - 4 \cos \theta}$ **29** $r = \dfrac{2}{1 - \sin \theta}$

31 $r = \dfrac{8}{5 + 2 \sin \theta}$ **33** $r = \dfrac{8}{1 + \sin \theta}$

35 (a) $\dfrac{3}{4}$ (b) $r = \dfrac{7}{4 - 3 \sin \theta}$

39 $e = \dfrac{r_{\text{aph}} - r_{\text{per}}}{r_{\text{aph}} + r_{\text{per}}}$, $a = \dfrac{r_{\text{aph}} + r_{\text{per}}}{2}$

CHAPTER 11 REVIEW EXERCISES

1 $V(0, 0)$; $F(16, 0)$ **2** $V(-2, 1)$; $F\left(-2, \dfrac{33}{32}\right)$

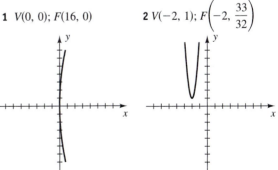

3 $V(0, \pm 4)$; $F\left(0, \pm\sqrt{7}\right)$

4 $V(0, \pm 4)$; $F(0, \pm 5)$

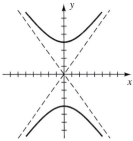

9 $V(-4 \pm 1, 5)$;
$\quad F\left(-4 \pm \dfrac{1}{3}\sqrt{10}, 5\right)$

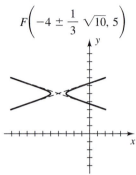

10 $V(-5, -2)$;
$\quad F\left(-\dfrac{39}{8}, -2\right)$

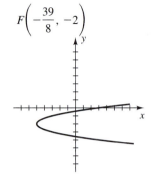

5 $V(\pm 2, 0)$; $F\left(\pm 2\sqrt{2}, 0\right)$

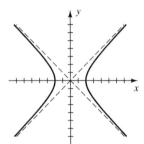

6 $V\left(\pm\dfrac{1}{5}, 0\right)$;
$\quad F\left(\pm\dfrac{1}{30}\sqrt{11}, 0\right)$

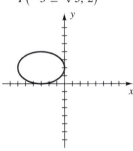

11 $V(-3 \pm 3, 2)$;
$\quad F\left(-3 \pm \sqrt{5}, 2\right)$

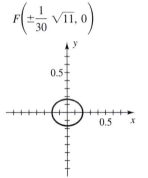

12 $V(5, -4 \pm 2)$;
$\quad F\left(5, -4 \pm \sqrt{5}\right)$

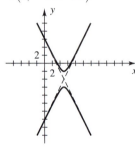

7 $V(0, 4)$; $F\left(0, -\dfrac{9}{4}\right)$

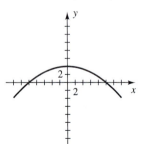

8 $V(3 \pm 2, -1)$;
$\quad F(3 \pm 1, -1)$

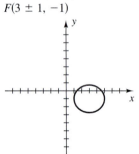

13 $V(2, -4)$; $F(4, -4)$

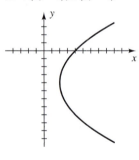

14 $V(3, -2 \pm 2)$;
$\quad F\left(3, -2 \pm \sqrt{3}\right)$

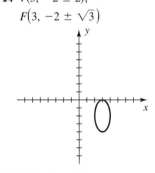

15 $V(-4 \pm 3, 0)$;
$\quad F\left(-4 \pm \sqrt{10}, 0\right)$

16 $V(2, -3 \pm 2)$;
$\quad F\left(2, -3 \pm \sqrt{6}\right)$

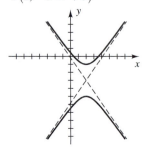

17 $y = 2(x + 7)^2 - 18$ **18** $y = -3(x + 4)^2 + 147$

19 $\dfrac{y^2}{49} - \dfrac{x^2}{9} = 1$ **20** $y^2 = -16x$ **21** $x^2 = -40y$

22 $x = 5y^2$ **23** $\dfrac{x^2}{75} + \dfrac{y^2}{100} = 1$ **24** $\dfrac{x^2}{25} - \dfrac{y^2}{75} = 1$

25 $\dfrac{y^2}{36} - \dfrac{x^2}{\frac{4}{9}} = 1$ **26** $\dfrac{x^2}{8} + \dfrac{y^2}{4} = 1$ **27** $\dfrac{x^2}{25} + \dfrac{y^2}{45} = 1$

28 $\dfrac{x^2}{256} + \dfrac{y^2}{112} = 1$ **29 (a)** $-\dfrac{7}{2}$ **(b)** Hyperbola

30 $A = \dfrac{4a^2b^2}{a^2 + b^2}$ **31** $x^2 + (y - 2)^2 = 4$

32 $2\sqrt{2}$ rad/sec ≈ 0.45 rev/sec

33 $x = 4y + 7$

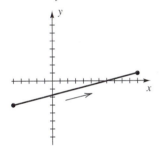

34 $y = x^4 - 4$

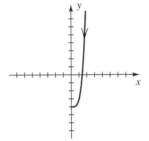

35 $(y - 1)^2 = -(x + 1)$

36 $y = 2^{-x^2}$

37 $y = \dfrac{2x^2 - 4x + 1}{x - 1}$

38 C_1

C_2

C_3

C_4

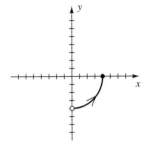

39 $20{,}480\sqrt{3}$; 9216 **40** $\left(-2, \dfrac{5\pi}{4}\right), \left(2, \dfrac{9\pi}{4}\right)$

41 $\left(\dfrac{5}{2}\sqrt{2}, -\dfrac{5}{2}\sqrt{2}\right)$ **42** $\left(4, \dfrac{11\pi}{6}\right)$

43 $r = 4 \cot \theta \csc \theta$ **44** $r = 3 \cos \theta - 4 \sin \theta$

45 $r(2 \cos \theta - 3 \sin \theta) = 8$ **46** $\theta = \dfrac{\pi}{4}$

47 $x^3 + xy^2 = y$ **48** $x^2 + y^2 = 2x + 3y$

49 $(x^2 + y^2)^2 = 8xy$ **50** $y = \left(\tan \sqrt{3}\right) x$

51 $8x^2 + 9y^2 + 10x - 25 = 0$ **52** $y^2 = 6 + x$

53

54

55

56

57

58

59

60

61

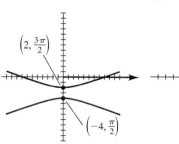

$\left(2, \frac{3\pi}{2}\right)$

$\left(-4, \frac{\pi}{2}\right)$

62

(3, 0)

63 $\frac{2}{3}$, ellipse

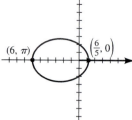

$(6, \pi)$ $\left(\frac{6}{5}, 0\right)$

64 $\frac{1}{2}$, ellipse

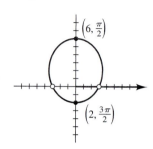

$\left(6, \frac{\pi}{2}\right)$

$\left(2, \frac{3\pi}{2}\right)$

CHAPTER 11 DISCUSSION EXERCISES

1 $w = 4|p|$
2 The circle goes through both foci and all four vertices of the auxiliary rectangle.
5 $\dfrac{(x-2)^2}{1} - \dfrac{y^2}{3} = 1, x \geq 3,$

or $x = 2 + \sqrt{1 + \dfrac{y^2}{3}}$

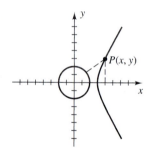

$P(x, y)$

6 $d = \dfrac{1}{4\sqrt{a^2 + b^2}}$ **7** 43.12°

9 $y = \pm\sqrt{\dfrac{1 \pm \sqrt{1 - x^2}}{2}}$

10 The graph of $r = f(\theta - \alpha)$ is the graph of $r = f(\theta)$ rotated counterclockwise through an angle α, whereas the graph of $r = f(\theta + \alpha)$ is rotated clockwise.
11 $(180/n)°$
12 $y = 2 \pm \sqrt{4 - x^2}, y = \pm\sqrt{4 - (x - 2)^2}$

INDEX OF APPLICATIONS

AGRICULTURE

Crop acreage, 577, 597
Crop growth, 306
Fencing, 71, 86, 194, 572
Fertilizer mixtures, 612
Grain storage structures, 68–69, 72, 246, 669–670
Orchard yield, 122
Watering fields, 580, 650

ANIMALS/WILDLIFE

Birds, 613, 652
Bobcats, 580
Cattle, 15
Competition for food, 572
Deer, 222, 274, 651
Elephants, 322
Elk, 296
Fish, 15, 26, 118, 264, 272, 307, 343, 344, 572, 598
Frogs, 194
German shepherds, 118
Livestock diet, 581
Moose, 598
Rabbits, 456
Whales, 26, 153, 307

ARTS AND ENTERTAINMENT

Backgammon moves, 725
Bridge hands, 709
Cards, 701, 705, 711–712, 713, 714, 724
Carnival game, 720–721
Coin toss, 710, 725
Compact disc rotation, 357
Craps, 722
Die toss, 700, 711, 712, 713, 720
DVD player costs, 663
Gambling survival formula, 275
Horserace results, 701
Human cannonball flight, 195
Lotteries, 708, 718, 721–722, 725
Maze dimensions, 670
Motorcycle daredevil's jump, 271
Movie attendance, 70
Movie frames, 15
Odds for dice, 714–715
Penrose tiles, 521
Phonograph record rotation, 433
Poker hands, 705, 709
Poster design, 86
Powerball, 726
Prize money, 670, 722, 726
Projection unit mounting, 438
Raffles, 725
Roulette, 721, 722
Slot machines, 720, 725–726
Stadium seating, 669
Ticket sales, 580, 589
Tightrope length, 170, 204
VCR taping, 582
Video games, 154–155, 428–429
Viewing angles for paintings, 495
Violin string vibration, 481

ASTRONOMY

Bode's sequence, 663
Brightness of stars, 322
Cassegrain telescope design, 761, 794
Comet's path, 745, 761
Earth
    distance to Venus from, 558
    orbit of, 749, 792
    radius of, 430
    rotation of, 357
Gravity simulation, 534
Jodrell Bank radio telescope, 737
Kepler's laws, 271, 791–792
Light year, 15
Mercury's orbit, 749
Milky Way galaxy, 15
Moon phases, 373
Period of a planet, 271
Reflecting telescope's mirror, 736
Satellite
    path of, 121, 737
    view from, 498
Telescope resolution, 373
Venus
    distance from sun of, 436
    elongation of, 429

BIOLOGY

Bacterial growth, 292, 297, 309, 343, 663, 678
Cell membrane dimensions, 121
Genetic mutation, 340
Genetic sequence, 670
Population growth curve, 338

BUSINESS/FINANCE

Annuities, 679
Apartment rentals, 122, 196
Appreciation, 208, 298
ATM codes, 701
Author's royalty rates, 184
Ben Franklin's legacy, 343
Billing for service, 70, 211
Business expenses, 111, 154
Compound interest, 293–295, 297, 299–300, 302, 303, 321, 340, 343, 678, 679
Consumer Price Index, 299
Daily savings, 673
Depreciation, 298, 677, 680
Effective yield, 307
Electricity rates, 184
Income tax rates, 158–159, 178, 184
Inflation comparisons, 299
Inventory levels, 589
Inventory value, 622–623
Investments, 64, 120, 580, 589, 620–621, 651
Land values, 307
Loans, 154, 299
Maximizing profit, 593, 597, 598, 609–610, 651
Minimizing cost, 594, 597, 598
Minimum wage growth, 295, 307
Mortgage payment, 298
Multiplier effect, 678
Office assignments, 708
Pareto's law for capitalist countries, 330
Payroll, 70, 204, 650
Price and demand, 87, 101, 328, 330, 582
Price calculations, 63

Property tax rates, 184
Quantity discount, 87, 196
Restaurant bills, 70
Salaries, 121, 346
Sales bonuses, 670
Savings accounts, 26, 70
Services swap, 581
Simple interest, 63–64
Starting work times, 721

CHEMISTRY

Acid solution, 65, 612
Avogadro's number, 15
Bactericide, 120
Copper-silver alloy, 70
Ideal gas law, 271
Mass
    of an electron, 15
    of a hydrogen atom, 15
pH calculation, 339
Salt dissolved in water, 297
Silver alloy, 581
Threshold response curve, 274

COMMUNICATION

Cable TV fees, 196
CB antenna length, 430
Communications satellite, 431
Express-mail rates, 651
First-class stamp rates, 299
Newspaper delivery, 73
Radio stations
    broadcasting ranges of, 140, 419, 420
    call letters of, 701
    number of, 287
Semaphore, 701
Telephone calls
    number of, 274
    rates for, 581
Telephone numbers, 701
Television transmitting tower, 429
Two-way radio range, 71, 86, 559

CONSTRUCTION

Bath, 121
Boxes, 82–83, 85, 86, 168, 221, 247
Buildings, 169, 623
Cages, 194
Conference table, 580

Elastic cylinder, 274
House, 72
Insulation, 70
Kennel, 121
Ladder, 101, 430, 667–668, 724, 740
Rain gutter, 190–191, 476
Ramp, 428
Sidewalk, 85
Solar heating panel, 120
Storage shelter, 208
Storage units, 598
Tent, 248
Tubing, 571
Wheelchair ramp, 208
Whispering gallery, 749
Window, 72, 589
Wire frame, 86, 196
Withdrawal resistance of nails, 101
Wooden brace placement, 495

EDUCATION

Class officers, 697
College budgets, 652
Committees, 707, 720
Fraternity designations, 701
Grade point average, 73, 265
Multiple-choice test, 701
Scheduling courses, 701
Scholarship selection, 708
Seat arrangement, 701
Teachers' eligibility for retirement, 120
Test question selection, 724
Test score average, 62, 70
True-or-false test, 701, 708, 719, 725

ELECTRICITY

Circuits and voltage, 51
Confocal parabolas, 736
Coulomb's law, 85, 271
Current in an electrical circuit, 321, 344, 454, 457, 552, 613, 716
Electrical condenser, 321
Electrical resistance, 270, 612
Electrical switches, 716
Heat production in an AC circuit, 476
Lissajous figure, 795
Ohm's law, 73, 111, 262
Resistors connected in parallel, 59, 111
Spotlight illuminance, 437

Voltage, 466, 552
Wind rotor power, 274
Windmill power, 101

ENGINEERING

Beams
    deflection of, 274
    strength of, 229
    support load of, 269
Bridges
    arches of, 748
    specifications for, 195
Crest vertical curves, 196
Cycloid, 768–769, 772
Drainage ditch, 72
Drawbridge, 428
Elliptical reflector, 749
Flexible cables, 304
Highways, 86, 195, 196, 456, 476
Jet fighter, 511–512, 560
Legendre polynomial, 221
Parabolic reflector, 736
Sag vertical curves, 196
Satellite TV antennas, 734, 736
Solar collector, 510
Sound receiving dish, 736
Stonehenge, 373, 536
Tunnel, 436
Water slide, 428

ENVIRONMENTAL SCIENCE

Atmospheric density, 85
Atmospheric pressure, 307, 344
Cable corrosion, 205
Cloud height, 73
Coal reserve depletion, 322
Daylight
    intensity of, 411, 412, 545
    length of, 409–410, 453–454
Earthquakes, 315, 321, 344, 346, 438, 494, 521
    frequency of, 344
Fires, 204, 432, 510
Forest growth, 590
Grassland growth, 590
Meteorological calculations, 400, 427
Ocean
    penetration of light in, 297, 336
    photic zone of, 340

salinity of, 153
tides of, 412
Oil-spill clean-up cost, 274
Ozone layer, 194, 211, 330, 341
Precipitation, 211, 264
Radioactive contamination, 170
Radon concentration, 272
River flow rate, 456
Sun's rays, 400, 456, 545
Temperature
and altitude, 73, 150–151
in a cloud, 73, 122
determination of, 221, 241, 274
in Fairbanks, 412, 456
and humidity, 392
in Ottawa, 439
in Paris, 154, 581
and precipitation, 590
variation in, 412
Tidal waves, 455
Tornados, 15, 357
Tree height, 340, 373
Tsunamis, 433
Urban heat island, 101, 154
Ventilation requirements, 287
Vertical wind shear, 155, 341
Water pollution, 330, 535
Wind velocity, 330, 530

FOOD

Cheese production, 154
Coffee bean mixtures, 613
Condiment selections, 708
Dietary planning, 598
Hospital food preparation, 120
Ice cream cone dimensions, 72
Ice cream selections, 708
Lunch possibilities, 701
Nut mixtures, 581, 611
Pizza toppings, 726
Pizza values, 358

GENERAL INTEREST

Aquariums, 121, 169, 568–569, 572
Balloons, 102, 201, 204
Book arrangement, 701
Chain weights, 613
Computer functions, 204, 545–546
Eiffel Tower, 435

Family makeup, 707, 720
Fibonacci sequence, 663
Flashlight mirror, 736
Forces on a Christmas ornament, 561
Garden fencing, 86
Genealogy, 678
Glottochronology (language dating), 298, 345
Great Pyramid, 436
Height
of a building, 430, 431, 437, 511, 558
of a flagpole, 362
of a kite, 427, 431
of a sign, 373
of a tower, 422–423, 430, 431
Key-ring arrangement, 707
Lawn-mowing rates, 72
Leaning Tower of Pisa, 510–511
Length
of a cable, 427, 515
of a telephone pole, 506–507, 510
Mine shaft rescue, 559
Pentagon, 429
Pools
chlorine levels in, 663
dimensions of, 121, 589
filling of, 73, 209
water in, 612
Reconnaissance sightings, 520
Sand piles, 101, 204
Searchlight reflector, 736
Stacks of logs, 669
Surveying, 427, 431, 438, 509, 510, 512, 520, 559
Wardrobe mix 'n' match, 700

GEOMETRY

Angles
of a box, 429, 520
of a triangle, 520
Area
of a conical cup, 102
of a parallelogram, 521
of a rectangle in a parabolic arch, 229
of a triangle, 170, 516–517, 518, 522
Center of a circle, 139
Depth of a conical cup, 436
Diagonal
of a cube, 204
of a parallelogram, 515, 521

Isoperimetric problem, 572
Moire pattern, 572
Sierpinski sieve, 679
Surface area
of a pyramid, 438
of a tank, 51
Volume
of a conical cup, 429–430
of a prism, 511
of a pyramid, 438

HEALTH/MEDICINE

Arsenic exposure and cancer, 721
Arterial bifurcation, 476
Bicep muscle, 533
Biorhythms, 411–412
Bone-height relationship, 73
Breathing analysis, 408
Calorie requirements, 40
Childhood growth, 154, 169, 307, 344
Children's weight, 321
Cholesterol levels, 323
Circadian rhythms, 439
Decreasing height, 111
Drugs
in the bloodstream, 297, 340, 678
concentration of, 71
dosage of, 154, 264
Electroencephalography, 411
Eye drops, 70
Fetal growth, 153
Force of a foot, 497
Glucose solution, 70
Heart action, 411
Heartbeats, 51
Infant growth, 194
Intelligence quotient, 70
Lifetime growth, 102
Limb dimensions, 271
Lithotripter operation, 746, 749
Memory, 341
Men's weight, 27
Minimum therapeutic level, 111, 116
Pill dimensions, 87, 229, 572
Poiseuille's law, 272
Protection from sunlight, 456
Radioactive tracer, 306
Radiotherapy, 304–305
Reaction to a stimulus, 330

Red blood cells, 51
Skin cancer, 342
Smoking deaths, 721
Surface area of the body, 26–27, 51
Threshold response curve, 274
Threshold weight, 271
Visual distinction, 498
Walking rates, 71, 322
Women's weight, 27

OPERATIONS MANAGEMENT

Adding salt water to a tank, 264
Cargo winch rotation, 358
Commercial fishing, 507–508
Computer chip production, 323, 721
Container production, 85, 209, 264, 589
Crayon production, 580
Employee productivity, 340, 341
Emptying a water tank, 73
Filling an extrusion bin, 120
Filling a storage tank, 69
Filtering water, 209
Fitting a box through a door, 196
Hydraulic conveyor belt, 429
Machine wheel speed, 355
Oil drum production, 87
Production capability, 612
Production planning, 581, 597, 651
Robotic movement, 392, 400, 535, 536, 559

PHYSICS/GENERAL SCIENCE

Acceleration
    of a ball, 581
    of a particle, 613
Adiabatic expansion, 51
Air pressure, 321
Alpha particle paths, 760
Astronaut's weight in space, 118
Boyle's law, 121
Carbon 14 dating, 345
Coulomb's law, 85, 271
Distance
    to a hot-air balloon, 170, 204, 430, 431, 510
    from lens to image, 107
    between points on Earth, 357, 373, 521, 558
    to a target, 71

traveled by a bouncing ball, 676, 678, 725
traveled by a falling object, 670
viewing, 26
Earth
    magnetic field of, 420
    surface area of, 70
Electron energy, 330
Elevation
    of a bluff, 438
    of a mountain, 431, 437, 438, 509
Energy-releasing events, 346
Flight of a projectile, 194, 766–768, 795
Flow rates, 580, 651
Gas pressure-volume proportionality, 267–268
Height
    of a cliff, 87
    of a projectile, 118, 194
    of a toy rocket, 73–74, 83, 210, 428
Hooke's law, 111, 270
Ideal gas law, 271
Illumination intensity, 270
Legendre polynomial, 221
Linear magnification, 111
Liquid pressure, 270
Lorentz contraction formula, 118
Magnetic pole drift, 358
Motion
    harmonic, 426, 432–433
    of a mass, 466
Newton's law of cooling, 297, 316
Nuclear explosion, 87
Particle stopping distance, 307
Path of a ball, 573
Pendulum swing, 358
Period of a pendulum, 121, 271, 678
Radioactive decay, 272, 293, 297, 307, 317, 321, 322, 343, 344
Range of a projectile, 271, 475, 795
Shadows, 546
Sound intensity, 321, 344, 467, 482
Speed
    of a particle, 118
    of sound, 121
Sun's elevation, 428
Temperature
    of boiling water, 85
    scales for, 58, 110, 155, 208
Vacuum pump, 677

Vapor pressure, 322
Velocity
    of a gas, 85
    laser measure of, 435
    of a rocket, 344
Vertical projection, 85, 581
Volume and decibels, 329
Water
    cooling of, 341
    cork in, 439
    vaporizing of, 101, 154
Weight-latitude relationship, 457
Work done
    in pulling, 533, 545
    in pushing, 542–545

POLITICS/URBAN ISSUES

City expansion, 86, 120
Cube rule, 102
Municipal funding, 70
Population density, 118, 264, 322
Population growth, 121, 303, 306, 316, 321, 345
Power plant location, 589
Skyline ordinance, 169
Urban canyons, 498
Water demand, 439

SPORTS

Aerobic power, 154
Baseball
    batting order in, 699
    buying equipment for, 586
    choosing a team for, 704–705
    distances in, 520
    series of games in, 725
    stats on, 153
Basketball
    leaps in, 210
    series of games in, 708
    standings in, 700
Bicycling, 670
Bowling, 722
Canoeing, 86
Discus throw, 208
Diving, 221–222
Football, 533, 707
Golf, 495
Golf club costs, 663

Hammer throw, 573
Handicapping weight lifters, 26
Rowing, 71
Running, 71, 439, 507, 520
Skateboard racecourse, 558
Tennis, 708
Track dimensions, 210, 439, 650
Track rankings, 708–709

STATISTICS AND PROBABILITY

Birthday probability, 722
Card and die experiment, 720, 725
Chebyshev polynomial, 221
Color arrangement, 701, 725
ESP experiment, 720
Letter and number experiment, 720
Numerical palindromes, 701
Probability demonstration, 721
Probability density function, 307

TRANSPORTATION

Airplanes
  distance between, 86
    distance to, 510
    landing speed of, 118
    path of, 427, 428, 432, 520, 533, 760
    propeller of, 435
    speed of, 430
    and windspeed, 120, 534, 535, 557
Airport runway distances, 171, 205
Antifreeze replacement, 66
Automobiles
  braking distance of, 118
  buying of, 111
  design of, 521–522
  distance between, 520
  gas mileage of, 71, 118, 120, 194, 208
  rental charges for, 184
  speed of, 120, 271
  tires on, 358
  trade-in value of, 297
  travel times of, 67–68, 121
  window wipers of, 357
Bicycle mechanics, 358
Boats
  course of, 423–424, 507–509, 520,
    535
    fuel consumption of, 120
    propeller for, 440
    speed of, 578, 580
    tracking of, 495
Cable car route, 509–510
Destination time, 171
Ferry routes, 99
Gasoline tax, 169
High-speed train travel, 120
Horsepower, 546
License plate numbers, 700
Maximizing passenger capacity, 598
Minimizing fuel cost, 598
Railroad route, 475
Shipping charges, 651
Ships
  bearings of, 424–425, 432, 521
  distance between, 165, 210, 520
  location of, 756, 761
  velocity of, 557
Snowplow's speed, 71
Spacecraft, 170, 210
Traffic flow, 184, 613
Tugboat force, 534

INDEX

A

Abscissa, 124
Absolute value, 10, 11, 21
  equations containing, 94
  graph of an equation containing, 180
  properties of, 108
  of a real number, 547
  system of inequalities containing, 585
  of a trigonometric function, 417
Absolute value function, 172
Addition
  of matrices, 614–615
  properties of, 4
  of vectors, 525, 526, 527
  of y-coordinates, 417
Addition formulas, 457, 459, 461, 462
Additive identity, 4
Additive inverse, 4, 615
Adjacent side, 359
Algebraic equation, 55
Algebraic expression, 28–38
Algebraic function, 198
Alternating infinite series, 676
Ambiguous case, 505, 514
Amplitude
  of a complex number, 548
  of a graph, 402
  of harmonic motion, 426
  of a trigonometric function, 402, 403,
    405, 406
Angle(s), 348–355
  acute, 350, 359
  central, 350, 354
  complementary, 350, 422
  coterminal, 348, 349, 393
  definition of, 348
  degree measure of, 348
  of depression, 422, 423
  of elevation, 422–423, 506–507
  initial side of, 348
  measures of, 350–352
  negative, 348
  obtuse, 350
  positive, 348
  quadrantal, 348, 370
  radian measure of, 350
  reference, 393, 394, 395
  right, 350

  standard position of, 348
  straight, 348
  subtended, 350
  supplementary, 350
  terminal side of, 348
  trigonometric functions of, 358–372
  between vectors, 538
  vertex of, 348
Angular speed, 355
Applied problems
  equations in, 61–69
  trigonometry in, 420–426
Approximately equal to ($\approx$), 3
Approximations, 13
Arc, of a circle, 350
Arc length, 483
Arccosine function, 486
Arcsine function, 483
Arctangent function, 488
Area
  of a circular sector, 354
  of a triangle, 126, 472–473, 516
Argand plane, 546
Argument, 178
  of a complex number, 548, 550
  of a function, 156
Arithmetic mean, 667
Arithmetic sequences, 664–668
Arrangements without repetitions, 696
Associative properties, 4
Astronomical unit (AU), 745
Asymptote
  horizontal, 251
  for a hyperbola, 751
  oblique, 261–262
  vertical, 250, 388, 412–413, 415, 485
Augmented coefficient matrix, 600
Augmented matrix, 600
Auxiliary rectangle, 751
Average, 667
Axis (axes)
  conjugate, 751
  coordinate, 124
  of an ellipse, 739
  of a hyperbola, 751
  imaginary, 546
  major, 739
  minor, 739

  of a parabola, 132, 729
  polar, 772
  real, 546
  transverse, 751

B

Back substitution, 599
Bacterial growth, 292
Base, 11, 16
  of an exponential function,
    287–288
  for exponential notation, 16
  logarithmic, 308–310, 331
Bearings, 424, 425, 517
Binomial(s), 28, 686
  multiplying, 30–31
Binomial coefficients, 688
Binomial expansion, 690–692
Binomial theorem, 686–693
Bounds for zeros, 236–238
Branches
  of a hyperbola, 752
  of the tangent, 385

C

Calculators, 12–13
  approximating functions values with,
    362–363, 396, 397, 398, 452–453
Cancellation of common factors, 41
Cardioid, 779, 780, 781
Cartesian coordinate system, 124–128
Catenary, 304
Center
  of a circle, 137
  of an ellipse, 737
  of a hyperbola, 750
Central angle, 350, 353
Change of base formula, 331
Circle, 728
  radius and center of, 137
  standard equation of, 135–136
  unit, 136
Circular arc, 353
Circular functions, 377
Circular sector, 354
Closed, definition of, 3
Closed curve, 762
Closed interval, 103

Coefficient, 16
  binomial, 688
Coefficient matrix, 600
Cofactor, 628–629
Cofunction, 459
Cofunction formulas, 460
Column, of a matrix, 600
Column matrix, 619
Column subscript, 600
Column transformation, 635, 636
Combination, 703, 704
Common denominator, 42
Common difference, 664
Common factors, 636–637
  cancellation of, 41
Common logarithms, 313
Common ratio, 671
Commutative properties, 4
Complement, of a set, 713
Complementary angles, 350, 422
Complete factorization theorem for
  polynomials, 230–231
Completing the square, 76, 136
Complex fraction, 44
Complex number(s), 87–93
  absolute value of, 547, 549
  addition of, 88–89
  amplitude of, 548
  argument of, 548, 549, 550
  conjugate of, 90–91
  difference of, 89
  equality of, 89
  imaginary part of, 88–89
  and imaginary unit i, 88
  modulus of, 548, 550
  multiplication of, 88–89
  multiplication of by a real number, 89
  multiplicative inverse of, 91
  nth root of, 554–555
  product of, 549
  quotient of, 91, 549
  real part of, 89
  trigonometric form of, 546, 548
Complex number system, 88
Complex plane, 546
Component(s)
  of **a** along **b**, 540, 541
  of a vector, 524
Composite function, 198–202
Compound interest, 293, 294
  formulas for, 294–295, 299–300
Conclusion, 9

Conditional equation, 55
Conic sections, 728
  polar equations of, 786–791
Conics. *See* Conic sections
Conjugate
  of a complex number, 90–91
  of an expression, 45
Conjugate axis, of a hyperbola, 751
Conjugate pair zeros of a polynomial,
  241
Consistent system of equations, 576
Constant(s), 27, 28
  of proportionality, 265
  sum of, 660
  of variation, 265
Constant force, 541, 542, 543
Constant function, 161
Constant multiple, of an equation, 574
Constant polynomials, 29
Constant term, 234
Constraints, of an objective function, 590
Continued inequality, 9, 106
Continuous functions, 214
Continuously compounded interest
  formulas, 300, 302–303
Converse, 9
Convex limaçon, 781
Coordinate, 7
Coordinate axes, 124
Coordinate line, 7
Coordinate plane, 124
Coordinate system, 7, 124
Correspondence
  one-to-one, 7
  between sets, 155–156
Cosecant function, 359, 384
Cosine function, 359, 384
  addition formula for, 459
  subtraction formula for, 457, 458, 460
Cosine wave, 382
Cotangent function, 359, 384, 386, 461
Cotangent identities, 364
Coterminal angles, 348, 349, 393
Cramer's rule, 638, 639, 640
Cube root function, 160
Cube roots, 20
  of unity, 93, 555
Cubic polynomials, 214
Cubing function, 160
Curve, 762, 763
  closed, 762
  endpoints of, 762

of least descent, 770
orientation of, 764
parametric equations for, 762
parametrized, 763
plane, 762
simple closed, 762
Cusp, 769
Cycle, 382
Cycloid, 768, 769, 770

D

Damped cosine wave, 418
Damped motion, 426
Damped sine wave, 418
Damping factor, 418
Decimal, 2, 3
Decreasing function, 161, 280
Definition
  of absolute value, 11
  of absolute value of a complex
    number, 547
  of addition of vectors, 525
  of arithmetic sequence, 664
  of combination, 703
  of common logarithm, 313
  of component of **a** along **b**, 540
  of composite function, 198
  of conjugate of a complex number, 90
  of determinant of a matrix, 628, 630,
    631
  of distance between points on a
    coordinate line, 11
  of dot product, 536
  of eccentricity, 744
  of ellipse, 737
  of equality and addition of matrices,
    614
  of event, 709
  of expected value, 717
  of function, 156, 166
  of geometric sequence, 671
  of graph of a function, 159
  of horizontal asymptote, 251
  of hyperbola, 750
  of **i** and **j**, 528
  of infinite sequence, 654
  of inverse cosine function, 486
  of inverse function, 280
  of inverse of a matrix, 623
  of inverse sine function, 483
  of inverse tangent function, 488
  of linear function, 163

of logarithm, 308
of magnitude of a vector, 525
of matrix, 601
of minors and cofactors, 628
of n factorial, 688
of nth root of a number, 20
of natural exponential function, 301
of natural logarithm, 314
of negative of a vector, 527
of odds of an event, 714
of one-to-one function, 278
of parabola, 728
of parallel and orthogonal vectors, 537
of parametric equations, 762
of periodic function, 380
of permutation, 698
of plane curve, 762
of polynomial, 29
of probability of an event, 710
of product of a real number and a matrix, 616
of product of two matrices, 617
of quadratic function, 185
of radian measure, 350
of rational exponents, 23
of reference angle, 393
of scalar multiple of a vector, 526
of simple harmonic motion, 425
of slope of a line, 140
of subtraction of vectors, 528
of trigonometric functions in terms of a unit circle, 377
of trigonometric functions of an acute angle of a right triangle, 359
of trigonometric functions of any angle, 368
of trigonometric functions of real numbers, 379
of vertical asymptote, 250
of work, 543
of zero vector, 527
Degenerate conic, 728
Degree
  as an angular measurement, 348
  of a polynomial, 29
  relation of to radian, 351, 352, 353
Delta, 141
De Moivre's theorem, 552–555
Denominator, 6
  least common, 42
  rationalizing, 22, 23, 45, 91

Dependent and consistent system, 576
Dependent variable, 165
Depressed equation, 227
Descartes, René, 124
Descartes' rule of signs, 234–236
Determinants, 628–632
  properties of, 634–640
Difference
  common, 664
  of complex numbers, 89
  of functions, 197
  of matrices, 616
  of real numbers, 6
  of two cubes, 35, 93
  of two squares, 35
Difference identity, 457
Difference quotient, 163
Digits, 13
Direct variation, 265
Directed line segment, 522
Direction, 424, 425
Directrix
  of a conic, 786
  of a parabola, 728, 730
Discriminant, 78
Displacement, 523
Distance, on a coordinate line, 11
Distance formula, 124, 125, 126
Distinguishable permutations, 702
Distributive property, 4
Divisible polynomial, 222
Division
  long, 222
  of polynomials, 222
  of real numbers, 6
  synthetic, 225–227, 236–237
Division algorithm, 223
Divisors, 2
Domain
  of an algebraic expression, 28
  of a composite function, 281
  of a function, 156
  implied, 157
  of a rational function, 249
  of a trigonometric function, 388
Dot product, 536–544
Double-angle formulas, 467
Double root, 75
Double subscript notation, 600
Doubling time, 316
Dyne, 542

E

e, the number, 301
Eccentricity, 744, 786
Echelon form, of a matrix, 603–606, 634
  reduced, 606
Element
  of a matrix, 601
  of a set, 27
Elementary row transformations, 602
Ellipse, 728, 737–746, 786
  center of, 737
  eccentricity of, 744
  foci of, 737
  major axis of, 739
  minor axis of, 739
  polar equations of, 788
  reflective property of, 746
  standard equation of, 739
  vertices of, 739
Ellipsoid, 746
End behavior, 131
Endpoints
  of a curve, 762
  of an interval, 103
Equal to ($=$), 27
Equality, 54
  of complex numbers, 89
  of functions, 157
  of matrices, 614
  of polynomials, 29
  properties of, 5
  of real numbers, 2
  of sequences, 655
  of sets, 27
  of vectors, 523
Equation(s), 54–59, 94–100
  algebraic, 55
  in applied problems, 61–69
  of a circle, 136
  conditional, 55
  depressed, 227
  of an ellipse, 739
  equivalent, 54
  exponential, 290, 330–334
  graphs of, 130–138
  of a half-ellipse, 741–743
  homogeneous, system of, 608
  of a hyperbola, 752
  identity, 55
  linear, 55–56, 146, 573–579
  of lines, 144–147
  logarithmic, 308, 323–328, 334–338

with no solutions, 57
of a parabola, 187–189, 731
of a perpendicular bisector, 149–150
quadratic, 73–83, 450
quadratic type, 97–98
root of, 133
solution of, 54
systems of, 564–570
theory of, 230
trigonometric, 447–454
in *x*, 54
in *x* and *y*, 130–131
Equivalent equation, 54
Equivalent inequalities, 103, 582
Equivalent matrices, 602
Equivalent notation, 28
Equivalent systems, 567, 574
Equivalent vectors, 522
Erg, 542
Euler's formula, 548
Even function, 171
Events, 709
independent, 715
mutually exclusive, 712
Expansion of a determinant, 631
Expected value, 717
Experiment, 709
Exponent(s), 11, 16–19
irrational, 24, 288
laws of, 17–18
negative, 16, 19
rational, 23
zero, 16
Exponential decay, 289
Exponential equation, 291, 330–334
Exponential form, 308, 548
Exponential function, 287–295
natural, 299–305
Exponential law of growth, 289
Exponential notation, 11, 16, 24
Extended principle of mathematical induction, 684–685
Extraneous root, 57
Extraneous solution, 57
Extremum, 215

F

Factor, 2, 33
Factor theorem, 224
Factorial form for a permutation, 699
Factorial notation, 688–689

Factoring, 33
formulas for, 34
by grouping, 37–38
method of, 74
in solving trigonometric equations, 449, 450, 451
with quadratic formula, 80–81
by trial and error, 36, 37
Feasible solutions, 590
First term of a sequence, 654
Focus (foci), 750
of a conic, 786
of an ellipse, 737
of a hyperbola, 750
of a parabola, 728, 730
of a paraboloid, 733
Force, 541
constant, 541, 542, 543
Force vector, 523
Formula
change of base, 331
compound interest, 294, 299–300
continuously compounded interest, 302–303
distance, 124, 125, 126
factoring, 34
law of growth (or decay), 303
midpoint, 127
for negatives, 382, 383–384
product, 32
simple interest, 63
special change of base, 332
Four-leafed rose, 782
Fractional expression, 40–47
Fractions, 6
adding, 42–43
complex, 44
partial, 642–647
Frequency, in harmonic motion, 426
Function(s)
absolute value, 172
algebraic, 198
alternative definition of, 166
amplitude of, 402
circular, 377
of a complex variable, 230
composite, 198–202
constant, 161
continuous, 214
cube root, 160
cubing, 160
decreasing, 161

defined on a set, 157–158
definition of, 156
difference of, 197
domain of, 281
equality of, 157
even, 171
existence of, 158–159
exponential, 287–295
extremum of, 215
graph of, 159, 160, 171–181
greatest integer, 179
growth, 289
hyperbolic cosine, 304
hyperbolic secant, 334
identity, 161
implied domain of, 157
increasing, 161
infinite sequence as, 654
inverse, 278–284
inverse trigonometric, 396, 482–492
linear, 163–165
logarithmic, 308–318
maximum value of, 187, 190–191
minimum value of, 187, 190
natural exponential, 301–305
natural logarithmic, 314
objective, 590
odd, 171
one-to-one, 278
operations on, 197–202
periodic, 380
piecewise-defined, 177–179
polynomial, 198
product of, 197
quadratic, 185–192
quotient of, 197
range of, 281
rational, 248–262
reciprocal, 252
square root, 160
squaring, 160
sum of, 197
test values for, 216
transcendental, 198
trigonometric, 359, 384, 385, 447, 448
undefined, 158
values of, 155–158
zeros of, 160, 216
Fundamental counting principle, 696
Fundamental identities, 363, 364, 365–366, 372

Fundamental theorem
  of algebra, 230
  of arithmetic, 2

G

Gauss, Carl Friedrich, 230
General form for equation of a line, 147
Geometric mean, 674
Geometric representation, 546
Geometric sequence, 671–676
Geometric series, 674
Graph(s)
  amplitude of, 402
  common, and their equations, 798–799
  definition of, 131
  of equations, 130–138
  of exponential functions, 289–292, 304
  of functions, 159–162, 171–181
  hole in, 256, 288
  horizontal compressions of, 176–177
  horizontal shifts of, 174–175
  horizontal stretches of, 176–177
  of inequalities, 103
  of linear equations, 147
  of logarithmic equations, 327
  of logarithmic functions, 311–313
  of a parametrized curve, 763, 765
  of a plane curve, 762
  of a polar equation, 776, 778, 780–783
  of polynomial functions, 217, 218
  of rational functions, 254, 255, 257–260, 262
  reflection of, 176, 177
  of a sequence, 655, 656
  of a set of ordered pairs, 130
  of a set of real numbers, 103
  summary of transformations of, 800–801
  symmetries of, 134–135
  of a system of inequalities, 582, 584, 586–587
  of trigonometric functions, 379, 386, 388, 400–410, 412–418, 802–804
  turning points of, 215
  vertical compressions of, 175, 176
  vertical shifts of, 173–174
  vertical stretches of, 175, 176
  x-intercepts of, 472, 479–480
Greater than ($>$), 8
Greater than or equal to ($\geq$), 9
Greatest common factor (gcf), 34, 47

Greatest integer function, 179
Grouping, solving equations using, 95
Growth function, 289
Guidelines
  for finding the echelon form of a matrix, 604
  for finding an element in a matrix product, 617
  for finding inverse functions, 282
  for finding partial fraction decompositions, 643
  for the method of substitution for two equations in two variables, 565
  for sketching the graph of an inequality in x and y, 583
  for sketching the graph of a rational function, 254
  for solving applied problems, 62
  for solving an equation containing rational expressions, 57
  for solving a linear programming problem, 592
  for solving variation problems, 267
  for synthetic division, 225

H

Half-angle formulas, 470, 471, 472
Half-angle identity, 469–470
Half-ellipse, equations for, 741–743
Half-life, 293
Half-open interval, 103
Half-plane, 584
Harmonic motion, 425, 426
Hemi-ellipsoid, 746
Heron's formula, 517, 518
Hole, in a graph, 256, 288
Homogeneous system of equations, 608
Horizontal asymptote, 251, 253
Horizontal compressions of graphs, 176–177
Horizontal line, 144
Horizontal line test, 279
Horizontal shifts of graphs, 174–175
Horizontal stretches of graphs, 176–177
Hyperbola, 252, 728, 750–758, 786
  asymptotes for, 751
  auxiliary rectangle for, 751
  branches of, 752
  center of, 750
  conjugate axes of, 751
  foci of, 750
  polar equations of, 789

  reflective property of, 757
  standard equation of, 752
  transverse axis of, 751
  vertices of, 751
Hyperbolic cosine function, 304
Hyperbolic secant function, 334
Hypotenuse, 359
Hypothesis, 9
  induction, 681

I

i, the complex number, 88
i, the vector, 528, 529
Identity
  additive, 4
  cotangent, 364
  equation as, 55
  multiplicative, 4
  Pythagorean, 364, 365
  reciprocal, 360, 364
  tangent, 364
  trigonometric, verifying, 367, 442–445
Identity function, 161
Identity matrix, 623, 624
Image, 156
Imaginary axis, 546
Imaginary number, 88
Imaginary part of a complex number, 88–89
Imaginary unit, 88
Implied domain, 157
Inconsistent system of equations, 576
Increasing function, 161, 280
Indefinite interval, 103
Independent events, 715
Independent variable, 165
Index
  of a radical, 20
  of summation, 658
Induction, mathematical, 680–685
Induction hypothesis, 681
Inequalities, 8, 102
  continued, 9, 106
  equivalent, 103
  graphs of, 103, 583–587
  linear, 583, 584
  nonstrict, 9
  properties of, 104, 117
  quadratic, 112, 113
  rational, 106
  solution of, 102–109

strict, 8
systems of, 582–587
Inequality signs, 8
Infinite geometric series, 674
Infinite interval, 103
Infinite repeating decimal, 675
Infinite sequence, 654
Infinite series, 675
Infinity (∞), 103, 250
Initial point of a vector, 522
Initial side of an angle, 348
Inner product, 536
Input variable, 165
Integers, 2
Integration, 477
Intercept, of a graph, 132–133
Intercept form, of a line, 153
Interest
compound, 293, 294
compounded continuously, 300, 302
simple, 63–64
Interest period, 299
Intermediate value theorem, 215, 216
Intersection ($\cap$) of sets, 108
Intervals, 103
Inverse
additive, 4, 615
of a matrix, 623–627
multiplicative, 91
Inverse cosine function, 486–487
Inverse functions, 278–284
Inverse method, 626
Inverse sine function, 483–485
Inverse tangent function, 488
Inverse trigonometric functions, 396, 482–492, 802–803
Inverse variation, 266, 267
Inversely proportional, definition of, 265
Invertible matrix, 624
Irrational exponents, 24
Irrational number, 3
Irreducible polynomial, 33
Isosceles triangle, 361, 472–473

J

j, the vector, 528, 529
Joint variation, 268
Joule, 542

K

Kepler, Johannes, 744–745

L

Law(s)
of cosines, 512, 513–515
of exponents, 17–18
of growth (or decay), 303
of logarithms, 323–325
of radicals, 21
of signs, 9
of sines, 502, 503, 504–506
of trichotomy, 8
Leading coefficient, 29
Least common denominator (lcd), 42
Length
of a circular arc, 353
of a line segment, 11
Less than ($<$), 8
Less than or equal to ($\leq$), 9
Limaçon(s), 780, 781
Line(s), 140–151
equation of, 147
general form of, 147
horizontal, 144
intercept form of, 152
parallel, 147
parametric equation of, 766
perpendicular, 148
point-slope form of, 145
polar equation of, 776
slope of, 140–143
slope-intercept form of, 146
vertical, 144
Linear combination
of **i** and **j**, 529, 530
of rows, 611
Linear equation, 55–56, 146
in more than two variables, 598–611
in two variables, 573–579
Linear function, 163–165
Linear inequality, 583
Linear programming, 590–596
Linear programming problem, 591
Linear speed, 355
Linearly related variables, 150
Lithotripter, 746
Logarithm(s)
base of, 308
change of base of, 331
common, 313
laws of, 323–325
natural, 314
properties of, 323–328
special changes of base of, 332

Logarithmic equation, 334–338
Logarithmic form, 308
Logarithmic functions, 308–318
Logistic curve, 338
Long division, of polynomials, 222
Lower bound, 236

M

Magnitude, of a vector, 522, 524–525
Major axis of an ellipse, 739
Maps, 156
Mathematical induction, 680–685
Mathematical model, 151
Matrix (matrices)
addition of, 614–615
additive inverse of, 615
algebra of, 614–621
augmented, 600
augmented coefficient, 600
coefficient, 600
column, 619
columns of, 600
definition of, 601
determinant of, 628, 631
double subscript notation for, 600
echelon form of, 603–606
element of, 601
elementary row transformations of, 602
equality of, 614
equivalent, 602
identity, 623, 624
inverse of, 623–627
linear combination of rows of, 611
main diagonal elements of, 601
of order n, 601
product of, 617
product of with a real number, 616
reduced echelon form of, 606
row, 619
row equivalent, 602
rows of, 600
size of, 601
square, 601
subtraction of, 616
of a system of equations, 600
zero, 615
Matrix invertibility, 632
Matrix row transformations, 602
Maximum value of a quadratic function, 187, 190–191

Mean
  arithmetic, 667
  geometric, 674
Method
  of completing the square, 76
  of elimination, 575, 578, 598
  of factoring, 74
  inverse, 627
  of substitution, 565–567
  of trial and error, 36
Midpoint, 128
Midpoint formula, 127
Minimum value of a quadratic function, 187, 190
Minor, 628–629
Minor axis of an ellipse, 739
Minus infinity ($-\infty$), 250
Minute, 350, 353
Mirror image, 134
Modulus, of a complex number, 548, 550
Monomial, 28
Motion, of a point, 764
Multiple-angle formulas, 467–473
Multiplication
  of matrices, 616–621
  properties of, 4
Multiplicative identity, 4
Multiplicative inverse
  of a complex number, 91
  of a real number, 4
Multiplicity of a zero, 232–233
Mutually exclusive events, 712

N

n factorial, 688
nth partial sum, 659, 672
nth power, 16
nth root, 19, 552, 554–555
  of unity, 555
nth term
  of an arithmetic sequence, 665
  of a geometric sequence, 671
  of a sequence, 654
  of a series, 675
Natural exponential function, 299–305
Natural logarithm, 314
Natural logarithmic function, 314
Natural numbers, 2
Negative(s)
  formulas for, 382–383
  of a real number, 4, 5

of a vector, 527
Negative angle, 348
Negative direction, 7
Negative exponents, 16, 19
Negative real numbers, 8
  square roots of, 92
Negative slope, 141
Newton, 542
Newton's law of cooling, 316
Nondistinguishable permutations, 702
Nonnegative integers, 2
Nonpolynomials, 30
Nonreal complex number, 88
Nonstrict inequalities, 9
Nontrivial factors, 33
Normal probability curve, 292
Number e, 301
Number(s)
  complex, 87–93
  imaginary, 88
  irrational, 3
  natural, 2
  negative real, 8
  nonreal complex, 88
  positive real, 8
  prime, 2
  pure imaginary, 88
  rational, 2
  real, 3
  unit real, 93
  whole, 2
Numerator, 6
  rationalizing, 45–46

O

Objective function, 590
Oblique asymptote, 261–262
Oblique triangle, 502, 503, 512
Obtuse angle, 350
Odd function, 171
Odds, 714
One-to-one correspondence, 7, 524
One-to-one function, 278, 289, 308
Open interval, 103
Opposite side, 359
Order of a matrix, 601
Ordered pair, 124, 130
Ordered r-tuple, 697
Ordered triple, 567
Ordering, 9
Ordinate, 124

Orientation, of a parametrized curve, 764
Origin, 7, 124, 739, 752, 772
Orthogonal vectors, 537, 539
Oscillation, 426
Outcome of an experiment, 709
Output variable, 165

P

Parabola(s), 132, 728–734, 786
  axis of, 729
  directrix of, 728
  focus of, 728
  polar equation of, 789–790
  reflective property of, 733
  standard equation of, 187–189, 730
  vertex of, 189, 190, 729
Paraboloid, 733
Parallel lines, 147
Parallel vectors, 537, 539
Parallelogram, diagonals of, 515
Parallelogram law, 524
Parameter, 762
Parametric equations, 762
  for a cycloid, 768–770
  for a line, 766
Parametrization, 763
Parametrized curve, 763, 765
Partial fraction, 642–647
Partial fraction decomposition, 642, 643–647
Partial sum, 659, 672
Pascal's triangle, 692
Path of a projectile, 766–768
Period, 380, 388, 402–403, 405, 406, 414
  of harmonic motion, 426
Periodic function, 380
Permutations, 695–700
  distinguishable, 702
  nondistinguishable, 702
Perpendicular bisector, 126, 127, 149–150
Perpendicular lines, 148
Phase shift, 405, 406, 414
Piecewise-defined functions, 177–179
Plane curve, 762
Plotting points, 124
Plus or minus ($\pm$), 23
Point
  motion of, 764
  on a unit circle corresponding to a real number, 376

Point-slope form, 145
Polar axis, 772
Polar coordinate system, 773
Polar coordinates, 772–784
    relationship of to rectangular
        coordinates, 774, 775, 776
Polar equation, 771–784
    of conics, 786–791
Polar form of a complex number, 548
Pole, 772
Pole values, 780
Polynomial(s), 28, 29
    adding and subtracting, 30
    bounds for zeros of, 236–238
    conjugate pair zeros of, 241
    constant, 29
    constant term of, 234
    cubic, 214
    degree of, 29
    dividing, 32, 222
    equal, 29
    factoring, 33, 34, 37, 224–225
    irreducible, 33
    leading coefficient of, 29
    in more than one variable, 31
    multiplying, 31
    prime, 33
    as a product of linear and quadratic
        factors, 244
    rational zeros of, 243, 244
    real zero of, 216
    term of, 29
    in x, 28, 29
    zero, 29
    zeros of, 229–238
Polynomial function, 198, 214–219
Position vector, 524
Positive angle, 348
Positive direction, 7
Positive integers, 2
Positive real numbers, 8
Positive slope, 141
Power functions, 265
Prime factorizations, 43
Prime number, 2
Prime polynomial, 33
Principal, 64
Principal nth root, 19
Principal square root, 20, 91
Principle of mathematical induction,
    681
Probability, 709–718

Product(s)
    of complex numbers, 88, 549
    of functions, 197
    involving zero, 5
    of matrices, 617–619
    of real numbers, 3
Product formulas, 32
Product-to-sum formulas, 477–478
Projectile, path of, 766–768
Projection, of **a** on **b**, 541
Properties
    of absolute values, 108
    of conjugates, 91
    of equality, 5
    of i, 88
    of inequalities, 104, 117
    of logarithms, 323–328
    of nth roots, 20
    of negatives, 5
    of quotients, 7
    of real numbers, 4
Proportionality
    constant of, 265
    direct, 265
    inverse, 265
    joint, 268
Pure imaginary number, 88
Pythagorean identities, 364, 365
Pythagorean theorem, 99, 125, 360, 361,
    490

Q

Quadrant(s), 124, 348, 371
Quadrantal angle, 348, 370
Quadratic equations, 73–83, 450
Quadratic formula, 77, 79–82
Quadratic functions, 185–192
Quadratic inequality, 112, 113
Quadratic type equation, 97–98
Quotient, 6, 7, 223
    of complex numbers, 91, 549
    difference, 163
    in division process, 223
    of factorials, 688, 689
    of functions, 197
    of real numbers, 6

R

$r\theta$-plane, 773
Radian, 350–353, 447
    relation of to degree, 351–352
Radian mode, 375, 376

Radical(s), 20–25
    combining, 25
    equations containing, 96–98
    laws of, 21
    removing factors from, 22–23
Radical sign, 20
Radicand, 20
Radioactive decay, 293
Radiotherapy, 304–305
Radius of a circle, 137
Range of a function, 156, 281, 388
Ratio, common, 671
Rational exponents, 23
    equations containing, 95
Rational expressions, 40
    equations containing, 57–58
    products and quotients of, 42
    simplified, 41
    sums and differences of, 43
Rational functions, 248–262
Rational inequality, 106
Rational number, 2
    infinite repeating decimal as, 675
Rational zeros of polynomials, 243, 244
Rationalizing denominators, 22–23, 45
Rationalizing numerators, 45
Rays, 348
Real axis, 546
Real line, 7
Real numbers, 2–13
    properties of, 4
Real part of a complex number, 89
Reciprocal, 4, 6
    notation for, 6
    of y-coordinates, 386
Reciprocal function, 252
Reciprocal identities, 360, 364
Rectangular coordinate system, 124–128
Rectangular coordinates, 124–128
    relation of to polar coordinates,
        774–776
Recursive definition, 657
Reduced echelon form, 606
Reduction formulas, 463
Reference angle, 393, 394, 395
Reflection of a graph, 134, 176, 177, 284
Reflective property
    of an ellipse, 746
    of a hyperbola, 757–758
    of a parabola, 733
Remainder, in division process, 223
Remainder theorem, 223, 224

Resultant force, 524
Resultant vector, 531
Resulting sign, 112
Richter scale, 315
Right angle, 350
Right triangle, 358, 420, 421, 422
Root(s)
    cube, 20
    double, 75
    of an equation, 54, 133
    existence of, 21
    extraneous, 57
    of multiplicity m, 232
    of multiplicity 2, 75
    nth, of complex numbers, 552–556
    principal nth, 19
    square, 3, 19, 91–92
    of unity, 93, 555
Row, of a matrix, 600
Row equivalent, 602
Row matrix, 619
Row subscript, 600
Row transformation of a matrix, 602,
    635, 636
Rule of 70, 317
Rule of 72, 317

S

Sample space, 709
Satisfying an equation, 54
Scalar, 522
Scalar multiple of a vector, 524, 526, 527
Scalar product, 536
Scalar quantity, 522
Scientific form, 12, 13
Secant function, 359, 384, 386
Second, 350, 353
Semicircle, 137
Sequence(s), 654
    arithmetic, 664–668
    equality of, 655
    geometric, 671–676
    graph of, 655, 656
    infinite, 654
    nth term of, 654
    of partial sums, 659
    recursively defined, 657
Series, 674, 675
Set(s), 27
    complement of, 713
    correspondence of, 155–156
    intersection of, 108

subsets of, 706
    union of, 108
Shifts of graphs, 173–175
Sign(s)
    laws of, 9
    of a real number, 9
    of trigonometric functions, 371
    variation of, 234
Sign chart, 112, 113
Sign diagram, 112, 113–116
Significant figures, 13
Simple closed curve, 762
Simple harmonic motion, 425
Simple interest, 63–64
Simplification
    of an exponential expression, 18, 19
    of a radical, 22
    of a rational expression, 41
Sine function, 359, 384, 447
    addition and subtraction formulas for,
    461
Sine wave, 382, 407
Sketching a graph, 103, 254
Slope(s)
    of a line, 140–143
    of parallel lines, 147
    of perpendicular lines, 148
Slope-intercept form, 146
Solution(s)
    bounds for, 236–238
    of an equation in x, 54
    of an equation in x and y, 130
    extraneous, 57
    feasible, 590
    of an inequality, 102
    of a polar equation, 776
    of a system of equations, 564, 567
    of a system of inequalities, 582, 584,
    585
    trivial, 608
Solving
    an equation, 54
    an inequality, 103, 104–106
    a system of equations, 564
    a triangle, 420
    for a variable, 58
Special change of base formulas, 332
Speed
    angular, 355
    linear, 355
Spiral of Archimedes, 782
Square matrix, 601

Square root, 3, 20, 91
    of negative numbers, 92
Square root function, 160
Squaring function, 160
Standard equation
    of a circle, 135
    of an ellipse, 739
    of a hyperbola, 752
    of a parabola, 730
Standard position, of an angle, 348
Straight angle, 348
Stretching of graphs, 176, 177
Strict inequalities, 8
Subset of a set, 27, 706
Subtended angle, 350
Subtraction
    of complex numbers, 90
    of matrices, 615, 616
    of real numbers, 6
Subtraction formulas, 457, 458, 461
Successive approximations, 216
Sum(s)
    of an arithmetic sequence, 666
    of complex numbers, 88
    of functions, 197
    of a geometric sequence, 672–673
    of an infinite geometric series, 674,
    675–676
    of matrices, 615
    partial, 659, 672
    of real numbers, 3
    of a series, 676
    theorem on, 661
    of trigonometric functions, 417, 463
    of two cubes, 35
    of vectors, 523
Sum identity, 457
Sum-to-product formulas, 478
Summation notation, 657, 668
Summation variable, 658
Supplementary angle, 350
Symmetries, of graphs
    of equations in x and y, 134–135
    of inverse functions, 284
    of polar equations, 783
    of trigonometric equations, 388
Synthetic division, 225–227, 236–237
Systems of equations, 564–570
    consistent, 576
    dependent and consistent, 576
    equivalent, 567, 574
    homogeneous, 608

inconsistent, 576
matrix of, 600
in more than two variables, 598–611
solution of, 564, 567
in two variables, 573–579
Systems of inequalities, 582–587

T

Tangent function, 359, 384, 385, 448
  addition and subtraction formulas for, 461
Tangent identity, 364, 461
Tangent line, to a parabola, 733
Term
  of a polynomial, 29
  of a sequence, 654, 655–656, 665, 671–672
  of a series, 675
Terminal point of a vector, 522
Terminal side of an angle, 348
Test point, 582
Test value, 112, 216
Tests for symmetry, 134
Theorem
  on amplitudes and periods, 402
  on amplitudes, periods, and phase shifts, 405
  binomial, 690
  on bounds for real zeros of polynomials, 237
  change of base, 331
  complete factorization, for polynomials, 230
  on conics, 786
  on conjugate pair zeros of a polynomial, 241
  on cosine of the angle between vectors, 538
  De Moivre's, 552–554
  on distinguishable permutations, 702, 703
  on dot product, 538
  on equivalent systems, 574
  on even and odd trigonometric functions, 384
  on exact number of zeros of a polynomial, 234
  on expansion of determinants, 631
  on expressing a polynomial as a product of linear and quadratic factors, 242
  factor, 224

fundamental, of algebra, 230
  on graph of the tangent function, 414
  on horizontal asymptotes, 253
  on identical rows, 635
  on independent events, 715
  intermediate value, for polynomial functions, 215
  on inverse functions, 281
  for locating the vertex of a parabola, 189
  on matrix invertibility, 632
  on matrix properties, 615, 616
  on matrix row transformations, 602
  on maximum or minimum value of a quadratic function, 190
  on maximum number of zeros of a polynomial, 231
  on mutually exclusive events, 712
  on negative exponents, 19
  on nth roots, 554
  on number of combinations, 704
  on number of different permutations, 698
  on one-to-one nature of exponential functions, 289
  on one-to-one nature of increasing or decreasing functions, 280
  on one-to-one nature of logarithmic functions, 310
  on orthogonal vectors, 539
  on polar equations of conics, 788
  on probability of the occurrence of either of two events, 713
  on products and quotients of complex numbers, 549
  on rational zeros of a polynomial, 243
  on reference angles, 395
  remainder, 223
  on repeated function values for sin and cos, 380
  on row and column transformations of a determinant, 634
  on a row of zeros, 632
  on slopes of parallel lines, 147
  on slopes of perpendicular lines, 148
  on sum of an arithmetic sequence, 666
  on sum of a constant, 660
  on sum of a geometric sequence, 672
  on sum of an infinite geometric series, 674
  on sum of a sequence, 661
  zero factor, 74

Theory of equations, 230
Tractrix, 336
Transcendental function, 198
Transformation
  of determinants, 634, 635
  of graphs, 181, 800–801
  of systems of equations, 574
Translations, 175
Transverse axis of the hyperbola, 751
Tree diagram, 695, 716–717
Trial and error, method of, 36
Triangle, 420
  area of, 125, 516
  isosceles, 361, 472–473
  oblique, 502, 503, 512
  right, 358, 420, 421, 422
  vertices of, 420
Triangle law, 523
Trichotomy law, 8
Trigonometric equation, 447–454
Trigonometric expression, 442
Trigonometric form of complex numbers, 546, 548
Trigonometric function(s), 359
  absolute value of, 417
  amplitude of, 402, 403, 405, 406
  of angles, 358–372
  and calculators, 396–398
  cofunction formulas for, 460
  domains of, 388
  double-angle formulas for, 467
  equations and inequalities involving, 388–389
  even and odd, 384
  graphs of, 379, 386, 388, 400–410, 412–418, 802–804
  half-angle formulas for, 470, 471
  half-angle identities for, 469, 470
  inverse, 396, 482–492
  multiple-angle formulas for, 467–473
  product-to-sum formulas for, 477–478
  of real numbers, 375–389
  and reference angles, 393–396
  signs of, 371
  special values of, 362, 378–379, 804
  subtraction formulas for, 457, 458, 461
  sum-to-product formulas for, 478
  in terms of right triangle, 358
  in terms of unit circle, 376–377
  values of, 360, 369–370, 393–398, 487, 489
Trigonometric identities, 442–445

Trigonometric substitution, 444, 445
Trinomial, 28
    factorization of, 36
Trivial solution, 608
Turning points, 215

U

Undefined function, 158
Union (∪), of sets, 108
Unit circle, 136, 376–377, 448
    arc length on, 483
Unit real number, 93
Unit vector, 528
Unity, roots of, 93, 555
Upper bounds, 236

V

Value
    of an expression, 28
    of a function, 156
    of trigonometric functions, 375,
        393–398
Variable, 27, 28
    dependent, 165
    directly proportional, 265
    independent, 165
    input, 165
    inversely proportional, 265
    linearly related, 150
    output, 165
    solving for, 58
    summation, 658
Variation, 265–269
    constant of, 265
    direct, 265
    inverse, 265
    joint, 268
    of sign, 234
Vector(s), 522–531
    addition of, 525, 526, 527
    angle between, 538

components along, 541
components of, 524
dot product of, 537
equal, 523
equivalent, 522
force, 523
horizontal component of, 529, 530
i, 528, 529
i, j form for, 529
initial point of, 522
inner product of, 536
j, 528, 529
linear combination of, 529, 530
magnitude of, 524–525
negative of, 527
one-to-one correspondence between,
    524
orthogonal, 537, 539
parallel, 537, 539
position, 524
projection of, 541
resultant, 531
scalar multiple of, 526, 527
scalar product of, 536
subtraction of, 528
sum of, 523
terminal point of, 522
unit, 528
velocity, 523
vertical component of, 529, 530
wind velocity as, 530
zero, 527
Velocity vector, 523
Verifying trigonometric identities,
    442–445
Vertex (vertices)
    of an angle, 348
    of an ellipse, 739
    of a hyperbola, 751
    of a parabola, 132, 189, 190, 729
    of a triangle, 420

Vertical asymptote, 250, 388, 412–413,
    415
Vertical compression of graphs, 175, 176
Vertical line, 144
Vertical line test, 159
Vertical shifts of graphs, 173, 405, 410
Vertical stretching of graphs, 175, 176

W

Wedge notation, 524
Whispering galleries, 746
Whole numbers, 2
Work, 541, 543–544

X

x-axis, 124
x-coordinate, 124
x-equation, 763
x-intercept, 132, 133, 388, 472, 479–480
xy-plane, 124

Y

y-axis, 124
y-coordinate, 124
y-equation, 763
y-intercept, 132, 133

Z

Zero, the number, 5, 7
Zero(s)
    of a function, 160, 216, 305
    of a graph, 132
    of multiplicity m, 232
    of a polynomial, 229–238
Zero exponent, 16
Zero factor theorem, 5, 74, 450, 451,
    452, 453
Zero matrix, 615
Zero polynomial, 29
Zero vector, 527

QUICK REFERENCE CARD

THE SWOKOWSKI • COLE SERIES

PRECALCULUS: FUNCTIONS AND GRAPHS
Tenth Edition

FUNDAMENTALS OF COLLEGE ALGEBRA
Eleventh Edition

ALGEBRA AND TRIGONOMETRY WITH ANALYTIC GEOMETRY
Classic Eleventh Edition

ALGEBRA AND TRIGONOMETRY WITH ANALYTIC GEOMETRY
Eleventh Edition

THOMSON™

BROOKS/COLE

FORMULAS FROM TRIGONOMETRY

FUNDAMENTAL IDENTITIES

$$\csc t = \frac{1}{\sin t}$$

$$\sec t = \frac{1}{\cos t}$$

$$\cot t = \frac{1}{\tan t}$$

$$\tan t = \frac{\sin t}{\cos t}$$

$$\cot t = \frac{\cos t}{\sin t}$$

$$\sin^2 t + \cos^2 t = 1$$

$$1 + \tan^2 t = \sec^2 t$$

$$1 + \cot^2 t = \csc^2 t$$

FORMULAS FOR NEGATIVES

$$\sin(-t) = -\sin t$$

$$\cos(-t) = \cos t$$

$$\tan(-t) = -\tan t$$

$$\cot(-t) = -\cot t$$

$$\sec(-t) = \sec t$$

$$\csc(-t) = -\csc t$$

ADDITION FORMULAS

$$\sin(u + v) = \sin u \cos v + \cos u \sin v$$

$$\cos(u + v) = \cos u \cos v - \sin u \sin v$$

$$\tan(u + v) = \frac{\tan u + \tan v}{1 - \tan u \tan v}$$

SUBTRACTION FORMULAS

$$\sin(u - v) = \sin u \cos v - \cos u \sin v$$

$$\cos(u - v) = \cos u \cos v + \sin u \sin v$$

$$\tan(u - v) = \frac{\tan u - \tan v}{1 + \tan u \tan v}$$

HALF-ANGLE FORMULAS

$$\sin \frac{u}{2} = \pm\sqrt{\frac{1 - \cos u}{2}}$$

$$\cos \frac{u}{2} = \pm\sqrt{\frac{1 + \cos u}{2}}$$

$$\tan \frac{u}{2} = \frac{1 - \cos u}{\sin u} = \frac{\sin u}{1 + \cos u}$$

DOUBLE-ANGLE FORMULAS

$$\sin 2u = 2 \sin u \cos u$$

$$\cos 2u = \cos^2 u - \sin^2 u$$

$$= 1 - 2\sin^2 u$$

$$= 2\cos^2 u - 1$$

$$\tan 2u = \frac{2 \tan u}{1 - \tan^2 u}$$

FORMULAS FROM TRIGONOMETRY

TRIGONOMETRIC FUNCTIONS

OF REAL NUMBERS

$$\sin t = y$$

$$\cos t = x$$

$$\tan t = \frac{y}{x}$$

$$\csc t = \frac{1}{y}$$

$$\sec t = \frac{1}{x}$$

$$\cot t = \frac{x}{y}$$

OF ACUTE ANGLES

$$\sin \theta = \frac{\text{opp}}{\text{hyp}}$$

$$\cos \theta = \frac{\text{adj}}{\text{hyp}}$$

$$\tan \theta = \frac{\text{opp}}{\text{adj}}$$

$$\csc \theta = \frac{\text{hyp}}{\text{opp}}$$

$$\sec \theta = \frac{\text{hyp}}{\text{adj}}$$

$$\cot \theta = \frac{\text{adj}}{\text{opp}}$$

LAW OF SINES

$$\frac{\sin \alpha}{a} = \frac{\sin \beta}{b} = \frac{\sin \gamma}{c}$$

LAW OF COSINES

$$a^2 = b^2 + c^2 - 2bc \cos \alpha$$

$$b^2 = a^2 + c^2 - 2ac \cos \beta$$

$$c^2 = a^2 + b^2 - 2ab \cos \gamma$$

OBLIQUE TRIANGLE

$y = \sin t, \; 0 \le t \le 2\pi$

$y = \cos t, \; 0 \le t \le 2\pi$

$y = \tan t$

area A perimeter P circumference C
volume V curved surface area S
altitude h radius r

RIGHT TRIANGLE

Pythagorean theorem:
$$c^2 = a^2 + b^2$$

TRIANGLE

$$A = \frac{1}{2}bh \qquad P = a + b + c$$

CIRCLE

$$A = \pi r^2 \qquad C = 2\pi r$$

SPHERE

$$V = \frac{4}{3}\pi r^3 \qquad S = 4\pi r^2$$

RIGHT CIRCULAR CYLINDER

$$V = \pi r^2 h \qquad S = 2\pi rh$$

RIGHT CIRCULAR CONE

$$V = \frac{1}{3}\pi r^2 h \qquad S = \pi r \cdot \sqrt{r^2 + h^2}$$

QUADRATIC FORMULA

If $a \neq 0$, the roots of $ax^2 + bx + c = 0$ are
$$x = \frac{-b \pm \sqrt{b^2 - 4ac}}{2a}$$

SPECIAL FACTORING FORMULAS

$$x^2 - y^2 = (x + y)(x - y)$$
$$x^2 + 2xy + y^2 = (x + y)^2$$
$$x^2 - 2xy + y^2 = (x - y)^2$$
$$x^3 - y^3 = (x - y)(x^2 + xy + y^2)$$
$$x^3 + y^3 = (x + y)(x^2 - xy + y^2)$$

EXPONENTIALS AND LOGARITHMS

$$y = \log_a x \quad \text{means} \quad a^y = x$$
$$\log_a xy = \log_a x + \log_a y$$
$$\log_a \frac{x}{y} = \log_a x - \log_a y$$
$$\log_a x^r = r \log_a x$$
$$a^{\log_a x} = x$$
$$\log_a a^x = x$$
$$\log_a a = 1$$
$$\log_a 1 = 0$$
$$\log x = \log_{10} x$$
$$\ln x = \log_e x$$
$$\log_b u = \frac{\log_a u}{\log_a b}$$

POINT-SLOPE FORM OF A LINE

$$y - y_1 = m(x - x_1)$$
m is the slope

SLOPE-INTERCEPT FORM OF A LINE

$$y = mx + b \quad m \text{ is the slope}$$

EXPONENTS AND RADICALS

$$a^m a^n = a^{m+n} \qquad a^{1/n} = \sqrt[n]{a}$$
$$(a^m)^n = a^{mn} \qquad a^{m/n} = \sqrt[n]{a^m}$$
$$(ab)^n = a^n b^n \qquad a^{m/n} = (\sqrt[n]{a})^m$$
$$\left(\frac{a}{b}\right)^n = \frac{a^n}{b^n} \qquad \sqrt[n]{ab} = \sqrt[n]{a}\sqrt[n]{b}$$
$$\frac{a^m}{a^n} = a^{m-n} \qquad \sqrt[n]{\frac{a}{b}} = \frac{\sqrt[n]{a}}{\sqrt[n]{b}}$$
$$a^{-n} = \frac{1}{a^n} \qquad \sqrt[m]{\sqrt[n]{a}} = \sqrt[mn]{a}$$

CIRCLE

$$(x - h)^2 + (y - k)^2 = r^2$$

PARABOLA

$$x^2 = 4py$$

ELLIPSE

$$\frac{x^2}{a^2} + \frac{y^2}{b^2} = 1 \quad \text{with} \quad a^2 = b^2 + c^2$$

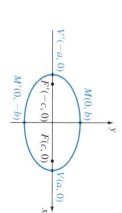

HYPERBOLA

$$\frac{x^2}{a^2} - \frac{y^2}{b^2} = 1 \quad \text{with} \quad c^2 = a^2 + b^2$$

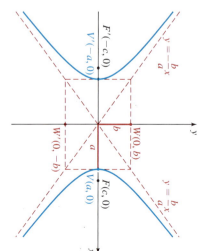

CONIC SECTIONS

PARABOLA

$$x^2 = 4py$$

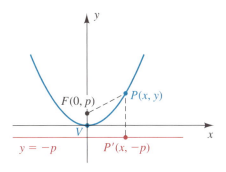

ELLIPSE

$$\frac{x^2}{a^2} + \frac{y^2}{b^2} = 1 \quad \text{with} \quad a^2 = b^2 + c^2$$

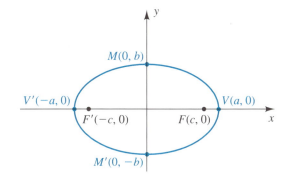

HYPERBOLA

$$\frac{x^2}{a^2} - \frac{y^2}{b^2} = 1 \quad \text{with} \quad c^2 = a^2 + b^2$$

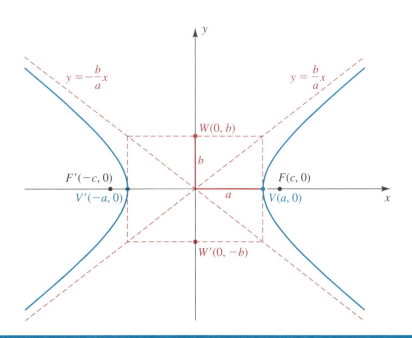

PLANE GEOMETRY

SIMILAR TRIANGLES

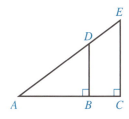

$$\frac{AB}{BD} = \frac{AC}{CE}$$

$$\frac{AB}{AD} = \frac{AC}{AE}$$

CONGRUENT ALTERNATE INTERIOR ANGLES

$l_1 \parallel l_2$

$\beta = 180° - \alpha$

TRIGONOMETRY

TRIGONOMETRIC FUNCTIONS

OF ACUTE ANGLES

$$\sin \theta = \frac{\text{opp}}{\text{hyp}} \qquad \csc \theta = \frac{\text{hyp}}{\text{opp}}$$

$$\cos \theta = \frac{\text{adj}}{\text{hyp}} \qquad \sec \theta = \frac{\text{hyp}}{\text{adj}}$$

$$\tan \theta = \frac{\text{opp}}{\text{adj}} \qquad \cot \theta = \frac{\text{adj}}{\text{opp}}$$

OF ARBITRARY ANGLES

$$\sin \theta = \frac{b}{r} \qquad \csc \theta = \frac{r}{b}$$

$$\cos \theta = \frac{a}{r} \qquad \sec \theta = \frac{r}{a}$$

$$\tan \theta = \frac{b}{a} \qquad \cot \theta = \frac{a}{b}$$

OF REAL NUMBERS

$$\sin t = y \qquad \csc t = \frac{1}{y}$$

$$\cos t = x \qquad \sec t = \frac{1}{x}$$

$$\tan t = \frac{y}{x} \qquad \cot t = \frac{x}{y}$$

SPECIAL RIGHT TRIANGLES

OBLIQUE TRIANGLE

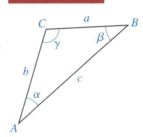

AREA

$$\mathcal{A} = \frac{1}{2}bc \sin \alpha$$

$$\mathcal{A} = \frac{1}{2}ac \sin \beta$$

$$\mathcal{A} = \frac{1}{2}ab \sin \gamma$$

$$\mathcal{A} = \sqrt{s(s-a)(s-b)(s-c)},$$

where $s = \frac{1}{2}(a+b+c)$ (*Heron's Formula*)

LAW OF COSINES

$$a^2 = b^2 + c^2 - 2bc \cos \alpha$$

$$b^2 = a^2 + c^2 - 2ac \cos \beta$$

$$c^2 = a^2 + b^2 - 2ab \cos \gamma$$

LAW OF SINES

$$\frac{\sin \alpha}{a} = \frac{\sin \beta}{b} = \frac{\sin \gamma}{c}$$

GREEK ALPHABET

Letter	Name	Letter	Name
A α	alpha	N ν	nu
B β	beta	Ξ ξ	xi
Γ γ	gamma	O o	omicron
Δ δ	delta	Π π	pi
E ϵ	epsilon	P ρ	rho
Z ζ	zeta	Σ σ	sigma
H η	eta	T τ	tau
Θ θ	theta	Υ υ	upsilon
I ι	iota	Φ ϕ (φ)	phi
K κ	kappa	X χ	chi
Λ λ	lambda	Ψ ψ	psi
M μ	mu	Ω ω	omega

SPECIAL VALUES OF TRIGONOMETRIC FUNCTIONS

θ (degrees)	θ (radians)	$\sin \theta$	$\cos \theta$	$\tan \theta$	$\cot \theta$	$\sec \theta$	$\csc \theta$
$0°$	0	0	1	0	—	1	—
$30°$	$\frac{\pi}{6}$	$\frac{1}{2}$	$\frac{\sqrt{3}}{2}$	$\frac{\sqrt{3}}{3}$	$\sqrt{3}$	$\frac{2\sqrt{3}}{3}$	2
$45°$	$\frac{\pi}{4}$	$\frac{\sqrt{2}}{2}$	$\frac{\sqrt{2}}{2}$	1	1	$\sqrt{2}$	$\sqrt{2}$
$60°$	$\frac{\pi}{3}$	$\frac{\sqrt{3}}{2}$	$\frac{1}{2}$	$\sqrt{3}$	$\frac{\sqrt{3}}{3}$	2	$\frac{2\sqrt{3}}{3}$
$90°$	$\frac{\pi}{2}$	1	0	—	0	—	1